Make Reform a Reality— with
HRW MATHEMATICS!

INTEGRATING

- **MATHEMATICS**
- **TECHNOLOGY**
- **EXPLORATIONS**
- **APPLICATIONS**
- **ASSESSMENT**

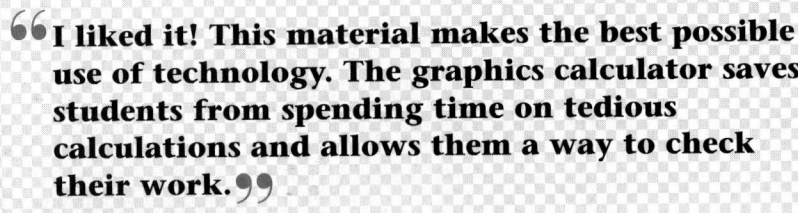

The Sign of Progress—
SATISFIED TEACHERS!

> **I liked it! This material makes the best possible use of technology. The graphics calculator saves students from spending time on tedious calculations and allows them a way to check their work.**
>
> Linda Craine
> Abilene, TX

> **I am impressed by the connection between mathematics and the real world. The examples are interesting and informative, and they motivate students to apply concepts.**
>
> Audrey Beres
> Bridgeport, CT

> **Your concrete use of algebra tiles is sound and well presented.**
>
> Robert E. Bauman
> Appleton, WI

> **The text captures the nature of modern-day mathematics. The calculator approach is wonderful and well-supported with written exercises.**
>
> Kenneth Dupuis
> Sterling Heights, MI

> **I really like the way you integrate material, especially cooperative-learning activities!**
>
> Dianne Hershey
> Jonesburg, GA

> **The real-life examples are meaningful to the students—so much, in fact, that many have even taken it upon themselves to discuss their work with their parents.**
>
> Melanie Gasperec
> Olympia Fields, IL

Making Progress– The Authors of
HRW MATHEMATICS!

JAMES E. SCHULTZ

"Technology has the capability of opening new worlds of mathematics to more students than ever."

Dr. Schultz is a co-author of the **NCTM Curriculum and Evaluation Standards for Mathematics** and **A Core Curriculum: Making Mathematics Count for Everyone.** He is especially well regarded for his inventive and skillful integration of mathematics and technology. Jim's dynamic vision of classroom reform recently earned him the prestigious Morton Chair at Ohio University.

KATHLEEN A. HOLLOWELL

"Mathematics classrooms should become laboratories of learning where excited students collect data, look for patterns, make and test conjectures, and explain their reasoning."

Dr. Hollowell's keen understanding of what takes place in the mathematics classroom recently helped her win a major NSF research grant to enhance mathematics teaching methods. She is particularly well-versed in the special challenge of motivating students and making the classroom a more dynamic place to learn.

WADE ELLIS, JR.

"Integration can cultivate an appreciation for the relevance of mathematics— provided you meaningfully unify material with an intuitive, common-sense approach. Otherwise, diversity fosters confusion and becomes a liability."

Professor Ellis has co-authored numerous books and articles on how to integrate technology realistically and meaningfully into the mathematics curriculum. He was a key contributor to the landmark study, **Everybody Counts: A Report to the Nation on the Future of Mathematics Education**.

Reform You Can Relate

"Get a head start teaching algebra!"

"Most math teachers say it's important to get your students on board as soon as possible or risk *'losing them'* for the year. Yet once school starts, it never fails—chapter one of your algebra text drags students through the obligatory prealgebra review of numbers, variables, expressions, and so on.

And suddenly, all of your back-to-school energy drifts back to summer."

To — HRW ALGEBRA!

"Break the pattern!"

"Unlike other programs, HRW doesn't tell students to forget about their world and pay attention to yours. Instead, from day one the book immerses students in their world through the use of patterns—the ones they see every day in their clothing, music, or TV.

Students immediately do what comes naturally—make sense of it all. In Chapter One, they collect data, organize it, and even make scatter plots to recognize relationships between patterns, formulas, and equations."

"Who couldn't function in this environment?"

"HRW makes another refreshing break when they preview functions in Chapter Two. My students explore exponential, quadratic, reciprocal, and other types of functions with real-world data.

Introducing functions early on gives my class the opportunity to handle interesting problems taken from real life. Try it and you'll wonder why nobody's taken this approach before. Maybe nobody's ever noticed how much realistic data enhances comprehension."

"This program doesn't mess around!"

"Data from real-world problems and situations can be pretty messy, and simple intuition tells you that people don't like messes. With this idea in mind, Chapter Seven does a convincing job of showing students how to use matrices to organize and analyze information.

By mid-year, HRW's coverage of matrices allows your students to apply their newfound knowledge in later chapters to organize data and solve equations."

"You see some amazing transformations."

"HRW's coverage of transformations in Chapter Nine gives students a genuine sense of how useful it can be to understand relationships between equations and their graphs. In conjunction, students relate lines and figures to what constitutes their shape, size, and position.

This integrated approach sure works a lot better than studying each element in isolation."

SO WHAT'S THE NEXT PROGRESSION?

"It's not unusual to see students build walls around themselves when it comes to learning math, and some textbooks help to build those walls. After all, how many books ever take you beyond the walls of your classroom—to the places where math really matters?

Well, no sooner do you open a new chapter in *HRW Algebra* than its applications take you outside your classroom. *Every chapter begins with a credible connection to the real world, including a lively portfolio activity.*"

Look for Some

"Turn snoring to exploring with Exploratory Lessons!"

"If you really want to get your students off to a good start, give them a job to do. That's exactly what HRW does with its *Exploratory Lessons*, which present concepts using a discovery approach.

For example, I have my students pretend they've been asked to play lead guitar for a music video, and if they want to look convincing, they'll need to explore the relationship between frequency and the length of a guitar string."

"Set an example with Expository Lessons"

"Another lesson format, *Expository Lessons,* provides step-by-step examples in a relevant, applied, or hands-on context. Students begin with applications or activities and stay involved in what they're doing. Sometimes they go on 'mini explorations' when a discovery approach is particularly useful."

Sure Signs of Progress!

"But first ask WHY?"

"People aren't born with genes for algebra, so our ability to ask 'why' is critical for understanding math. Unfortunately, it hasn't always come so naturally to math textbooks. HRW is a big exception. Each lesson begins from a student's point of view, asking *'Why should I learn this?'*

The same spirit of inquiry continues throughout HRW's concept development with examples and explorations that integrate higher-order *Critical Thinking* sections, and *Try This* practice. In fact, I like the way questions are interwoven throughout the book, for ongoing and self-assessment. And how can you miss them—they're all highlighted in yellow?"

"It's great exercise!"

"**HRW Algebra** pumps up math comprehension in four sessions of the best, no-nonsense math workout I've ever come across. In *Communicate*, students discuss, explain, or write about math to exercise one of the most powerful and underdeveloped problem-solving muscles—the logic of language.

In the next session, they break out in a healthy sweat with some robust *Practice and Apply* problems and applications. And when it's time to wind down, they can *Look Back* and review what they've learned, and *Look Beyond* to prepare for future workouts."

"Expect to see some healthy changes."

"When it's time to take a deep breath and measure individual or group progress, students stretch their minds with *portfolio activities*, long-term *Chapter Projects*, and *Eyewitness Math* activities. And they can further examine their progress with *Chapter Reviews* and *Chapter* and *Cumulative Assessments*."

SO WHAT'S THE NEXT PROGRESSION?

Plug Into Math with

PUT SOME ELECTRICITY IN THE AIR!

"My classroom really comes to life whenever we plug into technology. With HRW, technology is more than a computational toy—it's one of the most serious instructional advances ever to hit mathematics education.

HRW seamlessly integrates calculators and software into the text at the right place and the right time with all the help you need. Students are motivated to *explore, communicate,* and *apply* technology in a reasonable and balanced progression.

Although the book only requires a scientific calculator, students have the opportunity to explore concepts much further if other technologies are available to you."

"Explore!"

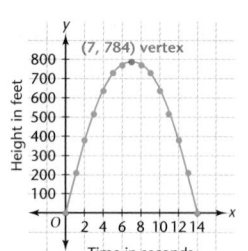

APPLICATION

Physics The following spreadsheet shows the data from the rocket problem in Lesson 1.2. The graph of the data is called a **parabola**.

seconds	height (ft)
0	0
1	208
2	384
3	528
4	640
5	720
6	768
7	784
8	768
9	720
10	640
11	528
12	384
13	208
14	0

Rocket Data

Use the data from the spreadsheet to draw a graph. The curve is a parabola that opens downward. You can see that the rocket reaches its maximum height, 784 feet, after 7 seconds.

A parabola may open either **downward** or **upward**. The point where the curve changes direction is the **vertex**. ❖

Exploration 2 Graphing Sequences

"HRW's use of technology allows me to go far beyond the surface of a math problem. Right now my class is exploring quadratic functions by using a spreadsheet to construct a table for finding the maximum height of a rocket."

"The spreadsheet helped us see the relationship between the time and height of a rocket."

TECHNOLOGY!

"*Communicate!*"

Exploration 2 Transformations of Quadratic Functions

1. Draw the graph of the parent function $y = x^2$.

2. Draw the graph of each of the following functions. Describe how the graph of each function has been transformed from the parent function.

   a. $y = 2x^2$ **b.** $y = -x^2$ **c.** $y = x^2 - 2$

3. Compare your graphs to those drawn in a-f in Exploration 1. How are they similar?

4. What must you do to the function rule for $y = x^2$ to make the new function rule result in a stretch? a reflection? a shift?

5. Explain how to use the parent function $y = x^2$ to grap $y = -2x^2 + 3$. ❖

GEOMETRY *Connection*

APPLICATION

Square photos are placed on a piece of cardboard backing. I special border that adds a total of 2 inches to each side of th area of a photo including its border is a function of the leng the photo. How can you model this application with a func

Method A Make a table.

...f the sides of the photo as they i...

"Any student who completes a project like painting the set of a rock concert can't wait to tell their friends about the experience—especially the time they saved through technology. HRW harnesses that excitement when it asks students to discuss and write about their discoveries.

For example, geometry gives students something to talk about and helps them explore an algebraic concept—transformations."

"Graphics calculators are cool—I can see the results of transforming a function so much easier and quicker."

"*Apply!*"

"Technology is incredibly useful, as my class discovered when they performed a weather simulation. Students used spreadsheet software to show the probability of rain over a two-day period. "

"I told my group not to waste money buying tickets for an outdoor concert, since the spreadsheet software showed it was probably going to rain!"

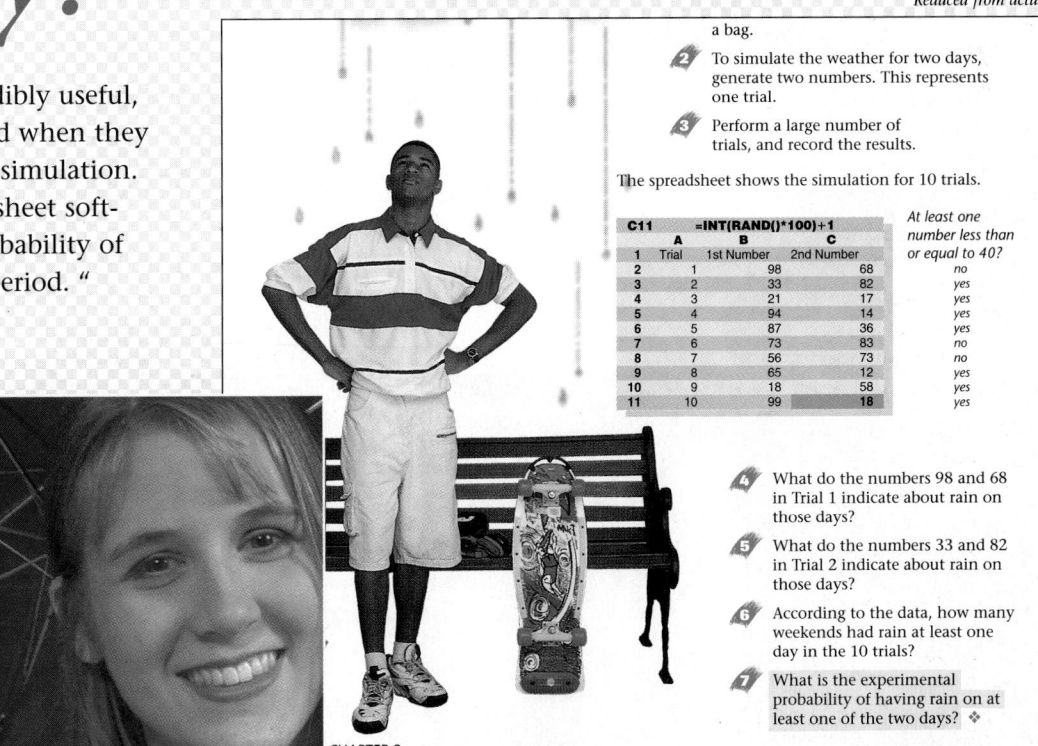

a bag.

2. To simulate the weather for two days, generate two numbers. This represents one trial.

3. Perform a large number of trials, and record the results.

The spreadsheet shows the simulation for 10 trials.

C11   =INT(RAND()*100)+1

	A	B	C	At least one number less than or equal to 40?
	Trial	1st Number	2nd Number	
2	1	98	68	no
3	2	33	82	yes
4	3	21	17	yes
5	4	94	14	yes
6	5	87	36	yes
7	6	73	83	no
8	7	56	73	no
9	8	65	12	yes
10	9	18	58	yes
11	10	99	18	yes

4. What do the numbers 98 and 68 in Trial 1 indicate about rain on those days?

5. What do the numbers 33 and 82 in Trial 2 indicate about rain on those days?

6. According to the data, how many weekends had rain at least one day in the 10 trials?

7. What is the experimental probability of having rain on at least one of the two days? ❖

CHAPTER 8

Look for exciting signs

"Don't miss the action!"

"It seems strange to me that math books tend to be so stationary because I've always thought of mathematics as an active and practical discipline—something you do.

That's probably why I'm so comfortable with *HRW Algebra*. Students learn by doing with thought-provoking materials, explorations, and hands-on activities. This approach really helps students to get interested in, and take responsibility for, their own work."

"Explore!"

"When I want my students to make discoveries, HRW gives them something worth exploring, like using algebra tiles for a visual representation to multiply binomials. My students discovered the results always relate to area. That came in handy when they calculated the area of a neighbor's lawn as a basis for charging a price to cut their grass."

Reduced from actual size

Exploration 1 — Modeling With Tiles

1. What product is shown in the diagram that models $(x)(-x)$?

2. Use tiles to model $(x)(x)$, $(-x)(x)$, and $(-x)(-x)$. What is the product for each?

3. The diagram models $(-x)(1)$ and $(1)(-x)$. What product is shown for each?

4. Use tiles to model $(x)(-1)$, $(x)(1)$, $(-x)(-1)$, $(1)(x)$, $(1)(-x)$, and $(-1)(-x)$. What is the product for each? ❖

Either tiles or the Distributive Property can be used to find the product of 3 and $2x + 1$.

Algebra-Tile Model

$2x + 1$

Distributive Property

$$3(2x + 1) = 3(2x) + 3(1)$$
$$= 6x + 3$$

"If I hadn't figured out the area of the lawn I was supposed to care for, I would have been working for far less than minimum wage!"

"Communicate!"

Reduced from actual size

LESSON 1.4

Order of Operations

Why *Calculators and computers have dramatically changed the way people work with numbers. New technology has made computation and mathematical exploration easier and more exciting. Even without technology, you need to understand the order for performing operations to get accurate results.*

Mary plans to buy carpet for two rooms. One room is 12 feet ▮ and the other room is 20 feet by 14 feet. She uses her calculat▮ number of square feet.

$$12 \boxed{\times} 15 \boxed{+} 20 \boxed{\times} 14 \boxed{=} \qquad 2800▮$$

Mary knows that 2800 square feet is way too much. Her entir▮ ▮ But calculators don't

"Ever notice how students seem to enjoy talking about movies as much as going to them? The same kind of thing happened in my class when students described steps for simplifying computations. In this case, they discussed plans to lay carpet at school."

"The calculator really helped me to explain that we had more than enough carpet to do the job."

"Apply!"

"You won't find busy work in *HRW Algebra.* Students always understand the purpose and relevance of what they're doing, and I think that's the secret to success in motivating students.

My class recently used number cubes to assess the probability of two absent-minded students showing up on time to meet each other for lunch. I told them their weekend plans to earn some money for concert tickets depended on that lunch meeting."

Reduced from actual size

Exploration 2 *Probability From a Number Cube*

1 The faces on an ordinary number cube are numbered 1 to 6. If the cube is rolled 10 times, guess how many times a 5 will appear on the top of the cube.

2 Roll one cube 10 times. Count how many times you get a 5.

3 Define an *event* and a *trial* in this experiment.

4 What is the experimental probability of getting a 5 in 10 trials in this experiment? ❖

CRITICAL Thinking

Tell how to find the experimental probability for each of the following.

- 3 on one roll of a number cube
- *heads* on a toss of a coin

"After using the number cubes, we decided to make back-up plans since Al and Zita probably wouldn't show up on time."

Put it All Together With

"Students climb out of the zip lock™ bag!"

"Sometimes it seems as if algebra lives in two worlds. In the everyday world, it touches everything from personal finances to fun and games. But in textbooks, algebraic concepts are all too often isolated from one another, not to mention the real world itself.

***HRW Algebra** helps my students bring those worlds together with seamlessly connected concepts, activities, applications, technology, disciplines, and cultures. Students make smooth, intuitive transitions as they explore, communicate, and apply."*

"It's a good thing we had 30 minutes to land—anything less and the tables showed we would have nosedived to the ground!"

"Explore!"

"I like to take my students to the places where math happens. With HRW, that means climbing into the cockpit of an airplane at 24,000 feet and figuring out the relationships between altitude, time, and distance so they can land safely.

This challenge certainly motivated my students to see how the slope of the rate of descent relates to linear functions. We just get so much more out of class when concepts are connected to real-world challenges."

LESSON 5.2

Exploring Graphs of Linear Functions

Why *A linear function can be used to model the descent of an airplane. When you model situations using the graph of a linear function, you need to know the slope and the point at which the line crosses the y-axis.*

A12

SEAMLESS INTEGRATION!

"Communicate!"

Spreadsheet

A calculator or spreadsheet will help you construct a table to find what the investment is worth after 10 years. Start with \$10,000 and multiply repeatedly by 1.08.

Will Jill have enough tuition for four years at State U? ❖

	A	B
1	Year	Cmp. interest
2	0	10,000.00
3	1	10,800.00
4	2	11,664.00
5	3	12,597.12
6	4	13,604.89
7	5	14,693.28
8	6	15,868.74
9	7	17,138.24
10	8	18,509.30
11	9	19,990.05
12	10	21,589.25

An equation such as $y = 2^x$ is an example of an **exponential function**. The base, 2, is the number that is repeatedly multiplied. The exponent, x, represents the number of times that the base, 2, occurs in the multiplications.

Exploration 2 · Growth and Decay

1. Copy and complete the table for $y = 2^x$.

x	1	2	3	4	5
y	2	4	8	?	?

2. Graph the function $y = 2^x$ for the values in Step 1.

3. What do you notice about the graph you made for $y = 2^x$

4. Make a table of x and y values for $y = \left(\frac{1}{2}\right)^x$. Use the same ⟨ you used for Step 1. Then graph the function.

5. Compare the graphs of $y = 2^x$ and $y = \left(\frac{1}{2}\right)^x$. How are they

Lesson 2.1 Previewing Expon

"Sometimes the most important connection students make is with one another, but you've got to have an interesting and relevant context to make that a rewarding experience. HRW comes up with connections you can build on—like managing money and saving for college—something students really would and should talk about.

My class uses spreadsheet software to present graphs that chart the compounding interest of a college tuition fund. They compare the fund's exponential growth to escalating college expenses."

"The payoff on the fund was unbelievable! In fact, my business class discussed the same topic, and I was able to show them what a huge difference compounding interest makes."

"Apply!"

"Integration works best when you have applications that stand at the intersection of concepts, disciplines, technology, or cultures, but most importantly, student interest. HRW does exactly that when, for example, students apply their knowledge of experimental probability as they simulate a Native American game of chance."

Discovering Experimental Probability

Cultural Connection: Americas One American Indian game, *shaymahkewuybinegunug*, involves tossing sticks. It is played by members of the Ojibwa, or Chippewa, tribe. This game uses five flat sticks that have carved pictures of snakes on one side and are plain on the other. Players take turns tossing the sticks to earn points.

Exploration 1 · Probability and the Coin Game

Games To play an adaptation of the Ojibwa game, use three coins in place of the sticks. Each person tosses the three coins 10 times. Each time the three coins are the same, that person gets a point. After the person completes 10 tosses, the person with the most points is the winner.

or

CHAPTER 8

"Before tossing the coins, I made a big deal about how lucky I was. But when my results matched everyone else's, I discovered the big deal was statistics."

Integrated Instruction

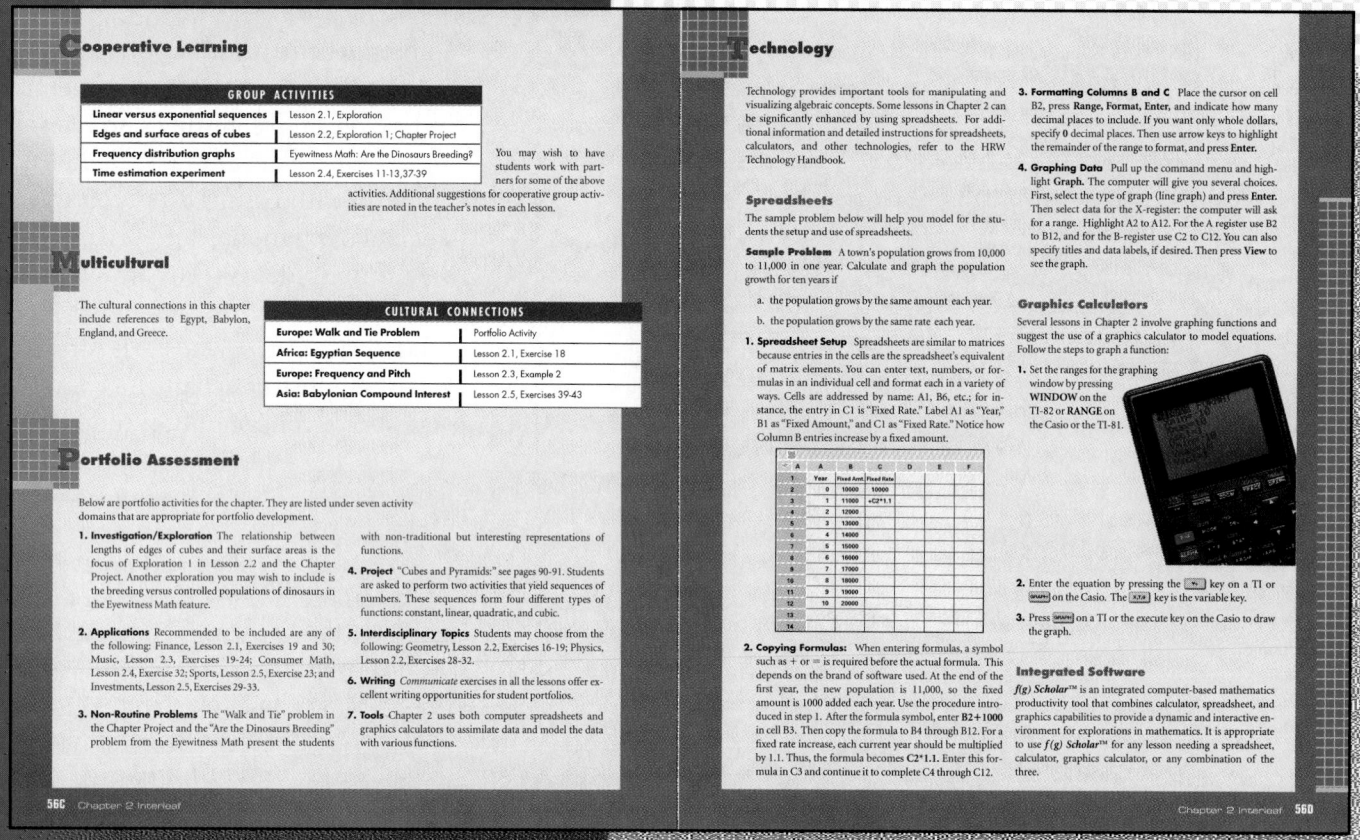

"Get full instructional support!"

"As a teacher, nothing is more important than the transfer of knowledge between me and my students. *HRW Algebra* gives teachers a full stream of instructional options and suggestions, so I can maintain a seamless relationship between me, the text, and my students.

HRW really supports instruction today rather than someone's vision of math instruction in the year 2010. I get useful information, like the *Alternative Teaching Strategies* you'll find in every lesson. Plus, I really appreciate HRW's easy-to-follow layout and organization."

"Before..."

"Before you begin a chapter, HRW's opening *Side Columns* on the *Chapter Openers* help you prepare to introduce new material by providing *Background Information*, *Chapter Objectives*, a list of available *Resources*, and more.

Then there's a series of *Interleaves*, which I find particularly useful. I read them before starting a chapter, then refer back to them whenever necessary. Interleaves include everything from a *Planning Guide* to *Reading*, *Visual*, and *Hands-on Strategies* for helping individual students.

You also get a quick-look reference to the chapter's *Cooperative Learning Activities*, *Cultural Connections*, and *Portfolio Assessment*, as well as a full page dedicated to *Technology* instruction."

begins with You!

Reduced from actual size

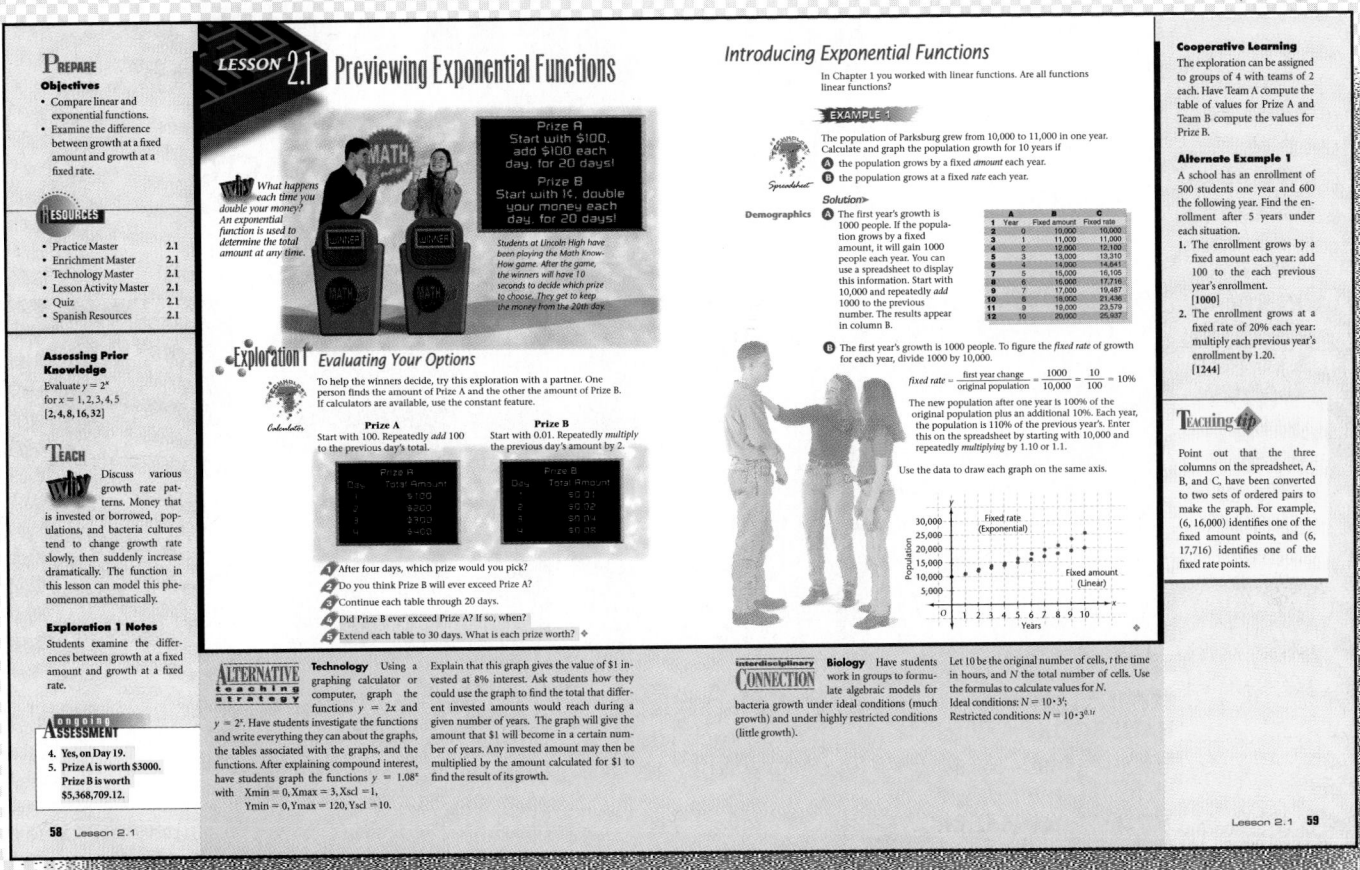

"During..."

"As you move through a lesson, you'll notice side column support is designed to lead you through content in a timely and logical progression.

Key sections include *PREPARE*, with *Objectives*, *Resources*, and *Prior Knowledge Assessment*, and *TEACH*, which features some substantial teaching strategies, notes, and tips, from *Critical Thinking* to *Alternative Teaching Strategies* to *Interdisciplinary Connections*."

"And After Your Lessons!"

"**HRW Algebra** gives you plenty of ways to measure the progress of your students with *Ongoing* and *Alternative Assessment*, and *Practice* and *Technology Masters*.

You'll also get lots of practical tips for the *Chapter Project*, which applies skills presented throughout the chapter. And finally, you'll find some great ideas for *Eyewitness Math*, a fun cooperative-learning activity that springs from today's headlines and stories."

Reform That's Based on What You Need!

HRW MATHEMATICS...
This *is* Progress

PRACTICE WORKBOOK
A full page of practice for each lesson helps students comprehend and review learned skills and concepts.

TECHNOLOGY HANDBOOK
Teacher's guide for using technology with ***HRW Mathematics.***

TEACHING RESOURCES
Support for each lesson in four convenient booklets.

TEACHING TRANSPARENCIES
Over 100 full-color visuals with suggested lesson plans for their use.

TEACHING TRANSPARENCIES DIRECTORY
A useful guide that makes it easy to view the quality and instructional value of each transparency.

TEST GENERATOR
A variety of assessment types delivered on user-friendly software. (Two versions available–Macintosh® and IBM®PC and compatibles.)

TEST GENERATOR: ASSESSMENT ITEM LISTING
Printout of all items included on the Test Generator.

SOLUTION KEY
Worked-out solutions for all *Exercises & Problems.*

SPANISH RESOURCES
Spanish translation of objectives, main ideas of lessons, and terminology.

LAB ACTIVITIES AND LONG-TERM PROJECTS
Hands-on activities and projects to be used before, during, or after each chapter.

EXPLORE • COMMUNICATE • APPLY

HRW ALGEBRA

TEACHER'S EDITION

explore

communicate

APPLY

Integrating

MATHEMATICS
TECHNOLOGY
EXPLORATIONS
APPLICATIONS
ASSESSMENT

HOLT, RINEHART AND WINSTON
Harcourt Brace & Company
Austin • New York • Orlando • Atlanta • San Francisco • Boston • Dallas • Toronto • London

A U T H O R S

James E. Schultz

Dr. Schultz is one of the math education community's most renowned mathematics educators and authors. He is especially well regarded for his inventive and skillful integration of mathematics and technology. He helped establish standards for mathematics instruction as a co-author of the NCTM "Curriculum and Evaluation Standards for Mathematics" and "A Core Curriculum, Making Mathematics Count for Everyone." Following over 25 years of successful experience teaching at the high school and college levels, his dynamic vision recently earned him the prestigious Robert L. Morton Mathematics Education Professorship at Ohio University.

Kathleen A. Hollowell

Dr. Hollowell is widely respected for her keen understanding of what takes place in the mathematics classroom. Her impressive credentials feature extensive experience as a high school mathematics and computer science teacher, making her particularly well-versed in the challenges associated with integrating math and technology. She currently serves as Associate Director of the Secondary Mathematics Inservice Program, Department of Mathematical Sciences, University of Delaware and is a past-president of the Delaware Council of Teachers of Mathematics.

Wade Ellis, Jr.

Professor Ellis has gained tremendous recognition for his reform-minded and visionary math publications. He has made invaluable contributions to teacher inservice training through a continual stream of hands-on workshops, practical tutorials, instructional videotapes, and a host of other insightful presentations, many focusing on how technology should be implemented in the classroom. He has been a member of the National Research Council's Mathematical Sciences Education Board, the MAA Committee on the Mathematical Education of teachers, and is a former Visiting Professor of Mathematics at West Point.

Printed in the United States of America
 3 4 5 6 7   041   00 99 98 97
ISBN: 0-03-097772-X

CONTRIBUTING AUTHOR

Irene "Sam" Jovell   An award winning teacher at Niskayuna High School, Niskayuna, New York, Ms. Jovell served on the writing team for the New York State Mathematics, Science, and Technology Framework. A popular presenter at state and national conferences, her workshops focus on technology-based innovative education.

PROGRAM CONCEPTUALIZER

Larry Hatfield   Dr. Hatfield is Department Head and Professor of Mathematics Education at the University of Georgia. He is a recipient of the Josiah T. Meigs Award for Excellence in Teaching, his university's highest recognition for teaching. He has served at the National Science Foundation and is Director of the NSF-funded Project LITMUS.

••

Editorial Director of Math
Richard Monnard

Executive Editor
Gary Standafer

Senior Editor
Ronald Huenerfauth

Project Editors
Stephen C. Johnson
Paula Jenniges

Design and Photo
Pun Nio
Diane Motz
Lori Male
Julie Ray
Rhonda Holcomb
Robin Bouvette
Katie Kwun
Paradigm Design
Mavournea Hay
Cindy Verheyden
Sam Dudgeon
Victoria Smith

Editorial Staff
Steve Oelenberger
Andrew Roberts
Joel Riemer
Ann Farrar
Editorial Permissions
Jane Gallion
Desktop Systems Operator
Jill Layson,
Department Secretary

Production
Donna Lewis
Amber Martin
Shirley Cantrell
Jenine Street

CONTENT CONSULTANT

Lloy Lizcano   Dr. Lizcano is a Mathematics Teacher and Department Chair at Bowie High School, Austin, Texas. She is also a Lecturer at the University of Texas at Austin, teaching Secondary Mathematics Methods and supervising student teachers. Her many areas of expertise include the integration of science and technology into the mathematics curriculum.

MULTICULTURAL CONSULTANT

Beatrice Lumpkin   A former high school teacher and associate professor of mathematics at Malcolm X College in Chicago, Professor Lumpkin is a consultant for many public schools for the enrichment of mathematics education through its multicultural connections. She served as a principal teacher-writer for the *Chicago Public Schools Algebra Framework* and has served as a contributing author to many other mathematics and science publications that include multicultural curriculum.

REVIEWERS

Toni Antonelli
St. Ignatius High School
Cleveland, Ohio

Judy B. Basara
St. Hubert High School
Philadelphia, Pennsylvania

Robert E. Bauman
Appleton East High School
Appleton, Wisconsin

Blanche Brownley
DC Public Schools
Washington, D.C.

Larry Clifford
Patterson Career Center
Dayton, Ohio

Carol Goehring
Dr. Phillips High School
Orlando, Florida

Michael Moloney
Snowden International High School
Boston, Massachusetts

Lisa Naxera
Westwood High School
Austin, Texas

Curt Perry
Plymouth-Salem High School
Canton, Michigan

Diane Young
Covington Middle School
Austin, Texas

TABLE OF CONTENTS

Technology T12
Exploration T14
Assessment T16
A Message From the Authors T19

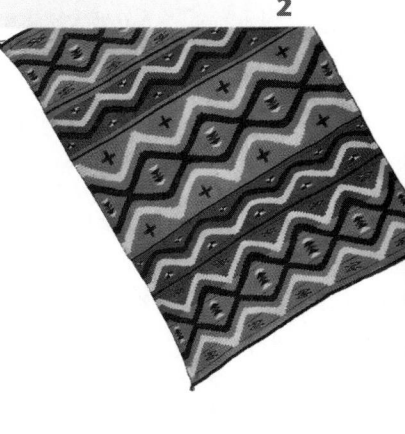

CHAPTER 1   **From Patterns to Algebra**   **2**

Chapter 1 Interleaf 2A-D
1.0   Mathematical Power 4
1.1   Representing Number Patterns 8
1.2   Using Differences to Identify Patterns 16
1.3   *Exploring* Variables and Equations 22
1.4   Order of Operations 28
1.5   Representing Linear Functions by Graphs 34
1.6   *Exploring* Scatter Plots and Correlation 40
1.7   Finding Lines of Best Fit 46
    Chapter Project   *Repeating Patterns* 50
    Chapter Review 52
    Chapter Assessment 55

CHAPTER 2   **A Preview of Functions**   **56**

Chapter 2 Interleaf 56A-D
2.1   Previewing Exponential Functions 58
2.2   *Exploring* Quadratic Functions 64
2.3   Previewing Reciprocal Functions 70
    Eyewitness Math   *Are the Dinosaurs Breeding?* 76
2.4   Previewing Other Functions 78
2.5   Identifying Types of Functions 84
    Chapter Project   *Cubes and Pyramids* 90
    Chapter Review 92
    Chapter Assessment 95

CUMULATIVE ASSESSMENT   **CHAPTERS 1–2** **96**

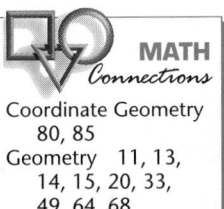

MATH Connections

Coordinate Geometry
80, 85
Geometry   11, 13,
14, 15, 20, 33,
49, 64, 68
Statistics   27, 33, 40,
44, 78

APPLICATIONS

Science
Ecology 24
Physics 18, 20, 66, 69,
84, 94
Physiology 37

Social Studies
Demographics 42, 59, 85
Geography 44
Music History 46

Psychological Experiment 83
Language Arts
Communicate 12, 19, 26,
32, 38, 43, 48, 62, 67,
74, 81, 87
Eyewitness Math 76

Business and Economics
Advertising 33
Cost of Living 43

Economics 49
Finance 60, 62, 63, 94
Fund-raising 27, 75
Investments 89

Life Skills
Consumer Economics 27,
89
Consumer Math 68, 82
Health 39

Home Economics 21,
82
Home Improvement 32

Sports and Leisure
Entertainment 27, 54
Music 74
Sports 45, 88
Other
Transportation 75, 89

Chapter 3 Interleaf 98A-D
3.1 Adding Integers 100
3.2 *Exploring* Integer Subtraction 107
3.3 Adding Expressions 112
3.4 Subtracting Expressions 118
3.5 Addition and Subtraction Equations 124
3.6 Solving Related Inequalities 131
Chapter Project *Find It Faster* 138
Chapter Review 140
Chapter Assessment 143

Chapter 4 Interleaf 144A-D
4.1 *Exploring* Integer Multiplication and Division 146
4.2 Multiplying and Dividing Expressions 151
Eyewitness Math *Is There Order in Chaos?* 158
4.3 Multiplication and Division Equations 160
4.4 Rational Numbers 167
4.5 Solving Problems Involving Percent 173
4.6 Writing and Solving Multistep Equations 180
4.7 *Exploring* Related Inequalities 186
4.8 Absolute Value Equalities and Inequalities 192
Chapter Project *Egyptian Equation Solving* 199
Chapter Review 200
Chapter Assessment 203

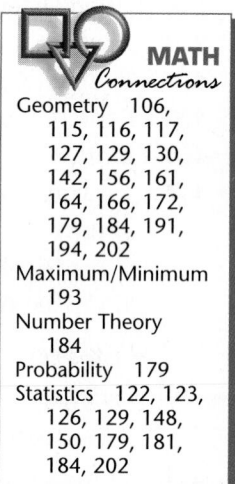

MATH *Connections*

Geometry 106, 115, 116, 117, 127, 129, 130, 142, 156, 161, 164, 166, 172, 179, 184, 191, 194, 202
Maximum/Minimum 193
Number Theory 184
Probability 179
Statistics 122, 123, 126, 129, 148, 150, 179, 181, 184, 202

CUMULATIVE ASSESSMENT **CHAPTERS 1–4** **204**

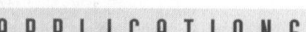

APPLICATIONS

Science
Physics 105, 130, 163
Temperature 102, 110, 136

Social Studies
Demographics 179
Government 198

Language Arts
Communicate 105, 110, 115, 121, 128, 135, 149, 155, 164, 171, 178, 184, 189, 197
Eyewitness Math 158

Business and Economics
Banking 150, 167

Business 106
Clothing Manufacturing 116
Discounts 176, 179
Finance 142
Fund-raising 122, 125, 172
Inventory 123
Investment 127, 168
Quality Control 192
Sales Tax 129, 177
Small Business 157
Wages 156

Life Skills
Carpentry 160, 165

Consumer Economics 172, 175, 185, 191
Health 197
Home Economics 129
Savings 171
Woodworking 195

Sports and Leisure
Ballet 189, 190
Hobbies 142
Picture Framing 116
Sports 102, 129, 136, 162, 171, 172, 202
Travel 129, 165, 185

CHAPTER **5** **Linear Functions** 206

Chapter 5 Interleaf 206A-D
5.1 Defining Slope 208
Eyewitness Math *Watch Your Step!* 216
5.2 *Exploring* Graphs of Linear Functions 218
5.3 The Slope-Intercept Formula 224
5.4 Other Forms for Equations of Lines 233
5.5 Vertical and Horizontal Lines 240
5.6 Parallel and Perpendicular Lines 246
Chapter Project *Diophantine Equations* 252
Chapter Review 254
Chapter Assessment 257

CHAPTER **6** **Systems of Equations and Inequalities** 258

Chapter 6 Interleaf 258A-D
6.1 Graphing Systems of Equations 260
6.2 *Exploring* Substitution Methods 267
6.3 The Elimination Method 272
6.4 Non-Unique Solutions 278
6.5 Graphing Linear Inequalities 284
6.6 Classic Applications 292
Chapter Project *Minimum Cost Maximum Profit* 298
Chapter Review 300
Chapter Assessment 303

CUMULATIVE ASSESSMENT **CHAPTERS 1-6** **304**

MATH
Connections

Coordinate Geometry
244, 261, 284
Geometry 213, 214,
232, 245, 247,
248, 250, 251,
270, 271, 276,
279, 282, 283,
289, 291, 302
Maximum/Minimum
288
Number Theory 283
Numeration 270,
294, 297
Statistics 211, 266

APPLICATIONS

Science
Chemistry 296, 302
Ecology 222
Physical Science 222, 232,
246
Physics 229
Science 209

Social Studies
Demographics 283
Social Studies 215, 239

Language Arts
Communicate 212, 221,
230, 237, 243, 250, 264,
270, 275, 281, 298, 296

Eyewitness Math 216
Language Arts 244

Business and Economics
Accounting 276
Equipment Rental 257
Finance 276, 302
Fund-raising 234, 238, 239,
271, 291
Income 276
Investment 268, 276
Money 293

Life Skills
Budget 244
Budgeting Time 291

Career Options 283
Consumer Mathematics 231

Sports and Leisure
Auto Racing 226
Recreation 213, 240, 245,
271, 296
Sports 214, 226, 266, 291
Travel 210, 256, 277, 294

Other
Age 295
Aviation 265, 270, 297
Mixture 292

CHAPTER **7** **Matrices** 306

Chapter 7 Interleaf 306A-D
7.1 Matrices to Display Data 308
7.2 Adding and Subtracting Matrices 315
     Eyewitness Math *Barely Enough Grizzlies?* 322
7.3 *Exploring* Matrix Multiplication 324
7.4 Multiplicative Inverse of a Matrix 331
7.5 Solving Matrix Equations 337
     Chapter Project *Secret Codes* 344
     Chapter Review 346
     Chapter Assessment 349

CHAPTER **8** **Probability and Statistics** 350

Chapter 8 Interleaf 350A-D
8.1 Experimental Probability 352
8.2 *Exploring* Simulations 360
     Eyewitness Math *Hot Hands or Hoop-la?* 366
8.3 Extending Statistics 368
8.4 *Exploring* The Addition Principle of Counting 375
8.5 Multiplication Principle of Counting 382
8.6 Theoretical Probability 388
8.7 Independent Events 394
     Chapter Project *Winning Ways* 400
     Chapter Review 402
     Chapter Assessment 405

CUMULATIVE ASSESSMENT **CHAPTERS 1–8** **406**

MATH *Connections*

Coordinate Geometry 336, 343, 365, 374, 387
Geometry 321, 330, 335, 342, 349, 370, 381, 384, 386, 387, 392, 394
Statistics 319, 321, 359, 378, 379, 380

APPLICATIONS

Science
Chemistry 343, 387
Meteorology 362, 372, 374
Science 336, 392
Space Exploration 313
Weather 364

Social Studies
Demographics 365
Geography 313, 392

Language Arts
Communicate 312, 318 328, 334, 341, 357, 364, 371, 379, 386, 391, 397
Eyewitness Math 322, 366
Language 387, 392

Business and Economics
Advertising 370, 373

Business 312, 348, 349
Fund Raising 325, 330
Inventory Control 316, 319, 324, 342, 384
Manufacturing 329
Marketing 404
Pricing 329
Sales Tax 329
Stocks 348
Work Schedule 336

Life Skills
Appointments 363, 394
Consumer Economics 337
Employment 343
Health 393
Job Opportunities 342
Membership Fees 359
Savings 381

Scheduling 382, 387, 399
Sports and Leisure
Contests 399
Games 325, 358, 378, 379, 385, 386, 390, 392 395, 396, 397, 398
Making Choices 388
Sports 320, 364, 368, 373, 381, 383
Ticket Sales 342
Travel 399

Other
Logic 376
Opinion Polls 380
Route Planning 398
Survey 391
Transportation 308, 313, 392

CHAPTER 9 Transformations

408

Chapter 9 Interleaf		408A-D
9.1	Functions and Relations	410
9.2	*Exploring* Transformations	418
9.3	Stretches	424
9.4	Reflections	431
9.5	Translations	436
9.6	Combining Transformations	443
	Chapter Project *Pick a Number*	448
	Chapter Review	450
	Chapter Assessment	453

CHAPTER 10 Exponents

454

Chapter 10 Interleaf		454A-D
10.1	*Exploring* Exponents	456
10.2	Multiplying and Dividing Monomials	463
10.3	Negative and Zero Exponents	469
	Eyewitness Math *All Mixed Up?*	474
10.4	Scientific Notation	476
10.5	Exponential Functions	483
10.6	Applications of Exponential Functions	489
	Chapter Project *Please Don't Sneeze*	496
	Chapter Review	498
	Chapter Assessment	501

CUMULATIVE ASSESSMENT **CHAPTERS 1–10** **502**

MATH Connections

Coordinate Geometry
462, 482
Geometry 420, 423,
435, 442, 452,
462, 467, 500
Maximum/Minimum
418
Number Theory 473
Probability 417, 430,
447, 462, 472,
491, 494, 500
Statistics 423, 430,
435, 439, 442,
462, 468, 482
Transformations 473

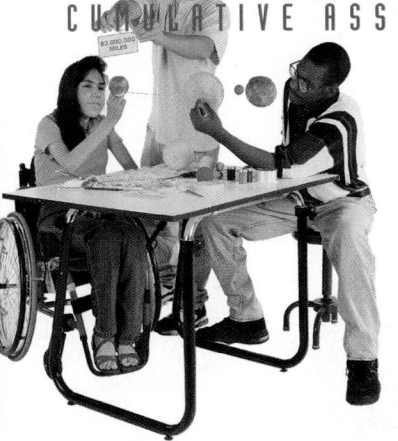

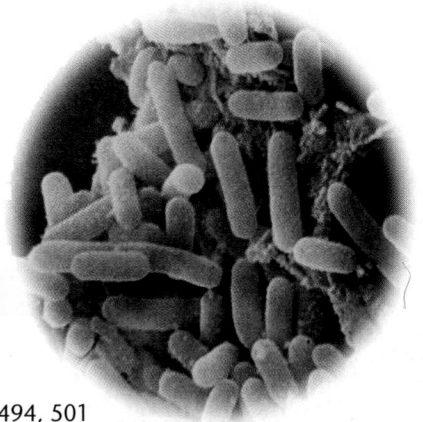

APPLICATIONS

Science
Astronomy 476, 481, 500
Bacteriology 459
Biology 461
Carbon Dating 494
Physics 435, 478
Science 417, 481, 489
Space Travel 477
Temperature 452

Social Studies
Demographics 488, 492, 494, 501

Social Studies 480, 483, 485
Taxes 501

Language Arts
Communicate 415, 421, 428
433, 440, 446, 460, 466,
471, 480, 487, 493
Eyewitness Math 474

Business and Economics
Business 452, 453
Economics 481
Finance 490

Packaging 447

Life Skills
Investment 491, 494, 501

Sports and Leisure
Art 429
Travel 443, 468

Other
Databases 416
Time 500

CHAPTER **11** **Polynomials and Factoring**   **504**

Chapter 11 Interleaf 504A-D
11.1 Adding and Subtracting Polynomials 506
11.2 *Exploring* Products and Factors 513
11.3 Multiplying Binomials 519
11.4 Common Factors 524
11.5 Factoring Special Polynomials 529
11.6 Factoring Trinomials 535
Chapter Project   *Powers, Pascal, and Probability* 540
Chapter Review 542
Chapter Assessment 545

CHAPTER **12** **Quadratic Functions**   **546**

Chapter 12 Interleaf 546A-D
12.1 *Exploring* Parabolas 548
12.2 Solving Equations of the Form $x^2 = k$ 552
12.3 Completing the Square 558
Eyewitness Math   *Rescue at 2000 feet* 564
12.4 Solving Equations of the form $x^2 + bx + c = 0$ 566
12.5 The Quadratic Formula 574
12.6 Graphing Quadratic Inequalities 581
Chapter Project   *What's the Difference?* 586
Chapter Review 588
Chapter Assessment 591

CUMULATIVE ASSESSMENT   **CHAPTERS 1–12** **592**

MATH *Connections*
Coordinate Geometry 518, 555
Geometry 509, 512, 515, 517, 518, 523, 526, 528, 529, 533, 534, 535, 544, 557, 563, 569, 572, 580
Maximum/Minimum 549, 560, 581
Probability 522, 528, 534, 563, 580
Statistics 534, 573

APPLICATIONS

Science
Biology 522
Optics 584
Physics 528, 551, 557, 562, 563, 573, 585
Language Arts
Communicate 510, 517, 522, 527, 532, 538, 550, 556, 561, 671, 579, 584
Eyewitness Math 564
Business and Economics
Accounting 580
Advertising 518
Discount Sales 518
Sales Tax 539
Small Business 521
Sports and Leisure
Crafts 544
Photography 572
Sports 590

CHAPTER 13 Radicals and Coordinate Geometry   594

Chapter 13 Interleaf 594A-D
13.1 *Exploring* Square Root Functions 596
13.2 Operations with Radicals 604
13.3 Solving Radical Equations 611
13.4 The "Pythagorean" Right-Triangle Theorem 619
13.5 The Distance Formula 625
13.6 *Exploring* Geometric Properties 632
13.6 Trigonometric Functions 638
    Chapter Project   *Radical Rectangles* 644
    Chapter Review 646
    Chapter Assessment 649

CHAPTER 14 Rational Functions   650

Chapter 14 Interleaf 650A-D
14.1 Rational Expressions 652
14.2 Inverse Variation 658
14.3 Simplifying Rational Expressions 664
14.4 Operations With Rational Expressions 670
    Eyewitness Math   *How Worried Should You Be?* 676
14.5 Solving Rational Equations 678
14.6 *Exploring* Proportions 684
14.7 Proof in Algebra 689
    Chapter Project   *A Different Dimension* 694
    Chapter Review 696
    Chapter Assessment 699

CUMULATIVE ASSESSMENT   CHAPTERS 1–14 700

MATH Connections

Coordinate Geometry
  626, 627, 628, 632
Geometry 601, 610,
  622, 623, 629,
  637, 643, 662,
  663, 667, 668,
  686, 688, 693, 698
Probability 669, 675
Statistics 693

APPLICATIONS

Science
Biology 668
Ecology 658, 674
Physics 611, 617, 618, 648, 660, 662

Social Studies
Politics 683

Language Arts
Communicate 601, 609, 616, 622, 629, 636, 641, 656, 662, 667, 673, 682 687, 692

Eyewitness Math 676

Business and Economics
Investment 661, 663

Life Skills
Auto Painting 683
Baking 688
Decorating 688
Fitness 698
House Construction 642
Landscaping 623
Savings 669
Surveying 634

Sports and Leisure
Boating Navigation 621
Model Airplanes 642
Music 663
Recreation 683
Sports 617, 623, 648, 678, 686
Travel 688, 698

Other
Emergency Service 625
Highways 642
Rescue Services 630
Transportation 656, 663, 688

Info Bank Table of Contents .... .703
Functions and Their Graphs .... .704
Table of Squares, Cubes, Square and Cube Roots ...... .708
Table of Random Digits .... .709
Glossary .... .710
Index .... .718
Credits .... .727
Additional Answers .... .729

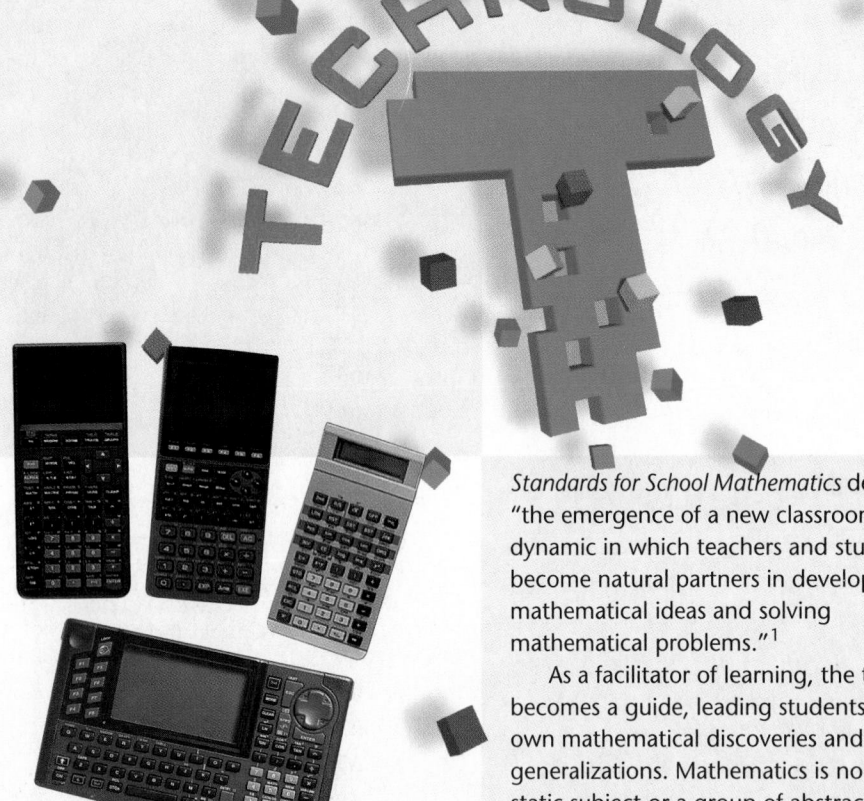

TECHNOLOGY

Interactive Learning

Students learn best if they are active participants in the learning process. Technology allows today's mathematics classroom to become a laboratory where students explore and experiment with mathematical concepts rather than just memorize isolated concepts. Students make generalizations and reach conclusions about mathematical concepts and relationships and apply them to real-life situations. *HRW Algebra* encourages instruction that utilizes technology with numerous explorations and examples. Where appropriate, students explore, work examples, solve exercises, make and test conjectures, and confirm mathematical ideas for themselves.

Role of the Teacher

Technology is changing the role of the teacher in the mathematics classroom. The National Council of Teachers of Mathematics' *Curriculum and Evaluation*

Standards for School Mathematics describes "the emergence of a new classroom dynamic in which teachers and students become natural partners in developing mathematical ideas and solving mathematical problems."[1]

As a facilitator of learning, the teacher becomes a guide, leading students to their own mathematical discoveries and generalizations. Mathematics is no longer a static subject or a group of abstract symbols. Mathematics becomes a dynamic field of related concepts that can be explored and experimented with in the same way that science concepts can be. The tedious arithmetic calculations that discouraged, if not prohibited, students from making statistical analysis of data are no longer a barrier between students and their understanding and application of real-life problems.

Technology allows students to quickly create visual representations of functions. Students concentrate on the effect of parameter changes on a function, rather than on calculating and plotting points. Students can more effectively explore and learn about the more complex algebraic concepts that require comparisons.

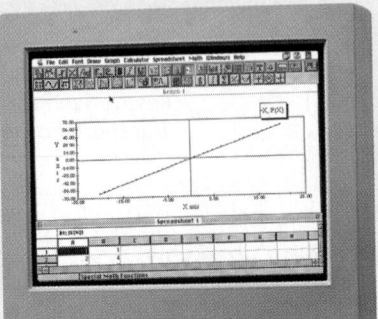

Changes in the Classroom

Teachers are finding that technology encourages cooperative learning. Cooperative-learning groups allow students to compare results, brainstorm, and reach conclusions based on group results. As in real life where scientists or financial analysts often consult each other, students in the technology-oriented classroom learn to communicate and consult with each other. Students learn that such consultations are not "cheating," but are rather a method of sharing information that will be used to solve a problem.

Another change in the classroom will be the role of the teacher from a dispenser of facts and formulas to a guide along the exploration trail. Teachers will become monitors who keep students headed in the right directions. As students gain confidence in their ability, they may ask more questions and even higher-level questions. Teachers no longer always need to have the "right" answer—teachers can suggest further exploration or research. In this way, teachers model real-life situations in which experts with the right answer are not always available or in which there is no right answer.

Real-World Applications

HRW Algebra integrates technology throughout the text. This use of technology allows students to explore real-world applications that would not be practical to explore without the use of such technology. Students are able to study realistic rather than contrived data. For instance, when calculating by hand, most students study only simple interest. The use of technology allows students to explore compound interest, which is the way that interest is usually calculated in the financial world today.

Real data about a student's local town or neighborhood can be used to find the average price or rental cost of a home. Students can study data that are more closely related to their own lives, such as the cost of a 4-year college degree based

[1]*National Council of Teachers of Mathematics. **Curriculum and Evaluation Standards for School Mathematics**. Reston, VA: The Council, 1989. (p. 128)*

on current costs and projections for the time students will actually be in college. The same can be done for projecting the cost of new cars when they graduate from high school.

	A	B
1	Year	Cmp. Interest
2	0	10,000.00
3	1	10,800.00
4	2	11,664.00
5	3	12,597.12
6	4	13,604.89
7	5	14,693.28
8	6	15,868.74
9	7	17,138.24
10	8	18,509.30
11	9	19,990.05
12	10	21,589.25

Exploratory Lessons

Exploratory lessons occur throughout *HRW Algebra*. Many of these exploratory lessons utilize the power of technology as a tool of exploration. For example, in Lesson 8.2 students explore simulations. Computers used as random number generators and spreadsheets are also utilized.

C11	= INT(RAND)()*100) + 1		
	A	B	C
1	Trial	1st Number	2nd Number
2	1	98	68
3	2	33	82
4	3	21	17
5	4	94	14
6	5	87	36
7	6	73	83
8	7	56	73
9	8	65	12
10	9	18	58
11	10	99	18

The use of technology in simulations allows students to concentrate on the simulation being studied rather than on the generation of random numbers.

Expository Lessons

The power of technology is often used to help students analyze functions that would be difficult or impossible to study without the use of technology. In Lesson 10.5,

students study the exponential graphs of $f(x) = 3x$ and $f(x) = 4x$.

In Lesson 12.3, a graphics calculator helps students study the maximum or minimum values of a parabola. In Lesson 12.4, calculators are also used to find the intersection of a line and a parabola.

Graphics Calculators

Graphics calculators perform many specialized and complex mathematical actions, including graphing. These calculators use certain keys to access full-screen menus that show additional operations. The scope of capabilities is wide, including traditional arithmetic calculations, matrix operations, statistics mode, function graphing, parametric graphing, polar graphing, sequence graphing, and more. Graphics calculators are capable of running user-defined programs. When using the graphing capabilities of the calculators, the use of a range-setting window adds greatly to the usefullness of the technology.

The trace and zoom functions allow students to find the coordinates of points on the curve and to enlarge sections of a graph.

Spreadsheets

A computer spreadsheet program is another valuable technological tool used in the mathematics classroom. By

utilizing a spreadsheet, students can enter, store, and sort the type of real-world data that would have been impractical in yesterday's classroom. Spreadsheets facilitate complex mathematics operations and data analysis. Students learn important mathematical concepts by entering formulas to generate simple or complex lists of data to solve problems. The spreadsheet shown is used to calculate the population growth of a small city for 10 years.

f(g) Scholar™

An effective, powerful yet easy to use tool in the classroom is *f(g) Scholar*™. This tool performs like a traditional graphics calculator, but it is used in conjunction with a personal computer. The advantages of *f(g) Scholar* are a larger screen for use with groups when individual calculators are not available and the ability to print out parameters, equations, and results of computations or graphs. This powerful tool has outstanding spreadsheet capabilities. The screen display offers the option of displaying the data from the calculator, the graph, and the spreadsheet all at once.

Explorations

Facilitator of Learning

An increasing number of educators are realizing that more learning takes place when students construct knowledge for themselves. This approach necessitates changes in both the form of instructional materials and the vehicles used to deliver instruction.

Instead of presenting students with rules, theorems, principles, and worked-out examples, this approach calls for students to be given questions to investigate, problems to explore, and conjectures to verify or disprove. A teacher is no longer the source of all information. Instead, the teacher acts as a facilitator, presenting the questions to explore and pointing to areas that need further discussion and clarification.

An activity-oriented approach gives teachers several advantages. Perhaps the most important of these is the ability to provide for learning styles other than verbal and visual. Students who need hands-on, tactile experiences—the kinesthetic learners—are no longer penalized by materials and instructional approaches that emphasize listening and recording information.

Students who need to work in collaboration with others to understand algebra concepts—the interpersonal learners—now have the opportunities that occur during small-group instruction. In fact, all students benefit from the increased written and verbal communication that occurs when active learning is implemented in the classroom.

So, what specific form should constructivist materials take in algebra? Although many variations are possible, materials suitable for active learning usually involve identification of patterns, making and testing conjectures, and exploring alternative approaches to problems. Students should be encouraged to try any approach that occurs to them as they work on a problem. Concrete models, graphic representations, and tabular methods such as those facilitated by spreadsheets all have an appropriate place in today's algebra classroom.

In many classroom situations, a key decision will be how much constructivism to use and when to use it. Some topics and situations will continue to require direct instruction; others will open opportunities for both students and teachers to experience the richer knowledge formation that results from students' active and personal involvement in their own learning.

New Tools

Throughout *HRW Algebra*, students are provided many opportunities for hands-on learning through the use of a wide variety of manipulatives. Probability simulations allow for the use of cards, number cubes, and marbles. Geometric concepts are enhanced by utilizing three-dimensional models. The use of algebra tiles enhances understanding of computation with integers as well as simplifying and factoring polynomials.

Rapid and continuing changes in technology provide teachers and students in the algebra classroom with more choices for strategies and procedures than in past years. Graphics calculators, spreadsheets, and computer programs are excellent tools for use in student explorations. These tools allow students to experiment with and explore many concepts that would otherwise be too tedious and time consuming.

Communicating In HRW Algebra

The exploratory approach in *HRW Algebra* provides many opportunities for students to communicate. While in a traditional classroom setting or in cooperative learning groups, students communicate orally as they complete

Real-World Applications

Teachers who implement activity-style learning can expect to see gradual improvement in long-term recall of information, clearer understanding of connections among concepts, more creative approaches to problem solving, and more sophisticated ways of using and communicating mathematical ideas.

In addition, teachers will be preparing students for future real-world problem-solving situations. Few, if any, of the problems and decisions that face people in their personal and professional lives come in tidy, textbook packaging. "Real" problems are vaguely defined and messy, with missing or unnecessary information, and require difficult compromises to be made. Students who have learned to tackle algebraic problems by actively exploring approaches, collaborating with others in group or team approaches, and comparing results to look for alternative strategies will be equipped to transfer these methodologies to personal and professional challenges later in their lives.

Activity-style instruction that makes frequent use of explorations and investigations gives students more than just a knowledge of algebraic concepts and skills. Careful and consistent implementation of these techniques will prepare all students to face and solve future challenges.

Explorations, answer Critical Thinking questions, and explain results of simulations and projects. Students communicate in writing as they complete written assignments, keep a journal, and build a portfolio of their work. Students communicate graphically as they make drawings to illustrate a concept or as they display data in a graph.

Active Learning in HRW Algebra

In the past, many algebra textbooks included little, if any, material appropriate for active-learning situations. Thus, teachers who realized a need for this type of approach had to search out or create special materials.

In marked contrast, *HRW Algebra* has been structured to include the frequent use of activity-style lessons and of activities. In lessons with the word Exploring, the entire instructional focus is centered around exploration and discovery. Also, many of the "non-exploration" lessons contain explorations that involve students in active-learning experiences. Most are appropriate for small-group work; many allow for use of alternative technologies; and all give students a chance to discover new ideas on their own.

As students begin a set of exercises at the end of a lesson, they may notice that they do not start with a set of simple-minded practice or drill problems. Instead, each exercise set starts with *Communicate*

questions. Here, students can talk or write about key ideas as a check on what they have learned from the lesson. Students may be asked to compare and contrast two different concepts, explain how they would solve a particular problem, or discuss why a particular answer is not reasonable for a given problem.

Students for whom active learning is a new approach may at first be confused by the way in which material is presented. They may have come to expect that they are simply to repeat whatever the teacher or the textbook has explained. But in *HRW Algebra* the textbook doesn't "give it all away." Students must explore and think a bit to get at the central concepts. After a few experiences, students will come to enjoy doing algebra in their own way, rather than being tied to someone else's thinking processes.

ASSESSMENT

Assessment Goals

An essential aspect of any learning environment, such as an algebra classroom, is the process of assessing or evaluating what students have learned. Informally this has been done using paper-and-pencil tests given by the teacher on a regular basis to measure students' performance against the material being studied. Formal evaluations using standardized tests are generally conducted over a period of years to establish performance records for both individuals and groups of students within a school or school district. Both types of tests are very good at measuring the ability of a student to use a particular mathematical skill or to recall a specific fact. They fall short, however, in evaluating other key goals of learning mathematics, such as being able to solve problems, to reason critically, to understand concepts, and to communicate mathematically, both verbally and in writing. Other techniques, usually referred to as alternative assessment, are needed to evaluate students' performance on these *process* goals of instruction.

The goals of an alternative assessment program are to provide a means of evaluating students' progress in non-skill areas of learning mathematics. Thus, the design and structure of alternative assessment techniques must be quite different from the skill-oriented, paper-and-pencil tests of the past.

Types of Alternative Assessment

In the world outside of school, a person's work is evaluated by what that person can do, that is, by the results the person achieves, and not by taking a test. For example, a musician may demonstrate skill by making music, a pilot by flying an air-plane, a writer by writing a book, and a surgeon by performing an operation. Students learn to think mathematically and to solve problems on a continuous basis over a long period of time as they study mathematics at many grade levels. Students, too, can demonstrate what they have learned and understand by collecting a representative sample of their best work in a **portfolio**. A portfolio should illustrate achievements in problem solving, critical thinking, writing mathematics, mathematical applications or connections, and any other activity that demonstrates an understanding of both concepts and skills.

Specific examples of the kinds of work that students can include in their portfolios are solutions to nonroutine problems, graphs, tables or charts, computer printouts, group reports or reports of individual research, simulations, and examples of artwork or models. Each entry should be dated and should be chosen to show the students' growth in mathematical competence and maturity.

A portfolio is just one way for students to demonstrate their performance on a mathematical task. Performance assessment can also be achieved in other ways, such as by asking students questions and evaluating their answers, by observing their work in cooperative learning groups, by having students give verbal presentations, by working on extended projects and investigations, and by keeping journals.

Peer assessment and self-evaluation are also valuable methods of assessing students' performance. Students should be able to critique their classmates' work and their own work against standards set by the teacher. In order to evaluate their own work, students need to know the teacher's goals of instruction and the criteria that have been established (scoring rubrics) for evaluating performance against the goals. Students can help to design their own self-assessment forms that they then fill out on a regular basis and give to the teacher. They can also help to construct test items that are incorporated into tests given to their classmates. This work is ideally done in small groups of four students. The teacher can then choose items from each group to construct the test for the entire class. Another alternative testing technique is to have students work on *take-home* tests that pose more open-ended and non routine questions and problems. Students can devote more time to such tests and, in so doing, demonstrate their understanding of concepts and skills and their ability to do mathematics independently.

Scoring

The use of alternative assessment techniques implies the need to have a set of standards against which students' work is judged. Numerical grades are no longer sufficient because growth in understanding and problem solving cannot be measured by a single number or letter grade. Instead, scoring rubrics or criteria can be devised that allow the teacher more flexibility to recognize and comment upon all aspects of a student's work, pointing out both strengths and weaknesses that need to be corrected.

A scoring rubric can be created for each major instructional goal, such as being able to solve problems or communicate mathematically. A rubric generally consists of four or five short descriptive paragraphs that can be used to evaluate a piece of work. For example, if a five-point paragraph scale is used, a rating of 5 may denote that the student has completed all aspects of the assignment and has a comprehensive understanding of problems. The content of paragraph 5 specifies the details of what constitutes the rating of highly satisfactory. On the other hand, a rating of 1 designates essentially an unsatisfactory performance, and paragraph 1 would detail what is unsatisfactory. The other three paragraphs provide an opportunity for the teacher to recognize significant accomplishments by the student and also aspects of the work that need improvement. Thus, scoring rubrics are a far more realistic and educationally substantive way to evaluate a student's performance than a single grade, which is usually determined by an answer being right or wrong.

The guide pictured, the Kentucky Mathematics Portfolio, was developed by the Kentucky Department of Education for use by school districts throughout that state. This guide is illustrated to show an example of an excellent and effective holistic scoring guide currently in use by math teachers who are practicing performance assessment in their classrooms. The scorer uses the Workspace/Annotations section of the guide to gather evidence about a student's mathematical ability. The top of the guide is then used to assign a single performance rating, based on an overall view of the full contents of the student's portfolio. The lower right-hand corner of the Holistic Scoring Guide lists the Types and Tools for the Breadth of Entries that are appropriate for a student to place in their portfolio. Within the interleaf pages that

Portfolio Holistic Scoring Guide

An individual portfolio is likely to be characterized by some, but not all, of the descriptors for a particular level. Therefore, the overall score should be the level at which the appropriate descriptors for a portfolio are clustered.

		NOVICE	APPRENTICE	PROFICIENT	DISTINGUISHED
PROBLEM SOLVING	Understanding/ Strategies	• Indicates a basic understanding of problems and uses strategies	• Indicates an understanding of problems and selects appropriate strategies	• Indicates a broad understanding of problems with alternate strategies	• Indicates a comprehensive understanding of problems with efficient, sophisticated strategies
	Execution/ Extensions	• Implements strategies with minor mathematical errors in the solution without observations or extensions	• Accurately implements strategies with solutions, with limited observations or extensions	• Accurately and efficiently implementes and analyzes strategies with correct solutions with extensions	• Accurately and efficiently implements and evaluates sophisticated strategies with correct solutions and includes analysis, justifications and extensions
REASONING		• Uses mathematical reasoning	• Uses appropriate mathematical reasoning	• Uses perceptive mathematical reasoning	• Uses perceptive, creative, and complex mathematical reasoning
MATHEMATICAL COMMUNICATION	Language	• Uses appropriate mathematical language some of the time	• Uses appropriate mathematical reasoning	• Uses precise and appropriate mathematical language most of the time	• Uses sophisticated, precise, and appropriate mathematical language
	Representations	• Uses few mathematical representations	• Uses a variety of mathematical representations accurately and appropriately	• Uses a wide variety of mathematical representations accurately and appropriately; uses multiple representations with some entries	• Uses a wide variety of mathematical representations accurately and appropriately uses multiple representations within entries and states their connections
UNDERSTANDING/ CONNECTING CORE CONCEPTS		• Indicates a basic understanding of core concepts	• Indicates an understanding of core concepts with limited connections	• Indicates a broad understanding of some core concepts with connections	• Indicates a comprehensive understanding of core concepts with connections throughout
TYPES AND TOOLS		• Includes few types; uses few tools	• Includes a variety of types; uses tools appropriately	• Includes a wide variety of types; uses a wide variety of tools appropriately	• Includes all types; uses a wide variety of tools appropriately and insightfully

PERFORMANCE DESCRIPTORS

PROBLEM SOLVING
- Understands the features of a problem (understands the question, restates the problem in own words)
- Explores (draws a diagram, constructs a model and/orchart, records data, looks for patterns)
- Selects an appropriate strategy (guesses and checks, makes an exhaustive list, solves a simpler but similar problem, works backward, estimates a solution)
- Solves (implements a strategy with an accurate solution)
- Reviews, revises, and extends (verifies, explores, analyzes, evaluates strategies/ solutions; formulates a rule)

REASONING
- Oberves data, records and recognizes pattern, makes mathematical conjectures (inductive reasoning)
- Validates mathematical conjectures through logical arguments or counter-examples; constructs valid arguments (deductive reasoning)

MATHEMATICAL COMMUNICATION
- Provides quality explanations and expresses concepts, ideas, and reflections clearly
- Uses appropriate mathematical notation and terminology
- Provides various mathematical representations (models, graphs, charts, diagrams, words, pictures, numerals, symbols, equations)

UNDERSTANDING/CONNECTING CORE CONCEPTS
- Demonstrates an understanding of core concepts
- Recognizes, makes, or applies the connections among the mathematical core concepts to other disciplines, and to the real world

WORKSPACE/ANNOTATIONS

PORTFOLIO CONTENTS
- Table of Contents
- Letter to Reviewer
- 5-7 Best Entries

BREADTH OF ENTRIES
TYPES
O INVESTIGATIONS/DISCOVERY
O APPLICATIONS
O NON-ROUTINE PROBLEMS
O PROJECTS
O INTERDISCIPLINARY
O WRITING

TOOLS
O CALCULATORS
O COMPUTER AND OTHER TECHNOLOGY
O MODELS MANIPULATIVES
O MEASUREMENT INSTRUMENTS
O OTHERS
O GROUP ENTRY

Place an X on each continuum to indicate the degree of understanding demonstrated for each core concept.

DEGREE OF UNDERSTANDING OF CORE CONCEPTS

Basic

NUMBER	
MATHEMATICAL PROCEDURES	
SPACE & DIMENSIONALITY	
MEASUREMENT	
CHANGE	
MATHEMATICAL STRUCTURE	
DATA: STATISTICS AND PROBABILITY	

precede each chapter of this Annotated Teachers Edition is a list of seven activity domains that correspond to the Breadth of Entries. Each of these activity domains is correlated to specific examples of activities in the pupil book that are appropriate for portfolio development.

Assessment and *HRW Algebra*

HRW Algebra provides many opportunities to employ alternative assessment techniques to evaluate students' performance. These opportunities are an integral part of the textbook itself and can be found in the Explorations, Try This, the interactive questions (which are highlighted), Critical Thinking questions, the Chapter Projects, and in the exercise and problem sets.

Throughout the textbook, students are asked to explain their work; describe what they are doing; compare and contrast different approaches; analyze a problem; make sketches, graphs, tables, and other models; hypothesize, conjecture, and look for counterexamples; and make and prove generalizations.

All of these activities, including the more traditional responses to routine prob-

lems, provide the teacher with a wealth of assessment opportunities to see how well students are progressing in their understanding and knowledge of algebra. The assessment task can be aligned with the major process goals of instruction and scoring rubrics established for each goal. For example, a teacher may decide to organize his or her assessment tasks in the following general areas of doing mathematics.

- Problem solving
- Reasoning
- Communication
- Connections

Within each of these areas, specific goals can be written and shared with students. In this way, the assessment process becomes an integral part, not only of evaluating students' progress, but also of the instructional process itself. The results of assessment can and should be used to modify the instructional approach to enhance learning for all students.

In addition to the many opportunities for performance assessment found in *HRW Algebra*, a variety of assessment types are integrated into the chapter-end material. The Chapter Review, the Chapter Assessment, and the Cumulative Assessment include both traditional and alternative assessment. All Chapter Tests include both

multiple choice and open-ended type questions. The Cumulative Reviews are formatted in the style of college preparatory exams. In addition to multiple-choice and free response questions, each Cumulative Assessment contains quantitative comparison questions which emphasize concepts of equalities, inequalities, and estimation. Other types of college entrance exam questions found in the Cumulative Review are student-produced response questions with gridded solutions. These Cumulative Assessments expose students to the new types of assessment that they will encounter when they take the latest form of college entrance examinations.

Eyewitness Math

Special two-page features called Eyewitness Math appear in almost every other chapter. These feature pages provide students with opportunities to read about current developments in mathematics and to solve real-life problems by working together in cooperative groups. Students' performance on Eyewitness Math can be assessed through group reports in writing or orally.

	Column A	Column B	Answers	
7.	$|-4.8|$	$|3.2|$	Ⓐ Ⓑ Ⓒ Ⓓ	[Lesson 6.2]
8.	INT(5.8)	INT(3.1)	Ⓐ Ⓑ Ⓒ Ⓓ	[Lesson 6.3]
9.	Perimeter	Area	Ⓐ Ⓑ Ⓒ Ⓓ	[Lesson 6.4]
10.	Play tickets: \$7 each for balcony seats, \$9 each for main floor seats. 9 balcony tickets	7 main-floor tickets	Ⓐ Ⓑ Ⓒ Ⓓ	[Lesson 6.5]

For row 9: $c \quad c = 4$, $c + 8$

A Message from the Authors

It is certainly a challenge in our rapidly changing world for textbooks to capture the essence of what students need and teachers can provide. Frequent visits to schools confirm that mathematics programs often continue to focus on skills - many which are diminishing in importance - even though there is an increasing need for students to be able to understand and apply concepts to solve problems in real-world settings using appropriate technology. To make matters worse, limited school budgets make it difficult to implement desired changes while teachers are faced with significant challenges which compete for their time and energy.

The authors are dedicated to the idea that mathematics programs should help all students gain mathematical power in a technological society. Based on careful examination of current recommendations and school mathematics programs, the authors have developed a program which strikes a balance in maintaining the strengths of former approaches while moving to mathematics content and methods of learning which are up-to-date and relevant to the present and future lives of students. In education, like so many other areas, even well-intended change should not be so rapid that students, teachers, and parents cannot cope with it. This program makes carefully chosen strides in the most vital areas, while staying within the comfort zones of students, teachers, and parents.

Our textbooks reflect a vision of mathematics instruction which includes three components:
- active students
- solving interesting and relevant problems
- using appropriate technology

For, example, in these books students use readily available technology (ordinary calculators). Thus, important problems that are traditionally attempted by 40% of high school juniors using advanced techniques, are now accessible to almost 100% of high school freshmen using simple, inexpensive technology. This earlier study of important topics makes mathematics more interesting and more relevant.

This program successfully solves a long-standing dilemma: textbooks don't feature the use of technology because schools don't have the equipment. And, schools don't have the equipment because it's not used in the textbooks. Thus, we have seen too many textbooks that do not reflect the needs of the students. In this series, technology is highly profiled, but only a limited amount is required. Courses that are strongly rooted in mathematical content can be enhanced by including the technology as it becomes available.

We wish you well in pursuing this timely, balanced approach!

1 From Patterns to Algebra

Meeting Individual Needs

1.1 Representing Number Patterns

Core Resources

Practice Master 1.1
Enrichment p. 10
Technology Master 1.1
Interdisciplinary
   Connection p. 9

[1 day]

Core Two-Year Resources

Inclusion Strategies p. 10
Reteaching the Lesson p. 11
Practice Master 1.1
Enrichment Master 1.1
Technology Master 1.1
Lesson Activity Master 1.1

[2 days]

1.4 Order of Operations

Core Resources

Practice Master 1.4
Enrichment p. 30
Technology Master 1.4
Mid-Chapter Assessment
   Master

[1 day]

Core Two-Year Resources

Inclusion Strategies p. 30
Reteaching the Lesson p. 31
Practice Master 1.4
Enrichment Master 1.4
Lesson Activity Master 2.4
Technology Master 1.4

[2 days]

1.2 Using Differences to Identify Patterns

Core Resources

Practice Master 1.2
Enrichment p 17
Technology Master 1.2

[1 day]

Core Two-Year Resources

Inclusion Strategies p. 17
Reteaching the Lesson p. 18
Practice Master 1.2
Enrichment Master p. 1.2
Technology Master 1.2
Lesson Activity Master 1.2

[2 days]

1.5 Representing Linear Functions by Graphs

Core Resources

Practice Master 1.5
Enrichment p. 36
Technology Master 1.5
Interdisciplinary
   Connection p. 35

[1 day]

Core Two-Year Resources

Inclusion Strategies p. 36
Reteaching the Lesson p. 37
Lesson Activity Master 1.5
Practice Master 1.5
Enrichment Master 1.5
Technology Master 1.5

[3 days]

1.3 Exploring Variables and Equations

Core Resources

Practice Master 1.3
Enrichment p. 24
Technology Master 1.3
Interdisciplinary
   Connection p. 23

[2 days]

Core Two-Year Resources

Inclusion Strategies p. 24
Reteaching the Lesson p. 25
Practice Master 1.3
Enrichment Master 1.3
Lesson Activity Master 1.3
Technology Master 1.3

[3 days]

1.6 Exploring Scatter Plots and Correlation

Core Resources

Practice Master 1.6
Enrichment Master 1.6
Technology Master 1.5
Interdisciplinary
   Connection p. 41

[1 day]

Core Two-Year Resources

Inclusion Strategies p. 42
Reteaching the Lesson p. 43
Practice Master 1.6
Enrichment p. 42
Lesson Activity Master 1.6
Technology Master 1.6

[2 days]

1.7 Finding Lines of Best Fit

Core Resources	Core Two-Year Resources
Practice Master 1.7	Inclusion Strategies p. 47
Enrichment p. 47	Reteaching the Lesson p. 48
Technology Master 1.7	Lesson Activity Master 1.7
	Practice Master 1.7
	Enrichment Master 1.7
	Technology Master 1.7
[1 day]	**[2 days]**

Chapter Summary

Core Resources	Core Two-Year Resources
Chapter 1 Project, pp. 50–51	Chapter 1 Project, pp. 50–51
Lab Activity	Lab Activity
Long-Term Project	Long-Term Project
Chapter Review, pp. 52–54	Chapter Review, pp. 52–54
Chapter Assessment p. 55	Chapter Assessment, p. 55
Chapter Assessment, A/B	Chapter Assessment, A/B
Alternative Assessment	Alternative Assessment
[3 days]	**[5 days]**

Reading Strategies

Having students keep a journal is a good way to help them assimilate new concepts. Have them respond to the following in their journals.

What is the goal of this lesson? What should you expect to learn? List or describe unfamiliar words, phrases, equations, or pictures. What do you know about these items?

Visual Strategies

As an introduction, have students spend time looking through their books. Ask students about the cover design and how the design reflects their expectations for this class. Ask them to point out different pictures in the book that interest them and to speculate about the contents of that chapter or lesson. Students can then take a closer look at Chapter 1, paying close attention to the pictures of patterns in the chapter opener. Ask students to look for patterns in the classroom and to identify numerical patterns or sequences that are familiar to them. Counting by multiples of five is one example. Explain that number patterns and sequences are a very important part of algebra, and working with patterns is the basis for Chapter One.

Hands-on Strategies

Attribute blocks come in varied shapes, colors, sizes, and thicknesses. They can be purchased, or students can make them in class using templates, scissors, different colored construction paper, glue, and cardboard of various thicknesses. One way to make attribute blocks of varying thickness is to glue two shapes of identical size and color onto opposite sides of an identically shaped piece of cardboard. Have the students construct several blocks in which no two are identical. Students can also complete sequences using attribute blocks. Have students complete sequences as they construct the attribute blocks. First, have them make conjectures about what it will take to complete a given sequence. Then have them describe the sequence in words.

Cooperative Learning

GROUP ACTIVITIES	
A Diminishing Sequence	Lesson 1.3, Exploration 2
Find a Correlation	Lesson 1.6, Exploration
Fitting a Line to a Set of Data	Lesson 1.7, Line of Best Fit
Repeating patterns	Chapter 1 Project

You may wish to have students work with partners for some of the above activities. Additional suggestions for cooperative group activities are noted in the teacher's notes in each lesson.

Multicultural

The cultural connections in this chapter include references to Africa and Europe.

CULTURAL CONNECTIONS	
Europe: Karl Fredrich Gauss **Rene Descartes**	Lesson 1.1, Exercises 25–27 Lesson 1.5
Africa: Square and triangular numbers	Lesson 1.3, Exercises 35–36

Portfolio Assessment

Below are portfolio activities for the chapter listed under seven activity domains which are appropriate for portfolio development.

1. **Investigation/Exploration** Two projects listed in the teacher notes ask students to investigate a relationship, compile data into a scatter plot, and analyze research data. They are the *Enrichment* activity on page 42 (Lesson 1.6) and the *Cooperative Learning* activity on page 47 (Lesson 1.7).

2. **Applications** Recommended to be included are any of the following: Home Economics (Lesson 1.2), Home Improvement (Lesson 1.4), and from the teacher notes *Enrichment* page 42 (Lesson 1.6), and *Alternative Assessment* p. 44 (Lesson 1.6).

3. **Non-Routine Problems** The Portfolio Activity on page 3 asks students to construct and extend patterns. This ties in with the Chapter Project on page 50.

4. **Project** The Chapter Project on page 50 asks students to create number patterns in the three Activities. The teacher notes includes an extension activity in which students may turn number patterns into picture patterns and draw these patterns on posterboard.

5. **Interdisciplinary Topics** Students may choose from the following: Physics (Lesson 1.2), Geometry (Lessons 1.1, 1.2), Consumer Economics (Lesson 1.3); and from the teacher notes, Biology on pages 9 and 35 (Lessons 1.1, 1.5), Aeronautics on page 23 (Lesson 1.3), Social Science on page 41 (Lesson 1.6)

6. **Writing** The interleaf material at the beginning of this chapter has information for journal writing. Also the Communicate questions at the beginning of the exercise sets can be used for writing assignments.

7. **Tools** Chapter 1 uses graphics calculators or graph paper for making scatter plots and finding the line of best fit. The technology masters are recommended to measure the individual student's proficiency.

Technology

Technology provides important tools for manipulating and visualizing algebraic concepts. Several lessons in Chapter 1 can be significantly enhanced by using appropriate technology. Scientific calculators make arithmetic calculations less tedious and help students to focus on the underlying algebraic concepts. Graphics calculators model functions visually, and students save the time it takes to create graphs by hand. For more information on the use of technology, refer to the *HRW Technology Handbook*.

Graphics Calculator

Graphing Linear Equations Several lessons in Chapter One and throughout the book suggest the use of a graphing calculator to model equations. Follow the steps to graph a linear equation.

1. The first step should be to adjust the ranges for the *x*- and *y*-axes in the graphing window. To make these adjustments, press the **WINDOW** key (TI-82) or the **RANGE** key (Casio, TI-81). Then enter the appropriate ranges for the graphing window using the arrows and number keys. Some automatic range configurations are available from the **ZOOM** menu. When you finish, press **ENTER** or **EXE**.

2. To enter the function, press the **Y=** key(TI) or the **GRAPH** key (Casio). Use the number and variable keys to enter the functions to be graphed.

3. Press **GRAPH** (TI) or **EXE** (Casio) to draw the graph.

Scatter Plots and Correlations

Lesson 1.7 suggests the use of a graphing calculator to show scatter plots and to determine lines of best fit. Follow the steps to create a scatter plot. The instructions listed below apply to the TI-82. For instructions

for using other calculators, please refer to the *Technology Manual for High School Mathematics*.

1. First clear the **Y=** function register, otherwise unwanted graphs may appear with your scatter plot. Select **STAT PLOT** to get the chart/graph function. This is the **2nd** and **Y=** key. Use the arrow keys and select **1:Plot 1** from the menu that appears on the screen. Then press **ENTER**.

2. With the arrow keys, move the cursor to **On** and press **ENTER**. Then, to choose the type of graph you want to make. Choose the scatter plot icon next to the word **Type**. The other icons are for line graphs, box plots, and histograms. **Xlist** and **Ylist** are the places where the (x,y) data will be stored. Choose **L1** for the **XList** and **L2** for the **Ylist**. **Xlist** and **Ylist** data must be stored in different locations.

3. Press **STAT** and select the **EDIT** mode at the top of the menu. You may now enter data for the *x*- and *y*-variables. Type in each number and press **ENTER**. When finished entering the *x*-variables, tab over to the next column using the arrow keys. The delete key **DEL** is used to correct mistakes.

4. When finished, press **STAT** again. This time select the **CALC** mode. When the new menu appears, select **5: LinReg(ax+b)** and press **ENTER**. When **LinReg(ax+b)** appears again, press **ENTER** to display the regression statistics. These include the *a*-value, *b*-value, and the correlation, *r*.

5. Press **GRAPH** to display graph. Press **CLEAR** to return to the statistics window.

Integrated Software

f(g) Scholar™ is an integrated computer-based mathematics productivity tool that combines calculator, spreadsheet, and graphics capabilities. It provides a dynamic and interactive environment for explorations in mathematics. It is appropriate to use *f(g) Scholar*™ for any lesson needing a spreadsheet, calculator, graphics calculator, or any combination of the three.

CHAPTER 1

From Patterns to Algebra

LESSONS

1.0 Making Connections in Algebra

1.1 Representing Number Patterns

1.2 Using Differences to Identify Patterns

1.3 *Exploring* Variables and Equations

1.4 Order of Operations

1.5 Representing Linear Functions by Graphs

1.6 *Exploring* Scatter Plots and Correlation

1.7 Finding Lines of Best Fit

Chapter Project
Repeating Patterns

ABOUT THE CHAPTER

Background Information

Patterns appear everywhere in human cultures and in nature. A wildflower field guide is one place to look for patterns that appear in nature.

CHAPTER RESOURCES

- Practice Masters
- Enrichment Masters
- Technology Masters
- Lesson Activity Masters
- Lab Activity Masters
- Long-Term Project Masters
- Assessment Masters
  Chapter Assessments, A/B
  Mid-Chapter Assessment
  Alternative Assessments, A/B
- Teaching Transparencies
- Spanish Resources

CHAPTER OBJECTIVES

- Find the pattern in a sequence.
- Generate and display sequences.
- Find and extend the pattern of a sequence using the difference method.
- Learn the basic concepts of variables, expressions, and equations.
- Evaluate an expression using substitution.
- Understand the order of operations and signs of inclusion for algebraic computation.
- Perform calculations in the proper order.

Throughout history, people from all parts of the world have been fascinated with patterns. The early native tribes of the American Southwest were aware of the patterns that influenced their lives. They recorded these patterns by weaving them into rugs.

In the technological world of modern society, patterns provide a basis for discoveries in science and engineering. Scientists use patterns to study, understand, and predict nature. Mathematicians look for regular patterns when investigating systems of numbers. Patterns can provide a powerful tool for solving problems.

Yearly Changes in Fish Population

(graph: Fish Population vs. Years)

In biology, a mathematical pattern can describe the changes in a region's fish population. The graph has a striking resemblance to the pattern shown on the Native American blanket.

ABOUT THE PHOTOS

The photos show similar mathematical patterns presented in a variety of ways. The pattern in the graph looks very much like the pattern in the blanket. Have students look for patterns in the classroom. Ask them how many varieties of the same pattern they can find.

- Identify and graph ordered pairs.
- Use a graph to find a solution.
- Understand the concept of a linear function.
- Interpret data in a scatter plot.
- Identify a correlation based on the orientation of points in a scatter plot.
- Find the line of best fit in a scatter plot.
- Understand and determine the closeness of fit.

The portfolio represents a profile of the student's work throughout the course in algebra. It should be the best that a student can do, but it can also show developmental progress. Portfolio Activities can provide a start, but more should be included. The reviewer should be able to see whether the student is willing to take risks, persevere to complete difficult problems, try various approaches, and continue beyond the answer with *What-if* questions or additional research.

PORTFOLIO ACTIVITY

Assign the Portfolio Activity as part of the assignment for Lesson 1.2 on page 21.

Students should be familiar with extending numerical sequences using a set rule. Here students will determine the rule for extending the visual sequence.

Students can determine the placement of the lines in the pattern using toothpicks. Have students explain how they arrived at their answers.

With toothpicks, have students see how far they can extend the pattern. Point out that many patterns may be extended infinitely.

Patterns are used in the structure and design of modern cities and the transportation networks that connect them. This student is using a computer to design a trestle bridge.

PORTFOLIO ACTIVITY

1. Construct the fourth and fifth dot patterns for this sequence of peaks. The fourth dot pattern should use 25 dots.

1   5   13

2. How many dots are in the pattern with 7 peaks?

ABOUT THE CHAPTER PROJECT

The Chapter 1 Project on pages 50–51, involves constructing picture patterns from numerical sequences. It would be helpful to have pictures of well-known mathematical patterns available, such as Pascal's Triangle. Books or documentaries on patterns in ancient art and architecture may help motivate students to create their own patterns.

PREPARE

Lesson 1.0 is a unique introductory lesson. It does not follow the same format as the other lessons in the text. This lesson does not contain instructional examples. It does, however, contain a set of exercises to be worked out by students in small groups.

One of the main objectives of this lesson is to introduce students to the many ways that mathematics is used in the real world. Another objective is to give students the opportunity to work in a cooperative group to explore a mathematical concept from a real-world setting.

LESSON 1.0 Mathematical Power

why *As you explore and investigate ideas from Algebra you will see how the power of mathematics is used to solve problems that occur in the everyday world. Put this power to work, and you will be successful in solving many problems.*

As you explore patterns and examine data in tables and graphs, you will learn to solve real problems for practical reasons and sometimes just for fun. Applying mathematics to the real world means interacting with science, economics, statistics, business, sports, entertainment, and music. As you explore mathematical concepts in the context of real world applications, math will seem more relevant and will answer the question, "Why do I have to learn this?"

As you explore, you will have the opportunity to see how technology can help you understand mathematics and process real world data. This technology takes the form of scientific calculators, graphics calculators, interactive dynamic graphics programs, and spreadsheets.

Algebra in Marketing

Statistics and probability are used to determine buying patterns, project future sales, determine packaging size and style, and create new products.

Algebra in Science

Mathematics has been referred to as the language of science. Whether it be in the field of biology, chemistry, physics, earth science, or ecology, scientists cannot investigate or represent their findings without understanding and using mathematics.

Advertising Spending

Category	Total ad spending (in billions)
Automotive	$5.9
Food	$3.5
Entertainment	$3.1
Travel & Hotels	$2.2
Snacks	$1.2

Algebra in Medicine

A person's heart rate is a function of many physiological factors such as age, diet, and heredity. A medical professional must understand these relationships to perform his or her job effectively.

The most often used expression heard from students that are new to algebra is, "Why do I need to learn this?" This lesson can be used to give your students an overview of how they will study algebra this year. Point out that each lesson will begin with a *Why* statement, and this statement usually reflects the utility of the mathematics presented in the lesson.

As the students proceed through the maze, discuss how the concepts learned in algebra are necessary or useful in a career that they might choose.

Ask students to work in groups of three of four to complete the exploration of page 7. This exploration is representative of how *HRW Algebra* presents mathematical concepts as patterns with numerical, graphical, and abstract representations.

Algebra in Mass Media

Today's information comes to us in many forms—newspapers, television, computer networks. Data is collected and stored in microseconds, making it possible to understand and predict in ways we never thought possible only a few years ago. As you explore mathematical concepts and relationships, you will learn how mathematics helps you make sense of real-world data.

Exploring with Algebra

As you explore, you may use various tools such as algebra tiles, integer squares, calculators, and other manipulative devices. As you use tools, make tables, and draw graphs, you will be given instructions to construct models. These models define and represent mathematical concepts you will use now and in the future. You will build your own mathematical power.

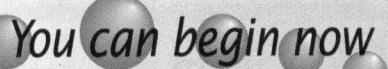
You can begin now . . . . .

Exploration — Graphing Changes

1 The table shows the distances covered by a mountain-bike racer on a level but rough road. Assuming a constant rate of speed, compute the values missing from the table.

Time	0	0.5	1.0	1.5	2.0	2.5	3.0
Distance	0	10	20	30	?	?	?

2 Write a sentence that tells how to find the distance if you know the time and rate of speed.

3 If the mountain bike-rider maintains the same speed, how far will the rider travel in 5 hours?

4 A graph is another way to show the relationship between time and distance. It can also be used to make predictions. Use the graph to predict how long it will take to cycle 100 miles.

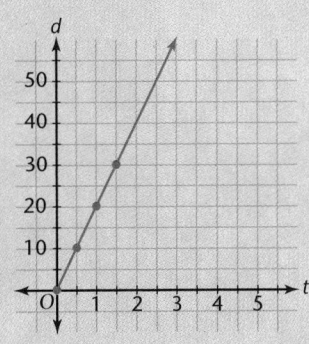

The graph in 4 describes the cyclist's movement traveling at a constant rate. Examine the graphs below.

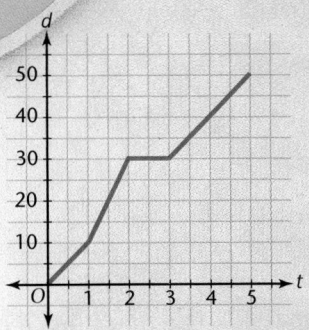
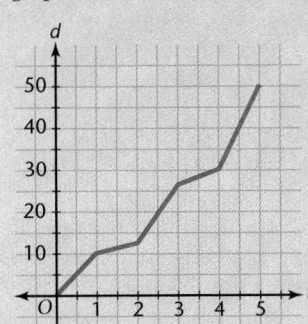
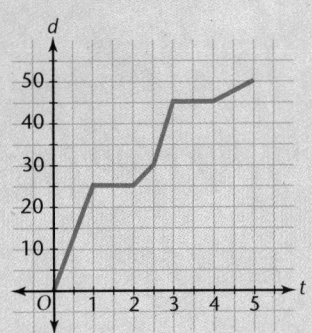

5 How does each graph describe the changes in the movement of a cyclist during each 5-hour trip?

1. The three missing distances are 40, 50, and 60 miles.

2. The distance traveled is the product of the rate times the time.

3. The biker will travel 100 miles in 5 hours.

4. From the pattern established, the graph indicates that it will take 5 hours to travel 100 miles. The line can be extended to show this.

5. Answers may vary.
   The first graph shows that for the first hour the biker travels at 10 miles per hour. For the next hour, the rate is 20 miles per hour, followed by an hour of rest. No additional distance is covered. The biker finishes the final 2 hours at 10 miles per hour.

   The second graph shows that after the 1 hour at 10 miles per hour the biker slows for an hour. This could indicate climbing a hill. For the third hour the biker travels faster, which could indicate going downhill. A slow interval followed by a fast one is repeated for the next 2 hours.

   The last graph shows a fast start at 25 miles per hour for the first hour, followed by a 1 hour rest. After the rest the biker travels at 10 miles per hour for a half hour, then speeds up. This is followed by another rest, perhaps for lunch, because the rate after the rest is slow.

Representing Number Patterns

PREPARE

Objectives

- Find the pattern in a sequence.
- Generate and display sequences.

RESOURCES

- Practice Master   1.1
- Enrichment Master   1.1
- Technology Master   1.1
- Lesson Activity Master   1.1
- Quiz   1.1
- Spanish Resources   1.1

Assessing Prior Knowledge

Complete each table.

1.

a	b
1	2
2	4
3	6
4	?
5	?

2.

a	b
1	1
2	4
3	9
4	?
5	?

[1. 8, 10;   2. 16, 25]

TEACH

What number patterns are useful for doing math? Discuss patterns in the multiplication tables, integers, perfect squares, and in cyclical number patterns such as measurements of time.

Patterns appear everywhere—in the architecture of buildings, in the designs of clothing, and in the advertisement of products. In music, sound patterns influence some people to like classical music and some to like rock and roll. Just as some people are fascinated by the patterns that occur in music, others are fascinated by the patterns formed by shapes and numbers.

Small portions of the fractal, when magnified, produce images similar to the entire fractal.

Suppose 10 people have just returned to school, and they all want to tell each other about what they did during summer break. Each person has just one conversation with each of the other people. Guess how many conversations will take place.

Using Problem-Solving Strategies

If you are not sure how to solve a problem directly, experiment with various problem-solving strategies. For example, you can draw a picture, think of a simpler problem, make a table, or look for a pattern.

Think of a Simpler Problem   Begin with the most basic situation for the problem, then make regular changes to build a pattern. Start with just 1 person. How many conversations can 1 person have if talking to oneself doesn't count? Organize the results, and make a table.

ALTERNATIVE teaching strategy

Using Visual Models

Explore the two-by-two conversations by using diagrams. Have students draw dots for the different numbers of students and represent the conversations by connecting the dots.

2 people
1 conversation

3 people
3 conversations

When you add a fourth person, D, the same three conversations are possible, but there are also three new conversations possible—A with D, B with D, and C with D, for a total of 6.

4 people
6 conversations

Look for a Pattern When a regular number pattern emerges, organize the information in a table. Are there other patterns that might help solve this problem?

People	1	2	3	4	5
Conversations	0	1	3	6	?
Increase		+1	+2	+3	

Once you discover how the pattern in the number of conversations increases, continue the pattern to find the number of conversations possible among 5 people.

Reason logically from the pattern When the fifth person joins, add 4 new conversations (one with each of the other four people) to the previous 6 conversations. Five people will have **6 + 4**, or **10**, conversations.

People	1	2	3	4	5
Conversations	0	1	3	6	10
Increase		+1	+2	+3	+4

When a sixth person joins, add 5 new conversations to the previous 10, and so on. Continue to build the pattern to 10 people.

People	1	2	3	4	5	6	7	8	9	10
Conversations	0	1	3	6	10	15	21	28	36	45
Increase		+1	+2	+3	+4	+5	+6	+7	+8	+9

interdisciplinary
CONNECTION

Biology Many examples of sequence patterns appear in nature. The Fibonacci sequence is one of these patterns. Originally developed to provide a mechanism to count a domestic rabbit population, the Fibonacci sequence appears in nature in such places as the petals of a daisy and the genealogical patterns of the drone honeybee. The first seven terms of the sequence are 1, 1, 2, 3, 5, 8, and 13. Each new term is derived by adding the previous two terms.

Cooperative Learning
Students can solve the conversation problem by role-playing. Have students act out the conversations in groups, with one student recording the results. Have students think of several methods for solving the same problem. Ask them which method works best for them and why.

TEACHING *tip*

Throughout the textbook, information necessary to work the examples may not appear in the written part of the example. Remind students to look for pertinent information in the art, photos, and captions.

Use Transparency 1

When mathematicians study number sequences for patterns, they often make *conjectures*. A **conjecture** is a statement about observations that they believe to be true. Mathematicians try to prove the conjecture or find a counterexample to show that the conjecture is not true.

The sum of conversations among any number of students is the sum of the numbers from 1 to one less than the number of students. Test this conjecture for 11, 12, 13, and 14 students.

Which figure completes the pattern?

Pattern

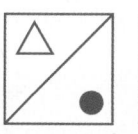 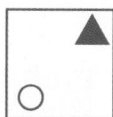  **?**

Choices

    A   B   C   D

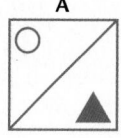  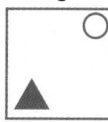 

Number Sequences

When a sequence shows a pattern, discover as much as you can about how the numbers relate to each other.

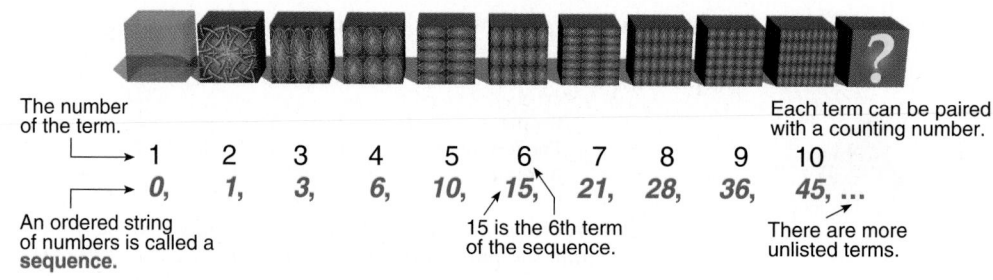

The number of the term.

| 1 | 2 | 3 | 4 | 5 | 6 | 7 | 8 | 9 | 10 |

An ordered string of numbers is called a **sequence**.

0,   1,   3,   6,   10,   15,   21,   28,   36,   45, ...

Each term can be paired with a counting number.

15 is the 6th term of the sequence.

There are more unlisted terms.

EXAMPLE 1

Predict the next three terms for each of the following sequences.

Ⓐ 15, 18, 21, 24, 27, . . .   Ⓑ 2, 6, 12, 20, 30, . . .

Solution▸

Ⓐ The first sequence begins with 15. Each of the following terms is 3 more than the one before. A calculator can be helpful when you explore sequences. A calculator with a *constant feature* can easily provide more terms of the sequence by repeating an operation.

Calculator

For example, the first sequence can be generated by entering

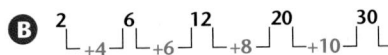

15 [+] 3 [=] [=] [=] [=].

Continue pressing the *equals key* three more times. The next three terms of the sequence are **30**, **33**, and **36**.

On some *graphics* calculators, enter

15 [ENTER] [+] 3 [ENTER] [ENTER] [ENTER] [ENTER].

Continue pressing the *enter key* three more times to generate **30**, **33**, and **36**.

B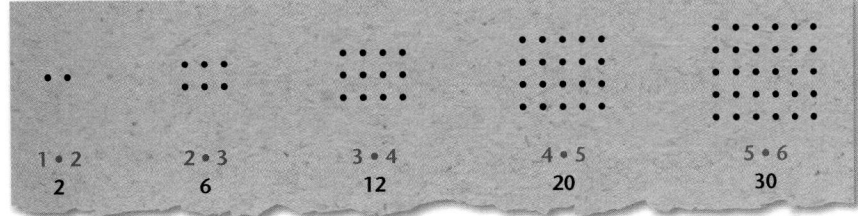

2 ⌐+4⌐ 6 ⌐+6⌐ 12 ⌐+8⌐ 20 ⌐+10⌐ 30

The next three terms are found by adding **12**, **14**, and **16**.

30 + **12** = **42**   **42** + **14** = **56**   **56** + **16** = **72** ❖

The sequence in Example 1B appears when counting the number of dots in the pattern below.

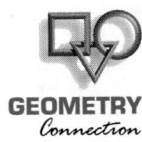

GEOMETRY
Connection

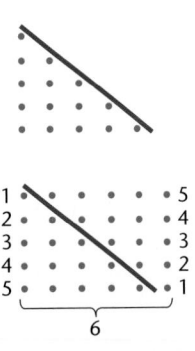

The next three terms are 6 · 7 = **42**, 7 · 8 = **56**, and 8 · 9 = **72**.

Why do you think the numbers in the sequence 2, 6, 12, 20, 30, . . . are called **rectangular numbers?**

EXAMPLE 2

Find the sum 1 + 2 + 3 + 4 + 5 using a geometric pattern.

Solution▶

There are many ways to find this sum. This method shows how geometric patterns can be used to find the sum of 1 + 2 + 3 + 4 + 5.

Represent the sum of 1 to 5 by a triangle of dots. Form a rectangle of dots by copying the triangle and rotating it.

The total number of dots is 5 · 6 = 30. There are half as many dots in the triangle. $\frac{5 \cdot 6}{2} = 15$.

The sum of the numbers from 1 to 5 is 15. ❖

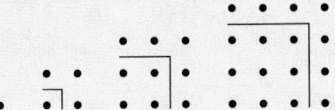

**Math Connection
Geometry**

The dot patterns utilize the concept of area to provide a visual image of the sums.

**Aongoing
ASSESSMENT**

The numbers are called rectangular numbers because each can be visualized with a rectangle one unit longer than it is wide.

Alternate Example 2

Use the pattern of squares below to find the sum of the first 4 odd numbers.

Ask students to find the sum of the first *n* odd numbers.
[**The sum of the first *n* odd numbers is n^2.**]

Technology can simplify the exploration of sequences and can provide tools to calculate sums of sequences. Try using a computer *spreadsheet* to add the numbers 1 to 9. A spreadsheet allows you to control lists or tables of data. Familiarize yourself with the layout of this spreadsheet.

Spreadsheet Title

Cell Reference (Shows where the results appear) — C2

Formula Bar — =SUM (A2:A10)

Sum of numbers

	A	B	C
1	Numbers		Sum
2	1		45
3	2		
4	3		
5	4		
6	5		
7	6		
8	7		
9	8		
10	9		
11			

Column Identifiers
Active Cell
Row
Cell
Row Identifiers **Column** **(Spreadsheet)**

Try This Find the sum $1 + 2 + 3 + 4 + 5 + 6 + 7 + 8 + 9$ using a geometric pattern. Then use your understanding of the number pattern to find a simple way to calculate the sum $1 + 2 + 3 + 4 + 5 + \cdots + 100$.

EXERCISES & PROBLEMS

Communicate

1. Explain how you would set up a table to solve the following problem. If each of the teams A, B, C, and D play each other, one of the games will be A versus B. Show how you would list all the games.

2. Describe how to find the sum of the numbers from 1 through 20.

Explain how you would find the pattern for the following sequences.

3. 3, 7, 11, 15, 19, 23, . . .   4. 1, 1, 2, 3, 5, 8, 13, . . .

5. Give four problem-solving strategies that are used in Lesson 1. Choose one strategy from your list, and explain how it helps in solving problems.

Practice & Apply

6. If each of the teams A, B, C, D, and E play each other, one of the games will be A versus B. List all of the games, and tell how many there are. 10

7. In 1993, after Penn State joined the Big Ten athletic conference, the Big Ten had 11 teams. If each of the 11 teams played each of the other teams, how many games would there be? Make a table similar to the conversation problem. 55

8. If an athletic conference has 12 teams and each of the teams play each of the other teams, how many games will there be? 66

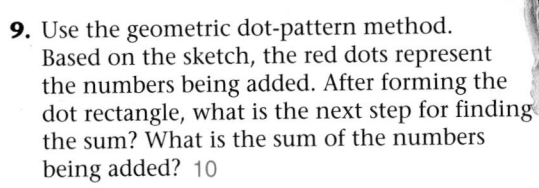

9. Use the geometric dot-pattern method. Based on the sketch, the red dots represent the numbers being added. After forming the dot rectangle, what is the next step for finding the sum? What is the sum of the numbers being added? 10

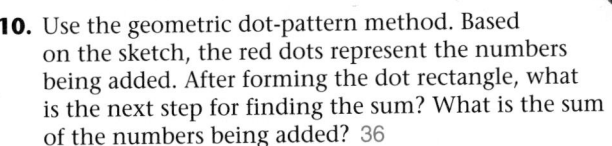

10. Use the geometric dot-pattern method. Based on the sketch, the red dots represent the numbers being added. After forming the dot rectangle, what is the next step for finding the sum? What is the sum of the numbers being added? 36

11. Use the geometric dot-pattern method to find the sum $1 + 2 + 3 + 4 + 5 + 6 + 7$. Include the sketch. 28

12. Find the sum $1 + 2 + 3 + 4 + 5 + \cdots + 40$. Think of the sketch without actually drawing it. 820

13. How can you find $1 + 2 + 3 + 4 + 5 + \cdots + 50$?
Form a rectangle 50 by 51. Divide by 2.

Find the next three terms in each sequence.
Then explain the pattern used to find the terms for each.

14. 4, 9, 14, 19, 24, . . .   **15.** 7, 16, 25, 34, 43, . . .   **16.** 9, 19, 29, 39, 49, . . .
29, 34, 39   52, 61, 70   59, 69, 79
17. 2, 4, 8, 16, 32, . . .   **18.** 5, 7, 9, 11, 13, . . .   **19.** 3, 9, 27, 81, 243, . . .
64, 128, 256   15, 17, 19   729, 2187, 6561
20. 8, 10, 12, 14, 16 . . .   **21.** 16, 8, 4, 2, 1, . . .   **22.** 5, 12, 19, 26, 33, . . .
18, 20, 22   40, 47, 54

23. Solve the conversation problem for 10 people by drawing a sketch. Draw 10 dots around a circle, and connect them with segments to represent conversations. How many segments did you draw? 45

24. 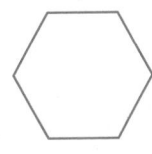 **Geometry** Diagonals are segments other than sides that connect the vertices of a polygon. Draw a hexagon and its diagonals. How many diagonals does a regular hexagon have? How does this problem relate to the conversation problem? 9

21. $\frac{1}{2}, \frac{1}{4}, \frac{1}{8}$

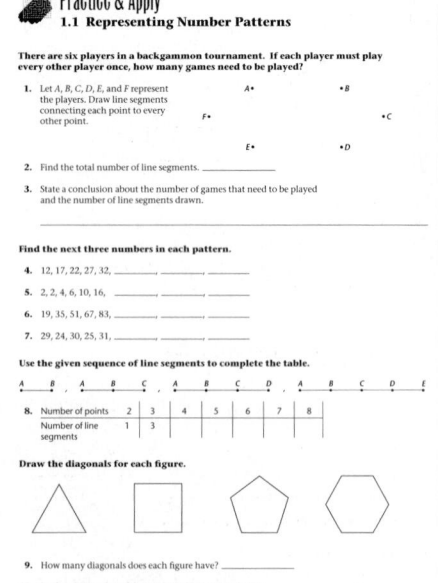

alternative ASSESSMENT

Performance Assessment

In groups, have students write a simple program for a computer or a graphics calculator that will generate numbers in the Fibonacci sequence.

Cultural Connection: Europe When Gauss was very young, a teacher told him to find the sum of all the whole numbers from 1 to 100, thinking this would keep him busy for a long time. But Gauss very cleverly paired the numbers and observed that $1 + 99 = 100$, $2 + 98 = 100$, $3 + 97 = 100$, and so on.

25. How many pairs could be formed in this way without repeating any numbers? 49

26. Which whole numbers from 1 to 100 would not appear in any of the pairs? 50 and 100

Karl Friedrich Gauss (1777–1855) was one of the greatest of all mathematicians.

27. Based on the previous two exercises, what is the sum of the numbers from 1 to 100? 5050

Geometry The numbers in each array are square numbers. The first four square numbers are 1, 4, 9, and 16.

28. Find the tenth square number. 100

29. Find the twentieth square number. 400

1	4	9	16

30. Suppose 8 cities are to be linked by phone lines, with 1 phone line between each pair of cities. How many phone lines will there be? 28

31. Use the geometric dot-pattern method to find the sum $1 + 3 + 5 + 7 + 9$. 25

Geometry The numbers in each array are triangular numbers. The first four triangular numbers are 1, 3, 6, and 10.

32. Find the tenth triangular number. 55

33. Find the twentieth triangular number. 210

1	3	6	10

34. Show visually that the square number 16 is the sum of two triangular numbers. Which two? 6 and 10

35. Show visually that 100 is the sum of two triangular numbers. Which two? 45 and 55

36. Copy and complete the following table.

Number	1	3	5	7	9	11	13	15	17
Sum	1	4	9	16	? 25	? 36	? 49	? 64	? 81

37. Examine the pattern in the table in Exercise 36. Guess the sum of the first 100 odd numbers. Explain your method.

38. Copy and complete the following table.

Number	1	2	3	4	5	6	7
Cube of the number	1	8	27	64	? 125	? 216	? 343
Sum of the cubes	1	9	36	100	? 225	? 441	? 784

39. Examine the pattern in the table in Exercise 38. Guess the sum of the first 10 cubes. Explain your method. 3025

37. $100^2 = 10,000$. The sum of the first n odd numbers is n^2.

Look Back

David and his 3 friends collected 196 aluminum cans.

40. They divided the cans among themselves equally into 4 bags. How many cans are in each bag? 49

41. The 4 friends take the cans to the recycling center where they can get $0.20 a pound for the cans. The total weight of the cans is 9 pounds. How much money will they get all together? $1.80

42. How much will each of the 4 friends get from recycling the cans if they split the money equally? $0.45

Recycled aluminum cans are compressed before shipping.

43. If a teacher collects $2.75 from each of 28 students for the field trip, how much does the teacher collect? $77

44. The citywide concert sold 6702 tickets to students and 3749 tickets to parents. How many tickets were sold all together? 10,451

45. **Geometry** What is the area of a square 23 centimeters on a side? 529

46. What are 3 fractions that are equivalent to $\frac{3}{5}$? $\frac{6}{10}, \frac{9}{15}, \frac{12}{20}, \cdots$

Find the value of the following expressions.

47. $6(9 + 4)$ 78

48. $\sqrt{225}$ 15

49. Kim had $4.75 and spent $3.12. How much money does she have left? $1.63

Look Beyond

50. The following pattern was first explored in China and Iran and was later called Pascal's Triangle. Extend the pattern by finding rows 6, 7, and 8.

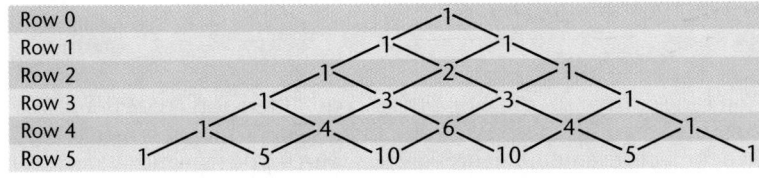

1, 6, 15, 20, 15, 6, 1

1, 7, 21, 35, 35, 21, 7, 1

1, 8, 28, 56, 70, 56, 28, 8, 1

51. Find the sum of the numbers in each row of Pascal's Triangle. 1, 2, 4, 8, 16, 32, 64, 128, 256

PREPARE

Objectives

• Find the pattern of a sequence using the difference method.

• Extend a sequence by working backward from the differences.

RESOURCES

• Practice Master	**1.2**
• Enrichment Master	**1.2**
• Technology Master	**1.2**
• Lesson Activity Master	**1.2**
• Quiz	**1.2**
• Spanish Resources	**1.2**

Assessing Prior Knowledge

Find the differences in each pair.

1. $5, 11$ **[6]**

2. $14, 9$ **[−5]**

Evaluate each of the following.

3. 7^2 **[49]**

4. 5^3 **[125]**

TEACH

 Discuss how various branches of science use mathematical patterns and equations to make predictions.

LESSON 1.2

Using Differences to Identify Patterns

A drum barometer keeps a running record of air pressure, one of the many factors important in the prediction of weather patterns.

why

People are always making predictions to answer questions they have about the future. We use data from the past and regular patterns in the present to make predictions about the future.

Recall from the previous lesson the problem of finding the number of possible conversations among 10 people. Examine the table for 5 people again. Look at the differences between consecutive terms in the sequence of numbers 0, 1, 3, 6, 10, . . .

People	1	2	3	4	5	
Conversations	0	1	3	6	10	
		1	2	3	4	← **First differences**

Extend the table. This time find the second differences.

People	1	2	3	4	5	
Conversations	0	1	3	6	10	
		1	2	3	4	← **First differences**
			1	1	1	← **Second differences constant**

You can see from the table that the second differences are *constant*. When the constant differences are greater than 0, the sequence is increasing. When the constant differences are less than 0, the sequence is decreasing.

If you know the differences, you can predict other terms of the sequence.

 **Technology** Use a graphics calculator or computer software to set up a simple program to perform repeated evaluations of expressions such as $2x + 5$, $3x − 1$, and $x + 2$. Using the calculator's table feature, the sequences can then be analyzed for patterns using the differences method described in the text.

EXAMPLE 1

Find the next two terms of each sequence.

A 80, 73, 66, 59, 52, . . .   **B** 1, 4, 9, 16, 25, . . .

Solution▸

A Find the first differences.

80 73 66 59 52
　 −7　 −7　 −7　 −7　 ◄──── **First differences are constant.**

Notice that each term decreases by 7. The first difference is represented by −7. Subtract 7 from the previous term to find each new term.

52 − 7 = 45   45 − 7 = 38

The next two terms are 45 and 38.

B Find the first differences.

1　 4　 9　 16　 25
　3　 5　 7　 9　 ◄──── **First differences are not constant.**

Since the first differences are not constant, find the second differences.

1　 4　 9　 16　 25
　3　 5　 7　 9
　　2　 2　 2　 ◄──── **Second differences are constant.**

Now work backward to find the first differences. To find each of the next first differences, add 2 to the previous first difference.

9 + 2 = 11   11 + 2 = 13

1　 4　 9　 16　 25
　3　 5　 7　 9　 11　 13　 ◄──── **First differences**
　　2　 2　 2　 2　 2 ◄──── **Second differences
are constant.**

Continue working backward to find the next two terms of the sequence. To find each of the next terms, add the first difference to the previous term.

25 + 11 = 36   36 + 13 = 49

1　 4　 9　 16　 25　 36　 49
　3　 5　 7　 9　 11　 13　 ◄──── **First differences**
　　2　 2　 2　 2　 2 ◄──── **Second differences
are constant.**

The next two terms are 36 and 49. ❖

Each successive term in the sequence is the square of the natural number with which it is paired.

Alternate Example 2

The data below represents the flight of a baseball thrown between two people 200 feet apart. The time is divided into tenths of a second.

Time (sec)	Height (feet)
0	0
0.1	5.7
0.2	10.8
0.3	15.3
0.4	19.2
. . .	. . .
2.0	0

Use the second differences to find the maximum height.
[30 feet]

Cooperative Learning

Technology Have students work in groups of two or three using a spreadsheet program to model sequences. Have them begin by generating simple constant difference sequences. Then have the students experiment with creating sequences that require computing three or four differences before arriving at a constant.

CRITICAL
Thinking

The numbers in the sequence 1, 4, 9, 16, . . . are called perfect square numbers. The first term is $1 \cdot 1 = 1^2$, or 1. The second term is $2 \cdot 2 = 2^2$, or 4. Make a conjecture about the term number and its square in the square number sequence.

EXAMPLE 2

Physics The data shown in the table is from the flight of a small rocket during the first 4 seconds of its flight. The flight of the rocket ends when it hits the ground 14 seconds after takeoff. Use the method of finding differences to find the maximum height during the rocket's flight.

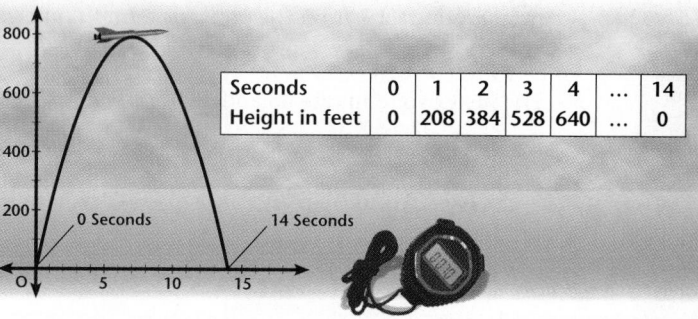

Seconds	0	1	2	3	4	...	14
Height in feet	0	208	384	528	640	...	0

Solution►

Look at the differences in the table to discover the pattern.

Seconds	0	1	2	3	4
Height in feet	0	208	384	528	640

208   176   144   112 ◄— **First differences**
   −32   −32   −32 ◄— **Second differences constant**

The second differences are each a constant −32 feet. This means the differences *decrease* by 32 feet each time. Use the strategy of **working backward** from the second differences to extend the table.

Seconds	0	1	2	3	4	5	6	7	8	9	10
Height in feet	0	208	384	528	640	720	768	784	768	720	640

208   176   144   112   80   48   16   −16   −48   −80
   −32   −32   −32   −32   −32   −32   −32   −32   −32

After 7 seconds, the numbers for the height begin to repeat in reverse order. By continuing the table to 14 seconds, the pattern shows points along the complete path of the rocket.

Seconds	11	12	13	14
Height in feet	528	384	208	0

The table shows the highest point reached is 784 feet. This occurs at 7 seconds. Why does this information seem reasonable? ❖

RETEACHING
the
lesson

Using Patterns Students can gain a fresh understanding of how sequences work by building them from the bottom up. This can be done by taking a common difference and adding to build another sequence, and then another. For example, using 3 as the first common difference and beginning each sequence with 1 produces the following pattern.

3   3   3   3   3   3
1   4   7   10   13   16   19
  1   5   12   22   35   51   70
    1   6   18   40   75   126   196

Students can then work back to see that what they have done is the reverse of the process of analysis presented in the lesson.

Try This   Suppose a rocket flight takes 15 seconds from launch to return to the Earth. After how many seconds do you think it will reach its maximum height?

Using a spreadsheet can extend your ability to use differences to discover sequence patterns, especially when the differences and calculations become more complicated.

Spreadsheet   Compute the first difference of 208 in cell C2 by subtracting the value in cell B2 from the value in cell B3. What instructions do you think are needed to compute the values in cells C3 and in D2 of the spreadsheet?

C2	=B3-B2			
	A	**B**	**C**	**D**
1	seconds	height	1st diff	2nd diff
2	0	0	**208**	−32
3	1	208	176	−32
4	2	384	144	−32
5	3	528	112	
6	4	640		
7	5	720		

EXERCISES & PROBLEMS

Communicate

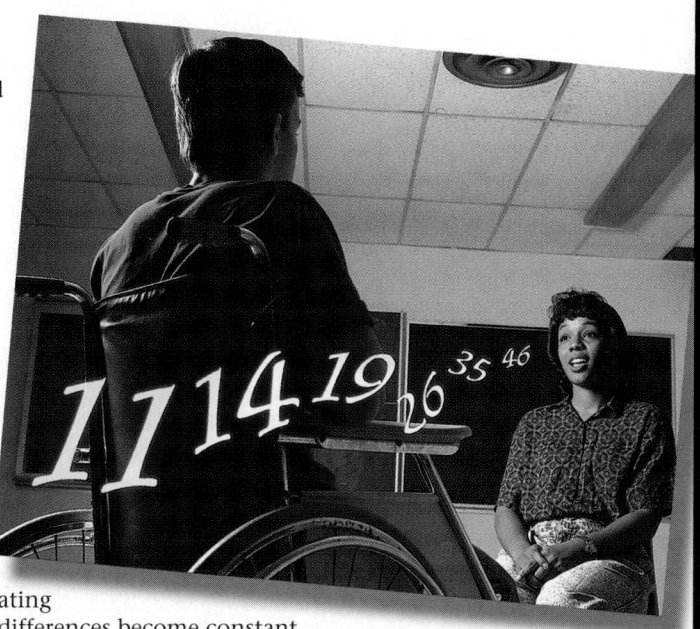

1. Explain how to find the first and second differences for the sequence that the students are discussing.

2. Describe the method for predicting the next two terms in the following sequence. 88, 76, 64, 52, 40, . . .

3. Suppose a rocket took 17 seconds to return to the Earth after launching. Explain how to find the time it takes the rocket to reach its maximum height.

4. **Technology** Explain how to find the entry in cell C4 on the spreadsheet above.

5. Explain why you can stop calculating additional differences when the differences become constant.

6. Explain how to work backward from a second difference to extend a sequence.

Practice & Apply

7. Find the first and second differences for the sequence
20, 27, 36, 47, 60.   7, 9, 11, 13
2, 2, 2

Find the next two terms of each sequence.

8. 18, 32, 46, 60, 74, . 88, 102

9. 33, 49, 65, 81, 97, . 113, 129

10. 20, 21, 26, 35, 48, 65, 86, 111

11. 30, 31, 35, 42, 52, 65, 81, 100

12. 100, 94, 88, 82, 76, 70, 64, 58

13. 44, 41, 38, 35, 32, 29, 26, 23

14. 12, 12, 18, 31, 53, 87, . . .   137, 208

15. 1, 7, 23, 50, 89, . . .   141, 207

Apply the method of finding differences until you get a constant. Which difference produces a constant? What is it?

16. 1, 2, 3, 4, 5, . . .   first; 1

17. $1^2, 2^2, 3^2, 4^2, 5^2,$ . . . HINT: This is the same as 1, 4, 9, 16, 25, . . .   second; 2

18. $1^3, 2^3, 3^3, 4^3, 5^3,$ . . . HINT: This is the same as 1, 8, 27, 64, 125, . . .   third; 6

19. Tell how many differences you will have to compute before you get a constant for $1^4, 2^4, 3^4, 4^4, 5^4, 6^4$ . . .   4

20. The first three terms of a sequence are 7, 11, and 16. Find the first and second differences. If the second differences are constant, what are the next three terms?   22, 29, 37

21. The first three terms of a sequence are 2, 6, and 12. The second differences are a constant 2. What are the next three terms of the sequence?   20, 30, 42

22. The third and fourth terms of a sequence are 15 and 23. If the second differences are a constant 2, what are the first 5 terms of the sequence?   5, 9, 15, 23, 33

23. If the second differences of a sequence are a constant 3, the first of the first differences is 7, and the first term is 2, find the first 5 terms of the sequence.
2, 9, 19, 32, 48

24. Physics If a rocket lands after 20 seconds of flight, at what time do you think it reached its maximum height? How can you tell?   10

25. Suppose a rocket took 24 seconds to reach its maximum height. After how many seconds do you think it hit the ground?   48

Technology Complete the spreadsheet that follows Example 2. What spreadsheet instruction formula should you place in the following cells?

26. C6 = B7 − B6

27. D5 = C6 − C5

28. Geometry Complete this table for the perimeter of a square with the length of a side given in centimeters.

Side	1	2	3	4	5	6	7	8
Perimeter	4	8	12	16	?	?	?	?

20, 24, 28, 32

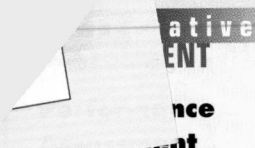

29. Apply the method of differences to find the next four terms of 4, 8, 12, 16, . . . Compare the results with the previous exercise. Do you get the same answer? Yes

30. Home Economics As the altitude increases, the boiling point of water decreases according to the table. For example, in Las Vegas, Nevada, at an altitude of 2180 feet, the boiling point of water is about 98° Celsius. Estimate the boiling point of water in Colorado Springs, Colorado, which is at an altitude of 6170 feet. ≈ 94° C

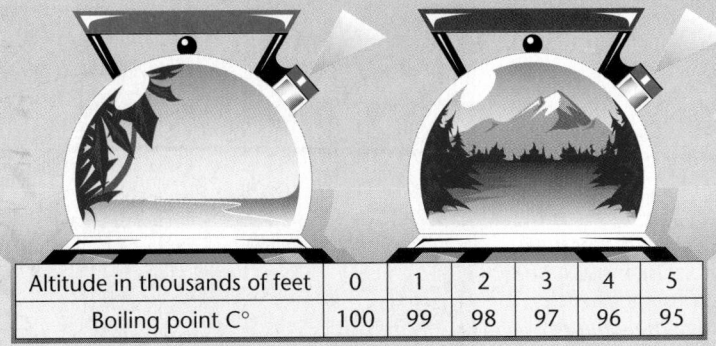

Altitude in thousands of feet	0	1	2	3	4	5
Boiling point C°	100	99	98	97	96	95

31. **Portfolio Activity** Complete the problem in the portfolio activity on page 3.

Look Back

32. If you read a 603-page novel at a rate of 24 pages a day, how many days will it take you to read it? about 25

33. If 675 people came to see the play and the admission was $1.60, how much money was taken in? $1080

34. There were 162 students who volunteered to help 3 organizations. How many students went to each of the organizations if each got the same number? 54

Perform each of the following operations.

35. $\sqrt{169}$ 13   **36.** 15^2 225

37. Find the sum of $1 + 2 + 3 + 4 + \cdots + 18$. **[Lesson 1.1]**
171

38. Find the next 3 numbers in the following pattern.
43, 49, 55, 61, 67, . . . **[Lesson 1.1]**
73, 79, 85

Look Beyond

The method of finding differences works only for certain kinds of patterns. Try it on the following patterns and tell what happens. Predict the next three terms of each pattern without using differences. Explain your method.
        64, 128, 256   100,000, 1,000,000, 10,000,000
39. 1, 2, 4, 8, 16, 32, . . .   **40.** 1, 10, 100, 1000, 10,000, . . .

31.

25 dots,

41 dots,

There are 85 dots in the pattern with 7 peaks.

Exploring Variables and Equations

PREPARE

Objectives

• Learn the basic concepts of variables, expressions, and equations.

• Evaluate an expression using substitution.

RESOURCES

• Practice Master   1.3
• Enrichment Master   1.3
• Technology Master   1.3
• Lesson Activity Master   1.3
• Quiz   1.3
• Spanish Resources   1.3

Assessing Prior Knowledge

Evaluate each with $x = 2$, and $x = 3$.

1. $3x - 1$ **[5, 8]**

2. $4(x + 3)$ **[20, 24]**

TEACH

Since the early grades, students have worked with placeholders for numbers and have had repeated experiences in evaluating expressions. Review how letters can be used to represent numbers.

Throughout the text, answers to exploration steps can be found in Additional Answers beginning on page 729.

You want to rent a bicycle, but you don't know how long you can rent it with the money you have. In mathematics, one technique used to solve this kind of problem begins when you represent the unknown number in the problem with a symbol. You then work with the symbol until the algebra reveals the symbol's value.

Exploration 1 *Defining a Variable*

For how many hours can you rent a bike if you have $15 to spend?

This relatively simple problem provides an example that can be used to explore the basic tools of algebra. Examine the pattern for the bike rental.

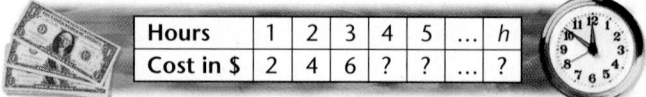

Hours	1	2	3	4	5	...	h
Cost in $	2	4	6	?	?	...	?

ALTERNATIVE teaching strategy

Technology Use the table function on a graphics calculator to emphasize how equations are useful for summarizing mathematical relationships. For example, have students calculate the total number of hours in a given number of days. Once students recognize the pattern
TOTAL HOURS = 24 · NUMBER OF DAYS, help them derive the equation, $y = 24x$. Students can enter the equation into the calculator and use the table to determine total hours for any number of days.

 What is the cost for 1 hour of bike rental?

 What is the cost for 2 hours of bike rental?

 What is the cost for 4 hours of bike rental?

 When you rent a bike for 5 hours, what number do you multiply by 5 to get the value for the *cost*?

 Let *h* stand for the number of hours. Follow the same pattern that you use to generate the values for *cost*. What number would you multiply times *h*?

 What should the *cost* entry be in this table for *h* hours?

 When *h* is 5, what is the value of 2*h* (which means 2 times *h*)? ❖

You can find the value of 2*h* when *h* is 6 by substitution. Replace *h* with 6, and simplify.

$$2h = 2 \cdot 6 = 12$$

Extend the table.

Hours (*h*)	1	2	3	4	5	6	7	8
Cost (*c*)	2	4	6	8	10	12	?	?

What is the cost for bike rental if the hours rented are halfway between 7 and 8 hours, and you can pay by the half-hour? For how many hours can you rent a bike at the park if you have $15?

If the cost for renting a bike is $3 an hour, how many hours can you rent a bicycle for $15? Try other values to see how they affect the hour and cost table.

Another way to represent the cost for renting the bicycle is to use the letter *c*. Since both *c* and 2*h* represent *cost*, they are equivalent. This is written as *c* = 2*h* or 2*h* = *c*.

In algebra, a **variable** is a letter or other symbol that can be replaced by any number or other expression. Variables combine with numbers and operations (addition, subtraction, multiplication, and division, for example) to form **expressions.** Two equivalent expressions, when separated by an equal sign, form an **equation.**

Examples of Variables	Examples of Expressions	Examples of Equations
a, *x*, *y*	$3x + 4$, $2h$,	$3x + 4 = 8$,
c, *h*, *d*, *t*	$-16t^2 + 4t$	$c = 2h$, $2h = c$

When you solve an equation, you find the value or values of the variable that make the equation a true statement. The solution for $2h = 15$ is $7\frac{1}{2}$ because $2 \cdot 7\frac{1}{2} = 15$.

 **interdisciplinary**
CONNECTION

Aeronautics The descent of an airplane can be described by a simple linear equation. Have students consider the following. Suppose a plane is descending at a rate of 500 feet per minute. If *y* represents the total number of feet descended and *x* represents the number of minutes, $y = 500x$ models the plane's descent. How many feet will the plane descend in 11 minutes? [**5500**] How long will it take for the plane to descend 10,000 feet? [**20 minutes**] Suppose the plane starts at 20,000 feet. Rewrite the equation so that *y* is the plane's altitude for a given time, *x*, of the plane's descent. [$y = -500x + 20,000$]

Exploration 1 Notes
The questions lead the students from the concrete numbers to the more abstract notation of equation and function. Large numbers can be used in this exploration. For example: During the course of one month, the park rents bicycles for 745 hours. How much money is received at $3 per hour? [$2235]

ongoing
ASSESSMENT

7. $2h = 2 \cdot 5 = 10$

Ecology   *The land area of Olympic National Park is 922,654 acres. How long would it take to lose forest land equal to the area of Olympic National Park if 57,600 acres were destroyed each day?*

Exploration 2   A Diminishing Sequence

Calculator

On a calculator with a constant feature, enter the area of Olympic National Park. Subtract 57,600 repeatedly, counting the number of times you subtract.

 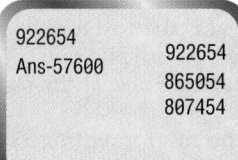 

As you continue to subtract, you can see how quickly the daily destruction diminishes the forest area. Continue repeating the subtraction until you have depleted the entire forest area.

To express this problem as an equation, you need to interpret and relate the information from the problem.

1. How many acres are in Olympic National Park?

2. How many acres are lost each day?

3. How would you express *57,600 fewer acres each day,* using the variable *d* to represent a day?

4. If you let *f* represent the number of days, then how much forest is left when *d* is 10?

5. If you let *f* represent the amount of Olympic National Park forest left, explain what it means when *f* = 0.

6. How many days does it take to completely destroy an area the size of Olympic National Park at the given rate? ❖

CRITICAL
Thinking

In one region there are 15,000 acres of forest. An average of 16 acres are destroyed every day from September 1 to October 30. Project the amount of forest lost after 30 more days. There are, in fact, 13,923 acres of the forest left. What could account for the difference between your projection and the actual amount?

APPLICATION

Danica places 3 photos on the cover of her picture album. On all of the other pages she places 4 photos. She has 178 photos in all. How many pages will she need if she uses all of her photos? This problem can be solved by writing an equation and using a guess-and-check strategy.

Let p equal the number of pages in Danica's picture album. To solve this problem, solve for p in the equation $4p + 3 = 178$. Guess the number of pages, substitute the number you guess for p, and check your guess. Keep track of your attempts in a table. Notice that in algebra, parentheses are used to show multiplication.

Guess 1: Try $p = 10$.
$$4p + 3 = 178$$
$$4(10) + 3 = 43$$

43 is too small. Try a larger number for p.

Guess 2: Try $p = 50$.
$$4p + 3 = 178$$
$$4(50) + 3 = 203$$

203 is too large. Try a smaller number for p.

Guess 3: Try $p = 40$.
$$4p + 3 = 178$$
$$4(40) + 3 = 163$$

163 is too small. You will find the number for p is between 43 and 44.

For 178 photos, 43 pages is not enough. The problem asks for the number of pages Danica will use, so the answer is 44 pages. One page more than 43 is needed for the extra photos even though the page will not be filled. ❖

Always be careful to interpret your answer in a way that makes sense. How many concert tickets can you buy for $15? The answer is 1, not $1\frac{7}{8}$. You cannot buy $\frac{7}{8}$ of a concert ticket, and you do not have enough money to buy 2 tickets.

RETEACHING
the
lesson

Have students make a table showing the relationship between a given number of $21 shirts and the total cost of the shirts. Help students derive the formula that describes the relationship algebraically. For example:

TABLE					EQUATION
Shirts	1	2	3	4	$C = 21s$
Cost	21	42	63	84	

CRITICAL
Thinking

In the 90 days involved (September 1 to October 1, plus 30) 16×90 acres could be destroyed. So, $15,000 - 1400$, or 13,560 acres would be left. Since there are actually 13,923 acres left, it would appear that about 12 acres per day, not 16, were being lost.

TEACHING tip

Introduce this application without referring to the text pages. Exploring the sequence using a guess-and-check strategy gives students experience with the algebraic expression for the sequence.

Assignment Guide

Core 1–8, 13–16, 19–39

Core Two-Year 1–38

Technology

A calculator will help students with Exercises 13–16.

Error Analysis

Exercises 27–31 emphasize that answers must be interpreted in context. Students often give an answer that is arithmetically correct, but not correct in context.

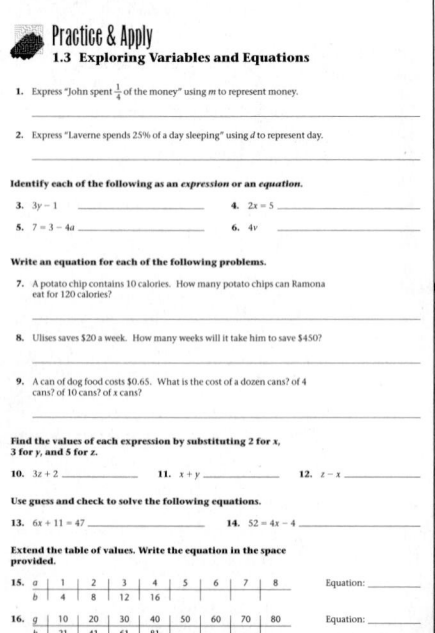
EXERCISES & PROBLEMS

Communicate

1. Define *variable*, *expression*, and *equation*.

2. Explain how to identify the unknown variable in the following problem: Jeff has 3 more apples than Phil, and together they have 9 apples. How many apples does Phil have?

3. Discuss how you would set up the equation for the following problem: If 12 pencils cost $1.92, find the cost of 1 pencil.

4. Tell how you would set up the equation for the following problem: If tickets for a concert cost $10 each, how many can you buy with $35?

5. Explain how you would use the guess-and-check strategy to solve the equation $10x + 3 = 513$.

Practice & Apply

6. Guess and check to solve the equation $4x + 3 = 49$. 11.5

7. Guess and check to solve the equation $49x + 3 = 4$. $\frac{1}{49}$

8. Is $14y$ an expression or a variable? expression

If pencils cost 20 cents each, find the cost of

9. 0 pencils. $0.00

10. 4 pencils. $0.80

11. 10 pencils. $2.00

12. p pencils. $0.20p$

Technology Tell how to generate each sequence on a calculator.

13. 6, 8, 10, 12, 14, . . . add 2

14. 15, 25, 35, 45, 55, . . . add 10

15. 100, 90, 80, 70, 60, . . . subtract 10

16. 52, 48, 44, 40, 36, . . . subtract 4

Make a table to show the substitutions of 1, 2, 3, 4, and 5 for the variables in the expressions below.

17. $4x$ 18. $5y$ 19. $7s + 4$ 20. $3n$

Beatrice solved the equation $7x = 14$ and got 2 for x.

21. Explain what Beatrice did to get this answer. Divide both sides by 2.

22. How can you determine if the answer is correct? Multiply: $7 \cdot 2$

23. If tickets for a concert cost $11 each, how many tickets can you buy with $126? 11

24. If tickets for a concert cost $9 each, how many can you buy with $135? 15

17.

x	1	2	3	4	5
4x	4	8	12	16	20

19.

s	1	2	3	4	5
7s + 4	11	18	25	32	39

18.

y	1	2	3	4	5
5y	5	10	15	20	25

20.

n	1	2	3	4	5
3n	3	6	9	12	15

25. Consumer Economics Write an equation that models the following situation. How many apples can you buy with 99 cents if apples cost 30 cents each? $30a = 99$

26. Solve the equation you wrote in Exercise 25. Does the exact solution to the equation answer the problem posed about the apples? Explain. $3\frac{3}{10}$; no

27. Entertainment How many $5 movie tickets can you buy with $32? 6

28. Fund–raising How many $5 raffle tickets must you sell to raise $32? 7

29. If 5 people split the cost of a $32 birthday gift equally, how much should each pay? $6.40

 **Statistics** The average number of people in a family in 1995 was 3.18.

30. How many families had exactly 3.18 people? none

31. How large would a typical family be? 3

 **Look Back**

32. How much of each ingredient should be used if the recipe is tripled?
3 times as much

Date and Nut Bars

1 Cup pecans or walnuts
1/2 Cup dates
3/4 Cup sifted all–purpose flour
3 eggs
1 1/2 Cups brown sugar, firmly packed
3/4 Teaspoon baking powder
1/4 Teaspoon salt

33. A can of a soft drink contains 355 milliliters of liquid. If 5 people split 2 cans evenly, how many milliliters will each get? 142

34. If each pizza is cut into 12 pieces, how many pieces will there be in $3\frac{1}{2}$ pizzas? 42

Cultural Connection: Africa Thousands of years ago Diophantus and Hypatia, who is the first known woman mathematician, investigated the relationship between square numbers and triangular numbers.

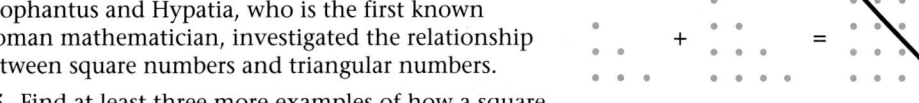

35. Find at least three more examples of how a square number can be represented as the sum of two triangular numbers. **[Lesson 1.1]** Examples: $4 = 1 + 3$; $9 = 6 + 3$; $25 = 10 + 15$

36. Are there any square numbers that can't be represented as the sum of two triangular numbers? **[Lesson 1.1]** yes, 1

37. In the first lesson you found that the sum $1 + 2 + 3 + 4 + 5$ is $\frac{5 \cdot 6}{2}$.

Examine the next two cases, the sum of the first 6 numbers and the sum of the first 7 numbers. Make a conjecture about the sum of the first n numbers.
[Lesson 1.1] $\frac{n(n + 1)}{2}$

38. The first differences of a sequence are 7, 9, and 11. The first term of the sequence is 5. What are the first eight terms of the sequence?
[Lesson 1.2] 5, 12, 21, 32, 45, 60, 77, 96

Look Beyond

39. If x^2 means x times x, then guess and check to solve for x in the equation $x^2 = 256$. 16

Objectives

- Understand the order of operations and signs of inclusion for algebraic computation.
- Perform calculations in the proper order.

RESOURCES

- Practice Master   1.4
- Enrichment Master   1.4
- Technology Master   1.4
- Lesson Activity Master   1.4
- Quiz   1.4
- Spanish Resources   1.4

Assessing Prior Knowledge

Evaluate the following:

1. $187 - 18.7$ [**168.3**]
2. $30 + 3.2 + 0.3$ [**33.5**]
3. 3^3 [**27**]

TEACH

why Students need to be familiar with the proper order of operations as a guideline for algebraic manipulation.

CRITICAL Thinking

460; The calculator multiplied 12 × 15, then added 20, then multiplied by 14.

LESSON 1.4 Order of Operations

why *Calculators and computers have dramatically changed the way people work with numbers. New technology has made computation and mathematical exploration easier and more exciting. Even without technology, you need to understand the order for performing operations to get accurate results.*

Mary plans to buy carpet for two rooms. One room is 12 feet by 15 feet, and the other room is 20 feet by 14 feet. She uses her calculator to find the number of square feet.

12 ☒ 15 ☐ 20 ☒ 14 ☐   2800

Mary knows that 2800 square feet is way too much. Her entire house is less than 2000 square feet. But calculators don't make mistakes. What is wrong?

CRITICAL Thinking

What is the correct answer for Mary's computation? What steps did her calculator use to get 2800?

To avoid misunderstandings and errors, mathematicians have agreed on certain rules for computation called the **order of operations.** One rule requires that multiplication be performed before addition. By following this rule, Mary can make the correct calculation.

$$12 \times 15 + 20 \times 14 =$$
$$180 + 280 =$$
$$460 \text{ square feet}$$

ALTERNATIVE teaching strategy

Cooperative Learning Divide the students into four to six groups. Give each group one of the following problems including the answer.

1. $100 - 4 \div 2 \div 2 = 99$
2. $[(12 + 3) - 3 \cdot 4] \div 3 = 1$
3. $8^2 + 4 \cdot 2 - 70 \div 2 = 37$
4. $[(3 + 2 \cdot 3) + 3] - 3^2 = 3$
5. $\dfrac{6 \cdot 3 + 2}{10 - 5} = 4$
6. $[(3^2 - 2^2) \cdot (10 + 2 \cdot 5)] \div 100 - 1 = 0$

Within their groups, students should conjecture how the answer was derived and write the steps down on a large sheet of paper. Have each group present the results to the class. The discussion should provide a basis for introducing the rules for the order of operations.

The Order of Operations

1. Perform all operations enclosed in symbols of inclusion (parentheses, brackets, braces, and bars) from innermost outward.
2. Perform all operations with exponents.
3. Perform all multiplications and divisions in order from left to right.
4. Perform all additions and subtractions in order from left to right.

Calculators that follow the order of operations use **algebraic logic.** You can use a computation like $2 + 3 \cdot 4$ to test for algebraic logic.

Key in 2 [+] 3 [×] 4 [=]. If the answer is 14, the calculator uses algebraic logic. On some calculators you might get 20. These calculators do not use algebraic logic, but you can use parentheses to get the correct answer.

2 [+] [( ] 3 [×] 4 [ )] [=]

You can also enter the multiplication first.

3 [×] 4 [+] 2 [=]

The examples you will see in this book assume that you have a calculator with algebraic logic.

EXAMPLE 1

How do you use a calculator with algebraic logic to work Mary's problem?

Solution▶

Key the numbers and operations just as they appear. 12 [×] 15 [+] 20 [×] 14 [ENTER]. The answer is 460. ❖

12*15+20*14
460

Using Inclusion Symbols

Symbols like parentheses, ( ), brackets, [ ], braces, { }, and the fraction bar are called **symbols of inclusion.** These symbols group numbers and variables. Treat any grouped numbers and variables as a single quantity. Operations should always be done within the innermost symbols of inclusion first. Then work outward.

Alternate Example 1

How do you use a calculator without algebraic logic to work Mary's problem? [**Use the order of operations and the memory function to calculate one part at a time or use parentheses.**]

TEACHING *tip*

The mnemonic device, **PEMDAS** (Parentheses, Exponents, Multiplication, Division, Addition, Subtraction), or the sentence Please Excuse My Dear Aunt Sally, is helpful to remember the order of operations.

EXAMPLE 2

Insert inclusion symbols to make $30 + 4 \div 2 - 1 = 16$ true.

Solution➤

Use parentheses to group 30 and 4 before dividing by 2. Begin with the operations in the innermost symbols of inclusion.

$$[(30 + 4) \div 2] - 1 \stackrel{?}{=} 16$$
$$[34 \div 2] - 1 \stackrel{?}{=} 16$$
$$17 - 1 \stackrel{?}{=} 16$$
$$16 = 16 \qquad \text{True} ❖$$

Try This   Insert inclusion symbols to make $5 + 30 \div 7 - 4 = 15$ true.

In algebra you are often asked to *evaluate* an expression. To do this, you replace the variables with the numbers that are assigned to those variables. Then proceed with the computation using the order of operations.

EXAMPLE 3

Evaluate $5x^2 + 7y$ when x is 3 and y is 2.

Solution➤

	$5x^2 + 7y$
Replace x with **3** and y with **2**.	$5 \cdot 3^2 + 7 \cdot 2$
First, square the 3.	$5 \cdot 9 + 7 \cdot 2$
Perform all of the multiplications.	$45 + 14$
Finally, add the results.	$45 + 14 = 59$ ❖

EXAMPLE 4

Show the keystrokes and the answer for $\dfrac{57 + 95}{16} + \dfrac{220}{88 + 104}$.

Solution➤

Since the entire quantity $57 + 95$ is divided by 16, place parentheses around $57 + 95$. Do the same for $88 + 104$.

$$(57 + 95) \div 16 + 220 \div (88 + 104)$$

Scientific Calculator

The keystrokes are shown.

(57 + 95) ÷ 16 + 220 ÷ (88 + 104) =

The answer is 10.646 to 3 decimal places. Is this reasonable? The first fraction is about $\frac{160}{16}$, or 10. The second fraction is about $\frac{200}{200}$, or 1. Thus, the estimated sum is about $10 + 1$, or 11. The answer 10.646 is reasonable. ❖

EXAMPLE 5

On Monday, John's father borrowed $3 and said that he would repay John double the amount plus $1 on Friday. The next Monday, John's father borrowed the amount he gave John the previous Friday and said that he would repay him double that amount plus $1 on Friday. The next week he did the same. On the third Friday, John figured his father owed him 2{2[2(3) + 1] + 1} + 1 dollars. How much money would John receive that Friday?

Solution►

When you simplify an expression that contains several pairs of inclusion symbols begin with the innermost pair and work outward.

$$2\{2[2(3) + 1] + 1\} + 1$$
$$2\{2[7] + 1\} + 1$$
$$2\{15\} + 1$$
$$31$$

If you use a calculator or a computer, the expression will usually contain only parentheses. The example would appear as 2(2(2(3) + 1) + 1) + 1. This does not affect the way you simplify the expression. ❖

If this pattern continues for one more week, how much money will John get on the fourth Friday?

Pattern Exploration and Technology

Calculator

Complete the pattern using a calculator.

$$1 \cdot 1 = 1$$
$$11 \cdot 11 = ?$$
$$111 \cdot 111 = ?$$

Predict the next three numbers in the pattern. Check your prediction with your calculator. What happens when you try to compute 111111 · 111111? Continue developing the pattern. What happens as the number of ones increases? What are the limitations of using a calculator to explore this pattern?

Using Algorithms Ask students to find as many values as they can for the expression 10 + 4 · 5 − 3 · 2. Then ask what problems result if we accept more than one value for this expression. Explain that to avoid such difficulties, mathematicians have agreed on a standard order of operations that allows everyone to get the same value for a given expression.

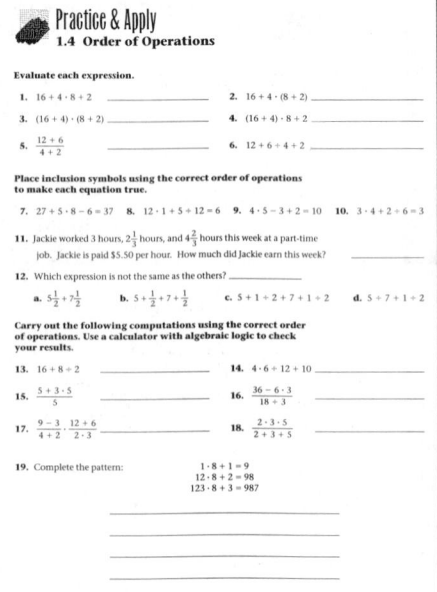

EXERCISES & PROBLEMS

Communicate

1. Explain how $3 + 2 \cdot 4$ can give two different answers.

2. Give two possible answers to the computation $20 \div 2 \cdot 5$.

3. Which answer to $20 \div 2 \cdot 5$ is correct? Why?

4. Describe the order of the steps for simplifying $\{[3(8-4)]^2 - 6\} \div (4-2)$.

5. Explain why rules called *the order of operations* are necessary for computation.

6. Explain why 3 is a reasonable estimate for $\frac{173 + 223}{151 - 21}$.

Practice & Apply

Place inclusion symbols using the correct order of operations to make each equation true. Tell how you would use a calculator to check your answers.

7. $28 - 2 \cdot 0 = 0$
   $(28 - 2) \cdot 0 = 0$

8. $59 - 4 \cdot 6 - 4 = 51$
   $59 - 4 \cdot (6 - 4) = 51$

Simplify using the correct order of operations. If answers are not exact, round them to the nearest thousandth.

9. $57 \cdot 29 + 89$ 1742

10. $7.2(9.8) + 1.2$ 71.76

11. $89 + 57 \cdot 29$ 1742

12. $0.3(1.5) + 9$ 9.45

13. $43 \cdot 32 + 91 \cdot 67$ 7473

14. $4.5(7.5) + 9.0(2.4)$ 55.35

15. $\frac{28 + 59}{97 - 17}$ 1.088

16. $\frac{9.6 - 1.7}{7.2 + 0.7}$ 1

17. $\frac{43 \cdot 91}{8 \cdot 25}$ 19.565

18. $157 - 29 + 23 \cdot 9$
    335

19. $91 \div 7 + 6$
    19

20. $187 - 34 \div 17$
    185

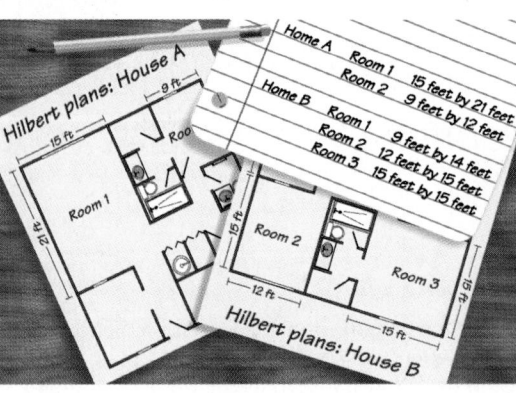

Hilbert plans: House A

Home A   Room 1   15 feet by 21 feet
         Room 2   9 feet by 12 feet
Home B   Room 1   9 feet by 12 feet
         Room 2   12 feet by 14 feet
         Room 3   15 feet by 15 feet

Hilbert plans: House B

Home Improvement Mr. Hilbert owns two rental properties, Home A and Home B. He wants to carpet two rooms in Home A and three rooms in Home B. He finds a carpet store selling carpet for $12.99 per square yard. The dimensions of the rooms in his homes are shown on the blueprint. HINT: $1\text{yd}^2 = 9\text{ft}^2$

21. Find the total number of square feet of carpet needed for Home A. 423

22. Find the total number of square feet of carpet needed for Home B. 531

23. Find the cost of carpeting the rooms in Home A. $610.53

24. Find the cost of carpeting the rooms in Home B. $766.41

Evaluate each expression.

25. $2(5 + 4) \div 9$ 2   **26.** $12 - 7 \cdot 3 + 9^2$ 72

27. $3 - 1 + 24 \div 6$ 6   **28.** $7 + 6 \div 2 \cdot 10$ 37

Given that a is 5, b is 3, and c is 4, evaluate each expression.

29. $a + b - c$ 4   **30.** $a^2 + b^2$ 34   **31.** $a^2 - b^2$ 16

32. $(a + b) \cdot c$ 32   **33.** $a^2 - b - c$ 18   **34.** $a^2 - (b + c)$ 18

35. Technology What answer do you get if you enter 2100.3636

$$32 \boxed{\times} 38 \boxed{\div} 11 \boxed{\times} 19 \boxed{=} ?$$

Explain how the calculator gets this answer.

36. Statistics
The teacher's grade
book shows the results
of an algebra quiz. Find
the average for the class.
Show your method.
     87.5

SCORE	100	90	80	70	60
NUMBER OF STUDENTS	4	12	7	0	1

Look Back

37. Advertising A photographer arranges cans of soup in a large triangle for
a supermarket ad. The top row contains 1 can, and each of the rows contains
1 can more than the row above it. If there are 10 rows of cans, how many cans
will the photographer need to form the triangle display? **[Lesson 1.1]**
     55

**Determine when the first constant difference appears, and find the
next 2 terms of each sequence. [Lesson 1.2]**

38. 2, 5, 8, 11, 14, . . .   **39.** 1, 6, 13, 22, 33, . . .   **40.** 2, 9, 20, 35, 54, 77, . . .
     17, 20   46, 61   104, 135

41. Geometry The perimeter of a rectangle is the distance around the border,
or the sum of twice the length and twice the width. The area is the product of
the length and width. For what length and width will the perimeter and the area
be the same number? 4

Look Beyond

An exponent indicates repeated multiplication. The 3^2 means $3 \cdot 3$.
Indicate whether each equation is true or false.

42. $(3 + 4)^2 = 3 + 4^2$ F   **43.** $(3 + 4)^2 = (3 + 4)(3 + 4)$ T

44. $(3 + 4)^2 = 3^2 + 2(3)(4) + 4^2$ T   **45.** $(3 + 4)^2 = 3^2 + 4^2$ F

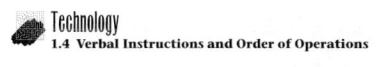

Objectives

- Identify and graph ordered pairs.
- Use a graph to find a solution.
- Understand the concept of a linear function.

RESOURCES

• Practice Master	1.5
• Enrichment Master	1.5
• Technology Master	1.5
• Lesson Activity Master	1.5
• Quiz	1.5
• Spanish Resources	1.5

Assessing Prior Knowledge

Have the students copy a number line (-3 to 3). Instruct the students to place points on the number line corresponding to each number.

$A\ (-3)\quad B\ (2.5)\quad C\left(-\frac{1}{3}\right)$

$$\left[\begin{array}{c} A \quad C \qquad B \\ \xleftarrow{\ \bullet\ \ \bullet\ \ \bullet\ \ \bullet\ \ \bullet\ \ \bullet\ \ \bullet\ } \\ -3\ -2\ -1\ 0\ 1\ 2\ 3 \end{array}\right]$$

TEACH

Lists, equations, and graphs can all be used to describe relationships between different sets of numbers. Point out how difficult it would be to interpret large amounts of data if it were not presented in a graph.

Aongoing ASSESSMENT

$\frac{1}{2}$ is halfway between 0 and 1.
$-\frac{1}{2}$ is halfway between -1 and 0.

Representing Linear Functions by Graphs

why *Sometimes it is difficult to recognize patterns in a numerical problem. It is often easier to see the pattern from a picture. In algebra, a graph is used to display a picture of a mathematical pattern.*

You can describe the location of your hometown on a road map by using the letters and numbers in the margins.

To find Northfield, Minnesota, on a map, for instance, you look up from the letter K on the bottom and across from the number 18 in the margin of a state road map. The region where these strips overlap contains Northfield.

Graphing With Rectangular Coordinates

Cultural Connection: Africa The idea of graphing has been around for over 4500 years. Around 2650 B.C.E., Egyptians sketched curves using pairs of numbers to locate points. This idea gave rise to the use of coordinates.

The set of real numbers can be represented on a number line.

- To construct the number line, mark a point on the line as zero.

- Mark off equal spaces to the right from zero, and number them 1, 2, 3, 4, . . .

- Then mark off corresponding spaces to the left from zero, and number them $-1, -2, -3, -4,$ . . .

$$\xleftarrow{\ \bullet\ \bullet\ \bullet\ \bullet\ \bullet\ \bullet\ \bullet\ \bullet\ \bullet\ }\rightarrow$$
$$-4\ -3\ -2\ -1\ \ 0\ \ 1\ \ 2\ \ 3\ \ 4$$

Where is $\frac{1}{2}$ on this number line? $-\frac{1}{2}$?

ALTERNATIVE teaching strategy

Technology Use a graphics calculator or computer to graph the equation $y = 3x - 1$. Find and write a set of ordered pairs (x, y) that satisfies the equation.

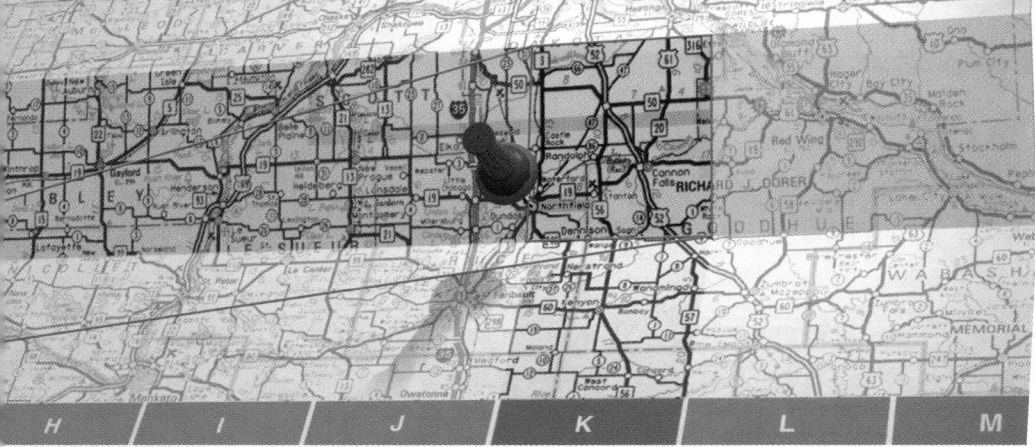

Cultural Connection: Europe In the seventeenth century, Rene Descartes, a French mathematician and philosopher, used a horizontal and vertical number line to divide a plane into four regions, called **quadrants**.

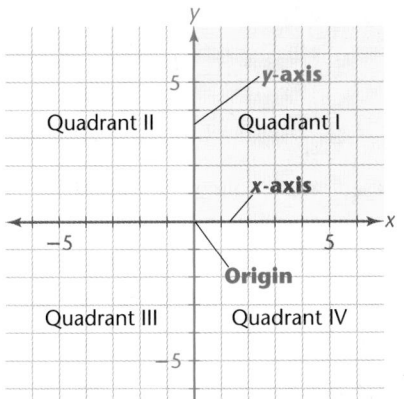

The horizontal line is called the **x-axis**, and points on this axis are called **x-coordinates**.

The vertical line is called the **y-axis**, and points on this axis are called **y-coordinates**.

The x- and y-axes intersect at the **origin**.

Letters other than x and y can represent the axes. For example, you might see the x-axis shown as the t-axis when t refers to time in a problem.

Graphing on the entire coordinate system will appear later in this book. For the rest of this chapter, however, the emphasis is on the first quadrant.

The rectangular, or Cartesian, coordinate system was named for Rene Descartes.

Coordinates give the address of a point. They are written as an **ordered pair**, indicated by two numbers in parentheses, (x, y). The number from the x-axis will appear first in an ordered pair.

To locate the point shown on the graph, start at the origin, (0, 0). Move to the right 10 units along the x-axis, then move up 5 units to the point. The ordered pair is (**10**, **5**).

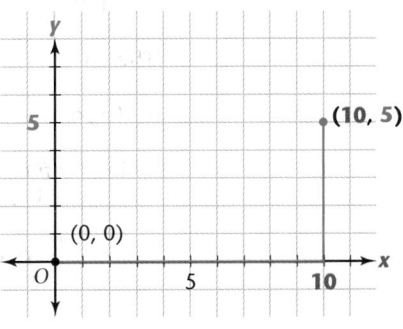

Find a way to locate a point on a plane without using the rectangular coordinate system.

Biology The heart-rate example is just one way that exercise physiologists use algebraic formulas to study how the human body responds to exercise stress. Students can interview an exercise physiologist to find out how math is used in the profession. See Example 2, page 37.

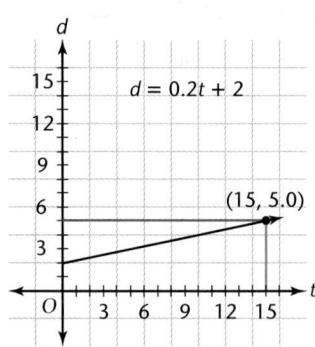

Sabrina normally rides her bicycle at a constant rate of 0.2 miles per minute.

Linear Functions

TEACHING *tip*

Students may be confused by the fact that Sabrina is already 2 miles from home when she starts her bicycle ride. Show the students how the problem can be broken down into two parts. First calculate the distance from school, then add the 2 miles from home.

Alternate Example 1

Jeremy drives an average of 50 mph. He leaves from a city 100 miles from home. Then he drives for $5\frac{1}{2}$ hours in a direction away from his home. How far away from home is Jeremy?
[**375 miles**]

TEACHING *tip*

For students who have difficulty with the table for the distance function, rewrite the ordered pairs in the more traditional table form.

Once you organize the data from a problem, you can form ordered pairs and graph the data.

EXAMPLE 1

On Wednesday, Sabrina left school and rode her bicycle for 15 minutes in a direction away from home. If Sabrina lives 2 miles from school, how far is Sabrina from home?

Solution▶

The equation that models Sabrina's distance from home with respect to time spent bicycling is based on the following relationship.

Distance = *rate · time* + miles from home to school.

Replace *distance* with *d*, *rate* with 0.2, *time* with *t*, and miles from home to school with 2.

The equation is now written as $d = 0.2t + 2$.

To make a table of values, substitute values for *t* in minutes, and find the corresponding values for *d* in miles. The ordered pairs are then plotted and connected by a line.

t	0.2t + 2.0	d	(t, d)
0	0.2(0) + 2.0	2.0	(0, 2.0)
1	0.2(1) + 2.0	2.2	(1, 2.2)
2	0.2(2) + 2.0	2.4	(2, 2.4)
3	0.2(3) + 2.0	2.6	(3, 2.6)
4	0.2(4) + 2.0	2.8	(4, 2.8)
5	0.2(5) + 2.0	3.0	(5, 3.0)
. . .			
15	0.2(15) + 2.0 =	5.0	(15, 5.0)

The table and the graph indicate that after 15 minutes, Sabrina is 5 miles from home. ❖

ENRICHMENT **Technology** On a graphics calculator, enter the equation $y = 224x - 16x^2$, using the following settings:
Xmin = 0, Xmax = 20, Xscl = 1;
Ymin = 0, Ymax = 800, Yscl = 50.
List several ordered pairs for this graph. What does the graph represent? Is the equation linear? Why or why not? [**Altitude/time relationship (rocket problem Lesson 1.2) No, the graph is curved.**]

INCLUSION strategies **Using Visual Models** Draw and label a graph for Example 2 on an overhead transparency. Show students the graph first, and ask them to find the appropriate heart rate for a given age. Help the students derive the formula for the line. Next, have students use graph paper and a straight edge to create their own lines and matching equations.

The graph of the equation $d = 0.2t + 2$ is a straight line. For this reason the equation is called a **linear equation.** This equation generates a set of ordered pairs, (t, d). We say that d is a function of t, or that d is dependent on t. Thus, $d = 0.2t + 2$ is a **linear function.** In the ordered pairs, the set of first values, t, is the **domain** of the function. The set of second values, d, is the **range** of the function.

In the function $d = 0.2t + 2$, t is called the independent variable. The d is called the dependent variable because its value depends on the value chosen for t.

Try This   How far would Sabrina be from home if she rides for 60 minutes?

EXAMPLE 2

Physiology   Represent the maximum heart rate by r and the age by a. Write a function to express the maximum heart rate in terms of a. Graph the function for ages 10 to 50.

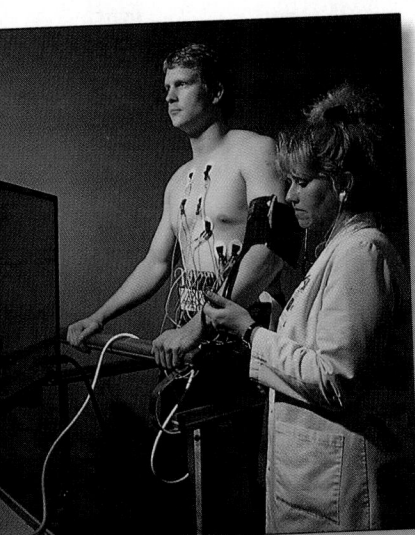

A reasonable estimate for the maximum heart rate during exercise should be no more than 220 beats per minute minus the person's age.

Solution▸

Since the rate, r, equals 220 beats per minute minus the age, write the equation $r = 220 - a$. Values for a are elements of the domain, and values for r are elements of the range. When graphing a function, it is customary to represent the domain on the x-axis and the range on the y-axis.

When the values 10, 20, 30, 40, and 50 are in the domain, the table shows how substitution allows you to find the corresponding values in the range.

To graph the function, enter **220 − x** after Y = on a graphics calculator. Set the limits of the domain and range in the range window. Then press GRAPH.

a	$220 - a$	r
10	$220 - 10$	210
20	$220 - 20$	200
30	$220 - 30$	190
40	$220 - 40$	180
50	$220 - 50$	170

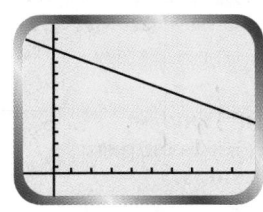

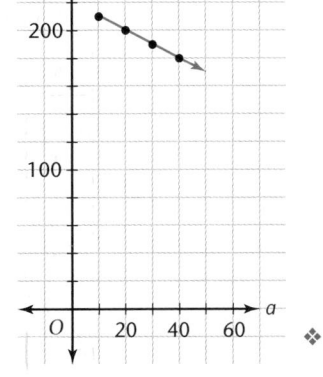

For some people the proper level of heart rate should not exceed 60% of the maximum found in this example. Individuals should consult a doctor if there is any doubt about their capability.

RETEACHING **the lesson**   **Using Patterns**   On graph paper, have students plot $(1, 3)$, $(4, 12)$, and $(5, 15)$. Explain that these numbers represent different points along a hiking trip. After 1 hour, the hiker has gone 3 miles, and after 4 hours, 12 miles. Have students connect the points to form the line and derive the linear function $y = 3x$. Extend the example by letting the hiker drive 7 miles before starting to hike. Ask students to derive the resulting equation, $y = 3x + 7$, and graph.

Try This

If Sabrina travels 12 miles per hour for 60 minutes, she travels 12 miles. Add the 2 initial miles to get 14 miles from home.

Alternate Example 2

A taxi ride costs $2.00 plus $0.25 per $\frac{1}{10}$ mile. Express the function letting x be the number of miles. Graph the function, and find the cost of a 1.5-mile trip.

$[y = 2 + 2.5x; \$5.75]$

TEACHING *tip*

Point out that often a letter is chosen for a variable because it is the first letter of a word, such as distance, rate, time, etc. However, x and y are usually the only variable choices available on a calculator.

TEACHING *tip*

Technology   A suitable calculator range for Example 2 is: Xmin = 0, Xmax = 100; Ymin = 0, Ymax = 250.

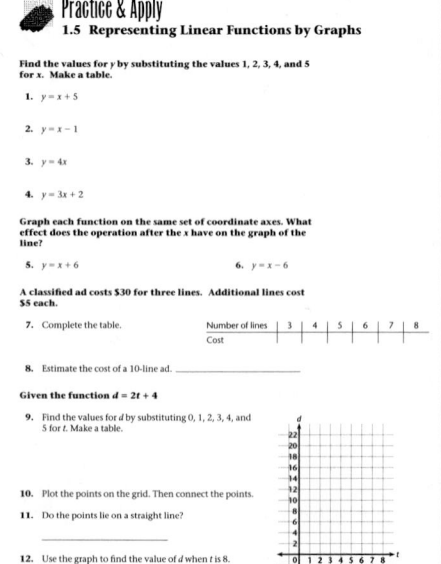

EXERCISES & PROBLEMS

Communicate

1. Explain how you find the coordinates of a given point from a graph.
2. Is (6, 7) the same point as (7, 6)? Explain.
3. Describe the steps for plotting the point with the coordinates (7, 3) on a graph.
4. Describe the relationship between the coordinates of an ordered pair and the domain and range of a function.
5. Compare the advantages of displaying data in a table and on a graph.
6. How would you make a table of the values for x and y for the equation $y = 2x + 5$?

Practice & Apply

Graph each list of ordered pairs. State whether they lie on a straight line.

7. (1, 3), (2, 6), (3, 9) yes
8. (1, 5), (2, 4), (3, 1) no
9. (1, 10), (2, 7), (3, 2) no
10. (1, −3), (2, −6), (3, −9) yes
11. (5, 2), (7, 2), (9, 2) yes
12. (4, 1), (4, 5), (4, 9) yes

What are the coordinates of the given points?

13. A   14. B   15. C   16. D
(0, 7)   (7, 7)   (5, 4)   (6, 0)

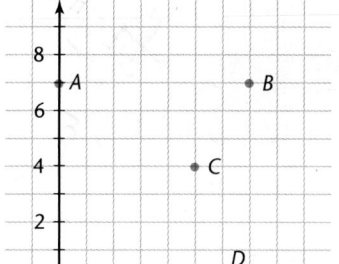

Graph and compare the following two linear functions. What effect does the operation before the x have on the graph of the line?

17. $y = 3 + x$ 18. $y = 3 − x$

Graph each of the following functions on the same set of coordinate axes, and compare the graphs. Explain your conclusions.

19. $y = x + 7$ 20. $y = x − 7$ 21. $y = 7 − x$ 22. $y = −7 − x$

Find the values for y by substituting 1, 2, 3, 4, and 5 for x. Make a table.

23. $y = x + 3$ 24. $y = x + 4$ 25. $y = 2x$ 26. $y = 2x + 5$

17–18.

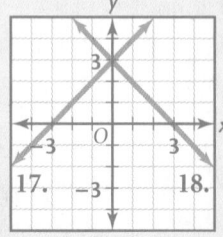

The operation changes the direction of the slope of the line.

The answers to Exercises 19–22 can be found in Additional Answers beginning on page 729.

23.

x	1	2	3	4	5
y	4	5	6	7	8

24.

x	1	2	3	4	5
y	5	6	7	8	9

25.

x	1	2	3	4	5
y	2	4	6	8	10

26.

x	1	2	3	4	5
y	7	9	11	13	15

28.

h	0	1	2	3
d	0	3	6	9

Don walks at a rate of 3 miles per hour. You can determine the distance that he walks by multiplying the rate times the number of hours that he walks.

27. Represent hours by h, and write an equation for the distance, d. $d = 3h$

28. Make a table to show how far Don walked in 0, 1, 2, and 3 hours.

29. Draw a graph of the function.

30. **Health** Using the equation given in Example 2, what should be the maximum heart rate for a 36-year-old person? 184

Suppose a T-shirt company charges a $3 handling fee per order.

31. How much does an order of 2 shirts cost? $19.00

32. How much does an order of 5 shirts cost? $43.00

33. Make a set of ordered pairs from the given information, and plot them as points on a graph. Do the points lie on a straight line? yes

34. Use your graph to see how many shirts you can buy with $75. 9

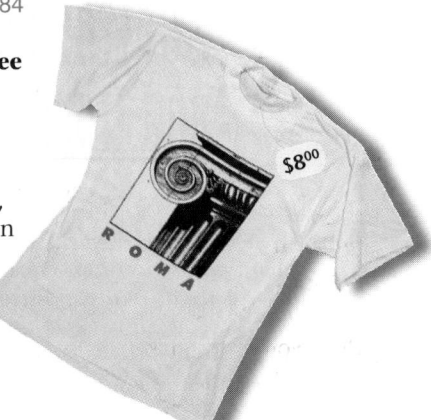

Look Back

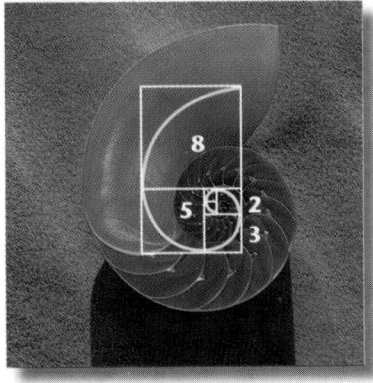

The Fibonacci sequence is found in many examples from nature.

35. What are the next 3 terms for the sequence 1, 1, 4, 10, 19 . . .? **[Lesson 1.1]** 31, 46, 64

36. How many differences are needed to reach a constant for the sequence 1, 2, 6, 15, 31, . . . ? What is the constant difference for the sequence? **[Lesson 1.2]** 3; 2

37. The sequence 1, 1, 2, 3, 5, 8, . . . is a famous sequence named for the Italian mathematician Fibonacci. Find the next two terms. Will the method of differences help? What kind of patterns do the differences show? 13, 21 **[Lesson 1.2]**

38. Guess the closest whole number value for x that satisfies the equation $3x + 406 = 421$. Test your guess using substitution. **[Lesson 1.3]** 5

39. Make a table to solve the equation $7x + 4 = 40$. $5\frac{1}{7}$ **[Lesson 1.3]**

Simplify. **[Lesson 1.4]**

40. $2 \cdot 14 \div 2 + 5$ 19

41. $6 + 12 \div 6 - 4$ 4

42. $4[(12 - 3) \cdot 2] \div 11$ $6\frac{6}{11}$

43. $[3(4) - 6] - [(15 - 7) \div 4]$ 4

Look Beyond

44. The graph of the path of a rocket is a parabola. The highest point of this curve is called the vertex. Graph each set of points, and determine which point is the vertex.

x	0	1	2	3	4	5	6	7	8	9	10
y	0	9	16	21	24	25	24	21	16	9	0

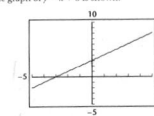
29.

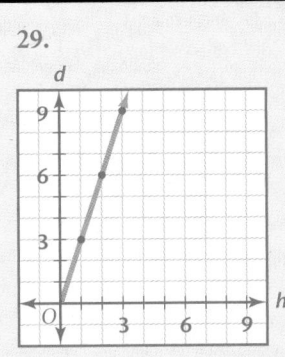

33.

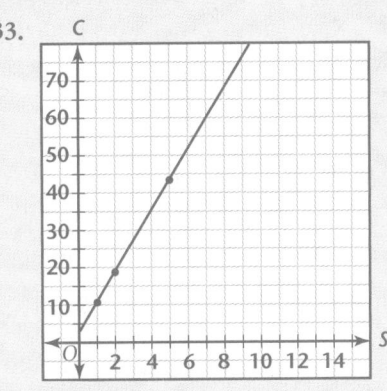

44.

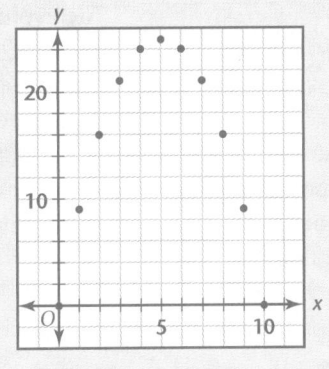

The vertex is (5, 25).

Exploring Scatter Plots and Correlation

PREPARE

Objectives

- Interpret data in a scatter plot.
- Identify a correlation based on the orientation of points in a scatter plot.

RESOURCES

- Practice Master   **1.6**
- Enrichment Master   **1.6**
- Technology Master   **1.6**
- Lesson Activity Master   **1.6**
- Quiz   **1.6**
- Spanish Resources   **1.6**

Assessing Prior Knowledge

Have students draw a coordinate plane and plot the following ordered pairs.

A. $(15, 700)$

B. $(10, 350)$

[Check student graphs.]

TEACH

To reinforce the concept of a relationship between different sets of data, brainstorm with students about different connections (for example, exercise and longevity).

Math Connection

Statistics The lesson introduces the concept of statistical correlation and emphasizes that a correlation between two variables does not necessarily represent a cause-and-effect relationship.

why *People frequently try to determine relationships between different factors such as math and science aptitude. How can you tell whether these factors are related? Correlations provide a way to measure numerically how well the sets of data are related.*

A newspaper article reported that students who do well in math also do well in science.

Scatter Plots

An effective way to see a relationship from data is to display the information as a **scatter plot**. Before you can use a scatter plot, however, you need to

- have clearly defined variables you can assign to the axes,
- form ordered pairs from the data you have, and
- locate each data point on the scatter plot the way you would graph a point on the coordinate plane.

You can then examine the patterns for clues to see how closely the variables are related.

•Exploration *Find a Correlation*

STATISTICS *Connection*

Ten students took a series of aptitude tests. The table shows the scores on two parts of the test. You want to know if students who do well in math also do well in science. Once you create a scatter plot, you can see the pattern for the relationship.

Student	A	B	C	D	E	F	G	H	I	J
Math	61	40	80	21	62	54	20	33	75	51
Science	70	38	92	50	68	41	38	20	73	48

ALTERNATIVE teaching strategy

Technology Use a graphing calculator to introduce the exploration. Enter the data into a list and graph as a scatter plot. Ask students to make a conjecture about the correlation. Use these calculator window or range settings:

Xmin = 0, Xmax = 100, Xscl = 10;
Ymin = 0, Ymax = 100, Xscl = 10.

1. Look at the scores of the students who did well in math. How well did they do in science?

2. Look at the scores of the students who did *not* do well in math. How well did they do in science?

3. Do the data points appear broadly scattered or do they seem to cluster along a line?

4. Does the line of the data rise or fall as you go from left to right?

5. What does the data pattern show about the relationship between the variables? ❖

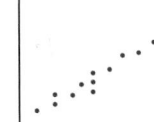

Math Scores

The answers to exploration steps can be found in Additional Answers beginning on page 729.

Exploration Notes

The exploration uses an established correlation between math aptitude and science aptitude. Tell students that statistics are generalized explanations for trends in large amounts of data. Individual exceptions will always occur.

Correlation

Scatter plots show a picture of how the variables relate to each other by displaying how well data points can fit the line.

A strong positive correlation As one variable increases, the other also tends to increase. This shows a *positive* correlation. When the points are nearly in a line, there is a *strong* correlation.

A strong negative correlation As one variable increases, the other tends to decrease. This shows a *negative* correlation. When the points are nearly in a line, there is a *strong* correlation.

Little or no correlation As one variable increases, you cannot tell if the other tends to increase or decrease. There is a weak correlation or none at all.

Aᵒⁿᵍᵒⁱⁿᵍ SSESSMENT

5. The data pattern shows a correlation between math and science scores. Students with high math scores tend to have high science scores, and vice versa.

Tᴇᴀᴄʜɪɴɢ tip

Some students may not immediately see that the procedure is the same for plotting data on a coordinate plane and a scatter plot.

Remind students that the procedure for plotting points is the same. Let them know, however, that it is not necessary to use the $(0, 0)$ origin when working with scatter plots. If the origin, $(0, 0)$, is used, a jagged symbol on the axis shows that the numbering on the axis is not continuous from zero.

APPLICATION

Compare the scatter plots with the statements.

1. The variables studied are *age* and *distance* from home to school.

2. The variables studied are *age* and the *time it takes to run a fixed distance.*

3. The variables studied are *age* and *height.*

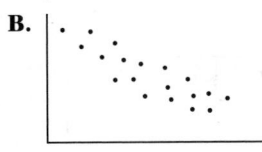

A.   B.   C.

Match the statements 1, 2, and 3 with the appropriate scatter plot.

interdisciplinary CONNECTION **Social Science** Correlation statistics are widely used in social science research. For example, the relationship between a student's math and science aptitude has been studied extensively by educational psychologists. Have students research various correlations studied in the social sciences.

Use Transparency ▶ 6

Scatter plot A

Scatter plot B

Scatter plot C

Scatter plot A shows a positive correlation. A positive correlation occurs when both variables increase. The correlation is strong. The data points are close to the shape of a straight line. This indicates that it is easy to predict the behavior of one variable by the behavior of the other. As a child's age increases, his or her height increases. This scatter plot would fit Statement 3.

Scatter plot B shows a negative correlation. A negative correlation occurs when one variable increases and the other variable decreases. The data points are close to the shape of a straight line, which is characteristic of a strong correlation. As a child gets older, bigger, and stronger, the time it takes the child to run a fixed distance decreases. This scatter plot would fit Statement 2.

Scatter plot C shows little or no correlation. The data points are randomly scattered. You cannot tell whether the correlation is positive or negative. The distance that a student lives from school has no relation to age. This scatter plot would fit Statement 1. ❖

Do not confuse a strong positive or negative correlation with a cause-and-effect relationship. Over the past few years there has been a strong positive correlation between annual consumption of diet soda and the number of traffic accidents reported in one Midwestern state. No one would claim that drinking diet soda would cause accidents, or that the trauma of being in an accident would cause people to drink more soda. Often a third variable can be responsible for the unexpectedly strong correlation between two otherwise unrelated variables. Can you think of what another variable might be in the Midwestern state?

◆ APPLICATION

Demographics   Make a scatter plot for the given data, showing the number of people working on farms for various years. Describe the correlation as strong positive, strong negative, or little to none.

A ongoing SSESSMENT

One answer is that an increase in population in the state would be responsible for an increase in soda consumption and accidents.

Applications appear in the Exploration lessons. They provide a practical example of the concepts presented in an exploration. The Application also reviews the concepts from the explorations. This can help students that have difficulty identifying or comprehending the key concepts.

Decline in Number of Farm Workers

Year	Number
1940	8995
1950	6858
1960	4132
1970	2881
1980	2818
1990	2864

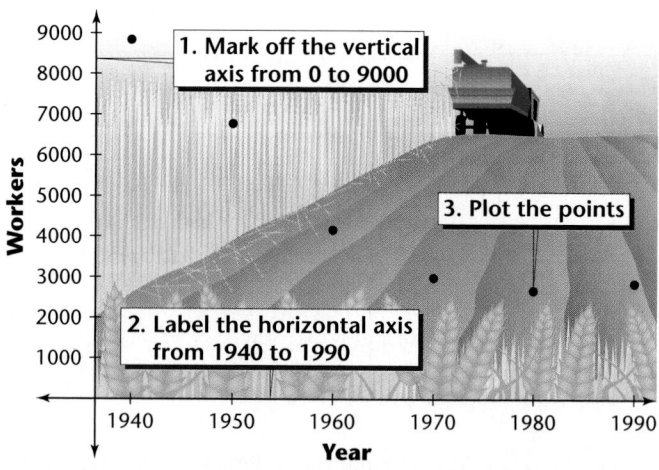

1. Mark off the vertical axis from 0 to 9000

2. Label the horizontal axis from 1940 to 1990

3. Plot the points

ENRICHMENT   **Technology**   Have students look up the records for some sport and write a table that relates the year to the record. Students should enter the data in a graphics calculator to obtain a scatter plot. Have the students guess the correlation before they calculate it.

INCLUSION strategies   **Using Discussion**   Students who want exact answers are likely to feel uncomfortable with the lesson. Make it clear that in the real world, especially in the social sciences and health professions, there are many correlations that cannot be precisely determined.

As the years increase, the number of workers decreases, so this is a strong negative correlation. However, you can notice a leveling off over the last 20 years. ❖

 Think of two examples of related data. Describe the kind of correlation you can expect from the data.

Exercises & Problems

Communicate

1. Use the scatter plot from the Exploration to explain how you would decide what type of correlation the math and science scores have.

2. How would you label the axes of a scatter plot involving the *cost* of notebooks and the *number of pages* each has? Which axis would you use for the cost? Which axis would you use for the number of pages?

3. Discuss how a correlation provides you with information about the variables.

Describe the correlation for the following scatter plots as strong positive, strong negative, or little to none. Explain the reason for your answer.

4.

5.

Cost of Living The following table shows the value of food items that cost $1.00 in 1982 has increased in cost over the years.

Year	1982	1985	1988	1991
Cost	1.00	1.06	1.18	1.36

6. Tell which axis you would use for the cost of food, and which you would use for the year. Why?

7. Tell how you would plot the points for the data in the table.

RETEACHING the lesson **Using Models** Review the connection between a scatter plot and correlation by giving a correlation and asking students to draw a corresponding scatter plot. Have students draw scatter plots for the following: strong positive correlation, strong negative correlation, and no correlation.

CRITICAL Thinking
Answers may vary. For example, car trip mileages and gasoline consumed per trip should show a high positive correlation. A person's height and the distance from the top of his/her head to the ceiling should have a perfect negative correlation.

ASSESS

Answers to all Communicate exercises can be found in Additional Answers beginning on page 729.

Selected Answers
Odd-numbered Exercises 9−31

Assignment Guide
Core 1−11, 16−34

Core Two-Year 1−32

Technology
All of the exercises exploring correlation can be done using a graphing calculator.

Error Analysis
When making scatter plots, students may have difficulty determining which variables should be represented by the axes and what scale to use. Discuss with the students how to set up the scatter plot for a few of the exercises.

Practice & Apply

Describe the correlation as strong positive, strong negative, or little to none.

8.
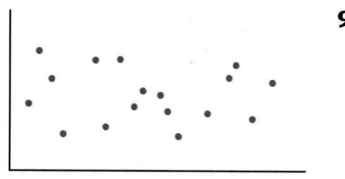

little or none

9.

strong positive

🔎 **Statistics** Use the data in the table for the following exercises.

Student	A	B	C	D	E	F	G	H	I	J
English	50	63	70	48	70	52	43	78	70	65
Math	61	40	80	21	62	54	20	33	75	51
Science	70	38	92	50	68	41	38	20	73	48

[34 ÷ 2] − 1 ? 16
17 − 1 ? 16
16 = 16
Read Chapters 8 & 9

10. Graph the student scores on math against the scores on English, and explain the correlation. little or none

11. Graph the student scores on science against the scores on English, and explain the correlation. little or none

Geography Use the following data to answer Exercises 12–15.

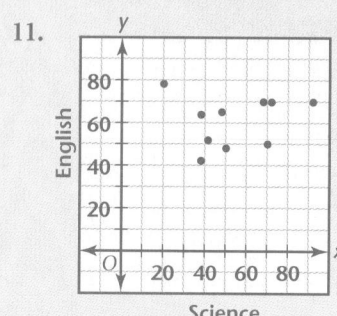

City	North Latitude	Elevation in Feet	Maximum Normal Temp. for Jan. (in °F)
Miami, FL	26°	7	75
Charleston, SC	33°	40	50
Washington, DC	39°	10	43
Boston, MA	42°	15	36
Portland, ME	44°	43	31

Look down the column for north latitude, which is arranged in increasing order. As north latitude increases, is there an obvious tendency in what happens to

12. elevation? no
13. temperature? yes
14. Describe the correlation between latitude and elevation. none
15. Describe the correlation between latitude and temperature. strong negative

 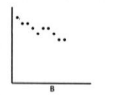
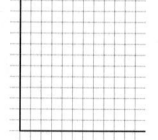
10.

English vs Math

11.

English vs Science

16. Sports Make a scatter plot of the following data for the women's 400-meter run in the Olympics.

Year	1964	1968	1972	1976	1980	1984	1988	1992
Seconds	52.00	52.00	51.08	49.29	48.88	48.83	48.65	48.83

17. Describe the correlation. strong negative

18. Draw these two coordinate axes on your graph paper. Plot the same points, (1, 2), (2, 3), and (3, 4), on each set of axes. Connect the points with a line.

a.

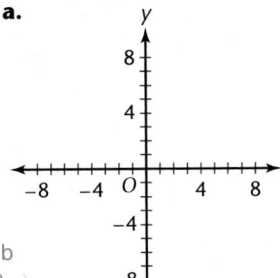

b.

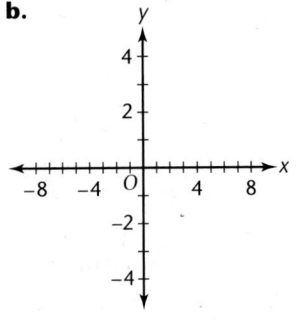

19. Which graph is steeper, a or b? b

20. What caused one graph to be steeper, even though the points are the same? Scale of b is double.

21. How does a change in scale on the axes affect the steepness of the line? *Scale* refers to the distance between units on an axis. Stretching increases steepness.

22. Does the scale of the graph affect a correlation? Explain your answer. no

Look Back

Simplify.

23. $\sqrt{144}$ 12   **24.** $\sqrt{7+2}$ 3   **25.** $\sqrt{171-2}$ 13   **26.** $2\sqrt{16}$ 8

27. What is the greatest common factor for 30, 60, and 15? 15

28. Pair the equivalent expressions.
  a. $2 \cdot 2 \cdot 2 \cdot 2$ **b.** $4(2)$ **c.** $2 + 2 + 2 + 2$ **d.** 2^4
  a and d;   b and c

Plot each point on the coordinate plane. [Lesson 1.5]

29. (5, 3)   **30.** (5, 5)   **31.** (3, 4)   **32.** (7, 2)

Look Beyond

33. Make a scatter plot from the data. Use a straightedge to draw a line that best fits the points on the scatter plot.

x	0	3	6	9	12
y	14	20	26	32	38

34. What is the relationship of the points to the line? very strong

16.

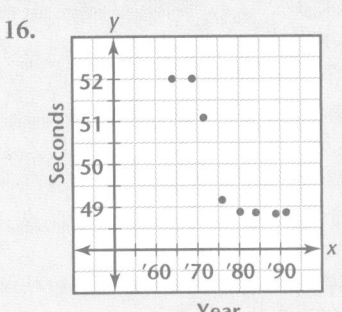

The answer to Exercise 18 can be found in Additional Answers beginning on page 729.

29.–32.

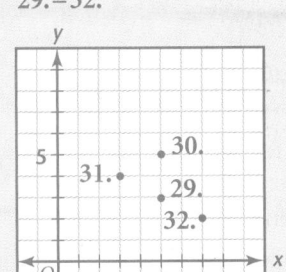

33.

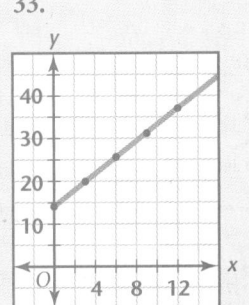

Look Beyond

Remind students that correlation depends, not upon the steepness of the line, but rather on how close the points are to the line.

Technology
1.6 Transposing Scatter Plot Data

The spreadsheet represents a table of values relating two variables *x* and *y*. Scatter plot I is the scatter plot for (*x*, *y*). Scatter plot II comes from transposing *y* and *x*.

Scatter plot I suggests a positive correlation between *x* and *y*.
Scatter plot II suggests a positive correlation between *y* and *x*.

In Exercises 1-8, make a scatter plot for each data set.

9. If *x* and *y* have a correlation, do you think that *y* and *x* have the same correlation?

Objectives

- Find the line of best fit on a scatter plot.
- Understand and determine closeness of fit.

RESOURCES

- Practice Master **1.7**
- Enrichment Master **1.7**
- Technology Master **1.7**
- Lesson Activity Master **1.7**
- Quiz **1.7**
- Spanish Resources **1.7**

Assessing Prior Knowledge

Make a scatter plot for the following set of data:

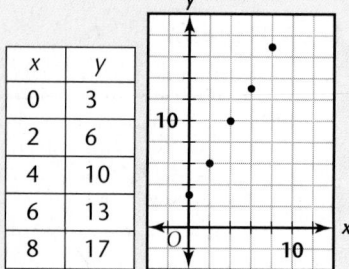

x	y
0	3
2	6
4	10
6	13
8	17

TEACH

Working with real data and with different kinds of scales will help students realize that linear relationships are applicable in many situations.

TEACHING tip

To help students understand the two sets of data, ask whether the number of compositions depends upon age or whether age depends upon the number of compositions.

LESSON 1.7 Finding Lines of Best Fit

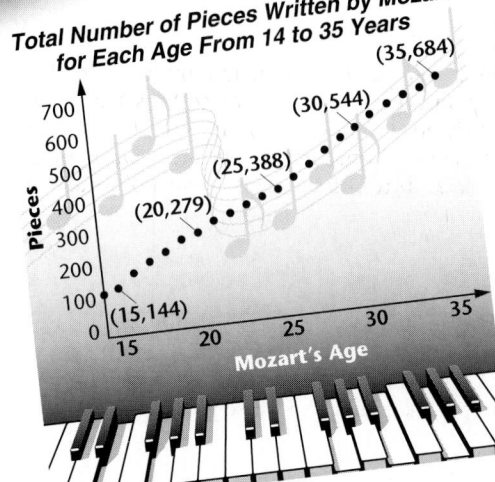

Total Number of Pieces Written by Mozart for Each Age From 14 to 35 Years

Why Scatter plots are used to show a trend in the data. This scatter plot presents an overall picture of Mozart's musical output. Since the points lie very nearly on a straight line, it tells you that Mozart wrote at an amazingly steady pace. When a trend is represented by a line that fits the data, you can study the line to see how the data behave in general.

Music History Wolfgang Amadeus Mozart (1756–1791) composed many musical pieces in his lifetime. The scatter plot shows the total number of musical pieces that Mozart had written as he reached different ages. For example, the point (35, 684) shows that by age 35 Mozart had written 684 pieces. The graph starts at (14, 121), indicating that by age 14 he had already written 121 musical pieces.

The Line of Best Fit

The line of best fit represents an approximation of the data on the scatter plot. The following steps should help you find this line.

Use a straightedge, such as a clear ruler, to model the line.

To fit the line to the points, choose a line that best matches the trend.

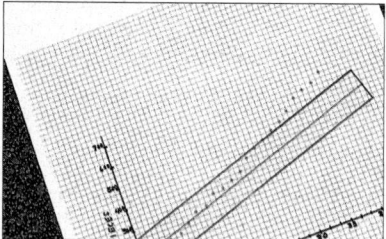
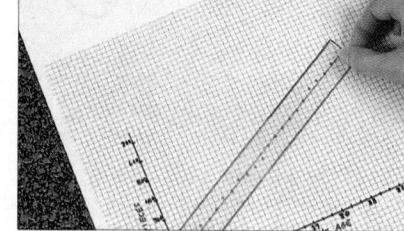

The line does not necessarily have to pass through any of the points.

ALTERNATIVE teaching strategy

Technology Use data from various sources to create tables. Enter the tables into a graphics calculator and graph the scatter plots. Graph the line of best fit, and have students conjecture about the relationship between the fit of the data around the line of best fit and the correlation coefficient.

If you look at the numbers along the *age*-axis, you see that there is a difference of 21 years between ages 14 and 35. How many pieces of music did Mozart write during that period of time? Form a ratio that compares the difference in the number of pieces to the difference in the number of years. What does this ratio tell you about Mozart's rate of composition?

Correlation and the Line of Best Fit

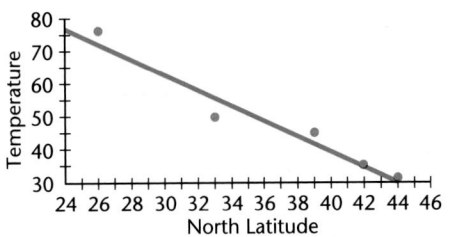

The scatter plot shows the maximum normal temperatures in January plotted for selected cities at different latitudes along the Atlantic coast. It also shows a computer-drawn line that best fits the data. In this first example, the points are close to the line.

The next scatter plot shows elevation in feet for selected cities along the Atlantic coast plotted against different average January temperatures. It shows a computer-drawn line that best fits the data. In this second example, the data points do not fit the line very closely.

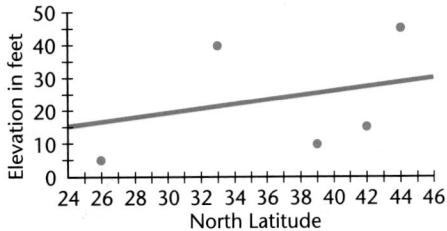

The measure of how closely a set of data points falls along a line is the **correlation coefficient,** *r*. The correlation coefficient can be computed automatically using a graphics calculator. The possible values for a correlation coefficient range from -1 to 1. The better the data fit the line, the closer the numerical value of the correlation coefficient is to *either* 1 or -1.

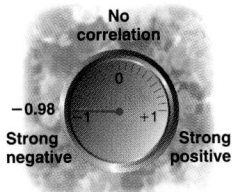

In the first scatter plot, the correlation coefficient computed on a calculator is -0.98, which is near -1. This indicates that the data are packed closely to a line that falls from left to right.

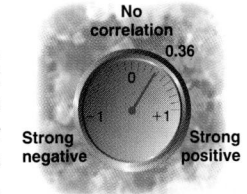

In the second scatter plot, the correlation coefficient from a calculator is 0.36, which is not close to 1 or -1. This indicates that the data do not cluster near the best fit line. Because the correlation is positive, the line rises from left to right.

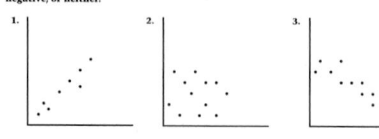
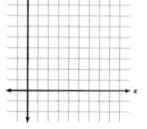

CRITICAL
Thinking

Think of several pairs of things that correlate well positively and several that correlate negatively. Try to identify the cause of the positive or negative correlation between the pairs.

EXAMPLE

Examine the scatter plots and the lines. Which line best fits the points, Line 1 or Line 2?

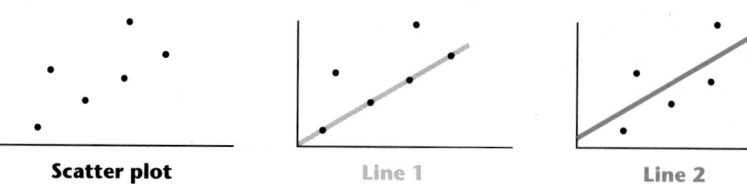

Scatter plot   Line 1   Line 2

A line of best fit should take all of the points into consideration.

Solution▸

Line 1 goes through four of the points but ignores the others.

Line 2 is closer to the four points than it is to the other two, but it takes all of the points into account. **Line 2** is a better fit, even though it doesn't go through any of the points. ❖

EXERCISES & PROBLEMS

Communicate

1. Explain how to determine the line of best fit between points using a piece of uncooked spaghetti for a line.

2. If the coefficient of correlation is 1 or − 1, what does this tell you about the data points and the line of best fit?

3. What does a correlation coefficient very close to + 1 tell you?

4. What does a correlation coefficient very close to − 1 tell you?

5. What does a correlation coefficient of 0.23 tell you?

Practice & Apply

Examine the airlines timetable spreadsheet.

6. What tends to happen to the minutes as the miles increase? *increase*

7. Is the correlation positive or negative? *positive*

8. The line of best fit rises and fits the points quite well. What does this tell you about the numerical value of the correlation? *close to 1*

City	miles	minutes
Jacksonville	270	64
St Louis	484	102
Pittsburgh	526	89
New York	765	120
Minneapolis	906	154
Boston	946	145
Denver	1208	190
Phoenix	1587	236
Los Angeles	1946	286
San Francisco	2139	312

Tell whether the given correlation coefficient describes a line of best fit that rises or falls. Also tell whether the line is a good fit or not.

9. 0.09 **10.** −0.92 **11.** 0.89 **12.** −0.45
rises; no   falls; yes   rises; yes   falls; no

Tell whether the correlation coefficient for the scatter plot is nearest to − 1, 0, or 1.

13. 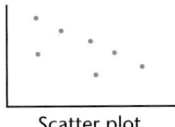 **14.**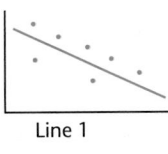

15. Given the scatter plot, which line best fits the points? Explain.
line 1

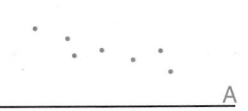

Scatter plot

Line 1

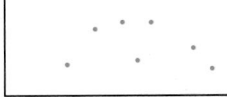

Line 2

Graph the points (1, 1), (1, 2), (2, 1), and (2, 2).

16. What shape do the points suggest? a square

17. According to one calculator, the line of best fit is a horizontal line $y = 1.5$. What other line fits equally well? $x = 1.5$

For Exercises 18–20, match the scatter plots to the correlation coefficients.   **A.** − 0.9   **B.** 0.2   **C.** 0.9

18. 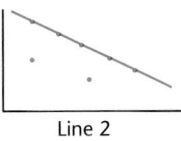A   **19.** C   **20.** B

21. Economics The chart shows how food items that cost $1.00 in 1982 have increased in cost over the years. Make a scatter plot, and use a straightedge to estimate the line of best fit.

Year	1982	1985	1988	1991
Cost	1.00	1.06	1.18	1.36

Look Back

What are the next 2 terms of each sequence? [Lesson 1.1]

22. 1, 4, 7, 10, 13, . . .
16, 19

23. 2, 5, 10, 17, 26, . . .
37, 50

24. 1, 2, 3, 4, 0, 1, 2, . . .
3, 4

25. Create two sequences, and an expression that generates the terms. **[Lesson 1.3]** Answers may vary.

26. Explain how to plot the point (5, 3) on a coordinate axis.   **[Lesson 1.5]**
Count 5 to the right, then 3 up.

27. **Geometry** The side of a square is 8 centimeters. What is its area?
64 cm²

Look Beyond

28. What value for x in $2x + 4 = 12$ will make the equality true? 4

29. What value for n in $32 − 2n = 10$ will make the equality true? 11

21.

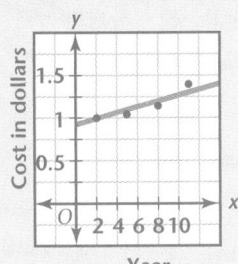

Year

Regression line: $y = 0.4x + 0.89$
(1980 is 0 on the x-axis)

25. One example is

$1, 4, 9, 16, \ . \ . \ . \ , n^2.$

Technology
1.7 Squares of Differences

Columns A and B of the spreadsheet shown give a data set involving two variables x and y. Suppose you wish to find out how well the equation $y_1 = 2x − 0.1$ fits the data. You could begin the comparison by entering the y_1-values into column C of the spreadsheet. Then in column D, enter the squares of the differences between the equation values and the y data values. The smaller that sum, the better the fit.

	A	B	C	D
1	X	Y	Y1	Y1 DIFF
2	0	0.4		
3	2	3.5		
4	3	6.4		
5	5	10.0		
6	6	12.1		
7	8	17.9		
8	9	18.4		
9			SUM >	

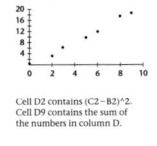

Cell D2 contains (C2−B2)^2.
Cell D9 contains the sum of the numbers in column D.

Use the spreadsheet to find the value in cell D9.

1. $y = 2x − 0.1$ _____   2. $y = 2.1x + 0.1$ _____
3. $y = 2.1x − 0.125$ _____   4. $y = 2x + 0.12$ _____

Which of the two equations given is the better line of best fit?

5. | x | y |
|---|---|
| 0 | 4.33 |
| 1 | 6.00 |
| 3 | 11.00 |
| 5 | 13.50 |
| 7 | 21.00 |
| 8 | 20.00 |

$y = 2.1x + 4.1$
$y = 2x + 4.23$

6. | x | y |
|---|---|
| 2 | 10.04 |
| 4 | 7.00 |
| 6 | 3.00 |
| 7 | 0.00 |
| 8 | −1.00 |
| 10 | −4.00 |

$y = −1.5x + 6$
$y = −1.8x + 13$

7. | x | y |
|---|---|
| 7 | 40 |
| 9 | 44 |
| 10 | 50 |
| 11 | 59 |
| 13 | 61 |
| 15 | 70 |

$y = 3x + 12$
$y = 3.9x + 12$

8. | x | y |
|---|---|
| 6 | 30 |
| 7 | 33 |
| 10 | 20 |
| 14 | 14 |
| 15 | 12 |
| 16 | 9 |

$y = −2.3x + 45$
$y = −2x + 45$

9. What does a sum of 0 in cell D9 tell you about the equation for a set of data and the data values?
_____

Repeating Patterns

Focus

The artistry of cultures over the centuries is often captured in the repetitive patterns woven or stitched in fabric. The art of the weaver or quilt maker can also be represented as mathematical patterns. Regularly repeating number patterns can produce spectacular designs. The project introduced here is an example of how this is accomplished.

Motivate

Invite a quilt maker or weaver to class to explain the art and show examples of the work. Explain also, that computers were historically related to advancements in the early weaving machinery.

Patterns in numbers are sometimes closely related to patterns in art. The work of a quilt maker, for example, is filled with mathematically intricate designs. This project will give you a method for creating artistic patterns from numbers.

Many patterns can be created by using the remainders from dividing numbers. Divide the whole numbers from 0 to 9 by 4, and examine the remainders.

Number	0	1	2	3	4	5	6	7	8	9
Remainder	0	1	2	3	0	1	2	3	0	1

```
8
4
0
        0
7 3  3        1   1 5  9
        2
2
6
```

Notice how the remainders repeat. A pattern that repeats is sometimes called a cycle. This is like a clock that cycles the hours from 1 to 12 and back to 1. In this case, the cycle goes from 0 to 3 before repeating. For example, 4 brings you back to 0, 5 brings you to 1, and so on.

Building a Design Table

On the left and top margins on a piece of graph paper, place the whole numbers from 0 to 9. Make a standard multiplication table. On another piece of graph paper, make a table of the remainders when each product is divided by 4.

Standard Multiplication Table

X	0	1	2	3	4	5	6	7	8	9
0	0	0	0	0	0	0	0	0	0	0
1	0	1	2	3	4	5	6	7	8	9
2	0	2	4	6	8	10	12	14	16	18
3	0	3	6	9	12	15	18	21	24	27
4	0	4	8	12	16	20	24	28	32	36
5	0	5	10	15	20	25	30	35	40	45
6	0	6	12	18	24	30	36	42	48	54
7	0	7	14	21	28	35	42	49	56	63
8	0	8	16	24	32	40	48	56	64	72
9	0	9	18	27	36	45	54	63	72	81

Remainder Table

X	0	1	2	3	4	5	6	7	8	9
0	0	0	0	0	0	0	0	0	0	0
1	0	1	2	3	0	1	2	3	0	1
2	0	2	0	2	0	2	0	2	0	2
3	0	3	2	1	0	3	2	1	0	3
4	0	0	0	0	0	0	0	0	0	0
5	0	1	2	3	0	1	2	3	0	1
6	0	2	0	2	0	2	0	2	0	2
7	0	3	2	1	0	3	2	1	0	3
8	0	0	0	0	0	0	0	0	0	0
9	0	1	2	3	0	1	2	3	0	1

Creating the Designs

1. Choose a color for each number in the remainder table.

2. Create an artistic design for each number in the remainder table.

3. On one grid, replace the numbers with the colored squares. On the second grid replace the numbers with the artistic designs.

| 0 | 1 | 2 | 3 |

| 0 | 1 | 2 | 3 |

Activity 1

Create a pattern graph using the remainders from division by 5. Be creative in your choice of cell designs to replace the remainders. Extend beyond nine numbers for the left and top margin numbers to enlarge the pattern.

Activity 2

On another piece of graph paper, create a pattern using a similar set of numbers greater than 5. Try an operation other than multiplication. Use addition or some combination of operations. Vary the colors. Be creative.

Activity 3

Compare the patterns. See what conclusions you can draw from comparing the patterns for addition with the patterns for multiplication. How do the patterns compare when you use more numbers? What other interesting information can you find from the patterns?

Cooperative Learning

Have groups of students design and create original numerical pattern posters like those described in the investigations. Use completed posters for classroom or hallway decorations. To extend, have students create patterns involving cycles of 12 or beyond.

Discuss

Look at patterns that appear in the art of different cultures, for example, the zigzag designs on Pueblo Indian pottery and the repeating patterns in quilts. Discuss the types of mathematical patterns that are related to the patterns in the pictures.

Chapter 1 Review

Vocabulary

conjecture	10	linear function	37	sequence	10
correlation coefficient	47	ordered pair	35	term	10
domain	37	quadrant	35	variable	23
equation	23	range	37	x- and y-axis	35
expression	23	scatter plot	40	x- and y-coordinates	35

Key Skills & Exercises

Lesson 1.1

➤ **Key Skills**

Generate the terms of a sequence.

Examine the sequence 6, 15, 24, 33, 42, . . . The first term is 6. Each of the terms is 9 more than the one before. To find the next three terms add 9.

$$42 + 9 = 51 \qquad 51 + 9 = 60 \qquad 60 + 9 = 69$$

➤ **Exercises**

Find the next three terms of the sequence, then explain the pattern used to find the terms for each.

1. 1, 4, 7, 10, 13, . . .
16, 19, 22

2. 1, 4, 16, 64, 256, . . .
1024, 4096, 16,384

3. 27, 9, 3, 1, $\frac{1}{3}$, . . .
$\frac{1}{9}, \frac{1}{27}, \frac{1}{81}$

Lesson 1.2

➤ **Key Skills**

Extend a sequence by working backward from the differences.

Apply the method of differences until you get a constant. Then use the differences to find the next two terms.

10, 13, 19, 28, 40, . . .

```
       10      13      19      28      40
1st        3       6       9      12
2nd          3       3       3
```

To find the next first differences, add 3 to the previous first differences.

$$12 + 3 = 15 \qquad 15 + 3 = 18$$

To find the next terms, add the first difference to the previous term.

$$40 + 15 = 55 \qquad 55 + 18 = 73$$

The next two terms are 55 and 73.

➤ **Exercises**

Find the next two terms of each sequence.

4. 1, 5, 11, 19, 29, . . . 41, 55

5. 1, 1, 6, 16, 31, . . . 51, 76

6. 90, 70, 54, 42, 34, . . . 30, 30

7. The fourth and fifth terms of a sequence are 21 and 32. If the second differences are a constant 3, what are the first 5 terms of the sequence? 6, 8, 13, 21, 32

8.

y	10y
1	10
2	20
3	30
4	40
5	50

9.

t	8t − 2
1	6
2	14
3	22
4	30
5	38

14.

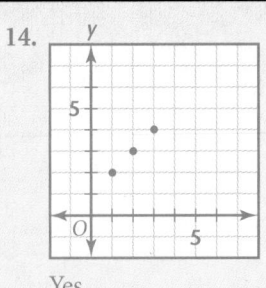

Yes

15.

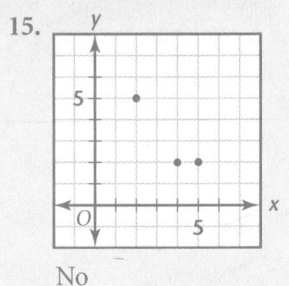

No

Lesson 1.3

> ### Key Skills

Evaluate an expression using substitution and use the results to solve an equation.

Make a table to show the substitutions of the values 1, 2, 3, 4, 5 for x in the expression $2x + 3$. Then use the table of values to solve the equation

$$2x + 3 = 9.$$

x	1	2	3	4	5
$2x + 3$	5	7	9	11	13

We can see from the table above that $2x + 3 = 9$ when $x = 3$.

> ### Exercises

Make a table to show the substitutions of the values 1, 2, 3, 4, 5 for the variable.

8. Find the values of $10y$ by substituting for y.

9. Find the values of $8t - 2$ by substituting for t.

10. Use the table of values in 9 to solve the equation $8t - 2 = 14$. $t = 2$

Lesson 1.4

> ### Key Skills

Perform calculations in the proper order.

To simplify the expression $5(7 - 4) - 6^2 \div 3 + 1$, follow the order of operations: parentheses, then exponents, next multiplication/division left to right, finally addition/subtraction left to right.

$$5(7 - 4) - 6^2 \div 3 + 1$$
$$5(3) - 6^2 \div 3 + 1$$
$$5(3) - 36 \div 3 + 1$$
$$15 - 12 + 1$$
$$4$$

> ### Exercises

Simplify.

11. $17 - 4 \cdot 3$ 5   **12.** $32 - 24 \div 6 - 4$ 24   **13.** $3 \cdot 4^2 - [24 \div (6 - 4)]$ 36

Lesson 1.5

> ### Key Skills

Graph points from the ordered pairs.

Graph the ordered pairs (7, 2), (1, 4), and (3, 5). Tell whether they lie on a straight line.

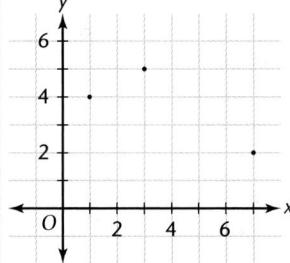

The points do not lie on a straight line.

> ### Exercises

Graph each list of ordered pairs. Tell whether they lie on a straight line.

14. (1, 2), (2, 3), (3, 4)   **15.** (2, 5), (4, 2), (5, 2)

16. (1, 3), (1, 4), (3, 5)   **17.** (0, 1), (1, 2), (3, 6)

18. Graph the linear function $y = x + 2$.

16.
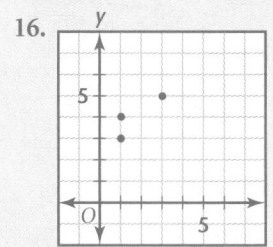
No

17.
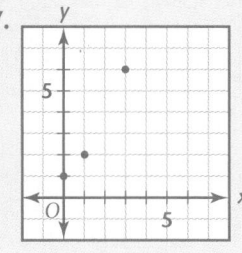
No

18.
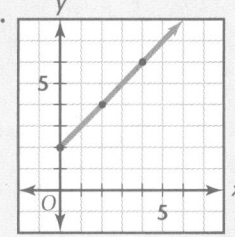

Lesson 1.6

➤ *Key Skills*

Identify the fit of data to a line.

Describe the correlation for the scatter plot as strong positive, strong negative, or little or none.

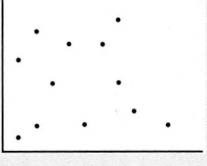

The correlation is little or none since the points do not cluster about a line.

➤ *Exercises*

Describe the correlation as strong positive, strong negative, or little or none.

19.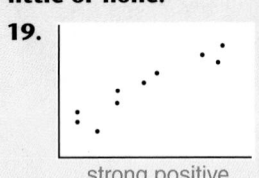
strong positive

20.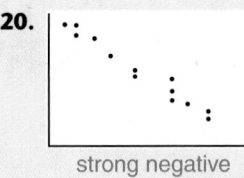
strong negative

Lesson 1.7

➤ *Key Skills*

Understand the relationship of the measure of correlation between the values − 1 and 1 to the closeness of the fit of data to the line.

A correlation coefficient of 0.99 describes a line of best fit which rises. It is a good fit because 0.99 is close to 1.

A correlation coefficient of − 0.98 describes a good fit, which is falling instead of rising.

➤ *Exercises*

Tell whether the given correlation coefficient describes a line of best fit which rises or falls. Also tell whether it is a good fit or not.

21. 0.02   rise; not good   **22.** − 0.95   falls; good

23. Tell whether the correlation coefficient for the scatter plot in 19 is nearest to − 1, 0, or 1. 1

24. Tell whether the correlation coefficient for the scatter plot in 20 is nearest to − 1, 0, or 1. − 1

Applications

25. Entertainment If it costs $13 per person to enter the amusement park, how many people can get in for $98? 7

Suppose the cost to order tickets to the ballet is $22 per person, plus a $5 handling charge per order (regardless of how many tickets are ordered).

26. How much does an order of 4 tickets cost? $93.00
27. How much does an order of 8 tickets cost? $181.00
28. Represent the number of tickets by *t*, and write an equation for the cost of an order of tickets. cost = 22*t* + 5

Chapter 1 Assessment

Find the next three terms of each sequence.

1. 4, 8, 16, 32, 64, . . . **2.** 49, 40, 32, 25, 19, . . . 14, 10, 7
128, 256, 512
3. If a softball league has 15 teams and each of the
teams play each of the other teams, how many
games will there be? 120
4. If the second differences of a sequence are a
constant 5, the second of the first differences is
11, and the second term is 12, find the first 5 terms of the sequence. 6, 12, 23, 39, 60
5. Use guess and check to solve the equation $3x + 4 = 16$. Find the value
of x. 4

If notebooks cost 59 cents, find the cost of the following.

6. 2 notebooks $1.18 **7.** 5 notebooks $2.95 **8.** 12 notebooks $7.08
9. Write an equation to model the situation. How many notebooks can you
get for $14.75? $14.75 = 0.59n$; 25

10. Find the values of $9x - 5$ by substituting 1, 2, 3, 4, 5 for x. 4, 13, 22, 31, 40
11. How much is $3 + 27 \div 3^2 - (7 - 5)$? 4

What are the coordinates of the given points?

12. A (0, 5)
13. B (2, 4)
14. C (5, 7)
15. D (7, 0)

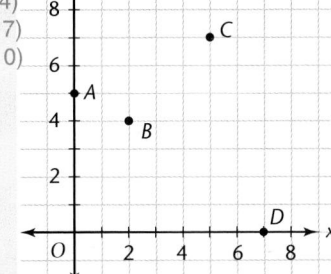

Teri drives at an average rate of 55 miles per hour.

16. Represent hours by h, and write an equation for the distance. $d = 55h$
17. Draw a graph of the function. What type of function is shown
by the graph?

18. Make a scatter plot of this data for the amount of time spent studying
and the score on the test the next day.

Time Studying in minutes	15	12	20	30	25	40	60	50	55	45
Test Score	65	68	75	80	79	85	97	80	88	73

19. Describe the correlation as strong positive, strong negative,
or little or none. strong positive
20. Tell whether the correlation coefficient for the scatter plot is nearest
to -1, 0 or 1. 1
21. Use a straightedge to estimate the line of best fit for this scatter plot.

17.
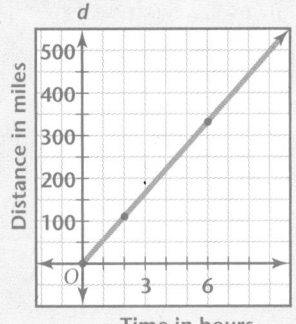
linear

18. and 21.
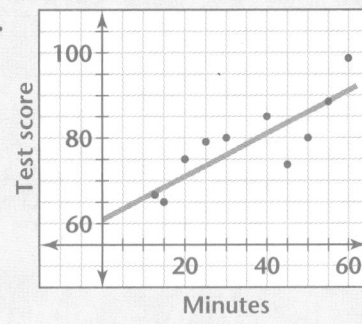
Regression Line: $y = 0.45x + 63.04$

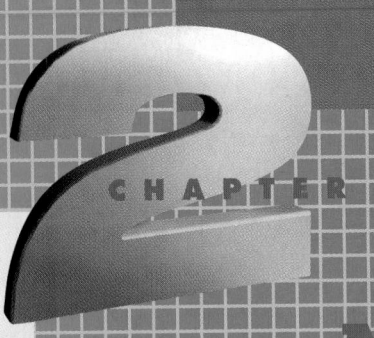

A Brief Introduction to Functions

CHAPTER 2

Meeting Individual Needs

2.1 Previewing Exponential Functions

Core Resources

Practice Master 2.1
Enrichment, p. 60
Technology Master 2.1
Interdisciplinary
    Connection, p. 59

[2 days]

Core Two-Year Resources

Inclusion Strategies, p. 60
Reteaching the Lesson, p. 61
Practice Master 2.1
Enrichment Master 2.1
Technology Master 2.1
Lesson Activity Master 2.1

[3 days]

2.2 Exploring Quadratic Functions

Core Resources

Practice Master 2.2
Enrichment Master 2.2
Technology Master 2.2
Interdisciplinary
    Connection, p. 65

[2 days]

Core Two-Year Resources

Inclusion Strategies, p. 66
Reteaching the Lesson, p. 67
Practice Master 2.2
Enrichment, p. 66
Technology Master 2.2
Lesson Activity Master 2.2

[3 days]

2.3 Previewing Reciprocal Functions

Core Resources

Practice Master 2.3
Enrichment Master 2.3
Technology Master 2.3
Mid-Chapter Assessment
    Master

[2 days]

Core Two-Year Resources

Inclusion Strategies, p. 72
Reteaching the Lesson, p. 73
Practice Master 2.3
Enrichment, p. 71
Technology Master 2.3
Lesson Activity Master 2.3
Mid-Chapter Assessment Master

[3 days]

2.4 Previewing Other Functions

Core Resources

Practice Master 2.4
Enrichment Master 2.4
Technology Master 2.4
Interdisciplinary
    Connection, p. 79

[2 days]

Core Two-Year Resources

Inclusion Strategies, p. 80
Reteaching the Lesson, p. 81
Practice Master 2.4
Enrichment, p. 80
Lesson Activity Master 2.4

[3 days]

2.5 Identifying Types of Functions

Core Resources

Practice Master 2.5
Enrichment, p. 86
Technology Master 2.5
Interdisciplinary
    Connection, p. 85

[1 day]

Core Two-Year Resources

Inclusion Strategies, p. 86
Reteaching the Lesson, p. 87
Practice Master 2.5
Enrichment Master 2.5
Technology Master 2.5
Lesson Activity Master 2.5

[3 days]

Chapter Summary

Core Resources

Chapter 2 Project,
    pp. 90–91
Lab Activity
Long-Term Project
Chapter Review,
    pp. 92–94
Chapter Assessment, p. 95
Chapter Assessment, A/B
Alternative Assessment
Cumulative Assessment,
    pp. 96–97

[3 days]

Core Two-Year Resources

Chapter 2 Project, pp.90–91
Lab Activity
Long-Term Project
Chapter Review, pp.92–94
Chapter Assessment, p. 95
Chapter Assessment, A/B
Alternative Assessment
Cumulative Assessment, pp. 96–97

[5 days]

Reading Strategies

Have students skim through each lesson and examine the pictures and figures before beginning a more careful reading. Ask students to tell what they think the lesson is about. The students should consider familiar situations such as acceleration, the path of a softball when thrown, rounding to whole numbers, and the like. Pre-reading could also include some hands-on activities, such as a scavenger hunt around school or home for items that look like or mimic the general shapes of the functions' graphs (e.g., stair steps for the integer function). During reading, have the students attach notes in the text to mark key concepts, such as the general form for an exponential function in Lesson 2.1, the reciprocal function in Lesson 2.3, the use of second differences to distinguish quadratic functions in Lesson 2.2, and how to represent and identify absolute value and greatest integer functions in Lessons 2.4 and 2.5. The students could also use journals to respond to the interactive questions throughout the chapter. After reading, have the students complete a writing assignment that compares and contrasts the different types of functions studied.

Visual Strategies

Direct the students to look at the pictures and the associated graphs in the Chapter Opener on page 56. Contrast the types of graphs shown there with the linear graphs from Chapter 1. How are they different? Are there any similarities? To get a visual representation of information in the various tables in each lesson, graph the data points on a Cartesian grid. To use the differences method on sequences of numbers, use individual grid blocks on graph paper to record the rows of computations.

Hands-On Strategies

Students can perform many scientific experiments in which data can be collected and modeled with functions from this chapter. Some examples are bacterial growth, length of a vibrating string versus its frequency or pitch, and acceleration due to gravity of a falling object. Problems can be set up to imitate some of the examples in the chapter. For example, each person could bring to class a package and could use scales to weigh it and compute the postal charges. The data could be collected for the entire class and graphed.

Cooperative Learning

GROUP ACTIVITIES	
Linear versus exponential sequences	Lesson 2.1, Exploration
Edges and surface areas of cubes	Lesson 2.2, Exploration 1; Chapter Project
Frequency distribution graphs	Eyewitness Math: Are the Dinosaurs Breeding?
Time estimation experiment	Lesson 2.4, Exercises 11-13, 37-39

You may wish to have students work with partners for some of the above activities. Additional suggestions for cooperative group activities are noted in the teacher's notes in each lesson.

Multicultural

The cultural connections in this chapter include references to Egypt, Babylon, England, and Greece.

CULTURAL CONNECTIONS	
Europe: Walk and Tie Problem	Portfolio Activity
Africa: Egyptian Sequence	Lesson 2.1, Exercise 18
Europe: Frequency and Pitch	Lesson 2.3, Example 2
Asia: Babylonian Compound Interest	Lesson 2.5, Exercises 39-43

Portfolio Assessment

Below are portfolio activities for the chapter. They are listed under seven activity domains that are appropriate for portfolio development.

1. **Investigation/Exploration** The relationship between lengths of edges of cubes and their surface areas is the focus of Exploration 1 in Lesson 2.2 and the Chapter Project. Another exploration you may wish to include is the breeding versus controlled populations of dinosaurs in the Eyewitness Math feature.

2. **Applications** Recommended to be included are any of the following: Finance, Lesson 2.1, Exercises 19 and 30; Music, Lesson 2.3, Exercises 19-24; Consumer Math, Lesson 2.4, Exercise 32; Sports, Lesson 2.5, Exercise 23; and Investments, Lesson 2.5, Exercises 29-33.

3. **Non-Routine Problems** The "Walk and Tie" problem in the Chapter Project and the "Are the Dinosaurs Breeding" problem from the Eyewitness Math present the students with non-traditional but interesting representations of functions.

4. **Project** "Cubes and Pyramids:" see pages 90-91. Students are asked to perform two activities that yield sequences of numbers. These sequences form four different types of functions: constant, linear, quadratic, and cubic.

5. **Interdisciplinary Topics** Students may choose from the following: Geometry, Lesson 2.2, Exercises 16-19; Physics, Lesson 2.2, Exercises 28-32.

6. **Writing** *Communicate* exercises in all the lessons offer excellent writing opportunities for student portfolios.

7. **Tools** Chapter 2 uses both computer spreadsheets and graphics calculators to assimilate data and model the data with various functions.

Technology

Technology provides important tools for manipulating and visualizing algebraic concepts. Some lessons in Chapter 2 can be significantly enhanced by using spreadsheets. For additional information and detailed instructions for spreadsheets, calculators, and other technologies, refer to the HRW Technology Handbook.

Spreadsheets

The sample problem below will help you model for the students the setup and use of spreadsheets.

Sample Problem A town's population grows from 10,000 to 11,000 in one year. Calculate and graph the population growth for ten years if

a. the population grows by the same amount each year.

b. the population grows by the same rate each year.

1. Spreadsheet Setup Spreadsheets are similar to matrices because entries in the cells are the spreadsheet's equivalent of matrix elements. You can enter text, numbers, or formulas in an individual cell and format each in a variety of ways. Cells are addressed by name: A1, B6, etc.; for instance, the entry in C1 is "Fixed Rate." Label A1 as "Year," B1 as "Fixed Amount," and C1 as "Fixed Rate." Notice how Column B entries increase by a fixed amount.

	A	B	C	D	E	F
1	Year	Fixed Amt.	Fixed Rate			
2	0	10000	10000			
3	1	11000	+C2*1.1			
4	2	12000				
5	3	13000				
6	4	14000				
7	5	15000				
8	6	16000				
9	7	17000				
10	8	18000				
11	9	19000				
12	10	20000				
13						
14						

2. Copying Formulas: When entering formulas, a symbol such as + or = is required before the actual formula. This depends on the brand of software used. At the end of the first year, the new population is 11,000, so the fixed amount is 1000 added each year. Use the procedure introduced in step 1. After the formula symbol, enter **B2+1000** in cell B3. Then copy the formula to B4 through B12. For a fixed rate increase, each current year should be multiplied by 1.1. Thus, the formula becomes **C2*1.1.** Enter this formula in C3 and continue it to complete C4 through C12.

3. Formatting Columns B and C Place the cursor on cell B2, press **Range, Format, Enter,** and indicate how many decimal places to include. If you want only whole dollars, specify **0** decimal places. Then use arrow keys to highlight the remainder of the range to format, and press **Enter.**

4. Graphing Data Pull up the command menu and highlight **Graph.** The computer will give you several choices. First, select the type of graph (line graph) and press **Enter.** Then select data for the X-register: the computer will ask for a range. Highlight A2 to A12. For the A register use B2 to B12, and for the B-register use C2 to C12. You can also specify titles and data labels, if desired. Then press **View** to see the graph.

Graphics Calculators

Several lessons in Chapter 2 involve graphing functions and suggest the use of a graphics calculator to model equations. Follow the steps to graph a function:

1. Set the ranges for the graphing window by pressing **WINDOW** on the TI-82 or **RANGE** on the Casio or the TI-81.

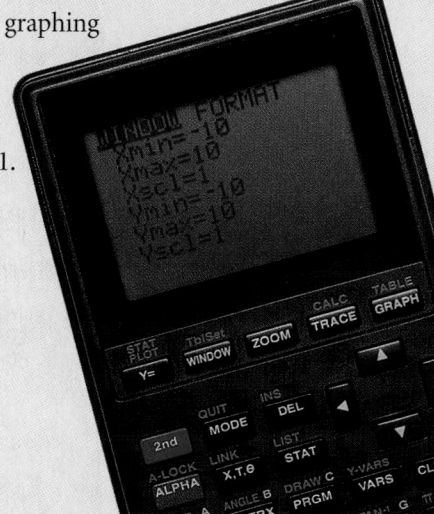

2. Enter the equation by pressing the [Y=] key on a TI or [GRAPH] on the Casio. The [X,T,θ] key is the variable key.

3. Press [GRAPH] on a TI or the execute key on the Casio to draw the graph.

Integrated Software

f(g) Scholar™ is an integrated computer-based mathematics productivity tool that combines calculator, spreadsheet, and graphics capabilities to provide a dynamic and interactive environment for explorations in mathematics. It is appropriate to use *f(g) Scholar*™ for any lesson needing a spreadsheet, calculator, graphics calculator, or any combination of the three.

2 A Preview of Functions

ABOUT THE CHAPTER

Background Information

Many real-world applications can be modeled by linear, exponential, quadratic, and reciprocal functions. The parent form of these functions is introduced through the multiple representations of tables, graphs, and equations. This chapter sets the scene for an in depth study of functions in later chapters.

CHAPTER RESOURCES

- Practice Masters
- Enrichment Masters
- Technology Masters
- Lesson Activity Masters
- Lab Activity Masters
- Assessment Masters
  Chapter Assessments, A/B
  Mid-Chapter Assessment
  Alternative Assessments, A/B
- Teaching Transparencies
- Cumulative Assessment
- Spanish Resources

CHAPTER OBJECTIVES

- Compare linear and exponential functions.
- Examine the difference between growth at a fixed amount and growth at a fixed rate.
- Examine, graph and define quadratic functions.
- Graph quadratic functions to identify characteristics of a parabola.
- Examine and define reciprocal functions.
- Graph and identify reciprocal functions.
- Examine and define absolute value and greatest integer functions.

CHAPTER 2

LESSONS

2.1 Previewing Exponential Functions

2.2 *Exploring* Quadratic Functions

2.3 Previewing Reciprocal Functions

2.4 Previewing Special Functions

2.5 Identifying Types of Functions

Chapter Project
Cubes and Pyramids

A Preview of Functions

Many activities that people take for granted are rich in mathematics. Functions are particularly valuable because they let you see the way numerical relationships behave.

Linear functions appeared in the first chapter. In this chapter, you will have the opportunity to examine other common functions that are part of the mathematician's toolbox. These functions include those that appear on standard scientific and graphics calculators. You will be able to study these functions in more detail in later chapters.

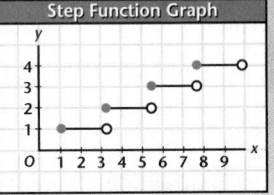

Step Function Graph

Paula walks into the post office to mail a package to her friend and notices that the cost for postage relates to the package's weight. A function shows this relationship.

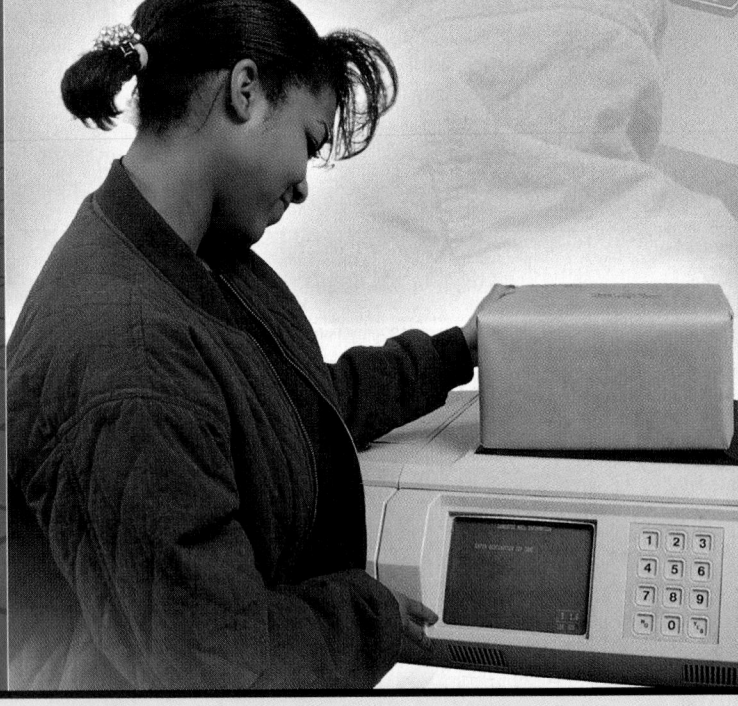

ABOUT THE PHOTOS

Playing or listening to a stringed instrument can be inspiring. A melody is achieved by varying the lengths of the individual strings to create different notes. When the string is shortened, more vibrations per second occur, and the pitch is higher. This is a classic example of a reciprocal function.

Functions can also be used to describe the events depicted in the other photos—a greatest integer function for the postal rates and an exponential function for the savings balance. Emphasize that much of mathematics is based on recognizing *different* patterns and finding models that represent those patterns.

Alex changes the length of the vibrating part of a guitar string by moving his fingers. A function shows how the number of vibrations per second relates to the length of the string.

Reciprocal Function Graph

PORTFOLIO ACTIVITY

"Walk-and-Tie" Problem

Years ago when people traveled on horseback, they might find themselves away from home without enough money for the return trip. A common solution for a pair of people with two horses was to sell one horse and take turns riding the other one home. Since a horse walks faster than a person, one person rode ahead for a while, tied up the horse, and continued walking toward home.

The second person would reach the point where the horse was tied and begin riding until he or she caught up with the first person. They would repeat this until they got home.

Suppose a horse travels 5 miles per hour and two people walk 3 miles per hour. How long will it take to make a 20-mile trip if the "walk-and-tie" method is used?

- Graph and identify absolute value and greatest integer functions.
- Graph data to identify functions.

PORTFOLIO ACTIVITY

Assign the Portfolio Activity as part of the assignment for Lesson 2.4 on page 83.

Try a class experiment. Use the length of the classroom as "two miles" and half the length as "one mile." Have two students start walking at the same time. One student should take regular strides, but the other should take "baby steps." They should each reverse strides upon reaching the "1-mile" mark. Drawing time lines will help students understand the amount of time needed for the journey. Have them decide what rates to assign to each phase of the journey and then generalize the results.

Walking:	Riding:
3 miles per hour	5 miles per hour
1 mile in 20 minutes	1 mile in 12 minutes

At the end of mile 2, the journey will have taken 32 minutes. There are ten 2-mile increments, so the journey will take 320 minutes. What would happen if three people had to share one horse? What about three people and two horses?

ABOUT THE CHAPTER PROJECT

In the Chapter Project on pages 90–91, students will perform two activities using small cubes to look for number patterns. The first activity consists of making a larger cube out of the smaller ones, and then counting and recording the number of small cubes that have exposed faces. The experiment is varied to include several sizes for the large cube (3×3, 4×4, 5×5, etc.).

The second activity also involves counting exposed faces on the small cubes, but this time the larger object formed is a pyramid. Successively larger bases are added to vary the experiment.

Students should look for ways to represent, with mathematical models, the patterns they find.

PREPARE

Objectives

- Compare linear and exponential functions.
- Examine the difference between growth at a fixed amount and growth at a fixed rate.

RESOURCES

- Practice Master   2.1
- Enrichment Master   2.1
- Technology Master   2.1
- Lesson Activity Master   2.1
- Quiz   2.1
- Spanish Resources   2.1

Assessing Prior Knowledge

Evaluate $y = 2^x$
for $x = 1, 2, 3, 4, 5$
$[2, 4, 8, 16, 32]$

TEACH

Discuss various growth rate patterns. Money that is invested or borrowed, populations, and bacteria cultures tend to change growth rate slowly, then suddenly increase dramatically. The function in this lesson can model this phenomenon mathematically.

Exploration 1 Notes

Students examine the differences between growth at a fixed amount and growth at a fixed rate.

A ongoing ASSESSMENT

4. Yes, on Day 19.
5. Prize A is worth $3000.
   Prize B is worth
   $5,368,709.12.

LESSON 2.1 Previewing Exponential Functions

Prize A
Start with $100,
add $100 each
day, for 20 days!

Prize B
Start with 1¢, double
your money each
day, for 20 days!

why What happens each time you double your money? An exponential function is used to determine the total amount at any time.

Students at Lincoln High have been playing the Math Know-How game. After the game, the winners will have 10 seconds to decide which prize to choose. They get to keep the money from the 20th day.

Exploration 1 Evaluating Your Options

Calculator

To help the winners decide, try this exploration with a partner. One person finds the amount of Prize A and the other the amount of Prize B. If calculators are available, use the constant feature.

Prize A
Start with 100. Repeatedly *add* 100 to the previous day's total.

Prize A	
Day	Total Amount
1	$100
2	$200
3	$300
4	$400

Prize B
Start with 0.01. Repeatedly *multiply* the previous day's amount by 2.

Prize B	
Day	Total Amount
1	$0.01
2	$0.02
3	$0.04
4	$0.08

1 After four days, which prize would you pick?

2 Do you think Prize B will ever exceed Prize A?

3 Continue each table through 20 days.

4 Did Prize B ever exceed Prize A? If so, when?

5 Extend each table to 30 days. What is each prize worth? ❖

ALTERNATIVE teaching strategy

Technology Using a graphing calculator or computer, graph the functions $y = 2x$ and $y = 2^x$. Have students investigate the functions and write everything they can about the graphs, the tables associated with the graphs, and the functions. After explaining compound interest, have students graph the functions $y = 1.08^x$ with $\text{Xmin} = 0, \text{Xmax} = 3, \text{Xscl} = 1,$ $\text{Ymin} = 0, \text{Ymax} = 120, \text{Yscl} = 10.$

Explain that this graph gives the value of $1 invested at 8% interest. Ask students how they could use the graph to find the total that different invested amounts would reach during a given number of years. The graph will give the amount that $1 will become in a certain number of years. Any invested amount may then be multiplied by the amount calculated for $1 to find the result of its growth.

Introducing Exponential Functions

In Chapter 1 you worked with linear functions. Are all functions linear functions?

EXAMPLE 1

Spreadsheet

The population of Parksburg grew from 10,000 to 11,000 in one year. Calculate and graph the population growth for 10 years if

A the population grows by a fixed *amount* each year.

B the population grows at a fixed *rate* each year.

Solution▸

Demographics

A The first year's growth is 1000 people. If the population grows by a fixed amount, it will gain 1000 people each year. You can use a spreadsheet to display this information. Start with 10,000 and repeatedly *add* 1000 to the previous number. The results appear in column B.

	A	B	C
	Year	Fixed amount	Fixed rate
1			
2	0	10,000	10,000
3	1	11,000	11,000
4	2	12,000	12,100
5	3	13,000	13,310
6	4	14,000	14,641
7	5	15,000	16,105
8	6	16,000	17,716
9	7	17,000	19,487
10	8	18,000	21,436
11	9	19,000	23,579
12	10	20,000	25,937

B The first year's growth is 1000 people. To figure the *fixed rate* of growth for each year, divide 1000 by 10,000.

$$fixed\ rate = \frac{first\ year\ change}{original\ population} = \frac{1000}{10,000} = \frac{10}{100} = 10\%$$

The new population after one year is 100% of the original population plus an additional 10%. Each year, the population is 110% of the previous year's. Enter this on the spreadsheet by starting with 10,000 and repeatedly *multiplying* by 1.10 or 1.1.

Use the data to draw each graph on the same axis.

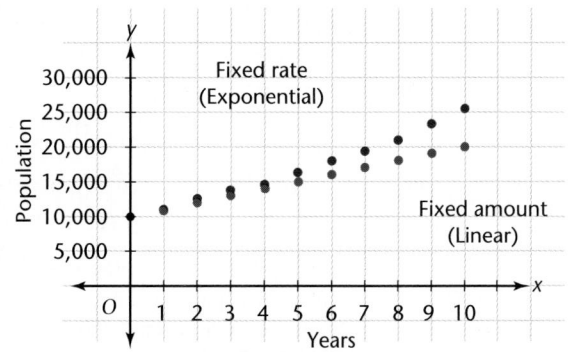

Cooperative Learning

The exploration can be assigned to groups of 4 with teams of 2 each. Have Team A compute the table of values for Prize A and Team B compute the values for Prize B.

Alternate Example 1

A school has an enrollment of 500 students one year and 600 the following year. Find the enrollment after 5 years under each situation.

1. The enrollment grows by a fixed amount each year: add 100 to the each previous year's enrollment.
   [1000]

2. The enrollment grows at a fixed rate of 20% each year: multiply each previous year's enrollment by 1.20.
   [1244]

TEACHING *tip*

Point out that the three columns on the spreadsheet, A, B, and C, have been converted to two sets of ordered pairs to make the graph. For example, (6, 16,000) identifies one of the fixed amount points, and (6, 17,716) identifies one of the fixed rate points.

interdisciplinary CONNECTION

Biology Have students work in groups to formulate algebraic models for bacteria growth under ideal conditions (much growth) and under highly restricted conditions (little growth).

Let 10 be the original number of cells, t the time in hours, and N the total number of cells. Use the formulas to calculate values for N.

Ideal conditions: $N = 10 \cdot 3^t$;

Restricted conditions: $N = 10 \cdot 3^{0.1t}$

Comparing Exponential Growth and Decay

When you found the population of Parksburg by repeatedly multiplying by 1.1, the population grew exponentially. The function modeled **exponential growth**.

EXAMPLE 2

Calculator

What would happen to Parksburg's population if you repeatedly multiplied it by 0.9?

Solution▶

If you use a calculator with a constant feature, enter 10,000. Now multiply by 0.9 repeatedly. The output is 9000, 8100, 7290, 6561, . . . The sequence decreases, as does the population. The function models **exponential decay**. ❖

Try This   Start with 100. First multiply repeatedly by 1.2. Write the first five terms of the sequence that results. Then multiply 100 repeatedly by 0.8, and write the first five terms of the sequence that results.

CRITICAL *Thinking*   What happens to the growth and decay of 100 as your multiplier gets closer and closer to 1?

Escalating costs are making it necessary to save for years to cover college expenses. Compound interest accounts earn increasing amounts of interest over long periods of time.

Compound Interest

Finance   The money paid or earned on a given amount of money often involves **compound interest**. Compound interest is an example of exponential growth.

EXAMPLE 3

When Jill was eight years old, her parents invested $10,000 in a college tuition fund. The interest rate is 8% compounded once at the end of each year. How much will Jill have for tuition after 10 years?

Solution▶

The interest for the *first year* is 8% of $10,000, which is

$$0.08 \cdot 10,000 = \$800.$$

After one year the investment is $10,000 + $800 = $10,800.

The interest for the *second year* is 8% of $10,800, which is

$$0.08 \cdot \$10,800 = \$864.$$

After two years the investment is $10,800 + $864 = $11,664.

Since you are adding 8% to 100%, multiply Jill's balance by 1.08.

Spreadsheet

A calculator or spreadsheet will help you construct a table to find what the investment is worth after 10 years. Start with $10,000 and multiply repeatedly by 1.08.

Will Jill have enough tuition for four years at State U? ❖

	A	B
1	Year	Cmp. interest
2	0	10,000.00
3	1	10,800.00
4	2	11,664.00
5	3	12,597.12
6	4	13,604.89
7	5	14,693.28
8	6	15,868.74
9	7	17,138.24
10	8	18,509.30
11	9	19,990.05
12	10	21,589.25

An equation such as $y = 2^x$ is an example of an **exponential function**. The base, 2, is the number that is repeatedly multiplied. The exponent, x, represents the number of times that the base, 2, occurs in the multiplications.

Exploration 2 Growth and Decay

 Copy and complete the table for $y = 2^x$.

x	1	2	3	4	5
y	2	4	8	?	?

 Graph the function $y = 2^x$ for the values in Step 1.

 What do you notice about the graph you made for $y = 2^x$?

4 Make a table of x and y values for $y = \left(\frac{1}{2}\right)^x$. Use the same x-values that you used for Step 1. Then graph the function.

 Compare the graphs of $y = 2^x$ and $y = \left(\frac{1}{2}\right)^x$. How are they different? ❖

Using Visual Models
The lesson may be introduced using a less dramatic but more visual image by comparing the growth of two trees. The first grows at a constant rate of 1 foot per year. The second grows at a rate of 1 inch the first year, 2 inches the second year, 4 inches the third year, and so on. Have students compare the heights of the trees after 2 years and after 10 years. After 2 years, the first tree will be 2 feet tall, and the second will be 3 inches; after 10 years, the first will be 10 feet tall, and the second will be 42.7 feet.

Assignment Guide

Core 1–8 , 12–19, 24–41

Core Two-Year 1–39

Error Analysis

Some students may have difficulty categorizing the sequence
in Exercise 12. The sequence is
exponential, but the multiplier
is 0.1. It is a sequence of exponential decay.

EXERCISES & PROBLEMS

Communicate

**Explain how to find the rule for each
of the following sequences.**

1. 10, 20, 40, 80, 160, . . . 2. 20, 10, 5, 2.5, 1.25, . . .

3. 10, 20, 30, 40, 50, . . . 4. 160, 140, 120, 100, 80, . . .

5. How can you tell if any of the sequences in Exercises 1–4
   are exponential?

6. Describe the characteristics of a linear sequence.
   Which of the sequences in Exercises 1–4 are linear?

7. Explain what is meant by *exponential decay*.
   Explain what is meant by *exponential growth*.

8. How do you determine the number that multiplies
   the principal when you solve the following problem?
   How much will Sabrina have to pay if she borrows $1000 at
   10% interest, compounded once each year for 4 years?

Practice & Apply

9. Write the first five terms of a sequence that starts with 5 and doubles the
   previous number. 5, 10, 20, 40, 80

What are the next three terms of each sequence?

10. 100, 300, 900, 2700, 8100, . . . 11. 100, 300, 500, 700, 900, . . .

12. 100, 10, 1, 0.1, 0.01, . . . 13. 40, 35, 30, 25, 20, . . .
    0.001, 0.0001, 0.00001   15, 10, 5

14. Which of the sequences in Exercises 10–13 show exponential growth? 10

15. Which of the sequences in Exercises 10–13 are linear? 11, 13

16. Is the sequence 64, 16, 4, 1, $\frac{1}{4}$, . . . linear or exponential? exponential

17. Generate a table for the function $y = 3^x$ when x is 1, 2, 3, 4, and 5.

18. **Cultural Connection: Africa** An ancient Egyptian papyrus of
    mathematical problems includes the statement, "Take $\frac{1}{2}$ to infinity."
    Explore the sequence you get by starting with 0.5 and
    repeatedly multiplying by 0.5. What happens to each successive
    product?

19. **Finance** Jacy invests $1000. His investment earns 5% interest per year
    and is compounded once each year. How long will it take Jacy to double
    his money? 15 years

10. 24,300, 72,900, 218,700

11. 1100, 1300, 1500

17.
x	1	2	3	4	5
y	3	9	27	81	243

18. The product gets smaller and smaller, but
    never equals zero.

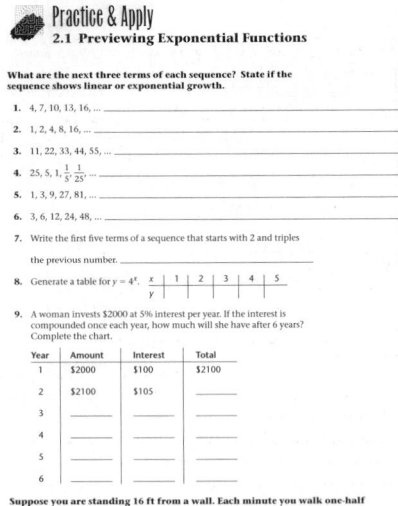

Practice Master

Practice & Apply
2.1 Previewing Exponential Functions

What are the next three terms of each sequence? State if the
sequence shows linear or exponential growth.

1. 4, 7, 10, 13, 16, ...
2. 1, 2, 4, 8, 16, ...
3. 11, 22, 33, 44, 55, ...
4. 25, 5, 1, $\frac{1}{5}$, $\frac{1}{25}$, ...
5. 1, 3, 9, 27, 81, ...
6. 3, 6, 12, 24, 48, ...
7. Write the first five terms of a sequence that starts with 2 and triples
   the previous number.
8. Generate a table for $y = 4^x$.

x	1	2	3	4	5
y					

9. A woman invests $2000 at 5% interest per year. If the interest is
   compounded once each year, how much will she have after 6 years?
   Complete the chart.

Year	Amount	Interest	Total
1	$2000	$100	$2100
2	$2100	$105	
3			
4			
5			
6			

Suppose you are standing 16 ft from a wall. Each minute you walk one-half
the distance to the wall. How far from the wall will you be after:

10. 1 min? _____ 11. 2 min? _____

12. 4 min? _____ 13. 6 min? _____

14. Juanita invests $1000. Her investment earns 8% interest per year and
    is compounded once each year. How long will it take Juanita to
    double her money?

Science Suppose a ball bounces $\frac{1}{2}$ its previous height on each bounce. Find how high the ball bounces on the   Answers in feet

20. second bounce. 5   **21.** third bounce. $2\frac{1}{2}$

22. fourth bounce. $1\frac{1}{4}$ **23.** fifth bounce. $\frac{5}{8}$

24. Write the sequence you get if you start with 1 and keep tripling the previous number. 1, 3, 9, 27, 81, . . .

25. Is the sequence   linear
100, 90, 80, 70, 60, . . .
linear or exponential?

26. Define *fixed rate*.

27. Define *fixed amount*.

28. Define *compound interest*.

29. Start with the number 8, and double the amount 5 times. What number do you get? 256

30. Finance Consider the terms for a First Money credit card. The company has a promotion where interest is not charged for the first 6 months. What is the balance to the nearest cent after the sixth payment on $1000 if only minimum payments are made? $735.08

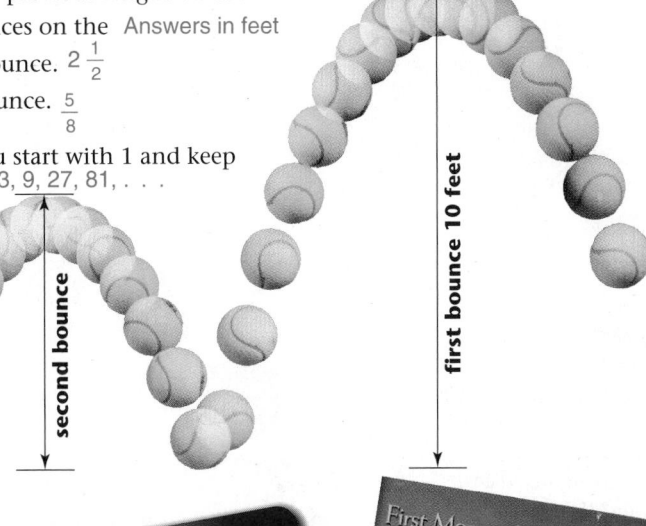

 **Look Back**

Write the rule used to find the next term in each pattern. Find the missing terms. [Lesson 1.1]

31. 3, 7, 11, 15, 19, __, __
add 4; 23, 27

32. 4, 2, 1, $\frac{1}{2}, \frac{1}{4}$, __, __
divide by 2; $\frac{1}{8}, \frac{1}{16}$

33. 33, 30, 27, 24, 21, __, __
subtract 3; 18, 15

Find the next two terms for each sequence. [Lesson 1.1]

34. 10, 21, 34, 49, 66, __, __
85, 106

35. 99, 78, 72, 51, 45, __, __
24, 18

Plot the following points on the coordinate plane. [Lesson 1.5]

36. $A(3, 5)$ **37.** $B(5, 6)$ **38.** $C(6, 5)$ **39.** $D(4, 5)$

Look Beyond

40. Technology Explore the sequence you find when you begin with 100 and repeatedly multiply by -2. Do the numbers get close to any particular number? Some calculators have a special key for "negative," such as $(-)$. On other calculators you enter 2 and then $+/-$ to get -2. no

41. Explore the sequence you find when you begin with 100 and repeatedly multiply by -0.5. Do the numbers get close to any particular number? close to 0

26. Each successive term is a constant multiple of the previous one.

27. A constant amount is added to or subtracted from each successive term.

28. Interest that is added on to money paid after each time period. Compound interest grows exponentially.

36.–39.

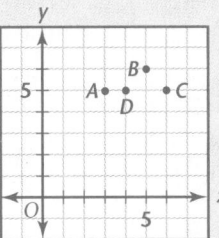

Performance Assessment

Have students do research and prepare reports on one of the following:

• an example of exponential growth or decay from science

• laws that require banks and lending institutions to reveal the total cost of any loan

Use Transparency 7

Look Beyond

The problems in this section foreshadow the notions of alternating sequences, limits, and asymptotes. Students will study exponential function concepts in more detail in Chapter 10.

Technology Master

Technology
2.1 Compound Interest Formulas

Because compound interest is interest given on all the money in an account at the time interest is paid, a compound interest problem generates a number pattern. If, for example, you deposit $1000 into an account that pays 6% interest compounded annually, you can write the following:

SAVINGS PASSBOOK			
DATE	DEPOSIT	INTEREST	BALANCE
1-2-96	$1000.00		$1000.00
1-2-97		6%	

(1) Starting Amount = 1000

(2) Amount at nth period = (1.05) × Amount at $(n − 1)$st period

With this in mind, you can represent the amount as a function of years with a spreadsheet.

	A	B
1	PERIOD	AMOUNT
2	1	1000
3	2	
4	3	
5	4	

Cell B3 contains 1.05*B2.

Use the FILL DOWN command to extend the formula.

Complete the first five rows of the spreadsheet for each initial deposit and annual rate of interest.

1. $250; 4% _____   2. $500; 4% _____

3. $3000; 3.5% _____   4. $3000; 7% _____

5. If you double the initial deposit as in Exercise 1, does the amount become double that in Exercise 2? _____

6. If you double the rate as in Exercise 4, does the amount become double that in Exercise 3? _____

7. In terms of interest rates, what does the following mean?
Amount at nth period = 2 × Amount at $(n − 1)$st period

8. Why is it that the formula shown cannot represent compound interest? Justify your answer with a table or graph.
Amount at nth period = 0.9 × Amount at $(n − 1)$st period

- Examine, graph, and define quadratic functions.
- Graph quadratic functions to identify characteristics of a parabola.

RESOURCES

- Practice Master   2.2
- Enrichment Master   2.2
- Technology Master   2.2
- Lesson Activity Master   2.2
- Quiz   2.2
- Spanish Resources   2.2

Assessing Prior Knowledge

1. Complete the sequence:
   3, 12, 27, 48, __, __.
   [**75, 108**]

TEACH

why Many real-world situations are modeled by quadratic functions including acceleration, projectile motion, and the curves of satellite dishes. The relationship between height and time for balls tossed in various sports is also represented by a quadratic function.

Math Connection
Geometry

The connection emphasizes the quadratic relationship between the length of the edge of a cube and its surface area.

LESSON 2.2
Exploring Quadratic Functions

why *The path of a softball, basketball, or volleyball is often a parabola. This path can be modeled by a quadratic function. In geometry, quadratic functions model problems that relate to area.*

• Exploration 1 *Edges and Surface Area*

GEOMETRY
Connection

Suppose you are going to paint all six faces of a cube. Guess how much paint will be needed if you decide to double the length of each edge. You might be surprised to find that it will take more than twice the amount to paint the faces of the larger cube.

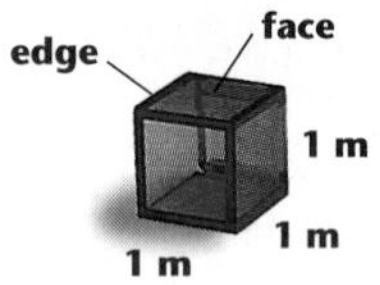

Suppose the edge of a cube is 1 meter. Since the area of each face is 1 square meter ($1m^2$), the surface area of the six faces is 6 m^2. Examine how the surface area of the six faces grows as the length of each edge is increased.

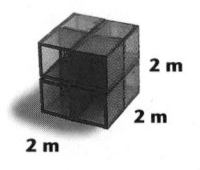

1 Compare the surface area of a 1-meter cube with the surface area of a 2-meter cube. What happens to the surface area of the six faces when the length of each edge doubles?

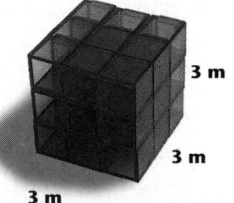

2 Examine a cube whose edge is triple the length of the first cube's edge. What happens to the surface area of the six faces when the length of each edge triples?

ALTERNATIVE teaching strategy

Cooperative Learning
Using nets for cubes of different sizes, first have students compute the area for one square on each net, and then have them compute the total surface area for each. Students should generalize the formula for edge length, L.

 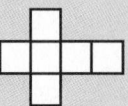

3 Examine the surface area patterns for the six faces of the 1-, 2-, and 3-meter cubes. What is the ratio of the surface area of a 2-meter cube to the surface area of a 3-meter cube? How does this ratio relate to the ratio for the lengths of the edges of the two cubes?

 4 Extend the following table to include a length of 5 meters for an edge. Notice the pattern.

Length of each edge in meters	1	2	3	4	5	...	e
Area of each face in m²	1	4	9	16	?	...	e^2
Formula for the total surface area of the cube	6(1) 6(1²)	6(4) 6(2²)	6(9) 6(3²)	6(16) 6(4²)	? ?	...	?
Total surface area of the cube in m²	6	24	54	96	?	...	

 5 Write a formula for the total area of the six faces of a cube based on the pattern you see in the table.

In the exploration the relationship involves the *square* of the length of an edge. Functions that involve squaring, such as $y = 6e^2$ or $y = x^2$, are called **quadratic functions**.

CRITICAL Thinking How can you use this pattern to find how much paint you need to paint cubes of different sizes?

These students are painting a portion of the set for a school play.

Exploration 1 Notes

Students examine surface areas of cubes in relation to the measures of the edges, and generalize for edge length, e.

Cooperative Learning

Students should work in groups of 3 or 4. For Exploration 1, try a hands-on activity, as suggested in Alternative Teaching Strategy. Exploration 2 can be done with graphics calculators.

Aongoing ASSESSMENT

5. **Surface Area** $= 6e^2$

CRITICAL Thinking

If you know the amount of paint needed for a particular size of cube, you can calculate the amount needed for areas that are multiples of that size.

Use Transparency ▶ 8

interdisciplinary CONNECTION

Physics For uniformly accelerated motion, distance is directly proportional to the square of the time $d = \frac{1}{2}at^2$. Have students tabulate and graph the following:

A race car accelerates at a uniform rate of 20 meters per second squared for 10 seconds. Find the distances traveled at 1 second intervals, and graph distance against time.

Students may use their graphing calculators to make a table. Have them enter the equation $Y_1 = 10x^2$. Use the Table key to find specific values for d.

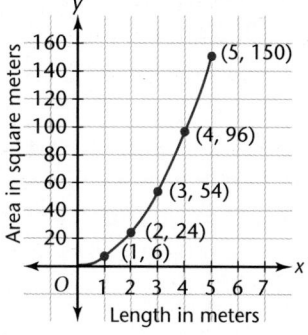

If you want to get a picture of the information in the table on page 65, you can draw a graph. Start by writing five ordered pairs that relate edge length to surface area (edge length, surface area).

These ordered pairs are (1, 6), (2, 24), (3, 54), (4, 96), and (5, 150). You can plot the points and connect them with a smooth curve like the one at the left.

Only part of the quadratic curve appears. Why does the graph show only the positive values of the quadratic function? ❖

APPLICATION

Physics   The following spreadsheet shows the data from the rocket problem in Lesson 1.2. The graph of the data is called a **parabola**.

seconds	height (ft)
0	0
1	208
2	384
3	528
4	640
5	720
6	768
7	784
8	768
9	720
10	640
11	528
12	384
13	208
14	0

Rocket Data

Use the data from the spreadsheet to draw a graph. The curve is a parabola that opens downward. You can see that the rocket reaches its maximum height, 784 feet, after 7 seconds.

A parabola may open either **downward** or **upward**. The point where the curve changes direction is the **vertex**. ❖

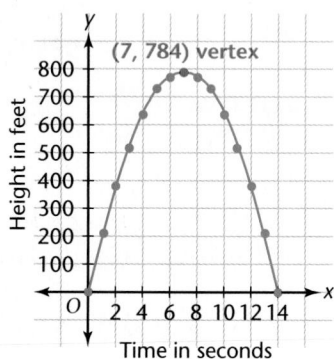

Exploration 2   Graphing Sequences

Use the method of differences (Lesson 1.2) to discover the next two terms in the sequence 1, 4, 9, 16, 25, . . .

1 Calculate the first differences. Are they constant? If not, what are the second differences?

2 Work backward using the pattern, and predict the missing terms.

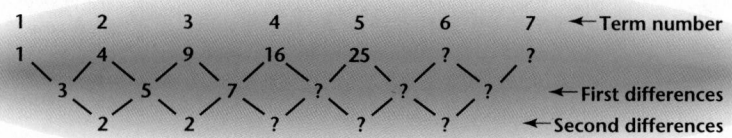

ENRICHMENT   Have students work in small groups to make a chart similar to the one in Exploration 2. Use the sequence 0, 1, 8, 27, 64, 125 . . . , and have students apply the differences method for sequences until they reach a constant difference. Discuss the results, and try to generalize any patterns.

INCLUSION strategies   **Hands-On Strategies** Some students have difficulty crossing from concrete to abstract thinking. Geoboards or graph paper will help these students see the areas of squares.

Aongoing **SSESSMENT**

Lengths of cube edges can never be negative.

TEACHING *tip*

Have the students examine the spreadsheet data and look for patterns. They should notice the duplicate heights for all entries but the vertex. Then have the students examine the graph and how the duplicate heights are plotted. Advanced students may notice the *symmetry*.

TEACHING *tip*

Technology   You may wish to have students plot the data using graphics calculators or spreadsheet software. See Inter-leaf "Technology," page 56D.

Exploration 2 Notes

Students should realize that the table for a quadratic function will take two rows to reach a constant difference.

Use Transparency ▶ 9

3 What do you notice about the second differences?

4 Graph the sequence. Use the term number for *x* and the value of the term for *y*. For example, the first three points will be (1, 1), (2, 4), and (3, 9). What kind of curve appears when you connect the points?

5 Find the first differences for the sequence 4, 8, 12, 16, 20. Are the differences constant?

6 Graph the sequence. For example, the first three points will be (1, 4), (2, 8), and (3, 12). What kind of curve appears when you connect the points?

7 How does the graph in Step 4 differ from the graph in Step 6? Which differences were constant for each sequence? ❖

TEACHING *tip*

For a linear relationship, the first row of differences will produce a constant difference.

Ongoing **A**SSESSMENT

7. The graph in Step 4 is quadratic. Its second difference is constant. The graph in Step 6 is linear. Its first difference is constant.

ASSESS

Selected Answers
Odd-numbered Exercises 11–41

Assignment Guide
Core 1–15, 16–20 even, 21–44

Core Two-Year 1–43

E**XERCISES** & P**ROBLEMS**

Communicate

1. Explain what happens to the area of a square if you double the length of each side.

2. What clues allow you to recognize a quadratic relationship? List at least two clues.

3. Explain how you would use a pattern to determine the type of relationship between the given pairs.

x	1	2	3	4	5
y	4	0	4	0	4

How would you decide which graph matches which relationship?

a. Linear
b. Exponential
c. Quadratic
d. None of these

4.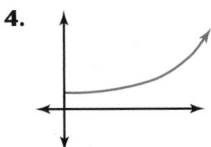

5.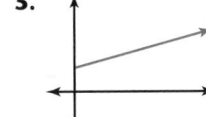

6.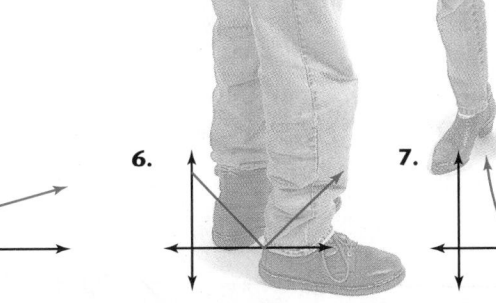

7.

R**ETEACHING**
the
l e s s o n

Using Tables Instead of using a cube, focus the discussion on the area of a square. Have students draw squares with sides of unit measure 1, 2, 3, 4, and 5. Then ask the students to set up a table showing the correspondence between the measure of sides and areas. The table will look something like the following:

side	1	2	3	4	5
area	1	4	9	16	25

Continue plotting the points to show the beginning of the parabola. When the relationship between side length and area (squaring) is clearly understood, return to the lesson and show that the pattern is nothing more than the sequence for squares with each term multiplied by 6.

Some students may not match the tables in Exercises 10–15 with the right relationship because they cannot distinguish an exponential relationship from a quadratic one.

Technology

A graphing calculator can be used for Exercises 10–15 to help students visualize the relationships represented by the tables. Students enter coordinates into a statistical register and then use the statistical plot to display the points.

8.

Explain how to determine the vertex in the parabola shown. Explain how to determine the vertex from a table.

9. Discuss how differences are used to determine sequences and how the differences relate to quadratic relationships. What is the importance of constant differences?

Practice & Apply

Match each table with the correct type of relationship.

a. Linear   **b.** Exponential   **c.** Quadratic   **d.** None of these

10.

x	1	2	3	4	5
y	3	6	11	18	27

c

11.

x	1	2	3	4	5
y	1	2	4	8	16

b

12.

x	1	2	3	4	5
y	9	8	7	6	5

a

13.

x	1	2	3	4	5
y	4	6	8	10	12

a

14.

x	1	2	3	4	5
y	10	17	26	37	50

c

15.

x	1	2	3	4	5
y	80	40	20	10	5

b

Geometry What happens to the area of a square if you

16. triple the length of each side? 9 times as large

17. multiply the length of each side by 10? 100 times as large

18. take $\frac{1}{3}$ the length of each side? $\frac{1}{9}$ as large

19. **Geometry** What is the area of each pizza shown? Use 3.14 for π.

20. **Consumer Math** Which pizza is the better buy? 15 inch pizza

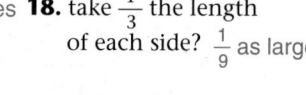

$d = 15''$
$9.00
A = 176.63 sq. in.

$d = 10''$
$4.50
A = 78.5 sq. in.

Match each relationship with its graph.

a. Linear   **b.** Exponential   **c.** Quadratic   **d.** None of these

21. 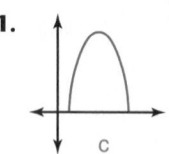 c

22. 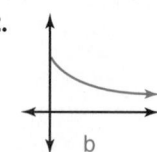 b

23. 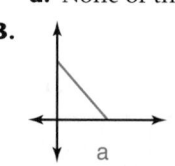 a

24. 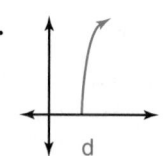 d

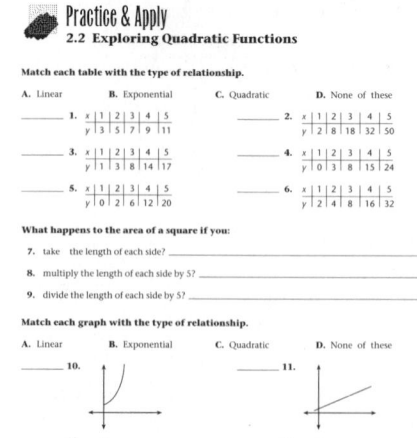

25. How many times larger is the side of the larger square than the side of the smaller square? twice

26. A parabola has a vertex at (8, − 2) and contains the point (5, 3). Give the (2, 34) coordinates of another point on the parabola.

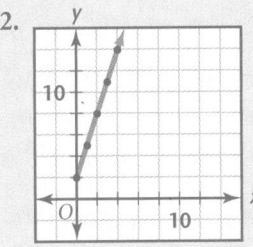
16 Square Inches

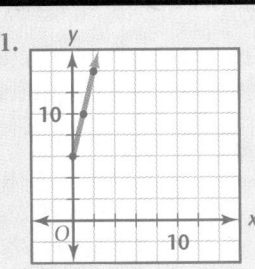
4 Square Inches

27. Tell whether the parabola opens upward or downward.
upward

Physics Find the height of each waterfall if the time the float takes to fall is

28. 1 second.   16 feet
29. 2 seconds.   64 feet
30. 5 seconds.   400 feet

31. Graph the points you found for the waterfalls in Exercises 28–30.

32. Use your graph to estimate how long it would take a float to fall from a waterfall with a height of 600 feet.   A little more than 6 seconds

Look Back

Write the next term in each pattern. Leave answers in fraction form.   [Lesson 1.1]

33. 11, 33, 99, 297, 891, ___   2673
34. 3125, 625, 125, 25, ___   5
35. $\frac{4}{5}, \frac{8}{10}, \frac{16}{20}, \frac{32}{40}$, ___ $\frac{64}{80}$

36. If each package of notebook paper costs $1.29, find the cost of 5 packages.   [Lesson 1.4]
$6.45

Evaluate each expression for the values $x = 2$, $y = 1$, and $z = 4$.   [Lesson 1.3]

37. $3xy$ 6
38. $4z$ 16
39. $21yz$ 84
40. xyz 8

Substitute values for x to find y. Form ordered pairs, and plot the points. [Lesson 1.5]

41. Graph the line for the equation $y = 4x + 6$.
42. Graph the line for the equation $y = 3x + 2$.

Look Beyond

Identify the coordinates for the vertex of each function.

43. $y = (x - 2)^2 + 7$ (2, 7)
44. $y = x^2 + 2x + 1$ (−1, 0)

h

The height, h, (in feet) of a waterfall can be determined by the time, t, (in seconds) that it takes a float in the water to fall from the top to the bottom, according to the formula h = 16t². For example, if the float takes 3 seconds to fall, the height is 16 · 3² = 144 feet.

alternative ASSESSMENT

Portfolio Assessment

Have students find an example of a quadratic relationship from another source. Students should write the equation, make a table of values, draw the graph, identify the vertex, explain what it means in the application, and point out whether the graph opens upward or downward.

Look Beyond

These exercises foreshadow further study of parabolas, their graphs, their equations, and their transformations in Chapters 9 and 12.

Technology Master

Technology
2.2 Symmetric Data

The spreadsheet shown provides a table of x-y values. Notice that if two values are the same distance from 0, then the y values that correspond to them are equal. You can think of the data as symmetric data.

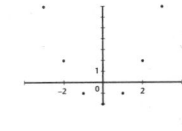

To get a picture of the data, you can use a spreadsheet chart. The graph of the spreadsheet suggests a pattern.

Graph each data set.

Describe the symmetry in each data set and graph.

5. Exercise 1 _____
6. Exercise 2 _____
7. Exercise 3 _____
8. Exercise 4 _____
9. Construct a data set like those above that is symmetric about x = 2.
10. Construct a data set like those above that is symmetric about x = −2.

31.
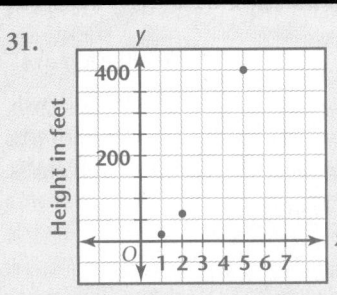
Height in feet / Time in seconds

41.

42.

- Examine and define reciprocal functions.
- Graph and identify reciprocal functions.

RESOURCES

- Practice Master   2.3
- Enrichment Master   2.3
- Technology Master   2.3
- Lesson Activity Master   2.3
- Quiz   2.3
- Spanish Resources   2.3

Assessing Prior Knowledge

Make a table of x and y values. Let x take on values from 1 to 5 for the function $y = 2x^2 + 5$.

x	1	2	3	4	5
y	7	13	23	37	55

A ongoing **SSESSMENT**

The results are the same.

TEACH

why This lesson introduces the concept of reciprocal functions to broaden students' experience in modeling mathematical applications.

TEACHING tip

An equation is a summarizing statement that will give the amount to be contributed per person as a function of the number of people.

LESSON 2.3   Previewing Reciprocal Functions

why How many cars must the students wash to meet their goal? If they decide to charge $4 per car, how many cars will they have to wash? As the price per car decreases, what happens to the number of cars they have to wash? A reciprocal function models this relationship.

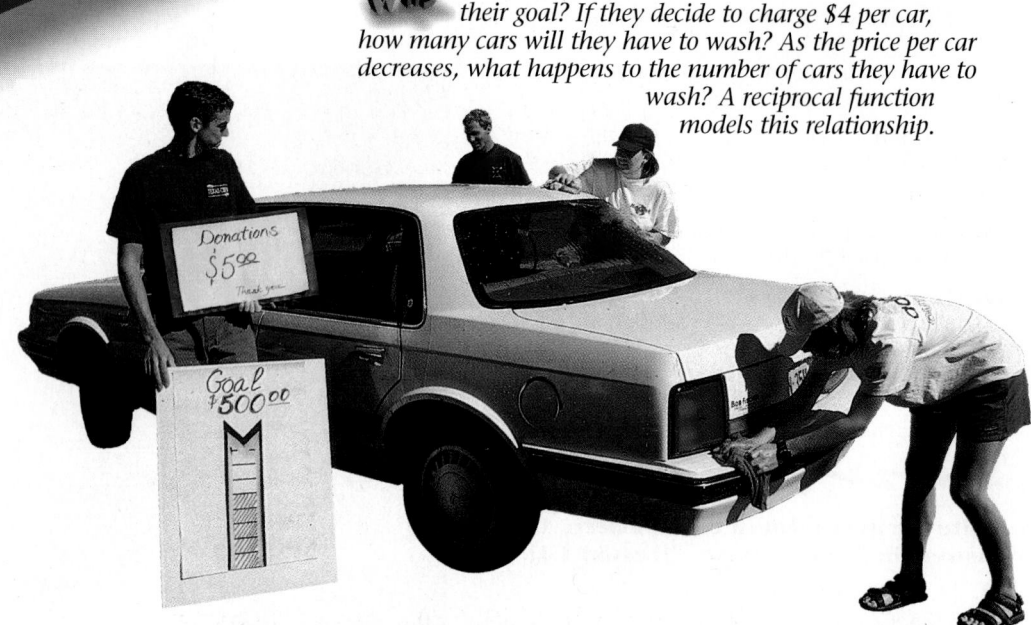

The carwash was successful. The students met their goal of $500, but the school needs $1000 more to help a needy family. If the students find 10 contributors, how much money will each person need to contribute to meet the goal?

For 10 contributors, divide the goal of $1000 by 10. Each of the 10 contributors will need to pay $100. Look at what happens when a greater number of contributors donate. For 20 contributors, divide $1000 by 20. What do you get? Calculate the amounts donated when there are 50, 100, and 1000 contributors. Compare your results with the data in the table.

Let n be the number of contributors	10	20	50	100	1000
Let d be the dollars given per person	100	50	20	10	1

Graphing Reciprocal Functions

To graph a function, substitute numbers into the equation to get ordered pairs. Then plot the points on the coordinate plane and connect the points to form a curve. You can save time by using a graphics calculator or computer to construct the graph directly from the function.

ALTERNATIVE teaching strategy   **Technology** Open the lesson by graphing the function $y = \frac{1}{x} + \frac{1}{2}$ in the first quadrant. Then have students introduce constants and investigate the resulting graphs and ordered pairs. For example, the functions $y = \frac{10}{x}$ and $y = \frac{1}{x} + 3$ will each produce a different set of ordered pairs and will slightly change the shape and/or position of the graph. A suitable calculator window setting is

$X\min = 0, X\max = 10, X\mathrm{scl} = 1;$
$Y\min = 0, Y\max = 10, Y\mathrm{scl} = 1.$

Plot the ordered pairs (10, 100), (20, 50), (50, 20), and (100, 10), and connect them with a curve. The point (1000, 1) is omitted because it would make the scale too large to see most of the other points. As the number of contributors increases, what happens to the number of dollars each contributor makes?

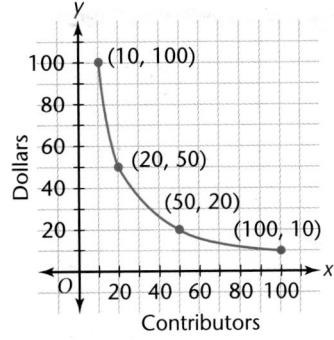

Examine the ordered pairs on the graph. Notice that in each case, the y-value is found by dividing 1000 by x. Functions such as $y = \frac{1000}{x}$ and $y = \frac{1}{x}$ are examples of **reciprocal functions**.

Try This Copy and complete the table which shows the amount, y, each contributor must make to raise $500 when x represents the number of contributors.

x	1	2	5	10	20	25	50	100	250	500
y	500	250	100	50	25	?	?	?	?	?

EXAMPLE 1

Make a table of the values for the reciprocal function $y = \frac{1}{x}$. Then graph the function.

Solution▶

Substitute values for x in the equation $y = \frac{1}{x}$ to make a table of ordered pairs. The y-values represent the reciprocals of the values that you choose for x.

x	1	2	4	5	$\frac{1}{2}$	$\frac{1}{4}$	$\frac{1}{5}$
y	1	$\frac{1}{2}$	$\frac{1}{4}$	$\frac{1}{5}$	2	4	5

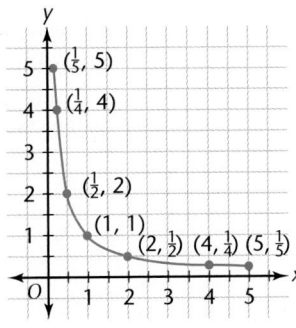

Plot the ordered pairs on the graph, and connect the points. Examine the curve formed from the values of the function. How can you find the reciprocal of $\frac{1}{8}$? ❖

Graphics Calculator

You can use a graphics calculator to display $\frac{1}{x}$. Enter x, and press the $\boxed{x^{-1}}$ key to enter the function after **Y=**. Draw the graph. Use the trace feature to find the reciprocal of $\frac{1}{8}$.

ENRICHMENT The graph of the reciprocal function has some unique qualities. Have students evaluate $y = \frac{1}{x}$ for $x = 1000$, $x = 10{,}000$, $x = 100{,}000$, $y = \frac{1}{10}$, $y = \frac{1}{1{,}000}$, and $y = \frac{1}{10{,}000}$. Ask students to speculate about whether the ends of the curve will ever touch the axes.

Ongoing ASSESSMENT

As the number of contributors increases, the number of dollars each needs to contribute decreases.

Cooperative Learning

Have each group graph several points of a reciprocal relationship and write the function for the relationship.

Ongoing ASSESSMENT

Try This

20, 10, 5, 2, 1

Alternate Example 1

Make a table of values, and draw a graph for $y = \frac{10}{x}$.

TEACHING tip

Technology A suitable range for the graphics calculator is Xmin = 0, Xmax = 10, Xscl = 1; Ymin = 0, Ymax = 10, Xscl = 1.

Ongoing ASSESSMENT

The reciprocal of $\frac{1}{8}$ is $\frac{1}{\frac{1}{8}} = 8$.

As the *x*-value comes closer to 0, the *y*-value gets larger and larger. The function is not defined when *x* is 0 because the *y*-values approach infinity. Notice that division by 0 is not defined.

Alternate Example 2

The area of a playground is to be 200 square feet. Show how the length and width of the playground may vary.

CRITICAL
Thinking

Examine the graph of $y = \frac{1}{x}$. What happens to the *y*-value of the function as the *x*-value decreases from 1 to 0? Why do you think the function is not defined when *x* is 0?

Reciprocals and Music

The frequency of the open string is 98 vps. Stopping the string halfway produces a frequency of 196 vps.

Cultural Connection: Europe A famous example of a reciprocal relationship occurs in music. Every culture that had stringed instruments probably discovered that the longer the string, the lower the frequency, or pitch, of the note it produced. By the time of the Greek scholar Aristotle, it was known that *doubling the length of a musical string would produce half the frequency* in vibrations per second.

EXAMPLE 2

Suppose a musical string 48 centimeters long produces a frequency of 440 vibrations per second (vps). This frequency produces the musical note A, the usual tuning note for an orchestra. What happens to the frequency when the length of the vibrating string is 96 centimeters?

Solution

To determine the frequency for the 96-centimeter string, find the relationship of the new length to the original 48-centimeter string.

$$96 = 2 \cdot 48$$

The string is 2 times as long as the 48-centimeter string. Based on the reciprocal relationship, *doubling the length halves the frequency.* The frequency of the 96-centimeter string is $\frac{1}{2}$ of the 440 vps you get from the 48-centimeter string.

$$\frac{1}{2} \cdot 440 = 220 \text{ vps} ❖$$

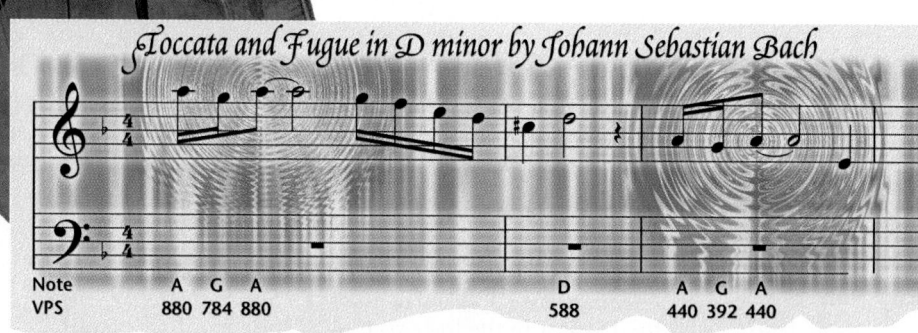
Toccata and Fugue in D minor by Johann Sebastian Bach

Note	A	G	A		D	A	G	A
VPS	880	784	880		588	440	392	440

INCLUSION **strategies** **Using Models** Use a guitar or borrow a large string instrument from the school orchestra, and demonstrate how shortening the length of a string with your finger results in a higher pitch.

EXAMPLE 3

Suppose a 15-inch organ pipe produces a frequency of 440 vps. What happens to the frequency when the length of the pipe is

Ⓐ 5 inches?

Ⓑ 60 inches?

The vibrations of air in organ pipes follow the same principle as vibrating strings.

Solution▶

Ⓐ To determine the frequency for a 5-inch pipe, find the relationship of the new length to the original 15-inch length.

$$5 = \frac{1}{3} \cdot 15$$

The pipe is $\frac{1}{3}$ as long as the 15-inch pipe.

Based on the reciprocal relation, $\frac{1}{3}$ of the length has 3 times the frequency. The frequency of the 5-inch pipe is 3 times the 440 vps you get from the 15-inch pipe.

$$3 \cdot 440 = 1320 \text{ vps}$$

Ⓑ To determine the frequency for a 60-inch pipe, find the relationship of the new length to the original 15-inch length.

$$60 = 4 \cdot 15$$

The pipe is 4 times as long as the 15-inch pipe. Based on the reciprocal relation, 4 times the length has $\frac{1}{4}$ of the frequency. The frequency of the 60-inch pipe is $\frac{1}{4}$ of the 440 vps you get from the 15-inch pipe.

$$\frac{1}{4} \cdot 440 = 110 \text{ vps} ❖$$

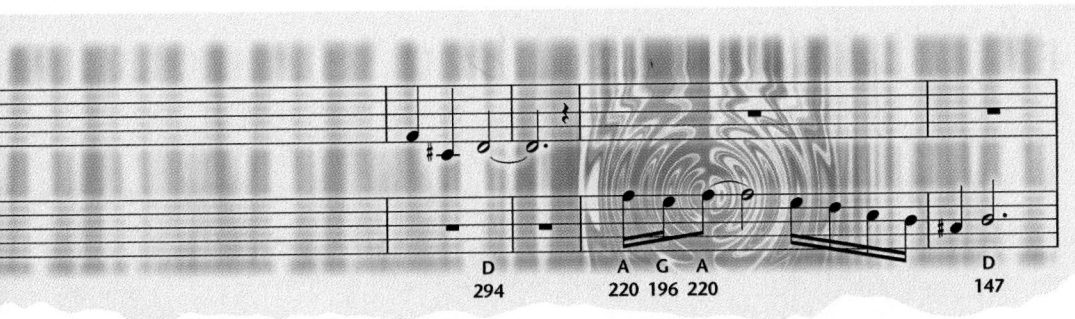

Alternate Example 3

Suppose a 2-foot organ pipe produces a pitch of 660 vibrations per second. What happens to the frequency when the length of the pipe is:

a. 6 inches?

b. 3 feet?

[2640], [440]

RETEACHING *the* **lesson**

Independent Learning The money-raising example can be explained by beginning with the multiplication $dn = 1000$, where d is the per-person contribution and n is the number of people contributing. Some students will more readily understand, if the relationship is multiplication rather than division.

EXERCISES & PROBLEMS

Communicate

1. Explain a reciprocal function using the terms *increasing* and *decreasing*. Use examples to describe the relationship.

Tell how to find the reciprocal of each number.

2. 5   **3.** 100   **4.** $\frac{1}{4}$ **5.** $\frac{1}{6}$

Discuss how you would find the amount each person would have to contribute to reach a goal of $1000 if there were

6. 4 people.   **7.** 200 people.

8. Describe how you would set up the following problem. Jane's co-workers are chipping in to buy her a birthday present. How much must each contribute if they expect to spend $30 and there are 5 co-workers?

Practice & Apply

x	6	5	3	1	$\frac{1}{3}$	$\frac{1}{5}$	$\frac{1}{6}$
y	$\frac{1}{6}$,	$\frac{1}{5}$,	$\frac{1}{3}$,	1,	3,	5,	6

9. Complete the table. Let the y-value for each x be the reciprocal of x. Graph the function.

10. Use variables to set up the following problem. Suppose a trip takes 4 hours if you average 45 miles per hour. How long will it take if you average 60 miles per hour? 3 hours

Find the reciprocal of each number.

11. 7 $\frac{1}{7}$ **12.** 5 $\frac{1}{5}$ **13.** 25 $\frac{1}{25}$ **14.** $\frac{1}{9}$ 9   **15.** $\frac{1}{10}$ 10   **16.** $\frac{1}{7}$ 7

17. Explain why zero has no reciprocal. Division by zero is not possible.

18. Explain why 1 is its own reciprocal. Because $1 \div 1 = 1$.

Suppose a string on a guitar is 60 centimeters long and produces a frequency of 300 vibrations per second.

Music   What frequencies will be produced by similar guitar strings with the following lengths?

19. 30 cm 600   **20.** 20 cm 900   **21.** 90 cm 200

Music   Given that a string 32 centimeters long produces a frequency of 660 vibrations per second, find the frequencies produced by similar strings with the following lengths.

22. 96 cm 220   **23.** 16 cm 1320   **24.** 24 cm 880

Plot several ordered pairs for each of the given rules, and join them using a smooth curve.

25. $y = \frac{12}{x}$ **26.** $y = \frac{20}{x}$ **27.** $y = \frac{24}{x}$

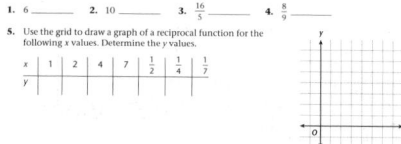
9.
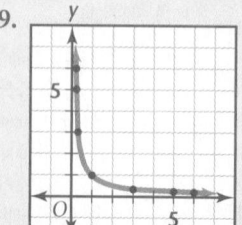

25.
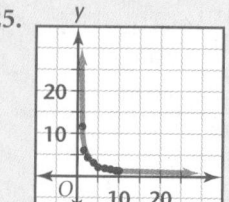

26.
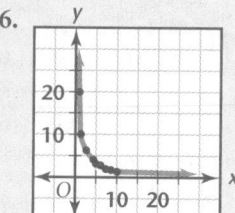

Fund-raising How much will each person have to contribute to reach a goal of $100 if there are

28. 10 people? $10   **29.** 20 people? $5   **30.** 100 people? $1   **31.** 1000 people? $0.10

Transportation Sixty students and teachers from two algebra classes are going on a field trip to a museum. How many cars are needed if each car holds the number of passengers shown?

32. 3 passengers 20   **33.** 4 passengers 15   **34.** 5 passengers 12

How fast would you have to drive to complete a 240-mile trip in

35. 6 hours? 40   **36.** 5 hours? 48   **37.** 4 hours? 60

38. 3 hours? 80   **39.** 1 hour? 240   **40.** $\frac{1}{2}$ hour? 480

41. Suppose a trip takes 4 hours if you average 50 mph. How long will it take if you average 40 mph? 5 hours

42. If other factors are held constant, the frequency of a musical note is a constant multiplied by the reciprocal of the length. If a musical string 48 centimeters long produces a frequency of 440 vps, find the constant.
21,120

 *Look Back*

43. Find the first differences for the following sequence. Are they constant? **[Lesson 1.2]** $-8, -7, -8, -6$
89, 81, 74, 66, 60 no

Guitar Picks
25¢

44. How many guitar picks can you buy with $2.25? Write an equation that models this problem.
[Lesson 1.3] $c = \frac{2.25}{0.25}$ 9 picks

45. How many $6 movie tickets can you buy with $21? **[Lesson 1.3]** 3

46. Make a table to solve the equation $3x + 4y = 12$.
[Lesson 1.5]

x	4	0
y	0	3

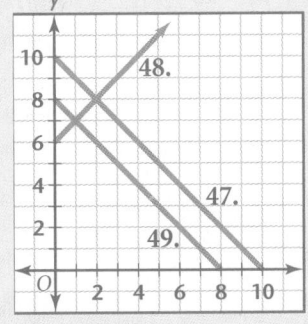

Graph each function on the same coordinate plane, and compare the graphs.   [Lesson 1.5]

47. $y = 10 - x$ **48.** $y = x + 6$ **49.** $x + y = 8$

Plot the following points on a graph.   [Lesson 1.5]

50. (1, 1), (2, 2), (6, 6), (7, 7)

51. (0, 0), (1, 4), (2, 6), (3, 2)

52. (0, 6), (2, 5), (3, 4), (4, 2)

Look Beyond

53. Graph $y = \frac{3}{x}$ for $x = 1, 2, 4, 5, \frac{1}{2}, \frac{1}{4}, \frac{1}{5}, -1, -2, -4, -5, -\frac{1}{2}, -\frac{1}{4}, -\frac{1}{5}$.

27.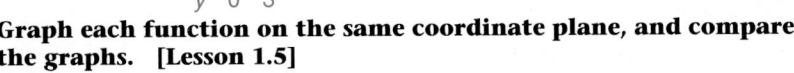

47–49.

The answers to Exercises 50–53 can be found in Additional Answers beginning on page 729.

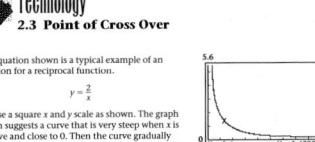

FOCUS

Use random numbers and frequency distribution tables to simulate animal population breeding.

MOTIVATE

1. After the first excerpt, discuss the graph and the story by asking questions like the following:

- What is this kind of graph called? [**frequency distribution**]
- What is this kind of distribution called? [**bell-shaped curve, normal curve, normal distribution, Poisson distribution**]
- Why does Dr. Malcolm think that the graph is worrisome. [**The park is a controlled environment, yet the graph is one you would expect from a natural enviroment.**]

2. After the second excerpt, compare the two graphs. Bring out these points:

- The first graph represents the actual population of the compys at Jurassic Park.
- Dr. Malcolm feels that the first graph would result from a breeding population and the second graph would result if the dinosaurs are not breeding.

EYEWITNESS MATH
Are the DINOSAURS Breeding?

In the novel *Jurassic Park*, dinosaurs created from preserved DNA roam an island theme park. New batches of these cloned creatures are added to the population every six months. During a pre-opening tour, Dr. Malcolm, a mathematician, and Mr. Gennaro, a lawyer, discuss a graph of the heights of one of the dinosaur species on the island.

Height Distribution: Procompsegnathide

"Yes," Malcolm said. "Look here. The basic event that has occurred in Jurassic Park is that the scientists and technicians have tried to make a new, complete biological world. And the scientists in the control room expect to see a natural world. As in the graph they just showed us. Even though a moment's thought reveals that nice, normal distribution is terribly worrisome on this island."

"It is?"

"Yes. Based on what Dr. Wu told us earlier, one should never see a population graph like that."

"Why not?" Gennaro said.

"Because that is a graph for a normal biological population. Which is precisely what Jurassic Park is not. Jurassic Park is not the real world. It is intended to be a controlled world that only imitates the natural world."

Later, Dr. Malcolm discusses the graph with one of the park's scientists, Dr. Wu. They debate whether the supposedly all-female population is somehow breeding.

"Notice anything about it?" Malcolm said.

"It's a Poisson distribution," Wu said. "Normal curve."

"But didn't you say you introduced the compys in three batches? At six-month intervals?"

"Yes . . ."

"Then you should get a graph with peaks for each of the three separate batches that were introduced," Malcolm said, tapping the keyboard. "Like this."

"But you didn't get this graph," Malcolm said.

"The graph you actually got is a graph of a breeding population. Your compys are breeding."

Height Distribution: Procompsegnathide

Wu shook his head. "I don't see how."

"You've got breeding dinosaurs out there, Henry."

"But they're all female," Wu said. "It's impossible."

Simulation Tables for both experiments can be found in Additional Answers beginning on page 729.

1. BREEDING FREQUENCY DISTRIBUTION

Range	Frequency
0–9	0
10–19	1
20–29	1
30–39	1
40–49	2
50–59	3
60–69	3
70–79	4
80–89	3
90–99	2

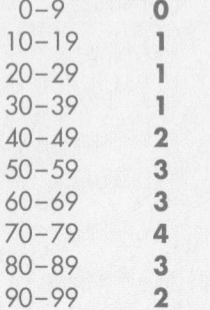

0- 10-20-30-40-50-60-70-80-90-
9 19 29 39 49 59 69 79 89 99

2. CONTROLLED POPULATION FREQUENCY DISTRIBUTION

Range	Frequency
0–9	0
10–19	0
20–29	2
30–39	5
40–49	3
50–59	1
60–69	2
70–79	4
80–89	3
90–99	0

Cooperative Learning

To help settle the breeding debate, you can model dinosaur populations with random numbers.

1. Model a breeding population by following these steps.
   a. Represent 15 baby dinosaurs with 15 letters A–O as shown.

A	B	C	D	E	F	G	H	I	J	K	L	M	N	O

   b. Use a random number table to generate digits, 0–9. Let each digit represent a week's growth for one dinosaur.

A	B	C	D	E	F	G	H	I	J	K	L	M	N	O	
8	1	0	1	7	4	9	0	2	7	7	9	0	3	1	(week 1)
5	0	9	1	2	0	9	3	9	9	2	3	5	0	1	(week 2)
2	2	6	4	2	6	3	0	8	1	0	8	1	9	1	(week 3)

   c. To simulate breeding population, add one new letter (for a new dinosaur) after every 3 weeks.

A	B	C	D	E	F	G	H	I	J	K	L	M	N	O	P	
8	1	0	1	7	4	9	0	2	7	7	9	0	3	1		(week 1)
5	0	9	1	2	0	9	3	9	9	2	3	5	0	1		(week 2)
2	2	6	4	2	6	3	0	8	1	0	8	1	9	1		(week 3)
8	9	4	2	0	6	7	8	0	0	5	5	1	3	7	5	(week 4)

   d. Run the simulation for about 15–20 weeks. Then add the digits for each dinosaur to find its height.
   e. Display the dinosaur heights in a frequency distribution graph. Which of the graphs on page 76 is shaped more like your graph?

2. Now model an artificially controlled population.
   a. Start with 10 dinosaurs labeled A–J, and run the simulation for 8 weeks. To model a population that is not breeding, do not add a new dinosaur every 3 weeks.
   b. After the first 8 weeks, introduce a new batch of 10 dinosaurs by adding letters K–T.
   c. Run the simulation for 8 more weeks. Now you will need 20 digits for each week.
   d. Find the height of each dinosaur. Display the heights on a frequency distribution graph. How is your graph like the bottom one on page 76? How is it different? Why does the difference make sense?

3. Do you agree with Dr. Malcolm or Dr. Wu? Why?

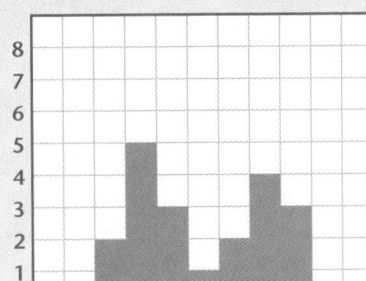

Sample responses:

1e. the bell-shaped graph on page 76

2d. Our graph has only 2 humps, whereas the one in the book has 3. We had only 2 batches of dinosaurs in our simulation.

3. Sample: Dr. Malcolm, because our simulation showed that with a breeding population you would get a graph closer to the single-peak graph on page 76.

Cooperative Learning

Activities 1–2

- Demonstrate *both* simulations by going through one or more "days" of each with the whole class before the groups work independently.
- To expedite simulation, one student in a group can read aloud from the random number table, while another student records the digits.
- You might review the term *frequency distribution.* Discuss how to make a graph from their data when they have finished filling in the growth charts for both simulations.

Discuss

Bring out these points:

- The random numbers are not intended to mimic actual dinosaur growth. (For example, baby dinosaurs are not 0 cm tall.) But the simulations offer a practical way to roughly compare the expected height distributions that would result from the two competing hypotheses – breeding versus not breeding.
- The first simulation includes only a single batch of dinosaurs.

Objectives

- Examine and define absolute value and greatest integer functions.
- Graph and identify absolute value and greatest integer functions.

RESOURCES

- Practice Master   2.4
- Enrichment Master   2.4
- Technology Master   2.4
- Lesson Activity Master   2.4
- Quiz   2.4
- Spanish Resources   2.4

Assessing Prior Knowledge

Find the coordinates of three points satisfying each function.

1. $y = x + 2$

2. $y = 2x - 1$

[Answers may vary. Sample answers follow.

1. $(0, 2), (1, 3), (2, 4)$

2. $(0, -1), (1, 1), (2, 3)$]

TEACH

 Discuss the mail function. For first-class mail, an amount up to one ounce requires one stamp; an amount of more than one ounce but less than two ounces requires two stamps.

Math Connection
Statistics

Students measure the actual error in the data from an experiment and calculate the absolute error.

LESSON 2.4 Previewing Other Functions

 *The absolute value function and greatest integer function model special relationships. The graph of each of these functions has a distinctive appearance. One looks like a* V, *and the other looks like a flight of steps.*

These students are participating in an experiment in which some of the students estimate elapsed time.

STATISTICS *Connection*

Students work in pairs. Half of the students close their eyes and raise their hands. They are told to put their hands down when they think a minute has passed. The other half of the students act as timers. Later they switch roles.

Data from the first group is entered in a spreadsheet with the students' names in column A and their estimates of the time elapsed in column B.

 *Spreadsheet*

Compute the error *in seconds* in column C by subtracting 60 from each student's time.

In column C, a negative number means the guess was *under* 1 minute, and a positive number means the guess was *over* 1 minute. The absolute error in column D shows only the *amount* of the error.

D2		=ABS(C2)		
	A	B	C	D
1	Student	Time	Error	Abs. error
2	Tricia	49	−11	11
3	Keira	59	−1	1
4	Tom	51	−9	9
5	Louise	65	5	5
6	James	68	8	8
7	Sakeenah	77	17	17
8	Hong	66	6	6
9	Louis	54	−6	6
10	Mary	67	7	7
11	Maria	46	−14	14
12	Marcus	62	2	2
13	Shamar	73	13	13
14	Lois	61	1	1
15	Dianne	53	−7	7
16	Suzanne	64	4	4

 **Hands-On Strategies** Begin the lesson by actually carrying out the experiment that is described. One student with a watch can easily monitor the two or three who are estimating.

 **Economics** Tax rates are often based on a step function. The sales tax table shows an example of this function. Have students create a graph from a sales tax table. Are tax amounts rounded up or down? Why?

For example, row 10 shows that Mary thought 1 minute had elapsed after 67 seconds. Her error was $67 - 60 = 7$, indicating that her answer was 7 seconds *over* 1 minute. Row 15 shows that Dianne's time was 53 seconds, so her error was $53 - 60 = -7$ or 7 seconds *under* 1 minute.

Look carefully at column D in the spreadsheet. Compare the numbers in column C with column D. What is the sign of each number in column D compared with the sign of the corresponding number in column C?

Absolute Value Function

The **absolute value** of a number can be shown on a number line. The absolute value of a number is its distance from zero.

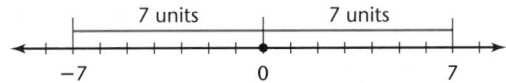

A function written in the form $y = |x|$ or $y = \text{ABS}(x)$ is an example of an **absolute value function**.

7 is 7 units from 0, so the absolute value of 7 is 7.

-7 is 7 units from 0, so the absolute value of -7 is 7.

Thus, $|7| = 7$.

Thus, $|-7| = 7$.

Exploration How Far From Zero?

1 Complete the table for $y = |x|$.

x	-4	-3	-2	-1	0	1	2	3	4
y	4	?	?	?	0	?	?	?	4

2 Graph the function $y = |x|$ for the values in Step 1.

3 Describe the graph for the absolute value function you drew in Step 2. ❖

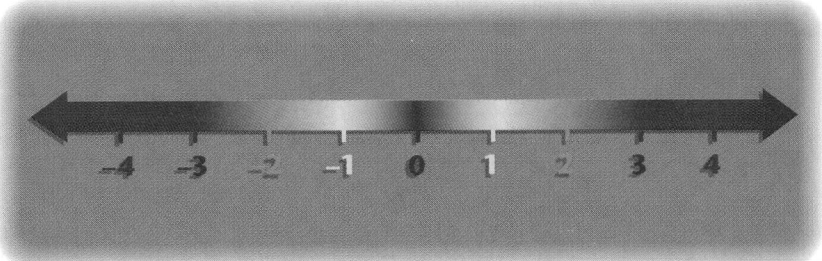

$7\frac{1}{2}$% Sales Tax Table					
Amount	**Tax**	**Amount**	**Tax**	**Amount**	**Tax**
0.01-0.06	0.00	0.74-0.85	0.06	1.54-1.66	0.12
0.07-0.19	0.01	0.87-0.90	0.07	1.67-1.79	0.13
0.28-0.33	0.02	1.00-1.13	0.08	1.80-1.93	0.14
0.34-0.46	0.03	1.14-1.26	0.09	1.94-2.06	0.15
0.47-0.53	0.04	1.27-1.30	0.10	2.07-2.19	0.16
0.60-0.73	0.05	1.40-1.53	0.11	2.20-2.33	0.17

Find the absolute value of each
of the following numbers:
12, 98, −75, 3.5, −1.1, $−\frac{1}{2}$, 0,
and 17.
[12, 98, 75, 3.5, 1.1, $\frac{1}{2}$, 0, 17]

Math Connection
Probability
Students graph or-dered pairs
to examine the characteristics of
an absolute value function.

CRITICAL
Thinking

Both have only positive *y*-val-
ues; both contain the points
(0, 0), (1, 1), and (−1, 1); and
both continue upward in the
first and second quadrants. One
graph is a parabola and the
other is V-shaped.

EXAMPLE 1

Graph the function $y = |x|$ from −3 to 3.

Solution▸

Make a table of ordered pairs. For every value of *x* in the table, write the
absolute value of *x* as the *y*-value. For example, if *x* is −2, then *y* is |−2|
or 2.

x	−3	−2	−1	0	1	2	3
y	3	2	1	0	1	2	3

**COORDINATE
GEOMETRY**
Connection

Plot the ordered pairs, and connect the
points.

The graph of the absolute value function
has a distinctive **V** shape. The tip of the
V shape is at the origin, (0, 0). ❖

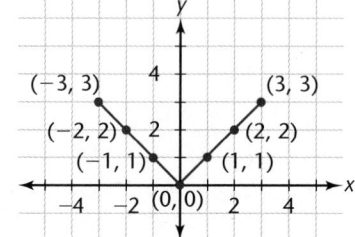

CRITICAL
Thinking

Compare
these graphs.
How are they
alike?
different?

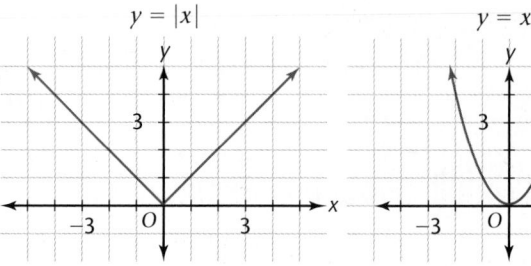

Integer Function

*Graphics
Calculator*

Another function that is especially important when using technology is the
integer function, also called the **greatest integer function**. It appears
on some calculators as a menu item (INT) or as a key . When given a
number, the integer function rounds down to the next integer. The
following are some examples.

INT (1) = 1   INT (2) = 2

INT $\left(\frac{5}{3}\right)$ = 1   INT (−2.1) = −3

INT (1.9) = 1   INT (−3.9) = −4

It is important to note that the integer function always rounds *down,* and
not *up,* to the nearest integer. When you use this function, you will see that
INT(1.9) is rounded *down* to 1, even though it is nearer to 2. You can also
represent these examples using ordered pairs, such as (1, 1), $\left(\frac{5}{3}, 1\right)$,
(1.9, 1), (2, 2), (−2.1, −3), and (−3.9, −4).

ENRICHMENT **Cooperative Learning** Have groups of
students design and make board games using
positive and negative integers and absolute values. When games are complete with rules and
game pieces, have students play them in class.

INCLUSION
strategies
Using Visual Models
Have each student draw a
number line and place on it
pairs of numbers with the same absolute value.
After this, make the connection with the coordinate system by having students draw the
y-axis through 0 on the number line, and then
draw the graph showing how each *x*-value and
its opposite correspond to the same *y*-value.

> ## EXAMPLE 2

If you have $8.00 and want to buy movie tickets that cost $3.00 each, how many movie tickets can you buy?

Solution▶

You can buy only 2 tickets, even though you have enough money for $2\frac{2}{3}$ tickets. You cannot buy $\frac{2}{3}$ of a ticket. In this situation, $2\frac{2}{3}$ tickets rounds down to 2 tickets. ❖

Try This How many movie tickets can you buy if you have $7.00? if you have $8.50? if you have $7.75? Remember, the integer function *rounds down* to the *next lower integer*.

> ## EXAMPLE 3

Amount	Number of tickets
$2.00	1
$2.50	1
$2.75	1
$3.50	2
$3.75	2
$4.00	2
$4.50	3
$5.00	3
$5.25	3
$6.00	4
$6.25	4
$6.50	4

Marilyn is ordering raffle tickets for her 12 friends. The tickets cost $1.50 each. Her friends have $2.00, $2.50, $2.75, $3.50, $3.75, $4.00, $4.50, $5.00, $5.25, $6.00, $6.25, and $6.50, respectively. Draw a graph representing the amount of money each has and the number of tickets that each can purchase.

Solution▶

Make a table to show the amount of money each has and the number of tickets that amount will purchase. Plot the pairs of numbers on a graph.

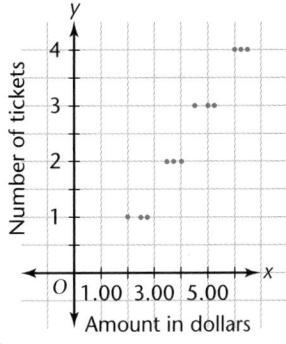

You can see that the *y*-value of the function stays level until the *x*-value reaches a certain point. Then the *y*-value *jumps*. This suggests that the graph is in steps. ❖

EXERCISES & PROBLEMS

Communicate ～～

Explain how to find the absolute value (ABS) of each.

1. 29   **2.** -34 **3.** 7.99   **4.** -3.44

Explain how to find the greatest integer (INT) of each.

5. 29   **6.** $\frac{34}{5}$ **7.** 7.99   **8.** 3.44

Selected Answers
Odd-numbered Exercises 11–37, 41–47.

Assignment Guide
Core 1–13, 14–30 even, 32–50

Core Two-Year 1–48

Technology
Exercises 37 and 38 involve using a spreadsheet. This can be done with pencil and paper, a computer, or a calculator with a table display.

Error Analysis
The most common error in finding the greatest integer occurs when students work with a negative number. They often give the integer that is actually greater than the number. For example, for -3.6, the answer -3 will be given. The number line is the best way to show why the answer is -4.

Practice Master

9. Describe the graph of an absolute value function.

10. Describe the graph of an integer function.

Practice & Apply

In the minute-timing experiment, one student's estimate of a minute was 52 seconds.

11. Was the student over or under the correct amount? under

12. What was the student's error? −8 seconds

13. What was the absolute value of the student's error? 8

Find the absolute value (ABS).

14. 17 17   15. −33 33
16. 8.67 8.67   17. −7.11 7.11

Evaluate.

18. |4.8| 4.8   19. |−3.2| 3.2   20. INT(5.8) 5
21. INT(11/2) 5   22. ABS(5.8) 5.8   23. ABS(11/2) 5.5

Find the greatest integer (INT).

24. 17 17   25. $\frac{33}{4}$ 8
26. 8.67 8   27. 7.11 7

Evaluate.

28. |8| 8   29. |−8| 8
30. −|8| −8   31. −|−8| −8

32. **Consumer Math** Another name for the integer function is the postal function. Paula often mails several packages and needs to know the cost for the various weights. Make a graph of the weight and cost from the table.

Home Economics Melissa is making place mats, each requiring 1 yard of fabric. How many place mats can she make from the following amounts of fabric?

33. $6\frac{3}{4}$ yards 6   34. $3\frac{1}{2}$ yards 3   35. $5\frac{1}{4}$ yards 5   36. $\frac{9}{10}$ yard 0

32. The answer to Exercise 32 can be found in Additional Answers beginning on page 729.

37. −1, 0, 11, −4, −2, 10, −2, 2, 8, 14, 10, 13, 5

38. 1, 0, 11, 4, 2, 10, 2, 2, 8, 14, 10, 13, 5

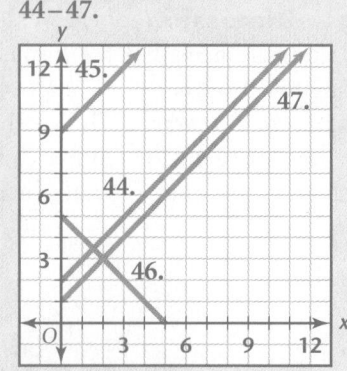

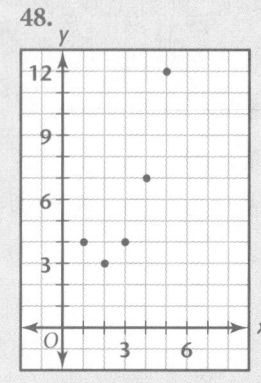

Technology After collecting the data shown for the minute-timing experiment, students switched roles. Only this time, the students estimating the time were distracted by having to carry on conversations at the same time. The spreadsheet shows the data for this group.

37. On your own paper, write the errors for column C.

38. Write the absolute errors for column D.

39. Psychological Experiment Perform both parts of the minute-timing experiment in your class. Give the estimate, the error, and the absolute value of the error for each student. Use a spreadsheet to display the results for the two groups. Answers may vary.

D3			=ABS(C3)	
	A	B	C	D
1	Student	Time	Error	Abs. error
2	Buster	57	−3	3
3	Tony	74	14	14
4	Charlotte	59		
5	Amy	60		
6	Gerti	71		
7	Nathan	56		
8	Jamie	58		
9	Marni	70		
10	Lee	58		
11	Wei	62		
12	Ned	68		
13	Debbie	74		
14	Wynton	70		
15	Bob	73		
16	Jill	65		

40. 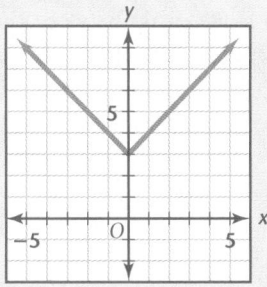 **Portfolio Activity** Complete the problem in the portfolio activity on page 57. 5 hours 20 minutes

Look Back

Find the next three numbers in each pattern.
[Lesson 1.1]

41. 2, 6, 10, 14, 18, __, __, __ 22, 26, 30
42. 27, 9, 3, 1, $\frac{1}{3}$, __, __, __ $\frac{1}{9}, \frac{1}{27}, \frac{1}{81}$
43. If Miguel has \$2, how many bookmarks can he buy if the bookmarks are \$0.38 each? **[Lesson 1.3]** 5

Graph each function on the same coordinate plane, and compare the graphs. [Lesson 1.5]

44. $y = x + 2$
45. $y = x + 9$
46. $y = 5 - x$
47. $y = x + 1$

48. Use this data to make a graph. Which point is the vertex?
[Lesson 2.2] (2, 3)

x	1	2	3	4	5
y	4	3	4	7	12

Look Beyond

Choose positive and negative values for x, and graph each of the following functions.

49. $y = |x| + 3$ **50.** $y = |x + 3|$

49.

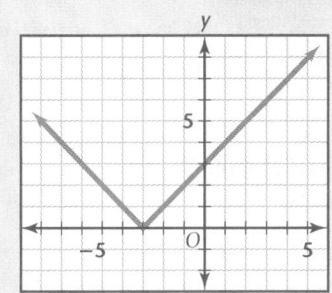

50.

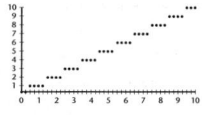

RESOURCES

• Practice Master   2.5
• Enrichment Master   2.5
• Technology Master   2.5
• Lesson Activity Master 2.5
• Quiz   2.5
• Spanish Resources   2.5

Assessing Prior Knowledge

For each equation, plot enough points to sketch a graph, and then discuss the type of function that is represented.

1. $y = 2|x|$ [**absolute value**]

2. $y = 3x$ [**linear**]

3. $y = \text{INT}(x)$ [**greatest integer**]

TEACH

This lesson continues the investigation of functions according to their basic types and curves.

Ongoing ASSESSMENT

Students should remember from Lesson 2.2 that for quadratic functions it takes two rows of differences to reach a constant.

LESSON 2.5 Identifying Types of Functions

Each second that an automobile accelerates increases the distance it travels. This relationship is not like a linear function.

Physics Have you ever heard the phrase "Zero to sixty in eight seconds"? The following table shows how time and distance are related.

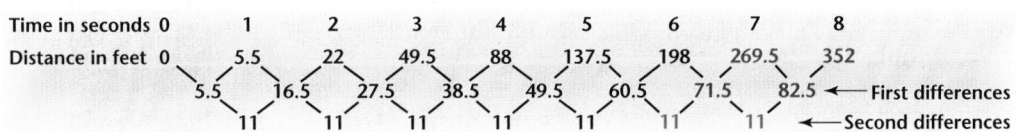

Time in seconds	0	1	2	3	4	5	6	7	8
Distance in feet	0	5.5	22	49.5	88	137.5	198		

How far do you think the car will have traveled after 8 seconds? Describe the graph of the data.

The method of differences can be used to find the distance for 7 and 8 seconds.

Time in seconds 0   1   2   3   4   5   6   7   8
Distance in feet 0   5.5   22   49.5   88   137.5 198   269.5 352
                    5.5   16.5 27.5   38.5 49.5   60.5   71.5   82.5 ← First differences
                      11   11   11   11   11   11   11   ← Second differences

The data indicate that after 8 seconds the car will have traveled 352 feet.

When you graph the ordered pairs, (time, distance), the results show that the relationship is quadratic. What was it about the constant differences that indicated that you should expect a quadratic function?

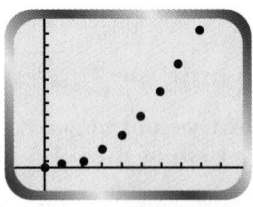

ALTERNATIVE teaching strategy

Using Cognitive Strategies If you wish to use a more familiar function, take one or several from the explorations or examples in previous lessons, and review them with special attention to the relationship between the variables. Because students will already be familiar with the specifics of the example, you can move to more abstract considerations stressing the shape of the curve and the connections between differences and equations.

EXAMPLE 1

The table shows the speed of a car with respect to the time it travels when accelerating at a constant rate of 11 feet per second per second. Find the speed of the car after 8 seconds.

Time in seconds	0	1	2	3	4
Speed in feet per second	0	11	22	33	44

COORDINATE GEOMETRY *Connection*

Solution▶

Plot the ordered pairs.

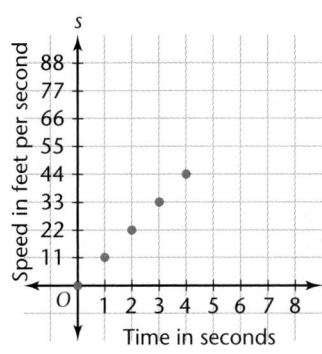

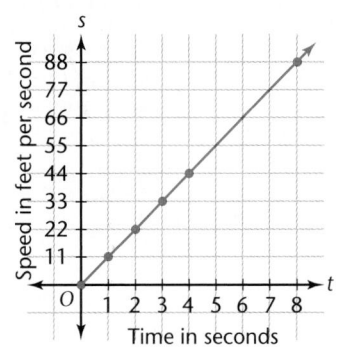

The points lie on a straight line. A linear function appears to be a good model. The graph indicates that after 8 seconds the car will be traveling 88 feet per second.

Notice that $\dfrac{60 \text{ miles}}{1 \text{ hour}} = \dfrac{60 \cdot 5280 \text{ feet}}{60 \cdot 60 \text{ seconds}} = \dfrac{88 \text{ feet}}{1 \text{ second}}$. ❖

CRITICAL *Thinking*

Once the driver reaches a speed of 88 feet per second, the speed limit, he or she stops accelerating, but doesn't decelerate. What happens to the graph in Example 1 after 8 seconds?

EXAMPLE 2

Demographics Draw a graph of the following United States population figures. Describe what pattern appears in the data as shown by the graph.

Year	1800	1820	1840	1860	1880	1900	1920	1940	1960	1980	1990
Population in millions	5	10	17	31	50	76	106	132	179	179	249

interdisciplinary **CONNECTION**

Science Sonar is often used to locate animals or objects in deep water. When a high frequency sound is bounced off an object, the time it takes for that sound to return indicates the location of the object.

In groups, have students design a graph and plot points that could simulate data taken by sonar. What measurements would be represented on the y-axis? (depth) What measurements would be represented on the x-axis? (time)

What type of function is the graph of your data? (linear)

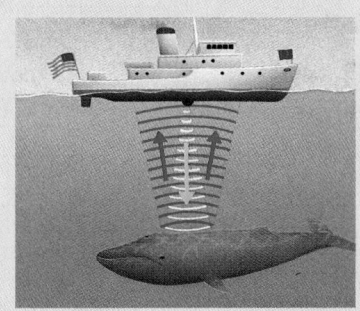

A group going on an outing travels at a rate of 40 miles per hour for 2 hours. They then stay at one place for 3 hours, and finally travel for another hour at a rate of 40 miles an hour. Construct a table and graph showing distance in miles as a function of time in hours.

Time	Distance
1	40
2	80
3	80
4	80
5	80
6	120

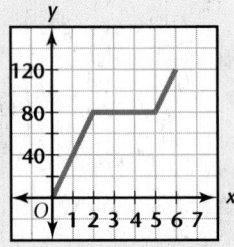

Aongoing**SSESSMENT**

Try This

The cost will be $65 if the car is driven 450 miles in one day.

Graphics Calculator

Solution▸

For changes in population, a good model is often an exponential function. A scatter plot drawn by a graphics calculator shows this model. The graphics calculator draws the exponential graph that best fits the points. According to the graphics calculator, the multiplier is 1.02, which means the growth is about 2% per year.

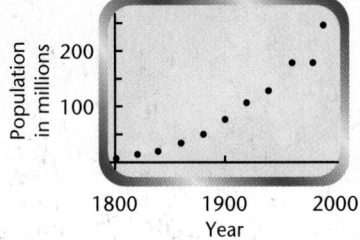
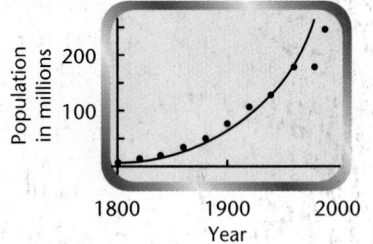

Piecewise Functions

In some cases, functions that are made up of pieces of other functions provide a model for the situation you are studying. These functions are called **piecewise functions**.

EXAMPLE 3

A car rental company charges $40 per day and allows 200 free miles. There is an additional charge of 10 cents per mile in excess of 200 miles. Graph the function that models this situation.

Solution▸

Make a table of values, and graph the ordered pairs.

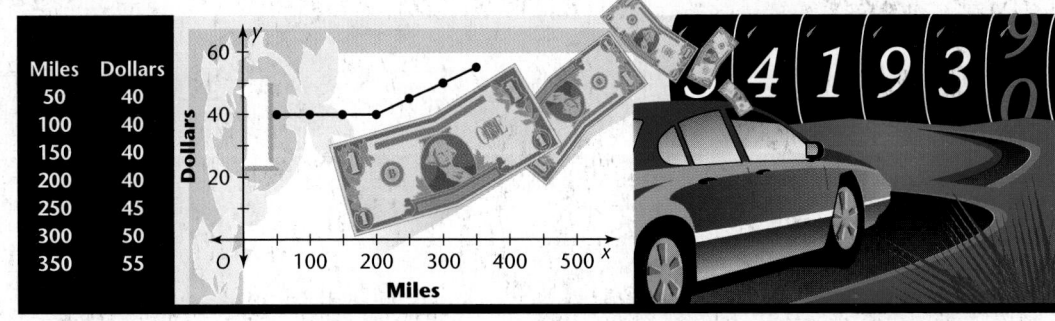

Miles	Dollars
50	40
100	40
150	40
200	40
250	45
300	50
350	55

Notice that the graph is constant until *x* is equal to 200. At that point it becomes a non-constant linear function because the cost increases by a fixed amount. ❖

Try This   What is the cost for 450 miles?

ENRICHMENT

x	1,	2 . . .
y	1,	4 . . .

Continue the table, adding three more pairs of numbers. Demonstrate (a) linear (b) quadratic functions. Possible answers:

a.

1	2	3	4	5
1	4	7	10	13

b.

1	2	3	4	5
1	4	9	16	25

INCLUSION **strategies**

Hands-On Strategies
Place grid paper over cork board and have students plot points using pushpins. Have them use different colors when graphing different types of functions.

Summary

The following is a summary of the different types of functions you have studied in this chapter. As you continue to study algebra, you will discover and use many properties of these functions.

Type	Growth Pattern	Graph	Typical Application
Linear	Fixed amount	Line ╱	Cost per item
Exponential	Fixed rate	Exp. curve ╯	Population growth
Quadratic	Up and down	Parabola ∩	Gravity
Reciprocal	Inverse	$\frac{1}{x}$ curve ⌡	Fund-raising
Absolute value	Linear pieces	V shape ∨	Absolute error
Integer	In steps	Steps ⌐	Postage

EXERCISES & PROBLEMS

Communicate

Match the graphs with the types of functions. Describe the characteristics that distinguish the functions.

a. Linear **b.** Exponential **c.** Quadratic

d. Reciprocal **e.** Absolute value **f.** Integer

1. 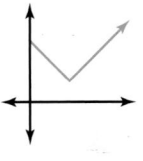 **2.** 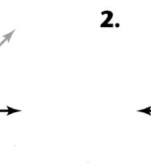 **3.**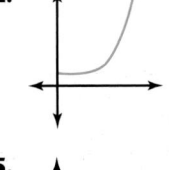

4. 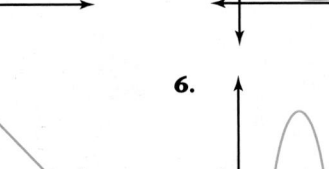 **5.** **6.**

Explain how to use the method of finite differences to determine whether the relationship is linear, quadratic, or neither.

7.

x	1	2	3	4	5
y	12	20	30	42	56

8.

x	1	2	3	4	5
y	3	4	12	24	48

9.

x	1	2	3	4	5
y	8.0	6.8	5.6	4.4	3.2

10.

x	1	2	3	4	5
y	2	4	6	8	10

Selected Answers
Odd-numbered Exercises 11–31, 35–43

Assignment Guide
Core 1–24, 29–45

Core Two-Year 1–43

Technology
A calculator may be used on Exercises 17–22, and 34–43. A graphics calculator may be used for 44–45.

Error Analysis
Students may have difficulty identifying the graph for a particular type of function because they are unfamiliar with the applications of the functions.

Practice & Apply

Match the graphs of the data points with the correct type of function.

a. Linear   **b.** Exponential   **c.** Quadratic

d. Reciprocal   **e.** Absolute value   **f.** Integer

11. a

12. c

13. b

14. d

15. f

16. e

Use the method of differences to determine whether the relationship is linear, quadratic, or neither.

17.

x	1	2	3	4	5
y	2	6	18	54	162

neither

18.

x	1	2	3	4	5
y	44	55	66	77	88

linear

19.

x	1	2	3	4	5
y	8	15	24	35	48

quadratic

20.

x	1	2	3	4	5
y	1	1	2	3	5

neither

21.

x	1	2	3	4	5
y	1	3	6	10	15

quadratic

22.

x	1	2	3	4	5
y	45	54	63	72	81

linear

23. Sports On graph paper, graph the data for the number of points scored (y-axis) versus

Player	Minutes	Points
Amy	23	14
Ling	18	10
Sarah	14	8
Trina	17	3
Latissa	10	6

the number of minutes played (x-axis) for a team's five players in a high school basketball game. Use spaghetti or thread to fit a straight line to the data. Write a sentence that explains what the fit tells you about the team's performance.

24. Describe the absolute value function as a piecewise function.

23.

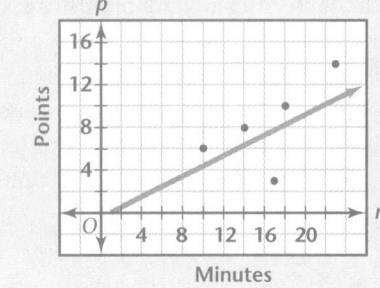

There is a positive correlation between time played and points scored.

24. The absolute value function can be described as two linear functions joined at a vertex.

Consumer Economics

25. Graph the function for the shipping charges.

26. What type of function is it?   integer

Transportation   A car rental charges $30 per day and allows 100 free miles. There is an additional charge of 5 cents per mile in excess of 100.

27. Graph the function that models the situation.

28. What kind of function is Exercise 27?   piecewise

Investments   The Rule of 72 is a method of estimating how long it will take to double your money at various interest rates. To make an estimate, divide 72 by the number of hundredths. For example, if the interest rate is 8%, it will take about $\frac{72}{8}$ or 9 years. Use the Rule of 72 to predict how long it will take to double an investment if the interest rate is

A catalog company charges $2 per pound (16 ounces), or fraction of a pound, for shipping its merchandise.

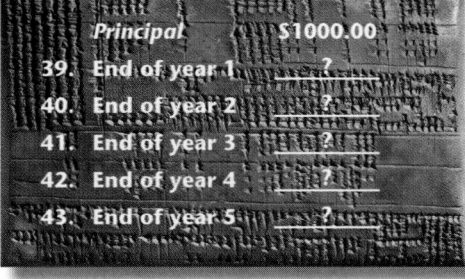

Biking Supplies

Item 108: Red Shorts and Jersey
Price: $56.00
Shipping Wt: 4 lb. 6 oz.

Item 109: Helmet and Gloves
Price: $98.00
Shipping Wt: 5 lb. 11 oz.

29. 4%.   18   **30.** 6%.   12   **31.** 9%.   8

32. Use the Rule of 72 to write an equation for the number of years (*t*) it will take to double an investment at *x*% interest. $t = \dfrac{72}{x}$

33. What kind of function did you write for Exercise 32?
        reciprocal

 *Look Back*

34. Find the sum of $0.1 + 0.2 + 0.3 + 0.4 + 0.5 + \cdots + 0.8$.   3.6
[Lesson 1.3]

Perform calculations in the proper order.   [Lesson 1.4]

35. $12 + 3^2 - 4 \div 2$ **36.** $74 - 10 \cdot 5 \div 2$ **37.** $5(1 \cdot 4 - 2) - 6 \div 3$ **38.** $60 \div \frac{6}{3} + 2^2 \cdot 9$
        19   49   8   66

Cultural Connection: Asia   Babylonian mathematicians 3800 years ago solved problems in compound interest. They asked how many years it would take to double money at 20% interest compounded annually. Today, bankers answer that question by looking it up in a table. The Babylonians did the same thing. Use your calculator to make a table that starts with $1000.00. Find how many years it would take to double the amount if it is compounded annually at 20% interest.   **[Lesson 2.1]**

	Principal	$1000.00
39.	End of year 1	?
40.	End of year 2	?
41.	End of year 3	?
42.	End of year 4	?
43.	End of year 5	?

Look Beyond

Graph the following two equations.

44. $y = 2x + 5$ **45.** $y = x + 6$

The answers to Exercises 25 and 27 can be found in Additional Answers beginning on page 729.

39. $1200.00
40. $1440.00
41. $1728.00
42. $2073.60
43. $2488.32

44.
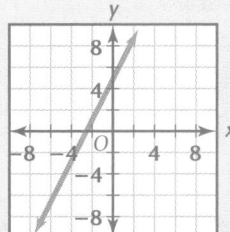

45.
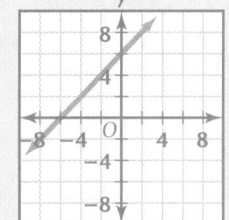

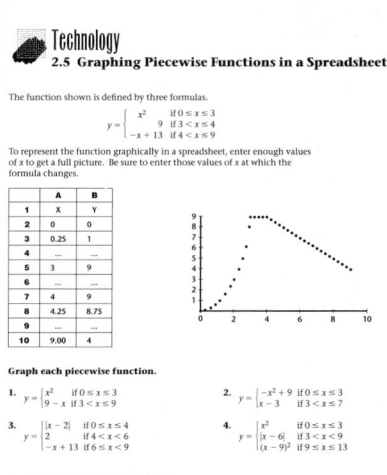

FOCUS

The focus should be on using problem-solving techniques to recognize different functions.

MOTIVATE

The number of exposed faces of two types of geometric solids yield different sequences of numbers. For both activities, students should complete the charts and look for the patterns.

Cooperative Learning

Have students work in pairs or small groups.

After students have completed the chart for the first activity, have them use the numbers from each individual column to make sequences. Have them answer questions like these:

- Start with 3 sides painted. What is the constant difference?
  [0]

- Can you predict what the table entry will be for side length n?
  [8]

- For two sides painted, what is the constant difference?
  [12]

- What difference level did you get to before reaching a constant difference?
  [the first difference]

- What does that tell you about what type of function it is?
  [It is linear.]

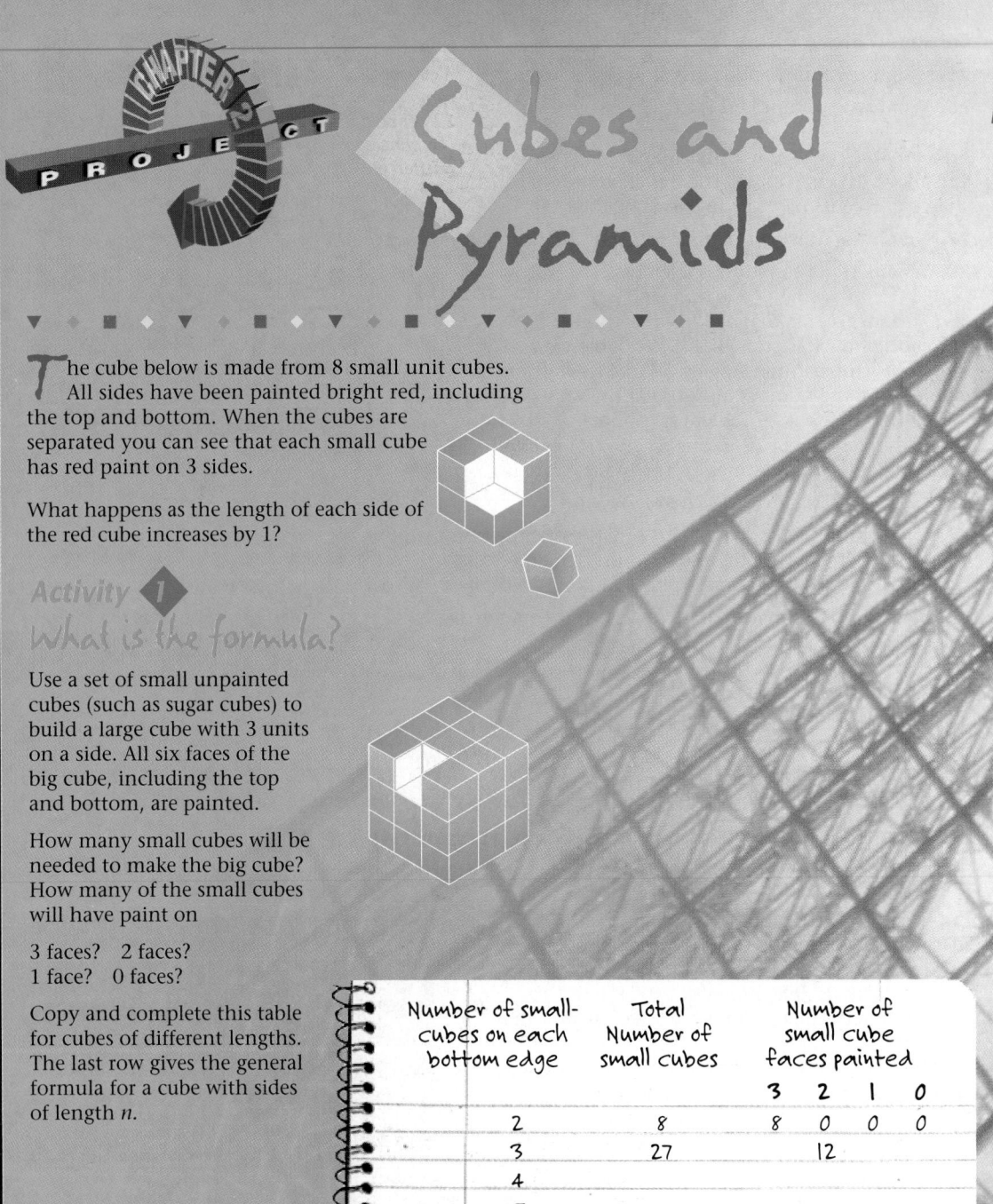

CHAPTER 2 PROJECT

Cubes and Pyramids

The cube below is made from 8 small unit cubes. All sides have been painted bright red, including the top and bottom. When the cubes are separated you can see that each small cube has red paint on 3 sides.

What happens as the length of each side of the red cube increases by 1?

Activity 1
What is the formula?

Use a set of small unpainted cubes (such as sugar cubes) to build a large cube with 3 units on a side. All six faces of the big cube, including the top and bottom, are painted.

How many small cubes will be needed to make the big cube? How many of the small cubes will have paint on

3 faces? 2 faces?
1 face? 0 faces?

Copy and complete this table for cubes of different lengths. The last row gives the general formula for a cube with sides of length n.

Number of small-cubes on each bottom edge	Total Number of small cubes	Number of small cube faces painted 3	2	1	0
2	8	8	0	0	0
3	27	12			
4					
5					
6					
n					

Activity 1:

Length of sides	Number of cubes	Number of sides painted 3	2	1	0
2	8	8	0	0	0
3	27	8	12	6	1
4	64	8	24	24	8
5	125	8	36	54	27
6	216	8	48	96	64
n	n^3	8	$12(n-2)$	$6(n-2)^2$	$(n-2)^3$

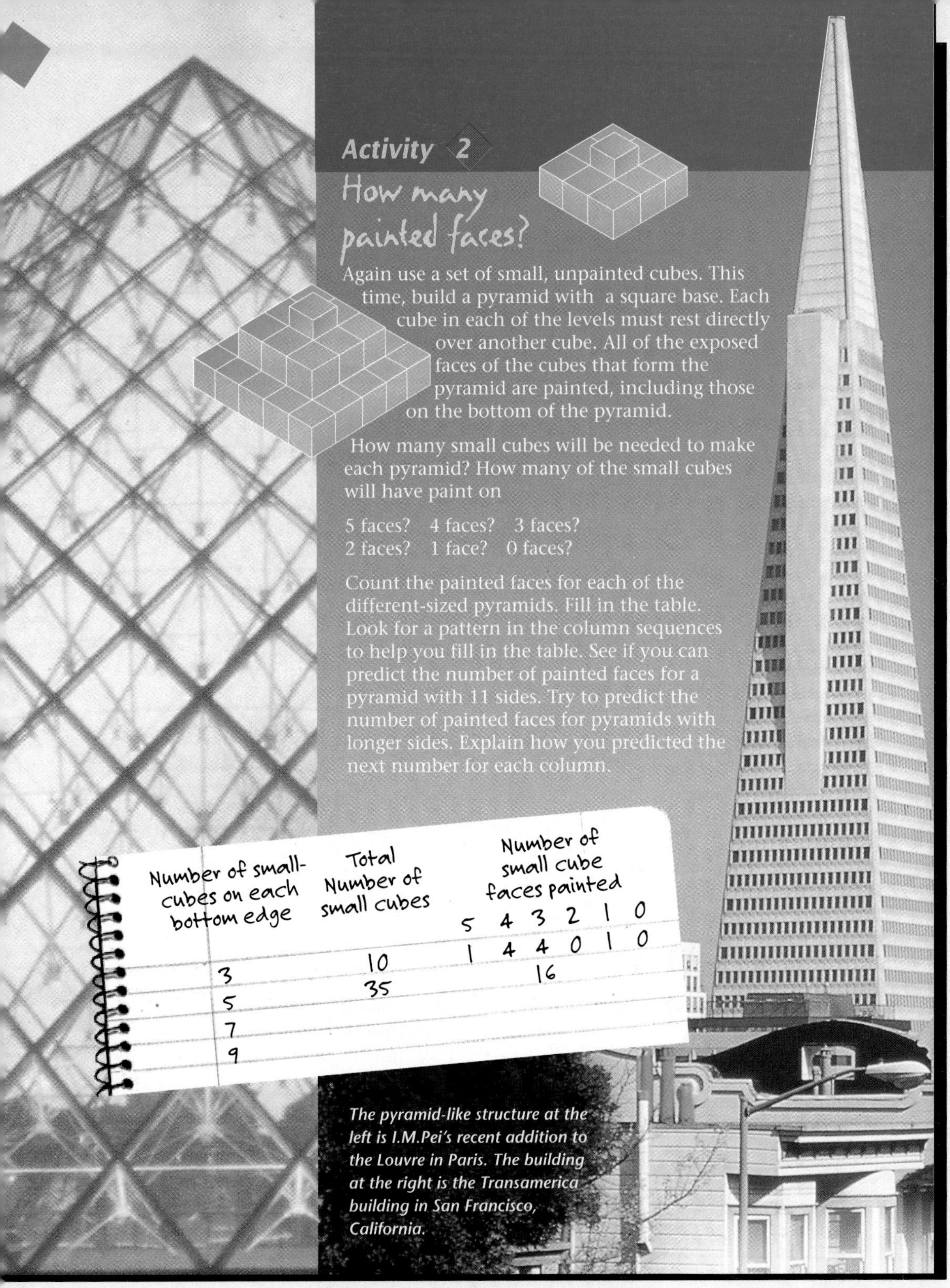

Activity 2

How many painted faces?

Again use a set of small, unpainted cubes. This time, build a pyramid with a square base. Each cube in each of the levels must rest directly over another cube. All of the exposed faces of the cubes that form the pyramid are painted, including those on the bottom of the pyramid.

How many small cubes will be needed to make each pyramid? How many of the small cubes will have paint on

5 faces?   4 faces?   3 faces?
2 faces?   1 face?   0 faces?

Count the painted faces for each of the different-sized pyramids. Fill in the table. Look for a pattern in the column sequences to help you fill in the table. See if you can predict the number of painted faces for a pyramid with 11 sides. Try to predict the number of painted faces for pyramids with longer sides. Explain how you predicted the next number for each column.

Number of small-cubes on each bottom edge	Total Number of small cubes	Number of small cube faces painted					
		5	4	3	2	1	0
3	10	1	4	4	0	1	0
5	35			16			
7							
9							

The pyramid-like structure at the left is I.M.Pei's recent addition to the Louvre in Paris. The building at the right is the Transamerica building in San Francisco, California.

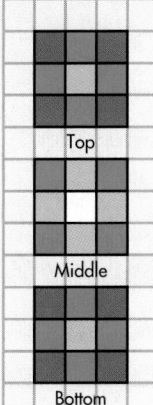

3 x 3 CUBE LAYERS

Top

Middle

Bottom

KEY
Purple: 3 faces exposed
Green: 2 faces exposed
Magenta: 1 face exposed
No color: 0 faces exposed

Discuss

In the second activity, students will again work with sequences that yield constant, linear, quadratic, and cubic functions. Have the students complete the chart.

Ask the class for their observations on the columns dealing with 5 and 4 sides painted. The 5 and 4 sides painted are constant functions because their entries do not change.

What about the sequence made by the number of cubes with 3 sides painted?

Technology

The explicit formula for most of these sequences can be determined by guessing and checking. Use the statistics functions on a graphics calculator and enter 1, 2, 3, 4, 5 in the first list and 0, 1, 10, 35, 84 in the second list. You may wish to have the students plot the data to see its pattern.

Activity 2:

Length of base	Number of cubes	Number of sides painted					
		5	4	3	2	1	0
3	10	1	4	4	0	1	0
5	35	1	4	16	4	9	1
7	84	1	4	28	16	25	10
9	165	1	4	40	36	49	35
11	286	1	4	52	64	81	84

Chapter 2 Project **91**

Chapter 2 Review

Vocabulary

absolute value function	79	integer function	80	quadratic function	65
compound interest	60	linear	60	reciprocal function	71
exponential function	60	parabola	66	vertex	66
exponential growth and decay	60	piecewise function	86		

Key Skills & Exercises

Lesson 2.1

➤ **Key Skills**

Identify linear and exponential functions.
Examine 4, 8, 16, 32, 64, . . . Each term is 2 times the previous term. To find the next three terms, multiply by 2. The next three terms are 64 · 2 = 128, 128 · 2 = 256, and 256 · 2 = 512. Since the sequence grows at the same rate, the sequence is exponential.

Identify an exponential function as showing growth or decay.
The sequence 50, 25, 12.5, 6.25, 3.125, . . . shows exponential decay. The multiplier is 0.5, and the terms decrease. When the multiplier is greater than 1, a sequence shows exponential growth.

➤ **Exercises**

Find the next three terms of each sequence. Then, identify each sequence, as linear or exponential.

1. 45, 43, 41, 39, 37, . . .   **2.** 9000, 900, 90, 9, 0.9, . . .   **3.** 3, 9, 27, 81, 243, . . .
4. 7, 12, 17, 22, 27, . . .   **5.** 2, 4, 6, 8, 10, . . .   **6.** 15625, 3125, 625, 125, 25, . . .

7. If the multiplier for an exponential function is 1.2, does the function show growth or decay? growth
8. Which of the exponential sequences in 1–6 show growth? decay?
growth: 3; decay: 2, 6

Lesson 2.2

➤ **Key Skills**

Graph basic quadratic functions.
You can use this data to make a graph and identify its vertex.

x	2	3	4	5	8
y	9	4	1	0	9

The vertex is (5, 0).

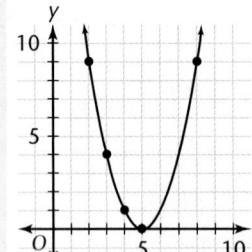

Use the differences method for sequences to identify quadratics.
Examine these sequences.

x	1	2	3	4	5
y	3	5	7	9	11

x	1	2	3	4	5
y	3	5	10	17	26

First differences for the first sequence are constant, so it is linear. Second differences for the second sequence are constant, so it is quadratic.

1. 35, 33, 31; linear

2. 0.09, 0.009, 0.0009; exponential

3. 729, 2187, 6561; exponential

4. 32, 37, 42; linear

5. 12, 14, 16; linear

6. 5, 1, 0.2; exponential

➤ **Exercises**

9. Which point is the vertex of this parabola? (4, 9)

10. Use the data to make a graph. Which point is the vertex? (6, 0)

x	5	6	7	8	9
y	1	0	1	4	9

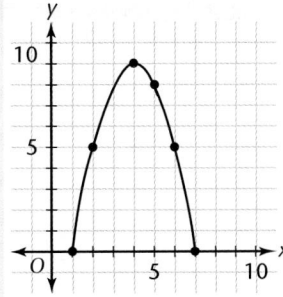

Identify each of the following sequences as quadratic or not quadratic. Explain your answer.

11.
x	1	2	3	4	5
y	2	5	10	17	26

12.
x	1	2	3	4	5
y	6	12	24	48	96

13.
x	1	2	3	4	5
y	3	12	27	48	75

Lesson 2.3

➤ *Key Skills*

Find the reciprocal of a number.

To find the reciprocal of 4, find a number which, when multiplied by 4, equals 1.

Since $4 \cdot \frac{1}{4} = 1$, $\frac{1}{4}$ is the reciprocal of 4.

Since $\frac{2}{3} \cdot \frac{3}{2} = 1$, $\frac{3}{2}$ is the reciprocal of $\frac{2}{3}$.

➤ **Exercises**

Find the reciprocal of each number.

14. 6 $\frac{1}{6}$ 15. $\frac{6}{7}$ $\frac{7}{6}$ 16. 3.5 $\frac{2}{7}$ 17. $\frac{1}{15}$ 15

Lesson 2.4

➤ *Key Skills*

Find the absolute value of a number.

The absolute value of a number is its distance from 0.

$|9.8| = 9.8$ $|-9.8| = 9.8$

9.8 and −9.8 are both 9.8 units from 0.

$\left|6\frac{1}{4}\right| = 6\frac{1}{4}$ $\left|-6\frac{1}{4}\right| = 6\frac{1}{4}$

$6\frac{1}{4}$ and $-6\frac{1}{4}$ are both $6\frac{1}{4}$ units from 0.

Find the greatest integer of a number.

To find INT $\left(3\frac{1}{2}\right)$, round $3\frac{1}{2}$ down to the next integer. So INT $\left(3\frac{1}{2}\right) = 3$.

➤ **Exercises**

Evaluate the following. $|x|$ means the same as ABS(x), the absolute value of x.

18. $|-3|$ 3   19. $|6.7|$ 6.7   20. $|-1.7|$ 1.7   21. ABS(-45)   45
22. INT(45)   45   23. INT(3.68)   3   24. INT(7.5)   7   25. INT(36)   36

10.
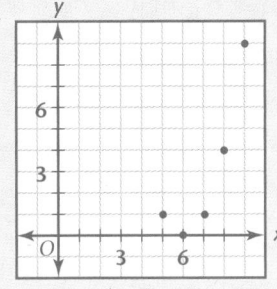

11. quadratic: the second differences are constant

12. not quadratic: neither first nor second differences are constant

13. quadratic: the second differences are constant

Lesson 2.5

➤ *Key Skills*

Identify specific types of functions by their graph.

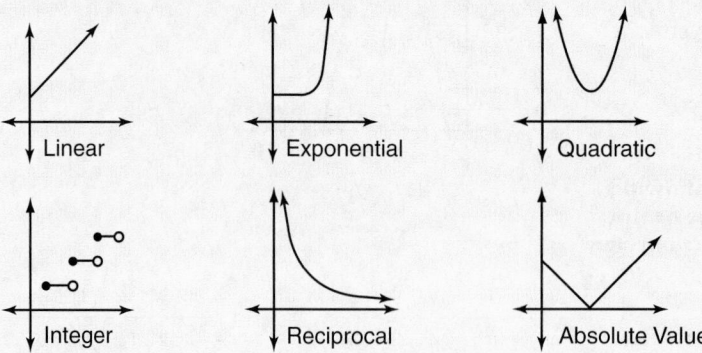

Linear   Exponential   Quadratic

Integer   Reciprocal   Absolute Value

➤ *Exercises*

What kind of function does each graph model?

26.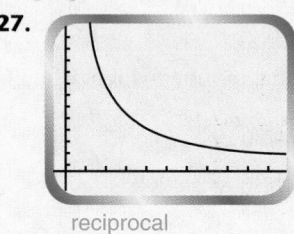

exponential

27.

reciprocal

28.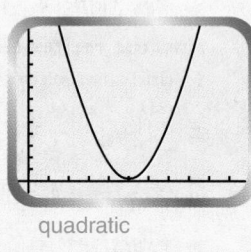

quadratic

Applications

29. **Finance** Roy invests $1000. His investment earns 7.9% interest per year, which is compounded once each year. How long will it take Roy to double his money? 10 years

30. **Physics** The distance, d, in meters that a race car travels when accelerating at a rate of 6.2 meters per second per second is determined by the time, t, in seconds that it travels, according to the formula $d = 3.1t^2$. For example, if the car travels for 10 seconds, the distance is $3.1 \cdot 10^2 = 310$ meters. Find the distance the car travels for 5 seconds, 8 seconds, and 20 seconds. 77.5 m, 198.4 m, 1240 m

31. Graph the points you found in Item 30. Use your graph to estimate how far the car would travel in 35 seconds.

32. How much would each person need to contribute toward a gift that costs $50 if 20 people are donating equal amounts? $2.50

Science One form of uranium has a half-life of 4.5 billion years. This means that after 4.5 billion years half of the atoms in any amount of uranium will have changed into a form of lead. After another 4.5 billion years, half of the remaining uranium atoms will have changed into lead, and so on.

33. What fraction of atoms would remain unchanged after 18 billion years? $\frac{1}{16}$

34. How long will it be before only $\frac{1}{64}$ of the atoms would be uranium? 27 billion years

31.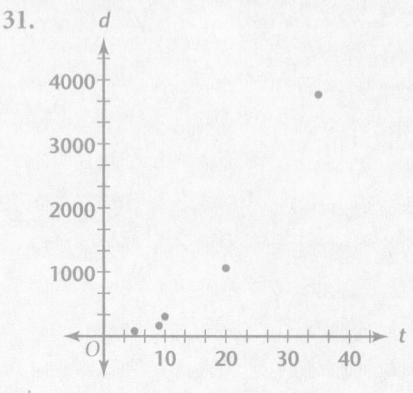

In 35 seconds, the car would travel about 4000 meters.

Chapter 2 Assessment

Find the next three terms of each sequence. Then identify each sequence as linear or exponential.

1. 10, 20, 40, 80, 160, . . .   **2.** 100, 96, 92, 88, 84, . . .

3. Does the exponential sequence 4000, 400, 40, 4, 0.4, . . . show growth or decay? Explain your answer.

4. Use the following data to make a graph. Which point is the vertex? (4, 2)

x	1	3	4	5	7
y	11	3	2	3	11

5. Mica's gymnastics team is raising money to buy equipment. How much does each team member need to raise toward a goal of $500 if there are 10 team members? $50.00

Find the reciprocal of each number.

6. 6 $\frac{1}{6}$ **7.** $\frac{1}{12}$ 12   **8.** 5.5 $\frac{2}{11}$

Evaluate.

9. $|-10|$ 10   **10.** ABS(7.9) 7.9   **11.** INT(9.75) 9

12. Tasha is planting a flower garden. She has $10 to spend. If each pack of flowers costs 49¢, how many packs of flowers can Tasha buy? 20

Determine whether each relationship is linear, quadratic, or neither.

13.

x	1	2	3	4	5
y	75	70	65	60	55

linear

14.

x	1	2	3	4	5
y	4	7	12	19	28

quadratic

Match the graphs with the type of function.

a. Linear   **b.** Exponential   **c.** Quadratic
d. Reciprocal   **e.** Absolute value   **f.** Integer

15. 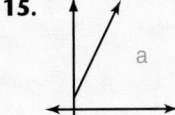   **16.** 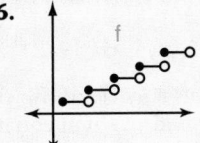   **17.**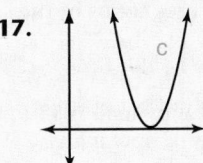

18.   **19.** 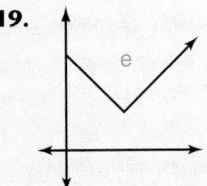   **20.**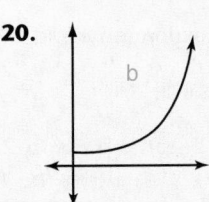

---

1. 320, 640, 1280; exponential

2. 80, 76, 72; linear

3. decay; The numbers are decreasing.

4.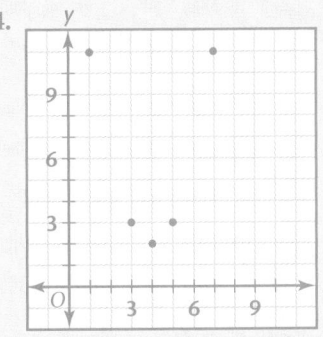

COLLEGE ENTRANCE-EXAM PRACTICE

Multiple-Choice and Quantitative Comparison Samples

The first half of the Cumulative Assessment contains two types of items found on standardized tests—multiple-choice questions and quantitative-comparison questions. Quantitative comparison items emphasize the concepts of quality, inequality, and estimation.

Free-Response Grid Samples

The second half of the Cumulative Assessment is a free-response section. A portion of this part of the Cumulative Assessment consists of student-produced response items commonly found on college entrance exams. These questions require the use of machine-scored answer grids. You may wish to have students practice answering these items in preparation for standardized tests.

Sample answer-grid masters are available in the *Chapter Teaching Resources Booklets*.

Chapter 1 – 2 Cumulative Assessment

College Entrance Exam Practice

Quantitative Comparison For Questions 1–4, write

A if the quantity in Column A is greater than the quantity in Column B;

B if the quantity in Column B is greater than the quantity in Column A;

C if the two quantities are equal; or

D if the relationship cannot be determined from the information given.

	Column A	Column B	Answers	
1.	$|-4.8|$	$|3.2|$	Ⓐ Ⓑ Ⓒ Ⓓ	**[Lesson 2.4}** A
2.	INT(5.8)	INT(3.1)	Ⓐ Ⓑ Ⓒ Ⓓ	**[Lesson 2.4]** A
3.	$c + 8$ ⬚ Perimeter	c $c = 4$ Area	Ⓐ Ⓑ Ⓒ Ⓓ	**[Lesson 1.3]** B
4.	Play tickets: $7 each for balcony seats, $9 each for main floor seats. 9 balcony tickets	7 main-floor tickets	Ⓐ Ⓑ Ⓒ Ⓓ	**[Lesson 1.3]** C

5. What are the next three terms of the sequence 5, 8, 11, 14, 17, . . . ? **[Lesson 1.1]** a

  a. 20, 23, 26   **b.** 14, 11, 8   **c.** 22, 27, 32   **d.** 19, 21, 23

6. The cost to order CDs by mail is $14 per CD plus a $7 handling charge per order (regardless of how many CDs are ordered). Which equation describes this situation? **[Lesson 1.5]** b

  a. $c = 7d + 14$ **b.** $c = 14d + 7$ **c.** $c = 14(d + 7)$ **d.** $c = 14d - 7$

7. Which sequence is exponential and shows growth? **[Lesson 2.1]** d

  a. 2, 4, 6, 8, 10, . . .   **b.** 20, 10, 5, 2.5, 1.25, . . .

  c. 20, 18, 16, 14, 12, . . .   **d.** 2, 4, 8, 16, 32, . . .

8. Which sequence is linear? **[Lesson 2.1]** b

  a. 400, 200, 100, 50, 25, . . .   **b.** 400, 350, 300, 250, 200, . . .

  c. 400, 390, 370, 340, 300, . . .   **d.** 400, 800, 1600, 3200, 6400, . . .

9. Which point is the vertex of this parabola? **[Lesson 2.2]** c
   a. (0, 4)   **b.** (0, 2)   **c.** (2, 0)   **d.** (4, 4)

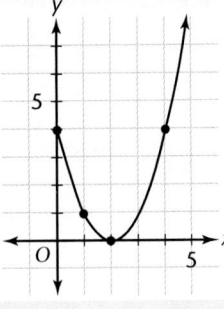

Name the graph shown. **[Lesson 2.5]**

10.

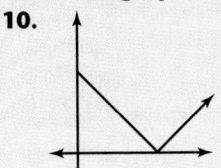

absolute value

11.

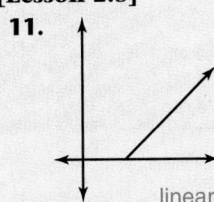

linear

12. Find the values for y by substituting the values 1, 2, 3, 4, and 5 for x in $y = 3x - 2$. **[Lesson 1.4]**
      1, 4, 7, 10, 13

13. Describe the correlation coefficient for the scatter plot at the right as strong positive, strong negative, or little to none. **[Lesson 1.6]** strong negative

14. Is the correlation for the scatter plot at the right nearest to -1, 0, or 1? **[Lesson 1.7]** -1

15. Find the next three terms of the sequence 3, 6, 12, 24, 48, . . . Is the sequence linear or exponential? **[Lesson 2.1]**

16. Use the data to make a graph. Which point is the vertex? **[Lesson 2.2]**

x	1	2	3	4	5
y	4	3	4	7	12

17. The first three terms of a sequence are 5, 7, and 13. The second differences are a constant 4. What are the next three terms? **[Lesson 1.2]** 23, 37, 55

Free-Response Grid The following questions may be answered using a free-response grid commonly used by standardized test services.

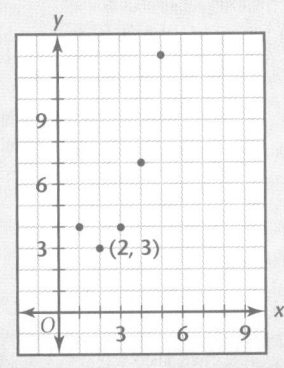

18. Misty's team is in a league with 12 other teams. If each of the teams plays each of the other teams once, how many games will be played? **[Lesson 1.1]** 78

19. Simplify $2(3 + 4) \div 2 + 3^2(4) - 6 \cdot 1$. **[Lesson 1.4]** 37

20. Sam is making bookshelves that are $1\frac{3}{4}$ feet long. How many shelves can he cut from a board that is 6 feet long? **[Lesson 2.4]** 3

15. 96, 192, 384; exponential

16.

CHAPTER 3
Addition and Subtraction in Algebra

Meeting Individual Needs

3.1 Adding Integers

Core Resources

Practice Master 3.1
Enrichment, p. 102
Technology Master 3.1
Interdisciplinary
   Connection, p. 101

[1 day]

Core Two-Year Resources

Inclusion Strategies, p. 103
Reteaching the Lesson, p. 104
Practice Master 3.1
Enrichment Master 3.1
Technology Master 3.1
Lesson Activity Master 3.1

[2 days]

3.2 Exploring Integer Subtraction

Core Resources

Practice Master 3.2
Enrichment, p. 108
Technology Master 3.2

[1 day]

Core Two-Year Resources

Inclusion Strategies, p. 108
Reteaching the Lesson, p. 109
Practice Master 3.2
Enrichment Master 3.2
Technology Master 3.2
Lesson Activity Master 3.2

[2 days]

3.3 Adding Expressions

Core Resources

Practice Master 3.3
Enrichment, p. 114
Technology Master 3.3
Interdisciplinary
   Connection, p. 113
Mid-Chapter Assessment
   Master

[1 day]

Core Two-Year Resources

Inclusion Strategies, p. 114
Reteaching the Lesson, p. 115
Practice Master 3.3
Enrichment Master 3.3
Technology Master 3.3
Lesson Activity Master 3.3

[2 days]

3.4 Subtracting Expressions

Core Resources

Practice Master 3.4
Enrichment, p. 118
Technology Master 3.4

[1 day]

Core Two-Year Resources

Inclusion Strategies, p. 119
Reteaching the Lesson, p. 120
Practice Master 3.4
Enrichment Master 3.4
Technology Master 3.4
Lesson Activity Master 3.4

[2 days]

3.5 Addition and Subtraction Equations

Core Resources

Practice Master 3.5
Enrichment, p. 126
Technology Master 3.5
Interdisciplinary
   Connection, p. 125

[2 days]

Core Two-Year Resources

Inclusion Strategies, p. 126
Reteaching the Lesson, p. 127
Practice Master 3.5
Enrichment Master 3.5
Technology Master 3.5
Lesson Activity Master 3.5

[3 days]

3.6 Solving Related Inequalities

Core Resources

Practice Master 3.6
Enrichment, p. 132
Technology Master 3.6
Interdisciplinary
   Connection, p. 132

[2 days]

Core Two-Year Resources

Inclusion Strategies, p. 132
Reteaching the Lesson, p. 134
Practice Master 3.6
Enrichment Master 3.6
Technology Master 3.6
Lesson Activity Master 3.6

[3 days]

Chapter Summary

Core Resources	Core Two-Year Resources
Chapter 3 Project, pp. 138–139	Chapter 3 Project, pp. 138–139
Lab Activity	Lab Activity
Long-Term Project	Long-Term Project
Chapter Review, pp. 140–142	Chapter Review, pp.140–142
Chapter Assessment, p. 143	Chapter Assessment, p. 143
Chapter Assessment, A/B	Chapter Assessment, A/B
Alternative Assessment	Alternative Assessment
[3 days]	**[5 days]**

Reading Strategies

Be sure to continually evaluate students' ability to read and comprehend the material. Assign a partner to help students who struggle with the reading. After students have read each lesson, have them respond to the following in their journals before they begin to complete the exercises provided.

• What is the main idea of this lesson?

• List and describe any words, phrases, equations, or pictures that you have not seen before. Why do you think they are important?

As you go through each lesson as a class, ask students to record their responses to the Exploration Labs, Critical Thinking, and Ongoing Assessment questions in their journals.

Visual Strategies

Have students look through this chapter and note any pictures of familiar objects. For example, most students have seen rockets, as shown in Lesson 3.1, or checkbooks, as in Lesson 3.2. Have students describe how they think the objects found relate to algebra. Encourage students to discuss their ideas with one another. After they have studied the lesson, have them compare their initial thoughts with the ways in which the objects were actually used.

Hands-On Strategies

Students can learn about adding and subtracting integers by using number lines and counters, tiles, and red and black sticks. Some examples throughout the chapter are illustrated by showing one or more of these manipulatives. Have students use the actual manipulatives to duplicate the process shown in the examples. Students may find it helpful to use manipulatives for the examples that do not illustrate their use. Allow students to use manipulatives as long as necessary, at the same time encouraging them to record their work using pencil and paper. If students are allowed enough time to work with manipulatives, they will begin to visualize the concept of addition and subtraction of integers and, with that understanding, will eventually feel comfortable working without the manipulatives.

Cooperative Learning

GROUP ACTIVITIES	
Adding integers using tiles	Lesson 3.1, Exploration
Subtracting integers using tiles	Lesson 3.2, Exploration
Distance on a grid	Lesson 3.3, Exercise 52
Number sequences in geometry	Lesson 3.6, Exercise 53
Creating a database	Chapter 3 Project

You may wish to have students work with partners for some of the above activities. Additional suggestions for cooperative group activities are noted in the teacher's notes in each lesson.

Multicultural

The cultural connections in this chapter include references to Asia and Africa.

CULTURAL CONNECTIONS	
Asia: Integers in China	Lesson 3.1
Africa: Hypatia	Lesson 3.2, 36–37
Adding and subtracting numbers	Lesson 3.4, 50–51

Portfolio Assessment

Below are portfolio activities for the chapter. They are listed under seven activity domains that are appropriate for portfolio development.

1. **Investigation/Exploration** Two projects listed in the teacher notes ask students to investigate a method and/or a relationship. They are the *Alternative Teaching Strategy* on page 118 (Lesson 3.4) and *Interdisciplinary Connection* on page 132 (Lesson 3.6). One project explores formulas used in another discipline. It is the *Interdisciplinary Connection* on page 125 (Lesson 3.5).

2. **Applications** Recommended to be included are any of the following: Picture Framing (Lesson 3.3), Fund-Raising (Lesson 3.4), Inventory (Lesson 3.4), Sports (Lessons 3.5 and 3.6), Home Economics (Lesson 3.5), Travel (Lesson 3.5), and, from the teacher notes, *Interdisciplinary Connection* on page 113 (Lesson 3.3).

3. **Non-Routine Problems** The Portfolio Activity, page 99, asks students to solve a problem using any method and then an algebraic method.

4. **Project** The Chapter Project on pages 138–139 asks students to create their own database using data that is of interest to them. They must also devise a method for sorting fields and retrieving records. The teacher notes include an extension activity in which students are asked to create their own database and provide a printout or a report about their database.

5. **Interdisciplinary Topics** Students may choose from the following: Physics (Lesson 3.1), Geometry (Lessons 3.3 and 3.5), Statistics (Lessons 3.4 and 3.5), Computer (Lesson 3.5), Consumer Math (Lesson 3.5), and, from the teacher notes, History on page 101 (Lesson 3.1) and Physics on page 125 (Lesson 3.5).

6. **Writing** The interleaf material at the beginning of this chapter has information for journal writing. Also recommended are the Communicate questions at the beginning of the exercise sets.

7. **Tools** Chapter 3 uses tiles, calculators, and computers for adding and subtracting integers and for solving equations. The technology masters are recommended to measure the individual student's proficiency.

Technology

Technology provides important tools for manipulating and visualizing algebraic concepts. Several lessons in Chapter 3 can be significantly enhanced by using appropriate technology. Scientific calculators make calculations with integers easier and help students focus on the underlying algebraic concepts. For more information on the use of technology, refer to the HRW Technology Handbook.

Graphics Calculator

Adding and Subtracting Integers

Lesson 3.1 suggests the use of a calculator to compute with integers. Either a graphics calculator or a scientific calculator can be used. Follow the steps given below to add or subtract integers on a graphics calculator. The instructions apply to the TI-82. For instructions for using other graphics and scientific calculators, please refer to the HRW Technology Handbook.

1. Press the [(-)] key *before* all negative integers.

2. Press the [+] key to add integers.

3. Press the [-] key to subtract integers.

4. When you are finished entering the expression to be evaluated, press the [ENTER] key. Press the [(] and [)] keys when parentheses are needed. It is not necessary to press the parentheses keys when they are shown around a single negative integer as in $3 + (-5)$.

5. Press [2nd] [ABS], then the integer to find the absolute value of that integer. Only the absolute value of the integer entered immediately after the [2nd] and [ABS] keys is calculated. Be sure to put parentheses around the entire expression that requires the absolute value.

For example, $3 + |-2| + 5$ can be entered as

3 [+] [2nd] [ABS] [(-)] 2 [+] 5 [ENTER],

but $3 + |-2 + 5|$ should be entered as

3 [+] [2nd] [ABS] [(] [(-)] 2 [+] 5 [)] [ENTER].

Integrated Software

$f(g)$ **Scholar**[TM] is an integrated computer-based mathematics productivity tool that combines calculator, spreadsheet, and graphics capabilities to provide a dynamic and interactive environment for explorations in mathematics. It is appropriate to use $f(g)$ **Scholar**[TM] for any lesson needing a spreadsheet, calculator, graphics calculator, or any combination of the three.

**Addition and
Subtraction
in Algebra**

ABOUT THE CHAPTER

Background Information

Balance plays an important role in many different activities. It must be achieved and maintained for the activity to work. The same holds true for balancing the expressions on each side of the equal side when solving an equation.

CHAPTER RESOURCES

- Practice Masters
- Enrichment Masters
- Technology Masters
- Lesson Activity Masters
- Lab Activity Masters
- Assessment Masters
    Chapter Assessments, A/B
    Mid-Chapter Assessment
    Alternative Assessment,
    A/B
- Teaching Transparencies
- Spanish Resources

CHAPTER OBJECTIVES

- Apply the concepts of absolute value and opposites to integers.
- Add two or more integers.
- Use neutral pairs of tiles to subtract integers.
- Subtract integers by adding the opposite.
- Represent and add expressions using tiles.
- Use the Distributive, Commutative, and Associative Properties of Addition to rearrange and group like terms.
- Simplify expressions with several variables by adding expressions with like terms.
- Use tiles to model subtraction of expressions as take away.

CHAPTER 3

LESSONS

3.1   Adding Integers

3.2   *Exploring* Integer
       Subtraction

3.3   Adding Expressions

3.4   Subtracting Expressions

3.5   Addition and
       Subtraction Equations

3.6   Solving Related
       Inequalities

Chapter Project
Find It Faster

Addition and Subtraction in Algebra

An equation is a statement that two mathematical expressions are equal. If each of these expressions were placed on a mathematical "scale," they would balance.

In this chapter you will study a number of mathematical properties. These properties maintain the balance in an equation as you add and subtract to find the value of the unknown.

ABOUT THE PHOTOS

The tight-rope-walker, the girl balancing on the beam, and the scale all show the need for balance when performing various activities. When solving equations using algebra, maintaining equality on the left and right sides of the equal sign is essential for finding the unknown value. Students can associate the need for maintaining equality in algebra with the need for balance by the performers in the opening photos.

- Subtract expressions by adding the opposite.
- Understand the basic concept of an equation in algebra.
- Represent and solve applications involving addition and subtraction equations using the four steps: organize, write, solve, and check.
- Solve literal equations that involve addition and subtraction.
- Define the concept and symbols of inequalities.
- Simplify and solve inequalities that involve addition and subtraction.

PORTFOLIO ACTIVITY

Assign the Portfolio Activity as part of the assignment for Lesson 3.5 on page 130.

Encourage students to discuss what it means for the dog to balance with the squirrels. It may be helpful for students to imagine actually placing them on a balance scale like the one shown in the photo.

Have students work in pairs or small groups and use manipulatives such as algebra tiles to help them solve the problem. After each group has agreed upon an answer, have them present their solution to the class. Encourage students to discuss the similarities and differences among the ways different groups solved the problem.

As an extension to this activity, have students determine how many cats will balance the dog.

PORTFOLIO ACTIVITY

A squirrel and a dog weigh the same as 2 cats. A dog weighs as much as 2 squirrels and a cat. Now how many squirrels will balance the dog?

Find a solution to the problem any way you can, then find an algebraic solution. Remember, if one quantity equals another, they can replace each other.

ABOUT THE CHAPTER PROJECT

The Chapter 3 project on pages 138–139, involves exploring the concept of balancing using weights of different animals. It would be helpful to have a real balance scale available with which students can experiment. Students may also find it useful to act out the problem in order to solve it. It may not be clear to students how to solve this problem algebraically until they have studied some of the lessons in this chapter.

- Apply the concepts of absolute value and opposites to integers.
- Add two or more integers.

RESOURCES

- Practice Master   **3.1**
- Enrichment Master   **3.1**
- Technology Master   **3.1**
- Lesson Activity Master   **3.1**
- Quiz   **3.1**
- Spanish Resources   **3.1**

Assessing Prior Knowledge

Find each sum.

1. $49 + 63$ **[112]**

2. $55 + 38$ **[93]**

3. $74 + 17 + 35$ **[126]**

4. $29 + 88 + 26$ **[143]**

TEACH

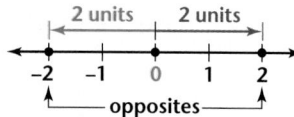

Integers are a part of everyday life. Encourage students to discuss ways in which they have used integers or have seen integers being used. Focus on situations in which integers are added.

LESSON 3.1 Adding Integers

 *"T minus 2 seconds and counting."* During the launch of the space shuttle, time is measured before and after the liftoff. Integers are needed to count units that occur on opposite sides of zero.

Numbers such as $+1$, $+31$, and $+53$ are examples of **positive integers**. However, they are usually written without the positive sign. Thus $+31$ and 31 represent the same number. Numbers such as -1, -31, and -53 are **negative integers**. The integers consist of the set of all positive integers, all negative integers, and **zero**. This set can be represented as points on a number line.

Negative integers   Positive integers

$$-5\ -4\ -3\ -2\ -1\quad 0\quad 1\ 2\ 3\ 4\ 5$$

Zero

Two numbers are **opposites** if they are

- on opposite sides of zero, and
- the same distance from zero.

2 units   2 units

$$-2\quad -1\quad 0\quad 1\quad 2$$

opposites

The opposite of 2 shown in symbols is
$$-(2) = -2.$$
The opposite of -2 shown in symbols is
$$-(-2) = 2.$$
The opposite of zero is zero itself,
$$-(0) = 0.$$

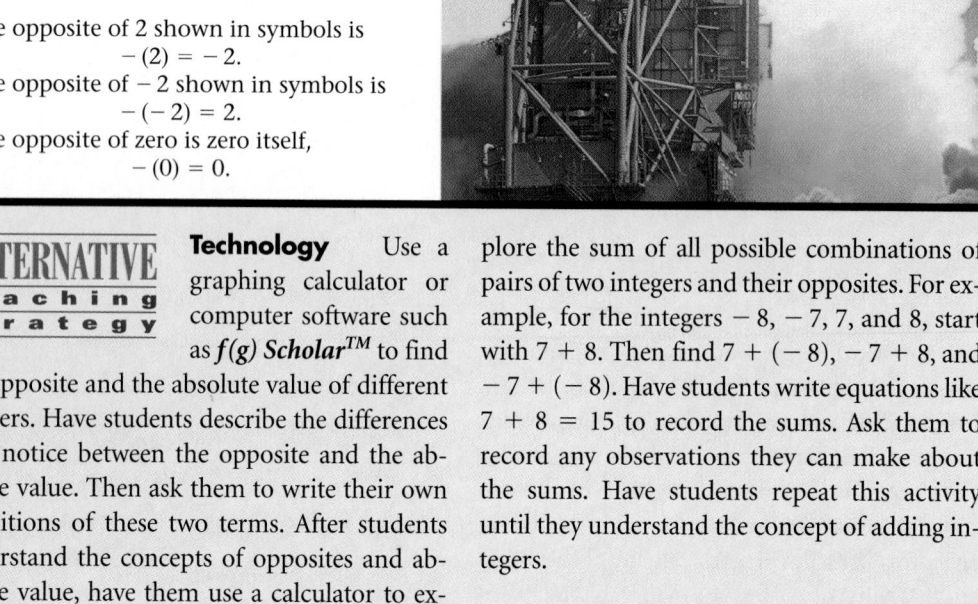

ALTERNATIVE
teaching strategy

Technology Use a graphing calculator or computer software such as *f(g) Scholar*™ to find the opposite and the absolute value of different integers. Have students describe the differences they notice between the opposite and the absolute value. Then ask them to write their own definitions of these two terms. After students understand the concepts of opposites and absolute value, have them use a calculator to ex-

plore the sum of all possible combinations of pairs of two integers and their opposites. For example, for the integers -8, -7, 7, and 8, start with $7 + 8$. Then find $7 + (-8)$, $-7 + 8$, and $-7 + (-8)$. Have students write equations like $7 + 8 = 15$ to record the sums. Ask them to record any observations they can make about the sums. Have students repeat this activity until they understand the concept of adding integers.

In Lesson 2.4 you were introduced to the absolute value function. The distance from zero to an integer is the absolute value of the integer. The absolute value of an integer less than 0 is the opposite of that integer. For example, $|-5| = -(-5) = 5$.

ABSOLUTE VALUE

For any number x,

if x is greater than or equal to 0, $|x| = x$, and

if x is less than 0, $|x| = -x$.

Using Models to Add Integers

Tiles can be used to model integers.

Represents 1

Represents -1

 *Patterns in Addition*

Tiles can also be used to model integer addition.

$3 + 2 = 5$ $-2 + -4 = -6$

1 Use tiles to model the addition shown, and solve each of the equations.

a. $7 + 3 = ?$ **b.** $4 + 5 = ?$
c. $-7 + (-3) = ?$ **d.** $-4 + (-5) = ?$

 Describe how to add two integers with like signs. ❖

---

interdisciplinary
CONNECTION
History The positive $(+)$ and negative $(-)$ symbols we use today made their first appearance in fifteenth-century German warehouses, where they were used to indicate a deviation from a standard measure. For example, a $(+)$ meant that the container weighed more than the standard weight and a $(-)$ meant that the container weighed less than the standard weight.

Use Transparency 11

Exploration Notes
Students should feel very comfortable adding positive integers. However, adding negative integers may be slightly more difficult. Encourage students to extend what they already know about adding positive integers to adding negative integers. If students still do not feel comfortable with these concepts after completing the exploration, encourage them to complete several more examples like those in Exercise 1.

Aongoing
ASSESSMENT

2. Add the absolute values of the integers, and use the sign of both integers.

Encourage students to use tiles to duplicate the procedure used in Example 1. Encourage students to use the tiles to complete Alternate Example 1.

Alternate Example 1

Suppose Chad was 2 seconds behind his competition after the first lap. He then gains 2 seconds on the second lap. What is his position in the race relative to his competition after completing the second lap?
[even]

Use Example 1 and Alternate Example 1 to show that addition of opposites is commutative.

Alternate Example 2

Dana recorded temperatures one cold December day. At 6 A.M. the temperature was −6°F. By noon the temperature had risen 11 degrees. What was the temperature at noon?
[5°F]

Teaching tip

Students can use a picture of a thermometer or a number line to help them visualize the situations in Example 2 and Alternate Example 2.

EXAMPLE 1

Sports *After the first lap, Chad was 2 seconds ahead of his competition.*

Chad loses 2 seconds of lead on the second lap. What is his position in the race relative to his competition after the second lap? Represent this problem using tiles.

Solution

Let one positive tile represent 1 second gain, and let one negative tile represent 1 second loss, relative to the competition. Then, 2 + (−2) represents Chad's position after the second lap.

Each pair of tiles consists of a positive tile and a negative tile. Since each positive tile is paired with a negative tile, the entire sum is 0. Chad is even with the competition after the second lap. ❖

2 + (−2) = 0

This example leads to an important property of integers.

> **PROPERTY OF OPPOSITES**
> For any number a, $-a$ is its opposite, and
> $$a + (-a) = 0.$$

The numbers a and $-a$ are also called **additive inverses.**

EXAMPLE 2

Temperature Anna recorded temperatures one cold day in January. At noon the temperature was 4°F. By 6 P.M. the temperature had dropped 7°F. What is the 6 P.M. temperature?

Solution A

The sum 4 + (−7) represents the temperature in the evening.

Each pair consisting of a positive tile and a negative tile is equal to 0, so three negative tiles remain. This represents −3°F.

$$4 + (-7) = -3$$

ENRICHMENT Have students do research to find out more about the countdown for the launch of a space shuttle. Have them answer such questions as "Why is T used?" and "Why are negative numbers used to represent the time before the launch?" Encourage students to share their findings with the rest of the class.

Solution B ➤

A thermometer is a vertical number line. To use the number line to model $4 + (-7)$, start at 4. Move 7 steps in the negative direction. You will end at -3.

$$4 + (-7) = -3 ❖$$

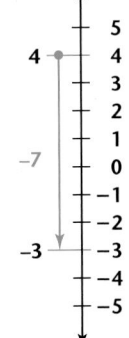

Cultural Connection: Asia   More than 2000 years ago, the Chinese used different colored rods to represent positive and negative numbers. To compute how much remains after a transaction involving a gain, $+23$, and a loss, -54, the Chinese combined sets of red and black rods.

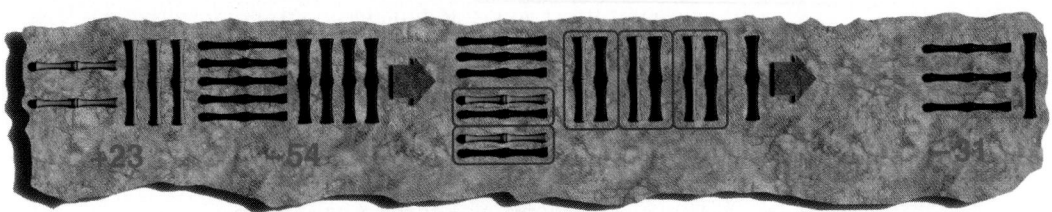

The corresponding red and black rods are eliminated. This leaves three horizontal black rods and one vertical black rod which represent -31.

There are only two conditions to consider when adding integers—either the signs are alike or the signs are different.

> **RULES FOR ADDING INTEGERS**
> **A. Like signs**
>   Find the **sum** of the absolute values.
>   Use the sign common to both integers.
> **B. Unlike signs**
>   Find the **difference** of the absolute values.
>   Use the sign of the integer with the greater absolute value.

EXAMPLE 3

Find each sum.   Ⓐ $-2 + (-8)$ Ⓑ $15 + (-23)$

Solution➤

Ⓐ If the terms have like signs, find the sum of the absolute values, then use the sign that is the same for both integers.

$$|-2| = 2 \qquad |-8| = 8 \qquad -2 + (-8) = -10$$

Ⓑ If the terms have unlike signs, find the difference of the absolute values, then use the sign of the integer with the greater absolute value.

$$|15| = 15 \qquad |-23| = 23 \qquad 15 + (-23) = -8 ❖$$

Try This   Find the sums.   **a.** $27 + (-16)$ **b.** $-27 + (-16)$

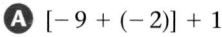

Adding More Than Two Integers

You will often need to add more than two integers.

EXAMPLE 4

Find each sum.   **A** $[-9 + (-2)] + 12$ **B** $-9 + [(-2) + 12]$

Solution➤

A Add the first pair of integers, and remove the brackets. Then add that sum and the remaining integer.

$$[-9 + (-2)] + 12 = -11 + 12 = 1$$

B The brackets instruct you to add the second pair of integers first. Next, remove the brackets. Then add the first integer to the sum.

$$-9 + [(-2) + 12] = -9 + 10 = 1 ❖$$

These examples illustrate the *Associative Property of Addition.* When adding integers, the grouping does not matter.

ASSOCIATIVE PROPERTY OF ADDITION

For all numbers *a*, *b* and *c*,

$$(a + b) + c = a + (b + c).$$

CRITICAL
Thinking

The Associative Property of Addition applies to adding 3 numbers. Examine several examples, and make a conjecture about adding 4 or more numbers.

TECHNOLOGY
Calculator

A calculator will save time when you need to add a series of numbers. Check your calculator to see how to key in a negative number. Try adding $-618 + 587 + (-10)$. Your calculator should show -41 as the answer. Explain how to use your calculator to find the sum of a loss of $26.46 followed by a second loss of $12.27 followed by a gain of $18.25.

Exercises & Problems

Communicate

1. Explain and give an example of what is meant by the opposite of a number.

2. Discuss and give examples of what is meant by the absolute value of a number.

3. Describe how to add two integers that have the same sign.

4. Describe how to add two integers that have unlike signs.

5. How would you add 25 and -13 using the ancient Chinese system? (Hint: Use colored toothpicks as counting sticks.)

6. Explain how to add -43, -9, and 128 in two different ways.

Practice & Apply

Find the opposite of each number.

7. 17 -17 8. -17 17   9. 0 0   10. $(12 - 5)$ -7

Use algebra tiles to find the following sums.

11. $-5 + (-2)$ -7 12. $-3 + 3$ 0   13. $2 + (-6)$ -4 14. $-2 + (-6)$ -8

Use a number line to find the following sums.

15. $8 + (-3)$ 5   16. $-4 + (-5)$ -9 17. $-1 + 5$ 4   18. $-1 + (-2)$ -3

Find each sum.

19. $-28 + 50$ 22   20. $17 + (-34)$ -17 21. $38 + (-72)$ -34

22. $14 + (-29) + (-12)$ -27 23. $-43 + 82 + |-19|$ 58   24. $-308 + |-24| + (-29)$
-313

Physics Electrons have a charge of -1 and protons have a charge of $+1$. The total charge of an atom is the sum of its electron charges and proton charges. Find the total charge of the following atoms.

25. 16 protons; 18 electrons -2 26. 10 protons; 10 electrons 0

27. The Lincoln High football team completes two series of downs with the following gains and losses. Find the total gain or loss. 6 yard gain

Down	1	2	3	4
Gain/Loss	8	-4	-3	15

Down	1	2	3	4
Gain/Loss	-3	-5	4	-6

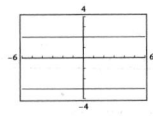

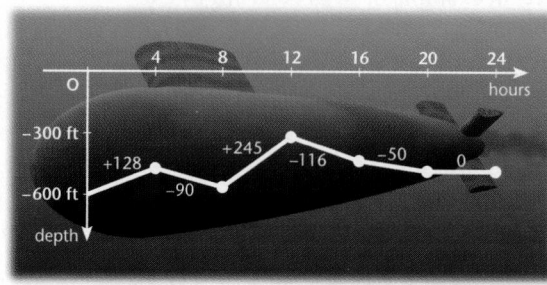

Alternative
ASSESSMENT

Performance Assessment

Ask students to write a problem for each sum in Exercises 15-18. Then have students write 3 problems that involve sums of integers of their choice.

Use Transparency ▶ 13

Look Beyond

Review how to find the opposite of a negative integer. Also review subtraction of two whole numbers, and explain that this is the same as subtracting two positive integers.

---

28. Business In business, a loss is recorded by placing the amount inside parentheses. What is the yearly profit or loss for the Family Shoe Store? $197.03 loss

1st Quarter	2nd Quarter	3rd Quarter	4th Quarter
(389.75)	(794.28)	1796.50	(809.50)

29. During one week Paul Phillips' stock had gains and losses. If the opening stock price was $34\frac{1}{2}$ on Monday, what was the price after the market closed on Friday? $33

M	T	W	Th	F
$+\frac{1}{2}$	$-\frac{3}{4}$	$+\frac{1}{4}$	0	$-1\frac{1}{2}$

30. The graph starts at -600 feet. It shows the change in depth of a submarine over six 4-hour intervals. What is the depth of the submarine after 24 hours? 483 feet below sea level

Let $a = 2, b = -3, c = 5$. Evaluate each expression.

31. $(a + b) + c$ 4

32. $|a + b| + c$ 6

33. $a + |b + c|$ 4

34. $a + (b + c)$ 4

35. $a + (c + b)$ 4

Look Back

36. Find the sum of $1 + 2 + 3 + \cdots + 40$. **[Lesson 1.1]** 820

Find the next two terms in each number sequence. [Lesson 1.3]

37. 40, 37, 34, 31, __, __   28, 25

38. $-5, -1, 3, 7,$ __, __   11, 15

39. 1, 3, 6, 10, __, __   15, 21

40. Geometry What is the perimeter of a square with a side of 3.5 centimeters? **[Lesson 1.3]** 14 cm

Calculate.   [Lesson 1.4]

41. $15 - 21 + 3 + 4$ 1

42. $[3(4 - 2)^2] + 7$ 19

43. $12 + 3^2 + (9 - 6)$ 24

Plot each point in the same coordinate plane.   [Lesson 1.5]

44. $(4, 4)$

45. $(5, 2)$

46. $(3, 4)$

47. $(6, 3)$

Look Beyond

48. You may recall from arithmetic that adding the same number repeatedly can be represented by multiplication.

$$2 + 2 + 2 + 2 + 2 = 5(2) = 10$$

With this in mind, represent $7(-2)$ as an addition problem, then find the value of $7(-2)$. -14

49. Recall that -14 is the opposite of 14, and 14 is the opposite of -14. Use this information and the method in Exercise 48 to find the product $-[7(-2)]$. $-(-14) = 14$

44-47.

LESSON 3.2

Exploring Integer Subtraction

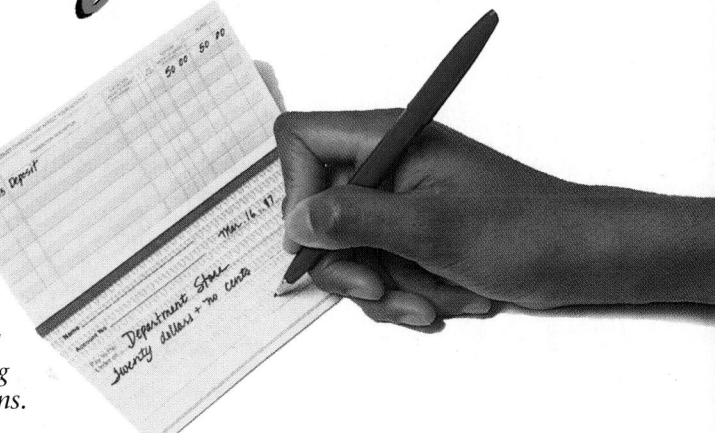

why *Each time Mark writes a check, the bank subtracts money from his account. When he deposits his paycheck, the bank adds money to his account. You use integers to model subtraction when working with financial transactions.*

Suppose Mark starts with $50 in his account and writes a check for $20. This transaction can be represented in two ways.

subtraction: $50 - 20$ or   addition: $50 + (-20)$

In either case, Mark will have $30 left.

$$50 - 20 = 30 \quad \text{or} \quad 50 + (-20) = 30$$

Using Tiles to Subtract Integers

You can model the subtraction of integers by taking away tiles.

6 − 4 = 2   **−4 − (−2) = −2**

In Lesson 3.1 you found the result of adding the same number of positive and negative tiles does not change the *value* of the tiles you have. The result is the same as adding 0. Why do you think the number 0 is called the **additive identity**?

ADDITION PROPERTY OF ZERO
For any number *a*,
$$a + 0 = a = 0 + a.$$

Technology Have students use a calculator, computer, or *f(g) Scholar*™ to explore the differences of 2 integers and their opposites. Use all combinations. For example, for the integers −9, −4, 4, and 9, start with 9 − 4. Then find 9 − (−4), −9 − 4, and −9 − (−4). Have students record the differences and any observations about the differences. Have them repeat this activity until they understand the concept of subtracting integers.

PREPARE

Objectives
• Use neutral pairs of tiles to subtract integers.
• Subtract integers by adding the opposite.

RESOURCES

• Practice Master   3.2
• Enrichment Master   3.2
• Technology Master   3.2
• Lesson Activity Master   3.2
• Quiz   3.2
• Spanish Resources   3.2

Assessing Prior Knowledge
Find each sum.
1. $-5 + 8$ [3]
2. $-13 + 6$ [−7]
3. $-25 + 19$ [−6]
4. $-19 + 28$ [9]

TEACH

why The opening example of using integer subtraction during financial transactions is a common one. There are many other situations in which integers are subtracted. Encourage students to discuss other ways in which they think subtraction of integers is used.

Ongoing ASSESSMENT

Adding 0 to any number results in a sum equal to the number itself.

Each pair of positive and negative tiles is sometimes called a *neutral pair* or a *zero pair* because the sum has a value of 0.

Cooperative Learning

Have students work in groups and use tiles to model subtraction of integers. Have them use tiles to find several differences. Encourage students to record all of their work and any observations they may make. Have each group share their observations with the class.

CRITICAL *Thinking*

Start with 2 negative tiles. Add 4 positive and 4 negative tiles. Take away 4 positive tiles. There are 6 negative tiles left, so the difference is -6. The equivalent addition expression is $-2 + (-4)$.

Exploration Notes

The method students will use to complete this exploration is similar to the method illustrated by the tile model at the top of the page. The fact that a positive 5 is being subtracted from a positive 3 is emphasized in the answer to question 3. Be sure students understand this.

Add enough neutral pairs so that you can take away the required number of positive or negative tiles.

The next activity makes use of the Addition Property of Zero. The tile model shows how it is possible to subtract, or *take away,* -4 from 2.

$$2 - (-4) = \underline{?}$$

1. Start with 2 positive titles.

2. Since you want to subtract -4, add 4 positive and 4 negative tiles. The total value of the tiles is still 2.

3. Now you can subtract, or *take away,* the 4 negative tiles and rewrite the original problem as $2 + 4$.

Thus, $2 - (-4) = 2 + 4 = 6$.

CRITICAL *Thinking*

Describe how you would use algebra tiles to calculate $-2 - 4$. What is the addition expression that is equivalent to $-2 - 4$?

•Exploration• *Subtraction With Tiles*

Use tiles to subtract 5 from 3.

1. Begin with 3 positive tiles.

2. The number you want to subtract is 5, so add 5 positive and 5 negative tiles. What value did you add to the expression when you added these tiles?

3. How many and what kind of tiles do you take away?

4. Remove the tiles. How many of each kind are left?

5. What addition expression is equivalent to $3 - 5$? What is the sum?

6. Use this procedure to calculate $-3 - (-4)$.

7. Explain how to subtract an amount with a greater absolute value from an amount with a lesser absolute value. ❖

ENRICHMENT Have students determine whether subtraction of integers is associative or commutative. Encourage them to experiment with many different examples before they come to a conclusion.

INCLUSION strategies **Using Symbols** Some students may be ready to extend the concept of subtracting integers to include the use of variables. For example, present students with the expression $a - b$, where both a and b represent integers. Ask them to describe when this difference is positive and when it is negative.

Relating Addition and Subtraction

When working with integers, you can use addition to calculate a difference. To subtract an integer, add its opposite. The same is true for any numbers.

> **THE DEFINITION OF SUBTRACTION**
> For all numbers a and b,
> $$a - b = a + (-b).$$

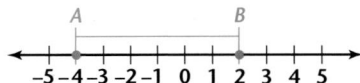 **EXTENSION**

You will sometimes need to determine the distance between points on the number line. One method uses absolute value.

To find the distance between points A and B, subtract the values, then find the absolute value of the difference.

The distance between -4 and 2 is $|2 - (-4)| = 6$, or $|(-4) - 2| = 6$.

In general, if you let a represent the value for the point A and let b represent the value for the point B, the distance between points on the number line is $|b - a|$ or $|a - b|$. ❖

What is the distance between -3 and 5 on the number line?

Calculator

On all calculators ⎓–⎓ is used for the subtraction operation. Recall from the previous lesson that some calculators use ⎓+/-⎓ or ⎓±⎓ to change the sign of a number. Other calculators use ⎓(-)⎓ to find the opposite of a number.

How does your calculator treat the difference between subtraction and the opposite of a number? What keystrokes on your calculator let you calculate $-35 - 27$?

 **R**ETEACHING *the* lesson

Using Visual Models Use the same method used in Lesson 3.1. Have students use different colored pencils and tally marks to represent positive and negative integers. Rewrite the difference $-4 - 6$ as a sum $-4 + (-6)$. Make 4 red marks and then 6 more red marks for a total of 10 red marks, which represent -10. To represent the difference $-4 - (-3)$, rewrite it as a sum, $-4 + 3$, and then add. Make 4 red marks

and 3 black marks. Cross out an equivalent number of black marks and red marks to determine the correct amount.

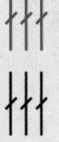

In this case, 3 black marks cross out 3 red marks, and there is 1 red mark left. The answer is negative 1.

Assignment Guide

Core 1–6, 11–18, 23–44

Core Two-Year 1–42

Technology

A calculator is helpful for exercises 7–26 and 38–40. You may want to have students complete the exercises using pencil and paper first, and then use the calculator to verify their answers.

Error Analysis

Many students struggle with the concept of when to add and when to subtract. Encourage these students to first write all differences as sums. They can then add the integers as they did in the last lesson.

EXERCISES & PROBLEMS

Communicate

1. What is the effect of adding 0 to a number?

2. How does adding pairs of positive and negative tiles help to model a subtraction problem?

3. Mark's account balance is $5. Explain what would happen if Mark were to write a check for $25 dollars.

4. For the problem $-4 - (-7)$, how many positive and negative tiles do you add to solve the problem? Why?

5. What is meant by adding the opposite?

6. Explain how addition can replace subtraction when you use integers.

Practice & Apply

Find the solutions of the following problems.

7. $67 - 3$ 64
8. $42 - (-9)$ 51
9. $-10 - (-21)$ 11
10. $-35 - 17$ -52

11. $33 - (-33)$ 66
12. $-78 + (-45)$ -123
13. $990 - (-155)$ 1145
14. $-97 - 88$ -185

15. $-43 + 23 + (-43)$ -63
16. $-77 - 77 + 5$ -149
17. $108 + (-18) - 8$ 82
18. $85 - (-12) - (-9)$ 106

Let $x = 5$, $y = -3$, and $z = -10$. Evaluate each expression.

19. $x - y$ 8
20. $x + y - z$ 12
21. $(x - z) - y$ 18
22. $x - (z - y)$ 12

23. $y - x$ -8
24. $y - x + z$ -18
25. $(x + y) - (x - y)$ -6
26. $y - y - y - y$ 6

Find the distance between the following pairs of points on the number line.

27. $4, 9$ 5 units
28. $-6, 15$ 21 units
29. $-47, -23$ 24 units
30. $-12, 74$ 86 units

Mike has a balance of $145 in his savings account.

31. If Mike withdraws $37, how much money does he have left in his account? $108

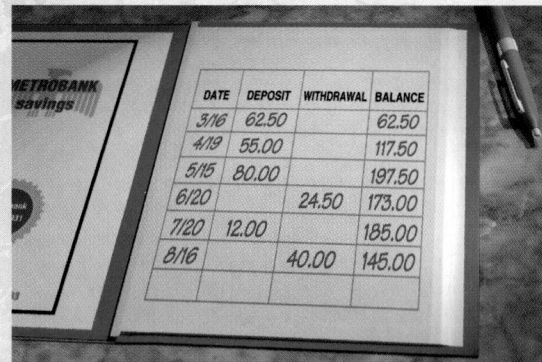

32. **Temperature** Theresa noticed that the temperature in her freezer was 5°F. She lowered the thermostat by 7°F. Later, she set the thermostat another 2°F lower. What temperature would you expect in the freezer after Theresa finished changing the thermostat? -4° F

This wind chill table depends on the air temperature and the speed of the wind.

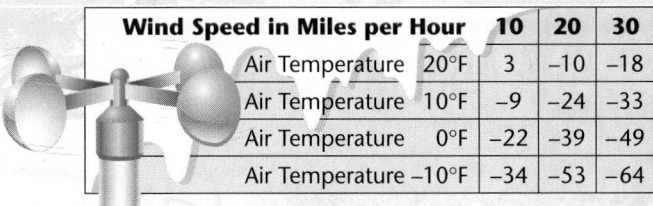

Wind Speed in Miles per Hour	10	20	30
Air Temperature 20°F	3	–10	–18
Air Temperature 10°F	–9	–24	–33
Air Temperature 0°F	–22	–39	–49
Air Temperature –10°F	–34	–53	–64

33. The air temperature is 20°F. The wind speed is 20 miles per hour. What is the wind-chill factor? − 10° F

34. The air temperature is − 10°F. How much colder does it feel when the wind speed is 20 miles per hour than when there is no wind? 43° F

35. The air temperature is − 10°F. The wind speed increases from 10 to 30 miles per hour. How many degrees does the temperature seem to drop? 30° F

 *Look Back*

36. Cultural Connection: Africa The earliest woman mathematician whose name we know is Hypatia. She was a professor at the University of Alexandria, Egypt, about 400 C.E. In her commentary on Diophantus, an earlier Egyptian mathematician, Hypatia spoke of pentagonal numbers.

The sequence for pentagonal numbers is 1, 5, 12, 22, 35, . . . Find how many differences it takes to reach a constant. What is the constant? **[Lesson 1.2]** 2 differences; 3

    •   1
    •
  • • •   5
  • • •
 • • • • 12
• • • •

37. What kind of function will generate the terms of the pentagonal numbers? **[Lesson 1.2]** quadratic

Simplify. [Lesson 1.4]

38. $36 − 12 ÷ 3 − 20$ 12

39. $3 · 5 + 7 ÷ 2$ 18.5

40. $28 ÷ 2 · 7 + 4$ 102

41. What function is associated with the graph of a parabola? **[Lesson 2.2]** quadratic

42. What kind of function will generate the values shown for y in the table? **[Lesson 2.3]** reciprocal

x	5	6	7	8	9
y	$\frac{1}{5}$	$\frac{1}{6}$	$\frac{1}{7}$	$\frac{1}{8}$	$\frac{1}{9}$

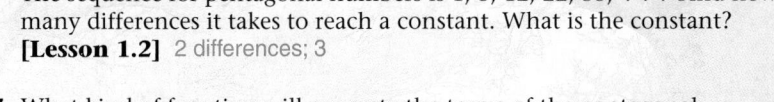 *Look Beyond*

43. One point on the number line is at 4. The location of the other point is not known. Let the unknown point be at x. If the distance between the points is 7, what is x? (HINT: Draw the number line, and try different values of x.) 11 or − 3

44. What is x if the distance between the points is 10? 14 or − 6

Objectives

• Represent and add expressions using tiles.
• Use the Distributive, Commutative, and Associative Properties of Addition to rearrange and group like terms.
• Simplify expressions with several variables by adding expressions with like terms.

RESOURCES

• Practice Master	3.3
• Enrichment Master	3.3
• Technology Master	3.3
• Lesson Activity Master	3.3
• Quiz	3.3
• Spanish Resources	3.3

Assessing Prior Knowledge

Find each sum.

1. $12 + -8$ [4]

2. $-6 + -5$ [-11]

3. $7 + 8$ [15]

4. $-11 + 7$ [-4]

TEACH

Students are introduced to a real-world situation that can be described using two algebraic expressions. Emphasize that each box contains the same number of comic books, so the variable b represents the number of comic books in each carton. Encourage students to discuss other situations that can be described by an algebraic expression.

LESSON 3.3 Adding Expressions

why *In Algebra you work with expressions that contain numbers and variables. When you add expressions, you need to write their sum in a simplified form.*

Juan is collecting comic books. He has 3 cartons of comics and 4 loose comics in his room. He has 2 more cartons of comics and 2 loose comics in the trunk of his car.

How can you write an expression that describes the total number of cartons and comics?

In his room	3 cartons of comics and 4 comics	$3x + 4$
In the car	2 cartons of comics and 2 comics	$2x + 2$
All together	5 cartons of comics and 6 comics	$5x + 6$

The quantity, $5x + 6$, is an example of an **algebraic expression.** Both $5x$ and 6 are **terms** of the expression $5x + 6$. The terms $3x$ and $2x$ are **like terms** because each contains the same form of the variable. The number 5 is a **factor** of $5x$. The 5 is also called the **coefficient** of x. The number 6, which represents a fixed amount, is often referred to as a **constant.**

Tiles can be used to model an expression containing a variable such as x.

Represents positive x Represents negative x

ALTERNATIVE teaching strategy

Technology Use a spreadsheet and set up a table to add the coefficients of like terms and constants. In the sample given below, $(5x - 6) + (2x + 4)$ is evaluated.

	A	B	C	D
	C4		=C2+C3	
1		x-coefficient	constant	
2		5	−6	
3		2	4	
4	sum	7	−2	
5				

The formula for cell B4 is = B2 + B3. The formula for cell C4 is = C2 + C3. Have students find the sum of many different expressions and record any ideas they have about adding algebraic expressions. Discuss the ideas, and guide students to the correct conclusions about adding algebraic expressions.

EXAMPLE 1

Simplify $(3x + 4) + (2x - 1)$ using tiles.

Solution

Represent each of the expressions using the appropriate tiles. Then combine like tiles.

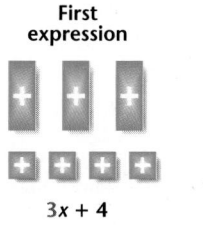

First expression	Second expression	Combination
$3x + 4$	$2x - 1$	$(3x + 4) + (2x - 1)$

Remember, a positive 1-tile and a negative 1-tile equal 0 when combined.

$5x + 3$

When you simplify the expression, the result is $5x + 3$. ❖

How would you model $(2x - 3) + (4x + 5)$? What is the result?

The Distributive Property

Combining like terms such as $3x$ and $2x$ uses an important property of numbers: multiplication can be distributed over addition.

The Distributive Property works in two directions. To multiply $3 \cdot (5 + 2)$, the multiplication is distributed over the addition.

$$3 \cdot (5 + 2) = 3 \cdot 5 + 3 \cdot 2 = 21$$

When used in reverse, a common factor can be removed from the terms.

$$3 \cdot 5 + 3 \cdot 2 = 3 \cdot (5 + 2) = 21$$

This can be shown on a number line.

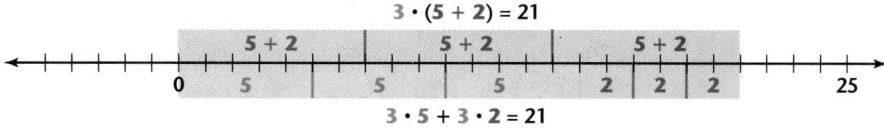

$3 \cdot (5 + 2) = 21$

$3 \cdot 5 + 3 \cdot 2 = 21$

> **DISTRIBUTIVE PROPERTY**
> For all numbers a, b, and c,
> $a(b + c) = ab + ac$ and $(b + c)a = ba + ca$.

interdisciplinary CONNECTION

Construction A builder needs to find the perimeter of the room to buy molding. The length of the room is 12 feet 2 inches, and the width is 15 feet 3 inches. To find the perimeter, he adds

```
  12 feet 2 inches
  12 feet 2 inches
  15 feet 3 inches
+ 15 feet 3 inches
  54 feet 10 inches
```

All of the lengths in feet are added in one column, and all of the inches are added in another. Students should see how adding measurements compares to adding expressions.

Some students may wonder about the doorways in the room. The builder would not need molding for the doorways, but he might buy 54 feet 10 inches of molding to be sure he has enough.

Cooperative Learning

Have students work in small groups. Ask each student in the group to write several algebraic expressions on a strip of paper. Have them use the same variable but different coefficients and constants. Place all of the strips in a pile and ask a student to choose two strips. Then have students work with their group to add the two expressions. Encourage the use of algebra tiles if available.

TEACHING *tip*

Be sure students understand that 1-tiles are the small square tiles and that x-tiles are the long tiles.

Alternate Example 1

Simplify $(4x - 5) + (3x + 2)$ using tiles. $[7x - 3]$

ongoing ASSESSMENT

Use two x-tiles and three negative 1-tiles to represent $2x - 3$ and use four x-tiles and five positive 1-tiles to represent $4x + 5$. The result is $6x + 2$.

CRITICAL *Thinking*

Starting at $0x$, draw an arrow to $3x$. Then from $3x$, draw another arrow $2x$ units long. The result is $5x$.

Some students may have difficulty with part (c) of Try This. Be sure students understand that the expressions can sometimes be added without changing the order of the terms.

A **ongoing**
SSESSMENT

Try This

  a. $[7b - 2]$

  b. $[3b + 4]$

  c. $[b + 7]$

The Distributive Property can be used to simplify an expression that contains like terms.

$$3x + 2x = (3 + 2)x$$
$$= 5x \; \diamond$$

CRITICAL *Thinking*  Explain how to add like terms such as $3x$ and $2x$ on the following number line.

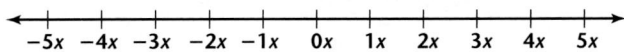

Rearranging Terms

Changing the order of two terms is called commuting the terms.

COMMUTATIVE PROPERTY FOR ADDITION
For any numbers a and b,
$a + b = b + a$.

Now the expressions in Example 1 can be added without using tiles for models.

$(3x + 4) + (2x - 1)$	Given
$3x + 4 + 2x + (-1)$	Definition of Subtraction
$3x + (4 + 2x) + (-1)$	**Associative Property**
$3x + (2x + 4) + (-1)$	**Commutative Property**
$(3x + 2x) + (4 + (-1))$	**Associative Property**
$5x + 3$	Combine like terms.

The Associative and Commutative Properties can be used to rearrange terms. For example, $(3B + 4) + (2 + 2B)$ can be rearranged as $(3B + 2B) + (4 + 2)$. Often the middle three steps of the process shown above are combined.

$(3x + 4) + (2x - 1)$	Given
$(3x + 2x) + (4 - 1)$	**Rearrange terms.**
$5x + 3$	Combine like terms.

To check the addition, replace the variable with a number that makes the expression easy to evaluate. For example, replace x with 10, and evaluate each expression.

$$(3x + 4) + (2x - 1) = 5x + 3$$
$$3(10) + 4 + 2(10) - 1 \stackrel{?}{=} 5(10) + 3$$
$$53 = 53 \qquad \text{True}$$

Try This Add each expression to $5b + 1$.
  a. $2b - 3$ **b.** $-2b + 3$ **c.** $6 - 4b$

 Ask students to write several problems that can be represented by adding two algebraic expressions.

INCLUSION
strategies

Using Cognitive Strategies Assign unusual words to variables, and then have students add. For Example 2 you could use

$(5 \text{ xebecs} + 1 \text{ yak}) +$
$(4 \text{ xebecs} + 8 \text{ yaks} + 8 \text{ zebras}) +$
$(9 \text{ xebecs} - 2 \text{ zebras}) =$
$18 \text{ xebecs} + 9 \text{ yaks} + 6 \text{ zebras}.$

Students may enjoy choosing their own words to assign to variables.

EXAMPLE 2

GEOMETRY
Connection

The lengths of the sides of triangle *ABC* are shown in the diagram. What is the perimeter of triangle *ABC*?

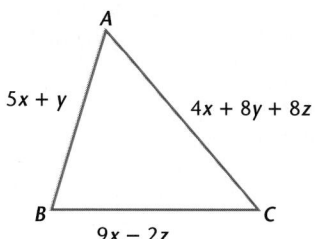

Solution ➤

To find the perimeter, add the lengths of the sides.

$$P = (5x + y) + (4x + 8y + 8z) + (9x - 2z)$$
$$= (5x + 4x + 9x) + (y + 8y) + (8z - 2z)$$
$$= 18x + 9y + 6z$$

Definition of Perimeter
Rearrange terms.
Combine like terms.

The perimeter is $18x + 9y + 6z$. ❖

EXERCISES & PROBLEMS

Communicate

1. Define an *algebraic expression*. Give two examples.

2. Discuss what is meant by a *coefficient*. Give two examples.

3. What is the Commutative Property for Addition? Give an example.

4. Describe the steps in using the Distributive Property to add $3x$ and $8x$.

5. Identify which of the following are like terms. Explain why.

$$7x, 2x, 3z, 5, 7y, 3x, -z, 3y, \text{ and } 23$$

6. Discuss how you would use algebra tiles to model $(5x + 2) + (3x - 4)$.

Practice & Apply

Add.

7. $(5a - 2) + (3a - 6)$
   $8a - 8$

8. $(2x + 3) + (7 - x)$
   $x + 10$

9. $(4x + 5y) + (x + 9y)$
   $5x + 14y$

10. $(1.1a + 1.2b) + (2a - 0.8b)$
   $3.1a + 0.4b$

11. $\left(\frac{x}{2} + 1\right) + \left(\frac{x}{3} - 1\right)$ $\frac{5}{6}x$

12. $\left(\frac{2m}{5} + \frac{1}{2}\right) + \left(\frac{m}{10} + \frac{5}{2}\right)$
   $\frac{1}{2}m + 3$

13. $(2a + 3b + 5c) + (7a - 3b + 5c)$
   $9a + 10c$

14. $(x + y + z) + (2w + 3y + 5)$
   $2w + x + 4y + z + 5$

15. If 6 cartons and 3 loose tiles are combined with 2 cartons and 1 loose tile, describe the result algebraically. $8c + 4$ or $8c + 4L$

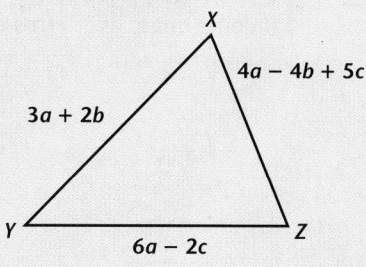

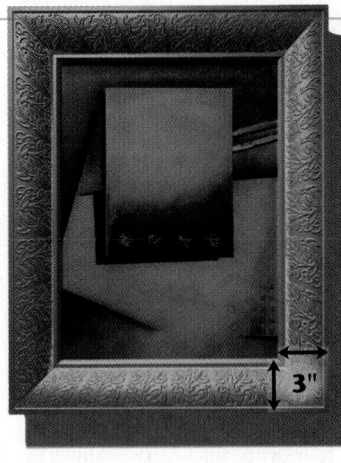

Picture Framing A picture is surrounded by a frame which is 3 inches wide. Find the total height of the frame and the picture if the height of the picture is

16. 12 inches. 18 inches

17. 20 inches. 26 inches

18. h inches. $h + 6$ inches

A lid 1 inch thick is placed on top of a box. Find the total height if the height of the box is

19. 10 inches. 11 inches

20. 27.5 inches. 28.5 inches

21. x inches. $x + 1$ inches

22. On Tuesday, John bought 2 large boxes of cassette tapes and 3 additional tapes. The next week he bought 1 large box and 7 additional tapes. Represent the boxes and tapes algebraically, and determine the expression for the sum. $3b + 10$

23. Each month for 7 months Don buys 2 boxes of cookies. In all, he let his friend have a total of 14 cookies. Let c represent the number of cookies in a box. Represent in an algebraic expression the total number of cookies Don has for himself during the 7 months. $14c - 14$

24. **Clothing Manufacturing** A piece of fabric is cut into 12 equal lengths with 3 inches left over. Define the variable, and write the length of the original piece as an expression. $12f + 3$

25. If m is an unknown integer, represent the next two consecutive integers in terms of m. $m + 1, m + 2$

26. If n is an unknown odd integer, represent the next two odd integers in terms of n. $n + 2, n + 4$

27. If apples cost 30 cents each, bananas cost 25 cents each, and plums cost 20 cents each, write an expression to find the total cost of a apples, b bananas, and p plums.
$0.30a + 0.25b + 0.20p$

Geometry Find the perimeter of each figure.

28.

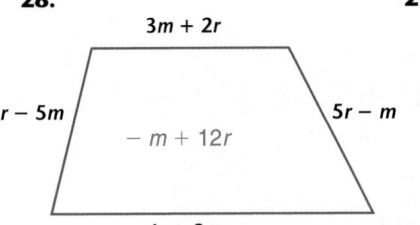

$3m + 2r$
$r - 5m$
$- m + 12r$
$5r - m$
$4r + 2m$

29.

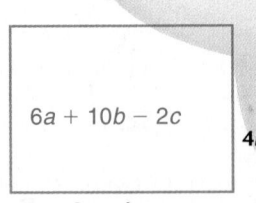

$6a + 10b - 2c$
$3a + b + c$

30.

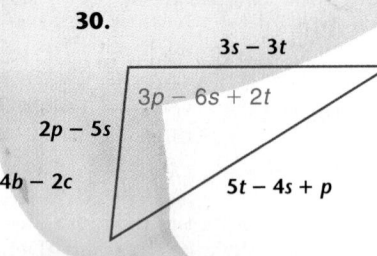

$3s - 3t$
$3p - 6s + 2t$
$2p - 5s$
$4b - 2c$
$5t - 4s + p$

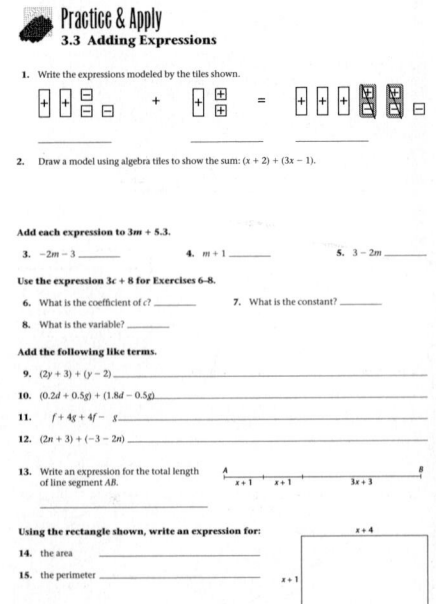

31. 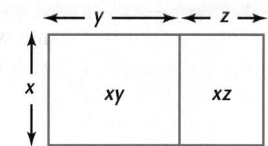 **Geometry** Write an expression for the area of the whole rectangle. Explain how the Distributive Property can be used to find this area. $xy + xz = x(y + z)$

Use the Distributive Property to write equivalent expressions for the following.

$6x - 6 \cdot 3$

32. $3(5 + 7)$ $3 \cdot 5 + 3 \cdot 7$ **33.** $6(x - 3)$ **34.** $(x + y)n$ $xn + yn$

35. $(4 \cdot 3 + 4 \cdot 5)$ **36.** $4x - 4 \cdot 7$ **37.** $az + bz$
$4(3 + 5)$ $4(x - 7)$ $z(a + b)$ or $(a + b)z$

38. Explain how you can use the Distributive Property to find $5 \cdot 12$.
$5 \cdot 12 = 5(2 + 10) = 5 \cdot 2 + 5 \cdot 10 = 60$

Look Back

39. You use the method of differences on a sequence, and a constant appears in the second difference. What type of function would you expect to find? **[Lesson 1.2]** quadratic

40. Find the next 3 terms in the sequence 10, 20, 40, 80, 160, ___, ___, ___.
[Lesson 1.2] 320, 640, 1280

41. Which ordered pair does not lie on the same line as the others?
$(3, 5), (4, 8), (6, 11), (4, 7)$ **[Lesson 1.5]** $(4, 8)$

Name the following type of function.

42. $y = 10x + 3$ **[Lesson 1.5]**
linear

43. $y = \dfrac{1}{x}$ **[Lesson 2.3]**
reciprocal

44. $y = x^2$ **[Lesson 2.2]**
quadratic

45. $y = |x|$ **[Lesson 2.4]**
absolute value

Simplify. [Lessons 3.1, 3.2]

46. $-2 + (-3)$ **47.** $-7 + 6$ **48.** $-5 - (-2)$ **49.** $-10 - 5$
-5 -1 -3 -15

Look Beyond

50. Find $(3A + 2B + 4P) - (5A + 3P)$. $-2A + 2B + P$

51. Draw a diagram to model the area represented by $(x + y)(x + y)$.

52. In the center of the City the blocks are in the shape of a grid. You are at the bank at A, and you want to walk to the post office at B. If all the different ways you take are exactly 7 blocks long, how many different ways are there to get from A to B?
35

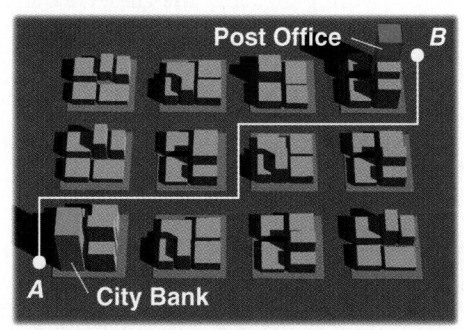

51.

	$(x + y)$	
	x	y
x	x^2	xy
y	xy	y^2

($(x + y)$ label on left side)

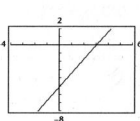

Objectives

- Use tiles to model the subtraction of expressions as take away.
- Subtract expressions by adding the opposite.

RESOURCES

- Practice Master — 3.4
- Enrichment Master — 3.4
- Technology Master — 3.4
- Lesson Activity Master — 3.4
- Quiz — 3.4
- Spanish Resources — 3.4

Assessing Prior Knowledge

Add.

1. $(3x + 5) + (-2x - 1)$
   $[x + 4]$

2. $(-7x + 3) + (4x - 2)$
   $[-3x + 1]$

3. $(2x - 3) + (5x - 2)$
   $[7x - 5]$

4. $(6x - 4) + (-3x + 1)$
   $[3x - 3]$

5. $(-3x - 1) + (-2x - 3)$
   $[-5x - 4]$

6. $(8x + 5) + (-2x - 3)$
   $[6x + 2]$

TEACH

 Use a real-world situation to illustrate subtraction of algebraic expressions.

In the computation, point out that both terms in the expression $3c + 4$ are subtracted from $5c + 6$. That is, $3c$ is subtracted from $5c$, *and* 4 is subtracted from 6.

why You can subtract one integer from another by adding its opposite. The same procedure can also be used to subtract expressions.

Bonnie bought 5 cartons of soda and 7 extra cans of soda for her party. Her guests drank 3 cartons and 4 cans of soda.

How do you write an algebraic expression that represents the total number of cans left after the party?

Before the party there are 5 cartons and 7 cans.

$$5x + 7$$

During the party the guests drank 3 cartons and 4 cans of soda.

$$3x + 4$$

After the party, the amount left is 2 cartons and 3 cans.

$$2x + 3$$

$$(5x + 7) - (3x + 4) = 2x + 3$$

If x represents the number of cans in a carton, then the algebraic expression $2x + 3$ represents the total number of cans left after the party.

ALTERNATIVE **teaching strategy**

Technology Use a spreadsheet to set up a table for subtracting the coefficients of like terms and constants. Evaluate $(3x - 4) - (5x + 6)$.

	A	B	C	D
		x-coefficient	constant	
1				
2		3	-4	
3		5	6	
4	difference	-2	-10	
5				

(C4 = C2-C3)

The formula for cell B4 is = B2 − B3. The formula for cell C4 is = C2 − C3. Have students find the difference for many different expressions. Discuss the ideas, and guide students to the correct conclusions about subtracting algebraic expressions.

ENRICHMENT Have students decide whether the expression $-(a + b)$ always represents a negative value. Then have the students explain their reasoning.

Modeling Subtraction With Tiles

Subtraction can be modeled with tiles.

EXAMPLE 1

Use algebra tiles to simplify $(2x + 3) - (4x - 1)$.

Solution➤

Start with 2 positive *x*-tiles and 3 positive 1-tiles to represent $2x + 3$.

2x + 3

Now subtract $(4x - 1)$.

You need to subtract 4 positive *x*-tiles and 1 negative 1-tile. Add 2 pair of positive and negative *x*-tiles and 1 pair of positive and negative 1-tiles. Now you can take away 4 positive *x*-tiles and 1 negative 1-tile.

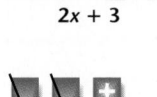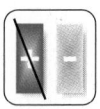

(2x + 3) − (4x − 1)

This leaves 2 negative *x*-tiles and 4 positive 1-tiles.

$(2x + 3) - (4x - 1) = -2x + 4$ ❖

−2x + 4

A tile pair that has opposite signs is sometimes referred to as a **neutral pair**. Explain how to use neutral pairs to subtract $6x + 2$ from $3x - 5$.

Neutral Pairs

Adding the Opposite of an Expression

You can subtract a number from another by adding its opposite. This is also true for expressions.

> **DEFINITION OF SUBTRACTION FOR EXPRESSIONS**
> To subtract an expression, add its opposite.

When you subtract a quantity by adding the opposite, first change the sign of *each term* in the quantity that you are subtracting. Then rearrange to group the like terms, and add.

$(2x + 3) - (4x - 1)$	Given
$(2x + 3) + (-4x + 1)$	Definition of Subtraction
$(2x - 4x) + (3 + 1)$	Rearrange terms.
$-2x + 4$	Combine like terms.

[It is *positive* for *a* and *b* both negative, for *a* positive and *b* negative with $|a| < |b|$, or for *a* negative and *b* positive with $|a| > |b|$.

It is *negative* for *a* and *b* both positive, when *a* is positive and *b* is negative with $|a| > |b|$, or for *a* negative and *b* positive with $|a| < |b|$.]

INCLUSION
strategies

Using Visual Models
Have students use red for all terms containing *x*, blue for all terms containing *y*, and green for all constant terms. Only terms that are the same color should be subtracted.

Cooperative Learning
Have students work in small groups. Ask each student to write a problem that can be solved by subtracting algebraic expressions. Have students exchange problems with other members of their group and solve them. Then have the members of the group discuss the solution to each problem and make any necessary corrections.

TEACHING *tip*

Have students use algebra tiles to duplicate the method used to subtract the expressions in Example 1.

Alternate Example 1
Use algebra tiles to simplify $(6x - 7) - (3x + 5)$.
$[3x - 12]$

Aongoing
SSESSMENT

Add 3 *x*-tile neutral pairs and 2 1-tile neutral pairs. Take away 3 positive *x*-tiles and 2 positive 1-tiles. The result is $-3x - 7$.

Alternate Example 2

Subtract.

a. $(8a - 9) - (3a - 7)$
$$[5a - 2]$$

b. $(6n + 5) - (10n + 8)$
$$[-4n - 3]$$

c. $(2x - 3) - (-5x - 6)$
$$[7x + 3]$$

Think $3x + 2x + (-4) +$
(-3), or $5x - 7$.

Alternate Example 3

Decide whether the expression
$-(-W)$ is positive or negative
when

a. $W = 3$.   [**positive**]

b. $W = -5$.   [**negative**]

c. $W = 0$.   [**neither**]

d. $W = n$.   [**cannot tell**]

CRITICAL
Thinking

It is the same as distributing a
-1 to each term.

EXAMPLE 2

Subtract.

A $(7m + 2) - (3m + 5)$ **B** $(10d - 3) - (4d + 1)$ **C** $(4x - 2) - (5x - 3)$

Solution

A
$(7m + 2) - (3m + 5)$	Given
$(7m + 2) + (-3m - 5)$	Definition of Subtraction
$(7m - 3m) + (2 - 5)$	Rearrange terms.
$4m - 3$	Combine like terms.

B
$(10d - 3) - (4d + 1)$	Given
$(10d - 3) + (-4d - 1)$	Definition of Subtraction
$(10d - 4d) + (-3 - 1)$	Rearrange terms.
$6d - 4$	Combine like terms.

C
$(4x - 2) - (5x - 3)$	Given
$(4x - 2) + (-5x + 3)$	Definition of Subtraction
$(4x - 5x) + (-2 + 3)$	Rearrange terms.
$-x + 1$	Combine like terms. ❖

Explain how to simplify $(3x - 4) - (-2x + 3)$ mentally.

A negative sign in front of a variable does not necessarily mean
its *value* is negative.

EXAMPLE 3

Decide whether the expression $-W$ is positive or negative when

A $W = 5$.   **B** $W = -7$.   **C** $W = 0$.   **D** $W = x$.

Solution

Substitute each value for W, and examine the value of $-W$.

A $-W = -(5) = -5$ The opposite of a positive 5 is negative 5. The
value is negative.

B $-W = -(-7) = 7$ The opposite of a negative 7 is positive 7. The
value is positive.

C $-W = -(0) = 0$ The value of 0 is neither positive nor negative.

D $-W = -(x)$ It is impossible to tell whether the expression is positive
or negative. Since x is a variable, the value will depend on the value
that x represents. ❖

CRITICAL
Thinking

Examine the expression $-(5x + 2y)$. The negative sign means to find the
opposite of *each term*. Explain why this might be considered a special case
of the Distributive Property.

RETEACHING
t h e
l e s s o n

Using Visual Models
Use a number line to
demonstrate how to find
the opposite of an inte-
ger. Show students how to start at 0 and draw an
arrow to show the integer. Then start at 0 and
move the same number of units in the opposite
direction to show the opposite number. In the
example that follows 3 and -3 are opposites.

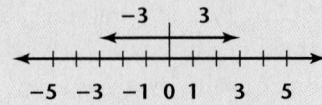

Try This If a is -4 and b is 3, find $-(a + b)$.

EXAMPLE 4

Simplify $(8x + 4y - z) - (6y + 3z - 5x)$.

Solution►

$(8x + 4y - z) - (6y + 3z - 5x)$	Given
$8x + 4y - z - 6y - 3z + 5x$	Definition of Subtraction
$(8x + 5x) + (-6y + 4y) + (-z - 3z)$	Rearrange terms.
$13x - 2y - 4z$	Combine like terms. ❖

EXERCISES & PROBLEMS

Communicate

Tell how to represent each subtraction using algebra tiles.

1. $4x - 3x$ **2.** $(3x + 2) - (2x + 1)$ **3.** $(5x + 3) - (2x + 4)$

4. For the following problem, assign a variable to represent the item. Express the information using algebraic expressions. Explain how to subtract the expressions. Then answer the question.

> Ms. Green had 5 reams of paper and 100 loose sheets. She gave Mr. Black 2 reams of paper and 50 loose sheets. How much did she have left?

5. Explain what is meant by adding the opposite when subtracting expressions.

6. Explain how to perform the following subtraction.
$$(7y + 9x + 3) - (3y + 4x - 1)$$

7. Describe the set of values for m that make $-6m$ positive.

Practice & Apply

Find the opposite of each expression.

8. 17 -17 **9.** -13 13 **10.** $2x$ $-2x$ **11.** -6 6 **12.** $9y + 2w$ $-9y - 2w$ **13.** $5a + 3b$ $-5a - 3b$

14. $2n - 3m$ $-2n + 3m$ **15.** $9c - 5d$ $-9c + 5d$ **16.** $-7x + 9$ $7x - 9$ **17.** $-7r + 4s$ $7r - 4s$ **18.** $-3p - q$ $3p + q$ **19.** $-j - k$ $j + k$

Perform the indicated operations.

20. $9x - 3x$ $6x$ **21.** $8y - 2y$ $6y$ **22.** $5c - (3 - 2c)$ $7c - 3$ **23.** $7d - (1 - d)$ $8d - 1$

24. $(7r + 2s) + (9r + 3s)$ $16r + 5s$ **25.** $(9k + 2k) + (11j - 2j)$ $11k + 9j$

26. $(2a - 1) - (5a - 5)$ $-3a + 4$ **27.** $(9v - 8w) - (8v - 9w)$ $v + w$

28. $(2x + 3) - (4x - 5) + (6x - 7)$ **29.** $(4y + 9) - (8y - 1) + (7 - y)$

28. $4x + 1$

29. $-5y + 17$

Tell whether each statement is true or false. Then explain why.

30. $x - (y + z) = x - y + z$

31. $x - (y - z) = x - y - z$

32. $-x + (y - z) = -x - y + z$

33. $-x$ is a negative number

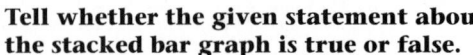 **Statistics** The graph is called a stacked bar graph. It shows the population of the United States by gender. For example, it shows that in 1970 there were approximately 100 million males and 100 million females in the United States.

34. What was the female population in 1950?
about 75 million

35. What was the male population in 1950?
about 75 million

36. How many people were there in 1950?
about 150 million

37. What was the female population in 1980?
about 110 million

38. What was the male population in 1980?
about 110 million

39. How many people were there in 1980?
about 220 million

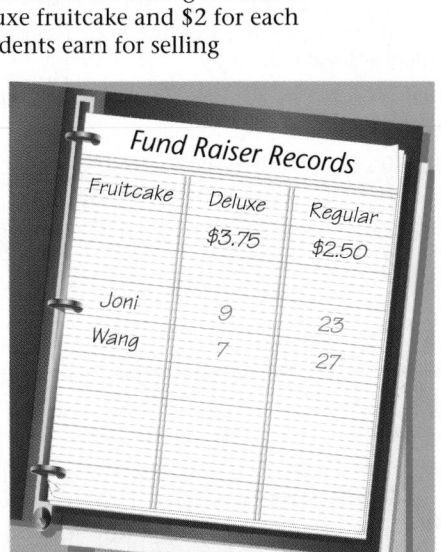

U.S. Population

Tell whether the given statement about the stacked bar graph is true or false.

40. The population rose during each 10-year period. T

41. The number of males and females were approximately equal during each 10-year period. T

42. The number of females rose much faster than the number of males. F

Fund-raising Students at Valley View High School are selling fruitcakes for a fund raiser. They earn $3 for each deluxe fruitcake and $2 for each regular fruitcake. How much would the students earn for selling

43. 89 deluxe fruitcakes and 234 regular fruitcakes? $735

44. d deluxe fruitcakes and r regular fruitcakes? $3d + 2r$

Suppose students earn $3.75 for each deluxe fruitcake and $2.50 for each regular fruitcake. How much would they earn for selling

45. 89 deluxe fruitcakes and 234 regular fruitcakes? $918.75

46. d deluxe fruitcakes and r regular fruitcakes? $3.75d + 2.50r$

Joni sold 9 deluxe fruitcakes and 23 regular fruitcakes.
Wang sold 7 deluxe fruitcakes and 27 regular fruitcakes.

47. How much did they earn together? $185

48. Who earned more? How much more? Wang; $2.50

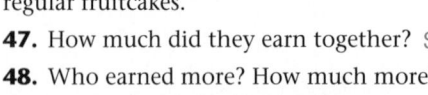

Fund Raiser Records		
Fruitcake	Deluxe	Regular
	$3.75	$2.50
Joni	9	23
Wang	7	27

30. False, it should be $x - y - z$.

31. False, it should be $x - y + z$.

32. False, it should be $-x + y - z$.

33. True, if $x > 0$ and false if $x < 0$.

49. Inventory The school kitchen had 11 cases of juice plus 3 extra cans of juice. After lunch they had 6 cases of juice and no extra cans. How much juice was distributed during lunch? 5 cases and 3 cans

Cultural Connection: Africa The idea of adding and subtracting numbers has been around for thousands of years. Early Egyptians represented addition by feet walking towards the number and subtracting by feet walking away from the number.

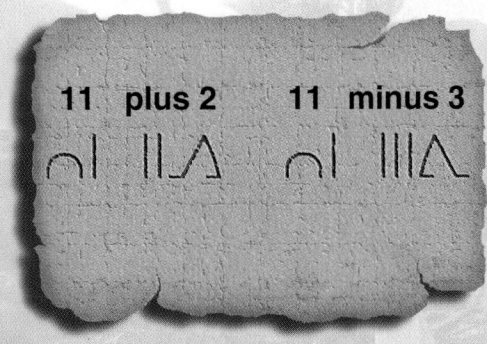
11 plus 2   11 minus 3

50. How would you represent $11 + 5$ using the early Egyptian notation if $| = 1$ and $\cap = 10$?

51. Represent $23 - 12$ using the early Egyptian notation.

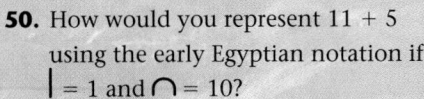 *Look Back*

52. Write 5 problem-solving strategies. **[Lesson 1.1]**

53. Create a sequence whose differences reach a constant at the third difference. **[Lesson 1.2]** Example: 0, 1, 18, 55, 116, 205, . . .

Place parentheses and brackets to make each equality true. [Lesson 1.4]

54. $28 \div 2 - 4 \cdot 1 = 10$ none needed

55. $16 \div 5 + 3 \div 2 = 1$ $16 \div (5 + 3) \div 2 = 1$

56. $40 \cdot 2 + 10 \cdot 4 = 1680$ $40[2 + (10 \cdot 4)] = 1680$

57. What is the value of $\text{INT}(x)$ for $x = \frac{7}{4}$? **[Lesson 2.4]** 1

58. Describe the graph of an integer function. **[Lesson 2.4]**

59. What is the shape of the graph for the absolute value function? **[Lesson 2.4]**

60. The first reading from a gauge is 100. The changes recorded at 1 hour intervals are $-4, +51, 0, +7, -12, -78, +2, -13, -1$. What is the current reading on the gauge? **[Lesson 3.1]** 52

Look Beyond

61. **Statistics** Fred has test scores of 87, 74, and 90. How many points does he need on the next test to have an average of 85? 89

62. Paul and Dan leave from the bandstand at one end of a lake and sail around the lake in the same direction. Paul sails at 12 miles per hour, while Dan sails at 10 miles per hour. The lake is 2 miles around, and they leave at noon. When will the two first meet again at the bandstand? 1 p.m.

50.

51.

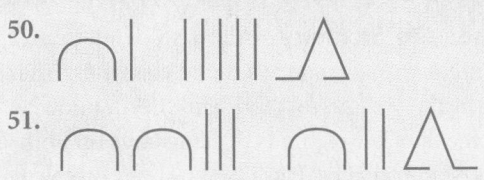

52. Think of a simpler problem, look for a pattern, reason logically from a pattern, make a table, or draw a diagram.

58. It has horizontal line segments with a filled-in dot on the left end and an open dot on the right end. The horizontal line segments appear as steps going up the graph diagonally to the right.

59. The absolute value graph is in the shape of a V.

Lesson 3.4 **123**

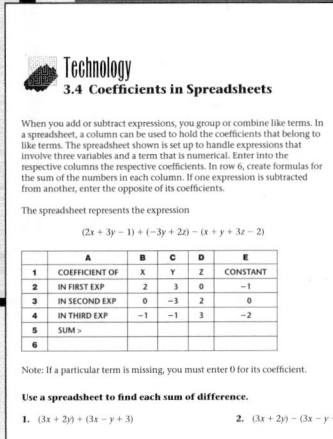

PREPARE

Objectives

- Understand the basic concept of an equation in algebra.
- Represent and solve applications involving addition and subtraction equations using the four steps: organize, write, solve, and check.
- Solve literal equations that involve addition and subtraction.

RESOURCES

- Practice Master   **3.5**
- Enrichment Master   **3.5**
- Technology Master   **3.5**
- Lesson Activity Master   **3.5**
- Quiz   **3.5**
- Spanish Resources   **3.5**

Assessing Prior Knowledge

Simplify.

1. $-5 - 6$ $[-11]$

2. $0.5 - 0.1$ $[0.4]$

3. $-\dfrac{3}{5} + \dfrac{7}{10}$ $\left[\dfrac{1}{10}\right]$

TEACH

why Many real-life problems can be solved using an equation. Students should notice that even though the problems involve different real-world situations, they can be modeled using the same equation.

CRITICAL Thinking

Draw 3 x-boxes and 4 tiles on one side, and 2 x-boxes and 10 tiles on the other side. Take away 2 x-boxes and 4 tiles from both sides. The result is $x = 6$.

why *Many problems can be modeled algebraically by writing an equation. A sequence of steps that isolates the unknown on one side of the equal sign provides the solution.*

How many degrees must the temperature rise to reach the freezing point, 32°F?

If Riva has $20, how much more money does Riva need to buy the boots?

Although each situation is different, each can be modeled by the same equation $x + 20 = 32$.

Tiles can be used to demonstrate the steps in solving the equation. Suppose the x-box represents an unknown number of tiles. The x-box and the tiles represent the equation $x + 20 = 32$.

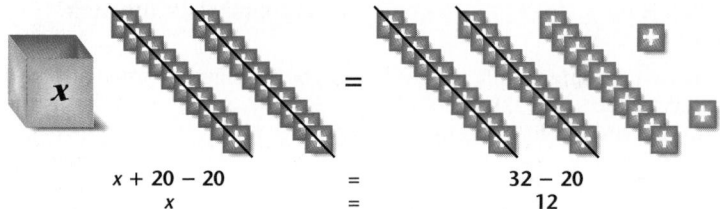

$x + 20 - 20 \qquad = \qquad 32 - 20$
$\qquad x \qquad\qquad = \qquad\quad 12$

After you take away 20 tiles from each side, the amounts on each side of the equal sign remain equal. The x-box is now alone on one side of the equal sign. Count the remaining tiles on the other side of the equal sign to find the number that the box represents.

The idea of subtracting the same number of tiles from each side of the equation is called the Subtraction Property of Equality.

CRITICAL Thinking

How can you use a model to represent and solve the equation $3x + 4 = 2x + 10$?

ALTERNATIVE teaching strategy

Using Visual Models Draw a balance scale on the chalkboard or overhead. Remind students that a balance scale is balanced as long as there are equal weights on both sides. Explain that an equation is like a balanced scale. If weight is added or subtracted from one side of the scale, the same must be done to the other side.

Hands-On Strategy Allow students to experiment with a balance scale in class. Disguise one heavy weight to represent the unknown. Use the lighter weights as the constants on both sides of the balance. Have the students discover the amount of the disguised weight by manipulating the lesser weights.

SUBTRACTION PROPERTY OF EQUALITY

If equal amounts are subtracted from the expressions on each side of an equation, the expressions remain equal.

EXAMPLE 1

Fund-raising The school band needs new uniforms. Write and solve an equation to find how much more money the band must raise.

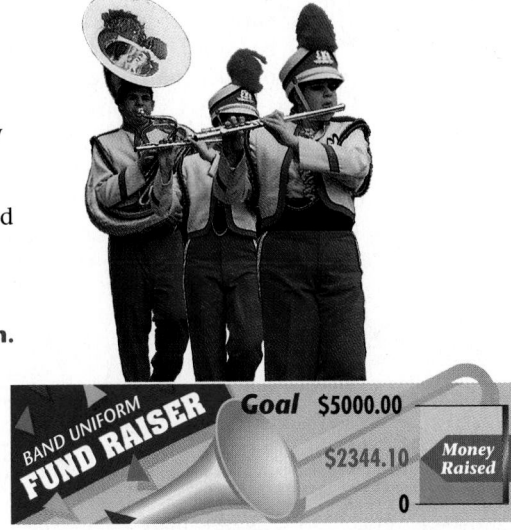

Solution▸

Organize the information.
Study the problem. Identify the known and unknown information. Then write a sentence using words and operation symbols.

(the amount to be raised) + (the amount they have) = (the goal)

Clearly identify a variable to represent the unknown value you are trying to find.

Let *x* represent the amount to be raised.

Write the equation.
Replace the words you wrote in the first step with the variable and the known information from the problem.

$$x + 2344.10 = 5000.00$$

Solve the equation.
Use the Subtraction Property of Equality.

$$x + 2344.10 - 2344.10 = 5000.000 - 2344.10$$
$$x = 2655.90$$

The equation is solved when the variable is isolated on one side of the equation.

Check the answer.
Replace the variable in the original equation with the value you found. See if it makes a true statement. Be sure that the solution is reasonable and that it answers the question in the problem.

$$2655.90 + 2344.10 \overset{?}{=} 5000.00$$
$$5000.00 = 5000.00 \quad \text{True}$$

The band must raise $2655.90. ❖

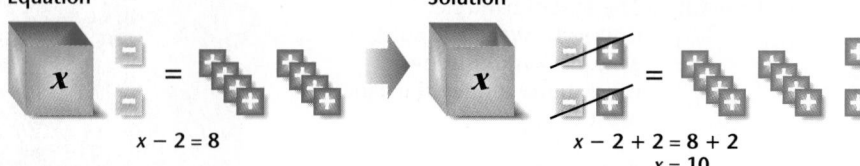

In the Try This students may write $x - (-3)$ as $x + 3$ before solving. Encourage them to solve the equation using both $x - (-3)$ and $x + 3$ so they can see that the results are the same.

Try This

a. $x = 9$

b. $x = 2$

Some students may have difficulty solving an equation with the variable on the right side. Encourage those students to rewrite the equation so that the variable is on the left side before solving.

Math Connections

Statistics

Review the definition of *range* as it is applied to statistical data. Ask students what other meaning they know for the term.

Alternate Example 2

Lea spent \$45.90 at the mall and has \$23.50 left. Write and solve an equation to find the amount she had at the beginning. $[x - 23.50 = 45.90; \$69.40]$

Some equations contain subtraction. To solve $x - 2 = 8$, use a procedure similar to that used for equations containing addition.

Suppose an x-box represents an unknown number of tiles. Represent the equation with tiles. To solve the equation, add 2 positive tiles to each side of the equation, and simplify. The unknown value is 10.

Equation

$$x - 2 = 8$$

Solution

$$x - 2 + 2 = 8 + 2$$
$$x = 10$$

Try This   Solve with tiles.   **a.** $x - 5 = 4$ **b.** $x - (-3) = 5$

ADDITION PROPERTY OF EQUALITY

If equal amounts are added to the expressions on each side of an equation, the expressions remain equal.

EXAMPLE 2

In a dart game, the range of scores is 47 points, and the lowest score is 52.

Write and solve an equation to find the highest score in the dart game.

Solution▶

Organize

In statistics, the *range* is the difference between the highest and lowest scores.

range = highest score − lowest score

Let H represent the highest score.

STATISTICS
Connection

Write

Replace the words in the range formula with the variable, H, and the known values from the problem.

$$47 = H - 52$$

Solve

$47 = H - 52$	Given
$47 + 52 = H - 52 + 52$	Addition Property of Equality
$99 = H$	Combine like terms.

Check

Substitute 99 for H.

Since $47 = 99 - 52$, the highest score is 99. ❖

Give students equations like $2y - 5 - y = 6$ and $4x + 7 - 3x = -10$ to solve. Tell them that there should only be one variable term before they start to solve. After they feel comfortable solving equations of this type, ask them to write a problem that can be solved using this type of equation.

English Language Development Be sure students understand the meaning of *addition, subtraction,* and the Addition and Subtraction Properties of Equality. Give examples.

$7 - 3 = 4$	$7 - 3 + 3 = 4 + 3$	$7 = 7$
$x - 3 = 4$	$x - 3 + 3 = 4 + 3$	$x = 7$
$2 + 3 = 5$	$2 + 3 - 3 = 5 - 3$	$2 = 2$
$x + 3 = 5$	$x + 3 - 3 = 5 - 3$	$x = 2$

Literal Equations

Scientists often express numerical relationships as **formulas**. For example, $C = \frac{5}{9}(F - 32)$ relates Fahrenheit and Celsius temperatures. Because formulas often contain a number of different letters that represent variables, they are called **literal equations**.

EXAMPLE 3

Investment   The formula $A = P + I$ shows that the total amount of money you receive from an investment equals the principal (the money you started with) and added interest. Write a formula for the interest, I, based on the principal and the amount.

Solution▶

To isolate I, subtract P from *both sides* of the equation.

$A = P + I$	Given
$A - P = P + I - P$	Subtraction Property of Equality
$A - P = I$	Group like terms and simplify.
$I = A - P$	Rewrite with I on the left. ❖

Literal equations that contain subtraction are treated in much the same way as equations that contain addition. In business, the equation $P = s - c$ is a formula for profit, where s represents the selling price and c represents the cost.

EXAMPLE 4

Solve the equation $P = s - c$ for s.

Solution▶

To solve the equation for s, add c to both sides of the equation.

$P = s - c$	Given
$P + c = s - c + c$	Addition Property of Equality
$P + c = s$	Combine like terms.
$s = P + c$	Rewrite with s on the left. ❖

Explain how you would solve $a = x - b$ for x.

Try This   The figure shows the sector of a circle with center O and radius R. The formula for the length h is $h = R - k$. Solve the equation for k.

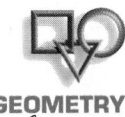

GEOMETRY
Connection

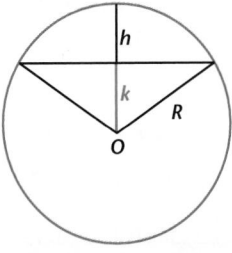

RETEACHING
t h e
l e s s o n

Using Cognitive Strategies   Begin with simple addition equations, such as $x + 3 = 6$. Encourage students to use questions such as, "What number plus 3 equals 6?" to describe the equation. Then have them state the solution and use the subtraction property to solve. Gradually increase the level of difficulty of the equation until students realize that the simplest way to find a solution is by using the Subtraction Property of Equality. Use a similar method for the Addition Property of Equality.

TEACHING *tip*▶

Have students highlight the variable for which they are solving.

Alternate Example 3
The formula $c = p + t$ gives the total cost (c) of an item, including tax. Write a formula for the amount of tax, t, based on the total cost, c, and the marked price, p, of the item. $[t = p - c]$

Alternate Example 4
Solve the equation $R = m + n$ for n. $[n = R - m]$

Aongoing
SSESSMENT

Add b to both sides of the equation. Since $-b + b = 0$, $a + b = x$.

Math Connection Geometry
Explain to the students that a sector of a circle is a a region shaped like a pie slice. Identify h and k as the lengths of the segments of the radius within the sector divided by the perpendicular segment, as shown in the diagram.

Aongoing
SSESSMENT

Try This

$k = R - h$

Use Transparency▶ 15

ASSESS

Selected Answers

Odd-numbered Exercises 7–57

Assignment Guide

Core 1–5, 6–26 even, 28–58

Core Two-Year 1–55

A ongoing ASSESSMENT

The computer will display −1.

Technology

For Exercises 46–47, permit the use of software that solves equations.

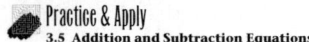

Technology that solves equations on a computer or a calculator already exists. For example, to solve the equation $23 = C + 14$ using the Maple™ software program, type in solve(23 = C + 14);. The computer will display 9.

Computer

To solve the equation $5 - 2r = 7 - (3r - 8)$, type in solve(5 − 2 * r = 7 − (3 * r − 8));. The computer will display 10. Note that 2 * r is entered for 2r. What do you think will result if you type in solve(4 + 3 * t = 6 + 5 * t);?

EXERCISES & PROBLEMS

Communicate

1. Compare the Addition Property of Equality with the Subtraction Property of Equality. Tell how to decide which one to use.

2. Give a situation that can be modeled by
   a. $x + 20 = 50$.   **b.** $x - 20 = 50$.

3. Tell how to solve each equation using algebra tiles.
   a. $x + 6 = 10$ **b.** $x - 6 = 10$

4. Explain how to solve $s + x = r$ for s.

5. Explain how to solve $m + b = n$ for m.

Practice & Apply

State which property you would use to solve each equation. Then solve.

6. $a - 16 = 15$ A; 31   7. $t + 29 = 11$ S; −18   8. $m + 54 = 36$ S; −18   9. $r - 10 = -80$
   A; −70

10. $l - 27 = 148$ A; 175   11. $b - 109 = 58$ 12. $y + 37 = -110$ S; −147   13. $396 = z + 256$
    A; 167   S; 140

14. $x + \frac{3}{4} = \frac{5}{4}$ S; $\frac{1}{2}$ 15. $7.4 + t = 5.2$ 16. $3r + 5.78 = 2r + 7$ 17. $\frac{3}{8} - x = \frac{3}{4}$
    S; −2.2   S; 1.22   A; S; $-\frac{3}{8}$

Solve each equation using algebra tiles.

18. $x + 9 = 6$ 19. $x - 7 = 3$ 20. $x - 10 = -4$ 21. $x + 6 = -4$
    −3   10   6   −10

Solve each equation for a.

22. $a + b = c$ 23. $a - b = c$ 24. $a + b = -c$ 25. $a - b = -c$
    $a = -b + c$ $a = b + c$ $a = -b - c$ $a = b - c$

26. If $a - \frac{2}{3} = 4$, what is the value of $3a$? 14

27. If $2.5 + s = 5.3$, what is the value of $2s - 3$? 2.6

Exercises 28-30 refer to the photo.

28. Solve equation a. 61

29. Solve equation b. 131

30. Compare the steps in solving the 2 equations.

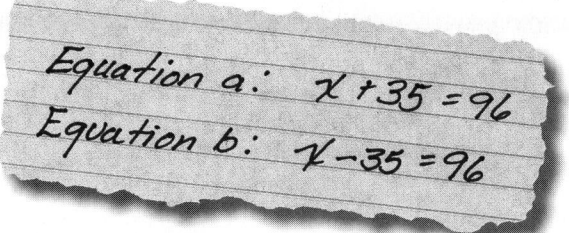

Equation a: $x + 35 = 96$

Equation b: $x - 35 = 96$

For Exercises 31-41, assign a variable and write an equation for each situation. Then solve the equation.

31. Sports The first- and second-string running backs on a football team ran for a total of 94 yards. If the first-string back ran 89 yards, how many yards did the other back run? $89 + y = 94; 5$

32. The first- and second-string running backs on a football team ran for a total of 89 yards. If the first-string back ran 94 yards, how many yards did the other back run? $94 + y = 89; -5$

33. The first-string running back on a football team ran for a total of 94 yards. If the second-string running back on a football team ran for a total of 89 yards, what was their combined yardage? $94 + 89 = c; 183$

34. Carl sees that the calendar indicates that his birthday, December 21, is the 355th day of the year and that the current day, October 15, is the 288th day of the year. In how many days is his birthday? $d + 288 = 355; 67$

35. Home Economics A recipe for turkey gravy says you should add water to the drippings (with most of the fat removed) to get $1\frac{1}{2}$ cups of liquid. If there is $\frac{7}{8}$ of a cup of drippings, how much water should be added? $w + \frac{7}{8} = 1\frac{1}{2}; \frac{5}{8}$

36. Travel If the odometer on Sarah's car registered 23,580 when she finished a vacation trip of 149 miles, what did it read when she started? $m + 149 = 23{,}580; 23{,}431$

37. If the odometer on Elizabeth's car registered 23,580 when she started a trip of 149 miles, what did it read when she finished her trip? $m - 149 = 23{,}580; 23{,}729$

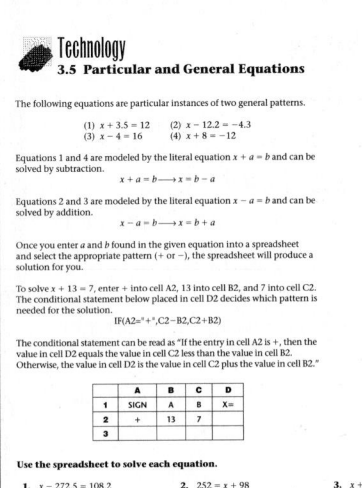

38. Sales tax Sam bought two items costing $12.98 and $14.95 plus tax. He got 39¢ change from $30. How much was the tax? $12.98 + 14.95 + t + 0.39 = 30; \$1.68$

39. Statistics The range of a set of scores was 28. If the highest score was 47, what was the lowest score? $28 = 47 - L; 19$

40. Geometry Supplementary angles are pairs of angles whose measures total 180 degrees. Find the measure of an angle supplementary to an angle with a measure of 92 degrees. $92 + s = 180; 88$

41. Two of the angles of a triangle each measure 50 degrees. Find the measure of the third angle. (HINT: The sum of the measures of the angles of a triangle is 180 degrees.) $2 \cdot 50 + a = 180; 80$

30. Both require the use of an equality property. Equation (a) needs the Subtraction Property of Equality while Equation (b) needs the Addition Property of Equality.

For Exercises 42 and 43, be sure students understand which angles are used. If necessary, encourage students to sketch the triangle on their paper and highlight the angles involved.

For Exercise 44, remind students that angles *RSP* and *PST* form a linear pair. They are therefore supplementary angles with a sum of 180°.

For Exercise 45, remind students that a right angle has a measure of 90°.

42.  **Geometry** The angles *SRU* and *SUR* of triangle *RSU* measure 70°. Write and solve an equation to find the measure of angle *SRT*. $x + 30 = 70; 40$

43. What would the measure of angle *SRT* be in triangle *SRU* if the measure of angle *TRU* were 18°?
    $x + 18 = 70; 52$

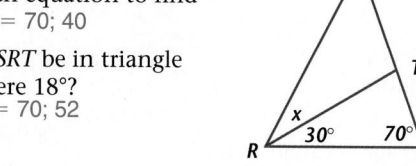

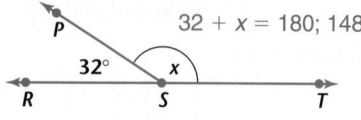 **Geometry**

44. Find *x* if *S* lies on line *RT*.

$32 + x = 180; 148$

45. Find *y* if angle *UVW* is a right angle.

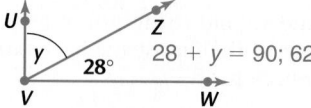 $28 + y = 90; 62$

Technology Some computer software packages, such as Maple™ can solve equations automatically, as shown in the following examples. The line after > is what you type in. The next line is what the computer types back when you hit ⌨ENTER.

> solve(x + 98765432123456789 = 444444444444444444);

345679012320987655

> solve(A = P + I,I);

A − P

Solve (A + B = C + D, D);

Give Maple™ commands to solve the following equations.

46. $x + 12345.6789 = 55555555$ 47. A + B = C + D (Solve for D.)
    Solve (x + 12345.6789 = 55555555);

48.  **Portfolio Activity** Complete the problem in the portfolio activity on page 99.   5 squirrels

 **Look Back**

49. State the pattern and find the next three numbers for the sequence 6, 0, 12, 6, 18, 12, __24__, __18__, __30__. **[Lesson 1.1]**   subtract 6, add 12

50. Find the perimeter of a square with sides 2.5 inches long.   **[Lesson 1.3]**
    10 inches

51. If poster board cost 39¢ for each piece, how many pieces of poster board can Shannon buy with $2.50?   **[Lesson 1.3]**   6

52. **Physics** Suppose a rocket took 23 seconds from take off to its return to Earth. Find the time it reached its maximum height.   **[Lesson 2.2]**
    11.5 seconds

Simplify.   [Lessons 3.3, 3.4]

53. $(3x − 7) − (x − 4)$ 54. $2(8 − 7y) + 5(2y)$ 55. $−4(n − 2) − (−3n + 4)$
    $2x − 3$ $−4y + 16$ $−n + 4$

Look Beyond

Solve these equations using algebra tiles.

56. $x − 3 = 8$ 11   57. $2x = 12$ 6   58. $2x + 5 = 11$ 3

Look Beyond

The last two problems provide a preview of equations that involve multiplication and two operations. Encourage students to experiment with algebra tiles to find the solutions and to record any findings.

LESSON 3.6 Solving Related Inequalities

 When the weather report says "The high temperature for today will be 75°," it implies that the temperatures throughout the day will be less than or equal to 75°F. This inequality describes a range of values. Inequalities can also be used to determine the items you can purchase when you have a given amount of money.

PREPARE

Objectives

- Define the concept and symbols of inequalities.
- Simplify and solve inequalities that involve addition and subtraction.

RESOURCES

- Practice Master   3.6
- Enrichment Master   3.6
- Technology Master   3.6
- Lesson Activity Master   3.6
- Quiz   3.6
- Spanish Resources   3.6

Michael can spend at most $3.10 for lunch. He buys a hamburger for $1.45. How much can he spend on dessert and stay within his spending limit?

Let *x* be the amount Michael spends on dessert. The total amount he spends is then $x + 1.45$. Michael's situation can be modeled by the statement $x + 1.45 \leq 3.10$. It is read "*x* plus 1.45 *is less than or equal to* 3.10." Michael will stay within his limit as long as $x + 1.45$ remains less than or equal to 3.10.

Suppose Michael buys a dish of fruit. Substitute 0.30 for *x*. The result is $0.30 + 1.45$, or 1.75. This is within the limit, since 1.75 is less than 3.10.

Suppose Michael buys a brownie with ice cream. Substitute 3.00 for *x*. The result is $3.00 + 1.45$, or 4.45. This is beyond Michael's limit because 4.45 is *not* less than or equal to 3.10.

Assessing Prior Knowledge

Graph each set of numbers on a number line.

1. $\{3, 5, 6, 8, 10\}$

   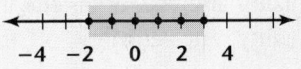

2. $\{-9, -6, -3, 0\}$

3. {integers between and including -2 and 3}

TEACH

 Discuss real-life situations that can be described by inequalities. Ask students what is implied by a weather report that says the low temperature for today will be 50°. [**The temperature will be greater than or equal to 50°.**]

ALTERNATIVE teaching strategy

Technology Use the TABLE function on a graphing calculator to find possible solutions to an inequality. For example, to find possible solutions to the inequality in the exploration, press the ⌷Y=⌷ key.

After Y1=, enter ⌷x⌷ ⌷+⌷ 1.25 . Press ⌷2nd⌷ ⌷TblSet⌷ and set TblMin = 0 and ΔTbl = 0.1. Then press ⌷2nd⌷ ⌷TABLE⌷ and examine the values listed under Y1. Scroll down the table of values until Y1 > 2.35. All of the X values listed before this point are in the solution set of the inequality.

Cooperative Learning

Have students work in small groups. Ask each student to write an inequality such as $x \geq 4$ on an index card and to draw the graph of their inequality on another index card. Each student should make at least 4 pairs of cards. Have each group put all of their cards together and shuffle them. Then have them place all of the cards facedown on a table or desk. Have students take turns turning over two cards at a time. If a student turns over an inequality and its graph, the student should pick up both cards and take another turn. Continue play until all of the cards have been picked up. The student with the most cards wins. As an extension, have each student write an inequality that has the solution on each pair of cards that he or she picked up. For example, $x + 3 > 7$ has the solution $x > 4$.

Exploration Notes

Emphasize that the only difference between the addition and subtraction properties of inequality and equality is the inequality and equality of the expressions involved. The method used to solve equalities and inequalities is the same.

Aongoing ASSESSMENT

3. Subtract 1.25 from both sides of the inequality.

For Michael, any amount for x that is less than or equal to 1.65 will make the inequality $x + 1.45 \leq 3.10$ a true statement. The *solution of an inequality* is the set of numbers that make the inequality statement true. There are several inequality statements used in algebra.

STATEMENTS OF INEQUALITY	
a is less than b	$a < b$
a is greater than b	$a > b$
a is less than or equal to b	$a \leq b$
a is greater than or equal to b	$a \geq b$
a is not equal to b	$a \neq b$

•Exploration *Solving an Inequality*

1. Tell whether each of the following values makes the inequality $x + 1.25 \leq 2.35$ true or false.
   a. 0.50
   b. 1.00
   c. 1.50
   d. 1.75

2. Continue guessing until you find the largest value of x which makes the statement true.

3. Give a method that does not use guessing to determine the largest value of x that makes the inequality statement true. ❖

The inequality $x + 1.45 \leq 3.10$ can be thought of as two statements.

a. $x + 1.45 < 3.10$ an inequality
b. $x + 1.45 = 3.10$ an equality

The equation $x + 1.45 = 3.10$ can be solved by using the Subtraction Property of Equality. Thus, $x + 1.45 = 3.10$ is true when $x = 1.65$. The inequality $x + 1.45 < 3.10$ is true when $x < 1.65$. This suggests that there are properties for solving inequalities similar to the ones for solving equations.

ADDITION PROPERTY OF INEQUALITY

If equal amounts are added to the expressions on each side of an inequality, the resulting inequality is still true.

SUBTRACTION PROPERTY OF INEQUALITY

If equal amounts are subtracted from the expressions on each side of an inequality, the resulting inequality is still true.

---

interdisciplinary CONNECTION **Geometry** Have students do research to find out about the Triangle Inequality Theorem. Ask them to write a report about their findings. Encourage students to include several examples in their report.

ENRICHMENT Give students inequalities like $3n + 6 - n > n + 1$ and $4g - 5 + g \leq 4g + 6$ to solve. Be sure they substitute values to test their solutions. Ask them to graph the solution sets.

English Language Development Some students may find it confusing to translate English phrases into mathematical in-

EXAMPLE 1

Solve the inequality $8m - 8 \geq 7m + 2$.

Solution ➤

One strategy is to arrange the terms with the variable on one side of the inequality sign and the numbers on the other side. Then combine like terms.

$8m - 8 \geq 7m + 2$	Given
$8m - 8 - 7m \geq 7m + 2 - 7m$	Subtraction Property of Inequality
$8m - 7m - 8 \geq 7m - 7m + 2$	Rearrange like terms.
$m - 8 \geq 2$	Combine like terms.
$m - 8 + 8 \geq 2 + 8$	Addition Property of Inequality
$m \geq 10$	Combine like terms.

The solution is $m \geq 10$. Now, substitute values into the original inequality to see if the result is true. For example, see if the values for m are greater than or equal to 10.

Substitute **10** for m.
$$8(\mathbf{10}) - 8 \geq 7(\mathbf{10}) + 2$$
$$80 - 8 \geq 70 + 2$$
$$72 \geq 72$$
True, because 72 equals itself.

Substitute **11** for m.
$$8(\mathbf{11}) - 8 \geq 7(\mathbf{11}) + 2$$
$$88 - 8 \geq 77 + 2$$
$$80 \geq 79$$
True, because 80 is greater than 79.

When m is 10 the inequality is true. When m is 11, the inequality is true. ❖

Using the Number Line to Represent an Inequality

EXAMPLE 2

Graph the solution to $x - 2 < 8$.

Solution ➤

First, solve the inequality for x.

$x - 2 < 8$	Given
$x - 2 + 2 < 8 + 2$	Addition Property of Inequality
$x < 10$	Combine like terms.

To graph the result on a number line, shade *all* points to the left of 10 on the number line. This includes all points whose coordinates are less than 10. To indicate that 10 *is not included,* put an *open dot* at 10.

$x < 10$
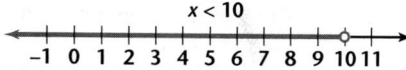

To check, substitute 9 for x.
$$9 - 2 = 7 \qquad \text{Since 7 is less than 8, 9 is a solution.}$$
Substitute 10 for x.
$$10 - 2 = 8 \qquad \text{Since 8 is not less than 8, 10 is not a solution.} ❖$$

equalities. Have students make a list of examples using key phrases and place the examples in a "> or ≥" list , a "< or ≤" list, or a "combination" list. Some examples are given below.

> or ≥

3 is greater than 2	$3 > 2$
x is greater than or equal to 4	$x \geq 4$
x is at least 2	$x \geq 2$
x exceeds 2	$x > 2$

< or ≤

-1 is less than 0	$-1 < 0$
x is less than or equal to 0	$x \leq 0$
x is at most 0	$x \leq 0$
values of x up to 0	$x < 0$

combination

x is between -1 and 3	$-1 < x < 3$
x is greater than -1 and less than 3	
	$-1 < x < 3$
x is from -1 to 3	$-1 < x < 3$

TEACHING *tip*

Encourage students to test several different values to check the solution to an inequality, both values that are in the solution set and values that are not in the solution set.

Alternate Example 1

Solve $9n + 3 < 8n - 5$.
$[n < -8]$

TEACHING *tip*

Be sure students understand that the opposite of an open dot is a solid dot and that a solid dot means that the value is included in the solution. The use of open dots and solid dots is summarized below.

open dot (○): $<$ $>$
solid dot (●): $\leq$ $\geq$

Alternate Example 2

Graph the solution to $x + 7 > 3$ for all numbers x.

all numbers greater than or equal to 3; $x \geq 3$

Alternate Example 3

Graph the points that satisfy both $x \leq 3$ and $x > -1$.

-5 -3 -1 0 1   3   5

Use Transparency ▶ 16

What numbers do you think are represented by the graph on this number line? Write an inequality that describes these points.

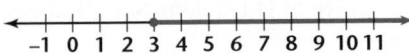

-1 0 1 2 3 4 5 6 7 8 9 10 11

Displaying Solutions

The graph of an inequality using integers is not the same as the graph using all numbers. Integers are graphed as dots (•). An inequality using all numbers is graphed as an interval, a ray, or a line.

$-2 \leq x < 1$

Using integers

-3 -2 -1 0 1 2 3

Using all numbers

-3 -2 -1 0 1 2 3

Remember, included endpoints are shown as dots, while endpoints not included are shown as open dots (○). It is assumed that all numbers are to be used unless otherwise noted.

The solution to an equation or inequality is the set of all numbers that makes the statement true. The solution might be listed, described, or graphed.

EXAMPLE 3

Graph the solution to $-4 \leq x$ and $x < 2$.

Solution ➤

The solution set must be true for both inequalities. This includes -4, since $-4 \leq 2$ and $-4 \leq -4$ are both true. It does not include 2, since $2 < 2$ is false.

-5 -4 -3 -2 -1 0 1 2 3 4 5 6 7

The statement **$-4 \leq x$ and $x < 2$** can also be written as **$-4 \leq x < 2$**. This means that x is between -4 and 2 and includes -4. ❖

Try This   Graph the solution to $-3 < x \leq 2$.

In everyday conversation, certain inequalities have important meanings. Some common examples are the following.

$x \geq 5$	$5 < x < 7$
x is at least 5	x is between 5 and 7
$x \leq 5$	$5 \leq x \leq 7$
x is at most 5	x is between 5 and 7 inclusive

What is the solution to $5 \geq x \geq 7$? Explain your answer.

ongoing
ASSESSMENT

Try This

-5 -3 -1 0 1   3   5

ongoing
ASSESSMENT

There is no solution. A number cannot be both less than 5 and greater than 7.

RETEACHING   **Using Manipulatives**
    t h e   Have students use alge-
l e s s o n   bra tiles to solve the
                    equation related to an in-
equality. For example, for $x + 5 \geq 7$, use algebra tiles to solve $x + 5 = 7$. Then replace the = sign with the $\geq$ in the final solution.

EXERCISES & PROBLEMS

Communicate

Explain how you would determine whether the following inequalities are true or false.

1. $5 \geq 2 + 3$ **2.** $4 \leq 2 + 3$

3. $6 < 7 - 3$ **4.** $8 > 10 - 2$

5. Tell how the properties for inequalities are similar to the properties for equality.

6. Tell the steps necessary to solve the inequality $3x - 4 \leq 2x + 1$. Name the property you would use for each step.

7. How do you find what values of the solution to $x - 4 \leq 9$ represent equality and what values represent inequality?

8. Explain how to draw the solution to $x + 3 < 7$ on a number line.

9. How do you write an inequality that describes the points that are shown on the graph?

10. Discuss the word *inclusive* and its meaning in mathematics. List the whole numbers between 5 and 10 inclusive.

Practice & Apply

State whether each inequality is true or false.

11. $8 > 9 - 1$ F   **12.** $-2 \leq 5 - 7$ T   **13.** $8 \leq 9 - 1$ T   **14.** $-2 > 5 - 7$ F

Solve each inequality.

15. $x + 8 > -1$ **16.** $x - 6 \leq 7$ **17.** $x + \frac{3}{4} < 1$ $x < \frac{1}{4}$
  $x > -9$ $x \leq 13$

18. $x + \frac{3}{4} \geq \frac{1}{2}$ **19.** $x + 0.04 > 0.6$ **20.** $x - 0.1 < 8$

21. Describe a real world situation modeled by $x + 10 < 100$.

22. Show the solution to $x - 4 \geq -1$ on a number line.

23. Show the solution to $x + 3 < 2$ on a number line.

---

18. $x \geq -\frac{1}{4}$

19. $x > 0.56$

20. $x < 8.1$

21. Answers will vary. For example, Rachel has already spent $10 and was told that she must stay under her $100 limit. How much does she have left to spend?

22.

23.

ASSESS

Selected Answers
Odd-numbered Exercises 11–19, 23–25, 29–47.

Assignment Guide
Core 1–10, 12–26 even, 27–53

Core Two-Year 1–48

Error Analysis
Caution students to be careful to use the correct symbol when writing the steps to solve an inequality. Students should also be careful not to switch the direction of the inequality sign or drop the equal line when using the $\leq$ or $\geq$ symbols.

Performance Assessment

Have students write 3 inequalities that have solutions that are $x \geq -4$. Then ask students to graph the inequality. Repeat this assessment with additional inequalities if necessary.

Write an inequality that describes the points on the following number lines.

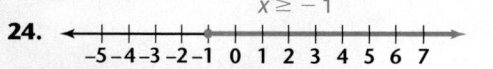

24. $x \geq -1$

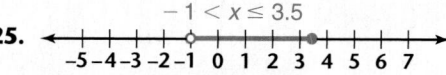

25. $-1 < x \leq 3.5$

26. List the whole numbers between 57 and 66.
   58, 59, 60, 61, 62, 63, 64, 65

Temperature Write an equation or an inequality to represent each situation. Use t to represent

27. all possible daytime temperatures where you live. depends on location

28. a high temperature of 66°F. $T \leq 66$

29. the low temperature was 54 degrees. $T \geq 54$

30. 54 degrees. $T = 54$; an equality results

31. If t represents the first reading, then after a rise of 5 degrees the temperature was between 70 and 80 degrees inclusive. What is the inequality? $65 \leq T \leq 75$

32. **Sports** A sports stadium holds 15,000 people. Everyone at the game has a seat, but the stadium is not full. Write an inequality that models this situation using P as the number of people. $1 \leq P < 15000$

33. The school auditorium can seat 450 people for graduation. Graduates will use 74 seats. Write an inequality to describe the number of others who can be seated in the auditorium. $0 \leq P \leq 376$

34. From 30 to 50 people attended a party. From 5 to 10 people left early. Write an inequality to represent the possible range for the number of people who did not leave early. $25 \leq P \leq 45$

35. A table is supposed to be 42.3 centimeters long. Write an inequality for M, the measure of the table, that allows a possible error of 0.5 centimeters. $41.8 \leq M \leq 42.8$

36. Students in Ms. Ambrose's algebra class earn an "A" if they average at least 90, and they earn a "B" if they average at least 80 but less than 90. This can be represented using the following inequalities.

$$A \geq 90 \qquad 80 \leq B < 90$$

Translate this inequality into words.

$$70 \leq C < 80$$

If a student's average is at least 70 but less than 80, he or she will earn a "C".

37. Represent the following statement using an inequality. $78 \leq B < 89$
Students earn a "B" if they average at least 78, but less than 89.

Practice Master

Practice & Apply
3.6 Solving Related Inequalities

For $x = -1$, determine whether each inequality is true or false.

1. $x - 1 > 0$ _____   2. $-x > 0$ _____

3. $x \leq 5 - 6$ _____   4. $x + 2 \geq 1$ _____

Wholesale Pet Supplies uses the table shown to determine shipping and handling charges.

SHIPPING AND HANDLING CHART RESIDENTIAL				
If your order is:	ZONE 1	ZONE 2	ZONE 3	
$25.00 to $50.00	$5.95	$7.35	$8.35	For orders over $145 there is no regular shipping charge.
$50.01 to $75.00	$6.40	$8.25	$9.95	
$75.01 to $145.00	$7.85	$9.95	$11.95	

5. Write an inequality to represent the possible amount A, of an order to be shipped to zone 3 if the shipping and handling cost is $9.95.
_____

6. Write an inequality that represents the minimum order M. _____

7. Find the cost of shipping an order totaling $48.56 to zone 1. _____

Write an inequality that describes the points on the number lines.

8. _____

9. _____

10. _____

Solve each inequality.

11. $x - 5 < 11$ _____   12. $3y \geq 2y - 8$ _____

13. $7n + 0.2 \leq 6n - 1.3$ _____   14. $\frac{4}{3} + 3m > 2m + \frac{4}{5}$ _____

15. Use the number line to graph the points that satisfy $n \geq -3$ and $n \leq 2$ when n is a real number.

38. State the pattern and find the next three numbers for the following sequence, 1, 1, 2, 3, 5, _8_, _13_, _21_. **[Lesson 1.1]**

39. If rulers cost 45¢, how many rulers can Mimi buy with $2.35? **[Lesson 1.3]** 5

40. If the points of a scatter plot lie near a line that rises from left to right, name the type of correlation. **[Lesson 1.6]** positive

41. If the points of a scatter plot lie near a line that falls from left to right, name the type of correlation. **[Lesson 1.6]** negative

42. A linear function grows by what kind of amount? **[Lesson 2.4]** constant

43. If a graph appears to be in steps, describe the type of function. **[Lesson 2.5]** integer

Find the values. [Lesson 3.1]

44. $|-4.5|$ 4.5

45. $|-9|$ 9

46. $|-3+4|$ 1

Solve for x. [Lesson 3.5]

47. $25x - 47 = 24x + 39$ 86

48. $4x + 110 = 5x + 17$ 93

Look Beyond

Solve each of the following inequalities. Try several values for x, and check the answers.

49. $2x < 8$ **50.** $4x + 5 \leq 16$ **51.** $8x - 3 > 33$
$x < 4$ $x \leq 2.75$ $x > 4.5$

52. How many ways can you make change for a quarter using pennies, nickels, and dimes? 12

53. Place dots on a circle, and connect the dots with line segments. Count the regions that are formed. Be sure that no three lines meet at the same point unless the point is on the circle. What is the maximum number of regions formed from 6 dots? 31

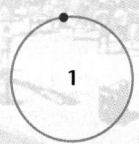
1 dot
1 region

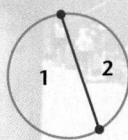
2 dots
2 regions

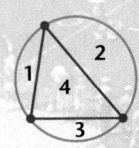
3 dots
4 regions

4 dots
8 regions

Databases are used today by many businesses to organize and manipulate important information. Lists of customers and potential customers can be put into a database. Lists of independent contractors a company uses can also be put into a database. What other information could possibly be put into a database?

MOTIVATE

Invite an owner of a small business to come and talk to your class about how his or her business uses databases. Ask students to write a report about the presentation.

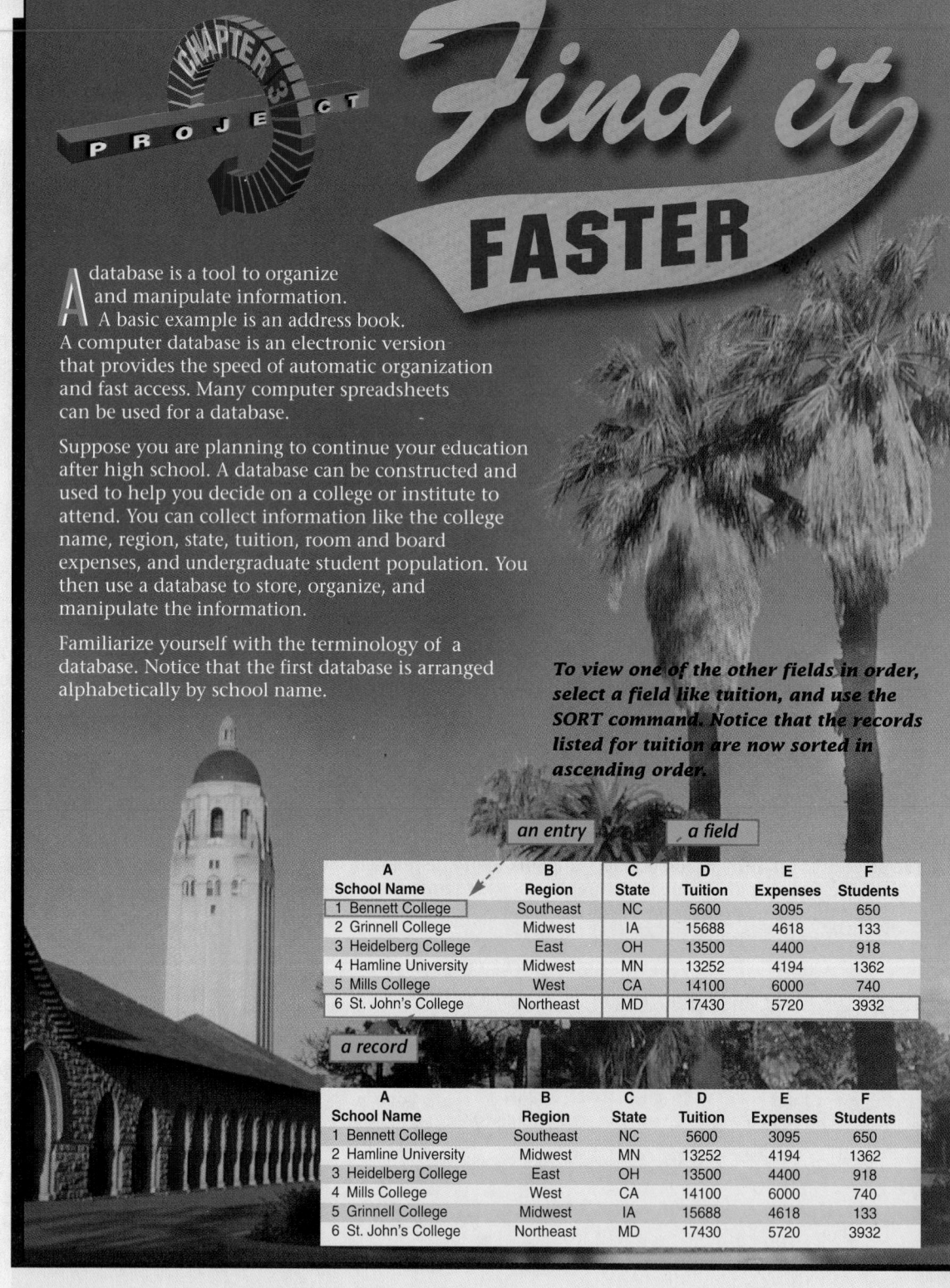

CHAPTER 3 PROJECT

Find it FASTER

A database is a tool to organize and manipulate information. A basic example is an address book. A computer database is an electronic version that provides the speed of automatic organization and fast access. Many computer spreadsheets can be used for a database.

Suppose you are planning to continue your education after high school. A database can be constructed and used to help you decide on a college or institute to attend. You can collect information like the college name, region, state, tuition, room and board expenses, and undergraduate student population. You then use a database to store, organize, and manipulate the information.

Familiarize yourself with the terminology of a database. Notice that the first database is arranged alphabetically by school name.

To view one of the other fields in order, select a field like tuition, and use the SORT command. Notice that the records listed for tuition are now sorted in ascending order.

an entry · a field

	A School Name	B Region	C State	D Tuition	E Expenses	F Students
1	Bennett College	Southeast	NC	5600	3095	650
2	Grinnell College	Midwest	IA	15688	4618	133
3	Heidelberg College	East	OH	13500	4400	918
4	Hamline University	Midwest	MN	13252	4194	1362
5	Mills College	West	CA	14100	6000	740
6	St. John's College	Northeast	MD	17430	5720	3932

a record

	A School Name	B Region	C State	D Tuition	E Expenses	F Students
1	Bennett College	Southeast	NC	5600	3095	650
2	Hamline University	Midwest	MN	13252	4194	1362
3	Heidelberg College	East	OH	13500	4400	918
4	Mills College	West	CA	14100	6000	740
5	Grinnell College	Midwest	IA	15688	4618	133
6	St. John's College	Northeast	MD	17430	5720	3932

Different spreadsheets and database software have different features for requesting various combinations of information. A request for information is called a **query**. Some common operations for making a query are comparisons that use AND, OR, and NOT.

A query using OR will retrieve all information from the records that fulfill any specification.

> **Query:** Schools in the **midwest OR** the **west**.
>
> **Output:** Hamline University
> Grinnell College
> Mills College

A query using AND will give only the information from the records that fulfill all the specifications.

> **Query:** Schools with **tuition less than $14,000 AND** schools in the **midwest**.
>
> **Output:** Hamline University

Different databases and spreadsheets treat AND and OR operations in different ways.

*A*ctivity 1

Alternative 1 (For those with a computer and a spreadsheet or database): Make a database from some data that is of interest to you. It should contain regular lists of information.

Alternative 2 (For those without a computer): Create your own database from index cards. As a group project, you can also act out the computer procedure for creating a database. This includes writing, storing, sorting, and retrieving information.

*A*ctivity 2

Once you have created your database, devise a strategy for sorting various fields and retrieving a record. Form questions that ask for specific combinations of the data. The search should contain a method for using OR and another using AND.

Discuss

Chapter 3 Review

Vocabulary

Addition Property of Equality	126	Distributive Property	113
Addition Property of Inequality	132	formula	127
algebraic expression	112	inequality	132
Associative Property	104	integer	100
coefficient	112	literal equation	127
Commutative Property	114	Subtraction Property of Equality	125
constant	112	Subtraction Property of Inequality	132

Key Skills & Exercises

Lesson 3.1

➤ **Key Skills**

Add two or more integers.

When the terms have like signs, find the sum of the absolute values, and use the common sign. When the terms have unlike signs, find the difference of the absolute values. Use the sign of the integer with the greater absolute value.

To add more than two integers, use the Associative Property of Addition.

$$(-5 + 5) + -3 = 0 + (-3) = -3$$
or
$$-5 + (5 + (-3)) = -5 + 2 = -3$$

$$-4 + (-7) = -11 \qquad 4 + 7 = 11$$
$$26 + (-9) = 17 \qquad -26 + 9 = -17$$

➤ **Exercises**

Find each sum.

1. $-17 + 6$ -11 **2.** $48 + (-15)$ 33   **3.** $-23 + (-25) + 3$ -45 **4.** $-39 + 68$ 29
5. $33 + (-55)$ -22 **6.** $-214 + 214$ 0   **7.** $6 + (-7) + (-9)$ -10 **8.** $-8 + 8 + (-12)$ -12

Lesson 3.2

➤ **Key Skills**

Subtract integers.

To subtract -8, add the opposite. $-16 - (-8) = -16 + 8 = -8$
To subtract 19, add the opposite. $-38 - 19 = -38 + (-19) = -57$

➤ **Exercises**

Find each difference.

9. $9 - (-15)$ 24   **10.** $48 - (-48)$ 96   **11.** $-13 - 28$ -41 **12.** $39 - (-18)$ 57
13. $-67 - (-42)$ **14.** $-23 - (-72)$ **15.** $-42 - (-42)$ -53 **16.** $8 - 14$ -27

13. -25

14. 49

15. -53

16. -33

Lesson 3.3

➤ **Key Skills**

Simplify expressions with several variables by adding like terms.

To simplify $(3r + 7t) + (4r - 8t)$, use this procedure.

$(3r + 7t) + (4r - 8t)$	Given
$(3r + 7t) + (4r + (-8t))$	Definition of Subtraction
$(3r + 4r) + (7t + (-8t))$	Rearrange terms.
$7r + (-1)t$	Combine like terms.
$7r - t$	Definition of Subtraction

➤ **Exercises**

Simplify.

17. $(6a - 1) + (5a - 4)$ $11a - 5$ **18.** $(7 - t) + (3t + 4)$ $2t + 11$ **19.** $\left(\frac{x}{3} - 2\right) + \left(\frac{x}{2} + 4\right)$ $\frac{5}{6}x + 2$

20. $(1.4m - 6.2n) + (2.4m - 5.5n)$ $3.8m - 11.7n$ **21.** $(3x + 2y + z) + (6x - 4y - 3z)$ $9x - 2y - 2z$

Lesson 3.4

➤ **Key Skills**

Use adding the opposite to subtract expressions.

To simplify $(7x + 4y - 2z) - (6x - 5y + z)$, use this procedure.

$(7x + 4y - 2z) - (6x - 5y + z)$	Given
$7x + 4y - 2z - 6x + 5y - z$	Definition of Subtraction
$7x - 6x + 4y + 5y - 2z - z$	Rearrange terms.
$x + 9y - 3z$	Combine like terms.

➤ **Exercises**

Simplify.

22. $3x - 5x$ $-2x$ **23.** $7y - (7 - 5y)$ $12y - 7$ **24.** $(8m - 4) - (6m - 3)$ $2m - 1$

25. $(6d + 3) - (4d - 7) + (3d - 5)$ $5d + 5$ **26.** $(4a - 3b - c) - (6a + 5b - 4c)$ $-2a - 8b + 3c$

Lesson 3.5

➤ **Key Skills**

Solve algebraic equations that contain addition and subtraction.

Solve $x + 15 = 11$.

$$x + 15 = 11$$
$$x + 15 - 15 = 11 - 15$$
$$x = -4$$

Check: $-4 + 15 = 11$ True

Solve $-8 = y - 14$.

$$-8 = y - 14$$
$$-8 + 14 = y - 14 + 14$$
$$6 = y$$

Check: $-8 = 6 - 14$ True

➤ **Exercises**

Solve.

27. $w + 16 = 25$ 9 **28.** $r + 26 = 16$ -10 **29.** $t + 7 = -5$ -12 **30.** $a + 1.5 = 3.6$ 2.1

31. $m + \frac{1}{2} = \frac{5}{6}$ $\frac{1}{3}$ **32.** $y - 13 = 12$ 25 **33.** $24 = x - 19$ 43 **34.** $-6 = g - 17$ 11

35. $h - \frac{1}{6} = \frac{2}{3}$ $\frac{5}{6}$ **36.** $7k - (6k + 5) = 7$ 12 **37.** $4 - (2 - 3z) = 6z - (4z + 3)$ -5

Lesson 3.6

➤ Key Skills

Solve inequalities, and show the solution on the number line.

For inequalities with addition and subtraction, solve as you would solve an equation. Then draw the solution on the number line.

$$5t - 6 \le 4t + 1$$
$$5t - 4t - 6 \le 4t - 4t + 1$$
$$t - 6 \le 1$$
$$t - 6 + 6 \le 1 + 6$$
$$t \le 7$$

0 1 2 3 4 5 6 7 8

➤ Exercises

Solve each inequality, and show the solution on a number line.

38. $x + 5 > 10$
$x > 5$

39. $n - 15 \le -3$
$n \le 12$

40. $y + 0.09 < 3.09$
$y < 3$

41. $d - \frac{2}{3} \ge \frac{3}{9}$
$d \ge 1$

Applications

42. Marsha is stocking the art supply cabinet at school. The paint brushes she needs to buy cost $1.75 each, and the paint she needs costs $2.45 per jar. Write an expression to show how much she will spend if she buys b paint brushes and j jars of paint. $1.75b + 2.45j$

43. Finance Lon's checking account is overdrawn by $30. How much does he need to deposit so that he has a balance of $150? $180

44. Toby bought a car for $9299. He paid $1500 as a down payment. How much money did he have to borrow to pay for the rest of the car? $7799

45. Hobbies Rita has $15 to spend on new baseball cards. She decides to buy a card that costs $3. What is the most she can spend on other baseball cards? $12

Geometry The formula for the perimeter, P, of a triangle with sides of length a, b, and c is $P = a + b + c$.

46. Use the formula to find P when $a = 3.5$, $b = 4.7$, and $c = 5.9$. 14.1

47. Solve this formula for b. $b = P - a - c$

48. Use the formula to find b when $P = 13.5$, $a = 4.5$, and $c = 4.5$. 4.5

49. Geometry Write a formula for the sum of the angles in triangle ABC. $x + y + z = 180$

50. Solve the formula for z. $z = 180 - x - y$

51. Find z when $x = 55$ and $y = 60$.
65

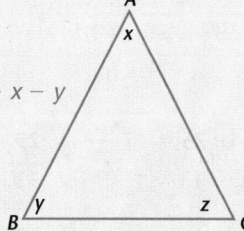

38.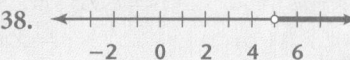
−2   0   2   4   6

41.
−4   −2   0   2   4

39.
−4 −2   0   2   4   6   8   10 12

40.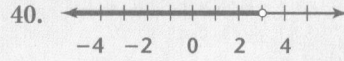
−4 −2   0   2   4

Chapter 3 Assessment

1. What are the opposites of 3, − 5, and 0? − 3; 5; 0
2. Find the absolute values of − 27, 82, and 0. 27; 82; 0
3. Draw algebra tiles to represent 3 + (− 5), and find the resulting sum. −2

Find each sum or difference.

4. − 7 + (− 5) − 12
5. 8 − (− 10) 18
6. 18 + (− 65) − 47
7. − 34 + 34 0
8. − 16 − (− 58) − 8 34
9. 42 + (− 78) + |− 9| − 27

10. Kylie had 13 tapes. Then she bought 6 tapes, gave 4 away, and bought 3 more. How many tapes does she have now? 18

Simplify by performing the indicated operations.

11. $(8n − 2) + (6n + 4)$ $14n + 2$
12. $5t − (9t − 6)$ $− 4t + 6$
13. $(8 − 9z) − (7z + 15)$ $− 16z − 7$

There are 5 boxes of books and 5 additional books in the book storage room.

14. If Mr. Weaver puts 2 boxes of books and 2 additional books in the book room, how many boxes and additional books are there now? 7 boxes, 7 books
15. If Mrs. Thompson *then* removes 3 boxes of books and 1 additional book, how many boxes and additional books are left in the room? 4 boxes, 6 books

Solve.

16. $c + 18 = 10$ $− 8$
17. $t − 36 = 19$ 55
18. $8 + y = − 14$ $− 22$
19. $34 = h − 4$ 38
20. $w + \frac{2}{5} = \frac{7}{10}$ $\frac{3}{10}$
21. $4x − (2x + 2) = 3 + x$ 5

22. Complementary angles are pairs of angles whose measures total 90°. Find the measure of an angle complementary to an angle with a measure of 48°. 42°
23. The relationship between total cost of an item, I, the cost without sales tax, C, and the amount of sales tax, T, is given by the formula $I = C + T$. Write a formula for C in terms of I and T. $C = I − T$
24. Using the formula $I = C + T$, find C when $I = \$25.19$ and $T = \$1.20$. $23.99

Solve each inequality.

25. $x + 6 \geq 4$ $x \geq − 2$
26. $y − 2.3 < 1.4$ $y < 3.7$
27. $t + \frac{1}{2} \geq 2$ $t \geq \frac{3}{2}$

28. Show the solution to $x − 8 > − 4$ on a number line.
29. Write an inequality that describes the points on the number line.

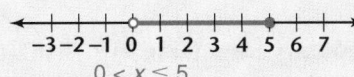

$0 < x \leq 5$

30. There is room for 40 people in an aerobics class. There are 25 people signed up for the class. Write an inequality to describe the number of other people that can still sign up for the class. $0 \leq P \leq 15$

3.
 =

−2

28.
−1 0 1 2 3 4 5 6 7

CHAPTER 4

Multiplication and Division in Algebra

Meeting Individual Needs

4.1 Exploring Integer Multiplication and Division

Core Resources

Practice Master 4.1
Enrichment, p. 148
Technology Master 4.1
Interdisciplinary
 Connection, p. 147

[1 day]

Core Two-Year Resources

Inclusion Strategies, p. 148
Reteaching the Lesson, p. 149
Practice Master 4.1
Enrichment Master 4.1
Technology Master 4.1
Lesson Activity Master 4.1

[3 days]

4.2 Multiplying and Dividing Expressions

Core Resources

Practice Master 4.2
Enrichment Master 4.2
Technology Master 4.2

[1 day]

Core Two-Year Resources

Inclusion Strategies, p. 152
Reteaching the Lesson, p. 154
Practice Master 4.2
Enrichment, p. 152
Technology Master 4.2
Lesson Activity Master 4.2

[3 days]

4.3 Multiplication and Division Equations

Core Resources

Practice Master 4.3
Enrichment, p. 161
Technology Master 4.3

[2 days]

Core Two-Year Resources

Inclusion Strategies, p. 161
Reteaching the Lesson, p. 162
Practice Master 4.3
Enrichment Master 4.3
Technology Master 4.3
Lesson Activity Master 4.3

[3 days]

4.4 Rational Numbers

Core Resources

Practice Master 4.4
Enrichment, p. 168
Technology Master 4.4
Interdisciplinary
 Connection, p. 168
Mid-Chapter Assessment
 Master

[1 day]

Core Two-Year Resources

Inclusion Strategies, p. 169
Reteaching the Lesson, p. 170
Practice Master 4.4
Enrichment Master 4.4
Technology Master 4.4
Lesson Activity Master 4.4
Mid-Chapter Assessment Master

[3 days]

4.5 Solving Problems Involving Percent

Core Resources

Practice Master 4.5
Enrichment, p. 174
Technology Master 4.5
Interdisciplinary
 Connection, p. 174

[1 day]

Core Two-Year Resources

Inclusion Strategies, p. 175
Reteaching the Lesson, p. 176
Practice Master 4.5
Enrichment Master 4.5
Technology Master 4.5
Lesson Activity Master 4.5

[4 days]

4.6 Writing and Solving Multistep Equations

Core Resources

Practice Master 4.6
Enrichment, p. 181
Technology Master 4.6

[2 days]

Core Two-Year Resources

Inclusion Strategies, p. 181
Reteaching the Lesson, p. 182
Practice Master 4.6
Enrichment Master 4.6
Technology Master 4.6
Lesson Activity Master 4.6

[4 days]

4.7 Exploring Related Inequalities

Core Resources

Practice Master 4.7
Enrichment, p. 188
Technology Master 4.7

[1 day]

Core Two-Year Resources

Inclusion Strategies, p. 188
Reteaching the Lesson, p. 189
Practice Master 4.7
Enrichment Master 4.7
Technology Master 4.7
Lesson Activity Master 4.7

[2 days]

4.8 Absolute Value Equalities and Inequalities

Core Resources

Practice Master 4.8
Enrichment, p. 194
Technology Master 4.8
Interdisciplinary
    Connection, p. 193

[2 days]

Core Two-Year Resources

Inclusion Strategies, p. 194
Reteaching the Lesson, p. 196
Practice Master 4.8
Enrichment Master 4.8
Lesson Activity Master 4.8

[4 days]

Chapter Summary

Core Resources

Eyewitness Math, pp.
    158–159
Chapter 4 Project, p.199
Lab Activity
Long-Term Project
Chapter Review,
    pp. 200–202
Chapter Assessment, p.
    203
Chapter Assessment, A/B
Alternative Assessment
Cumulative Assessment,
    pp. 204–205

[3 days]

Core Two-Year Resources

Eyewitness Math, pp. 158–159
Chapter 4 Project, p. 199
Lab Activity
Long-Term Project
Chapter Review, pp. 200–202
Chapter Assessment, p. 203
Alternative Assessment
Cumulative Assessment, pp.
    204–205

[5 days]

Reading Strategies

Encourage students to jot down key concepts in journals like Properties of Zero in Lesson 1, the Multiplicative Property of −1 in Lesson 2, and the Multiplication and Division Properties of Inequalities in Lesson 7. Students can then use their journals to respond to interactive questions. They can also make brief presentations that explain the everyday use of algebraic expressions and equations that involve multiplication and division.

Visual Strategies

Encourage students to use algebra tiles or drawings of algebra tiles to model the multiplication and division of expressions. For percents, encourage visual learners to sketch their own percent bars as an aid to completing the exercises. Visual learners may also benefit from sketching number lines as they work the exercises.

Hands-on Strategies

Many manipulatives exist to provide hands-on practice. *Algebra tiles* provide a hands-on method used to explore the concepts of multiplication and division for integers and algebraic expressions. *Coins* can be used to model many multiplication situations with integers. *Number cubes* provide a way of generating data for multiplication. *Rulers* can be used to make physical number lines, to find errors in measurement, and to study precision and tolerance.

Cooperative Learning

GROUP ACTIVITIES	
Working with bank transactions	Lesson 4.1, Opening Discussion
Developing spreadsheets	Lesson 4.2, Opening Discussion
Iteration formulas	Eyewitness Math
Solving formulas	Lesson 4.3, Example 5
Solving equations	Lesson 4.6, Example 4

You may wish to have students work with partners or in small groups to complete any of these activities. Additional suggestions for cooperative group activities can be found in the teacher's notes in the lessons.

Multicultural

The cultural connections in this chapter include references to Egypt and Russia.

CULTURAL CONNECTIONS	
Africa: Ancient Age Problem	Portfolio Activity
Africa: Egyptian Measurements	Lesson 4.6, Exercise 40
Europe: Russian Prime Problem	Lesson 4.7, Exercises 40–43

Portfolio Assessment

Below are portfolio activities for the chapter listed under seven activity domains which are appropriate for portfolio development.

1. **Investigation/Exploration** The relationship between the sign of the product or quotient of two or more integers is the focus of Explorations 1 and 2 in Lesson 4.1 and the Exploration in Lesson 4.2. Another Exploration appears in the Eyewitness Math feature. It examines some aspects of the science of chaos.

2. **Applications** Recommended to be included are any of the following: Banking, Lesson 4.1, Exercises 35–37; Carpentry, Lesson 4.3, Exercises 47–51; Consumer Economics, Lesson 4.4, Exercise 36; Demographics, Lesson 4.5, Exercise 53; Travel, Lesson 4.6, Exercises 37–39; Government, Lesson 4.8, Exercises 35–37.

3. **Non-Routine Problems** The Ancient Age Problem found in the Portfolio Activity provides an opportunity for algebraic thinking and logical reasoning.

4. **Project** Egyptian Equation Solving appears on page 199.

Students are asked to study the methods used by Egyptians over 3500 years ago and to use these methods to solve equations.

5. **Interdisciplinary Topics** Students may choose from the following: Statistics, Lesson 4.5, Exercises 47, 49–52; Number Theory, Lesson 4.6, Exercise 33; Health, Lesson 4.8, Exercises 22–24. Geometry, Lesson 4.2, Exercises 54–55; Lesson 4.3, Exercises 58–64; Lesson 4.6, Exercise 35; Lesson 4.7, Exercises 44–45.

6. **Writing** The *Communicate* exercises offer excellent possibilities for writing assignments that can be included in the portfolio.

7. **Tools** Chapter 4 uses both computer spreadsheets and graphics calculators to work with integers and rational numbers.

Technology

Technology provides important tools for manipulating and visualizing arithmetic and algebraic concepts. Spreadsheets help students organize and manipulate data. Scientific calculators perform arithmetic calculations quickly and help students focus on underlying algebraic concepts. Calculators can also be used to explore and verify hypotheses when solving problems such as finding the product or quotient of two integers or finding the inverse of a particular rational number. Graphics calculators provide an alternative method of solving equations in one variable. They also allow more efficient use of time for creating graphs. Detailed instructions for spreadsheets, calculators, and other technologies can be found in the *HRW Technology Handbook*.

Spreadsheets

Spreadsheets can be used to efficiently record and compute employee wages, such as suggested in Lesson 4.2. To expand on this concept, have students develop a spreadsheet that will keep track of five employees' hours and wages for the current month.

Employee A works 8 hours per day, Monday through Friday, at $8 per hour.

Employee B works 4 hours on Thursday, Friday, and Saturday at $5 per hour.

Employee C works between 1 and 6 hours on Monday through Saturday at $12 per hour. (Students can roll a number cube or use a random number generator to find how many hours Employee C works each day.)

Employee D works 7.5 hours on Monday through Friday at $5.80 per hour.

Employee E works 4 hours on Monday and Tuesday and between 1 and 6 hours on Thursday and Friday. He earns $5.40 per hour.

Students' spreadsheets should also calculate the total number of hours worked during the month as well as the total wages for each employee. As an extension, students can have their spreadsheet calculate the total hours worked and total wages for all employees together. Have students discuss possible ways to set up this spreadsheet.

After students have chosen a way to organize the data, show them how to format the various columns. Format the required columns or rows by selecting the appropriate cells, entering **Range** and **Format** from the menu, and specifying how many decimal places to include.

Graphics Calculator

Students learn about the use of the reciprocal key in Lesson 4.4. Be sure students understand that they must use parentheses when finding the reciprocal of rational numbers in fractional form. If they correctly enter $\frac{4}{5}$ as a fraction on some calculators, the calculator automatically puts in the parentheses when the x^{-1} key is pressed. Have students experiment with their calculators to find the correct procedures. Students may find a method of using the reciprocal key that works for unit fractions but will not work for other fractions. This is because

$$\left(\frac{1}{x}\right)^{-1} = \frac{1}{x^{-1}}, \text{ but } \left(\frac{z}{x}\right)^{-1} \neq \frac{z}{x^{-1}}.$$

In Lesson 4.6, students learn to solve an equation with variables on each side of the equality by graphing each side separately and then finding the point of intersection. Be sure students reset the original values in the calculator to the default values. After graphing the equations, students will find it necessary to Zoom Out, Trace to find the intersection, then Zoom In and Trace again to find the solution.

Integrated Software

f(g) Scholar™ is an integrated computer-based mathematics productivity tool that combines calculator, spreadsheet, and graphics capabilities to provide a dynamic and interactive environment for explorations in mathematics. It is appropriate to use *f(g) Scholar*™ for any lesson needing a spreadsheet, calculator, graphics calculator, or any combination of the three.

B11		=B10*8					
	A	B	C	D	E	F	G
1		Employee A	Employee B	Employee C	Employee D	Employee E	
2	Monday	8		3	7.5	4	
3	Tuesday	8		6	7.5	4	
4	Wednesday	8		2	7.5	5	
5	Thursday	8	4	5	7.5	3	
6	Friday	8	4	4	7.5		
7	Saturday		4	1			
8	Sunday						
9							
10	Total hours	40	12	21	37.5	16	
11	Pay	$320	$60	$252	$218	$86	

4 Multiplication and Division in Algebra

ABOUT THE CHAPTER

Background Information

The previous chapter focused on how to solve equations that involve addition and subtraction. This chapter extends the procedures to equations that involve multiplication and division. The central idea is still to maintain the balance or equality when performing operations to solve the equation.

CHAPTER RESOURCES

- Practice Masters
- Enrichment Masters
- Technology Masters
- Lesson Activity Masters
- Assessment Masters:
   Chapter Assessments, A/B
   Mid-Chapter Assessment
   Alternative Assessments,
   A/B
- Teaching Transparencies
- Cumulative Assessment
- Spanish Resources

CHAPTER OBJECTIVES

- Multiply and divide negative and positive integers.
- Define properties of zero.
- Evaluate algebraic expressions using multiplication and division.
- Multiply and divide algebraic expressions.
- Use the Distributive Property Over Subtraction to simplify expressions.
- Solve multiplication and division equations.
- Define rational numbers and display them on the number line.
- Find reciprocals of rational numbers.

CHAPTER 4

Multiplication and Division in Algebra

LESSONS

4.1 *Exploring* Integer Multiplication and Division

4.2 Multiplying and Dividing Expressions

4.3 Multiplication and Division Equations

4.4 Rational Numbers

4.5 Solving Problems Involving Percent

4.6 Writing and Solving Multistep Equations

4.7 *Exploring* Related Inequalities

4.8 Absolute Value Equalities and Inequalities

Chapter Project
Egyptian Equation Solving

Multiplication and Division in Algebra

Chapter 4 continues the development of solving linear equations. It extends the concept of maintaining balance in an equation when you multiply and divide to find the value of the unknown.

This chapter blends traditional methods with the latest technology. It begins with a very old problem written about Diophantus, who is often called the father of algebra. Applications of linear functions to business are stressed. Although modern applications of mathematics to commerce and business are often rather complex, we still use linear relationships to compute percentages, simple interest, and discounts.

ABOUT THE PHOTOS

The balances show how scaling can affect objects that are weighed without affecting the balance during multiplication and division.

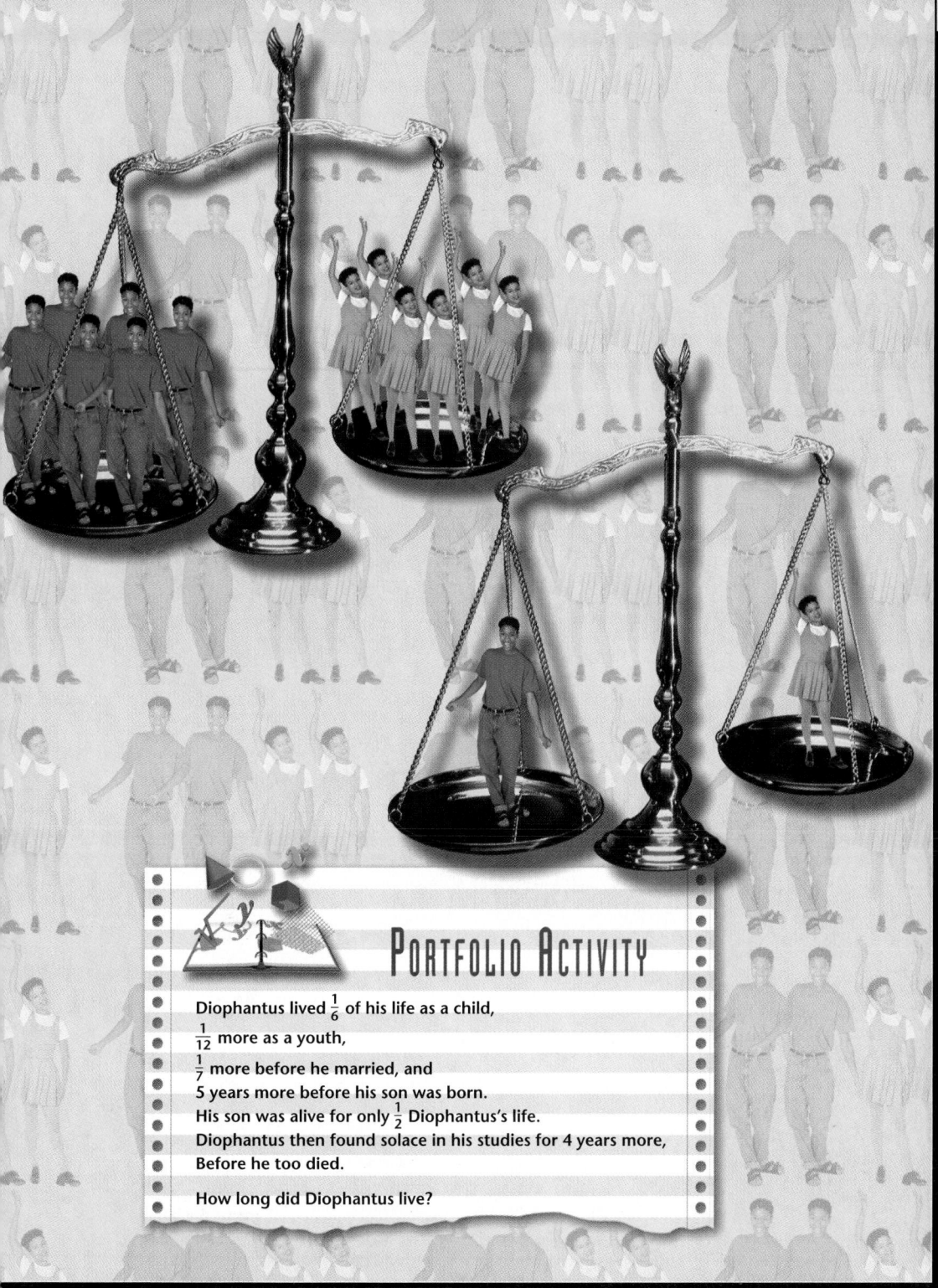

- Use the Reciprocal Property to solve equations.
- Find equivalencies between fractions, decimals, and percents.
- Estimate with percents.
- Solve problems involving percent.
- Write equations that represent problem situations.
- Solve multistep equations.
- Investigate the Multiplication and Division Properties of Inequality.
- Solve inequalities in one variable.
- Solve absolute value equations.
- Solve absolute value inequalities.

PORTFOLIO ACTIVITY

Assign the Portfolio Activity as part of the assignment for Lesson 4.8 on page 198.

Students may be interested to know that this puzzle was actually written over 3500 years ago. Diophantus was an Egyptian mathematician who was interested in linear equations as well as mathematical recreations.

Have students work in small groups. Encourage a variety of approaches to the problem. Some possible approaches might be: write an equation, draw a time line, or use guess-and-check. Each group is to clearly describe what strategy they used and how they arrived at their answer.

PORTFOLIO ACTIVITY

Diophantus lived $\frac{1}{6}$ of his life as a child,
$\frac{1}{12}$ more as a youth,
$\frac{1}{7}$ more before he married, and
5 years more before his son was born.
His son was alive for only $\frac{1}{2}$ Diophantus's life.
Diophantus then found solace in his studies for 4 years more,
Before he too died.

How long did Diophantus live?

ABOUT THE CHAPTER PROJECT

In the Chapter Project on page 199 students will investigate the guess-and-check methods the Egyptians used for solving equations.

PREPARE

Objectives

- Multiply and divide negative and positive integers.
- Define properties of zero.

RESOURCES

- Practice Master   4.1
- Enrichment Master   4.1
- Technology Master   4.1
- Lesson Activity Master   4.1
- Quiz   4.1
- Spanish Resources   4.1

Assessing Prior Knowledge

Find each product or quotient.

1. $32(4)$ **[128]**

2. $\dfrac{144}{6}$ **[24]**

3. $\dfrac{6(7)(4)}{12}$ **[14]**

4. $1.8 \div 90$ **[0.02]**

Aongoing ASSESSMENT

A deposit is a positive transaction. A withdrawal is a negative transaction. A removal of a deposit is a negative transaction. A removal of a withdrawal is a positive transaction.

TEACH

When students follow a class schedule, they are following a pattern. Mathematicians follow patterns to develop the rules for algebraic activities. By following the multiplication patterns on this page, students can learn the rules for multiplication of integers.

LESSON 4.1

Exploring Integer Multiplication and Division

why *Measures of gain and loss in financial affairs, time before and after blastoff, and distance above and below sea level are reported using integers. Calculations using this data often involve multiplication and division. When calculating with positive and negative numbers, it is important to know what happens to the sign of the result.*

Examine what happens to an account when transactions are made. Explain how the signs of the integers relate to the transactions.

Transaction	Representation	Result
Add 3 *deposits* of $100	$(3)(100) = 300$	Increase $300
Add 4 *withdrawals* of $20	$(4)(-20) = -80$	Decrease $80
Remove 5 *deposits* of $50	$(-5)(50) = -250$	Decrease $250
Remove 2 *withdrawals* of $10	$(-2)(-10) = 20$	Increase $20

 Exploration 1 *Multiplying Integers*

1 Complete each pattern.

a.
$(2)(3) = 6$
$(2)(2) = 4$
$(2)(1) = 2$
$(2)(0) = 0$
$(2)(-1) = \underline{?}$
$(2)(-2) = \underline{?}$
$(2)(-3) = \underline{?}$

b.
$(3)(3) = 9$
$(2)(3) = 6$
$(1)(3) = 3$
$(0)(3) = 0$
$(-1)(3) = \underline{?}$
$(-2)(3) = \underline{?}$
$(-3)(3) = \underline{?}$

c.
$(3)(-3) = -9$
$(2)(-3) = -6$
$(1)(-3) = -3$
$(0)(-3) = 0$
$(-1)(-3) = \underline{?}$
$(-2)(-3) = \underline{?}$
$(-3)(-3) = \underline{?}$

 **ALTERNATIVE teaching strategy**

Using Patterns Give students a grid like the one shown. Have students study how the upper-right section has been completed using multiplication. Have students look at the 2 row and note the pattern. Have students continue that pattern to complete the 2 row in the upper-left section. Notice that all answers are negative. Have students complete all the rows of grid.

					X	0	1	2	3	4
					4	0	4	8	12	16
					3	0	3	6	9	12
					2	0	2	4	6	8
					1	0	1	2	3	4
					0	0	0	0	0	0
-4	-3	-2	-1	X	0	1	2	3	4
				-1					
				-2					
				-3					
				-4					

Identify the sign for each statement.
  a. (positive) • (positive) = ____?____
  b. (positive) • (negative) = ____?____
  c. (negative) • (positive) = ____?____
  d. (negative) • (negative) = ____?____

 Find the products.
  a. (12)(4) = _?_   **b.** (13)(− 3) = _?_
  c. (− 81)(6) = _?_   **d.** (− 16)(− 5) = _?_

 To multiply two integers, first multiply their absolute values.
  a. What is the sign of the product when two integers have *like signs*?
  b. What is the sign of the product when two integers have *unlike signs*? ❖

Is the product positive or negative when an even number of negative numbers are multiplied? Is the product positive or negative when an odd number of negative numbers are multiplied? Explain your answers.

Multiplication and division are related.

Multiplication Fact

6 • 5 = 30

Related Division Facts

a. 30 ÷ 5 = 6

b. 30 ÷ 6 = 5

Exploration 2 Dividing Integers

 Complete each multiplication. Examine all the signs.
  a. (8)(7) = _?_   **b.** (5)(− 3) = _?_
  c. (− 4)(2) = _?_   **d.** (− 8)(− 1) = _?_

 Write the related division facts for a, b, c, and d. Compare the signs of the completed multiplications with the signs of the related division facts.

 Identify the sign for each statement.
  a. (positive) ÷ (positive) = ____?____
  b. (positive) ÷ (negative) = ____?____
  c. (negative) ÷ (positive) = ____?____
  d. (negative) ÷ (negative) = ____?____

 Find the quotients.
  a. (48) ÷ (2) = _?_   **b.** (18) ÷ (− 2) = _?_
  c. (− 24) ÷ (− 3) = _?_   **d.** (− 49) ÷ (7) = _?_

 To divide two integers, first divide their absolute values.
  a. What is the sign of the quotient when the two integers have *like signs*?
  b. What is the sign of the quotient when the two integers have *unlike signs*? ❖

interdisciplinary CONNECTION

Economics Economists study changes in the cost of living. For example, a gallon of gasoline cost $1.70 in 1975. Between 1975 and 1995, the average cost of a gallon of gasoline has changed by − $0.025 per year. According to that change, what was the cost of a gallon of gasoline in 1995?   [**$1.20**]

During that same time, the monthly rent in one town has had an average change of + $11 per year. If the average monthly rent was $434 in 1995, what was it in 1975?   [**$214**]

Zero is also an integer. As you know, zero has several special properties.

PROPERTIES OF ZERO

Let a represent any number.

1. **The product of any number and zero is zero.**
   $a \cdot 0 = 0$ and $0 \cdot a = 0$

2. **Zero divided by any nonzero number is zero.**
   $\frac{0}{a} = 0, \, a \neq 0$

3. **A number divided by zero is undefined.**
   That is, NEVER DIVIDE BY ZERO.

APPLICATION

STATISTICS *Connection*

What is the average score for the bowling team? Is a guess of 133 reasonable?

To see how accurate the guess is, find the differences between each actual score and the guess of 133. Add the differences and divide the total by 5.

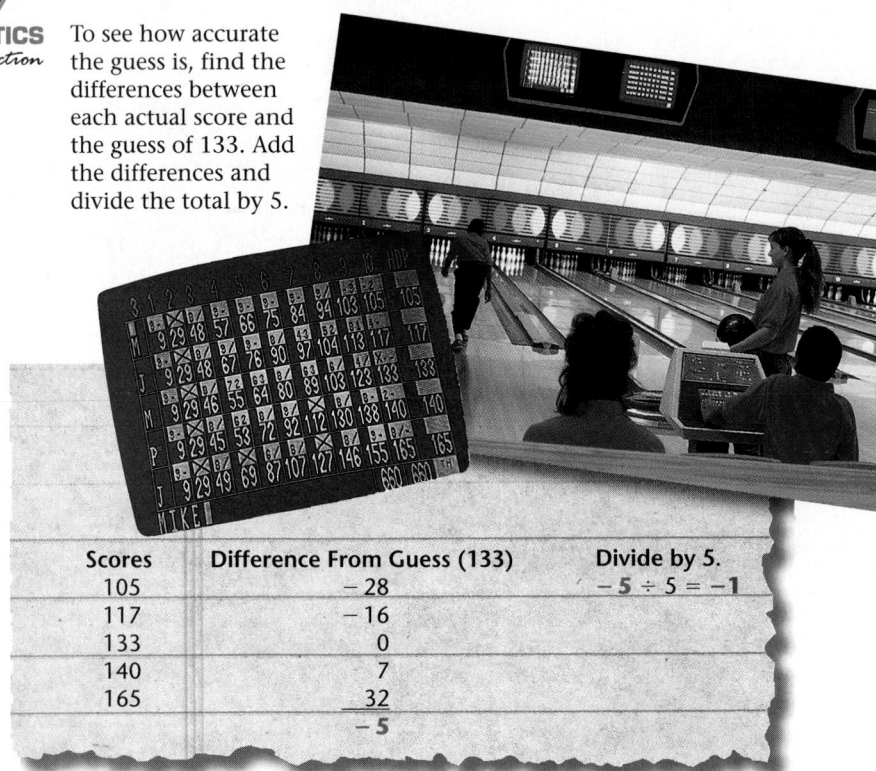

Scores	Difference From Guess (133)	Divide by 5.
105	−28	$-5 \div 5 = -1$
117	−16	
133	0	
140	7	
165	32	
	−5	

The average difference is -1. Add -1 to the guess of 133. The average bowling score is $133 + (-1)$ or 132. The guess is reasonable.

You can check the answer by adding all of the scores and dividing by 5. ❖

ENRICHMENT Have students identify occurrences of integer multiplication and division in the real world. Students should then show examples and the numerical representation.

INCLUSION **strategies** **Using Manipulatives** Find or create several manipulatives such as counters, tiles, number lines, ladders, graph paper, rulers, and others. Arrange the class in groups. One person in the group thinks of a multiplication or division problem. The other students each use a different manipulative to represent the given problem.

EXERCISES & PROBLEMS

Communicate

Discuss what would happen to the balance in a savings account if you make the following transactions. Complete Exercises 1-2.

1. Add 4 withdrawals of $25.

2. Subtract 3 withdrawals of $10.

3. Explain how to write a bank transaction modeling deposits or withdrawals from a bank account for the representation $(3)(-5)$.

4. Discuss how to write two division problems and their answers based on the multiplication problem $(3)(-5) = -15$.

5. Explain how to find the average of 95, 119, 110, 130, 141, and 155 using differences. Is a guess of 120 reasonable?

Practice & Apply

Evaluate.

6. $(-12)(-6)$ 72
7. $(-12) - (-6)$ -6
8. $(-8) - (-2)$ -6
9. $(-6.6) \div (3)$ -2.2
10. $(-0.8)(-2)$ 1.6
11. $(-8) \div (-2)$ 4
12. $(-8) + (-2)$ -10
13. $(-22) \div (-1)$ 22
14. $(-12) + (-6)$ -18
15. $(-1.2) \div (-6)$ 0.2
16. $(-5)[6 + (-6)]$ 0
17. $(7)(4)(-6)$ -168
18. $(-9) - [8 + (-3)]$ -14
19. $(-54)(-115)$ 6210
20. $(-8)(-2)(-3)$ -48
21. $(-225) \div (-5)$ 45
22. $(-47)(23)$ -1081
23. $(-2108) \div (124)$ -17
24. $(-6942) \div (-78)$ 89
25. $(-90) \div (-15)(-3)$ -2
26. $\dfrac{(7)(-1)}{-7}$ 1
27. $\dfrac{(-6)(-12)}{3}$ 24
28. $\dfrac{(-100)(-5)}{-25}$ -20
29. $\dfrac{(-1)(10)(-80)}{-4}$ -200

Tell whether the following statements are true or false.

30. The sum of two negatives is negative. T
31. The difference of two negatives is always negative. F
32. The product of two negatives is negative. F
33. The quotient of two negatives is negative. F
34. The average of a set of negative numbers is negative. T

RETEACHING the lesson

Using Manipulatives
Begin with several counters or tiles with positive on one side and negative on the other. Form one pile of positive counters and another pile of negative counters. Count the tiles in each group. Then separate each pile into groups that contain the same amount. Explain how this represents division. Show how division of a negative number by a lesser negative can be demonstrated by first separating the negative pile of tiles into groups, and then by turning the tiles in each group over to show the opposite side.

Multiplication can be demonstrated by conducting the division in reverse. Separate groups of positive or negative tiles are consolidated into a pile. Negative times a negative is represented by consolidating and turning over the result to show the opposite side.

ASSESS

Selected Answers
Odd-numbered problems 7–53

Assignment Guide
Core 1–5, 6–28 even, 30–58

Core Two-Year 1–54

Technology
A calculator, although not necessary, might be helpful for completing Exercises 6–29.

Error Analysis
In the section with Exercises 6–25, students may become confused and write the sum of two negative numbers as a positive number. Review addition and subtraction of integers to avoid this error.

Technology Master

Technology
4.1 Closure Under Division

You can easily, perhaps mentally, compute the quotient of $\frac{-8}{2}$. Clearly, the quotient is 4 with a negative sign affixed to it. Suppose, however, you choose two integers at random, except for 0 as a divisor. What do you think will happen? Will you always get a quotient that is another integer? To explore the question, you can use a spreadsheet. Keep in mind that a decimal represents an integer when all digits to the right of the decimal point are 0.

In column A of the spreadsheet shown, enter the integers from -20 to 20. In column B, enter a particular integer k, 2. The cells in column C contain the quotients. Cell C2, for example, contains A2/B2.

	A	B	C
	NUMERATOR	DENOMINATOR	QUOTIENT
2	-20	2	
3	-19	2	
4	-18	2	
...	...	2	
18	18	2	
19	19	2	
20	20	2	

Use a spreadsheet to find the quotients when the denominator is the given number k. Then identify which numerators give a quotient that is another integer.

1. $k = 1$ 2. $k = 2$ 3. $k = 3$ 4. $k = 4$

5. $k = 5$ 6. $k = 6$ 7. $k = 7$ 8. $k = 8$

Describe the number pattern for those numerators that give integer quotients for each exercise.

9. Exercise 3

10. Exercise 5

11. Compare the number pattern in Exercise 9 with that in Exercise 10.

Banking Juan opened a savings account with a $20 deposit. He made a total of 4 more deposits of $20 and withdrawals of $10, $15, and $25.

35. What is the total increase in Juan's account *after* his initial deposit? $80

36. What is the total decrease in Juan's account *since* his initial deposit? $50

37. What is the total amount currently in Juan's account? $50

Statistics A running back in football ran 12 times for the following yardage: $-2, -2, -1, 0, 0, 2, 2, 3, 3, 3, 6, 16$.

38. What is his total *net* yardage? HINT: To get the net yardage, subtract the total losses from the total gains. 30

39. What is the *median* (middle value) number of yards? HINT: Order the yardage from smallest to largest to find the middle value. 2

40. What is the *mode* (most frequent value) for the number of yards? 3

41. What is the *mean* (average) number of yards? 2.5

42. What is his longest run? 16

43. Find the average of 95, 104, 87, 120, 102, 100, and 99 using differences. Is a guess of 100 reasonable? 101; yes

44. **Statistics** Five people were asked to estimate the number of beans in a jar. Their errors were $-135, -43, -22, 38,$ and 111. What is the average error? Make a guess and use differences to find the average. -10.2

Look Back

Evaluate. [Lessons 3.1, 3.2]

45. $-2 + (-7)$ -9 **46.** $-8 - (-3)$ -5 **47.** $|-7|$ 7   **48.** $|13|$ 13

Solve the following equations.   [Lesson 3.5]

49. $3x - 5 = 2x + 12$ 17   **50.** $4(y - 7) = 3(y + 2)$ 34   **51.** $2(5y + 7) = 11y + 8$ 6

Solve the following inequalities.   [Lesson 3.6]

52. $3p - 9 > 16 + 2p$ $p > 25$ **53.** $15 - 3x \le 2x$ $x \ge 3$ **54.** $-3c + 4 > -2c$ $c < 4$

Look Beyond

Simplify.

55. $\frac{6x + 12}{2}$ $3x + 6$ **56.** $\frac{-5y - 25}{-5}$ $y + 5$ **57.** $\frac{3w + 24}{-3}$ $-w - 8$ **58.** $\frac{5y - 45}{-5}$ $-y + 9$

Multiplying and Dividing Expressions

Businesses use spreadsheets to
compute employees' earnings.
The spreadsheet automatically multiplies
and divides any integers needed.

D2	=6*C2			
	A	**B**	**C**	**D**
1	Month	Day	Hours	Earnings
2	October	1	3	**18**
3	October	2	2	12
4	October	3	4	24
5	October	4	0	0
6	October	5	6	36

*Richard has a part-time
job making $6 an hour.
His supervisor keeps a
record of his earnings on
a spreadsheet.*

Spreadsheet

The formula **=6*C2** at the top of the spreadsheet computes
the earnings for October 1. The value in Cell D2 is 6 times the
value in Cell C2.

Just as a spreadsheet uses **6*C2** to mean *6 times the value in
Cell C2*, you can write $6 \cdot h$, or simply $6h$, to represent 6 times
a particular quantity h.

Word Expression	**Algebraic Expression**
six times the number of hours	$6h$

The algebraic expression $6h$ can be used to find how much
Richard earned on any day.

To find what Richard earned on October 1, evaluate the
expression $6h$ for h equal to 3. That is, replace h with 3, and
then multiply 6 times 3.

$$6h = 6(3) = 18$$

Richard earned $18 on October 1. What expression represents Richard's
earnings for 5 hours? What are Richard's earnings for 5 hours? for 5.5
hours?

Alternate Example 1

Have students evaluate $5m - 3n$ when $m = 3$ and $n = 5$ and then explain why the answer is zero.

TEACHING *tip*

Allow students to use algebra tiles to model the situations on this page.

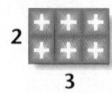

EXAMPLE 1

Evaluate $5m - 3n$ when m is 6 and n is -2.

Solution▸

$5m - 3n$ means five times m minus 3 times n. Replace m by **6** and n by -2, and then simplify.

$$5m - 3n = 5(6) - 3(-2)$$
$$= 30 - (-6)$$
$$= 30 + 6$$
$$= 36 \ \diamondsuit$$

Multiplying Expressions

To model $2 \cdot 3$, use 1-tiles. Think of $2 \cdot 3$ as the area of a rectangle with dimensions 2 by 3.

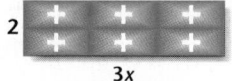

The area of the rectangle is $2 \cdot 3 = 6$.

To model $2 \cdot 3x$, use x-tiles. Think of $2 \cdot 3x$ as the area of a rectangle with dimensions 2 by $3x$.

The area of the rectangle is $2 \cdot 3x = 6x$.

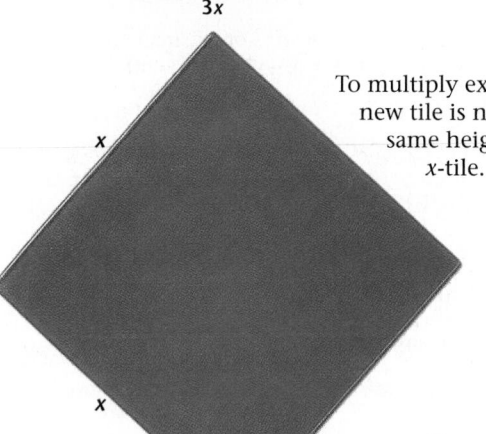

To multiply expressions, each of which contains x, a new tile is needed. Think of a tile that has the same height and length as the longer side of an x-tile. The area of the x^2-tile is x^2.

To model $2x \cdot 3x$, use x^2-tiles. Think of $2x \cdot 3x$ as the area of a rectangle with dimensions $2x$ by $3x$.

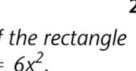

The area of the rectangle is $2x \cdot 3x = 6x^2$.

To check the multiplications, substitute 10 for x.

$2 \cdot 3x = 2 \cdot 3(10) = 60$ $2x \cdot 3x = 2(10) \cdot 3(10) = 600$
$6x = 6(10) = 60$ $6(x^2) = 6(10^2) = 6(100) = 600$

Thus, $2 \cdot 3x = 6x$ checks.   Thus, $2x \cdot 3x = 6x^2$ checks.

ENRICHMENT Provide each group of four students with a bag and six number cubes (three of one color representing positive and three of another color representing negative). Each student, in turn, draws two cubes from the bag, rolls them, and finds the product. If the product is incorrect, another member can challenge. The student giving the correct product scores that many positive points. The student with the greatest total after three rounds is the winner.

INCLUSION **strategies** **Inviting Participation** Helping students to overcome math anxiety is very important. Always provide positive reinforcement. Students should feel that they can ask questions to learn what they want to know, not to show what they do not know.

Exploration — *Multiplying by −1*

Copy the following table.

Scientific Calculator

1. Choose seven values for *n*. Make three positive, three negative, and one 0. Include at least two fractions and two decimals. List your choices in the first row of the table.

2. Use your calculator to multiply each value in the first row by −1, and record the answer in the corresponding space in the second row.

3. What do you notice about the entries in the second row?

4. In relation to the original number, what do you call the numbers that have been multiplied by −1?

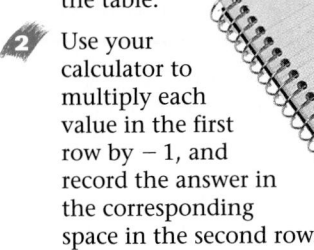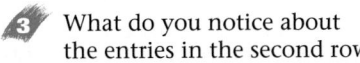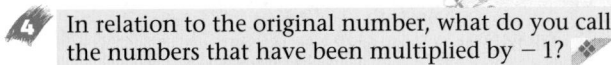

To multiply an expression by −1, find the opposite of the expression.

MULTIPLICATIVE PROPERTY OF −1

For all numbers *a*,

$$-1(a) = -a.$$

The Distributive Property Over Subtraction

Which expression below is equal to $100 - 2(12 - 9)$?

a. $100 - 2(12) - 9$ b. $100 - 2(12) + 9$
c. $100 - 2(12) - 2(9)$ d. $100 - 2(12) + 2(9)$

Since $100 - 2(12) + 2(9) = 94$, and $100 - 2(12 - 9) = 94$, **d** is the correct choice. Multiplication can be distributed over subtraction. The sign of the number being distributed must be distributed to every term.

DISTRIBUTIVE PROPERTY OVER SUBTRACTION

For all numbers *a*, *b*, and *c*,

$$a(b - c) = ab - ac, \text{ and } (b - c)a = ba - ca.$$

EXAMPLE 2

Simplify $-2(5a - 4)$.

Solution ➤

An expression is simplified when no more operations can be performed.

$(-2)(5a - 4) = (-2)(5a) - (-2)(4)$ Distributive Property
$= -10a + 8$ Simplify.   ❖

You can distribute a negative sign across the expressions you are subtracting. That is, if a subtraction sign is in front of an expression, change the sign to addition, and then distribute the negative number to every term of the quantity.

EXAMPLE 3

Simplify $(5x + 3y - 7) - 3(2x - y)$.

Solution ➤

Use the definition of subtraction to write an addition problem.
$(5x + 3y - 7) - 3(2x - y) = (5x + 3y - 7) + (-3)(2x - y)$

Distribute -3 over $(2x - y)$.

$(5x + 3y - 7) + (-3)(2x - y) = 5x + 3y - 7 + [(-3)(2x) - (-3)(y)]$
$= 5x + 3y - 7 + (-6x + 3y)$

Complete the addition.
$= 5x + 3y - 7 - 6x + 3y$
$= (5x - 6x) + (3y + 3y) - 7$
$= -x + 6y - 7$

You can check the answer by substituting 2 for x and 3 for y. ❖

Try This   Simplify $(2b - 5c) - 4(3b + c)$.

Dividing Expressions

The division problem $\frac{2x + 6}{2}$ can be modeled with tiles.

 $= 2x + 6$

$= \dfrac{2x + 6}{2}$

Since each set contains one x-tile and three 1-tiles, $\frac{2x + 6}{2} = x + 3$.

Try This Simplify. **a.** $\dfrac{16x - 4}{-4}$ **b.** $\dfrac{5x + 7}{5}$

Notice that when you divide an expression by a number, *each term* is divided by that same number.

DIVIDING AN EXPRESSION
For all numbers a, b, and c, $c \neq 0$,
$$\dfrac{a + b}{c} = \dfrac{a}{c} + \dfrac{b}{c}, \text{ and } \dfrac{a - b}{c} = \dfrac{a}{c} - \dfrac{b}{c}.$$

EXAMPLE 4

Simplify $\dfrac{8 - 4x}{-4}$.

Solution ➤

Remember that since the quantity $8 - 4x$ is being divided by -4, each term must be divided by -4.

$$\dfrac{8 - 4x}{-4} = \dfrac{8}{-4} - \dfrac{4x}{-4}$$
$$= \dfrac{8}{-4} + \dfrac{-4x}{-4}$$
$$= -2 + x \quad \diamond$$

 Explain why $\dfrac{-4x^2}{-4} = x^2$.

Exercises & Problems

Communicate

Explain how to write an expression that shows Jan's earnings if she earns $6 an hour and works

1. 4 hours. **2.** 2.5 hours. **3.** h hours.

4. Discuss how to evaluate $2j + 3$ when j is 10.

5. Explain how to multiply the expression $3y \cdot 4y$.

6. Discuss how to simplify $-2(4m + 5)$.
What does it mean when an expression is simplified?

7. Describe how to simplify the expression $\dfrac{5p - 15}{-5}$.

ongoing

ASSESSMENT

Try This

a. $-4x + 1$

b. $x + \dfrac{7}{5}$

Alternate Example 4

Simplify $\dfrac{9 - 6x}{-3}$. $[-3 + 2x]$

CRITICAL
Thinking

Students should recognize that the phrase can be rewritten as $\dfrac{4}{4}x^2$ and that any number divided by itself is equal to one.

Assess

Selected Answers
Odd-numbered Exercises 9 −69

Assignment Guide
Core 1 −7, 8 −48 even, 49 −74

Core Two-Year 1 −72

Students often fail to completely distribute a quantity, particularly -1, over all terms of a factor. Alternate Example 3 on page 154 provides many chances for this error. Encourage students to write each step in the distribution to avoid this error.

Practice & Apply

Evaluate 13r for the following values of r.

8. -4 -52 **9.** 1.5 19.5 **10.** 7 91 **11.** $\frac{1}{2}$ $\frac{13}{2}$

Evaluate 2t + 1 for the following values of t.

12. 10 21 **13.** 8.5 18 **14.** -6.2 -11.4 **15.** $\frac{3}{4}$ $\frac{5}{2}$

Simplify the following expressions.

16. $2 \cdot 6x$ $12x$ **17.** $-6x \cdot 2$ $-12x$ **18.** $6x \cdot 2x$ $12x^2$ **19.** $-66x \div 2$ $-33x$

20. $12x \cdot 3x$ $36x^2$ **21.** $-2(6x + 3)$ $-12x - 6$ **22.** $-1.2x \cdot 3x$ $-3.6x^2$ **23.** $-12x \div 3$ $-4x$

24. $7x - (3 - x)$ $8x - 3$ **25.** $-3(7x - 3)$ $-21x + 9$ **26.** $-2(4x - 1) - 8x + 2$ **27.** $3x \cdot 5 + 2x \cdot 2$ $19x$

28. $-2 \cdot 8x$ $-16x$ **29.** $8x \cdot 2x$ $16x^2$ **30.** $-8x \div 2$ $-4x$ **31.** $-21x \cdot 3x$ $-63x^2$

32. $2.1x \cdot 3x$ $6.3x^2$ **33.** $-21x \div 7$ $-3x$ **34.** $8x - (2 - 5x)$ $13x - 2$ **35.** $8(x + 1) - (2 - 5x)$ $13x + 6$

36. $(3x + 2y - 9) - 2(x - y)$ $x + 4y - 9$ **37.** $4(x - 5y) - (2x - 2y)$ $2x - 18y$ **38.** $\frac{5w + 15}{-5} - w - 3$

39. $8(x - y) + 5(3x - 3y)$ $23x - 23y$ **40.** $(3x + 5y - 9) - 7(x - y)$ $-4x + 12y - 9$ **41.** $\frac{8 + 16w}{8}$ $1 + 2w$

42. $\frac{-90x + 2.7}{-9}$ $10x - 0.3$ **43.** $-(x - y) - 4(x - y + 9)$ $-5x + 5y - 36$ **44.** $9(2x + y) - 3(3x + 3y)$ $9x$

45. $\frac{11 - 33y}{11}$ $1 - 3y$ **46.** $\frac{-10x + 35}{5}$ $-2x + 7$ **47.** $-6(4x - y) - 4(2x - 3y)$ $-32x + 18y$

48. A computer program shows the formula A=$-5*$B+B. What is the value of A when B is 3? -12

Nicole has a part-time job 3 days after school and on Saturdays.

Wages If Nicole earns $5.25 per hour, find her earnings for the following 4 days.

49. 4 hours on Friday $21   **50.** 6.5 hours on Saturday $34.13

51. h hours on Monday $5.25h$ **52.** $(h + 1)$ hours on Tuesday $5.25(h + 1)$

53. Write an expression in simplified form that shows the total number of hours Nicole worked for the 4 days.

54. **Geometry** The formula for area is $A = lw$. Find the area of the rectangle by evaluating the formula for the given length and width. 24 cm^2

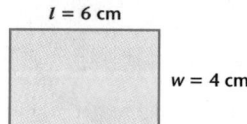
$l = 6$ cm $w = 4$ cm

55. **Geometry** Evaluate the volume formula $V = lwh$ of a rectangular prism with the given dimensions. 96 in.^{3}

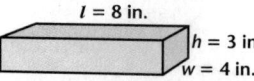
$l = 8$ in. $h = 3$ in. $w = 4$ in.

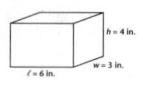
53. $11.5 + 2h$

Charlotte is a plumber who charges $35 an hour, plus a fixed service fee of $20 per job.

Small Business How much does Charlotte charge for a job that takes

56. 1 hour?
$55

57. 3 hours?
$125

58. h hours?
$35h + 20$

On Sundays, Charlotte doubles her charges. How much does she charge for a job that takes

59. 1 hour?
$110

60. 3 hours?
$250

61. h hours?
$70h + 40$

Look Back

Add the like terms. [Lesson 3.3]

62. $(3x + 2y) + (3x - 2y) - (3x - 2y)$ $3x + 2y$

63. $(x^2 + 2y + 4) - (2x - 3y - 2)$ $x^2 - 2x + 5y + 6$

Solve each inequality. [Lesson 3.6]

64. $x - 4 \leq 3$
$x \leq 7$

65. $x + 5 \geq -2$
$x \geq -7$

66. $x < -3 + 4$
$x < 1$

67. Find the average of 2, 3, 4, 5, and 6. [Lesson 4.1] 4

Simplify. [Lesson 4.2]

68. $2(-3 + c)$
$-6 + 2c$

69. $-6(-4 + x)$
$24 - 6x$

70. $4(2x + 4)$
$8x + 16$

Look Beyond

Simplify.

71. $\dfrac{2x^2 + 4x}{4x}$ $\dfrac{x}{2} + 1$

72. $\dfrac{5y^2 - 15y}{-5y}$ $-y + 3$

73. In a bowling alley, the pins are set up as shown. Move exactly three of the pins to form a triangle exactly like the original one, but pointing away from you.

74. Sixty-four teams are picked to play in the NCAA tournament. Each team plays until it loses one game. How many games are played in order to have a champion? 63

73.

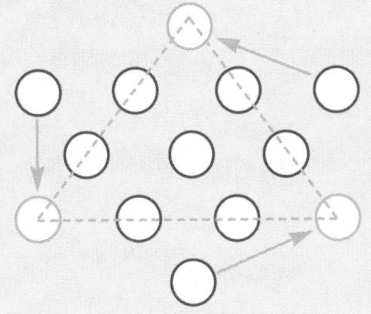

Technology

To solve Exercises 56-60, students could develop and use a simple spreadsheet program.

Alternative ASSESSMENT

Performance Assessment

Have students write a paragraph explaining how using the Distributive Property and dividing an expression are alike. [Dividing an expression by a quantity is the same as multiplying by the reciprocal of that quantity.]

Look Beyond

Exercises 71 and 72 involve simplifying quadratic expressions. Exercises 73-74 encourage students to apply logical reasoning to solve a problem. Other strategies might include using models.

Technology Master

Technology
4.2 The Expression $nx(nx - 1)$

The expression $nx(nx - 1)$ gives a product for each value of n. You can assume that n is an integer. What happens if the expression is evaluated for a set of x-values and a particular value of n? To explore the question, a spreadsheet will be valuable. Let x take on each value in the set $\{-10, -9, -8, ..., 8, 9, 10\}$. Column B contains $n = 1$ and column D contains the differences between successive numbers in column C.

	A	B	C	D
1	X	N=	VALUES	DIFFERENCES
2	-10	1		
3	-9	1		
4	-8	1		
...	...	1		
19	8	1		
20	9	1		
21	10	1		

Write an expression using spreadsheet terminology for the contents of each cell.

1. Cell C2 _____ 2. Cell D3 _____

3. Describe the number pattern in column D. Use a verbal description or a pictorial description.

4. Add column E to include differences between successive numbers in column D. Describe the pattern.

5. Let $n = 2$. Describe the number pattern in column D. Use a verbal description or a pictorial description.

6. Describe the pattern found in column E. _____

7. What do you think will happen if $n = 3$? Justify your conclusion.

8. Do the values in column D for $n = 2$ have a relationship with those in column D when $n = 3$? Explain.

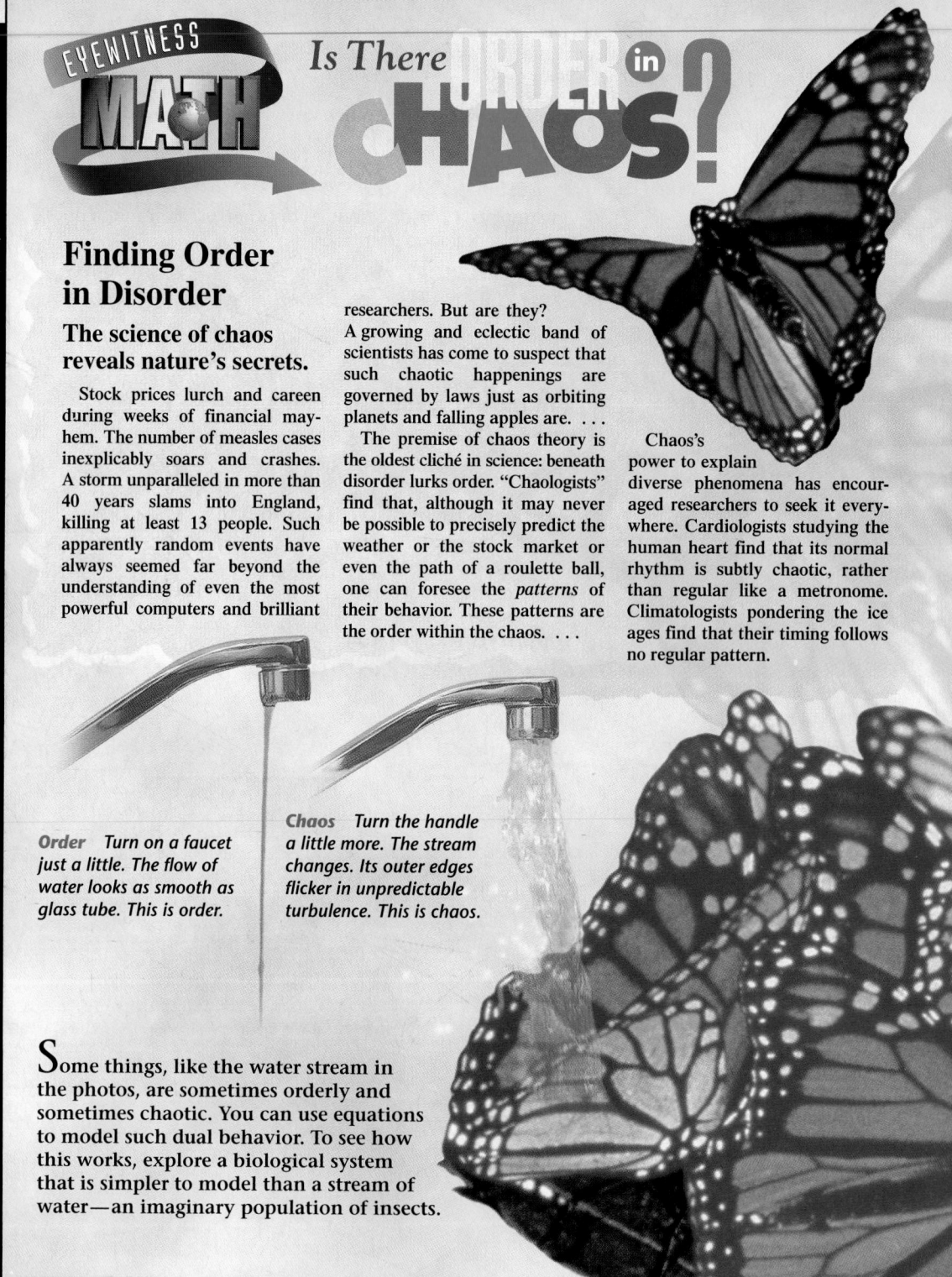

EYEWITNESS MATH
Is There ORDER in CHAOS?

Finding Order in Disorder

The science of chaos reveals nature's secrets.

Stock prices lurch and career during weeks of financial mayhem. The number of measles cases inexplicably soars and crashes. A storm unparalleled in more than 40 years slams into England, killing at least 13 people. Such apparently random events have always seemed far beyond the understanding of even the most powerful computers and brilliant researchers. But are they?

A growing and eclectic band of scientists has come to suspect that such chaotic happenings are governed by laws just as orbiting planets and falling apples are. . . .

The premise of chaos theory is the oldest cliché in science: beneath disorder lurks order. "Chaologists" find that, although it may never be possible to precisely predict the weather or the stock market or even the path of a roulette ball, one can foresee the *patterns* of their behavior. These patterns are the order within the chaos. . . .

Chaos's power to explain diverse phenomena has encouraged researchers to seek it everywhere. Cardiologists studying the human heart find that its normal rhythm is subtly chaotic, rather than regular like a metronome. Climatologists pondering the ice ages find that their timing follows no regular pattern.

Order   Turn on a faucet just a little. The flow of water looks as smooth as glass tube. This is order.

Chaos   Turn the handle a little more. The stream changes. Its outer edges flicker in unpredictable turbulence. This is chaos.

Some things, like the water stream in the photos, are sometimes orderly and sometimes chaotic. You can use equations to model such dual behavior. To see how this works, explore a biological system that is simpler to model than a stream of water—an imaginary population of insects.

The insects hatch in the spring and die after laying eggs in the fall. The population each year depends on the population of the prior year. The more insects there are, the more eggs there are to hatch next spring. Other factors, such as limited food supply, keep the population from becoming infinitely large.

If you know the size of the insect population one year, you can find the size the next year by using the following equation.

$$P_2 = r \cdot P_1(1 - P_1)$$

P_1 is the size of the population in year 1, P_2 is the size of the population in year 2, and r is some constant.

To find the population in year 3, use $P_3 = r \cdot P_2(1 - P_2)$.

What equation would you use for year 4?

In these equations, all values of P must stay in the range 0 to 1. A value of 0 represents no insects, and a value of 1 represents the final population size.

Now you are ready to see how these equations can model behavior that is sometimes orderly and predictable and sometimes chaotic and unpredictable.

Cooperative Learning

1. Copy and complete the chart for $r = 2$ and $r = 4$.

   a. To start, choose any value for P_1 where $0 < P_1 < 1$. Round each answer to three decimal places.

   b. Do you see a pattern in your results for each value of r? Explain.

   c. Compare your results with the results of groups that started with a different value for P_1.

   d. For which values of r can you make a good prediction about the size of the population in year 6 even if your year 1 data were a little off?

2. Copy and complete the chart for $r = 4$. This time start with a value for P_1 that is 0.001 less or 0.001 greater than the value you originally used.

3. In chaos, a tiny difference at the beginning can make a big difference later on. For which value of r did the equation make chaos? Explain.

$r = 2$	
P_1	
P_2	
P_3	
P_4	
P_5	
P_6	

$r = 4$	
P_1	
P_2	
P_3	
P_4	
P_5	
P_6	
P_7	
P_8	

Cooperative Learning

Activity 1 Work in pairs. Have each pair of students choose a different value for P_1. For better comparison later in part 1c, make sure there is a wide range of values in the class (for example: 0.1, 0.4, 0.75, 0.9). Let students perform their calculations and then discuss questions 1b-1d before going on to part 2.

Activities 2 and 3 Point out that in these activities each pair of students will do 2 sets of calculations (with slightly different values for P_1).

Discuss

You may wish to explore how the population equation behaves when other values of r are used. [It turns out that different ranges of r produce different behavior patterns.]

1a. When r is 2 regardless of the values used for P_1, the population will eventually reach 0.5. When r is 4, the population never stabilizes.

1b. There is a pattern when r is 2, but no pattern when r is 4.

1c. All groups should get the same general result, no matter what value between 0 and 1 is chosen for P_1.

1d. When r is 2, you can make a good prediction.

2-3. When r is 4 even miniscule changes in P_1, cause varied results.

PREPARE

Objectives

- Solve multiplication and division equations.

RESOURCES

- Practice Master **4.3**
- Enrichment Master **4.3**
- Technology Master **4.3**
- Lesson Activity Master **4.3**
- Quiz **4.3**
- Spanish Resources **4.3**

Assessing Prior Knowledge

Tell whether each statement is true or false.

1. $a + 3 = a - 3$ **[false]**

2. $17 - b = b - 17$ **[false]**

3. $c + 17 = 17 + c$ **[true]**

4. $(16)(4) = (64)(1)$ **[true]**

5. $\left(\dfrac{11}{8}\right)(8) = 11$ **[true]**

6. $\left(\dfrac{a}{b}\right)b = a$ **[true]**

TEACH

 Students should find specific mathematical problems that a carpenter or other professional might have to solve. Encourage students to make a display of such problems.

---

why *Carpenters use problem-solving strategies, estimation, and mental computation in their work. They write and solve equations to calculate measurements and to compute the cost of materials and labor.*

Alice needs to purchase 20, 8-foot reinforcement beams. What is the cost per foot?

Multiplication Equations

Carpentry To find the cost per foot, you can draw a picture or use an equation.

Draw a picture of a beam divided into 8, 1-foot pieces.

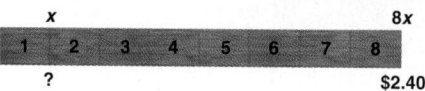

To find the cost of one piece, divide $2.40 by 8. $\dfrac{2.40}{8} = 0.30$

The cost per foot is $0.30.

If the cost per foot is represented by x, the equation that models the problem is $8x = 2.40$. To find x, divide both sides of the equation by 8.

$$8x = 2.40 \qquad \text{Given}$$
$$\frac{8x}{8} = \frac{2.40}{8} \qquad \text{Divide both sides by 8.}$$
$$x = 0.30 \qquad \text{Simplify.}$$

Each piece costs $2.40 \div 8$, or $0.30.

Technology Have students name actual situations in which knowing the unit price might be helpful. Use newspaper advertisements to find different costs of different sizes of products. Have students use a calculator to determine the unit price.

For example, 8 oz can for $1.99 (unit cost $0.25) and 20 oz can for $4.99 (unit cost $0.25). Point out that in real life, as in this example, the larger size is not always the most economical. Have students do research to see if they can find an example in their local stores.

You can solve multiplication equations by using division.

DIVISION PROPERTY OF EQUALITY
If each side of an equation is divided by equal nonzero numbers, the results are equal.

EXAMPLE 1

GEOMETRY
Connection

In a regular pentagon all of the angles have equal measures. If the sum of the measures is 540°, find the measure of each angle.

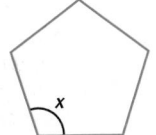

Regular Pentagon

Solution ➤

Organize
Find the information needed to write an equation. The total number of angles is 5. The sum of the measures of the angles is 540°.

Write
Let x be the measure of each angle.
Then $5x$ is the sum of the measures of all 5 angles.
Thus, $5x = 540$.

Solve

$5x = 540$	Given
$\dfrac{5x}{5} = \dfrac{540}{5}$	Division Property of Equality
$x = 108$	Simplify.

Check
$5(108) \stackrel{?}{=} 540$

$540 = 540$ True

Each angle has a measure of 108°. ❖

EXAMPLE 2

GEOMETRY
Connection

The formula for the circumference of a circle is $C = \pi d$. Solve for d.

Solution ➤

Divide both sides of the equation by π.

$C = \pi d$	Given
$\dfrac{C}{\pi} = \dfrac{\pi d}{\pi}$	Division Property of Equality
$\dfrac{C}{\pi} = d$ or $d = \dfrac{C}{\pi}$	Simplify. ❖

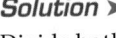

$C = \pi d$

Try This   The area of a rectangle is $A = lw$. Solve for w.

Have students work in small groups. Each group should do the following:

a. Brainstorm to come up with as many different formulas as possible.

b. Choose three or four formulas and explain how to solve for each of the variables.

c. Write a real-world problem that can be solved using one of their formulas.

You may also want groups to exchange and solve the problems.

Alternate Example 3

Solve the equation.

$$\frac{x}{2.4} = 0.25 \quad [\mathbf{0.6}]$$

TEACHING *tip*

Students should note that the properties apply to all rational numbers, not just integers.

Division Equations

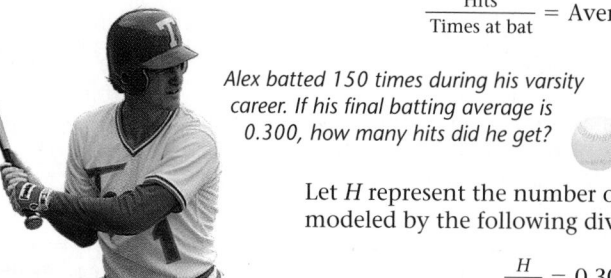

Sports Division is used to find batting averages. The number of hits is divided by the number of times at bat. To find the number of hits, a division equation can be used as a model.

$$\frac{\text{Hits}}{\text{Times at bat}} = \text{Average}$$

Alex batted 150 times during his varsity career. If his final batting average is 0.300, how many hits did he get?

Let H represent the number of hits. The problem can be modeled by the following division equation.

$$\frac{H}{150} = 0.300$$

To solve the division equation, multiply both sides of the equation by the divisor of H.

$$150\left(\frac{H}{150}\right) = 150(0.300)$$
$$H = 45$$

Alex had 45 hits in his varsity career.

The solution to the batting average problem leads to another important property of equality.

MULTIPLICATION PROPERTY OF EQUALITY
If each side of an equation is multiplied by equal numbers, the results are equal.

EXAMPLE 3

Solve the equation $\frac{x}{-3} = -6$.

Solution ➤

Use the Multiplication Property of Equality.

$$\frac{x}{-3} = -6 \qquad \text{Given}$$
$$-3\left(\frac{x}{-3}\right) = -3(-6) \qquad \text{Multiplication Property of Equality}$$
$$x = 18 \qquad \text{Simplify.}$$

Check the solution.

$$\frac{18}{-3} \overset{?}{=} -6$$
$$-6 = -6 \qquad \text{True} ❖$$

RETEACHING
t h e
l e s s o n

Have students review the relationship of multiplication and division. Then have students solve simple equations such as $2x = 6$. Ask how they found the answer. Have them apply the same

techniques to solve equations with larger numbers, decimals, and other rational numbers such as the following:

a. $324n = 81,324 \quad [n = 251]$

b. $-\frac{2}{3}p = -12 \quad [p = 18]$

CRITICAL
Thinking

Compare the Division Property of Equality with the Multiplication Property of Equality. How are they alike? different?

You now have tools to solve addition, subtraction, multiplication, or division equations. Make sure you apply the correct property to solve an equation.

EXAMPLE 4

Solve. **A** $y + \frac{1}{3} = 6$ **B** $\frac{x}{0.7} = -56$

Solution ➤

A Since this is an addition equation, subtract $\frac{1}{3}$ from each side.

$$y + \frac{1}{3} = 6$$

$$\left(y + \frac{1}{3}\right) - \frac{1}{3} = 6 - \frac{1}{3}$$

$$y = 5\frac{2}{3}$$

Check $5\frac{2}{3} + \frac{1}{3} \stackrel{?}{=} 6$

$6 = 6$ True

B This is a division equation. Multiply each side by 0.7.

$$\frac{x}{0.7} = -56$$

$$0.7\left(\frac{x}{0.7}\right) = (0.7)(-56)$$

$$x = -39.2$$

Check $\frac{-39.2}{0.7} \stackrel{?}{=} -56$

$-56 = -56$ True ❖

Why is it important to identify what type of equation you are solving before you apply one of the properties of equality?

Try This   Solve.   **a.** $d - \frac{3}{4} = 8$ **b.** $\frac{m}{-1.5} = -6$

EXAMPLE 5

Physics   In science, **density** is the mass of a material divided by its volume. If the formula for density is $\frac{m}{v} = d$, find the mass, m, in terms of the density, d, and the volume, v.

Solution ➤

Multiply each side of the equation by v.

$$\frac{m}{v} = d$$ Given

$$v\left(\frac{m}{v}\right) = v(d)$$ Multiplication Property of Equality

$$m = vd$$ Simplify. ❖

Try This   Solve $m = \frac{F}{a}$ for F.

CRITICAL
Thinking

Both sides of the equation are equal if both sides are multiplied or divided by equal numbers. Furthermore, multiplication and division are opposite operations. Students should note that to solve a multiplication equation, you use division, and to solve a division equation, you use multiplication.

Alternate Example 4

Solve.

a. $2m + 1 = 0.5$

$[m = -0.25]$

b. $\frac{1}{4} - \frac{1}{2}n = 1\frac{1}{4}$

$[n = -2]$

Aongoing
SSESSMENT

You must know the type of equation you are solving in order to choose and apply the correct property of equality.

Aongoing
SSESSMENT

Try This

a. $d = 8\frac{3}{4}$

b. $m = 9$

Alternate Example 5

In the formula $\frac{m}{v} = d$, find the formula for volume, v, in terms of the density, d, and the mass, m. $\left[v = \frac{m}{d}\right]$

Aongoing
SSESSMENT

Try This

$[F = ma]$

SUMMARY FOR SOLVING EQUATIONS	
Type of Equation	**Operation**
Addition	Subtract equal amounts from each side.
Subtraction	Add equal amounts to each side.
Multiplication	Divide each side by equal nonzero amounts.
Division	Multiply each side by equal amounts.

EXERCISES & PROBLEMS

Communicate

1. Make up a real world situation modeled by each equation.
   a. $5x = 100$
   b. $\frac{x}{2} = 10$

Explain how to solve for x.

2. $592x = 812$

3. $x - 246 = 528$

4. $5x = \frac{1}{10}$

5. $x + 10 = 5$

6. **Geometry** From the formula for the circumference of a circle, explain how to find the radius, r, in terms of the circumference, C, and π.

Write a real-world problem modeled by each equation.

7. $4x = 40$

8. $\frac{x}{4} = 20$

9. $2.50p = 15$

10. $\frac{x}{20} = 4$

Practice & Apply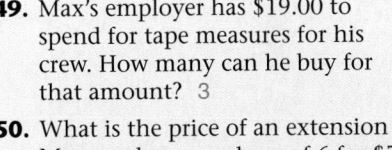

Give the property needed to solve each equation. Do not solve.

11. $\frac{y}{3} = -13$ M

12. $\frac{x}{27} = -26$ M

13. $x - \frac{1}{3} = 2$ A

14. $7x = 56$ D

15. $\frac{b}{-9} = 6$ M

16. $-12y = 84$ D

17. $x + \frac{2}{3} = 2$ S

18. $777x = -888$ D

19. $5.6v = 7$ D

20. $7 = -56w$ D

21. $\frac{x}{-7} = -1.4$ M

22. $3x = 2$ D

Solve each equation.

23. $888x = 777$ $\frac{7}{8}$

24. $y + \frac{3}{4} = 12$ $11\frac{1}{4}$

25. $x - 888 = 777$ 1665

26. $-3f = 15$ -5

27. $\frac{y}{7} = -8$ -56

28. $\frac{x}{0.5} = 6$ 3

29. $w + 0 = -22$ -22

30. $-4x = -3228$ 807

31. $4b = -15$ $-\frac{15}{4}$

32. $\frac{p}{111} = -10$ -1110

33. $\frac{s}{-1} = -40$ 40

34. $c + 7 = 63$ 56

35. $r - \frac{1}{5} = 2$ $2\frac{1}{5}$

36. $2a = 13$ $\frac{13}{2}$

37. $\frac{x}{10} = -1.9$ -19

38. $d - 7 = 35$ 42

39. $\frac{w}{-1.2} = 10$ -12

40. $7e = -14$ -2

41. $\frac{m}{-9} = 0$ 0

42. $\frac{b}{15} = 1$ 15

43. $0.55x = 0.55$ 1

44. $\frac{p}{-9} = 0.9$ -8.1

45. $-3x = -4215$ 1405

46. $p + 2300 = 890$ -1410

Carpentry Max is apprenticing as a carpenter during his summer vacation. Write and solve an equation to describe each situation in Exercises 47–51.

Masking Tape $1.15
Hammers $7.99
Tape Measure $4.95

47. How many rolls of masking tape can Max buy for $6.00? 5

48. Max finds masking tape in packages of 4 rolls for $4.32. What is the cost of one roll in the package? Is the cost of a single roll of masking tape more than one packaged roll? $108; yes

49. Max's employer has $19.00 to spend for tape measures for his crew. How many can he buy for that amount? 3

50. What is the price of an extension cord if Max can buy a package of 6 for $7.26? $1.21

51. Max finds that a package of 4 AA-type batteries costs $2.52. What is the price for each battery? $0.63

52. Travel Natalie's family wants to make a 400-mile trip in 10 hours, but they figure that they will stop a total of 2 hours along the way. What speed must they average for the trip? 50

53. Travel Derrick averaged 50 miles per hour during three hours of driving, but he still has 75 miles to go. How many miles does he drive in all? 225

54. Maria wants to drive 320 miles in 8 hours. What speed should she average for the trip? 40

Solve each formula for the variable indicated.

55. $r = \dfrac{d}{t}$ for d
$rt = d$

56. $A = bh$ for b
$\dfrac{A}{h} = b$

57. $V = lwh$ for w
$\dfrac{V}{lh} = w$

Geometry In a regular polygon with n sides, the measure, m, of each interior angle is given by the formula $m = \dfrac{180(n-2)}{n}$. Find the measure of an interior angle of the following regular polygons. HINT: Use the number of sides of each polygon for n.

58. equilateral triangle 60°
59. square 90°
60. pentagon 108°
61. octagon 135°

62. What effect does the *number* of angles in each regular polygon have on the *size* of the angles? The greater the number of angles, the larger the size of the angle.

63. **Geometry** Solve the formula for the area of a rhombus for the variable n. $n = \dfrac{2A}{m}$

64. The diagonals of the rhombus are m and n. If the area is 21 square inches and the longer diagonal, m, is 7 inches, what is the length of the other diagonal? 6

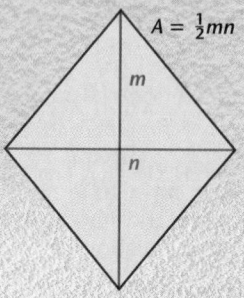

$A = \tfrac{1}{2}mn$

m

n

Look Back

65. Write a sequence of 5 terms, starting with 3 and doubling to get the next number. **[Lesson 1.1]** 3, 6, 12, 24, 48

Evaluate the following if $a = 3$, $b = 2$, and $c = 0$. [Lesson 1.4]

66. $\dfrac{a + b \cdot c}{3}$ 1

67. $\dfrac{a \cdot b}{b + c}$ 3

68. $\dfrac{a + b}{b + c}$ $\dfrac{5}{2}$

69. Place parentheses to make $30 \cdot 7 - 4 \cdot 10 \div 2 \div 2 = 200$ true.
[Lesson 1.4] $30 \cdot 7 - (4 \cdot 10 \div 2 \div 2) = 200$

Find the absolute value for each number. [Lesson 3.1]

70. $|0|$ 0

71. $|-5|$ 5

72. $\left|-\dfrac{1}{6}\right|$ $\dfrac{1}{6}$

73. $|9|$ 9

Look Beyond

Solve each equation.

74. $3y = 10 + 5y$ -5

75. $4x + \dfrac{1}{2} = 5x$ $\dfrac{1}{2}$

76. $x + 4 = 3x - 2$ 3

77. $6x - 2 = 14 - 2x$ 2

Look Beyond

Exercises 74–77 anticipate solving equations with variables on each side of the equality. Allow students to brainstorm and exchange ideas on how to solve these exercises. Remind students to always check their solutions in the original equations.

LESSON 4.4 Rational Numbers

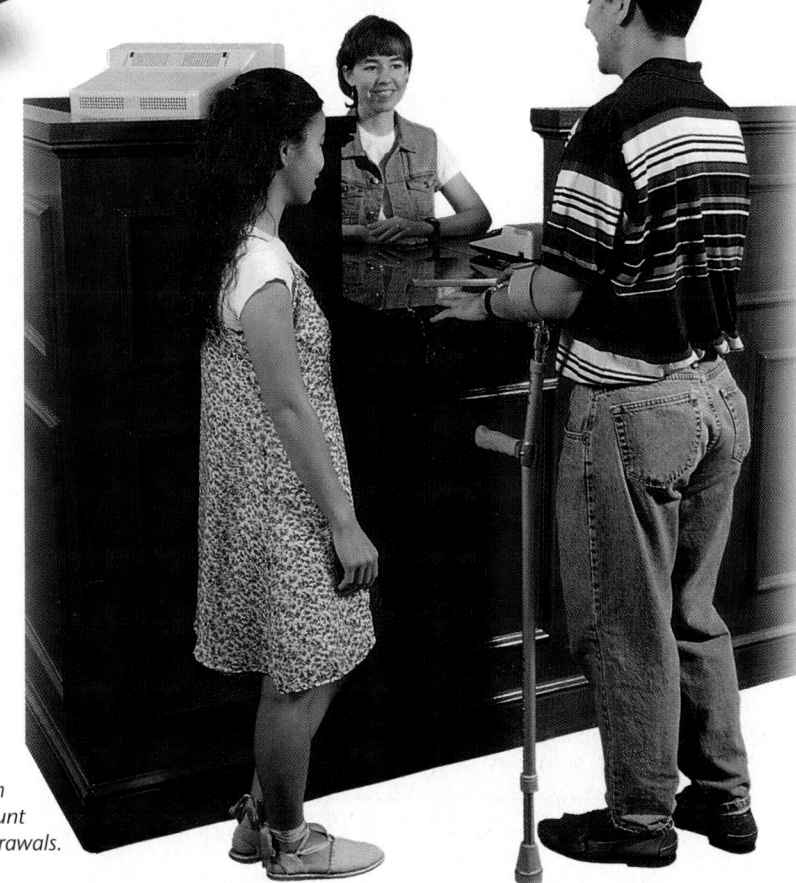

Geno decides to borrow $25 from his savings account in 2 equal withdrawals.

 *In Lesson 4.3 you solved equations involving fractions and decimals. Although fractions and decimals are not integers, the properties for solving equations still apply.*

Banking How can Geno use negative numbers to express the amount of withdrawal?

Since Geno is withdrawing $25, use -25 to represent the total withdrawal.

Let x represent each withdrawal. Then $2x$ represents the 2 equal withdrawals. The equation to model the problem is

$$2x = -25.$$

Apply the Division Property of Equality.

$$\frac{2x}{2} = \frac{-25}{2}$$

$$x = -12\frac{1}{2}$$

The result is $x = -12\frac{1}{2}$ or -12.50. Geno should withdraw $12.50 each time.

ALTERNATIVE teaching strategy

Using Manipulatives Have students use rulers to model positive rational numbers by locating rational numbers and modeling addition and subtraction on the ruler.

Have students place another ruler upside down and to the left of the original ruler. The upside-down ruler indicates negative rational numbers. Now have students locate numbers and perform operations using this "number line."

PREPARE

Objectives

- Define rational numbers and display them on a number line.
- Find reciprocals of rational numbers.
- Use the Reciprocal Property to solve equations.

RESOURCES

- Practice Master   4.4
- Enrichment Master   4.4
- Technology Master   4.4
- Lesson Activity Master   4.4
- Quiz   4.4
- Spanish Resources   4.4

Assessing Prior Knowledge

Solve each equation.

1. $6k = -30$ $[k = -5]$
2. $25n = 25$ $[n = 1]$
3. $\frac{r}{8} = 6$ $[r = 48]$
4. $3x = 1$ $\left[x = \frac{1}{3}\right]$
5. $\frac{w}{-3} = 1$ $[w = -3]$

TEACH

 In real life, fractions can be both positive and negative. Students need to learn how to deal with negative fractions for amounts of money, temperatures, distances, or other occasions when amounts are not exact whole numbers.

Zero and -2 are integers by definition. Since both can be expressed as the ratio of two integers, both are also rational numbers.

For instance, $0 = \frac{0}{-3}$ and $-2 = \frac{-4}{2}$.

Alternate Example 1

The value of a share of stock decreased by $\frac{1}{4}$ of a point on Monday morning and increased by $1\frac{1}{8}$ points on Monday afternoon. What was the net change for that share of stock on that Monday?

$$\left[\frac{7}{8} \text{ point gain} \right]$$

CRITICAL
Thinking

The shape of the graphs are the same. The graph for $y = \frac{1}{x}$ with positive values for x lies in the first quadrant. The graph for $y = \frac{1}{x}$ with negative values for x lies in the third quadrant.

Numbers such as -12.5, $-12\frac{1}{2}$, or $\frac{-25}{2}$ are called negative **rational numbers.** Other examples of rational numbers include $\frac{1}{2}$, $-\frac{1}{2}$, 0, and 2. A rational number is a number that can be expressed as the ratio of two integers with 0 excluded from the denominator.

Rational numbers can be graphed on a number line.

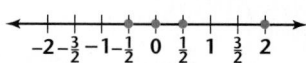

CRITICAL
Thinking

How can you show that the integers 0 and -2 are also rational numbers?

 **EXAMPLE 1**

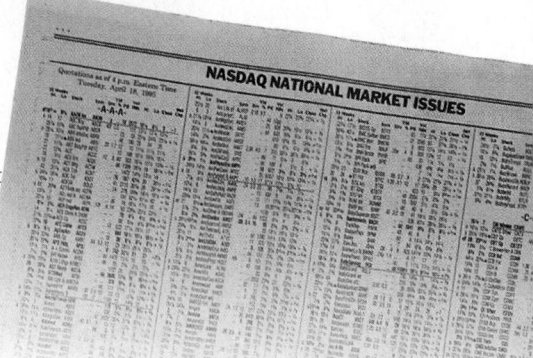

Investment A share of stock increased by $\frac{1}{4}$ of a point on Monday and decreased by $\frac{5}{8}$ of a point on Tuesday. What is the net effect on the stock?

Solution A ➤

Use a number line to model this problem.

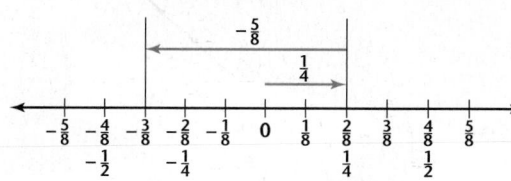

Solution B ➤

Use a common denominator to solve this problem.

$$\frac{1}{4} + \left(-\frac{5}{8} \right) = \frac{2}{8} + \left(-\frac{5}{8} \right)$$

$$= \frac{2 + (-5)}{8}$$

$$= -\frac{3}{8}$$

The net effect is a decrease of $\frac{3}{8}$ of a point. ❖

The set of rational numbers shares all the properties of the set of integers. However, there is one additional rational number property not shared by the set of integers.

RECIPROCAL PROPERTY

For any nonzero number r, there is a number $\frac{1}{r}$ such that

$$r \cdot \frac{1}{r} = 1.$$

CRITICAL
Thinking

In Lesson 2.3 you graphed the reciprocal function $y = \frac{1}{x}$ for positive values of x. Now draw the graph for the negative values of x. How do the two graphs compare?

---

interdisciplinary
CONNECTION

Home Economics Have students look in various cookbooks to find how rational numbers are used in them. Challenge students to make up real world problems that involve cooking. Choose a recipe and notice how many servings it makes. How much of each ingredient would you need if you wanted to make enough of this recipe to serve your entire class? your entire school?

ENRICHMENT

Have students study the patterns below:

$$\frac{1}{8} + \frac{1}{8} = \frac{1}{4}; \frac{1}{10} + \frac{1}{10} = \frac{1}{5}; \frac{1}{6} + \frac{1}{6} = \frac{1}{3}$$

a. Describe the pattern. [**The numerators are all one. The denominator of the sum is one-half the denominator of the addends.**]

EXAMPLE 2

Write the reciprocal.

A $\frac{2}{3}$ **B** 7   **C** $-\frac{4}{5}$ **D** $-1\frac{1}{2}$

Solution ➤

A The reciprocal of $\frac{2}{3}$ is $\frac{3}{2}$ $\left(\text{or } 1\frac{1}{2}\right)$ because $\frac{2}{3} \cdot \frac{3}{2} = \frac{6}{6} = 1$.

B The reciprocal of 7 is $\frac{1}{7}$ because $7 \cdot \frac{1}{7} = \frac{7}{7} = 1$.

C The reciprocal of $-\frac{4}{5}$ is $-\frac{5}{4}$ $\left(\text{or } -1\frac{1}{4}\right)$ because $-\frac{4}{5} \cdot -\frac{5}{4} = \frac{20}{20} = 1$.

D The reciprocal of $-1\frac{1}{2}$ $\left(\text{or } -\frac{3}{2}\right)$ is $-\frac{2}{3}$ because $-\frac{3}{2} \cdot -\frac{2}{3} = \frac{6}{6} = 1$. ❖

Why is -1 its own reciprocal?

Calculator

In Lesson 2.3 you used a calculator to find reciprocals. The reciprocal of a number is also called its **multiplicative inverse** and the $\boxed{x^{-1}}$ key is called an inverse key. How can you use the inverse key to find the reciprocal of $-\frac{2}{5}$?

The Reciprocal Property can be used to solve an equation when the coefficient of a variable is a rational number.

EXAMPLE 3

Find the coefficient of the variable for each of the following.

A $\frac{3x}{4}$ **B** $-\frac{5}{6}k$

C $\frac{-y}{8}$ **D** $-\frac{p}{7}$

Solution ➤

A The coefficient of $\frac{3x}{4}$ is $\frac{3}{4}$ because $\frac{3x}{4} = \frac{3}{4}x$.

B The coefficient of $-\frac{5}{6}k$ is $-\frac{5}{6}$.

C The coefficient of $\frac{-y}{8}$ is $-\frac{1}{8}$ because $\frac{-y}{8} = -\frac{1}{8}y$.

D The coefficient of $-\frac{p}{7}$ is $-\frac{1}{7}$ because $-\frac{p}{7} = -\frac{1}{7}p$. ❖

How does the placement of a negative sign affect the coefficient of a variable?

b. Explain why it is true. Hint: Use multiplication. [**Adding the same number twice is the same as multiplying by 2.** $2\left(\frac{1}{8}\right) = \frac{1}{4}$]

Using Visual Models Simple fraction models made from circles split into halves, thirds, fourths, and so on could be used to assist students who have difficulty with the number-line model. Negative fractions can be shaded with one color, and positive fractions with a different color. Have students match equivalent positive and negative values that equal zero and exchange equivalent fractions (for example, two-fourths for one-half) until the sum is simplified.

EXAMPLE 4

Solve. **A** $\frac{2}{3}x = 8$ **B** $\frac{-x}{5} = -3$

Solution ➤

Multiply each side of the equation by the reciprocal of the coefficient of x. The coefficient of $\frac{2}{3}x$ is $\frac{2}{3}$. The coefficient of $\frac{-x}{5}$ is $-\frac{1}{5}$.

A

$$\frac{2}{3}x = 8$$
$$\frac{3}{2}\left(\frac{2}{3}x\right) = \frac{3}{2}(8)$$
$$x = \frac{24}{2}$$
$$x = 12$$

B

$$\frac{-x}{5} = -3$$
$$-\frac{1}{5}x = -3$$
$$-5\left(-\frac{1}{5}x\right) = -5(-3)$$
$$x = 15 ❖$$

What is the reciprocal for the coefficient of x in the expression $\frac{2x}{-3}$?

Try This   Solve.   a. $\frac{3}{4}y = -15$ b. $\frac{-w}{9} = 8$

Remember, a proportion states that two ratios are equal. Example 5 shows two different methods to solve a proportion.

EXAMPLE 5

Solve the proportion $\frac{x}{-5} = \frac{2}{3}$.

Solution ➤

Method A	Method B
Multiply both sides by the reciprocal of the coefficient of x.	Multiply both sides by the least common denominator.

Method A

$$\frac{x}{-5} = \frac{2}{3}$$
$$-5\left(\frac{x}{-5}\right) = -5\left(\frac{2}{3}\right)$$
$$x = \frac{-10}{3}$$
$$x = -3\frac{1}{3}$$

Method B

$$\frac{x}{-5} = \frac{2}{3}$$
$$-15\left(\frac{x}{-5}\right) = -15\left(\frac{2}{3}\right)$$
$$3x = -10$$
$$\frac{3x}{3} = \frac{-10}{3}$$
$$x = \frac{-10}{3}$$
$$x = -3\frac{1}{3}$$

❖

CRITICAL *Thinking*

Compare these methods to Example 4. How are they alike? How are they different?

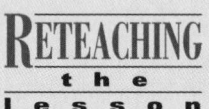**RETEACHING** **the lesson**

Using Algorithms
Help students to see that rational number equations involving only multiplication and division can be rewritten as proportions.

Using Example 4a: $\frac{2}{3}x = 8$ can be written as $\frac{2x}{3} = \frac{8}{1}$. Then students can solve the equation by finding the cross products $2x = 24$. Dividing by 2 then leaves $x = 12$.

Using Example 4b: $\frac{-x}{5} = -3$ can be written as $\frac{-x}{5} = \frac{-3}{1}$. Using cross products gives $-x = -15$, and multiplying by -1 leaves $x = 15$.

EXERCISES & PROBLEMS

Communicate

1. Define a rational number. Give three examples.
2. Explain why an integer is also a rational number.
3. Describe how to model $\frac{1}{2} + (-3)$ on the number line.
4. How can you find $\frac{1}{2} - \frac{5}{6}$ using the number line?
5. State the Reciprocal Property. Use the number -3 to give an example of the Reciprocal Property.
6. Give two ways to solve the proportion $\frac{y}{8} = \frac{-2}{3}$.

Practice & Apply

Write the reciprocal for each number.

7. -2 $-\frac{1}{2}$ **8.** $\frac{-1}{9}$ -9 **9.** $4\frac{1}{5}$ $\frac{5}{21}$ **10.** $\frac{10}{3}$ $\frac{3}{10}$ **11.** 1 1

Solve each equation.

12. $\frac{3}{5}x = 3$ 5 **13.** $\frac{1}{-7}y = -6$ 42 **14.** $\frac{-5}{8}q = 10$ -16 **15.** $-\frac{3}{4}x = -9$ 12

16. $\frac{-y}{7} = 11$ -77 **17.** $\frac{p}{4} = -1.2$ -4.8 **18.** $\frac{p}{-11} = -5$ 55 **19.** $-\frac{t}{4} = -5$ 20

20. $-\frac{w}{8} = 9$ -72 **21.** $\frac{-m}{-12} = -10$ -120

Solve each proportion.

22. $\frac{x}{5} = \frac{3}{4}$ $3\frac{3}{4}$ **23.** $\frac{-x}{3} = \frac{2}{5}$ $-1\frac{1}{5}$ **24.** $\frac{x}{4} = \frac{-3}{6}$ -2 **25.** $\frac{p}{-3} = \frac{-4}{7}$ $1\frac{5}{7}$

26. $\frac{x}{16} = \frac{-3}{8}$ -6 **27.** $\frac{x}{-25} = \frac{3}{-5}$ 15 **28.** $\frac{-x}{6} = \frac{5}{3}$ -10 **29.** $\frac{w}{-14} = \frac{6}{7}$ -12

30. $\frac{-x}{6} = \frac{3}{7}$ $-2\frac{4}{7}$ **31.** $\frac{-y}{7.5} = \frac{-2}{4}$ $3\frac{3}{4}$

32. **Savings** Miguel plans to mow lawns during summer vacation. He spends $\frac{2}{3}$ of his savings to buy a lawn mower. If Miguel's savings totaled $210, how much of his savings did he use to buy the lawn mower? $140

33. **Sports** If Alicia shoots $\frac{3}{4}$ of her free throws by noon, how many practice free throws has she taken? 45

Alicia spent Saturday shooting free throws for basketball. Her coach asked her to practice shooting 60 baskets.

Performance Assessment

A senior class sold $\frac{2}{3}$ of their spirit ribbons on the first day.

Set up a proportion and solve each problem.

34. Fund-raising If they had 450 ribbons left, how many ribbons did the senior class sell on the first day? 1350 ribbons

35. Sports Mary can run a 50-yard race in 9 seconds. How many seconds will it take Mary to run 75 yards at the same speed? 13.5 seconds

36. Consumer Economics Miguel went to the store to buy dog food. Six cans of dog food sell for $1.74. He has $5 for his purchase. Does he have enough money to buy 16 cans? $4.64; yes

37. Cultural Connection: Americas If 9 crowns were exchanged for 7 pesos, how many pesos were exchanged for 63 crowns? 49 pesos

In the days of colonial Mexico, crowns and pesos were forms of money.

38. Geometry At a certain time of day a tree casts a shadow 12 meters long. If a meter stick (1 meter long) casts a shadow of 60 centimeters at the same time of day, how tall is the tree? 20 meters

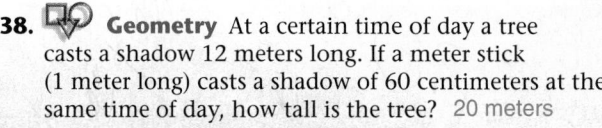 **Look Back**

39. Evaluate $6p - 3q + r$ if $p = 3$, $q = 4$, and $r = 10$. **[Lesson 1.4]** 16

Evaluate. [Lesson 2.4]

40. INT(1.6)　　**41.** INT(6.35)　　**42.** INT(9)
　　　1　　　　　　　6　　　　　　　9

Simplify. [Lesson 3.3]

43. $(3x - 2) + (4x + 7)$　　**44.** $2(4q + 3) - (q - 2)$　　**45.** $(2a + 3b - 1) - 2(a - 2b - 1)$
　　　$7x + 5$　　　　　　　　　$7q + 8$　　　　　　　　　　$7b + 1$

Solve each inequality. [Lesson 3.6]

46. $x + 4 > -12$　　**47.** $y - 0.05 \leq 10.5$　　**48.** $w + 1.4 \geq 10.2$
　　　$x > -16$　　　　　　$y \leq 10.55$　　　　　　$w \geq 8.8$

Look Beyond

Find a pair of numbers (x, y) that satisfies both equations.

49. $3x + 2y = 14$ and $6x - 4 = 8$ (2, 4)

LESSON 4.5 Solving Problems Involving Percent

 *The next time you read a newspaper, notice how many times a percent is used. Percents are found in tables, charts, and graphs. They are also in ads, on display signs, and in financial statements.*

THE WEATHER PAGE

WARM FRONT

Palm Beach, FL
Temperature 30°C
Relative Humidity 50%

PREPARE

Objectives

- Find equivalencies between fractions, decimals, and percents.
- Estimate with percents.
- Solve problems involving percent.

RESOURCES

- Practice Master   4.5
- Enrichment Master   4.5
- Technology Master   4.5
- Lesson Activity Master   4.5
- Quiz   4.5
- Spanish Resources   4.5

The relative humidity is expressed as a percent. For example, at 30°C, if 1 cubic meter of air holds 42 liters of water vapor, the air is saturated, and the relative humidity is 100%. If the air contains 21 liters of water vapor or $\frac{1}{2}$ the saturated amount, the relative humidity is 50%.

Fractions, Decimals, and Percent

A **percent** is a *ratio* that compares a number with 100. 50% is the same as $\frac{50}{100}$ or $\frac{1}{2}$. 100% is the same as $\frac{100}{100}$ or 1. A percent can be expressed in many ways. For example, 25% can be written as a decimal or a fraction. Sometimes you may wish to write the fraction in lowest terms.

$$25\% = \frac{25}{100} = 0.25 \qquad \frac{25}{100} = \frac{1}{4}$$

Suppose the air contains $10\frac{1}{2}$ liters of water vapor. Then the relative humidity is $\frac{10\frac{1}{2}}{42}$ or $\frac{1}{4}$ of 100%. The relative humidity is 25%.

Assessing Prior Knowledge

Write each fraction as a decimal.

1. $\frac{1}{4}$ **[0.25]**

2. $\frac{3}{8}$ **[0.375]**

3. $\frac{7}{200}$ **[0.035]**

4. $1\frac{4}{5}$ **[1.8]**

Solve for x.

5. $\frac{3}{100} = \frac{x}{75}$ **[$x = 2.25$]**

6. $\frac{x}{100} = 0.105$ **[$x = 10.5$]**

TEACH

 Students need to learn to use percents so they can understand how taxes, sales items, and salary increases or decreases are calculated. Grades and class rank are also often based on percents.

ALTERNATIVE teaching strategy

Using Visual Models Use 10-by-10 grid paper to make models. Have students shade one row and then write a fraction, a decimal, and a percent that represent that picture.
$$\left[\frac{1}{10}, 0.1, 10\%\right]$$

Have students randomly shade parts of their grids. Then have them exchange papers with one another, writing a fraction, a decimal, and a percent that represent each picture.

Write each percent as a decimal and as a fraction.

a. 80% $\left[0.8, \frac{4}{5}\right]$

b. 450% $\left[4.50, 4\frac{1}{2}\right]$

c. 9.99% $\left[0.0999, \frac{999}{10,000}\right]$

ongoing ASSESSMENT

8%

TEACHING tip

Review common decimal/fraction/percent equivalences, such as

$\frac{1}{2} = 0.5 = 50\%$,

$\frac{1}{4} = 0.25 = 25\%$,

$\frac{1}{8} = 0.125 = 12.5\%$.

TEACHING tip

Some students write 0.1 as 1%. Remind students of the meaning of percent as parts per hundred. 0.1 means 1 in 10, 1% means 1 in 100.

Use Transparency ▶ 21

EXAMPLE 1

Write each percent as a decimal and as a fraction.

Ⓐ 75%   Ⓑ 110%   Ⓒ 4.4%

Solution ➤

Ⓐ $75\% = \frac{75}{100} = 0.75$ $\frac{75}{100} = \frac{3}{4}$

Ⓑ $110\% = \frac{110}{100} = 1.10$ $\frac{110}{100} = \frac{11}{10}$

Ⓒ $4.4\% = \frac{4.4}{100} = 0.044$ $\frac{4.4}{100} = \frac{44}{1000} = \frac{11}{250}$ ❖

What percent would you write for the decimal 0.08?

Writing and Solving Percent Problems

In a percent equation such as 25% of 40 = 10, 25% is the percent, 40 is the base, and 10 is the percentage. How can you solve an equation if one of the parts is unknown? Explain how each equation is solved using basic properties.

1. 25% of $40 = x \longrightarrow \frac{1}{4} \cdot 40 = x \longrightarrow 10 = x$

2. 25% of $x = 10 \longrightarrow \frac{1}{4} \cdot x = 10 \longrightarrow x = 40$

3. $x\%$ of $40 = 10 \longrightarrow \frac{x}{100} \cdot 40 = 10$

$$x \cdot \frac{40}{100} = 10$$

$$x = 10 \cdot \frac{100}{40}$$

$$x = 25$$

You can also visualize a percent bar and form a proportion.

1. The percentage is unknown. 25% of $40 = x$

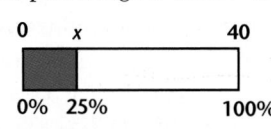

$$\frac{x}{25} = \frac{40}{100}$$

$$100\left(\frac{x}{25}\right) = 100\left(\frac{40}{100}\right)$$

$$4x = 40$$

$$x = 10$$

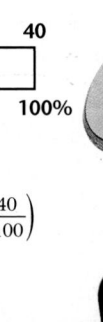

---

interdisciplinary CONNECTION

Consumer Awareness Have students interview adults to see how they use percents in their lives. Have students make a display to show the various ways percents are used. Suggest that students contact some professional people like insurance salespeople, bank officials, real-estate brokers, school principals, department store owners, and farmers. If possible, have students bring examples of work or data involving percents.

ENRICHMENT Have students solve the problem that follows: A woman started a business. During the first year she made a profit of $100,000. During the second year, her profits increased 100%. During the third year, her profits again increased by 100%. During the fourth year, her profits decreased by 100%. What were her profits during the second, third, and fourth years? [$200,000; $400,000; $0]

2. The base is unknown. 25% of $x = 10$

0	10		x

0% 25% 100%

$$\frac{10}{25} = \frac{x}{100} \rightarrow 100\left(\frac{10}{25}\right) = 100\left(\frac{x}{100}\right)$$
$$40 = x$$

3. The percent is unknown. x% of $40 = 10$

0	10		40

0% x% 100%

$$\frac{10}{x} = \frac{40}{100} \text{ or } \frac{x}{10} = \frac{100}{40}$$
$$10\left(\frac{x}{10}\right) = 10\left(\frac{100}{40}\right)$$
$$x = 25$$

CRITICAL *Thinking*

Explain how to use one step to solve for each variable in the equation a% of $b = c$.

EXAMPLE 2

Consumer Economics

Tom plans to buy a CD player for his mother. He sees that the Royal Appliance Store is having a sale. How much will he save at the sale if he buys a $120 CD player?

TV's 30% OFF

VCR's 40% OFF

CD PLAYERS 25% OFF

Solution ➤

The CD player Tom is buying has an original price of $120. CD players are on sale for 25% off. You want to find the amount of savings. Let x represent the amount of savings.

Write a percent statement for the problem, and then solve the equation.

$$25\% \text{ of } \$120 \text{ is } \underline{\ ?\ }.$$
$$0.25 \cdot 120 = x$$
$$30 = x$$

Tom will save $30 on the CD player. ❖

CRITICAL *Thinking*

a% of $b = c$ can be written as an equation in the form $\frac{a}{100}(b) = c$ which is solved for c. To find b, solve $\frac{100c}{a}$. To find a, solve $\frac{100c}{b}$.

Alternate Example 2

Arejay's Sports Shop has baseball card sets that normally sell for $24.80 each. During a sale, these sets were sold for 37.5% off the regular price. What was the cost of the baseball card sets during the sale? **[$15.50]**

Watch for students who answer $200,000; $300,000; $200,000. They are using $100,000 as the increase and decrease, not 100%.

INCLUSION **strategies** **English Language Development** Students whose primary language is not English or students who have difficulty reading may need to tackle percent problems one sentence or phrase at a time until they understand the entire problem. Encourage these students to rewrite the problem using their own words. Students who struggle with the interpretation of application problems dealing with percents should be encouraged to draw a picture or simple diagram of the given information.

Alternate Example 3

Second Chance Used Video/CD Store sells used videos and CDs at a 75% discount off the original cost. If you bought your current favorite CD at Second Chance, how much would it cost? [**Answers will vary depending on the original cost of the CD chosen by students.**]

EXAMPLE 3

Discounts   The freshman class is sponsoring a trip to see a play. If a group of 20 people buy tickets, there is a 30% discount. How much will each ticket cost at the discounted price?

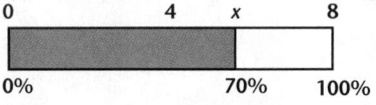

Students $8.00
Group discounts available

The Rialto Theater
Matinee Wednesday 3:00pm
Daily 8:30pm

Solution A ➤

Proportion method   You can visualize the problem using a percent bar. The regular cost, $8, is the length of the bar. Since the tickets will be 30% off, the discounted price will be 70% of the original price. Mark off 70% of the bar, and use x to represent the discounted price.

```
0        4    x      8
┌────────────┬────────┐
│            │        │
└────────────┴────────┘
0%          70%    100%
```

The figure indicates that the price is between $4 and $8. Write the proportion $\frac{x}{70} = \frac{8}{100}$ from the percent bar. Apply the Multiplication Property of Equality, and simplify.

$$70\left(\frac{x}{70}\right) = 70\left(\frac{8}{100}\right)$$

$$x = \frac{560}{100}$$

$$x = 5.60$$

The discounted price is $5.60, which agrees with the estimate.

Solution B ➤

Equation method   Use x to represent the discounted price.

The discounted price, x, is 70% of the full price, $8. Model this statement by an equation.

$$x = \frac{70}{100} \cdot 8$$

$$x = 5.60$$

The discounted price is $5.60. ❖

Try This   Solve the following percent statements.

a. _?_% of 80 is 15.   **b.** 115% of 200 is _?_.   **c.** 35% of _?_ is 45.

Aongoing ASSESSMENT

Try This

a. 18.75   **b.** 230

c. about 128.6

RETEACHING the lesson

Using Estimation   For students having difficulty solving percent problems, suggest that they estimate the answers before solving the problems. Review these relationships between common fractions and percents.

$\frac{1}{2} = 50\%$	$\frac{1}{4} = 25\%$
$\frac{3}{4} = 75\%$	$\frac{1}{3} \approx 33\%$
$\frac{2}{3} \approx 67\%$	$\frac{1}{5} = 20\%$
$\frac{2}{5} = 40\%$	$\frac{3}{5} = 60\%$
$\frac{4}{5} = 80\%$	$\frac{6}{5} = 120\%$

A VCR is on sale for $239.40.

Alternate Example 4

Mario bought a coat for $45. This was 20% less than the regular price. Find the regular price of the coat.   [**$56.25**]

EXAMPLE 4

What was the original price of the VCR before the sale?

Solution ➤

Use the equation method. Let w represent the original price, and note that $239.40 is 60% of the original price. Model this statement by an equation.

$239.40 = 0.60\,w$

Apply the Division Property of Equality, and simplify.

$$\frac{239.40}{0.60} = \frac{0.60w}{0.60}$$

$$399 = w$$

The original price was $399. ❖

EXAMPLE 5

Sales Tax Some states have a sales tax for certain items. Linda paid $47.74 for a $45.25 item. Find the percent of tax or the tax *rate*.

Solution ➤

Use the proportion method. First, find the amount of tax. Subtract the amount of the item from the total amount paid.

$$\$47.74 - \$45.25 = \$2.49$$

The tax was $2.49. Use the percent bar to show the proportion.

Let p represent percent.

Use the Multiplication Property of Equality to solve the proportion.

$0 ⟍ $2.49 $45.25

0% ⟍ p%   100%

$$\frac{p}{2.49} = \frac{100}{45.25}$$

$$2.49\left(\frac{p}{2.49}\right) = 2.49\left(\frac{100}{45.25}\right)$$

$$p = \frac{249}{45.25} \approx 5.5\% ❖$$

What is a percent statement for Example 5?

Alternate Example 5

A video tape is marked $14.97. The clerk asks for $16.09. What is the sales tax rate?   [**7.5%**]

Aongoing **SSESSMENT**

5.5% of $45.25 is $2.49

ASSESS

Selected Answers
Odd-numbered Exercises 11–63

Assignment Guide
Core 1–10, 12–44 even, 45–66

Core Two-Year 1–64

Error Analysis
If students become confused when trying to set up percent problems, review the meaning of percent and encourage the use of a visual model, such as a percent bar.

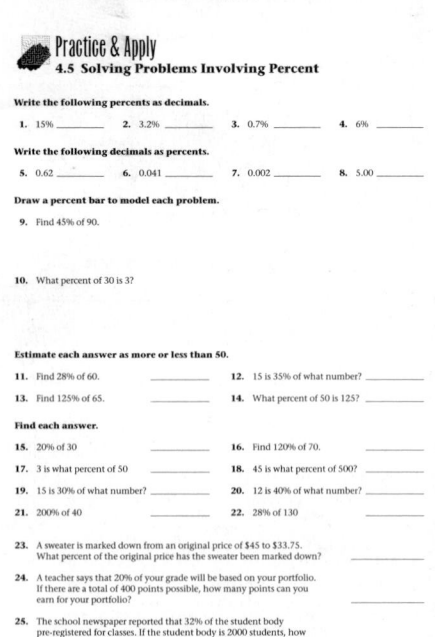

EXERCISES & PROBLEMS

Communicate

1. Explain the procedure for changing a percent to a fraction.

Explain how to draw a percent bar to model each problem.

2. 40% of 50   3. 200% of 50

4. 30 is 50% of what number?

5. What percent of 80 is 60?

Estimate each answer as more or less than 50. Explain how you made your estimate.

6. 40% of 50   7. 200% of 50

8. 30 is 50% of what number?

9. What percent of 80 is 60?

Describe how to set up the equation for the following problem.

10. How many ounces of pretzels would you have to eat to get 100% of the U.S. RDA of iron?

Each ounce of pretzels contains 10% of the U.S. Recommended Daily Allowance (U.S. RDA) of iron.

Practice & Apply

Write the following percents as decimals.

11. 55%   12. 1.2%   13. 8%   14. 145%   15. 0.5%
    0.55   0.012   0.08   1.45   0.005

Write the following decimals as percents.

16. 0.47   17. 0.019   18. 8.11   19. 0.001   20. 9.00
    47%   1.9%   811%   0.1%   900%

Draw a percent bar to model each problem.

21. 35% of 80   22. 5 is what percent of 25?   23. What number is 10% of 8?

24. 150% of 40   25. What percent of 90 is 40?   26. 18 is 20% of what number?

Estimate each answer as more or less than 50 or 50%.

27. 5% of 80 less   28. 18 is 20% of what number? more   29. 150% of 40 more

30. 2.5% of 100 less   31. What percent of 90 is 40? less   32. What percent of 60 is 120? more

Find each answer.

33. 40% of 50 20   34. 125% of what number is 45? 36   35. 8 is 20% of what number? 40

36. 200% of 50 100   37. 30 is 60% of what number? 50   38. What percent of 80 is 10? 12.5%

39. 35% of 80 28   40. What number is 3.5% of 120? 4.2   41. 3 is what percent of 3000? 0.1%

42. 75% of 900 675   43. 72 is 9% of what number? 800   44. What percent of 60 is 24? 40%

The answers to Exercises 21-26 can be found in Additional Answers beginning on page 729.

49.
CLINTON 43.24%
BUSH   37.73%
PEROT   19.02%

50.
CLINTON 68.77%
BUSH   31.23%
PEROT   0%

51.

1992 Presidential Popular Vote

Clinton	43.3%
Bush	37.7%
Perot	19.0%

(Percentages have been rounded.)

 **Probability** According to the National Highway Safety
Administration, a 0.1% blood alcohol concentration increases the odds of a
car accident 7 times, and even one-half of that amount impairs reflex time
and depth perception.

45. Write the percent that is one-half of 0.1%. 0.05%

46. Write a percent statement using "of" and "is" for Exercise 45.
$\frac{1}{2}$ of 0.1% is 0.05%

47. **Statistics** In a sample of 50 seniors, 52% opposed the school's
new driving policy. How many students in the survey opposed the
policy? 26

48. Discounts Troy buys a tennis racket for 30% off the original price of
$66. What is the sale price of the racket? $46.20

 **Statistics**

49. What percent of the popular
vote did each candidate get?

50. What percent of the electoral
vote did each candidate get?

51. Make a pie graph of the popular
vote for all three candidates.

52. Make a pie graph of the electoral
vote for all three candidates.

Candidate	Popular Vote	Electoral Vote
Clinton (Democrat)	43,682,624	370
Bush (Republican)	38,117,331	168
Perot (Independent)	19,217,213	0

53. Demographics The newspaper says that only 42% of the registered
voters voted to determine a school bond election. If 11,960 people
voted, how many registered voters are there? 28,476

54.  **Geometry** The dimensions of a rectangle are increased by 10%.
By what percent is the perimeter increased? 10%

55. Mr. Arnes assigns 100 points out of 600 points for a group project. What
percent is this? about 16.7%

Look Back

Plot the following points on a coordinate plane. [Lesson 1.5]

56. $A(7, 8)$ **57.** $B(5, 2)$ **58.** $C(3, 3)$ **59.** $D(5, 7)$

60. If x is an even integer, represent the next two even integers in terms of x.
[Lesson 3.3] $x + 2$; $x + 4$

Simplify. [Lesson 4.2]

61. $-2 \cdot 3x$ **62.** $-8x \div 4$ **63.** $\frac{10x + 25}{-5}$ **64.** $\frac{22 - 2y}{2}$
$-6x$ $-2x$ $-2x - 5$ $11 - y$

Look Beyond

65. Solve $4x + 6 = 2x - 2$. -4 **66.** Solve $4x + \frac{1}{2} = 5x - \frac{3}{4}$. $1\frac{1}{4}$

52.

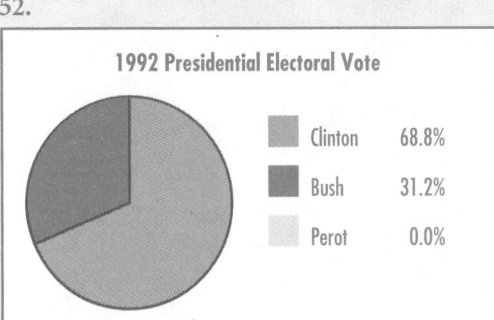

1992 Presidential Electoral Vote

Clinton	68.8%
Bush	31.2%
Perot	0.0%

56–59.

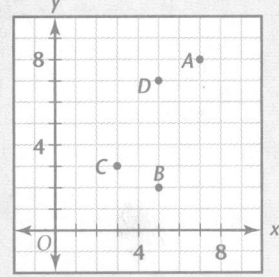

Objectives

• Write equations that represent problem situations.
• Solve multistep equations.

RESOURCES

• Practice Master	4.6
• Enrichment Master	4.6
• Technology Master	4.6
• Lesson Activity Master	4.6
• Quiz	4.6
• Spanish Resources	4.6

Assessing Prior Knowledge

Solve each equation.

1. $64 + k = 58$ $\quad [k = -6]$

2. $86n = 17.2$ $\quad [n = 0.2]$

3. $r - 5.6 = 5.6$ $\quad [r = 11.2]$

4. $\frac{x}{4} = \frac{9}{24}$ $\qquad [x = 1.5]$

TEACH

 Real life involves multistep problem solving. Have students give other examples of problems that involve many steps, for example, finding batting averages (add and then divide) or finding how much each person should pay if the cost of a pizza with extra toppings is divided evenly among those eating the pizza.

LESSON 4.6 **Writing and Solving Multistep Equations**

Jim bought 6 glasses and 6 mugs.

why *It usually takes more than one step to solve a problem about money. To find the sales tax, you first add the item prices, then multiply by the tax rate. Solving multistep problems involves solving multistep equations.*

Jim knew that one mug and one glass would cost between $9 and $10. Six of each should cost between $54 and $60, so he knew that the amount the salesperson rang up was too low. Jim believed that the salesperson had entered the correct amount for the mugs and wanted to find the number of glasses he had been charged for.

The cost of the mugs is $5.98 · 6 or $35.88. Let g represent the number of glasses. The cost of the glasses is $2.99g$. The total cost for mugs and glasses is $47.84.

Write the equation that models the problem.

$$\text{Cost of mugs} + \text{Cost of glasses} = \text{Total cost}$$
$$35.88 \quad + \quad 2.99g \quad = 47.84$$

Solve for g.

$35.88 + 2.99g = 47.84$	Given
$35.88 + 2.99g - 35.88 = 47.84 - 35.88$	Subtraction Property of Equality
$2.99g = 11.96$	Simplify.
$\frac{2.99g}{2.99} = \frac{11.96}{2.99}$	Division Property of Equality
$g = 4$	Simplify.

Jim thought the salesperson accidentally entered 4 glasses instead of 6.

ALTERNATIVE teaching strategy

Using Models Make a function machine for the overhead projector. Have students name the output for various input values. Then reverse the process and have students name the input for various output values.

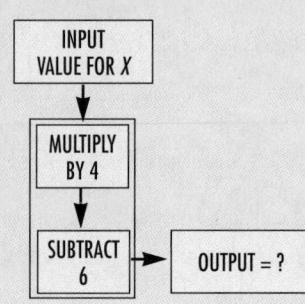

INPUT VALUE FOR X
↓
MULTIPLY BY 4
↓
SUBTRACT 6 → OUTPUT = ?

When the salesperson returned with the register receipt and the purchase, the receipt indicated that the salesperson had indeed miscalculated the bill, but she had charged Jim for 6 glasses and only 5 mugs. How was this possible?

Computer

There are computer programs that solve equations like the one in the previous problem. For example, if you type in

solve(35.94+3.49*n=46.41); ENTER

using a certain Maple™ software program, the computer will immediately display the answer, 3. As equations become more complicated, you can use technology to solve them.

Why is it important to know how to write and solve equations without relying on technology?

EXAMPLE 1

Heather is waiting to take her third test.

What score does Heather need so that her average will be 80?

STATISTICS
Connection

Solution ➤

Let s represent the score Heather needs on her third test. The average after 3 tests can be modeled by $\frac{72 + 79 + s}{3}$. She wants her average to equal 80.

$$\frac{72 + 79 + s}{3} = 80 \qquad \text{Given}$$
$$\frac{151 + s}{3} = 80 \qquad \text{Simplify.}$$
$$3\left(\frac{151 + s}{3}\right) = (3)(80) \qquad \text{Multiplication Property of Equality}$$
$$151 + s = 240 \qquad \text{Simplify.}$$
$$151 + s - 151 = 240 - 151 \qquad \text{Subtraction Property of Equality}$$
$$s = 89 \qquad \text{Simplify.}$$

Heather needs 89 on her third test.

Check
$$\frac{72 + 79 + (89)}{3} \stackrel{?}{=} 80$$
$$80 = 80 \qquad \text{True} ❖$$

How do you determine the number to divide by to find an average?

CRITICAL
Thinking

Do you think Heather will receive an 80 for her average if she scores 88 on her third test? Why or why not?

Solve $-2x-3 = 2x+9$.

$$[-2x - 3 - 2x = 2x + 9 - 2x$$
$$-4x - 3 = 9$$
$$-4x - 3 + 3 = 9 + 3$$
$$-4x = 12$$
$$\frac{-4x}{-4} = \frac{12}{-4}$$
$$x = -3]$$

A ongoing SSESSMENT

Try This

$$y = -\frac{1}{2}$$

Alternate Example 3

Solve $2x + 1 = -3x + 2$ using a graphics calculator. *Hint:* Set x-min and y-min at -2 and x-max and y-max at 2. $[x = 0.2]$

Technology

For Example 3, set x-min, y-min, x-max, y-max, and scale at default values. Then show students how to use ZOOM to find the solution to the equation.

EXAMPLE 2

Solve $5x + 6 = 2x + 18$.

Solution ➤

Isolate x on one side of the equation. Begin by subtracting $2x$ from both sides.

$5x + 6 = 2x + 18$	Given
$5x + 6 - 2x = 2x + 18 - 2x$	Subtraction Property of Equality
$3x + 6 = 18$	Simplify.
$3x + 6 - 6 = 18 - 6$	Subtraction Property of Equality
$3x = 12$	Simplify.
$\frac{3x}{3} = \frac{12}{3}$	Division Property of Equality
$x = 4$	Simplify.

Check the solution.

$$5(4) + 6 \stackrel{?}{=} 2(4) + 18$$
$$26 = 26 \qquad \text{True} \diamond$$

Try This   Solve $4y - 3 = 6y - 2$.

The next example shows how technology can be used to solve an equation graphically.

EXAMPLE 3

Graphics Calculator

Solve $5x + 6 = 2x + 18$ using a graphics calculator.

Solution ➤

On a graphics calculator, let $Y_1 = 5x + 6$, and let $Y_2 = 2x + 18$. Graph the functions.

Start *with the standard window.*

Zoom out. *Trace to find the intersection.*

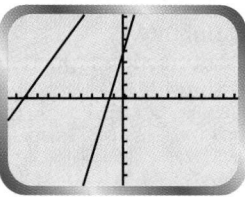
X=3.8947368 Y=25.473684

Zoom in. *Trace to find the coordinates.*

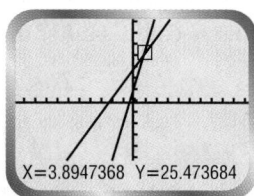
X=4   Y=26

The value of the x-coordinate is 4. $\diamond$

Sometimes, there are many steps needed to find a solution. In the next example all four properties of equality are used.

RETEACHING the lesson

Using Manipulatives

Have students use algebra tiles to solve equations. Start with simple equations with integral coefficients and constants. The variable, however, can be on both sides of the equation. Explain to the students that an equation is like a balance scale, and that whatever is done to one side of the equation must also be done to the other side of the equation.

EXAMPLE 4

Solve $x - \frac{1}{4} = \frac{x}{3} + 7\frac{3}{4}$.

Solution ➤

Step 1

Clear the equation of fractions by multiplying both sides by the least common denominator, 12.

$x - \frac{1}{4} = \frac{x}{3} + 7\frac{3}{4}$	Given
$12\left(x - \frac{1}{4}\right) = 12\left(\frac{x}{3} + 7\frac{3}{4}\right)$	Multiplication Property of Equality
$12x - 12\left(\frac{1}{4}\right) = 12\left(\frac{x}{3}\right) + 12\left(7\frac{3}{4}\right)$	Distributive Property
$12x - 3 = 4x + 93$	Simplify.

Step 2

Isolate x on one side of the equation.

$12x - 3 - 4x = 4x + 93 - 4x$	Subtraction Property of Equality
$8x - 3 = 93$	Simplify.

Step 3

Collect the constant terms on the other side of the equation, and divide both sides by 8 to find x.

$8x - 3 + 3 = 93 + 3$	Addition Property of Equality
$8x = 96$	Simplify.
$\frac{8x}{8} = \frac{96}{8}$	Division Property of Equality
$x = 12$	Simplify.

Step 4

Check your result by substituting 12 for x in the original equation.

$$12 - \frac{1}{4} \overset{?}{=} \frac{12}{3} + 7\frac{3}{4}$$

$$11\frac{3}{4} = 11\frac{3}{4} \qquad \text{True} \quad ❖$$

Solve $-\frac{x}{3} + \frac{4}{5} = 2x - \frac{5}{6}$.

$$\left[30\left(-\frac{x}{3} + \frac{4}{5}\right) = 30\left(2x - \frac{5}{6}\right)\right.$$
$$-10x + 24 = 60x - 25$$
$$-10x - 60x = -25 - 24$$
$$-70x = -49$$
$$\left. x = \frac{7}{10}\right]$$

Cooperative Learning

Have students work in groups of five. Each student is to be an "expert" on one of the Equality Properties or the Distributive Property. Provide each group with the following equations to be solved:

$$\frac{x}{8} + 4 = -2x - \frac{1}{4}$$
$$[x = -2]$$

$$\frac{3x}{5} - (-2) = -\frac{x}{2} - \frac{3}{4}$$

$$\left[x = \frac{-5}{2}\right]$$

Each member is to solve the part of the equation that requires his or her expertise. Then each student in the group should check to see if the solution is correct.

ASSESS

Selected Answers

Odd-numbered Exercises 7–49

Assignment Guide

Core 1–5, 6–32 even,
33–54

Core Two-Year 1–54

Technology

Use of a graphics calculator is appropriate for Exercises 18–26, 36, and 39.

Error Analysis

In Exercises such as 15–17 and 24–26, students may make errors when clearing the fractions from the equation. Some students only multiply fractions by the least common denominator (LCD) and forget to multiply any whole number or constant parts by the LCD. Be sure students understand that every term must be multiplied by the LCD.

EXERCISES & PROBLEMS

Communicate

Without actually solving the equation, explain the steps needed to solve the problem.

1. $23p + 57 = p + 984$ **2.** $\frac{x}{24} + 33 = 84$

3. $5z - 5 = -2z + 19$

4. **Statistics** Brittany scored 85 on her first 2 tests. Explain how to find what score she needs on her third test to average 90 for all 3 tests.

5. Explain how to solve $2x + 3 = 8 - 2x$ using a graphics calculator.

Practice & Apply

Solve each equation.

6. $2m + 5 = 17$ 6 **7.** $9p + 11 = -7$ -2

8. $5x + 9 = 39$ 6 **9.** $3 + 2x = 21$ 9

10. $6 - 8d = -42$ 6 **11.** $2s + 9 = -13$ -11

12. $2(x + 3) = 14$ 4 **13.** $4(x - 20) = 16$ 24 **14.** $2(5 - x) = -25$ $17\frac{1}{2}$

15. $\frac{z}{5} = 22 - 20$ 10 **16.** $\frac{v}{4} + 8 = 11$ 12 **17.** $\frac{x}{10} - 1 = -31$ -300

18. $5x - 7 = 2x + 2$ 3 **19.** $4x + 1 = 12 - 8x$ $\frac{11}{12}$ **20.** $3x - 8 = 5x - 20$ 6

21. $1 - 3x = 2x + 8$ $-1\frac{2}{5}$ **22.** $15 - 4x = 12 - 8x$ $\frac{3}{4}$ **23.** $2(x - 3) = 3(x - 4)$ 6

24. $3v - 8 = \frac{v}{2} + 2$ 4 **25.** $\frac{w}{2} + 7 = \frac{w}{3} + 9$ 12 **26.** $6 - \frac{t}{4} = 8 + \frac{t}{2}$ $-2\frac{2}{3}$

Solve each equation for the indicated variable.

27. $3t = r$ for t **28.** $53 + 9s = P$ for s **29.** $ma = q$ for a

30. $al + r = 7$ for l **31.** $2a + 2b = c$ for b **32.** $y = mx + b$ for x

33. **Number Theory** The sum of three consecutive whole numbers is 48. Write an equation, and solve to find the numbers. 15; 16; 17

34. Lynn has scores of 95, 91, and 88 on three tests. Write an equation and solve to find a fourth score to average 90 on all four tests. 86

35. **Geometry** If the perimeter of the rectangle is 180 feet, write and solve an equation to find the length and width. 75 and 15

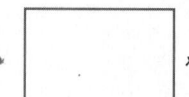

x

$x + 60$

27. $t = \dfrac{r}{3}$

28. $s = \dfrac{P - 53}{9}$

29. $a = \dfrac{q}{m}$

30. $l = \dfrac{7 - r}{a}$

31. $b = \dfrac{c - 2a}{2}$

32. $x = \dfrac{y - b}{m}$

$15.95 per disc, plus a shipping and handling charge of $2.95 per order.

Stone Age	House	Faith	King
The Tiger A25647	Best Of A46578	Once To A84950	Shadow A48593

Yes, Please Send Me____ CDs Now!

7.	8.	9.
1.	2.	3.

Advance Bonus Offer:

4.	5.	6.	10.	11.	12.

36. Consumer Economics If the total cost, C, is $98.65, how many compact discs did Kara and a friend buy? 6

Travel You can estimate the time T (in hours) it takes to fly non-stop a distance D (in miles) by the equation $T = \frac{D}{500} + \frac{1}{2}$. For example, a flight of 1000 miles would take about $2\frac{1}{2}$ hours.

37. Use the formula to estimate the time it takes to fly 1300 miles. $3\frac{1}{10}$ hours

38. Solve the formula for D. $D = 500\ T - 250$

39. Use the new formula to find how many miles you can travel in 4 hours.
1750 miles

40. Cultural Connection: Africa Find the answer to this problem from ancient Egypt about 3800 years ago: Fill a large basket $1\frac{1}{2}$ times. Then add 4 *hekats* (a *hekat* is about half a bushel). The total is 10 *hekats*. How many *hekats* does the basket hold? 4

Look Back

41. Find the area of a square with a side of 6 inches. **[Lesson 2.2]**
36 square inches

Write the opposite for each expression. [Lesson 3.4]

42. $6c - 3d$
$-6c + 3d$

43. $-s + t$
$s - t$

44. $-a - c$
$a + c$

Solve. [Lesson 3.6]

45. $x + 4 = 10$ 6

46. $x - 6 = 23$ 29

47. $x - 13 = 10$ 23

48. $5 - x = -12$
17

49. $2.1 + x = -8.3$
-10.4

50. $-\frac{2}{3} + x = -\frac{1}{6}$
$\frac{1}{2}$

Look Beyond

Solve each equation for x.

51. $x^2 = 4$ $2, -2$

52. $x^2 = 169$ $13, -13$

53. $x^2 = 24$ $\pm 2\sqrt{6}$

54. Paul, Pete, and Pam each took a different type of fruit to school for lunch. Paul does not like bananas and Pete does not like to peel fruit. If the fruits are an apple, an orange, and a banana, who brought the apple? Pete

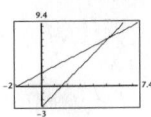

Objectives

• Investigate the Multiplication and Division Properties of Inequality.

• Solve multistep inequalities in one variable.

RESOURCES

• Practice Master	**4.7**
• Enrichment Master	**4.7**
• Technology Master	**4.7**
• Lesson Activity Master	**4.7**
• Quiz	**4.7**
• Spanish Resources	**4.7**

Assessing Prior Knowledge

Use > or < to make each sentence true.

1. $6 + 3$ ___ 8 [>]

2. 4 ___ 2 ___ $1/3$ [>]

3. $8(-3)$ ___ 16 [<]

4. $-3 + 17$ ___ $(-3)(-6)$ [<]

5. $-7 + (-8)$ ___ $-18 - (-2)$ [>]

TEACH

Often in the real world there is more than one answer to a problem, but some solutions are better than others. For example, to cross-country ski, you need snow, but for *maximum* enjoyment 5 inches is a *minimum*.

LESSON 4.7 Exploring Related Inequalities

Why Many real-life situations are not solved by finding a single answer, but by finding a minimum, maximum, or range of answers that satisfy the conditions. For problems of this type, you often need to solve an inequality.

This ballet company must maintain a certain minimum average attendance to stay in business. Fire codes and health regulations establish a certain maximum number of people allowed in the theater at one time.

• Exploration 1 Multiplying and Dividing Inequalities

Numbers graphed on the number line are in order from left to right.

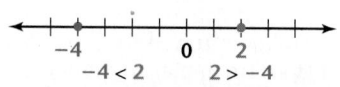

$-4 < 2$ $2 > -4$

ALTERNATIVE teaching strategy

Using Manipulatives

Allow students to work in pairs. Each group is to have 12 coins, with heads representing positive integers and tails representing negative integers.

1. Student A chooses from 1 to 3 coins and places them with all heads or all tails showing. Student B then does the same, but cannot pick the same number.

2. The students now compare the number and signs of the coins to determine which set has the greater value. Students should write down each inequality they create.

3. Now each student takes more coins and models multiplying his or her number of coins by 2. Then students compare sets again and write down the new inequality. [**The same student still should have the greater value.**]

Complete the pattern by filling in $<$, $=$, or $>$.

a. $-4 < 2$, so $-4 \cdot 2 < 2 \cdot 2$
b. $-4 < 2$, so $-4 \cdot 1 < 2 \cdot 1$
c. $-4 < 2$, so $-4 \cdot 0 \underline{\ ?\ } 2 \cdot 0$
d. $-4 < 2$, so $-4 \cdot -1 \underline{\ ?\ } 2 \cdot -1$
e. $-4 < 2$, so $-4 \cdot -2 \underline{\ ?\ } 2 \cdot -2$

a. $-4 < 2$, so $-4 \div 2 < 2 \div 2$
b. $-4 < 2$, so $-4 \div 1 < 2 \div 1$
c. $-4 < 2$, so $-4 \div -1 \underline{\ ?\ } 2 \div -1$
d. $-4 < 2$, so $-4 \div -2 \underline{\ ?\ } 2 \div -2$

3 What happens to an inequality when each side is multiplied or divided by the same positive number?

4 What happens to an inequality when each side is multiplied or divided by 0?

5 What happens to an inequality when each side is multiplied or divided by the same negative number?

6 **a.** Write a rule to explain what you must do to the inequality when you multiply or divide each side by the same positive number.
b. Write a rule to explain what you must do to the inequality when you multiply or divide each side by the same negative number. ❖

•Exploration 2 *Multiplying Inequalities*

To visualize what happens when you multiply each side of an equality by the same number, complete this exploration.

1 Start with a set of 3 positive tiles and a set of 5 positive tiles. The $<$ sign is used

$3 < 5$

2 To multiply both sides of the inequality by 2, think of doubling each set of tiles.

Fill in $<$, $=$, or $>$.

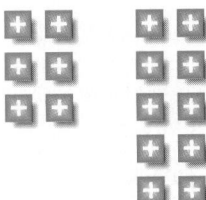

$2 \cdot 3 \underline{\ ?\ } 2 \cdot 5$

3 To multiply both sides of the inequality by -2, think of doubling each set *and* changing the sign of each tile.

Fill in $<$, $=$, or $>$.

$-2 \cdot 3 \underline{\ ?\ } -2 \cdot 5$

4 Draw a picture to show what happens to $-3 > -4$ when each side of the inequality is multiplied by **a.** 2. **b.** -2. ❖

4. Students start again with their first set of coins. This time they are modeling multiplying by a negative. This is done by taking twice as many coins, but turning each one over. Students compare sets again and write down the new inequality. [**Now the opposite student should have the greater value.**] After several rounds, have students look at their charts and write the rule for multiplying by negative numbers in their own words.

5. The same procedure can be used for division, with students starting with more coins and separating them into equal parts.

Exploration 3 *Dividing Inequalities*

To visualize what happens when you divide each side by the same number, complete this exploration.

1 Start with a set of 4 positive tiles and a set of 8 positive tiles. The < sign is used for the inequality.

4 < 8

2 To divide both sides of the inequality by 4, think of dividing each set of tiles by 4.

Fill in <, =, or >.

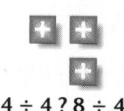

4 ÷ 4 ? 8 ÷ 4

3 To divide both sides of the inequality by −4, think of dividing each set by 4 *and* changing the sign of each tile.

Fill in <, =, or >.

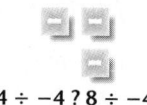

4 ÷ −4 ? 8 ÷ −4

4 Draw a picture to show what happens when each side of the inequality −4 < 2 is divided by   **a.** 2.   **b.** −2. ❖

The properties for multiplying and dividing inequalities express the results of multiplying or dividing an inequality by a positive or negative number.

MULTIPLICATION AND DIVISION PROPERTY OF INEQUALITIES

If each side of an inequality is multiplied or divided by the same positive number, the resulting inequality has the same solution.

If each side of an inequality is multiplied or divided by the same negative number and the inequality sign is reversed, the resulting inequality has the same solution.

CRITICAL *Thinking*

A famous inequality states that $|a| + |b| \geq |a + b|$. Explain why this inequality is true for all numbers a and b.

4. a. 2 negative tiles are fewer than 1 positive tile.

   b. 2 positive tiles are more than 1 negative tile.

CRITICAL *Thinking*

Let $|a| = 2$ and $|b| = 3$. Consider the four cases.

1. $a > 0$ and $b > 0$.
   $|2| + |3| \geq |5|$
   $5 \geq 5$ **True**

2. $a < 0$ and $b < 0$.
   $|-2| + |-3| \geq |-5|$
   $5 \geq 5$ **True**

3. $a > 0$ and $b < 0$.
   $|2| + |-3| \geq |-1|$
   $5 \geq 1$ **True**

4. $a < 0$ and $b > 0$.
   $|-2| + |3| \geq |1|$
   $5 \geq 1$ **True**

ENRICHMENT Have students try to solve these inequalities and show the solution on a number line.

1. $0 < 2x < 4$

   $0 < x < 2$

2. $-6 < 2x < -2$

   $-3 < x < -1$

3. $-6 < -2x < -2$

   $1 < x < 3$

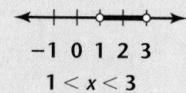 **Using Symbols** Students with learning disabilities may have difficulty remembering and deciphering inequality symbols. You may suggest that they use colored pencils or markers to differentiate between the symbols.

Special women's ballet slippers cost about $50 a pair and wear out very quickly.

Ballet A dancer rehearsing and performing the role of Clara during one run of *The Nutcracker* can typically use 15 to 25 pairs of these ballet slippers. If a ballet company has at most $1000 to spend, how many pairs of these special ballet slippers can the company buy?

Use an inequality to model the problem.

Let x equal the number of ballet slippers the company can buy. Then, $50x$ equals the amount they can spend.

The amount must be less than or equal to 1000. Thus, $50x \leq 1000$.

Divide both sides of the inequality by 50.

$$\frac{50x}{50} \leq \frac{1000}{50}$$

$$x \leq 20$$

The company can buy at most 20 pairs of special ballet slippers. ❖

EXERCISES & PROBLEMS

Communicate

Tell whether each statement is true or false. Explain your reason.

1. $7 < 8$ **2.** $7 < 7$ **3.** $7 \leq 7$ **4.** $7 \neq 7$

Describe the steps needed to solve each inequality.

5. $x + 1 > 4$ **6.** $x - 3 \leq 13$ **7.** $-3p < 12$ **8.** $4x - 2 \geq 2x + 3$

The inequality symbols may confuse some students. You might suggest that they replace the symbol with an equal sign and solve that related equation. They should then replace the inequality symbol, being sure to use the correct one. Tell them to check their solutions by choosing a value for the variable and putting it in the original inequality. They should make sure this solution has the correct sign.

Practice & Apply

Write an inequality that corresponds to each statement.

9. L is greater than W. $L > W$

10. r is greater than or equal to 4. $r \geq 4$

11. V is between 3.1 and 3.2 inclusive. $3.1 \leq V \leq 3.2$

12. x cannot equal 0. $x \neq 0$

13. m is positive. $m > 0$

14. y is non-negative. $y \geq 0$

Tell whether each statement is true or false.

15. $4.2 \geq 4.2$ T

16. $9.22 \leq 9.22$ T

17. $3.1 < 3.01$ F

18. $8.55 > 8.505$ T

19. $\frac{1}{7} \geq \frac{1}{6}$ F

20. $\frac{3}{4} \leq \frac{4}{5}$ T

21. $-8 < -4$ T

22. $0 \geq -3$ T

Solve each inequality.

23. $x + 8 \geq 11$ $x \geq 3$

24. $x - 11 < -20$ $x < -9$

25. $G - 6 \leq 9$ $G \leq 15$

26. $8 - H > 9$ $H < -1$

27. $6 - x > -1$ $x < 7$

28. $5 - y \leq 2$ $y \geq 3$

29. $\frac{x}{8} < 1$ $x < 8$

30. $\frac{u}{-3} \geq 21$ $u \leq -63$

31. $5b > 3$ $b > \frac{3}{5}$

32. $9c \geq -21$ $c \geq -2\frac{1}{3}$

33. $2d + 1 < 5$ $d < 2$

34. $5x - 2 > 2x + 9$ $x > 3\frac{2}{3}$

35. $\frac{x}{3} + 4 < 10$ $x < 18$

36. $\frac{-x}{5} - 1 < 3$ $x > -20$

37. $8x - 3 \leq 9$ $x \leq 1\frac{1}{2}$

38. $15 + \frac{y}{4} > 10$ $y > -20$

39. Ballet Express the cost, C, of shoes used in a year (for one man) as an inequality. $255 \leq C \leq 300$

A man uses about 15 pairs of ballet shoes a year, and the price ranges from $17 to $20 per pair.

Cultural Connection: Europe A Russian mathematician named Chebyshev (1821–1894) proved that the next prime after p is less than $2p$. For example, 7 is a prime, so according to Chebyshev, the next prime after 7 is less than 14, which is true since 11 is the next prime and $11 < 14$.

Substitute the following numbers for p to show that Chebyshev's statement is true.

40. 2 $3 < 4$

41. 3 $5 < 6$

42. 5 $7 < 10$

43. 89 $97 < 178$

Geometry The length, *l*, of a rectangle is to be at least 5 centimeters more than the width, *w*.

44. Express the statement as an inequality. $l \geq w + 5$

45. Write an inequality for the length if the width is 20 centimeters.
$l \geq 25$

All Tapes On Sale
$14.99

BIG City
Snow Mountain
Paradise Shore

Big City
Snow Mountain
Paradise Shore

Sun River
Nobodys Home
The Painter

Sun River
Nobodys Home
The Painter

$3.95
handling charge
per order.
Fill out the order form
on the reverse side.

Consumer Economics Robin has $50 to order videos on sale from a catalog.

46. Write *two* inequalities that express the possible number of videos she can buy.
   a. the maximum number $14.99v + 3.95 \leq 50$
   b. the minimum number $v \geq 0$

47. List all the possibilities for the number of videos Robin can buy. 0, 1, 2, or 3

48. How many videos can Robin and two friends buy if they have $80 to spend? 5 or less

~~~ *Look Back* ~~~

Evaluate. [Lesson 3.2]

49. $89 - (-14)$ **50.** $400 - (-111)$ **51.** $-16 - (-3)$ **52.** $-674 - 9(-900)$
       103   511 -13 7426

State if the product is negative or positive. [Lesson 4.1]

53. $(-3)(-3)(-1)(-1)$ P   **54.** $(-1)(1)(-1)(-1)$ N   **55.** $\dfrac{-22}{2}$ N   **56.** $\dfrac{-16}{-4}$ P

Use the Distributive Property to simplify the following expressions. [Lesson 4.2]

57. $(3x - 2y + 1) - 3(x + 2y - 1)$ **58.** $3(a + b) - 2(a - b)$
       $-8y + 4$ $a + 5b$

Look Beyond ~~~

Sometimes inequalities are used to indicate values that make sense in an equation, such as in these examples.

If $y = \dfrac{1}{x}$, then $x \neq 0$.   If $y = \sqrt{x}$, then $x \geq 0$.

Complete each of the following.

59. If $y = \dfrac{1}{x - 2}$, then $x \neq \underline{\ ?\ }$. 2   **60.** If $y = \sqrt{x + 3}$, then $x \geq \underline{\ ?\ }$. -3

Absolute Value Equalities and Inequalities

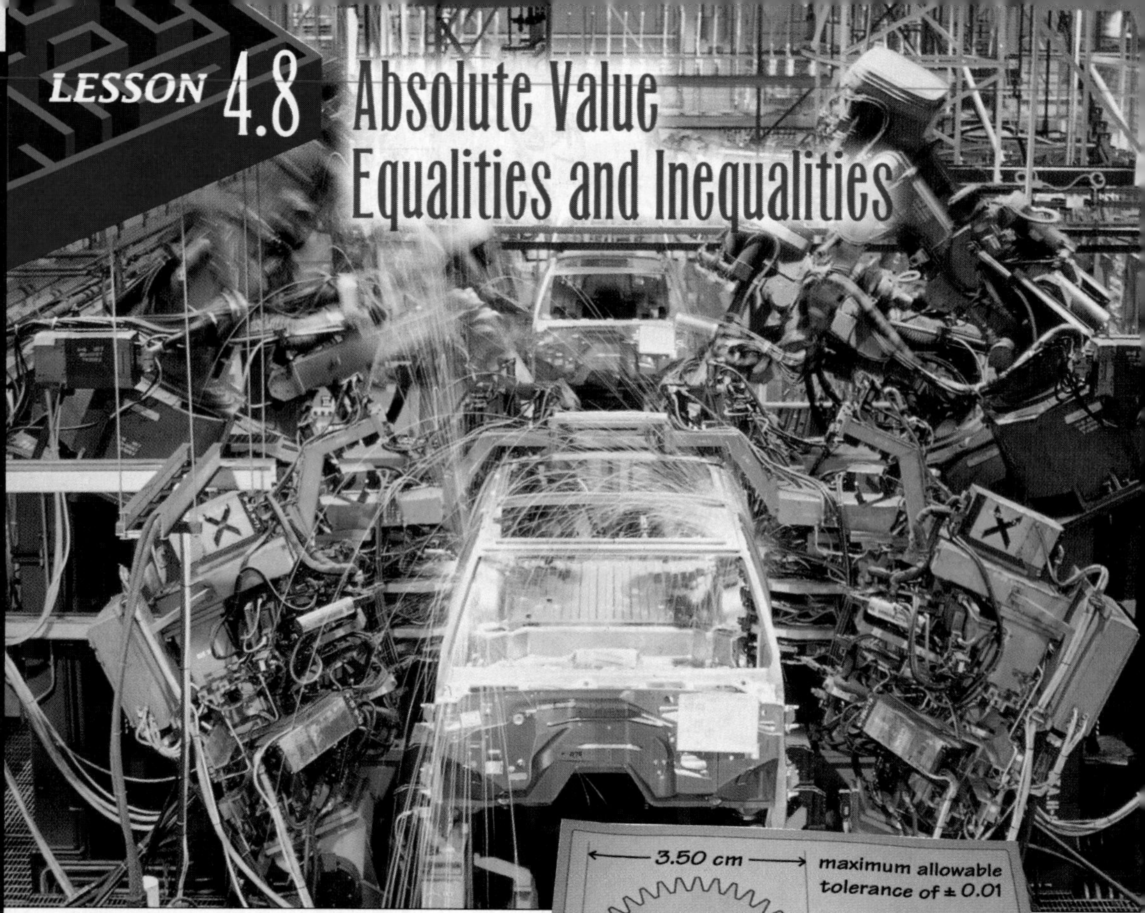

PREPARE

Objectives

- Solve absolute value equations.
- Solve absolute value inequalities.

RESOURCES

- Practice Master   4.8
- Enrichment Master   4.8
- Technology Master   4.8
- Lesson Activity Master   4.8
- Quiz   4.8
- Spanish Resources   4.8

Assessing Prior Knowledge

Solve each equation.

1. $3x - 2 = 7$ $\quad [x = 3]$

2. $-(3x - 2) = 7$ $\quad \left[x = -\frac{5}{3}\right]$

TEACH

 Discuss with students whether they think anything can ever be measured exactly. What does an exact measurement mean? To an engineer, *tolerance* is the amount something is allowed to differ from the requirement. Compare this definition to the usual meaning of the word, which implies acceptance of things that differ from the norm.

why *Measurement is important in science, engineering, economics, industry, medicine, and many other fields. Measurements and the errors they contain can be expressed using absolute value.*

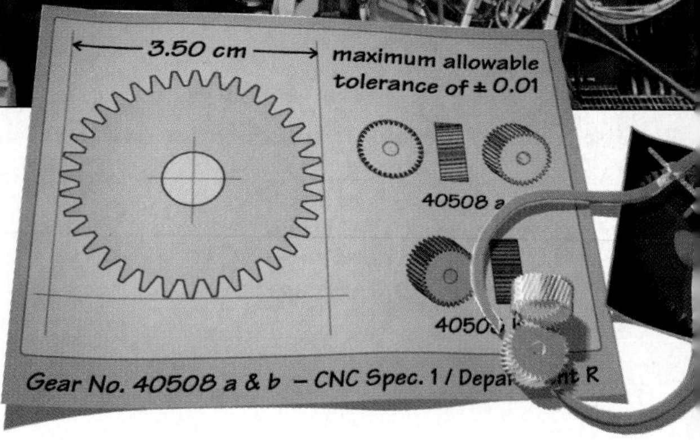

Gear No. 40508 a & b – CNC Spec. 1 / Department R

3.50 cm — maximum allowable tolerance of ± 0.01

40508 a

4050_ _

Error and Absolute Value

A company manufactures a small gear for a car. If the gear is made too large, it will not fit. If it is made too small, the car will not run properly. How accurate is close enough?

Quality Control A gear is designed with a specification of 3.50 centimeters for the diameter. It will work if it is within ± 0.01 centimeter of the specified measurement. The **absolute error** is the absolute value of the difference between the actual measure, x, and the specified measure 3.50 cm. This is written $|x - 3.50|$.

**ALTERNATIVE teaching strategy** **Using Models** Provide students with an object with a circular hole in the center, such as a nut that goes on a bolt. Have students measure the diameter of the hole. Then have students exchange nuts with other students and measure again. Make a list of all the students' results on the board. Which measurement was largest? Which was smallest? What was the range? Were all the measurements within a "tolerable" range? Ask students to decide what the range should be.

MAXIMUM MINIMUM
Connection

If the maximum error permitted is 0.01 centimeters, the acceptable diameters can be shown with an absolute value inequality.

$$|x - 3.50| \leq 0.01$$

To find the maximum and minimum diameters, solve the equation part of the inequality.

Recall the definition for absolute value.

$$|x| = x \text{ if } x \geq 0$$
$$|x| = -x \text{ if } x < 0$$

To solve absolute value equations you must consider two cases.

Case 1
Consider the quantity within the absolute value sign to be positive.

$$|x - 3.50| = 0.01$$
$$x - 3.50 = 0.01$$
$$x = 3.50 + 0.01$$
$$x = 3.51$$

Case 2
Consider the quantity within the absolute value sign to be negative.

$$|x - 3.50| = 0.01$$
$$-(x - 3.50) = 0.01$$
$$-x = -3.50 + 0.01$$
$$x = 3.49$$

The maximum and minimum allowable lengths for the gear's diameter are 3.51 centimeters and 3.49 centimeters.

EXAMPLE 1

Solve $|3x - 2| = 10$.

Solution ➤

Case 1
Consider the quantity within the absolute value sign to be positive.

$$|3x - 2| = 10$$
$$3x - 2 = 10$$
$$3x = 2 + 10$$
$$3x = 12$$
$$x = 4$$

Check

$$|3x - 2| \stackrel{?}{=} 10$$
$$|3(4) - 2| \stackrel{?}{=} 10$$
$$|12 - 2| \stackrel{?}{=} 10$$
$$|10| = 10 \quad \text{True}$$

Case 2
Consider the quantity within the absolute value sign to be negative.

$$|3x - 2| = 10$$
$$-(3x - 2) = 10$$
$$-3x + 2 = 10$$
$$-3x = -2 + 10$$
$$-3x = 8$$
$$x = -\frac{8}{3}$$

Check

$$|3x - 2| \stackrel{?}{=} 10$$
$$\left|3\left(-\frac{8}{3}\right) - 2\right| \stackrel{?}{=} 10$$
$$|-8 - 2| \stackrel{?}{=} 10$$
$$|-10| = 10 \quad \text{True} \quad ❖$$

Try This Solve $|2x - 4| = 8$.

Math Connections

Geometry Before beginning a discussion of page 194, draw a large circle on the board or overhead with a compass. Then draw a line through the center of the circle that extends beyond the circle in each direction. Have students discuss how far the points of intersection of that line and the circle are from the center of the circle. [**the same distance**] Have students try drawing the same situation on their own. Is the distance always the same? [**yes**] Why? [**In drawing a circle, the radius does not change.**]

CRITICAL *Thinking*

The absolute value of a number is defined as the distance of that number from zero, and distance is always positive.

Distance and Absolute Value

GEOMETRY *Connection*

Absolute value can be used to describe the distance between any two points on a number line. The number line and geometry can help you visualize the meaning of the absolute value equation and its solution.

What values of x are 2 units from 5?

Geometric solution On a number line locate the center of a circle at the point representing 5. Draw an arc 2 units from the center on each side. The arc intersects the number line at 3 and 7.

Algebraic solution Express this relationship as an absolute value equation. The expression $|x - 5|$ represents the distance between 5 and each of the two points marked on the number line.

$$|x - 5| = 2$$
Center ↗   ↖ Radius

Then solve the absolute value equation $|x - 5| = 2$.

Case 1
The quantity within the absolute value sign is positive.

$$|x - 5| = 2$$
$$x - 5 = 2$$
$$x = 7$$

Case 2
The quantity within the absolute value sign is negative.

$$|x - 5| = 2$$
$$-(x - 5) = 2$$
$$-x + 5 = 2$$
$$-x = -3$$
$$x = 3$$

These are the points where the arcs of the circle cross the number line. They are each two units from 5.

CRITICAL *Thinking*

Explain why absolute value is always a non-negative number. Why is absolute value a good method for finding distance?

ENRICHMENT Challenge students to solve $|x - 3| > -2$ and explain how they arrived at their solution. Some students may intuitively see that any number will be a solution, since the absolute value is always positive and therefore always greater than -2.

INCLUSION strategies

Using Visual Models Students solving inequalities involving absolute values may solve the separate cases correctly and then not know how to interpret the individual answers as one combined response. Encourage students to draw a number line encompassing both endpoint values. They should then draw the solution to the first case above the number line, draw the solution to the second case below

Absolute Values and Inequalities

An artist must fit a square piece of metal in a 48.00 centimeter by 48.00 centimeter space in a wood block.

EXAMPLE 2

Woodworking

The metal square will fit properly in the block if the amount of error in the length of the sides is within 0.05 centimeter. Use absolute value to express the amount of error allowed for the length of the metal square's side.

Solution ➤

The absolute error can be expressed as $|x - 48.00|$. The allowable error is ± 0.05 centimeter. Since the allowable error includes all the lengths within 0.05 centimeters of the specified measure, use an inequality.

Solve $|x - 48.00| \le 0.05$.

Case 1
Consider the quantity within the absolute value sign to be positive.

$$|x - 48.00| \le 0.05$$
$$x - 48.00 \le 0.05$$
$$x \le 48.00 + 0.05$$
$$x \le 48.05$$

Case 2
Consider the quantity within the absolute value sign to be negative.

$$|x - 48.00| \le 0.05$$
$$-(x - 48.00) \le 0.05$$
$$-x + 48.00 \le 0.05$$
$$-x \le -48.00 + 0.05$$
$$x \ge 47.95$$

In case 2, recall that the inequality reverses when you multiply both sides of the inequality by a negative value.

The allowable lengths for the sides of the stained-glass square are all the measures between 48.05 centimeters and 47.95 centimeters. Write $47.95 \le x \le 48.05$. ❖

the number line, and draw the union or intersection of the two cases on the number line.

Case 1: $x \le 7$

Solution: $3 \le x \le 7$

0 1 2 3 4 5 6 7

Case 2: $3 \le x$

EXAMPLE 3

Alternate Example 3

Solve $|x + 4| < 5$.
$[-9 < x < 1]$

Alternate Example 4

Solve $|x + 4| > 5$.
$[x > 1 \text{ or } x < -9]$

TEACHING *tip*

Have students graph the solutions to Alternate Example 3 and Alternate Example 4 on the same number line, but using different colors for each. What do they notice about the graphs? [**Together, the graphs cover the entire number line, except for 1 and −9. The graphs do not overlap.**]

CRITICAL *Thinking*

There can be no solution because absolute values cannot be less than a negative number.

Practice Master

Practice & Apply
4.8 Absolute Value Equalities and Inequalities

Find the values of x that solve each absolute value equation. Check your answer.

1. $|x - 2.75| = 0.05$ 2. $|x - 7| = 4$ 3. $|x - 3| = 5$

4. $|4x - 2| = 6$ 5. $|3x + 5| = 11$ 6. $|-4 + x| = 7$

Find the values of x that solve each absolute value inequality. Graph the answer on the number line. Check your answer.

7. $|x + 2| > 7$

8. $|-2 - x| \geq 4$

9. $|x + 1| \leq 4$

10. $|x - 3| > 2$

11. $|4 - x| \geq 5$

12. $|x + 2| > 2$

The distance between x and 3 is 2.

13. Draw a number line diagram to illustrate the given sentence.

14. Translate the given sentence into an absolute value equation.

196 Lesson 4.8

What values of x are less than or equal to 2 units from 5? Solve the inequality $|x - 5| \leq 2$.

Solution ➤

Case 1
The quantity within the absolute value sign is positive.

$$|x - 5| \leq 2$$
$$x - 5 \leq 2$$
$$x \leq 7$$

Case 2
The quantity within the absolute value sign is negative.

$$|x - 5| \leq 2$$
$$-(x - 5) \leq 2$$
$$-x + 5 \leq 2$$
$$-x \leq -3$$
$$x \geq 3$$

The inequality is true when x is less than or equal to 7 and greater than or equal to 3. We write $x \leq 7$ *and* $x \geq 3$, or $3 \leq x \leq 7$.

This can be represented on the number line.

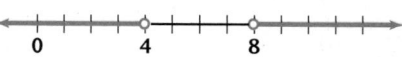

$3 \leq x \leq 7$

EXAMPLE 4

Solve $|x - 6| > 2$.

Solution ➤
Consider two cases.

Case 1
The quantity $x - 6$ is positive.

$$|x - 6| > 2$$
$$x - 6 > 2$$
$$x > 8$$

Case 2
The quantity $x - 6$ is negative.

$$|x - 6| > 2$$
$$-(x - 6) > 2$$
$$-x + 6 > 2$$
$$-x > -4$$
$$x < 4$$

The inequality is true when $x > 8$. The inequality is also true when $x < 4$. We write $x > 8$ *or* $x < 4$.

Check by testing numbers from the solution in the original inequality. ❖

CRITICAL *Thinking*

What problem occurs when you try to solve $|x - 5| < -1$? Test some values to see. Why does this happen?

RETEACHING the lesson

Using Visual Models
Use a number line to show absolute value. Emphasize that the absolute value is the distance of that number from zero. Distance cannot be negative.

Have students find the absolute value of several positive and negative numbers by locating them on the number line. Then have students generalize: $|x| = x$ if $x > 0$, and $|x| = -x$ if $x < 0$.

EXERCISES & PROBLEMS

Communicate

1. What is the meaning of the specification 45 ± 0.001 centimeters?

2. Explain how to write an absolute value inequality that can be used to express 45 ± 0.001 centimeters.

3. Why must you consider two cases when you solve absolute value equations and inequalities?

4. Describe how to use an absolute value inequality to represent all the values on the number line that are within 3 units of the number -7.

5. Explain how to check the values of an absolute value inequality to see if they are inside or outside the boundary.

Practice & Apply

Find the values of x that solve each absolute value equation. Check your answer.

6. $|x - 5| = 3$ 8, 2   7. $|x - 1| = 6$ 7, -5 8. $|x - 2| = 4$ 6, -2 9. $|x - 8| = 5$ 13, 3

10. $|5x - 1| = 4$ 1, $-\frac{3}{5}$ 11. $|2x + 4| = 7$ $1\frac{1}{2}$, $-5\frac{1}{2}$ 12. $|4x + 5| = 1$ -1, $-1\frac{1}{2}$ 13. $|-1 + x| = 3$ 4, -2

Find the values of x that solve each absolute value inequality. Graph the answer on a number line. Check your answer.

14. $|x - 3| < 7$ 15. $|x + 4| > 8$ 16. $|x - 8| \le 4$ 17. $|x - 5| \ge 2$

18. $|x - 2| > 6$ 19. $|x - 2| \le 10$ 20. $|x + 1| < 5$ 21. $|x - 4| > 2$

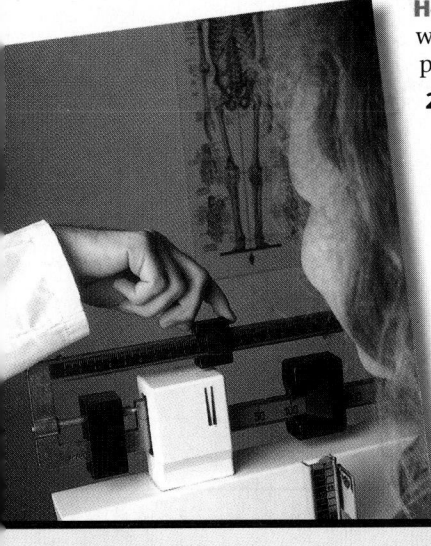

Health According to a height and weight chart, Margo's ideal weight is 118 pounds. She will be satisfied if she stays within 5 pounds of her ideal weight.

22. Draw a diagram. Identify boundary values, acceptable weights, and unacceptable weights on your diagram.

23. Write an absolute value equation to describe the boundary values. What is the solution of your equation?   113, 123

24. Write an absolute value inequality to describe the acceptable weights. What is the solution of your inequality? $113 \le w \le 123$ $|x - 118| \le 5$

The distance between x and 2 is 7.

25. Draw a number line diagram to illustrate the given sentence.

26. Translate the given sentence into an absolute value equation.

The answers to Exercises 14–21 can be found in Additional Answers begining on page 729.

22.   Unacceptable Wts.   Acceptable Wts.   Unacceptable Wts.

106 110 114 118 122 126 130

25.
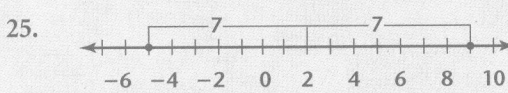

-6 -4 -2 0   2   4   6   8   10

26. $|x - 2| = 7$

ASSESS

Selected Answers

Odd-numbered Exercises 7–49

Assignment Guide

Core   1–5, 6–20 even, 22–52

Core Two-Year   1–50

Technology

Students with graphing calculators may want to experiment with the absolute value key. First have them choose expressions for which they know the answer, such as ABS (5).

Error Analysis

Students may become confused and think that $|-8| = 8$ and $|8| = -8$. Reinforce the idea that absolute value refers to distance and cannot be negative.

Technology
4.8 Exploring Absolute Value Inequalities

In Lesson 4.8 you learned how to solve absolute value inequalities like $|x - 3| < 2$ and $|x - 3| > 2$. These inequalities are significantly different from one another.

To explore the difference, you can use a spreadsheet like the one outlined below.

In column A, enter various values of x and evaluate $|x - 3|$ for those values of x in column B.

    Cell A1 contains -8. Then fill column A with integers from -8 to 8.
    Cell B1 contains the value of ABS(A1$-$3) Then FILL DOWN column B.

In columns C and D, test whether a value of column B is less than or greater than 2, that is, whether it is a solution of each inequality.

    Cell C1 contains the formula IF(B1 < 2,1,0). Then FILL DOWN column C.
    Cell D1 contains the formula IF(B1 > 2,1,0). Then FILL DOWN column D.

By studying the results of the spreadsheet, you can discover two important properties of absolute value inequalities.

1. Create a spreadsheet like the one shown.

Use your spreadsheet to determine the solution set of each inequality in the given pair of inequalities.

2. $|x - 3| < 2$;   3. $|x + 3| < 3$;   4. $|2x| < 4$;
   $|x - 3| > 2$ $|x + 3| > 3$ $|2x| > 4$

5. $|2x - 1| < 3$;   6. $|3x + 6| < 12$;   7. $|2x + 1| < 3$;
   $|2x - 1| > 3$ $|3x + 6| > 12$ $|2x + 1| > 3$

8. Refer to Exercise 2. What do the values in columns C and D tell you about the solutions of $|x - 3| < 2$ and $|x - 3| > 2$?

9. Use algebra to solve $|ax + b| = c$ for x where c is positive and a is not 0. Use your solution of $|ax + b| = c$ and your results from Exercises 2–7 to describe all solutions of the inequality $|ax + b| < c$ and the inequality $|ax + b| > c$.

Performance Assessment

Have students write a problem that can be solved using absolute value. Then have students exchange problems and solve them.

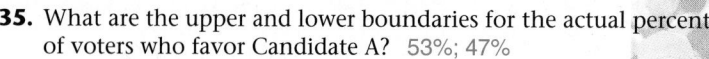
Let |*x* − 3| < 4.

27. Write the absolute value inequality as a sentence that begins, "The distance between . . ." *x* and 3 is less than 4.

28. Draw a number line solution for the inequality.

29. Describe the solution in words. any number between but not including −1 and 7

30. Give 5 specific numbers that satisfy the inequality.
0, 1, 2, 3, and 4

Let |*x* − 3| > 4.

31. Write the absolute value inequality as a sentence that begins, "The distance between . . ." *x* and 3 is greater than 4.

32. Draw a number line solution for the inequality.

33. Describe the solution in words. all numbers greater than 7 or less than −1

34. Give 5 specific numbers that satisfy the inequality.
−2, −3, −4, 8, and 9

Government In a recent voter preference poll between two candidates, respondents gave Candidate A 50% of the vote. The polling technique used gives results accurate to within 3 percentage points.

35. What are the upper and lower boundaries for the actual percent of voters who favor Candidate A? 53%; 47%

36. Describe the upper and lower boundaries using an absolute value equation. |*A* − 50| ≤ 3

37. Is it possible that Candidate A will lose the election? Explain. yes

Technology Use technology to compare each set of two graphs.

38. $y = 2x - 6$
$y = |2x - 6|$

39. $y = -x + 5$
$y = |-x + 5|$

40.  **Portfolio Activity** Complete the problem in the portfolio activity on page 145. 84 years

Look Back

Evaluate the following if *p* = 4, *q* = 1, and *r* = 2. [Lesson 1.4]

41. $pqr - q$ 7   **42.** $\dfrac{pq}{r}$ 2   **43.** $\dfrac{pqr}{q} + pqr$ 16

Simplify. [Lesson 3.5]

44. $-5(8c + 3)$ **45.** $9(7b + 2)$ **46.** $-4(-5k + 8)$
−40c − 15   63b + 18   20k − 32

Solve the following inequalities. [Lesson 3.6]

47. $4x + 5 \le 25$ **48.** $6y - 10 > 5$ $y > 2\frac{1}{2}$ **49.** $9m - 8 < 4 + 8m$
x < 5   m < 12

50. What percent of 480 is 60? [Lesson 4.5]
$12\frac{1}{2}$%

Look Beyond

Look Beyond

Exercises 51–52 anticipate working with polynomials. Allow students to solve these using common sense. Discuss the correct solutions, and assure students that they will learn more about solving this type of exercise later in the course.

51. Subtract $x^2 + 4$ from $x^2 - x + 5$.   −x + 1

52. Subtract $x^2 + 2x + 3$ from $x^2 + 10$.   −2x + 7

28. −1 < *x* < 7

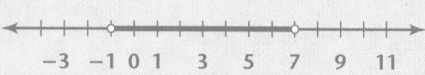

32. *x* < −1 or *x* > 7

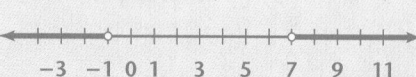

38. The function $y = 2x - 6$ is a line. The function $y = |2x - 6|$ is a V that opens upward and whose lowest point, or vertex, is at (3, 0). The portion of the line $y = 2x - 6$ that lies above the *x*-axis is the right half of the V.

39. The function $y = -x + 5$ is a line. The function $y = |-x + 5|$ is a V that opens upward and whose lowest point (vertex) is at (5, 0). The portion of the line $y = -x + 5$ that lies above the *x*-axis is the left side of the V.

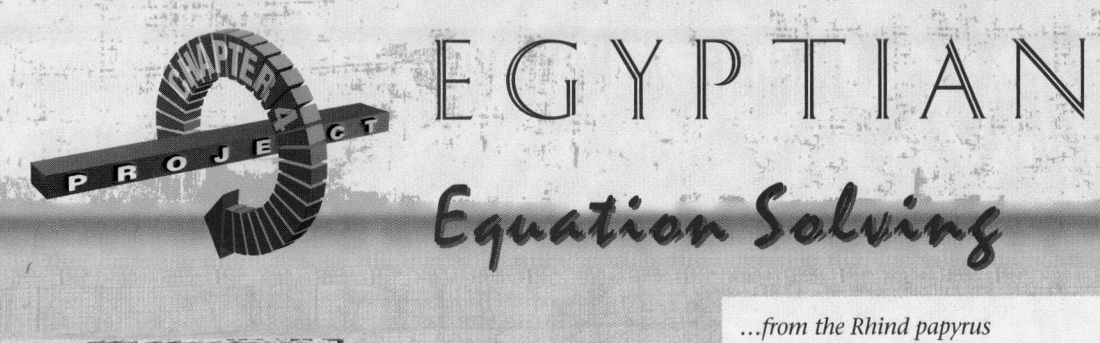

EGYPTIAN
Equation Solving

...from the Rhind papyrus

tp-ḥśb, *accurate reckoning*
n, *of*
hʹ·t , *entering*
m, *into*
ḫ·t, *things*
rḫ, *knowledge*
nt·t, *of existing things*
nb·t, *all*

About 4000 years ago, Egyptian mathematicians wrote texts to prepare students for jobs in commerce. In 1858, a Scotsman named Rhind bought one of these texts which had been copied by a scribe named Ahmes about 1650 B.C.E.

One problem of the Ahmes Papyrus asks, "What is the amount of meal in each loaf of bread if 2520 *ro* of meal is made into 100 loaves?" In the form of an equation: **$100m = 2520$**

The Egyptians solved the equation using the methods of make-a-table and guess-and-check.

They first substituted key values, like different powers of ten, for the variable to find the value of the left side of the equation.

They then used multiples of those values to see how close they could get to 2520. Since $2520 = 2000 + 500 + 20$, they asked what value of *m* would produce 20 for $100m$.

For the equation $100m = 2520$, they needed to compute $100(20) + 100(5) + 100\left(\frac{1}{5}\right)$ on the left side to get 2520 on the right side.

So, *m* was $20 + 5 + \frac{1}{5}$ or $25\frac{1}{5}$ *ro*.

m	100m
1	100
10	1000
20	2000
5	500
$\frac{1}{5}$	20

Activity

Solve these equations as the Egyptians might have.

a. $40m = 4800$ **c.** $60j = 930$

b. $24p = 252$ **d.** $10w = 2432$

Make up and solve some equations of your own.

> HINT:
> For problem **a** complete this table.
>
> | m | 40m |
> |---|-----|
> | 1 | 40 |
> | 100 | — |
> | 20 | — |

FOCUS

This project involves solving an equation by using a guess-and-check strategy much like the Egyptians did almost 4000 years ago.

MOTIVATE

About 1850 B.C.E., Amenemhat III ruled Egypt. During his rule an extensive system of irrigation was built. This required the knowledge of leveling, surveying, and measuring. These needs forced Egyptians to develop more sophisticated mathematical methods, such as solving equations.

About 1650 B.C.E., a scribe named Ahmes wrote or copied an older work on the mathematics known to the Egyptians at that time. The Ahmes Papyrus is not a textbook, but rather a practical handbook. Much of what we know about Egyptian mathematics at that time comes from this handbook.

Cooperative Learning

Have students work in pairs to solve the given equations in the same way the Egyptians might have solved them. Point out that Egyptians only used *unit fractions*—that is, fractions with a numerator of one. Thus, only unit fractions should be used in the solution of the equations.

DISCUSS

You may wish to discuss with students how the quantity $\frac{7}{8}$ could be written as a sum of unit fractions. $\left[\frac{1}{2} + \frac{1}{4} + \frac{1}{8}\right]$

---

a. m	40m	**b.** j	60j	**c.** p	24p	**d.** w	10w
1	40	1	60	1	24	1	10
10	400	10	600	10	240	10	100
100	4000	5	300	$\frac{1}{2}$	12	100	1000
20	800	$\frac{1}{2}$	30			200	2000
						40	400
						3	30
						$\frac{1}{5}$	2

$m = 100 + 20$
$m = 120$

$j = 10 + 5 + \frac{1}{2}$
$j = 15\frac{1}{2}$

$p = 10 + \frac{1}{2}$
$p = 10\frac{1}{2}$

$w = 200 + 40 + 3 + \frac{1}{5}$
$w = 243\frac{1}{5}$

Chapter 4 Review

Vocabulary

Distributive Property Over Subtraction	153	Multiplication Property of Inequality	188
Dividing an Expression	155	percent	173
Division Property of Equality	161	percent bar	174
Division Property of Inequality	188	Properties of Zero	148
equation method	176	proportion method	176
multiplicative Inverse	169	rational numbers	168
Multiplicative Property of − 1	153	reciprocal	168
Multiplication Property of Equality	162	Reciprocal Property	168

Key Skills & Exercises

Lesson 4.1

➤ *Key Skills*

Use the rules for multiplying and dividing integers to evaluate the following expressions.

a. $(-14)(-3) = 42$ **b.** $(-72) \div (24) = -3$ **c.** $(-8)[(-7) + 7] = (-8)(0) = 0$

➤ *Exercises*

Evaluate.

1. $(-12)(-5)$ 60   **2.** $(-0.9)(3)$ -2.7 **3.** $(54) \div (-18)$ -3
4. $(-121) \div (-11)$ 11   **5.** $(-2)(-2)(-2)(-2)$ 16   **6.** $(-6)(-3) \div (-9)$ -2
7. $45[8 + (-8)]$ 0   **8.** $(-4)(4) \div (-1)$ 16   **9.** $(-5)(-5)(-1)(1)$ -25

Lesson 4.2

➤ *Key Skills*

Use the rules for multiplying and dividing expressions.

a. $-4(5x - 6) = (-4)(5x) - (-4)(6)$
$\qquad\qquad\quad = -20x + 24$

b. $(2a - 3b + 1) - 4(a + 6b - 3) = 2a - 3b + 1 - 4a - 24b + 12$
$\qquad\qquad\qquad\qquad\qquad\qquad = (2a - 4a) + (-3b - 24b) + 13$
$\qquad\qquad\qquad\qquad\qquad\qquad = -2a - 27b + 13$

c. $\frac{12 - 8y}{-4} = \frac{12}{-4} + \left(\frac{-8y}{-4}\right) = -3 + 2y$

➤ *Exercises*

Multiply and divide expressions.

10. $3 \cdot 9x$ 27x   **11.** $-33x \div 3$ $-11x$ **12.** $-2(7y - 2)$ $-14y + 4$
13. $-2.4x \cdot 2x$ $-4.8x^2$ **14.** $\frac{-30v + 3.6}{-3}$ $10v - 1.2$ **15.** $9r^2 - 8(4 - 3r^2)$ $33r^2 - 32$

Lesson 4.3

➤ Key Skills

Solve multiplication equations.

Solve $-9k = 108$.

$-9k = 108$	Given
$\dfrac{-9k}{-9} = \dfrac{108}{-9}$	Division Property of Equality
$k = -12$	Simplify.

Solve division equations.

Solve $\dfrac{w}{-5} = -2.2$.

$\dfrac{w}{-5} = -2.2$	Given
$-5\left(\dfrac{w}{-5}\right) = -5(-2.2)$	Multiplication Property of Equality
$w = 11$	Simplify.

➤ Exercises

Solve each equation.

16. $17x = -85$ $\;-5$ **17.** $-4g = -56$ $\;14$ **18.** $-2.2h = 33$ $\;-15$

19. $24f = 150$ $\;6\frac{1}{4}$ **20.** $\dfrac{w}{-8} = 0.5$ $\;-4$ **21.** $\dfrac{y}{-2.4} = -10$ $\;24$

Lesson 4.4

➤ Key Skills

Solve an equation using the Reciprocal Property.

Solve $\dfrac{-2}{5}t = 14$.

One way to solve this equation is to multiply both sides by $\dfrac{5}{-2}$.

$$\dfrac{-2}{5}t = 14$$

$$\dfrac{5}{-2}\left(\dfrac{-2}{5}t\right) = \dfrac{5}{-2}(14)$$

$$t = -35$$

➤ Exercises

Solve each equation.

22. $\dfrac{4}{7}x = 4$ $\;7$ **23.** $\dfrac{1}{-9}y = 2.5$ $\;-22.5$ **24.** $\dfrac{-j}{7} = -8$ $\;56$ **25.** $\dfrac{w}{-8} = \dfrac{3}{-4}$ $\;6$ **26.** $-\dfrac{3}{5}m = \dfrac{8}{15}$ $\;-\dfrac{8}{9}$

Lesson 4.5

➤ Key Skills

Solve percent problems.

a. Find 30% of 15. You can use the equation method.

$$0.30 \cdot 15 = x$$
$$4.5 = x$$

b. What percent of 50 is 75? You can use a percent bar or an equation.

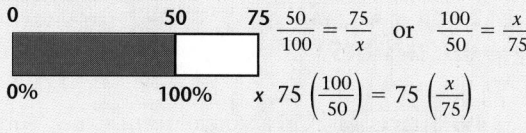

$$\dfrac{50}{100} = \dfrac{75}{x} \quad \text{or} \quad \dfrac{100}{50} = \dfrac{x}{75}$$

$$75\left(\dfrac{100}{50}\right) = 75\left(\dfrac{x}{75}\right)$$

$$150 = x$$

The solution is 150%.

➤ Exercises

Find each answer.

27. 55% of 60 $\;33$ **28.** 28 is 70% of what number? $\;40$ **29.** 200% of 40 $\;80$

30. 35% of 140 $\;49$ **31.** What percent of 90 is 40.5? $\;45\%$ **32.** 4 is what percent of 50? $\;8\%$

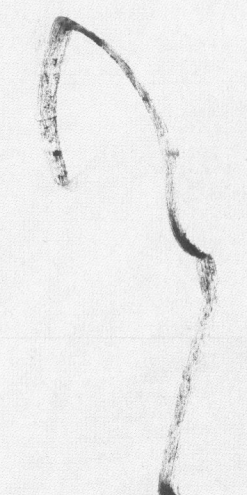

Lessons 4.6–4.7

➤ Key Skills

Solve multistep equations.

Solve $8c - 9 = 15$.

$8c - 9 = 15$	Given
$8c - 9 + 9 = 15 + 9$	Addition Property of Equality
$8c = 24$	Simplify.
$\dfrac{8c}{8} = \dfrac{24}{8}$	Division Property of Equality
$c = 3$	Simplify.

Solve inequalities.

Solve $14 - x \le -6$.

$14 - x \le -6$	Given
$14 - 14 - x \le -6 - 14$	Subtraction Property of Inequality
$-x \le -20$	Simplify.
$-1(-x) \ge -1(-20)$	Multiplication Property of Inequality
$x \ge 20$	Simplify.

➤ Exercises

Solve.

33. $3a + 7 = 31$ 8

34. $-2x + 10 = -4$ 7

35. $n - 6 = 2n - 14$ 8

36. $\dfrac{y}{5} + 4 = 4.05$ 0.25

37. $7k - 2(k + 6) = -2$ 2

38. $2q + 5 = \dfrac{q}{5} - 4$ −5

39. $x + 4 < 6$ $x < 2$

40. $8 - y \ge 7$ $y \le 1$

41. $5r > -60$ $r > -12$

42. $\dfrac{-p}{8} \le -3$ $p \ge 24$

43. $x + 3 \ge 9 - x$ $x \ge 3$

44. $t + \dfrac{1}{2} < \dfrac{t}{4} + 2$ $t < 2$

Lesson 4.8

➤ Key Skills

Solve absolute value equations and inequalities.

Solve $|x - 2| = 3$.

Case 1
The quantity $x - 2$ is positive.

$$|x - 2| = 3$$
$$x - 2 = 3$$
$$x = 5$$

Case 2
The quantity $x - 2$ is negative.

$$|x - 2| = 3$$
$$-(x - 2) = 3$$
$$-x + 2 = 3$$
$$-x = 1$$
$$x = -1$$

Solve $|x - 2| < 3$.

Case 1
The quantity $x - 2$ is positive.

$$|x - 2| < 3$$
$$x - 2 < 3$$
$$x < 5$$

Case 2
The quantity $x - 2$ is negative.

$$|x - 2| < 3$$
$$-(x - 2) < 3$$
$$-x + 2 < 3$$
$$-x < 1$$
$$x > -1$$

$$-1 < x < 5$$

➤ Exercises

Solve each equation or inequality. Graph the inequalities on a number line.

45. $|x - 4| = 8$ $-4, 12$

46. $|3x + 2| = 6$ $1\tfrac{1}{3}, -2\tfrac{2}{3}$

47. $|x - 3| \le 1$ $2 \le x \le 4$

48. $|x - 6| < 2$ $4 < x < 8$

Applications

Solve.

49. Sports Mark's basketball team scored 66, 74, 70, 82, and 58 points in their last 5 games. What is the average score for the team? Is a guess of 72 reasonable? 70; yes

50. Geometry The area of a triangle with a base of 6 centimeters is 21 square centimeters. What is the height of the triangle? (HINT: $A = \dfrac{1}{2} bh$) 7 cm

51. Statistics Suppose that of all of the potatoes grown in the United States, 22% are made into French fries. Out of 50 pounds of potatoes, how many pounds are made into French fries? 11

47.

$-4 \quad -2 \quad 0 \quad 2 \quad 4$

48.

$0 \quad 2 \quad 4 \quad 6 \quad 8$

Chapter 4 Assessment

Evaluate.

1. $(-25)(-4)$ 100   **2.** $(-3.6) \div (18)$ -0.2

3. $(-6)[(-3) + 3]$ 0   **4.** $(-80) \div (-4)(-5)$ -4

5. Explain how to write a bank transaction modeling deposits or withdrawals from a bank account for the representation $(-6)(-30)$.

6. Are all of the following expressions equivalent? If not, which are different?   a, b, and d

   a. $(8x - 32) \div 4$ **b.** $(8x - 32)(0.25)$ **c.** $\dfrac{32 - 8x}{4}$ **d.** $\dfrac{-32 + 8x}{4}$

Multiply or divide the following expressions.

7. $-5x \cdot 7$ **8.** $-3(4 - 11y)$ **9.** $\dfrac{-28q}{-14}$ **10.** $\dfrac{-10t + 35}{-5}$

   $-35x$ $-12 + 33y$ $2q$ $2t - 7$

Solve each equation.

11. $|x - 4| = 11$ **12.** $\dfrac{r}{5} = -22$ **13.** $-10f = 0.55$ **14.** $\dfrac{z}{-3.5} = -7$

   $-7, 15$ -110 -0.055 24.5

15. Farah needs to make a 360 mile trip in 9 hours. What speed should she average for the trip?   40 mph

16. Are all rational numbers integers? Explain your answer.

17. What is the only rational number that does not have a reciprocal? Explain why it does not have a reciprocal.

Solve each equation or proportion.

18. $\dfrac{2}{3}y = -8$ -12 **19.** $\dfrac{-5}{11}p = -15$ 33   **20.** $\dfrac{-c}{14} = -\dfrac{1}{7}$ 2

21. $\dfrac{x}{3} = \dfrac{-2}{5}$ $-1\dfrac{1}{5}$ **22.** $\dfrac{-a}{6} = \dfrac{4}{-9}$ $2\dfrac{2}{3}$ **23.** $\dfrac{r}{2.8} = -\dfrac{5}{7}$ -2

24. Paula made $350 in 4 weeks. At this rate, how much will she make in 10 weeks? $875

Find each answer.

25. 35% of 90   31.5   **26.** 3 is what percent of 15?   20%   **27.** 150% of 28   42

28. Jamie made 80% of his free throws in his game last night. If he attempted 15 free throws, how many free throws did he make?   12

Solve each equation.

29. $3w + 5 = -16$ **30.** $\dfrac{m}{2} - 1 = 11$ **31.** $4x + 9 = 2x + 6$

   -7 24 $-1\dfrac{1}{2}$

32. The formula for the area of a triangle is $A = \dfrac{1}{2}bh$. If the area of a triangle is 336 square centimeters, and the base is 24 centimeters, what is the height of the triangle?   28 cm

Solve each inequality.

33. $|x - 8| \geq 14$ **34.** $-9n < 42$ **35.** $\dfrac{j}{4} + 3 \leq -10$ **36.** $|x - 7| < 9$

---

5. Since the 6 is negative, an amount will be withdrawn from the bank account. Since the 30 is also negative, the amount being withdrawn is also negative. Therefore, the transaction is to remove 6 withdrawals of $30 each. This restores $180 to the account.

16. All rational numbers are not integers, because rational numbers include the fractions between the integers.

17. 0; By definition a number is the reciprocal of another if their product is equal to 1. There is no number that when multiplied by 0 will yield a product of 1.

33. $x \leq -6$ OR $x \geq 22$

34. $n > -4\dfrac{2}{3}$

35. $j \leq -52$

36. $-2 < x < 16$

Chapters 1 – 4
Cumulative Assessment

College Entrance Exam Practice

Quantitative Comparison   For Questions 1–5, write

A if the quantity in Column A is greater than the quantity in Column B;
B if the quantity in Column B is greater than the quantity in Column A;
C if the two quantities are equal; or
D if the relationship cannot be determined from the information given.

	Column A	Column B	Answers
1.	$2x + 1$ $x = 6$	$4x - 3$	B Ⓐ Ⓑ Ⓒ Ⓓ **[Lesson 1.3]**
2.	The number of \$14 CDs you can buy for \$86.	The number of \$12 CDs you can buy for \$75.	C Ⓐ Ⓑ Ⓒ Ⓓ **[Lesson 3.5]**
3.	$|-12.4|$	$|12.4|$	C Ⓐ Ⓑ Ⓒ Ⓓ **[Lesson 2.4]**
4.	INT(4.75)	INT(5.2)	B Ⓐ Ⓑ Ⓒ Ⓓ **[Lesson 2.4]**
5.	$(3y + 2) + (2y - 5)$	$(2y + 6) - (y + 7)$	D Ⓐ Ⓑ Ⓒ Ⓓ **[Lesson 3.4]**

6. Which expression is equal to 19?   **[Lesson 1.4]** d
   a. $15 \cdot (6 + 4) - 57$ **b.** $2^3 - 6 \cdot (3 + 2)$ **c.** $4^2 - 5^2 \div 5 + 2(0)$ **d.** $9 \div 3 \cdot 2^3 - 5$

7. Which decimal number is the equivalent of 143%?   **[Lesson 4.5]** b
   a. 0.143   **b.** 1.43   **c.** 143.0   **d.** 14.3

8. Which percent is equivalent to 0.5?   **[Lesson 4.5]** b
   a. 0.5%   **b.** 50%   **c.** 5%   **d.** 0.05%

9. The correlation between two variables is described as strongly positive. What is the relationship between the two variables?   **[Lesson 1.6]** a
   a. As one variable increases, the other also tends to increase.
   b. As one variable increases, you cannot tell if the other tends to increase or decrease.
   c. As one variable increases, the other tends to decrease.

10. Which sequence is linear?   **[Lesson 2.1]** d
    a. 60, 30, 15, 7.5, 3.75, . . .   **b.** 6, 12, 24, 48, 96, . . .
    c. 6, 9, 13.5, 20.25, 30.375, . . .   **d.** none

11. What is the value of $\frac{5^2 + 11}{9 \cdot 3 + 9}$? **[Lesson 1.4]** c

    a. 36   **b.** 0   **c.** 1   **d.** $\frac{4}{9}$

12. Which expression is *not* equivalent to the others? **[Lesson 4.2]** c

    a. $3x - 2(x - 3)$ **b.** $3x + 2(3 - x)$ **c.** $3x - 2(x + 3)$ **d.** $3x - 2(-3 + x)$

13. What is the solution to the equation $\frac{q}{139.2} = -58$? **[Lesson 4.3]** b

    a. -2.4 **b.** -8073.6 **c.** -0.417 **d.** -7551.6

14. 60 is 75% of what number? **[Lesson 4.5]** d

    a. 0.45   **b.** 0.8   **c.** 45   **d.** 80

15. Which is the value of x for the absolute value equation $|x - 6| = 10$? d
    [Lesson 4.8]

    a. -16 **b.** 4   **c.** 10   **d.** -4

16. Is $18t$ a variable or an expression? Explain your answer. **[Lesson 1.3]**

What are the coordinates of the given points? [Lesson 1.5]

17. X (1, 3)   **18.** Y (4, 0)   **19.** Z (5, 5)

20. Solve the absolute value inequality $|x - 4| > 11$.
    [Lesson 4.8] $x < -7$ or $x > 15$

21. Determine whether the relationship is linear, quadratic,
    or neither. **[Lesson 2.1]** quadratic

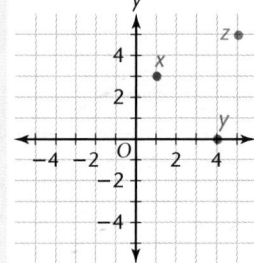

x	1	2	3	4	5
y	4	6	10	16	24

Simplify the following expressions. [Lesson 4.2]

22. $(9w - 5) - (3w - 4)$ $6w - 1$ **23.** $(-54s) \div (6s)$ -9 **24.** $4x^2 - 4(3x^2 + 7)$
                                                 $-8x^2 - 28$

25. Solve the inequality $y + 4.5 \geq 6$. **[Lesson 3.6]**
    $y \geq 1.5$

26. Jeff is on the 14th floor of a medical building. He rides the elevator up 3
    floors, down 5 floors, and then up 4 floors. What floor is Jeff on now?
    [Lesson 1.1] 16th

Evaluate. [Lesson 3.1]

27. $[-16 + (-4)] + 6$ **28.** $-8 + [(-23) + 23]$ **29.** $-48 + 8 - 3 + 2$
    -14 -8 -41

30. Solve the proportion $\frac{g}{12} = \frac{-3}{20}$. **[Lesson 4.4]** $-1\frac{4}{5}$

Free-Response Grid The following questions may be answered
using a free-response grid commonly used by standardized test
services.

31. What is the solution to the equation
    $3r + 2(r - 4) = 12$? **[Lesson 4.6]** 4

32. Hannah got 84% of the questions correct on a test. If
    there were 50 questions on the test, how many did she
    answer correctly? **[Lesson 4.5]** 42

33. Maurice has scores of 88, 90, and 80 on three tests.
    What score must he make on his next test to have a
    test average of 85? **[Lesson 4.1]** 82

16. It is an *expression* because it contains a
    number, 18, a variable t, and a mathemati-
    cal operator, multiplication.

Lines and Slopes

CHAPTER 5

Meeting Individual Needs

5.1 Defining Slope

Core Resources	Core Two-Year Resources
Practice Master 5.1	Inclusion Strategies, p. 210
Enrichment, p. 210	Reteaching the Lesson, p. 211
Technology Master 5.1	Practice Master 5.1
Interdisciplinary	Enrichment Master 5.1
Connection p. 209	Technology Master 5.1
	Lesson Activity Master 5.1
[1 day]	**[2 days]**

5.2 Exploring Graphs of Linear Functions

Core Resources	Core Two-Year Resources
Practice Master 5.2	Inclusion Strategies, p. 220
Enrichment Master 5.2	Reteaching the Lesson, p. 221
Technology Master 5.2	Practice Master 5.2
Interdisciplinary	Enrichment, p. 220
Connection, p. 219	Technology Master 5.2
	Lesson Activity Master 5.2
[1 day]	**[2 days]**

5.3 The Slope-Intercept Formula

Core Resources	Core Two-Year Resources
Practice Master 5.3	Inclusion Strategies, p. 226
Enrichment, p. 226	Reteaching the Lesson, p. 227
Technology Master 5.3	Practice Master 5.3
Interdisciplinary	Enrichment Master 5.3
Connection, p. 225	Lesson Activity Master 5.3
Mid-Chapter Assessment	Mid-Chapter Assessment Master
Master	
[1 day]	**[2 days]**

5.4 Other Forms for Equations of Lines

Core Resources	Core Two-Year Resources
Practice Master 5.4	Inclusion Strategies, p. 235
Enrichment, p. 235	Reteaching the Lesson, p. 236
Technology Master 5.4	Practice Master 5.4
Interdisciplinary	Enrichment Master 5.4
Connection, p. 234	Technology Master 5.4
	Lesson Activity Master 5.4
[1 day]	**[2 days]**

5.5 Vertical and Horizontal Lines

Core Resources	Core Two-Year Resources
Practice Master 5.5	Inclusion Strategies, p. 241
Enrichment, p. 241	Reteaching the Lesson, p. 242
Technology Master 5.5	Enrichment Master 5.5
	Practice Master 5.5
	Technology Master 5.5
	Lesson Activity Master 5.5
[1 day]	**[2 days]**

5.6 Parallel and Perpendicular Lines

Core Resources	Core Two-Year Resources
Practice Master 5.6	Inclusion Strategies, p. 248
Enrichment, p. 248	Reteaching the Lesson, p. 249
Technology Master 5.6	Practice Master 5.6
Interdisciplinary	Enrichment Master 5.6
Connection, p. 247	Technology Master 5.6
	Lesson Activity Master 5.6
[1 day]	**[3 days]**

Chapter Summary

Core Resources	Core Two-Year Resources
Eyewitness Math, pp. 216–217	Eyewitness Math, pp. 216–217
Chapter 5 Project, pp. 252–253	Chapter 5 Project, pp. 252–253
Lab Activity	Lab Activity
Long Term Project	Long-Term Project
Chapter Review, pp. 254–256	Chapter Review, pp. 254–256
Chapter Assessment, p. 257	Chapter Assessment, p. 257
Chapter Assessment, A/B	Chapter Assessment, A/B
Alternative Assessment	Alternative Assessment
[3 days]	**[5 days]**

Reading Strategies

Continually evaluate the students' ability to read and comprehend the material. Ask students to record responses to the Exploration Labs, Critical Thinking, and Ongoing Assessment questions in their journals. These features provide additional insight and information about the concepts taught in each lesson. After discussing each lesson, have the students review their notes on each lesson in groups and make any necessary corrections. Students can further review their comprehension of the concepts presented in the chapter through class discussion of the Communicate exercises. You may want to have students write explanations of the answers to some of the Communicate exercises in their journals.

Visual Strategies

Have students look through the different graphs in this chapter. Ask them to describe situations in which they have seen similar kinds of graphs. For example, most students have seen linear graphs in newspapers and magazines. Have students describe how they think the graphs they see in this chapter can be used in algebra. Encourage students to discuss their ideas with one another. After they have studied the lesson, have them compare their initial thoughts about graphs with the ways in which the graphs are actually used.

Hands-on Strategies

Students can learn about lines and slopes by using wire, string, and pieces of hard spaghetti to represent lines, and by using graph paper, colored pencils, transparencies, and markers to draw lines. Some Explorations throughout the chapter are to be completed using one or more of these manipulatives.

Cooperative Learning

GROUP ACTIVITIES	
Applying slope	Chapter Opener, The Portfolio Activity
Steps and slope	Eyewitness Math
Exploring $y = mx$	Lesson 5.2, Exploration 1, 2, and 3
Regression lines	Lesson 5.3, Example 3 and Exercise 47
Slopes and perpendicular lines	Lesson 5.6, Exploration 1 and 2
Using Diophantine equations	Chapter 5 Project

You may wish to have students work with partners for some of the above activities. Additional suggestions for cooperative group activities are noted in the teacher's notes in each lesson.

Multicultural

The cultural connections in this chapter include references to Africa and North America.

CULTURAL CONNECTIONS	
Africa: slope and Egyptian pyramids	Chapter Opener and Portfolio Activity, 206–207
Africa: slope and sekeds	Lesson 5.2, 223
Africa: Diophantine Equations	Chapter Project, 252–253
Americas: slope and Mexican Pyramids	Lesson 5.1, 214

Portfolio Assessment

Below are portfolio activities for the chapter, listed under seven activity domains that are appropriate for portfolio development.

1. **Investigation/Exploration** Two projects listed in the teacher notes ask students to investigate a method and/or a relationship. They are the *Alternative Teaching Strategy* on page 233 (Lesson 5.4) and *Interdisciplinary Connection* on page 219 (Lesson 5.2). One project, *Interdisciplinary Connection* on page 247 (Lesson 5.6), explores the use of parallel and perpendicular lines in an application.

2. **Applications** Recommended to be included are any of the following: Travel (Lesson 5.1), Recreation (Lessons 5.1 and 5.5), Sports (Lessons 5.1 and 5.3), Ecology (Lesson 5.2), Auto Racing (Lesson 5.3), Fund-raising (Lesson 5.4), Budget (Lesson 5.5), and, from the teacher notes, *Construction* on page 209 (Lesson 5.1), *Sports* on page 225 (Lesson 5.3), and *Surveying* on page 246 (Lesson 5.6).

3. **Non-Routine Problems** The Portfolio Activity on page 207 asks students to calculate slope using two different units of measure and then compare the results.

4. **Project** The Chapter Project on pages 252–253 asks students to use special equations called Diophantine equations to solve several different problems.

5. **Interdisciplinary Topics** Students may choose from the following: Science (Lessons 5.1 and 5.6), Geometry (Lessons 5.1, 5.3, 5.5, and 5.6), Statistics (Lesson 5.1), Social Studies (Lessons 5.1 and 5.4), Consumer Mathematics (Lesson 5.3), Language Arts (Lesson 5.5), and, from the teacher notes, *Physics* on page 219 (Lesson 5.2) and *Biology* on page 234 (Lesson 5.4).

6. **Writing** The interleaf material at the beginning of this chapter has information for journal writing. Also recommended are the *Communicate* questions at the beginning of the exercise sets.

7. **Tools** Chapter 5 uses calculators, computers, and graph paper for graphing lines and finding slopes.

Technology

Technology provides important tools for manipulating and visualizing linear equations and lines. Several lessons in Chapter 5 can be significantly enhanced by using appropriate technology. Graphics calculators make comparing the slope and y-intercept of different linear equations quick and easy. For more information on the use of technology, refer to the *HRW Technology Handbook*.

Graphics Calculator

Graphing Scatter Plots and Regression Lines

Lesson 5.3 gives general instructions on using a graphics calculator to create a scatter plot. Then find the equation and graph a regression line for the data. Specific instructions are given below for the TI-82. Please refer to the *HRW Technology Handbook* for instructions on using other graphics calculators.

1. Press the ⌊STAT⌋ key, choose **1: Edit,** and press ⌊ENTER⌋. If there are entries in any of the lists, press ⌊2nd⌋ ⌊QUIT⌋ to return to the Home screen. Then press ⌊STAT⌋ and choose **4: ClrList. ClrList** will appear on the Home screen, followed by the cursor. Press ⌊2nd⌋ ⌊L₁⌋ ⌊ENTER⌋. **Done** will appear on the Home screen, indicating that the entries in **L1** have been cleared. Repeat the process for each additional list that contains entries.

2. Press the ⌊STAT⌋ key and choose **1: Edit** and press ⌊ENTER⌋ again. All entries in all lists should now be cleared. Enter the x-values into **L1**, press the ⌊▶⌋ to move to the **L2** column, and enter the y-values into the **L2** column.

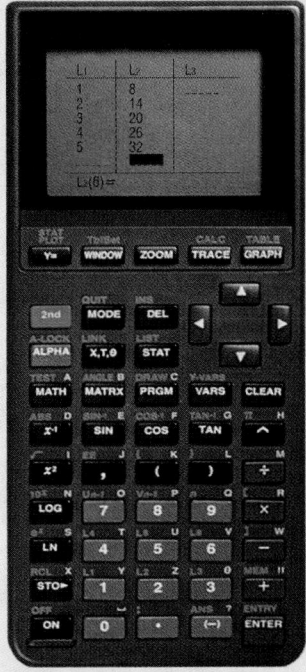

3. After entering all values, press the ⌊STAT⌋ key and choose the **5:LinReg(ax+b)** from the **CALC** menu. **LinReg(ax+b)** will appear on the Home screen. Press ⌊ENTER⌋ and the calculator will show values for a, b, and r. Substitute a and b in the equation $y = ax + b$ to get the equation of the line of best fit.

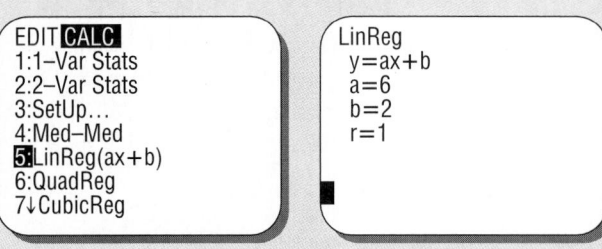

4. To draw the scatter plot, press ⌊2nd⌋ ⌊STAT PLOT⌋ and choose **On**, the scatter plot for Type: (which is the first choice), **L1** for **Xlist:,** **L2** for **Ylist:,** and the mark of your choice for Mark. Press ⌊ENTER⌋ and the scatter plot will appear. Then press ⌊Y=⌋ and enter the equation for the line of best fit. Press ⌊GRAPH⌋ and the line will be shown with the scatter plot.

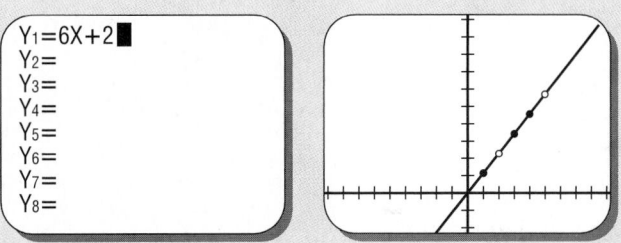

Integrated Software

f(g) Scholar[TM] is an integrated computer-based mathematics productivity tool that combines calculator, spreadsheet, and graphics capabilities to provide a dynamic and interactive environment for explorations in mathematics. It is appropriate to use *f(g) Scholar*[TM] for any lesson needing a spreadsheet, calculator, graphics calculator, or any combination of the three.

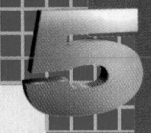

ABOUT THE CHAPTER

Background Information

The slope of each side of a pyramid is determined by the height of the pyramid and half the length of the pyramid's base, or the rise over the run.

CHAPTER RESOURCES

- Practice Masters
- Enrichment Masters
- Technology Masters
- Lesson Activity Masters
- Lab Activity Masters
- Assessment Masters
    Chapter Assessments, A/B
    Mid-Chapter Assessment
    Alternative Assessment,
    A/B
- Teaching Transparencies
- Spanish Resources

CHAPTER OBJECTIVES

- Calculate slope of a line using the concept of rise and run.
- Find the slope of the line of best fit from a set of data.
- Calculate slope of a line from the ratio of the differences in y- and x-coordinates.
- Identify how a change of slope affects a graph and the equation of a line.
- Identify how changing where the line crosses the y-axis affects a graph and the equation of a line.
- Define and explain the components of the Slope-Intercept Form of a line.
- Define and use the Slope-Intercept Form of the equation of a line.
- Solve problems of direct variation.

CHAPTER 5

LESSONS

5.1 Defining Slope

5.2 *Exploring* Graphs of Linear Functions

5.3 The Slope-Intercept Formula

5.4 Other Forms for Equations of Lines

5.5 Vertical and Horizontal Lines

5.6 Parallel and Perpendicular Lines

Chapter Project
Diophantine Equations

Linear Functions

Pyramids are part of the heritage of people who trace their roots to Africa or to Central America. The great pyramid of Gizeh in Egypt was built over a 30-year period around 2900 B.C.E. The builders used blocks averaging 2.5 tons, but some weighed as much as 54 tons. The blocks were transported from a quarry 600 miles away. The pyramid has a square base, and the error in constructing the right angles of the base was only one part in 14,000.

In order to build the pyramids, people needed to solve problems about slope. One problem asks to find the *seked* (a form of slope) of a pyramid with a base of 360 cubits (618 feet) on a side and a height of 250 cubits (429 feet).

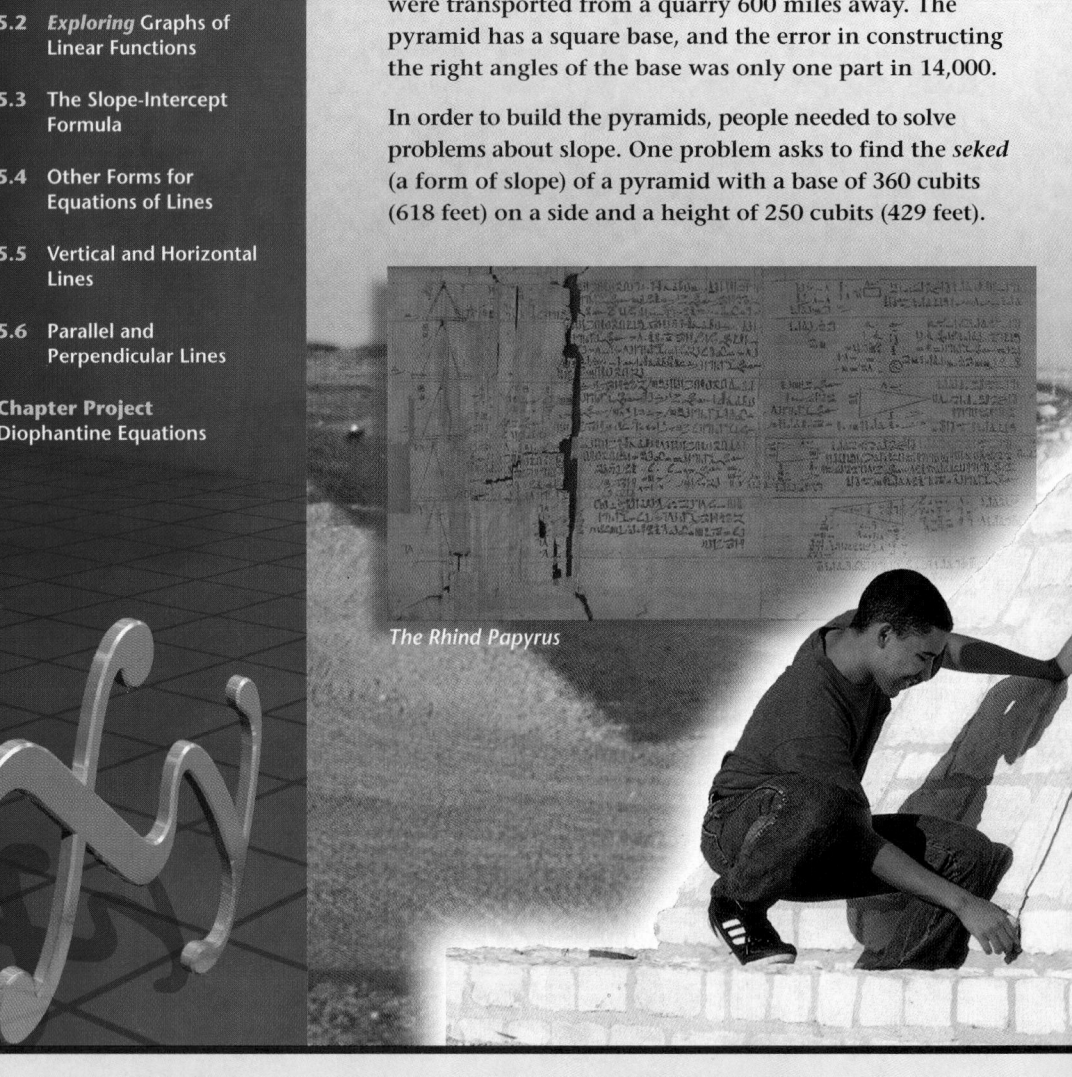

The Rhind Papyrus

ABOUT THE PHOTOS

There are several pyramids pictured in the chapter opener. Have students study these pyramids and discuss the similarities and differences. Focus their attention on the slope of the sides. Also ask students how they think mathematics may have been used in the building of these pyramids.

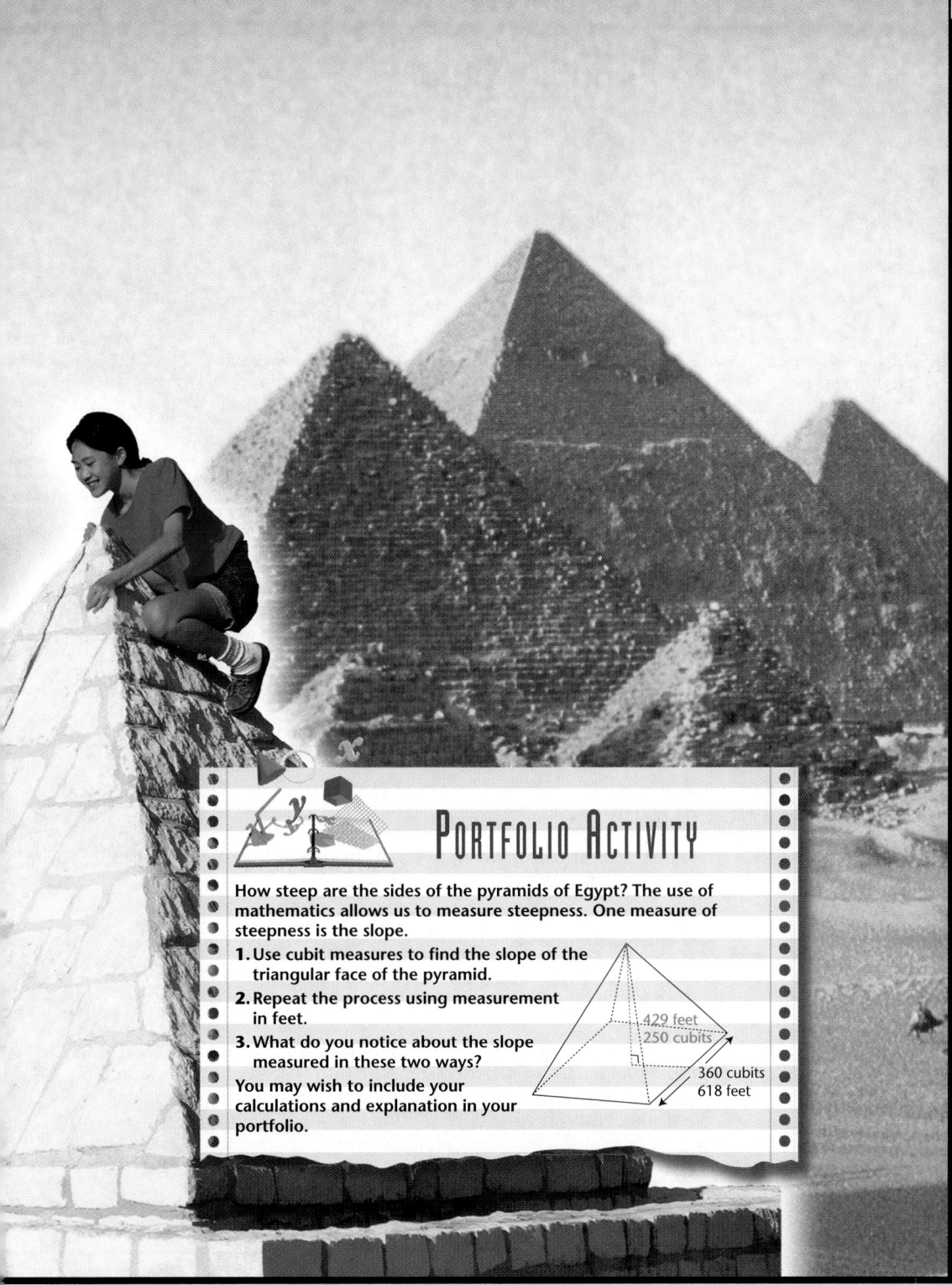

- Define and use the Standard Form of the equation of a line.
- Define the Point-Slope Form of the equation of a line.
- Investigate the function $y = a$ and the relation $x = b$ as examples of horizontal and vertical lines.
- Relate a slope of 0 to the graph of a horizontal line, and relate an undefined slope to the graph of a vertical line.
- Identify parallel lines as lines with the same slope.
- Identify perpendicular lines as lines with slopes that are the negative reciprocal of each other.
- Write equations of lines that are parallel and perpendicular to other given lines.

PORTFOLIO ACTIVITY

Assign the Portfolio Activity as part of the assignment for Lesson 5.2 on page 223.

Have students work in small groups to answer the questions in the portfolio activity. It may be helpful for them to draw and label a sketch of the pyramid. The labels should include both cubits and feet. This will further emphasize that the slope of a side is the same no matter how it is measured.

After students have found the slope using both units of measure, ask them to think about how they could build a small model of the pyramid. Have them work in small groups to carry out their plan. Conduct a class discussion of the plans used to construct the models, emphasizing the similarities and differences.

PORTFOLIO ACTIVITY

How steep are the sides of the pyramids of Egypt? The use of mathematics allows us to measure steepness. One measure of steepness is the slope.

1. Use cubit measures to find the slope of the triangular face of the pyramid.
2. Repeat the process using measurement in feet.
3. What do you notice about the slope measured in these two ways?

You may wish to include your calculations and explanation in your portfolio.

429 feet
250 cubits

360 cubits
618 feet

ABOUT THE CHAPTER PROJECT

The Chapter 5 Project on pages 252–253 introduces students to Diophantine equations. Such equations have only integers for solutions. Solutions involving both tables and graphs are presented. After students have used these equations to solve two problems, they summarize their work by completing three activities.

PREPARE

Objectives

- Calculate slope of a line using the concept of rise and run.
- Find the slope of the line of best fit from a set of data.
- Calculate the slope of a line from the ratio of the differences in *y*- and *x*-coordinates.

RESOURCES

- Practice Master   **5.1**
- Enrichment Master   **5.1**
- Technology Master   **5.1**
- Lesson Activity Master   **5.1**
- Quiz   **5.1**
- Spanish Resources   **5.1**

Assessing Prior Knowledge

Graph the pair of points. Then draw a line that contains both points.

1. $A(1,5), B(-2,3)$

2. $M(0,-6), N(-3,1)$

3. $X(2,5), Y(3,-1)$

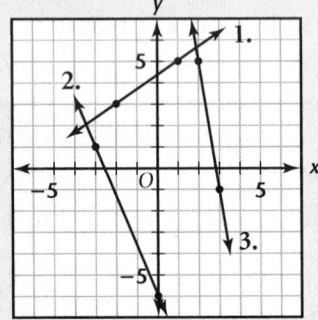

TEACH

Be sure students understand that the slope of this trail would be the same no matter whose step they use. Ask students to discuss and make sketches of examples of these different types of hills.

208 Lesson 5.1

why *Kara and Ricardo hike up the steep slope of a mountain trail. They remember how easy it was walking along the nearly flat streets back home. If you want to compare the steepness of a mountain trail with a level street, you can compare their slopes. You can also measure the steepness of lines on a coordinate plane the same way.*

Finding the Slope

Slope is the ratio of the vertical rise to the horizontal run. The slope of the hill is the same for both Kara and Ricardo.

EXAMPLE 1

What is the slope of the hill?

Solution ➤

To calculate the slope, find the ratio of rise to run for each hiker. The horizontal length of Ricardo's step (the run) is 35 centimeters, and the vertical length (the rise) is 14 centimeters. The run for Kara's step is 30 centimeters and the rise is 12 centimeters.

Ricardo	**Kara**
$slope = \dfrac{rise}{run} = \dfrac{14}{35} = \dfrac{2}{5}$	$slope = \dfrac{rise}{run} = \dfrac{12}{30} = \dfrac{2}{5}$

The slope of the hill is $\dfrac{2}{5}$. ❖

ALTERNATIVE teaching strategy

Hands-On Strategies

Have groups of students draw axes on graph paper and graph the line that passes through the points (1, 2) and (3, 5). Have each group draw another line on the piece of acetate. Ask each group to place the line over the original line they drew on the graph paper. Then have students slide the acetate line so that it is parallel to the original line. Ask students to record any observations they can make about

the lines, especially about their slopes. Then have them move the line to examine other lines parallel to the original line.

Have students place the line so that it passes only through the point (1, 2). Have them move the line to show several additional lines that pass through (1, 2). In each case, ask students to record their observations, especially about the slopes. This activity can be repeated with different lines.

Try This   Find the slope of a hill if the rise is 12 and the run is 24.

> **GEOMETRIC OR GRAPHIC INTERPRETATION OF SLOPE**
>
> **Slope** measures the steepness of a line by the formula
>
> $$\text{slope} = \frac{\text{rise}}{\text{run}}.$$

Sign of the Slope

The slope of the mountain trail was found from an upward rise left to right and calculated to be positive $\frac{2}{5}$. In Example 2, you will investigate a situation in which a line slopes downward from left to right.

EXAMPLE 2

Science   Maya and her class are studying the effects of friction. To model the concept, they lean a meter stick against a wall. The object is to see how steep an incline is necessary to keep the ruler from slipping. They find that the friction prevents the ruler from slipping when the top of the ruler is 80 centimeters from the table top and the base is 60 centimeters from the wall. The students record the results on a graph. The wall is the *y*-axis, and the table top is the *x*-axis. What is the slope of the line formed by the meter stick?

Solution ►

To measure the slope, start at the point where the line touches the *y*-axis. Measure the run, then the rise, and end at the point where the line touches the *x*-axis.

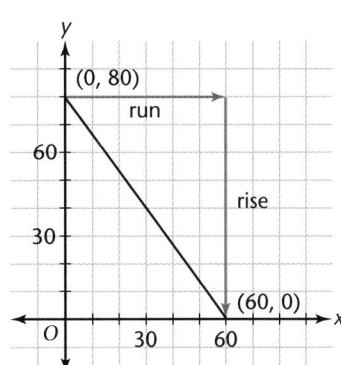

The distance toward the *right* is positive. Thus, the run is 60. The distance moving *down* is negative. Thus, the rise is -80.

$$\text{slope} = \frac{\text{rise}}{\text{run}} = \frac{-80}{60} \text{ or } -\frac{4}{3} \text{ ❖}$$

What makes the slope of a line negative?

interdisciplinary CONNECTION   **Construction**   There are many examples of the use of slope in construction. The use of slope in the construction of stairs is discussed in Eyewitness Math on pages 216 and 217. Have students do research to find other examples. Ask them to write a report about their findings, including diagrams as necessary. Have students share their reports with the class.

What is the slope of the line shown on the graph paper?

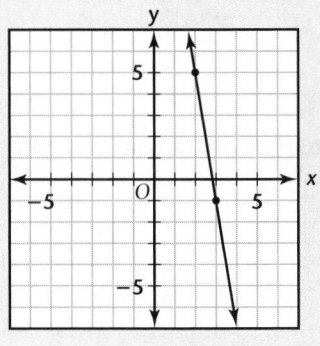

[−6]

Positive slope: $\frac{+}{+}$ or $\frac{-}{-}$

Negative slope: $\frac{+}{-}$ or $\frac{-}{+}$

Alternate Example 4

Find the average slope of the highway between Kansas City, Kansas, and Goodland, Kansas. [6.8 feet per mile]

Use Transparency 23

DIRECTIONS AND SIGNS

The *positive* directions (+) on the graph are toward the right and up. The *negative* directions (−) on the graph are toward the left and down.

EXAMPLE 3

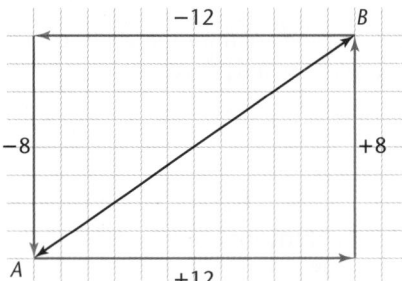

What is the slope of line *AB*?

Solution ➤

The run from *A* to *B* is 12, and the rise is 8.

$$\text{slope} = \frac{\text{rise}}{\text{run}} = \frac{8}{12} = \frac{2}{3}$$

The run from *B* to *A* is −12, and the rise is −8.

$$\text{slope} = \frac{\text{rise}}{\text{run}} = \frac{-8}{-12} = \frac{2}{3} \; ❖$$

What combinations of positive and negative values for rise and run produce positive slope? negative slope?

EXAMPLE 4

Travel Jim drives from Kansas City, Kansas, to Denver, Colorado, along Interstate 70. Although the land seems flat, the drive is uphill much of the time. The table shows the elevation for several cities along the way, and how many miles the cities are from Kansas City. What is the average slope of the highway between Kansas City and Denver?

CITY	MILES from Kansas City	ELEVATION in feet
Kansas City, KS	0	744
Lawrence, KS	29	850
Topeka, KS	61	951
Salina, KS	171	1220
Hays, KS	264	1997
Goodland, KS	407	3683
Burlington, CO	437	4160
Denver, CO	600	5280

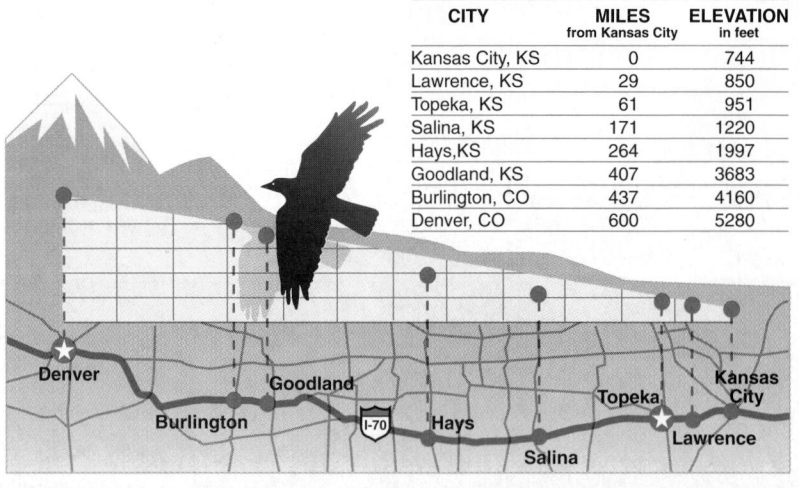

ENRICHMENT Have students find the slope between each pair of cities in the table in Example 4. Then have them compare the slopes and use these slopes to draw a graph that shows a side view of the road from Kansas City to Denver.

INCLUSION strategies **Using Visual Models** Have students graph the point (2, 2). Then have them draw several lines with positive and negative slopes through the point. Ask students to find the actual slope of each line, first by counting squares and then by using (2, 2) and another point on each line in the slope formula. Have them record any observations.

Solution ➤

To determine the slope, the average increase in elevation, the miles listed in the table are assumed to be horizontal.

STATISTICS
Connection

Use the data in the table to make a scatter plot. Draw a line that best fits the data points. Choose two convenient points on the line, the points (100, 1200) and (500, 4375), for example.

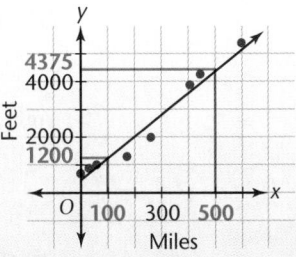

The *x*-coordinate changes from 100 to 500, an increase of 400 miles. The *y*-coordinate changes from 1200 to 4375, an increase of 3175 feet.

$$\text{slope} = \frac{\text{rise}}{\text{run}} = \frac{4375 - 1200}{500 - 100} = \frac{3175}{400} \approx 7.9$$

On the average, the elevation increases by about 7.9 feet per mile. ❖

EXAMPLE 5

Burlington, Colorado, elevation 4160 feet, is on Interstate 70 near the border of Kansas and Colorado. The horizontal distance from Denver to Burlington is 163 miles. Is the average slope of the highway steeper in Kansas or in eastern Colorado?

Solution ➤

Use the information from the table in Example 4.

Find the slope from Kansas City to Burlington.

$$\text{slope} = \frac{\text{rise}}{\text{run}} = \frac{4160 - 744}{437 - 0} = \frac{3416}{437} \approx 7.8$$

Find the slope from Burlington to Denver.

$$\text{slope} = \frac{\text{rise}}{\text{run}} = \frac{5280 - 4160}{600 - 437} = \frac{1120}{163} \approx 6.9$$

The slope is steeper in Kansas than in eastern Colorado. ❖

In the last two examples slope was determined by finding the differences in the *y*-coordinates and *x*-coordinates. This gives another interpretation of slope.

SLOPE AS A DIFFERENCE RATIO

If the coordinates of two points on a line are given, the slope, *m*, is the ratio $\frac{\text{difference in } y\text{-coordinates}}{\text{difference in } x\text{-coordinates}}$.

Hands-On Strategies
Draw several right triangles on the chalkboard, each with the right angle in the lower left-hand corner of the triangle. Ask students to study the slant of the hypotenuse of each right triangle and record any observations. Then have students study the lengths of the legs of each triangle and compare those lengths with their observations about the slant of the hypotenuse. Show students how the legs of a right triangle represent the rise and run of the slope of the hypotenuse. Then have students draw several of the triangles on a coordinate plane to verify this. Use the slope formula and two points to calculate the slope of each hypotenuse.

EXAMPLE 6

Find the slope of the line containing points $M(3, 5)$ and $N(-2, -6)$.

Solution ➤

$$\text{slope} = \frac{\text{rise}}{\text{run}} = \frac{\text{difference in } y\text{-coordinates}}{\text{difference in } x\text{-coordinates}}$$

You must remember to subtract the x-coordinates in the same order as the corresponding y-coordinates.

$$\text{slope} = \frac{5 - (-6)}{3 - (-2)} = \frac{11}{5} \quad \text{or} \quad \text{slope} = \frac{(-6) - 5}{(-2) - 3} = \frac{-11}{-5} = \frac{11}{5} \; ❖$$

What is the effect on the slope if the coordinates are not entered in the same order?

EXERCISES & PROBLEMS

Communicate

1. Explain how to draw a line with a rise of 4 and a run of 3.

2. Describe the slope of a line with negative rise and positive run.

3. Explain how to find the slope of line AC.

4. Explain why the slope of line AC is the same as the slope of line AB.

5. What is the slope of a line with rise s and run t? Compare this line with a line that has a rise $-s$ and run t. Which is steeper?

6. Point $R(5, -3)$ and point $N(9, -4)$ are on line k. Explain how to find the slope of line k.

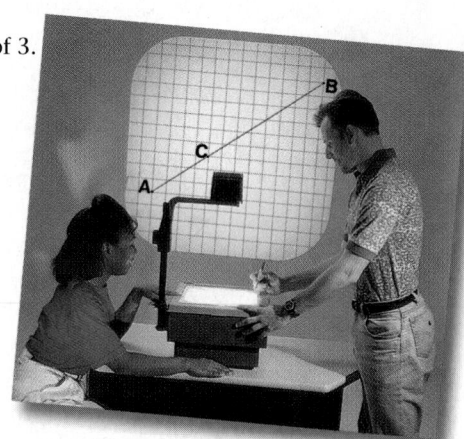

Practice & Apply

Examine the graphs below. Which lines have positive slope? Which have negative slope? Which have neither?

7. positive

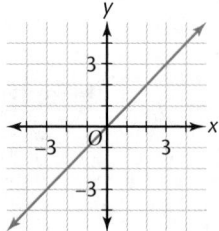

8. negative

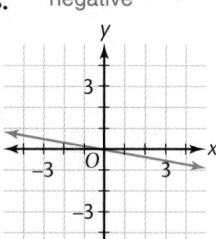

9. neither

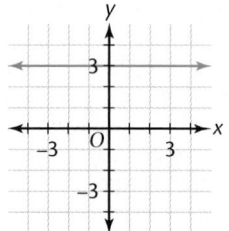

10. positive

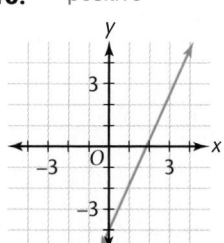

11. Answers may vary, for example: beginner slope, $-\frac{1}{10}$, and advanced slope, $-\frac{1}{2}$.

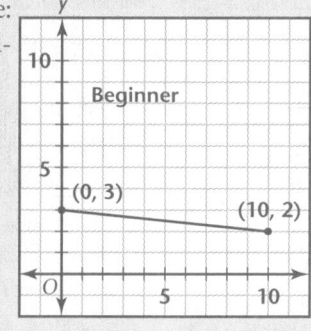

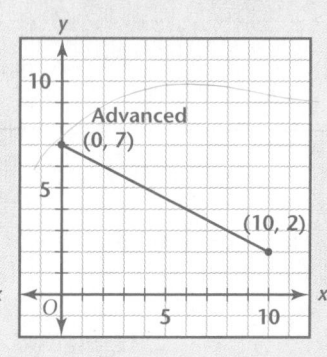

11. **Recreation** Suppose the slopes of two hills are negative. On graph paper draw a line to represent a beginner's ski hill and another to represent an advanced ski hill. Indicate each slope.

Find the slope for each of the given lines.

12. rise 6, run 2 3

13. rise 1, run 7 $\frac{1}{7}$

14. rise -1, run 7 $-\frac{1}{7}$

15. rise 0, run 5 0

16. rise $5-3$, run -3 $-1\frac{1}{2}$

17. rise $-7 + 2$, run $3 - 1$ $-\frac{5}{2}$

18. Draw coordinate axes on graph paper. Place a point somewhere in the upper left of your graph. From that point, move 6 spaces to the right, then down 4. Place a point there. Move again 6 spaces right, then down 4. Repeat this at least four times. What shape appears when you connect the points? What is the slope of the line connecting the points? Try this exercise with patterns of your own. The shape is a line with slope $-\frac{2}{3}$.

Each pair of points is on a line. What is the slope of the line?

19. $M(9, 6)$, $N(1, 4)$ $\frac{1}{4}$

20. $M(1, 3)$, $N(1, 5)$ none

21. $M(-3, 1)$, $N(2, 6)$ 1

22. $M(0, 2)$, $N(3, 0)$ $-\frac{2}{3}$

23. $M(-3, 5)$, $N(7, 6)$ $\frac{1}{10}$

24. $M(2, -4)$, $N(5, 9)$ $\frac{13}{3}$

25. $M(3.2, 8.9)$, $N(9.1, 7.2)$ $-\frac{17}{59}$

26. $M(8, 10)$, $N(8, 7)$ none

27. $M(10, 4)$, $N(7, 4)$ 0

28. $M(-3.2, -9.0)$, $N(1.8, 8.2)$ $\frac{172}{50}$

29. $M(0, 7)$, $N(0, 0)$ none

30. $M(-1, 1)$, $N(-3, -3)$ 2

31. Use the information in the table to find the average amount you climb per mile from Kansas City to Denver. 7.56 feet per mile

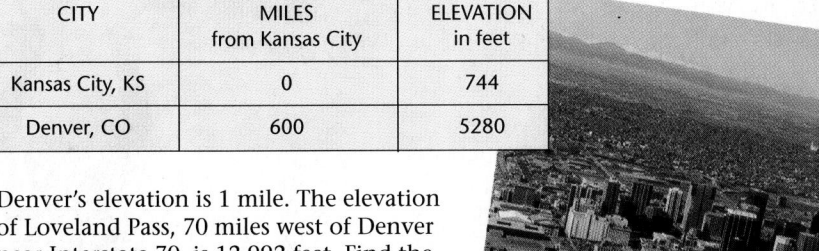

Denver, Colorado, elevation 1 mile

CITY	MILES from Kansas City	ELEVATION in feet
Kansas City, KS	0	744
Denver, CO	600	5280

32. Denver's elevation is 1 mile. The elevation of Loveland Pass, 70 miles west of Denver near Interstate 70, is 12,992 feet. Find the average slope of the land from Denver to Loveland Pass. 110.2 feet per mile

33. Find the difference for several pairs of x-coordinates and the difference for corresponding pairs of y-coordinates. Next, plot the values on a graph, and find the slope. Explain the relationships that you find.

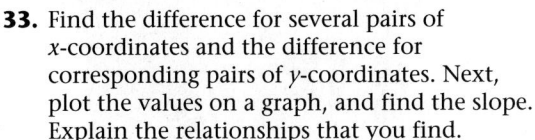

x	1	2	3	4	8
y	3	6	9	12	24

34.  **Geometry** Find at least 4 pictures of houses that have roofs with different slopes. Use a protractor to measure the angle for the slope of each roof. Use a ruler to find the slopes. Make a table to match the measure of each angle to each slope.

33. The change between coordinate pairs is constant, 3 for y and 1 for x, which means the graph is linear. The ratio $\frac{y}{x}$ is a constant $\frac{3}{1}$. This is the same as the slope of the line.

34. Answers may vary. The slope of the roof for the house in the picture is about 0.5. The angle is about 26°.

Error Analysis

Watch for students who write the ratio for the slope as $\frac{run}{rise}$ instead of $\frac{rise}{run}$. Also be sure that students are subtracting the coordinates in the same order in the numerator and denominator. Students can graph the line on a coordinate plane and check to be sure the slope they have found for the line makes sense.

Portfolio Assessment

Have students find several objects that involve slope, such as ramps or the roof of a dog house. Ask them to actually take the necessary measurements to find the slope of each object. Have them graph the line whose slope represents the slope of each object. Ask each student to summarize all of this information in a report that they can place in their portfolio.

There are 91 steps in each of the 4 stairways on the Mayan/Toltec pyramid at Chichen Itza, in the Yucatan.

Cultural Connection: Americas
Mexican pyramids are somewhat different in concept from African pyramids. They have stairways on each side.

35. If the height (rise) of the stairway is 27 meters and the length (run) is 25 meters, what is the slope of the stairway of the pyramid? 1.08

36. The rise of the average step is 29.7 centimeters, and the run is 27.5 centimeters. Find the slope of a step, and compare it with the slope of the pyramid's stairway. 1.08

37. **Geometry** Use a string to measure several round objects. Record their diameters and circumferences in a table. Then graph the data with diameters on the *x*-axis, and fit it with a straight line. How well does the line fit the points? Find the slope of the line. What conjectures can you make from this information? about 3.14

38. **Sports** On graph paper, graph the following data for the minutes played (*x*-axis) versus the number of points scored (*y*-axis) for the last game of the 1993 NBA playoffs. Use a clear ruler to fit a straight line to the data. Estimate the slope, giving the units. Write a sentence that explains what the slope tells you about the players' performance. about $\frac{1}{2}$

Player	Minutes	Points
Pippen	43	23
Armstrong	41	18
Jordan	44	33
Paxson	22	8
S. Williams	22	5
Tucker	7	9
King	2	0

The answers to Exercises 38, 52, and 53 can be found in Additional Answers beginning on page 729.

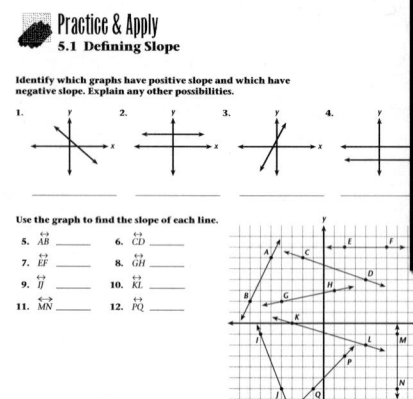

Practice Master

Practice & Apply
5.1 Defining Slope

Identify which graphs have positive slope and which have negative slope. Explain any other possibilities.

1. 2. 3. 4.

Use the graph to find the slope of each line.

5. $\overleftrightarrow{AB}$ _____   6. $\overleftrightarrow{CD}$ _____
7. $\overleftrightarrow{EF}$ _____   8. $\overleftrightarrow{GH}$ _____
9. $\overleftrightarrow{IJ}$ _____   10. $\overleftrightarrow{KL}$ _____
11. $\overleftrightarrow{MN}$ _____   12. $\overleftrightarrow{PQ}$ _____

Find the slope of the given lines.

13. rise −5, run −5   14. rise 2, run 3   15. rise 1, run −6

Each pair of points is on a line. What is the slope of each line?

16. $A(3, 9), B(1, 5)$ _____   17. $A(7, 5), B(2, 4)$ _____
18. $A(-5, 10), B(-5, -4)$ _____   19. $A(5, 2), B(2, -1)$ _____
20. $A(3, -2), B(-1, 3)$ _____   21. $A(-1, 3), B(5, 3)$ _____

22. If the height (rise) of a stairway is 15 m, and the length (run) is 18 m, what is the slope of the stairway?

39. Social Studies Graph the data in the table for the populations and number of representatives in the U.S. Congress for all the states starting with M. Fit the data to a line and find its slope. Explain what information the slope gives us about the 1990 data.

The number of representatives to Congress depends on the population of the state.

State	Representatives	Population
Maine	2	1,227,928
Maryland	8	4,781,468
Massachusetts	10	6,016,425
Michigan	16	9,295,297
Minnesota	8	4,375,099
Mississippi	5	2,573,216
Missouri	9	5,117,073
Montana	1	799,065

40. A line passes through the origin $O(0, 0)$ and point $P(c, d)$. What is the slope of the line? $\frac{d}{c}$

41. A line passes through point $M(a, b)$ and point $N(c, d)$. What is the slope of the line? $\frac{d - b}{c - a}$ or $\frac{b - d}{a - c}$

Look Back

Name the type of correlation indicated by the given conditions. As one variable increases, **[Lesson 1.6]**

42. the other variable tends to decrease. negative

43. you cannot tell if the other variable tends to increase or decrease. little or no

44. the other variable tends to increase. positive

Evaluate. **[Lesson 3.1]**

45. $|-7|$ 7 **46.** $|50|$ 50 **47.** $-|-9|$ -9 **48.** $-|99|$ -99

49. If there are 8 people contributing to a charity, how much would each person have to contribute to raise $98? **[Lesson 4.3]** $12.25

Solve for x. **[Lesson 4.6]**

50. $3x + 16 = 19$ 1 **51.** $28 = -4 + 4x$ 8

Look Beyond

52. Graph $y = 4x - 2$, $y = 4x$, and $y = 4x + 3$ on the same axes. Find the slope of each function. What do you notice about the lines?

53. On graph paper, plot the point (0, 0). From the point, move right 5 and up 3 to plot the second point. Then move right 5 and up 6 to plot the third. Next move right 5 and up 9 for the fourth point. Continue until you have plotted 6 points. Connect the points with a smooth curve. Have you seen a curve like this before? If so, where?

39. The slope is approximately $\frac{8 - 4}{14 - 7} = \frac{4}{7}$ or about 0.57. These are approximately 570,000 people for each representative.

Focus

An architect's mission to make stairways safer stirs a debate centering on the tradeoff between safety and cost. Students learn the math behind stairway design in order to take part in the debate.

Motivate

As students read the news article, they should refer to the diagrammed staircase on the page to help them with any unfamiliar terms, such as *tread* and *riser*.

After reading the first article, discuss the issues raised. Ask: *Why does the architect want building codes changed?* (He feels new specifications will make for safer stairs.) *Why might the National Association of Home Builders not want to change building codes?* (They may feel the change will unnecessarily add to the cost of a house.)

In discussion, bring out the notion that the tradeoff between safety and cost is an issue because it may not always be practical to make something as safe as possible. There may be a point of diminishing returns—by spending a lot of money to make something a little bit safer, you may be raising the cost so much that the item becomes too expensive to buy.

Tell the students that in order to understand the issues more fully, they will be reading another article—one that describes in more detail the mathematics of stairway design.

Again, have students refer to the diagram on page 216 as they read the article, especially the middle part that describes a sample stairway calculation.

EYEWITNESS MATH — Watch Your Step!

Atlanta Architect Steps Up Quest for Safe Staircases

Numerous Accidents, Deaths Push Researcher into Career

By Lauran Neergaard
THE ASSOCIATED PRESS

ATLANTA – ...Every year, 1 million Americans seek medical treatment for falls on staircases. About 50,000 are hospitalized and 4,000 die.

Templer's interest began as a student at Columbia University, when someone asked him why people were always falling down the steps outside Lincoln Center. Templer discovered there was no research on stair safety. Nobody even counted falls.

Curious, he visited Lincoln Center with his family, and his sister-in-law tripped on the steps.

Templer called his mentor to propose the topic as a thesis and learned that the man had just broken his leg falling down a stairwell of a subway station.

A career was born.

"All stairs are dangerous, it's a matter of degree," Templer said. "There are ways to mitigate that danger, if we could get that message to people."

Stairs evolved from ladders and first were used for defense. Narrow winding staircases, for instance, hampered intruders. Europeans wrought stairs into works of art, building grand palace staircases that gradually steepened, forcing visitors into a slow, stately pace as they approached royalty.

Most stairs now have 9-inch treads and $8\frac{1}{2}$-inch risers, a size determined around 1850, Templer said. But people today have bigger feet that hang over the edges of stairs, throwing them off balance, he said.

Stairs also are too high, Templer concluded after experiments in which he forced volunteers to trip on collapsible stairs. They were harnessed so they didn't tumble all the way to the floor, but Templer used videos to simulate how they would have landed.

He wants building codes revised for stairs with 11-inch treads and 7-inch risers. His proposal prompted a lobbying blitz from the National Association of Home Builders, which contends that larger stairs would add at least 150 square feet and $1,500 in costs to a house.

NAHB's Richard Meyer dismissed Templer's work, saying people fall when stairs are improperly lighted, have loose carpeting or have objects placed in the way.

That's true, too, Templer said. But he said his experiments, funded by the National Science Foundation, prove stair shape is a large problem.

Cutaway Diagram of Steps of Stairway

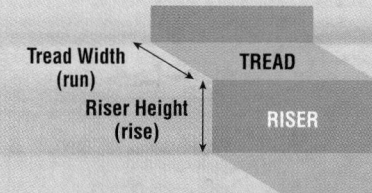

Tread Width (run) — TREAD
Riser Height (rise) — RISER

Count Your Way to Stair Success

By Karol V. Menzie and Randy Johnson

Baltimore Sun

The riser height (rise) and the tread width (run) determine how comfortable the stairs will be to use. If the rise and run are too great, the stairs will strain your legs and be hard to climb. If the rise and run are too small, you may whack your toe on the back of each step.

Over the years, carpenters have determined that tread width times riser height should equal somewhere between 72 to 75 when the measurements are in inches.

On the main stair, the maximum rise should be no more than $8\frac{1}{4}$ inches and the minimum run should be no less than nine inches.

To determine how many steps, or treads, you need, measure from the top of the finished floor on the lower level to the top of the finished floor on the upper level.

To figure the rise and run in a house with eight-foot ceilings, for instance, start by figuring the total vertical rise. By the time you add floor joists, subfloor and finish floor, the total is about 105 inches.

A standard number of treads in a stair between first and second floors is 14. One hundred five divided by 14 equals $7\frac{1}{2}$. That means the distance from the top of each step to the top of the next step will be $7\frac{1}{2}$ inches.

With a riser height of $7\frac{1}{2}$ inches, tread width (run) should be at least nine inches. Ten inches is a more comfortable run; when you multiply $7\frac{1}{2}$ inches by 10 inches, you get 75–within the conventional *ratio* of 72 to 75. With fourteen 10-inch treads, the total run of the stairs will be 140 inches. In other words, the entire stair will be 105 inches tall and 140 inches deep.

You can alter the rise and run to some extent. If you used 15 risers instead of 14, for instance, the rise would be 7 inches, and the tread width would be $10\frac{1}{2}$ inches (7 times $10\frac{1}{2}$ equals 73.5, within the rise and run guidelines).

Cooperative Learning

1. In the news article above, is the underlined term *ratio* used correctly? Explain.

2. Find the angle of each of the following staircases. Draw the triangle and measure the angle.

   a. the 105-inch by 140-inch staircase described in the article above

   b. the same staircase changed as described in the article to include 15 risers instead of 14

   c. the same staircase changed to meet Templer's preference for 7-inch risers and 11-inch treads

3. Which of the three staircases in activity 2 would take up the most room? Explain.

4. Suppose you did not want the building code for stairs to change. Write an argument (with numbers and a diagram) to show that larger stairs would add at least 150 square feet to a typical house. Make up your own estimates of the size of a typical house.

Project

Measure the rise and run on staircases in your school or home. Do they match the information in the two news articles about standard stairs?

1. No; 72 to 75 is actually the range of a product (tread width • riser height). In the subsequent paragraph, the article refers to this product more appropriatley as the "rise and run guidelines."

2a. about 37°

2b. about 34°

2c. about 32.5°

3. Stairway c; because it has the smallest slope, it will extend farthest into the room.

4. Answers may vary. For example, the stairway in (c) is about 2 feet longer than the stairway in (a). To make up for that lost space, the room would need to be about 2 feet longer. In a house 40 feet wide, that's an additional 80 square feet. Since placing the wall 2 feet further out also affects the second story, that's another 80 additional square feet, for a total extra area of 160 square feet.

PREPARE

Objectives

- Identify how a change of slope affects a graph and the equation of a line.
- Identify how changing where the line crosses the *y*-axis affects a graph and the equation of a line.

RESOURCES

- Practice Master   5.2
- Enrichment Master   5.2
- Technology Master   5.2
- Lesson Activity Master   5.2
- Quiz   5.2
- Spanish Resources   5.2

Assessing Prior Knowledge

Graph the ordered pairs. Tell whether the points lie on a straight line.

$(2, 0), (0, 2), (1, 1), (3, -1)$

[Yes]

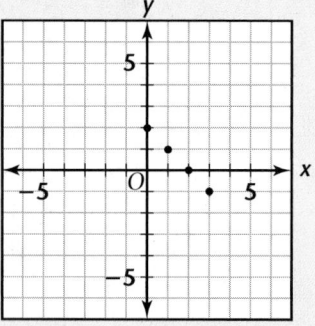

TEACH

why The descent of an airplane is a real-world example of the concept of slope. Ask students to give other examples of this concept.

Exploring Graphs of Linear Functions

why *A linear function can be used to model the descent of an airplane. When you model situations using the graph of a linear function, you need to know the slope and the point at which the line crosses the y-axis.*

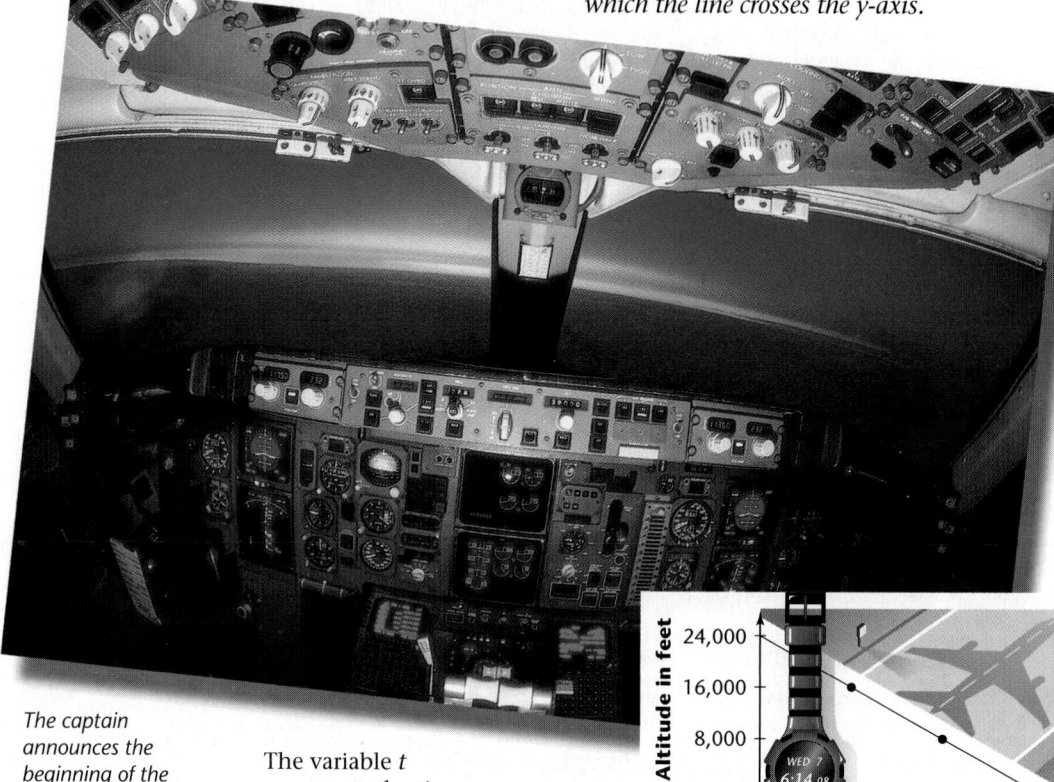

The captain announces the beginning of the descent from 24,000 feet for a landing in Detroit in 30 minutes.

The variable *t* represents the time in minutes after the airplane begins its descent. The starting condition is 24,000 feet. The constant rate of descent is −800 feet per minute because the plane must descend 24,000 feet in 30 minutes. The function $y = 24{,}000 - 800t$ models the descent.

Recall from Lesson 1.5 that the equation $y = 24{,}000 - 800t$ is a linear function because its graph is a line. In fact, any equation of the form $y = mx + b$ is a linear function. When $b = 0$, $y = mx$ is a line passing through the origin.

ALTERNATIVE teaching strategy

Using Visual Models
Have students work in small groups. Provide each group with grid paper and different colored pencils. Ask them to explore the general linear equation $y = mx$ by choosing different values to substitute for *m*. Students should make a table like the one shown. Have the students place values they choose for *m* in the first column. They should then substitute the values above each column for *x* to find the values of *y*. Fill in the last 5 columns accordingly.

$y = mx$	-2	-1	0	1	2
$y = \_\_\_ x$					
$y = \_\_\_ x$					
$y = \_\_\_ x$					

Then ask each group to plot the points and draw each line in a different color. Have stu-

Graph the point $A(3, 6)$ on a coordinate plane. Start with the equation $y = mx$. Guess and check numbers for m by graphing the equations until you find a line that intersects point A.

If you use a graphics calculator, enter point $A(3, 6)$, and let your first guess for m be 4. Enter **Y = 4X**, then graph the function. See if the line passes through $A(3, 6)$.

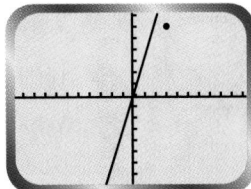

If you use graph paper, substitute values for x in the equation $y = 4x$ to locate points like $(0, 0)$ and $(2, 8)$. Then draw the line that connects the points. See if the line passes through $A(3, 6)$.

When m is 4, the line representing $y = 4x$ is too steep. It does not go through point A.

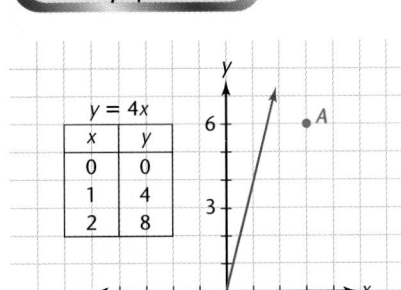

$y = 4x$	
x	y
0	0
1	4
2	8

Repeat the process with different values for m until you find an equation for the line that intersects point A.

Exploration 1 Fitting the Line to a Point

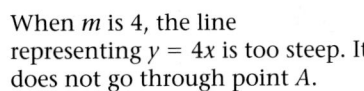

Repeat the process above for points B through F.

$B(2, 8)$ $C(6, 3)$ $D(3, -6)$ $E(2, -8)$ $F(4, 7)$

What is the connection between the coordinates of the given target point and the value of m? ❖

Exploration 2 Changing the Slope

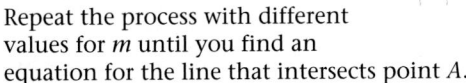

Draw the graph of each equation on the same axes. Compare the graphs you get from varying the values of m.

$y = x$ $y = 5x$ $y = -2x$ $y = \frac{1}{2}x$ $y = -\frac{1}{3}x$

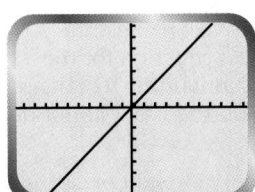

1   What stays the same when the lines are drawn?

2   What is different?

3   Guess what the graph of $y = 3x$ looks like.

4   Check your guess by drawing the graph.

5   Make a conjecture about how the coefficient of x affects the graph of $y = mx$. ❖

---

dents compare the lines and record any observations. Then have each group complete a similar activity for the equation $y = mx + b$.

interdisciplinary
CONNECTION

Physics The equation $y = mx$ can be used in physics to describe the relationship between the distance an object travels and the amount of time it travels. The slope of the line, m, gives the speed of the object, which is constant. The value of x represents the time, and the value of y represents the distance. Have students use this information to write equations describing the motion of a car traveling at different constant speeds. Then have them graph the equations and record any observations.

Before students complete this exploration, ask them to look at the four equations they will use to answer questions 1–5 and compare them with the four equations they will use to answer questions 6–10. They should notice that only the value of m changes in the equations for 1–5, and that only the value of b changes for 6–10.

Aongoing
SSESSMENT

5. Adding 3 raises the graph so that the y-intercept is at $(0, 3)$.

10. The value of b is where the line crosses the y-axis.

Aongoing
SSESSMENT

The value of m is the slope, and the value of b is the y-intercept of the line.

TEACHING *tip*

For students who find it difficult to comprehend the relationship between $y = mx + b$ and the graph of the line, continue to provide experience in graphing related equations like those presented in Exploration 3. Working with a partner who understands the concept may be beneficial for such students.

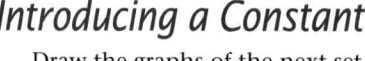

Exploration 3 *Introducing a Constant*

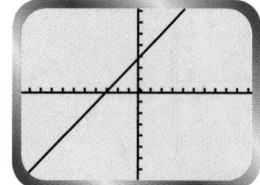

Draw the graphs of the next set of equations on the same axes.

$y = x + 3$ $y = 2x + 3$ $y = 5x + 3$ $y = -2x + 3$

1 What stays the same when the lines are drawn?

2 What is different?

3 Guess what the graph of $y = 4x + 3$ looks like.

4 Check your guess by graphing.

5 Make a conjecture about how adding 3 affects the graph of $y = mx$.

Draw the graphs of the next set of equations on the same axes.

$y = 2x$ $y = 2x + 3$ $y = 2x + 5$ $y = 2x - 3$

6 What stays the same when the lines are drawn?

7 What is different?

8 Guess what the graph of $y = 2x - 4$ looks like.

9 Check your guess by graphing.

10 Make a conjecture about how the value of b affects the graph of $y = 2x + b$. ❖

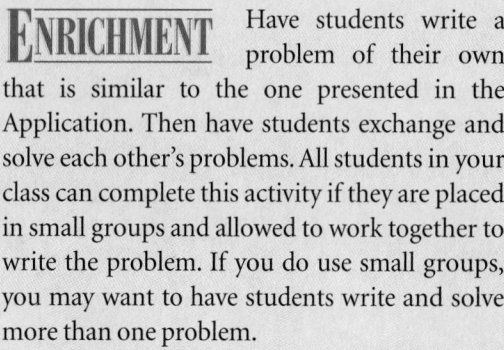

Picture the graph of each equation in your mind. Check the accuracy of your pictures by graphing each equation.

$y = -3x$ $y = 5x - 6$

$y = 5 + 6x$ $y = \frac{1}{2}x + 3$

How do the values of m and b affect the graph of $y = mx + b$? Discuss the results of what you have discovered from the explorations.

APPLICATION

An orange carton weighs 3 pounds when empty, and each orange you place in it weighs an average of half a pound. If you know the number of oranges, what is the total weight of the carton and the oranges? Let n equal the number of oranges. Write an equation for the total weight, w, of a carton containing n oranges. Graph the equation with n on the x-axis and w on the y-axis.

$$\text{total weight} = \text{weight of oranges} + \text{weight of carton}$$
$$w \qquad = \qquad \frac{1}{2}n \qquad + \qquad 3$$

ENRICHMENT   Have students write a problem of their own that is similar to the one presented in the Application. Then have students exchange and solve each other's problems. All students in your class can complete this activity if they are placed in small groups and allowed to work together to write the problem. If you do use small groups, you may want to have students write and solve more than one problem.

INCLUSION
strategies

English Language Development   The concepts presented in this lesson provide a way for students who are weak in the English language to practice communicating their ideas. Be sure students can use a graph to visually demonstrate the concepts presented and that they understand the basic language associated with these concepts. Encourage them to discuss the concepts with one another using their own words.

The weight and the number of oranges cannot be negative. Thus, the graph will include only points where n and w are greater than or equal to 0.

Try $n = 0$. When there are 0 oranges, the weight is only that of the carton, or 3 pounds. Plot the first point $(0, 3)$. Each time you add an orange, the weight increases by $\frac{1}{2}$ pound. This is the **rate of change of the function** $w = \frac{1}{2}n + 3$. How does the rate of change compare with the slope of a line?

You can add new points to the graph by moving to the right 1 unit and up $\frac{1}{2}$ unit from each point. If you draw a line through the points, the line will have a slope of $\frac{1}{2}$.

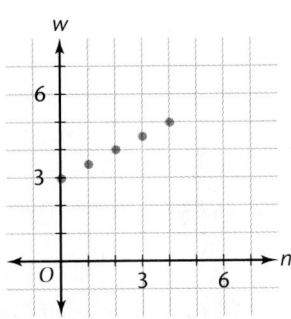

The individual points represent the weight of whole oranges.

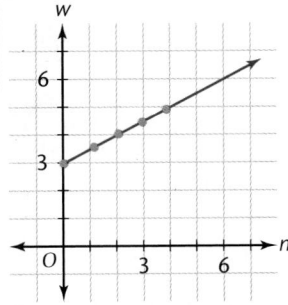

The solid line also includes the weight of partial oranges. ❖

CRITICAL Thinking

What happens to the function that models the airplane descent if the starting altitude or rate of descent changes? Find the new equation if the starting condition is 30,000 feet and the rate of change is -1000 feet per minute.

EXERCISES & PROBLEMS

Communicate

1. Summarize your discoveries from each of the explorations.

Guess the equation of the line through the origin that intercepts each of the following points. Check by drawing the graph.

2. $G(8, 4)$ 3. $H(4, 8)$ 4. $I(4, 9)$ 5. $J(6, -4)$

6. Line l passes through the origin $(0, 0)$ and $A(3, 6)$. Explain how to write the equation for line l.

7. If the slope of a line through the origin and through a given point is a, what is the equation of the line?

Aongoing ASSESSMENT

They are the same.

CRITICAL Thinking

If the starting altitude changes, then the value for the constant, 24,000, changes. If the rate changes, then the value of the slope, -800, changes. The new equation would be

$$y = 30,000 - 1000t.$$

Assess

Selected Answers
Odd-numbered Exercises 9–45

Assignment Guide
Core 1–7, 8–22 even, 23–47

Core Two-Year 1–46

Technology
A graphics calculator or computer software such as *f(g) Scholar*™ is helpful for Exercises 8–13, 21, 22, 26–37, and 47.

Practice & Apply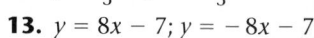

Draw each *pair* of graphs on the same axes. In each case, tell what is the same and what is different.

8. $y = 6x$; $y = -6x$

9. $y = 8x$; $y = -8x$

10. $y = \frac{1}{2}x$; $y = -\frac{1}{2}x$

11. $y = \frac{4}{3}x$; $y = -\frac{4}{3}x$

12. $y = 4x - 7$; $y = -4x - 7$

13. $y = 8x - 7$; $y = -8x - 7$

14. What effect does changing $y = ax$ to $y = -ax$ have on the line that is graphed?

15. Physical Science From the information shown, write an equation for the total mass, T, of a beaker containing m milliliters of water. Substitute values for m, and graph the equation.
$$T = m + 30$$

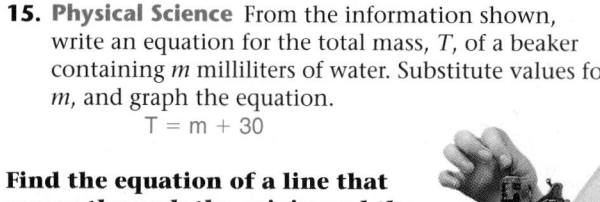

Empty beaker 30 grams

1 milliliter of water = 1 gram

Find the equation of a line that passes through the origin and the following points.

16. $(2, 5)$ **17.** $(5, 8)$ **18.** $(1, 9)$

19. $(4, 2)$ **20.** $(7, 3)$

Draw each *pair* of graphs on the same axes. In each case, tell what is the same and what is different.

21. $y = \frac{1}{4}x$ and $y = \frac{1}{4}x + 6$

22. $y = \frac{1}{4}x + 8$ and $y = \frac{1}{4}x - 2$

23. Describe the effect on the lines that are graphed if only the values of b change in the equation $y = ax + b$.

The normal water level of the river is 32 feet.

24. Ecology The water level of a river is rising 0.8 foot per day. Write an equation for the water level after d days. Graph the equation.

25. Suppose the water level of the river is 34 feet and is receding 0.5 foot per day. Write an equation for the water level after d days. Graph the equation. In how many days will the water level be 26 feet?

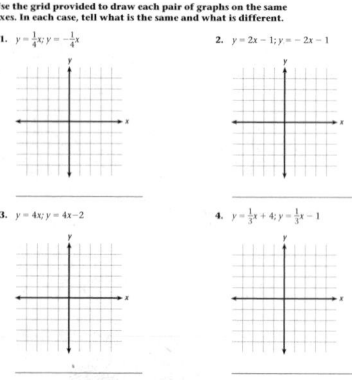
The answers to Exercises 8–14, 16–20, and 26–38 can be found in Additional Answers beginning on page 729.

21. same slope; different y-intercepts

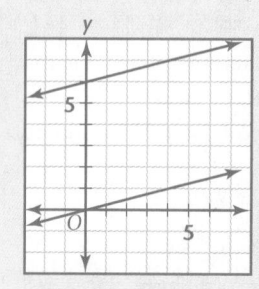

22. same slope; different y-intercepts

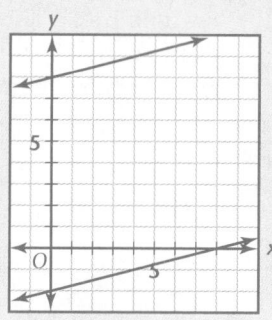

Guess what each line will look like when it is graphed. Then check your guess by graphing.

26. $y = -5x$

27. $y = -6x + 3$

28. $y = 7$

29. $y = -6$

30. $y = -5x - 1$

31. $y = -2$

32. $y = -x - 3$

33. $y = 2x + 3$

34. $y = 3x + 7$

35. $y = 7 - x$

36. $y = -3 + \frac{1}{2}x$

37. $y = x + 4$

38. 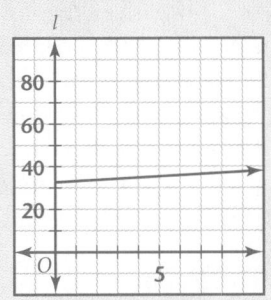 **Portfolio Activity** Complete the problem in the portfolio activity on page 207. 1.4; 1.4

Look Back

39. If Calvin earns $8 an hour, find his earnings for 33 hours. Write an equation and find the solution. **[Lesson 4.1]** $264

40. Dan earns 15% of the price of each $20 box of candles he sells. How many boxes will he have to sell to earn at least $100? **[Lesson 4.5]** 34

41. Solve the equation $3x + 4y = 24$ for y. **[Lesson 4.6]** $y = -\frac{3}{4}x + 6$

42. Suppose that in the orange carton example on page 220 the total weight is 51 pounds. How many oranges are in the carton? **[Lesson 4.6]** 96

Is the slope of the line negative, positive, or neither? [Lesson 5.1]

43.

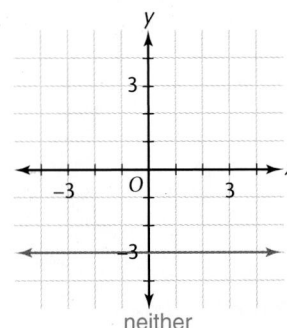

neither

44.

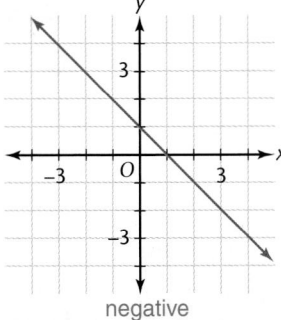

negative

45.

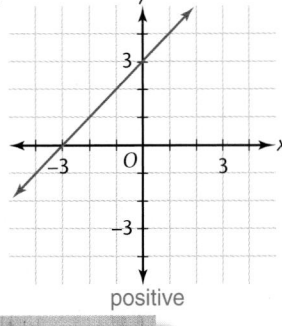

positive

46. Cultural Connection: Africa The Rhind papyrus that describes the mathematics of building the ancient pyramids of Egypt refers to a *seked*. Consider a square pyramid with a base edge of 360 cubits and a height of 250 cubits. Scribes took half the base as the run and the height as the rise. They calculated the *seked* to be $\frac{180}{250}$. Compare the calculation of a *seked* with the calculation of the slope for the same pyramid. Explain the relationship between *seked* and slope. **[Lesson 5.1]**

slope $= \frac{250}{180}$; seked $= \frac{180}{250}$; reciprocals

Look Beyond

47. If you graph the lines $y = 0.5x + 3$ and $y = \frac{3}{6}x + 4$, will the graphs of the lines ever cross? no

23. The line will have the same slope but different y-intercepts.

24. $l = 32 + 0.8d$

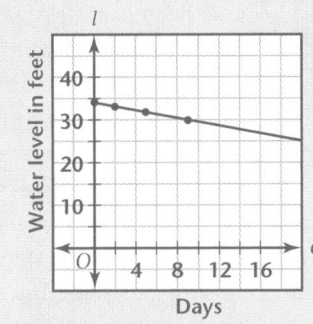

25. $l = 34 - 0.5d$; 16 days

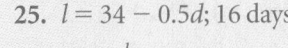

Days

Look Beyond

The graphs of the two equations are parallel because they have the same slope. This concept is formally presented in Lesson 5.6.

Technology Master

Technology
5.2 Collinearity

When you make a scatter plot of data and find the line of best fit for the data, you will most often find that the line passes through some but not all of the points. If the line does pass through all the points, then the set of points is called collinear. Rarely does this happen.

You can use a spreadsheet to check for collinearity by computing the slopes of the line segments that connect the data points. The spreadsheet shown contains the x-coordinates of the set of points in column A and the y-coordinates of the points in column B. Column C contains the slopes of the line segments. To use the spreadsheet, you must first arrange the points so that the x-coordinates increase.

	A	B	C
1	X	Y	SLOPE
2			
3			
4			
5			
6			
7			
8			
9			

Use a spreadsheet to determine whether each set of points is collinear. If it is, tell the slope of the line through them.

1.

X	Y
1	1.2
3	3.6
5	6.0
6	7.2

2.

X	Y
1	2.6
3	7.6
5	13.0
6	15.0

3.

X	Y
7.2	-18.72
8.2	-21.32
9.2	-23.92
10.2	-26.52

4.

X	Y
11	39.7
0	6.7
8	30.7
3	15.7

5.

X	Y
-1	3.7
3	6.0
4	10.2
5	15.0

6.

X	Y
2	12.7
-3	-2.3
5	21.7
-6	-11.3

7. Suppose that the slopes of the line segments that join a set of data points are 0.55, 0.54, 0.56, 0.55, 0.56, 0.55, 0.54, and 0.55. What can you say about the set of data points?

Objectives
- Define and explain the components of the Slope-Intercept of a line.
- Use the Slope-Intercept Form of the equation of a line.
- Solve problems of direct variation.

RESOURCES

- Practice Master   5.3
- Enrichment Master   5.3
- Technology Master   5.3
- Lesson Activity Master   5.3
- Quiz   5.3
- Spanish Resources   5.3

Assessing Prior Knowledge

Find the pattern. Write the next three numbers.

1. $1, 5, 9, 13, 17, \ldots$
   [21, 25, 29]

2. $-8, -1, 6, 13, 20, \ldots$
   [27, 34, 41]

3. $-1.5, 1, 3.5, 6, 8.5 \ldots$
   [11, 13.5, 16]

Find the slope of the line containing the two points given.

4. $A(3, 12), B(1, 6)$ [3]

5. $X(5, 9), Y(1, 8)$ $\left[\frac{1}{4}\right]$

6. $M(-1, -2), N(2, 0)$ $\left[\frac{2}{3}\right]$

TEACH

 Remind students that a function is a special relationship between two sets of data that can be represented by a table of values, an equation, or a graph. Remind them that one of the simplest functions is the linear function.

LESSON 5.3 The Slope-Intercept Formula

Skaters practice several hours a week. Kim pays $50 for club fees and $3 an hour to practice.

why *Once you understand how to interpret the components of a linear function $y = mx + b$, you can sketch the graph. When you can identify the slope and the y-intercept of a line, you can write its equation.*

Kim can use a graph to see the pattern of her skating expenses. How would you draw this graph?

Examine the data. Let x be the number of hours Kim practices. Let y be the total cost of practice time. Calculate the first differences to determine the slope from a table of x and y values.

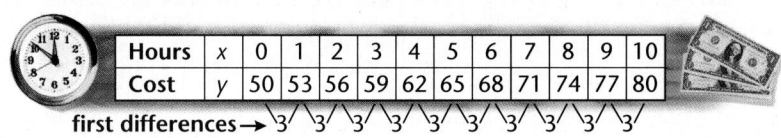

Hours	x	0	1	2	3	4	5	6	7	8	9	10
Cost	y	50	53	56	59	62	65	68	71	74	77	80

first differences → 3 3 3 3 3 3 3 3 3 3

Since the first differences are 3, the table represents a linear function in the form $y = mx + b$. How can you find the slope of the line that represents the function? The table shows that the change in the y-variable is 3. Since the x-values are consecutive, the change in the x-variable is 1.

$$\text{slope} = m = \frac{\text{change in } y}{\text{change in } x} = \frac{3}{1} = 3$$

The slope indicates that Kim pays an additional $3 for every hour she practices.

ALTERNATIVE teaching strategy

Visual Models Discuss the meaning of m and b in the equation $y = mx + b$. Then display the graph of the equation $y = 3x + 50$ from Example 1. To demonstrate the rise and run as given by the slope, m, ask students to think of stair steps. To go up stairs, you lift your foot up and then move it forward to place it on the step. The same is true when using the slope on a graph. Use a blue marker to represent moving up 30 units for the rise and a green marker to represent moving right 10 units for the run. To represent a negative slope, ask students to think about walking down steps. First you move your foot forward (representing a move to the right on a graph), and then down (representing a move down on a graph) to reach the next step. Use a red marker to represent moving down and a green marker to represent moving to the right.

The table also shows that when x is 0, y is 50. Thus, (0, 50) is on the graph. The y-value, 50, of the point where the line crosses the y-axis, is the **y-intercept**. It means that Kim pays a $50 fixed fee in addition to the hourly rate.

Recall from Lesson 5.2 that the graph of an equation in the form $y = mx + b$ has slope m and crosses the y-axis at b. If you substitute 3 for m and 50 for b, the equation for Kim's practice costs becomes $y = 3x + 50$.

SLOPE-INTERCEPT FORM

The slope-intercept formula or form for a line with slope m and y-intercept b is $y = mx + b$.

EXAMPLE 1

The equation of the line that represents Kim's skating costs is $y = 3x + 50$. How can you use the slope-intercept formula to construct the graph?

Solution ➤

The formula is $y = mx + b$. Since b is 50, measure 50 units up from the origin (0, 0) on the y-axis and graph a point.

From that point, measure the run and then the rise of the slope to locate a second point.

The slope, m, is 3. You can use any equivalent of $\frac{3}{1}$ to measure the run and rise. The axes are in intervals of 10 units, so use $\frac{30}{10}$. Move right 10, then up 30. Graph the second point. Draw a line through the points. ❖

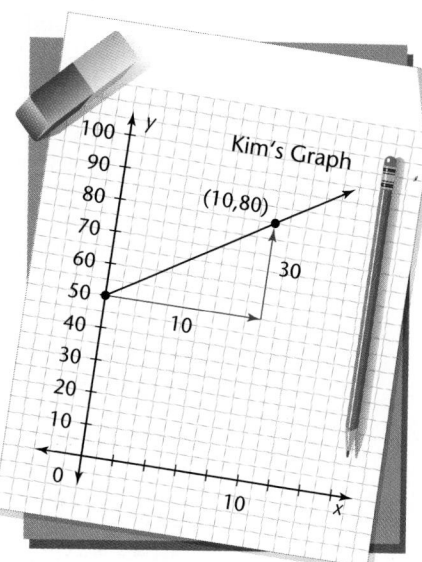

Kim's Graph

(10,80)

CRITICAL *Thinking*

Why do you think the line in Kim's graph does not extend past the y-axis? How far do you think the line would realistically extend in the positive direction?

From Two Points to an Equation

When you know 2 points on a line, you can write the equation for that line. First, calculate the slope, m, using the slope formula. Then calculate b from the slope-intercept formula and one of the points.

interdisciplinary
CONNECTION
Sports Have students do research to find data about their favorite sport that can be represented using a linear function. Ask students to determine whether their data represent a true linear function or whether they need to find a line of best fit. Suggest that they use a graphics calculator to help them make this determination. Make a display of students' data and graphs.

Alternate Example 1

Brian bought a new car for $12,500. Each year the value of the car decreases by $1500. For example, after he has owned the car for one year, it will be worth $11,000, after two years it will be worth $9,500, and so on. The equation to determine the value of Brian's car for a particular year is $y = 12,500 - 1500x$.

How do you use the slope-intercept formula to construct the graph? [**Locate 12,500 units up from the origin (0, 0) and plot the point. From that point, move a run of 1 and a rise of −1500 to locate a second point, (1, 11,000). Draw a line through the points.**]

CRITICAL *Thinking*

To the left of the y-axis, the x-values are negative, and Kim cannot skate a negative amount of time. The graph would probably extend to the point where she skated as many hours as she could skate and afford the cost.

Susan kept a record of the value of her car each year. She found that after 3 years the car would be worth $9,500 and after 8 years the car would be worth $2,000. Write an equation of a line based on this information.
[$y = 14,000 - 1500x$]

Aongoing
SSESSMENT

$(7, 71)$ and $(5, 65)$ are points on the same line as the line for the equation in Example 1.

Aongoing
SSESSMENT

Try This

$y = 2x - 3$

Alternate Example 3

Since 1928, the winning times in the Women's 100-meter run in the Olympics have been faster. Find the equation for the line of best fit that shows the trend in winning times.

Year	Time (sec.)	Year	Time (sec.)
1928	12.2	1968	11.00
1932	11.9	1972	11.07
1936	11.5	1976	11.08
1948	11.9	1980	11.60
1952	11.5	1984	10.97
1956	11.5	1988	10.54
1960	11.0	1992	10.82
1964	11.4		

[$y = -0.0187x + 47.96$]

SLOPE FORMULA

Given two points with coordinates (x_1, y_1) and (x_2, y_2), the formula for the slope is

$$m = \frac{\text{change in } y}{\text{change in } x} = \frac{\text{difference in } y}{\text{difference in } x} = \frac{y_2 - y_1}{x_2 - x_1}.$$

EXAMPLE 2

Sports   Kim's notes show that after 7 hours of practice the total cost for her skating is $71. Earlier in her notes it shows that after 5 hours of practice the total cost was $65. How can you write an equation for a line knowing only this information?

Solution ➤

You can represent the data as points by writing the hours and cost as ordered pairs, $(5, 65)$ and $(7, 71)$. To write the equation of a line from these two points, substitute the values into the slope formula.

$$m = \frac{\text{difference in } y}{\text{difference in } x}$$
$$m = \frac{71 - 65}{7 - 5} = \frac{6}{2} = 3$$

Substitute 3 for m in $y = mx + b$.

$$y = 3x + b$$

Next, choose either point, and substitute the coordinates for x and y into the equation. If you use the point $(5, 65)$, substitute 5 for x and 65 for y. Then, solve for b.

$$y = 3x + b$$
$$65 = 3(5) + b$$
$$65 = 15 + b$$
$$50 = b$$

Now substitute 3 for m and 50 for b in $y = mx + b$. The equation for the line is $y = 3x + 50$. ❖

Why is the equation in Example 2 the same as the equation in Example 1?

Try This   Write an equation for a line passing through points $(3, 3)$ and $(5, 7)$.

EXAMPLE 3

Auto Racing   For decades the Indianapolis 500 automobile race has attracted the attention of millions of enthusiasts. Since 1911 the average speed for the race has increased from 74.59 miles per hour to 185.987 in 1990. How steadily has this average increased? How can you find an equation for the line of best fit that shows the trend in average speed? Use the data given for a sample of speed averages from 1915 to 1975 in 5-year intervals. Consider 1900 as year 0.

ENRICHMENT   Have students do research to find out more about the use of linear regression in statistical analysis. Ask them to write a report about their findings, including appropriate data, graphs, and equations. Have students share their reports with the class.

INCLUSION **strategies**   **Technology**   Have students make a scatter plot of the data from Example 3 on graph paper, including carefully labeled axes. Then have them use a graphics calculator to find the equation of the regression line and draw the line on their scatter plot. Ask them to compare their graph with the graph shown by the graphics calculator.

Year	Average Speed in MPH	Year	Average Speed in MPH	Year	Average Speed in MPH	Year	Average Speed in MPH
1915	89.84	1935	106.240	1955	128.209	1975	149.213
1920	88.62	1940	114.277	1960	138.767	1980	?
1925	101.13	1945	none	1965	150.686	1985	?
1930	100.448	1950	124.002	1970	155.749	1990	?

Solution ▶

One way to find the equation for the line of best fit is to plot the points on a graph and estimate the location of the line using a clear ruler.

Find the slope and y-intercept. Then substitute the numbers into the formula for a line $y = mx + b$.

Graphics Calculator

The developments in calculator technology have made graphing a scatter plot much easier. On a graphics calculator, the line of best fit is called the **regression line**. The calculator will automatically calculate the slope, the y-intercept, and the equation, based on the line of best fit.

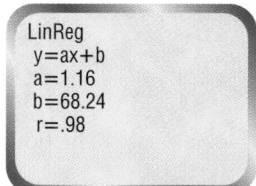

L1	L2	L3
15.00	89.84	------
20.00	88.62	
25.00	101.13	
30.00	100.45	
35.00	106.24	
40.00	114.28	
50.00	124.00	

L2={89.84,88.62...

*Select the **Stat** key of your calculator. Place the information from the table into a 2-variable (x, y) data table.*

LinReg
y=ax+b
a=1.16
b=68.24
r=.98

*Locate and select the **Reg** or **LinReg** for linear regression. Here, the slope for the line of best fit is a, and the y-intercept is b. The r is the correlation coefficient.*

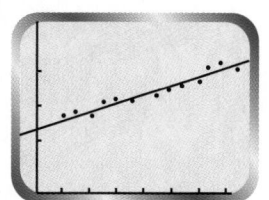

*The calculator will graph the scatter plot and draw the line of best fit when you enter the equation for the regression line after **Y =**.*

The rate of increase in average speed is the slope, about 1.16 miles per hour per year. The line of best fit (regression line) has the approximate equation $y = 1.16x + 68.24$. ❖

> Estimate the average speeds that would account for the missing data in 1980, 1985, 1990, and 1995. Compare your estimates with the actual speeds achieved in those years. What might explain the variation?

RETEACHING the lesson

Provide students with several different tables of values that exhibit a constant rate of change for each variable. Ask students to use each table of values to graph the functions. Then ask them to calculate the slope and y-intercept of each function from the values in the table and write an equation in slope-intercept form. Have them compare the equation to the graph they have already drawn, and ask if they think the equation matches the graph. Have them check their answer by locating the y-intercept and checking the slope.

Jay works as an electrician and charges by the hour. His records show the hours and pay he receives for various jobs.

Hours	Pay
2	$31.00
5	$77.50
8	$124.00
10	$155.00
12	$186.00

Write an equation that will calculate the hourly wage that Jay earns.   [$p = 15.5h$]

TEACHING *tip*

Have students do research to find out more about Hook's law and its history. Ask them to report their findings to the class.

Direct Variation

Springs are used in science to measure mass. Some springs are thin, and you can stretch them easily. Others are much stiffer, and you cannot seem to stretch them at all. The relationship between the amount of stretch and the load on the spring is defined by a law of physics called Hook's law.

Hook's law is an example of direct variation. It states that the distance a spring stretches varies directly as the force applied.

EXAMPLE 4

Jill works as a cable technician and charges by the hour. Her records show the hours and pay she receives for various jobs.

Hours on the job	3	5	7	10	15
Pay for the job	$28.80	$48	$67.20	$96	$144

Write an equation that will calculate the hourly wage that Jill earns.

Solution ➤

In looking at Jill's records, as the number of hours she works increases, the pay she earns also increases. This is an example of *direct variation*. The pay that Jill earns varies directly as the number of hours she works. The hourly wage that Jill earns is determined by

$$\frac{\text{Pay for the job}}{\text{Hours on the job}} = \text{Hourly wage, so } \frac{48}{5} = \$9.60.$$

The hourly wage, $9.60, is a constant.

Let p = the pay for the job. Let h = the hours on the job.

$$\frac{p}{h} = 9.6$$

If you solve for p, you find that $p = 9.6h$. ❖

The equation $\frac{p}{h} = 9.6$ models direct variation. The expression $\frac{p}{h}$ is a ratio, and 9.6 is called the constant of variation.

DIRECT VARIATION

If y varies directly as x, then $\frac{y}{x} = k$ or $y = kx$. The k is called the **constant of variation.**

The direct variation equation written in the form $y = kx$ is a linear function. To find the constant of variation, find the ratio $\frac{y}{x}$. How does this compare with finding the slope of a line?

> **EXAMPLE 5**

Physics   A force of 5 pounds stretches a spring a distance of 17.5 inches. How far will a spring stretch if an 8-pound weight is applied?

Solution ➤

The distance of stretch varies directly as the force applied.

Let d = the distance the spring stretches.

Let f = the force of the weight applied.

$\frac{d}{f} = k$ From the definition of direct variation, identify the value of k, the constant of variation.

$\frac{17.5}{5} = k$ The ratio of 17.5 inches to a 5-pound weight simplifies to the constant $k = 3.5$. The constant k is the constant of variation.

17.5 in.

$\frac{d}{f} = 3.5$ Once you know the constant of variation, you can find the distance of a given stretch, d.

$d = 3.5f$ Solve for d to form a linear function in terms of f.
$d = 3.5(8)$ Substitute 8 for the weight.

$d = 3.5(8)$ The distance the spring stretches is 28 inches.❖
    $= 28$ inches

If you graph a direct variation equation, you will find that direct variation represents a linear function in the form $y = mx + b$. The y-intercept is 0 and the slope is k, the constant of variation.

A ongoing
ASSESSMENT

The ratio is $\frac{y}{x}$, which is the same ratio as the slope of a line through the origin.

Alternate Example 5

A force of 8 pounds stretches a spring a distance of 19.5 inches. How far will a spring stretch if a 10-pound weight is applied? [24.375 inches]

TEACHING *tip*

Students can also solve Example 5 using a proportion. Since both $\frac{17.5}{5}$ and $\frac{d}{8}$ are equal to k, they are equal to each other. So the proportion $\frac{17.5}{5} = \frac{d}{8}$ can be solved to find the distance the spring stretches.

Use Transparency ▶ **24**

ASSESS

Selected Answers
Odd-numbered Exercises 9–45, 49–55

Assignment Guide
Core 1–16, 18–40 even, 41–58

Core Two-Year 1–58

EXERCISES & PROBLEMS

Communicate

1. Explain what effect a change in *b* represents on the graph of the equation $y = mx + b$.

2. Explain what effect a change in *m* represents on the graph of the equation $y = mx + b$.

3. Describe how to find the slope, *m*, of a line passing through the points (3, 5) and (7, 2).

4. How do you determine the *y*-intercept for the equation $y = 5x - 1$?

5. How do you determine the slope for the equation $y = 10 - x$?

6. Tell how to graph the equation $y = 2x - 3$ without plotting points.

7. Explain how to write the equation of a line passing through points $A(-2, 4)$ and $B(1, 5)$.

Practice & Apply

Make a table from the given data points, and find the first differences of the *x*- and *y*-values.

(0, 4), (2, 10), (4, 16), (6, 22), (8, 28), (10, 34)

8. According to the differences, what is the slope? 3

9. At what point does the line cross the *y*-axis? How do you know?
(0, 4)

10. Write the equation, and graph the line.

Give the coordinates where the line for each equation crosses the *y*-axis.

11. $y = 4x + 5$ (0, 5)   12. $y = 8x - 1$ (0, −1)

13. $y = -3x + 7$ (0, 7)   14. $y = -5x - 9$
(0, −9)

Examine the lines on the graph.

15. Which line has a *y*-intercept of −2? *k*

16. Which line has a positive slope? *n*

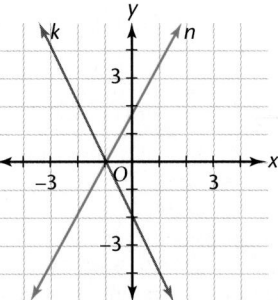

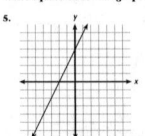

Practice & Apply
5.3 The Slope-Intercept Formula

Give the coordinates where the line for each equation crosses the *y*-axis.

1. $y = 3x + 4$ _____   2. $y = 2x - 3$ _____

3. $y = \frac{1}{2}x$ _____   4. $y = 2 - x$ _____

Write equations for the graphs of the following lines.

5. 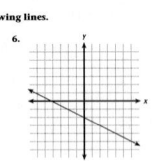   6.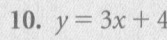

Orlando sings in the school chorus. He has to travel 42 miles round trip to the nearest college to take special lessons. Let x be the number of trips. Let y be the total miles he will travel.

7. Make a table and find the first differences of the x- and y-values.

8. Find the slope of the line that represents the function. _____

9. Write the equation and graph the line. _____

Find the slope of the lines passing through the following points.

10. (3, 8), (2, 6) _____   11. (0, −6), (−3, 3) _____

12. (−2, −4), (5, −1) _____   13. (−1, −2), (−3, −4) _____

Write the equation for a line

14. with slope 2 and y-intercept 5. _____

15. with slope −3 and y-intercept 1. _____

10. $y = 3x + 4$

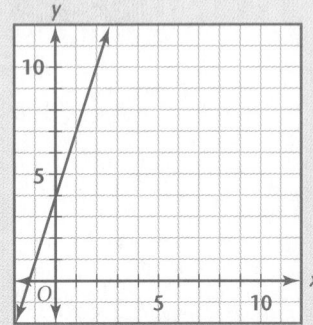

Find the slope of the lines passing through the following points.

17. $(0, 6), (5, 0)$ $-\frac{6}{5}$ **18.** $(3, 4), (-1, -2)$ $\frac{3}{2}$ **19.** $(1, 7), (5, 3)$ -1

20. $(7, 2), (-4, -2)$ $\frac{4}{11}$ **21.** $(-5, 0), (8, -4)$ $-\frac{4}{13}$ **22.** $(7, -7), (-4, -3)$ $-\frac{4}{11}$

23. $(-4, -3), (-2, -6)$ $-\frac{3}{2}$ **24.** $(6, 6), (-2, -2)$ 1 **25.** $(-1, 1), (5, -7)$ $-\frac{4}{3}$

Write the equation for a line

26. with slope -1 and with y-intercept 0.   **27.** through $(0, -4)$ and with slope -4.

28. with slope 11 and with y-intercept 15.   **29.** with slope -5 and y-intercept 7.

30. through $(0, 5)$ and with slope 1.   **31.** with slope -3 and y-intercept -1.

32. through $(0,3)$ and with slope 3.   **33.** with slope $\frac{2}{3}$ and y-intercept 2.

Write the equation for a line passing through the following points.

34. $(7, 9), (3, -2)$ **35.** $(-1, 0), (0, 3)$ **36.** $(4, 5), (-1, -2)$

37. $(3, 3), (-2, -6)$ **38.** $(-8, -3), (6, -2)$ **39.** $(-1, -4), (-3, -5)$

40. Consumer Mathematics A catalog company charges a fixed charge of $2 per order plus $0.50 per pound for shipping. Write an equation for the total shipping cost, c, in terms of the number of pounds, p. What is the form of your equation? $c = 0.50p + 2$; linear

Write equations for the graphs of the following lines.

41.
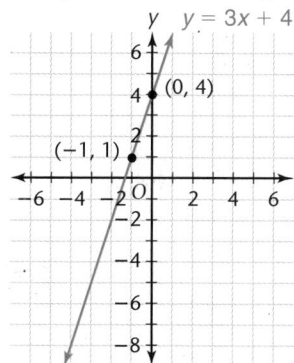
$y = 3x + 4$
$(0, 4)$
$(-1, 1)$

42.
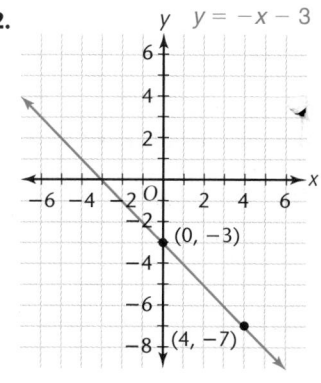
$y = -x - 3$
$(0, -3)$
$(4, -7)$

Josh sold 20 animal figures. The cost of materials was $43.80.

43. Write an equation for the amount of money Josh will have after each sale. $y = 6.50x - 43.80$

44. How many figures must Josh sell before the cost of materials is less than the income from sales? 7

45. Graph the equation for Josh's business.

$ 6.50 each

46. The distance a car travels at a constant rate of 65 miles per hour varies directly as the time in hours. Write an equation to express this direct variation. Then find the distance if the time elapsed is 3 hours.

26. $y = -x$

27. $y = -4x - 4$

28. $y = 11x + 15$

29. $y = -5x + 7$

30. $y = x + 5$

31. $y = -3x - 1$

32. $y = 3x + 3$

33. $y = \frac{2}{3}x + 2$

34. $y = \frac{11}{4}x - \frac{41}{4}$

35. $y = 3x + 3$

36. $y = \frac{7}{5}x - \frac{3}{5}$

37. $y = \frac{9}{5}x - \frac{12}{5}$

38. $y = \frac{1}{14}x - \frac{17}{7}$

39. $y = \frac{1}{2}x - \frac{7}{2}$

45.
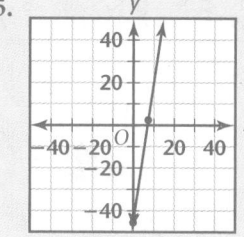

46. $\frac{d}{t} = k; \frac{d}{3} = 65$; 195 miles

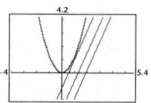

alternative
ASSESSMENT

Performance Assessment

Have students work in small groups to find an example of data that can be analyzed using a linear equation. They should write a report about the data and why it can be described using a linear equation. Their report should also include a table of values, an equation, and a graph. Have them make a poster that shows the results of their work.

Technology

A graphics calculator or computer software such as $f(g)$ Scholar™ may be used for Exercises 49 and 51.

Look Beyond

These exercises will help prepare students for their work with vertical and horizontal lines in Lesson 5.5.

47. Physical Science Find a sturdy rubber band or spring and suspend it so that you can measure its length. Measure the length without any weight on it. Find several weights that the rubber band or spring will hold without breaking. Hang the weights on the rubber band or spring and measure the length of the stretch. Make a table of the weights and the stretch lengths of the rubber band or spring. Make a scatter plot and find the line of best fit. Use a graphics calculator if possible to find the equation of the regression line. Write a description of your findings.

48. 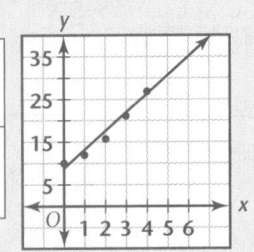 **Geometry** If the base area of a container is kept constant, the volume of the container varies directly as the height. The table shows the volumes and heights. Find the constant of variation. Then find the volume if the height is 12 centimeters. 9; 108 cubic cm

V	27	36	45	54
h	3	4	5	6

Look Back

49. Technology The population of Cooperstown is 12,345 and is growing by 678 people per year. Make a table or use the repeating operations feature of a calculator to determine when the population will reach 20,000. **[Lesson 1.3]** between 11 and 12 years

50. Write an equation for the population, p, in terms of y, the number of years the population increases at this rate. **[Lesson 1.3]** $p = 12{,}345 + 678y$

51. Technology Plot a graph (try using a graphics calculator) to find when the population will reach 20,000. **[Lesson 1.3]**

Perform the indicated operations.

52. $(3a - 2) + (2a - 2)$ **[Lesson 3.3]**
$5a - 4$

53. $(4x + 3y) - (3x - 2y)$ **[Lesson 3.4]**
$x + 5y$

Solve each inequality. **[Lesson 3.6]**

54. $x + 7 \leq -2$
$x \leq -9$

55. $x - 4 \geq 6$
$x \geq 10$

56. $x - 5 < 21$
$x < 26$

Look Beyond

57. If a line is horizontal, what is the rise when you compute the slope? 0

58. If a line is vertical, what is the run when you compute the slope? What does this indicate about the slope? 0; undefined

47. Answers may vary. For example:

Weight in Newtons	0	1	2	3	4
Length in Centimeters	10	12	15.5	21	27

Regression Line:
$y = 4.3x + 8.5$

51.

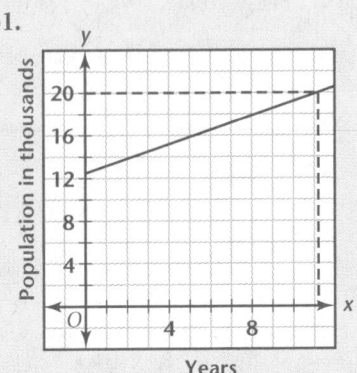

LESSON 5.4

Other Forms for Equations of Lines

Wynton Marsalis studied at the Julliard School of Music and played trumpet with the New Orleans Philharmonic Orchestra before gaining fame as a jazz musician. He was born in New Orleans on October 16, 1961.

 *The equation of a line can be written in different forms. If you know which form to use, you can often save yourself time and effort.*

You can write the birth date of Wynton Marsalis many ways. Each way has its advantages. The long form is easy to read. The short form is quick and easy to write. The computer form is easy to sort by year, month, and day.

<table>
<tr><td>October 16, 1961</td><td>10-16-61</td><td>611016</td></tr>
<tr><td>long form</td><td>short form</td><td>computer form</td></tr>
</table>

You have been using the slope-intercept form of a line, $y = mx + b$, to solve problems. As with the dates, there are other forms for the equation of a line. Each form has its advantages. The standard form provides a simple method for graphing the equation of a line.

Standard Form

Consider an equation written in the form $Ax + By = C$. After values for A, B, and C are determined, a solution to this equation is any ordered pair of numbers (x, y) that makes the equation true. When these ordered pairs are graphed, they form a straight line.

> ### STANDARD FORM
> An equation in the form **Ax + By = C** is in standard form when
> - **A**, **B**, and **C** are integers,
> - **A** and **B** are not both zero, and
> - **A** is not negative.

ALTERNATIVE
teaching
strategy

Using Visual Models
Have students graph the equations listed below.

1. $2x + 3y = 9$ **2.** $y = -\frac{2}{3}x + 3$

3. $y - 1 = -\frac{2}{3}(x - 3)$

Ask them to record their observations, including the slope and y-intercept of each graph. For equation 3, ask students whether the point

(3, 1) is on the graph and how that point is represented in the equation. Students should discover that all of the equations given are equations for the same line. Inform students that equation 1 is said to be in *standard form*, $Ax + By = C$; equation 2 is in *slope-intercept form*, $y = mx + b$; and equation 3 is in *point-slope form*, $y - y_1 = m(x - x_1)$. Discuss the advantages and disadvantages of each form. Have students write their own equation in three forms.

PREPARE

Objectives
- Define and use the Standard Form of the equation of a line.
- Define the Point-Slope Form of the equation of a line.

RESOURCES

- Practice Master   5.4
- Enrichment Master   5.4
- Technology Master   5.4
- Lesson Activity Master 5.4
- Quiz   5.4
- Spanish Resources   5.4

Assessing Prior Knowledge

Solve each equation for the indicated variable.

1. $C = 2\pi r$, for r
$$\left[r = \frac{C}{2\pi}\right]$$

2. $P = 2l + 2w$, for l
$$\left[l = \frac{P - 2w}{2}\right]$$

3. $C = \frac{5}{9}(F - 32)$, for F
$$\left[F = \frac{9}{5}C + 32\right]$$

4. $y = mx + b$, for x
$$\left[x = \frac{y - b}{m}\right]$$

TEACH

why Discuss the different ways that the same data can be written. Emphasize the fact that although the forms may be different, the data remain the same. For example, the date that Wynton Marsalis was born remains the same no matter how the date is written. Have students write their own birth dates according to the forms shown in the text.

EXAMPLE 1

Alternate Example 1

Mrs. McCarthy works part time at the bakery for $5 an hour and part time at the supermarket for $6 an hour. She wants to earn $195 a week.

a. Write an equation in standard form that models this situation.
   $[5x + 6y = 195]$

b. Graph the equation.

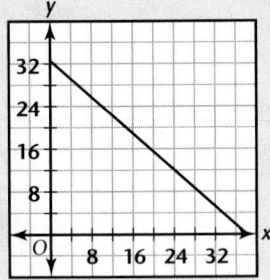

Fund-raising Jackie is in charge of selling tickets for the school jazz concert. She hopes the total ticket sales will be about $588. This will cover expenses and make a modest profit.

A Write an equation in standard form that models this situation.

B Graph the equation.

Solution ➤

A The equation must show that $4 times the number of adult tickets added to $2 times the number of student tickets is $588.

Let x represent the number of adult tickets sold. Let y represent the number of student tickets sold. The equation that models the problem is $4x + 2y = 588$. Notice that the equation is in standard form $Ax + By = C$.

B Graph the adult tickets, x, on the horizontal axis and student tickets, y, on the vertical axis. The points (x, y) represent (adult tickets, student tickets).

To graph an equation in standard form, substitute 0 for x, and solve for y. Then substitute 0 for y, and solve for x. This gives the intercepts $(0, y)$ and $(x, 0)$, which are the points where the line crosses each axis. Draw the line connecting these 2 points.

Start with the equation $4x + 2y = 588$.

First let $x = 0$.
$$4x + 2y = 588$$
$$0 + 2y = 588$$
$$y = 294$$
(0, 294) is on the graph.

Then let $y = 0$.
$$4x + 2y = 588$$
$$4x + 0 = 588$$
$$x = 147$$
(147, 0) is on the graph

Draw the line connecting the points. The coordinates of any point on the line will solve the equation. ❖

To solve the jazz concert ticket problem, however, the values for x and y must be whole numbers.

What does the point (97,100) represent with respect to the problem?

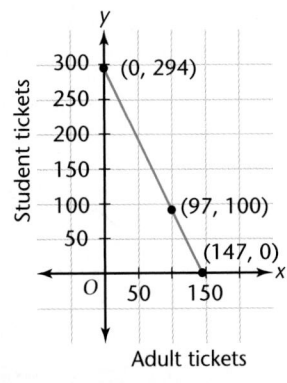

Adult tickets

interdisciplinary **Biology** Biologists
CONNECTION have discovered that temperature and the number of chirps some crickets make in one minute is a linear relationship. When it is 68°F, crickets chirp 124 times in one minute. When it is 80°F, crickets chirp 172 times in one minute.

a. Write an equation that models this situation, and graph the equation.
$$\left[(y - 68) = \tfrac{1}{4}(x - 124)\right]$$

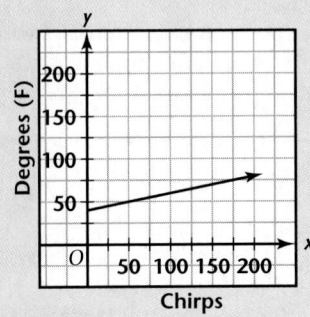

Chirps

b. Suppose you hear 100 chirps. Predict the temperature. [62°F]

 **EXAMPLE 2**

Graphics Calculator

Use a graphics calculator to graph the equation $4x + 2y = 588$.

Solution ➤

To enter an equation in a graphics calculator, first rewrite the equation in slope-intercept form $y = mx + b$.

$$4x + 2y = 588$$
$$2y = 588 - 4x$$
$$y = 294 - 2x$$
$$y = -2x + 294$$

The equation can now be entered into the calculator.

$$\mathbf{Y = -2X + 294} \; \diamond$$

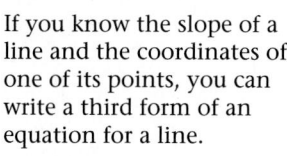
CRITICAL *Thinking*

If the equation of a line is written in standard form $Ax + By = C$, the slope is $-\frac{A}{B}$ and the y-intercept is $\frac{C}{B}$. Explain why.

Point-Slope Form

If you know the slope of a line and the coordinates of one of its points, you can write a third form of an equation for a line.

Remember that the value for the slope m can be calculated from any two points (x_1, y_1) and (x_2, y_2), and the formula $m = \frac{y_2 - y_1}{x_2 - x_1}$.

Multiply both sides of the slope formula by $x_2 - x_1$, and simplify.

If you replace the specific point (x_2, y_2) with a general point on the line, represented by (x, y), the result is $y - y_1 = m(x - x_1)$.

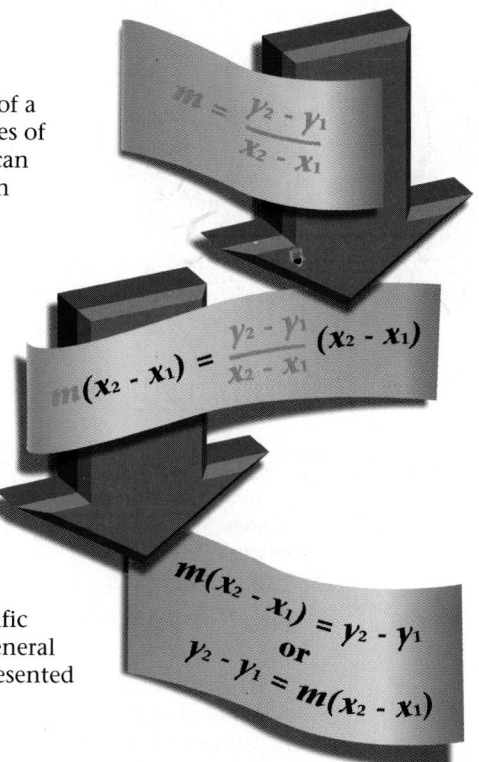

$$m = \frac{y_2 - y_1}{x_2 - x_1}$$

$$m(x_2 - x_1) = \frac{y_2 - y_1}{x_2 - x_1}(x_2 - x_1)$$

$$m(x_2 - x_1) = y_2 - y_1$$
or
$$y_2 - y_1 = m(x_2 - x_1)$$

> **POINT-SLOPE FORM**
>
> The form $y - y_1 = m(x - x_1)$ is the **point-slope form** for the equation of a line. The coordinates x_1 and y_1 are taken from a given point (x_1, y_1), and the slope is m.

Alternate Example 3

Write the point-slope form of a line with slope -2 that passes through point $(3, -1)$.
$[y + 1 = -2(x - 3)]$

Cooperative Learning

Have students work in groups of four. One student begins by writing an equation of a line in any one of the three forms. Two other students each write the given equation in the other two forms. The fourth student draws the graph of the original equation. Then students check to be sure each equation matches the graph.

Alternate Example 4

The monthly cost of Jon's checking account is $0.20 a check, plus a monthly service charge. Last month, Jon wrote 12 checks and the total cost of the account for the month was $5.40. Write the point-slope equation that models the cost of Jon's account last month.
$[y - 5.40 = 0.20(x - 12)]$

Alternate Example 5

Change $y - 5.40 = 0.20(x - 12)$ to an equation

a. in slope-intercept form.
$[y = 0.20x + 3]$

b. in standard form.
$[2x - 10y = -30]$

EXAMPLE 3

A line with slope 3 passes through point $(2, 7)$. Write the equation of the line in point-slope form.

Solution ➤

Let $m = 3$, $x_1 = 2$, and $y_1 = 7$. Substitute the given values into the point-slope equation.

$$y - y_1 = m(x - x_1)$$
$$y - 7 = 3(x - 2)$$ ❖

EXAMPLE 4

Janet's class is ordering T-shirts. Write the point-slope equation that models the information on the order form.

T-Shirts $9.00 ea.
* Plus Shipping and Handling

COLOR	S	M	L	TOTAL
Red	2			2
Blue	1	4	5	10
White		1	3	4
TOTAL				16
TOTAL COST				$149.00

Solution ➤

Each time a shirt is ordered, the cost increases by $9, so 9 is the rate of change or slope. Since 16 shirts cost $149, $(16, 149)$ represents a point on the graph. Substitute 9 for m, 16 for x_1, and 149 for y_1 into $y - y_1 = m(x - x_1)$. The result is
$y - 149 = 9(x - 16)$. ❖

EXAMPLE 5

Change $y - 149 = 9(x - 16)$ to an equation

Ⓐ in slope-intercept form. Ⓑ in standard form.

Solution ➤

Ⓐ Begin with $y - 149 = 9(x - 16)$ Given
$y - 149 = 9x - 144$ Distributive Property
$y = 9x + 5$ Addition Property of Equality

The equation is now in slope-intercept form. Note that the slope represents $9, the cost per shirt. The y-intercept represents $5, the shipping and handling charge.

Ⓑ Next, change the slope-intercept form to standard form, $Ax + By = C$.

$y = 9x + 5$ Given
$y - 9x = 5$ Subtraction Property of Equality
$9x - y = -5$ Multiply each side by -1 and rearrange terms. ❖

R**ETEACHING**
t h e
l e s s o n

Give students the coordinates of two points. Have them plot the points and draw a line through them. Ask students to determine the slope and y-intercept from the graph. Then have them write an equation for the line in slope-intercept form. Ask students to check their equation by substituting the two original points into the equation. Finally, have students write the equation in standard form and point-slope form.

EXAMPLE 6

Find an equation for the graph of a line that passes through the points $(-1, 10)$ and $(5, 8)$.

Solution ➤

The point-slope form simplifies the way this problem is solved. Let $(x_1, y_1) = (-1, 10)$ and $(x_2, y_2) = (5, 8)$. Find the slope.

$$\text{Slope} = \frac{y_2 - y_1}{x_2 - x_1} = \frac{8 - 10}{5 - (-1)} = \frac{-2}{6} = -\frac{1}{3}$$

Use the point $(-1, 10)$ for (x_1, y_1). Substitute the slope and the coordinates of the point into the equation $y - y_1 = m(x - x_1)$. Then $y - 10 = -\frac{1}{3}(x - (-1))$ or $y - 10 = -\frac{1}{3}(x + 1)$. ❖

Try This Use the point-slope form to find the equation of the line from the points $(5, 65)$ and $(7, 71)$.

SUMMARY OF THE FORMS OF A STRAIGHT LINE

Name	Form	Example
Slope-intercept	$y = mx + b$	$y = 3x + 5$
Standard	$Ax + By = C$	$3x - y = -5$
Point-slope	$y - y_1 = m(x - x_1)$	$y - 11 = 3(x - 2)$

EXERCISES & PROBLEMS

Communicate

1. Explain how to write the equation $5y - 2 = -3x$ in standard form.
2. Tell how you would find the intercepts for the equation $3x + 6y = 18$.
3. Describe how to graph $2x + 3y = 12$ by finding the intercepts.
4. Explain how to change $x - 3y = 9$ into slope-intercept form.
5. How would you use the point-slope formula to write the equation of a line through points $(-2, 4)$ and $(4, -8)$?
6. The equation of a given line is $5x + 2y = 40$. Tell how you would find the slope of the line.

Practice & Apply

Write the following equation in standard form.

7. $4x = -3y + 24$ **8.** $7y = -5x - 35$ **9.** $6x + 4y + 12 = 0$ **10.** $2x = 4y$

11. $6x - 8 = 2y + 6$ **12.** $x = \frac{2}{3}y + 6$ **13.** $2 + 7x + 14y = 3x - 10$ **14.** $5 = y - x$

Find the intercepts for the following equations.

15. $x + y = 10$
(0, 10), (10, 0)

16. $3x - 2y = 12$
(0, -6), (4, 0)

17. $5x + 4y = 20$
(0, 5), (4, 0)

18. $x = 2y$
(0, 0)

19. Draw a graph of a line that intercepts the axes at (3, 0) and (0, 7).

20. Graph the equation $2x + 6y = 18$ by finding the intercepts.

21. The equation of a given line is $6x + 2y = 40$. What is the slope of the line? -3

22. Compare the graphs of $4a + 2s = 588$ and $2a + s = 294$. What do you find? the same line

23. **Fund-raising** Suppose that in Example 1 on page 234, 50 is selected for y in the equation $4x + 2y = 588$. What is x? What is the maximum number of adult tickets you can buy for $588? What is the maximum number of student tickets you can buy for $588? 122; 147; 294

Write the equation of a line

24. through (5, 2) and with slope -8. $8x + y = 42$

25. that crosses the x-axis at $x = 7$ and the y-axis at $y = 2$. $y = -\frac{2}{7}x + 2$

26. through (5, 2) and with slope 8. $8x - y = 38$

27. through (5, 2) and with slope 0. $y = 2$

28. that crosses the x-axis at $x = 1$ and the y-axis at $y = \pi$. $y = -\pi x + \pi$

29. through (2, 3) and (8, -3). $x + y = 5$

30. through (5, 9) and (10, 9). $y = 9$

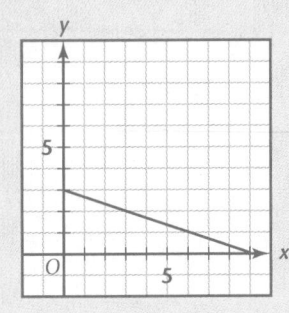

31. **Technology** On some graphing calculators, equations must be in function form, $y = f(x)$, to draw graphs. The symbol $f(x)$ indicates a function. Which of the three forms for the equation of a line given in the summary at the end of the lesson is/are in $y = f(x)$ form?
slope-intercept form; $y = mx + b$

Rewrite each of the following in the function form $y = f(x)$, where the equation is solved for y.

32. $2x + 3y = 12$ **33.** $y - 2 = 3(x - 7)$ **34.** $2x + 2y + 5 = 0$

35. $3x + 4y = 6 + 5y$ **36.** $\frac{x}{2} + \frac{y}{3} = 18$ **37.** $\frac{5x - y}{7} = 14$

38. Write an equation that shows the value of n nickels and d dimes is $5. HINT: Be careful about the units. $5.00 = 0.05n + 0.10d$

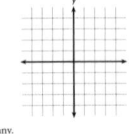
7. $4x + 3y = 24$

8. $5x + 7y = -35$

9. $6x + 4y = -12$

10. $2x - 4y = 0$

11. $6x - 2y = 14$

12. $3x - 2y = 18$

13. $4x + 14y = -12$

14. $x - y = -5$

19.

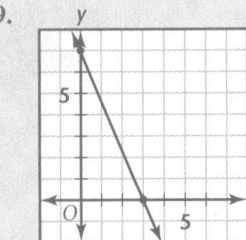

20.

Intercepts: (0, 3), (9, 0)

Fund-raising Write an expression for the sales in dollars of

39. 40 adult tickets. 5(40)

40. 20 student tickets and 37 adult tickets. 3(20) + 5(37)

41. *s* student tickets and *a* adult tickets. 3s + 5a

42. Write an equation that says the total sale of *s* student tickets and *a* adult tickets is $700. Which of the given forms did you choose for your equation? 3s + 5a = 700

Solve the equation in Exercise 42 when

43. 90 student tickets are sold. 86

44. 80 adult tickets are sold. 100

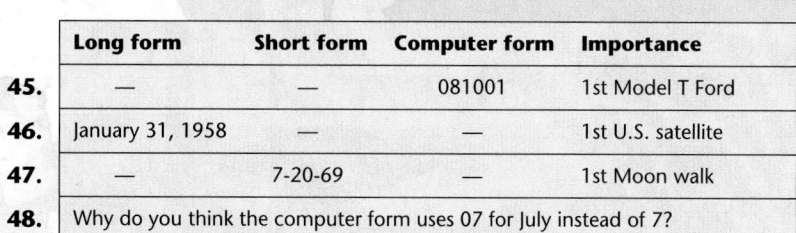

Social Studies Complete this table showing various forms of the same date.

	Long form	Short form	Computer form	Importance
45.	—	—	081001	1st Model T Ford
46.	January 31, 1958	—	—	1st U.S. satellite
47.	—	7-20-69	—	1st Moon walk
48.	Why do you think the computer form uses 07 for July instead of 7?			

Look Back

Evaluate the following if *x* = 1, *y* = 1, and *z* = 2, [Lesson 1.3]

49. $x^2 + y + z^2$ 6

50. $x - y + z$ 2

51. $x + y - z$ 0

52. $-(x + y + z)$ − 4

53. Describe the scatter plot of a strong negative correlation. **[Lesson 1.6]**
Points are close to the line of best fit, sloping downward left to right.

Solve the following equations. [Lesson 4.3]

54. $-5y = 30$ − 6

55. $3x = 420$ 140

56. $\frac{y}{9} = 36$ 324

57. $\frac{x}{2} = 108$ 216

Find the slope of the line containing the origin (0, 0) and the given point. Then give the equation of the line passing through the origin and the given point. [Lesson 5.1]

58. $A(3, 6)$
2; y = 2x

59. $B(2, 8)$
4; y = 4x

60. $C(6, 3)$
$\frac{1}{2}$; $y = \frac{1}{2}x$

61. $D(-5, -7)$
$\frac{7}{5}$; $y = \frac{7}{5}x$

Look Beyond

62. If two lines intersect, how many points do they have in common? 1

63. If two lines are parallel, how many points do they have in common? 0

64. Michael has a 5-liter and a 3-liter bottle. Neither have markings. He has a supply of water. How can he measure 1 liter exactly?

32. $y = -\frac{2}{3}x + 4$

33. $y = 3x - 19$

34. $y = -x - \frac{5}{2}$

35. $y = 3x - 6$

36. $y = \frac{3}{2}x + 54$

37. $y = 5x - 98$

The answers to Exercises 45–48 can be found in Additional Answers beginning on page 729.

64. Fill the 3-liter bottle. Pour the contents into the 5-liter bottle. Fill the 3-liter bottle again. Pour water from it into the 5-liter bottle. It takes 2 more liters to be full. One liter will remain in the 3-liter bottle.

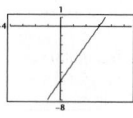

LESSON 5.5   Vertical and Horizontal Lines

WHY *A line along the diving tower, seen from the side is horizontal. A line down the tower is vertical. The slopes and equations for such horizontal or vertical lines have a special significance in algebra.*

A pool offers two options. Residents can either pay \$2 per visit or \$25 for a season pass. How many times do you have to use the pool to make the season pass the better option?

Recreation   The data points showing the number of visits to the pool, x, and the total cost, y, can be modeled for both options. The graphs show only the first quadrant because the number of visits and the amount of money spent will always be 0 or greater.

Per-visit option   For 0 visits the cost is \$0. Thus, the y-intercept is 0, and (0, 0) is one point on the graph. Since the rate of increase is \$2 per visit, the slope is 2.

The first graph is a line drawn through **(0, 0)** with slope 2. The line represents a typical equation in slope-intercept form, $y = 2x + 0$. You can check this by seeing that it contains the point **(10, 20)**. Ten visits cost \$20.

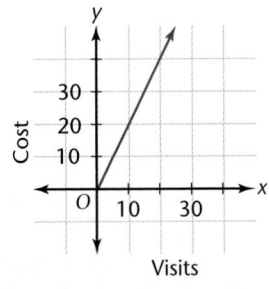

Season-pass option The cost is a constant $25. If you go to the pool twice, the cost is $25. Thus, **(2, 25)** is a point on the graph. If you go to the pool 40 times, the cost is still $25. The point **(40, 25)** is on the graph.

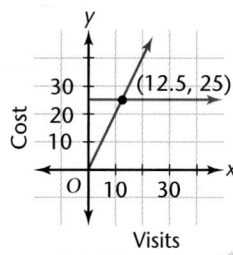

The second graph is a horizontal line consisting of all y-values equal to 25. The line drawn through the two points is a special equation in slope-intercept form, $y = 0x + 25$ or $y = 25$.

Now draw both graphs on one coordinate plane.

The graph shows that for $12\frac{1}{2}$ visits the cost is $25. Since you cannot make $\frac{1}{2}$ visit, the season-pass option is better for 13 or more visits.

Notice that the graph of $y = 25$ is a horizontal line. The graph of $y = b$ is a horizontal line that crosses the y-axis at the point $(0, b)$. The equation $y = b$ is a **constant function** where b is the **constant**. It can be written as $y = mx + b$ where the slope is zero. In standard form the equation is written $0x + 1y = b$.

EXAMPLE 1

Compare the graphs of $y = 4$ and $x = 4$.

Solution ➤

The graph $y = 4$ consists of all points with a y-coordinate of 4. The graph $x = 4$ consists of all points with an x-coordinate of 4.

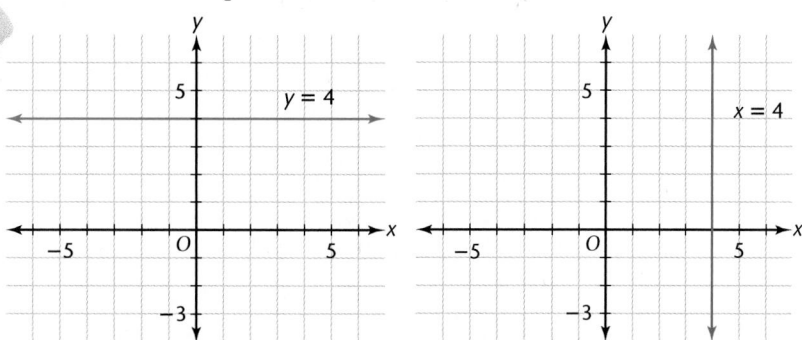

The equation $y = 4$ is a horizontal line 4 units above the origin.

The equation $x = 4$ is a vertical line 4 units to the right of the origin. ❖

Try This Compare the graphs of $y = -3$ and $x = -3$.

Alternate Example 1

Compare the graphs of $y = 1$ and $x = 1$. [**The graph $y = 1$ consists of all points with a y-coordinate of 1. The graph $x = 1$ consists of all points with an x-coordinate of 1.**]

Aongoing
SSESSMENT

Try This

The graph $y = -3$ consists of all points with a y-coordinate of -3. The graph of $x = -3$ consists of all points with an x-coordinate of -3.

TEACHING *tip*

Stress that there are an infinite number of points that can be chosen for either of the equations in Example 2. You may want to emphasize this by asking students to complete the example again with two different points on each line.

Alternate Example 2

Find the slope of
  a. $y = -2$. [0]

  b. $x = 3$. [undefined]

Alternate Example 3

Compare the graphs of $y = 2x$ and $x = 2y$. [The graph of the equation $y = 2x$ is a line with a slope of 2 and a y-intercept of 0. The equation $x = 2y$ is written $y = \frac{1}{2}x + 0$ in slope-intercept form. Its graph is a line with a slope of $\frac{1}{2}$ and y-intercept of 0.]

$y = 2x$:

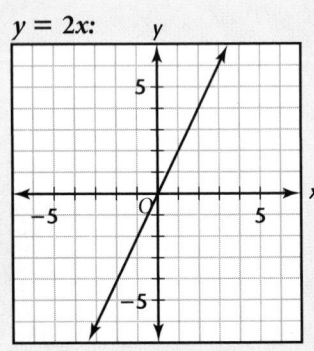

$x = 2y$:

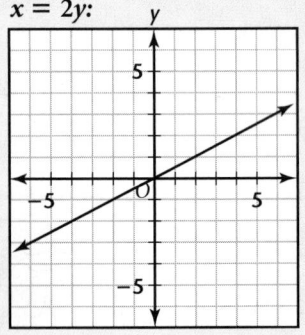

EXAMPLE 2

Find the slope of **A** $y = 5$. **B** $x = -3$.

Solution ➤

A The graph of $y = 5$ has the same value of y for any value of x. Use any two points with a y-coordinate of 5 to find the slope. The points $(3, 5)$ and $(7, 5)$ will work.

$$\text{slope} = \frac{y_2 - y_1}{x_2 - x_1} = \frac{5 - 5}{7 - 3} = \frac{0}{4} = 0$$

B The graph of $x = -3$ has the same value of x for any value of y. Use any two points with an x-coordinate of -3 to find the slope. The points $(-3, 3)$ and $(-3, 7)$ will work.

$$\text{slope} = \frac{y_2 - y_1}{x_2 - x_1} = \frac{7 - 3}{(-3) - (-3)} = \frac{4}{0}$$

The run is 0. Since dividing by 0 is impossible, the *slope* is *undefined*. ❖

The graph of an equation in the form $x = a$ is a vertical line that crosses the x-axis where x equals a. The equation cannot be written in slope-intercept form, but it can be written in standard form.

$$1x + 0y = a$$

HORIZONTAL AND VERTICAL LINES

The equation for a horizontal line is written in the form $y = b$. The slope is 0.

The equation for a vertical line is written in the form $x = a$. The slope is undefined.

EXAMPLE 3

Compare the graphs of $y = 3x$ and $x = 3y$.

Solution ➤

Write the equations in slope-intercept form. Graph the lines.

The equation $y = 3x$ is in slope-intercept form, and its graph is a line with slope 3 and y-intercept 0.

Change $x = 3y$ to slope-intercept form by solving for y. Since $x = 3y$ is equivalent to $3y = x$,

$$3y = x \rightarrow y = \frac{x}{3} \rightarrow y = \frac{1}{3}x + 0.$$

The equation $x = 3y$ is written $y = \frac{1}{3}x + 0$ in slope-intercept form. Its graph is a line with slope $\frac{1}{3}$ and y-intercept 0.

RETEACHING the lesson

Cooperative Learning Have students work in small groups to solve the following problem. The monthly commuter pass for a bridge is $30. Without the pass, each bridge crossing costs $4.

1. Write an equation to represent each plan.
   [$y = 30; y = 4x$]

2. Graph both equations on the same coordinate plane.

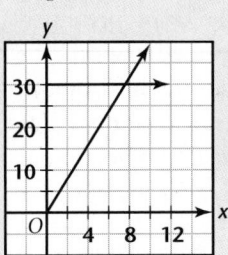

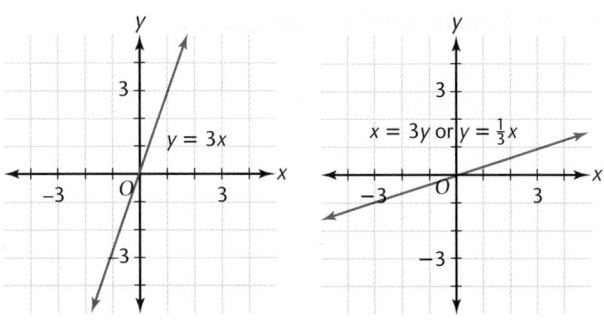

Recall that numbers are reciprocals when their product equals 1. Since $3 \cdot \frac{1}{3} = 1$, the slopes of $y = 3x$ and $y = \frac{1}{3}x$ are reciprocals.

The graphs of the original equations $y = 3x$ and $x = 3y$ are lines with reciprocal slopes and a common point at the origin, (0, 0). ❖

CRITICAL
Thinking

Draw the graph of $y = x$. On the same set of axes, draw the graphs of $y = 2x$ and $x = 2y$. Try other pairs of lines in the form $y = mx$ and $x = my$ with different m values. What relationships do you see?

EXERCISES & PROBLEMS

Communicate

1. Describe how to find the slope of $y = 3$. Is the line vertical or horizontal?

2. Explain how to find the slope of $x = 1$. Is the line vertical or horizontal?

3. Discuss how to find the slope of $y = 5x$. Choose your own points to graph the equation. At what point does the line cross the y-axis?

4. Tell how to find the slope of $x = 5y$. Choose your own points to graph the equation. At what point does the line cross the y-axis?

5. Explain why the graph of $y = 15$ is a constant function.

6. Tell why the slope for a vertical line is unusual.

7. Tell why the slope for a horizontal line is 0.

3. How many times would you need to cross the bridge to make the commuter pass the better option? Explain how you know.   [8; **When you have crossed the bridge 8 times, you have spent \$32, which is more than \$30. Up until that point, you have spent less than \$30.**]

CRITICAL
Thinking

The lines have reciprocal slopes and pass through a common point (0, 0).

ASSESS

Selected Answers
Odd-numbered Exercises 9–47

Assignment Guide
Core   1–7, 8–12 even, 13–19, 20–28 even, 29–50

Core Two-Year   1–50

Practice & Apply

Find the slope for each equation. Plot points, graph a line for each equation, and indicate if the line is vertical or horizontal.

8. $y = 5$ **9.** $x = 7$ **10.** $y = 9x$ **11.** $x = 5y$ **12.** $y = \frac{1}{3}x$

13. Language Arts Where do you think the word *horizontal* comes from? from the word horizon

Match each equation with the appropriate description.

14. $x + y = 9$ E   **A.** a horizontal line 9 units above the origin
15. $xy = 9$ F   **B.** a vertical line 9 units to the right of the origin
16. $x = 9$ B   **C.** a line through the origin with slope 9
17. $y = 9$ A   **D.** a line through the origin with slope $\frac{1}{9}$
18. $y = 9x$ C   **E.** a line with slope -1 and y-intercept 9
19. $x = 9y$ D   **F.** something other than a straight line

Coordinate Geometry The equation $x = 4$ cannot be written in slope-intercept form because the slope is undefined. It can, however, be written in standard form as $1x + 0y = 4$. Copy and complete the table. Write an equivalent form of each equation, writing "undefined slope" when appropriate.

Given	Slope-intercept Form	Standard Form
20. $x = 1$	?	?
21. $y = 4$	?	?
22. $x + y = 5$	?	?
23. $y = 4x$	?	?
24. $x = 4y$	?	?

Determine whether the following lines are vertical or horizontal.

25. $y = 8$ **26.** $x = -2$ **27.** $y = -4$ **28.** $x = 9$
horizontal   vertical   horizontal   vertical

29. Technology Find how to graph a vertical line on a graphics calculator. Answer depends on calculator.

30. Find the slope for the equations $y = 6x$ and $x = 6y$. What point do these lines have in common? $6; \frac{1}{6}; (0, 0)$

31. Write the reciprocal of 0.25. 4

32. Write the reciprocal of $-4\frac{1}{2}$. $-\frac{2}{9}$

33. Budget Saul's father is thinking of buying his son a six-month movie pass for $40. If matinees are $3.00 each, how many times must Saul attend before it would benefit his father to buy the pass? more than 13 times

8–9.
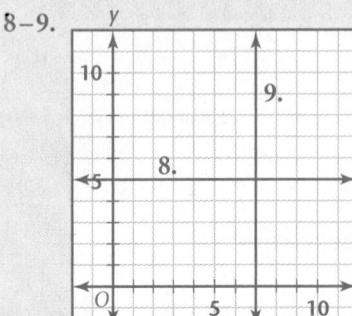

10–12.
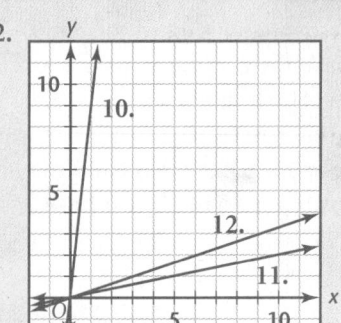

10. $m = 9$
11. $m = \frac{1}{5}$
12. $m = \frac{1}{3}$

8. $m = 0$, horizontal
9. slope is undefined, vertical

In the swimming pool example from the beginning of the lesson, which option is cheaper if you visit the pool

34. 8 times?
per-visit

35. 29 times?
season-pass

36. Recreation Repeat the swimming pool problem, but change the pool costs to those shown at the right. After how many visits does it pay to buy a season pass?
12 or more

37. Technically, should the graphs of the problem consist of points or lines? For example, does it make sense to visit the pool $4\frac{2}{3}$ times? For ease in making and reading graphs, often lines are used instead of points. What is another example where this might happen?

Season Opening
at the
Community Pool!
Please note our new rates

	old	new
Daily Pass	$ 2.00	$ 3.00
Season Pass	$25.00	$35.00

Lifeguards always on duty!
hours: 9:00am to 9:00pm
Bring the whole family!

Look Back

Use the order of operations to simplify the following expressions. [Lesson 1.4]

38. $30 \cdot 2 \div 1 + 10$
70

39. $[2(105 - 5)] \div 2$
100

40. $\frac{(3 + 9)^2 - 4}{2^2 \cdot 15 - 10}$
$2\frac{4}{5}$

Simplify. [Lesson 3.3]

41. $(6a + 3) + (2a + 6)$
$8a + 9$

42. $(x + y + z) + (3x + y + 4z)$
$4x + 2y + 5z$

43. $-7 + 3p - 9 - 7p$
$-16 - 4p$

44. Mary needs $1\frac{1}{3}$ yards of fabric to make a skirt. A yard of fabric costs $4.58. How many skirts can she make from 5 yards of fabric? 3
[Lesson 4.3]

Lunch cost $15.50. There is a 7% sales tax, and the customary tip is 15%. [Lesson 4.5]

45. How much is the tax on the lunch? $1.09

46. How much is the tip? $2.33

47. Given points (3, 5) and (−4, −2), write the slope-intercept equation for the line. **[Lesson 5.3]** $y = x + 2$

Look Beyond

48. Geometry What kind of angle is formed when a horizontal and a vertical line intersect? right angle

49. Will two distinct vertical lines ever intersect? no

50. Will two distinct horizontal lines ever intersect? no

	Slope-intercept	Standard
20. $x = 1$	undefined slope	$1x + 0y = 1$
21. $y = 4$	$y = 4$	$0x + 1y = 4$
22. $x + y = 5$	$y = -x + 5$	$1x + 1y = 5$
23. $y = 4x$	$y = 4x$	$4x - 1y = 0$
24. $x = 4y$	$y = \frac{1}{4}x$	$1x - 4y = 0$

37. points; You cannot visit the pool a fraction of times. Another example could be population growth. To indicate a constant population growth during the year for a community, you would show the growth with the graph of a line, not points. In reality, you could not indicate a partial person.

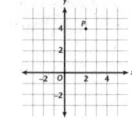

Objectives

- Identify parallel lines as lines with the same slope.
- Identify perpendicular lines as lines with slopes that are the negative reciprocal of each other.
- Write equations of lines parallel and perpendicular to other given lines.

RESOURCES

- Practice Master   5.6
- Enrichment Master   5.6
- Technology Master   5.6
- Lesson Activity Master 5.6
- Quiz   5.6
- Spanish Resources   5.6

Assessing Prior Knowledge

Write each equation in slope-intercept form.

1. $2x - y = -3$
   $[y = 2x + 3]$

2. $3x + y = -5$
   $[y = -3x - 5]$

3. $x + 2y = 2$
   $\left[y = -\dfrac{1}{2}x + 1 \right]$

4. $4x - 3y = 12$
   $\left[y = \dfrac{4}{3}x - 4 \right]$

TEACH

 Discuss the three different temperature scales discussed in the text. Students should be familiar with the Fahrenheit and Celsius scales. The Kelvin temperature scale is used by scientists when they need to measure very low temperatures.

LESSON 5.6 Parallel and Perpendicular Lines

whY *The graphs of parallel and perpendicular lines are related by their slopes. Algebra can help you solve problems in geometry.*

Parallel lines often look like they meet, but they never do.

Physical Science Examine the three different thermometer readings. It is possible to derive a linear equation for changing degrees in one scale to degrees in another. Notice how slope is used in each solution.

EXAMPLE 1

Use the data in the table to find the formula that changes

Ⓐ degrees Celsius (°C) to degrees Fahrenheit (°F).

Ⓑ kelvins to degrees Fahrenheit.

Reference	°F	°C	K
Water boils	212	100	373
Water freezes	32	0	273
Absolute 0	−460	−273	0

ALTERNATIVE **teaching strategy**

Using Visual Models Have students draw the graph of the equation $y = 2x$. Then ask them to draw several lines in red that are parallel to this line and determine the equations. Ask them to compare the equations, note any similarities, and share their observations with the class.

Then have students draw several lines in blue that are perpendicular to $y = 2x$. Have them use the graph of each line to determine the equations. Ask them to compare the equations and note the relationships between the slopes of these lines and the slope of $y = 2x$ and all of the red lines on the graph. Have the students share their observations with the class.

Solution ➤

Ⓐ The relationship between any two of the temperature scales is linear. Compare the Fahrenheit and Celsius scales. In the equation $y = mx + b$, let the y-values be Fahrenheit and the x-values be Celsius. Use the data from the table to calculate the slope and y-intercept.

$$\text{slope} = \frac{\text{change in Fahrenheit}}{\text{change in Celsius}} \quad \rightarrow \quad m = \frac{212 - 32}{100 - 0} = \frac{180}{100} = \frac{9}{5}$$

Notice that when C is 0, F is 32. The y-intercept, b, is 32. Find the equation that changes Celsius to Fahrenheit by substituting $\frac{9}{5}$ for m and 32 for b in $y = mx + b$. The equation $y = \frac{9}{5}x + 32$ gives the formula $F = \frac{9}{5}C + 32$.

Ⓑ You can change from kelvins to degrees Fahrenheit in much the same way. The equation for this change is $y = \frac{9}{5}x - 460$. The formula is $F = \frac{9}{5}K - 460$. ❖

Parallel Lines

Graphics Calculator

Graph the equations $F = \frac{9}{5}C + 32$ and $F = \frac{9}{5}K - 460$. Let y represent degrees in Fahrenheit and x represent either degrees in Celsius or kelvins. On a graphics calculator, enter the Celsius formula as the function **Y = (9/5)X + 32**. Then enter the Kelvin formula as the function **Y = (9/5)X − 460**. Place both graphs on the same axes.

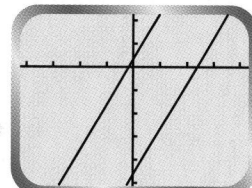

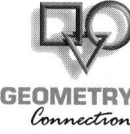
GEOMETRY Connection

If you use graph paper, use the y-intercept to locate the first point for one equation. Then use the run and rise of the slope for the second point. Once you have determined 2 or 3 points for the equation, use a straightedge to connect the points. Repeat the process for the second equation.

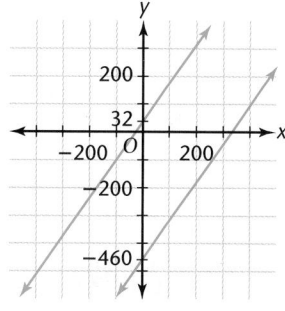

How are the slopes of the two lines related?

PARALLEL LINES

If two different lines have the same slope, the lines are **parallel**.

If two non-vertical lines are parallel, they have the same slope. Two parallel, vertical lines have undefined slopes.

---

interdisciplinary **CONNECTION** **Surveying** Have students do research to find how surveyors use parallel and perpendicular lines in their work. Be sure students include information about slope. Have students summarize the information in a presentation for the class.

Alternate Example 1

Use the data in the table to find the formula that changes

a. degrees Fahrenheit (°F) to degrees Celsius (°C).
$$\left[C = \frac{5}{9}F - \frac{160}{9} \right]$$

b. degrees Fahrenheit to kelvins.
$$\left[K = \frac{5}{9}F + \frac{2300}{9} \right]$$

Cooperative Learning

Have students work in pairs or small groups to complete the explorations on page 248 and 249.

Math Connection Geometry

Review the definition of parallel lines with students. Emphasize the fact that parallel lines never intersect.

ongoing ASSESSMENT

The slopes are the same.

Exploration 1 Notes

It may be helpful to place coordinate axes so that the origin is at point *A*. However, emphasize that students should concentrate on the slopes of the lines and how they compare.

Aongoing ASSESSMENT

4. **The slope of the perpendicular line is the reciprocal slope of the original line and has the opposite sign.**

5. **The slope of the original line is $-\frac{3}{5}$. The slope of the perpendicular line is $\frac{5}{3}$. The slope of the perpendicular line is the reciprocal of the original slope and has the opposite sign.**

Exploration 2 Notes

Suggest that students use their results from Exploration 1 to help them complete their work in Exploration 2. After they have completed this exploration, students should be sure that the results of both explorations lead them to the same conclusions about the slopes of perpendicular lines.

Perpendicular Lines

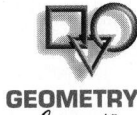

GEOMETRY
Connection

In geometry, **perpendicular lines** form right angles. In the next exploration, you can discover the algebraic relationship between two lines that are perpendicular. The symbol for perpendicular is ⊥.

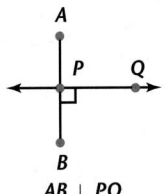

$AB \perp PQ$

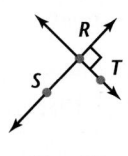

$SR \perp RT$

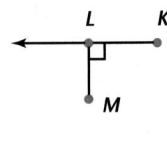

$LM \perp KL$

•Exploration 1 Slopes and Perpendicular Lines

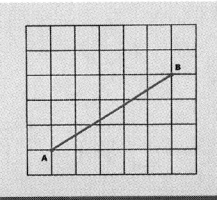

On a piece of graph paper, draw a line that has a run of 5 squares and a rise of 3 squares, as shown on the grid. Make a copy on clear plastic or tracing paper. Rotate the copy of the graph 90 degrees clockwise, as shown. When one graph is placed on the other, you can see the relationship of the two lines.

The lines are **perpendicular**, because the line on the second graph was rotated 90 degrees. Where the lines on the two graphs intersect, they will form right angles.

1. What is the slope of the original line?

2. Find the slope of the rotated line. Count the squares to find the rise over the run. Notice, for example, that the new line shows a run of 3. Consider the direction of the line.

3. What is the sign for the slope of the perpendicular line?

4. What is the relationship of the slope of the perpendicular line to the original line?

5. What happens when this exploration is repeated with a negative slope? ❖

•Exploration 2 Perpendicular Slope on a Calculator

Graphics Calculator

You can use a graphics calculator to find the equation of a line perpendicular to a given line. Be sure to select a *square window* when you set the range on your graphics calculator. Then the right angles will have the correct appearance on the screen.

ENRICHMENT Have students work in groups and experiment with rectangles on the coordinate plane. Have them begin by drawing a rectangle such that none of the sides are vertical or horizontal. Then ask students to use the coordinates of the vertices to determine the equation of one of the sides. Have them use the relationships discussed in this lesson to find the equations of the other three sides of the rectangle. Ask each group to present their results to the class. You may want to use the results of this activity in a bulletin board display.

INCLUSION strategies **Using Cognitive Strategies** Have students draw the graphs of several equations for parallel lines in blue pencil. Then have them make a table of the slopes for the parallel equations in blue pencil. Have students draw the graphs of the equations of perpendicular lines in red pencil and write a table of the slopes

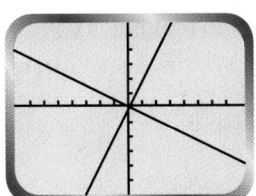

 Enter the graph for the line **Y = 2X**.

 On the same axes, graph new lines in the form $y = mx$ by trying different values for m until the new line looks perpendicular to the original line. Keep a record of your guesses.

 Decide on your best guess for the slope. Write an equation for a line perpendicular to $y = 2x$ from the slope you find. ❖

On the same set of axes graph $y = 2x$ and the line perpendicular to $y = 2x$ that passes through (0, 0). Find the slope of each line. Compare the signs. Compare the absolute values. Multiply the slopes. What is the result?

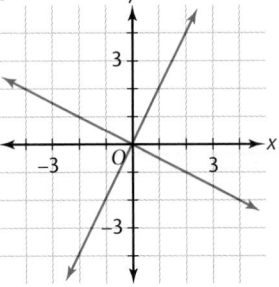

Suppose a given line has positive slope m. Make a conjecture about the product of m and the slope of a line perpendicular to the given line.

PERPENDICULAR LINES

If the slopes of two lines are m and $-\frac{1}{m}$, the lines are perpendicular.

If the slope of a line is m, then the slope of a line perpendicular to it is $-\frac{1}{m}$.

EXAMPLE 2

Find an equation for the line that contains point (4, 5) and is

Ⓐ parallel to the line $2x + 3y = 7$.

Ⓑ perpendicular to the line $2x + 3y = 7$.

Solution ➤

Ⓐ First, write the equation of the given line in slope-intercept form.

$$2x + 3y = 7$$
$$3y = -2x + 7$$
$$y = -\frac{2}{3}x + \frac{7}{3}$$

The slope of the line is $-\frac{2}{3}$. Any line parallel to the original line must also have slope $-\frac{2}{3}$. The coordinates of the given point are (4, 5), and the slope is $-\frac{2}{3}$. Now use the *point-slope form* to write an equation for the parallel line through the given point. Substitute $-\frac{2}{3}$ for m and (4, 5) for (x, y) into the point-slope form of the equation, $y - y_1 = m(x - x_1)$.

$$y - 5 = -\frac{2}{3}(x - 4)$$

for the perpendicular equations in red. Then have students examine the results. Ask them for their conclusions.

 **Independent Learning** Have students draw several lines on a coordinate plane that have a rise of 3 and a run of 2. Ask students to record any observations they can make about the lines they have drawn. They should find that all of the

lines have the same slope and are parallel. Ask students to write an equation for each line. Then have students use another coordinate plane and draw two lines, one with a rise of 2 and a run of 5, and one with a rise of − 5 and a run of 2. Ask students to record any observations they can make about the lines they have drawn. They should find that the pair of lines have slopes that are negative reciprocals of each other and that the lines are perpendicular.

ASSESSMENT

3. $y = -\frac{1}{2}x$

---

ongoing
ASSESSMENT

The slopes are 2 and $-\frac{1}{2}$. The signs are opposite. The absolute values are reciprocals. When the slopes are multiplied they equal -1.

CRITICAL
Thinking

A line perpendicular to a line with slope m has a slope of $-\frac{1}{m}$.

The product of the two slopes is -1.

T**EACHING** *tip*

Ask students to write the equation given as the answer to part (a) of Example 2 in slope-intercept form.   [$2x + 3y = 23$]

Alternate Example 2

Find an equation for the line that contains point (2, 3) and is

a. parallel to the line
$x - 2y = 5$.
$$\left[y - 3 = \frac{1}{2}(x - 2)\right]$$

b. perpendicular to the line
$x - 2y = 5$.
$$[y = -2x + 7]$$

Review the definition of perpendicular lines with students. Emphasize the fact that perpendicular lines always form four right angles.

Aongoing ASSESSMENT

Try This

$y - 1 = -\frac{1}{2}(x - 4)$ or

$y = -\frac{1}{2}x + 3$

ASSESS

Selected Answers
Odd numbered Exercises 9–37

Assignment Guide
Core 1–14, 16–28 even, 29–39

Core Two-Year 1–39

Technology
A calculator is helpful for Exercises 32–34.

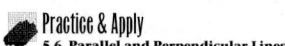

Practice Master

Practice & Apply
5.6 Parallel and Perpendicular Lines

Write the slope of a line that is parallel to the following lines.

1. $y = 2x - 5$ _____   2. $y = -x + 2$ _____
3. $3x + y = 10$ _____   4. $5x - y = 11$ _____
5. $x + 2y = 6$ _____   6. $2x - 3y = 9$ _____

Write the slope of a line that is perpendicular to the following lines.

7. $y = 4x + 6$ _____   8. $y = -\frac{1}{5}x - 3$ _____
9. $x + y = 7$ _____   10. $6x - y = 14$ _____
11. $x + 7y = -21$ _____   12. $5x - 4y = 12$ _____

Write an equation for a line containing the point (6,−2) and

13. parallel to the line $2x + y = 5$. _____
14. perpendicular to the line $y = -3x + 4$. _____

Write an equation for a line containing the point (−6, 5) and

15. parallel to the line $x + 2y = 6$. _____
16. perpendicular to the line $3x - 4y = -8$. _____

Write an equation for a line containing the point (−3, 2) and

17. parallel to the line $y = -4$. _____
18. perpendicular to the line $y = -4$. _____

State whether the graphs of these pairs of equations are parallel, perpendicular, or neither.

19. $y = 5x + 7$ 20. $y = x - 6$
    $y = -5x - 4$ _____ $y = -x$ _____
21. $4x + y = 3$ 22. $x + 2y = 14$
    $8x + 2y = 10$ _____ $y = -\frac{1}{2}x + 6$ _____
23. $2x + 3y = 9$ 24. $y = \frac{1}{3}x + 2$
    $3x - 2y = -8$ _____ $2y = -2x - 8$ _____

B When a line has slope $-\frac{2}{3}$, any line perpendicular to that line has slope $\frac{3}{2}$. Since the line also contains the point (4, 5), you can substitute the coordinates into $y - y_1 = m(x - x_1)$.

$$y - 5 = \frac{3}{2}(x - 4)$$

Graphics Calculator

As a check, change the equations $y - 5 = -\frac{2}{3}(x - 4)$ and $y - 5 = \frac{3}{2}(x - 4)$ to slope-intercept form. Graph the lines using a square window. The lines will appear perpendicular. Both lines will also contain the point (4, 5) where they intersect. ◆

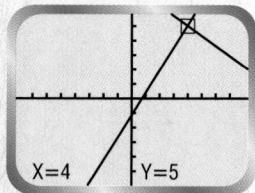

X=4   Y=5

Try This Find an equation for a line containing the point (4, 1) and perpendicular to $y = 2x - 7$.

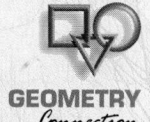

GEOMETRY Connection

If line l has slope $\frac{2}{5}$, and line m has slope $-\frac{5}{2}$, then $l \perp m$.
If $r \perp s$ and r has slope $\frac{2}{5}$, then s has slope $-\frac{5}{2}$.

EXERCISES & PROBLEMS

Communicate

1. Explain how to write an equation for a line parallel to $y = 4x + 3$.

2. The slope of a line is $\frac{3}{2}$. Explain how to find the slope of a line that is perpendicular to that line.

3. How would you find the slope of a line perpendicular to the line $y = \frac{1}{3}x + 2$?

4. Describe how to find an equation for a line perpendicular to $y = 4x + 3$.

5. Explain how to write the standard form of an equation for a line with y-intercept -4 that is parallel to a line with slope 3.

6. Tell how to write an equation in standard form for a line perpendicular to the line with $m = -6$ and $b = 12$.

7. Discuss how to write the equation for a line that passes through the point (0, 0) and is perpendicular to $x - 5y = 15$.

8. Answers may vary. The slope must be 5 to be parallel, but the y-intercept can vary. For example, $y = 5x - 6$.

9. Answers may vary. The slope must be $-\frac{1}{5}$ to be perpendicular, but the y-intercept can vary. For example, $y = -\frac{1}{5}x$.

10. One line has a slope of 3 and one has a slope of -3. Both lines have a y-intercept of 2. They are not perpendicular.

Practice & Apply

Graph the line $y = 5x$.

8. Sketch a graph that is parallel to $y = 5x$. Write its equation.

9. Sketch a graph that is perpendicular to $y = 5x$. Write its equation.

10. What relationship is there between the graph of $y = 3x + 2$ and $y = -3x + 2$? Are these lines perpendicular?

What is the slope of a line

11. parallel to a horizontal line? 0   **12.** perpendicular to a horizontal line? undefined

13. parallel to a vertical line? undefined   **14.** perpendicular to a vertical line? 0

Write an equation for a line according to the instructions.

	Contains	Is parallel to		Contains	Is perpendicular to
15.	$(3, -5)$	$5x - 2y = 10$	**16.**	$(3, -5)$	$5x - 2y = 10$
17.	$(-2, 7)$	$y = 3x - 4$	**18.**	$(-2, 7)$	$y = 3x - 4$
19.	$(2, 4)$	$y = 7$	**20.**	$(2, 4)$	$y = 7$

Write the slope of a line that is parallel to the following lines.

21. $y = 3x + 14$ 3   **22.** $2x + y = 6$ -2 **23.** $8 = -4x + 2y$ 2   **24.** $-2x + \frac{1}{2}y = 16$ 4

Write the slope of a line that is perpendicular to the following lines.

25. $y = -\frac{1}{3}x + 10$ 3   **26.** $-\frac{1}{2}x - y = 20$ 2   **27.** $13 = -x + y$ -1 **28.** $3x + 12y = 12$ 4

Geometry Write equations of 4 lines that meet to form a square whose sides are

29. *parallel* to the axes.   **30.** are *not parallel* to the axes.

Look Back

31. Place parentheses to make $2 \cdot 7 + 35 \div 7 - 10 = 2$ true. **[Lesson 1.4]**
$2(7 + 35) \div 7 - 10 = 2$

Evaluate the following. [Lesson 3.1]

32. $-4 + (-3) + 1$ -6 **33.** $-2 + 3 + (-7) + 3$ -3 **34.** $-12 + 4 - (-4)$ -4

Add the like terms. [Lesson 3.3]

35. $2x^2 + 3y + 4y + 3x^2$ $5x^2 + 7y$ **36.** $3x + 2 + 4y + 2 + 3y$ $3x + 7y + 4$ **37.** $2x + 3xy + 5x^2 + 7xy$ $5x^2 + 10xy + 2x$

Look Beyond

How many ordered pairs will satisfy both equations simultaneously, if two linear equations have graphs

38. that are parallel?   none   **39.** that are perpendicular?   one

---

15. $5x - 2y = 25$

16. $2x + 5y = -19$

17. $3x - y = -13$

18. $x + 3y = 19$

19. $y = 4$

20. $x = 2$

29. Answers may vary. For example, $x = 2$, $x = 4, y = 2, y = 4$.

30. Answers may vary. For example, $y = x$, $y = -x, y = x - 2, y = -x + 2$.

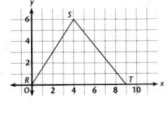

FOCUS

Algebra has been used for centuries to solve problems. The Diophantine equations are a perfect example of how ancient civilizations used algebraic equations. As students work through this project, they will learn about the usefulness of these equations today.

MOTIVATE

Have students either act out or draw a diagram to solve the problem about the horses and the riders. Ask students to consider what the minimum number of legs could be. **[6]**

Have students discuss how this information can be used to solve the problem.

Diophantine
EQUATIONS

Donna sees only the legs of horses and the legs of the riders walking the horses. She counts 22 legs in all. How many horses and how many riders does she see?

This is a version of a classic algebra problem. The numbers of legs must be whole numbers, so the number of solutions is restricted.

Equations that require the condition that *the solutions must be integers* are called **Diophantine equations.** The equations are named for Diophantus, who lived about 2000 years ago in the city of Alexandria in Egypt.

Using a Table Solution

Find all the solutions to the horse-and-rider problem.

Let h be the number of horses and r be the number of riders. Since each horse has 4 legs and each rider has two legs, $4h + 2r = 22$. Notice that all the coefficients are even numbers, so you can divide both sides of $4h + 2r = 22$ by 2 to get $2h + r = 11$.

Make a table to find the solutions. Remember, all solutions must be whole numbers. The first solution can be found by letting $h = 0$. When you substitute, you will find that $r = 11$. To find other solutions, note that each time there is one more horse, there are 4 more legs. That means there are 2 fewer riders. The table shows the solutions.

Horses	0	1	2	3	4	5
Riders	11	9	7	5	3	1

It is impossible to have more than 5 horses, because there would then be more than 22 legs.

---

1. There are two possibilities: four 2-pound bags plus two 5-pound bags, or nine 2-pound bags and no 5-pound bags. These are the only non-negative integer coordinates that lie on the graph.

2. $9a + 8b = 100$; Ms. Smiley should give 4 points for each question in part I and 8 points for each question in part II. This is the only whole number solution to the equation.

$9a + 8b = 200$

a	8	16
b	16	7

Using a Graph Solution

Find all of the whole-number solutions to the equation $4h + 2r = 22$.

If $r = 0$, the equation becomes $4h = 22$, which has no whole-number solution. If $r = 1$, the equation becomes $4h + 2 = 22$, or $4h = 20$, which does have a solution, $h = 5$. So $(1, 5)$ is one solution to the Diophantine equation. Now look at the Diophantine equation in slope-intercept form. Solve for h.

$4h + 2r = 22$ becomes $h = -\frac{2}{4}r + \frac{22}{4}$.

The slope of the equation in lowest terms is $-\frac{1}{2}$.

Start with $(1, 5)$. Use a run of 2 and a rise of -1 to find any other solutions. The second solution is $(3, 4)$. After $(11, 0)$, another run of 2 and rise of -1 would place you at $(13, -1)$. But -1 is not a whole number.

How can you be sure that no other whole-number solutions exist between $(1, 5)$ and $(11, 0)$?

Try whole-number values for r between 1 and 11, to be sure.

Notice that if you had used $-\frac{2}{4}$ for the slope from $(1, 5)$, you would have missed the point $(3, 4)$. Why must the slope be in lowest terms?

Activity 1

If flour comes in 2-pound bags and in 5-pound bags, list all the ways to buy exactly 18 pounds of flour. Draw a graph, and locate the solutions on the graph. What can you say about the solutions that appear on the graph?

Activity 2

Ms. Smiley has written a math exam with two parts. Part I has 9 questions, and part II has 8 questions. She wants to know how to assign points to each part so that the total will be exactly 100 points. Find a Diophantine equation that represents this situation and solves Ms. Smiley's problem.

Make a table to represent tests that have 9 questions in part I, 8 questions in part II, and 200 points in all.

Make tables to represent tests that have different numbers of questions but have 100 total points.

Activity 3

How can you tell that $4x + 6y = 125$ has no whole-number solution?

How can you tell that $5x + 10y = 112$ has no whole-number solution?

How can you tell whether $6x + 9y = 100$ has whole-number solutions?

Cooperative Learning

Have students work in small groups to complete the activities. Suggest that each member of each group attempt to solve the problem in Activity 1 and then work together to finalize their solution. For Activity 2, have students work together to complete the first two parts and then work individually to make tables with different numbers. Have each group compare and discuss their individual tables. Groups should then complete the project by working together to answer the questions in Activity 3.

Discuss

Have each group present their results of each activity to the class. As a class, discuss exactly how each group used Diophantine equations to solve the problems. As an extension, have students work together in their groups to write problems that can be solved using Diophantine equations.

3. $4x + 6y = 125$ has no whole-number solution because the left side is always an even number and 125 is not a whole-number multiple of 2.

$5x + 10y = 112$ has no whole-number solution because the left side is always a multiple of 5 and 112 is not a whole-number multiple of 5.

$6x + 9y = 100$ has no whole-number solution because the left side is always a multiple of 3 and 100 is not a whole-number multiple of 3.

Chapter 5 Review

Vocabulary

constant	241	perpendicular lines	248	run	209
constant function	241	point-slope form	235	slope	209
constant of variation	228	rate of change	221	slope-intercept form	225
direct variation	228	regression line	227	standard form	233
parallel lines	247	rise	209	y-intercept	225

Key Skills & Exercises

Lesson 5.1

➤ **Key Skills**

Find the slope of a line passing through two points.

Find the slope of a line passing through the points $A(-2, 4)$ and $B(3, 5)$.

$$m = \text{slope} = \frac{\text{difference in } y\text{-values}}{\text{difference in } x\text{-values}}$$

$$= \frac{5 - 4}{3 - (-2)} = \frac{1}{5}$$

➤ **Exercises**

If each pair of points is on a separate line, what is the slope of each line?

1. $A(-3, 2), B(2, 3)$ $\frac{1}{5}$ **2.** $A(2, 5), B(4, 1)$ -2

3. $A(-5, 4), B(1, 4)$ 0 **4.** $A(-3, -1), B(-3, 3)$ undefined

Lesson 5.2

➤ **Key Skills**

Find the equation of a line passing through the origin and a given point.

Find the equation of a line passing through the origin and the point $(2, 6)$.

Since the slope is $\frac{6}{2}$ or 3, the equation is $y = 3x$.

➤ **Exercises**

Find the equation of a line passing through the origin and the following points.

5. $(-3, 2)$ $y = -\frac{2}{3}x$ **6.** $(2, 3)$ $y = \frac{3}{2}x$

7. $(-2, -5)$ $y = \frac{5}{2}x$ **8.** $(-5, 4)$ $y = -\frac{4}{5}x$

9. $y = 2x + 1$

10. $y = -\frac{1}{2}x + 5$

11. $y = 8x - 4$

12. $y = -6x + 26$

13. $y = -1$

14. $y = -\frac{1}{3}x - 2$

Answers to Exercises 15–18 can be found in Additional Answers beginning on page 729.

Lesson 5.3

➤ Key Skills

Find the equation of a line in slope-intercept form.

Write the equation for a line

a. with slope 3 and y-intercept 5.
b. passing through $(-2, 3)$ and $(-1, 5)$.

a. Substitute 3 for m and 5 for b.
$$y = mx + b \qquad y = 3x + 5$$

b. Slope $= m = \dfrac{5-3}{-1-(-2)} = \dfrac{2}{1} = 2$

Substitute 2 for m, -2 for x, and 3 for y in the equation $y = mx + b$. Then solve for b.
$$3 = 2(-2) + b$$
$$3 = -4 + b$$
$$7 = b$$

Substitute 2 for m and 7 for b.
$$y = mx + b \qquad y = 2x + 7$$

Graph an equation in slope-intercept form.

Graph the equation $y = \dfrac{1}{2}x + 5$.

The slope is $\dfrac{1}{2}$ and the y-intercept is 5.

Locate $(0, 5)$. From that point move 2 to the right and up 1 to $(2, 6)$. Draw the line.

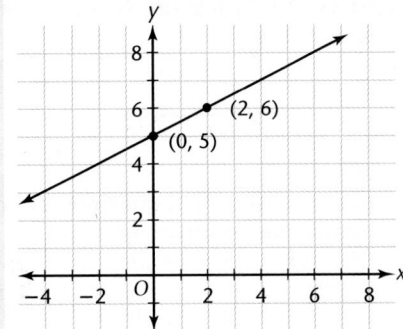

➤ Exercises

Write the equation in slope-intercept form

9. with slope 2 and y-intercept 1.
10. with slope $-\dfrac{1}{2}$ and y-intercept 5.
11. through $(0, -4)$ and with slope 8.
12. through $(4, 2)$ and $(5, -4)$.
13. through $(2, -1)$ and $(-3, -1)$.
14. through $(-6, 0)$ and $(0, -2)$.

Graph each equation.

15. $y = -\dfrac{1}{2}x - 2$ **16.** $y = -x + 6$ **17.** $y = -\dfrac{3}{4}x + 4$ **18.** $y = 4x - 1$

Lesson 5.4

➤ Key Skills

Write equations in standard form.

Write $y + 10 = -8x - 22$ in standard form.
$$y + 10 = -8x - 22$$
$$8x + y + 10 = -22$$
$$8x + y = -32$$

Write equations in point-slope form.

Write an equation for the line passing through the point $(4, 5)$ and $(2, 1)$.
$$m = \dfrac{1-5}{2-4} = \dfrac{-4}{-2} = 2$$

Use the point $(4, 5)$ for (x_1, y_1).
$$y - 5 = 2(x - 4)$$

➤ Exercises

Write the following equations in standard form.

19. $y + 9 = 4x - 8$ **20.** $y - 4 = -x + 1$ **21.** $y - 13 = 2x + 4$ **22.** $3x + y + 6 = 9$

Write the equations in point-slope form

23. with slope of 2 and y-intercept of -1.
24. through $(0, 4)$ and $(1, 2)$.
25. through $(3, 5)$ and with a slope of 4.
26. that crosses the x-axis at $x = -2$ and the y-axis at $y = 1$.

---

19. $4x - y = 17$

20. $x + y = 5$

21. $2x - y = -17$

22. $3x + y = 3$

23. $y + 1 = 2(x - 0)$

24. $y - 2 = -2(x - 1)$ or $y - 4 = -2(x - 0)$

25. $y - 5 = 4(x - 3)$

26. $y - 1 = \dfrac{1}{2}(x - 0)$ or $y - 0 = \dfrac{1}{2}(x + 2)$

Lesson 5.5

➤ **Key Skills**

Recognize and find the slope of vertical and horizontal lines.

Identify each of the following lines as vertical or horizontal. State the slope for each.

a. $y = -8$ b. $x = 7$

a. The line is horizontal and the slope is 0.

b. The line is vertical and the slope is undefined.

➤ **Exercises**

Identify the following lines as vertical or horizontal, and state the slope.

27. $y = -6$
horizontal; 0

28. $x = 2$
vertical; undefined

29. $y = \frac{1}{2}$
horizontal; 0

30. $x = -\frac{1}{3}$
vertical; undefined

Lesson 5.6

➤ **Key Skills**

Recognize and write equations of lines parallel and perpendicular to given lines.

Write equations in slope-intercept form containing the point (0, 4) that are parallel and perpendicular to the line $2x + 3y = 5$.

$2x + 3y = 5$ in slope-intercept form is $y = -\frac{2}{3}x + \frac{5}{3}$. The slope is $-\frac{2}{3}$. Since (0, 4) tells us the y-intercept is 4, the equation of the parallel line is $y = -\frac{2}{3}x + 4$.

The slope of a perpendicular line is the negative reciprocal of $-\frac{2}{3}$, or $\frac{3}{2}$. Its y-intercept is also 4. The equation of the perpendicular line is $y = \frac{3}{2}x + 4$.

➤ **Exercises**

Write the slope of one line parallel to and one line perpendicular to each of the following lines.

31. $y = \frac{2}{3}x + 4$ $\frac{2}{3}; -\frac{3}{2}$

32. $y = -7x - 1$ $-7; \frac{1}{7}$

Write an equation for a line that contains the point (1, 5) and is

33. parallel to $2x + y = -1$.
$y = -2x + 7$

34. perpendicular to $2x + y = -1$.
$y = \frac{1}{2}x + \frac{9}{2}$

Applications

Travel   Marie had driven 110 miles when she stopped at a gas station. After she left the station, she recorded the hours that she drove. If she averaged 50 miles per hour for the rest of the trip, write an equation for the total distance she drove after

35. 1.6 hours.   **36.** 2.9 hours.   **37.** 3.3 hours.

38. Suppose that during a flood the water level, L, of a river is 38 feet on Monday and is receding by 2 feet a day. Write and solve an equation of the water level in d days. Graph the equation. In how many days will the water level be 26 feet? 6

35. $d = 50(1.6) + 110$

36. $d = 50(2.9) + 110$

37. $d = 50(3.3) + 110$

Chapter 5 Assessment

1. Copy and complete the table for $y = 2x + 3$.

x	-2	-1	0	1	2
y	-1	1	3	5	7

Write the following equations in standard form.

2. $y + 3 = -5x + 6$
$5x + y = 3$

3. $4x + 2y + 8 = 10$
$4x + 2y = 2$

Write the following equations in slope-intercept form.

4. $2x + y = 10 - 3$
$y = -2x + 7$

5. $3x - y = 2y - 5$
$y = x + \frac{5}{3}$

Find the slope and the y-intercept for the following equations.

6. $3x + 6y = 12$ $\quad -\frac{1}{2}, 2$

7. $2x + 3y = -3$ $\quad -\frac{2}{3}, -1$

8. On a sheet of graph paper, sketch the graph $y = 3x + 1$.

9. Find the slope of a line passing through the points $(5, 3)$ and $(8, -2)$.
$-\frac{5}{3}$

Write the slope-intercept form of an equation for a line that contains the point (2, 4) and is

10. parallel to the graph of $y = -3x - 7$. $\quad y = -3x + 10$

11. perpendicular to the graph of $2x + y = 13$. $\quad y = \frac{1}{2}x + 3$

12. On a sheet of graph paper draw a graph of a line that has slope -4 and y-intercept 1.

13. Write the equation in slope-intercept form for a line with slope $-\frac{2}{3}$ and y-intercept 5. $\quad y = -\frac{2}{3}x + 5$

14. Write the equation in point-slope form for a line that has a slope of -3 and passes through the point $(2, -3)$. $\quad y + 3 = -3(x - 2)$

15. Write the equation in slope-intercept form for a line that passes through the points $(-5, -1)$ and $(1, 5)$. $\quad y = x + 4$

16. What is the slope-intercept form of the equation for the graph at the right? $\quad y = \frac{3}{5}x - 3$

17. Give the slope of a line drawn through the points $(1, 8)$ and $(-2, 8)$. $\quad 0$

18. Equipment Rental Steve can rent a small lawn mower for a daily rate of \$15, or he can rent a mower for 10 days at a flat rate of \$100. How many days must Steve use the mower before it would benefit him to pay the flat rate? $\quad$ 7 or more

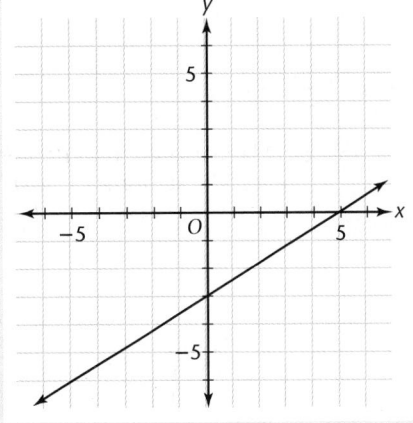

8.

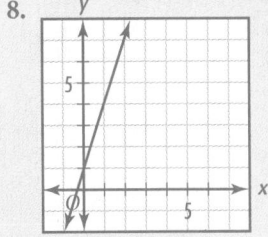

12.

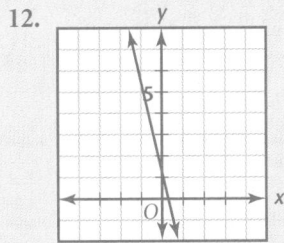

CHAPTER 6 Systems of Equations and Inequalities

Meeting Individual Needs

6.1 Graphing Systems of Equations

Core Resources

Practice Master 6.1
Enrichment, p. 261
Technology Master 6.1

Core Two-Year Resources

Inclusion Strategies, p. 262
Reteaching the Lesson, p. 263
Practice Master 6.1
Enrichment Master 6.1
Technology Master 6.1
Lesson Activity Master 6.1

[1 day]   [2 days]

6.2 Exploring Substitution Methods

Core Resources

Practice Master 6.2
Enrichment Master 6.2
Technology Master 6.2
Interdisciplinary
    Connection, p. 268

Core Two-Year Resources

Inclusion Strategies, p. 269
Reteaching the Lesson, p. 270
Practice Master 6.2
Enrichment, p. 268
Technology Master 6.2
Lesson Activity Master 6.2

[1 day]   [2 days]

6.3 The Elimination Method

Core Resources

Practice Master 6.3
Enrichment, p. 274
Technology Master 6.3
Interdisciplinary
    Connection, p. 273
Mid-Chapter Assessment
    Master

Core Two-Year Resources

Inclusion Strategies, p. 274
Reteaching the Lesson, p. 275
Practice Master 6.3
Enrichment Master 6.3
Technology Master 6.3
Lesson Activity Master 6.3
Mid-Chapter Assessment Master

[1 day]   [2 days]

6.4 Non-Unique Solutions

Core Resources

Practice Master 6.4
Enrichment, p. 280
Technology Master 6.4
Interdisciplinary
    Connection, p. 279

Core Two-Year Resources

Inclusion Strategies, p. 280
Reteaching the Lesson, p. 281
Practice Master 6.4
Enrichment Master 6.4
Technology Master 6.4
Lesson Activity Master 6.4

[1 day]   [2 days]

6.5 Graphing Linear Inequalities

Core Resources

Practice Master 6.5
Enrichment Master 6.5
Technology Master 6.5

Core Two-Year Resources

Inclusion Strategies, p. 286
Reteaching the Lesson, p. 287
Practice Master 6.5
Enrichment, p. 285
Technology Master 6.5
Lesson Activity Master 6.5

[2 days]   [3 days]

6.6 Classic Applications

Core Resources

Practice Master 6.6
Enrichment, p. 293
Technology Master 6.6

Core Two-Year Resources

Inclusion Strategies, p. 294
Reteaching the Lesson, p. 295
Practice Master 6.6
Enrichment Master 6.6
Lesson Activity Master 6.6

[2 days]   [3 days]

Chapter Summary

Core Resources	Core Two-Year Resources
Chapter 6 Project, pp. 298–299	Chapter 6 Project, pp. 298–299
Lab Activity	Lab Activity
Long-Term Project	Long-Term Project
Chapter Review, pp. 300–302	Chapter Review, pp. 300–302
Chapter Assessment, p. 303	Chapter Assessment, p. 303
Chapter Assessment A/B	Chapter Assessment A/B
Alternative Assessment	Alternative Assessment
Cumulative Review, pp. 304–305	Cumulative Assessment, pp. 304–305
[3 days]	**[5 days]**

Reading Strategies

After students have read each lesson, have them answer questions like the following:

- What is the main idea of this lesson?

- List and describe any words, phrases, equations, or graphs that you have not seen before. Why do you think they are important?

- What new methods are presented in this lesson? Why do you think these methods are important?

Responses to the questions may be included in student journals. Also have students write answers to selected Communicate exercises in their journals.

Visual Strategies

One of the best ways to present mathematical data is to use a graph. Graphs enable students to see a visual representation of the relationship contained in a collection of data. Linear equations can be quickly and easily graphed using a graphics calculator or computer software such as $f(g)$ $Scholar^{TM}$. Encourage students to use a graph as often as they feel it is necessary. The linear programming application presented in the chapter project shows students how useful graphs are when solving practical problems.

Hands-on Strategies

Suggest to students that they use manipulatives whenever they feel it will help them understand a concept in the text. Encourage students to be creative and to devise their own ways to use these manipulatives. You may want to have students demonstrate to the class any of their own ways of using these manipulatives.

Cooperative Learning

GROUP ACTIVITIES	
Using guess-and-check	Chapter Opener, The Portfolio Activity
Graphing and substituting	Lesson 6.2, Explorations 1, 2, 3
Solutions to different systems	Lesson 6.4, Explorations 1, 2
Graphing linear inequalities	Lesson 6.5, Exploration
Linear programming	Chapter 6 Project

You may wish to have students work with partners for some of the above activities. Additional suggestions for cooperative group activities are noted in the teacher's notes in each lesson.

Multicultural

The cultural connections in this chapter include references to Asia and Africa.

CULTURAL CONNECTIONS	
Asia: Chinese problem	Lesson 6.2, page 271
Asia: Chinese numbers	Lesson 6.4, page 281
Asia: Aryabhata problem	Lesson 6.6, page 297
Africa: Rhind papyrus by Ahmes	Lesson 6.3, page 277

Portfolio Assessment

Below are portfolio activities for the chapter, listed under seven activity domains that are appropriate for portfolio development.

1. **Investigation/Exploration** Two projects listed in the teacher notes ask students to investigate a method and/or a relationship. They are the *Enrichment* on page 285 (Lesson 6.5) and *Inclusion Strategies* on page 269 (Lesson 6.2). One project explores writing a system of equations for an application. It is the *Enrichment* on page 261 (Lesson 6.1).

2. **Applications** Recommended to be included are any of the following: Sports (Lessons 6.1 and 6.5), Investment (Lessons 6.2 and 6.3), Fund-raising (Lesson 6.2), Recreation (Lessons 6.2 and 6.6), Finance (Lesson 6.3), Income (Lesson 6.3), Travel (Lesson 6.3), Career options (Lesson 6.4), Budgeting time (Lesson 6.5), Numeration (Lesson 6.6), and, from the teacher notes, *Cooking* on page 268 (Lesson 6.2) and *Sports* on page 279 (Lesson 6.4).

3. **Non-Routine Problems** The Portfolio Activity on page 259 asks students to use the information given to find the cost of each of two different types of seats at a concert.

Students are asked to solve this problem using different methods throughout the chapter.

4. **Project** The Chapter Project on pages 298–299 leads students through the process of writing a set of linear inequalities and using them to find a minimum cost for a mixture of raisins and nuts. Students are then asked to use a similar procedure to find the maximum profit for a manufacturer of motorcycles.

5. **Interdisciplinary Topics** Students may choose from Accounting (Lesson 6.3) in the student text and *Business* on page 273 (Lesson 6.3) in the teacher notes.

6. **Writing** The interleaf material at the beginning of this chapter has information for journal writing. Also recommended are the *Communicate* questions at the beginning of the exercise sets.

7. **Tools** Chapter 6 uses calculators, computers, and graph paper for graphing systems of equations and inequalities. The technology masters are recommended to measure the individual student's proficiency.

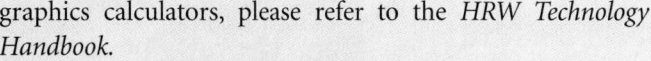

Technology enables students to graph a system of equations or inequalities quickly so that their solutions can be found or checked. Students will appreciate the capabilities of graphics calculators and computer software such as *f(g) Scholar*^{TM} even more after they have graphed several systems by hand. Encourage students to check all of their work using the available technology. For more information on the use of technology, refer to the *HRW Technology Handbook*.

Graphics Calculator

Graphing Systems of Linear Inequalities In Lesson 6.5, students are introduced to systems of linear inequalities. The solution set of a system of inequalities can be shown using a graph. A graphics calculator can be used to draw such graphs. Specific instructions are given below for the TI-82. For instructions for using other graphics calculators, please refer to the *HRW Technology Handbook*.

1. Press 2nd DRAW and select **7: Shade (** to have **Shade (** appear on the Home screen.

2. Enter the function above which you want to shade, followed by a comma. For example, if you want to shade above $y = 3x - 2$, enter 3 X,T,θ − 2.

3. Enter the function below which you want to shade after the comma. For example, if you want to shade below $y = 2x + 1$, enter 2 X,T,θ + 1 , .

4. If you want, you can specify a shading resolution for the graph. Do this by entering a comma after the second expression and choosing an integer from 1 to 9. The 1 gives the lightest shading and 9 the heaviest.

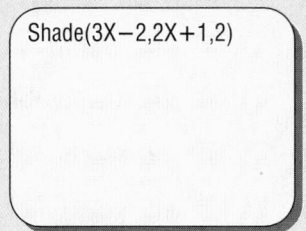

Shade(3X−2,2X+1,2)

5. Press ENTER to draw the graph.

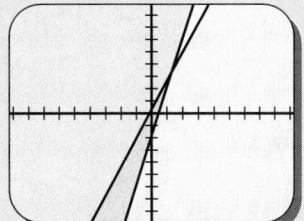

Integrated Software

f(g) Scholar^{TM} is an integrated computer-based mathematics productivity tool that combines calculator, spreadsheet, and graphics capabilities to provide a dynamic and interactive environment for explorations in mathematics. It is appropriate to use *f(g) Scholar*^{TM} for any lesson needing a spreadsheet, calculator, graphics calculator, or any combination of the three.

ABOUT THE CHAPTER

Background Information

A rock concert is something with which most students are familiar. Encourage those students who have attended such concerts to discuss the expenses involved. Also discuss other events that might be ticketed in a similar manner, such as plays or sporting events.

CHAPTER RESOURCES

- Practice Masters
- Enrichment Masters
- Technology Masters
- Lesson Activity Masters
- Lab Activity Masters
- Assessment Masters:
  Chapter Assessments, A/B
  Mid-Chapter Assessment
  Alternative Assessment, A/B
- Teaching Transparencies
- Cumulative Assessment
- Spanish Resources

CHAPTER OBJECTIVES

- Create tables, and graph systems of equations.
- Find an approximate solution to a system by inspecting a graph.
- Graph a system of equations, and estimate a solution from a graph by inspection.
- Find an exact solution to a system of equations by the substitution method.
- Find an exact solution to a system of equations by the elimination method.

CHAPTER 6

LESSONS

6.1 Graphing Systems of Equations

6.2 *Exploring* Substitution Methods

6.3 The Elimination Method

6.4 Non-Unique Solutions

6.5 Graphing Linear Inequalities

6.6 Classic Applications

Chapter Project
  Minimum Cost
  Maximum Profit

Systems of Equations and Inequalities

The atmosphere is electric! It's minutes before your favorite rock star begins her huge benefit concert. You can't believe you're here. You had to work hard to make the money and time to attend, but you love her music, and the money goes to charity. Thousands of people like you crowd the hall, eager for the concert to start. A hush passes over the crowd. The curtain begins to rise. The crowd cheers. It has begun!

ABOUT THE PHOTOS

A concert is a source for many applications of mathematics. The price of tickets, the cost of putting on a performance, the number of people in attendance, the time it takes people to enter and depart, and the amount contributed to charity are a few examples that provide the numbers and the problems that mathematics can solve. The popularity of concerts also provides motivation because of its relevance for many students. In the Portfolio Activity, a concert provides the basis for a problem that can be solved by systems of equations.

- Identify consistent and inconsistent systems of equations.
- Compare slopes and intercepts of independent and dependent systems.
- Determine the boundary line that solves a linear inequality.
- Graph the solution of a linear inequality.
- Graph the solution for a system of linear inequalities.
- Solve traditional application problems.

PORTFOLIO ACTIVITY

Assign the Portfolio Activity as part of the assignment for Lesson 6.1 on page 226.

The problem presented in the first paragraph is a typical problem that can be solved using a system of linear equations. The problem is analyzed in the second paragraph.

Have students work in pairs or small groups to write the equations described. Ask them to first write out each equation in words and then substitute mathematical symbols. Suggest that students attempt to solve the problem using guess-and-check. Be sure they record their work. After students have found the solution, ask them to discuss the difficulties they encountered using this method of solution. Inform students that they will learn more efficient ways to solve this problem in this chapter.

To give students more practice writing systems of equations, give them the same problem or similar problems that contain different numbers from those given in the text. Ask students to write two equations for each problem.

PORTFOLIO ACTIVITY

You are so excited about the concert that you ask all of your friends to join you. The tickets are sold at two different prices, with the center section costing $10 more than the balcony seats. You buy 7 center seats and 5 balcony seats for a total of $310. What is the cost of each type of seat?

Use c as the cost of a center section ticket and b as the cost of a balcony ticket. Write two equations using c and b that show the relationships in this problem. Save them. As you progress through the chapter, you will be asked to solve this problem using various methods. You may wish to use these exercises as part of your portfolio.

ABOUT THE CHAPTER PROJECT

The Chapter 6 project on pages 298-299 presents students with two excellent applications of everything they have learned in this chapter.

Students will find that using a graphics calculator or computer software such as *f(g) Scholar™* makes some of their work easier and allows them more time to concentrate on the application of the graph.

Assessing Prior Knowledge

Write an equation for each of the following statements.

1. The cost of renting a car is $35 a day plus $0.12 a mile.
   $[c = 0.12m + 35]$

2. Aaron spent a total of $65 for tapes at $5 each and CDs at $8 each.
   $[5t + 8c = 65]$

3. During the summer Joe earned $600 and received an allowance of $10 a week.
   $[A = 10w + 600]$

TEACH

 Graphs of linear equations help to show a trend over a period of time and are often used in business to convey this information. If a situation consists of two conditions, then two equations are used to model it. Have students discuss situations that might involve two conditions.

One of the best ways to present mathematical data is to use a graph. For business decisions and customer choices, graphs can represent costs and show comparisons. When you graph two linear equations on the same coordinate system, you can see where the lines intersect.

If the Smiths decide to buy the VCR, they plan to spend $20 per month to rent 8 tapes. Should the Smiths buy the VCR or install cable TV?

VCRs on sale NOW
Starting at
$185.00
12:00
12:00

Installation $45.⁰⁰
Movies
Sports
Only $34.⁰⁰ a month!
News
EAST COAST CABLE

One way to help the Smiths decide on the right purchase is to compare the cost of each purchase over several months. An equation can be written to model each purchase.

First, define two variables.

Let c = the total costs for a system.
Let m = the number of months.

Next, write an equation in words to represent the costs. Then substitute variables and numbers for the words.

Cable TV costs = installation + cost for m months
c =   45   + $34m$ (or $c = 34m + 45$)

VCR and tape costs = machine + tape rental for m months
c =   185   + $20m$ (or $c = 20m + 185$)

 **ALTERNATIVE teaching strategy**

Hands-On Strategies
Provide small groups of students with three transparencies, one containing a coordinate plane and the other two containing lines that are different colors. Then

have different groups of students demonstrate on an overhead projector the solution to various examples like the ones provided in the text. Ask the other students to record the work shown on the overhead.

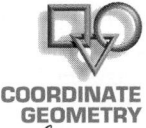

Create a table of values from the equations to compare the data. Then graph each equation on one coordinate plane.

	Table of Costs	
	Cable	VCR
m	$34m + 45$	$20m + 185$
1	79	205
2	113	225
3	147	245
4	181	265
5	215	285
6	249	305
7	283	325
8	317	345
9	351	365
10	385	385

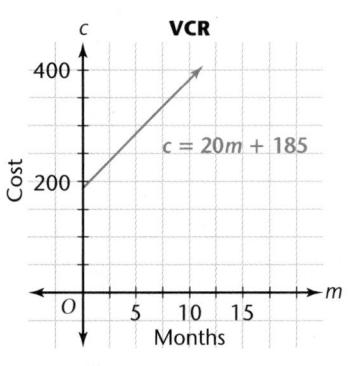

Cable TV

$c = 34m + 45$

VCR

$c = 20m + 185$

Cable TV versus VCR

$c = 20m + 185$

$c = 34m + 45$

The table and the graphs show that each cost continues to increase, but one increases at a faster rate than the other. At what point do the graphs intersect?

At 10 months the value for both equations is $385. When the value is the same for both equations, the equations have a **common solution**.

CRITICAL
Thinking

Use the information shown in the graph to explain which decision is best for the Smith family.

A System of Equations

Two or more equations in two or more variables is called a **system of equations**. When a system of two equations in two variables has a single solution, the solution is the ordered pair, (x, y), that satisfies both equations. The solution to the system used in the Smith family problem is $(10, 385)$.

Math Connection
Coordinate Geometry Students use coordinate geometry methods to solve systems of equations.

TEACHING *tip*

Technology The table of costs given in the text can be generated using spreadsheet software. Many spreadsheet programs can then draw the graph directly from the data.

Use Transparency 27

ongoing
ASSESSMENT

$(10, 385)$

CRITICAL
Thinking

A VCR would be the best choice because after 10 months the cost would be less.

ENRICHMENT Have students find out about the cable costs in their area. They should find the cost for installation and the monthly rental fee. They should find the cost of the VCR they would like to have and the cost of tape rental at a store near their home. Then have students use a system of equations to determine which option to choose according to their needs.

Ask students to think of one other example of a system of equations that could help them make a decision. Ask them to do any necessary research, write a problem about their example, and then use a system of equations to solve the problem.

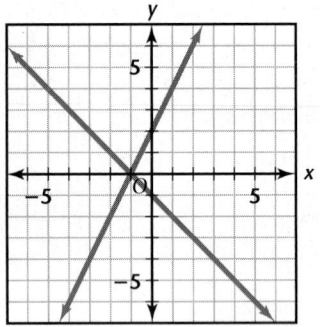

TEACHING tip

Technology Students can use the TRACE or TABLE feature of a graphics calculator to find the point of intersection. Such equations can also be graphed using computer software such as $f(g)$ *Scholar*™.

Aongoing ASSESSMENT

Try This

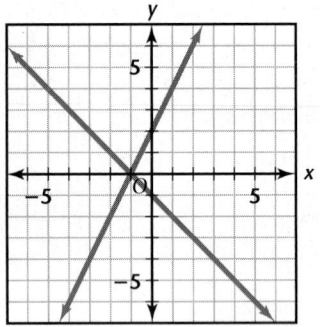

Graph of $y = 2x + 2$ and $y = -x - 1$ [$(-1, 0)$]

EXAMPLE 1

Use a graphics calculator or graph paper to graph the equations $3x + y = 11$ and $x - 2y = 6$. Find the common solution by examining the graph.

Graphics Calculator

Solution ▶

Change each equation to slope-intercept form, $y = mx + b$. Graph the two equations to find a common solution.

$$3x + y = 11 \qquad\qquad x - 2y = 6$$
$$y = -3x + 11 \qquad\qquad -2y = -x + 6$$
$$y = \frac{1}{2}x - 3$$

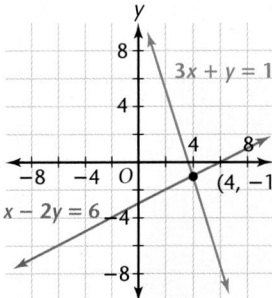

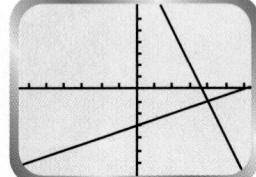

The graphs show that the lines intersect at exactly one point. To find the common solution, read the coordinates for the point of intersection from the x- and y-axes. It appears from the graph that the point of intersection is $(4, -1)$.

To check the solution, substitute 4 for x and -1 for y in each equation.

$$3x + y = 11 \qquad\qquad x - 2y = 6$$
$$3(4) + (-1) \stackrel{?}{=} 11 \qquad\qquad 4 - 2(-1) \stackrel{?}{=} 6$$
$$11 = 11 \quad \text{True} \qquad\qquad 6 = 6 \quad \text{True}$$

Since the substitution is true for each equation, $(4, -1)$ is the common solution. ❖

Try This Graph the equations $y = 2x + 2$ and $y = -x - 1$. Find a common solution by examining the graph. Check your answer.

EXAMPLE 2

Ms. Alyward gives a science midterm with 200 possible points. There are a total of 38 questions on the test. How many of each type of question are on the test?

Solution ➤

First define two variables. Then write equations to model the problem.

Organize

Let x = the number of 4-point multiple-choice questions.
Let y = the number of 20-point essay questions.

Write

number of 4-point items + number of 20-point items = 38 items
$$x \quad + \quad y \quad = \quad 38$$

points for multiple choice + points for essay = 200 points
$$4x \quad + \quad 20y \quad = \quad 200$$

Solve

To find how many of each type question appear on the test, solve the system of equations for (x, y). Make a table, or graph the two equations to find the point of intersection for a common solution. Change the equations to slope-intercept form to find values for the variables.

$$x + y = 38 \qquad\qquad 4x + 20y = 200$$
$$y = -x + 38 \qquad\qquad 20y = -4x + 200$$
$$y = -\frac{1}{5}x + 10$$

Use the table or the graph to find the point of intersection.

Table of Values

Multiple choice	Essay	Essay
x	$-x + 38$	$-\frac{1}{5}x + 10$
10	28	8
15	23	7
20	18	6
25	13	5
30	8	4
✔ 35	3	3

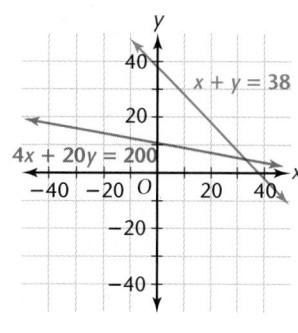

The solution is (35, 3).

There will be 35 multiple-choice questions and 3 essay questions on Ms. Alyward's science midterm.

Check

$$35 + 3 \stackrel{?}{=} 38 \qquad\qquad 4(35) + 20(3) \stackrel{?}{=} 200$$
$$38 = 38 \quad \text{True} \qquad\qquad 140 + 60 \stackrel{?}{=} 200$$
$$200 = 200 \quad \text{True} ❖$$

Why were the equations written in slope-intercept form before they were graphed?

Hands-On Strategies
Have small groups of students experiment with the graphs of two lines. Have them draw any two lines on a coordinate plane and determine whether they intersect. If the lines do intersect, have them find the coordinates of the point of intersection. Be sure students understand that the point of intersection

is the only point that lies on both lines. To verify that the point is correct, have them find the equation of both lines and check the coordinates by substituting the values into both equations. If two true statements result, then the point is indeed a solution to the system. Ask students to repeat this activity with several different pairs of lines.

Cooperative Learning
Solving systems of equations is ideal for cooperative-learning groups because of the multiple tasks involved. Assign each member of each group one of the tasks required to solve each problem. For example, one student can write the equations, another can graph the equations, and a third can find the point of intersection. Have students perform different tasks with different systems so that each member gets a chance to perform each task on at least one problem.

Aongoing
SSESSMENT

Slope-intercept form allows you to substitute values for x to find values for y. These (x, y) values can be used as ordered pairs to graph the equations. If you use a graphics calculator, the equations are entered in slope-intercept form.

Approximate Solution

It is sometimes difficult to find an exact solution from a graph. A reasonable estimate for a point of intersection is an **approximate solution** for a system of equations.

EXAMPLE 3

Solve by graphing. $\begin{cases} 4x + 3y = 6 \\ 2y = x + 2 \end{cases}$

The brace { is used to indicate that the equations form a system.

Solution ➤

Graphics Calculator

Write the equations in slope-intercept form.

$$4x + 3y = 6 \qquad\qquad 2y = x + 2$$
$$3y = -4x + 6 \qquad\qquad y = \frac{1}{2}x + 1$$
$$y = -\frac{4}{3}x + 2$$

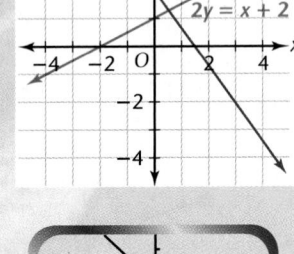

From the graph, locate the point of intersection. Use a graphics calculator, or make a reasonable estimate by inspecting the graph to approximate a common solution. An approximate solution is (0.55, 1.27).

Check the solution by substituting 0.55 for x and 1.27 for y in the original equations.

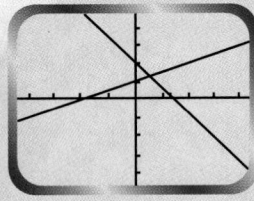

$$4x + 3y = 6 \qquad\qquad 2y = x + 2$$
$$4(0.55) + 3(1.27) \overset{?}{=} 6 \qquad 2(1.27) \overset{?}{=} 0.55 + 2$$
$$2.2 + 3.81 \overset{?}{=} 6 \qquad\qquad 2.54 \approx 2.55$$
$$6.01 \approx 6$$

A check shows the approximate solution is reasonable. ❖

EXERCISES & PROBLEMS

Communicate

Exercises 1–3 refer to the following system. $\begin{cases} 2x - 3y = 4 \\ x + 4y = -9 \end{cases}$

1. How do you write each equation in slope-intercept form? Explain why this form is used to find values for the variables.

2. Explain how to create a table of values for the system of equations. Discuss how to find the point of intersection from a table.

3. Discuss how to graph the system of equations from the table in Exercise 2. Tell how to find a common solution from the graph.

4. Discuss how to graph the system at the right. Tell how to make a reasonable estimate of the solution by inspecting the graph.

$$\begin{cases} x + y = 3 \\ x - y = 4 \end{cases}$$

5. Why is it important to check your estimate?

Practice & Apply

Solve by graphing. Round approximate solutions to the nearest tenth. Check algebraically.

6. $\begin{cases} 2x + 10y = -5 \\ 6x + 4y = 2 \end{cases}$
7. $\begin{cases} x - 2y = 60 \\ 3y = 30 - 2x \end{cases}$
8. $\begin{cases} 3y = 4x - 1 \\ 2x + 3y = 2 \end{cases}$
9. $\begin{cases} x + 3y = 6 \\ -6y = 2x + 12 \end{cases}$

10. $\begin{cases} 5x + 6y = 14 \\ 3x + 5y = 7 \end{cases}$
11. $\begin{cases} x = 400 - 2y \\ x - 100 = y \end{cases}$
12. $\begin{cases} 3x + 5y = 12 \\ 7x - 5y = 8 \end{cases}$
13. $\begin{cases} x = 2 \\ 2y = 4x + 2 \end{cases}$

Algebraically determine whether the point (2, 10) is a solution for each pair of equations.

14. $\begin{cases} y = 2x - 4 \\ y = x + 8 \end{cases}$ no
15. $\begin{cases} y = -x + 12 \\ y = -3x + 16 \end{cases}$ yes
16. $\begin{cases} y = x + 8 \\ y = -3x + 16 \end{cases}$ yes
17. $\begin{cases} x + 3y = 6 \\ -6y = 2x + 12 \end{cases}$ no

18. Graph the equations for Exercises 14-17 to check the results.

19. Find the point of intersection of $y = -x + 8$ and $y = -3x + 16$. (4, 4)

20. Use a graph to find the approximate point of intersection of $y = -x + 12$ and $y = 2x - 4$. Estimate the exact fractional value of these decimals. Algebraically check your estimate.

Write a system of equations. Then graph the equations to determine the solution.

21. Aviation The plane at 3,800 feet is descending at a rate of 120 feet per minute, and the other plane is climbing at a rate of 40 feet per minute. In how many minutes will they be at the same altitude?

$y = -120x + 3800$
20.5 minutes $y = 40x + 520$

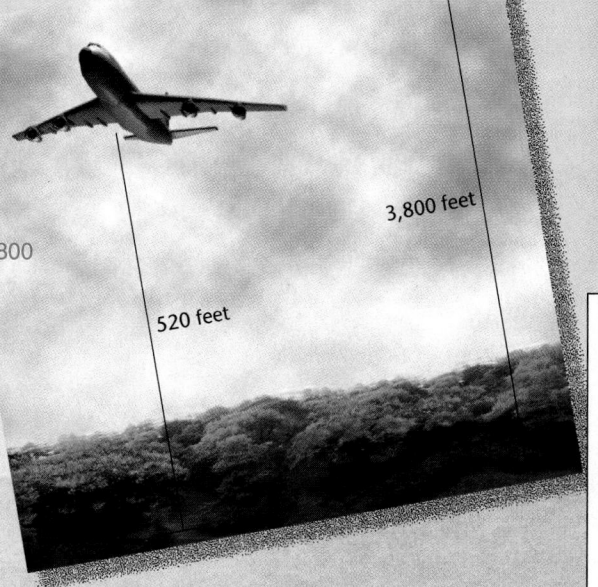

3,800 feet

520 feet

Refer to points A(3, 5), B(4, −1), and C(9, 3).Two lines meet at point A.

22. Write an equation for line AB.

23. Write an equation for line AC.

24. Check your equations by graphing.

25. Use the point of intersection to check your equations algebraically.

6. $(0.8, -0.7)$ **7.** $(34.3, -12.9)$

8. $(0.5, 0.3)$ **9.** no solution

10. $(4, -1)$ **11.** $(200, 100)$

12. $(2, 1.2)$ **13.** $(2, 5)$

18. The graphs of the systems in Exercises 15 and 16 intersect at $(2, 10)$.

20. $\left(\frac{16}{3}, \frac{20}{3}\right)$

22. $y = -6x + 23$ **23.** $y = -\frac{1}{3}x + 6$

24.

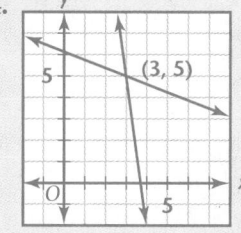

25. $(3, 5)$ satisfies both equations.

Technology Master

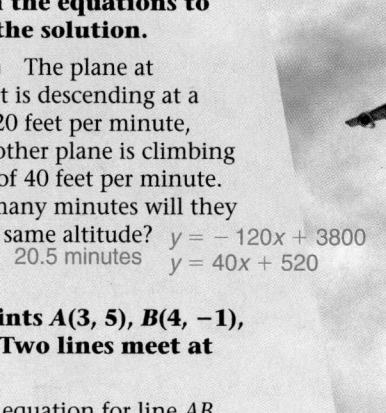

Technology
6.1 Parallel and Intersecting Lines

From geometry you know that two lines are parallel if they do not intersect. Thus, two lines are parallel if the vertical distance between them is a nonzero constant. As a result, you can conclude that two lines intersect if the vertical distance between them is not constant.

Suppose you have $y = 2.5x - 3.2$ and $y = 1.3x + 4.1$. By using a spreadsheet, you can tell if the lines intersect. Column A in the spreadsheet below contains a range for x, say the integers from −4 to 4. Use the FILL DOWN command to complete columns B, C, and D.

	A	B	C	D	
1	X	Y1	Y2	DIFFERENCE	
2	−4				Cell B2 contains 2.5*A2−3.2.
3	−3				Cell C2 contains 1.3*A2+4.1.
4	...				Cell D2 contains B2−C2.
9	3				
10	4				

Use a spreadsheet to tell whether the graphs of the given equations are intersecting lines or parallel lines.

1. $y = -1.2x + 4$ and $y = -1.2x + 3$

2. $y = -1.2x + 4$ and $y = 1.2x + 4$

3. $y = -3.3$ and $y = 3.3$

4. $y = 1.2x + 4$ and $y = 1.3x - 3$

5. $2x - y = 3$ and $3x - 2y = 3$

6. $3x + 3y = 3$ and $3x + 3y = 4$

7. What can you say about the two given lines if the difference column gives a list of decreasing positive numbers?

8. What can you say about the two given lines if the difference column gives a list of increasing positive numbers?

9. Is it possible for the difference column to include decreasing positive numbers followed by numbers that are the same? Explain.

Portfolio Assessment

Have students collect real-world data that can be described using a system of equations. Suggest that they look for data in newspapers, magazines, and almanacs. Have students write one problem about the data they collect. Then have students exchange and solve each other's problems.

Statistics Use the following data to solve Exercises 26–30.

26. **Technology** Use a graphics calculator to plot the data as ordered pairs (year, degrees). Use the statistical functions of your calculator to show a scatter plot of the data.

27. Fit lines to this data.

28. Find the approximate point of intersection of the two models. About how many years beyond the given data will it be before the number of education degrees and the engineering degrees conferred are the same?

29. If you had to make a career choice based on this data, which career would you choose and why?

30. Estimate the point of intersection for $y = -\frac{1}{2}x + 5$ and $y = \frac{1}{4}x - 5$.

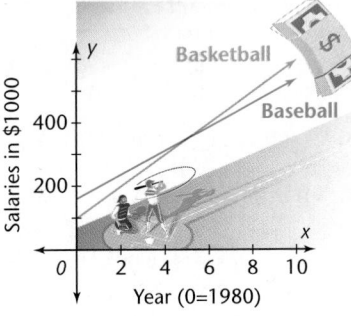

Degrees Conferred for Five Consecutive Years

Year	1	2	3	4	5
Education degrees	7473	7151	7110	6909	6544
Engineering degrees	2981	3230	3410	3820	4191

Sports From 1980 to 1990, there was a shift in salaries for basketball players and baseball players. The graph reflects the average salaries for the players.

31. Estimate the year in which their salaries became equal.

32. Estimate the salary.

33. **Portfolio Activity** Complete the problem in the portfolio activity on page 259 using graphs.

Look Back

34. Place parentheses and brackets to make $21 \div 3 - 4 \cdot 0 + 6(-3) = -39$ true. **[Lesson 1.4]** $[(21 \div 3) - (4 \cdot 0) + 6] \cdot (-3) = -39$

35. If data points are randomly scattered, what type of correlation exists? **[Lesson 1.6]** zero

36. Graph $-3 < x$ and $x < 5$ on the number line. State the solution set clearly. **[Lesson 3.6]** $-3 < x < 5$

37. Solve $3(5 - 2x) - (8 - 6x) = -9 + 2(3x + 4) - 10$. **[Lesson 4.6]** 3

Find the slope of the following lines. **[Lesson 5.3]**

38. $3x + 2y = 12$ 39. $8y = -6x + 12$ 40. $2y = \frac{1}{2}x + 6$ 41. $y = -3x - 18$
$-\frac{3}{2}$ $-\frac{3}{4}$ $\frac{1}{4}$ -3

Look Beyond

Look Beyond

These exercises review the concept of substitution, which will be used in the next lesson.

Substitute each value of x into the equation $-2x + 4y = 12$, and solve for y. EXAMPLE: Let $x = 2$. $-2(2) + 4y = 12$ $y = 4$

42. Let $x = -6$. 0   43. Let $x = 8y$. -1 44. Let $x = y - 3$. 3   45. Let $x = 3y + 1$. -7

The answers to Exercises 26 and 27 can be found in Additional Answers beginning on page 729.

28. $\approx$ 5 years

29. Answers may vary.
   For example, choose engineering because the graph shows an increase in degrees for engineers. Engineering is a growing field.

30. Answers may vary. $\approx \left(13\frac{1}{3}, -1\frac{2}{3}\right)$

31. Answers may vary. $\approx$ 1985

32. Answers may vary. $\approx$ $360,000

33. Floor section, $30 each; Balcony, $20 each

36.
$-3 < x < 5$

LESSON 6.2
Exploring Substitution Methods

why *Sometimes graphing can be used to find the solution to a system of equations. However, when an exact solution cannot be determined from a graph, other methods are needed to find the solution.*

Sue paid a total of $27 for 3 shares of Allied Sports and 4 shares of Best Design.

Best Design, Ltd.
4 Shares Common Stock

Allied Sports, Inc.
3 Shares Common Stock
@
$4.00 per Share

Exploration 1 What Is the Cost?

This problem can be solved by graphing.

Let a = price of one share of Allied Sports.
Let b = price of one share of Best Design.

Then, $\begin{cases} 3a + 4b = 27 \\ a = 4 \end{cases}$ is the system
that models the problem.

1 Examine the graph. Guess the value of b at the point of intersection.

2 Explain why your guess is an approximation.

3 Substitute 4 for a in $3a + 4b = 27$, and solve the resulting equation for b.

4 How can you find the solution to a system if you know the value of one variable?

5 Give two methods to solve the system. $\begin{cases} 8x + 2y = 19 \\ x = 3 \end{cases}$

6 Which method results in an exact solution? Use what you have discovered to solve the system in Step 5. ❖

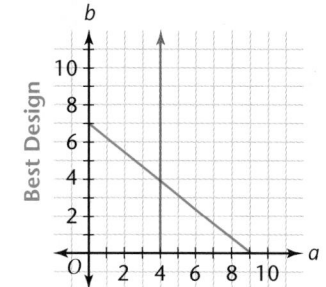

b

Best Design

10
8
6
4
2

O   2   4   6   8   10   *a*

Allied Sports

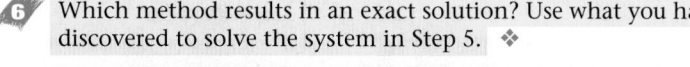

Technology Give students several systems of equations like those in the explorations. Have them use a graphics calculator or computer software such as *f(g) Scholar*™ to graph the system and determine a solution, either approximate or exact. Students will need to write the equations in slope-intercept form to use technology to graph these equations. Since the right-hand sides of both equations in each system are now equal to *y*, students can set the expressions on the right-hand sides equal to each other. Have them solve the equation for *x* and then substitute to find the value of *y*. Have students compare this solution with the solution they found by graphing. Encourage students to discuss the advantages and disadvantages of both methods. Emphasize that even when the substitution method is used to solve a system, a graph can be used to check the solution.

Two ways to use the substitution method to solve the system are given. Emphasize that the coefficients of both x and y in the second equation are 1.

4. Solve for either variable. Substitute the value of one variable in terms of the other variable into the other equation.

Exploration 3 Notes

Discuss why a coefficient of 1 is significant when solving a system using substitution. Emphasize that a system can be solved by substitution even if one of the variables does not have a coefficient of 1, but fractions may result.

3. Solve for the variable whose coefficient is 1. Substitute this expression into the other equation to solve for the value of the unknown variable. Substitute this value into the equation to determine the value of the other variable.

CRITICAL
Thinking

Each exploration involves solving an equation for a variable whose coefficient is 1 and using this value to solve for the other variable. The known value is then used to determine the value of the unknown variable whose coefficient is 1.

 Exploration 2 Substituting Expressions

Examine the system. $\begin{cases} 3x + 2y = 4 \\ x + y = 1 \end{cases}$

1 Solve $x + y = 1$ for x.
   a. Substitute this expression for x in $3x + 2y = 4$.
   b. What values of x and y solve the system?

2 Solve $x + y = 1$ for y.
   a. Substitute this expression for y in $3x + 2y = 4$.
   b. What values for x and y solve the system?

3 How do the methods of Step 1 and Step 2 compare?

4 What substitutions can be made to solve a system if each coefficient in one equation is equal to 1? ❖

 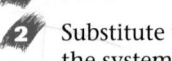 **Exploration 3** Finding a Value

Examine the system. $\begin{cases} 15x - 5y = 30 \\ -2x + y = -6 \end{cases}$

1 Solve $-2x + y = -6$ for y.

2 Substitute this expression for y in the first equation, and then solve the system for x.

3 Explain how to use substitution to solve a system if at least one variable has a coefficient of 1. ❖

 **CRITICAL** *Thinking*

How are Explorations 2 and 3 alike? Summarize what you have learned.

APPLICATION

Investment Each share of New Sport Times stock costs $2 more than a share of Modern Design stock. Find the cost for one share of each stock.

A student in a social studies class purchased 2 shares of Modern Design and 3 shares of New Sport Times for $96.

Define the variables.
   Let m = the cost of a share of Modern Design stock.
   Let n = the cost of a share of New Sport Times Stock.

Write the equations as a system. $\begin{cases} 2m + 3n = 96 \\ n = m + 2 \end{cases}$

 **interdisciplinary** **CONNECTION**

Cooking Have students use substitution to solve the problem.

A recipe for punch calls for 2 parts orange juice to 3 parts ginger ale. Mark wants to make 6 gallons, or 24 quarts, of punch. How many quarts each of orange juice and ginger ale will Mark need?

[9.6 quarts of orange juice and 14.4 quarts of ginger ale]

ENRICHMENT Provide students with a list of different systems of equations like those shown below. Ask them to determine which systems can be easily solved using substitution and which equations would be easier to solve using a different method, such as graphing. Be sure students can justify their answer.

$\begin{cases} 2x + 5y = 19 \\ x = 3y + 2 \end{cases}$ [yes] $\begin{cases} y = 2x - 1 \\ 3x + 2y = 9 \end{cases}$ [yes]

Use the substitution method to solve the system of linear equations. Substitute $m + 2$ for n in the equation $2m + 3n = 96$.

$$2m + 3n = 96$$
$$2m + 3(m + 2) = 96$$
$$2m + 3m + 6 = 96$$
$$5m + 6 = 96$$
$$5m = 90$$
$$m = 18$$

Now substitute 18 for m in the equation $n = m + 2$.

$$n = m + 2$$
$$n = 18 + 2$$
$$n = 20$$

The solution is (18, 20).

Check your answer to see that the values satisfy the original equations. The cost of each share of Modern Design is $18, and the cost of each share of New Sport Times is $20. ❖

APPLICATION

One share of Disposable Inc. and 2 shares of Software Today are worth $65. What is the cost for one share of each stock?

A student in a social studies class purchased 2 shares of Disposable Inc. and 3 shares of Software Today for $105.

Define the variables.

Let d = the cost of each share of Disposable Inc.

Let s = the cost of each share of Software Today.

Write the system. $\begin{cases} 2d + 3s = 105 \\ d + 2s = 65 \end{cases}$

To use the substitution method, first solve $d + 2s = 65$ for d.

$$d = -2s + 65$$

Now substitute $-2s + 65$ for d in the equation $2d + 3s = 105$.

1. Solve the equation $2(-2s + 65) + 3s = 105$ for s.

2. Use your value of s to find the value of d in the equation $d = -2s + 65$.

3. Is (15, 25) your solution?

The cost of one share of Disposable Inc. is $15. The cost of one share of Software Today is $25. ❖

Cooperative Learning

Have students work in small groups to complete the explorations. Have each student complete the equation-solving process individually. Have them compare their solutions and answer the assessment questions.

TEACHING *tip*

Emphasize that after solving the system in the second application, students should check their answers in both of the original equations.

$\begin{cases} 6x + 9y = 20 \\ 3x + 2y = 8 \end{cases}$ **[no]** $\begin{cases} x + 4y = 3 \\ 3x + 2y = 1 \end{cases}$ **[yes]** $\begin{cases} x + 4y = 3 \\ 3x + 2y = 1 \end{cases}$, students could write each x in red. The x is the variable with a coefficient of 1, so it is used for the substitution in the second equation.

INCLUSION **strategies** — **Using Visual Models** For a system of equations, have students use color when they write the variable they will use to substitute. For example, in the system

This procedure should help students keep better track of the variable for which they are initially solving and substituting.

Assignment Guide

Core   1–3, 4–18 even, 20–43

Core Two-Year   1–39

Technology

A graphing calculator or computer software such as *f(g) Scholar*™ is helpful for Exercises 12–19 and 37–39.

Error Analysis

Be sure students complete each step of the procedure carefully. Following the steps as outlined in Reteaching the Lesson on page 270 can be very helpful to students who are struggling with the substitution method.

EXERCISES & PROBLEMS

Communicate

1. If you know $y = 42$, explain how to use substitution to solve $y = 2x + 8$.

2. Given the equations $-4x + y = 2$ and $2x + 3y = 34$, tell how to find the expression to substitute and tell how to find the common solution.

3. Explain how to use substitution to solve the system.
$$\begin{cases} x - 2y = 8 \\ 2x + 3y = 23 \end{cases}$$

Practice & Apply

Solve and check by the substitution method.

4. $\begin{cases} 2x + 8y = 1 \\ x = 2y \end{cases}$ $\left(\frac{1}{6}, \frac{1}{12}\right)$

5. $\begin{cases} x + y = 7 \\ 2x + y = 5 \end{cases}$ $(-2, 9)$

6. $\begin{cases} 3x + y = 5 \\ 2x - y = 10 \end{cases}$ $(3, -4)$

7. $\begin{cases} y = 5 - x \\ 1 = 4x + 3y \end{cases}$ $(-14, 19)$

8. $\begin{cases} 2x + y = -92 \\ 2x + 2y = -98 \end{cases}$ $(-43, -6)$

9. $\begin{cases} 4x + 3y = 13 \\ x + y = 4 \end{cases}$ $(1, 3)$

10. $\begin{cases} 6y = x + 18 \\ 2y - x = 6 \end{cases}$ $(0, 3)$

11. $\begin{cases} 2x + y = 1 \\ 10x - 4y = 2 \end{cases}$ $\left(\frac{1}{3}, \frac{1}{3}\right)$

First graph each system and estimate the solution. Then use the substitution method to get an exact solution.

12. $\begin{cases} 5x - y = 1 \\ 3x + y = 1 \end{cases}$ $\left(\frac{1}{4}, \frac{1}{4}\right)$

13. $\begin{cases} 2x + y = 1 \\ 10x = 4y + 2 \end{cases}$ $\left(\frac{1}{3}, \frac{1}{3}\right)$

14. $\begin{cases} 5x = 3y + 12 \\ x = y \end{cases}$ $(6, 6)$

15. $\begin{cases} 3x - 2y = 2 \\ y = 2x + 8 \end{cases}$ $(-18, -28)$

✗ 16. $\begin{cases} 2x + 3y = 7 \\ x + 4y = 9 \end{cases}$ $\left(\frac{1}{5}, \frac{11}{5}\right)$

17. $\begin{cases} 4x - y = 15 \\ -2x + 3y = 12 \end{cases}$ $\left(\frac{57}{10}, \frac{78}{10}\right)$

18. $\begin{cases} 2y + x = 4 \\ y - x = -7 \end{cases}$ $(6, -1)$

19. $\begin{cases} 4y - x = 4 \\ y + x = 6 \end{cases}$ $(4, 2)$

Write a system of equations for each problem, and solve for an exact solution. Show checks.

20. **Aviation** A hot-air balloon is rising at the rate of 4 feet every second. A small aircraft in the vicinity at 7452 feet is losing altitude at a rate of 30 feet every second. In how many seconds will both be at the same altitude?   197 seconds

754 ft

21. **Numeration** The sum of two numbers is 27. The larger is 3 more than the smaller. Find the two numbers.   15 and 12

22. **Numeration** One number is 4 less than 3 times a second number. If 3 more than twice the first number is decreased by twice the second, the result is 11. Find both numbers.   8 and 4

23. **Geometry** Find the dimensions of this rectangle if its length is twice its width.   about 34.7 by 69.4

Perimeter is 208 meters.

RETEACHING the lesson

Using Algorithms
Guide students through the step-by-step procedure outlined below for using the substitution method to solve systems of equations. Have students complete several examples using these steps.
$$\begin{cases} 3x + 5y = 19 \\ x + 2y = 7 \end{cases}$$

1. Solve one of the equations in terms of one of the variables.
$$x + 2y = 7$$
$$x = 7 - 2y$$

2. Substitute the variable you solved for in step 1 into the other equation.
$$3(7 - 2y) + 5y = 19$$

Fund-raising At the Boy Scout "all you can eat" Spaghetti dinner, the troop served 210 people and raised $935. **Complete Exercises 24–27.**

24. Write an equation for the *total amount raised* from adult and child dinners served. $6.00a + 3.50c = 935$

25. Write an equation for the *total number* of adult and child dinners served. $a + c = 210$

26. Solve the system of equations from Exercises 24 and 25. How many adult and child dinners were served? 80; 130

27. Name two methods of solving systems of equations that can be used to solve Exercise 26. Why would you choose one over another?
graphing; substitution

28. **Geometry** The sum of the measures of angle A and angle B is 90°. If angle A is 30° less than twice angle B, then find the degree measures for each angle.
50° and 40°

29. **Cultural Connection: Asia** A Chinese problem states that several people pooled their money to buy a farm tool to share. If each person paid 8 coins, 3 coins were left over. If each paid 7 coins, they were 4 coins short. How many people were there, and what was the price of the farm tool? 7; 53

30. **Portfolio Activity** Complete the problem in the portfolio activity on page 259 using substitution. $30; $20

 *Look Back*

31. Recreation In a foot race Sam finished 20 feet in front of Joe. Joe was 5 feet behind Mark, and Mark was 10 feet behind Rob. Tom was 15 feet ahead of Rob. In what order did they finish? **[Lesson 3.1]**
Tom; Sam; Rob; Mark; Joe

Solve each equation for x. [Lesson 4.4]

32. $\frac{x}{15} = 3$ 45

33. $\frac{3}{x} = 15$ $\frac{1}{5}$

34. $\frac{15}{x} = 3$ 5

35. $\frac{x}{3} = 15$
45

36. The number 12.6 is 42% of what number? **[Lesson 4.5]** 30

Tell if the following equations represent lines that are parallel, perpendicular, or neither. [Lesson 5.6]

37. $3y = 2x - 15$
$3x + 2y = 24$
perpendicular

38. $2y = x - 12$
$2y - x = 12$
parallel

39. $y = x - 1$
$y = -x + 3$
perpendicular

Look Beyond

Examine the following equations. Use substitution to solve for x and y.

40. $(4, -1)$
$$\begin{cases} x + 2y + 3z = 8 \\ y + 2z = 3 \\ z = 2 \end{cases}$$

41. $(-9, 14)$
$$\begin{cases} 2x + 3y + 5z = 44 \\ 2y - 6z = 4 \\ z = 4 \end{cases}$$

42. $(-3b + a, 3b)$
$$\begin{cases} x + y = a \\ 2y = 6b \end{cases}$$

43. $(-d, -2d)$
$$\begin{cases} x - y - d = 0 \\ 6x - 4y = 2d \end{cases}$$

3. Solve the resulting equation.
$$21 - 6y + 5y = 19$$
$$21 - y = 19$$
$$y = 3$$

4. Find the value of the other variable by substituting the result from step 3 into either of the original equations.
$$x + 2 \cdot 3 = 7$$
$$x = 1$$

5. Check the solution in both of the original equations.
$$3 \cdot 3 + 5 \cdot 2 \stackrel{?}{=} 19$$
$$19 = 19 \quad \text{True}$$

$$3 + 2 \cdot 2 \stackrel{?}{=} 7$$
$$7 = 7 \quad \text{True}$$

Objectives

• Find an exact solution to a system of equations by the elimination method.

RESOURCES

• Practice Master   **6.3**
• Enrichment Master   **6.3**
• Technology Master   **6.3**
• Lesson Activity Master   **6.3**
• Quiz   **6.3**
• Spanish Resources   **6.3**

Assessing Prior Knowledge

Simplify each expression.

1. $3(2x)$ $[6x]$

2. $-2(4x)$ $[-8x]$

3. $-4(-3x)$ $[12x]$

4. $-1(-10x)$ $[10x]$

TEACH

why Not all systems of equations can be solved easily using the substitution method. If there are no coefficients of 1 or -1, the substitution method can be awkward to use. An alternative method, *the elimination method*, provides another way of solving such systems.

Aongoing
ASSESSMENT

Addition of opposites allows one of the variables to be eliminated when the Addition Property of Equality is used. This determines the value of the other variable which is then substituted to determine the value of the first variable.

LESSON 6.3   The Elimination Method

why

The graphing method and the substitution method both provide a way to solve systems of equations. Another method used to solve systems is the elimination method.

To find the daily fee as well as the per-mile cost, you can set up and solve a system of equations by graphing or by substitution. However, the system can also be solved by another method called *elimination*. The **elimination method** for solving equations eliminates one variable by adding or subtracting opposites.

Airport ✈ Rent-A-Car
Bill No.   001
2 Days Rental
Please pay $95.75
Bill No.   002
4 Days Rental
Please pay $226.50

Jason drove 125 miles on the 2-day trip. He drove 350 miles on the 4-day trip.

• Exploration   Using Opposites

1 What is true about the sum of any expression and its opposite?

2 Consider the system of equations. $\begin{cases} 3x + 2y = 7 \\ 5x - 2y = 9 \end{cases}$

   a. Which terms of the equations are opposites?
   b. Use the Addition Property of Equality to add the terms on each side of the equal signs.
   c. Solve the resulting equation for x, and then solve for y.
   d. Check the solution in each equation of the system.

3 Consider the system of equations. $\begin{cases} 2a + b = 9 & \text{Equation 1} \\ 3a - 4b = 8 & \text{Equation 2} \end{cases}$

   a. Multiply both sides of Equation 1 by 4.
   b. Use the Addition Property of Equality. Solve for a, and then solve for b.
   c. Check the solution in each equation of the system.

4 How are opposites used to solve a system of equations? ❖

ALTERNATIVE
teaching strategy

Technology Give small groups of students several systems of equations like those in the exploration and the examples. Have them use a graphics calculator or computer software such as *f(g) Scholar*™ to graph each system and determine a solution, either approximate or exact.

Then introduce students to the elimination method for solving systems of equations. After students have completed their work, have them compare their solutions from graphing with the results from using the elimination method. They should record any observations. Each group should then share the results of their work with the class.

The Exploration used the method of elimination to solve a system of equations. This method used opposites to eliminate one of the variables. Examine how this method is used to solve the rental car problem.

Define the variables.

Let x = the cost per day. Let y = the cost per mile.

Write the equations in standard form.

$$\begin{cases} 2x + 125y = 95.75 & \leftarrow \text{Bill No. 001} \\ 4x + 350y = 226.50 & \leftarrow \text{Bill No. 002} \end{cases}$$

Multiply each term of the first equation by -2.

$$-2(2x) + (-2)(125)y = -2(95.75)$$
$$4x \quad + \quad 350y \quad = \quad 226.50$$

Explain why x rather than y was chosen to be eliminated.

Simplify by using the Addition Property of Equality.

$$\begin{aligned} -4x + (-250)y &= -191.50 \\ \underline{4x + \quad 350y} &= \underline{\quad 226.50} \\ 100y &= \quad 35 \\ y &= \quad 0.35 \end{aligned}$$

Substitute (0.35) in the first equation to find x.

$$\begin{aligned} 2x + 125(0.35) &= 95.75 \\ 2x + \quad 43.75 &= 95.75 \\ 2x &= 52 \\ x &= 26 \end{aligned}$$

The solution is (26, 0.35). A check proves the solution correct. Jason rented the car for $26 a day, and the cost per mile was $0.35.

EXAMPLE 1

Solve. $\begin{cases} 2x + 3y = 1 & \textbf{Equation 1} \\ 5x + 7y = 3 & \textbf{Equation 2} \end{cases}$

Solution ➤

Multiply **Equation 1** by 5 and **Equation 2** by -2.

$$\begin{aligned} (5)2x + (5)3y = (5)1 &\quad \rightarrow \quad 10x + 15y = 5 \\ (-2)5x + (-2)7y = (-2)3 &\quad \rightarrow \quad -10x - 14y = -6 \end{aligned}$$

Use the Addition Property of Equality.
$$15y + (-14y) = 5 + (-6)$$
$$y = -1$$

Then solve for x by substituting.
$$10x + 15y = 5$$
$$10x + 15(-1) = 5$$
$$10x = 20$$
$$x = 2$$

Check each value by substituting the solutions in the original equations. ❖

Try This Solve by elimination. $\begin{cases} 3a - 2b = 6 \\ 5a + 7b = 41 \end{cases}$

Exploration Notes

Be sure students work carefully through all of the steps. You may want to have students work individually to complete steps 1–3 and then work in pairs or small groups to answer the question in step 4.

A_{ongoing} SSESSMENT

The x can be eliminated with fewer steps than y. 125 does not divide into 350 exactly, but 2 divides into 4 exactly.

Alternate Example 1

Solve. $\begin{cases} 2x - 7y = 3 \\ 5x - 4y = -6 \end{cases}$

$[(-2, -1)]$

A_{ongoing} SSESSMENT

Try This

$a = 4$ and $b = 3$

Business Have students write a system of equations to solve the following problem.

A hotel is considering the two special weekend plans given below.

Plan 1: 3 nights and 4 meals for $233
Plan 2: 3 nights and 3 meals for $226.50

If the hotel wants the cost per night and the cost per meal to be the same for both plans, what would the costs be?
[$69 per night, $6.50 per meal]

Have students contact several major hotels in your area to find out about their weekend specials. Have them write a problem using the information they find. Then ask students to exchange and solve each other's problems.

Cooperative Learning

Have students work in pairs or small groups to solve several systems of equations using the steps as outlined in the summary. Be sure they number the steps and follow them carefully.

Alternate Example 2

Choose a method for solving the following systems of equations. Explain why you choose each method.

System 1 $\begin{cases} x + 2y = 6 \\ 5x - 2y = 30 \end{cases}$

[Use the elimination method because $2y$ and $-2y$ are opposites.]

System 2 $\begin{cases} 2x + 3y = 7 \\ 3x - y = 5 \end{cases}$

[Use the substitution method because y has a coefficient of -1.]

System 3 $\begin{cases} 5x - 2y = 8 \\ 2x + 7y = 1 \end{cases}$

[Use the elimination method because no terms have 1 or -1 as coefficients.]

System 4

$\begin{cases} 12x - 1.2y = 3.024 \\ 1.34x - 1.2y = 6.928 \end{cases}$

[Use the elimination method because the coefficients of y can easily be made opposites, and use technology to combine the decimals.]

CRITICAL Thinking

AND; the solution must work in both equations, not just in one or the other.

Aongoing ASSESSMENT

Try This

Each student's choice of method may vary. [(5, 2)]

SUMMARY OF THE ELIMINATION METHOD

1. Write both equations in standard form.
2. Use the Multiplication Property of Equality to write two equations in which the coefficients of one of the variables are opposites.
3. Use the Addition Property of Equality to solve for one of the variables.
4. Substitute the value of that variable into one of the original equations. Solve for the remaining variable.
5. State a solution clearly and check.

You now have several methods for solving a system of equations. Graphing is used as a visual model for a problem involving a system of equations. You can use graphing to check the reliability of a calculated solution or to approximate a solution. The substitution method and the elimination method are used to find an exact solution. Substitution may be the more efficient method if either of the given equations has at least one coefficient of 1 or -1.

EXAMPLE 2

Choose a method for solving the following systems of equations. Explain why you chose each method.

System ❶ $\begin{cases} y = 10 - 6x \\ y = 3x - 6 \end{cases}$ System ❷ $\begin{cases} 2x - 5y = -20 \\ 4x + 5y = 14 \end{cases}$

System ❸ $\begin{cases} 9a - 2b = -11 \\ 8a - 7b = 25 \end{cases}$ System ❹ $\begin{cases} 324p + 456t = 225 \\ 178p - 245t = 150 \end{cases}$

Solution ➤

There may be more than one correct answer for each set of equations.

❶ Use the substitution method. Because y has a coefficient of 1, the equivalent expression can be substituted.

❷ Use the elimination method. $5y$ and $-5y$ are opposites.

❸ Use the elimination method. No term has coefficient 1 or -1.

❹ Use technology. Because the coefficients are such large numbers, ordinary algebra or manual graphing techniques would be cumbersome. ❖

Try This Choose any method to solve the following system of equations. Explain why you chose that method.

$$\begin{cases} y + 2x = 12 \\ 3y = 5x - 19 \end{cases}$$

CRITICAL Thinking

Which word, **AND** or **OR**, connects the equations in a system? How does this help to explain why a solution to a system of equations must be checked in *both* equations?

ENRICHMENT Have students use the elimination method to solve the following system of equations.
$\begin{cases} 3x - y = -3 \\ 2x - y = -12 \end{cases}$

$[x = 9, y = 30]$

INCLUSION strategies **Using Visual Models** Have students use colored pencils to write each equation in a system. Then have them write that equation using the same color throughout the equation-solving process no matter what is done to the equation. This will help students to "see" that even though both sides of an equation are multiplied by the same number, the equation is still equivalent to the original equation.

EXERCISES & PROBLEMS

Communicate

In the following systems, which terms are opposites? How would you solve each system?

1. $\begin{cases} x + y = 13 \\ x - y = 5 \end{cases}$

2. $\begin{cases} 2x - 3y = 8 \\ 5x + 3y = 20 \end{cases}$

3. $\begin{cases} 2a + b = 6 \\ -2a - 3b = 8 \end{cases}$

How would you use opposites to solve each system?

4. $\begin{cases} 4x + 3y = 8 \\ x - 2y = 13 \end{cases}$

5. $\begin{cases} a - 2b = 7 \\ 4b + 2a = 15 \end{cases}$

6. $\begin{cases} 3m - 5n = 11 \\ 2m - 3n = 1 \end{cases}$

Explain how you would use elimination to solve each system.

7. $\begin{cases} 2x + 3y = 9 \\ 3x + 6y = 7 \end{cases}$

8. $\begin{cases} 2x - 5y = 1 \\ 3x - 4y = -2 \end{cases}$

9. $\begin{cases} 9a + 2b = 2 \\ 21a + 6b = 4 \end{cases}$

10. The sum of two numbers is 156. Their difference is 6. Write a system to model the problem. Tell what method you would choose to solve the system of equations. Tell why you chose that method.

Practice & Apply

Solve each system of equations by elimination and check.

11. $\begin{cases} -x + 2y = 12 \\ x + 6y = 20 \end{cases}$

12. $\begin{cases} 2x + 3y = 18 \\ 5x - y = 11 \end{cases}$

13. $\begin{cases} -4x + 3y = -1 \\ 8x + 6y = 10 \end{cases}$

14. $\begin{cases} 2x - 3y = 5 \\ 5x - 3y = 11 \end{cases}$

15. $\begin{cases} 6x - 5y = 3 \\ -12x + 8y = 5 \end{cases}$

16. $\begin{cases} 4x + 3y = 6 \\ -2x + 6y = 7 \end{cases}$

17. Mrs. Jones is celebrating Gauss' birthday by having a pizza party for her two algebra classes. She orders 3 pizzas and 3 bottles of soda for her 2nd period class and 4 pizzas and 6 bottles of soda for her 7th period class. How much does each pizza and bottle of soda cost?

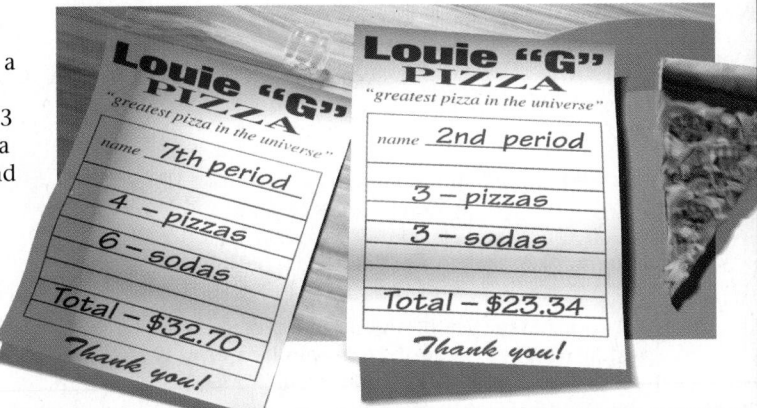

Louie "G" PIZZA
"greatest pizza in the universe"
name 7th period
4 – pizzas
6 – sodas
Total – $32.70
Thank you!

Louie "G" PIZZA
"greatest pizza in the universe"
name 2nd period
3 – pizzas
3 – sodas
Total – $23.34
Thank you!

Error Analysis

Be sure students align terms that contain the same variables when preparing to solve a system by elimination. Students will need to be especially careful about this when working with the systems in Exercises 18–23.

Solve each system of equations using any method.

18. $\begin{cases} 2m = 2 - 9n \\ 21n = 4 - 6m \end{cases}$

19. $\begin{cases} x - 7 = 3y \\ 6y = 2x - 14 \end{cases}$

20. $\begin{cases} 2x = 3 - y \\ y = 3x - 12 \end{cases}$

21. $\begin{cases} 3b = -6a - 3 \\ b = 2a - 1 \end{cases}$

22. $\begin{cases} y = 1.5x + 4 \\ 0.5x + y = -2 \end{cases}$

23. $\begin{cases} 2x = 3y - 12 \\ \frac{1}{3}x = 4y + 5 \end{cases}$

Write a system of equations for each problem, and choose the best method to solve the system. Solve and show checks. **Complete Exercises 24–30.**

At Highview Towers, a one-time deposit is required with the first month's rent.

24. **Finance** Roberto paid $900 the first month and a total of $6950 during the first year. Find the monthly rent and the deposit.

25. **Income** As a parking attendant at the Saratoga Raceway, Juan earns a fixed salary for the first 15 hours worked each week and then additional pay for overtime. During the first week, Juan worked 25 hours and earned $240, while the second week he worked 22.5 hours and earned $213.75. What is Juan's weekly salary for 15 hours and his overtime pay per hour?

26. Shopping at Super Sale Days, Martha buys her children 3 shirts and 2 pairs of pants for $85.50. She returns during the sale and buys 4 more shirts and 3 more pairs of pants for $123. What is the sale price of the shirts and pants?

Bill's grandfather sells his tractor for $6000.

27. **Investment** With the money from the sale of his tractor, Bill's grandfather makes two investments, one at a bank paying 5% per year and the rest in stocks yielding 9% per year. If he makes a total of $380 in interest in the first year, how much is invested at each rate?

28. **Accounting** At a local music store, single-recording tapes sell for $6.99 and concert tapes sell for $10.99. The total number of these tapes sold on Monday was 25. If the sales total for Monday is $230.75 for these two items, find the number of each type sold.

29. **Geometry** The perimeter of a rectangle is 24 centimeters, and its length is 3 times its width. Find the length and width of this rectangle.

18. $\left(-\dfrac{1}{2}, \dfrac{1}{3}\right)$

19. infinite number of solutions

20. $(3, -3)$

21. $(0, -1)$

22. $\left(-3, -\dfrac{1}{2}\right)$

23. $(-9, -2)$

24. $d = \$350,\ r = \$550$

25. $p = \$10.50,\ s = \$135$

26. $p = \$27.00,\ s = \$10.50$

27. $b = \$4000,\ s = \$2000$

28. $c = 14,\ s = 11$

29. $l = 9,\ w = 3$

30. Travel At the rates shown, what is the daily room rate and cost of a meal at the Shamrock Inn?

Use the equations $y = x - 2$ and $y = 2x$ for Exercises 31–33.

31. Graphically find the common solution.

32. Algebraically solve the system.

33. What method did you use to solve Exercise 32?

34. 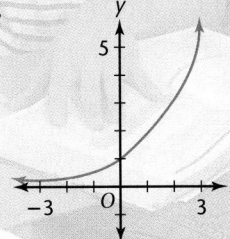 **Portfolio Activity** Complete the problem in the portfolio activity on page 259 using the elimination method.

The Shamrock Inn offers two special holiday weekends.

Look Back

35. Find the pattern and give the next 3 terms of the sequence. **[Lesson 1.1]** 20, 23, 26

5, 8, 11, 14, 17, ___, ___, ___

Evaluate the following. [Lesson 2.4]

36. INT $\left(31\frac{4}{5}\right)$ 31 **37.** INT (31.4) 31 **38.** INT (6.91) 6

39.

This graph represents a function that would be best described as

a. absolute value.

b. quadratic.

c. exponential.

d. linear. **[Lesson 2.5]** c

40. Cultural Connection: Africa The following problem is in the Rhind papyrus by Ahmes. A bag contains equal weights of gold, silver, and lead purchased for 84 *sha'ty*. What is the weight (in *deben*) of each metal if the price for 1 *deben* weight is 6 *sha'ty* for silver, 12 *sha'ty* for gold, and 3 *sha'ty* for lead? **[Lesson 4.6]** 4 deben

Find the slope for the following lines. [Lesson 5.3]

41. $7x - 3y = 22$ $\frac{7}{3}$ **42.** $3x + 2y = 6$ $-\frac{3}{2}$ **43.** $y = 2x$ 2

44. Write the equation of the line through $(-3, 8)$ and parallel to $y = 0.8x - 7$. **[Lesson 5.6]** $y = 0.8x + 10.4$

Look Beyond

Technology Graph the system of equations on the same set of axes, and describe your graph. Use a graphics calculator if you have one.

45. $\begin{cases} 2x - 3y = 6 \\ 4x - 6y = 18 \end{cases}$ The lines are parallel.

30. $m = \$17.50, r = \$67.50$

31 and **32.** $(-2, -4)$

33. Answers may vary. Use substitution.

34. Floor, $30 each; Balcony, $20 each

PREPARE

Objectives

- Identify consistent and inconsistent systems of equations.
- Compare slopes and intercepts of independent and dependent systems.

RESOURCES

- Practice Master **6.4**
- Enrichment Master **6.4**
- Technology Master **6.4**
- Lesson Activity Master **6.4**
- Quiz **6.4**
- Spanish Resources **6.4**

Assessing Prior Knowledge

Graph the systems of equations.

$$\begin{cases} x - y = 2 \\ x + y = 3 \end{cases}$$

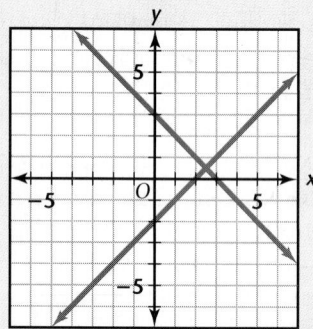

TEACH

Two lines in a plane will either intersect, be parallel, or be the same (concurrent). Have students discuss what results when you try to find the intersection of two lines in a plane that are parallel or concurrent.

Sara's brother is getting married in a few weeks. While reading a magazine, the information in the chart catches her eye.

20.0	21.9	20.8	22.7	21.8	23.6	23.0	24.8	23.9	25.8
1970		1975		1980		1985		1990	

Average Ages for First Marriages

why You have studied systems of equations that have exactly one solution. There are also systems that have an infinite number of solutions or no solution at all. A comparison of systems can be made by studying the graphs of their lines and their solutions.

In the chart you can see that the average age for first marriages is increasing for both men and women. If the ages continue to increase at the same rate for both men and women, will the average age ever become the same?

Inconsistent Systems

The data in the chart can be compared by making a graph.

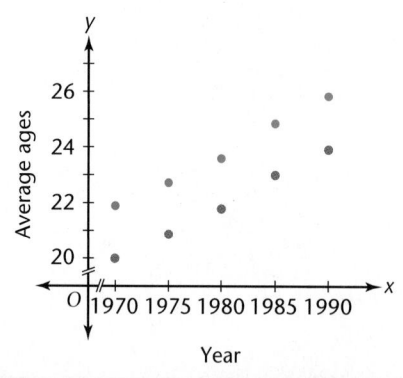

Year

Let x = the year in 5-year intervals.
Let y = the average age for first marriages.

On the graph the years begin with 1970 and the ages begin with 20. What pattern do you see? A line of best fit can be used to model each data set.

The data for men can be modeled by $y = 0.2x - 368$.
The data for women can be modeled by $y = 0.2x - 374$.

Since the slopes of the equations are equal, the lines are parallel. If the trend continues, the lines never intersect, and the mean ages for first marriages will never be equal.

ALTERNATIVE teaching strategy

Technology Provide students with one of each of the three different types of systems of equations. Have them use a graphics calculator or computer software such as *f(g) Scholar*™ to graph each system. After they have studied the graph to determine the type of system, ask students to use the table feature to study the coordinates of the points on each line. Ask them to record any observations.

Now consider the system formed by the lines of best fit. $\begin{cases} y = 0.2x - 374 \\ y = 0.2x - 368 \end{cases}$

Solve the system by substituting $0.2x - 368$ for y in the first equation.

$$0.2x - 368 = 0.2x - 374$$
$$-368 = -374 \qquad \text{False}$$

Notice that both variables have been eliminated and 2 unequal numbers remain. This means there is no possible solution for this system. Since the graphs of the two equations are parallel, this system is called an *inconsistent system*.

An **inconsistent system** has no ordered pair that satisfies both of the original equations. Inconsistent systems have no common solutions. Their graphs are parallel and have the same slope but different y-intercepts.

The graph for mean ages for marriages between 1970 and 1990 is a good example of an inconsistent system. Give another real life problem that might be modeled by parallel lines.

Consistent Systems

When a system has one or more solutions, the system is **consistent**.

 Dependent Systems

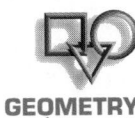
GEOMETRY
Connection

Line AB passes through $A(3, 2)$ and $B(-2, -8)$.
Line CD passes through $C(1, -2)$ and $D(7, 10)$.

1. Draw line AB and line CD on the same coordinate plane. What do you notice about the geometric relationship of the lines?

2. Use the point-slope formula, $y - y_1 = m(x - x_1)$, to write the equation for line AB and line CD. Compare your equations. What do you notice about the slopes of the two equations?

3. Solve the two equations as a system.

4. What happens to the variables? Is the statement of equality true? Do the two lines have any points in common? Is more than one solution possible? How many? ❖

It may be helpful for students to write the equations of the two lines described in slope-intercept form. The equation for line 1 is $y = m_1x + b_1$, and the equation for line 2 is then $y = m_2x + b_2$. As students work through the steps for this exploration, they may find it helpful to draw an example of each pair of lines described before attempting to answer the questions.

Use Transparency ▶ 29

Aongoing **SSESSMENT**

Independent systems have lines with different slopes. Both dependent and inconsistent systems have lines with the same slope.

Independent systems have lines that intersect at one point, dependent systems have only one line, and inconsistent systems have parallel lines.

Independent systems have one solution, dependent systems have an infinite number of solutions, and inconsistent systems have no solutions.

Independent systems have the same or different y-intercepts, dependent systems have the same y-intercept, and inconsistent systems have different y-intercepts.

•Exploration 2 *Other Systems*

Let line 1 have slope m_1, y-intercept b_1, and be represented by equation 1.
Let line 2 have slope m_2, y-intercept b_2, and be represented by equation 2.

Let Equation 1 and Equation 2 form the following system. $\begin{cases} \text{Equation 1} \\ \text{Equation 2} \end{cases}$

1 Suppose line 1 and line 2 intersect in exactly 1 point. What can you say about
   a. m_1 and m_2?
   b. b_1 and b_2?
   c. the solution to the system?

2 Suppose line 1 and line 2 are parallel. What can you say about
   a. m_1 and m_2?
   b. b_1 and b_2?
   c. the solution to the system?

3 Suppose line 1 and line 2 are the same line. What can you say about
   a. m_1 and m_2?
   b. b_1 and b_2?
   c. the solution to the system? ❖

Study the systems shown below.

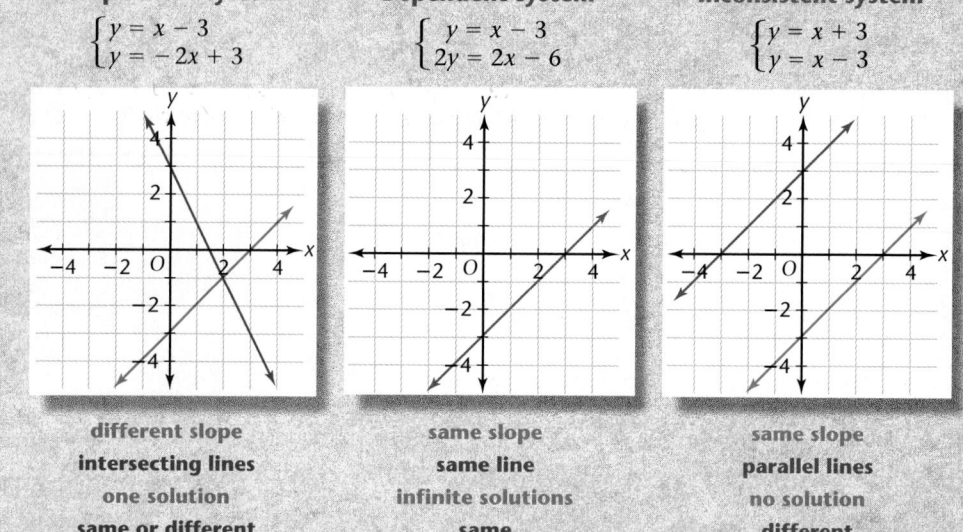

Independent system	Dependent system	Inconsistent system
$\begin{cases} y = x - 3 \\ y = -2x + 3 \end{cases}$	$\begin{cases} y = x - 3 \\ 2y = 2x - 6 \end{cases}$	$\begin{cases} y = x + 3 \\ y = x - 3 \end{cases}$
different slope	same slope	same slope
intersecting lines	same line	parallel lines
one solution	infinite solutions	no solution
same or different y-intercepts	same y-intercepts	different y-intercepts
consistent	consistent	inconsistent

What characteristics are the same or different for each pair of systems?

ENRICHMENT Have students write several true-or-false questions about the concepts presented in this lesson. Then have students exchange and answer each other's questions. You may also want to use students' questions to compile a set of exercises for the entire class to complete.

INCLUSION **strategies** **English Language Development** Ask students to think about ways in which they have heard the terms *inconsistent, consistent, dependent,* and *independent* used in everyday life. Discuss those meanings and how they apply to the mathematical meanings of the terms. You may want to have each student make his or her own list of the terms and definitions.

Cultural Connection: Asia In B.C.E. 213, a Chinese ruler ordered all books to be burned. All of the written knowledge of mathematics was lost. Four hundred fifty years later, Liu Hui came to the rescue. He was able to redo the lost mathematics because people still used rod numerals. The rod numerals were used to represent a decimal system with place value, much like the one we use today. Recall from Chapter 3 that red rods were used for positive numbers and black rods for negative numbers. Vertical and horizontal numerals were used alternately to keep the place values separate.

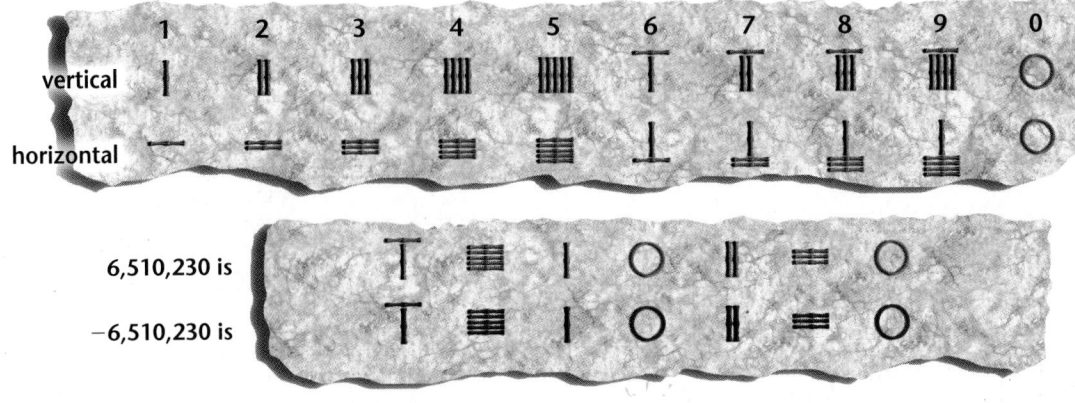

6,510,230 is

−6,510,230 is

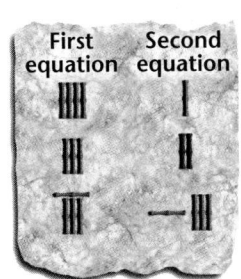

First equation   Second equation

To represent a system of equations, the counting rods were arranged in such a way that one column was assigned to each equation of the system.

The figure at the left represents the following system.
$$\begin{cases} 4x + 3y = 8 \\ x - 2y = 13 \end{cases}$$

One row was assigned to the coefficients of each unknown in the equations. The elements of the last row consisted of the entries on the right-hand side of each equation. Red rods were used to represent positive *(cheng)* coefficients and black rods for negative *(fu)* coefficients.

EXERCISES & PROBLEMS

Communicate

Discuss how to identify the following systems of equations as inconsistent, dependent, or independent. Give reasons.

1. $\begin{cases} y = 2x - 3 \\ 3y = 6x - 9 \end{cases}$
2. $\begin{cases} y = -3x + 2 \\ y = 2x + 2 \end{cases}$
3. $\begin{cases} x + y = 4 \\ x + y = 5 \end{cases}$

4. Explain how to write an equation that would form a dependent system with the equation $y = 2x + 3$. Name what parts of the original equation you would use to write a new equation.

ASSESS

Selected Answers
Odd-numbered Exercises 7–23, 29–43

Assignment Guide
Core   1–6, 8–20 even, 22–45

Core Two-Year   1–44

Technology
A graphing calculator or computer software such as *f(g) Scholar*™ is useful for working Exercises 16–21.

5. If two lines are parallel, discuss how you would write another equation that would form an inconsistent system with the equation $y = -3x + 4$.

6. If two equations in a system intersect at only one point, describe their slopes and y-intercepts. Tell how you determined your answer.

Practice & Apply

Algebraically solve each system. State your solution set clearly.

7. $\begin{cases} x + y = 7 \\ x + y = -5 \end{cases}$ none

8. $\begin{cases} 3y = 3x - 6 \\ y = x - 2 \end{cases}$ infinite

9. $\begin{cases} 3y = 2x - 24 \\ 4y = 3x - 3 \end{cases}$ $(-87, -66)$

10. $\begin{cases} y = -2x - 4 \\ 2x + y = 6 \end{cases}$ none

11. $\begin{cases} 2x + 3y = 11 \\ x - y = -7 \end{cases}$ $(-2, 5)$

12. $\begin{cases} 4x = y + 5 \\ 6x + 4y = -9 \end{cases}$ $\left(\frac{1}{2}, -3\right)$

13. $\begin{cases} 4x + y = 8 \\ y = 4 - 2x \end{cases}$ $(2, 0)$

14. $\begin{cases} y = \frac{3}{2}x + 4 \\ 2y - 8 = 3x \end{cases}$ infinite

15. $\begin{cases} y = \frac{1}{2}x + 9 \\ 2y - x = 1 \end{cases}$ none

Determine which of the following systems are dependent, independent, or inconsistent.

16. $\begin{cases} 2y + x = 8 \\ y = 2x + 4 \end{cases}$ indep

17. $\begin{cases} 3x + y = 6 \\ y = 3x + 9 \end{cases}$ indep

18. $\begin{cases} y + 2 = 5x \\ y = -3x - 2 \end{cases}$ indep

19. $\begin{cases} y + 6x = 8 \\ y = -6x + 8 \end{cases}$ dep

20. $\begin{cases} x - 5y = 10 \\ -5y = -x + 6 \end{cases}$ inc

21. $\begin{cases} y + 4x = 12 \\ y = 5x + 12 \end{cases}$ indep

Use the graph to answer Exercises 22–27.

22. Write the equations of the lines AB and CD.

23. What kind of system do lines AB and CD represent? Write a complete sentence that states at least two reasons for your answer.

24. Write the equation of a line that would form an independent system with line AB and that contains the point A. Graph your line.

25. Write the equation of a line that would form a dependent system with line CD and that contains the point D. Graph your line.

26. What is the point of intersection of $y = 6$ with line AB? with line CD?

27. Write the equation of a line that would form an inconsistent system with $y = 6$.

28. ✏️ **Geometry** Triangle ABC has a perimeter of 18 centimeters while rectangle $PQRS$ has a perimeter of 24 centimeters. Write a system of equations for the figures. How many solutions are possible?

29. The equations for the lines AB and BC form a dependent system. If the slope of the line BC is -3 and the points $A(-3, p)$ and $B(5, 2 - p)$ are given, find the value of p. 13

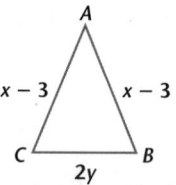

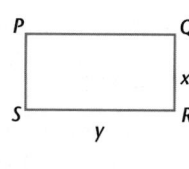

22. $AB: y = \dfrac{-3}{2}x + 10$;

$CD: y = \dfrac{-3}{2}x + 3$

23. inconsistent; The lines are parallel.

24. Answers may vary. For example, $y = \dfrac{2}{3}x + \dfrac{4}{3}$.

25. Answers may vary. For example, $2y = -3x + 6$.

26. $\left(\dfrac{8}{3}, 6\right)$; $(-2, 6)$

27. Answers may vary. For example, $y = 8$.

28. $\begin{cases} 2x + 2y = 24 \\ 2x + 2y = 24 \end{cases}$; infinite solutions

30. Choose the first job. The salary will always be $100 greater than the second job's salary.

33. Center City: $y = 500x + 40,070$
Bay City: $y = 500x + 43,750$

34. Never, the system is inconsistent.

30. Career Options After you work for 10 years, you are offered the choice of two different jobs. The first has a starting salary of $2500 per month with a $750 per month expense account and an annual $800 bonus. The other job pays $3250 per month and has two semi-annual bonuses of $350 each. Write a short paragraph that explains which of the two job offers you would take and why.

31. 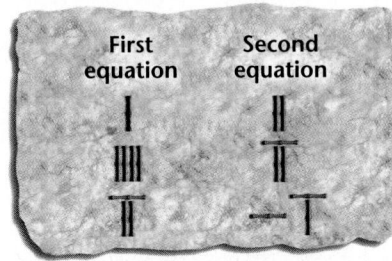 **Geometry** The sum of two angles is 180°. What is the measure of each angle if one angle is 30° more than twice the other? 50° and 130°

32. Number Theory One number is 24 more than another number. If the sum of the numbers is 260, what are the numbers? 142 and 118

Demography The chart shows the populations of Center City and Bay City from 1970 to 1990. Complete Exercises 33–34.

	Center City	Bay City
1970	40,070	43,750
1975	42,570	46,250
1980	45,070	48,750
1985	47,570	51,250
1990	50,070	53,750

33. Write a system of equations that can be used to predict the population (y) in years (x) for each city.

34. In how many years will the population of Center City equal that of Bay City?

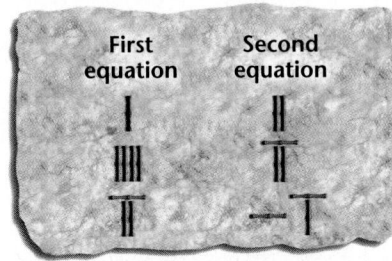

First equation Second equation

35. Cultural Connection: Asia Change these equations from the Chinese counting-rod system to standard algebraic equations. Then solve the equations.

Look Back

36. Make a table of x and y values, and graph the equation $2x + 5y = 7$. **[Lesson 1.5]**

37. At what point does a parabola change direction? **[Lesson 2.2]**
at the vertex

38. Make a table of ordered pairs for the reciprocal function $y = \frac{1}{x}$, when $x = 1, 3, 6, 7, \frac{1}{3}, \frac{1}{6}, \frac{1}{7}$. **[Lesson 2.3]**

39. What is the value of the integer function, $INT\left(7\frac{2}{3}\right)$? **[Lesson 2.4]** 7

40. Write the y values for ordered pairs for the absolute value function of x for integers from -5 to 5. **[Lesson 2.4]**

Solve for x. [Lesson 4.6]

41. $2x - 5 = 15$
10

42. $-3x + 4 = -14$
6

43. $4x + 7 = -5$
-3

44. $ax + b = c$
$\frac{c - b}{a}$

Look Beyond

45. To get to work, Charley travels down Interstate 87 for 15 miles at 60 miles per hour and then 12 miles down Central Avenue at 40 miles per hour. How many minutes should it take Charley to get to work? 33 minutes

35. $\begin{cases} -x + 4y = 7 \\ 2x + 7y = 16 \end{cases}; (1, 2)$

36.
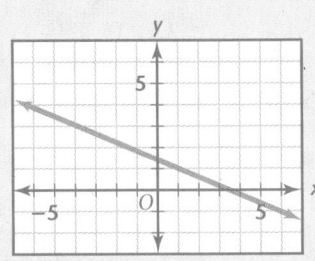

38.

x	1	3	6	7	$\frac{1}{3}$	$\frac{1}{6}$	$\frac{1}{7}$
y	1	$\frac{1}{3}$	$\frac{1}{6}$	$\frac{1}{7}$	3	6	7

40.

x	-5	-4	-3	-2	-1	0	1	2	3	4	5
y	5	4	3	2	1	0	1	2	3	4	5

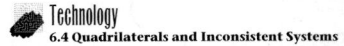
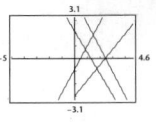

PREPARE

Objectives

- Determine the boundary line that solves a linear inequality.
- Graph the solution of a linear inequality.
- Graph the solution for a system of linear inequalities.

RESOURCES

- Practice Master   6.5
- Enrichment Master   6.5
- Technology Master   6.5
- Lesson Activity Master   6.5
- Quiz   6.5
- Spanish Resources   6.5

Assessing Prior Knowledge

Write the inequality for each.
1. Barbara spent at least $12.
   $[b \geq 12]$
2. Sam made more than two dozen cookies. $[c > 24]$

TEACH

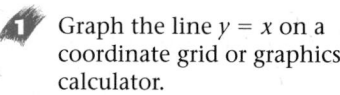 Many expressions that we use every day suggest inequalities. Discuss the situation presented in the text. Ask students to share any others.

Exploration Notes

Have students plot additional points on the *P* side of the line, then on the *Q* side, and confirm the responses to steps 2 and 3.

ongoing ASSESSMENT

2. The *y*-coordinate is always greater than the *x*-coordinate. $y > x$.
3. The *y*-coordinate is always less than the *x*-coordinate. $y < x$.

LESSON 6.5 Graphing Linear Inequalities

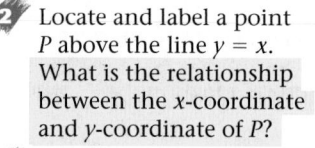 *A tape cassette player costs more than you have saved. Your grade on the last test was less than you expected. These conditions create inequalities. Sometimes unequal conditions can be reversed. Two inequalities form a system that can be solved by graphing.*

Linear Inequalities

Graphing on a coordinate plane allows you to find solutions to linear inequalities that contain more than one variable.

•Exploration• Graphing Linear Inequalities

1 Graph the line $y = x$ on a coordinate grid or graphics calculator.

2 Locate and label a point *P* above the line $y = x$. What is the relationship between the *x*-coordinate and *y*-coordinate of *P*?

3 Locate and label a point *Q* below the line $y = x$. What is the relationship between the *x*-coordinate and the *y*-coordinate of *Q*? ❖

 **COORDINATE GEOMETRY** *Connection*

The graph of a linear equation such as $y = x$ divides a coordinate plane into two half-planes. The line graphed for $y = x$ is a **boundary line** for the two half-planes. A solution for the linear inequality $y \geq x$ includes *all ordered pairs* that make the inequality true. The shaded area, plus the line $y = x$ contains all possible solutions for $y \geq x$. Test the coordinates for points *R* and *S*. Which point has coordinates that satisfy the inequality $y \geq x$?

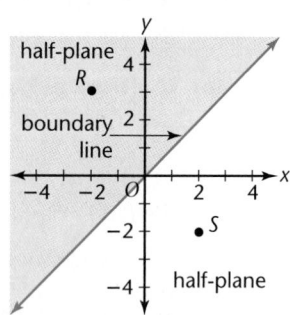

 **ALTERNATIVE teaching strategy**

Technology Have students work in groups of three. One member of each group uses a graphing calculator or computer software such as *f(g) Scholar*™ to graph the boundary line of a linear inequality. Another member of each group draws the line on a piece of graph paper, using a dashed or solid line as necessary. The third student shades the correct half-plane. Then ask each member of the group to choose a point in the shaded region and check to be sure it is in the solution set. Instruct students how to use the available technology to graph a system of inequalities. Provide students with several systems, and have them work in their groups to graph the system. Ask students to record any observations they make as they work. Discuss these observations as a class.

When the inequality symbol is ≤ or ≥, the points on the boundary line *are* included. In this case a solid line is used to represent the boundary. When the inequality symbol is < or >, the points on the boundary line *are not* included. In this case a dashed line is used to represent the boundary.

EXAMPLE 1

Graph the following inequality to find the solution.

$$x - 2y < 4$$

Solution ➤

Graph the boundary line $x - 2y = 4$.

Let $x = 0$.	Let $y = 0$.
$0 - 2y = 4$	$x - 2(0) = 4$
$-2y = 4$	$x - 0 = 4$
$y = -2$	$x = 4$

The points $(0, -2)$ and $(4, 0)$ are on the boundary line. Since the inequality symbol does not include the equal sign, use a dashed line to connect the points.

Try a point such as $(0, 0)$ to see if it satisfies the inequality. Substitute 0 for x and 0 for y.

$$x - 2y < 4$$
$$0 - 2(0) < 4$$
$$0 < 4 \quad \text{True}$$

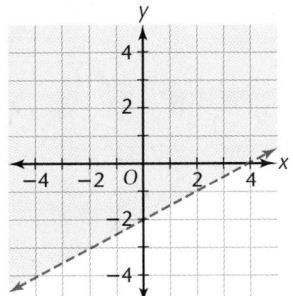

Since the substitution makes the inequality true, $(0, 0)$ is a point in the solution. Shade the half-plane that contains that point. ❖

Systems of Linear Inequalities

For systems of linear inequalities, the solution for the system is the **intersection** of the solution for each inequality. The intersection on the graph is that portion of the plane where the solutions for each inequality overlap. Every point in the intersection satisfies the system.

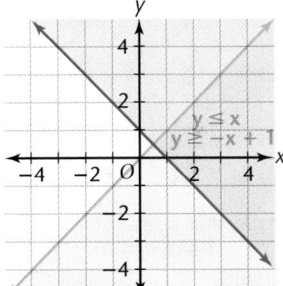
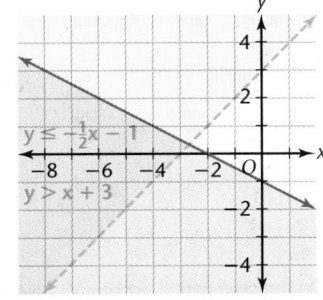
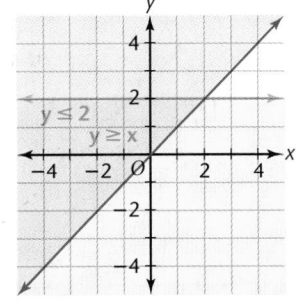

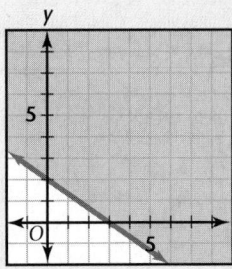
Use Transparency ▶ 30

ongoing ASSESSMENT

point R

TEACHING tip

Show students the graph of $y \leq x$, and ask them to compare it with the graph of $y \geq x$. Use these two graphs to emphasize how to determine when to shade above the boundary line and when to shade below the boundary line.

Cooperative Learning

Have students work in small groups to solve systems of inequalities. Assign a specific step for each student in each group to complete. For example, ask the first student to draw the boundary line for the first inequality. Then ask the second student to shade the correct half-plane for the first inequality. Ask the third and fourth students to graph the second inequality in a similar manner. Then have students switch roles to graph a different system.

Alternate Example 1

Graph the following inequality.
$$2x + 3y \geq 6$$

ENRICHMENT

Give students several systems of three or four inequalities to graph. Two possible systems and their graphs are given below.

1. $\begin{cases} x > 0 \\ y > 0 \\ x + y < 6 \end{cases}$

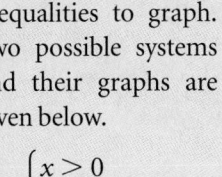

2. $\begin{cases} x > 0 \\ y > 0 \\ x < 4 \\ y < 4 \end{cases}$

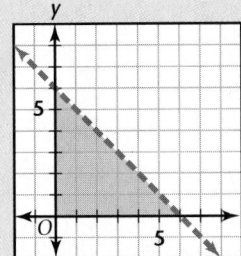
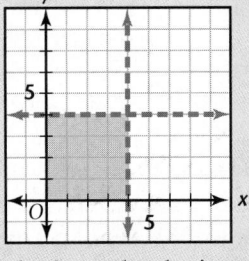

Then have students write several systems of their own. Then have them exchange and solve each other's systems.

Alternate Example 2

Solve by graphing.

$$\begin{cases} x + y \leq 2 \\ 3x + y > 4 \end{cases}$$

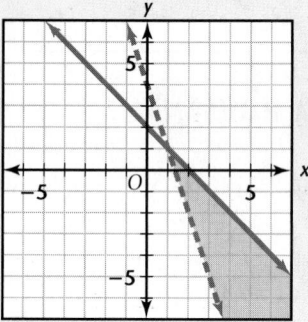

EXAMPLE 2

Solve by graphing. $\begin{cases} x + y \geq -1 \\ -2x + y < 3 \end{cases}$

Solution ➤

Graph the boundary line
$x + y = -1$.

Let $x = 0$.   Let $y = 0$.
$0 + y = -1$ $x + 0 = -1$
   $y = -1$ $x = -1$

The points $(0, -1)$ and $(-1, 0)$ are on the boundary line. Since the inequality symbol includes the equal sign, draw a solid line to connect the points.

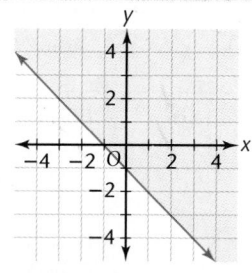

Try a point $(0, 0)$ to see if it satisfies the inequality.

$$0 + 0 \geq -1$$
$$0 \geq -1 \quad \text{True}$$

Since the substitution is true, shade the half-plane that contains $(0, 0)$.

Graph the boundary line
$-2x + y = 3$.

Let $x = 0$.   Let $y = 0$.
$-2(0) + y = 3$ $-2x + 0 = 3$
     $y = 3$ $x = -\frac{3}{2}$

The points $(0, 3)$ and $(-\frac{3}{2}, 0)$ are on the boundary line. Since the inequality symbol does not include the equal sign, draw a dashed line to connect the points.

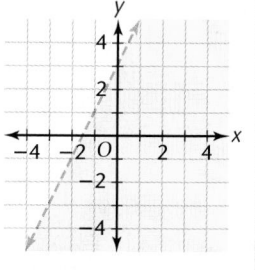

Try a point $(0, 0)$ to see if it satisfies the inequality.

$$-2(0) + 0 < 3$$
$$0 < 3 \quad \text{True}$$

Since the substitution is true, shade the half-plane that contains $(0, 0)$.

The graph shows the solution for the two inequalities. Try a point such as $(1, 1)$ from the intersecting region to test both inequalities.

$$x + y \geq -1$$
$$1 + 1 \geq -1$$
$$2 \geq -1 \quad \text{True}$$

$$-2x + y < 3$$
$$-2 + 1 < 3$$
$$-1 < 3 \quad \text{True} \, \diamond$$

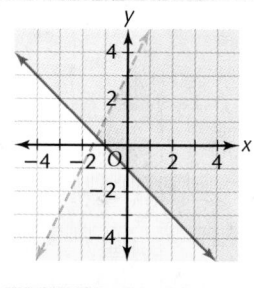

Why is $(1, 1)$ a good point to use as a check for both inequalities?

Aongoing
ASSESSMENT

(1, 1) is clearly in the overlapping shaded region, and 1 is an easy number to use to evaluate an equation.

INCLUSION strategies

English Language Development Take time to go over the translation of words to mathematical symbols. For example, the symbol $\leq$ is used for the phrase "no more than," and the symbol $\geq$ is used for the phrase "at least." Have students use these and any other phrases they can think of in several of their own

statements. Then ask students to write their statements using mathematical symbols.

Using Visual Models Students may benefit from using different colors to graph each inequality in a system. The intersection of the two graphs will then be easy to find.

Try This Solve by graphing. $\begin{cases} y < -2x + 4 \\ y > 3x - 4 \end{cases}$

Kwame gets one job mowing lawns at $4.50 an hour and a second job at the library for $6.00 an hour.

EXAMPLE 3

During the summer, Kwame wants to make at least $90 a week working part-time. He can work no more than 30 hours per week. Write a system of inequalities that represents all of the combinations of hours and jobs that Kwame can work. Graph this solution.

Solution ▶

Define the variables.

Let m = the number of hours mowing.
Let l = the number of hours at the library.

Write the inequalities that satisfy the conditions in the problem. Use the correct inequality symbols.

Kwame can work *no more* than 30 hours a week.	Kwame wants to make *at least* $90 a week.
$m + l \le 30$	$4.5m + 6l \ge 90$
Graph $m + l = 30$.	Graph $4.5m + 6l = 90$.

Graph $m + l = 30$.

Let $l = 0$.	Let $m = 0$.
$m + 0 = 30$	$0 + l = 30$
$m = 30$	$l = 30$

Graph $4.5m + 6l = 90$.

Let $l = 0$.	Let $m = 0$.
$4.5m + 0 = 90$	$4.5(0) + 6l = 90$
$4.5m = 90$	$0 + 6l = 90$
$m = 20$	$l = 15$

The points (30, 0) and (0, 30) are on the boundary line. Since the inequality symbol includes the equal sign, use a solid line to connect the points.

The points (20, 0) and (0, 15) are on the boundary line. Since the inequality symbol includes the equal sign, use a solid line to connect the points.

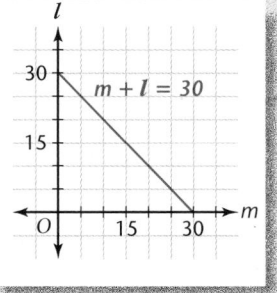

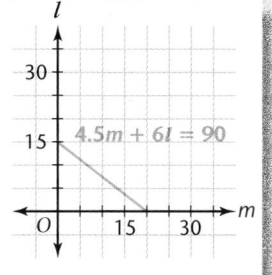

Why do we exclude the points to the left of the *l*-axis and below the *m*-axis?

Using Symbols Begin by asking students to graph several linear inequalities that are written in slope-intercept form. After students have completed these graphs, ask them to study the relationship of the values of y and the inequality symbol to the half-plane that is shaded. Students should notice that if y is greater than the other side of the inequality, the half-plane

above the boundary line should be shaded. If y is less than the other side of the inequality, the half-plane below the boundary line should be shaded.

Provide groups of students with several systems of linear inequalities to graph. Instruct students to use what they have learned about graphing an inequality to graph each system.

Aongoing
ASSESSMENT

Try This

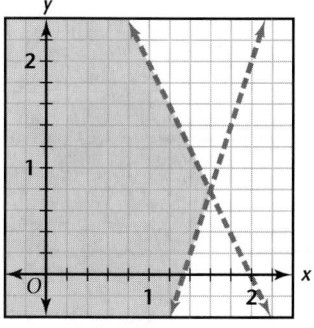

Alternate Example 3

Marilyn has at most $2000 to invest. She plans to invest some of the money in a long-term CD at 4% and some of it in a short-term CD at 2%. She wants to earn at least $50 interest a year. Write a system of inequalities that represents all of the combinations of investments that can be made in each CD. Graph this solution.

$[0.04x + 0.02y \ge 50;$
$x + y \le 2000;$

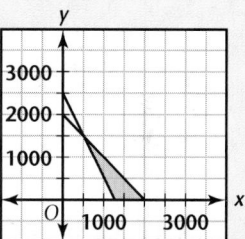

Aongoing
ASSESSMENT

Kwame cannot work a negative number of hours.

The point (0, 0) makes the inequality true, so shade below.

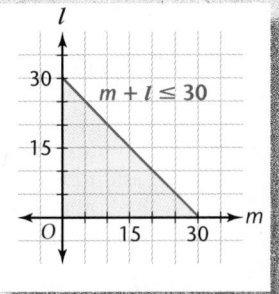

The shaded area shows the greatest number of hours Kwame can work at each job.

The point (15, 5) makes the inequality true, so shade above.

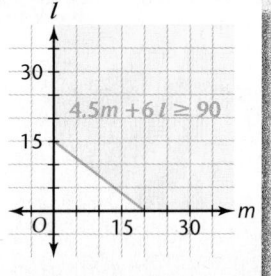

The shaded area shows the least number of hours Kwame works to make at least $90.

The following graph represents the solution for the two inequalities.

Any of the points in the intersecting region will satisfy the conditions for both the inequalities.

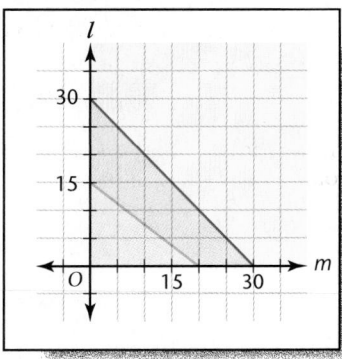

TEACHING tip

It may be helpful to draw just the part of the graph that represents the solution without any additional shading.

The total hours for each set of points is less than or equal to 30. The total number of dollars Kwame earns for each set of points is $90 or more. ❖

MAXIMUM MINIMUM *Connection*

Explain how you would determine how many hours Kwame can work at each job.

To check, try two points from the intersecting region. For example (15, 5) and (14, 10) show that:

• Kwame can work 15 hours mowing and 5 hours at the library.
• Kwame can work 14 hours mowing and 10 hours at the library.

CRITICAL *Thinking*

What is the least number of hours Kwame can work to earn at least $90? What is the greatest amount of money he can make and still work 30 hours or less?

CRITICAL *Thinking*

0 hours of lawn mowing and 15 hours of library work; $90

30 hours of library work and 0 hours of lawn mowing; $180

EXERCISES & PROBLEMS

Communicate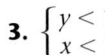

1. Explain how you would decide whether to draw a solid or a dashed line for an inequality.

2. Tell how to choose the area to shade for the inequality $x + 2y < 2$. What is a good point to test in the shaded area?

Discuss how to graph the intersection for the following inequalities. Describe the process for determining the shaded region that satisfies both inequalities in each pair.

3. $\begin{cases} y < 1 \\ x < 1 \end{cases}$ 4. $\begin{cases} 8x + 4y > 12 \\ y < 3 \end{cases}$ 5. $\begin{cases} x \leq 3 \\ x - 2y \geq 2 \end{cases}$ 6. $\begin{cases} x + y \leq -2 \\ x + y > -2 \end{cases}$

Explain how to write a system of inequalities defined by the problem. Describe how to graph the solution for values that make sense. Show your graph.

7. Each spring Marta and her family add flowers to their garden. They plan on buying no more than 20 new perennials at $5.00 per pot and a number of annuals at $1.50 per pot. Marta knows they have budgeted at least $30 for the plants. What combination of each type of plant can they buy?

Practice & Apply

Graph the common solution of each system. Choose a point from the solution. Check both inequalities.

8. $\begin{cases} 2x - 3y > 6 \\ 5x + 4y < 12 \end{cases}$ 9. $\begin{cases} x - 4y \leq 12 \\ 4y + x \leq 12 \end{cases}$ 10. $\begin{cases} y < x - 5 \\ y \leq 3 \end{cases}$

11. $\begin{cases} 2x + y \leq 4 \\ 2y + x \geq 8 \end{cases}$ 12. $\begin{cases} 5y < 2x - 5 \\ 4x + 3y \leq 9 \end{cases}$ 13. $\begin{cases} 4y < 3x + 8 \\ y \leq 1 \end{cases}$

14. $\begin{cases} x \geq 4 \\ 0 < y \end{cases}$ 15. $\begin{cases} x - 2y < 8 \\ x + y \geq 5 \end{cases}$ 16. $\begin{cases} 3x - y > -2 \\ x - y > -1 \end{cases}$

17. 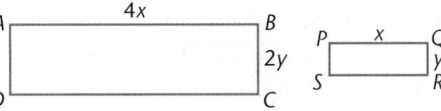 **Geometry** Write the system of inequalities defined by each perimeter, and graph the common solution. Make sure that the solution represents real life possibilities.

Rectangle ABCD's perimeter is at most 30 centimeters, while rectangle PQRS's perimeter is at least 12 centimeters.

The answers to Exercises 8–16 can be found in Additional Answers beginning on page 729.

17.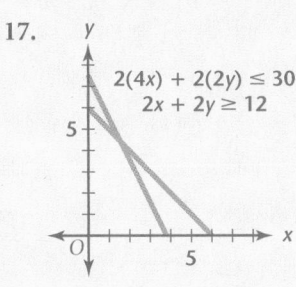

$2(4x) + 2(2y) \leq 30$
$2x + 2y \geq 12$

ASSESS

Selected Answers
Odd-numbered Exercises 9–19, 23–25, 29–37

Assignment Guide
Core   1–7, 8–16 even, 17–40

Core Two-Year   1–37

Technology
A graphing calculator or computer software such as *f(g) Scholar*™ can be useful for Exercises 8–16.

You and a friend are going to have lunch at Al's Burgers.

18. Between the two of you, you have $8.00 and each of you wants at least one Alburger and an order of fries. Can either of you have seconds?

Yes, one person can have a second order of fries.

Match the set of inequalities to the graph that represents the solution.

19. $y < -2x + 6$ and $y \leq 3x - 4$ b

20. $y \leq -2x + 6$ and $y > 3x - 4$ a

a.

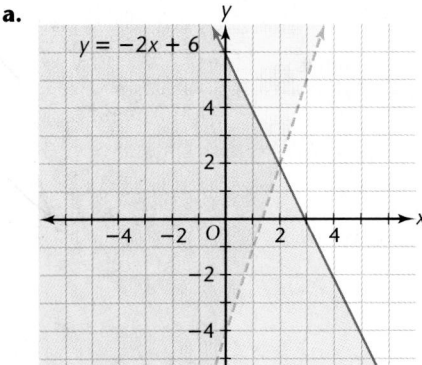

b.

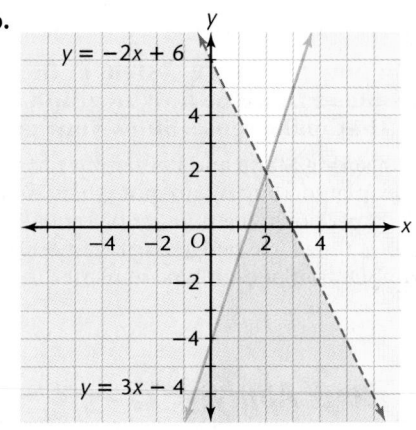

21. Write the ordered pairs for the points $A, B, C, D.$ $A(3, 8)$; $B(9, 5)$; $C(8, 2)$; $D(1, 2)$

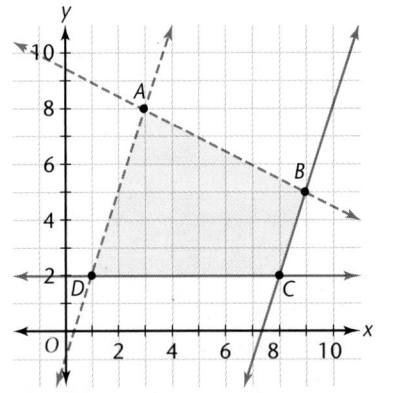

22. Write the system of inequalities that represents the shaded region. Use the points shown to find equations for the boundary lines.

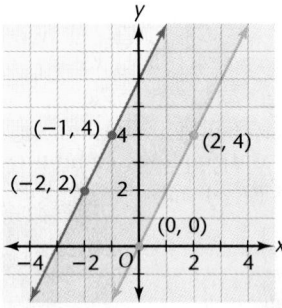

23.

Baskets	0	1	2	3	4	5	6	7	8
Free Throws	16	14	12	10	8	6	4	2	0

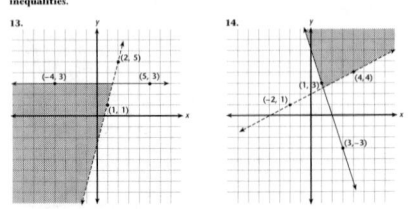
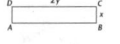

23. Sports You usually average no more than 16 points in a game. Find the combinations of baskets and free throws you could have made.

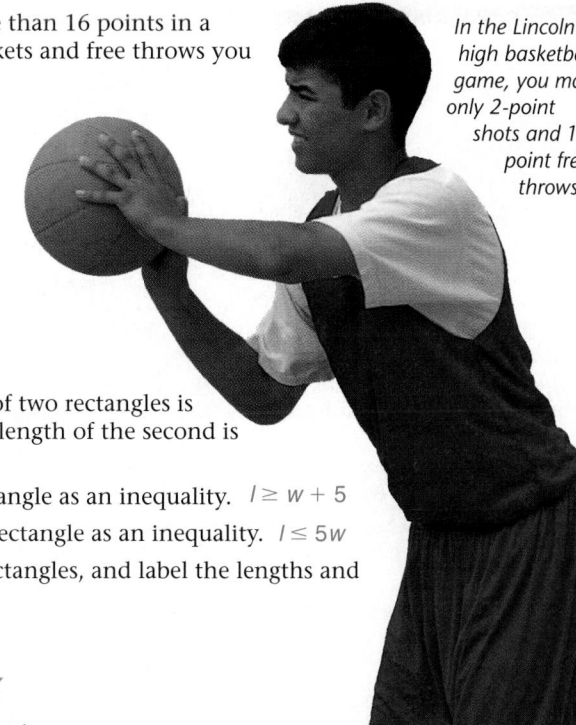

In the Lincoln high basketball game, you made only 2-point shots and 1-point free throws.

24. Budgeting Time On most days Abel spends no more than 2 hours on his math and science homework. If math always takes about twice as long as science, what is the maximum time that Abel usually spends on science? $\frac{2}{3}$ hours

Geometry The length of the first of two rectangles is at least 5 more than its width, while the length of the second is no more than 5 times its width.

25. Represent the length of the first rectangle as an inequality. $l \geq w + 5$

26. Represent the length of the second rectangle as an inequality. $l \leq 5w$

27. Draw figures to represent the two rectangles, and label the lengths and widths as inequalities.

Look Back

28. Find the first differences for the following sequence.
[Lesson 1.2]   12; 20; 28; 36
    5 17 37 65 101, . . .

29. Find the second differences for the following sequence.
[Lesson 1.2]   8; 8; 8
    5 17 37 65 101, . . .

30. Simplify $-8 - 2\{3 + 6\left(5 - \frac{4}{2}\right) + 7\}$.   **[Lesson 1.4]** -64

31. Fund-raising How much would each person contribute toward a goal of $1200 if there were 20 people contributing equally?
[Lesson 2.3] $60

Add the following like terms.   [Lesson 3.3]

32. $2x^2 + 4 + 3y - x^2 + 2$ **33.** $3a + a + z + 4a + 2b^2 - b$
    $x^2 + 3y + 6$ $8a + 2b^2 - b + z$

Use the slopes to tell whether the following lines are parallel, perpendicular, or neither.   [Lesson 5.6]

34. $x + y = 7$ **35.** $-2x + y = -5$ **36.** $-x + 2y = 6$ **37.** $x - 2y = 4$
    $x + 3y = 9$ $-2x - y = 5$ $-x + 2y = -3$ $2x + y = 1$
    neither   neither   parallel   perpendicular

Look Beyond

For $f(x) = 2x + 4$, evaluate the following.

38. $f(3)$ 10   **39.** $f(-2)$ 0   **40.** $f\left(-\frac{1}{2}\right)$ 3

27.

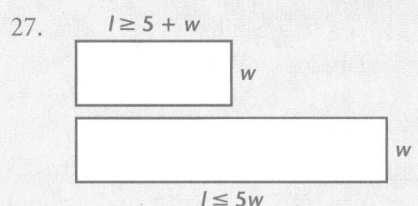

$l \geq 5 + w$

w

$l \leq 5w$

w

Technology
6.5 Discovering an Area Theorem

Suppose that the sum of the length (x) and the width (y) of a rectangle is at most 15 units. This condition is expressed symbolically as $x + y \leq 15$ and graphically as the region shown.

Of all the points in the shaded region, which pair (length, width) gives the maximum area of the rectangle?

If the dimensions of the rectangle are different, they differ by some positive amount a.

Area $= x(x - a) = x^2 - ax < x^2$

This means that the area of a rectangle with unequal length and width is less than the area of a rectangle with equal length and width. Consequently, the maximum area comes from a point on the line $y = x$.

By using a spreadsheet like the one shown, you can find the pair (length, width) that gives the maximum area. Columns A and B contain numbers that are equal. These numbers are the dimensions to be checked. From column D, choose the greatest value under the condition that the corresponding value in column C is no more than 15.

	A	B	C	D
	X	Y	SUM	PRODUCT
1				
2	0	0	0	0
3	1	1	2	1
4	2	2	4	4
5	3	3	6	9
6	...	...	...	...

Use a spreadsheet to find the maximum area of the rectangle whose length and width satisfy the given inequality.

1. $x + y \leq 15$ 2. $x + y \leq 20$

3. $x + y \leq 30$ 4. $x + y \leq 36$

5. $x + y \leq 10.5$ 6. $x + y \leq 50.5$

7. Find the maximum area of the rectangle whose length and width satisfy $x + y \leq s$, where $s > 0$.

Objectives

• Solve traditional application problems.

RESOURCES

• Practice Master   **6.6**
• Enrichment Master   **6.6**
• Technology Master   **6.6**
• Lesson Activity Master **6.6**
• Quiz   **6.6**
• Spanish Resources   **6.6**

Assessing Prior Knowledge

Write an equation to describe each problem. You may use either one or two variables.

1. The dimes in Joan's pocket are worth $2.50.
   [$0.10d = 2.50$]

2. The dimes and nickels in Sam's pocket are worth $3.00.
   [$0.05n + 0.10d = 3$]

3. Aaron has 8 ounces of peanuts and cashew nuts.
   [$p + c = 8$]

4. The sum of the digits of a two-digit number is 6.
   [$t + u = 6$]

5. At r miles per hour, Eileen drove 150 miles in 3 hours.
   [$150 = 3r$]

TEACH

Some types of problems have been studied by mathematics students for many decades. In this lesson, students will study some classic applications of systems of equations. Students should find it helpful to use a table to organize the information from a problem.

LESSON 6.6 Classic Applications

why Have you ever seen a classic car? Classic cars usually represent a model that was revolutionary for its time because of either style or technology. Mathematics has classic problems involving mixtures, money, digits, rates, and age.

Many people collect, restore, and display classic cars that were revolutionary for their day. Only time will tell whether new models will become classics.

Each classic problem from algebra can usually be solved by setting up the equations in special ways.

EXAMPLE 1

Mixture This classic application uses percent to express how much of a chemical is in a solution.

How many ounces of a 15% acid solution should be mixed with a 40% acid solution to produce 60 ounces of a 25% acid solution?

Solution ➤

Draw a diagram and make a table to help solve this problem.

Define the variables.

Let x = the number of ounces of 15% acid solution.

Let y = the number of ounces of 40% acid solution.

ALTERNATIVE **teaching strategy**

Technology After students have written the system of equations that can be used to solve the problems given in each example, have them use a graphics calculator or computer software such as $f(g)$ *Scholar*™ to graph the system and estimate the solution. Students should then use an algebraic method to solve the system and check the answer against the estimate. Students can then use technology to check their solutions in each of the original equations. One benefit of graphing the systems for problems like these is that students can gain additional insight about the situation described in the problem.

Place the information in a chart to help organize the facts.

	% acid	Amount of solution	Amount of acid
First solution	15%	x	$0.15x$
Second solution	40%	y	$0.40y$
Final solution	25%	60	$0.25(60)$

From the table, write two equations.

$$\begin{cases} x + y = 60 \\ 0.15x + 0.40y = 0.25(60) \end{cases}$$

Solving the system gives the solution (36, 24).

To get the required 60 ounces of the 25% acid solution, mix 36 ounces of the 15% acid solution with 24 ounces of the 40% acid solution. ❖

EXAMPLE 2

Money In classic coin problems you can usually set up two equations—one involving the number of coins and the other involving the value of the coins.

In a coin bank there are 250 dimes and quarters worth a total of $39.25. Find how many of each kind of coin are in the bank.

Solution ➤

Create a chart to organize the information.

Coin type	Number	Coin value	Value in cents
Quarters	q	25¢	$25q$
Dimes	d	10¢	$10d$
Total	250		3925

From the table, write two equations based on the totals.

$$\begin{cases} q + d = 250 \\ 25q + 10d = 3925 \end{cases}$$

Solving the system, you find that q is 95 and p is 155.

The coin bank has 95 quarters and 155 dimes. A check indicates that the solution is correct. ❖

Alternate Example 1

How many ounces of a 20% acid solution should be mixed with a 70% acid solution to produce 50 ounces of a 40% solution? [**30 ounces of the 20% solution and 20 ounces of the 70% solution**]

Cooperative Learning

Have students work in small groups to find classic word problems like those presented in this lesson. Have students write each problem on one side of an index card and write the solution to the problem on the other side. They should also write the type of problem—either mixture, money, numeration, travel, or age—in the upper left-hand corner of the card. Place the dividers for each type of problem in a card box. As students complete their index cards, they should put them in the correct place in the card file.

Alternate Example 2

In a coin bank there are 125 nickels and dimes worth a total of $10.50. Find how many of each kind of coin are in the bank. [**85 dimes and 40 nickels**]

ENRICHMENT

Have students work in small groups to create problems of their own. Suggest that they start with the answer and then work backward. For example, to write a motion problem, students could decide to write a problem about a plane that is traveling 350 miles per hour with a wind of 50 miles per hour for 3 hours. The distance the plane travels can be found using $d = rt$, so $d = 1200$ miles. Using the same distance, they can determine that the time for the return trip is 4 hours. Then students can create a problem with any one of the pieces of information missing. Encourage students to be creative while writing these problems. It may be helpful for them to use a calculator.

EXAMPLE 3

Alternate Example 3

The sum of the digits of a two-digit number is 6. The original two-digit number is six less than twice the number with its digits reversed. Find the original two-digit number. **[42]**

Numeration

Other examples of classic problems are digit problems. In most digit problems the trick is to write the value of a number in expanded form. You can write 52 as 5(10) + 2. If you reverse the digits in 52, you get 25, and can write it as 2(10) + 5.

The sum of the digits of a two-digit number is 7. The original two-digit number is 3 less than 4 times the number with its digits reversed. Find the original two-digit number.

Solution ➤

Define the variables.

Let t = the ten's digit of the original number.
Let u = the unit's digit of the original number.

The two-digit number expressed in expanded notation is
$$10t + u.$$

The number with its digits reversed in expanded notation is
$$10u + t.$$

The problem says the sum of the digits is 7.
$$t + u = 7$$

The original 2-digit number is 3 less than 4 times the number with its digits reversed.
$$10t + u = -3 + 4[10u + t]$$

Solving the system $\begin{cases} t + u = 7 \\ 10t + u = -3 + 4[10u + t] \end{cases}$ gives $u = 1$ and $t = 6$.

If t is 6 and u is 1, then the original number is 61. The number 61 is a two-digit number which is 3 less than 4 times itself with its digits reversed. A check proves the solution is correct. ❖

Try This

[75]

Try This   A two-digit number whose ten's digit is 2 more than the units digit is 3 more than 6 times the sum of its digits. Find the original number.

EXAMPLE 4

Alternate Example 4

A boat went upstream (against the current) a distance of 90 miles in 4.5 hours. The boat went the same distance downstream (with the current) in 3 hours. Find the rate of the current and the rate of the boat with no current. **[The rate of the current is 5 mph, and the rate of the boat with no current is 25 mph.]**

Travel   Classic motion problems have some basic elements. Travel by plane, train, or automobile involves speed, distance, and time. Create charts, draw diagrams, and organize the information to help solve the following problem.

A plane leaves New York City and heads for Chicago, which is 750 miles away. The plane, flying against the wind, takes 2.5 hours to reach Chicago. After refueling the plane returns to New York City, traveling with the wind, in 2 hours. Find the rate of the wind and the rate of the plane with no wind.

 **Using Visual Models**
Many times, diagrams can be drawn to show the situation described in a classic problem. Encourage students to draw diagrams as needed to help them understand the problem. Working on these diagrams with other students may also be helpful.

Solution ➤

Define the variables. Chart the information.

Let x = the rate of the plane without any wind.
Let y = the rate of the wind.

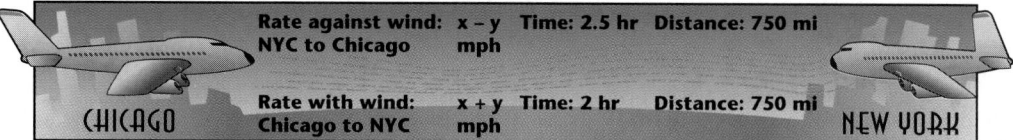

Rate against wind: $x - y$ mph   Time: 2.5 hr   Distance: 750 mi
NYC to Chicago

Rate with wind: $x + y$ mph   Time: 2 hr   Distance: 750 mi
Chicago to NYC

CHICAGO   NEW YORK

To write the system of equations, recall the relationship between rate, time, and distance.

	Rate	× Time	=	Distance
For Chicago to NYC:	$(x - y)$	(2.5)	=	750
For NYC to Chicago:	$(x + y)$	(2)	=	750

To solve, divide to get simpler equations.

$$(x - y)(2.5) = 750 \longrightarrow x - y = 300$$
$$(x + y)(2) = 750 \longrightarrow x + y = 375$$

Solving the system gives $x = 337.5$ and $y = 37.5$.

The rate of the plane is 337.5 miles per hour and the rate of the wind is 37.5 miles per hour. A check proves the solution is correct. ❖

CRITICAL *Thinking*

How would the rate of the plane be affected if there were no wind?

EXAMPLE 5

Age

A father is 32 years older than his son.

Age problems are classic problems that can usually be solved by writing the two related ages at two points in time. In 4 years the father will be 5 times as old as his son. How old is each now?

Solution ➤

Write two equations about the age of the father and son at two different times. Define the variables.

Let f = the father's age.   Let s = the son's age.

The father is 32 years older than his son.
$$f = s + 32$$

In 4 years the father will be 5 times the son's age. Add 4 years to each age.
$$f + 4 = 5(s + 4)$$

Solving the system gives $s = 4$ and $f = 36$.

At the present time the father is 36 years old, and his son is 4 years old. A check proves the solution is correct. ❖

CRITICAL *Thinking*

It would not be affected.

Alternate Example 5

In 2 years Sue will be 4 times as old as her daughter. Right now, Sue is 30 years older than her daughter. How old is each now? **[Sue is 38 years old, and her daughter is 8 years old.]**

R**ETEACHING** t h e l e s s o n

Inviting Participation Have students use the problems that they created in the Cooperative Learning activity on page 293. Ask small groups of students to choose a problem card from one of the problem types and work together to solve it. Continue to have them choose cards from that problem type until they feel comfortable solving that type of problem. Have students move to the next problem type and repeat the procedure. They should continue in this manner. Be sure to allow students ample time to work on each problem type.

ASSESS

Selected Answers

Odd-numbered Exercises 7–23

Assignment Guide

Core 1–26

Core Two-Year 1–24

Technology

A graphics calculator or computer software such as *f(g) Scholar™* is useful for Exercises 25 and 26.

EXERCISES & PROBLEMS

Communicate

Discuss the problem-solving strategy you would use to solve the following problems, and explain how to *set up* the equations.

1. When Jim cleaned out the reflecting pool at the library, he found 20 nickels and quarters. The mixture of nickels and quarters totaled $2.60. How many quarters did Jim find?

2. **Chemistry** You are running a chemistry experiment that requires you mix a 25% alcohol solution with a 5% alcohol solution. This mixture produces 20 liters of a solution that contains 2.6 liters of pure alcohol. How many liters of the 25% alcohol solution must you use?

A bird can fly with the wind 3 times as fast as it can fly against the wind. How can you represent

3. the rate of the bird in still air (no wind) and the rate of the wind.

4. the rate of the bird flying with the wind and against the wind.

5. Discuss how the rate of the bird with the wind compares with the rate of the bird against the wind.

Describe how to set up Exercise 6, and tell how to check your answer after discussing how to solve the problem.

6. The sum of the digits of a two-digit number is 8. If 16 is added to the original number, the result is 3 times the original number with its digits reversed. Find the original number.

Practice & Apply

7. **Recreation** Ramona turns her canoe around and returns downstream to her cabin in exactly 1 hour. What is the rate of the creek's current and the rate of Ramona's paddling in still water?
   1 mph and 5 mph

Ramona is staying at a vacation resort called Stony Cabins. She rents a canoe and takes 1.5 hours to paddle her canoe upstream.

8. The coin box of the hospital's vending machine contains 6 times as many quarters as dimes. If the total amount of money at the end of the day was $28.80, how many of each coin was in the box?
   18 d and 108 q

9. Numeration Find the two-digit number whose ten's digit is 4 less than the unit's digit and the original number is 2 more than 3 times the sum of the digits. 26

With a tailwind, a jet flew 2000 miles in 8 hours, but the return trip against the same wind required 10 hours.

10. Aviation Find the jet's speed and the wind speed. 225 mph and 25 mph

11. In 15 years, Maya will be twice as old as David is now. In 15 years, David will be as old as Maya will be 10 years from now. How old are they now? 25 and 20

12. A 4% salt solution is mixed with a 16% salt solution. How many milliliters of each solution are needed to obtain 600 milliliters of a 10% solution? 300 ml each

13. Wymon is 20 years older than Sabrina. In 8 years, Wymon will be twice as old as Sabrina. How old is each today? 12 and 32

14. Cultural Connection: Asia Apply the problem-solving strategy of working backwards (inversion) to a problem supplied by Aryabhata, circa 500 C.E.

"O beautiful maiden with beaming eyes, tell me, since you understand the method of inversion:
What number multiplied by 3, then increased by $\frac{3}{4}$ of the product, then divided by 7, then diminished by $\frac{1}{3}$ of the result, then multiplied by itself, then diminished by 52, whose square root is then extracted before 8 is added, and then divided by 10, gives the final result of 2?" 28

Look Back

Evaluate the following expressions if $x = 2$, $y = 1$, and $z = 1$. [Lesson 1.3]

15. $2x + 4y - 3z$ 5 **16.** $\frac{2x - z}{5}$ $\frac{3}{5}$ **17.** $\frac{x + y}{z}$ 3 **18.** $3xyz$ 6

19. Simplify $(29 + 1) \div 3 \cdot 2 - 4$. **[Lesson 1.4]** 16

Multiply. [Lesson 4.2]

20. $-[2(a + b)]$ **21.** $-[-11(x - y)]$ **22.** $2(x - y) + 3[-2(x + y)]$
$-2a - 2b$ $\quad 11x - 11y$ $\quad\quad -4x - 8y$

Find the slope of the following equations. [Lesson 5.4]

23. $3x + 4y = 12$ **24.** $5y = 10x - 20$
$-\frac{3}{4}$ $\quad\quad\quad 2$

Look Beyond

Find each square root.

25. $\sqrt{\dfrac{25}{625}}$ $\frac{1}{5}$ **26.** $\sqrt{a^2}$
$\quad\quad\quad\quad a \geq 0, \sqrt{a^2} = a$
$\quad\quad\quad\quad a < 0, \sqrt{a^2} = -a$

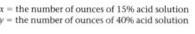
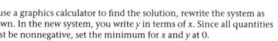
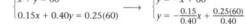

Minimum Cost Maximum Profit

Finding the Right Mixture

For a hike you want to make a mixture of nuts and raisins to provide all of the needed nutrients but at the lowest cost. The chart shows the nutritional information and cost per ounce.

You have a weight limit for your pack and can allow no more than 18 ounces for the mix. The snack should supply at least 1800 calories with no more than 117 grams of fat. How many ounces each of raisins and nuts should you buy to minimize the costs?

	Calories	Fat	Price/ounce
Nuts	150	13g	$0.40
Raisins	90	0g	$0.10

1 Define the variables for the unknowns.
 Let n = the number of ounces of nuts.
 Let r = the number of ounces of raisins.

2 Write a linear equation, called the optimization equation, to express the relationship between the highest and lowest cost.
 Thus, $0.40n$ equals the cost of nuts.
 Thus, $0.10r$ equals the cost of raisins.
 Total cost $0.40n + 0.10r$.

FOCUS

Linear programming is one of the most common applications of the concepts presented in this chapter. The graph of the system of inequalities can be used to describe a situation visually and to provide valuable insight into the relationship.

MOTIVATE

Think about the first problem. When have you or someone you know encountered a similar problem? Think about how you might use the skills you have learned in this chapter to solve this problem. Share your ideas with the class.

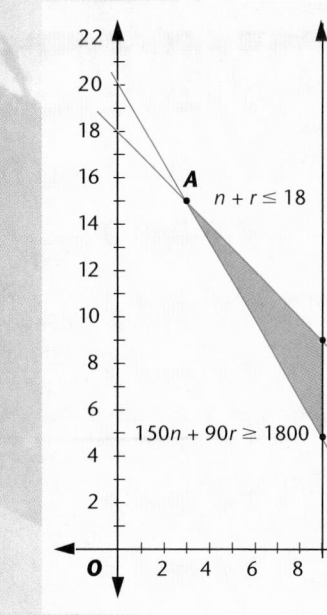

$$13n \le 117$$

$$n + r \le 18$$

$$150n + 90r \ge 1800$$

3 Write a set of inequalities that define the conditions for the problem.

The snack should supply at least 1800 calories.

$$150n + 90r \ge 1800$$

The snack should supply no more than 117 grams of fat. $13n \le 117$

You can carry no more than 18 ounces of the mix $n + r \le 18$

The conditions defined by the inequalities that must be satisfied are called the **constraints**.

4 Sketch or use graphing technology to graph each of the three inequalities. Shade the region that represents the solution set.

The shaded region, *ABC*, represents the solution set. This region, determined by the linear inequalities that contains the possible solutions, is called the **feasibility region**.

5 Find the coordinates of the vertices of *ABC*. They are $A(3, 15)$, $B(9, 9)$, and $C(9, 5)$.

6 Use each vertex of the shaded region to find values for the optimization equation. Then select the equation with the lowest cost.

Substitute $A(3, 15)$.

    Total cost = $0.40n + $0.10r$

    Total cost = $0.40(3) + $0.10(15)$

    Total cost = $1.20 + $1.50 = $2.70

Substitute $B(9, 9)$.

    Total cost = $3.60 + $0.90 = $4.70

Substitute $C(9, 5)$.

    Total cost = $3.60 + $0.50 = $4.10

Three ounces of nuts and 15 ounces of raisins gives a minimum cost of $2.70.

Activity
Solve Using the Method of Linear Programming

The maker of recreational motorcycles lists two models, the Mountain Climber and the Dune Crawler. The chart shows some of the assembly information.

	Labor Hours	Maximum produced	Profit
Mountain Crawler	150	60	$120
Dune Crawler	200	45	$180

If the company has no more than 12,000 hours of labor available for production of motorcycles, find the number of each model that should be built to give the maximum profit.

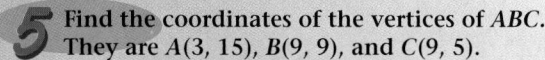

20 Mountain Climbers; 45 Dune Crawlers

Chapter 6 Review

Vocabulary

approximate solution	264	dependent system	280	linear inequalities	284
boundary line	284	elimination method	272	substitution method	267
common solution	261	inconsistent system	280	system of equations	261
consistent system	280	independent system	280		

Key Skills & Exercises

Lesson 6.1

➤ **Key Skills**

Find a solution to a system of equations by graphing the equations and finding the point of intersection.

Solve by graphing. $\begin{cases} x + 3y = 12 \\ x - y = 4 \end{cases}$

Write the equations in slope-intercept form.

$$x + 3y = 12 \qquad x - y = 4$$
$$y = -\frac{x}{3} + 4 \qquad y = x - 4$$

Graph the lines on graph paper or with a graphics calculator. The solution is (6, 2). Check by substituting.

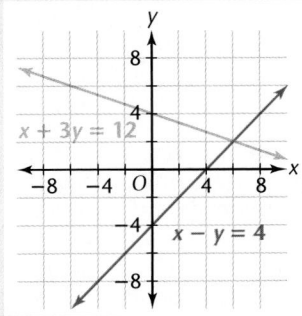

➤ **Exercises**

Solve each system by graphing.

1. $\begin{cases} x - y = 6 \\ x + 2y = -9 \end{cases}$
2. $\begin{cases} y = -x \\ y = 2x \end{cases}$
3. $\begin{cases} x + 6y = 3 \\ 3x + y = -8 \end{cases}$
4. $\begin{cases} 3x + y = 6 \\ y + 2 = x \end{cases}$

Lesson 6.2

➤ **Key Skills**

Find a solution to a system of equations by the substitution method.

Solve. $\begin{cases} c + 4d = 1 \\ 2c + 7d = 6 \end{cases}$

Solve $c + 4d = 1$ for c to get $c = 1 - 4d$.
Then substitute in $2c + 7d = 6$.

$$2(1 - 4d) + 7d = 6$$
$$d = -4$$

Now substitute -4 for d in the equation $c = 1 - 4d$.

$$c = 1 - 4(-4)$$
$$c = 17$$

The solution is $(17, -4)$.

1. $(1, -5)$

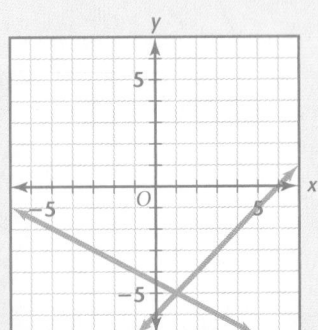

2. $(0, 0)$

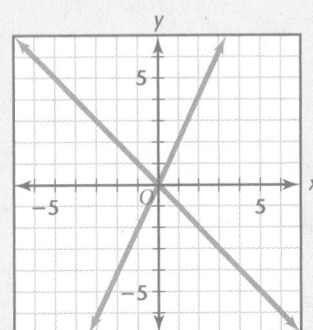

3. $(-3, 1)$

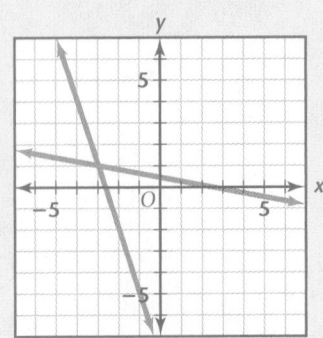

4. $(2, 0)$

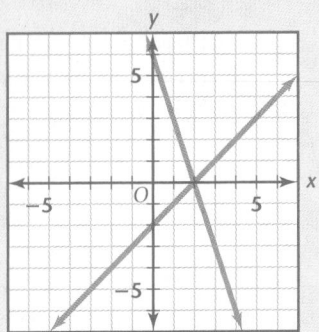

> ### Exercises

Solve each system by the substitution method.

5. $\begin{cases} 2x + y = 1 \\ x + y = 2 \end{cases}$
$(-1, 3)$

6. $\begin{cases} x - y = 6 \\ 2x - 4y = 28 \end{cases}$
$(-2, -8)$

7. $\begin{cases} x + 2y = 1 \\ 2x - 8y = -1 \end{cases}$
$\left(\frac{1}{2}, \frac{1}{4}\right)$

8. $\begin{cases} 4x = 3y + 44 \\ x + y = -3 \end{cases}$
$(5, -8)$

Lesson 6.3

> ### Key Skills

Find a solution to a system of equations by the elimination method.

Solve by the elimination method.

$$\begin{cases} x + y = -3 \\ 2x = 3y - 11 \end{cases}$$

Write the equations in standard form, and compare the coefficients.

$\begin{array}{ll} x + y = -3 & \text{Equation 1} \\ 2x - 3y = -11 & \text{Equation 2} \end{array}$

Choose a variable to eliminate. Multiply both sides of one of the equations by the same number, and solve for the remaining variable.

$$\begin{aligned} 3(x + y) = 3(-3) &\rightarrow 3x + 3y = -9 \\ 2x - 3y = -11 &\rightarrow \underline{2x - 3y = -11} \\ & \quad\quad\quad 5x \quad\quad = -20 \\ & \quad\quad\quad\quad x = -4 \end{aligned}$$

Substitute this value of x in $x + y = -3$ to find the value of y.

$$\begin{aligned} -4 + y &= -3 \\ y &= 1 \end{aligned}$$

The solution is $(-4, 1)$.

> ### Exercises

Solve each system by the elimination method.

9. $\begin{cases} 4x + 5y = 3 \\ 2x + 5y = -11 \end{cases}$
$(7, -5)$

10. $\begin{cases} 2x + 3y = -6 \\ -5x - 9y = 14 \end{cases}$
$\left(-4, \frac{2}{3}\right)$

11. $\begin{cases} 0.5x + y = 0 \\ 0.9x - 0.2y = -2 \end{cases}$
$(-2, 1)$

12. $\begin{cases} -2x + 4y = 12 \\ 3x - 2y = -10 \end{cases}$
$(-2, 2)$

Lesson 6.4

> ### Key Skills

Identify a system of equations as independent, dependent, or inconsistent by its solution.

Solve the system algebraically.

$$\begin{cases} 2x + y = 4 \\ 2x + y = 1 \end{cases}$$

Use the elimination method.

$$\begin{aligned} 2x + y &= 4 \\ -1(2x + y) &= -1(1) \end{aligned}$$

$$\begin{aligned} 2x + y &= 4 \\ \underline{-2x - y} &= \underline{-1} \\ 0 &\neq 3 \end{aligned}$$

All variables are eliminated, and you are left with two unequal numbers. There is no possible solution for this system of equations. The system is inconsistent.

> ### Exercises

Solve each system. State your solution set clearly, and identify the system as independent, dependent, or inconsistent.

13. $\begin{cases} 3x - 2y = 7 \\ 4y = -14 + 6x \end{cases}$

14. $\begin{cases} 4y = 2x + 20 \\ 3x - y = -20 \end{cases}$

15. $\begin{cases} 2x + 5y = 7 \\ 3y = 2x + 17 \end{cases}$

16. $\begin{cases} x + y = 7 \\ 28 - 2y = 2x \end{cases}$

13. Infinite number of solutions; dependent system

14. $(-6, 2)$; independent system

15. $(-4, 3)$; independent system

16. no solution; inconsistent system

Lesson 6.5

➤ **Key Skills**

Graph the solution for a system of linear inequalities.

Graph the solution set for the system.
$$\begin{cases} y < -x - 1 \\ y \leq x + 1 \end{cases}$$

Graph $y < -x - 1$ as a dashed line and $y \leq x + 1$ as a solid line. Use substitution to determine which half-plane to shade for each inequality. The intersection of the solution sets for the two inequalities is the solution set for the system.

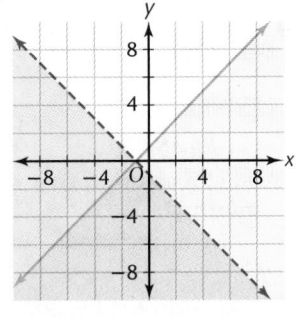

➤ **Exercises**

Graph the common solution. Choose a point to check.

17. $\begin{cases} y \geq 4 \\ y \leq x + 1 \end{cases}$ **18.** $\begin{cases} 4x - y > 1 \\ 2x + y > -2 \end{cases}$ **19.** $\begin{cases} y - 1 \leq x \\ 4y - 2x \geq 2 \end{cases}$ **20.** $\begin{cases} 2x - y < 3 \\ 2y - 2 \leq 6x \end{cases}$

Lesson 6.6

➤ **Key Skills**

Use a system of equations to solve classic problems.

Finance Mr. Carver invested $5000, part at 5% interest and the rest at 3.9% interest. If he earned $233.50 in interest the first year, how much did he invest at each rate?

Make a table to organize the information.

Amount	Rate	Interest
x	5%	$0.05x$
y	3.9%	$0.039y$
Total 5000		233.50

From the table, write two equations.

$$x + y = 5000$$
$$0.05x + 0.039y = 233.50$$

Solving the system gives $x = 3500$ and $y = 1500$.

Mr. Carver invested $3500 at 5% and $1500 at 3.9%.

➤ **Exercises**

Use a system of equations to solve classic problems.

21. Samantha has 43 quarters and nickels in her wallet. If her change is worth $7.75, how many of each kind of coin does Samantha have? 15 nickels; 28 quarters

22. Finance Tina invested $600 of her bonus at 4.3% and the rest at 9% interest. If she earned $97.80 in interest in one year, how much was her bonus? $1,400

Applications

23. Chemistry How many ounces of a 25% acid solution should be mixed with a 50% acid solution to produce 100 ounces of a 40% acid solution? 40 ounces

24. Geometry The sum of the measures of angles x, y, and z of triangle XYZ is 180°. Angle x has a measure of 60°. The measure of angle y is 15 more than twice the measure of the angle z. What are the measures for angles y and z? 85° and 35°

17.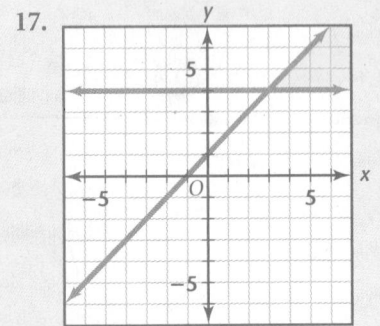

18.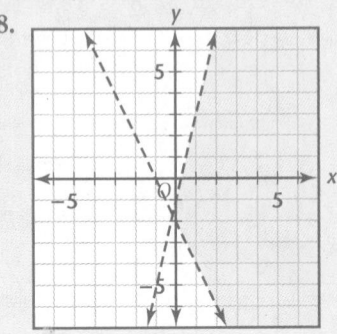

19.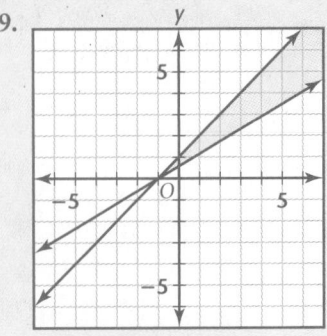

20.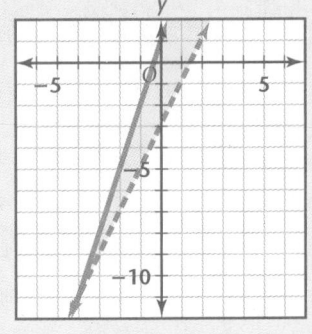

Chapter 6 Assessment

Solve each system by graphing.

1. $\begin{cases} 2x = y + 1 \\ y = 3x + 2 \end{cases}$

2. $\begin{cases} 4y - 2x = -2 \\ 4y - x = 0 \end{cases}$

3. $\begin{cases} 3y - 12x = 18 \\ 3x + y = -1 \end{cases}$

Solve each system by the substitution method.

4. $\begin{cases} y = x + 3 \\ 2x - 4y = -12 \end{cases}$
(0, 3)

5. $\begin{cases} 2x - 2y = -6 \\ x + 2y = -9 \end{cases}$
(−5, −2)

6. $\begin{cases} 2x - 3y = -2 \\ 4x + y = 1 \end{cases}$
$\left(\dfrac{1}{14}, \dfrac{5}{7}\right)$

7. One number is 4 greater than another number. Twice the larger number is 2 less than 3 times the smaller number. Find both numbers. 10 and 14

Solve each system by the elimination method.

8. $\begin{cases} 2x + 3y = 3 \\ 4x - 3y = 3 \end{cases}$
$\left(1, \dfrac{1}{3}\right)$

9. $\begin{cases} 3x + 4y = -2 \\ x - 2y = 6 \end{cases}$
(2, −2)

10. $\begin{cases} 5y = 3x - 18 \\ 2x = 3y + 12 \end{cases}$
(6, 0)

11. The length of a rectangle is 3 inches more than its width. Find the length and width of this rectangle. 7 × 4

Perimeter is 22 inches.

12. Marla bought 12 books at a garage sale. Some of them were hardback and the rest were paperback. She paid 50¢ for each paperback book and 75¢ for each hardback book. If she spent $6.75, how many of each type of book did she buy? 3 hardback and 9 paperback

Solve each system algebraically. State your solution clearly.

13. $\begin{cases} y + 5 = x \\ y = x + 2 \end{cases}$
none

14. $\begin{cases} 5y - 2x = -15 \\ 4x = 10y + 30 \end{cases}$
infinite

15. $\begin{cases} 4x + 5y = 14 \\ 5y = 8x + 2 \end{cases}$
(1, 2)

Graph the common solution of each system. Choose a point from the solution, and check both inequalities.

16. $\begin{cases} y > x \\ y - 2x \le 2 \end{cases}$

17. $\begin{cases} y \ge 3x - 1 \\ y < 5 \end{cases}$

18. $\begin{cases} 2y + 10x > -12 \\ y + 5 < 4x \end{cases}$

19. Yvette has $16.30 in dimes and quarters. If she has a total of 88 coins, how many of each type of coin does she have? 38 dimes and 50 quarters

20. Henry divides Hannah's current age by 2 and then adds that result to her current age. The result is Henry's current age. Henry notices that in 24 years, his age will be 3 times Hannah's current age. How old are Henry and Hannah? 16 and 24

21. Tamara drove 810 miles in 14 hours over 2 days. If she drove at an average rate of 55 miles per hour the first day and an average rate of 60 miles per hour the second day, how far and how long did she drive each day? 6 hours; 330 miles and 8 hours; 480 miles

1.
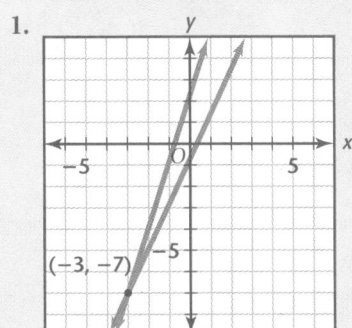

2.
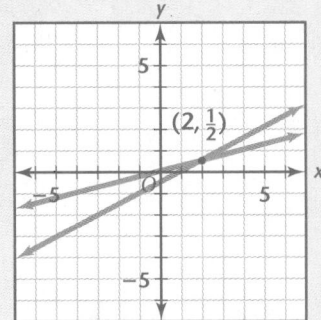

3.
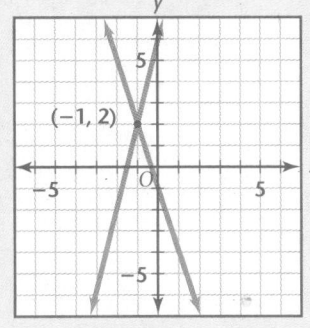

The answers to items 16–18 can be found in Additional Answers beginning on page 729.

**Multiple-Choice and
Quantitative-
Comparison Samples**

The first half of the Cumulative Assessment contains two types of items found on standardized tests—multiple-choice questions and quantitative-comparison questions. Quantitative-comparison items emphasize the concepts of equalities, inequalities, and estimation.

Free-Response Grid Samples

The second half of the Cumulative Assessment is a free-response section. A portion of this part of the Cumulative Assessment consists of student-produced response items commonly found on college entrance exams. These questions require the use of machine-scored answer grids. You may wish to have students practice answering these items in preparation for standardized tests.

Sample answer-grid masters are available in the *Chapter Teaching Resources Booklets.*

Chapters 1-6 Cumulative Assessment

College Entrance Exam Practice

Quantitative Comparison For Questions 1–5, write

A if the quantity in Column A is greater than the quantity in Column B;
B if the quantity in Column B is greater than the quantity in Column A;
C if the two quantities are equal; or
D if the relationship cannot be determined from the information given.

	Column A	Column B	Answers
1.	$\begin{cases} 3x + 2y = 6 \\ x + y = 0 \end{cases}$ x	y	A Ⓐ Ⓑ Ⓒ Ⓓ **[Lesson 6.2]**
2.	0.37	$\dfrac{37}{10}$	B Ⓐ Ⓑ Ⓒ Ⓓ **[Lesson 4.5]**
3.	The eighth term of the sequence 2, 5, 8, 11, 14, . . .	The eighth term of the sequence 32, 30, 28, 26, 24, . . .	A Ⓐ Ⓑ Ⓒ Ⓓ **[Lesson 1.1]**
4.	The solution to $x + 3 \geq 4$	The solution to $x - 7 < 10$	D Ⓐ Ⓑ Ⓒ Ⓓ **[Lesson 4.7]**
5.	The slope of the line passing through the points (1, 2) and (0, 3).	The slope of the line $4x + 4y = 12$.	C Ⓐ Ⓑ Ⓒ Ⓓ **[Lesson 5.1]**

6. Which expression is the solution to the equation $-4g = -64$? **[Lesson 4.2]** c

    a. $\dfrac{-64}{4}$ **b.** $-64 \cdot (-4)$ **c.** $\dfrac{-64}{-4}$ **d.** -16

7. What is the reciprocal of $-\dfrac{1}{6}$? **[Lesson 2.3]** d

    a. 1.6   **b.** -1.6 **c.** 6   **d.** -6

8. What are the first three terms of a sequence if the fourth and fifth terms are 27 and 39 and the second differences are a constant 2? **[Lesson 1.2]** b

    a. 21, 23, 25   **b.** 3, 9, 17   **c.** 6, 8, 10   **d.** 13, 15, 17

9. If a basketball league has 8 teams and each of the teams plays each of the other teams, how many games will be played? **[Lesson 1.1]** a

    a. 28   **b.** 36   **c.** 72   **d.** 56

10. What is ABS(8.6)? **[Lesson 2.4]** c

    a. 8   **b.** -8 **c.** 8.6   **d.** -8.6

11. Which point is the vertex of the graph of the data in the table? **[Lesson 2.2]** a

x	1	2	3	4	5
y	1	0	1	4	9

    a. $(2, 0)$ **b.** $(1, 1)$ **c.** $(3, 1)$ **d.** $(4, 4)$

12. Which number line shows the solution set to the inequality $x - 6 > -4$? **[Lesson 4.8]** d

a.
-5 -4 -3 -2 -1 0 1 2 3 4 5

b.
-5 -4 -3 -2 -1 0 1 2 3 4 5

c.
-5 -4 -3 -2 -1 0 1 2 3 4 5

d.
-5 -4 -3 -2 -1 0 1 2 3 4 5

13. What is the solution to $\frac{3}{5}p = -3$? **[Lesson 4.2]** c

    a. $-\frac{9}{5}$ **b.** $-\frac{5}{9}$ **c.** -5 **d.** $-\frac{1}{5}$

14. What is 65% of 120? **[Lesson 4.5]** a
    a. 78   **b.** 185   **c.** 7800   **d.** 780

15. Which of the following lines is vertical? **[Lesson 5.5]** b
    a. $y = -5$ **b.** $x = 3$ **c.** $y = -3x + 1$ **d.** $y = 2x - 2$

Evaluate the following. [Lesson 2.4]
16. $|-2.2|$ 2.2   **17.** ABS(7) 7   **18.** INT(9.6) 9
19. Does the exponential sequence 4, 8, 16, 32, 64 show growth or decay? Explain. **[Lesson 2.2]** growth

Find each sum. [Lesson 3.1]
20. $-25 + (-4)$ **21.** $-36 + 6 + |-6|$ **22.** $452 + (-452)$
    -29 -24 0

Solve each equation. [Lesson 4.7]
23. $t + \frac{1}{6} = \frac{1}{3}$ $\frac{1}{6}$ **24.** $2a - 5 = 11$ 8   **25.** $\frac{r}{4} + 6 = -4$ -40

26. What is the y-intercept of a line that is parallel to the line $2x + 3y = 4$ and contains the point $(3, -1)$? **[Lesson 5.6]** 1

Solve each system. [Lessons 6.1, 6.2, 6.3]
27. $\begin{cases} y = 7x \\ 2x - y = -10 \end{cases}$ $(2, 14)$ **28.** $\begin{cases} 5x + 2y = 8 \\ x + y = 1 \end{cases}$ $(2, -1)$

Free-Response Grid The following questions may be answered using a free-response grid commonly used by standardized test services.

29. The sum of the digits of a 2-digit number is 7. The tens digit is 1 less than the units digit. Find the number. 34

30. What is the solution to the proportion $\frac{c}{16} = \frac{3}{8}$? **[Lesson 4.5]** 6

31. A suit is marked down from an original price of $75 to $63.75. By what percent of the original price has the sweater been marked down? **[Lesson 4.5]** 15%

7 CHAPTER

Matrices

Meeting Individual Needs

7.1 Matrices to Display Data

Core Resources	Core Two-Year Resources
Practice Master 7.1	Inclusion Strategies, p. 310
Enrichment, p. 310	Reteaching the Lesson, p. 311
Technology Master 7.1	Practice Master 7.1
Interdisciplinary	Enrichment Master 7.1
Connection, p. 309	Technology Master 7.1
	Lesson Activity Master 7.1
[1 day]	[2 days]

7.2 Adding and Subtracting Matrices

Core Resources	Core Two-Year Resources
Practice Master 7.2	Inclusion Strategies, p. 317
Enrichment, p. 316	Reteaching the Lesson, p. 318
Technology Master 7.2	Practice Master 7.2
	Enrichment Master 7.2
	Technology Master 7.2
	Lesson Activity Master 7.2
[1 day]	[2 days]

7.3 Exploring Matrix Multiplication

Core Resources	Core Two-Year Resources
Practice Master 7.3	Inclusion Strategies, p. 326
Enrichment, p. 326	Reteaching the Lesson, p. 327
Technology Master 7.3	Practice Master 7.3
Interdisciplinary	Technology Master 7.3
Connection, p. 325	Lesson Activity Master 7.3
Mid-Chapter Assessment	Mid-Chapter Assessment Master
Master	
[2 days]	[3 days]

7.4 Multiplicative Inverse of a Matrix

Core Resources	Core Two-Year Resources
Practice Master 7.4	Inclusion Strategies, p. 333
Enrichment, p. 332	Reteaching the Lesson, p. 333
Technology Master 7.4	Practice Master 7.4
	Lesson Activity Master 7.4
	Technology Master 7.4
[2 days]	[3 days]

7.5 Solving Matrix Equations

Core Resources	Core Two-Year Resources
Practice Master 7.5	Inclusion Strategies, p. 339
Enrichment, p. 338	Reteaching the Lesson, p. 340
Technology Master 7.5	Practice Master 7.5
	Lesson Activity Master 7.5
	Technology Master 7.5
[2 days]	[3 days]

Chapter Summary

Core Resources	Core Two-Year Resources
Chapter 7 Project, pp. 344–345	Chapter 7 Project, pp. 344–345
Lab Activity	Lab Activity
Long-Term Project	Long-Term Project
Chapter Review, pp. 346–348	Chapter Review, pp. 346–348
Chapter Assessment, p. 349	Chapter Assessment, p. 349
Chapter Assessment, A/B	Chapter Assessment, A/B
Alternative Assessment	Alternative Assessment
[3 days]	[5 days]

Reading Strategies

Be sure to continually evaluate students' ability to read and comprehend the material. Assign a partner to help students who struggle with the reading. After students have read each lesson, have them respond to the following questions in their journals before they begin to complete the exercises provided:

• What is the main idea of this lesson?

• List and describe any words, phrases, equations, or graphs that you have not seen before. Why do you think they are important?

• Give at least one application of this lesson's main idea that is applicable to everyday life.

As you go through each lesson as a class, ask students to record their responses to the Explorations, Critical Thinking, and Ongoing Assessment questions in their journals.

Visual Strategies

Matrices are used every day to organize and store data. To find data in a matrix, simply look in the correct row and column. Students will find that organizing data in a matrix is similar to using a table to organize data. In Lesson 7.1, students are shown data that is organized in a

table before it is written in a matrix. You may want students to continue to use a table before putting data in a matrix until they understand the concept. Progress to writing the matrices with labels. Only when students feel completely comfortable with matrices should they be encouraged to write matrices without labels. The more comfortable students become with matrices now, the easier their work with matrices will be in the future.

Hands-on Strategies

As students will quickly discover, graphics calculators and computer software such as $f(g)$ Scholar™ simplify the task of computation with matrices. However, students should still be able to recognize when matrices can be added, subtracted, or multiplied. Matrices that can be added or subtracted need only have the same dimensions, so they should be easily recognized. Recognizing matrices that can be multiplied can be slightly more difficult for students. To help students grasp this concept, have students circle the first row of the first matrix and record the number of elements in that row. Then have them circle the first column of the second matrix and record the number of elements in that column. If these numbers match, the matrices can be multiplied. An example illustrating this process is given below.

$$\begin{bmatrix} 3 & 3 & -4 \\ 1 & -2 & 5 \end{bmatrix} \begin{bmatrix} 7 & 8 & 1 \\ 3 & 6 & -2 \\ -4 & 3 & -1 \end{bmatrix}$$

Since $3 = 3$, these matrices can be multiplied.

Cooperative Learning

GROUP ACTIVITIES	
Using matrices to organize information	Chapter Opener, The Portfolio Activity
Organizing and analyzing data	Eyewitness Math
Addition and subtraction properties	Lesson 7.2 Exploration
Multiplying matrices	Lesson 7.3 Explorations 1, 2, and 3
Coding and decoding messages	Chapter 7 Project

You may wish to have students work with partners for some of the above activities. Additional suggestions for cooperative group activities are noted in the teacher's notes in each lesson.

Multicultural

The Cultural Connections in this chapter include references to the Americas and Asia.

CULTURAL CONNECTIONS	
Americas: Spanish problem	Lesson 7.1, page 314
Asia: Chinese method for solving systems in matrix form	Lesson 7.5, page 341

Portfolio Assessment

Below are portfolio activities for the chapter listed. The activities are listed in seven activity domains that are appropriate for portfolio development.

1. **Investigation/Exploration** Three projects listed in the teacher notes ask students to investigate a method and/or a relationship. They are the *Enrichment* on pages 316 (Lesson 7.2), 326 (Lesson 7.3), and 332 (Lesson 7.4). One project explores the history of mathematics. It is the *Enrichment* on page 310 (Lesson 7.1).

2. **Applications** Recommended to be included are any of the following: Transportation (Lesson 7.1), Geography (Lesson 7.1), Space exploration (Lesson 7.1), Inventory control (Lessons 7.2, 7.3, and 7.5), Sports (Lesson 7.2), Fund-raiser (Lesson 7.3), Manufacturing (Lesson 7.3), Pricing (Lesson 7.3), Sales tax (Lesson 7.3), Consumer economics (Lesson 7.5), Job opportunities (Lesson 7.5), and Ticket sales (Lesson 7.5); and from the teacher notes, *Sports* on page 309 (Lesson 7.1).

3. **Non-Routine Problems** The Portfolio Activity on page 307 asks students to organize given information into matrix form to solve a problem. Initially, this problem may seem to be too confusing to students. However, as they begin to think about organizing the information into ma-

trix form, it should begin to make sense and at the same time allow students to experience the advantages matrices provide in problem solving.

4. **Project** The Chapter Project on pages 344–345 introduces students to the use of matrices in the science of cryptography. Most students will find this project intriguing and very motivating.

5. **Interdisciplinary Topics** Students may choose from Business (Lesson 7.1), Geography (Lesson 7.1), and Chemistry (Lesson 7.5) in the student text, and from *Business* on page 325 (Lesson 7.3) in the teacher notes.

6. **Writing** The interleaf material at the beginning of this chapter has information for journal writing. Also recommended are the Communicate questions at the beginning of the exercise sets.

7. **Tools** Chapter 7 uses calculators or computer software such as *f(g) Scholar*™. The technology masters are recommended to measure the individual student's proficiency.

Technology

Graphics calculators and computer software such as *f(g) Scholar*™ provide students with an opportunity to explore matrices and their sums, differences, products, and inverses without performing extensive calculations with pencil and paper. Included among the advantages to using such technology is the reduction of some of the frustration students often experience when working with matrices and the opportunity to focus on the use of matrices in problem solving. For more information on the use of technology, refer to the *HRW Technology Handbook*.

Graphics Calculator

Entering and Multiplying Matrices Using a graphics calculator to multiply matrices enables students to concentrate on using the matrices rather than the tedious and time-consuming calculations that can be involved. General instructions for using a graphics calculator are given in individual lessons for the concepts covered in that lesson. Specific instructions are given below for the TI-82. For instructions for using other graphics calculators, please refer to the *HRW Technology Handbook*.

1. Press the MATRX key. To enter the elements of matrix A, Choose EDIT 1: and press ENTER.

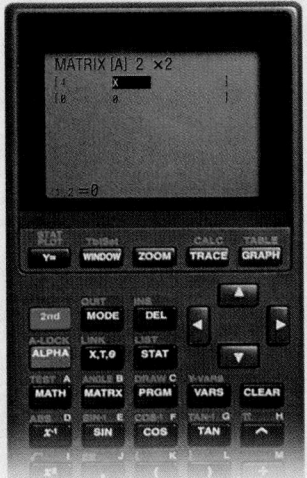

2. Enter the appropriate dimensions for matrix A.

3. To enter the elements for matrix A, start with a_{11} and enter the elements by rows. Notice that the row and column of the element being entered are given at the bottom of the screen in the form **row, column = entry,** for example, **1,1=4.** Repeat steps 1–3 to enter the elements for matrix B, except choose EDIT 2: in step 1.5. To multiply matrix A and matrix B, press 2nd QUIT to return to the Home screen. Then press MATRX, choose NAMES 1:, and press ENTER. The Home screen appears showing [A]. Press ×. Then press MATRX, choose NAMES 2:, and press ENTER. The Home screen appears showing [A]*[B]. Press ENTER and the product of the matrices will appear on the Home screen.

Integrated Software

f(g) Scholar™ is an integrated computer-based mathematics productivity tool that combines calculator, spreadsheet, and graphics capabilities to provide a dynamic and interactive environment for explorations in mathematics. It is appropriate to use *f(g) Scholar*™ for any lesson needing a spreadsheet, calculator, graphics calculator, or any combination of the three.

ABOUT THE CHAPTER

Background Information

Mathematicians, scientists, and engineers have known for many decades that matrices provide an excellent way to organize data. Too much information, however, can be a problem. For example, a large matrix with 512 rows and 171 columns requires powerful technology to handle even well-organized data.

CHAPTER RESOURCES

- Practice Masters
- Enrichment Masters
- Technology Masters
- Lesson Activity Masters
- Lab Activity Masters
- Long-Term Project Masters
- Assessment Masters
  Chapter Assessments, A/B
  Mid-Chapter Assessment
  Alternative Assessments, A/B
- Teaching Transparencies
- Spanish Resources

CHAPTER OBJECTIVES

- Interpret data from given information, and write the data in table or matrix form.
- Determine the dimensions and addresses of a matrix.
- Determine whether two matrices are equal.
- Add and subtract matrices.
- Determine whether the Identity Property for Addition is true for matrices.

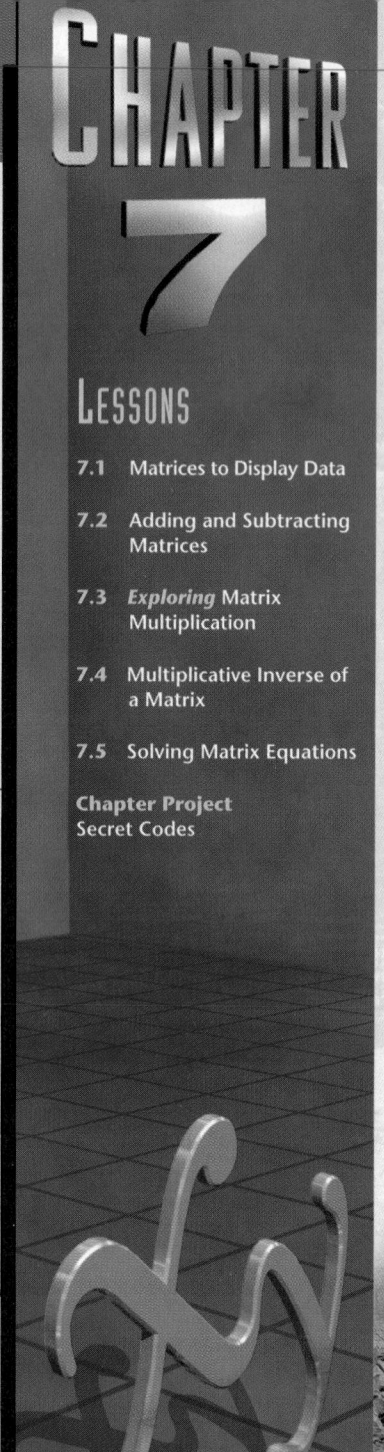

CHAPTER 7 Matrices

LESSONS

7.1 Matrices to Display Data

7.2 Adding and Subtracting Matrices

7.3 *Exploring* Matrix Multiplication

7.4 Multiplicative Inverse of a Matrix

7.5 Solving Matrix Equations

Chapter Project
Secret Codes

Ancient Nubians of Sudan left spectacular pyramids and tombs containing precious artifacts. They left us a puzzle as well. In 1960, the building of the Aswan High Dam on the Nile River meant that the tombs would soon be flooded. Archaeologists from all countries rushed in to excavate the sites and save the artifacts.

Millions of artifacts were recovered, representing civilizations from the Paleolithic to the Middle Ages. But what story do they tell? Researchers used a matrix to organize and classify the treasures at the Fadrus site. The rows and columns of the matrix quickly filled with numbers. These numbers represented the amount of artifacts with similar characteristics found at burial sites. Was it a heart scarab from Fadrus or spiral earrings from Kerma? It was all recorded.

There was a problem with the size of the matrix, 512 rows and 171 columns. The matrix did an excellent job of organizing the data, but in the 1960s, no computer was powerful enough to do the mathematical analysis. For years the artifacts remained unstudied in museum basements. It was not until the 1990s that a computer was available that could handle the analysis and begin to reveal the secrets of these ancient civilizations.

ABOUT THE PHOTOS

The background for the Chapter Opener is an aerial photograph of the Aswan Dam area before the construction of the dam. In the foreground are archaeologists excavating at the dam sight. The other photos are of relics unearthed from various Mideast ruins: a likeness of a human head which is the lid of a canopic urn, two views of a heart-shaped scarab, a primitive rendering of a hawk, and a comb decorated with a lion.

- Multiply a matrix by a scalar, and multiply matrices.
- Determine the identity matrix for multiplication.
- Determine the inverse matrix for multiplication.
- Use technology to find the inverse of a matrix.
- Solve a system of equations using matrices.
- Use technology to solve a system of equations using matrices.

PORTFOLIO ACTIVITY

Assign the Portfolio Activity as part of the assignment for Lesson 7.5 on page 343.

Begin this activity by asking students to organize the information given in tables. You may want to provide at least one table with the headings in place to get students started. Then have students progress to writing the matrices.

Have students work in small groups. Ask each group to work together to organize the information into matrix form. Be sure all students participate in the work. Have each group present their final matrices to the class, describing any strategies they used to get to the final matrix.

As an extension to this activity, have each group do research to find data that can be organized using matrices.

PORTFOLIO ACTIVITY

The many sites on the Aswan High Dam flood plain were divided into three different culture groups: C-Group, Pan-grave, and Transitional. Excitement mounted when excavators realized that some early C-Group sites were royal burial sites. The kings buried at these sites may have been the first pharaohs. Thousands of artifacts needed to be classified. The initial categories were tomb materials, pottery, and nonceramic pieces.

The typical C-Group site might contain 5 pieces of tomb materials, 8 pieces of decorated pottery, and 15 non-ceramic artifacts. These nonceramic pieces were often decorative hair clasps, the most identifying feature of the C-Group sites. The average Pan-grave site would contain 9 tomb materials, 5 pottery pieces, and 12 nonceramic artifacts. The Pan-grave sites were the only places where nonceramic Kohl pots were found. At the Transitional sites, an average of 18 tomb materials, 3 bits of pottery, and 3 nonceramic artifacts were found. The Transitional sites seem to be characterized by burial rituals.

At the end of this phase of digging, the total number of artifacts in each of the categories was 175 tomb materials, 104 pottery pieces, and 195 nonceramic artifacts.

Organize this information into matrix form, and find how many of each type of site were included in this part of the study.

ABOUT THE CHAPTER PROJECT

The Chapter 7 project on pages 344–345 asks students to use the skills they have acquired in this lesson to code and decode messages. This can be an interesting and enjoyable application of matrix multiplication. The final activity of the project gives students the opportunity to encode and send a short message of their own.

- Interpret data from given information, and write the data in table or matrix form.
- Determine the dimensions and addresses of a matrix.
- Determine whether two matrices are equal.

CHAPTER RESOURCES

• Practice Master	**7.1**
• Enrichment Master	**7.1**
• Technology Master	**7.1**
• Lesson Activity Master	**7.1**
• Quiz	**7.1**
• Spanish Resources	**7.1**

Assessing Prior Knowledge

Solve each equation.

1. $11x = 66$ [6]

2. $0.3x = 27$ [90]

3. $4x + 3 = -17$ [-5]

4. $4(x - 2) = -6$ [0.5]

TEACH

Ask students to think about why it is important to have both their possessions and their thoughts well organized. In both cases, poor organization can result in the inability to find what one needs. Then ask students to describe ways they have organized information in this mathematics course. Use this discussion to lead into the introduction of matrices.

LESSON 7.1 Matrices to Display Data

why *Do you know exactly where all your papers and projects are? You can easily find what you are looking for when you are organized. You have already seen how tables of values, graphs, and spreadsheets are used in problem solving. Now you will use a closely related and very important organizational tool called a matrix.*

Organizing Data

Imagine you are writing a paper on the effect that automobiles have had on society. You want to include the following data as evidence.

Transportation

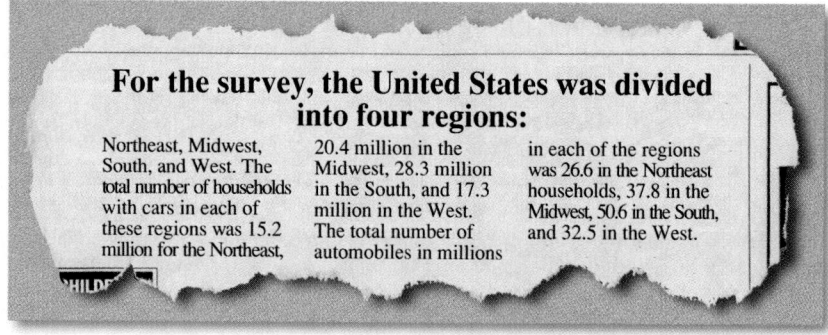

For the survey, the United States was divided into four regions:

Northeast, Midwest, South, and West. The total number of households with cars in each of these regions was 15.2 million for the Northeast, 20.4 million in the Midwest, 28.3 million in the South, and 17.3 million in the West. The total number of automobiles in millions in each of the regions was 26.6 in the Northeast households, 37.8 in the Midwest, 50.6 in the South, and 32.5 in the West.

The amount of information in the paragraph is difficult to interpret. However, when organized in a table the same information is easier to read and understand.

Region	Number of households in millions	Number of vehicles in millions
Northeast	15.2	26.6
Midwest	20.4	37.8
South	28.3	50.6
West	17.3	32.5

$$C = \begin{bmatrix} 15.2 & 26.6 \\ 20.4 & 37.8 \\ 28.3 & 50.6 \\ 17.3 & 32.5 \end{bmatrix}$$

A matrix is named with a capital letter.

When data are arranged in a table of rows and columns and enclosed by brackets, [ ], the structure is a **matrix** (plural, matrices).

Labels describe the information in the rows and columns, but are not part of the matrix.

Matrix C has 4 rows and 2 columns. The number of rows by the number of columns describes the *row by column* dimensions of the matrix. This is sometimes written 4×2 or $C_{4 \times 2}$.

The information that appears in each position of the matrix is called an **entry**. Each entry in the matrix can be located by its matrix **address**.

Use Transparency 31

	Number of households in millions	Number of vehicles in millions			Addresses for matrix C Column 1	Column 2
Northeast	15.2	26.6		Row 1	c_{11}	c_{12}
Midwest	20.4	37.8	$= C$	Row 2	c_{21}	c_{22}
South	28.3	50.6		Row 3	c_{31}	c_{32}
West	17.3	32.5		Row 4	c_{41}	c_{42}

*4 rows by 2 columns
The dimension is 4×2.*

The address c_{32} represents the entry in row 3 and column 2 of matrix C. You will find 50.6 at this address. The labels show that c_{32} represents the fact that in the *South* there are 50.6 *million vehicles*.

What is the address in matrix C for the entry 37.8?

interdisciplinary  **Sports** Matrices are often used to keep track of statistics for different sporting events. Have students do research to find data about a sporting event that they think can be written in matrix form. Suggest that stu-

dents look in newspapers, magazines, almanacs, or use an on-line service. Have students make a poster showing their data in matrix form. Be sure they include row and column headings, as well as a title for the matrix.

Alternative Example 1

Information about temperatures is in table form. This is easily changed to the format of a matrix. Answer the following questions about the location of the information in the temperature matrix.

Predicted Temperatures for May 10 (in °F)

$$T = \begin{array}{c} \\ \text{Amsterdam} \\ \text{Beijing} \\ \text{Brussels} \\ \text{Dublin} \\ \text{Istanbul} \\ \text{Rome} \\ \text{Tokyo} \end{array} \begin{array}{cc} \text{Hi} & \text{Lo} \\ \begin{bmatrix} 57 & 45 \\ 72 & 46 \\ 54 & 46 \\ 57 & 37 \\ 70 & 55 \\ 73 & 52 \\ 75 & 59 \end{bmatrix} \end{array}$$

a. What do the entries in matrix T represent? [**the high or low temperature in a different city around the world**]

b. What are the row and column dimensions of T? [**7 by 2**]

c. What was the high temperature in Istanbul on this date? What is the matrix address of this entry? [**70°F; t_{51}**]

d. What does the entry t_{32} represent? [**t_{32} is 46, the low temperature predicted for Brussels on May 10.**]

e. What does the entry t_{71} represent? [**t_{71} is 75, the high temperature predicted for Tokyo on May 10.**]

Aongoing ASSESSMENT

Try This

4139; d_{25}, d_{52}

EXAMPLE 1

Information in an atlas is in table form. This is easily changed to the format of a matrix. Answer the following questions about the location of the information in the matrix.

Airline Distances in Kilometers

	London	Mexico City	New York	Paris	San Francisco	Tokyo
London	0	8944	5583	344	8637	9590
Mexico City	8944	0	3363	9213	3037	11,321
New York	5583	3363	0	5851	4139	10,874
Paris	344	9213	5851	0	8975	9741
San Francisco	8637	3037	4139	8975	0	8288
Tokyo	9590	11,321	10,874	9741	8288	0

$= D$

A What do the entries in matrix D represent?

B What are the row and column dimensions of D?

C How many kilometers is it from Paris to Mexico City? What is the matrix address of this entry?

D What does the entry d_{36} represent?

E What does the entry d_{22} represent?

Solution ➤

A Each entry in the matrix represents the distance measured in kilometers between the cities whose names appear in the row and column headings.

B D has dimensions 6 by 6. A matrix is called a **square matrix** if its row and column dimensions are equal.

C It is 9213 kilometers between Paris and Mexico City. This information is located at d_{42} and d_{24}.

D The entry at d_{36} is 10,874. It represents the number of kilometers between New York and Tokyo.

E The entry at d_{22} is 0, and it shows that there are 0 kilometers between Mexico City and Mexico City. ❖

Try This What is the entry at d_{35}? What are the addresses for the entry 3037?

CRITICAL Thinking — Identify at least 3 reasons why matrices are a good way to organize data.

ENRICHMENT Have students do research to find out about the history of matrices. Ask them to write a paper about their findings. Have students share a summary of their findings with the class.

INCLUSION strategies **Using Visual Models** Provide large grids of different sizes, labeling each cell with the address of the cell. Make the cells large enough so that you can place small index cards on top of each cell without overlapping. Then write the actual entries for a matrix on index cards, and place them in their appropriate cell on the grid. Students can then lift each card up to see the address of the entry underneath. After students become more comfortable with the concept of giving the address of the entries of a matrix, ask them to place the cards at addresses you specify. Save these grids to use in the *Alternative Teaching Strategy* in Lesson 7.2.

Matrix Equality

Two matrices are equal when their dimensions are the same and their corresponding entries are equal.

$$A = \begin{bmatrix} 1 & 2 & 3 \\ 4 & 5 & 6 \end{bmatrix} \qquad B = \begin{bmatrix} 1 & 3 & 5 \\ 2 & 4 & 6 \end{bmatrix} \qquad C = \begin{bmatrix} 1 & 4 \\ 2 & 5 \\ 3 & 6 \end{bmatrix}$$

Matrix A and matrix B have the same dimension, but unequal entries, so $A \neq B$. Matrix B and matrix C do not have the same dimensions, so $B \neq C$.

EXAMPLE 2

Does $S = T$? Explain why or why not.

$$S = \begin{bmatrix} 2^2 & (12 - 3) \\ \sqrt{49} & 11 \\ 3(-9) & \frac{24}{3} \end{bmatrix} \qquad T = \begin{bmatrix} 2(2) & \sqrt{81} \\ 7 & \frac{44}{4} \\ -(30 - 3) & 2^3 \end{bmatrix}$$

Solution ➤

Recall the characteristics that make two matrices equal.
  a. The matrices have the same dimensions. In this example, both matrices have dimensions 3 by 2.
  b. The matrices have equal entries. Each matrix simplifies to the matrix at the right.

$$\begin{bmatrix} 4 & 9 \\ 7 & 11 \\ -27 & 8 \end{bmatrix}$$

Since the dimensions and corresponding entries of each matrix are equal, $S = T$. ❖

EXAMPLE 3

Matrices U and V are equal. Find the values for x, y, and z.

$$U = \begin{bmatrix} 3x + 2 & x + 2 \\ 2xy & 3y \end{bmatrix} \qquad V = \begin{bmatrix} 14 & 2y \\ 3z & -4y + 21 \end{bmatrix}$$

Solution ➤

Since $U = V$, the corresponding entries are equal. You can form individual equations from the equivalent entries.

$$3x + 2 = 14 \qquad x + 2 = 2y$$
$$2xy = 3z \qquad 3y = -4y + 21$$

When you solve $3x + 2 = 14$, you find that $x = 4$. You can then substitute **4** for x in any of the other equations.

If you substitute 4 for x in the equation $x + 2 = 2y$ and solve, you get $4 + 2 = 2y$, or $y = 3$. You can then use $3y = -4y + 21$ to check the y-value. ❖

How do you find the value for z?

Using Models Have students work in pairs. Give students several matrices containing blanks where entries are to be placed. One student in each pair chooses one matrix, names it, and fills in the blanks with entries of his or her choice. The other student names either the entry or the address that the first student requests. Then have students switch roles. Repeat this activity as many times as necessary. Have each pair of students look at all of the matrices they have written and identify which matrices are equal.

ASSESS

Selected Answers

Odd-numbered Exercises 5–27

Assignment Guide

Core 1–9, 15–33

Core Two-Year 1–33

Technology

A graphing calculator or computer software such as *f(g) Scholar™* should be used for Exercises 29 and 30, and may be helpful for Exercise 28.

EXERCISES & PROBLEMS

Communicate

1. The paragraph contains information that would be more efficiently represented in matrix form. Explain how to create two matrices to organize this data. Place descriptive labels in the left and top margins to identify the rows and columns.

2. Discuss how to determine the dimensions of each matrix. Select an entry of your own choice from each matrix, and give its address.

> The retail business relies on data to predict the buying habits of the public. During the years 1985, 1989, and 1993, the Retailers' News records show that the sales of men's suits was 14.6, 10.8, and 11.5 units respectively. For all apparel, a sales unit represents 1000 items. For the same years, the sales of women's suits were 17.4, 12.3, and 8.6 units. Shirt sales for women were 25.6, 21.3, and 16.2 units and jeans sales for women were 98.2, 90.1, and 80.3 units. Men's jeans sales were 242.7, 210.5, and 186.9 units. Shirt sales of 16.7, 16.2, and 17.3 units were posted for men.
> ...knowing last years numbers. **xt season.**

$$P = \begin{bmatrix} x & y \\ -11 & 2 \\ -10 & 6 \\ -9 & 4 \\ -8 & 6 \\ -7 & 2 \end{bmatrix} \quad Q = \begin{bmatrix} x & y \\ -5 & 2 \\ -3 & 6 \\ -1 & 2 \\ -2 & 4 \\ -4 & 4 \end{bmatrix} \quad R = \begin{bmatrix} x & y \\ 3 & 2 \\ 3 & 6 \\ 1 & 6 \\ 5 & 6 \end{bmatrix} \quad T = \begin{bmatrix} x & y \\ 7 & 2 \\ 7 & 6 \\ 7 & 4 \\ 9 & 4 \\ 9 & 6 \\ 9 & 2 \end{bmatrix}$$

3. Discuss how you would tell which matrix or matrices are equal to matrix *G*.

$$G = \begin{bmatrix} |-5| & \sqrt{4} \\ -(5)^2 & (-2)^3 \end{bmatrix}$$

$$A = \begin{bmatrix} 5 & 2 \\ -25 & -8 \end{bmatrix} \quad B = \begin{bmatrix} -5 & 2 \\ 25 & -8 \end{bmatrix} \quad C = \begin{bmatrix} 8-3 & (\sqrt{2})^2 \\ 20-45 & -20+12 \end{bmatrix} \quad D = \begin{bmatrix} |5| & -2 \\ -\dfrac{100}{4} & -|8| \end{bmatrix}$$

Practice & Apply

4. Matrices *M* and *N* are equal. Find the values of *a*, *b*, *c*, *d*, *g*, and *k*.

$$M = \begin{bmatrix} 2(a+4) & 77 & \frac{1}{3}c \\ -5d-1 & 30 & -\frac{1}{2}k \end{bmatrix} \quad N = \begin{bmatrix} -12 & 11b & 5 \\ -(3-d) & 0.4g & \frac{3}{4}k-3 \end{bmatrix}$$

Business The Ace Auto Repair shop keeps a record of employee absences. Use this record for Exercises 5–8.

	M	T	W	Th	F
Managers	2	1	0	1	4
Mechanics	2	0	2	1	0	= A
Secretaries	3	2	1	0	1

4. $a = -10; b = 7; c = 15; d = \dfrac{1}{3};$
$g = 75; k = \dfrac{12}{5} \text{ or } 2\dfrac{2}{5}$

5. What is the entry in the a_{24} address of the auto repair shop matrix? What does this entry represent? 1; One mechanic absent on Thursday.

6. On which day were the most employees absent? What reason would you give for this high number of absences? Monday

7. Which group of employees has the least number of absences this week? mechanics

8. If there is a total of 30 employees at Ace Auto Repair, what percent of the employees were absent on Friday? 16.7%

$$\begin{bmatrix} 0 & 2 & 4 & 8 \\ 4 & 0 & 2 & 4 \\ 2 & 5 & 0 & 2 \\ 1 & 3 & 5 & 0 \end{bmatrix}$$

9. **Transportation** Four cities, A, B, C, and D, and the air routes between them are represented by the diagram. The arrows show the only directions of travel allowed. From the diagram you can see there are 2 routes from A to B. In the matrix there is the number 2 for the entry AB. The same number is shown for BC. Complete the given matrix to show the number of allowable air routes between the cities. Use 0 if there is no direct path between cities.

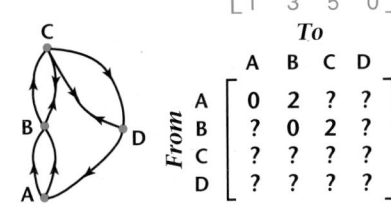

$$\begin{array}{c} \text{To} \\ \begin{array}{cccc} A & B & C & D \end{array} \\ \text{From} \begin{array}{c} A \\ B \\ C \\ D \end{array} \begin{bmatrix} 0 & 2 & ? & ? \\ ? & 0 & 2 & ? \\ ? & ? & ? & ? \\ ? & ? & ? & ? \end{bmatrix} \end{array}$$

Date	Shuttle	Date	Shuttle
9/88	Discovery	10/90	Discovery
12/88	Atlantis	11/90	Atlantis
3/89	Discovery	12/90	Columbia
5/89	Atlantis	3/91	Discovery
8/89	Columbia	4/91	Atlantis
10/89	Atlantis	5/91	Columbia
11/89	Discovery	7/91	Discovery
1/90	Columbia	8/91	Atlantis
2/90	Atlantis	11/91	Discovery
4/90	Discovery	12/91	Atlantis

Space Exploration The table shows the space shuttle names and launch dates for the years 1988–1991.

10. Form matrix L to display the number of shuttle launches in a given year. Begin with 1988, and label the rows by year. Label the columns with the shuttle names in alphabetical order.

11. What are the dimensions of your matrix? 4 by 3

12. What does the matrix address l_{42} represent in matrix L? 1 launch of Columbia in 1991

13. Create another column to represent total launches for the year. In which year did the most launches occur? 1991

14. Over the 4-year period, which year had the least number of launches? What historical event resulted in so few launches that year?

Geography The population of Chula Vista, California and Abilene, Texas for the years 1980 and 1990 is represented in the matrix.

15. If the rate of growth was assumed to be constant for both cities over this 10-year period, what was each city's population increase per year?

16. Using the assumed population increase for each city, create another matrix that shows each city's population for **each** year from 1980 to 1990.

17. In what year were the populations equal? 1984

18. Write equations that could be used to predict the population for each city. Graphically or algebraically predict the cities' populations in 2000.

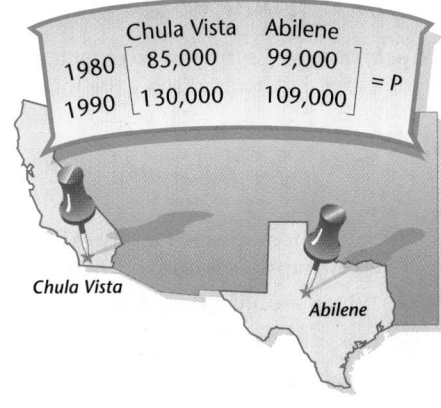

$$\begin{array}{c} \begin{array}{cc} \text{Chula Vista} & \text{Abilene} \end{array} \\ \begin{array}{c} 1980 \\ 1990 \end{array} \begin{bmatrix} 85{,}000 & 99{,}000 \\ 130{,}000 & 109{,}000 \end{bmatrix} = P \end{array}$$

Chula Vista

Abilene

14. 1988; The launching of the *Discovery* space shuttle on September 29, 1988, was the first launching since the explosion of the *Challenger* space shuttle on January 28, 1986.

15. Chula Vista: 4,500; Abilene: 1,000

The answer to Exercise 16 can be found in Additional Answers beginning on page 729.

18. Let $x = 0$ for 1980 and 1 for 1981, etc. Chula Vista: $P = 85{,}000 + 4500x$; Abilene: $P = 99{,}000 + 1{,}000x$; 175,000; 119,000

Portfolio Assessment

Have students write several problems like Example 3 and Exercise 4. Then ask them to exchange and solve each other's problems.

19. Is the sequence 55, 65, 75, 85, . . . linear or exponential? **[Lesson 2.1]**
linear

Match the name of the property to the example that best illustrates it. [Lessons 3.1–3.3]

Examples	Properties
20. $5 + 0 = 5$ c	**a.** Property of Opposites
21. $-8 + 8 = 0$ a	**b.** Associative for Addition
22. $4 + 9 = 9 + 4$ d	**c.** Addition Property of Zero
23. $12 + 18 = 6(2 + 3)$ e	**d.** Commutative for Addition
24. $7 + (11 + 4) = (7 + 11) + 4$ b	**e.** Distributive Property

25. At the end of the season sale at Well-Dressed Women, Ms. Sanchez bought a suit for 40% off the ticket price. She paid $96. What was the original price? **[Lesson 4.5]** $160

26. Cultural Connection: Americas
The earliest math texts in the Americas were in Spanish. Solve this problem from Guatemala written by D. Juan Joseph de Padilla in 1732. **[Lesson 4.6]**
10; 20; 90

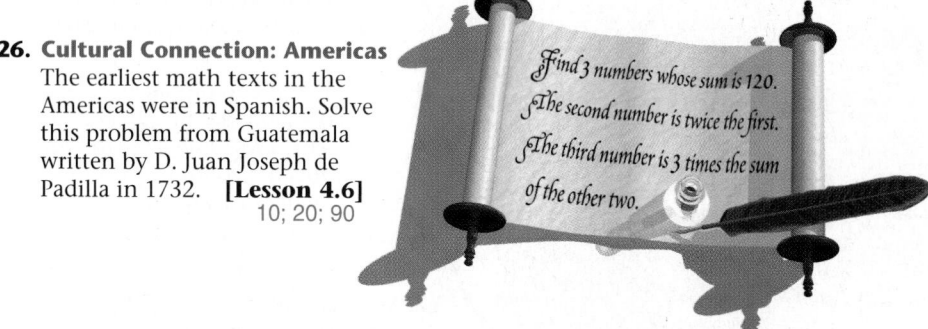

*Find 3 numbers whose sum is 120.
The second number is twice the first.
The third number is 3 times the sum of the other two.*

27. Solve and graph the solution set on the number line. **[Lesson 4.7]**
$$3x - (5 - 2x) \le x - 21 \quad \text{and} \quad -7x - 12 \le 16$$
$x \le -4$ and $x \ge -4$

28. Graph the equation $3x + 4y = 12$. **[Lesson 5.4]**

Look Beyond ~~~~

Look Beyond

Exercises 29 and 30 deal with technology and matrices. Exercises 31–33 introduce students to the use of matrices for analyzing data.

Technology Learn how to enter a matrix into a graphics calculator or computer. Enter and display the following matrices. Refer to the calculator manual.

29. $M = \begin{bmatrix} 1 & -5 & 7 \\ 0.5 & 0 & -12 \end{bmatrix}$ **30.** $N = \begin{bmatrix} 4 & 0 \\ 0 & -1 \end{bmatrix}$

The matrix displays the number of people that helped in an environmental clean-up project for 3 areas of a city.

31. Which area had the greatest number of helpers in week 1? 3

32. How many helpers were in Areas 2 and 3 in week 3? 87

33. What was the total number of helpers in week 2? 87

	Area 1	Area 2	Area 3
Week 1	25	35	50
Week 2	15	40	32
Week 3	18	42	45
Week 4	21	53	46

27. The graph is a single point at -4.

29–30. Answers may vary.

28.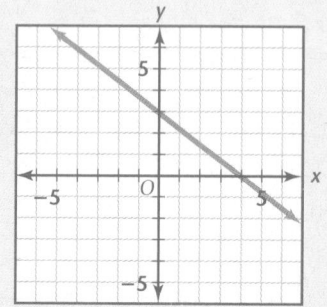

LESSON 7.2 Adding and Subtracting Matrices

 *A matrix is an effective tool for storing data in a way that makes sense visually. What if the data changes? If you keep getting new data tables, you will need a way to combine the information in the old and new matrices without disturbing the organization. You need to know how to add and subtract matrices.*

Your last look at the stock inventory for the Super Sight and Sound Shop was on April 1. Since that time, business has been good, and much of the stock has been sold. The matrices below represent the *stock* available on April 1 and the *sales* for April.

Stock [April 1]:

Videos	DATs
Games	CDs
Laser disks	Tapes

$$A = \begin{bmatrix} 3000 & 1000 \\ 800 & 3000 \\ 1500 & 1000 \end{bmatrix}$$

Sales [April]:

Videos	DATs
Games	CDs
Laser disks	Tapes

$$B = \begin{bmatrix} 2254 & 952 \\ 675 & 1325 \\ 1187 & 548 \end{bmatrix}$$

PREPARE

Objectives

- Add and subtract matrices.
- Determine whether the Identity Property for Addition is true for matrices.

RESOURCES

- Practice Master   7.2
- Enrichment Master   7.2
- Technology Master   7.2
- Lesson Activity Master   7.2
- Quiz   7.2
- Spanish Resources   7.2

Assessing Prior Knowledge

Add or subtract.

1. $5 + (-3.7)$ **[1.3]**
2. $-7.2 - 8.8$ **[-16]**
3. $\dfrac{3}{8} + \left(-\dfrac{3}{4}\right)$ $\left[-\dfrac{3}{8}\right]$
4. $0 + (-0.09)$ **[-0.09]**

TEACH

Why Point out to students that matrices greatly simplify the handling of certain repetitive additions and subtractions for arrays of numbers that occur in real-life situations. Ask students to suggest situations in which data stored in matrices may need to be added or subtracted.

ALTERNATIVE teaching strategy

Using Visual Models Use the grids suggested on page 310 in Lesson 7.1 under **Inclusion Strategies.** Use two grids that have the same dimensions. If necessary, write numbers on additional index cards. Be sure to include positive and negative numbers, decimals, and fractions. Have students work in pairs and choose enough cards to fill up each of two matrix grids that have the same dimensions. Then have students pick up the cards containing the corresponding entries from each matrix and add the numbers. Repeat this procedure until all corresponding entries have been added. Use a similar procedure for subtraction. Be sure that students subtract the numbers in the correct order.

Suppose that the stock was not replenished in May and that the matrix for May sales is given in matrix M below.

$$M = \begin{bmatrix} 600 & 34 \\ 93 & 423 \\ 197 & 9 \end{bmatrix}$$

What is the available stock on June 1?

$$\begin{bmatrix} 146 & 14 \\ 32 & 1252 \\ 116 & 443 \end{bmatrix}$$

Alternative Example 2

Refer to Alternative Example 1. Suppose the June delivery arrives on June 2 with more stock. The shipment includes an invoice matrix to let the store know what is being delivered.

$$J = \begin{bmatrix} 3189 & 1642 \\ 600 & 2560 \\ 1980 & 954 \end{bmatrix}$$

How would you update your stock matrix to show the increase in available stock?

$$\begin{bmatrix} 3335 & 1656 \\ 632 & 3812 \\ 2096 & 1397 \end{bmatrix}$$

Cooperative Learning

In the exploration on page 317, have students work in pairs. Have one student complete the work by hand while the other uses a calculator or computer. Have students reverse roles for each of the required additions in step 1. Repeat the same procedure for the subtractions in step 3. Have each pair work together to complete the rest of the exploration.

EXAMPLE 1

Inventory Control The sales matrix shows how many of each item were sold in April. Since the original stock has decreased, how would you determine the available stock for May?

Solution ➤

Subtract each entry in B from the corresponding entry in A. The result will be a new matrix, C, that shows the stock available May 1. For example, the video stock available **May 1** is $3000 - 2254$ or 746. Place 746 in address c_{11}.

$$
\begin{array}{ccccc}
A & - & B & = & C \\
\textbf{Stock [April 1]} - & & \textbf{Sales [April]} & = & \textbf{Stock [May 1]}
\end{array}
$$

$$\begin{bmatrix} 3000 & 1000 \\ 800 & 3000 \\ 1500 & 1000 \end{bmatrix} - \begin{bmatrix} 2254 & 952 \\ 675 & 1325 \\ 1187 & 548 \end{bmatrix} = \begin{bmatrix} 746 & \\ & \\ & \end{bmatrix}$$

Matrix C is completed by subtracting corresponding entries.

Stock [May 1]:

	Videos	DATs
	Games	CDs
	Laser disks	Tapes

$$C = \begin{bmatrix} 746 & 48 \\ 125 & 1675 \\ 313 & 452 \end{bmatrix}$$ ❖

EXAMPLE 2

The next week, the May delivery arrives with more stock. The shipment includes an invoice matrix D to let the store know what is being delivered.

Delivery [May]:

	Videos	DATs
	Games	CDs
	Laser disks	Tapes

$$D = \begin{bmatrix} 2500 & 1500 \\ 625 & 775 \\ 1500 & 400 \end{bmatrix}$$

How would you update your stock matrix to show the increase in stock?

Solution ➤

If you add each entry in C to the corresponding entry in D, you will create a new stock matrix, E, for the available stock for May.

$$
\begin{array}{ccccc}
C & + & D & = & E \\
\textbf{Stock [May 1]} & + & \textbf{Delivery [May]} & = & \textbf{Stock [May]}
\end{array}
$$

For example, video stock in the new matrix **Stock [May]** is $746 + 2500$ or 3246 at address e_{11}.

Stock [May]:

	Videos	DATs
	Games	CDs
	Laser disks	Tapes

$$E = \begin{bmatrix} 3246 & 1548 \\ 750 & 2450 \\ 1813 & 852 \end{bmatrix}$$ ❖

ENRICHMENT Have students investigate to determine whether the Associative Property holds true for subtraction of matrices. Begin by asking students to recall the relationship between subtraction and the Associative Property. Ask them to use their answer to make a conjecture about the relationship between subtraction of matrices and the Associative Property. Then have them use several examples to test their conjecture. After students come to the conclusion that subtraction of matrices is not associative, ask them if they can think of any matrices that are an exception to the rule. One example of matrices with which subtraction is associative is three equal matrices.

Try This Let $X = \begin{bmatrix} 5 & -3 \\ 2 & 0 \end{bmatrix}$ and $Y = \begin{bmatrix} -1 & 7 \\ 0 & 3 \end{bmatrix}$. Find $X + Y$.

CRITICAL *Thinking*

In a large business, how often does this process of updating the stock matrices take place? If there are thousands of different items in the stock of a large business, how does a store manage its inventory and manipulate its matrices?

•Exploration• *Addition and Subtraction Properties*

Use the following matrices to explore and develop conjectures about the algebra of adding and subtracting matrices. This is a good opportunity to use a calculator or computer.

$$A = \begin{bmatrix} 6 & -3 \\ -4 & \frac{9}{2} \end{bmatrix} \quad B = \begin{bmatrix} -4.2 & 1 \\ 0 & -6.6 \end{bmatrix} \quad C = \begin{bmatrix} \frac{3}{4} & -10 \\ 9 & -2 \end{bmatrix} \quad D = \begin{bmatrix} -2 & 11 & 7 \\ 5 & 9 & -12 \end{bmatrix}$$

 Add.
  a. $A + C$ **b.** $C + A$ **c.** $B + D$
  d. $A + (B + C)$ **e.** $(A + B) + C$ **f.** $C + D$

  Describe the procedure for adding matrices.

2 Compare and contrast these additions. Are there any restrictions on whether the matrices can be added? Write a conjecture.

 Subtract.
  a. $C - A$ **b.** $B - B$ **c.** $B - A$
  d. $A - C$ **e.** $B - C$ **f.** $C - D$

  Describe the procedure for subtracting matrices.

4 Compare and contrast these subtractions. Are there any restrictions on whether the matrices can be subtracted? Write a conjecture. ❖

The **identity matrix** for addition is a matrix filled with zeros. It is also called the **zero matrix.** The zero matrix for a 2 by 2 matrix is shown here.

$$\begin{bmatrix} 0 & 0 \\ 0 & 0 \end{bmatrix} = \mathbf{0}_{2 \times 2}$$

Adding the zero matrix to another matrix does not change the original matrix.

$$\begin{bmatrix} 0 & 0 \\ 0 & 0 \end{bmatrix} + \begin{bmatrix} -1 & 5 \\ 7 & 0 \end{bmatrix} = \begin{bmatrix} -1 & 5 \\ 7 & 0 \end{bmatrix}$$
$$\begin{bmatrix} -1 & 5 \\ 7 & 0 \end{bmatrix} + \begin{bmatrix} 0 & 0 \\ 0 & 0 \end{bmatrix} = \begin{bmatrix} -1 & 5 \\ 7 & 0 \end{bmatrix}$$

Consider matrix C and matrix D. Form the zero matrix that can be added to C. Then form the zero matrix that can be added to D.

$$? + C = \begin{bmatrix} \frac{3}{4} & -10 \\ 9 & -2 \end{bmatrix} = C + ? \qquad ? + D = \begin{bmatrix} -2 & 11 & 7 \\ 5 & 9 & -12 \end{bmatrix} = D + ?$$

How do the zero matrices for C and D compare?

$$L + M = \begin{bmatrix} 6 & 8 \\ 4 & -1 \end{bmatrix}$$

$$M + L = \begin{bmatrix} 6 & 8 \\ 4 & -1 \end{bmatrix}$$

$$L + M = M + L$$

$$(L + M) + N = \begin{bmatrix} 5 & 12 \\ 11 & -6 \end{bmatrix}$$

$$L + (M + N) = \begin{bmatrix} 5 & 12 \\ 11 & -6 \end{bmatrix}$$

$$(L + M) + N = L + (M + N)$$

ASSESS

Selected Answers

Odd-numbered Exercises 9–39

Assignment Guide

Core 1–7, 8–18 even, 19–41

Core Two-Year 1–39

Technology

A graphing calculator or computer software such as *f(g) Scholar*™ may be helpful for Exercises 1–4 and 8–17.

$$L = \begin{bmatrix} 2 & 3 \\ 6 & -1 \end{bmatrix}$$

$$M = \begin{bmatrix} 4 & 5 \\ -2 & 0 \end{bmatrix}$$

$$N = \begin{bmatrix} -1 & 4 \\ 7 & -5 \end{bmatrix}$$

When you add numbers, you know that certain properties are true.

The *Commutative Property for Addition* states that for numbers a and b, $a + b = b + a$.

The *Associative Property of Addition* states that for numbers a, b, and c, $a + (b + c) = (a + b) + c$.

Use the given matrices L, M, and N to show that matrix addition is also commutative and associative.

EXERCISES & PROBLEMS

Communicate

Explain how to perform each of the matrix operations. If a solution is not possible, explain why.

1. $\begin{bmatrix} 10 & 6 \\ 4 & -5 \end{bmatrix} + \begin{bmatrix} -1 & 6 \\ 9 & -7 \end{bmatrix}$

2. $\begin{bmatrix} -11 & 6 \\ 13 & 8 \\ 17 & -9 \end{bmatrix} - \begin{bmatrix} -2 & 10 \\ -16 & 12 \\ 4 & -3 \end{bmatrix}$

3. $\begin{bmatrix} \frac{5}{6} \\ \frac{7}{12} \end{bmatrix} + \begin{bmatrix} -\frac{5}{4} & \frac{2}{3} \end{bmatrix}$

4. $\begin{bmatrix} \frac{1}{2} & \frac{1}{4} & \frac{-1}{2} \\ \frac{2}{3} & \frac{4}{5} & \frac{1}{4} \end{bmatrix} + \begin{bmatrix} \frac{1}{2} & \frac{1}{4} & \frac{-1}{2} \\ \frac{2}{3} & \frac{4}{5} & \frac{1}{4} \end{bmatrix}$

5. Is matrix subtraction commutative? Explain.

6. Is matrix subtraction associative? Explain.

7. Is the zero matrix always a square matrix? Explain why or why not.

Practice & Apply

Perform each of the matrix operations. If an operation is not possible, explain why.

8. $\begin{bmatrix} 23 & 46 \\ 4 & -35 \end{bmatrix} + \begin{bmatrix} -51 & 16 \\ 29 & -7 \end{bmatrix}$

9. $\begin{bmatrix} 4 & 2 \\ 6 & -7 \\ 3 & 9 \end{bmatrix} + \begin{bmatrix} -5 & 7 \\ -3 & 9 \\ 3 & -6 \end{bmatrix} - \begin{bmatrix} -11 & 3 \\ 8 & -15 \\ -7 & -2 \end{bmatrix}$

10. $\begin{bmatrix} -6 & 1 \\ -1 & 7 \end{bmatrix} - \begin{bmatrix} -3 & 0 \\ 9 & -2 \end{bmatrix}$

11. $\begin{bmatrix} -1.4 & 4.3 & -9.6 \\ 15.8 & 1 & -3.5 \end{bmatrix} - \begin{bmatrix} 6.2 & 3.2 \\ -9.6 & 7.1 \\ 2.6 & -8.5 \end{bmatrix}$

RETEACHING
the lesson

Cooperative Learning

Have students work in small groups to find an example of a local business or organization that from time to time needs to order new supplies. Suggest that they think about such possibilities as a textbook storage center or a Red Cross facility. Have students gather data about recent stocking activities at the facilities that can be written in matrix form.

8. $\begin{bmatrix} -28 & 62 \\ 33 & -42 \end{bmatrix}$

9. $\begin{bmatrix} 10 & 6 \\ -5 & 17 \\ 13 & 5 \end{bmatrix}$

10. $\begin{bmatrix} -3 & 1 \\ -10 & 9 \end{bmatrix}$

11. These 2 matrices cannot be subtracted because the first one has 2 rows and 3 columns, and the second one has 3 rows and 2 columns.

12. $\begin{bmatrix} 3.2 & \frac{3}{5} \\ \frac{-8}{5} & -4.2 \end{bmatrix} - \begin{bmatrix} \frac{-4}{5} & 5.6 \\ -6.7 & \frac{3}{5} \end{bmatrix}$

13. $\begin{bmatrix} 2.3 & -9.3 & 1.8 \\ -8.2 & -4.1 & 10 \\ 2.1 & 6 & -0.16 \end{bmatrix} + \begin{bmatrix} -1.2 & -2.1 & 5.3 \\ 0.2 & -4 & -2.3 \\ -3.4 & 2.4 & 6 \end{bmatrix}$

14. $\begin{bmatrix} \frac{-2}{3} & \frac{4}{5} \\ \frac{1}{15} & \frac{7}{30} \end{bmatrix} - \begin{bmatrix} \frac{9}{15} & \frac{-4}{3} \\ \frac{-2}{5} & \frac{-4}{15} \end{bmatrix}$

15. $\begin{bmatrix} 0.4 & -1.5 & 0.9 \\ 2.6 & 6.9 & 3.7 \end{bmatrix} + \begin{bmatrix} -4.7 & 2.6 & 6.9 \\ -7.3 & 9.8 & -5.5 \end{bmatrix}$

16. $\begin{bmatrix} -3 & 4 \\ -4 & 0 \\ 2 & -5 \end{bmatrix} + \begin{bmatrix} 0 & -1 \\ -6 & 9 \\ -7 & 5 \end{bmatrix}$

17. $\begin{bmatrix} -25 & 32 & 14 \\ 36 & -42 & -45 \\ -71 & 65 & 29 \end{bmatrix} - \begin{bmatrix} 16 & -34 & -55 \\ 21 & 11 & 22 \\ -43 & -67 & -44 \end{bmatrix} + \begin{bmatrix} 57 & 79 & 64 \\ -38 & -22 & -48 \\ -56 & 88 & 26 \end{bmatrix}$

18. Inventory Control At the Super Sight and Sound Shop, the May sales matrix and June delivery invoice have just appeared on your desk. Make a new matrix to update the May stock matrix to reflect the May sales and June delivery. Label your answer matrix **Stock [June]**.

Stock [May 1]

Videos	DATs		3246	1548
Games	CDs		750	2450
Laser disks	Tapes		1813	852

Sales [May]

Videos	DATs		2748	1081
Games	CDs		702	1456
Laser disks	Tapes		1679	622

Delivery [June]

Videos	DATs		3000	1500
Games	CDs		750	1000
Laser disks	Tapes		1800	500

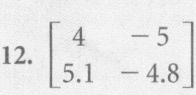 **Statistics** In the Woodlake public school system, there are two junior high schools, Glenn and Kelly. The enrollment for the electives music (Mu), art (Ar), technology (Te), and health (He) in each of the schools appears in the matrices below.

Glenn

	Mu	Ar	Te	He
Boys	447	199	514	389
Girls	498	352	432	399

Kelly

	Mu	Ar	Te	He
Boys	387	276	489	367
Girls	505	392	387	437

19. How many girls are enrolled in art? How many boys are taking health? 744; 756

20. Create a matrix to show the total enrollment in each elective.

21. How many students will take technology while in junior high school? 1822

12. $\begin{bmatrix} 4 & -5 \\ 5.1 & -4.8 \end{bmatrix}$

15. $\begin{bmatrix} -4.3 & 1.1 & 7.8 \\ -4.7 & 16.7 & -1.8 \end{bmatrix}$

13. $\begin{bmatrix} 1.1 & -11.4 & 7.1 \\ -8 & -8.1 & 7.7 \\ -1.3 & 8.4 & 5.84 \end{bmatrix}$

16. $\begin{bmatrix} -3 & 3 \\ -10 & 9 \\ -5 & 0 \end{bmatrix}$

14. $\begin{bmatrix} \frac{-19}{15} & \frac{32}{15} \\ \frac{7}{15} & \frac{1}{2} \end{bmatrix}$ or $\begin{bmatrix} -1.267 & 2.133 \\ 0.467 & 0.5 \end{bmatrix}$

The answers to Exercises 17, 18, and 20 can be found in Additional Answers beginning on page 729.

Error Analysis
Watch for students who do not add or subtract corresponding entries. Emphasize that corresponding entries are in the same position in each of the two matrices that are being added or subtracted.

Portfolio Assessment

Have students write a short lesson on adding and subtracting matrices. Be sure they include several examples and explanations of the properties that apply to each operation. Have students share their lessons with the class.

Sports You keep all of the statistics for the girls' basketball team in a matrix, a portion of which is shown below. You record the number of 3-point baskets, 2-point baskets, 1-point free throws (FTs), and the total points. The matrices show the statistics for the end of the regular season and the total points scored after the playoff series. Use these matrices to complete Exercises 22–24.

	Points scored at end of regular season						**Points scored at end of playoff series**			
	3pts.	2pts.	FTs	Total			3pts.	2pts.	FTs	Total
Waters	3	15	12	51		Waters	5	25	15	80
Riley	1	20	16	59		Riley	2	22	20	70
Sharp	3	17	13	56		Sharp	5	20	15	70
Evans	4	6	9	33		Evans	5	12	12	51
Jones	0	14	18	46		Jones	0	17	22	56

22. Create a matrix to show how many points were scored by each player in the playoff series.

23. Which player should get the Most Valuable Player award for the playoff series based on points?   Waters

24. What is the total number of free throws made during the playoffs?   16

25. Find the values of w, x, y, and z by solving the matrix equation $A + B = C$. $-2; 6; 1; -15$

$$A = \begin{bmatrix} -5w + 2 & 2x - 4 \\ 8y - 1 & \frac{1}{5}z \end{bmatrix} \qquad B = \begin{bmatrix} -13 & 5 - x \\ -8 & z + 9 \end{bmatrix} \qquad C = \begin{bmatrix} 2w + 3 & 7 \\ 3y - 4 & -9 \end{bmatrix}$$

26. Use the values you found for w, x, y, and z to evaluate matrices A, B, and C. Solve the matrix equation $A + B = C$ numerically to check your solutions.

The symbol that represents the additive inverse of a matrix is $-A$. An additive inverse occurs if $-A + A = 0$ (where 0 is the identity matrix for addition).

27. Is there one matrix that can be used as the identity matrix for A, B, and C in Exercise 25? Why? $\begin{bmatrix} 0 & 0 \\ 0 & 0 \end{bmatrix}$

28. Describe the additive identity matrix that will fit the matrix below.

$$\begin{bmatrix} 1 & 4 & -3 \\ 7 & 12 & 0 \end{bmatrix} \qquad \begin{bmatrix} 0 & 0 & 0 \\ 0 & 0 & 0 \end{bmatrix}$$

29. Form the matrices $-A$, $-B$, and $-C$ for matrices A, B, C in Exercise 25.

---

22. Points Scored in the Playoff Series

	Total
Waters	29
Riley	11
Sharp	14
Evans	18
Jones	10

26. $A = \begin{bmatrix} 12 & 8 \\ 7 & -3 \end{bmatrix}$ $B = \begin{bmatrix} -13 & -1 \\ -8 & -6 \end{bmatrix}$

$C = \begin{bmatrix} -1 & 7 \\ -1 & -9 \end{bmatrix}$

29. $-A = \begin{bmatrix} -12 & -8 \\ -7 & 3 \end{bmatrix}$

$-B = \begin{bmatrix} 13 & 1 \\ 8 & 6 \end{bmatrix}$ $-C = \begin{bmatrix} 1 & -7 \\ 1 & 9 \end{bmatrix}$

Statistics Students collect data by measuring the number of inches around the thickest part of their thumb and then around their wrist. The data for 5 students appear in the matrix below.

	Thumb	Wrist
Samir	3.7	7.3
Betty	2.8	4.7
Ruth	3.1	6.2
Todd	3.6	7.1
Han	3.3	6.6

30. Plot the data and describe the correlation. **[Lesson 1.7]**

31. Graph $y = |x|$ and $y = 9$. Find their common solutions. **[Lesson 2.4]**

Write the opposite of each expression.
[Lesson 3.4] $2x - 3y$

32. $-2x + 3y$ **33.** $-4a - 2b$ $\;4a + 2b$

34. $-x + 3$ $\;x - 3$ **35.** $4(-4x - 7)$ $\;4(4x + 7)$

36. Graph $-3 < x$ or $x \le -5$ on the number line. State the solution set clearly. **[Lesson 3.6]**

37. **Geometry** You have both a square and a rectangle. The rectangle is drawn so that its length is 3 times the side of the square and its width is twice the side of the square. Algebraically express the perimeter and area of the rectangle. **[Lesson 4.6]** $\;10s; 6s^2$

38. Write the equation $5x = 2y - 12$ in standard form, slope-intercept form, and point-slope form. **[Lessons 5.3, 5.4]**

39. Solve the system by elimination and check. $\begin{cases} 3x = 12 + 4y \\ 2y - x = -5 \end{cases}$ **[Lesson 6.3]** $\left(2, -\frac{3}{2}\right)$

Look Beyond

40. **Geometry** The matrix T represents the coordinates of the vertices of triangle ABC graphed on a coordinate plane. Matrix M represents a translation matrix.

$$T = \begin{bmatrix} x & y \\ 4 & 3 \\ 7 & -2 \\ 10 & 6 \end{bmatrix} \quad M = \begin{bmatrix} x & y \\ -7 & 2 \\ -7 & 2 \\ -7 & 2 \end{bmatrix}$$

Write the matrix equal to $T + M$. Plot the original points and the translated points on a graph.

---

30. Regression line: $y = 2.70x - 2.54$; Strong Positive

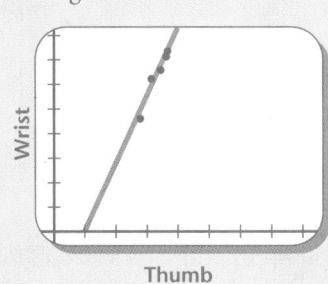

31. The answer to Exercise 31 can be found in Additional Answers beginning on page 729.

36. $x > -3$ or $x \le -5$

number line from $-9\; -7\; -5\; -3\; -1\; 0\; 1$

38. $5x - 2y = -12$; $y = \frac{5}{2}x + 6$; Answers may vary. For example, $y - 11 = \frac{5}{2}(x - 2)$.

40. $\begin{bmatrix} -3 & 5 \\ 0 & 0 \\ 3 & 8 \end{bmatrix}$

Look Beyond

These exercises use matrices to preview translations, which are presented in Chapter 9.

Technology Master

Technology
7.2 Matrix Addition and Translations

A translation of the plane is a function according to which (1) you add a fixed number r to each x-coordinate of a point and (2) you add a fixed number s to each y-coordinate of that point P. If you let a point (x, y) be represented by a matrix $\begin{bmatrix} x \\ y \end{bmatrix}$, then the matrix sum shown represents the translation.

$$\begin{bmatrix} x \\ y \end{bmatrix} + \begin{bmatrix} r \\ s \end{bmatrix} = \begin{bmatrix} x + r \\ y + s \end{bmatrix}$$

If you need to translate many points in a plane, you can use a spreadsheet to make the work go quickly. Enter the x-coordinates and the y-coordinates into columns A and B. The new coordinates are calculated in columns C and D. If, for example, $r = 3$ and $s = 2$, then cell C2 contains A2+3 and cell D2 contains B2+2.

	A	B	C	D
1	X	Y	NEW X	NEW Y
2				
3				
4				
5				

Let $x = 0, 1, 2, 3, 4, 5, 6, 7, 8,$ and 9 and $y = 2x^2 - 3x$. Let $r = -7.5$ and $s = 13.9$. Use a spreadsheet to find the translation of the point corresponding to each x-value.

1. $x = 2$ _____   **2.** $x = 6$ _____

3. $x = 8$ _____   **4.** $x = 9$ _____

Let $y = 2x^3 - 7x^2 - x$ and let $r = 7.5$ and $s = -13.9$. Use a spreadsheet to find the translation of the point corresponding to each x-value.

5. $x = 2$ _____   **6.** $x = 6$ _____

7. $x = 8$ _____   **8.** $x = 9$ _____

9. Does $\begin{bmatrix} x \\ y \end{bmatrix} - \begin{bmatrix} r \\ s \end{bmatrix}$, where r and s are positive, represent a translation? If so, describe it.

Barely Enough GRIZZLIES?

Counting Big Bears

Government Wants Them Off the Endangered List; How Many is Enough?

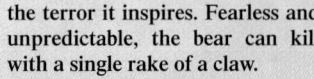

By Marj Charlier
Staff Reporter of The Wall Street Journal

KALISPELL, Mont.—The solitary monarch of the wild, the grizzly rules the rugged high country. Scientists call it *Ursus arctus horribilus,* and the name suggests the terror it inspires. Fearless and unpredictable, the bear can kill with a single rake of a claw.

Man is the grizzly's only natural enemy, but the clash almost proved fatal to the species. Unrestricted hunting and development of the animal's habitat decimated the bear population over the last century. By 1975 when the grizzly was placed on the federal government's Endangered Species List, only 1,000 were thought to be alive.

Now, federal officials say the grizzly population has rebounded, though they concede there is no hard evidence to support that conclusion, and they are proposing to take the bear off the list.

A Death Sentence?

Many environmentalists, however, believe that even if the number of bears has increased, the grizzly population is still too small to strip the animal of federal protection. Such a plan, they say, would be a death sentence for the species. "If the grizzly bear is delisted, it will be strangled by development and go extinct quite promptly," says Lance Olsen, president of the nonprofit Great Bear Foundation.

Population Unknown

Complicating the issue is the difficulty of determining how many grizzlies roam the Rockies and their foothills. Everyone concedes that the 1,000 figure for 1975 was only a guess. Attempts to take a bear census in the years since then have proved difficult, and no one knows how many grizzlies there are now.

Richard Mace, a wildlife biologist for the state of Montana, knows how tough it is to count grizzlies. Displaying a battered steel box with a jagged rip down one side, he explains that the box once held a tree-mounted camera that was rigged to snap photos when it sensed body heat from a large animal. Biologists had placed some 43 cameras in a section of Montana's Swan Mountains, hoping to get a more accurate bear count.

Apparently, Mr. Mace says, when the camera flashed, the grizzly slashed. Its two-inch long claws ripped through the steel casing, which is a bit thicker than a coffee can. The bear dug out the camera and chewed it to pieces.

1a-c. Answers may vary. The following show the results from a batch of 450 objects with 50 marked.

Sample	Total Number of Objects in Sample	Number of Marked Objects in Sample
1	30	2
2	30	4
3	30	3
4	30	4

1e. Sample calculation based on table follow.

$$M_s = (2 + 4 + 3 + 4)/4 = 13/4 = 3.25$$
$$N_s = 30$$
$$M_p = 50$$

$$\frac{N_p}{N_s} = \frac{M_p}{M_s} \rightarrow \frac{N_p}{30} = \frac{50}{3.25}$$

$$N_p = 462$$

1f. Answer will depend on estimate and actual number of objects in batch.

Cooperative Learning

Counting animals in the wild is not easy. How do you know if you have counted some more than once? How can you possibly find them all? Wildlife managers sometimes use a technique called *tag and recapture*. To see how the *tag and recapture* method works, you can try it with objects instead of animals.

You will need:

- a batch of several hundred objects (macaroni, beans, crumpled pieces of paper) in a paper bag or other container
- a marker (or some other way to *tag*, or mark, the objects)

1. Follow steps a–e to estimate the number of objects in your batch.

   a. Remove a handful of objects from the batch and mark them. Record how many you have marked.

   b. Return the marked objects to the batch. Mix thoroughly.

   c. Take 30 objects from the batch at random. Count the number of marked objects in your sample. Copy the chart and fill in the first row. Repeat for each group member.

Sample	Total Number of Objects in Sample (Ns)	Number of Marked Objects in Sample (Ms)
1	30	
2	30	

   d. How do you think the fraction of marked objects in your samples compares with the fraction of marked objects in the whole batch?

   e. Use the proportion to estimate the size of the whole batch. (Average your samples for M_s and N_s.)

   $$\frac{M_p}{N_p} = \frac{M_s}{N_s}$$

   M_p is the number of <u>marked objects</u> in the batch.
   N_p is the <u>total number of objects</u> in the batch.
   M_s is the number of <u>marked objects</u> in the sample.
   N_s is the <u>total number of objects</u> in the sample.

   f. Check your estimate by counting. Within what percent of the actual total was your estimate?

2. Describe how the *tag and recapture* method could be used to estimate an animal population.

3. Do you think the *tag and recapture* method would work for a grizzly bear population? Why or why not?

2. Answers may vary. Capture a number of the animals you are studying, tag or mark them in some way, and return them to the wild. After some time passes, capture a sample of the animals and count how many in the sample are marked. Set up the equation used in the activity and solve for N_p to get an estimate of the number of animals in that whole population.

3. Answers may vary. No, because it would be too hard to capture enough grizzlies without harming them.

Cooperative Learning

Activity 1

Discuss how the batch of beans (or whatever objects students are using) represents a population of animals the size of which needs to be estimated.

To make sure groups stay on track, you might work with all the groups at the same time. Have all the groups complete one part of the activity before any group goes on to the next.

1a A *handful* should contain roughly 10% of the objects. If you are using very small objects, like popcorn or peas, students should take a very small handful.

1c Before proceeding, students should make a copy of the chart, with enough room to record one sample per group member.

1d Before going on to Part e, help students see that the samples should be representative of the batch.

1e Students find the mean of the data in the third column of their chart for M_s, but they need not calculate a mean for N_s if all the samples for the group are the same size.

1f You might use the term *percent error,* because students are finding the percent error of their estimate.

Discuss

You may wish to explore how the tag-and-recapture estimate would be affected by the traveling patterns of the animals, the social groupings of the animals, or the size of the group captured or recaptured.

Exploring Matrix Multiplication

PREPARE

Objectives

• Multiply a matrix by a scalar, and multiply matrices.

• Determine the identity matrix for multiplication.

RESOURCES

• Practice Master	**7.3**
• Enrichment Master	**7.3**
• Technology Master	**7.3**
• Lesson Activity Master	**7.3**
• Quiz	**7.3**
• Spanish Resources	**7.3**

Assessing Prior Knowledge

Simplify each expression.

1. $-2(8) + 3.1(7)$ **[5.7]**
2. $12(4) + 10(6)$ **[108]**
3. $4(1) + 7(0)$ **[4]**
4. $4(x) + (-5)(x - 1)$
           [$-x + 5$]

TEACH

Scalar multiplication is often easy for students to understand. Matrix multiplication is usually quite a bit more challenging for students, but the availability of calculators and computers today makes the task much easier. However, students should still understand how to complete matrix multiplication manually.

why *If the entries in a matrix are doubled or tripled, you need a way to multiply the data in the matrix by a constant. This is scalar multiplication, and it is related to addition. Another operation called matrix multiplication will enable you to find the product of 2 matrices.*

The Teen Boutique is planning a Back-to-School sale, with cotton sweaters featured. The current stock of sweaters is represented in matrix S.

$$\text{Stock:}\quad \begin{array}{c} \\ \text{Small} \\ \text{Medium} \\ \text{Large} \end{array} \begin{array}{ccc} \text{Blue} & \text{Red} & \text{Black} \end{array} \\ \begin{bmatrix} 16 & 18 & 15 \\ 14 & 22 & 17 \\ 21 & 20 & 19 \end{bmatrix} = S$$

Inventory Control In anticipation of the sale, the boutique orders three times the present stock. How many of each item are ordered? Can you represent the order in matrix form?

If you want three times the stock, find the sum of 3 identical matrices.

$$\begin{array}{ccccccc} S & + & S & + & S & = & T \end{array}$$
$$\begin{bmatrix} 16 & 18 & 15 \\ 14 & 22 & 17 \\ 21 & 20 & 19 \end{bmatrix} + \begin{bmatrix} 16 & 18 & 15 \\ 14 & 22 & 17 \\ 21 & 20 & 19 \end{bmatrix} + \begin{bmatrix} 16 & 18 & 15 \\ 14 & 22 & 17 \\ 21 & 20 & 19 \end{bmatrix} = \begin{bmatrix} 48 & 54 & 45 \\ 42 & 66 & 51 \\ 63 & 60 & 57 \end{bmatrix}$$

An easier way to do this is to multiply each entry by 3.

$$\begin{array}{ccc} 3S & = & T \end{array}$$
$$3\begin{bmatrix} 16 & 18 & 15 \\ 14 & 22 & 17 \\ 21 & 20 & 19 \end{bmatrix} = \begin{bmatrix} 48 & 54 & 45 \\ 42 & 66 & 51 \\ 63 & 60 & 57 \end{bmatrix}$$

This type of multiplication is called **scalar multiplication**. The **scalar** is the number that multiplies each entry in the matrix. In this case, the scalar is 3.

ALTERNATIVE teaching strategy

Technology Begin by asking students to study the opening situation. Have them add matrix S three times using a graphics calculator or computer software such as *f(g) Scholar™*. Then ask students to perform the scalar multiplication $3S$ and compare the results to the sum of $S + S + S$. Continue this lesson based on the situation described in Exploration 1. Ask students to study the two 2×2 matrices given and decide what Sam and Kim will need to know in order to make the cars and trucks. Have students work in small groups to determine how the entries in the product matrix were derived and have them experiment until they find the correct procedure.

Multiplying two matrices creates a **product matrix**. This matrix combines part of the information from each of the original matrices. Pay close attention to the procedure for multiplying matrices. It is different from scalar multiplication.

Exploration 1   *Understanding the Multiplication of Matrices*

Fund-raising   Each year Sam and Kim make toy cars and trucks for the service club fair. The following matrix examples show how many cars and trucks they plan to make and the number of wheels and nails they will need to construct each of them.

Before you can multiply the two matrices, the *left matrix column labels* (*Cars and Trucks*) must match the *right matrix row labels* (*Cars and Trucks*).

	Cars	Trucks
Sam	8	15
Kim	12	10

	Wheels	Nails
Cars	4	7
Trucks	6	9

Sam builds 8 cars with 4 wheels each. He will need 8(4), or 32 wheels. He is also building 15 trucks with 6 wheels each. He will need 15(6), or 90 more wheels. The total number of wheels will be 8(4) + 15(6), or 122.

Place this information in the Sam/Wheels address for the product matrix.

	Wheels	**Nails**
Sam	8(4) + 15(6) 32 + 90 122	
Kim		

The matching column and row labels disappear after you multiply. The product matrix will no longer show labels for (***Cars and Trucks***). ***Sam and Kim*** and (***Wheels and Nails***) become the labels of the product matrix.

Answer the questions to complete the table.

 How many wheels will Kim need to complete her work?

 How many nails will Sam need to complete his work?

 How many nails will Kim need to complete her work?

 What matrix represents the product? ❖

interdisciplinary
CONNECTION

Business Present the following problem to students to discuss and solve. Two restaurants receive daily delivery of bread from a bakery. The matrix below shows the number of loaves of each type of bread each restaurant receives.

	wheat	white	rye
Bob's	15	10	8
Betty's	12	16	9

The matrix below shows the cost of one loaf of each type of bread.

	Cost
wheat	0.75
white	0.60
rye	0.85

How much does each restaurant spend per day on bread?

	Cost
Bob's	24.05
Betty's	26.25

Cooperative Learning

Have students work in small groups to complete the explorations in this lesson. Encourage each group to thoroughly discuss each step of each exploration before deciding upon a final answer to record. Ask volunteers from each group to share their results with the class.

Exploration 1 Notes

The fact that the labels "Cars" and "Trucks" disappear after multiplying should help students to use the correct combinations of rows and columns. Suggest that students think of "turning" the first matrix on its side to match the "Cars" and "Trucks" labels. Then 8 goes with 4 and 15 goes with 6. Likewise, 12 goes with 7 and 10 goes with 9. You can create a visual example of this by writing the matrices on index cards and then actually turning the cards to match up the correct entries.

ongoing
ASSESSMENT

4. $\begin{bmatrix} 122 & 191 \\ 108 & 174 \end{bmatrix}$

Have students write the general product matrix containing just the addresses of the entries before they actually complete the matrix multiplication. Then students can look at the numbers in the address to determine the row and column to use to obtain that entry.

4. To find the entry at the address p_{ab}, multiply the elements in row a of Matrix A by the corresponding elements in column b of Matrix B. Then find the sum of the products. For example, in Exploration 2 to find the entry at the address p_{11} in AB, multiply $(3)(-2)$ and $(5)(6)$. Then find the sum, $-6 + 30$, or 24. Repeat the process for row 1 of A and column 2 of B to get $(3)(-7) + (5)(5)$, or 4 for address p_{12} of the product. Continue the process with row 2 of A and column 1 of B for p_{21} of AB and with row 2 of A and column 2 of B for p_{22} of AB

Use Transparency ▶ 33

Exploration 2 *The Procedure for Multiplying Matrices*

From Exploration 1, it is possible to create a procedure for multiplying matrices. First, check to see if the number of left matrix columns matches the number of right matrix rows. Then follow the pattern to see how to multiply the matrices.

Let $A = \begin{bmatrix} 3 & 5 \\ 4 & -1 \end{bmatrix}$ and $B = \begin{bmatrix} -2 & -7 \\ 6 & 5 \end{bmatrix}$. Find the product, $P = AB$.

First, find the products of each entry in row 1 of matrix A and each corresponding entry in column 1 of matrix B. Then place the sum of the products at the address p_{11} in the product matrix, P.

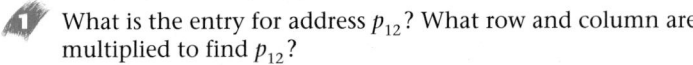

Matrix A   Matrix B

$\begin{bmatrix} 3 & 5 \\ 4 & -1 \end{bmatrix}$ times $\begin{bmatrix} -2 & -7 \\ 6 & 5 \end{bmatrix} \longrightarrow 3(-2) + 5(6) \longrightarrow P = \begin{bmatrix} 24 & \\ & \end{bmatrix}$

The entry in p_{11} is the sum of [**row 1** entries times **column 1** entries].

Continue this pattern to complete the remaining entries.

1 What is the entry for address p_{12}? What row and column are multiplied to find p_{12}?

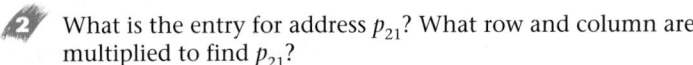

Matrix A   Matrix B

$\begin{bmatrix} 3 & 5 \\ 4 & -1 \end{bmatrix}$ times $\begin{bmatrix} -2 & -7 \\ 6 & 5 \end{bmatrix} \longrightarrow 3(?) + 5(?) \longrightarrow P = \begin{bmatrix} 24 & ? \\ & \end{bmatrix}$

The entry in p_{12} is the sum of [**row _?_** entries times **column _?_** entries].

2 What is the entry for address p_{21}? What row and column are multiplied to find p_{21}?

Matrix A   Matrix B

$\begin{bmatrix} 3 & 5 \\ 4 & -1 \end{bmatrix}$ times $\begin{bmatrix} -2 & -7 \\ 6 & 5 \end{bmatrix} \longrightarrow ?(-2) + ?(6) \longrightarrow P = \begin{bmatrix} 24 & p_{12} \\ ? & \end{bmatrix}$

The entry in p_{21} is the sum of [**row _?_** entries times **column _?_** entries].

3 What is the last entry for matrix P? What row and column are multiplied to find the last entry? What is the address for the last entry in matrix P?

Matrix A   Matrix B

$\begin{bmatrix} 3 & 5 \\ 4 & -1 \end{bmatrix}$ times $\begin{bmatrix} -2 & -7 \\ 6 & 5 \end{bmatrix} \longrightarrow ?(?) + ?(?) \longrightarrow P = \begin{bmatrix} 24 & p_{12} \\ p_{21} & ? \end{bmatrix}$

The entry in p_{22} is the sum of [**row _?_** entries times **column _?_** entries].

4 Summarize in your own words the procedure for multiplying matrices. ❖

Have students find a formula for each entry in the product matrix C for matrices A and B as shown below.

$A = \begin{bmatrix} a_{11} & a_{12} \\ a_{21} & a_{22} \end{bmatrix}$ $B = \begin{bmatrix} b_{11} & b_{12} \\ b_{21} & b_{22} \end{bmatrix}$

$[A \cdot B = C =$

$\begin{bmatrix} a_{11}b_{11} + a_{12}b_{21} & a_{11}b_{12} + a_{12}b_{22} \\ a_{21}b_{11} + a_{22}b_{21} & a_{21}b_{12} + a_{22}b_{22} \end{bmatrix}]$

Kinesthetic Learners Use strips of paper to write down the entries in each row of the first matrix to be multiplied and the entries in each column of the second matrix to be multiplied. As students multiply the matrices, have them pick up the correct row from the first matrix and match it to the correct column of the second matrix. Then have them calculate the entry for the product matrix.

Mary places two orders for new tableware. Mary's first order is 1 cup, 4 saucers, and 7 plates. Her second order consists of 3 cups, 6 saucers, and 8 plates. What is the cost of each of the two orders?

Let matrix C represent the cost for each saucer, cup, and plate. Let matrix D represent the number of pieces in her first order and the number of pieces in her second order.

$$\begin{array}{c} \quad\quad \textbf{Cups} \quad \textbf{Saucers} \quad \textbf{Plates} \\ \textbf{Cost} \quad [2 \quad\quad 5 \quad\quad\quad 9] = C \end{array}$$

$$\begin{array}{c} \quad\quad\quad \textbf{First} \quad \textbf{Second} \\ \quad\quad\quad \text{order} \quad \text{order} \\ \begin{array}{c}\textbf{Cups}\\\textbf{Saucers}\\\textbf{Plates}\end{array} \begin{bmatrix} 1 & 3 \\ 4 & 6 \\ 7 & 8 \end{bmatrix} = D \end{array}$$

Check the row and column dimensions of each given matrix to see if multiplication is possible. Then find the product matrix.

Since the number of columns in C matches the number of rows in D, multiplication is possible.

C (1 by 3) times D (3 by 2) $\rightarrow P$ (1 by 2)

$$\textbf{Matrix } C \;\cdot\; \textbf{Matrix } D$$

$$[2 \quad 5 \quad 9] \begin{bmatrix} 1 & 3 \\ 4 & 6 \\ 7 & 8 \end{bmatrix} = P$$

$CD = [2(1) + 5(4) + 9(7) \quad\quad 2(3) + 5(6) + 9(8)] = [85 \quad 108] = P$

The cost for the first order is $85, and the cost for the second order is $108. ❖

Exploration 3 The Multiplicative Identity Matrix

1 Let $L = \begin{bmatrix} 3 & -7 \\ 16 & 9 \end{bmatrix}$, $M = \begin{bmatrix} 1 & 0 \\ 0 & 1 \end{bmatrix}$, and $N = \begin{bmatrix} 12 & 9 \\ -7 & -4 \end{bmatrix}$. Multiply.

a. $\begin{bmatrix} 1 & 0 \\ 0 & 1 \end{bmatrix}\begin{bmatrix} 3 & -7 \\ 16 & 9 \end{bmatrix}$ **b.** $\begin{bmatrix} 12 & 9 \\ -7 & -4 \end{bmatrix}\begin{bmatrix} 1 & 0 \\ 0 & 1 \end{bmatrix}$ **c.** $\begin{bmatrix} 1 & 0 \\ 0 & 1 \end{bmatrix}\begin{bmatrix} 12 & 9 \\ -7 & -4 \end{bmatrix}$

2 What do you notice about the products?

3 Notice that $N \cdot M = M \cdot N$. Why is M called an identity matrix?

4 Find $L \cdot N$ and $N \cdot L$. Make a conjecture about the commutative property and matrix multiplication.

5 Explain why a multiplication identity matrix must be a square matrix. ❖

Using Algorithms

Have students work in pairs. Have students write their own exercises. These should include some that involve scalar multiplication, several pairs of matrices that can be multiplied, several pairs of matrices that cannot be multiplied, and several that involve multiplying by the identity matrix. Have different pairs of students exchange papers and complete the exercises, when possible.

Exploration Notes 3

Emphasize that the 2 × 2 identity matrix is the same for all 2 × 2 matrices.

A ongoing SSESSMENT

5. The identity matrix must work on both sides. Thus, the dimensions of the original matrix and its identity matrix must be equal.

EXERCISES & PROBLEMS

Communicate

Discuss how to identify the row and column dimensions of the following matrices.

1. $\begin{bmatrix} 1 \\ -5 \\ 3 \end{bmatrix}$ 2. $\begin{bmatrix} 1 & 3 \\ 0 & 5 \\ 12 & -4 \end{bmatrix}$ 3. $\begin{bmatrix} 1 & 4 & 3 & 108 \end{bmatrix}$

Discuss how to recognize if two matrices can be multiplied. Explain how to perform the multiplication with the given matrices.

4. $3\begin{bmatrix} -6 & 1 \\ 4 & -1 \end{bmatrix}$ 5. $\begin{bmatrix} 2 & -4 \\ 1 & 0 \end{bmatrix}\begin{bmatrix} 5 & -1 \\ 6 & 5 \end{bmatrix}$ 6. $\begin{bmatrix} 1 & 2 \\ 1 & -3 \\ 3 & 5 \end{bmatrix}\begin{bmatrix} 4 & 1 & 3 \\ 2 & 1 & 6 \end{bmatrix}$

Practice & Apply

If you have technology available to assist you with the matrix operations, you should make use of it as directed by your teacher.

The dimensions of two matrices are given. If the two matrices can be multiplied, indicate the dimensions of the product matrix. If they cannot, write *Not Possible*.

	A	B	AB
7.	2 by 4	3 by 5	___
8.	1 by 3	3 by 5	___
9.	4 by 2	2 by 1	___
10.	3 by 3	3 by 2	___
11.	1 by 2	5 by 2	___

Use the given matrices to evaluate the following operations. Write *Not Possible* when appropriate.

$A = \begin{bmatrix} 2 & 0 \\ -1 & 5 \end{bmatrix}$ $B = \begin{bmatrix} 1 & 3 \\ -2 & -3 \\ 4 & 0 \end{bmatrix}$ $C = \begin{bmatrix} 6 & -4 \\ -1 & 12 \\ -2 & 5 \end{bmatrix}$ $D = \begin{bmatrix} -3 & 1 & 4 \\ -2 & 2 & 6 \end{bmatrix}$

12. $B + C$ 13. $-5D$ 14. $3B - 4C$

15. AB 16. BD 17. DA

18. DC 19. $-BC$ 20. CA

Find the missing element(s) in each product matrix.

21. $-3\begin{bmatrix} -1 & 8 \\ 0 & 3 \end{bmatrix} = \begin{bmatrix} 3 & ? \\ 0 & -9 \end{bmatrix} \begin{matrix} \\ -24 \end{matrix}$

22. $\begin{bmatrix} -1 & 8 \\ 0 & 3 \end{bmatrix}\begin{bmatrix} -5 & 6 \\ -9 & 8 \end{bmatrix} = \begin{bmatrix} ? & 58 \\ ? & 24 \end{bmatrix} \begin{matrix} -67 \\ -27 \end{matrix}$

7. Not Possible

8. 1 by 5

9. 4 by 1

10. 3 by 2

11. Not Possible

12. $\begin{bmatrix} 7 & -1 \\ -3 & 9 \\ 2 & 5 \end{bmatrix}$

13. $\begin{bmatrix} 15 & -5 & -20 \\ 10 & -10 & -30 \end{bmatrix}$

14. $\begin{bmatrix} -21 & 25 \\ -2 & -57 \\ 20 & -20 \end{bmatrix}$

15. Not Possible

16. $\begin{bmatrix} -9 & 7 & 22 \\ 12 & -8 & -26 \\ -12 & 4 & 16 \end{bmatrix}$

17. Not Possible

18. $\begin{bmatrix} -27 & 44 \\ -26 & 62 \end{bmatrix}$

The answers to Exercises 19–20 can be found in Additional Answers beginning on page 729.

23. $\begin{bmatrix} 0 & 8 \\ -5 & 1 \\ -2 & 3 \end{bmatrix} \begin{bmatrix} 0 & 2 & -1 \\ 4 & 6 & -3 \end{bmatrix} = \begin{bmatrix} 32 & 48 & ? \\ 4 & ? & 2 \\ ? & 14 & -7 \end{bmatrix} \begin{matrix} -24 \\ -4 \\ 12 \end{matrix}$ **24.** $5.7 \begin{bmatrix} 0.03 & -2.3 \\ 1.8 & 0.4 \end{bmatrix} = \begin{bmatrix} 0.171 & ? \\ 10.26 & ? \end{bmatrix} \begin{matrix} -13.11 \\ 2.28 \end{matrix}$

Use the given matrices for the following questions about the properties of matrix multiplication. Complete Exercises 25–29.

$$A = \begin{bmatrix} 0 & 4 \\ -5 & 1 \\ -2 & 3 \end{bmatrix} \qquad B = \begin{bmatrix} 2 & -1 \\ 4 & -3 \end{bmatrix} \qquad C = \begin{bmatrix} -1 \\ 2 \end{bmatrix} \qquad D = \begin{bmatrix} 1 & -2 \\ 3 & -4 \end{bmatrix}$$

25. What are the dimensions of the matrices A, B, and C? Are the dimensions compatible for multiplication? 3×2; 2×2; 2×1; AB, AC, and BC are possible.

26. Does $A(BC) = (AB)C$? If they are equal, what is the product?

27. What is your conjecture about the Associative Property for Multiplication? Find 3 more matrices that confirm your conjecture, or find 3 more matrices that show a counterexample.

28. What are the row and column dimensions of matrices B and D? Are these dimensions compatible for multiplication?

29. Does $BD = DB$? Make a conjecture about the Commutative Property for multiplying matrices. Is this true for all matrices?

30. What characteristics of matrices are necessary to make multiplication possible?

31. If $X = \begin{bmatrix} 1 & 2 & 3 \end{bmatrix}$ and $Y = \begin{bmatrix} 3 \\ 2 \\ 5 \end{bmatrix}$, what are the products XY and YX?

32. Manufacturing At the fair, toy cars and trucks must be sold at a price that will generate a profit. Use the matrices below to calculate how much it will cost to make a car and how much it will cost to make a truck. Express your answer as a labeled matrix.

Wooden Toy Craftsman
Denis Grant
764 Harmony Lane
Maplewood, WI 54321
Phone: (123) 456-7891

	Supplies		
Type:	Wheels	Nails	Blocks
Cars	4	7	2
Trucks	6	9	3

Supplies:		Cost in cents
	Wheels	25
	Nails	3
	Blocks	45

Cost of supplies:

wheels	$0.25
Nails	$0.03
Blocks	$0.45

33. Pricing Use the given *Stock by Type* matrix to determine how much it will cost Sam and Kim to produce all of the toy cars and trucks. Express your answer as a labeled matrix in dollars.

		Type	
Stock:		cars	trucks
	Sam	8	15
	Kim	12	10

34. Sam and Kim decide to charge $5 per car and $6.50 per truck. Create your selling price matrix, and determine how much money Sam and Kim will collect at the fair if all of the cars and trucks are sold.

35. Sales Tax If you make a purchase, the total price you pay is the retail price plus sales tax. If sales tax is 6% in Sam and Kim's town, create the matrix that represents the amount of sales tax for cars and trucks sold at the fair.

36. How much profit will Sam and Kim each make after paying cost and sales tax?

26. Yes; $\begin{bmatrix} -40 \\ 10 \\ -22 \end{bmatrix}$

27. The Associative Property of Multiplication is true for numbers and matrices.

28. Both are 2 by 2; yes

29. No; The Commutative Property of Multiplication does not apply to matrices; No

30. The number of columns in the first matrix must equal the number of rows in the second matrix.

31. $[22]$; $\begin{bmatrix} 3 & 6 & 9 \\ 2 & 4 & 6 \\ 5 & 10 & 15 \end{bmatrix}$

32. Cars $\begin{bmatrix} 2.11 \\ 3.12 \end{bmatrix}$
Trucks

33. Sam $\begin{bmatrix} 63.68 \\ 56.52 \end{bmatrix}$
Kim $\begin{matrix} \text{Cost} \end{matrix}$

34. Cars $\begin{bmatrix} 137.50 \\ 125.00 \end{bmatrix}$
Trucks $\begin{matrix} \text{Sales} \end{matrix}$

35. Cars $\begin{bmatrix} 8.25 \\ 7.50 \end{bmatrix}$
Trucks $\begin{matrix} \text{Sales Tax} \end{matrix}$

36. Cars $\begin{bmatrix} 65.57 \\ 60.98 \end{bmatrix}$
Trucks $\begin{matrix} \text{Profit} \end{matrix}$

Technology Master

Technology
7.3 Matrix Multiplication and Closure

You know that the set of integers is closed under multiplication. That is, when you multiply any two integers, you get another integer.

Consider the matrix $\begin{bmatrix} a & 0 \\ 0 & a \end{bmatrix}$. Its form indicates that the entries on the main diagonal (addresses p_{11} and p_{22}) are a, and its entries elsewhere are 0. Do you think that the product of two matrices of this form will be another matrix of this form?

Use a graphics calculator to find each product.

1. $\begin{bmatrix} 2 & 0 \\ 0 & 2 \end{bmatrix} \begin{bmatrix} 3 & 0 \\ 0 & 3 \end{bmatrix}$ **2.** $\begin{bmatrix} -2 & 0 \\ 0 & -2 \end{bmatrix} \begin{bmatrix} 1.5 & 0 \\ 0 & 1.5 \end{bmatrix}$

3. $\begin{bmatrix} \sqrt{2} & 0 \\ 0 & \sqrt{2} \end{bmatrix} \begin{bmatrix} \sqrt{2} & 0 \\ 0 & \sqrt{2} \end{bmatrix}$ **4.** $\begin{bmatrix} 6.2 & 0 \\ 0 & 6.2 \end{bmatrix} \begin{bmatrix} \frac{1}{2} & 0 \\ 0 & \frac{1}{2} \end{bmatrix}$

5. Without multiplying, predict $\begin{bmatrix} a & 0 \\ 0 & a \end{bmatrix} \begin{bmatrix} b & 0 \\ 0 & b \end{bmatrix}$. Is the set of matrices of the form described above closed under multiplication?

Use a graphics calculator to find each product.

6. $\begin{bmatrix} 1 & 3 \\ 3 & 1 \end{bmatrix} \begin{bmatrix} 1 & 3 \\ 0 & 1 \end{bmatrix}$ **7.** $\begin{bmatrix} 1 & -2.5 \\ 0 & 1 \end{bmatrix} \begin{bmatrix} 1 & 6.5 \\ 0 & 1 \end{bmatrix}$

8. Without multiplying, predict $\begin{bmatrix} 1 & a \\ 0 & 1 \end{bmatrix} \begin{bmatrix} 1 & b \\ 0 & 1 \end{bmatrix}$. Is the set of matrices of the form shown here closed under multiplication?

alternative ASSESSMENT

Portfolio Assessment

Have students write a report explaining how to multiply by a scalar, how to multiply two matrices, and how to multiply by the identity matrix. The report should be written for a person who has never performed these tasks before. Be sure they include information about how to determine whether two matrices can be multiplied. Ask students to give at least one real-world application as a part of their report.

Look Beyond

These exercises preview the concept of the multiplicative inverse of a matrix, which is presented in the next lesson.

Fund-raising The first matrix shows the number of flags and buttons sold by each class. The second gives the purchase cost and selling price of the flags and buttons.

	Flags	Buttons
Freshmen	125	200
Sophomore	100	200
Junior	50	150
Senior	200	250

	Purchase cost	Selling price
Flags	0.75	2.00
Buttons	0.35	1.00

37. Create the matrix that represents the money both spent and made by each of the classes.

38. Add a column to the matrix to represent the profit made by each class.

39. Why do you think the number of flags and buttons sold decreases each year after the freshman year, but increases again during senior year?

At Central City High School, each class raises funds by selling class flags and buttons.

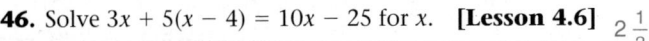

Look Back

40. Find the sum of $1 + 2 + 3 + \cdots + 15$. **[Lesson 1.1]** 120

41. Simplify $5 - 2y\{3 - 2[x + 6y - (3y + x) + 8]\}$. **[Lesson 1.4]**
$12y^2 + 26y + 5$

Find the value of y if x is equal to 5. [Lessons 3.5, 3.6]

42. $2x + y = 70$ **43.** $2x - y = 61$ **44.** $2x + 2y = 100$
  60   -51 45

45. **Geometry** The area of trapezoid is defined by the formula $A = \frac{1}{2}h(b_1 + b_2)$ where h is the height and b_1 and b_2 are the bases. Solve the equation for h. **[Lesson 4.3]** $\frac{2A}{b_1 + b_2}$

46. Solve $3x + 5(x - 4) = 10x - 25$ for x. **[Lesson 4.6]** $2\frac{1}{2}$

47. Graph the equation of the line that is perpendicular to $-3x + 9y = 27$ and contains the point $(-6, 7)$. **[Lesson 5.6]**

48. Solve by the elimination method. $\begin{cases} 2x - 5y = 20 \\ 3x - 4y = 37 \end{cases}$ **[Lesson 6.3]** (15, 2)

49. You and a friend can paddle your canoe 6 miles upstream in 3 hours. It takes the two of you 2 hours to return the same distance downstream. Find the rate of the current in the stream. **[Lesson 6.6]** $\frac{1}{2}$ mph

Look Beyond

Use matrices A and B to complete Exercises 50–51.

$$A = \begin{bmatrix} 2 & -1 \\ 4 & -3 \end{bmatrix} \qquad B = \begin{bmatrix} 1.5 & -0.5 \\ 2 & -1 \end{bmatrix}$$

50. Find AB. Label the resulting matrix C.

51. Why is matrix B called the inverse of matrix A?

37.

	Cost	Sale Price
Freshman	163.75	450.00
Sophomore	145.00	400.00
Junior	90.00	250.00
Senior	237.50	650.00

38. Profit

$$\begin{bmatrix} 286.25 \\ 255.00 \\ 160.00 \\ 412.50 \end{bmatrix}$$

39. Answers may vary. Seniors have graduation activities and trips that require more funds to be raised.

The answers to Exercises 47, 50, and 51 can be found in Additional Answers beginning on page 729.

LESSON 7.4 Multiplicative Inverse of a Matrix

PREPARE

Objectives
- Determine the inverse matrix for multiplication.
- Use technology to find the inverse of a matrix.

RESOURCES

- Practice Master   7.4
- Enrichment Master   7.4
- Technology Master   7.4
- Lesson Activity Master   7.4
- Quiz   7.4
- Spanish Resources   7.4

 *You have used the Reciprocal Property to solve multiplication equations. When you solve a matrix equation, you use the inverse of a matrix in much the same way.*

You know that $r \cdot \frac{1}{r} = 1$, when $r \neq 0$. Since their product is the multiplicative identity, r and $\frac{1}{r}$ are called reciprocals or multiplicative inverses. A similar situation occurs with matrices. If the product of two matrices equals an identity matrix, the matrices are inverses. The symbol A^{-1} is used to represent the inverse of matrix A.

Examine the matrix multiplication.

$$A \quad \cdot \quad B \quad = \quad I \quad = \quad B \quad \cdot \quad A$$

$$\begin{bmatrix} 3 & -1 \\ 4 & 2 \end{bmatrix} \begin{bmatrix} 0.2 & 0.1 \\ -0.4 & 0.3 \end{bmatrix} = \begin{bmatrix} 1 & 0 \\ 0 & 1 \end{bmatrix} = \begin{bmatrix} 0.2 & 0.1 \\ -0.4 & 0.3 \end{bmatrix} \begin{bmatrix} 3 & -1 \\ 4 & 2 \end{bmatrix}$$

The matrices $\begin{bmatrix} 3 & -1 \\ 4 & 2 \end{bmatrix}$ and $\begin{bmatrix} 0.2 & 0.1 \\ -0.4 & 0.3 \end{bmatrix}$ are **inverses.** The product of A

and B is the **identity matrix**, $\begin{bmatrix} 1 & 0 \\ 0 & 1 \end{bmatrix}$. Thus, $B = A^{-1}$.

MULTIPLICATIVE INVERSE MATRIX
The matrix A has an inverse matrix, A^{-1}, if
$$AA^{-1} = I = A^{-1}A,$$
where I is the identity matrix.

Assessing Prior Knowledge

Find the multiplicative inverse, or reciprocal, of each number.

1. $-2 \quad \left[-\frac{1}{2}\right]$

2. $\frac{2}{3} \quad \left[\frac{3}{2}\right]$

3. $-\frac{4}{5} \quad \left[-\frac{5}{4}\right]$

4. $3\frac{1}{3} \quad \left[\frac{3}{10}\right]$

5. $-1\frac{3}{5} \quad \left[-\frac{5}{8}\right]$

TEACH

 The importance of the multiplicative inverse of a matrix will become more evident to students when they study matrix equations in the next lesson.

ALTERNATIVE teaching strategy

Technology Have students use a graphics calculator or computer software such as *f(g) Scholar*™ to find the inverse of matrix A shown on page 331. To find the inverse on a TI-82, enter the matrix, press [2nd] [QUIT] to return to the Home screen, press [MATRX] and choose A:, and then press [x⁻¹] [ENTER]. The inverse matrix will appear on the screen. Have students compare the matrix on the screen to matrix B given on page 331. Students should find that the matrices are the same. Then have students use technology to multiply A and A^{-1}. They should find the result to be the identity matrix. Have students complete several more examples.

Technology provides a quick and easy way to find the multiplicative inverse of a matrix, but students should successfully complete several examples of the manual method first.

Cooperative Learning

Have students work in pairs to study the examples. After students have studied the solution presented in the text, have them attempt to find the inverse of the same matrix on their own. Then have them work together to write the equations. Have one student work to find the values of two of the variables, while the other student finds the values of the other two variables. Have them reverse roles as they work through the second example.

Alternative Example 1

If matrix $B = \begin{bmatrix} 2 & 6 \\ -2 & 4 \end{bmatrix}$, find B^{-1}.

$[B^{-1} = \begin{bmatrix} 0.2 & -0.3 \\ 0.1 & 0.1 \end{bmatrix}]$

Finding the Inverse Matrix

EXAMPLE 1

If matrix $A = \begin{bmatrix} 3 & 5 \\ 4 & 6 \end{bmatrix}$, find A^{-1}.

Solution ➤

You know from the definition that $AA^{-1} = I = A^{-1}A$. Since A is a 2 by 2 matrix, A^{-1} must be a 2 by 2 matrix. Let $\begin{bmatrix} a & b \\ c & d \end{bmatrix}$ represent A^{-1}, the inverse matrix. Then substitute the matrices in $AA^{-1} = I$.

$$A \quad \cdot \quad A^{-1} \quad = \quad I$$

$$\begin{bmatrix} 3 & 5 \\ 4 & 6 \end{bmatrix}\begin{bmatrix} a & b \\ c & d \end{bmatrix} = \begin{bmatrix} 1 & 0 \\ 0 & 1 \end{bmatrix}$$

Multiply.

$$AA^{-1} \qquad = \quad I$$

$$\begin{bmatrix} 3a + 5c & 3b + 5d \\ 4a + 6c & 4b + 6d \end{bmatrix} = \begin{bmatrix} 1 & 0 \\ 0 & 1 \end{bmatrix}$$

Equal matrices have equal, corresponding matrix entries. Write the equation for each of the corresponding entries. Pair the equations that have the same variables. Then solve the systems of equations.

$$\begin{cases} 3a + 5c = 1 \\ 4a + 6c = 0 \end{cases} \qquad \begin{cases} 3b + 5d = 0 \\ 4b + 6d = 1 \end{cases}$$

Use elimination to solve for a and b.

$$\begin{array}{ll} 18a + 30c = 6 & 18b + 30d = 0 \\ \underline{-20a - 30c =} & \underline{-20b - 30d = -5} \\ \quad -2a \quad = 6 & \quad -2b \quad = -5 \\ \quad\quad a = -3 & \quad\quad b = 2.5 \end{array}$$

Substitute and solve for c and d.

$$\begin{array}{ll} 3(-3) + 5c = 1 & 3(2.5) + 5d = 0 \\ -9 + 5c = 1 & 7.5 + 5d = 0 \\ 5c = 10 & 5d = -7.5 \\ c = 2 & d = -1.5 \end{array}$$

Replace the variables in $\begin{bmatrix} a & b \\ c & d \end{bmatrix}$ with the values you found. The inverse matrix A^{-1} is $\begin{bmatrix} -3 & 2.5 \\ 2 & -1.5 \end{bmatrix}$.

$$A \quad \cdot \quad A^{-1} \quad = \quad I \quad = \quad A^{-1} \quad \cdot \quad A$$

Check $\begin{bmatrix} 3 & 5 \\ 4 & 6 \end{bmatrix}\begin{bmatrix} -3 & 2.5 \\ 2 & -1.5 \end{bmatrix} = \begin{bmatrix} 1 & 0 \\ 0 & 1 \end{bmatrix} = \begin{bmatrix} -3 & 2.5 \\ 2 & -1.5 \end{bmatrix}\begin{bmatrix} 3 & 5 \\ 4 & 6 \end{bmatrix}$ ❖

CRITICAL Thinking

If a matrix has an inverse, what are the requirements for row and column dimensions of the inverse matrix? Why is this the case?

EXAMPLE 2

Find the inverse of $B = \begin{bmatrix} 4 & 6 \\ 2 & 3 \end{bmatrix}$.

Solution ➤

Let $\begin{bmatrix} a & b \\ c & d \end{bmatrix}$ represent the inverse matrix B^{-1}.

Since $BB^{-1} = I$, write the matrix equation.

$$B \quad \cdot \quad B^{-1} \quad = \quad I$$

$$\begin{bmatrix} 4 & 6 \\ 2 & 3 \end{bmatrix} \begin{bmatrix} a & b \\ c & d \end{bmatrix} = \begin{bmatrix} 1 & 0 \\ 0 & 1 \end{bmatrix}$$

Show the matrix multiplication.

$$BB^{-1} \qquad = \quad I$$

$$\begin{bmatrix} 4a + 6c & 4b + 6d \\ 2a + 3c & 2b + 3d \end{bmatrix} = \begin{bmatrix} 1 & 0 \\ 0 & 1 \end{bmatrix}$$

Write the equations from corresponding matrix entries.

$$\begin{cases} 4a + 6c = 1 \\ 2a + 3c = 0 \end{cases} \qquad \begin{cases} 4b + 6d = 0 \\ 2b + 3d = 1 \end{cases}$$

You can use any of the methods you learned in the previous chapter to solve the systems of equations. If you use the elimination method, the result is the following.

$$\begin{array}{ll} 4a + 6c = 1 \\ \underline{-4a - 6c = 0} \\ \quad\quad 0 = 1 \text{ Not true} \end{array} \qquad \begin{array}{ll} 4b + 6d = 0 \\ \underline{-4b - 6d = -2} \\ \quad\quad 0 = -2 \text{ Not true} \end{array}$$

In this case, you find 0 equal to 1 and 0 equal to -2, which cannot be true.

There is *no solution* for either system. The matrix $\begin{bmatrix} 4 & 6 \\ 2 & 3 \end{bmatrix}$ *has no inverse.* ❖

Carefully study the entries in the rows of the matrix that has no inverse. How is one row related to the other? Make a conjecture about a matrix that has no inverse.

Try This Find the inverse of $\begin{bmatrix} 0 & 1 \\ -6 & 2 \end{bmatrix}$.

ASSESS

Selected Answers
Odd-numbered Exercises 7–39

Assignment Guide
Core 1–6, 8–20, even 22–41

Core Two-Year 1–39

Technology
A graphing calculator or computer software such as *f(g) Scholar™* should be used for Exercises 19–21, 24–26, and 30.

Matrix Technology

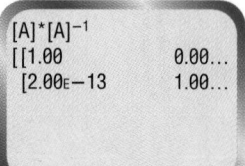

```
ERR:DIVIDE BY 0
1:Goto
2:Quit
```

```
[A]*[A]⁻¹
 [[1.00      0.00…
  [2.00ᴇ-13   1.00…
```

Working with matrices can be tedious because of the large number of arithmetic calculations. As you continue studying matrices, you will soon encounter larger matrices. The formation of an inverse matrix with dimensions beyond 2 by 2 takes many calculations and a lot of time unless you use technology. Computers and calculators perform these calculations instantaneously, freeing you to explore the more interesting aspects of matrices.

When you use technology, however, you should be aware of unusual situations that cause calculators and computers to display error messages. For example, you can expect an error message if a matrix does not have an inverse or if you try to divide by 0.

Another difficulty could appear when you encounter rounding errors. You might see unusual expressions such as 2E − 13 for the number 0.

EXERCISES & PROBLEMS

Communicate

Explain how to find the inverse of the following matrices, if possible.

1. $\begin{bmatrix} -1 & 6 \\ 0 & 3 \end{bmatrix}$ **2.** $\begin{bmatrix} 1 & 2 \\ 2 & 4 \end{bmatrix}$ **3.** $\begin{bmatrix} 2 & -5 & 7 \end{bmatrix}$

4. What conditions would cause a matrix not to have an inverse?

5. How is the inverse of a matrix used?

6. What occurs when a number is multiplied by its inverse?

Practice & Apply

Use the definition of inverse matrices to find the inverse matrix for each of the given matrices.

7. $\begin{bmatrix} 5 & -7 \\ -2 & 3 \end{bmatrix}$ **8.** $\begin{bmatrix} 3 & 4 \\ 5 & 5 \end{bmatrix}$ **9.** $\begin{bmatrix} -1 & 0.04 \\ -0.6 & 1 \end{bmatrix}$

10. $\begin{bmatrix} 1 & 0 \\ 0 & -1 \end{bmatrix}$ **11.** $\begin{bmatrix} -1 & -2 \\ 2 & 0 \end{bmatrix}$ **12.** $\begin{bmatrix} -\frac{4}{5} & 1 \\ \frac{9}{5} & -2 \end{bmatrix}$

7. $\begin{bmatrix} 3 & 7 \\ 2 & 5 \end{bmatrix}$

8. $\begin{bmatrix} -1 & 0.8 \\ 1 & -0.6 \end{bmatrix}$

9. $\begin{bmatrix} -1.025 & 0.041 \\ -0.615 & 1.025 \end{bmatrix}$ or $\begin{bmatrix} -\frac{125}{122} & \frac{5}{122} \\ -\frac{75}{122} & \frac{125}{122} \end{bmatrix}$

10. $\begin{bmatrix} 1 & 0 \\ 0 & -1 \end{bmatrix}$

11. $\begin{bmatrix} 0 & 0.5 \\ -0.5 & -0.25 \end{bmatrix}$

20. $\begin{bmatrix} 0.833 & 1 & 0.167 \\ -0.133 & -0.2 & 0.133 \\ 0.083 & 0 & -0.083 \end{bmatrix}$

21. $\begin{bmatrix} 0 & -0.2 & 0.2 \\ -0.2 & 0 & 0.2 \\ 1.2 & 1.2 & -1.4 \end{bmatrix}$

The answer to Exercise 12 can be found in Additional Answers beginning on page 729.

Does the matrix have an inverse?

13. $\begin{bmatrix} 5 & -3 \\ 2 & -4 \end{bmatrix}$ yes

14. $\begin{bmatrix} 6 & 8 \\ 3 & 4 \end{bmatrix}$ no

15. $\begin{bmatrix} -1 & 2 & -3 \\ -4 & 5 & -6 \end{bmatrix}$ no

16. $\begin{bmatrix} \frac{1}{2} & -1 \\ 1 & -2 \end{bmatrix}$ no

17. $\begin{bmatrix} -0.2 & 1 \\ 0.3 & -1.5 \end{bmatrix}$ no

18. $\begin{bmatrix} 5 \\ -6 \\ 7 \end{bmatrix}$ no

Technology Use technology to find the inverse if it exists.

19. $\begin{bmatrix} 2 & -4 & 6 \\ 5 & 0 & -8 \\ 3 & -6 & 9 \end{bmatrix}$ none

20. $\begin{bmatrix} 1 & 5 & 10 \\ 0 & -5 & -8 \\ 1 & 5 & -2 \end{bmatrix}$

21. $\begin{bmatrix} 6 & 1 & 1 \\ 1 & 6 & 1 \\ 6 & 6 & 1 \end{bmatrix}$

22. A general 2 by 2 matrix can be written $\begin{bmatrix} a & b \\ c & d \end{bmatrix}$. Calculate the products ad and bc for each of the matrices $\begin{bmatrix} 3 & 5 \\ 4 & 6 \end{bmatrix}$ and $\begin{bmatrix} 4 & 6 \\ 2 & 3 \end{bmatrix}$. Do you find anything significant about these products? $18 \neq 20$ and $12 = 12$. $\begin{bmatrix} 4 & 6 \\ 2 & 3 \end{bmatrix}$ has no inverse.

23. Choose at least three of the 2 by 2 matrices that *do not have inverses* from the exercises. Again, for each of these matrices, find the value of the products ad and bc. Do you find anything significant about these products? Yes, the products are equal.

For each matrix below find the products ad and bc. Write and test a conjecture about the matrices. Use technology to check your conjecture.

24. $\begin{bmatrix} -6 & 60 \\ 0.5 & -5 \end{bmatrix}$
$30 = 30$
no inverse

25. $\begin{bmatrix} 7 & -4 \\ 14 & -7 \end{bmatrix}$
$-49 \neq -56$
has inverse

26. $\begin{bmatrix} -8 & 3 \\ 7 & 2 \end{bmatrix}$
$-16 \neq 21$
has inverse

Geometry The matrices represent the endpoints of two line segments AB and CD.

$$\begin{array}{c} \\ A \\ B \end{array} \begin{array}{cc} x & y \\ \begin{bmatrix} 0 & 6 \\ 4 & 12 \end{bmatrix} \end{array} \qquad \begin{array}{c} \\ C \\ D \end{array} \begin{array}{cc} x & y \\ \begin{bmatrix} 6 & 0 \\ 12 & 4 \end{bmatrix} \end{array}$$

27. Graph both line segments. Describe how the line segment slopes are related to each other. They are reciprocals.

28. If $\begin{bmatrix} 0 & 6 \\ 4 & 12 \end{bmatrix} \cdot T = \begin{bmatrix} 6 & 0 \\ 12 & 4 \end{bmatrix}$, find the matrix T.

29. Draw a line segment of your own choosing, and write a matrix to represent the endpoints. Multiply your matrix by T, and graph the solution. Compare your results to segments AB and CD.

30. Technology Use technology to find the inverse of matrix A.

$$A = \begin{bmatrix} 1 & 2 & 3 \\ 2 & 4 & 5 \\ 3 & 5 & 6 \end{bmatrix}$$

The answer to Exercise 27 can be found in Additional Answers beginning on page 729.

28. $\begin{bmatrix} 0 & 1 \\ 1 & 0 \end{bmatrix}$

29. Answers may vary. For example,
$$\begin{bmatrix} 4 & 2 \\ 6 & 7 \end{bmatrix} \cdot T = \begin{bmatrix} 2 & 4 \\ 7 & 6 \end{bmatrix}.$$
The slope of each line segment is the reciprocal of the other.

30. Answers may vary. For example, enter matrix A as a 3×3 matrix. Use the inverse key on a calculator to determine the inverse,
$$\begin{bmatrix} 1 & -3 & 2 \\ -3 & 3 & -1 \\ 2 & -1 & 0 \end{bmatrix}.$$

Technology Master

Technology
7.4 Sequences and Matrices

There is something special about matrix A.

$$A = \begin{bmatrix} 1 & 2 \\ 3 & 4 \end{bmatrix}$$

If you read its entries in order from left to right and from row 1 to row 2, you get an arithmetic sequence.

1, 2, 3, 4

When you try to find the inverse of A by using a system of equations, you will get this matrix.

$$\begin{bmatrix} -2 & 1 \\ 1.5 & 0.5 \end{bmatrix}$$

To use a graphics calculator to find A^{-1}, enter the numbers into a matrix and then use the inverse key. By either method, you will see that A has an inverse.

Is it always true that a matrix whose entries form an arithmetic sequence with nonzero common difference will have an inverse? You can use the exercises that follow to find out.

Use a graphics calculator to find the inverse of each matrix, if the inverse exists.

1. $\begin{bmatrix} 2 & 3 \\ 4 & 5 \end{bmatrix}$

2. $\begin{bmatrix} 3 & 6 \\ 9 & 12 \end{bmatrix}$

3. $\begin{bmatrix} 0 & 4 \\ 8 & 12 \end{bmatrix}$

4. $\begin{bmatrix} -1 & -2 \\ -3 & -4 \end{bmatrix}$

5. $\begin{bmatrix} -2 & 0 \\ 2 & 4 \end{bmatrix}$

6. $\begin{bmatrix} 1.2 & 1.7 \\ 2.2 & 2.7 \end{bmatrix}$

7. Based on your results from Exercises 1–6, do you think that a 2×2 matrix whose entries form an arithmetic sequence with nonzero common difference will have an inverse?

8. Test your answer to Exercise 7 by making your own matrix with entries that form an arithmetic sequence.

31. Science You want to analyze an experiment in which you have collected data that describes the height in centimeters and body temperature in degrees Fahrenheit for 20 people. The data is collected, and a scatter plot is drawn. Describe the function that would best fit this data. **[Lesson 1.7]** linear

Make a table to evaluate each of the following functions when x is -3, -2, -1, 0, 1, 2, and 3. Are any values undefined? **[Lessons 2.1, 2.3, 2.4]**

32. $h(x) = x^2 - 2x + 3$ **33.** $f(x) = \dfrac{2}{x - 2}$ **34.** $g(x) = |x|$

Identify the property of equality that you would use to solve each of the following equations. Then solve each equation. **[Lessons 3.6, 4.3]**

35. $y - 11 = 67$ **36.** $\dfrac{2}{3}x = 16$ **37.** $\dfrac{z}{-3} = 20$
addition, 78   multiplication, 24   multiplication, -60

38. **Coordinate Geometry** A line has slope $\dfrac{3}{4}$ and passes through the point $P(2, -4)$. Write an equation in standard form for a line that is perpendicular to the original line and that passes through the common point, P. **[Lesson 5.6]**
$4x + 3y = -4$

39. Work Schedule Rick works 2 jobs. He keeps track of his hours using matrices for each day of a five-day workweek. Matrices A and B represent 2 weeks of work. Create a matrix, C, that shows the total hours worked each day and at each job for the 2 weeks he worked. **[Lesson 7.2]**

Hours Worked

Week 1: Job 1 $\begin{bmatrix} 3 & 4 & 2 & 3 & 3 \\ 1 & 6 & 5 & 5 & 4 \end{bmatrix}$
   Job 2

M T W Th F

Week 2: Job 1 $\begin{bmatrix} 5 & 0 & 2 & 5 & 3 \\ 3 & 6 & 3 & 5 & 1 \end{bmatrix}$
   Job 2

Look Beyond

The given matrix equation represents a system of equations.

$$\begin{bmatrix} -2 & 5 \\ 4 & -3 \end{bmatrix} \begin{bmatrix} x \\ y \end{bmatrix} = \begin{bmatrix} 10 \\ 1 \end{bmatrix}$$

40. Using the algebra of matrices, write the system of equations in standard form.

41. Solve the system and use matrix multiplication to check your solution.

32.

x	-3	-2	-1	0	1	2	3
$h(x)$	18	11	6	3	2	3	6

No

33.

x	-3	-2	-1	0	1	2	3
$f(x)$	$-\dfrac{2}{5}$	$-\dfrac{1}{2}$	$-\dfrac{2}{3}$	-1	-2	undefined	2

Yes

34.

x	-3	-2	-1	0	1	2	3
$g(x)$	3	2	1	0	1	2	3

No

39.

	M	T	W	Th	F
$C =$ Job 1	8	4	4	8	6
Job 2	4	12	8	10	5

40. $\begin{cases} -2x + 5y = 10 \\ 4x - 3y = 1 \end{cases}$ **41.** $\left(\dfrac{5}{2}, 3\right)$

Solving Matrix Equations

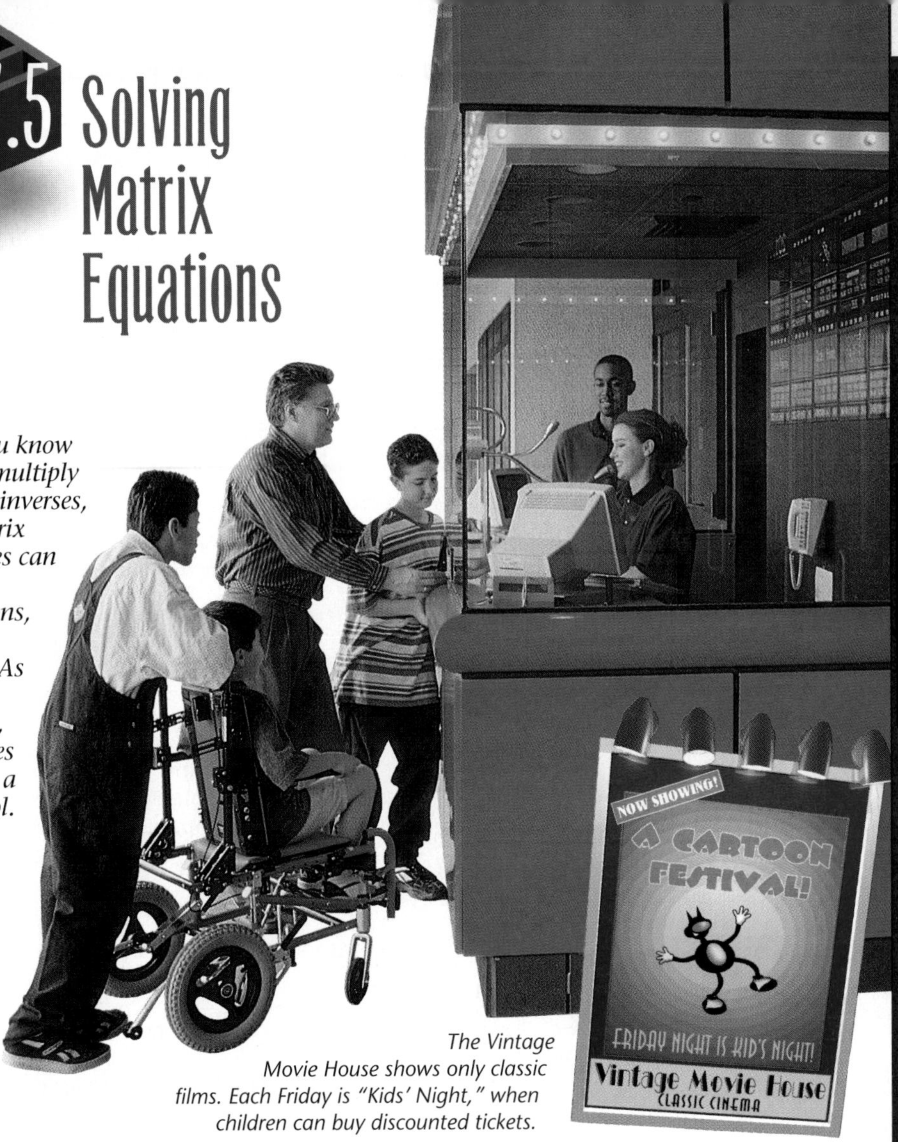

 *Once you know how to multiply matrices and find inverses, you can solve matrix equations. Matrices can be used to solve systems of equations, even those with many unknowns. As the size of the matrices increases, technology becomes more necessary as a computational tool.*

The Vintage Movie House shows only classic films. Each Friday is "Kids' Night," when children can buy discounted tickets.

NOW SHOWING!
A CARTOON FESTIVAL!
FRIDAY NIGHT IS KID'S NIGHT!
Vintage Movie House
CLASSIC CINEMA

Consumer Economics
One Friday, a group of 3 parents and 5 children went to see *Gone with the Wind*, and the cost was $34. A week later, the group that went to *A Cartoon Festival* consisted of 4 parents and 6 children. The cost was $43. Find the cost of the regular and discounted tickets.

Let r = the regular cost of a parent's ticket.
Let d = the discounted cost of a child's ticket.
Write a system of equations.

3 parents and 5 children cost $34. $\begin{cases} 3r + 5d = 34 \\ 4r + 6d = 43 \end{cases}$
4 parents and 6 children cost $43.

This system can be solved by graphing, elimination, and substitution. It can also be solved by using matrices.

ALTERNATIVE teaching strategy

Visual Models Write
$\begin{cases} 3x + 5y = 34 \\ 4x + 6y = 43 \end{cases}$
on the chalkboard have students write this system using one color for the coefficients of x and y, a second color for the variables, and a third color for the constants. Then ask students to write "Coefficient matrix:" in the first color,

"Variable matrix:" in the second color, and "Constant matrix:" in the third color. Show students how to write the coefficient matrix, the variable matrix, and the constant matrix. Have them write each matrix on their paper in the apropriate color. Then name the matrices A, X, and B, and give students the matrix equation $AX = B$.

PREPARE

Objectives

- Solve a system of equations using matrices.
- Use technology to solve a system of equations using matrices.

RESOURCES

- Practice Master   7.5
- Enrichment Master   7.5
- Technology Master   7.5
- Lesson Activity Master 7.5
- Quiz   7.5
- Spanish Resources   7.5

Assessing Prior Knowledge

Solve each equation.

1. $8x = 24$ $[3]$
2. $-14x = 7$ $\left[-\dfrac{1}{2}\right]$
3. $-5x = -36$ $\left[7\dfrac{1}{5}\right]$
4. $\dfrac{x}{7} = 4$ $[28]$
5. $\dfrac{2}{3}x = 12$ $[18]$

TEACH

Remind students that they learned how to solve systems of equations using graphing, substitution, and elimination in Chapter 6. Stress that while all of these methods are good ways to solve a system of equations, using matrix equations to solve such systems has many advantages. Ask students to speculate about what some of those advantages might be.

The Matrix Solution for Systems of Equations

The system of equations $\begin{cases} 3r + 5d = 34 \\ 4r + 6d = 43 \end{cases}$ can be written in **matrix equation** form using the following matrices.

Coefficient matrix	Variable matrix	Constant matrix
$A = \begin{bmatrix} 3 & 5 \\ 4 & 6 \end{bmatrix}$	$X = \begin{bmatrix} r \\ d \end{bmatrix}$	$B = \begin{bmatrix} 34 \\ 43 \end{bmatrix}$

Notice that together these three matrices form the matrix equation $AX = B$.

$$\begin{array}{ccccc} A & \cdot X & = & B \\ \begin{bmatrix} 3 & 5 \\ 4 & 6 \end{bmatrix} & \begin{bmatrix} r \\ d \end{bmatrix} & = & \begin{bmatrix} 34 \\ 43 \end{bmatrix} \end{array}$$

The product matrix, AX, is $\begin{bmatrix} 3r + 5d \\ 4r + 6d \end{bmatrix}$. When AX is set equal to the constant matrix, B, the result is the original system of equations.

To solve the matrix equation, $AX = B$, multiply **both sides** of the matrix equation by the inverse A^{-1}. Be careful to multiply both sides of the matrix equation with A^{-1} on the left. Matrix multiplication is not generally commutative. Recall that $A^{-1}A = I$ and $IX = X$.

$$\begin{aligned} AX &= B \\ A^{-1}AX &= A^{-1}B \\ IX &= A^{-1}B \\ X &= A^{-1}B \end{aligned}$$

Use this idea to solve the equation $\begin{array}{ccccc} A & \cdot X & = & B \\ \begin{bmatrix} 3 & 5 \\ 4 & 6 \end{bmatrix} & \begin{bmatrix} r \\ d \end{bmatrix} & = & \begin{bmatrix} 34 \\ 43 \end{bmatrix} \end{array}$.

In Lesson 7.4, you found that the inverse of $\begin{bmatrix} 3 & 5 \\ 4 & 6 \end{bmatrix}$ is $\begin{bmatrix} -3 & 2.5 \\ 2 & -1.5 \end{bmatrix}$, or A^{-1}. Multiply both sides of $AX = B$ by A^{-1}.

Make sure both multiplications are on the left of A and B.

$$\begin{array}{ccccccc} A^{-1} & \cdot & A & \cdot X & = & A^{-1} & \cdot & B \\ \begin{bmatrix} -3 & 2.5 \\ 2 & -1.5 \end{bmatrix} & & \begin{bmatrix} 3 & 5 \\ 4 & 6 \end{bmatrix} & \begin{bmatrix} r \\ d \end{bmatrix} & = & \begin{bmatrix} -3 & 2.5 \\ 2 & -1.5 \end{bmatrix} & & \begin{bmatrix} 34 \\ 43 \end{bmatrix} \end{array}$$

$$\begin{array}{ccc} I & \cdot X & = & A^{-1}B \\ \begin{bmatrix} 1 & 0 \\ 0 & 1 \end{bmatrix} \begin{bmatrix} r \\ d \end{bmatrix} & = & \begin{bmatrix} -3(34) + 2.5(43) \\ 2(34) - 1.5(43) \end{bmatrix} \end{array}$$

$$\begin{array}{ccc} X & = & A^{-1}B \\ \begin{bmatrix} r \\ d \end{bmatrix} & = & \begin{bmatrix} 5.50 \\ 3.50 \end{bmatrix} \end{array} \quad \text{so} \quad \begin{array}{l} r = 5.50 \\ d = 3.50 \end{array}$$

The cost of the regular ticket, r, for the parents is \$5.50, and the cost of the discounted ticket, d, for the children is \$3.50.

Why must the inverse matrix be multiplied to the left of both the coefficient and constant matrices? How does this procedure compare to solving the linear equation $ax = b$?

Solving Matrix Equations With Technology

Once you know the inverse of matrix A, you can solve the matrix equation $AX = B$. In Lesson 7.4 you used a system of equations to find A^{-1}. If you have a calculator or computer that works with matrices, you can find A^{-1} directly and solve the equation easily.

EXAMPLE 1

Use matrices to solve. $\begin{cases} 6x - y = 3 \\ 2x - 7y = 1 \end{cases}$

Solution ➤

Write the matrices in the following form.

Coefficient matrix	Variable matrix	Constant matrix
$A = \begin{bmatrix} 6 & -1 \\ 2 & -7 \end{bmatrix}$	$X = \begin{bmatrix} x \\ y \end{bmatrix}$	$B = \begin{bmatrix} 3 \\ 1 \end{bmatrix}$

Substitute the matrices into $AX = B$.

$$\begin{matrix} A & \cdot & X & = & B \end{matrix}$$
$$\begin{bmatrix} 6 & -1 \\ 2 & -7 \end{bmatrix} \begin{bmatrix} x \\ y \end{bmatrix} = \begin{bmatrix} 3 \\ 1 \end{bmatrix}$$

Graphics Calculator

Find the inverse for the coefficient matrix. This step can be simplified by using technology. Enter the original matrix into the calculator, and name it A. Place A on the screen, and press the inverse key $\boxed{x^{-1}}$ to get the inverse matrix. You can now solve the matrix equation. Multiply both sides of the matrix equation by the inverse matrix.

$$\begin{matrix} A^{-1} & \cdot & A & \cdot & X & = & A^{-1} & \cdot & B \end{matrix}$$
$$\begin{bmatrix} 0.175 & -0.025 \\ 0.05 & -0.15 \end{bmatrix} \begin{bmatrix} 6 & -1 \\ 2 & -7 \end{bmatrix} \begin{bmatrix} x \\ y \end{bmatrix} = \begin{bmatrix} 0.175 & -0.025 \\ 0.05 & -0.15 \end{bmatrix} \begin{bmatrix} 3 \\ 1 \end{bmatrix}$$

$$\begin{matrix} I & \cdot & X & = & A^{-1}B \end{matrix}$$
$$\begin{bmatrix} 1 & 0 \\ 0 & 1 \end{bmatrix} \begin{bmatrix} x \\ y \end{bmatrix} = \begin{bmatrix} 0.5 \\ 0 \end{bmatrix}$$

```
MATRIX[A] 2 ×2
[6      −1           ]
[2      −7           ]
```

```
[A]⁻¹
       [ [.175 −.025]
         [.050 −.150] ]
```

$$\begin{matrix} X & = & A^{-1}B \end{matrix}$$
$$\begin{bmatrix} x \\ y \end{bmatrix} = \begin{bmatrix} 0.5 \\ 0 \end{bmatrix}$$

The solution to the system is (0.5, 0); x is 0.5, and y is 0. You can check the results by substituting the values for x and y into the original equation. ❖

Try This   Use matrices to solve this system. $\begin{cases} -x + y = 1 \\ -2x + 3y = 0 \end{cases}$

Alternative Example 2

Two clubs bought food for their banquets at the same store. One club served 12 chicken breasts and 10 pieces of fish. They figured the total cost was $47.50. The second club served 15 chicken breasts and 9 pieces of fish. It cost them $53.25. What was the price for each chicken breast and each piece of fish? **[Each chicken breast was $2.50, and each piece of fish was $1.75.]**

On the first weekend, the Woodwards served 14 hamburgers and 12 hot dogs. They figured the total cost was $49.50.

Once you obtain the matrix equation, you can solve the system of equations quickly and easily with technology.

EXAMPLE 2

Two families buy food at the same store. They decide to have cookouts on consecutive weekends. The next week with prices still the same, the Tates had a cookout and served 10 hamburgers and 7 hot dogs. It cost the Tates $33.00. What was the price for each hamburger and each hot dog?

Solution ➤

Organize the information. Define the variables. Let h = the cost of each hamburger. Let d = the cost of each hot dog.

Write the equations.

$$14h + 12d = 49.50$$
$$10h + 7d = 33.00$$

Solve the system of equations. Form the matrices for $AX = B$.

$$A = \begin{bmatrix} 14 & 12 \\ 10 & 7 \end{bmatrix} \quad X = \begin{bmatrix} h \\ d \end{bmatrix} \quad B = \begin{bmatrix} 49.50 \\ 33 \end{bmatrix}$$

$$A \quad \cdot X \quad = \quad B$$
$$\begin{bmatrix} 14 & 12 \\ 10 & 7 \end{bmatrix} \begin{bmatrix} h \\ d \end{bmatrix} = \begin{bmatrix} 49.50 \\ 33 \end{bmatrix}$$

Graphics Calculator

The numbers in the coefficient matrix make it difficult to find an inverse. A graphics calculator saves valuable time.

Enter the coefficient matrix, $\begin{bmatrix} 14 & 12 \\ 10 & 7 \end{bmatrix}$, for A.

Enter the constant matrix, $\begin{bmatrix} 49.50 \\ 33 \end{bmatrix}$, for B.

Select A to recall the coefficient matrix, and take the inverse to get A^{-1}. Then multiply A^{-1} by B. This will give you $\begin{bmatrix} 2.25 \\ 1.50 \end{bmatrix}$.

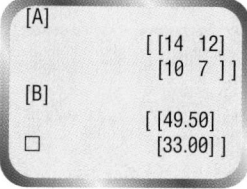

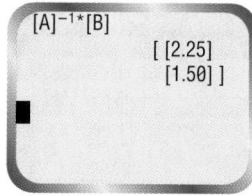

Each hamburger was $2.25, and each hot dog was $1.50.

Check the answer.

$$14(2.25) + 12(1.50) = 49.50$$
$$10(2.25) + 7(1.50) = 33.00 \; ❖$$

RETEACHING the lesson

Using Algorithms

Have students go back to Chapter 6 and choose several systems of equations from Lessons 6.2 and 6.3. Ask students to work in pairs to solve each system using matrix equations. As students work, have them record any questions they may have about solving matrix equations. After students have completed their work, have them present their questions to the class for discussion.

Cultural Connection: Asia Over 2000 years ago, the Chinese developed a way to solve systems of linear equations written in matrix form.

Given the equations: $\begin{cases} 4x + 5y = 22 \\ 2x + 3y = 13 \end{cases}$ The Chinese would form the matrices: $\begin{bmatrix} 4 & 5 & 22 \\ 2 & 3 & 13 \end{bmatrix}$

Apply scalar multiplication, $\dfrac{1}{2}\begin{bmatrix} 4 & 5 & 22 \\ 2 & 3 & 13 \end{bmatrix}$, to produce $\begin{bmatrix} 4 & 5 & 22 \\ 4 & 6 & 26 \end{bmatrix}$.

Subtract row 1 from row 2. $[0 \quad 1 \quad 4] \longrightarrow y = 4$

Finally, substitute into the first equation to get $x = \dfrac{1}{2}$.

Explain how to use the Chinese method to solve the system.

$\begin{cases} 3x + y = 0 \\ 2x + 3y = 7 \end{cases}$

EXERCISES & PROBLEMS

Communicate

Explain how to represent each system in matrix equation form $AX = B$.

1. $\begin{cases} 3a - 6b = 9 \\ 5a + 2b = 9 \end{cases}$ **2.** $\begin{cases} 12x + 5y = 30 \\ 4x - y = 6 \end{cases}$

Explain the method for solving the matrix equation for x and y.

3. $\begin{bmatrix} -2 & -1 \\ 1 & 0 \end{bmatrix}\begin{bmatrix} x \\ y \end{bmatrix} = \begin{bmatrix} -9 \\ -9 \end{bmatrix}$ **4.** $\begin{bmatrix} 1 & 1 \\ 1 & -1 \end{bmatrix}\begin{bmatrix} x \\ y \end{bmatrix} = \begin{bmatrix} 6 \\ 2 \end{bmatrix}$

Practice & Apply

Represent the systems of equations in matrix equation form, $AX = B$.

5. $\begin{cases} 4y = 6x - 12 \\ 5x = 2y + 10 \end{cases}$ **6.** $\begin{cases} 6a - b - 3c = 2 \\ -3a + b - 3c = 1 \\ -2a + 3b + c = -6 \end{cases}$ **7.** $\begin{cases} 2x + 3y + 4z = 8 \\ 6x + 12y + 16z = 31 \\ 4x + 9y + 8z = 17 \end{cases}$

Solve the systems of equations using matrices if possible.

8. $\begin{cases} 2x + y = 4 \\ x + y = 3 \end{cases}$ **9.** $\begin{cases} 2x = 4 \\ 3x + y = 6 \end{cases}$ **10.** $\begin{cases} 2x - 3y = 4 \\ -4x + 6y = 1 \end{cases}$
$(1, 2)$ $(2, 0)$ none

11. Which of the following ordered pairs represents the solution to $\begin{bmatrix} 3 & \frac{1}{2} \\ -2 & \frac{1}{3} \end{bmatrix}\begin{bmatrix} x \\ y \end{bmatrix} = \begin{bmatrix} -6 \\ 12 \end{bmatrix}$?

   a. $(-4, 6)$ **b.** $(4, 12)$ **c.** $(-4, 12)$ **d.** $(4, -6)$
c

5. $\begin{bmatrix} -6 & 4 \\ 5 & -2 \end{bmatrix}\begin{bmatrix} x \\ y \end{bmatrix} = \begin{bmatrix} -12 \\ 10 \end{bmatrix}$

7. $\begin{bmatrix} 2 & 3 & 4 \\ 6 & 12 & 16 \\ 4 & 9 & 8 \end{bmatrix}\begin{bmatrix} x \\ y \\ z \end{bmatrix} = \begin{bmatrix} 8 \\ 31 \\ 17 \end{bmatrix}$

6. $\begin{bmatrix} 6 & -1 & -3 \\ -3 & 1 & -3 \\ -2 & 3 & 1 \end{bmatrix}\begin{bmatrix} a \\ b \\ c \end{bmatrix} = \begin{bmatrix} 2 \\ 1 \\ -6 \end{bmatrix}$

Job Opportunities The potential for growth in your chosen career is an important consideration. The matrix shows the job growth in two related fields.

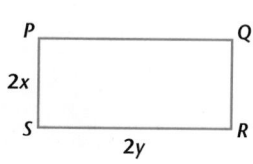
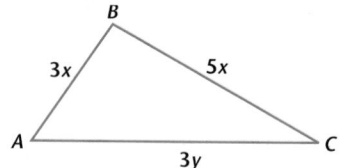

Occupation	U.S. Jobs (in 1000s)	
	1990	2000
Electrical engineers	440	615
Electrical assemblers	235	135

12. What is the projected average increase per year in the engineer's jobs?

13. What is the projected average decline per year in the assembler's jobs?

14. Write the linear equations that best fit this data.

15. In what year were the numbers of these two jobs equal? Use matrices to solve, but check by graphing.

16. **Geometry** The perimeter of rectangle *PQRS* is 30. The perimeter of triangle *ABC* is 40. Find the value of *x* and *y*. 3.5; 4

17. At Sid's Burger Barn, after the Friday concert, Kisha and her friend Maya ordered 3 Sid burgers and 2 super shakes. They were joined by Jason and Steve, who ordered 4 Sid burgers and 4 super shakes. When the checks arrived, the girls' check was $14, and the boys' check was $21. What was the cost of a Sid burger, and what was the cost of a super shake?
$3.50; $1.75

Inventory Control The matrices below indicate the sweater sales for three days at two branches of the Stylish Woman Sportswear Center.

Store #6 sales

	Cashmere	Cotton	Wool
Day 1	4	8	3
Day 2	3	10	5
Day 3	5	6	8

Store #8 sales

	Cashmere	Cotton	Wool
Day 1	2	5	3
Day 2	4	7	6
Day 3	3	8	8

The total sales for both stores for each day is

Day 1	$465
Day 2	$650
Day 3	$730

18. What was the total number of each type of sweater sold each day?

19. What was the selling price of each type of sweater?

20. **Ticket Sales** During Central City High School's spirit weekend, there was a concert on Friday night and an alumni football game on Saturday. For each event, adult and student tickets were sold. What was the price for an adult ticket and for a student ticket?
$2.50; $1.50

12. 17.5 thousand per year
13. 10 thousand per year
14. $y = 17.5x + 440$
    $y = -10x + 235$
15. 1982

18.

	Cashmere	Cotton	Wool
Day 1	6	13	6
Day 2	7	17	11
Day 3	8	14	16

19. Cashmere: $25.00; Cotton: $15.00; Wool: $20.00

Employment Kevin and Doug have part time jobs at the local library. One week Kevin worked 14 hours, and Doug worked 16 hours. Together they took home $149.50. The next week they each worked 12 hours and had a combined take-home pay of $120.

21. How much does each boy earn per hour? Kevin, $5.25
Doug, $4.75

22. How much did Kevin earn during the two weeks?
$136.50

23. Cultural Connection: Asia Solve the system using the Chinese method from the end of the lesson. $(-1, -4)$ $\begin{cases} 2x + 5y = -22 \\ 4x - 3y = 8 \end{cases}$

24. What method of solving linear equations does the Chinese method most resemble? elimination

25.  **Portfolio Activity**
Complete the problem in the portfolio activity on page 307.
8; 5; 5

Look Back

$y = -0.15x^2 + 2x + 2$

26. A ball is tossed between two players. If x represents the distance from the player tossing the ball, y represents the height of the ball. Would a third player between 6 and 7 feet from the player tossing the ball be able to intercept the toss? Explain your answer. **[Lesson 2.2]**

27. The given matrix represents 5 ordered pairs of a function. Use differences to show that the function represented by these points is quadratic. **[Lesson 2.2]**

Ordered Pairs

$$\begin{array}{c} \\ x \\ y \end{array} \begin{array}{ccccc} 1 & 2 & 3 & 4 & 5 \\ \begin{bmatrix} 3 & 4 & 5 & 6 & 7 \\ 0 & 2 & 6 & 12 & 20 \end{bmatrix} \end{array}$$

28. If the blue jeans you just bought cost $26.70, including the 7% sales tax, what was the ticket price? **[Lesson 4.5]** $24.95

Coordinate Geometry Write the equation of the line through the point $(-2, 5)$ and

29. parallel to the x-axis. $y = 5$

30. perpendicular to the x-axis. **[Lesson 5.6]** $x = -2$

31. Chemistry How many ounces of water should be added to a 75% acid solution to produce 30 ounces of 55% acid solution? **[Lesson 6.6]** 8

Look Beyond

Technology On a calculator, graph $y = \left| \dfrac{4}{3} x + 8 \right|$. Find the points where this graph intersects the graphs of the following functions.

32. $y = x$ **33.** $y = x + 5$ **34.** $y = 4$

26. Yes; The third player can if he or she can jump high enough so that his or her hands reach 8.65 feet which is the maximum height that the ball will reach.

27. The first differences for the y-values are 2, 4, 6, and 8, and the second differences are 2, 2, and 2. Since the second differences are constant, the function is quadratic.

32. There are no intersection points.

33. There are no intersection points.

34. $(-3, 4)$ and $(-9, 4)$

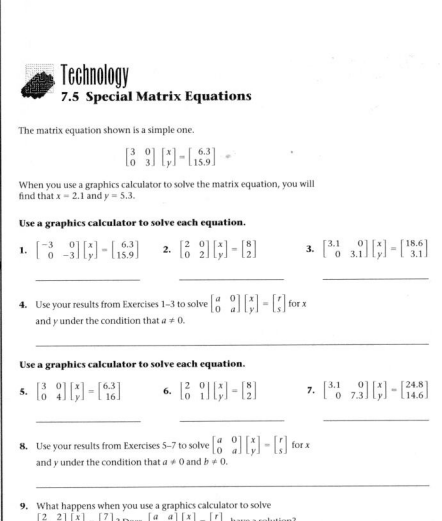

FOCUS

Students have already learned about many applications of matrices. This project introduces them to yet another very interesting application.

MOTIVATE

Have you ever wondered how codes are derived? Coding and decoding information is actually considered a science and has been given a name. It is called cryptography. Why do you think that name was chosen for this branch of science?

SECRET CODES

Navajo code talkers, one of the best kept secrets of World War II.

Cryptography, the science of coding and decoding information, is an interesting application of the matrix techniques that you have learned.

So, XRXEPENZ!

A	B	C	D	E	F	G	H	I	J	K	L	M	N
1	2	3	4	5	6	7	8	9	10	11	12	13	14

O	P	Q	R	S	T	U	V	W	X	Y	Z	*	
15	16	17	18	19	20	21	22	23	24	25	26	27	

Use asterisk * for space.

1 Use this table to switch to and from the number code.
The coding matrix that you will use is $\begin{bmatrix} 1 & -1 \\ -3 & 4 \end{bmatrix}$

The secret message you want to send is ADIOS AMIGO. Change the letters to numbers.

A D I O S * A M I G O * → 1 4 9 15 19 27 1 13 9 7 15 27
Form a matrix with these numbers. It must be compatible with the 2 by 2 coding matrix so that they can be multiplied. This message matrix is a compatible 2 by 6 matrix with the last position filled with a space, 27.

(Message Matrix) $\begin{bmatrix} 1 & 4 & 9 & 15 & 19 & 27 \\ 1 & 13 & 9 & 7 & 15 & 27 \end{bmatrix}$

2 Multiply the message matrix by the coding matrix.

$$\begin{bmatrix} 1 & -1 \\ -3 & 4 \end{bmatrix}\begin{bmatrix} 1 & 4 & 9 & 15 & 19 & 27 \\ 1 & 13 & 9 & 7 & 15 & 27 \end{bmatrix} = \begin{bmatrix} 0 & -9 & 0 & 8 & 4 & 0 \\ 1 & 40 & 9 & -17 & 3 & 27 \end{bmatrix}$$

3 Some of the numbers in the coded message matrix are not in the alphabet code table because they are either greater than 27 or less than 1. If a number is less than 1, keep adding 27 until the number corresponds to a value on the table. If the number is greater than 27, keep subtracting 27 until the result corresponds to a value in the table.

0	−9	0	8	4	0	1	40	9	−17	3	27
27	18	27	8	4	27	1	13	9	10	3	27

4 Use the alphabet-code table to change the numbers into letters and symbols.

27 18 27 8 4 27 1 13 9 10 3 27 → * R * H D * A M I J C *

5 Send the coded message.

6 Receiver gets **coded message**, and uses the alphabet code table to form the coded-message matrix.

* R * H D * A M I J C * → 27 18 27 8 4 27 1 13 9 10 3 27

$$\begin{bmatrix} 27 & 18 & 27 & 8 & 4 & 27 \\ 1 & 13 & 9 & 10 & 3 & 27 \end{bmatrix}$$

7 Multiply the secret matrix by a **decoding** matrix.

If C (coding) is $\begin{bmatrix} 1 & -1 \\ -3 & 4 \end{bmatrix}$, then C^{-1} is $\begin{bmatrix} 4 & 1 \\ 3 & 1 \end{bmatrix}$.

Use C^{-1} to decode the coded message matrix.

$$\begin{bmatrix} 4 & 1 \\ 3 & 1 \end{bmatrix}\begin{bmatrix} 27 & 18 & 27 & 8 & 4 & 27 \\ 1 & 13 & 9 & 10 & 3 & 27 \end{bmatrix} = \begin{bmatrix} 109 & 85 & 117 & 42 & 19 & 135 \\ 82 & 67 & 90 & 34 & 15 & 108 \end{bmatrix}$$

8 Add or subtract 27 as explained in step 3 to find the numbers that fit the table.

109	85	117	42	19	135	82	67	90	34	15	108
1	4	9	15	19	27	1	13	9	7	15	27

9 Translate from numbers to letters using the alphabet-code table.

A D I O S * A M I G O S *

Activity 1

The message in the introduction of this lesson, XRXEPENZ, was coded with the matrix $\begin{bmatrix} 3 & 7 \\ 2 & 5 \end{bmatrix}$. Decode the message.

Activity 2

Use available technology to code, send, and decode the message **BEWARE FEARLESS LEADER**. Use the coding matrix $\begin{bmatrix} -1 & 3 \\ -2 & 4 \end{bmatrix}$.

Activity 3

$\begin{bmatrix} 2 & 3 \\ 1 & 2 \end{bmatrix}$ Write your own short message of 15 characters or less. Use the given coding matrix to form the secret message. Give the secret message and the coding matrix to a classmate to decode.

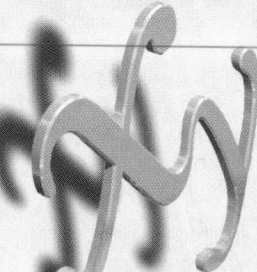

Chapter 7 Review

Vocabulary

coefficient matrix	338	matrix equality	311	scalar multiplication	324
constant matrix	338	matrix equation	338	square matrix	310
entry	309	matrix inverse	331	variable matrix	338
identity matrix	327	product matrix	325	zero matrix	317
matrix	309	scalar	324		

Key Skills & Exercises

Lesson 7.1

➤ **Key Skills**

Determine the dimensions and addresses of a matrix.

a. What are the dimensions of each matrix?
b. Find the entry at m_{32} for matrix M.
c. Find the entry at n_{21} for matrix N.
d. Can the matrices be equal? Why or why not?

Let $M = \begin{bmatrix} -5 & 6 \\ 7 & -2 \\ 1 & 0 \end{bmatrix}$ and $N = \begin{bmatrix} -5 & 6 & 7 \\ -2 & 1 & 0 \end{bmatrix}$.

a. Matrix M has dimensions 3×2. Matrix N has dimensions 2×3.
b. 0
c. -2
d. No, the dimensions are unequal.

➤ **Exercises**

Let $R = \begin{bmatrix} -4 & 1 \\ 2 & -3 \\ 2 & 0 \\ 8 & -5 \end{bmatrix}$.

1. What are the dimensions of matrix R?
2. Find the entry at r_{32}.
3. Find the entry at r_{41}.
4. What is the address of -5 in the matrix?
5. Give a matrix that is equal to matrix R.

Lesson 7.2

➤ **Key Skills**

Add matrices.

Evaluate the matrix operation.

$$\begin{bmatrix} -2.1 & 3.5 & 6.7 \\ 8.1 & -0.4 & -9.9 \\ 2.1 & 5.4 & -4.7 \end{bmatrix} + \begin{bmatrix} -6.4 & 3.2 & 3.5 \\ 6.5 & 7.6 & 1.2 \\ -9.3 & -4.2 & 0.5 \end{bmatrix}$$

Add entries with corresponding addresses.

$$\begin{bmatrix} -2.1 + (-6.4) & 3.5 + 3.2 & 6.7 + 3.5 \\ 8.1 + 6.5 & -0.4 + 7.6 & -9.9 + 1.2 \\ 2.1 + (-9.3) & 5.4 + (-4.2) & -4.7 + 0.5 \end{bmatrix}$$

$$= \begin{bmatrix} -8.5 & 6.7 & 10.2 \\ 14.6 & 7.2 & -8.7 \\ -7.2 & 1.2 & -4.2 \end{bmatrix}$$

Subtract matrices.

Evaluate the matrix operation.

$$\begin{bmatrix} -2.1 & 3.5 & 6.7 \\ 8.1 & -0.4 & -9.9 \\ 2.1 & 5.4 & -4.7 \end{bmatrix} - \begin{bmatrix} -6.4 & 3.2 & 3.5 \\ 6.5 & 7.6 & 1.2 \\ -9.3 & -4.2 & 0.5 \end{bmatrix}$$

Subtract entries with corresponding addresses.

$$\begin{bmatrix} -2.1 - (-6.4) & 3.5 - 3.2 & 6.7 - 3.5 \\ 8.1 - 6.5 & -0.4 - 7.6 & -9.9 - 1.2 \\ 2.1 - (-9.3) & 5.4 - (-4.2) & -4.7 - 0.5 \end{bmatrix}$$

$$= \begin{bmatrix} 4.3 & 0.3 & 3.2 \\ 1.6 & -8 & -11.1 \\ 11.4 & 9.6 & -5.2 \end{bmatrix}$$

1. 4 rows and 2 columns
2. 0
3. 8
4. r_{42}

5. Answers may vary.

$$\begin{bmatrix} 2(-2) & 3 - 2 \\ 4\left(\dfrac{1}{2}\right) & 3 - 6 \\ 16 \div 8 & \dfrac{0}{6} \\ 1 + 7 & -25\left(\dfrac{1}{5}\right) \end{bmatrix}$$

➤ **Exercises**

Evaluate each matrix operation.

6. $\begin{bmatrix} 4 & 3 & 5 \\ 2 & 1 & 2 \\ 5 & 3 & 1 \end{bmatrix} + \begin{bmatrix} -3 & 1 & 3 \\ -4 & -3 & 1 \\ -1 & 5 & 9 \end{bmatrix}$
7. $\begin{bmatrix} 5.6 & 6.5 \\ 4.3 & 4.6 \end{bmatrix} + \begin{bmatrix} 7.1 & 3.4 \\ -6.7 & 1.3 \end{bmatrix}$

8. $\begin{bmatrix} -6 & 1 \\ -1 & 2 \\ -2 & 7 \end{bmatrix} - \begin{bmatrix} 5 & -7 \\ -5 & 1 \\ -2 & 4 \end{bmatrix}$
9. $\begin{bmatrix} -9.9 & 8.8 \\ -2.3 & 1.2 \end{bmatrix} - \begin{bmatrix} 8.3 & 7.3 \\ -9.6 & -5.5 \end{bmatrix}$

Lesson 7.3

➤ *Key Skills*

Multiply a matrix by a scalar.

Evaluate the matrix operation.

$$4\begin{bmatrix} -4 & 8 \\ 9 & -5 \end{bmatrix}$$

Multiply each entry of the matrix by 4.

$$\begin{bmatrix} 4 \cdot -4 & 4 \cdot 8 \\ 4 \cdot 9 & 4 \cdot -5 \end{bmatrix} = \begin{bmatrix} -16 & 32 \\ 36 & -20 \end{bmatrix}$$

Multiply two matrices to form a product matrix.

Evaluate the matrix operation.

$$[4 \quad 1 \quad -6]\begin{bmatrix} -1 & 4 \\ 2 & 7 \\ -5 & 3 \end{bmatrix}$$

The product can be found either by the matrix multiplication procedure or by technology. The product is [28 5].

➤ *Exercises*

Use the given matrices to evaluate the following operations. Write *Not Possible* when appropriate.

$$A = \begin{bmatrix} -2 & -1 \\ 1 & 3 \end{bmatrix} \quad B = \begin{bmatrix} 1 & -3 & -4 \\ 3 & 5 & -1 \end{bmatrix} \quad C = \begin{bmatrix} 5 & 4 & 7 \\ 2 & 1 & 7 \\ 3 & 2 & 6 \end{bmatrix} \quad D = \begin{bmatrix} -1 & -3 \\ 4 & 5 \\ 6 & 1 \end{bmatrix}$$

10. $3A$ **11.** $-4D$ **12.** AB **13.** BC **14.** BA **15.** $-DA$

Lesson 7.4

➤ *Key Skills*

Determine the inverse matrix for multiplication.

Find the inverse of $V = \begin{bmatrix} 4 & 3 \\ -1 & -2 \end{bmatrix}$.

By definition, $AA^{-1} = I$.

$$\begin{bmatrix} 4 & 3 \\ -1 & -2 \end{bmatrix}\begin{bmatrix} a & b \\ c & d \end{bmatrix} = \begin{bmatrix} 1 & 0 \\ 0 & 1 \end{bmatrix}$$

The inverse can be found either by the matrix multiplication procedure or by technology.

The inverse is $\begin{bmatrix} 0.4 & 0.6 \\ -0.2 & -0.8 \end{bmatrix}$.

➤ *Exercises*

Find the inverse matrix if it exists for the given matrices.

16. $\begin{bmatrix} -4 & 1 \\ 3 & -3 \end{bmatrix}$ **17.** $\begin{bmatrix} 0 & -1 \\ -1 & -1 \end{bmatrix}$ **18.** $\begin{bmatrix} 0.1 & 0.2 \\ 1 & 2 \end{bmatrix}$

Chapter Review

15. $\begin{bmatrix} 1 & 8 \\ 3 & -11 \\ 11 & 3 \end{bmatrix}$

16. $\begin{bmatrix} -0.33 & -0.11 \\ -0.33 & -0.44 \end{bmatrix}$

or

$\begin{bmatrix} \frac{-1}{3} & \frac{-1}{9} \\ \frac{-1}{3} & \frac{-4}{9} \end{bmatrix}$

17. $\begin{bmatrix} 1 & -1 \\ -1 & 0 \end{bmatrix}$

18. The inverse for this matrix does not exist.

6. $\begin{bmatrix} 1 & 4 & 8 \\ -2 & -2 & 3 \\ 4 & 8 & 10 \end{bmatrix}$

7. $\begin{bmatrix} 12.7 & 9.9 \\ -2.4 & 5.9 \end{bmatrix}$

8. $\begin{bmatrix} -11 & 8 \\ 4 & 1 \\ 0 & 3 \end{bmatrix}$

9. $\begin{bmatrix} -18.2 & 1.5 \\ 7.3 & 6.7 \end{bmatrix}$

10. $\begin{bmatrix} -6 & -3 \\ 3 & 9 \end{bmatrix}$

11. $\begin{bmatrix} 4 & 12 \\ -16 & -20 \\ -24 & -4 \end{bmatrix}$

12. $\begin{bmatrix} -5 & 1 & 9 \\ 10 & 12 & -7 \end{bmatrix}$

13. $\begin{bmatrix} -13 & -7 & -38 \\ 22 & 15 & 50 \end{bmatrix}$

14. Not Possible

Lesson 7.5

> ### Key Skills

Solve a system of equations using matrices.

Use matrices to solve the system.

$$\begin{cases} 3x + 2y = 16 \\ 2x + 3y = 19 \end{cases}$$

Prepare the matrices so that the matrix equation form is $AX = B$.

$$\begin{bmatrix} 3 & 2 \\ 2 & 3 \end{bmatrix} \begin{bmatrix} x \\ y \end{bmatrix} = \begin{bmatrix} 16 \\ 19 \end{bmatrix}$$

To solve the matrix equation, multiply both sides by the inverse of A which is

$$\begin{bmatrix} 0.6 & -0.4 \\ -0.4 & 0.6 \end{bmatrix}.$$

Since $A^{-1}AX = A^{-1}B$ simplifies to $X = A^{-1}B$, find $A^{-1}B$ to solve the system.

$$A^{-1}B = \begin{bmatrix} 2 \\ 5 \end{bmatrix}, \text{ so the solution is } (2, 5).$$

> ### Exercises

Solve the systems of equations using matrices if possible.

19. $\begin{cases} 2x + 5y = 12 \\ x + 3y = 7 \end{cases}$

20. $\begin{cases} -3x + y = -5 \\ x + 2y = 10 \end{cases}$

21. $\begin{cases} 4x + 3y = 6 \\ x - 2y = -15 \end{cases}$

22. $\begin{cases} x - 7y = 5 \\ -4x + 5y = -20 \end{cases}$

23. $\begin{cases} -2x - 6y = 3 \\ 4x + 12y = -6 \end{cases}$

Applications

24. Business Mr. Tunney is giving four of his employees an 8% pay raise. Their current hourly wages are $5.50, $4.95, $6.35, and $5.75. Use scalar-matrix multiplication to find the new hourly wage for each employee.

25. Stocks The matrix shows the number of shares of three different stocks that are owned by three different people.

	Julie	Kira	Tony
Stock: 1	14	12	10
2	9	4	8
3	17	20	2

Last Friday, stock 1 gained $5.50 a share, stock 2 lost $3.00 a share, and stock 3 lost $4.50 a share. How much did each person gain or lose last Friday?

26. At Frank's Gym, a basic membership costs $250 a year, a silver membership costs $350 a year, and a gold membership costs $500 a year. During the first week of a 4-week sales promotion, 4 basic, 3 silver, and 2 gold memberships were sold. During the second week, 3 basic, 4 silver, and 3 gold memberships were sold. During the third week, 5 basic, 1 silver, and 2 gold memberships were sold. During the fourth week, 1 basic, 1 silver, and 3 gold memberships were sold. What was the total number of memberships sold for each week of the sales promotion?

19. $\begin{bmatrix} 1 \\ 2 \end{bmatrix}$

20. $\begin{bmatrix} \frac{20}{7} \\ \frac{25}{7} \end{bmatrix}$

21. $\begin{bmatrix} -3 \\ 6 \end{bmatrix}$

22. $\begin{bmatrix} 5 \\ 0 \end{bmatrix}$

23. This system of equations has an infinite number of solutions.

24. $5.94; $5.35; $6.86; $6.21

25. Julie: $-$26.50; Kira: $-$36.00; Tony: $22.00

26. Memberships Sold

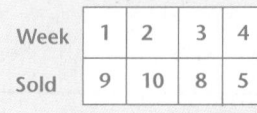

Week	1	2	3	4
Sold	9	10	8	5

Chapter 7 Assessment

Use the given matrices to answer items 1–21. Write *Not Possible* when appropriate.

$$M = \begin{bmatrix} -2 & -1 \\ 1 & 3 \end{bmatrix} \quad N = \begin{bmatrix} 1 & -3 & -4 \\ 3 & 5 & -1 \end{bmatrix} \quad P = \begin{bmatrix} 5 & 4 & 7 \\ 2 & 1 & 7 \end{bmatrix} \quad Q = \begin{bmatrix} -1 & -3 \\ 4 & 5 \\ 6 & 1 \end{bmatrix}$$

1. Give the dimensions of each matrix. $2 \times 2; 2 \times 3; \ 2 \times 3; 3 \times 2$

2. Find the entry at n_{23} for matrix N. -1

3. Find the entry at q_{12} for matrix Q. -3

4. What is the address of 4 in matrix P? P_{12}

5. What is the address of 3 in matrix M? M_{22}

6. Discuss how you would determine whether a matrix is equal to matrix N.

7. Discuss how to recognize whether two matrices can be multiplied.

8. $N + P$ **9.** $Q - P$ **10.** $P - N$

11. $2M$ **12.** $-4N$ **13.** MN

14. QP **15.** $-PQ$ **16.** NP

17. M^{-1} **18.** MM^{-1} **19.** P^{-1}

20. Let *I* be the identity matrix for addition. Evaluate $N + I$.

21. Let *I* be the identity matrix for multiplication. Evaluate MI.

22. If it is possible to multiply matrix *A* times matrix *B*, is it also possible to multiply matrix *B* times matrix *A*? Explain your answer.

23. Business   Kyle and Kodie both work at the Ice Cream Shop and the Burger Palace. Kyle makes \$4.50 per hour at the Ice Cream Shop and \$5.00 per hour at the Burger Palace. Kodie makes \$4.75 per hour at the Ice Cream Shop and \$4.50 per hour at the Burger Palace. Last week, they each worked 15 hours at the Ice Cream Shop and 12 hours at the Burger Palace. How much did each earn last week?   \$127.50; \$125.25

Solve the systems of equations using matrices if possible.

24. $\begin{cases} -2x - 3y = 5 \\ x + 2y = -2 \end{cases} \begin{bmatrix} -4 \\ 1 \end{bmatrix}$ **25.** $\begin{cases} -5x + 2y = -3 \\ 10x - 4y = 6 \end{cases}$ infinite

26. $\begin{cases} -2x + 7y = 0 \\ x + 3y = 1 \end{cases} \begin{bmatrix} \frac{7}{13} \\ \frac{2}{13} \end{bmatrix}$ **27.** $\begin{cases} 4x - 2y = -2 \\ 2x - 4y = 10 \end{cases} \begin{bmatrix} -\frac{7}{3} \\ -\frac{11}{3} \end{bmatrix}$

28. Geometry   Recall that two supplementary angles have measures whose sum is 180 and that two complementary angles have measures whose sum is 90. If two supplementary angles have measures of $3x$ and $3y$ and two complementary angles have measures of $4x$ and y, what are the measures of the four angles?   30°; 150°; 40°; 50°

13. $\begin{bmatrix} -5 & 1 & 9 \\ 10 & 12 & -7 \end{bmatrix}$

14. $\begin{bmatrix} -11 & -7 & -28 \\ 30 & 21 & 63 \\ 32 & 25 & 49 \end{bmatrix}$

15. $\begin{bmatrix} -53 & -12 \\ -44 & -6 \end{bmatrix}$

16. Not possible.

17. $\begin{bmatrix} -0.6 & -0.2 \\ 0.2 & 0.4 \end{bmatrix}$

18. $\begin{bmatrix} 1 & 0 \\ 0 & 1 \end{bmatrix}$

19. Not possible.

20. $\begin{bmatrix} 1 & -3 & -4 \\ 3 & 5 & -1 \end{bmatrix}$

21. $\begin{bmatrix} -2 & -1 \\ 1 & 3 \end{bmatrix}$

22. Sometimes, if matrix *A* and matrix *B* are square matrices.

6. A matrix equal to matrix *N* would need to have 2 rows and 3 columns and entries that are equal to each corresponding entry in matrix *N*.

7. The number of columns in the first matrix must equal the number of rows in the second matrix.

8. $\begin{bmatrix} 6 & 1 & 3 \\ 5 & 6 & 6 \end{bmatrix}$

9. Not possible.

10. $\begin{bmatrix} 4 & 7 & 11 \\ -1 & -4 & 8 \end{bmatrix}$

11. $\begin{bmatrix} -4 & -2 \\ 2 & 6 \end{bmatrix}$

12. $\begin{bmatrix} -4 & 12 & 16 \\ -12 & -20 & 4 \end{bmatrix}$

CHAPTER 8 Probability and Statistics

Meeting Individual Needs

8.1 Experimental Probability

Core Resources

Practice Master 8.1
Enrichment, p. 354
Technology Master 8.1
Interdisciplinary
   Connections, p. 353

[1 day]

Core Two-Year Resources

Inclusion Strategies, p. 354
Reteaching the Lesson, p. 356
Practice Master 8.1
Enrichment Master 8.1
Technology Master 8.1
Lesson Activity Master 8.1

[3 days]

8.2 Exploring Simulations

Core Resources

Practice Master 8.2
Enrichment, p. 362
Technology Master 8.2
Interdisciplinary
   Connections, p. 361

[1 day]

Core Two-Year Resources

Inclusion Strategies, p. 362
Reteaching the Lesson, p. 363
Practice Master 8.2
Enrichment Master 8.2
Technology Master 8.2
Lesson Activity Master 8.2

[3 days]

8.3 Extending Statistics

Core Resources

Practice Master 8.3
Enrichment, p. 369
Technology Master 8.3

[2 days]

Core Two-Year Resources

Inclusion Strategies, p. 369
Reteaching the Lesson, p. 370
Practice Master 8.3
Enrichment Master 8.3
Technology Master 8.3
Lesson Activity Master 8.3

[3 days]

8.4 Exploring the Addition Principle of Counting

Core Resources

Practice Master 8.4
Enrichment, p. 376
Technology Master 8.4
Mid-Chapter Assessment
   Master

[2 days]

Core Two-Year Resources

Inclusion Strategies, p. 377
Reteaching the Lesson, p. 378
Practice Master 8.4
Enrichment Master 8.4
Technology Master 8.4
Lesson Activity Master 8.4
Mid-Chapter Assessment Master

[3 days]

8.5 Multiplication Principle of Counting

Core Resources

Practice Master 8.5
Enrichment, p. 383
Technology Master 8.5

[2 days]

Core Two-Year Resources

Inclusion Strategies, p. 383
Reteaching the Lesson, p. 384
Practice Master 8.5
Enrichment Master 8.5
Technology Master 8.5
Lesson Activity Master 8.5

[4 days]

8.6 Theoretical Probability

Core Resources

Practice Master 8.6
Enrichment, p. 390
Technology Master 8.6
Interdisciplinary
   Connections, p. 389

[2 days]

Core Two-Year Resources

Inclusion Strategies, p. 390
Reteaching the Lesson, p. 391
Practice Master 8.6
Enrichment Master 8.6
Technology Master 8.6
Lesson Activity Master 8.6

[4 days]

8.7 Independent Events

Core Resources

Practice Master 8.7
Enrichment, p. 395
Technology Master 8.7

[1 day]

Core Two-Year Resources

Inclusion Strategies, p. 395
Reteaching the Lesson, p. 396
Practice Master 8.7
Enrichment Master 8.7
Technology Master 8.7
Lesson Activity Master 8.7

[2 days]

Chapter Summary

Core Resources

Eyewitness Math,
    pp. 366–367
Chapter 8 Project,
    pp. 400–401
Lab Activity
Long-Term Project
Chapter Review,
    pp. 402–404
Chapter Assessment,
    p. 405
Chapter Assessment, A/B
Alternative Assessment
Cumulative Assessment,
    pp. 405–407

[3 days]

Core Two-Year Resources

Eyewitness Math, pp. 366–367
Chapter 8 Project, pp. 400–401
Lab Activity
Long-Term Project
Chapter Review, pp. 402–404
Chapter Assessment, p. 405
Chapter Assessment, A/B
Alternative Assessment
Cumulative Assessment,
    pp. 406–407

[5 days]

Visual Strategies

Tree diagrams provide one visual strategy that allows students to see all the possible outcomes of two or more events involved in a probability exercise. For some problems, you may want to suggest that students solve a simpler problem and use the visual strategy of drawing a tree diagram.

Using Venn diagrams is another visual strategy. Students can see the intersection of two sets. This is particularly useful when teaching the Addition Principle of Counting, in which the number of common elements must be subtracted from the sum of the elements of the two parts.

Hands-on Strategies

Chapter 8 requires the use of many manipulatives, such as counters, coins, dice, cards, and so on. The Explorations found throughout this chapter provide for active learning by students and require hands-on participation. Be sure students understand the difference between single events and compound events. A single event might be finding the probability of getting heads when you flip a coin. A compound event might be tossing dice, then drawing a marble from a bag; a compound event can be dependent or independent. This depends on whether the first event has an effect on the second.

Cooperative Learning

Probability games/experiments	Lesson 8.1, Explorations 1, 2, and 3
Simulations	Lesson 8.2, Explorations 1 and 2
Scatter plots	Lesson 8.3, Example 2
Theoretical probability	Lesson 8.6, Opening Discussion

You may wish to have students work with partners or in small groups to complete any of these activities. Additional suggestions for cooperative group activities can be found in the teacher's notes in the lessons.

Multicultural

The cultural connections in this chapter include references to the Ojibwa, or Chippewa, American Indian Tribe and to Egypt.

CULTURAL CONNECTIONS	
Americas: Indian Games	Lesson 8.1, Exploration 1
Africa: Ancient Games	Lesson 8.1, Exploration 3

Portfolio Assessment

Below are portfolio activities for the chapter. They are listed under seven activity domains that are appropriate for portfolio development.

1. **Investigation/Exploration** The methods used to find experimental probability in Lessons 8.1 and 8.2 provide interesting mathematical concepts that can be included. Another interesting investigation involves the theory of "Hot Hands" found in the Eyewitness Math feature.

2. **Applications** Recommended to be included are any of the following: Games, Lesson 8.1, Exercises 16–23; Sports, Lesson 8.2, Exercises 7–10; Meteorology, Lesson 8.3, Exercises 22–25; Games, Lesson 8.4, Exercises 13–15; Language, Lesson 8.5, Exercises 13–20; Geography, Lesson 8.6, Exercise 16.

3. **Non-Routine Problems** The "Hot Hand" phenomenon studied in the Eyewitness Math feature provides an application of probability to a non-routine problem.

4. **Project** Permutations and Combinations: see pages 400–401. Students study the similarities and differences between permutations and combinations.

5. **Interdisciplinary Topics** Students may choose from the following: Statistics, Lesson 8.4, Application on Page 378 and Exercises 5–8. Geometry is closely related to many topics in this chapter and may be studied in any of the following: Lesson 8.3, discussion on page 370; Lesson 8.5, Example 3 and Exercises 6 and 21; Lesson 8.6, Exercise 25; Lesson 8.7, opening discussion.

6. **Writing** *Communicate* exercises that ask students to describe and explain procedures involved in solving probability problems offer excellent writing selections for the portfolio.

7. **Tools** Chapter 8 uses both computer spreadsheets and graphics calculators to work with probability, simulations, and random numbers.

Technology

Spreadsheets

Students use spreadsheets in Lesson 8.1 to generate random numbers using the RAND (random) and INT (integer) commands. The formula used is

$$INT(RAND()*K)+A.$$

In the same lesson, students can see how to use a spreadsheet to display the results of an experiment in tabular form. The use of spreadsheets to generate and record random numbers continues in Lesson 8.2. In the spreadsheet on page 361, the desired results (four in a row) are printed in bold type. The spreadsheet does not do this. Suggest that students use a highlight marker to show desired outcomes on their spreadsheets. In Lesson 8.4, data bases are studied and displayed on a spreadsheet.

A6	=RAND()	
	A	B
1	0.30261695	
2	0.08310659	
3	0.77245826	
4	0.70603577	
5	0.34317643	
6	0.66527653	

Spreadsheet

Rand	
	.0078387869
	.9351587791
	.1080114624
	.0062633066
	.5489861799
	.8555803143

Graphics calculator

C11	=INT(RAND()*100)+1	
	A	B	C
1	Trial	1st Number	2nd Number
2	1	98	68
3	2	33	82
4	3	21	17
5	4	94	14
6	5	87	36
7	6	73	83
8	7	56	73
9	8	65	12
10	9	18	58
11	10	99	18

Graphics Calculator

In Lesson 8.1, students learn to use a graphics calculator to generate random numbers from a list of K consecutive integers beginning with A by using the INT and RAND keys. The use of a graphics calculator to generate random numbers is continued in Lesson 8.2 as students study and design simulations.

Making scatter plots and finding the correlation were originally studied in Lesson 1.6 and are now related to measures of central tendency in Lesson 8.3. The graphics calculator is used to generate the line of best fit and correlation in this lesson.

When working with simulations, students may use data that are not reasonable. This may be a result of incorrect use of technology to generate random numbers. To generate random numbers on a graphics calculator, students should consult the manual for their particular calculator, since some require different keystrokes. Students should also check to see that data that are generated fit the data requirements and yield a reasonable answer. For instance, a weight or height of zero for a child is not reasonable, although it might be included in a line of best fit.

An extension of the chapter project on permutations and combinations lends itself nicely to the use of a graphics calculator to find the value of a permutation or combination.

Integrated Software

f(g) Scholar[TM] is an integrated computer-based mathematics productivity tool that combines calculator, spreadsheet, and graphics capabilities to provide a dynamic and interactive environment for explorations in mathematics. It is appropriate to use *f(g) Scholar*[TM] for any lesson needing a spreadsheet, calculator, graphics calculator, or any combination of the three.

Probability and Statistics

ABOUT THE CHAPTER

Background Information

Probability is involved in many facets of everyday life. This chapter will help students become more aware of how probability affects their lives and will explore the mathematics of probability.

CHAPTER RESOURCES

- Practice Masters
- Enrichment Masters
- Technology Masters
- Lesson Activity Masters
- Lab Activity Masters
- Long-Term Project Masters
- Assessment Masters
    Chapter Assessments, A/B
    Mid-Chapter Assessment
    Alternative Assessment, A/B
- Teaching Transparencies
- Cumulative Assessment
- Spanish Resources

CHAPTER OBJECTIVES

- Find the experimental probability that an event will occur.
- Use random numbers in probability situations.
- Design probability simulations.
- Use simulations to find experimental probability.
- Find the mean, median, mode, and range for a set of data.
- Find the line of best fit for a set of data.
- Recognize deceptive uses of statistics.
- Identify the margin of error for a set of data.

CHAPTER 8

Probability and Statistics

LESSONS

8.1 Experimental Probability

8.2 *Exploring* Simulations

8.3 Extending Statistics

8.4 *Exploring* the Addition Principle of Counting

8.5 Multiplication Principle of Counting

8.6 Theoretical Probability

8.7 Independent Events

Chapter Project
Winning Ways

Probability touches your life in many ways. Games, insurance, matters of health and diet, and countless other situations involve an element of chance. Statistics makes extensive use of probability to study, predict, and draw conclusions from data. Sometimes probabilities are found experimentally by looking at data. Other times, probabilities are found theoretically by considering all the possibilities. In this chapter, you can learn about both kinds of probability and explore examples of probability in action.

ABOUT THE PHOTOS

Have students look at the photographs on these pages and identify as many of the following as possible: coins, dice, and other objects associated with games of chance. Ask students why a dartboard would be included on this page. Have students discuss how the recording of data is shown on this page. Tell students that all of the materials shown here are involved with probability and that they will learn more about such tools throughout this chapter.

- Draw and interpret Venn diagrams.
- Use the Addition Principle of Counting.
- Make a tree diagram to show possible outcomes.
- Use the Multiplication Principle of Counting to find the number of possible outcomes.
- Find the probability that an event will occur.
- Use counting principles to find the probability that an event will occur.
- Define and give examples of independent events.
- Find the probability of independent events.

PORTFOLIO ACTIVITY

Assign the Portfolio Activity as part of the assignment for Lesson 8.7 on page 399.

To introduce the activity, ask students if they think they can simply guess at the answers to a multiple-choice test or a true-false test and get a passing grade. Allow students to discuss their responses. Allow students to work alone or in cooperative groups to complete this activity. Challenge students to find the probability of guessing the correct answer 70% of the time on a multiple-choice test that has four possible answers for each question.

PORTFOLIO ACTIVITY

Bob doesn't study as much as he should. On a true-false quiz in science class he guesses at all five answers. Use experimental or theoretical methods to find the probability that he gets at least 80% on the quiz.

ABOUT THE CHAPTER PROJECT

In the Chapter Project on pages 400–401 students investigate the basic concepts of permutations and combinations. These are related to the Multiplication Principle of Counting, which is studied in this chapter.

Objectives

- Find the experimental probability that an event will occur.
- Use random numbers in probability situations.

RESOURCES

- Practice Master   **8.1**
- Enrichment Master   **8.1**
- Technology Master   **8.1**
- Lesson Activity Master   **8.1**
- Quiz   **8.1**
- Spanish Resources   **8.1**

Assessing Prior Knowledge

Write each fraction in simplest form.

1. $\frac{4}{20}$ $\left[\frac{1}{5}\right]$

2. $\frac{15}{18}$ $\left[\frac{5}{6}\right]$

3. $\frac{14}{20}$ $\left[\frac{7}{10}\right]$

Evaluate each of the following.

4. $3 \cdot 2 + 4$ [10]
5. $8 \cdot 10 + 1$ [81]
6. $0.56 \cdot 100 + 1$ [57]
7. $0.081 \cdot 100 + 1$ [9.1]

TEACH

Probability and chance are central to many games and other experiences in real life. Have students discuss their experiences with events that involve probability. Sample answers may include games, contests, or sports.

LESSON 8.1 **Experimental Probability**

why *People in all cultures seem to enjoy games. Many games involve an element of luck or chance. Chance is often introduced by number cubes or spinners. You can use experimental probability to determine a number that represents the chance that an event will happen.*

Discovering Experimental Probability

Cultural Connection: Americas One American Indian game, *shaymahkewuybinegunug*, involves tossing sticks. It is played by members of the Ojibwa, or Chippewa, tribe. This game uses five flat sticks that have carved pictures of snakes on one side and are plain on the other. Players take turns tossing the sticks to earn points.

Exploration 1 **Probability and the Coin Game**

Games To play an adaptation of the Ojibwa game, use three coins in place of the sticks. Each person tosses the three coins 10 times. Each time the three coins are the same, that person gets a point. After the person completes 10 tosses, the person with the most points is the winner.

or

ALTERNATIVE teaching strategy

Hands-On Strategies
Begin class with a discussion of various experiments that students have participated in or watched. Have students discuss what the experiment was and what each outcome was. Students should also note whether the outcomes were successful or not. Finally, have students decide whether chance or probability was involved in the experiments.

1 Play the game. How many times did all 3 of your coins come up alike?

2 If you make 20 tosses, how many times do you think all 3 coins will come up alike?

3 Perform an experiment to check your guess. Toss 3 coins 20 times, and record how often the 3 coins come up alike.

Experimental probability is represented as a fraction usually written in lowest terms. The numerator is the number of times all 3 coins come up alike. The denominator is the total number of tosses, 20. Probabilities are sometimes expressed as decimals or percents.

4 Give the experimental probability of your result in Step 3. How does your result compare with results from other groups? Notice that there can be many different experimental probabilities.

5 Pool the data for the class. What is the experimental probability for the class? Is your result close to the result for the class?

6 In this *experiment,* each toss of 3 coins is a *trial.* Let *t* represent the number of trials. When all 3 coins come up alike it is considered a *successful event.* Let *f* represent the number of times a successful event occurs. Express the experimental probability, *P*, in a formula using *f* and *t*. ❖

•Exploration 2 *Probability From a Number Cube*

1 The faces on an ordinary number cube are numbered 1 to 6. If the cube is rolled 10 times, guess how many times a 5 will appear on the top of the cube.

2 Roll one cube 10 times. Count how many times you get a 5.

3 Define an *event* and a *trial* in this experiment.

4 What is the experimental probability of getting a 5 in 10 trials in this experiment? ❖

CRITICAL *Thinking*

Tell how to find the experimental probability for each of the following.

- 3 on one roll of a number cube
- *heads* on a toss of a coin

---

interdisciplinary CONNECTION

Social Studies Have students complete research to find why the height of the Nile River flood would have been important enough to the people of Egypt to keep records as mentioned on page 354. **[The height of the flood often determined the type of growing season as well as the growing area for raising crops for food.]**

Exploration 1 Notes

The game in Exploration 1 is to be played by pairs of students. Two-colored counters can be used in place of coins. Students should write the expression for the experimental probability as $P = \frac{f}{t}$.

ongoing ASSESSMENT

4. The answer will depend on individual rolls of the number cube. It should be a fraction with a denominator of 20 and a numerator from 0 to 20, depending on the actual data.

6. $P = \frac{f}{t}$

Exploration 2 Notes

Some students may say that the experimental probability of rolling a 5 is $\frac{1}{6}$. Emphasize that experimental probability is based on the results of the experiment, not on theoretical probability.

ongoing ASSESSMENT

4. Answer will depend on individual rolls of the number cube. It should be a fraction with a denominator of 10 and a numerator from 0 to 10 depending on the actual data.

CRITICAL *Thinking*

Have each student in class roll one cube. Use a ratio to compare the number of 3s to the number of students. The coin toss can be done in like manner. The *experimental probability* is the given ratio.

Lesson 8.1 **353**

Have students work in pairs and play the game several times. Then have them write the experimental probability for their results. Collect data regarding the results for the entire class. Find the experimental probability for those results. Have students use those results to answer the question in part 3.

1. **Player A has a better chance. Player A can get a point 24 ways. Player B has only 12 ways to get a point. The game is not fair.**

Cooperative Learning

You may wish to have students work in small groups and make up two new games—one that is fair and one that is not fair. Then let students actually try playing these new games to see if they are fair or unfair.

Experimental probability varies when an experiment is conducted several times.

EXPERIMENTAL PROBABILITY

Let t be the number of trials in the experiment. Let f be the number of times a successful event occurs.

The **experimental probability**, P, of the event is given by

$$P = \frac{f}{t}.$$

Exploration 3 A Fair Game With Two Number Cubes

1 A two-player game involves rolling two number cubes and finding the sum. If the sum is 5, 6, 7, 8, or 9, Player A gets a point. If the sum is 2, 3, 4, 10, 11, or 12, Player B gets a point. Guess which player has a better chance of winning. Explain your answer.

2 When a game is fair, each player has the same chance of winning. Design and conduct a fair game using two number cubes. ❖

Cultural Connection: Asia
Even in prehistoric times people played games of chance and strategy. They must have wondered what the probability of winning would be if they followed a certain strategy. But one of the earliest records of the collection of statistical data had geographical implications.

About 2100 B.C.E., a record was engraved on a clay tablet calculating the surface area of a terrain at Umma, Mesopotamia (Iraq).

ENRICHMENT Students could conduct an experiment to determine the probability that an event will *not* occur. They could roll a pair of dice for 20 trials and record trials as either a 6 or not a 6. Repeat the experiment a number of times. Have students write a statement summarizing their conclusions about determining the probability that an event will not occur.

INCLUSION
strategies

Hands-On Strategies Experimental probability is ideal for the hands-on learner. Some students may want to do more research on this, or other ancient games. Allow students to make their own sticks or other game materials and present the game to the class. Then have students discuss how probability is involved in the game. Allow students to play the game to find the experimental probability of a particular outcome or of winning.

Random Numbers

A key ingredient in experimental probability is the use of random numbers. In games, you may have used simple random number generators such as spinners. For making selections, you may have tossed a coin or drawn a number from a hat. There are also books that contain tables of random numbers. Some calculators and computers are useful because they can generate random numbers that can be adapted to fit specific conditions.

Spreadsheet

Spreadsheet software and calculators typically use a command like RAND() or RAND to generate random numbers. The basic random numbers are decimals from 0 to 1, including 0 but not including 1. On some graphics calculators, as you continue to press ENTER you get more random numbers.

A6	=RAND()	
	A	**B**
1	0.30261695	
2	0.08310659	
3	0.77245826	
4	0.70603577	
5	0.34317643	
6	0.66527653	

Spreadsheet

Rand	
	.0078387869
	.9351587791
	.1080114624
	.0062633066
	.5489861799
	.8555803143

Graphics calculator

A6	=RAND()*5	
	A	**B**
1	1.40817388	
2	2.17844473	
3	3.12062745	
4	3.38549538	
5	0.02076914	
6	4.06690849	

Rand*5	
	4.885421232
	1.391544132
	1.376071474
	.6089506775
	.2629490302
	3.611896579

If each random number is multiplied by 5, the result is six new random numbers. The new random numbers range in value from 0 to 5. Can these new random numbers ever include 5? Explain.

The integer value function INT applied to the random number generator produces random integers from 0 through 4.

A6	=INT(RAND()*5)	
	A	**B**
1	1	
2	0	
3	2	
4	3	
5	4	
6	0	

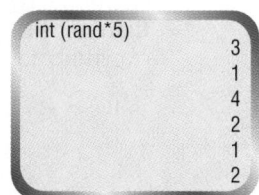

int (rand*5)	
	3
	1
	4
	2
	1
	2

Thus the spreadsheet command INT(RAND()*5) gives random integers from 0 through 4. Most software uses a symbol like = or @ to signal a formula.

You can adapt the output of a random number generator to your needs by changing the command.

> **GENERATING RANDOM INTEGERS**
> The command INT(RAND() * K) + A generates random integers from a list of K consecutive integers beginning with A.

EXAMPLE 1

Determine the possible numbers that can appear at random when you use the spreadsheet command.

$$INT(RAND()*2) + 3$$

Spreadsheet

Solution ➤

Examine the function in steps from the inside out.

Ⓐ **RAND()** results in a decimal value from 0 to 1 excluding 1.

Ⓑ **RAND()*2** results in a decimal value from 0 to 2 excluding 2.

Ⓒ **INT(RAND()*2)** results in the integer 0 or 1.

Ⓓ **INT(RAND()*2) + 3** results in the integer 3 or 4. ❖

Try This Generate sets of random numbers from 3 different commands. Describe the numbers that can possibly appear for each.

EXAMPLE 2

Suppose you are given two random numbers from 1 to 100. Find the experimental probability that at least one of them is less than or equal to 40.

Solution ➤

Generate 10 pairs of random numbers from 1 to 100. Count how many times at least one of each pair is less than or equal to 40.

Spreadsheet

This spreadsheet shows 10 pairs of random integers from 1 to 100 generated by the command INT(RAND()*100)+1.

In 7 of the 10 trials, at least one of the two numbers is less than or equal to 40. So the experimental probability is $\frac{7}{10}$. ❖

C11	=INT(RAND()*100)+1	
	A	**B**	**C**
1	Trial	1st Number	2nd Number
2	1	98	68
3	2	33	82
4	3	21	17
5	4	94	14
6	5	87	36
7	6	73	83
8	7	56	73
9	8	65	12
10	9	18	58
11	10	99	18

CRITICAL
Thinking

Will you get the same experimental probability if you use a greater number of trials? How does the number of trials affect the probability?

EXERCISES & PROBLEMS

Communicate

1. Explain what is meant by experimental probability.

2. Describe an experiment to find the experimental probability of getting at least 3 *heads* on a toss of 4 coins.

3. Is it possible for two separate groups to conduct the same experiment to determine a probability and get different results? Explain or give an example.

4. Is it possible for someone to conduct the same experiment twice to determine a probability and get different results? Explain or give an example.

5. If RAND generates a number from 0 to 1 (including 0, but not 1), describe the numbers you get from INT(RAND*7) + 1.

6. Look at the spreadsheet data for Example 2. Tell how to find the experimental probability that both numbers are less than 50.

Practice & Apply

7. Describe an experiment to find the experimental probability that if 4 coins are flipped, there will be either 4 *heads* or 4 *tails*.

8. Describe an experiment to find the experimental probability that if 2 number cubes are rolled, at least one of them will show a 6.

9. To determine an experimental probability, Fred conducted 15 trials and Ted conducted 16 trials. Is it possible that they arrived at the same experimental probability? Explain.

Two coins were flipped 20 times with the following results.

Trial	1	2	3	4	5	6	7	8	9	10	11	12	13	14	15	16	17	18	19	20
Coin 1	H	H	H	H	T	H	T	T	T	H	H	H	T	T	H	T	T	T	H	H	H
Coin 2	T	H	H	H	T	T	T	T	H	T	T	T	H	H	H	H	H	H	H	T	H

According to the data, find the following experimental probabilities.

10. Both coins are alike. $\frac{7}{20}$
11. Both coins are *heads*. $\frac{1}{4}$
12. At least one coin is *heads*. $\frac{9}{10}$
13. Neither coin is *heads*. $\frac{1}{10}$

Look at the spreadsheet data for Example 2. Find the experimental probability that

14. both numbers are less than 80. $\frac{2}{5}$
15. the first number is greater than the second number. $\frac{3}{5}$

7. Toss 4 coins, recording the result for each coin. One toss of all coins represents 1 trial. Repeat for 10 trials. Count the number of trials where 4 heads or 4 tails show up. Divide this number by 10. The quotient represents the experimental probability.

8. Roll 2 number cubes. Record the results. This represents 1 trial. Repeat for 10 trials. Count the number of trials where at least one cube shows a 6. Divide this number by 100. The quotient represents the experimental probability.

9. Yes, both Fred and Ted will have the same result if all trials were successful events or if all trials were unsuccessful events.

$$\frac{15}{15} = \frac{16}{16} \text{ or } \frac{0}{15} = \frac{0}{16}$$

Games Design and conduct experiments to determine the following experimental probabilities. Describe each experiment carefully and give your results for

16. getting *tails* when a coin is flipped.

17. getting 2 *tails* when a coin is flipped twice.

18. getting at least 3 when a number cube is rolled.

19. getting a multiple of 3 when a number cube is rolled.

20. getting an odd sum when two number cubes are rolled.

21. getting "doubles" (the same number on both cubes) when two number cubes are rolled.

22. getting the same result at least 3 times in a row when a coin is tossed 5 times.

23. not getting the same result twice in a row when a coin is tossed 5 times.

Technology The command RAND generates a decimal value from 0 to 1, including 0, but not 1. Describe the output of

24. RAND*2.

25. INT(RAND*2).

26. INT(RAND*2)+1.

27. 100*(INT(RAND*2)+1).

Before graphics calculators were used, engineers and scientists used books of tables to find data such as random numbers. A table of random digits can be found on page 709.

Look at the command INT(RAND*5)+10, where RAND generates a number from 0 to 1, including 0, but not 1.

28. How many different numbers are there? 5

29. What is the least number? 10

30. What is the greatest number? 14

Technology Write commands to generate random numbers from each of the following lists, where RAND generates a number from 0 to 1, including 0, but not 1. If you can, check your results using technology. Adapt the command to suit the computer or calculator you are using.

31. 0, 1, 2

32. 0, 1, 2, 3, 4, 5, 6, 7, 8, 9

33. 1, 2, 3, 4, 5, 6

34. 1, 2, 3, 4, 5, 6, 7, 8, 9, 10

Answers may vary for Exercises 16–23.

16. Toss a coin. Record the results. Repeat for 10 trials. Divide the number of trials where tails show by 10.

17. Toss a coin twice, recording the result for each toss. Repeat for 10 trials. Divide the number of trials where 2 tails show up by 10.

18. Roll a number cube; record the result. Repeat for 20 trials. Divide the number of trials where 3, 4, 5, or 6 show up by 20.

19. Roll a number cube; record the result. Repeat for 20 trials. Divide the number of trials where 3 or 6 show up by 20.

20. Roll 2 number cubes; record the result. Repeat for 40 trials. Divide the number of trials where an odd sum shows up by 40.

21. Roll 2 number cubes; record the result. Repeat for 40 trials. Divide the number of trials where the same numbers show up on both cubes by 40.

35. Answers may vary; correlation should be zero.

35. 🔺 **Statistics** Use a random number generator to make two sets
of 10 numbers. Pair the numbers from the two sets, and make a scatter
plot. What kind of correlation does the set of random number pairs
show? **[Lesson 1.6]**

36. Write the 6 values of the function $y = 5x + 2$ for x-values of 0, 1, 2, 3, 4,
and 5. Examine the differences. Explain what you notice about the
differences, and name the function. **[Lesson 2.1]**

> 2, 7, 12, 17, 22; The differences are the same; linear

Technology Find the values for the following functions. **[Lesson 2.4]**

37. INT(32.573) 32 **38.** INT(34.747) 34 **39.** ABS(−33.725) 33.725

Solve each equation for t. [Lesson 4.6]

40. $7(t - r) = 2r - 3(4r + t)$

$-\dfrac{3r}{10}$

41. $\dfrac{3(4t - 2)}{2} = \dfrac{3}{2} - (5 - t)$

$-\dfrac{1}{10}$

42. Use the data in the table to make a scatter plot. Find the slope of the line
that best fits the data. **[Lesson 5.1]** 0.375

x	2	5	6	10	12	15
y	4.750	5.875	6.250	7.750	8.500	9.625

43. Membership Fees Jenni pays a membership fee of \$47, plus \$4 per
hour at a tennis club. Kari pays a flat fee of \$275. For how many hours
can Jenni play tennis before her cost equals Kari's? **[Lesson 6.2]** 57

44. Solve the system of equations. **[Lesson 6.4]**

$\begin{cases} 3x + 5y = 2 \\ x - 3y = -4 \end{cases}$ $(-1, 1)$

45. Write the system of equations in Exercise 44 as a matrix equation, and
solve using matrix methods. **[Lesson 7.5]**

$$\begin{bmatrix} 3 & 5 \\ 1 & -3 \end{bmatrix} \begin{bmatrix} x \\ y \end{bmatrix} = \begin{bmatrix} 2 \\ -4 \end{bmatrix}; \begin{bmatrix} -1 \\ 1 \end{bmatrix}$$

Look Beyond ~~~

Look Beyond

Exercises 46–48 foreshadow
simulations in Lesson 8.2.

46. How many pairs of numbers are possible from one red cube and one
green cube if each has faces numbered 1 through 6? 36

47. How many of the pairs in Exercise 46 form a sum of 3? 2

48. What is the calculator command that will generate the numbers from
tossing one number cube? int (rand*6) + 1

Complete Exercises 49–54.

49. $1^2 - 0^2$ 1 **50.** $2^2 - 1^2$ 3 **51.** $3^2 - 2^2$ 5

52. $4^2 - 3^2$ 7 **53.** $5^2 - 4^2$ 9 **54.** $6^2 - 5^2$ 11

Use the pattern from Exercises 49–54 to solve.

55. $100^2 - 99^2$ 199 **56.** $a^2 - (a - 1)^2$ 2a − 1

22. Toss a coin 5 times, recording the result
after each toss. Repeat for 40 trials. Divide
the number of trials where at least 3 heads
or at least 3 tails appear in a row by 40.

23. Toss a coin 5 times, recording the result
after each toss. Repeat for 40 trials. Divide
the number of trials where heads and tails
alternate by 40.

24. numbers between 0 and 2, including 0,
but not 2

25. integers 0 and 1

26. integers 1 and 2

27. integers 100 and 200

31. INT(RAND*3)

32. INT(RAND*10)

33. INT(RAND*6)+1

34. INT(RAND*10)+1

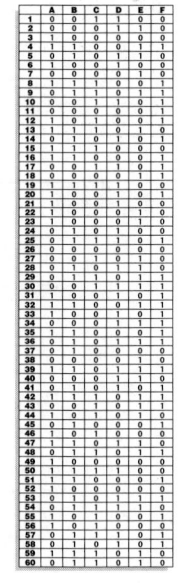

Technology
8.1 Larger and Larger Experiments

If you let 1 represent heads and 0 represent tails,
you can create a spreadsheet like the one shown to
model the flipping of a coin many times. In the
spreadsheet, each cell contains INT(2*RAND()).

Use the spreadsheet to answer the exercises.

1. The sum of the values in column A tells the
number of heads that occur in the first 60
tosses. Write spreadsheet formulas to find the
probability of flipping heads in each of the six
60-toss experiments for cells A61, B61, C61,
D61, E61, and F61.

2. Write a formula for cell G61 to find the mean of
the sums in cells A61, B61, C61, D61, E61, and
F61. From the spreadsheet, find the mean and
compare it with $\frac{1}{2} = 0.5$.

3. With your eyes closed, select twenty cells with
numerical values at random. Find the sum of
those values and divide the sum by 20. How
does the quotient compare to 0.5?

4. From the spreadsheet, find the probability of
tossing heads from all 360 flips. Use the result
to predict the number of heads in 10,000 tosses
of a coin.

5. Create your own spreadsheet like the one
shown. Use your spreadsheet to rework
Exercises 1–4 and compare your answers to
those you already found.

PREPARE

Objectives

- Design probability simulations.
- Use simulations to find experimental probability.

RESOURCES

- Practice Master 8.2
- Enrichment Master 8.2
- Technology Master 8.2
- Lesson Activity Master 8.2
- Quiz 8.2
- Spanish Resources 8.2

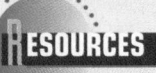

Assessing Prior Knowledge

Write each percent as a fraction in simplest form.

1. 50% $\left[\frac{1}{2}\right]$ **2.** 25% $\left[\frac{1}{4}\right]$

3. 80% $\left[\frac{4}{5}\right]$

4. 64% $\left[\frac{16}{25}\right]$

Solve.

5. You flipped a coin 10 times and it landed heads 3 times. What is the experimental probability of tossing tails? $\left[\frac{7}{10}\right]$

TEACH

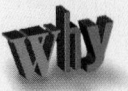

why Discuss any simulations that students have used. These might include science class or video games. Ask students to describe what a simulation is and how it is used.

LESSON 8.2 Exploring Simulations

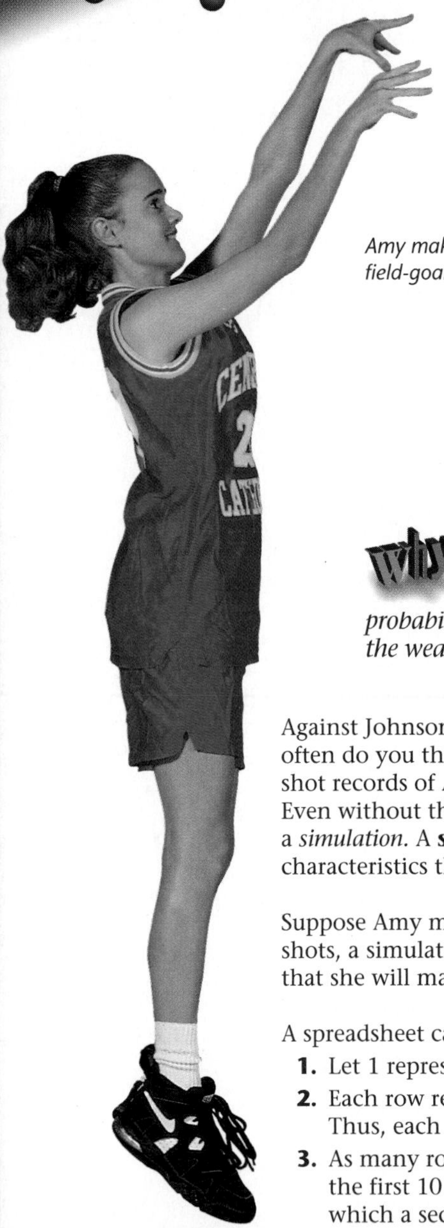

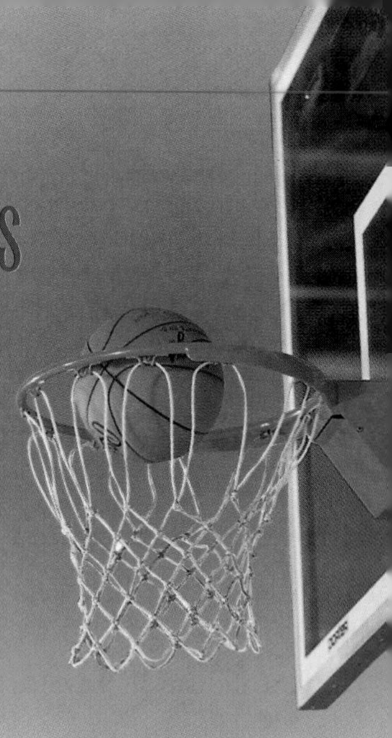

Amy makes 50% of her field-goal attempts.

why *Probabilities derived from coin, number cube, or technology experiments can be used to simulate outcomes when you want to find experimental probabilities. Simulations are used in studying games, sports, the weather, political elections, and many other situations.*

Against Johnson High School, Amy made four consecutive shots. How often do you think Amy makes four in a row? If you had the actual shot-by-shot records of Amy's performance, you could see how often this happens. Even without the records, you can find the experimental probability using a *simulation.* A **simulation** is an experiment with mathematical characteristics that are similar to the actual event.

Suppose Amy makes 50% of the shots she takes in a game. If Amy takes 20 shots, a simulation can be designed to find the experimental probability that she will make four in a row.

A spreadsheet can be used to generate random numbers.

1. Let 1 represent each shot made, and let 0 represent each shot missed.
2. Each row represents a trial. Generate 20 random numbers for each row. Thus, each row will have 20 columns.
3. As many rows as needed can be generated. For this simulation examine the first 10 rows or trials as shown. Count the number of trials in which a sequence of at least four consecutive 1s appear.

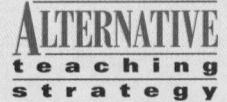

ALTERNATIVE teaching strategy

Using Manipulatives Allow students to use the flip of a coin as the random number generator. Let heads = 1 and tails = 0. To simulate the situation, students make 20 flips of the coin for each trial. There will be 10 such trials. Enter the results in a hand-drawn table similar to the one on page 361. Have students find the experimental probability of at least four consecutive ones based on the results entered in their table. Then have students compare their results with those shown on page 361.

	A	B	C	D	E
1	Trial	Results			
2	1	1	1	1	0

In the first trial, the beginning of the spreadsheet shows that Amy made her first three shots and missed the next one.

At the end of the first trial, Amy makes 4 in a row.

	R	S	T	U
	1	1	1	1

U11   **=INT(RAND()*2)**

	A	B	C	D	E	F	G	H	I	J	K	L	M	N	O	P	Q	R	S	T	U
1	Trial	Results																			
2	1	1	1	1	0	1	1	1	0	1	0	0	0	0	1	1	0	1	1	1	1
3	2	1	0	0	0	1	0	0	0	0	0	0	0	0	1	1	1	0	1	1	0
4	3	0	0	1	1	1	1	1	0	1	1	1	1	0	1	0	0	0	0	0	1
5	4	1	1	1	0	1	0	0	1	0	1	1	0	0	0	1	1	0	0	0	1
6	5	1	1	0	1	0	1	0	1	1	0	1	1	0	0	0	1	0	1	1	1
7	6	0	0	1	0	1	0	0	1	0	0	1	1	1	0	1	1	1	1	1	0
8	7	1	0	0	0	1	0	0	0	1	1	1	1	0	1	0	0	0	0	0	0
9	8	1	0	1	0	1	0	1	1	1	0	1	1	1	0	0	1	0	1	0	0
10	9	1	0	1	0	1	1	1	1	1	0	0	1	0	0	1	0	0	0	0	0
11	10	1	1	1	1	1	0	1	1	1	0	1	1	0	0	0	1	0	1	1	1

Of the 10 trials, there were at least four consecutive 1s recorded in six of the trials: 1, 3, 6, 7, 9, and 10. The experimental probability that Amy will make 4 shots in a row is $\frac{6}{10}$. ❖

 **Exploration 1** *Basketball Simulation*

Toss a coin to simulate a 50% chance of making a shot.

1. Toss a coin 20 times, and record the sequence of heads and tails.

2. Does the sequence have at least four consecutive heads?

3. Pool the results of the class.

4. Give the experimental probability based on the class experiment. ❖

DESIGNING A SIMULATION

1. Choose a random number generator such as a coin toss, number-cube roll, calculator, computer, or other method that will be used to simulate a situation. Describe what each result represents.
2. Plan how to perform the experiment to simulate one trial.
3. Perform a large number of trials, and record the results.

 **interdisciplinary** **CONNECTION**

Physical Education Have students investigate the actual statistics for their school's boys or girls basketball team. Find a player who has averaged making about 50% of his or her free-throw attempts and have that player attempt 20 free throws to find the experimental probability of making 4 shots in a row. Have students compare the results with those found in the simulation on page 361.

TEACHING tip

Technology Ask students whether they think using a spreadsheet program is practical for recording the results of simulations.
[**Answers may vary. If computers are not readily available, students may say they are impractical.**]

Simulations are easily done using the integrated software *f(g) Scholar™*, if available.

Exploration 1 Notes

Ask students why it is necessary to pool the class results before stating the experimental probability. [**The larger the number of trials, the more accurate the probability should be.**]

ongoing ASSESSMENT

4. **Experimental probability will depend on actual data, with the denominator equal to the total number of class tosses.**

TEACHING *tip*

Technology In Exploration 2 some students may be able to design a spreadsheet that includes another column in which the computer indicates yes or no. Challenge students to write a formula for such a column.

Aongoing
ASSESSMENT

7. $\frac{7}{10}$

Exploration 2 Weather Simulation

Spreadsheet

Meteorology Perform a simulation to find the experimental probability that it rains at least one of the two days.

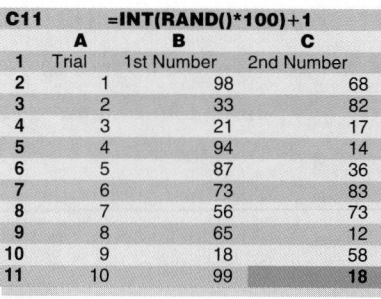

The weather report predicts that the probability of rain is 40% on Saturday and 40% on Sunday.

① Simulate a 40% chance of rain. If you use technology, generate a sequence of random integers from 1 to 100. A number less than or equal to 40 means it will rain. A number greater than 40 means it will not rain.

You can also use slips of paper or chips. Make 4 slips that show *rain* and 6 that show *no rain*. Then draw at random from a bag.

② To simulate the weather for two days, generate two numbers. This represents one trial.

③ Perform a large number of trials, and record the results.

The spreadsheet shows the simulation for 10 trials.

C11	=INT(RAND()*100)+1			At least one number less than or equal to 40?
	A	**B**	**C**	
1	Trial	1st Number	2nd Number	
2	1	98	68	no
3	2	33	82	yes
4	3	21	17	yes
5	4	94	14	yes
6	5	87	36	yes
7	6	73	83	no
8	7	56	73	no
9	8	65	12	yes
10	9	18	58	yes
11	10	99	**18**	yes

④ What do the numbers 98 and 68 in Trial 1 indicate about rain on those days?

⑤ What do the numbers 33 and 82 in Trial 2 indicate about rain on those days?

⑥ According to the data, how many weekends had rain at least one day in the 10 trials?

⑦ What is the experimental probability of having rain on at least one of the two days? ❖

ENRICHMENT Students could use a random-number generator to simulate a large number of coin tosses. Have students determine how to interpret the results. Then have them find the experimental probability of rolling a number less than 3.

INCLUSION **strategies** **Using Visual Models** Provide students with several copies of a sheet with 100 squares (a 10-by-10 grid numbered 1 through 100. Ask students to start at the top of the large square and shade 40% of the area of the entire large square. Ask what numbers are shaded. [**1 through 40**] Have students shade squares to show other percentages and relate the percent shaded to a corresponding number range. Relate these explorations to probability simulations, such as Exploration 2, to help students visualize why using the numbers 1–40 can be used to simulate a 40% probability.

The probability that Allessandro shows up on time is 60%. The probability that Zita shows up on time is 80%.

Appointments Allessandro and Zita are absent-minded students who plan to meet for lunch. A simulation can be designed to find the experimental probability that both Allessandro and Zita show up on time.

Design 1: Using number cubes

1. Roll two number cubes, a red one and a green one, for example. Let a 1, 2, or 3 on the red cube mean Allessandro is on time, and a 4 or 5 mean that he is late. Roll again if you get a 6. Note that 60% of the rolls that count means he is on time and 40% means he is late. Let a 1, 2, 3, or 4 on the green cube mean Zita is on time and a 5 mean she is late. Roll again if you get a 6.

2. Roll the two number cubes to represent one trial, which represents one lunch together.

3. Roll the cubes 10 times. Disregard any rolls that include a 6. Determine how many times, out of the 10 attempts, they meet on time for lunch. How many times does the red cube show a 1, 2, or 3 and the green one show a 1, 2, 3, or 4?

4. Express the experimental probability as a fraction with *the number of times the friends are on time* over *the total number of trials*. This represents the experimental probability that both friends are on time for lunch.

Spreadsheet

Design 2: Using technology

1. Use a calculator or computer to generate a pair of random integers from 1 to 100. If the first number is less than or equal to 60, it means Allessandro is on time. If the second number is less than or equal to 80, it means Zita is on time.

2. Each pair of numbers represents one trial, or meeting.

3. Generate 10 pairs. Determine how many times out of 10 that they meet on time.

4. Write the fraction with the number of successful trials over the total number of trials. This represents the experimental probability that both friends are on time for lunch. ❖

CRITICAL
Thinking

Make a list of real life situations that can be simulated with a coin flip and with a toss of a number cube. Compare the characteristics of the different types of situations.

RETEACHING
the
lesson

Hands-On Strategies
Have students complete Exploration 1 a second time, but use a different random-number generator. Again, pool the responses to find the experimental probability for the class. Ask whether the experimental proba-

bility for the class was the same as the first time they did the exploration. Have students discuss the results, particularly if the results were significantly different. Were the results different because of a different random-number generator or for other reasons?

TEACHING *tip*

Pose the following situation to students: Some member of your family is to pick you up at 4:30. How many of you will be sure to be ready before 4:30? How many of you will still be waiting at 4:40? Ask students why they answered as they did. Is there a pattern for certain members of their family. Are some members usually late or sometimes early? Relate this discussion to the Application on page 363.

CRITICAL
Thinking

Situations will vary. Sample answers: chance of a baby being a boy or a girl; chances of it raining, based on a weather forecast.

Assignment Guide

Core 1–6, 7–11, 12–22 even, 23–33

Core Two-Year 1–32

Technology

For Exercises 13–17, you may wish to require students to show how to use technology to simulate one or more of the events.

Error Analysis

In working with simulations, students may set up the situation correctly, but interpret the probability incorrectly. Caution students to be careful when deciding what they are to find and to make sure their results are reasonable.

Practice Master

Practice & Apply
8.2 Exploring Simulations

Suppose the Denver Nuggets and the New York Knicks are in the basketball championship finals and each team has an equal chance of winning a given game. The finals end when one team wins four games. Use simulation to play the finals for at least 10 trials.

1. Describe the three steps of the simulation.

2. Complete the chart to give game-by-game results of each of the 10 trials. Use random numbers generated by a coin toss or technology. Use W for a Knicks win, and L for a Knicks loss.

Trial	1	2	3	4	5	6	7
1							
2							
3							
4							
5							
6							
7							
8							
9							
10							

Use your results of the basketball championship simulation to find the experimental probability of each of the following.

3. The Nuggets win the finals. _____

4. The Knicks win the finals in 4 games. _____

5. The finals last 7 games. _____

6. A team wins the finals after losing the first two games. _____

EXERCISES & PROBLEMS

Communicate

1. What are the three steps for designing a simulation?

2. Assume that the chance of making a shot is 50%. Give three different ways to simulate making or missing a shot.

3. Suppose you want to simulate that it rains when there is a 30% chance of rain. Tell how to do this using random numbers from 1 to 10.

4. Name three things that could be simulated by flipping a coin once.

5. Tell how you might simulate guessing correctly a multiple-choice item with 5 possible responses.

6. Suppose 90% of the flights for an airline are on time. Explain how to design a simulation to find the experimental probability that three consecutive flights are on time.

Practice & Apply

Sports Look at Amy's basketball game at the beginning of the lesson.

7. How many shots did Amy make in the first trial? 13

8. In how many trials did Amy make more than 50% of her shots? 6

9. Give the value in cell Q4. What formula was used to generate the value?

10. What does the value in cell Q4 tell you? 0; INT(RAND()*2)
Amy did not make her 16th shot in trial 3.

Weather Suppose you want to simulate two days during which there is an 80% chance of rain and it rains at least one of the two days. Design the simulation and explain how you would perform it using

11. random numbers from 1 to 10. 12. a six-sided number cube.

Describe one way to simulate selecting at random

13. a day of the week. 14. a day of the year.

15. 1 student out of 24 students. 16. which of 8 team captains gets to choose first.

Describe a simulation for Exercises 17–22.

17. You are given two random numbers from 1 to 100. Find the experimental probability that both numbers are less than or equal to 20.

18. There are 10 ways to make a choice. There are 3 possible winning choices. Find the experimental probability that a person will win 2 out of 5 choices.

The answers to Exercises 13–16, 17, 18, 21, 22, 24, and 27 can be found in the Teacher's Answer Key beginning on page 729.

11. Let the numbers 1 to 8 represent the number of occurrences of rain, and the number 9 and 10 mean no rain. Generate 2 random numbers, let the results represent the weather on day 1 and day 2, respectively. Repeat for 10 trials. Divide the number of trials where rain occurred on at least one day by 10. The quotient is the experimental probability.

12. Let the numbers 1 to 4 represent the number of occurrences of rain, let 5 represent no rain, and let 6 mean "roll the number cube again." Toss the number cube twice with each toss representing 1 day. Record the result. Repeat for 10 trials. Divide the number of trials where rain occurred on at least one day by 10. The quotient is the experimental probability.

19. Demographics Find the experimental probability that a family with 4 children includes 2 boys and 2 girls. Assume that births of boys and girls are equally likely.

20. Find the experimental probability that a student correctly guesses all 3 questions on a true-false quiz.

21. Find the experimental probability of guessing the correct answer at least 70% of the time on a 4-question multiple choice quiz, if each item has 5 possible responses.

22. Repeat the previous problem for a 6-question quiz.

Look Back

23. If the variable n represents the number of the term, write an expression based on n that will generate any term of the following sequence. 4, 8, 12, 16, 20, 24, . . . **[Lesson 1.3]** term $n = 4n$

24. Draw examples, and explain the difference in the graphs of a quadratic and an exponential function. **[Lessons 2.1, 2.2]**

25. Write an expression to represent the length of a piece of fabric that is 74 centimeters longer than the width, w. **[Lesson 3.3]** $l = w + 74$

26. Solve the previous problem for the width, w, if the length is 86 centimeters. **[Lesson 3.5]** 12 cm

27. Represent the inequality $-7 < n < 8$ on a number line. **[Lesson 3.6]**

28. Solve $ac = p$ for c. **[Lesson 4.6]** $c = \dfrac{p}{a}$

29. **Coordinate Geometry** A line passes through the origin and the point $A(3, -2)$. Write the equation for the line that passes through point A and is perpendicular to the original line. **[Lesson 5.6]** $3x - 2y = 13$

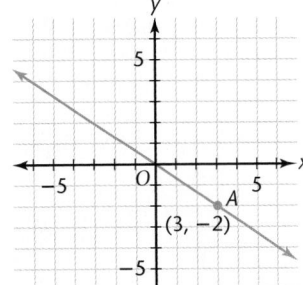

30. Multiply the two pairs of matrices. Are the results the same for each pair? **[Lesson 7.3]** no

$$\begin{bmatrix} -1 & 3 \\ 5 & 2 \end{bmatrix}\begin{bmatrix} 2 & -4 \\ 1 & -3 \end{bmatrix} \quad \text{and} \quad \begin{bmatrix} 2 & -4 \\ 1 & -3 \end{bmatrix}\begin{bmatrix} -1 & 3 \\ 5 & 2 \end{bmatrix}$$

31. Multiply the two pairs of matrices. Are the results the same for each pair? **[Lesson 7.3]** yes

$$\begin{bmatrix} -1 & 3 \\ 3 & -1 \end{bmatrix}\begin{bmatrix} 2 & 1 \\ 1 & 2 \end{bmatrix} \quad \text{and} \quad \begin{bmatrix} 2 & 1 \\ 1 & 2 \end{bmatrix}\begin{bmatrix} -1 & 3 \\ 3 & -1 \end{bmatrix}$$

32. What do Exercises 30-31 show about the commutativity of matrix multiplication? Matrix multiplication is not always commutative.

Look Beyond

33. Find the *theoretical* probability that a student correctly guesses all 3 questions on a true-false quiz. $\dfrac{1}{8}$

19. Answers may vary. Use a calculator to generate the numbers 1 and 2. Let 1 represent a boy, and 2 represent a girl. Generate 4 numbers; record the results. Repeat for 10 trials. Divide the number of trials where there are 2 boys and 2 girls by 10.

20. Answers may vary. Use a calculator to generate the numbers 0 and 1. Let 0 represent an incorrect response, and 1 represent a correct response. Generate 3 numbers; record the results. Repeat for 10 trials. Divide the number of trials where all three responses are correct by 10.

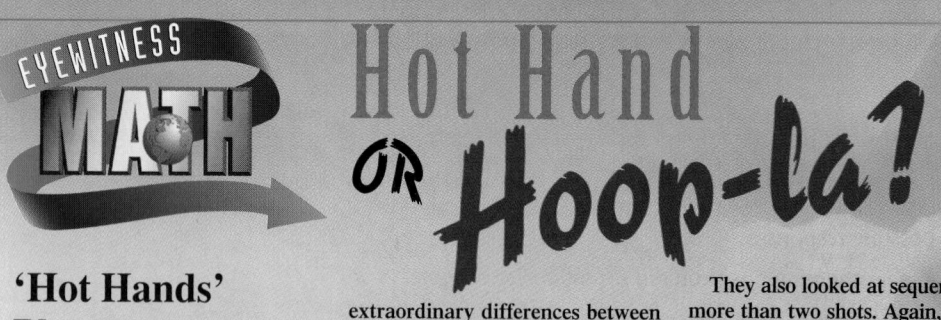

Hot Hand or Hoop-la?

'Hot Hands' Phenomenon: A Myth?

The gulf between science and sports may never loom wider than in the case of the hot hands.

Those who play, coach or otherwise follow basketball believe almost universally that a player who has successfully made his last shot or last few shots—a player with hot hands—is more likely to make his next shot. An exhaustively statistical analysis led by a Stanford University psychologist, examining thousands of shots in actual games found otherwise: the probability of a successful shot depends not at all on the shots that came before.

To the psychologist, Amos Tversky, the discrepancy between reality and belief highlights the extraordinary differences between events that are random and events that people perceive as random. When events come in clusters and streaks, people look for explanations; they refuse to believe they are random, even though clusters and streaks do occur in random data.

To test the theory, researchers got the records of every shot taken from the field by the Philadelphia 76ers over a full season and a half. When they looked at every sequence of two shots by the same player—hit–hit, hit–miss, miss–hit or miss–miss, they found that a hit followed by a miss was actually a tiny bit likelier than a hit followed by a hit.

They also looked at sequences of more than two shots. Again, the number of long streaks was no greater than would have been expected in a random set of data, with every event independent of its predecessor.

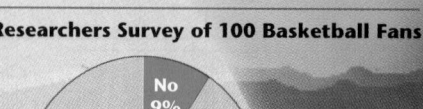

Researchers Survey of 100 Basketball Fans

No 9%
Yes 91%

Does a player have a better chance of making a shot after having just made his last two or three shots than he does after having just missed his last two or three shots?

No 16%
Yes 84%

Is it important to pass the ball to someone who has just made several (2, 3, or 4) shots in a row?

Discuss Do you agree with the majority on the two survey questions? Why or why not?

FOCUS

A study of player's shooting performances questions the widely held belief in the "hot hand" in basketball. Students tabulate some of the researchers' raw data to help decide for themselves if players actually "get in a groove" after making a few shots in a row or whether such streaks are merely the result of chance.

MOTIVATE

Ask students their opinions about hot streaks and cold streaks in sports. Ask what is meant by a "hot hand" in basketball and how you might go about deciding if there really is such a thing, or if it's just a matter of chance that there will be frequent streaks of hits and of misses.

1a. Player A

After a Hit	
No. of times next shot is made	No. of times next shot is missed
17	19

Player B

After a Hit	
No. of times next shot is made	No. of times next shot is missed
16	18

1b. Player A hit followed by hit: $\frac{17}{36} \approx 47.2\%$ or about 47%

hit followed by miss: $\frac{19}{36} \approx 52.8\%$ or about 53%

Player B hit followed by hit: $\frac{16}{34} \approx 47.1\%$ or about 47%

hit followed by miss: $\frac{18}{34} \approx 52.9\%$ or about 53%

1c. About 50% for each because whether the player just hit or missed, the chances of making the next shot would be equal.

Before proceeding, make sure students understand that they will be analyzing some of the actual data from the study to help reach a more informed opinion and that the actual data included many more players and many more games.

Go over the sample tabulation in the margin. Explain that spaces between batches of 1s and 0s mean a new game; why a string of *four* 1s in a single game counts as *three* instances of "a hit followed by a hit"; that the first shot in a game is not counted because the preceding shot was in a different game and thus would not be part of a streak due to a hot or cold hand. To help students answer question **c**, have them keep track of a series of coin tosses to get an idea of what a set of random data would look like and imagine a coin being tossed. Ask: *Does the coin remember whether it landed heads or tails on the previous toss?*

Discuss

Have students collect their own shooting performance data by watching an NBA, college, or local team play. They can compare these records to "random" data obtained by tossing a coin or programming a computer. If they use a set of coin data, they should compare it to the shooting data for a 50% shooter.

Cooperative Learning

To form your own opinion about the hot hand study, it helps to know how the researchers tabulated their data. For simplicity, you will use a portion of their data, which shows field goal attempts by two players for several games. Both players usually make about half of their shots from the field.

A—11100010 01110111010010 000011101 0100000101 010101001110 1001111 01001010001 011110110

B—10101000100 1110000110011 0101010111001111 111001001101 0001101010 10010001111

(1 = hit   0 = miss   Space = new game)

1. First, look at what happens after a shot is made within each game.

   a. Copy and complete Table 1, using the data for either Player A or Player B.

   b. What percentage of the times was a hit followed by another hit? What percentage of the times was a hit followed by a miss?

   c. What would you expect your results in **b** would be if there no such thing as a *hot hand*, that is, if the shots were hit or missed just by chance? Explain. (Assume the player makes 50% of his shots on average.)

2. Now look at what happens after a shot is missed within each game.

   a. Copy and complete Table 2. Use the same player you used for Table 1.

   b. What percentage of the times was a miss followed by a hit? What percentage of the times was a miss followed by another miss?

   c. What would you expect your results in **b** would be if the shots were hit or missed just by chance?

TABLE 1

After a Hit	
Number of times next shot is made	Number of times next shot is missed

TABLE 2

After a Miss	
Number of times next shot is missed	Number of times next shot is made

3. Do you think there is such a thing as a hot hand in basketball? Why or why not? How does your belief affect the strategy you would use if you were coaching a basketball team?

2a. Player A

After a Miss	
No. of times next shot is missed	No. of times next shot is made
15	21

Player B

After a Miss	
No. of times next shot is missed	No. of times next shot is made
16	18

2b. **Player A** miss followed by miss: $\frac{15}{36} \approx$ 41.7% or about 42%

miss followed by hit: $\frac{21}{36} \approx 58.3\%$ or about 58%

Player B miss followed by miss: $\frac{16}{34} \approx$ 47.1% or about 47%

miss followed by hit: $\frac{18}{34} \approx 52.9\%$ or about 53%

2c. About 50% for each because whether the player just hit or missed, the chances of making the next shot would be even.

3. Answers vary. The study showed that the shooting records of players is about what you would get if they hit or missed at random. If I were coaching, I would try to get the ball to the player who has the highest overall shooting percentage.

Objectives

- Find the mean, median, mode, and range for a set of data.
- Find the line of best fit.
- Recognize deceptive uses of statistics.
- Identify the margin of error for a set of data.

RESOURCES

- Practice Master   8.3
- Enrichment Master   8.3
- Technology Master   8.3
- Lesson Activity Master   8.3
- Quiz   8.3
- Spanish Resources   8.3

Assessing Prior Knowledge

Have students find the mean, median, and mode.

1. 3, 5, 2, 5, 6, 2, 1, 1, 5, 4
   [3.4, 3.5, 5]
2. 0.25, 0.5, 0.25, 0.25, 0.75
   [0.4, 0.25, 0.25]
3. $\frac{5}{8}, \frac{7}{8}, \frac{7}{4}, \frac{3}{8}, \frac{1}{2}, \frac{1}{8}$

   $\left[\frac{11}{24}, \frac{3}{16}, \text{no mode}\right]$

TEACH

 Have students discuss how statistics are important to their lives. Have them give examples from school, health, and consumer situations.

Alternate Example 1

Have students find the mean, median, mode, and range of the following data: Math Test Scores 90, 88, 64, 88, 96, 75, 78, 63, 90, 88, 92, 91, 82, 64, 78, 85.
[mean: 82, median: 86.5, mode: 88, range: 33]

LESSON 8.3 Extending Statistics

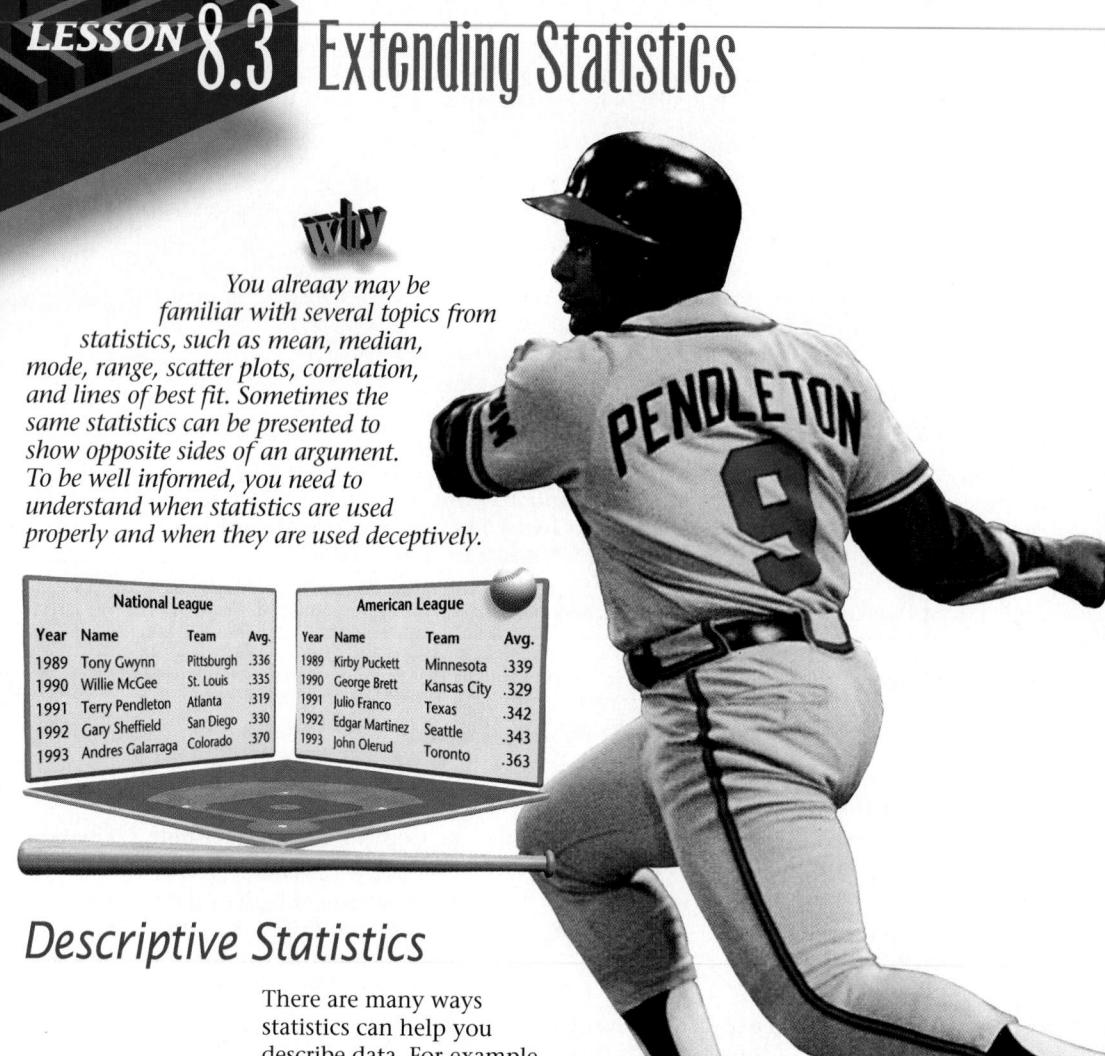

why

You already may be familiar with several topics from statistics, such as mean, median, mode, range, scatter plots, correlation, and lines of best fit. Sometimes the same statistics can be presented to show opposite sides of an argument. To be well informed, you need to understand when statistics are used properly and when they are used deceptively.

National League

Year	Name	Team	Avg.
1989	Tony Gwynn	Pittsburgh	.336
1990	Willie McGee	St. Louis	.335
1991	Terry Pendleton	Atlanta	.319
1992	Gary Sheffield	San Diego	.330
1993	Andres Galarraga	Colorado	.370

American League

Year	Name	Team	Avg.
1989	Kirby Puckett	Minnesota	.339
1990	George Brett	Kansas City	.329
1991	Julio Franco	Texas	.342
1992	Edgar Martinez	Seattle	.343
1993	John Olerud	Toronto	.363

Descriptive Statistics

There are many ways statistics can help you describe data. For example, *mean, median,* and *mode* indicate the central tendency of data. The *range* provides information about the spread of the data.

EXAMPLE 1

Give the mean, median, mode, and range for the batting averages (avg.) in each league. Use these statistics to write a summary statement.

Solution ➤

Sports To find the *mean,* add the batting averages for each player in each league. Then divide each sum by 5. The means are .338 and .3432, respectively. The .3432 is customarily rounded to .343.

Using Cooperative Learning Have students work in small groups to research statistics relevant to their own experience, such as grade-point average and number of hours spent watching television in one week, or number of minutes spent studying for a test and the grade received on the test. Each group should collect at least 6 sets of data and find the mean, median, mode, and range for each set. Then use the data to make a scatter plot and find the line of best fit.

To find the *median,* arrange the batting averages in order, then choose the middle number. If the number of batting averages is even, the median is the mean of the two middle numbers. The National League batting averages in order from least to greatest are .319, .330, .335, .336, and .370. The median is .335. The median for the American League is .342.

To find the *mode,* choose the batting average that occurs the most. Since no number appears more than once in either set of data, the mode is not reported.

To find the *range,* determine the difference between the highest and lowest batting averages. The range of the National League is .370 − .319 = .051. The range of the American League is .363 − .329 = .034.

In summary, the mean and median batting averages were higher in the American League. The batting averages in the National League had a greater spread, or range. ❖

EXAMPLE 2

Make a scatter plot of the batting averages for each year for the National League. Use a calculator or computer to find the correlation and the line of best fit.

Solution ➤

Graphics Calculator

A graphics calculator will help locate the data points for a scatter plot and find the line of best fit.

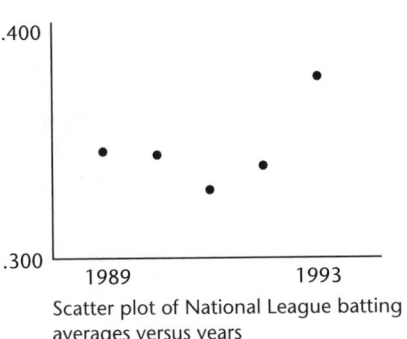

Scatter plot of National League batting averages versus years

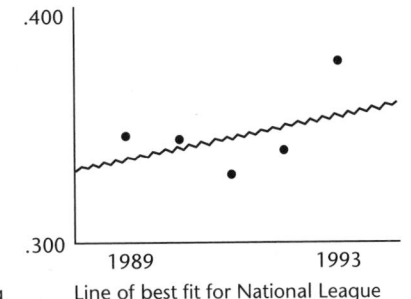

Line of best fit for National League batting averages versus years

The variables show a moderate relationship. The correlation is about 0.52, and the line of best fit has a slope of about 0.006. What does this trend suggest about the batting averages over the five-year period? ❖

Try This

Fans Interviewed
24 38 26 31 31
30 29 29 37 32
39 40 29 35 45

Find the mean, median, and mode of the number of fans interviewed as they entered the stadium during a 15-game study.

Alternate Example 2

Technology Use a calculator or computer to find the correlation and line of best fit for these data.

Year	'90	'91	'92	'93	'94	'95
Sales	84	84	81	78	76	76

TEACHING *tip*

If available, the integrated software *f(g) Scholar*™ would be useful for demonstrations of scatter plots and lines of best fit such as Example 2.

Aongoing ASSESSMENT

This trend suggests that the batting averages increased slightly over the five-year period.

Aongoing ASSESSMENT

Try This

mean, 33; median, 31; mode, 29

INCLUSION strategies

Guided Research Students can be actively involved in collecting data for scatter plots by having them measure each other with a centimeter ruler or tape measure. Have students find each of the following measurements in centimeters: height, length of arm from fingertips to tip of elbow, circumference of head, circumference of neck, length of foot, length of leg from bottom of heel to middle of knee, etc. Then have students plot various pairs of mea-surements on large grids to make scatter plots. Students can use thumbtacks and string to show the line of best fit (if there is one) for various pairs of data, such as circumference of head and height.

ENRICHMENT Have students find at least three newspaper graphs that give misleading information. Have the student write a report that includes a way to correct the graphs.

Math Connection
Geometry

Circle graphs can also be distorted by drawing the circle as an ellipse. If the desired sector is to seem larger, it is displayed in the foreground (lower part) of the ellipse. If it is to seem smaller, it is displayed in the background (upper part) of the ellipse.

CRITICAL
Thinking

Remind students that the formula for volume is $V = lwh$. Thus, doubling the dimensions of box A yields a volume that is 8 times as great.

Deceptive Statistics

Advertising While statistics are very powerful and useful, they are sometimes used deceptively. Some graphs can be deliberately misleading because of the way the scale is selected.

% of trucks still in use

Brand C has the most trucks in use by far.

It appears at a glance that Model C, the commercial's sponsor, has far more trucks still in use than the other models. It looks like Model C has about twice as many trucks still in use as Model D. According to the scale, however, Model C differs from Model D by only a small amount, 3%.

GEOMETRY
Connection

Graphs can also be deliberately distorted by using area or volume to compare amounts. The fact that twice as many doctors recommended Brand B can be represented by a 3-dimensional box. A box twice as high with the same width and depth would fairly represent the difference. But a box that has twice the height, twice the width and twice the depth makes Brand B appear much greater.

CRITICAL
Thinking

If box B has twice the dimensions of box A, how many times as great is its volume?

Surveys are used to sample people's opinions about everything from politics to new cosmetics. Major decisions rest on how a sample of the population responds. There is often a question of how certain anyone can be with the results. A phrase used when discussing the results of a survey is *margin of error*.

RETEACHING
the
lesson

Using Visual Models
Alicia had test scores of 80, 81, 83, 84, and 85. Challenge students to make two line graphs of these scores: one that shows the actual improvement in scores and one that makes the improvement in scores look more impressive. **[For the second graph, students should start the grid near 80 and have a large space between 80 and 85.]**

Exploration 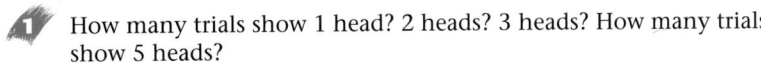 *Margin of Error*

Each trial represents the number of heads thrown in 10 tosses of a coin. The table shows the results after 20 trials.

1 How many trials show 1 head? 2 heads? 3 heads? How many trials show 5 heads?

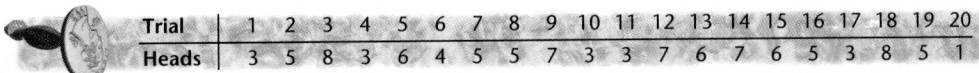

Trial	1	2	3	4	5	6	7	8	9	10	11	12	13	14	15	16	17	18	19	20
Heads	3	5	8	3	6	4	5	5	7	3	3	7	6	7	6	5	3	8	5	1

You can see the distribution if you make a tally of the trials for the number of heads thrown.

2 What is the mean number of heads thrown for this set of 20 trials?

3 What percent of the 20 trials show from 2 to 8 heads?

4 How many trials are contained within 3 heads of the mean?

5 In general, what do you expect the mean to be for flipping a coin?

6 According to the data you can be 95% certain that the results are within 3 heads on either side of 5 for the 20 trials. The 3 heads on either side of the mean represent the *margin of error.*

7 What percent of all the trials show only 1 head? Is this inside or outside the margin of error? ❖

EXERCISES & PROBLEMS

Communicate

1. Explain the difference between the mean, median, and mode.
2. In Example 1, tell how the median was found for the American League.
3. Bob thought the median for the data 10, 20, 8, 5, and 17 was 8. Explain what he did wrong.
4. Tell how to find the median for a set of six pieces of data.

Exploration Notes

Have students use 5 coins and complete this experiment. Have them graph the distribution in the same way as shown on page 371. Then have each student determine how many of their results were outside the margin of error.

Aongoing
ASSESSMENT

7. 5%; outside the margin of error

Assignment Guide

Core 1–7, 12–15, 22–40 even, 42–51

Core Two-Year 1–48

Technology

A standard calculator may be used by students to complete Exercises 22–25. For Exercises 26–31, it is recommended that students use a graphics calculator.

Error Analysis

Students may not understand how using only part of a scale on the side of a graph affects the interpretation of the information contained on a bar or line graph. For Exercises 34–39, have students draw the bar graph when the scale on the left starts at zero and goes to 21. Then have students complete the exercises.

Practice Master

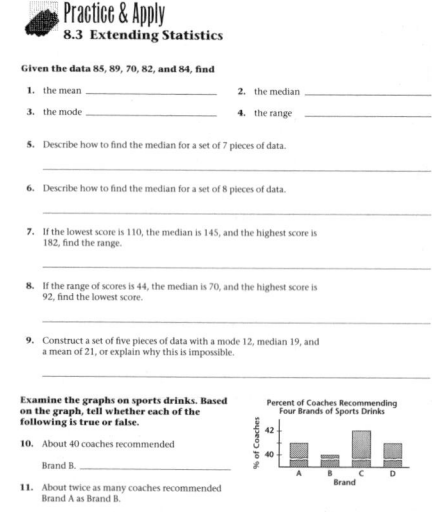

5. In an ad to show that its lawn products were preferred 2 to 1, a company showed a lawn that its product covered, twice as long and twice as wide as another lawn. This resulted in an area how many times as large? Explain.

6. In the margin-of-error exploration, how confident can you be that the true mean is within 2 of the mean for the total number trials?

7. Describe two ways statistics can be used deceptively.

Practice & Apply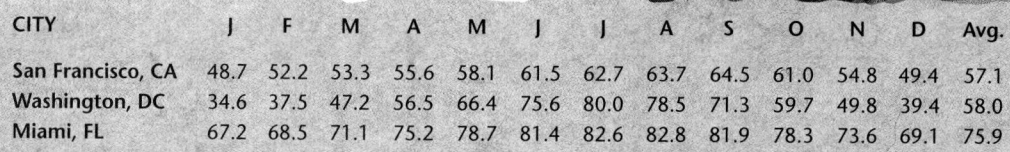

Given the data 24, 26, 19, 20, and 33, find

8. the mean.	**9.** the median.	**10.** the mode.	**11.** the range.
24.4	24	none	14

Given the data 12.4, 14.1, 14.1, 14.6, 15.0, and 15.3, find

12. the mean	**13.** the median.	**14.** the mode.	**15.** the range.
14.25	14.35	14.1	2.9

16. Tell how to find the median for a set of 9 pieces of data.

17. Tell how to find the median for a set of 10 pieces of data.

18. If the lowest score is 62, the median score is 76, and the highest score is 98, find the range. 36

19. If the range of scores is 32, the median score is 80, and the highest score is 100, find the lowest score. 68

20. Construct a set of five pieces of data with mode 18, median 20, and mean 21, or explain why this is impossible. one possibility: 18, 18, 20, 24, 25

21. Construct a set of five pieces of data with mode 18, median 20, and mean 19, or explain why this is impossible. impossible

Meteorology The monthly normal mean temperatures for three cities are shown below.

CITY	J	F	M	A	M	J	J	A	S	O	N	D	Avg.
San Francisco, CA	48.7	52.2	53.3	55.6	58.1	61.5	62.7	63.7	64.5	61.0	54.8	49.4	57.1
Washington, DC	34.6	37.5	47.2	56.5	66.4	75.6	80.0	78.5	71.3	59.7	49.8	39.4	58.0
Miami, FL	67.2	68.5	71.1	75.2	78.7	81.4	82.6	82.8	81.9	78.3	73.6	69.1	75.9

22. Compute the range of temperatures for each city. 15.8°; 45.4°; 15.6°

23. Which city stands out because of its average temperature? Miami

24. Which city stands out because of its temperature range? Washington

25. Write several sentences comparing the temperatures of the cities without using any numbers.

16. Arrange in order, then find the middle value.

17. Arrange in order, then find the mean or average of the 2 middle values.

25. Answers may vary. The average temperatures for San Francisco and Washington are almost equivalent. The range of temperatures for San Francisco and Miami are almost equivalent. The summer temperatures for Miami and Washington are almost equivalent.

National League			
Year	Name	Team	Avg.
1989	Tony Gwynn	Pittsburgh	.336
1990	Willie McGee	St. Louis	.335
1991	Terry Pendleton	Atlanta	.319
1992	Gary Sheffield	San Diego	.330
1993	Andres Galarraga	Colorado	.370

American League			
Year	Name	Team	Avg.
1989	Kirby Puckett	Minnesota	.339
1990	George Brett	Kansas City	.329
1991	Julio Franco	Texas	.342
1992	Edgar Martinez	Seattle	.343
1993	John Olerud	Toronto	.363

26. **Sports** Make a scatter plot of the batting averages versus years for the American League.

27. Describe the trend for the American League. Are the batting averages getting better or worse? How much is the change per year?

28. **Technology** Find the correlation for the data in Exercise 26. 0.79

29. Compare the correlation in Exercise 28 with the correlation you found for the National League in Example 2. Is the relationship stronger or weaker? National League is weaker.

30. Use an almanac or another source to find data for more recent National League batting champions. Analyze the data, and compare the results with the data given in Example 1. Depends on year used.

31. Repeat Exercise 30 for the American League. Depends on year used.

32. **Advertising** Suppose a lawn-care company used an ad to show that its products were preferred 3 to 1 by showing that its product covered a lawn three times as long and three times as wide as a lawn of its competition. This would result in an area how many times as large? 9

33. Suppose a cereal manufacturer used an ad to show that 20% more people preferred its products by showing a cereal box 20% longer, 20% wider, and 20% taller than its competition. This would result in a box with a volume how many times as large as the competition? 1.7

Advertising Examine the graph on pain relievers. Based on the graph, tell whether each of the following is true or false.

34. About 20 doctors recommended Brand A. T

35. About twice as many doctors recommended Brand X as Brand A. F

36. The majority of doctors recommended Brand X. F

37. What percent of doctors recommended Brand C? 20.5%

38. If there were 200 doctors in the survey, how many recommended Brand X? 42

39. If there were 200 doctors in the survey, how many more recommended Brand X than Brand C? 1

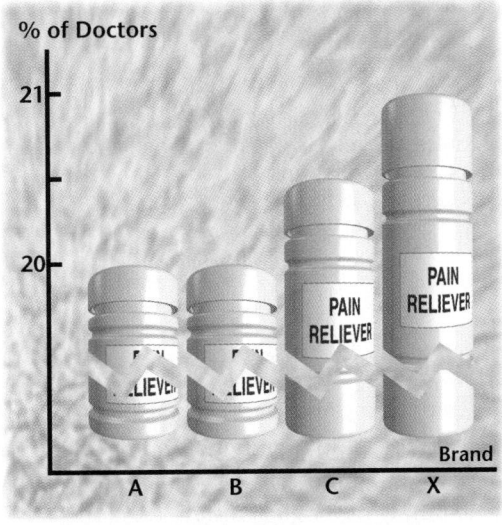

% of Doctors

26.

American League Averages (y-axis): 0.365, 0.355, 0.345, 0.335, 0.325

Year (x-axis): '89 '90 '91 '92 '93

27. The averages dropped the first year, then improved. The averages improved by a mean of 0.006.

Find the margin of error for each situation.

40. The mean is 78, and 95% of the scores are between 75 and 81, inclusive. 3

41. The mean is 8.4, and 95% of the scores are between 7.8 and 9.0 0.6 inclusive.

Look Back

42. The sequence 1, 6, 13, 22, 33, 46, ... is represented by what kind of function? How do you know?
[Lesson 2.2] quadratic; second differences—constant 2

43. **Meteorology** The lowest temperature on record for each of several cities is shown in the table. What are the differences in temperature? **[Lesson 3.1]**

H-J:15°
H-D: 32°
J-D: 17°

CITY	RECORD LOW
Juneau, AK	−22
Duluth, MN	−39
Houston, TX	−7

44. A radio is marked $63. What is the actual price of the radio if you also pay a sales tax of 6%? **[Lesson 4.5]** $66.78

45. An automatic coin counter totals Denise's collection of dimes and nickels at $36.45. Earlier, she counted 231 nickels. Write and solve an equation to find the number of dimes. **[Lesson 4.6]**
$0.05(231) + 0.1d = 36.45$; 249

46. **Coordinate Geometry** What figure is shown when you connect the points $A(-1, -2)$, $B(1, -1)$, $C(3, 0)$, $D(1, 2)$, $E(0, 0)$, and then A again? What is the equation for a line containing points A and D?
[Lesson 5.4] triangle; $y = 2x$

47. In Exercise 46 what is the slope of a line parallel to a line containing points D and C? **[Lesson 5.6]** −1

48. Solve the matrix equation. Then write the matrix equation as a system of linear equations, and solve the system. Are your answers the same? **[Lesson 7.5]**

$$\begin{bmatrix} 9 & -4 \\ -3 & 2 \end{bmatrix} \begin{bmatrix} x \\ y \end{bmatrix} = \begin{bmatrix} 61 \\ -23 \end{bmatrix} \quad \begin{bmatrix} 5 \\ -4 \end{bmatrix}; (5, -4); \text{yes}$$

Look Beyond

Exercises 49–51 extend students' understanding of statistics, which will be studied in future sections.

Look Beyond

49. Find the mean of the numbers .336, .335, .319, .330, and .370. 0.338

50. Find the mean of the whole numbers 336, 335, 319, 330, and 370. 338

51. If you multiply each piece of data by 1000, by what value does this multiply the original mean? 1000

Exploring the Addition Principle of Counting

why

The process of counting entries in lists such as databases can often be tedious. Once you learn the shortcuts to counting, you save yourself time and effort. You can also discover many useful and interesting mathematical properties.

Databases display records that have a selected entry in one or more fields. For example, you can print the records of all the people in the database that live in Texas.

White	Marsalis	TX	Yes
Clinton	John and Ann	TX	No
Johnson	Merrill	TX	No

Similarly, if you print the records of all the people who are relatives, you will get the following.

Schmidt	Hans, Gretchn	WI	Yes
Post	Alex	OH	Yes
White	Marsalis	TX	Yes
Childress	Andrea	OH	Yes

Meg keeps a data base of addresses of her friends and relatives. Each line, like the information about Alex Post, is a *record*. Each category, like State is a *field*.

ALTERNATIVE teaching strategy

Using Tables On the board or overhead, complete this table for members of the class.

Last Name	First Name	Hair Color	Eye Color

After completing the table, tell students that this is one type of data base. Each line is a *record*, and each category, like Hair Color, is a *field*. Use the data base to make various Venn diagrams, such as one that shows students who have black hair and brown eyes.

PREPARE

Objectives
- Draw and interpret Venn diagrams.
- Use the Addition Principle of Counting.

RESOURCES

- Practice Master   8.4
- Enrichment Master   8.4
- Technology Master   8.4
- Lesson Activity Master   8.4
- Quiz   8.4
- Spanish Resources   8.4

Assessing Prior Knowledge

List five multiples of each number. Answers may vary for 1–2. Sample answers are given.

1. 4: $[8, 12, 16, 20, 24]$

2. 11: $[22, 33, 44, 55, 66]$

3. What is the meaning of the symbol $>$?   [**is greater than**]

4. What is the meaning of $\geq$? [**is greater than or equal to**]

5. Is 4 a solution to $x > 4$? [**no**]

6. Is 4 a solution to $x \geq 4$? [**yes**]

TEACH

Ask students if any of them have an address book. An address book is a data base. Ask how students use such a data base. One answer might be to have a phone number or address available when it is needed.

Databases can also combine information from two different fields. When you request combined information, you will need to use the instructions AND and OR. The words **AND** and **OR** have special meaning when working with databases. For example, if you print the records of all the people who are living in Texas OR who are relatives, you will get the following.

Schmidt	Hans, Gretchn	WI	Yes
Post	Alex	OH	Yes
White	Marsalis	TX	Yes
Clinton	John, Ann	TX	No
Johnson	Merrill	TX	No
Childress	Andrea	OH	Yes

If you print the records of all the people who are living in Texas AND who are relatives, you will get the following.

White	Marsalis	TX	Yes

Logic The relationship between AND and OR can be illustrated with a *Venn diagram*.

Name and Address Database

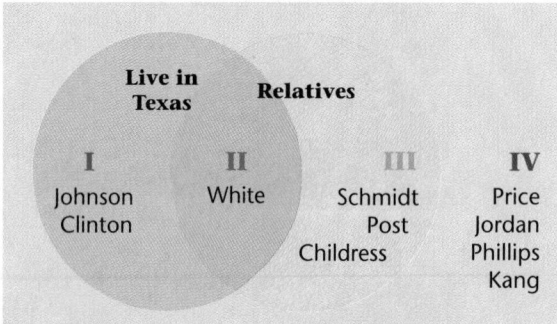

Region I *contains people who live in Texas and who are not relatives.*

Region II *contains people who live in Texas and are relatives.*

Region III *contains people who are relatives and who do not live in Texas.*

Region IV *contains people who do not live in Texas and who are not relatives.*

Region II represents the overlapping regions, or **intersection**, of the people who are in the first set **AND** the second set. Regions I, II, and III together represent the combined regions or **union** of the two sets of people. These people are in the first set **OR** the second set.

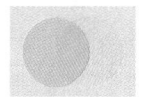

intersection

When two sets have no members in common, the sets are called **disjoint**. For example, the set of people who live in Ohio and the set of people who live in Texas are disjoint.

union

 **CRITICAL** *Thinking*

What is the relationship between set *A* and set *B* if *A* is contained in *B* and *B* is contained in *A*?

disjoint

CRITICAL *Thinking*

If *A* is contained in *B* and *B* is contained in *A*, then *A* and *B* are the same sets and contain all the same elements.

ENRICHMENT Suppose Clinton, White and Post have red hair. Include this new data with the Name and Address Database above. Have students draw a Venn diagram involving the intersection of three circles. Fill in the circles and explain the meaning of the Venn diagram.

Original Database

Last name	First name	Street	City	State	Zip	Relative?
Schmidt	Hans, Gretchen	1025 Union Ave.	Sheboygan	WI	53081	Yes
Post	Alex	555 Weber Rd.	Columbus	OH	43202	Yes
White	Marsalis	33 S. 5th St.	Edinburg	TX	78539	Yes
Phillips	Jane	6122 Douglas St.	Greeley	CO	80634	No
Price	Latisha	144 5th Ave.	Harrisburg	PA	17126	No
Clinton	John, Ann	223 Houston Ave.	Huntsville	TX	77341	No
Jordan	Cornelius	815 S. 1st Ave.	Ashland	OH	44906	No
Johnson	Merrill	9935 Lincoln Rd.	Austin	TX	78746	No
Kang	Hong	1332 Hartford St.	Pensacola	FL	32514	No
Childress	Andrea	RR # 3	Berea	OH	44017	Yes

Exploration 1 *Interpreting a Venn Diagram*

Use the original database that shows all 10 records.

1 Draw a Venn diagram using the last names of people living in Ohio and people living in Texas as your sets.

2 How many of the people live in Ohio?

3 How many of the people live in Texas?

4 How many of the people live in either Ohio OR Texas?

5 How many of the people live in both Ohio AND Texas?

6 Recall the number of people that answers the question in Step 4. Then find the sum of the numbers that answers Steps 2 and 3. What do you notice when you compare these two amounts? ❖

Exploration 2 *Counting from a Venn Diagram*

1 Draw a Venn diagram that represents the people who live in Ohio and who are relatives.

2 How many of the people live in Ohio?

3 How many of the people are relatives?

4 How many of the people either live in Ohio OR are relatives?

5 How many of the people live in Ohio AND are relatives?

6 Does your answer to Step 4 equal the sum of the answers to Steps 2 and 3? Explain.

7 Show how to produce the answer to Step 4 based on the numbers from Steps 2, 3, and 5. ❖

The next application illustrates the use of a counting technique in statistics.

INCLUSION strategies

Using Visual Models

Venn diagrams can greatly aid visual learners. Have students make a Venn diagram for the Application on page 378. (Note: The Venn diagram needed to answer the questions will show the number of boys and the number of students who favor the new rule.) Then relate the elements of the Venn diagram to the Addition Principle of Counting.

Exploration 1 Notes

Students should note that the sum of Steps 2 and 3 yields the same answer as for Step 4. Ask students how many total people are involved in Steps 2 and 3? [6] Have students also note that the answer to Step 5 is zero. You may wish to ask students to conjecture about the total number of people involved if the answer to Step 5 is not zero. This will be discussed in Exploration 2.

Aongoing **SSESSMENT**

6. The amounts are the same.

Exploration 2 Notes

Students should realize that some people are relatives AND live in Ohio. Thus, that number of people (2) should be shown in the intersection of the two circles in the Venn diagram. For Step 7, students should add the answers to Steps 2 and 3 and subtract the answer to Step 5 from that sum. (3 + 4 − 2 = 5) The result (5) is the same as the answer to Step 4.

Aongoing **SSESSMENT**

7. Ohio + Relatives − Relatives in Ohio = Total [3 + 4 − 2 = 5]

Have students discuss how statistics influence daily life on many different levels. They may make a personal choice of what to buy based on their previous experience. Those earlier experiences are part of the database they use to make choices.

	Favor rule	Oppose rule	Total
Boys	4	9	13
Girls	7	10	17
Total	11	19	30

Mark took a class survey to get student opinions about a new rule concerning students driving to school.

STATISTICS *Connection*

After reviewing the data, the following are some of the questions that were asked. Notice that you need to know what is meant by AND and OR.

Q: How many students sampled are boys?
A: The total for the first row is 13, the total number of boys.

Q: How many students favor the rule?
A: The total for the first column is 11, the total number who favor the rule.

Q: How many students are boys AND favor the rule?
A: If you look under *Favor rule* and across from *Boys,* the intersection shows the number 4. This is the number of students who favor the rule and are boys.

Q: How many students are boys OR favor the rule?
A: The answer is the number of boys plus the number favoring the rule, except that some have been counted twice. Subtract the number who are boys AND favor the rule. If you use the information from the responses to the previous questions, the answer is $13 + 11 - 4 = 20$. ❖

The response to these questions leads to an important principle used in counting.

> ### ADDITION PRINCIPLE OF COUNTING
> Suppose there are m ways to make a first choice, n ways to make a second choice, and t ways that have been counted twice. Then there are $m + n - t$ ways to make the first choice OR the second choice.

Games In an old card game called *Crazy Eights,* if the queen of hearts is showing, you can play either a queen, a heart, or an eight. How many different kinds of cards can be played? Keep in mind that an ordinary deck of cards has 4 queens, 13 hearts, and 4 eights.

You do not count the queen of hearts, since it has already been played. There are 3 other queens, 12 other hearts, and 4 eights, for a total of 19 cards. The eight of hearts has been counted twice, so subtract 1. Thus, $3 + 12 + 4 - 1$ totals 18 cards that can be played. ❖

RETEACHING *the lesson*

Using Manipulatives Provide each student with a yellow acetate circle and a blue acetate circle. Using washable markers, have students write the names of all the boys in the class on the left half of the yellow circle and all the names of students who wear glasses on the right half of the blue circle. Using the overhead projector, project each color circle. Overlap the two circles (yellow on the left and blue on the right). The center section should appear green. Have students identify any name that appears on both circles, erase that name from the yellow and blue sections, and write it in the green section. Those names are in the intersection of the two sets. Continue with other Venn diagrams for different pairs of characters of students.

EXERCISES & PROBLEMS

Communicate

1. Explain the difference between a field and a record in a database.

2. In the original data base for this lesson, name another entry beside the Schmidts that is in the intersection of the set of relatives and the set of people who are not from Texas.

3. In the game of *Crazy Eights* application, explain how the 8 of hearts is counted twice.

4. What do you know about the sets if, in the union of the sets, you do not have to subtract?

Statistics   A survey of participation in school music programs has the results shown in the table. Explain how to find each of the following.

5. How many students are girls?

6. How many students participated?

7. How many students are girls AND participated?

8. How many students are girls OR participated?

	Participated	Did not participate
Boys	43	59
Girls	49	57

Practice & Apply

List the integers from 1 to 10, inclusive, which are

9. even. 2, 4, 6, 8, 10

10. multiples of 3. 3, 6, 9

11. even AND multiples of 3. 6

12. even OR multiples of 3.
2, 3, 4, 6, 8, 9, 10

Games   Tell how many of each of the following are in an ordinary deck of 52 cards.

13. aces 4

14. red cards 26

15. cards that are red OR aces 28

In the first application on page 378, how many students

16. are girls? 17

17. oppose the rule? 19

18. are girls AND oppose the rule? 10

19. are girls OR oppose the rule? 26

20. When does the number of members in the union of two sets equal the sum of the number of members in each set? When they do not intersect.

ASSESS

Selected Answers
Odd-numbered Exercises 9–45

Assignment Guide
Core   1–20, 22–28 even, 40–50

Core Two-Year 1–46, 50

Technology

If students have difficulty using spreadsheets allow more time to work in order to become familiar with the vocabulary as well as the techniques involved in using a spreadsheet.

Error Analysis

Students may become confused about the words *AND* and *OR* as they are used in conjunction with statistics. Remind students that *AND* means "in both only," referring to the intersection of the sets. *OR* means "in either" or "in both," referring to the union of the sets.

Performance Assessment

Have students use the information to solve the problems, and explain their method.

KING HIGH SCHOOL STUDENTS		
	Girls	Boys
Walk To School	145	155
Do Not Walk To School	230	205

1. How many girls are there in King High School? **[375]**

2. How many students in all are there in King High School? **[735]**

3. How many students are girls AND walk to school? **[145]**

4. How many students are girls OR walk to school? **[530]**

List the integers from 1 to 20, inclusive, that are multiples

21. of 5. 5, 10, 15, 20

22. of 3. 3, 6, 9, 12, 15, 18

23. of 5 AND multiples of 3. 15

24. of 5 OR multiples of 3. 3, 5, 6, 9, 10, 12, 15, 18, 20

25. People waiting to hear a concert were given numbers. At one point they let people enter with tickets numbered 100 to 150. How many people entered? 51

26. If it takes 15 minutes to saw a log into 3 pieces, how long does it take to saw it into 4 pieces? HINT: The answer is *not* 20 minutes. 22.5

27. A spreadsheet has 25 rows of data beginning in row 5. Which row has the last row of data? 29

Statistics In a class of 28 students, 17 have brown eyes and 13 have black hair. This includes 10 who have both.

28. Show this in a Venn diagram. Include all 28 students.

29. How many students have either brown eyes OR black hair? 20

30. In a poll, 13 students said they had taken music lessons and 5 said they had taken dance lessons. How many students had taken either music lessons OR dance lessons? What is the most? What is the least? HINT: There is more than one answer possible. The most is 18; the least is 13.

A survey of participation in school team sports had the results shown in the table.

31. How many students are girls? 361

32. How many students participated? 235

33. How many students are girls AND participated? 111

34. How many students are girls OR participated? 485

	Participated	Did not participate
Boys	124	239
Girls	111	250

Opinion Polls The survey in the first application on page 378 had the following results for the entire school.

Grade	Favor rule	Oppose rule	Total
9	2	97	99
10	34	61	95
11	57	40	97
12	81	7	88

35. How many 9th graders opposed the rule? 97

36. How many students were either in 9th or 10th grade AND opposed the rule? 158

37. How many students were either in 9th or 10th grade OR opposed the rule? 241

38. How many non-seniors favored the rule? 93

39. How many non-seniors opposed the rule? 198

28.

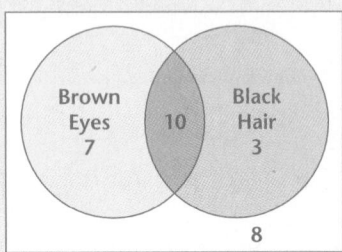

Brown Eyes 7   10   Black Hair 3   8

40. Savings   Chou deposits $150 in a bank that pays 5% interest per year compounded annually. How much money will she have in this account after 5 years?   **[Lesson 2.1]** $191.44

Simplify.

41. $-(-3 + (-7)) + 23 - 2(7 - 5)$ **[Lesson 3.3]** 29   **42.** $4(a + 2b) - 5(b - 3a)$ **[Lesson 3.4]**

$19a + 3b$

43. 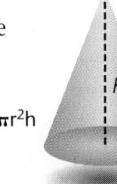 **Geometry**   The volume of a cone is $V = \frac{1}{3}\pi r^2 h$. Solve the equation for h.
[Lesson 4.3] $h = \dfrac{3V}{\pi r^2}$

$V = \frac{1}{3}\pi r^2 h$

Solve.

44. $3(x + 5) - 23 = 2x - 47$
[Lesson 4.6] -39

45. $\frac{x}{3} = -3x + 5$ **[Lesson 4.6]** $\frac{3}{2}$

46. Sports   A major league baseball player has a .300 batting average. Use a simulation to estimate the probability that the batter gets at least two hits in his next four at bats. Describe the 3 steps.
[Lesson 8.2]

Look Beyond

Of the 28 equally-likely ways of choosing a student, there are 17 ways of choosing a student with brown eyes.

In an earlier exercise there was a class of 28 students. Of those, 17 had brown eyes and 13 had black hair. There were 10 who had both. If a student is chosen at random, the (theoretical) probability that the student has brown eyes is $\frac{17}{28}$. Find the experimental probability that a student chosen at random has

47. brown eyes AND black hair. $\frac{5}{14}$

48. black hair. $\frac{13}{28}$

49. brown eyes OR black hair. $\frac{5}{7}$

50. A book has pages numbered from 1 to 203. How many of the page numbers contain at least one 7? 38

46. Answers may vary. Use a calculator to randomly generate the numbers 1 to 10. Let the numbers, 1, 2, and 3 represent a hit; let the numbers 4, 5, 6, 7, 8, 9, and 10 represent a miss. Generate 4 numbers; record the results. Repeat for 10 trials. Divide the number of trials where two hits are made by 10. This is the experimental probability.

PREPARE

Objectives

- Make a tree diagram to show possible outcomes.
- Use the Multiplication Principle of Counting to find the number of possible outcomes.

RESOURCES

- Practice Master   **8.5**
- Enrichment Master   **8.5**
- Technology Master   **8.5**
- Lesson Activity Master   **8.5**
- Quiz   **8.5**
- Spanish Resources   **8.5**

Assessing Prior Knowledge

Find each product.

1. 5 · 3   **[15]**
2. 2 · 3 · 7   **[42]**
3. 20 · 19   **[380]**
4. 8 · 3 · 5   **[120]**
5. A line contains points B, J, and W. Using two letters, list each possible name for this line.   **[BJ, BW, JB, JW, WB, WJ]**

TEACH

 Ask if any students have ever tried to list all of the possibilities for the results of an event, only to give up in frustration when they realize how difficult it is. Ask how a person can find all the different ways three people can be elected president, vice-president and secretary of an organization.

LESSON 8.5 Multiplication Principle of Counting

why *In the previous lesson you learned how to count in situations in which you made one choice OR another. It is also important to know how to handle situations where you make one choice AND another.*

Jordanna is scheduling her classes for next year. After she registers for her required courses, Jordanna may choose one elective from cluster 1 and one from cluster 2.

Scheduling   One way to find all the possible selections is to make a **tree diagram.**

The tree diagram shows that for each of the 3 ways to choose a music elective there are 2 ways to choose a cluster 2 elective.

You can see by following along the branches of the tree diagram that there are 3 · 2, or 6 possible choices.

Cluster 1	Cluster 2
band	home economics / woodworking
orchestra	home economics / woodworking
chorus	home economics / woodworking

- band and home economics
- orchestra and home economics
- chorus and home economics

- band and woodworking
- orchestra and woodworking
- chorus and woodworking

Jordanna has six possible selections.

 **ALTERNATIVE teaching strategy**

Using Models   Pose the problem that you have 3 different vegetables and 2 different fruits. Display 1 can of each different type. Ask students how many different possibilities there are if you want to serve one of these vegetables and one fruit at a meal. Have students show how to model a list by physically placing a can of vegetables and a can of fruit together on the desk. Continue putting pairs on the desk until all possibilities are shown. Students should note that there are 6 such possibilities.

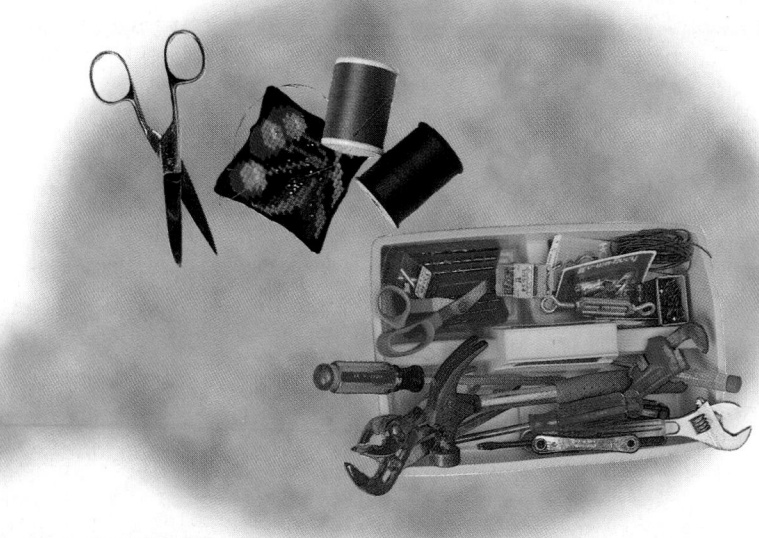

Alternate Example 1

A restaurant serves four kinds of sandwiches: cheese, roast beef, chicken, and peanut butter. With each sandwich, a free beverage is offered: coffee, milk, or tea. Make a tree diagram to show the possible combinations of a sandwich and free beverage. How many possible outcomes are there? [12]

EXAMPLE 1

Sports   There are 2 volleyball teams, each with 6 members. One player is to be selected at random. Flip a coin to select the team, then roll a number cube to select a player from that team. How many pairings of outcomes are possible?

Solution ➤

Make a tree diagram.

For each of the 2 outcomes from the coin, there are 6 outcomes from the number cube. That is, $2 \cdot 6 = 12$ pairings. ❖

What if there were 3 teams instead of 2? How would you find how many outcomes would be possible?

These examples lead to another useful principle used in counting.

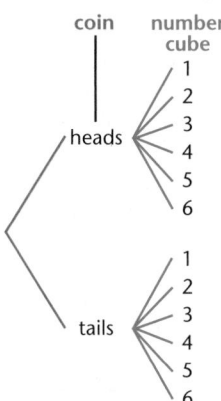

```
coin      number
          cube
            1
            2
            3
 heads      4
            5
            6

            1
            2
            3
 tails      4
            5
            6
```

TEACHING *tip*

Point out that in Example 1 a different tree diagram could be made if each number from the number cube were listed and followed by the result of each coin toss. For example, 1H, 1T, 2H, 2T, etc. Ask students whether the change in order affects the number of possible outcomes. [No]

MULTIPLICATION PRINCIPLE OF COUNTING

If there are *m* ways to make a first choice and *n* ways to make a second choice, then there are $m \cdot n$ ways to make a first choice AND a second choice.

The Multiplication Principle of Counting is often used instead of a tree diagram, especially when the answers to counting problems are large. Even when the original numbers are small, some multiplications can produce numbers too inconvenient for a tree diagram.

ENRICHMENT   Have students gather information about different choices that can be made when buying a new car. Have students make a list of these different options and then make a tree diagram to show the number of choices. If there are too many possibilities, have students explain how they can use the Multiplication Principle of Counting to determine how many possibilities there are.

INCLUSION  **strategies**   **Using Manipulatives**
Provide students with colored centimeter cubes, paper clips, chips, or other objects that come in various colors. Students will need 6 each of 6 different colors. Pose situations such as a red ball and a blue ball in one bag and a yellow ball, a green ball, and a white ball in another bag. Have students use the colored objects to show the possible outcomes.

EXAMPLE 2

Alternate Example 2

A print shop offers custom-made personalized stationery. The customer can choose the color of the paper (white, pink, or blue), the type of border (one line all around, double lines all around, double lines on top and bottom only, or wavy lines all around), and the color of ink used to print the border (black, blue, red, brown, purple, or green). If each choice contains one selection from each of the 3 characteristics, paper, border, and ink, how many different choices of stationary are there? [3 · 4 · 6 = 72]

Aongoing ASSESSMENT

Try This

The tree diagram should show the number of choices, 2, 4, and 2. The Multiplication Principle of Counting says that there are 2 · 4 · 2, or 16 possible outcomes.

Alternate Example 3

Each slip of paper in a bag has one of the following letters: O, P, S, T. Four slips are chosen, one at a time, without replacing the paper slips. Make a tree diagram to show the possible outcomes. **[There should be 24 different outcomes.]** How many of the outcomes are real words? [6] How can the Multiplication Principle of Counting be used to solve this problem? [4 · 3 · 2 = 24]

Math Connections
Geometry

In Example 3, ask students how many points are needed to name a line. [2] Students should note that once a letter is chosen, it cannot be chosen again as the second letter.

Inventory

20 styles!
Each in 8 Sizes
and 3 Colors!

A shoe store wants to stock 20 styles of shoes in various sizes and colors. How many pairs of shoes must the store carry to have one pair of each combination?

Solution ➤

For each of the 20 styles there are 8 sizes. This means there are 20 · 8, or 160 possible combinations of styles and sizes. For each of these 160 combinations there are 3 colors, so there are 160 · 3, or 480 combinations in all. You can also perform the multiplications together.

20 · 8 · 3 = 480

As you can see, the Multiplication Principle of Counting can be extended beyond two choices. ❖

Try This Tammy is selecting colored pens. The first characteristic is the kind—felt tip or ballpoint. The second characteristic is the color—red, blue, green, or black. The third characteristic is the brand—Able or Blakely. Use a tree diagram to show the possible selections there are for a pen. Explain how this relates to the Multiplication Principle of Counting.

The next example shows how counting problems occur in geometry.

EXAMPLE 3

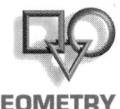

GEOMETRY
Connection

Points *A*, *B*, and *C* lie on the same line. How many ways are there to name the line using two of the letters?

<——•——————•——————•——>
    A   *B*   *C*

Solution A ➤
Make a tree diagram.

There are 6 ways to name the line.

AB, AC, BA, BC, CA, and *CB*

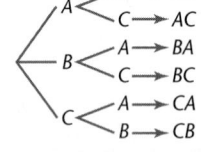

RETEACHING
the
lesson

Have students work in small groups and draw tree diagrams of possible outcomes of events. By working together, students can share insights and understanding with each other. The tree diagrams help students to visualize results and relate to the multiplication concepts.

Solution B ➤

Use the Multiplication Principle of Counting. There are 3 ways to choose the first letter. Once this is done, there are 2 ways remaining to choose the second letter. There are 3 · 2 = 6 ways. ❖

CRITICAL Thinking

What are the advantages of using a tree diagram, and what are the advantages of using the Multiplication Principle of Counting?

The addition principle and the multiplication principle are very useful methods to simplify counting. Nevertheless, you must be careful to understand when and where to use each principle.

EXAMPLE 4

Games   A regular deck of 52 playing cards contains 26 red cards and 26 black cards. The deck contains 4 aces, 2 black and 2 red.

A One card is drawn. How many ways are there to draw a black ace?

B One card is drawn. How many ways are there to draw a black card or an ace?

C Two cards are drawn. The first card is replaced before drawing the second card. How many ways are there to draw a black card followed by an ace?

Solution ➤

A Two aces are black, the ace of spades and the ace of clubs. Thus, there are 2 ways to draw a black ace.

B There are 26 black cards and 4 aces, but the 2 black aces are counted twice. According to the addition principle, there are 26 + 4 − 2 = 28 ways to draw a black card OR an ace.

C There are 26 ways to draw a black card. No matter which card is drawn, there are 4 ways to draw an ace as the second card. Using the multiplication principle, there are 26 · 4 = 104 ways to draw a black card AND an ace in two draws. ❖

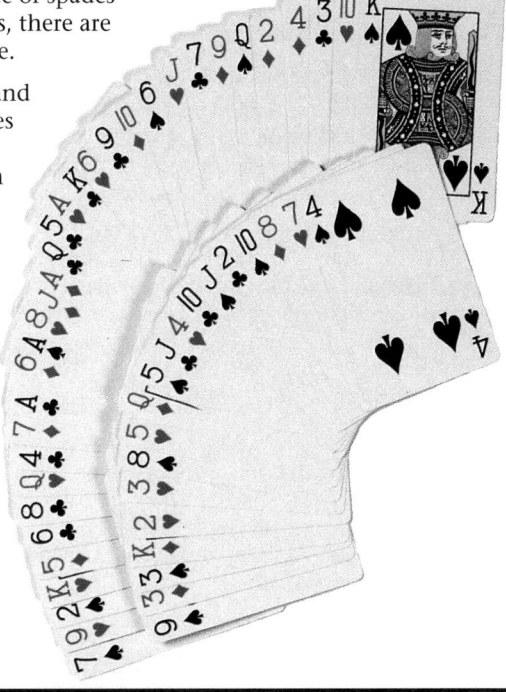

CRITICAL Thinking

Answers may vary. Sample answer: An advantage of a tree diagram is that it shows all the possible outcomes. An advantage of the Multiplication Principle of Counting is that it is shorter and faster to complete, especially when there are many possible outcomes.

Alternate Example 4

Jacques took these clothes to camp: 3 pairs of shorts (white, blue, and black), 4 pull-over shirts (white, blue, red, and black), and 2 pairs of tennis shoes (1 white and 1 black). Each outfit is considered to be a pair of shorts, a shirt, and shoes.

1. How many different outfits can Jacques wear?   [24]
2. How many outfits can Jacques wear that will have black shorts and black shoes?   [4]
3. How many outfits can Jacques wear that will have three different colors?   [9]
4. How many outfits can Jacques wear that will be all the same color?   [2]

ASSESS

Selected Answers

Odd-numbered Exercises 7–27

Assignment Guide

Core 1–6, 7–11 odd, 13–21, 22–32

Core Two-Year 1–28

Technology

Exercises 13–20 deal with anagrams. Suggest that students use a spreadsheet program to create all possible anagrams of a word with five different letters, such as PARTS. The computer can list all the possible combinations and then students can determine which are really words. [**There are 120 combinations. Anagrams include TRAPS, STRAP, and TARPS.**]

EXERCISES & PROBLEMS

Communicate

1. Explain the difference between these two sentences and explain the special meaning of AND and OR.
   "I am going to the mall AND I am going to study."
   "I am going to the mall OR I am going to study."

Suppose in the opening example on page 382, there were four cluster 1 electives instead of three for Jordanna. Explain how to determine how many choices there are if Jordanna were to select

2. one cluster 1 elective OR one cluster 2 elective.

3. one cluster 1 elective AND one cluster 2 elective.

4. Suppose there are 2 volleyball teams, and each team had 8 players. One player is to be selected at random. How can you find the number of branches the corresponding tree diagram has?

5. Two cards are drawn. The first card is replaced before drawing the second card. How can you find the number of ways there are to draw a three followed by a red card?

6. **Geometry** To name a ray, always start with the endpoint. In this case it is *A*. How many ways are there to use two of the letters to name the ray?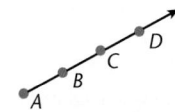

Practice & Apply

7. The Raholls are ordering a new car. There are 9 exterior colors from which to choose. For each of these there are 4 interior colors. How many color combinations are possible? 36

8. **Games** In chess, at the start of the game White moves first and has 20 possible moves. For each of these possibilities, Black has 20 possible countermoves. How many ways are there for the game to start with each player making one move? 400

9. Tricia bought 2 CDs and 3 tapes. How many recordings did she buy? 5

10. Ed is trying to choose 1 CD out of 2 that he likes and 1 tape out of 3 that he likes. In how many ways can he do this? 6

A menu contains 4 appetizers, 3 salads, 5 main courses, and 6 desserts.

11. How many ways are there to choose one of each? 360

12. How many ways include a chocolate sundae, one of the 6 desserts? 60

Language A word which is obtained by rearranging the letters of another word is called an *anagram*. For example, *art* is an anagram of *rat*.

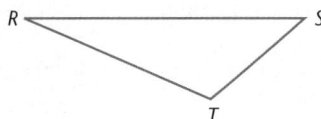

13. Find another anagram of *rat*. tar

14. How many ways are there to arrange the letters of the word *rat*, using each letter exactly once? 6

15. How many ways are there to arrange the five letters of the word *stake*? HINT: There are 5 ways to choose which letter is first. For each of these 5 ways, there are 4 ways to choose which letter is second. 120

16. Write two or three anagrams of *stake*. steak; skate

17. How many ways are there to arrange the six letters of the word *foster*? 720

18. Find two anagrams for *foster*. softer; fortes

19. What anagrams can you find within the sentence, "Ed cares that no races on acres in the country scare the birds"? cares; races; acres; scare

20. Make up at least 3 anagrams of your own.
example: live, evil, vile

21. **Geometry** How many ways are there to name the given triangle using 3 letters? 6

R ——————— S

T

Look Back

22. Hexagonal numbers are in a sequence 1, 6, 15, 28, 45, 66, 91, . . . Find the next two hexagonal numbers. **[Lesson 1.2]** 120, 153

Solve. **[Lesson 4.4]**

23. $\frac{3}{2}x = 6$ 4   **24.** $\frac{x}{-4} = -\frac{5}{8}$ 2.5   **25.** $\frac{x}{5} = 10$ 50

26. **Coordinate Geometry** The points $P(7, -4)$ and $Q(-2, 5)$ are on a line. What is the slope of that line? **[Lesson 5.1]** -1

27. If lines on a graph are parallel, how are their slopes related? **[Lesson 5.6]** same

28. **Chemistry** How many milliliters of a 90% acid solution are needed to mix with 450 milliliters of a 16% acid solution to produce a 50% acid solution? **[Lesson 6.6]** 382.5 ml

Look Beyond

Scheduling In the opening to this lesson, the probability that Jordanna chooses band and home economics is $\frac{1}{6}$ because making that choice is 1 of 6 equally likely possibilities. Find the following probabilities.

29. She chooses band for her cluster 1 elective. $\frac{1}{3}$

30. She chooses woodworking for her cluster 2 elective. $\frac{1}{2}$

31. She chooses either band or woodworking (or both). $\frac{2}{3}$

32. She chooses band and woodworking. $\frac{1}{6}$

- Find the probability that an event will occur.
- Use counting principles to find the probability that an event will occur.

RESOURCES

- Practice Master   **8.6**
- Enrichment Master   **8.6**
- Technology Master   **8.6**
- Lesson Activity Master   **8.6**
- Quiz   **8.6**
- Spanish Resources   **8.6**

Assessing Prior Knowledge

Write each fraction as a percent. (Round to the nearest whole percent, if necessary.)

1. $\frac{4}{25}$ **[16%]**

2. $\frac{4}{28}$ **[14%]**

3. $\frac{7}{7}$ **[100%]**

4. $\frac{0}{50}$ **[0%]**

5. How many possible outcomes are there for drawing a black card and then a red card from a standard deck of cards?

  [26 · 26 = 676]

TEACH

Have students make a list of the ways in which they use probability and how it affects their lives. Ask why knowing a probability is important. Ask students to speculate on the difference between experimental probability and theoretical probability.

LESSON 8.6 Theoretical Probability

In the first lesson of this chapter you had the chance to explore experimental probability. Another way to find the numerical measure of chance is to use theoretical probability. Many probabilities can be found more quickly and exactly using theoretical probability.

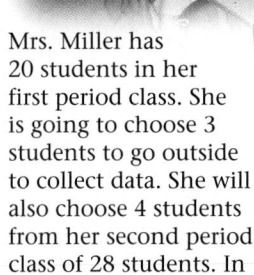

Making Choices Mrs. Miller has 20 students in her first period class. She is going to choose 3 students to go outside to collect data. She will also choose 4 students from her second period class of 28 students. In which class is the chance of being selected greater?

More students will be chosen from Mrs. Miller's second period class, but does this mean that the chances of being selected are greater? There are also more students from which to choose. To make it easier to compare, find the fraction that represents the number to be chosen out of the total number for each class.

First period (3 out of 20) The chances are $\frac{3}{20}$, or 15%.

Second period (4 out of 28) The chances are $\frac{4}{28}$, or about 14%.

The chances are slightly greater that a certain person will be chosen in the first period class even though fewer students will be chosen from that class.

The *event* is *to be chosen to go outside*. Since each of the students has an equal chance of being chosen, the number of possible outcomes is the number of students in the class. The number of successful outcomes in the event is the number of students to be chosen.

ALTERNATIVE
teaching strategy

Using Models Show students spinners that fit each of these descriptions. Each spinner is a circle that is separated into A: halves, B: thirds, C: four unequal sections, and D: three unequal sections. Have students determine which spinners have equally-likely outcomes. **[A and B]** Continue by shading one section of spinner A and one section of spinner B. Have students find the theoretical probability of each spinner stopping in the shaded section. $\left[A: \frac{1}{2}, B: \frac{1}{4}\right]$

THEORETICAL PROBABILITY

Let n be the number of *equally-likely* outcomes in the event. Let s be the number of successful outcomes in the event. Then, the **theoretical probability** that an event will occur is $P = \frac{s}{n}$.

The condition that the outcomes are *equally likely* is important. It means that each outcome has the same probability of happening.

Consider a simple example. A 5 on a toss of an ordinary six-sided number cube is considered a success. You might think that the probability is $\frac{1}{2}$ since only two things can happen—either you will roll a 5 or you will not. Experimentally, however, 5 does not usually come up half the time, so you know this is not correct.

The problem is that the two possibilities are *not equally likely*. The correct probability must take into account six *equally-likely outcomes*. Notice that you can roll either a 1, 2, 3, 4, 5, or 6, but only one of these is a 5. Each of the numbers has 1 chance in 6 that it will appear. The probability of rolling a 5 is $\frac{1}{6}$, not $\frac{1}{2}$.

Always be sure to consider *equally-likely outcomes* when determining the theoretical probability. Note that when you are asked to find probability, it will be the theoretical probability unless otherwise specified.

EXAMPLE 1

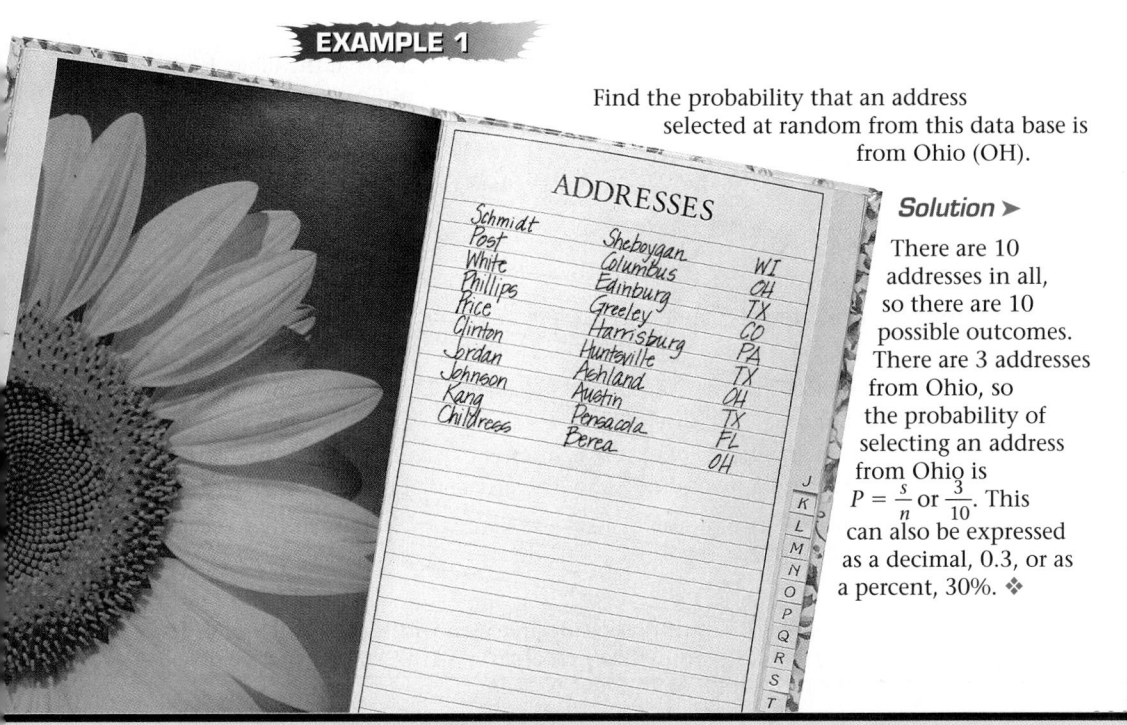

Find the probability that an address selected at random from this data base is from Ohio (OH).

ADDRESSES

Schmidt	Sheboygan	WI
Post	Columbus	OH
White	Edinburg	TX
Phillips	Greeley	CO
Price	Harrisburg	PA
Clinton	Huntsville	TX
Jordan	Ashland	OH
Johnson	Austin	TX
Kang	Pensacola	FL
Childress	Berea	OH

Solution ➤

There are 10 addresses in all, so there are 10 possible outcomes. There are 3 addresses from Ohio, so the probability of selecting an address from Ohio is $P = \frac{s}{n}$ or $\frac{3}{10}$. This can also be expressed as a decimal, 0.3, or as a percent, 30%. ❖

Alternate Example 1

Each box contains eight different colors of pencils. One of these colors is red. Find the theoretical probability of choosing red from the box without looking. $\left[\frac{1}{8}, \text{ or } 12.5\% \right]$

interdisciplinary CONNECTION **Biology** The science of heredity, known as genetics, relies on probability. One example is that crossing a round-seeded plant (*RR*) with a wrinkled-seeded plant (*rr*) theoretically results in three round-seeded plants for every wrinkled-seeded plant. What is the theoretical probability that a plant bred from the two original plants will be round-seeded? $\left[\frac{1}{2} \right]$

Roll a number cube once. What is the theoretical probability that the result is greater than 3? $\left[\frac{1}{2}\right]$

Roll the number cubes. What is the theoretical probability that the sum is 8? $\left[\frac{5}{36}\right]$

Aongoing
SSESSMENT

There are 36 sums in all; four are 5.

Use Transparency ▶ 37

Aongoing
SSESSMENT

Try This

The probability of getting a sum of 9 is $\frac{4}{36}$, or $\frac{1}{9}$.

CRITICAL
Thinking

Events will vary. For 0, the event can never occur. For 1, the event must always occur. The range of values for theoretical probability is from 0 to 1, inclusive.

390 Lesson 8.6

EXAMPLE 2

Games Roll a number cube once. What is the probability that the result is less than 3?

Solution ➤

The equally-likely outcomes are 1, 2, 3, 4, 5, or 6. Two of these outcomes (the event 1 appears or the event 2 appears) are successes, so the probability is $\frac{2}{6}$, or $\frac{1}{3}$. ❖

EXAMPLE 3

Roll two number cubes. What is the probability that the sum is 5?

Solution ➤

There are 6 equally-likely outcomes for the first number cube. For each of these, there are 6 equally-likely outcomes for the second number cube. By the Multiplication Principle of Counting, there are 6 · 6, or 36, ways to get a number on the first cube AND a number on the second cube. These are summarized in the table.

	Second number cube					
	1	2	3	4	5	6
1	2	3	4	5	6	7
2	3	4	5	6	7	8
3	4	5	6	7	8	9
4	5	6	7	8	9	10
5	6	7	8	9	10	11
6	7	8	9	10	11	12

First number cube

Look at the table. How many sums are there in all? How many sums of 5 are there? The probability of rolling a sum of 5 is $\frac{4}{36}$, or $\frac{1}{9}$. ❖

Try This Find the probability that a sum of 9 is tossed on one roll of two dice.

Notice that there are 11 possible sums: 2, 3, 4, 5, 6, 7, 8, 9, 10, 11, and 12. But the probability of getting a sum of 7 is not $\frac{1}{11}$. These sums are *not* equally likely. For example, a sum of 7 can happen 6 ways, while a sum of 12 can happen only 1 way. The probability of a sum of 7 is $\frac{6}{36}$, or $\frac{1}{6}$. The possibility of a sum of 12 is $\frac{1}{36}$. A sum of 7 is more likely to occur than a sum of 12.

CRITICAL
Thinking

Describe an event that results in a sum with a probability of 0. Then describe an event with a probability of 1. What is the range of values for the probability of an event occuring?

ENRICHMENT Have students consider a deck of 52 playing cards. The deck consists of "4 suits" (hearts, spades, trumps, and diamonds) of 13 cards each. The suit contains cards numbered 2–10 and face cards (jack, queen, king, and ace). Ask students to name events about drawing cards from a deck with theoretical probabilities of $\frac{1}{4}, \frac{1}{13}, \frac{1}{26}$, and $\frac{5}{52}$.

INCLUSION
strategies
Using Cognitive Strategies Pose variations on typical probability problems, and ask students whether the probability remains the same or changes. For example: Two number cubes are numbered from 11 to 16. Is the probability that the sum will be 16 the same as the probability that sum will be 6 on a standard pair of dice? [Yes] If the bull's-eye in Exercise 26 on page 392 is not in the center, is the probability of hitting it the same? [Yes]

EXAMPLE 4

Statistical Survey The table shows the results of a class survey about a new rule concerning students driving to school. A student who participated in the survey is chosen at random. What is the probability that the student

	Favor rule	Oppose rule	Total
Boys	4	9	13
Girls	7	10	17
Total	11	19	30

Ⓐ is a boy?

Ⓑ favors the rule?

Ⓒ is a boy AND favors the rule?

Ⓓ is a boy OR favors the rule?

Solution ➤

The denominator of the fraction is the number of *equally-likely* possibilities. In this case, it is the number of students in the survey, 30. You can use counting principles to find the values for the numerator.

Ⓐ 13 students are boys. The probability is $\frac{13}{30}$.

Ⓑ 11 students favor the rule. The probability is $\frac{11}{30}$.

Ⓒ 4 students are boys who favor the rule. The probability is $\frac{4}{30}$, or $\frac{2}{15}$.

Ⓓ 13 students are boys and 11 favor the rule, but this counts 4 students twice. By the Additional Principle of Counting there are $13 + 11 - 4$, or 20, students who are boys OR who favor the rule. The probability is $\frac{20}{30}$, or $\frac{2}{3}$. ❖

EXERCISES & PROBLEMS

Communicate

1. Explain the difference between experimental probability and theoretical probability.

2. John said he had a 50-50 chance of getting all 10 questions right on a quiz because either he would or he would not get each one right. Is he correct? Explain.

3. Based on his record so far this season, the probability that Nat will make a free throw is 40%. Is this theoretical probability or experimental probability? Explain how you can tell.

4. In a two-player game, four coins are flipped. The first player wins if the coins come up all heads or all tails. Otherwise the second player wins. Is this game fair? Explain.

5. When finding theoretical probabilities, why must the possibilities be equally likely?

Ask students: How many boys are in this class? How many girls are in this class? How many students are in this class? Have students use that data to answer these questions:

1. What is the probability that I will be chosen if one student in the class is chosen?

2. What is the probability that I will be chosen if one girl in the class is chosen?

3. What is the probability that I will be chosen if one boy in the class is chosen?

[Answers will vary, depending upon class size.]

Every student should get an answer of 0 for either Question 2 OR 3. Have students discuss why this must be true. Can any student answer 0 for both Questions 2 AND 3?   **[No]**

Example 4 on page 391 builds upon the survey used in Lesson 8.4 on page 378. Each part of Example 4 utilizes the questions and answers found in the Application.

Alternate Example 4

Assorted screws are sold in packages. The contents are listed as follows:

	1-inch	2-inch
Round head	36	24
Flat head	36	24
Phillips	48	32

You choose one screw from the package without looking. What is the probability that you choose

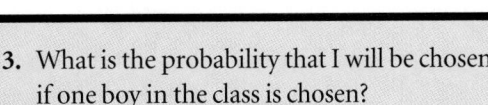

a. a 1-inch screw? $\left[\frac{120}{200}, \text{or } \frac{3}{5}\right]$

b. a 2-inch Phillips-head screw? $\left[\frac{32}{200}, \text{or } \frac{4}{25}\right]$

c. a 1-inch screw OR a flat-head screw? $\left[120 + 60 - 36 = 144; \frac{144}{200}, \text{or } \frac{18}{25}\right]$

Assignment Guide

Core 1–9, 10–24 even, 26–41

Core Two-Year 1–41

Error Analysis

For Exercises like 7 and 8 on page 392, students may think the probability of getting an even number or an odd number is $\frac{1}{2}$, or 50%, since they are equally-likely events. This is true only when the database contains an equal number of even and odd numbers. Point out that in the set of integers from 1 to 25, there are 12 even numbers and 13 odd numbers.

Practice Master

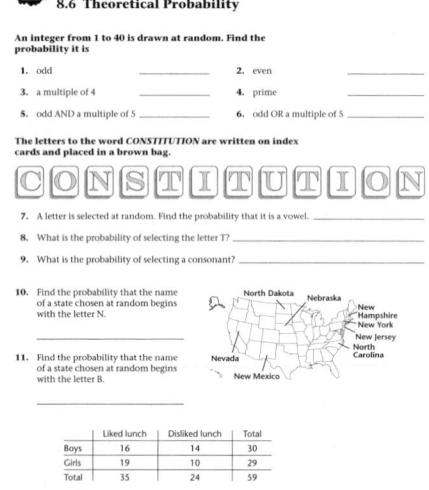

Practice & Apply
8.6 Theoretical Probability

An integer from 1 to 40 is drawn at random. Find the probability it is

1. odd _____   2. even _____

3. a multiple of 4 _____   4. prime _____

5. odd AND a multiple of 5 _____   6. odd OR a multiple of 5 _____

The letters to the word *CONSTITUTION* are written on index cards and placed in a brown bag.

⟦C⟧⟦O⟧⟦N⟧⟦S⟧⟦T⟧⟦I⟧⟦T⟧⟦U⟧⟦T⟧⟦I⟧⟦O⟧⟦N⟧

7. A letter is selected at random. Find the probability that it is a vowel. _____

8. What is the probability of selecting the letter T? _____

9. What is the probability of selecting a consonant? _____

10. Find the probability that the name of a state chosen at random begins with the letter N.
11. Find the probability that the name of a state chosen at random begins with the letter B.

	Liked lunch	Disliked lunch	Total
Boys	16	14	30
Girls	19	10	29
Total	35	24	59

A student in the grade surveyed is chosen at random. What is the theoretical probability that the student

12. is a girl? _____   13. likes the lunch? _____

14. is a girl AND likes the lunch? _____   15. is a girl OR likes the lunch? _____

Practice & Apply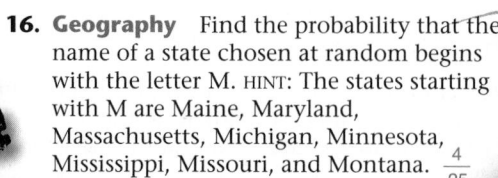

Games An integer from 1 to 25 is drawn at random. Find the probability it is

6. even. $\frac{12}{25}$ **7.** odd. $\frac{13}{25}$ **8.** a multiple of 3. $\frac{8}{25}$

9. a prime. $\frac{9}{25}$ **10.** even AND a multiple of 3. $\frac{4}{25}$ **11.** even OR a multiple of 3. $\frac{16}{25}$

Language Select a letter of the alphabet at random. Find the probability it is

12. r. $\frac{1}{26}$ **13.** a vowel (a, e, i, o, u, or y). $\frac{3}{13}$ **14.** a consonant. $\frac{10}{13}$

15. one of the letters of the word *mathematics*. $\frac{4}{13}$

16. Geography Find the probability that the name of a state chosen at random begins with the letter M. HINT: The states starting with M are Maine, Maryland, Massachusetts, Michigan, Minnesota, Mississippi, Missouri, and Montana. $\frac{4}{25}$

17. Science Find the probability that a planet in our solar system chosen at random is closer to the sun than Earth is. HINT: Of the 9 planets, only Mercury and Venus are closer. $\frac{2}{9}$

Transportation According to the National Transportation Safety Board there were about 0.243 fatal crashes for each 100,000 departures of commuter planes in 1992.

18. Express this ratio as a fraction with a numerator of 1. $\frac{1}{411,523}$

19. Is this experimental probability or theoretical probability? experimental

20. For larger planes (carrying over 30 passengers), the number of fatal crashes was 0.05 per 100,000. Express this ratio as a fraction with a numerator of 1. $\frac{1}{2,000,000}$

21. According to these figures, a commuter plane is how many times more likely to have a fatal crash than a larger plane? about 5 times

22. Language If the letters of the word *tap* are rearranged, find the probability that an anagram is formed. $\frac{1}{2}$ with tap

23. John is trying to solve the equation $2x - 1 = 15$ on a quiz in algebra. He doesn't remember how to do it, but he hopes that the answer is a positive one-digit number and makes a guess. Find the probability that he guesses the correct solution. $\frac{1}{9}$

24. Repeat Exercise 23 if the equation is $1 - 2x = 15$. 0

25. ▽ **Geometry** A dart is thrown at a circular target. If it is equally likely to land anywhere in the large circle, what is the probability that it hits the bull's-eye, the dark region? $\frac{1}{16}$

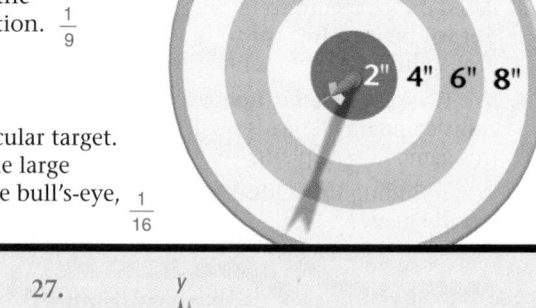

2" 4" 6" 8"

26.

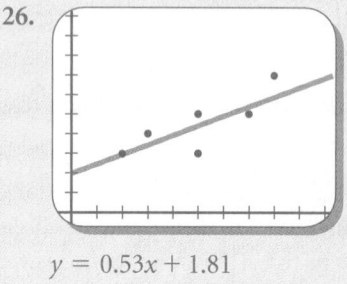

$y = 0.53x + 1.81$

27.

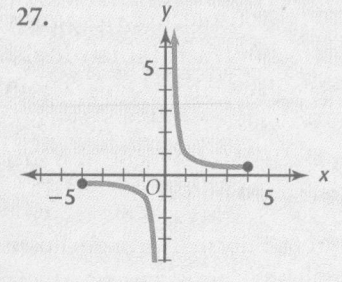

28.

-4 0 4 8 12 16

26. Technology   Make a scatter plot for the points (2, 3), (3, 4), (5, 3), (5, 5), (7, 5), and (8, 7), and find a line of best fit.   **[Lesson 1.7]**

27. Draw the graph of the function $y = \frac{1}{x}$ for values $0 < x \le 4$. What do you notice about the graph?   **[Lesson 2.3]**

28. If $-6 \le n - 4 < 10$, show the solution set on a number line when the values of n are integers.   **[Lesson 3.6]**

29. Write an explanation of why you cannot divide by zero.   **[Lesson 4.1]**

30.   Health   A study released at the 1991 annual meeting of the American Heart Association found that taller people have less risk of heart attack. It was found that every inch of height reduced the chance of heart attack by about 3%. Compare the risk of someone 5 feet tall with that of someone 6 feet tall.   **[Lesson 4.5]**

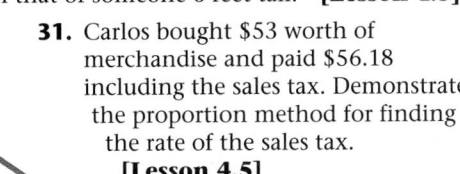

31. Carlos bought $53 worth of merchandise and paid $56.18 including the sales tax. Demonstrate the proportion method for finding the rate of the sales tax.
[Lesson 4.5]

32. Solve the equation $y = mx + b$ for m.   **[Lesson 4.6]** $m = \dfrac{y - b}{x}$

33. A coin is improperly balanced and shows the probability of *heads* to be 0.64. If Suzanne flips the coin 50 times, how many times would you expect *heads* to appear?   **[Lesson 8.1]** 32

Karl's typing rates on 5 tests were 19, 21, 26, 21, and 27 words per minute.

34. What is the mode for Karl's typing rates?   **[Lesson 8.3]** 21

35. What is the median for the typing rates?   **[Lesson 8.3]** 21

Look Beyond

The symbol 3! is 3 *factorial* and means the product of the whole numbers from 3 to 1, inclusive. Thus, $3! = 3 \cdot 2 \cdot 1 = 6$.

Find the following values.

36. 4! 24   **37.** 5! 120   **38.** 6! 720   **39.** $\frac{3!}{2!}$ 3   **40.** $\frac{7!}{5!}$ 42   **41.** $\frac{10!}{6!}$ 5,040

---

29. If $b = \frac{a}{0}$ and $a \ne 0$, then there is no value of b such that $b \cdot 0 = a$.

30. A 6 foot tall person is 12 inches taller than a 5 foot tall person. The 12 inches would reduce the risk by 36%, or over $\frac{1}{3}$, for the taller person.

31.
$$\frac{3.18}{53.00} = \frac{x}{100}$$
$$x = \frac{3.18}{53.00}(100)$$
$$= 6$$
The sales tax is 6%.

Technology
8.6 Electronic Dartboards

The dartboard shown consists of a set of circles with the same center. Each section of the dartboard is a circular annulus, the region bounded by two circles with the same center.

Define the probability of a dart landing in a certain annulus to be the area of an annulus divided by the area of the entire dartboard.

To study the probabilities, create a spreadsheet like the one shown. Column A contains the name of the region.

	A	B	C
1	RADIUS	AREA	PROB
2	1		
3	2		
4	3		
5	4		
6	5		
7	6		
8	7		

Cell B2 contains 3.14159*A2^2.
Cell B3 contains 3.14159*A3^2−3.14159*A2^2.
Cell C2 contains B2/(3.14159*A8^2).

Make a spreadsheet to find the probability of a dart landing in each annulus for each set of radii.

1. 1, 2, 3, 4, 5, 6, 7   **2.** 1, 2, 4, 8, 16, 32, 64   **3.** 12, 24, 34, 42, 48, 52, 54

4. What can you say about the probability of a dart landing in the regions of the dartboards you examined in Exercises 1–3?

Make a spreadsheet to find the probability of a dart landing in each square annulus for each set of half lengths.

5. 1, 2, 3, 4, 5, 6, 7   **6.** 1, 2, 4, 8, 16, 32, 64

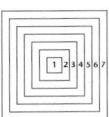

7. What can you say about the probability of a dart landing in the regions of the dartboards you examined in Exercises 5 and 6?

- Define and give examples of independent events.
- Find the probability of independent events.

RESOURCES

• Practice Master	**8.7**
• Enrichment Master	**8.7**
• Technology Master	**8.7**
• Lesson Activity Master	**8.7**
• Quiz	**8.7**
• Spanish Resources	**8.7**

Assessing Prior Knowledge

A card is drawn from a standard deck. Find the probability that it is

1. the 5 of hearts. $\left[\frac{1}{52}\right]$

2. a 5. $\left[\frac{1}{13}\right]$

3. a heart. $\left[\frac{1}{4}\right]$

4. a 5 or a heart. $\left[\frac{16}{52}, \text{ or } \frac{4}{13}\right]$

5. not a heart. $\left[\frac{3}{4}\right]$

TEACH

Have students discuss factors involved in being able to go to a concert. Such factors might include getting permission to go, transportation, money for the tickets, and getting tickets. Use the information to consider independent and dependent combinations of events.

LESSON 8.7 Independent Events

Many situations involve more than one event. Sometimes the occurrence of a previous event will affect the probability of the events that follow. As probability becomes more difficult to determine, geometry can provide assistance.

In Lesson 8.5, Jordanna is selecting one elective from cluster 1, band, orchestra, or chorus, and one elective from cluster 2, home economics or woodworking. If all of the choices are equally likely, find the probability that she chooses band and woodworking.

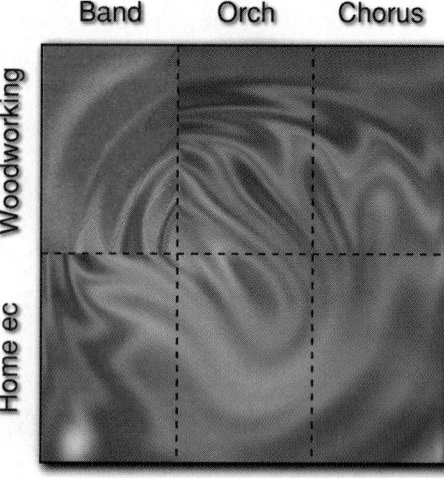

Cluster 1
Band   Orch   Chorus

Cluster 2

Woodworking

Home ec

GEOMETRY *Connection*

Use 2 vertical lines to divide a square into three equal regions, one for each cluster 1 elective. Then divide it again with a horizontal line, so that there is one region for each cluster 2 elective. The six regions of the square represent the six equally-likely possibilities or outcomes. One region represents the pair, *band* and *woodworking*. The probability for this choice is $\frac{1}{6}$.

EXAMPLE 1

Appointments

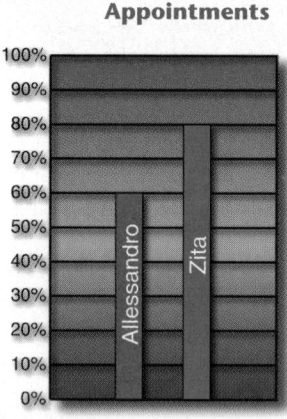

100%
90%
80%
70%
60%
50%
40%
30%
20%
10%
0%

Allessandro   Zita

■ Zita

■ Allessandro

Recall that Allessandro and Zita are absent-minded students who plan to meet for lunch. Use an area model to find the probability that both Allessandro and Zita show up on time.

ALTERNATIVE teaching strategy

Using Cognitive Strategies Have students decide whether each of the following events is dependent or independent. Students should also explain the reasoning used to determine each answer.

- A representative from each class in the school is chosen to be on a committee. [**independent**]

- From a group of four people, two are chosen to be on a committee. [**dependent**]
- One card from a deck is chosen. The card is not replaced. Then another card is chosen. [**dependent**]
- One card from a deck is chosen. The card is replaced. Then another card is chosen. [**independent**]

Solution ➤

On graph paper, divide a 10 by 10 square into two parts with a vertical line as shown in the first figure. This shows a probability of 60% that Allessandro is on time. Divide the square into two parts with a horizontal line as shown in the second figure. This shows a probability of 80% that Zita is on time.

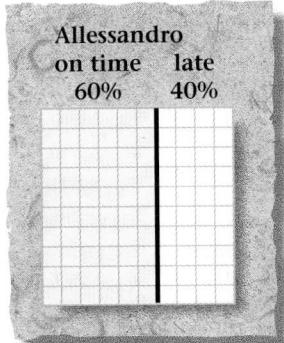

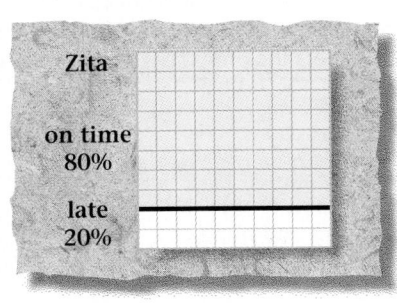

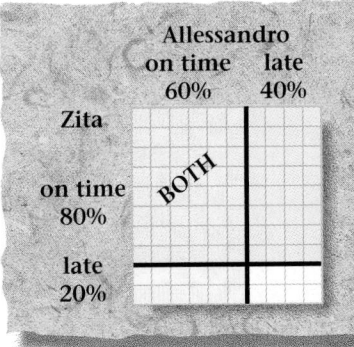

When the squares for Allessandro and Zita are combined, they divide the square into four regions. The part marked **BOTH** represents the probability that both are on time. It contains 8 · 6 or 48 out of 100 possible squares. The probability is $\frac{48}{100} = 0.48$, or 48%. ❖

Try This Use the combined figure to determine the probability that Allessandro is on time and Zita is not.

Sometimes you can solve a problem without using a model.

EXAMPLE 2

Games Sarah draws a card, then draws another from a regular 52-card deck. Find the probability that the 2 cards she holds are black. In a regular deck, half of the cards are black.

Solution ➤

There are 52 ways of drawing the first card. After Sarah draws the first card, there are 51 ways of drawing the second card. Since she draws a first card and a second card, apply the Multiplication Principle of Counting. There are 52 · 51 = 2652 equally-likely ways of drawing the two cards when the first card is not replaced.

A success occurs if Sarah draws two black cards. There are 26 ways to draw the first black card. After the first is drawn, there are 25 ways to draw the second card. Apply the Multiplication Principle of Counting. There are 26 · 25 = 650 ways to draw two black cards.

The probability of drawing two black cards is then $\frac{650}{2652}$, or $\frac{25}{102}$. ❖

 **ENRICHMENT** Encourage students to draw a Venn diagram of two independent events. [**Circles should not intersect.**]

INCLUSION strategies **Using Listening Skills** For auditory learners, orally describe the visual diagrams. For Example 1, reinforce the diagram by de-

scribing a large square made of 100 smaller squares. It is separated with a horizontal line to show 80% and 20%. This represents Zita's probabilities of being on time. It is also separated with a vertical line to show 60% an 40%. This represents Allesandro's probabilities of being on time. Together there are four sections. Have students describe what each section represents.

Alternate Example 1

In a close baseball game, Crystal is at bat and Karen is on deck (next to bat). Crystal gets on base 65% of the time she bats. Karen gets on base 40% of the time she bats. What is the probability that both girls will get on base? [**26%**] What is the probability that neither girl will get on base? [**21%**] What is the probability that exactly one of the girls will get on base? [**53%**]

ongoing ASSESSMENT

Try This

The probability is 12%.

TEACHING tip

In Example 2, the probability of each event can be found and the results multiplied. First black card = 26/52, or 1/2. Second black card = 25/51. Two black cards drawn = 1/2 · 25/51, or 25/102. Students should note that Example 2 shows events that are dependent, not independent.

Alternate Example 2

Each person in your class writes his or her name on a slip of paper and places it in a box. What is the probability that two boys are chosen in succession? [**Answers will vary depending upon the number of boys and girls in the class.**]

Use Transparency ▷ 38

Alternate Example 3

Six marbles (1 red, 2 green, and 3 blue) are placed in a bag. One marble is drawn, then replaced. Then another marble is drawn. Find the probability of drawing a green marble twice. $\left[\frac{1}{3}\cdot\frac{1}{3}=\frac{1}{9}\right]$

Aongoing
SSESSMENT

The probability of both occurring is the product of the probabilities of each occuring.

Aongoing
SSESSMENT

Draw a face card from a regular deck, and then without replacement, draw another face card.

Probability of Independent Events

Independence has a special significance in probability. In Jordanna's problem she first selects an elective from cluster 1 and then selects an elective from cluster 2. The outcome of her first selection has no effect on her second selection. The two events are independent.

> **INDEPENDENT EVENTS**
> Two events are **independent** if the occurrence of the first event does *not* affect the probability of the second event occurring.

EXAMPLE 3

What is the probability that the red cube shows an odd number and the green cube shows a number greater than or equal to 5? Are the events independent?

Todd rolls a red number cube and records the number shown. He then rolls a green number cube, and records the number.

Solution ➤

Apply the Multiplication Principle of Counting. There are 6 · 6, or 36, equally-likely ways that two numbers can appear in two rolls. On the first roll, 3 of the 6 numbers are odd. On the second roll 2 of the 6 numbers are greater than or equal to 5. There are 3 · 2, or 6 ways of rolling a success on both rolls.

The probability of success is $\frac{6}{36}$, or $\frac{1}{6}$. Since the result of the first roll does not affect the second roll, the events are independent. ❖

Examine the probabilities of success on the first roll, the second roll, and both rolls.

The probability of rolling an odd number on the red cube is $\frac{1}{2}$.

The probability of rolling 5 or more on the green cube is $\frac{1}{3}$.

The probability of rolling both an odd number and 5 or more is $\frac{1}{2}\cdot\frac{1}{3}$, or $\frac{1}{6}$.

What pattern do you see?

Events that are not independent are called **dependent**. Give an example of a dependent event.

Finding Probabilities With the Complement

In Lesson 8.6, you determined theoretical probability in three steps.

1. Find the denominator, the number of equally-likely possibilities.

2. Find the numerator, the number of successes.

3. Express the probability as a fraction.

RETEACHING
the
l e s s o n

Place eight colored objects, such as marbles, in a box. Use 1 green, 2 red, 3 blue, and 2 white. Ask students to determine the probability of each of these events:

A: drawing a blue marble $\left[\frac{3}{8}\right]$

B: drawing a red marble $\left[\frac{1}{4}\right]$

C: drawing a blue marble, putting it back, then drawing a green marble $\left[\frac{3}{8}\cdot\frac{1}{8}=\frac{3}{64}\right]$

D: drawing a blue marble, **not** putting it back, then drawing a green marble $\left[\frac{3}{8}\cdot\frac{1}{7}=\frac{3}{56}\right]$

E: drawing a blue marble, putting it back, then drawing a blue marble $\left[\frac{3}{8}\cdot\frac{3}{8}=\frac{9}{64}\right]$

F: drawing a blue marble, **not** putting it back, then drawing a blue marble $\left[\frac{3}{8}\cdot\frac{2}{7}=\frac{3}{28}\right]$

Another useful technique in solving probability problems is to count the ways something *does not* happen. In some cases this provides an easier way to solve the problem.

EXAMPLE 4

Games Two different cards are drawn from an ordinary deck of 52 cards. Find the probability that at least one of them is red.

Solution ➤

The event *at least one card is red* includes every case that can happen when two cards are drawn *except* getting two black cards. Thus, out of the 2652 ways of drawing two cards, the event includes all except the 650 ways of getting two black cards. This amounts to $2652 - 650$, or 2002, ways. The probability of getting at least one red card is $\frac{2002}{2652}$, or $\frac{1001}{1326}$. ❖

When you find the number of possible ways that an event occurs by considering the number of ways that it does not occur, you are using the **complement**.

Try This Use the complement to determine the following theoretical probability. When two number cubes are tossed, what is the probability that the sum is at least 3?

 **CRITICAL** *Thinking* A box contains 7 quarters, 6 dimes, 5 nickels, and 8 pennies. One coin is drawn from the box. Explain how to find the probability that the coin is worth at least ten cents.

EXERCISES  PROBLEMS

Communicate

1. Name two ways to represent independent events using a diagram.
2. Explain what it means for two events to be independent.
3. When you find probabilities of independent events, are you usually using the Addition Principle of Counting or the Multiplication Principle of Counting?
4. How many paths are there through this tree diagram?
5. Explain the method of finding a probability by finding the complement.

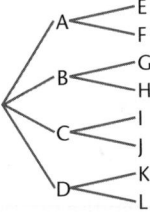

TEACHING *tip*

In Example 4, students should simplify the probability to $\frac{77}{102}$.

Alternate Example 4

Each person in your class writes his or her name on a slip of paper and places it in a box. What is the probability that at least one of two slips chosen at random would be a girl?   [**The answer may vary depending on the size of the class.**]

Aongoing **SSESSMENT**

Try This

The probability that the sum is less than 3 is $\frac{1}{36}$. So the probability that the sum is at least 3 is $1 - \frac{1}{36}$, or $\frac{35}{36}$.

CRITICAL *Thinking*

There are 26 coins. 13 of these coins are worth ten cents or more. The probability is $\frac{13}{26}$ or $\frac{1}{2}$.

Odd–numbered Exercises 11–39

ASSESS

Assignment Guide

Core 1–9, 10–14, 16–30 even, 31–40

Core Two-Year 1–39

Technology

Have students write a simulation that would use random numbers to complete an experiment like those in Exercises 15–18.

Error Analysis

Students may confuse independent and dependent events. Remind students that if the second event is affected by what happened in the first event, then it is a dependent event. For instance, if a marble is drawn and not replaced, then the drawing of a second marble is a dependent event.

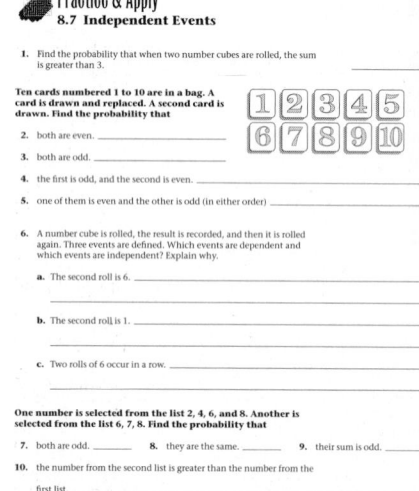

In this area model, the small squares to the left of the vertical red line represent that event A occurs; the small squares above the horizontal blue line represent that event C occurs; etc. Tell how to find the following probabilities.

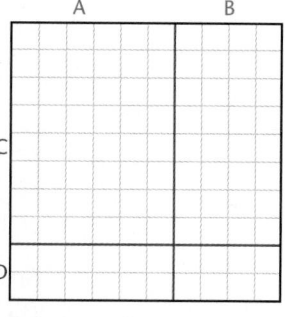

6. A occurs.

7. C occurs.

8. A occurs and C occurs.

9. A occurs or C occurs.

Practice & Apply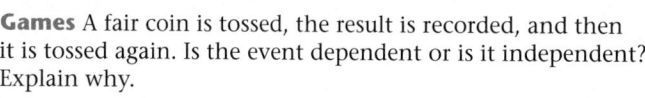

10. Find the probability that when two number cubes are thrown, the sum is less than 12. $\frac{35}{36}$

Route Planning John and Jim are each hiking from Wapitu Falls to Songbird Lake using the trails shown on the map.

11. Find the probability that John takes Trail 101 from Wapitu Falls to Lookout Rock. $\frac{1}{3}$

12. Find the probability that John takes Trail 201 from Lookout Rock to Songbird Lake. $\frac{1}{4}$

13. Find the probability that John takes Trail 101 from Wapitu Falls to Lookout Rock AND he takes Trail 201 from Lookout Rock to Songbird Lake. $\frac{1}{12}$

14. Find the probability that John takes Trail 101 from Wapitu Falls to Lookout Rock OR he takes Trail 201 from Lookout Rock to Songbird Lake. $\frac{1}{2}$

Games Five chips numbered 1 through 5 are in a bag. A chip is drawn and replaced. Then a second chip is drawn. Find the probability that

15. both are even. $\frac{4}{25}$

16. both are odd. $\frac{9}{25}$

17. the first is even and the second is odd. $\frac{6}{25}$

18. one is even and the other is odd. $\frac{12}{25}$

Games A fair coin is tossed, the result is recorded, and then it is tossed again. Is the event dependent or is it independent? Explain why.

19. Second toss is heads. I

20. Second toss is tails. I

21. Two heads occur in a row. D

Games One number is selected from the list 1, 3, 5, 7. Another is selected from the list 5, 6, 7. Find the probability that

22. both are even. 0

23. they are the same. $\frac{1}{6}$

24. their sum is even. $\frac{2}{3}$

25. the number from the second list is greater than the number from the first list. $\frac{2}{3}$

30. Answers may vary for experimental probability.

Experimental Probability
Randomly generate 0 or 1. Let 0 represent an incorrect response and 1 represent a correct response. Generate 5 numbers, and record the results. This represents 1 trial. Repeat for 30 trials. Divide the number of trials where 4 or 5 questions were answered correctly by 30. This is the experimental probability.

Theoretical Probability
Each question has 1 out 2 of chances of being correct or a probability of $\frac{1}{2}$. To get at least 80% on the 5 question test, you must get 4 or 5 questions correct.
$$P = P(4) + P(5)$$
$$= \left(\frac{1}{2}\right)^4 + \left(\frac{1}{2}\right)^5 = 0.094$$

The probability is 9.4%.

Use the grid to find the probability that

26. A occurs. $\frac{1}{3}$

27. C occurs. $\frac{1}{6}$

28. A occurs AND C occurs. $\frac{1}{18}$

29. A occurs OR C occurs. $\frac{4}{9}$

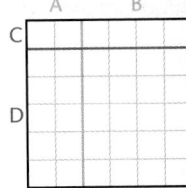

30. **Portfolio Activity** Complete the problem in the portfolio activity on page 351.

Look Back

Find the y-values of the following functions for x = −2, −1, 0, 1, and 2. Then identify the function.
[Lesson 2.5]

31. $y = \frac{4}{x}$

32. $y = 4x^2$

33. $y = |4x|$

34. INT(4x)

35. Find the product. $-6(2)\left(\frac{5}{-3}\right)$ **[Lesson 4.1]** 20

36. Tim leaves home on his bike and rides at a rate of 12 miles per hour. Two hours after he leaves, his mother remembers he has a doctor's appointment. She sets out after him in her car at 40 miles per hour. How long does it take her to catch up? **[Lesson 6.6]** $\frac{6}{7}$ hours

37. **Scheduling** At one school, 48 students signed up to take Spanish, and 23 signed up to take French. If there were 12 that signed up for both, how many signed up for these foreign languages? **[Lesson 8.4]** 59

38. If Rich has 5 books, how many ways can he arrange them on a shelf of his bookcase? **[Lesson 8.5]** 120

39. **Travel** There are 3 highways from Birchville to Pine Springs. From Pine Springs to Clearwater there are 5 roads. How many possible ways are there to travel from Birchville to Clearwater? **[Lesson 8.5]** 15

Look Beyond

40. **Contest** Two radio stations are having a contest. At the first station, the prize is $1000, and your probability of winning is $\frac{1}{100,000}$. At the second station, the prize is $25, and your probability of winning is $\frac{1}{2000}$. Which provides a better return for the listener's time? Compare your chances of winning the two contests. HINT: What could you expect to happen if you played each contest 100,000 times?

31.

x	−2	−1	0	1	2
y	−2	−4	undefined	4	2

reciprocal function

32.

x	−2	−1	0	1	2
y	16	4	0	4	16

quadratic function

33.

x	−2	−1	0	1	2
y	8	4	0	4	8

absolute value function

34.

x	−2	−1	0	1	2
y	−8	−4	0	4	8

integer function

The answer to Exercise 40 can be found in Additional Answers on page 729.

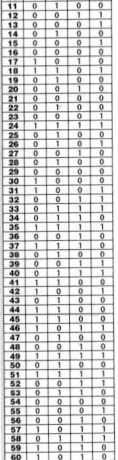

Focus

This project involves finding the number of permutations or combinations involved in a specific situation.

Motivate

Remind students of the Multiplication Principle of Counting and how it was used in probability. Ask if students think this can be extended if there are more than two or three items to be selected for a simulation. [**Yes**] Tell students that this project will show them a way to handle situations that involve large numbers of possible choices.

Chapter 8 Project

Winning Ways

Only four out of ten students will be selected to compete against an academic team from another school. If the order in which the students are chosen to appear matters, how many ways can the four students be selected?

If you use the Multiplication Principle of Counting, there are 10 ways to make the first choice. For each of these choices, there are 9 ways to make the second choice.

First choice	Second Choice	Third Choice	Fourth Choice
10	9		

That means that there are 10 • 9 ways to make the first two choices. If you continue in the same way, there are 10 • 9 • 8 • 7 or 5040 ways to make the 4 choices.

First choice	Second Choice	Third Choice	Fourth Choice
10	9	8	7

When r objects are chosen and the order matters, the result is called a permutation on n things taken r at a time. The number of possible **permutations** is written $_nP_r$. The example, $_{10}P_4 = 10 • 9 • 8 • 7 = 5040$.

Other examples are:
a. The permutations of 5 letters taken 2 at a time is $_5P_2 = 5 • 4 = 20$.
b. The permutation of 7 digits taken 3 at a time is $_7P_3 = 7 • 6 • 5 = 210$.

Of the 5040 permutations, some of the choices contain the same students. For example, the choice Andy, Bob, Cardi, and Donna is the same as Cardi, Bob, Donna, and Andy when order does not matter. There are 4 • 3 • 2 • 1 or 24 permutations for each group of 4 students. Thus, if the total number of permutations, 5040, is divided by 24, the result is 210.

Each of these 210 sets of 4 students is called a **combination**. So, while there are 5040 permutations of 10 objects taken 4 at a time, there are only 210 combinations of 10 objects taken 4 at a time.

Thus, when r objects are chosen from n objects and the order does not matter, it is called the **combination** of n things taken r at a time. This is written as $_nC_r$. In the example, $_{10}C_4 = \dfrac{10 \cdot 9 \cdot 8 \cdot 7}{4 \cdot 3 \cdot 2 \cdot 1} = \dfrac{5040}{24} = 210$.

Other examples are:
a. The combination of 5 letters taken 2 at a time is
$_5C_2 = \dfrac{5 \cdot 4}{2 \cdot 1} = \dfrac{20}{2} = 10$.
b. The combination of 7 digits taken 3 at a time is
$_7C_3 = \dfrac{7 \cdot 6 \cdot 5}{3 \cdot 2 \cdot 1} = \dfrac{210}{6} = 35$.

Activity 1

Use a tree diagram to show the number of permutations of the letters A, B, C, and D taken 2 at a time. Explain how the result can be used to find the number of combinations of 4 objects taken 2 at a time.

Activity 2

Find the number of arrangements consisting of 1, 2, 3, 4, and 5 letters that can be made from the letters in the word *point*. What is the sum of all possible arrangements?

Show all the 3 letter arrangements, and determine which ones form English words. How many English words did you find? What is the probability that a 3-letter word can be formed from the letters in the word *point*?

Cooperative Learning

Have students work in small groups to complete this lesson. Group discussion should include deciding whether a permutation or combination is involved in a specific situation.

Technology

As an extension, allow students to experiment and discover how to use their graphics calculator to solve problems involving permutations and combinations.

Discuss

Discuss the difference between permutations and combinations. When order is important, a permutation is needed. When order is not important, a combination is involved. For instance, the letters of the word MANY have many permutations—MANY, MNAY, YANM, etc., but only one combination—AMNY—if listed alphabetically. Have students see if $_nP_r$ and $_nC_r$ work for that situation.

$$\left[\text{Yes: } _4P_4 = 4 \cdot 3 \cdot 2 \cdot 1 = 24 \right.$$
$$\left. \text{and } _4C_4 = \dfrac{4 \cdot 3 \cdot 2 \cdot 1}{4 \cdot 3 \cdot 2 \cdot 1} = 1 \right]$$

Chapter Review

Chapter 8 Review

Vocabulary

Addition Principle of Counting	378	intersection	376	successful event	353
dependent events	396	margin of error	371	theoretical probability	389
disjoint	376	Multiplication Principle of Counting	383	tree diagram	382
equally likely outcomes	389	RAND	355	trial	353
experimental probability	354	range	369	union	376
independent events	396	simulation	360	Venn diagram	376

Key Skills & Exercises

Lesson 8.1

➤ **Key Skills**

Calculate experimental probability.

Two coins are flipped 5 times with the following results.

Trial	1	2	3	4	5
Coin 1	H	T	T	T	H
Coin 2	T	T	H	H	H

The experimental probability that both coins are alike is $\frac{2}{5}$.

Use technology to generate random numbers.

If RAND generates a number from 0 to 1 (including 0, but not 1), then the output of the command INT(RAND( ) * 4) is 0, 1, 2, 3, and the output of the command INT(RAND( ) * 4) + 2 is 2, 3, 4, 5.

In general, the function INT(RAND( ) * K) + A generates random integers from a list of K consecutive integers beginning with A.

➤ **Exercises**

Use the above data to find the following experimental probabilities.

1. At least one coin is heads. $\frac{4}{5}$ **2.** Both coins are tails. $\frac{1}{5}$

Write commands to generate random numbers from the following lists.

3. 4, 5, 6, 7, 8
   INT (RAND( )*5) + 4

4. 0, 1, 2, 3, 4, 5, 6, 7, 8, 9, 10
   INT (RAND( )*11)

Lesson 8.2

➤ **Key Skills**

Design a simulation to determine experimental probability.

Rolling a number cube can be used to simulate choosing at random the correct response to a multiple-choice question. Suppose there are 5 possible choices. Let 1, 2, 3, 4, or 5 represent one of the possible responses, with a 1 representing a correct response. Roll again if you get a 6. One roll represents one trial. The results are as follows.

Trial	1	2	3	4	5	6	7	8	9	10
Roll	1	4	2	1	3	6	1	4	3	5

➤ **Exercises**

**Suppose you want to simulate answering all of the questions on a
10-question quiz correctly. Design a simulation using**

5. random numbers.   **6.** coins.

Lesson 8.3

➤ *Key Skills*

Identify deceptive uses of statistics.

A graph advertising the amount of iron in
Cereal X is shown at the right. It appears at a
glance that Cereal Z has about twice as much
iron as Cereals X and Y. According to the
scale, however, Cereal Z differs from Cereals X
and Y by only 10%.

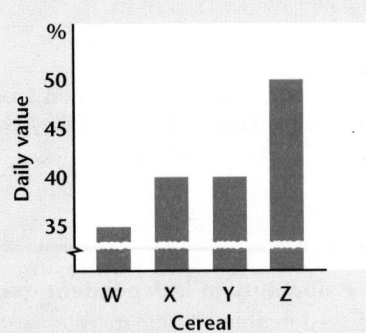

➤ *Exercises*

Refer to the graph for Cereals W, X, Y, and Z above.

7. How much more iron does there appear to be in Cereal Z than in
Cereal W? 3 or 4 times as much

8. By what percent does Cereal Z actually differ from Cereal W?
15% more

Lesson 8.4

➤ *Key Skills*

**Use the Addition Principle of Counting to determine the number
of elements in the union of sets.**

In a music class, 11 students play the clarinet and 15 play the flute,
including 8 who play both. You can use the Addition Principle of
Counting to determine the number of students in the music class. There
are 11 students who play the clarinet, 15 who play the flute, and 8 who
have been counted twice. So, there are $11 + 15 - 8 = 18$ students.

➤ *Exercises*

Consider an ordinary deck of 52 playing cards.

9. Draw a Venn diagram showing the cards that are black OR face cards.
10. How many cards are black OR face cards? 32
11. How many cards are red OR numbered cards? 46

Lesson 8.5

➤ *Key Skills*

**Use the Multiplication Principle of Counting to find the number of
possible ways to make multiple choices.**

A craft store stocks 5 styles of T-shirts in 4 sizes and in 10 different colors.
The craft store stocks $5 \cdot 4 \cdot 10 = 200$ T-shirts.

➤ *Exercises*

12. How many ways are there to arrange 4 different letters? 24

13. A dress shop stocks 6 styles of blouses. Each of these comes
in 7 sizes, and each in 5 colors. How many different selections
are possible? 210

5–6. Answers may vary.

9.

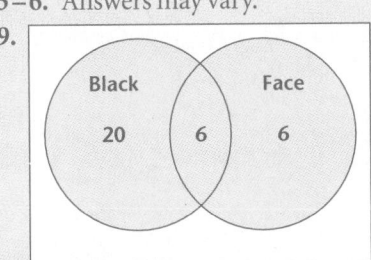

Lesson 8.6

➤ *Key Skills*

Find the theoretical probability.

Suppose you roll a number cube once. The probability that you will roll a number greater than 4 is $\frac{2}{6}$, or $\frac{1}{3}$.

➤ *Exercises*

An integer from −10 to 10 is drawn from a bag containing numbered chips. Find the probability that it is

14. negative. $\frac{10}{21}$ **15.** zero. $\frac{1}{21}$ **16.** even. $\frac{11}{21}$

Lesson 8.7

➤ *Key Skills*

Find the probability of independent events.

There are 2 red marbles, 4 blue marbles, and 3 white marbles in a bag. If you select a marble, replace it, and then select another marble, the probability of selecting a red marble *and* a blue marble is $\frac{2}{9} \cdot \frac{4}{9} = \frac{8}{81}$.

➤ *Exercises*

Ten cards numbered 1 through 10 are placed in a box. One card is drawn and replaced. Then another card is drawn. Find the probability that both are

17. less than 5. $\frac{4}{25}$ **18.** multiples of 3. $\frac{9}{100}$ **19.** the same. $\frac{1}{10}$

Applications

Marketing The Hamburger Hut is giving away free food during their grand opening. They are distributing 1,000 cards, of which 100 are for a free hamburger, 150 are for free fries, and 200 are for a free soft drink. What is the probability of winning

20. a hamburger? $\frac{1}{10}$ **21.** fries? $\frac{3}{20}$ **22.** a soft drink? $\frac{1}{5}$

23. The Hamburger Hut advertises 150 different ways to order their hamburgers, 2 different types of fries, and 8 different flavors of soft drinks. How many ways are there to choose one of each? 2400

24. The probability that a household in the United States will have some kind of pet is about $\frac{6}{10}$. The probability that a household will have at least one child is $\frac{1}{3}$. What is the probability of a household in the United States having a pet and a child? $\frac{1}{5}$

25. Tim, Mary, and Paul have applied for a loan to buy a used car. The probability that Tim's application is approved is 75%. The probability that Mary's application is approved is 80%. The probability that Paul's application is approved is 62.5%. What is the probability that all three loans will be approved? 37.5%

26. In Item 25, what is the probability that only Mary's application will be approved? 7.5%

Chapter 8 Assessment

If RAND( ) generates a random number from 0 to 1 (including 0, but not 1), then describe the output of the following commands.

1. RAND( ) * 5   **2.** INT(RAND( ) * 5)   **3.** INT(RAND( ) * 5) + 4

4. Write a command using RAND( ) to generate random numbers from the list 1, 2, 3, 4, 5, 6, 7, 8. (RAND( )*8) + 1

5. Show how to simulate randomly picking a phone number from a list of 6 phone numbers.

6. If the range for a set of scores is 45, the mean is 75, and the highest score is 98, find the lowest score. 53

Suppose the manager of a health club used an ad to show that people preferred her club 4 to 1 by showing a locker four times as long and four times as wide as the competition.

7. This would result in an area how many times as large? 16 times

8. Would this accurately show the comparison of the preference of health clubs? Explain. No; greater than 4 times

In an ordinary deck of 52 cards tell how many of the cards are

9. numbered. 36   **10.** red. 26

11. red AND numbered. 18   **12.** red OR numbered. 44

13. There are 16 girls on a softball team and 11 girls on a track team. If there are 5 girls on both teams, how many girls are on the softball team and the track team? 22

14. The Tiberi's are ordering a new couch. There are 5 styles from which to choose. For each of these there are 15 fabrics from which to choose. How many combinations are possible? 75

15. What does it mean to say that outcomes are equally likely? Why is this important?

Two number cubes are rolled. Find the probability that the sum is

16. 10. $\frac{1}{12}$ **17.** less than 5. $\frac{1}{6}$

18. a multiple of 2 AND 3. $\frac{1}{6}$ **19.** a multiple of 2 OR 3. $\frac{2}{3}$

An integer from 1 to 10 is selected. Then a letter from the alphabet is selected. Find the probability that

20. the integer is odd AND the letter is a vowel (a, e, i, o, u, or y). $\frac{3}{26}$

21. the integer is a multiple of 5 AND the letter is a consonant. $\frac{2}{13}$

22. the first letter of the name of the integer chosen is t AND the letter chosen is t. $\frac{3}{260}$

A box contains 3 blue stickers, 4 red stickers, and 11 yellow stickers. Maria chooses one sticker and then another. Find the probability that she chooses

23. a blue sticker, puts it back, and then chooses a yellow sticker. $\frac{11}{108}$

24. a blue sticker, does *not* put it back, and then chooses a yellow sticker. $\frac{11}{102}$

25. two red stickers without putting the first one back. $\frac{2}{51}$

26. Explain the difference between independent and dependent events.

1. decimal numbers greater than or equal to 0, but less than 5

2. 0, 1, 2, 3, and 4

3. 4, 5, 6, 7, and 8

5. Answers may vary. Assign a number from a number cube to each of the 6 phone numbers. Roll the number cube. Record the phone number that corresponds to the number cube roll.

15. The outcomes have the same chance of occurring. This is important because it means the event is fair.

26. Independent event: event that is not influenced by the outcome of another event. Dependent event: event that is affected by the outcome of another event.

Chapters 1 – 8 Cumulative Assessment

College Entrance Exam Practice

Multiple-Choice and Quantitative-Comparison Samples

The first half of the Cumulative Assessment contains two types of items found on standardized tests—multiple-choice questions and quantitative-comparison questions. Quantitative-comparison items emphasize the concepts of equalities, inequalities, and estimation.

Free-Response Grid Samples

The second half of the Cumulative Assessment is a free-response section. A portion of this part of the Cumulative Assessment consists of student-produced response items commonly found on college entrance exams. These questions require the use of machine-scored answer grids. You may wish to have students practice answering these items in preparation for standardized tests.

Sample answer grid masters are available in the *Chapter Teaching Resources Booklets*.

Quantitative Comparison For Questions 1–4, write

A if the quantity in Column A is greater than the quantity in Column B;
B if the quantity in Column B is greater than the quantity in Column A;
C if the two quantities are equal; or
D if the relationship cannot be determined from the information given.

	Column A	Column B	Answers
1.	reciprocal of -2	reciprocal of -3	A B C D **[Lesson 2.3]** B
2.	$-18 + 4$	$-10 + (-4)$	A B C D **[Lesson 3.1]** C
3.	$|-5.2|$	$|4.9|$	A B C D **[Lesson 2.4]** A
4.	45% of 16	60% of 12	A B C D **[Lesson 4.5]** C

5. What is INT(4.5)? **[Lesson 2.4]** a
  a. 4 **b.** 4.5 **c.** -4.5 **d.** -5

6. What are the next three terms of the sequence 6, 12, 24, 48, . . .?
  [Lesson 1.1] b
  a. 72, 96, 120 **b.** 96, 192, 384 **c.** 86, 162, 240 **d.** 50, 52, 54

7. What is the solution to $16y = -120$? **[Lesson 4.3]** d
  a. 1920 **b.** -1920 **c.** 7.5 **d.** -7.5

8. What is the equation of a line passing through $(-1, 2)$ and parallel to $y = 2x - 1$? **[Lesson 5.6]** c
  a. $y = -\frac{x}{2} + 4$ **b.** $y = -\frac{x}{2} + 2$ **c.** $y = 2x + 4$ **d.** $y = -x - 1$

9. Find a solution to the system. $\begin{cases} 2x + 3y = 9 \\ x - 4y = -23 \end{cases}$
  [Lessons 6.1, 6.2, 6.3] b
  a. (3, 5) **b.** $(-3, 5)$ **c.** $(3, -5)$ **d.** $(-3, -5)$

10. A television is marked $198. What is the actual cost of the television if the sales tax is 7.5%? **[Lesson 4.5]** a
  a. $212.85 **b.** $14.85 **c.** $183.15 **d.** $205.50

11. Graph the list of ordered pairs. Tell whether they lie on a straight line.
$(-2, 2), (1, 4), (-4, 0)$ **[Lesson 1.5]** no

12. Use this data to make a graph. Which point is the vertex? **[Lesson 2.2]** $(0, -1)$

x	-2	-1	0	1	2
y	3	0	-1	0	3

13. Is the graph of the equation $x = -4$ vertical or horizontal? What is the slope of the line? **[Lesson 5.5]** vertical; undefined

14. Solve $4 - 3t \geq 19$. Show the solution set on a number line.
[Lesson 3.6] $t \leq -5$

Let $X = \begin{bmatrix} 2 & -1 & 5 \\ 1 & 0 & -2 \end{bmatrix}$ **for Items 15–17.**

15. What are the dimensions of matrix X? **[Lesson 7.1]** 2 by 3

16. Find the entry at x_{23}. **[Lesson 7.1]** -2

17. Find $3X$. **[Lesson 7.3]** $\begin{bmatrix} 6 & -3 & 15 \\ 3 & 0 & -6 \end{bmatrix}$

18. A jacket has been marked down from an original price of $85 to $59.50. What percent from the original price has the jacket been marked down? **[Lesson 4.5]** 30%

19. Mark has 3 nickels, 5 dimes, and 7 quarters in his pocket. He chooses one coin and then another. If he does not replace the first coin, what is the probability that he chose a dime first and then a quarter?
[Lesson 8.1] $\frac{1}{6}$

20. The perimeter of a rectangle is 52 and its width is 6 less than its length. Find the length and width of this rectangle. **[Lesson 3.3]** 16; 10

Free-Response Grid The following questions may be answered using a free-response grid commonly used by standardized test services.

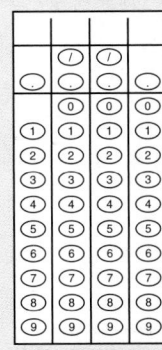

21. Simplify $4^2 \div 8 + 5(8 - 2) \cdot 2$. **[Lesson 1.4]** 62

22. How much would each of 16 people need to contribute toward a gift that costs $136? **[Lesson 2.3]** $8.50

23. Solve $7(p + 4) = 49$. **[Lesson 4.6]** 3

24. Find the slope of the line that passes through the origin and $(4, 2)$. **[Lesson 5.3]** $\frac{1}{2}$

25. What is the value of $54 - 68 + |-80|$? **[Lesson 3.1]** 66

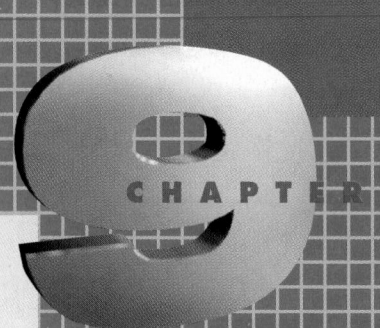

CHAPTER 9

Transformations

Meeting Individual Needs

9.1 Functions and Relations

Core Resources

Practice Master 9.1
Enrichment, p. 412
Technology Master 9.1
Interdisciplinary
　Connection, p. 411

[1 day]

Core Two-Year Resources

Inclusion Strategies, p. 413
Reteaching the Lesson, p. 414
Practice Master 9.1
Enrichment Master 9.1
Technology Master 9.1
Lesson Activity Master 9.1

[2 days]

9.2 Exploring Transformations

Core Resources

Practice Master 9.2
Enrichment, p. 419
Technology Master 9.2

[1 day]

Core Two-Year Resources

Inclusion Strategies, p. 419
Reteaching the Lesson, p. 421
Practice Master 9.2
Enrichment Master 9.2
Technology Master 9.2
Lesson Activity Master 9.2

[3 days]

9.3 Stretches

Core Resources

Practice Master 9.3
Enrichment, p. 426
Technology Master 9.3
Interdisciplinary
　Connection, p. 425
Mid-Chapter Assessment

[2 days]

Core Two-Year Resources

Inclusion Strategies, p. 426
Reteaching the Lesson, p. 427
Practice Master 9.3
Enrichment Master 9.3
Technology Master 9.3
Lesson Activity Master 9.3
Mid-Chapter Assessment

[3 days]

9.4 Reflections

Core Resources

Practice Master 9.4
Enrichment, p. 432
Technology Master 9.4

[2 days]

Core Two-Year Resources

Inclusion Strategies, p. 432
Reteaching the Lesson, p. 433
Practice Master 9.4
Enrichment Master 9.4
Technology Master 9.4
Lesson Activity Master 9.4

[3 days]

9.5 Translations

Core Resources

Practice Master 9.5
Enrichment, p. 438
Technology Master 9.5
Interdisciplinary
　Connection, p. 437

[2 days]

Core Two-Year Resources

Inclusion Strategies, p. 438
Reteaching the Lesson, p. 439
Practice Master 9.5
Enrichment Master 9.5
Technology Master 9.5
Lesson Activity Master 9.5

[4 days]

9.6 Combining Transformations

Core Resources

Practice Master 9.6
Enrichment, p. 444
Technology Master 9.6

[2 days]

Core Two-Year Resources

Inclusion Strategies, p. 444
Reteaching the Lesson, p. 445
Practice Master 9.6
Enrichment Master 9.6
Technology Master 9.6
Lesson Activity Master 9.6

[4 days]

Chapter Summary

Core Resources

Chapter 9 Project,
   pp. 448–449
Lab Activity
Long-Term Project
Chapter Review,
   pp. 450–452
Chapter Assessment,
   p. 453
Chapter Assessment, A/B
Alternative Assessment

[3 days]

Core Two-Year Resources

Chapter 9 Project, pp. 448–449
Lab Activity
Long-Term Project
Chapter Review, pp. 450–452
Chapter Assessment, p. 453
Chapter Assessment, A/B
Alternative Assessment

[5 days]

Reading Strategies

Are students familiar with terms from the chapter, like *functions*, *reflections*, or *transformations*? As they read, suggest to students that they outline the chapter to keep track of key concepts. These include the definitions of *relation*, *function*, and *transformation*. Students might also use journals to respond to the interactive questions throughout the chapter.

Visual Strategies

The visual learner will benefit from the use of a graphics calculator. Using the *f(g) Scholar*™ is even more beneficial, in that the screen is larger and the values of the graphed data points are displayed as the calculator is working in the Graph, Trace, and Zoom modes. When graphing a function, the equation is also shown on the screen. Another advantage is that the graphs can be printed out for inclusion in a portfolio or as part of an assessment activity.

Encourage students to make sure their graphs are accurately drawn and accurately represent the function. You may wish to have students sketch graphs by hand and then use a calculator or *f(g) Scholar*™ for comparison.

Hands-on Strategies

The main manipulatives used in Chapter 9 are pencil and grid paper. Hands-on graphing of equations is the best way to learn about graphing, but graphics calculators allow students to avoid unnecessarily tedious procedures. It is recommended that students graph some functions by hand. Completing a table of values locating the points on a grid, and sketching a smooth curve through the points provides the basis for understanding transformations.

Cooperative Learning

GROUP ACTIVITIES	
Graphing transformations	Lesson 9.2, Explorations 1 and 2
Experiments with stretches	Lesson 9.3, Opening Discussion
Mirror experiments	Lesson 9.4, Exploration
Combining transformations	Lesson 9.6, Example 1

You may wish to have students work with partners or in small groups to complete any of these activities. Additional suggestions for cooperative group activities can be found in the teacher's notes in the lessons.

Multicultural

The cultural connections in this chapter include references to American Indians, to the Han dynasty of China, and to Africa.

CULTURAL CONNECTIONS	
Americas: Indian Dwellings	Lesson 9.1, Opening Discussion
Asia: Ancient Tombs	Lesson 9.4, Exercise 24
Africa: African Art	Lesson 9.5, Discussion, p. 440

Portfolio Assessment

Below are portfolio activities for the chapter. They are listed under seven activity domains that are appropriate for portfolio development.

1. **Investigation/Exploration** The methods used to sketch graphs and recognize graphs of functions used throughout the chapter, especially in Lessons 9.2 and 9.4, are appropriate for inclusion.

2. **Applications** Recommended for inclusion are any of the following: Databases, Lesson 9.1, Exercises 22–25; Science, Lesson 9.1, Exercises 49 and 50 and Lesson 9.3, Exercises 1–6; Art, Lesson 9.3, Exercise 24; and Travel, Lesson 9.6, opening discussion.

3. **Non-Routine Problems** The statistics experiment in Lesson 9.5, Exercises 34–36 provides an opportunity for non-routine problem solving. Any of the following other non-routine problems are also appropriate for inclusion: Combination problem, Lesson 9.3, Exercise 38; Cartooning problem, Lesson 9.3, Exercise, 24; and Playing games, Lesson 9.1, Exercise 58.

4. **Project** Composite Functions: see pages 448–449. Students explore composite functions through arithmetic and algebraic expressions and a graphics calculator.

5. **Interdisciplinary Topics** Students may choose from the following interdisciplinary topics: Computer Science, Lesson 9.1, Exercises 22–25; Statistics, Lesson 9.3, Exercises 29–31 and Lesson 9.5, Exercises 34–35; and Geometry, Lesson 9.2, Exercises 32–34 and Lesson 9.5, Exercises 38–39.

6. **Writing** The *Communicate* exercises that ask students to describe and explain procedures involved in graphing functions offer excellent writing selections for the portfolio.

7. **Tools** Chapter 9 makes extensive use of the graphics calculator. Also useful is the *f(g) Scholar*[TM]. To measure the individual student's proficiency, it is recommended that he/she complete selected worksheets from the Technology Masters.

Technology

Graphics Calculator

Graphics calculators provide a quick and accurate way to graph functions on a coordinate plane. Another valuable tool for use with this chapter is the *f(g) Scholar*™. This computer program works like a standard graphics calculator. One advantage of utilizing *f(g) Scholar*™ is that all graphs, drawings, and spreadsheets can be printed on paper directly from the program.

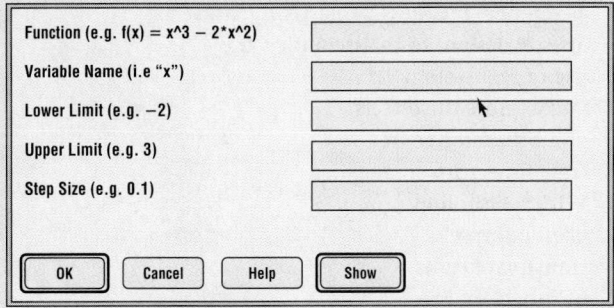

To graph a function on *f(g) Scholar*™, the **graph function** is utilized. This template will be displayed.

The following must be specified.
Function: Specify in the form $f(x) = x^2$.
Variable: Name the variable, such as x.
Lower Limit: Specify the lower limit of the x-axis.
Upper Limit: Specify the upper limit of the x-axis.
Step Size: Specify the interval between plotted points.

The graph is then created on the window below.

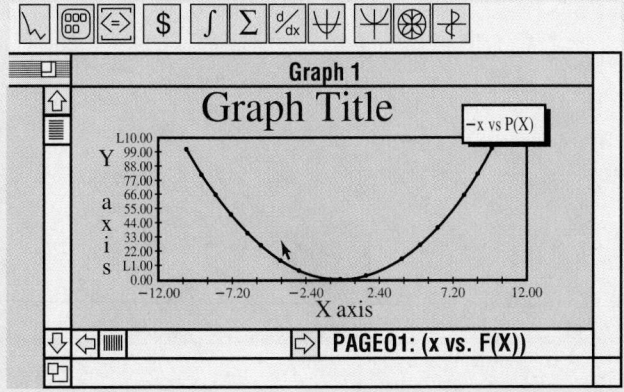

To go to a full-screen graph window, press **ALT+W** to open the Windows menu. Then choose **Active Screen** and select (**4**) **Graph.** Alternatively you can press **Shift-F4.**

TRACE The trace feature of a graphics calculator allows students to trace the graph and see the values associated with points on the graph. To do this, locate the pixel or cursor on the graph, press TRACE, and use the < or > keys. The values of x and y will be displayed. However, these may not be "nice" integral values. Students may have to interpolate to find integral values.

ZOOM The Zoom feature can be used to magnify part of the graph. After pressing ZOOM on a graphics calculator, a display similar to this may be shown:

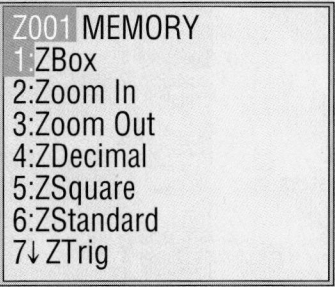

To use the Zoom Box on the TI-82, Select ZBox, move the cursor to any corner of the box you want to define, and press ENTER. Move the cursor to the diagonal corner of the box you want to define. When the box is defined as you want it, press ENTER to replot the portion of the graph that is inside the box.

ZOOM IN Magnifies the graph around the cursor location. ZOOM OUT displays a greater portion of the graph, centered on the cursor location.

More capable students may wish to experiment with the Zoom Factors and Zoom Memory.

To leave the ZOOM menu, press 2ND [QUIT] to return to the Home screen.

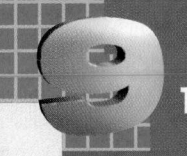

Transformations

CHAPTER 9

ABOUT THE CHAPTER

Background Information

Students may be familiar with transformations of geometric figures such as triangles. In this chapter, students will study the transformation of functions. The use of technology will help students focus their attention on the transformations.

CHAPTER RESOURCES

- Practice Masters
- Enrichment Masters
- Technology Masters
- Lesson Activity Masters
- Lab Activity Masters
- Long-Term Project Masters
- Assessment Masters
    Chapter Assessments, A/B
    Mid-Chapter Assessment
    Alternative Assessment,
    A/B
- Teaching Transparencies
- Spanish Resources

CHAPTER OBJECTIVES

- Understand the definitions of *relation* and *function.*
- Test a graph to identify a function using the vertical-line test.
- Use the *f(x)* function notation to represent and evaluate a function.
- Identify basic transformations of parent functions.
- Identify stretch, reflect, and shift as they apply to the graphs of functions.
- Describe the effect of a stretch on the graph of a function when the coefficient is greater than or equal to 1, and when it is between 0 and 1.

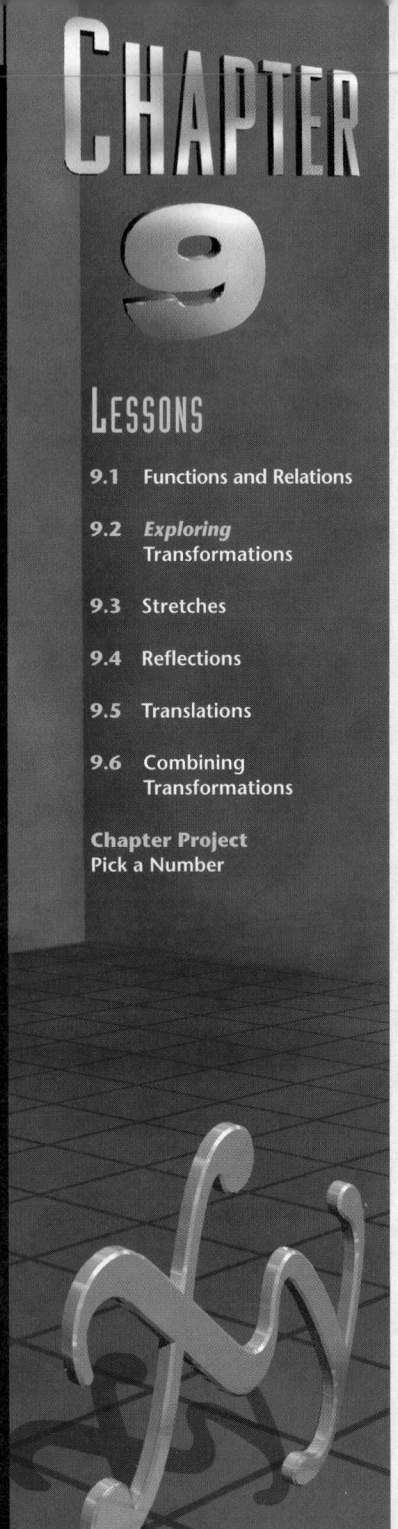

LESSONS

9.1 Functions and Relations

9.2 *Exploring* Transformations

9.3 Stretches

9.4 Reflections

9.5 Translations

9.6 Combining Transformations

Chapter Project
Pick a Number

Transformations

Once a basic function and its graph are understood, it is possible to create many adaptations that help fit the function to specific needs. Adaptations, known as transformations in mathematics, appear in such fields as science, business, advertising, and artistic design. In mathematics, the use of technology lets you explore the effects of a broad range of transformations. With technology, even complex transformations of functions are now much easier to perform.

ABOUT THE PHOTOS

Have students look at the photographs of these pages and identify as many transformations as possible. Remind students of the different types of transformations—slides, reflections, and stretches. Have students discuss other occurrences of transformations, for example, a person on roller skates, a reflection in a pond, or putting a picture on clay and then stretching the clay.

- Identify the coefficient to determine the amount of stretch.
- Describe the effect of a reflection on the graph of a function.
- Identify the minimum or maximum value for absolute value and quadratic functions.
- Describe the effect of a translation on the graph of a function.
- Identify the relationship between the translation of a graph and the addition or subtraction of a constant to the function.
- Identify the parent function in a transformation.
- Understand the effect of order on combining transformations.

PORTFOLIO ACTIVITY

Draw coordinate axes on a piece of graph paper. Leave most of the space to the right of the *y*-axis. Then design a figure in the first quadrant like the one shown in the example. Reflect the figure through the *x*-axis. Translate the reflection to the right far enough so the translation does not overlap the reflection. Then reflect the translation through the *x*-axis. Continue the procedure to create a repeating pattern.

Create a new figure, and perform several different reflections and translations to create other repeating patterns.

Identify one point on your original design by its coordinates. List the coordinates of that point after each successive transformation.

PORTFOLIO ACTIVITY

Assign the Portfolio Activity as part of the assignment for Lesson 9.6 on page 447.

Start the activity by having students discuss two transformations—reflections and slides. Have students discuss the picture on page 409.

Allow students to work independently or in cooperative groups to complete this activity.

Challenge students to repeat their transformations by first sliding the figure and then reflecting the figure. If the distance of the slide is the same, and the same line of reflection is used, are the results the same as their original transformations? **[Yes]**

ABOUT THE CHAPTER PROJECT

In the Chapter Project, on pages 448–449, students investigate number games and composite functions. Technology is used to help students complete and analyze composite functions.

PREPARE

Objectives

- Understand the definitions of *relation* and *function*.
- Test a graph to identify a function using the vertical line test.
- Use the $f(x)$ function notation to represent and evaluate a function.

RESOURCES

- Practice Master   **9.1**
- Enrichment Master   **9.1**
- Technology Master   **9.1**
- Lesson Activity Master   **9.1**
- Quiz   **9.1**
- Spanish Resources   **9.1**

Assessing Prior Knowledge

Graph each equation on a coordinate plane.

1. $x + y = 3$
2. $2x - 4 = y$

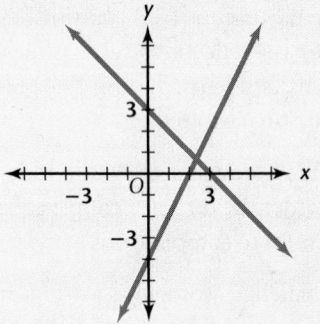

TEACH

Have students discuss relations and functions that occur in their lives. Sample answers might include the following: your shoe size is a function of the length of your foot; your earnings are a function of the number of hours you work.

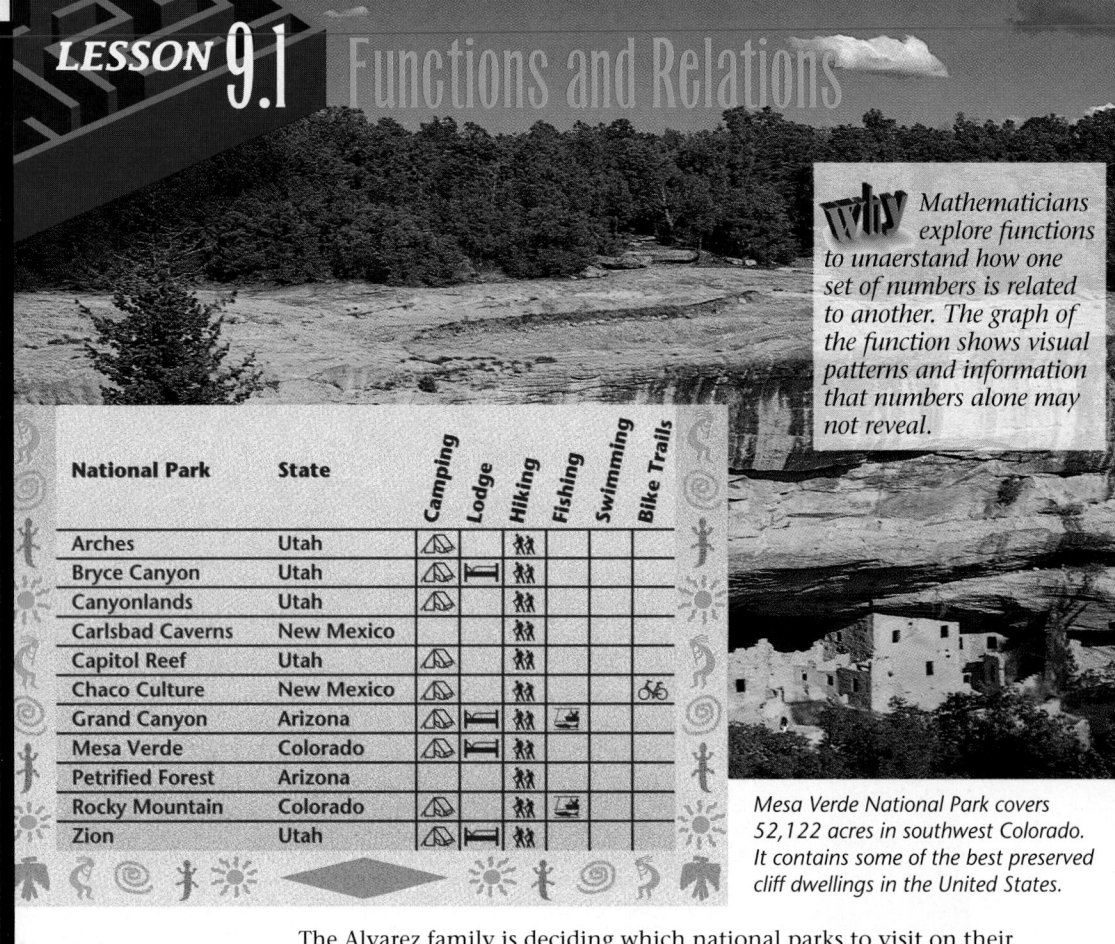

Mathematicians explore functions to understand how one set of numbers is related to another. The graph of the function shows visual patterns and information that numbers alone may not reveal.

National Park	State	Camping	Lodge	Hiking	Fishing	Swimming	Bike Trails
Arches	Utah	⛺		🚶			
Bryce Canyon	Utah	⛺	🛏	🚶			
Canyonlands	Utah	⛺		🚶			
Carlsbad Caverns	New Mexico			🚶			
Capitol Reef	Utah	⛺		🚶			
Chaco Culture	New Mexico	⛺		🚶			🚲
Grand Canyon	Arizona	⛺	🛏	🚶	🎣		
Mesa Verde	Colorado	⛺	🛏	🚶			
Petrified Forest	Arizona			🚶			
Rocky Mountain	Colorado	⛺		🚶	🎣		
Zion	Utah	⛺	🛏	🚶			

Mesa Verde National Park covers 52,122 acres in southwest Colorado. It contains some of the best preserved cliff dwellings in the United States.

The Alvarez family is deciding which national parks to visit on their summer vacation. In their road atlas they have a chart that gives the camping features at each park.

Defining Relation and Function

Two different pairings appear in the table. The first pairs a park with the features available at that park. For example, Mesa Verde is paired with camping, a lodge, and hiking trails. This pairing is an example of a **relation**.

Mesa Verde — Camping, Lodge, Hiking Trails

A **relation** pairs elements from one set with elements of another. In other words, a relation is a set of ordered pairs.

The second pairs each park with *exactly one* state. For example, Mesa Verde is paired only with Colorado. This pairing is an example of a **function**. A function is a special relation.

Mesa Verde — Colorado

ALTERNATIVE teaching strategy

Using Discussion Use the equation $y = x^2 - 3$. Ask students to find the value of y if $x = 2$ [1], if $x = -2$ [1], if $x = 10$ [97], if $x = -10$ [97]. Tell students that another way to show a relationship is to use *function notation*. Write $f(x) = x^2 - 3$ on the board or overhead projector. Have students discuss how the notation is similar to the equation $y = x^2 - 3$. Have students find the value of $f(x)$ when $x = 2, -2, 10,$ and -10. Allow students to discuss the process. Tell students that finding the value of $f(x)$ for particular values of x is called *evaluating the function* and can be written as $f(2) = 1$ and $f(-10) = 97$.

TEACHING tip

Provide students with sets of ordered pairs and have students determine whether each set is a function. Have students refer to the definition of a function on page 411. If the set is not a function, have them explain why.

1. $(2, -3)$ $(2, 0)$, $(2, -1)$
   [**not a function; 2 equal values for x with different y-values**]
2. $(0, 2)$, $(1, 2)$, $(2, 2)$ [**is a function**]
3. $(0, 0)$, $(-1, -1)$, $(2, 2)$
   [**is a function**]
4. $(-223, 132)$, $(132, -223)$, $(132, 223)$ [**not a function; 2 equal values for x with different y-values**]

FUNCTION
A **function** is a set of ordered pairs for which there is exactly one second coordinate for each first coordinate.

You have already been introduced to several elementary functions and have examined their graphs by plotting ordered pairs. Remember, to graph a function use the coordinates of ordered pairs such as (x, y) to locate a set of points on a coordinate plane.

The graph shows speed as a function of time. The points on the graph form a set of ordered pairs (time, speed). There is exactly one speed for any moment in time.

The graph shows the relation between

- the speed in miles per hour at which the Alvarez's car travels and

- each moment of time during the first 25 minutes of their trip.

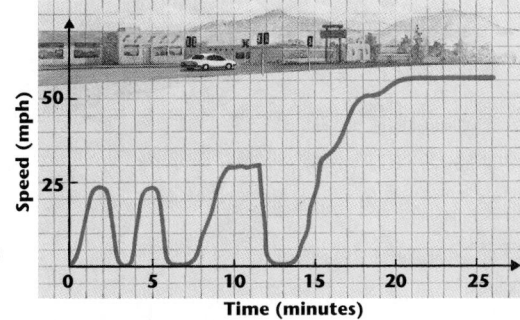

CRITICAL Thinking

Answers may vary. For example, they began the trip in a residential area and had to stop for stoplights, stop signs, or traffic. They may have stopped to buy gas 6 minutes from home. Their speed levels off at 55 mph around 17 minutes because they reached a highway.

CRITICAL *Thinking*

Use the graph to determine what might have happened during the first 25 minutes of the trip to cause the changes in the graph.

interdisciplinary
CONNECTION

Geometry Algebra is used to describe many geometric concepts. Algebraic equations are often used when finding the length of a line segment.

Provide students with grid paper that contains several sets of axes. Have students follow your directions and determine whether the line they have drawn shows a function.

1. Draw a horizontal line through 2 on the *y*-axis. [**is a function**]
2. Draw a vertical line through −1 on the *x*-axis. [**is not a function**]
3. Draw a diagonal line that passes through the origin. [**is a function**]
4. Draw a circle that crosses both the *x*- and *y*-axes. [**not a function**]
5. Draw a circle that does not cross either of the axes. [**not a function**]
6. Draw a wavy line of any type. [**Answers will depend upon the line drawn.**]

Testing for a Function

Once a relation is graphed, a **vertical line test** will show whether the relation is a function. A relation is a function *if any vertical line intersects the graph of the relation no more than once.*

EXAMPLE 1

Which of the following graphs represent functions?

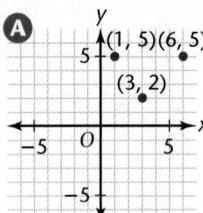
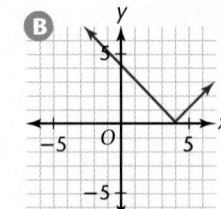
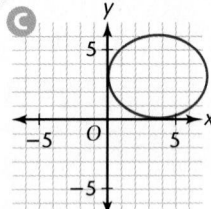

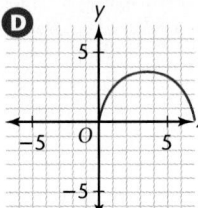
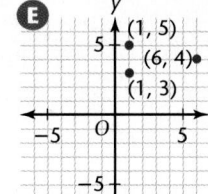
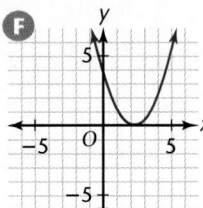

Solution ➤

Any vertical line will intersect graphs A, B, D, and F *at most once,* so these graphs represent functions. The vertical lines on graphs C and E intersect the graph of the relation twice. These graphs do *not* represent functions.

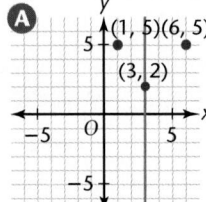
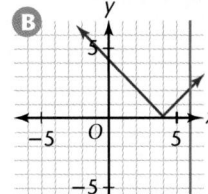
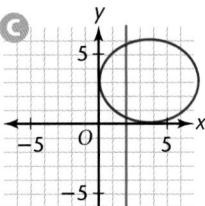

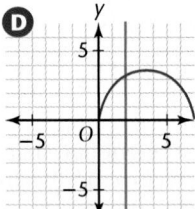
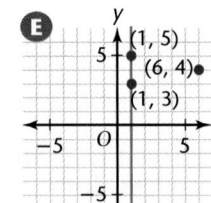
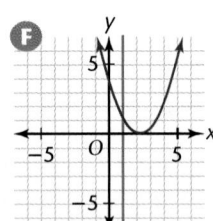

ENRICHMENT   Show students a set of capital letters in block printing on a coordinate grid, such as ABCDEFGHIJKLMNOPQRSTUVWXYZ. Ask students which letters could be represented by functions according to the definition of a function. [**V and W**]

For any function, the set of *x*-values is the **domain** and the set of *y*-values is the **range**.

The domain contains only the numbers that can be used in the function. The range contains only the numbers that the function can produce. For example, division by 0 is not permitted, so 0 cannot be in the domain of $y = \frac{1}{x}$. How do you know that there are no negative numbers in the range of the function, $y = |x|$?

Examine the graphs in Example 1. Can you identify the domain and range of the functions A, B, D, and F?

Example	Domain (x-values)	Range (y-values)
A	1, 3, and 6	2 and 5
B	all numbers	all numbers greater than or equal to zero
D	$0 \le x \le 7$	$0 \le y \le 3.5$
F	all numbers	all numbers greater than or equal to zero

Function Notation

You have used the equation $y = x^2$ to represent a quadratic function. Another way to express $y = x^2$ is to use *function notation*. Function notation uses a symbol such as $f(x)$ to write $y = x^2$ as $f(x) = x^2$.

This is read "*f* of *x* equals *x* squared." The variable in parentheses is the *replacement* variable. The expression on the right side of the equal sign is the *function rule*.

To evaluate a function, replace the variable *x* with a value from the domain. Then perform the operations given in the rule.

For example, to evaluate the function $f(x) = x^2$ for $x = 3$, substitute **3** for *x* and simplify.

$$f(\mathbf{3}) = \mathbf{3}^2 = 9$$

In algebra it is customary to use *f*, *g*, and *h* to represent functions. The domain of a function is sometimes called the input values. The range is sometimes called the output values.

input value ⟶ function rule ⟶ output value

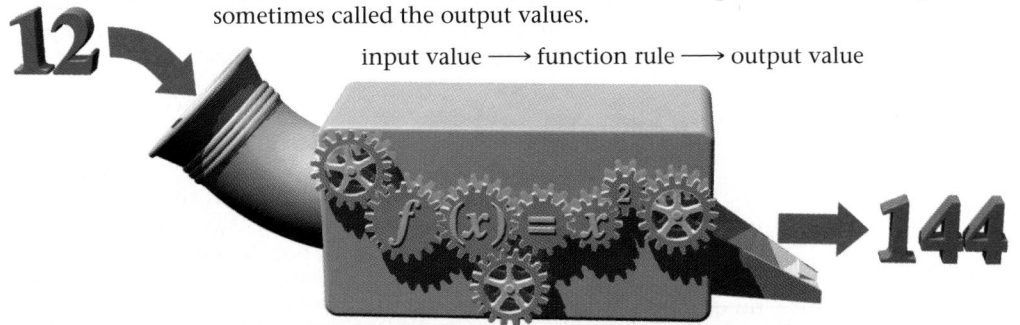

Hands-On Strategies Use the overhead to project relations like those shown on page 412. Allow students to go to the screen or board and use a yard or meter stick to show why particular relations are not functions. Have them show that it is possible to draw a vertical line through more than one point of the graph of a relation.

INCLUSION strategies

Cooperative Learning

Have students discuss situations involving the domain and range of relations and functions.

1. graph of $y = 3$ [**domain, all numbers; range, 3**]
2. graph of $y = -|x|$ [**domain, all numbers; range, all numbers ≤ 0**]
3. line segment connecting points $(-2, -3)$ and $(3, 2)$ [**domain, all numbers from -2 to 3, inclusive; range, all numbers from -3 to 2, inclusive**]

Domain: all numbers; Range: all numbers. Evaluate each function.

1. $f(x) = x + 2$ for $f(3)$ [5]

2. $g(x) = x^2 - x$ for $g\left(\frac{1}{4}\right)$

$$\left[-\frac{3}{16}\right]$$

3. $h(x) = 3x$ for $h(2)$ [9]
4. $j(x) = \frac{1}{x}$ for $j(0.5)$ [2]

Aongoing
SSESSMENT

Try This

[90]

EXAMPLE 2

Suppose h is a function and $h(x) = x^2 - 2x + 1$. The domain of the function is *all of the integers*. Evaluate $h(4)$.

Solution ➤

The notation $h(4)$ indicates that you substitute 4 for each x in the rule $x^2 - 2x + 1$. Since 4 is in the domain, simplify to find the value of the function.

$$h(x) = x^2 - 2x + 1$$
$$h(4) = 4^2 - 2(4) + 1$$
$$h(4) = 9 \; ❖$$

The h identifies or names the function. The input values for x in Example 2 are all integers. The *function rule* is the expression $x^2 - 2x + 1$. The values you get after substituting each input value and simplifying the expression are the output values.

The values shown in $h(4) = 9$ represent the ordered pair (4, 9). Notice that 4 is an element of the domain, and 9 is an element of the range. If you graph the rule, you can see that it satisfies the vertical line test for a function.

Try This   For the function $d(t) = 10t^2$ evaluate $d(-3)$.

The distance between two points on a number line is always a nonnegative number. This makes the absolute value function a good representation for distance on the number line. For example, the absolute value function can represent the distance between a given number, such as 5, and some other number, x.

RETEACHING
the
lesson

Using Manipulatives Write "+ 3" on the sides of a large box. Display a set of number cards for the numbers 1 through 20. Tell students to choose a card between 1 and 10 and put it in the box. If the box represents a function machine that adds 3 to any number, what will the output be for the chosen card.   [**3 more than the chosen card**]   Have students repeat the process. Write $f(x) = x + 3$ on the board and have students relate them to each other.   [$f(x)$ **denotes the output of the function machine that adds 3 to each input number.**]

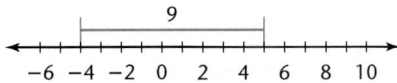

EXAMPLE 3

Determine a function for the distance between any number and the fixed point, 5. Find the distance between −4 and 5 and between 0 and 5.

Solution ➤

The function $g(x) = |x - 5|$, where the domain is all numbers, gives the distance between any number and 5. To find the distance between −4 and 5, evaluate $g(-4)$.

The distance between x and 5 when $x = -4$ is $g(-4) = |-4 - 5| = 9$.

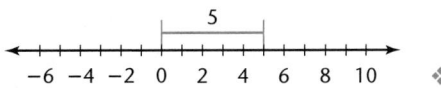

To find the distance between 0 and 5, evaluate $g(0)$.

The distance between x and 5 when $x = 0$ is $g(0) = |0 - 5| = 5$.

EXERCISES & PROBLEMS

Communicate

1. What is the difference between a function and a relation?

2. Examine the table of national parks at the beginning of the lesson. If national parks and states are paired, they form a function. Explain why this is a function.

3. Describe the relationship between the definition of a function and what the vertical line test shows.

4. Explain the procedure for evaluating the function $f(x) = 2x^2 - 5x$ if the domain consists of integers from −2 to 2.

5. You are given an ordered pair (6, −46) from a function $h(x)$. How do you know which element is from the domain and which is from the range?

6. Explain how to change the information $h(-24) = 16$ to an ordered pair so that it can be graphed.

Alternate Example 3

Tell students that the function $d(x) = |x - (-3)|$ will give the distance between any number and − 3. Have students evaluate $d(6)$ to find the distance between 6 and − 3. Then have students draw a number line to check whether their answer is correct.

ASSESS

Selected Answers

Odd-numbered Exercises 7–59

Assignment Guide

Core 1–6, 8–50 even, 51–62

Core Two-Year 1–60

Practice & Apply

Which of the following are functions? Explain.

7. {(3, 4), (4, 4), (5, 4)}
yes

8. {(1, 1)}
yes

9.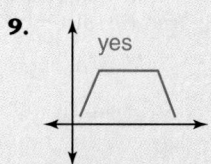
yes

10.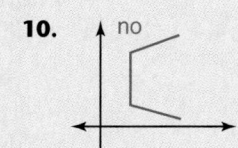
no

Find the domain and range of each of the following functions.

11. {(3, 4), (4, 4), (5, 4)}
D: 3,4,5
R: 4

12.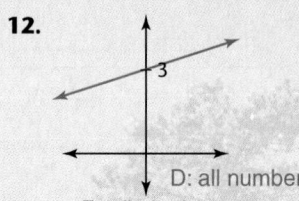
D: all numbers
R: all numbers

13.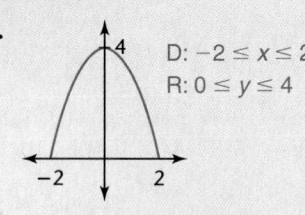
D: $-2 \leq x \leq 2$
R: $0 \leq y \leq 4$

14. Given $y = \frac{1}{x}$, find the second coordinate of an ordered pair if the first coordinate is 3. $\frac{1}{3}$

15. If $h(x) = x + 3$, find $h(7)$. 10

Answer these questions about the national parks trip.

16. Which parks have fishing?

17. What are the features at Canyonlands?

18. Which parks have bike trails?

19. Which have swimming?

20. Which have a lodge or fishing?

21. Which have a lodge and fishing?

Databases Enter the national parks information in a data base. Sort the database to list the parks

22. in Utah.

23. with camping.

24. that do *not* have a lodge.

25. that have a lodge and fishing.

Which of the following are functions?

26. {(9, 5), (9, −5)} no

27. {(5, 9), (−5, 9)} yes

28.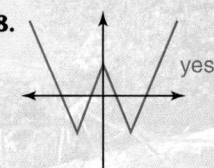
yes

29.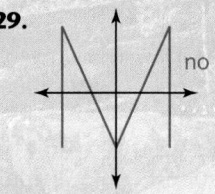
no

Find the domain and range of each of the following functions.

30. $\left\{ \left(1, \frac{1}{2}\right), \left(2, \frac{1}{2}\right), \left(3, \frac{1}{3}\right), \left(4, \frac{1}{4}\right) \right\}$

31. {(0.1, 1), (0.2, 2), (0.3, 3)}

32.

33.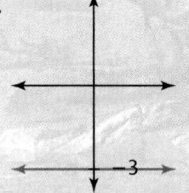

16. Grand Canyon and Rocky Mountain
17. Camping and hiking
18. Chaco Culture
19. None
20. Bryce Canyon, Grand Canyon, Mesa Verde, Rocky Mountain, and Zion
21. Grand Canyon
22. Arches, Bryce Canyon, Canyonlands, Capitol Reef, and Zion
23. Grand Canyon, Mesa Verde, Rocky Mountain, Chaco Culture, Arches, Bryce

Canyon, Canyonlands, Capitol Reef, and Zion
24. Petrified Forest, Carlsbad Caverns, Rocky Mountain, Chaco Culture, Arches, Canyonlands, and Capitol Reef
25. Grand Canyon
30. D: 1, 2, 3, 4   R: $\frac{1}{2}, \frac{1}{3}, \frac{1}{4}$
31. D: 0.1, 0.2, 0.3   R: 1, 2, 3
32. D: all numbers   R: − 3
33. D: all numbers   R: − 1 ≤ y ≤ 1

If $f(x) = 5x$, find the following.

34. $f(3)$ 15   **35.** $f(0)$ 0   **36.** $f(-2)$ −10   **37.** $f(-6)$ −30

Evaluate the functions for $x = 3$.

38. $g(x) = x^2$ 9   **39.** $f(x) = 2^x$ 8   **40.** $h(x) = |x|$ 3   **41.** $k(x) = \frac{1}{x}$ $\frac{1}{3}$

42. **Technology** Tell how to graph $|y| = x$ on a graphics calculator. HINT: Graph the top half using $Y_1 =$ and the bottom half using $Y_2 =$.
   Let $y_1 = x$ and $y_2 = -x$. Let $x \geq 0$.

Which of these points satisfy $|x| + |y| = 8$?

43. (5, 3)   **44.** (5, −3)   **45.** (−5, 3)   **46.** (−5, −3)   **47.** (7, 1)
   yes   yes   yes   yes   yes

48. Graph $|x| + |y| = 8$. Is the graph a function? Explain.

49. **Science** If you know the Celsius temperature, can you necessarily know the Fahrenheit temperature? Is Fahrenheit temperature a function of Celsius temperature? yes; yes

50. If you know the Fahrenheit temperature, can you necessarily know the Celsius temperature? Is Celsius temperature a function of Fahrenheit temperature? yes; yes

 *Look Back*

Solve the following equations.   [Lesson 4.6]

51. $3r + 4 = -2 + 6r$ 2   **52.** $8(x - 7) = 3x + 4$ 12

53. Solve $A = \frac{h}{2}(B + b)$ for B. **[Lesson 4.6]** $B = \frac{2A}{h} - b$

54. Graph the linear equation $4x - 2y = 24$. **[Lesson 5.4]**

55. A line crosses the y-axis at 5 and is perpendicular to $y = -\frac{3}{5}x$. Write the equation in standard form for that line. **[Lesson 5.6]** $5x - 3y = -15$

56. Find the product of $\begin{bmatrix} 1 & 3 \\ -5 & 7 \end{bmatrix}\begin{bmatrix} 6 & 0 \\ 1 & -2 \end{bmatrix}$. **[Lesson 7.3]** $\begin{bmatrix} 9 & -6 \\ -23 & -14 \end{bmatrix}$

57. Find the inverse of $A = \begin{bmatrix} 3 & -5 \\ -1 & 2 \end{bmatrix}$. **[Lesson 7.4]** $\begin{bmatrix} 2 & 5 \\ 1 & 3 \end{bmatrix}$

58. There are 5 friends that play tennis. How many games must occur for each player to play all the others once? **[Lesson 8.5]** 10

Probability There are 8 red, 4 green, 9 yellow, and 3 white tickets in a jar. The tickets are thoroughly mixed. **[Lesson 8.6]**

59. Jacob draws one ticket from the jar without looking. What is the probability that he will draw a red ticket? $\frac{1}{3}$

60. Jacob puts the ticket back and draws another without looking. What is the probability that he will now draw a yellow or a green ticket? $\frac{13}{24}$

Look Beyond

Graph each on a graphics calculator. HINT: **Graph the top half using $Y_1 =$ and the bottom half using $Y_2 =$.**

61. $x = y^2$ **62.** $x^2 + y^2 = 25$

48.

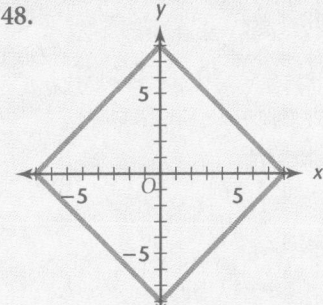

54.

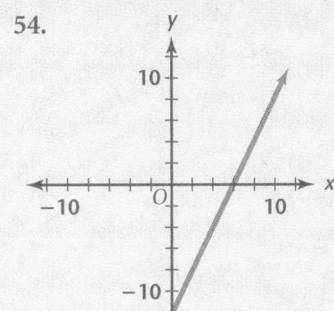

The answers to Exercises 61–62 can be found in Additional Answers beginning on page 729.

It is not a function. (5,3) and (5,−3).

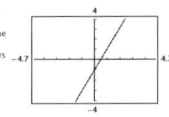

Objectives

- Identify basic transformations of parent functions.
- Identify stretch, reflect, and shift as they apply to the graphs of functions.

- Practice Master   **9.2**
- Enrichment Master   **9.2**
- Technology Master   **9.2**
- Lesson Activity Master   **9.2**
- Quiz   **9.2**
- Spanish Resources   **9.2**

Assessing Prior Knowledge

Graph each function.

1. $y = |x + 1|$
2. $y = x^2 - 1$
3. $y = 3x^2$

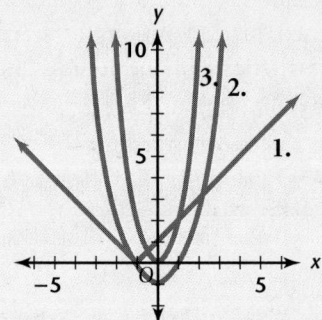

TEACH

 Ask the students whether they have ever visited a carnival and seen themselves in a distorted mirror that stretches their image. Have them describe what they saw. Tell students that these stretches represent a type of transformation when an image is on a coordinate plane. In this lesson, they will study transformations of this kind.

LESSON 9.2
Exploring Transformations

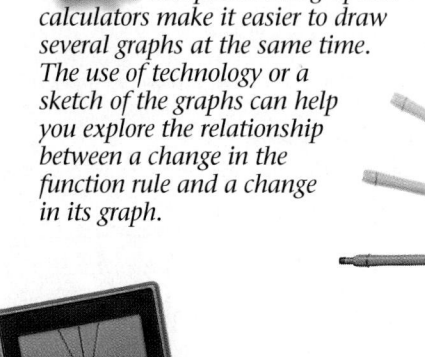 *Computers and graphics calculators make it easier to draw several graphs at the same time. The use of technology or a sketch of the graphs can help you explore the relationship between a change in the function rule and a change in its graph.*

In Chapter 2 you learned how to draw the graphs of several functions. One of those is the absolute value function, $f(x) = |x|$ or $y = |x|$. The *vertex* of this graph is where the V comes to a point. This vertex has its lowest, or minimum, point at the origin (0, 0).

**MAXIMUM
MINIMUM**
Connection

This graph is also *symmetric* with respect to the *y*-axis. The axis that passes vertically through the vertex of this function is the *axis of symmetry*. If you fold the graph along this axis, the left and right halves of the graph will match exactly.

x	|x|
−3	3
−2	2
−1	1
0	0
1	1
2	2
3	3

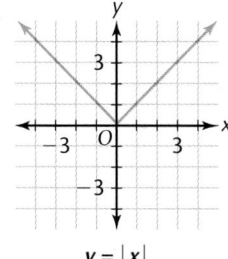

$y = |x|$

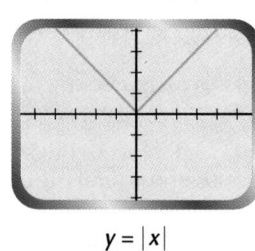

$y = |x|$

The original function $y = |x|$ is an example of a **parent function**. A parent function is the most basic of a family of functions. A variation such as a stretch, reflection, or shift is a **transformation** of the parent function.

ALTERNATIVE
**t e a c h i n g
s t r a t e g y**

Using Visual Models

On an overhead-grid transparency, draw the graph of $y = |x|$ in red. Draw several transformations of the graph in other colors on the same axes. Have students describe the changes that have taken place. Encourage students to use the terms *slide*, *flip*, and *stretch* in their descriptions. Repeat with $y = x^2$ and $y = \frac{1}{x}$.

There are various ways to stretch, reflect, and shift the graph of $y = |x|$
Examine the graphs in a-f. The function $y = |x|$ has been

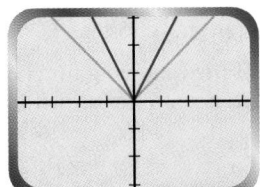
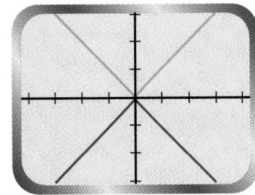

a. *stretched vertically by 2.* **b.** *shifted vertically by −2.* **c.** *reflected through the x-axis.*

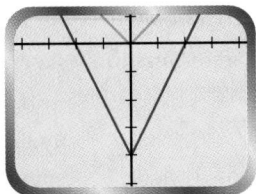
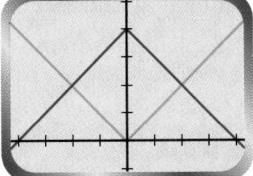
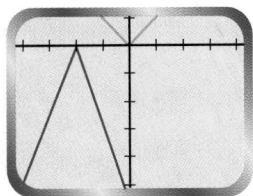

d. *stretched vertically by 2 and shifted vertically by −4.* **e.** *reflected through the x-axis and shifted vertically by 4.* **f.** *stretched vertically by 3, reflected through the x-axis, and shifted horizontally by −2.*

•Exploration 1 *Transformations of Absolute Value Functions*

Graphics Calculator

1 Draw the graph of the parent function, $y = |x|$. If you have a graphics calculator, enter ABS X or ABS(X) to select the function.

2 Draw the graph of each of the following functions. Describe how the graph has been transformed from the parent function.

   a. $y = -|x|$ **b.** $y = 2|x|$ **c.** $y = |x| - 2$

   d. $y = -|x| + 4$ **e.** $y = 2|x| - 4$ **f.** $y = -3|x + 2|$

3 Compare your graphs a-f from Step 2 with the graphs a-f above. Match the functions and the graphs.

4 What must you do to the function rule for $y = |x|$ to make the new function rule represent a stretch? a reflection? a shift?

5 How can you tell if the graph of the absolute value has been stretched, reflected, or shifted?

6 Explain how to use the parent function $y = |x|$ to graph

   a. $y = 3|x|$.   **b.** $y = -5|x|$.   **c.** $y = |x| + 2$.

7 Make a conjecture about the graph of $y = |x + 1|$ as a transformation of $y = |x|$.

8 Test your conjecture by graphing

   a. $y = |x + 3|$.   **b.** $y = |x - 3|$. ❖

Math Connection
Geometry

Students should be familiar with the axis of symmetry of common geometric figures. In this case, the axis of symmetry is the *x*- or *y*-axis.

TEACHING *tip*

Technology Remind students to clear the memory and to check the range set on their calculators before beginning Exploration 1.

Exploration 1 Notes

One of the advantages of using a graphics calculator is that students can quickly and accurately create graphs to test hypotheses. Encourage the students to explore "What if . . ." situations beyond those presented in the text.

Aongoing ASSESSMENT

5. If it has been stretched, then the sides of the graph would not be made up of the portions of the lines $y = x$ and $y = -x$ that lie above the x-axis. If it has been reflected, then it would open downward. If it has been shifted, then the vertex of the graph would not be at $(0, 0)$.

ENRICHMENT Students could conduct research to examine one or more cultures that have applied transformations to the mathematics of tiling. Possibilities include Islamic and Arabic designs, the Chinese lattice pattern, sand drawings of the Tchokwe people of Angola and the Shongo children of Zaire, Celtic knotworks found on Scottish stones, and designs found on American Indian pottery.

INCLUSION strategies **Hands-On Strategies** Have students draw and then cut out a simple figure. Ask them to move the figure to several different positions on a grid using stretches, reflections, and shifts while writing the directions for each move. Students could exchange figures and directions, and then trace the figure after each move to demonstrate the directions described.

Exploration 2 Notes

After students have graphed $y = x^2$, have them discuss how this graph is different from the graphs in Exploration 1.
[The graph of $y = x^2$ is a curve. The graphs in the exploration are straight lines.]

ongoing ASSESSMENT

5. **Stretch the graph vertically by 2, reflect it through the x-axis and shift it vertically by 3.**

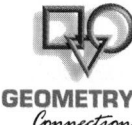

Exploration 2 Transformations of Quadratic Functions

1. Draw the graph of the parent function $y = x^2$.
2. Draw the graph of each of the following functions. Describe how the graph of each function has been transformed from the parent function.

   a. $y = 2x^2$ **b.** $y = -x^2$ **c.** $y = x^2 - 2$

3. Compare your graphs to those drawn in a-f in Exploration 1. How are they similar?
4. What must you do to the function rule for $y = x^2$ to make the new function rule result in a stretch? a reflection? a shift?
5. Explain how to use the parent function $y = x^2$ to graph $y = -2x^2 + 3$. ❖

APPLICATION

GEOMETRY *Connection*

Square photos are placed on a piece of cardboard backing. Each photo has a special border that adds a total of 2 inches to each side of the photo. The area of a photo including its border is a function of the length of a side of the photo. How can you model this application with a function?

Method A Make a table.

Measure the lengths of the sides of the photo as they increase by 1 unit. Then find the corresponding areas.

Length of photo	x	0	1	2	3	4	5	6	x
Area with border	$A(x)$	4	9	16	25	36	49	64	$(x + 2)^2$

Notice that 2 is added to x and then squared. The function rule is $(x + 2)^2$.

TECHNOLOGY *Graphics Calculator*

Method B Make a graph.

Graph the ordered pairs in the table.

X	Y_1
0	4
1	9
2	16
3	25
4	36
5	49
6	64

X=0

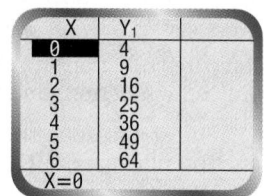

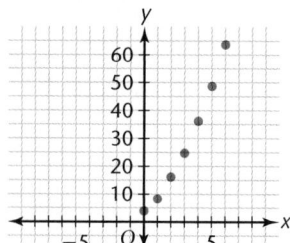

When a graphics calculator is used, the entire parabola is shown. Remember to match the graph with the original information in the problem.

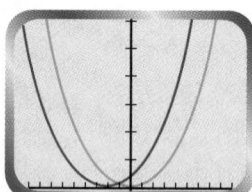

Compare the graph with the parent function $y = x^2$. Try various transformations until one fits. Notice that the x^2 function is shifted 2 units to the left. Thus, the function rule for the application is $(x + 2)^2$ and $A(x) = (x + 2)^2$ is the function that models the application data. ❖

The following is a summary of some parent functions from Chapter 2.

SUMMARY OF PARENT FUNCTIONS

FUNCTION	NAME	PAGE
$y = x$	Linear	60
$y = 2^x$ and $y = 10^x$	Exponential	60
$y = x^2$	Quadratic (parabola)	65
$y = \dfrac{1}{x}$	Reciprocal	70
$y = |x|$	Absolute value	79

In the following lessons you will examine some transformations of these parent functions. A guide to the parent functions and some of their transformations can be found on pages 704-707.

EXERCISES & PROBLEMS

Communicate

1. Explain how to identify the vertex and axis of symmetry for the graph of the function $y = x^2$.

2. Describe how the graph changes when the graph of a parent function is stretched.

3. Describe how the graph changes when the graph of a parent function is reflected.

4. Describe how the graph changes when the graph of a parent function is shifted.

5. Explain how the vertex of $y = |x|$ and $y = x^2$ changes when the graph of the function is shifted.

6. Review the Summary of Parent Functions. How do you determine which of the functions have a vertex and which have symmetry through the y-axis?

TEACHING tip

Technology Have students create each of the parent functions on their graphics calculator and sketch them on paper. The sketches can be included in their portfolio for future reference.

ASSESS

Selected Answers
Odd-numbered Exercises 7–43

Assignment Guide
Core 1–6, 8–24 even, 26–47

Core Two-Year 1–44

Technology
Students should be allowed to use their graphics calculators for Exercises 10–21. Remind students that analysis statements are required to complete each exercise.

Error Analysis
Students may confuse the graphs of $y = |x| + 2$ and $y = |x + 2|$. Have students graph each on a graphics calculator. A number added or subtracted within the absolute-value symbol moves the graph left or right while a number outside the absolute-value symbol shifts the graph up or down.

Practice & Apply

Give the parent function for each graph. Then describe the transformation using the terms stretch, reflect, and shift.

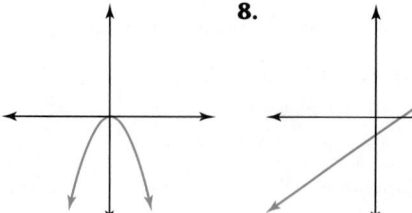

7.   8.   9.

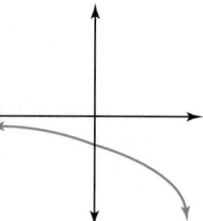

Graph each function. Then tell whether the graph of $y = |x|$ is stretched, reflected, or shifted when you graph the following functions.

10. $y = 4|x|$ **11.** $y = 4 + |x|$ **12.** $y = -|x|$

13. $y = 0.5|x|$ **14.** $y = |x - 5|$ **15.** $y = |x| - 3$

Graph each function. Then tell whether the graph of $y = \dfrac{1}{x}$ is stretched, reflected, or shifted, or give the combination of transformations when you graph the following functions.

16. $y = 5 + \dfrac{1}{x}$ **17.** $y = \dfrac{4}{x}$ **18.** $y = -\dfrac{1}{x}$

19. $y = \dfrac{1}{x} + \dfrac{1}{3}$ **20.** $y = \dfrac{1}{x + 2}$ **21.** $y = 3 - \dfrac{1}{x}$

Identify the parent function. Tell what transformation is applied to each, and then draw a carefully labeled graph.

22. $y = 5x^2$ **23.** $y = \sqrt{10}\,x$ **24.** $y = \mathrm{INT}\,\dfrac{x}{2}$ **25.** $y = -|x - 2|$

A square picture is to be framed with a 2-inch mat around it. The area of the picture without the mat is $A = x^2$, where each side of the picture is x inches. The area of the picture including the mat is $A = (x + 4)^2$.

26. Make a table of values for $A = x^2$.

27. Make a table of values for $A = (x + 4)^2$.

28. Graph the two functions on the same axes.

29. Use transformations to tell how the graphs compare.

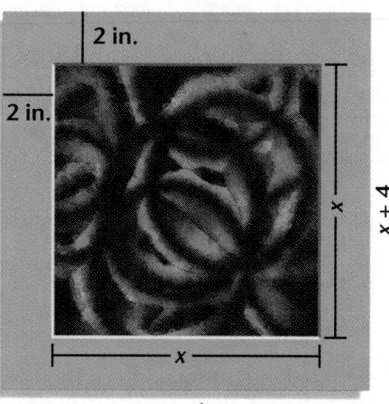

2 in.

2 in.

x + 4

x

x

x + 4

30. A group of students are weighed at the beginning of the year. What transformation of the data takes place if each student gains 3 pounds?

31. A teacher decides to double every student's score on a quiz because she used the wrong grading scale. What kind of transformation is this?

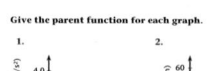
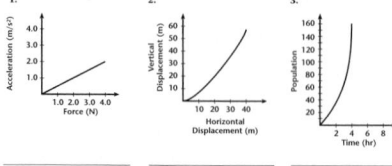
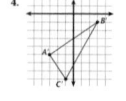 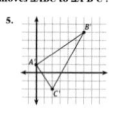 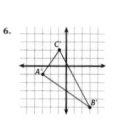
The graphs for Exercises 10–25 can be found in Additional Answers beginning on page 729.

7. $y = x^2$; reflection through the x-axis
8. $y = x$; shifted vertically downward
9. $y = 2^x$, reflection through the x-axis
10. stretched vertically by 4
11. shifted vertically by 4
12. reflected through the x-axis
13. stretched vertically by 0.5
14. shifted horizontally by 5

15. shifted vertically by -3
16. shifted vertically by 5
17. stretched vertically by 4
18. reflected through the x-axis
19. shifted vertically by $\dfrac{1}{3}$
20. shifted horizontally by -2
21. reflected through the x-axis and shifted vertically by 3
22. $y = x^2$; stretched vertically by 5

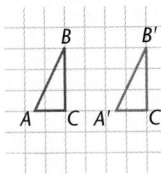 **Geometry** Tell what kind of transformation moves $\triangle ABC$ to $\triangle A'B'C'$, which is read *A* prime, *B* prime, *C* prime.

32.

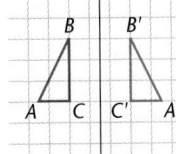

horizontal shift

33.

B B'

A C C' A'

reflected through a vertical line

34. Find the equation of the axis of symmetry for this graph.

$x = 3$

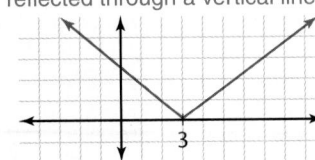

3

Look Back

35. What are the missing numbers in the sequence? **[Lesson 1.2]**
3, 6, 11, 18, 27, _?_, _?_, 66, _?_   38; 51; 83

36. Calculate the value of $6 + 36 \div 3 - 1$. **[Lesson 1.4]**   17

37. The cost of lunch was $6.09. The total amount for the food was $5.80. What percent of the total was the sales tax? **[Lesson 4.5]**   5%

Graph the solution of each inequality on a number line.
[Lesson 4.8]

38. $3x - 5 < 5x - 17$ **39.** $7x + 4 \geq -10$ and $2x + 3 < 13$

40. Find an equation of a line that contains the points $(3, -7)$ and $(-4, 3)$.
[Lessons 5.3, 5.4] $10x + 7y = -19$

Solve the following systems of equations if possible. If the system has no solution, explain why. **[Lessons 6.2, 6.3]**

41. $\begin{cases} 8x - 3y = -1 \\ 2x + y = 5 \end{cases}$ **42.** $\begin{cases} 4x - 3y = 6 \\ x = 12 \end{cases}$ **43.** $\begin{cases} 4x - 6y = 3 \\ 2x - 3y = 4 \end{cases}$
$(1, 3)$ $(12, 14)$ none; parallel lines

44. Given the following matrices, find the matrix products for both YA and XA. **[Lesson 7.4]**

$A = \begin{bmatrix} -6 & 3 \\ -8 & 7 \end{bmatrix}$ $Y = \begin{bmatrix} -1 & 0 \\ 0 & 1 \end{bmatrix}$ $X = \begin{bmatrix} 1 & 0 \\ 0 & -1 \end{bmatrix}$

$YA = \begin{bmatrix} 6 & -3 \\ -8 & 7 \end{bmatrix}$ $XA = \begin{bmatrix} -6 & 3 \\ 8 & -7 \end{bmatrix}$

Look Beyond

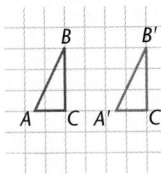 **Statistics** Tell what happens to the mean of a set of data if each data point is
increased by 5   doubles   opposite
45. increased by 5.   **46.** doubled.   **47.** replaced by its opposite.

23. $y = x$; stretched vertically by $\sqrt{10}$
24. $y = \text{INT}(x)$; stretched horizontally by 2
25. $y = |x|$; reflected through the *x*-axis and shifted horizontally by 2

The answers to Exercises 28–31 and 38–39 can be found in Additional Answers beginning on page 729.

26.

x	$A = x^2$
2	4
1	1
0	0
-1	1
-2	4

27.

x	$A = (x + 4)^2$
0	16
-1	9
-2	4
-3	1
-4	0
-5	1
-6	4

Look Beyond

Exercises 45–47 foreshadow statistical transformations that will be studied in Lesson 9.3.

- Describe the effect of a stretch on the graph of a function when the coefficient is greater or equal to 1, and when it is between 0 and 1.
- Identify the coefficient to determine the amount of stretch.

RESOURCES

- Practice Master   **9.3**
- Enrichment Master   **9.3**
- Technology Master   **9.3**
- Lesson Activity Master   **9.3**
- Quiz   **9.3**
- Spanish Resources   **9.3**

Assessing Prior Knowledge

Evaluate each of the following for $f(4)$:

1. $f(x) = x^2$ **[16]**
2. $f(x) = x^2 - 12$ **[4]**
3. $f(x) = 3x^2$ **[48]**
4. $f(x) = 3x^2 - 12$ **[36]**
5. $f(x) = 3(x^2 - 12)$ **[12]**
6. $f(x) = 3(x - 12)^2$ **[192]**

TEACH

 Ask students what is meant by the scale of a map. Suppose one map has a scale of 1 cm : 1 km and another map of the exact same region has a scale of 4 cm : 1 km. How are the two maps alike and how are they different?

Aongoing ASSESSMENT

256 feet

LESSON 9.3 Stretches

why *When you evaluate a function and plot the points, the function will have a distinctive shape. Changes in scale will change the shape of that graph. One such change is a vertical stretch. Once you know the effects of a change in scale on a parent function, it is easier to sketch or visualize the graph.*

The height of a waterfall can be approximated if you know the time it takes the water to fall. This relationship can be modeled by the function $S = 16t^2$, where S is the distance in feet and t is the time in seconds. For example, if it takes the water 3 seconds to fall, the height is about $16(3)^2$ or 144 feet. What is the height if the time is 4 seconds?

The table shows the values of the function $S = 16t^2$ for times from 0 to 3 seconds.

How do these values compare to the values of the parent function $S = t^2$? That is, compare the functions $f(x) = x^2$ and $g(x) = 16x^2$.

$S = 16t^2$		
t	$16t^2$	S
0	$16(0)^2$	0
1	$16(1)^2$	16
2	$16(2)^2$	64
3	$16(3)^2$	144

ALTERNATIVE teaching strategy

Using Patterns On an overhead projector, show a graph of each of the following: $f(x) = x^2$, $f(x) = 4x^2$, and $f(x) = 16x^2$. Have students describe patterns and relationships they observe between the graphs and their related equations. Repeat the procedure for these graphs: $f(x) = \frac{1}{x}$, $f(x) = 6\left(\frac{1}{x}\right)$, and $f(x) = 12\left(\frac{1}{x}\right)$.

$16t^2$

What is the value of the parent function $f(x) = x^2$ when x is 5?

If $x = 5$, then $f(5) = 5^2 = 25$.

What is the value of the transformed function $g(x) = 16x^2$ when x is 5?

If $x = 5$, then $g(5) = 16(5)^2 = 400$.

The values of the function $y = 16x^2$ are 16 times greater than the corresponding values of the parent function $y = x^2$. The coefficient, 16, is called the **scale factor** of the function.

What does $g(5)$ represent in the waterfall problem?

Can you visualize the graph of $y = 16x^2$? Can you sketch it quickly without plotting the points? To do this you must understand how the scale factor affects the parent function.

Compare the graphs.

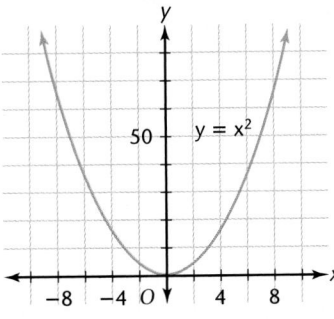

The graph of $y = x^2$ is *stretched* vertically to become the graph $y = 16x^2$.

When you apply the scale factor of 16, the point (2, 4) on the parent graph becomes the point (2, 64) on the transformed graph.

$$16(4) = 64$$

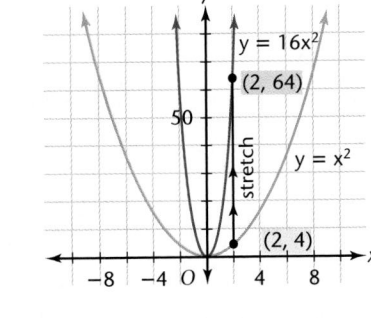

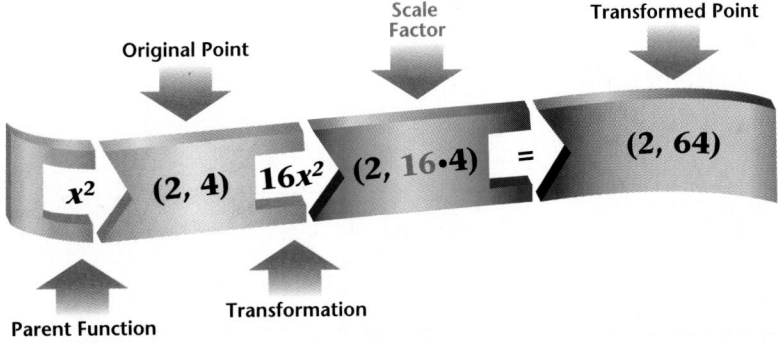

Original Point · Scale Factor · Transformed Point

Parent Function · Transformation

x^2 · (2, 4) · $16x^2$ · (2, 16·4) · = · (2, 64)

Science Have students discuss any science formulas they are familiar with. Do any of these formulas involve transformations?

interdisciplinary **CONNECTION**

TEACHING tip

If it takes 4 seconds for the water to fall, the height is $16(4)^2$ or 256 feet.

TEACHING tip

Students should realize that values of $g(x)$ are 16 times as great as values of $f(x)$ for any particular value of x.

ongoing ASSESSMENT

In the waterfall problem, $g(5)$ represents the height of the waterfall if it takes the water 5 seconds to fall from the top of the waterfall to the bottom.

TEACHING tip

Technology Have students graph both $y = x^2$ and $y = 16x^2$ on their graphics calculator. For these graphs have students set the range of their calculator at

Xmin = −10, Xmax = 10, Xscl = 1, Ymin = −1, Ymax = 100, Yscl = 10.

Have students use the ZOOM and TRACE keys to find the value of $f(x)$ when $x = 2$ on each graph. [**4 and 64**]

Alternate Example 1

Use the parent graph to sketch the graph of $y = \frac{1}{4}|x|$.

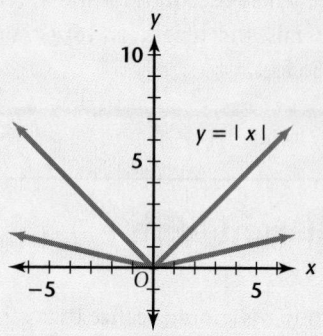

Aongoing SSESSMENT

Try This

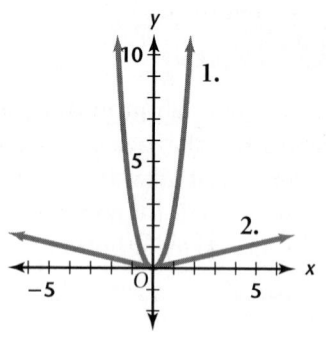

1. $f(x) = 4x^2$

2. $g(x) = \dfrac{|x|}{4}$

Alternate Example 2

Sketch the graph of $y = \frac{1}{4}x$ and describe how it compares to its parent function. [**Each point is stretched to $\frac{1}{4}$ the height of the parent function.**]

EXAMPLE 1

Use the parent function to sketch the graph of $y = \frac{1}{2}|x|$.

Solution ➤

The parent function is $y = |x|$. The transformation has a scale factor of $\frac{1}{2}$. Each of the y-values is $\frac{1}{2}$ the corresponding value of the parent function.

The graph will be stretched vertically to only $\frac{1}{2}$ the height of $y = |x|$, so the graph of $y = \frac{1}{2}|x|$ is between the parent function and the x-axis.

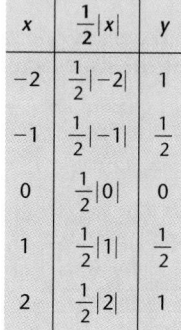

x	$\frac{1}{2}\lvert x\rvert$	y
−2	$\frac{1}{2}\lvert-2\rvert$	1
−1	$\frac{1}{2}\lvert-1\rvert$	$\frac{1}{2}$
0	$\frac{1}{2}\lvert 0\rvert$	0
1	$\frac{1}{2}\lvert 1\rvert$	$\frac{1}{2}$
2	$\frac{1}{2}\lvert 2\rvert$	1

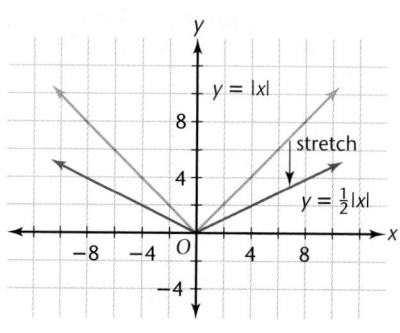

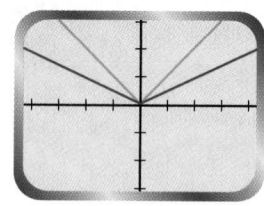

Try This Sketch the functions $f(x) = 4x^2$ and $g(x) = \frac{|x|}{4}$.

The graph of $y = 16x^2$ and $y = \frac{1}{2}|x|$ have been stretched from the graph of their parent functions. When $a > 0$, $y = a \cdot f(x)$ is stretched vertically from the graph of its parent function, $y = f(x)$, by the scale factor a.

EXAMPLE 2

Sketch the graph of $y = \frac{6}{x}$. How does it compare to its parent function?

Solution ➤

The function can be written $y = 6\left(\frac{1}{x}\right)$. Its parent function is the reciprocal function, $y = \frac{1}{x}$. Each point of the parent graph is stretched vertically by a scale factor of 6.

x	$6\left(\frac{1}{x}\right)$	y
−3	$6\left(\frac{1}{-3}\right)$	−2
−2	$6\left(\frac{1}{-2}\right)$	−3
1	$6\left(\frac{1}{1}\right)$	6
2	$6\left(\frac{1}{2}\right)$	3
3	$6\left(\frac{1}{3}\right)$	2

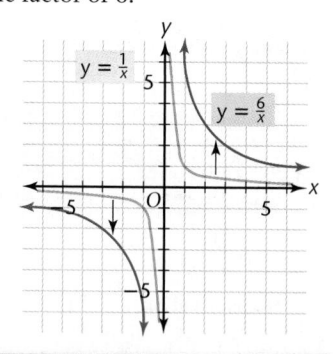

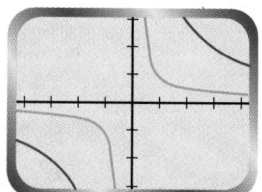

ENRICHMENT

Have students use science books as sources for equations and formulas. Have students examine various formulas and determine the parent function and any appropriate transformations.

INCLUSION strategies

Guided Analysis Have students work in pairs. One student explains and orally analyzes each of the following functions while the other uses a graphics calculator to check that the analysis is correct.

$$f(x) = 5x^2 + 1, \quad g(x) = \frac{4}{x}, \quad h(x) = x^2 - \frac{5}{2}$$

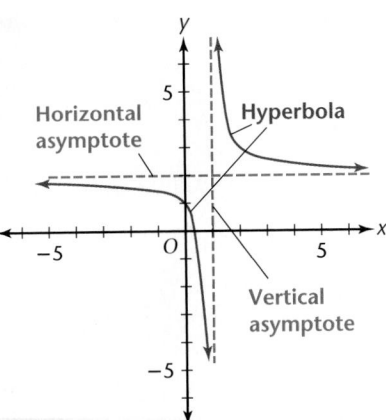

The graph of the reciprocal function forms a curve called a **hyperbola**. Notice that the graph of $y = \frac{1}{x}$ gets closer and closer to the x- and y-axes, but will never touch them. The lines that a hyperbola will approach, but not touch, are called **asymptotes**. The intersection of the asymptotes provides a reference point when transforming reciprocal functions. At what point do the asymptotes intersect in the graph at the left?

How does the change in the graph of $y = ax^2$ differ from the change in the graph of $y = a\left(\frac{1}{x}\right)$ as the scale factor a increases when $a > 0$?

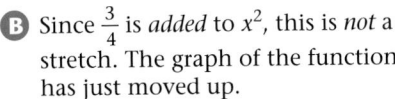

EXAMPLE 3

Which functions result in stretches of the parent graph?

A $y = \frac{x^2}{3}$ **B** $y = x^2 + \frac{3}{4}$ **C** $y = 5|x|$

Graphics Calculator

Solution ➤

A Since x^2 is multiplied by $\frac{1}{3}$, this is a vertical stretch by the scale factor $\frac{1}{3}$. The y-values will be only $\frac{1}{3}$ the y-values of the parent function.

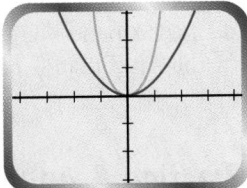

B Since $\frac{3}{4}$ is *added* to x^2, this is *not* a stretch. The graph of the function has just moved up.

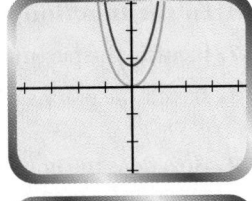

C Since $|x|$ is multiplied by the scale factor, 5, each value of the parent function stretches vertically by 5. ❖

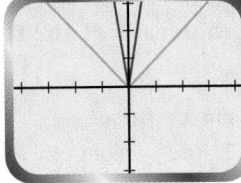

SUMMARY

To graph a vertical stretch, you should

1. identify the parent function and
2. determine the scale factor.

The graph of the parent function will be stretched vertically by the amount of the scale factor.

CRITICAL Thinking

In $y = ax^2$, as a increases, the graph gets closer to the y-axis. In $y = a\left(\frac{1}{x}\right)$, as a increases, the graph becomes flatter as it approaches the x-axis.

Alternate Example 3

Which of these functions result in stretches of the parent graph?

1. $y = 0.4x^2$
   [**involves a stretch**]
2. $y = \frac{x^2}{0.4}$
   [**involves a stretch**]
3. $y = x^2 - 0.4$
   [**does not involve a stretch**]

RETEACHING **the** **lesson**

Have each student graph each of these points on a large coordinate grid: $m = (16, 8)$, $n = (-4, 4)$, and $p = (0, -12)$. Ask students where they think the graph of $\frac{1}{4} m$ would be located.

$\left[(4, 2) \text{ Each coordinate is multiplied by} \frac{1}{4}.\right]$

Ask students where they think the graph of $2n$ would be located.   [$(-8, 8)$ **Each coordinate is multiplied by 2.**]   Repeat for $\frac{1}{4} p$ and $2p$. [$(0, -3)$ **and** $(0, -24)$]   Have students refer to these concepts as they graph stretches of functions in the lesson.

Assess

Selected Answers

Odd-numbered Exercises 7–41

Assignment Guide

Core 1–11, 13–18, 23–26, 29–44

Core Two-Year 1–41

Technology

You may wish to allow students to use their graphics calculators for Exercises 8, 10–12, 16, and 42–44.

Error Analysis

For Exercises 22 and 23, students may simply write the equation of the parent graph, $y = x^2$. Remind students that the point or points shown on the graph must make the equation true. For Exercise 22, $8 \neq 2^2$, so $y = x^2$ is not the correct equation. $[y = 2x^2]$

The answers to Exercises 8, 10–12, and 17–21 can be found in Additional Answers beginning on page 729.

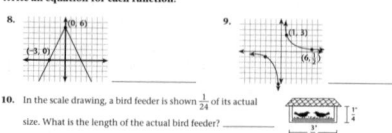
EXERCISES & PROBLEMS

Communicate

1. Explain the pattern of changes to the graph of $y = ax^2$ for different values of a from 1 to 4.

2. Explain the pattern of changes to the graph of $y = ax^2$ for different values of a from 1 to 0.

3. Describe what is meant by vertical stretch.

4. Explain the pattern of changes to the graph of $y = b\left(\frac{1}{x}\right)$ for different values of b from 1 to 4.

5. Explain the pattern of changes to the graph of $y = b\left(\frac{1}{x}\right)$ for different values of b, such as $b = \frac{1}{2}, \frac{1}{3},$ and $\frac{1}{4}$.

6. If E is kinetic energy, m is mass, and v is velocity, explain the effect that $\frac{m}{2}$ has on the graph of $E = v^2$.

The formula for kinetic energy is $E = \frac{m}{2} v^2$.

Practice & Apply

Given the function $y = 5x^2$,

7. identify the parent function. $y = x^2$

8. sketch the graphs of the function and the parent function on the same axes.

9. How does the graph of $y = 5x^2$ compare to the graph of the parent function? vertical stretch of $y = x^2$ by a factor of 5

Graph each of the following.

10. $y = 2\,|x|$ 11. $y = 0.5\,x^2$ 12. $y = 10\,\text{INT}(x)$

True or false?

13. The y-values of $y = 5\,|x|$ are five times as large as the y-values of $y = |x|$. T

14. The x-values of $y = 5\,|x|$ can be $\frac{1}{5}$ as large as the x-values of $y = |x|$ in order to get the same y-values. T

Given the function $y = \dfrac{1}{2x}$,

15. identify the parent function. $y = \frac{1}{x}$

16. sketch the graphs of the function and the parent function on the same axes.

17. tell how the graph of the parent function was changed by the 2.

16.

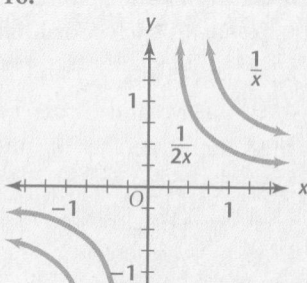

26. $y = \text{ABS}(x)$

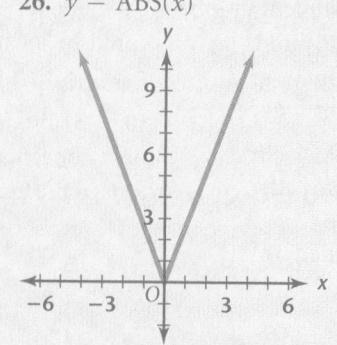

The graph of $y = \text{ABS}(3x)$ has been stretched by a factor 3.

Which functions result in stretches of the parent function?

18. $y = 3x^2$
stretch; 3

19. $y = |x| + 1$
no stretch

20. $y = \dfrac{x^2}{5}$
stretch; $\frac{1}{5}$

21. $y = \dfrac{5}{x}$
stretch; 5

Write an equation for each function.

22.

$y = 2x^2$

23.

$y = \frac{1}{4}x^2$

24. Art Pictures can be enlarged by using a grid. Cut a cartoon or other figure from a newspaper or magazine. Draw a grid on your picture. Make a larger grid off to the side and use it to make an enlargement of the picture.   Answers may vary. Check student's work.

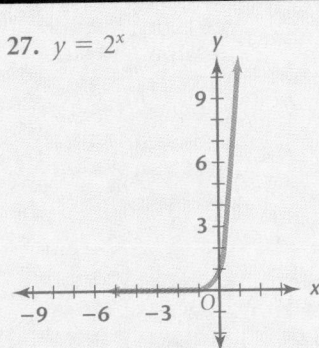

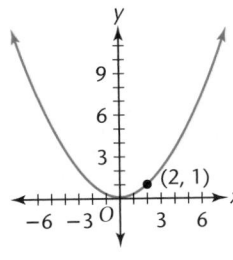

25. The workers in an electronics plant got an "across the board" raise of 5%. Find the scale factor, the amount their salaries were "stretched."
HINT: Be careful. It isn't 0.05.   1.05

For each function, identify the parent function, draw the graph, and tell how the graph of the parent function is transformed by the 3.

26. $y = ABS(3x)$

27. $y = 2^{3x}$

28. $y = \left(\dfrac{x}{3}\right)^2$

27. $y = 2^x$

28. $y = x^2$

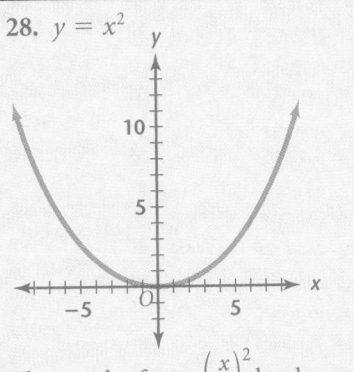

The graph of $y = 2^{3x}$ has been stretched by a factor of 3.

The graph of $y = \left(\dfrac{x}{3}\right)^2$ has been stretched by a factor of $\frac{1}{9}$.

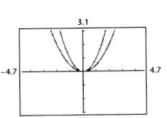

TEST SCORE DISTRIBUTION

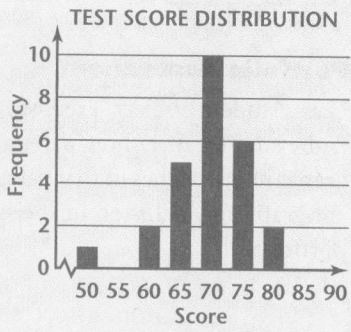

29. **Statistics** Draw a bar graph for this frequency distribution of Ms. Thompson's test.

Score	50	55	60	65	70	75	80	85	90	95	100
Frequency	1	0	2	5	10	6	2	0	0	0	0

30. Because the students didn't have enough time and the scores were low, Ms. Thompson decided to add 20 points to every student's score. Draw the new bar graph, and explain how it compares with the first graph.

31. Ms. Thompson also considered multiplying every student's original score by $\frac{5}{4}$. Draw the new bar graph, and explain how it compares with the graph you drew in Exercise 30.

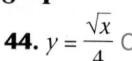 **Look Back**

32. Which number indicates the strongest correlation? **[Lesson 1.6]** d
   a. 0.57   **b.** 0.86   **c.** −0.01   **d.** −0.91

33. The cost for 2 assistants both working 12 hours of part-time work was $129.60. One was paid $0.30 more per hour because of experience. What is the hourly pay for each of the workers? **[Lesson 4.6]** $5.25; $5.55

34. Solve the equation $5x - (7x + 4) = 3(x + 5) + 4(5 - 2x)$. **[Lesson 4.6]** 13

Solve the systems of linear equations. **[Lessons 6.2, 6.3]**

35. $\begin{cases} 3x + 7y = -6 \\ x - 2y = 11 \end{cases}$ **36.** $\begin{cases} 5y - 3x = -31 \\ 4y = 16 \end{cases}$ **37.** $\begin{cases} 1.25x + 2y = 5 \\ 3.75x + 6y = 15 \end{cases}$

   (5, −3)   (17, 4)   infinite

38. The menu of a restaurant has 4 different burgers, 6 different soft drinks, and 2 extras, French fries or onion rings. How many different combinations of lunch can you select if each lunch consists of a burger, drink, and 1 extra? **[Lesson 8.5]** 48

Probability A spinner on a hexagonal base has 1-6 as possible numbers. After the spinner determines the first number, a coin is tossed. **Complete Exercises 39-40.**

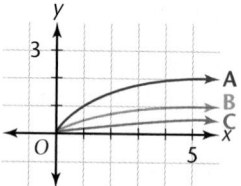

39. What are the possible combinations of numbers and letters, H and T, that could appear? **[Lesson 8.5]**
   1H, 1T, 2H, 2T, 3H, 3T, 4H, 4T, 5H, 5T, 6H, 6T

40. What is the probability of selecting any one of the pairs? **[Lesson 8.6]** $\frac{1}{12}$

41. Two sets of numbers form the ordered pairs (3, 7), (5, −1), (1, 0), (2, 5), (16, −6), (0, 15), (1, −1), and (−2, 5). Is this relation a function? Why? **[Lesson 9.1]** no; 1 is paired with 0 and −1.

Look Beyond

Exercises 42−44 foreshadow working with radical equations, which will be studied in subsequent chapters.

Look Beyond

Match each equation with its graph.

42. $y = \sqrt{x}$ A   **43.** $y = \sqrt{\dfrac{x}{4}}$ B   **44.** $y = \dfrac{\sqrt{x}}{4}$ C

30.

TEST SCORE DISTRIBUTION

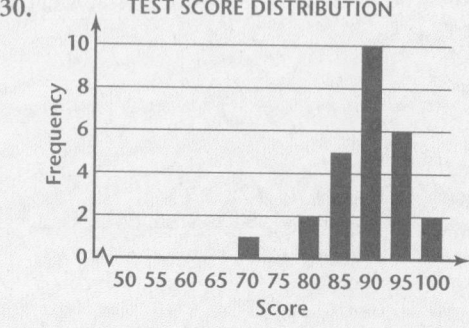

31.

TEST SCORE DISTRIBUTION

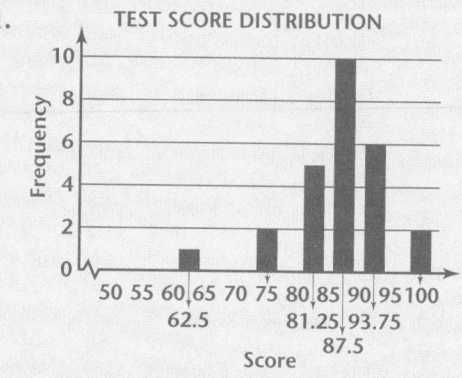

LESSON 9.4 Reflections

  A mirror reflects a figure as its image on the other side. The graph of a function can be reflected through an axis or a given line. This reflection can be represented by a function rule.

Exploration A Reflection

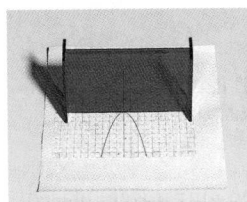

1 Use the values -2, -1, 0, 1, and 2 for x in this exploration. Make a table of values, and graph the parent function $f(x) = x^2$.

2 Reflect f through the x-axis. That is, create a mirror image of f as if the mirror were placed on the x-axis. Let g represent the function for this reflected image.

3 Make a table of values for the reflected function g. Use the same first coordinates or x-values that you used to graph f.

4 Compare the tables for the two graphs. The point $(2, 4)$ is on the parent graph. What is the y-value on the reflected graph when x is 2? If $(-3, 9)$ is a point on the parent graph, what is the corresponding y-value on the reflected graph?

5 Evaluate $g(x)$ for the given values of x.

   a. $g(3) = \underline{?}$ **b.** $g(0) = \underline{?}$ **c.** $g(-3) = \underline{?}$ **d.** $g(a) = \underline{?}$

6 Follow the same procedure for the following functions.

   a. $f(x) = |x|$

   b. $f(x) = 2^x$

   c. $f(x) = -2x^2$

Examine the graphs, and describe in general the relationship between the graph of the function and the graph of its reflection.

7 Suppose the graph of $y = f(x)$ is reflected through the x-axis. What is the function for the reflected graph? ❖

ALTERNATIVE teaching strategy

Using Models Provide students with a full sheet of grid paper and a piece of tracing paper. Have students draw axes on the paper so that the origin is near the center. Tell students to draw a figure in the first quadrant. Have students carefully fold the grid paper along the x-axis. Now place the tracing paper over the grid, so that the bottom aligns with the x-axis, and trace the y-axis and the figure they drew. Have students carefully fold the grid paper, keeping the tracing paper in place. Now unfold the paper, but with the tracing paper on the bottom half of the grid. If the tracing paper slips out, have students try again. The image on the tracing paper now shows the reflection of the figure over the x-axis.

PREPARE

Objectives

- Describe the effect of a reflection on a graph of a function.
- Identify the minimum or maximum value for absolute value and quadratic functions.

RESOURCES

- Practice Master   **9.4**
- Enrichment Master   **9.4**
- Technology Master   **9.4**
- Lesson Activity Master   **9.4**
- Quiz   **9.4**
- Spanish Resources   **9.4**

Assessing Prior Knowledge

Graph.

   1. $f(x) = (x^2 + 1)$
   2. $g(x) = -(x^2 + 1)$

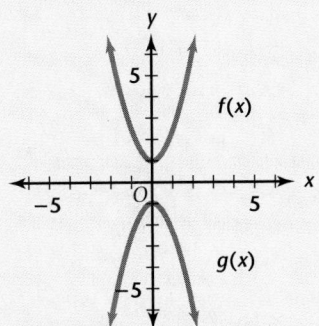

Exploration 1 Notes

Make sure students include y-values from 9 to -9 on their graph. You may want to supply groups of students with Miras® or hand mirrors so that they may experiment before actually drawing the reflected images.

Ongoing ASSESSMENT

   7. $y = -f(x)$

CRITICAL
Thinking

Compare the graphs of $y = x$ and $y = -x$. Describe the transformation. How does the graph compare with $y = |x|$?

EXAMPLE 1

Describe each graph. **A** $y = -|x|$ **B** $y = -3x^2$

Graphics Calculator

Solution ➤

A The parent function is $y = |x|$. The *y*-value of every point on the graph of the parent function changes to its opposite. Thus, the graph of $y = -|x|$ is a reflection of the graph of $y = |x|$ through the *x*-axis.

Notice that the vertex of the graph of $y = -|x|$ is now a maximum.

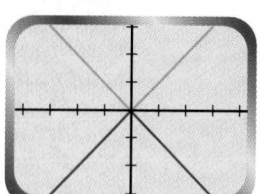

B The parent function is $y = x^2$. First, every point on the graph of the parent function is stretched vertically by a scale factor of 3. Then the negative sign in $y = -3x^2$ indicates a reflection of the graph $y = 3x^2$ through the *x*-axis.

The parabola for $y = -3x^2$ opens downward, so the vertex of the parabola is the maximum point on the graph. ❖

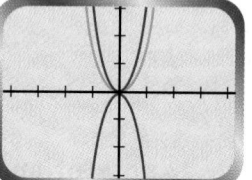

A **vertical reflection** through the *x*-axis results when each point (a, b) of the original graph is replaced by the point $(a, -b)$. Thus, the function $y = f(x)$ is transformed to $y = -f(x)$.

Let $f(x) = x + 2$. If a mirror image of *f* is created as if the mirror were placed on the *y*-axis, *g* would represent the function for the reflected image.

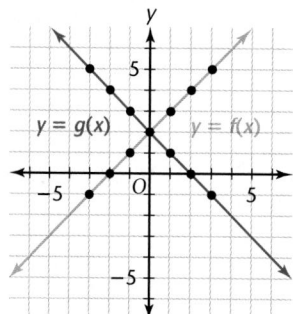

x	y = f(x)
3	5
2	4
1	3
0	2
−1	1
−2	0
−3	−1

x	y = g(x)
−3	5
−2	4
−1	3
0	2
1	1
2	0
3	−1

$(3, 5)$ is replaced by $(-3, 5)$.

A **horizontal reflection** through the *y*-axis results when each point (a, b) of the original graph is replaced by the point $(-a, b)$. Thus, the function $y = f(x)$ is transformed to $y = f(-x)$.

EXAMPLE 2

Identify which graphs are vertical reflections of parent graphs.

Ⓐ $y = -2^x$ **Ⓑ** $y = x^2 - 2$ **Ⓒ** $y = -\frac{1}{x}$

Solution ➤

Ⓐ The values of -2^x are opposite those for 2^x. This is a reflection through the x-axis of the parent graph $y = 2^x$.

Ⓑ This is not a reflection because -2 is *added* to x^2.

Ⓒ The y-values of $y = -\frac{1}{x}$ are the opposite of the y-values of the function $y = \frac{1}{x}$. This is a reflection through the x-axis of $y = \frac{1}{x}$. ❖

Remember, a reflection occurs when the values of the function are replaced by their opposites. For this reason, look for reflections when negative signs affect the function values. Plot a few points to check.

SUMMARY

To graph a vertical reflection

1. identify the parent function and
2. replace the rule with its opposite.

The graph of the parent function will be reflected through the x-axis.

EXERCISES & PROBLEMS

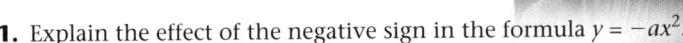

Communicate

1. Explain the effect of the negative sign in the formula $y = -ax^2$.

2. If $h(x) = a\frac{1}{x}$ is reflected through the x-axis, how would you change the original function to express the reflection?

3. Describe how to graph the reflection of the function $f(x) = 2x$ through the x-axis.

4. How does a reflection of $g(x)$ through the x-axis change the formula of the function?

5. What is the effect of reflecting the function through the y-axis when the function is symmetric about the y-axis? Give an example.

6. Identify the axis through which the graph $y = -(x)^2$ and the graph $y = (-x)^2$ are reflected. What effect does the location of the negative sign have on the reflection of the parent function?

7. Why are left and right reversed in a mirror image?

Error Analysis

For exercises like Exercise 13, students graph $y = \text{INT}(-x)$ rather than $y = -\text{INT}(x)$. Have students graph pairs of functions, such as those given, to see that the graphs are different. Caution students to carefully enter the exact function desired.

The answers to Exercises 12–18 and 21 can be found in Additional Answers beginning on page 729.

Practice & Apply

8. A vertical reflection is through which axis?   x-axis   y-axis
9. A horizontal reflection is a reflection through which axis?
10. If you see the reflection of a mountain in a lake, is this a vertical reflection or a horizontal reflection?   vertical
11. If the right half of a fir tree is the mirror image of the left half, is this a vertical reflection or a horizontal reflection?   horizontal

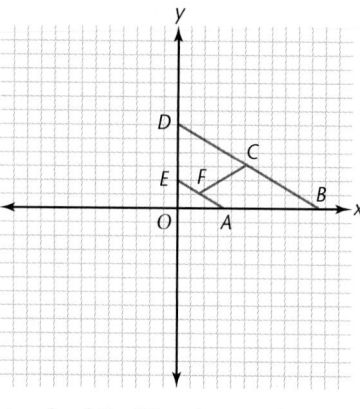

Graph each of the following.

12. $y = -x^2$ **13.** $y = -\text{INT}(x)$ **14.** $y = -2^x$
15. Compare the graphs of $y = -\dfrac{1}{x}$ and $y = \dfrac{1}{-x}$. Explain why they are the same or different.
16. What happens to the graph of $y = |x|$ when you reflect it through the x-axis? through the y-axis? Explain.
17. Graph $f(x) = 2x + 1$.   **18.** For $f(x) = 2x + 1$, find $f(3)$.   7

In each of the following, find $f(3)$, and use this information to tell which kind of reflection (horizontal or vertical) is applied to $y = 2x + 1$.

19. $y = -(2x + 1)$.
−7; vertical
20. $y = 2(-x) + 1$
−5; horizontal
21. Show the result when a parabola is reflected vertically, then horizontally.
22. How can you describe the two reflections in Exercises 19 and 20 as a single transformation?   a rotation of 180°
23. Draw a sketch that reflects (2, 5) into (5, 2), (10, 3) into (3, 10), etc. What happens to (a, b)?   It becomes (b, a).

Stone figure in Kong Woods, family cemetery of Confucius, a famous Chinese philosopher

24. Cultural Connection: Asia

On a piece of graph paper, construct the line segments shown in bold. Point E is (0, 2), and D is (0, 6). Segment EA is twice the length of segment OE, and DB is twice OD. The segment FC connects the midpoints of the segments EA and DB. Reflect these lines through both the x- and y-axes into all 4 quadrants to complete the pattern. This basic pattern appears in a carving found in a tomb of the Han dynasty in China (206 B.C.E. to 222 C.E.). When the segments are reflected through both axes the shape is said to have *bilateral symmetry*.

25. Explain how the symmetry of a parabola can help you graph $y = x^2$ using fewer points.

Practice & Apply
9.4 Reflections

1. Draw $\triangle A'B'C'$, the reflection of $\triangle ABC$ through the y-axis, on the same grid.

Write an equation for each graph, which is a reflection of the given parent function.

2. parent: $y = |x|$ **3.** parent: $y = 2^x$
(-4, 16)
(0,1)

4. parent: $y = \text{INT}(x)$ **5.** parent: $y = x^2$

6. An eyeglass lens bends the rays of light entering the eye. The ability to bend light rays is measured in dioters d, and depends on the focal length f, of the lens. Write the function that is modeled by the given set of data.

d	400	200	100	50	25	2.5	6.25
f	0.0025	0.005	0.01	0.02	0.04	0.08	0.16

24.

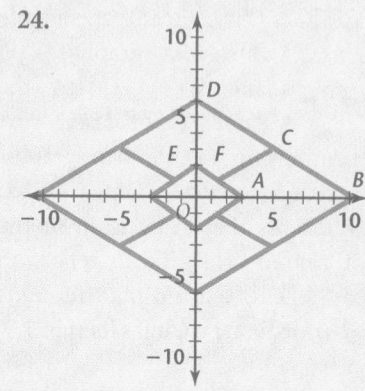

A (3.46, 0)
B (10.39, 0)
C (4.90, 3)

D (0, 6)
E (0, 2)
F (1.73, 1)

25. Since the y-values are the same for a given x on either side of the y-axis, plot a y-value for a positive x and use that same y-value for negative x.

Look Back

26. Physics If the speed of sound through air is approximately 330 meters per second, how far away did the lightning strike? **[Lesson 2.3]**
2310 meters

27. Geometry The length of a rectangular garden needs to be 3 times the width. If the perimeter of the rectangular garden is 72 meters, what is the area? **[Lesson 4.6]**
243m²

Solve for the unknown. [Lesson 4.6]

28. $6(3 - 2r) = -(5r + 3)$ 3   **29.** $\frac{x}{2} + 5 = \frac{3x - 2}{4}$ 22   **30.** $3x = -9(x - 4)$ 3

31. Graph the solution set for $x \geq 3$ or $x < -1$ on a number line. **[Lesson 4.7]**

32. **Statistics** Quality control measures 5 random samples of glass rods in a 15-minute production run. What is the average length of the glass rods tested in this run based on the following sample measurements? **[Lesson 8.3]** 23.74 cm

Sample number	1	2	3	4	5
Measurement in cm	23.5	23.9	23.6	24.0	23.7

33. A number cube and a number tetrahedron (4 faces) are used for a probability experiment. How many possible outcomes are there when both are rolled once? **[Lesson 8.6]** 24

Look Beyond

Match each of these functions with the appropriate graphs.

34. $y = \sqrt{x}$ B   **35.** $y = \sqrt{-x}$ A   **36.** $y = -\sqrt{x}$ D   **37.** $y = -\sqrt{-x}$ C

A.
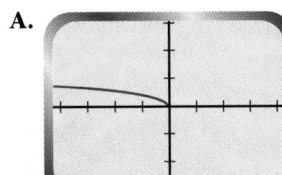

B.
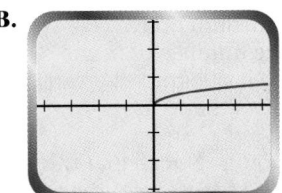

C.
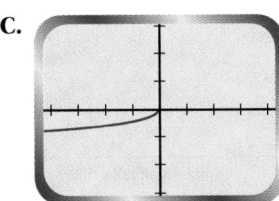

D.
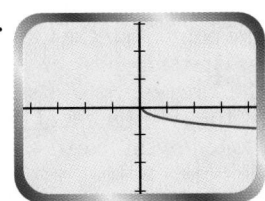

38. Graph $y = 5 - x^2$.

31.
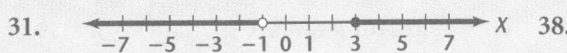

38.
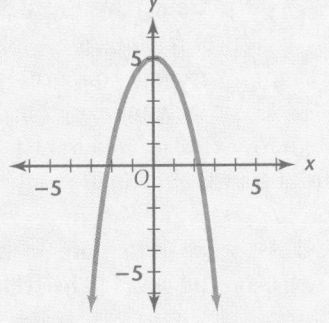

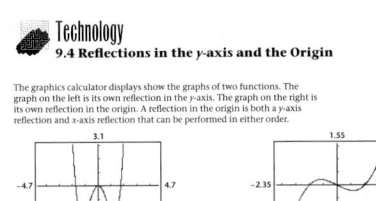

- Describe the effect of a translation on the graph of a function.
- Identify the relationship between the translation of a graph and the addition or subtraction of a constant to the function.

RESOURCES

- Practice Master   **9.5**
- Enrichment Master   **9.5**
- Technology Master   **9.5**
- Lesson Activity Master   **9.5**
- Quiz   **9.5**
- Spanish Resources   **9.5**

Assessing Prior Knowledge

Graph the function and sketch the graph of its reflection through the *x*-axis.
$f(x) = 2x^2$

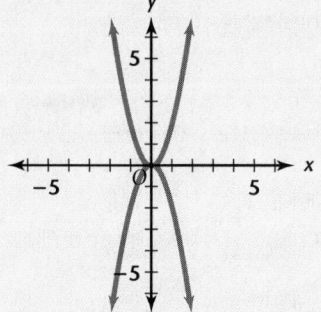

TEACH

A translation preserves the shape of an original drawing. Ask students to discuss other transformations that preserve shape.

LESSON 9.5 Translations

In 1960 Escher designed this tiled facade for the hall of a school in The Hague. It was executed in concrete in two colors.

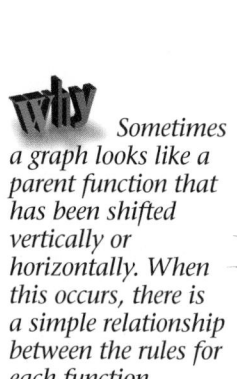

Why *Sometimes a graph looks like a parent function that has been shifted vertically or horizontally. When this occurs, there is a simple relationship between the rules for each function.*

A transformation that shifts the graph of a function horizontally or vertically is called a **translation**. In the Escher facade point *A* is translated to point *B* by shifting point *A* to the right.

Compare the graph of $E(x) = |x - 10|$ with the graph of its parent function.

Solution ➤

To graph the function $E(x) = |x - 10|$, start with a table of values. Include the number 10 in the table, as well as numbers greater than and less than 10.

Plot the points. Notice that the function is a variation of the absolute value function $y = |x|$.

Connect the points to make the familiar V-shaped graph. The graph of the function, $E(x) = |x - 10|$, is the graph of its parent function shifted 10 units to the right. ❖

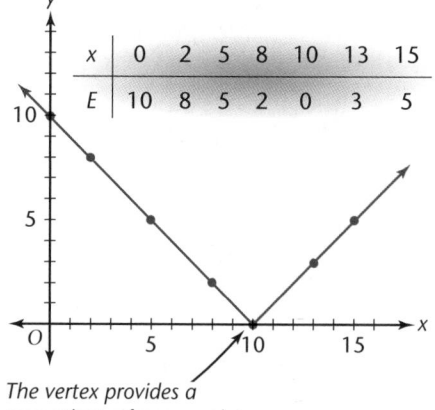

x	0	2	5	8	10	13	15
E	10	8	5	2	0	3	5

The vertex provides a convenient reference point.

ALTERNATIVE **teaching strategy**

Using Manipulatives
Ask students to place a triangle on grid paper with the base on the *x*-axis and the third vertex in Quadrant I. Have students slide or translate the figure so that one side is still on the *x*-axis, but the third vertex is in Quadrant II. Ask students to compare the triangles and their coordinates. [**The triangles** are congruent and have the same orientation or position, but have a different location. The *x*-coordinates change, but the *y*-coordinates remain the same.]

Repeat the procedure, but this time the side on the *y*-axis stays in position. [**The x-coordinate stays the same and the y-coordinate changes.**]

Vertical Translation

A transformation that shifts the graph of a function up (positive direction) or down (negative direction) is a **vertical translation**.

EXAMPLE 2

Compare the graph $y = x^2 - 3$ with the graph of the parent function $y = x^2$. In what direction does the translation shift the vertex? How far?

Solution ➤

Evaluate $h(x) = x^2 - 3$ for several values of x, and plot the graph with $h(x)$ on the y-axis.

$h(x) = x^2 - 3$		
x	$(x)^2 - 3$	$h(x)$
-2	$(-2)^2 - 3$	1
-1	$(-1)^2 - 3$	-2
0	$(0)^2 - 3$	-3
1	$(1)^2 - 3$	-2
2	$(2)^2 - 3$	1

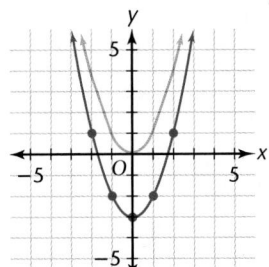 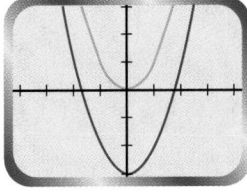

The vertex is vertically shifted downward 3 units. ❖

Try This Sketch the graph of the function $y = |x| + 4$.

VERTICAL TRANSLATION

The graph of $y = f(x) + k$ is translated vertically by k units from the graph of $y = f(x)$.

The following translations show how the formula applies to basic parent functions. The graph of a function shifts vertically when you add a positive or negative constant to the function.

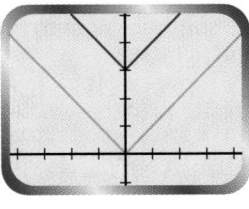

$y = |x| + 3, (k = 3)$

The graph is shifted up 3 units from the graph of $y = |x|$.

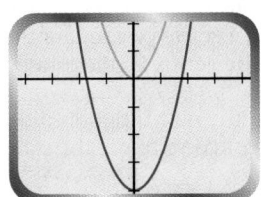

$y = x^2 - 4, (k = -4)$

The graph is shifted down 4 units from the graph of $y = x^2$.

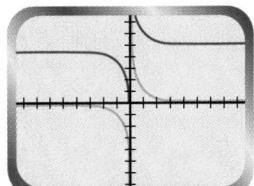

$y = \frac{1}{x} + 5, (k = 5)$

The graph is shifted up 5 units from the graph of $y = \frac{1}{x}$.

---

interdisciplinary

CONNECTION

Art The Dutch artist Mauritz Escher (1898–1972) utilized transformations in much of his artwork. Display Escher prints, if they are available. Have students discuss what transformations are used. Have students find other examples of art that involve transformations and bring examples to class if possible.

Alternate Example 1

Use a graphics calculator to compare these graphs.

1. $f(x) = |x + 1|$
2. $g(x) = |x + 2|$
3. $h(x) = |x + 3|$
4. $k(x) = |x + 4|$

[The graphs are the same, except that they are moved to the left.]

Alternate Example 2

Graph $f(x) = -x^2$ and $g(x) = -x^2 + 2$ and describe the graphs. [The second graph is the same as the first, except that it is translated up 2 units.]

Aongoing
ASSESSMENT

Try This

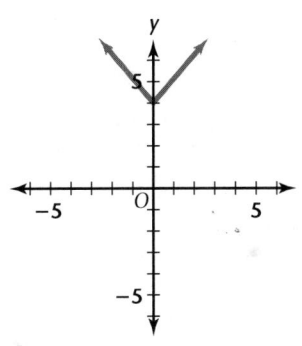

TEACHING *tip*

Technology To duplicate the graphs at the bottom of page 437, students will have to change the range values on their calculators. First graph:
Xmin $= -1$, Xmax $= 5$,
Ymin $= -5$, Ymax $= 5$
Second graph:
Xmin $= -5$, Xmax $= 5$,
Ymin $= -5$, Ymax $= 2$
Third graph:
Xmin $= -15$, Xmax $= 15$,
Ymin $= -10$, Ymax $= 10$
(All have Xscl and Yscl $= 1$.)

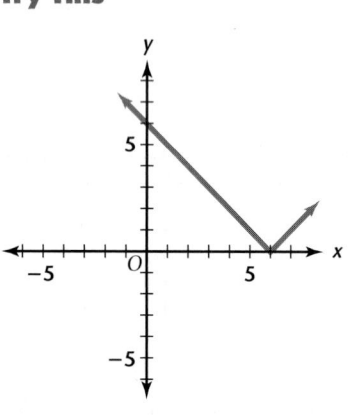
Horizontal Translation

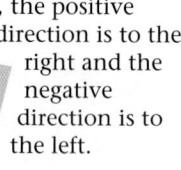

Horizontal translation is the transformation that shifts the function left or right. Remember, the positive direction is to the right and the negative direction is to the left.

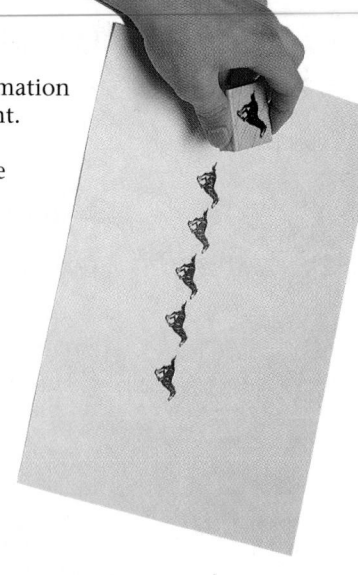

EXAMPLE 3

Compare the graph of $y = (x - 3)^2$ with the graph of its parent function. Which direction does the translation shift the vertex and by how much?

Solution ➤

Graph the parent function $y = x^2$. The transformed function, $y = (x - 3)^2$, can be interpreted, "Subtract 3 from x, then square the quantity." The table shows some of the values of this function. Sketch this function and the parent function on the same axes.

$y = (x - 3)^2$		
x	$(x - 3)^2$	y
-2	$(-2 - 3)^2$	25
-1	$(-1 - 3)^2$	16
0	$(0 - 3)^2$	9
1	$(1 - 3)^2$	4
2	$(2 - 3)^2$	1
3	$(3 - 3)^2$	0
4	$(4 - 3)^2$	1

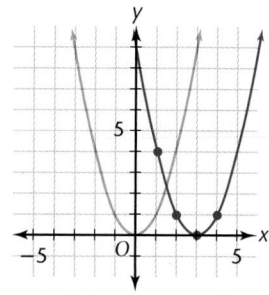

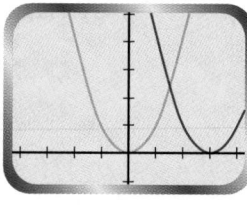

Notice that a translation does not change the shape of the parabola. If you subtract 3 from x *before* you square the quantity, you shift the graph to the right 3 units. The vertex is horizontally shifted to the right 3 units.

The x-values in the new function must be 3 units greater than the x-values for the parent function to get the same y-value. ❖

Try This Sketch the graph of $y = |x - 6|$.

HORIZONTAL TRANSLATION

The graph of $y = f(x - \mathbf{h})$ is translated horizontally by h units from the graph of $y = f(x)$.

438 Lesson 9.5

The following examples show how the graphs and the formulas represent the horizontal translation of various parent functions.

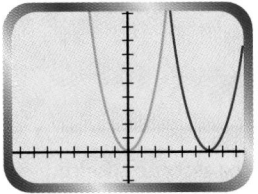

$y = (x - 6)^2,\ (h = 6)$

The graph shifts right 6 units from the graph of $y = x^2$.

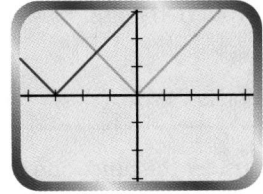

$y = |x + 3|,\ (h = -3)$

The graph shifts left 3 units from the graph of $y = |x|$.

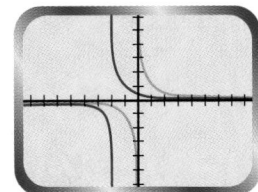

$y = \dfrac{1}{x + 2},\ (h = -2)$

The graph shifts left 2 units from the graph of $y = \dfrac{1}{x}$.

EXAMPLE 4

Which of the functions represents translations of the parent graph?

A $y = x^2 + 3$ **B** $y = \dfrac{3}{4} + 5^x$ **C** $y = \dfrac{|x|}{9}$ **D** $y = \dfrac{1}{x - 10}$

Solution ▶

A Since 3 is added to x^2, the parent function, $y = x^2$, is translated upward by 3 units.

B Since $\dfrac{3}{4}$ is added to 5^x, the parent function is translated upward by $\dfrac{3}{4}$ of a unit.

C Since $|x|$ is divided by 9, this is a stretch. It is not a translation.

D Since 10 is subtracted from x in $\dfrac{1}{x - 10}$, this is a horizontal translation of $y = \dfrac{1}{x}$ to the right by 10 units. ❖

EXAMPLE 5

STATISTICS
Connection

Three salespeople have salaries of $21,000, $24,000, and $30,000. What happens to the average salary if each gets a $3000 raise?

Solution ▶

The average salary before the raise is
$\dfrac{21,000 + 24,000 + 30,000}{3} = 25,000.$

If each gets a $3000 raise, the average will be $\dfrac{24,000 + 27,000 + 33,000}{3} = 28,000.$

If each salary increases by $3000, the average will increase by $3000. In general, if each salary is translated by a number, k, the average is translated by the same number, k. ❖

Technology For the graphs at the top of Page 439, set these range values:
First graph:
Xmin $= -10,$ Xmax $= 10,$
Xscl $= 2$
Ymin $= -2,$ Ymax $= 10,$
Yscl $= 2$
Second and third graphs:
Xmin $= -5,$ Xmax $= 5,$
Xscl $= 1$
Ymin $= -3,$ Ymax $= 3,$
Yscl $= 1.$

Alternate Example 4
Which of the functions represents a translation of the parent graph? Answer yes or no.

a. $y = 3^x$ [No]
b. $y = x^2 - 2$ [Yes]
c. $y = \dfrac{1}{x - 2}$ [Yes]
d. $y = \dfrac{1}{4} - x^2$ [No]

Alternate Example 5
The sticker prices of four new cars are $12,000, $13,500, $10,500, and $16,000. What happens to the average sticker price if all prices are raised $750? [**The average is also raised $750.**]

The bar chart (Statistics Connection) shows values:
35,000 / 30,000 / 25,000 / 20,000 / 15,000 / 10,000 / 5,000 / 0 for Bill, Alan, Rene.

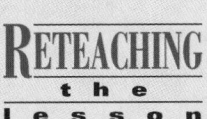

RETEACHING **the lesson**

Have students graph $x = y$ and $x = y - 2$ on the same set of axes. Ask how the y-intercept is affected. [**It moves down 2 units.**] Ask how the x-intercept is affected. [**It moves right 2 units.**] Have students discuss the similarities of the changes in the x- and y-coordinates and what happens in the vertical or horizontal translation of a parabola. [**If h and k are positive, then in $y = f(x) + k$, the y-coordinate shifts down k units and the x-coordinate in $y = f(x - h)$ shifts right h units.**]

Provide each group with two different coins and a standard number cube.

Coin 1: heads = vertical shift, tails = horizontal shift

Coin 2: heads = x^2 function, tails = $|x|$ function

The number cube determines the number of units in the shift. Each player, in turn, flips both coins and rolls the number cube. Then each student will write an equation based on the results and sketch the graph of the equation.

Score:
Each correct equation is worth 1 point. Each correct graph is worth 1 point. The student with the most points at the end of 5 rounds wins.

ongoing
ASSESSMENT

The entire graph moves down by 5.

Practice Master

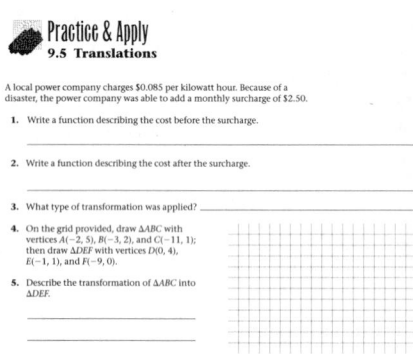

Practice & Apply
9.5 Translations

A local power company charges $0.085 per kilowatt hour. Because of a disaster, the power company was able to add a monthly surcharge of $2.50.

1. Write a function describing the cost before the surcharge.

2. Write a function describing the cost after the surcharge.

3. What type of transformation was applied? _____

4. On the grid provided, draw △ABC with vertices A(−2, 5), B(−3, 2), and C(−11, 1); then draw △DEF with vertices D(0, 4), E(−1, 1), and F(−9, 0).

5. Describe the transformation of △ABC into △DEF.

Note that the point (1, 2) is on the graph of the parent function $f(x) = 2^x$. Use this fact to tell which way the parent function is translated in each of the following.

6. $f(x) = 2^x − 1$ _____   7. $f(x) = 2^x + 4$ _____

8. $f(x) = 2^{(x−1)}$ _____   9. $f(x) = 2^{(x+2)}$ _____

10. Scores on a national exam were adjusted according to the given table. Write an equation and describe the transformation modeled by the data.

Original Score	30	35	40	45	50	55	60
New Score	38	43	48	53	58	63	68

11. Write an equation for the given graph.

SUMMARY

A translation of the parent function $y = f(x)$ occurs if a constant is added or subtracted from:

1. the variable x. $y = f(x − h)$

2. the function value $f(x)$. $y = f(x) + k$

For example, consider the function $y = (x − 5)^2$. The parent function is $y = x^2$. When x is 5, $(x − 5)^2$ is 0. Thus, the vertex (0, 0) of $y = x^2$ has been shifted to the right 5 units. What happens to the vertex of $y = x^2$ when $y = x^2 − 5$?

Cultural Connection: Africa The designs of African art for pottery and fabric show the use of translation and reflection for decoration. Such examples show that geometric transformations have existed for centuries in this as well as in many other cultures. Algebra has created a way to represent these geometric features by using symbols and expressions.

Contrast the two earthenware vessels. The one in the background dates from the early Twentieth Century, the other from 1200 C.E.

EXERCISES & PROBLEMS

Communicate

1. What is meant by a *translation* in mathematics?

2. A transformation of the parent quadratic function is $f(x) = (x − 3)^2 + 5$. What value tells you the amount and direction of the *horizontal* translation?

3. A transformation of the parent absolute value function is $g(x) = |x − 3| + 5$. What value tells you the amount and direction of the *vertical* translation?

4. Explain why the graph of $f(x) = (x + h)^2$ is translated horizontally when a constant, h, is added to x before the quantity is squared.

5. Explain why the graph of $f(x) = x^2 + k$ is translated vertically when a constant, k, is added after the variable is squared.

6. What is the relationship of the numbers 4 and 3 in the formula, and what is the location of the vertex of the graph for $y = (x − 4)^2 + 3$?

Practice & Apply

Identify the parent function. Tell what type of transformation is applied to each, then draw a carefully labeled graph. HINT: Remember to plot a few points to help see what happens to the parent function.

7. $y = -x^2$ **8.** $y = (-x)^2$ **9.** $y = x^2 + 3$ **10.** $y = (x + 3)^2$

11. $y = \text{INT}\left(\dfrac{x}{2}\right)$ **12.** $y = \text{INT}(-x)$ **13.** $y = \text{ABS}(x) - 1$ **14.** $y = \text{ABS}(x - 1)$

Consider the parent function $f(x) = x^2$. Note that $f(3) = 9$, so the point $(3, 9)$ is on the graph. In each of the following, use the given fact to tell which way the parent function is translated.

15. $f(x) = x^2 + 5$ contains the point $(3, 14)$.   **16.** $f(x) = x^2 - 5$ contains the point $(3, 4)$.

17. $f(x) = (x + 5)^2$ contains the point $(-2, 9)$.   **18.** $f(x) = (x - 5)^2$ contains the point $(8, 9)$.

The point $(5, 8)$ is on the graph of $f(x)$. Tell what happens to $(5, 8)$ when each of the following transformations are applied to the function.

19. vertical translation by 6 $(5, 14)$ **20.** vertical translation by -2 $(5, 6)$

21. vertical translation by -10 $(5, -2)$ **22.** vertical stretch by 3 $(5, 24)$

23. horizontal translation by 3 $(8, 8)$ **24.** horizontal translation by -1 $(4, 8)$

25. horizontal translation by -12 $(-7, 8)$ **26.** vertical stretch by 10 $(5, 80)$

Tell whether the given graph is stretched or translated from the graph of the parent function.

27. $y = \dfrac{2}{3}|x|$ stretched   **28.** $y = \dfrac{2}{3} + |x|$ translated

29. What happens to the average salary of the workers in Example 5 if all the salaries were increased by 10%?   increase by 10% to $27,500

30. Suppose one ball is thrown from level ground in a parabolic path as shown. A second ball is thrown in an identical path, but from a point on top of a cliff to a point on another cliff. What transformation relates the higher path to the lower one?

The graphs for Exercises 7–14 can be found in Additional Answers beginning on page 729.

7. $y = x^2$; vertical reflection
8. $y = x^2$; horizontal reflection
9. $y = x^2$; translation up 3 units
10. $y = x^2$; translation left 3 units
11. $y = \text{INT}(x)$; stretched horizontally by 2.
12. $y = \text{INT}(x)$; horizontal reflection through y-axis

13. $y = \text{ABS}(x)$; translation down 1 unit
14. $y = \text{ABS}(x)$; translation right 1 unit
15. translated up 5 units
16. translated down 5 units
17. translated left 5 units
18. translated right 5 units
30. vertical translation

alternative ASSESSMENT

Performance Assessment

Provide each student with a large coordinate grid and a cardboard cutout of the function $y = x^2$ that fits that grid. (See INCLUSION STRATEGIES, page 438.) Write each function on the board and have students show the correct location of the parent graph.

$y = -x^2 + 3$
[shifts 3 units up]
$y = (x - 2)^2$
[shifts 2 units right]
$y = (x + 4)^2$
[shifts 4 units down]
$y = x^2 - 1$
[shifts 1 unit down]

Look Beyond

Exercises 40 and 41 anticipate the graphing of higher degree equations.

Look Back

31. Simplify the expression $-3(8 - 4x) - 5(x + 2)$. **[Lesson 4.2]** $7x - 34$

32. 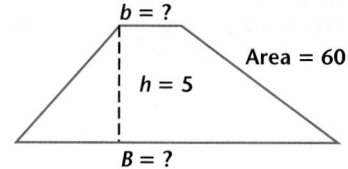 **Geometry** The area of a trapezoid can be determined by the formula $A = \frac{h(B + b)}{2}$. The area, A, of a given trapezoid is 60. The altitude, h, is 5, and the larger base, B, is twice the smaller base, b. What is the length of each base?
[Lesson 4.6] 8; 16

33. Solve for x. $7x - (24 + 3x) = 0$ **[Lesson 4.6]** 6

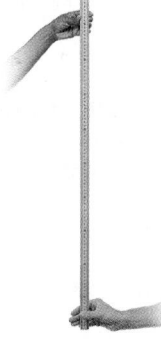

There should be a 4-centimeter distance between thumb and index finger with the ruler centered between them.

Statistics Investigate differences in reaction time. Place a ruler with the zero point between a person's thumb and index finger. Release the ruler. The person should catch the ruler as quickly as possible. Determine the point at which the ruler was caught. **[Lesson 8.3]**

34. Find the average distance along the ruler where most people catch it.

35. Compare the average catch distance along the ruler for groups of males and females. Which group showed the shortest average catch distance? What does this say about the reaction time of each group?

36. Sketch the graph of $y = -\frac{10}{x}$. **[Lesson 9.3]**

37. Determine a formula for the function $f(x)$ that will model the following data. Give the steps you used for finding the formula.
[Lesson 9.3]

x	0	1	2	3	4	5	6	7
f(x)	0	3	12	27	48	75	108	147

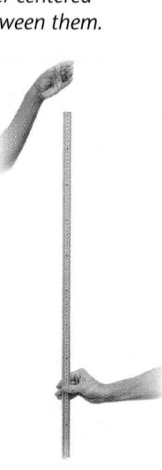

Geometry Notice the triangle on the coordinate graph with vertices at points $A(2, 1)$, $B(4, 4)$, and $C(7, 2)$. **[Lesson 9.4]**

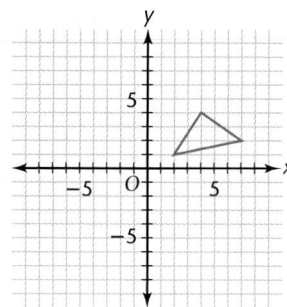

38. Reflect the triangle through the x-axis. What are the coordinates for the vertices of the transformed triangle?

39. Reflect the original triangle through the y-axis, and give the new coordinates.

Look Beyond

Graph the parent function and the following transformations. How is each changed from the basic cubic function $y = x^3$?

40. $y = -x^3$ **41.** $y = (-x)^3$

34–35. Answers may vary, depending on the outcome of the experiment.

36.

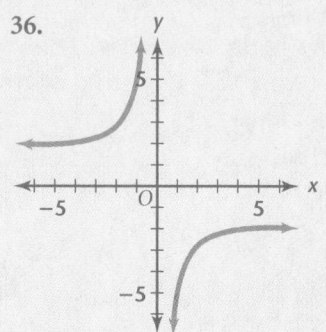

37. $y = 3x^2$; Answers may vary. For example, from examining the first 2 points, y could be $3x$. The third point, however, makes this impossible. Divide both the third and fourth y-values by 3. What remains is the x-value multiplied by itself. Hence, y could be $3x^2$. The fifth and following points confirm that the answer is correct.

The answers to Exercises 38–41 can be found in Additional Answers beginning on page 729.

LESSON 9.6 Combining Transformations

Bob and his friend allow a
total of 2 hours for rest breaks.

 Once
a function
is recognized as
related to a known
parent function,
the graph can be
sketched by
identifying and
performing several
transformations.

Travel Bob and his friends are planning a 100-mile bike ride. How fast must they ride to complete the whole trip in 6 hours? 7 hours? h hours?

To finish the trip in 6 hours, the boys expect to ride 4 hours with 2 hours rest (6 − 2 = 4). That means they would have to ride 100 miles in 4 hours.

$$\text{rate} = \frac{100 \text{ miles}}{4 \text{ hours}} = \frac{25 \text{ miles}}{1 \text{ hour}}, \text{ or } 25 \text{ mph}$$

To finish the trip in 7 hours, they would need to ride for 5 hours and rest for 2 hours. They would have to ride 100 miles in 5 hours.

$$\text{rate} = \frac{100 \text{ miles}}{5 \text{ hours}} = \frac{20 \text{ miles}}{1 \text{ hour}}, \text{ or } 20 \text{ mph}$$

To finish the trip in h hours, they would need to ride h − 2 hours. They would have to ride 100 miles in h − 2 hours.

$$\text{rate} = \frac{100 \text{ miles}}{h - 2 \text{ hours}}, \text{ or } \frac{100}{h - 2} \text{ mph}$$

If R is the rate, and h is the total number of hours for the trip, the rate the boys would have to travel is expressed as

$$R = \frac{100}{h - 2}.$$

ALTERNATIVE teaching strategy **Cooperative Learning** Have students work in small groups. Write the quadratic equation $y = -4(x - 3)^2 + 2$ on the board. Each group should try to graph the equation without a graphics calculator. Each group should record the method they tried and explain how it worked, or why it didn't work. After all groups are finished, have the class discuss the problem as a whole.

PREPARE

Objectives

- Identify the parent function in a transformation.
- Understand the effect of order on combining transformations.

RESOURCES

• Practice Master	9.6
• Enrichment Master	9.6
• Technology Master	9.6
• Lesson Activity Master	9.6
• Quiz	9.6
• Spanish Resources	9.6

Assessing Prior Knowledge

Ask students to describe the second graph, in terms of the first graph, for each pair of equations.

1. $y = x^2$ and $y = x^2 - 3$
   [**The second graph is shifted down 3 units.**]
2. $y = |x|$ and $y = |x + 2|$
   [**The second graph is shifted up 2 units.**]
3. $y = \frac{1}{x}$ and $y = \frac{1}{x - 3}$
   [**The second graph is shifted right 3 units.**]
4. $y = x^2$ and $y = (x - 2)^2$
   [**The second graph is shifted right 2 units.**]

TEACH

 Ask students to look again at some Escher artwork. Ask: Which transformation was used to make the pattern? Students should reply that more than one transformation was used. Tell students that they will learn about combining transformations in this lesson.

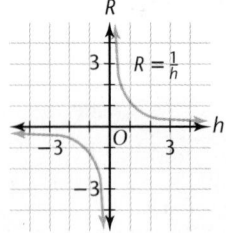
To sketch the graph of a transformed function, begin with the parent function, then perform the individual transformations one at a time.

Sketch the graph of $R = \frac{100}{h-2}$.

First determine the parent function. Since h is in the denominator, you should first consider the parent reciprocal function $R = \frac{1}{h}$. Start with $R = \frac{1}{h}$ and find the transformations needed to end with $R = \frac{100}{h-2}$.

Next determine whether the function has been stretched or reflected. A graph has been stretched when its parent function is multiplied by a nonzero constant. If the constant is negative, the graph is also reflected.

Examine $R = \frac{100}{h}$. Since $R = \frac{100}{h}$ can be written as $100\left(\frac{1}{h}\right)$, the parent function is stretched vertically by 100 and there is no reflection. Notice on the blue and red graphs that the y-axis is labeled in hundreds. Although the graphs of $R = \frac{1}{h}$ and $R = \frac{100}{h}$ look alike, each y-value in the graph of $R = \frac{100}{h}$ is 100 times the amount of the y-value in $R = \frac{100}{h}$.

Then determine whether the parent function has been translated. Remember that the graph is translated when you add or subtract a constant. Since 2 is subtracted from x in the function $y = \frac{100}{h-2}$, the x coordinate needs to be 2 units greater to get the same value as $y = \frac{100}{h}$. This indicates a horizontal translation of 2 to the right.

To summarize, the parent graph, $R = \frac{1}{h}$ is transformed by a vertical stretch of 100 followed by a translation of 2 to the right. The point $\left(3, \frac{1}{3}\right)$ on the original graph, is transformed to the point $\left(5, \frac{100}{3}\right)$.

From the function, determine how fast they must travel to finish in 8 hours.

CRITICAL Thinking

Suppose the function $y = x^2$ represents the area of a square with side x. Interpret what the function $y = 3x^2 + 5$ would indicate about the change in the area of the original square.

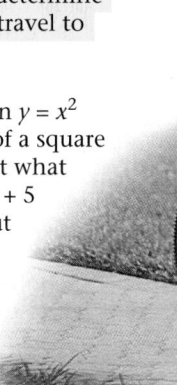

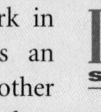

EXAMPLE 1

Sketch the graph of $y = 2|x + 4|$.

Graphics Calculator

Solution ➤

1. The parent graph is $y = |x|$.

2. Since the scale factor is 2, the y-values will be twice as great as the parent function. This is a vertical stretch by 2.

3. Because 4 is added to the variable, the x-values in the transformed function need to be 4 units less to produce the same y-values that the stretched parent function produces. The result is a translation to the left by 4.

The parent graph $y = |x|$ is stretched vertically by 2 and translated to the left by 4.

The vertex is at -4 on the x-axis. Since $y = 2|0| = 0$, $(-4, 0)$ is the new position of the vertex.❖

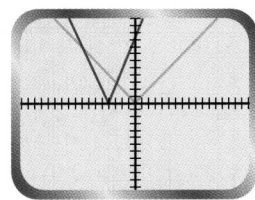

The graphics calculator can plot the parent function and transformation quickly.

Try This Check several points for $y = 2|x + 4|$ to see how steep the V of the graph is. Is the graph symmetric? If so, what is the axis of symmetry?

Reflection   **Horizontal translation**

Summary

For many functions, the graph can be sketched directly from the information that appears in the function rule.

$$y = -a \cdot f(x - h) + k$$

Vertical stretch   **Vertical translation**

EXAMPLE 2

Use the information from the Summary to sketch the graph of the function $y = -(x + 3)^2 - 2$.

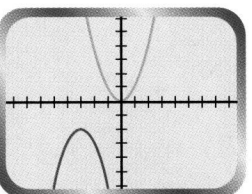

Solution ➤

The parent function is $y = x^2$, so the graph is a parabola. Since the sign is negative, the parabola is reflected through the x-axis. There is no constant following the negative sign. Thus, there is no vertical stretch. The value of h must be -3, since $x - (-3) = x + 3$. This indicates a horizontal translation to the left 3 spaces. Finally, $k = -2$, so the parabola is translated vertically downward 2 spaces. ❖

RETEACHING the lesson

Have students use a geoboard to experiment with different graphs. First, have students use rubber bands to identify the x- and y-axes. Then have them model the function $y = |x|$. Have students translate this graph 1 unit to the right and determine its equation. $[y = |x - 1|]$ Next, have students translate the graph 1 unit to the left and determine its equation. $[y = |x + 1|]$

Repeat this activity for a vertical translation (up and down 1 unit), a reflection over the x-axis and down 1 unit, and a vertical stretch up 1 unit.

$[y = |x| + 1;$
$y = |x| - 1;$
$y = -|x|;$
$y = -|x| + 1;$
$y = 2|x|]$

Alternate Example 1

Sketch the graph of $y = 3x^2 - 2$. Describe the process. [**The parent graph is $y = x^2$. It is stretched by a multiple of 3 and moved 2 units down.**]

A ongoing SSESSMENT

Try This

The graph is symmetric with respect to $x = -4$.

Alternate Example 2

Describe how transformations can be used to sketch the graph of $y = \dfrac{-3}{x - 2} - 1$. [**The parent graph is $\frac{1}{x}$, so start with the graph of a hyperbola. Since the coefficient (-3) is negative, the hyperbola is reflected through the x-axis. The graph is then stretched by a factor of 3. The value of h is 2, so the graph has a horizontal translation of 2 units to the right. Finally, $k = -1$, so the hyperbola is translated vertically downward 1 unit.**]

ASSESS

Selected Answers

Odd-numbered Exercises 7–31

Assignment Guide

Core 1–6, 8–24 even, 25–36

Core Two-Year 1–32

Technology

The use of a graphics calculator is left to the discretion of the teacher for Exercises 7–15. For Exercises 35 and 36, encourage students to use a graphics calculator with the range set from -10 to 10 for both x and y.

Error Analysis

Students may forget reflections or stretches or may confuse vertical and horizontal transformations. Make and display a poster of the summary for $y = \pm af(x - h) + k$.

Have students copy the information onto a notecard to use when they are studying.

Practice Master

Practice & Apply
9.6 Combining Transformations

Tell what happens to the point (3, 8) on $y = 2^x$ when each of the following pairs of transformation are applied in the order given.

1. A vertical stretch by 2 followed by a vertical translation up by 2.
2. A vertical translation up by 2 followed by a vertical stretch by 2.
3. A horizontal translation right by 5 followed by a vertical stretch by 3.
4. A vertical stretch by 3 followed by a horizontal translation right by 5.

In a chain discount, the second discount is taken on the amount left after the first discount is applied. Suppose the price of a watch is $120.

5. Find the price of the watch after a discount of 25% is followed by a discount of 20%.
6. Find the price of the watch after a single discount of 45%.
7. Describe each transformation.
8. Find the single discount that gives the same result as a chain discount of 25% followed by 20%.

Graph each function on the grid provided.

9. $y = \text{INT}(x - 2)$
10. $y = -x + 1$

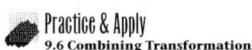

EXERCISES & PROBLEMS

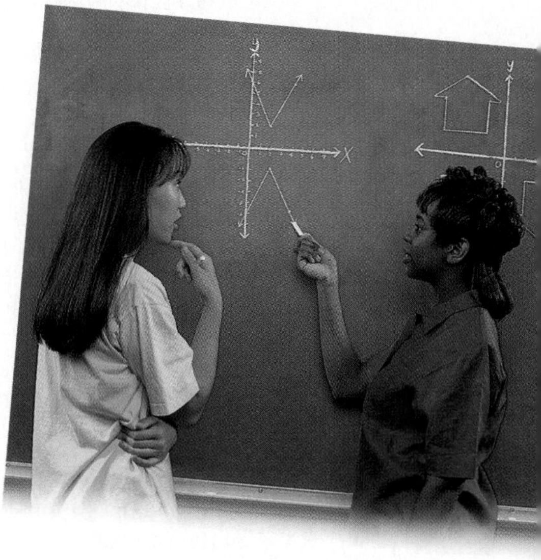

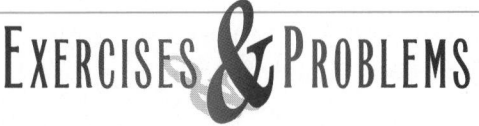

Communicate

1. In the 100-mile bike ride problem, explain how to identify which number in the function represents the translation.
2. In the 100-mile bike ride problem, explain how to identify which number in the function represents a stretch of the graph.
3. Discuss the characteristics of various parent functions that will help you to identify them from their transformed functions.
4. Describe how to write the function for a translation of 4 units to the left if the parent function is the reciprocal function.
5. Given the function $y = -a(x - h)^2 + k$, identify and explain the effect of each transformation from the letters and symbols in the formula.
6. Explain the steps you would perform to sketch the graph of the function $y = |x + 7| - 3$.

Practice & Apply

Sketch the graph of each function.

7. $j(x) = 2(x + 5)^2$
8. $m(x) = 2x^2 + 5$
9. $t(x) = |x + 3| - 4$
10. $v(x) = 10|x - 4|$
11. $f(x) = \frac{3}{x - 2}$
12. $g(x) = \frac{3}{x} - 2$
13. $h(x) = 0.5(2^x) - 1$
14. $z(x) = \frac{3}{4} + \frac{2}{x}$
15. $p(x) = \frac{3}{4} + \frac{2}{4 + x}$

Tell what happens to the point (4, 10) on $y = \frac{5}{2}x$ when each of the following pairs of transformations are applied in the order given.
HINT: If you are not sure, draw a sketch.

16. a vertical stretch by 2, followed by a vertical translation by 3 (4, 23)
17. a vertical reflection, followed by a vertical stretch by 2 (4, -20)
18. a vertical reflection, followed by a vertical translation by -3 (4, -13)
19. a vertical stretch by -3, followed by a horizontal translation by 5 (9, -30)
20. a vertical reflection, then a horizontal translation by -5, followed by a vertical stretch by $\frac{1}{2}$ (-1, -5)
21. Simplify the formula $y = 3(x^2 + 5)$ so that it can be sketched from the clues in the function rule. $3x^2 + 15$

The answers to Exercises 7–15, 24–25, 33, 35–36 can be found in Additional Answers beginning on page 729.

22. $A'(5, -3); A''(-4, -3);$ *yes;* $(x, y)'' = (x - 9, -y)$

26. $0.75b + 0.35 = 32$; 42 boxes

27.

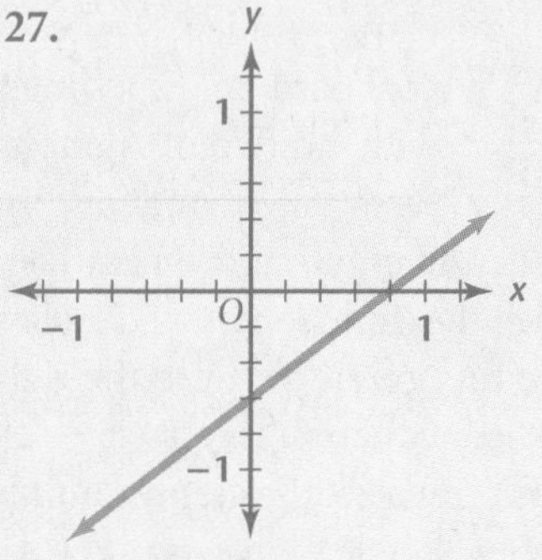

22. In the figure at the right, an irregular triangle is shown in the first quadrant. Its reflection and translation are also shown. If the coordinates of A are $(5, 3)$, find the coordinates of A' and A''. Do the other vertices change using the same rule?

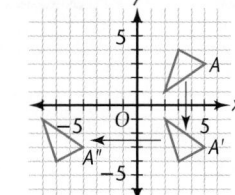

23. Use the information in the table to determine the transformation of a parent function that fits the data. $y = -5x + 6$

x	-3	-2	-1	0	1	2	3	4	5	6	7
y	21	16	11	6	1	-4	-9	-14	-19	-24	-29

24. **Portfolio Activity** Complete the problem in the portfolio activity on page 409.

Look Back

25. Sketch a scatter plot that could have a correlation of -0.95. **[Lesson 1.6]**

26. **Packaging** A roll of tape contains 32 meters of tape. Each box requires 75 centimeters of tape to seal for delivery. If 35 centimeters of tape must be left for sealing other items, write and solve an equaton that shows how many boxes can be sealed with the available tape. **[Lesson 4.6]**

27. Graph the function $4x - 5y = 3$. Describe the procedure for graphing this equation. **[Lesson 5.4]**

28. If the graph of a line contains the points $A(-3, 7)$ and $B(2, -9)$, what is the equation for the line perpendicular to the given line at point A? **[Lesson 5.6]** $5x - 16y = -127$

Solve the system of equations. $\begin{cases} 2x - 3y = 8 \\ 6x + 5y = -4 \end{cases}$

29. by graphing **[Lesson 6.1]** $(1, -2)$ **30.** by elimination **[Lesson 6.3]** $(1, -2)$

31. by matrices **[Lesson 7.5]** $(1, -2)$

32. **Probability** What is the theoretical probability of throwing 7 on a number cube? **[Lesson 8.6]** 0

Look Beyond

33. The rocket function given in Chapter 2 can be written as $y = -16(x - 7)^2 + 784$. Graph the function.

34. Describe each transformation of the parent function in Exercise 33.

Plot the following graphs.

35. $y = 3 + \sqrt{x + 4}$

36. $y = -7 + 10\sqrt{x}$

27. First write the equation in slope-intercept form giving $y = \frac{4}{5}x - \frac{3}{5}$. Plot the y-intercept of $\left(0, -\frac{3}{5}\right)$. Apply the slope of $\frac{4}{5}$ (up 4 units and right 5 units) to this point to generate a second point $\left(5, \frac{17}{5}\right)$. Draw a line through these two points.

34. The parent function is $y = x^2$. The negative sign in front of the 16 denotes a vertical reflection. The 16 represents a stretching of the y-values by 16. The -7 denotes a shift to the right by 7 units. The 784 is a shift up of 784 units.

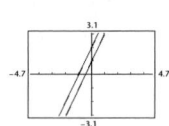

FOCUS

This project involves working with number games and composite functions.

MOTIVATE

Start by playing this game with students. Tell students, "Choose a number between 10 and 20. Double the number. Add 12. Divide by 2. Subtract the number you started with. What is your number now?" All the students should say, "Six." Tell students they will learn how these types of games work as they complete this project.

CHAPTER 9 PROJECT

Pick a Number

Try this with a classmate. Ask your classmate to

pick a number from 1 through 9, triple the number and add 6, then divide the result by 3.

Ask for the number. If you mentally subtract 2, you can tell your classmate the original number.

Try other numbers. What do you notice?

Algebra can be used to see why this happens. Let *x* be the number chosen.

	Arithmetic	Algebra
Pick a number	5	x
Triple the number and add 6	15 + 6 = 21	3x + 6
Divide by 3	21 ÷ 3 = 7	x + 2

The algebra shows you how to go directly to the result. This is an example of finding a function of a function. In terms of a function machine, the output from the first machine is used as the input for the second machine.

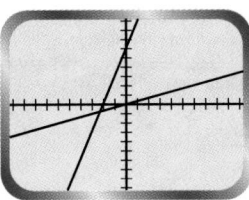

When you find a function of a function, you are forming a **composite function**. If f is one function and g is another function, write $g(f(x))$ to indicate the value of the composite function for any x value.

If $f(x) = 3x + 6$ and $g(x) = \frac{x}{3}$, then

$$f(g(x)) = \frac{f(x)}{3} = \frac{3x + 6}{3} = x + 2.$$

The composition of two functions produces a new function with a new graph.

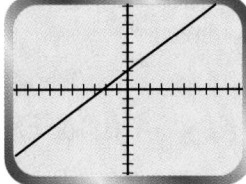

Composition of functions is not a commutative operation. $g(f(x)) = x + 2$, but

$$f(g(x)) = 3(g(x)) + 6 = 3\left(\frac{x}{3}\right) + 6 = x + 6.$$

Activity 1
Make up a two step computation game like that in the opener of the project. Show how the domain and range are related through the steps. Then show the simplified composition of the functions that lets you perform the calculation easily. Try the game on your friends.

Activity 2
Use what you have learned about composing functions to graph the composition of two functions. Begin by selecting a parent function from those you studied in Chapter 2. Replace the x in the function rule with a function, and graph the composition.

a. Identify the domains and ranges of the first function and the composite function.

b. Make a table of the input values, the intermediate values for input in the second step, and the final composite values.

c. Use a graphics calculator or draw the graph of the new composite function, and comment on the graph.

Cooperative Learning
Have students work in small groups to complete this lesson. Have students discuss composite functions and write a definition, including an example, of a composite function.

Technology
When exploring the composite function at the top of page 449, students should set the range of x and y from -5 to 5.

Discuss

Discuss why the domain of $g(x)$ cannot include zero. [**Division by zero is not defined.**]

Have students discover the function used to play the game in the **MOTIVATE** section.

$$\left[f(x) = \frac{2x + 12}{2} - x \right]$$

Chapter 9 Review

Vocabulary

asymptotes	427	parent function	418	symmetric	418
axis of symmetry	418	range	413	transformation	418
domain	413	reflection	431	translation	436
function	411	relation	410	vertex	418
hyperbola	427	stretch	425	vertical line test	412

Key Skills & Exercises

Lesson 9.1

➤ **Key Skills**

Test a graph to identify a function using the vertical line test.

Any vertical line will intersect the graph at the right no more then once, so the graph is a function.

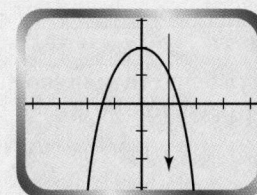

Evaluate functions.

To evaluate the function $f(x) = x^2 + 1$ for $f(-3)$, substitute -3 for x, and simplify.

$$f(x) = x^2 + 1$$
$$f(-3) = (-3)^2 + 1$$
$$f(-3) = 10$$

➤ **Exercises**

Which of the following are functions?

1. 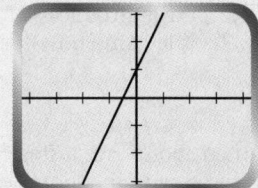 yes

2. 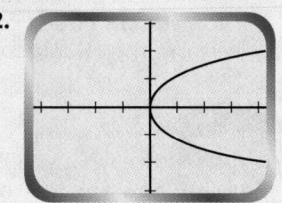 no

3. 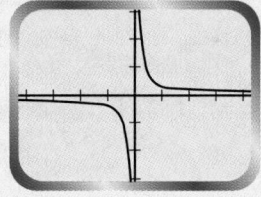 yes

Evaluate the following functions when $x = 4$.

4. $f(x) = x - 8$ -4 **5.** $g(x) = |x + 2|$ 6   **6.** $h(x) = 3^x$ 81

Lesson 9.2

➤ **Key Skills**

Identify which transformation, stretch, reflect, or shift, was applied to a parent function.

The parent function for the graph at the right is $y = x^2$. It has been reflected through the x-axis and shifted vertically.

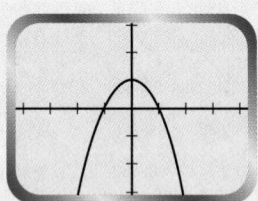

➤ **Exercises**

Give the parent function for each graph. Then tell what kind of transformation was applied.

7.
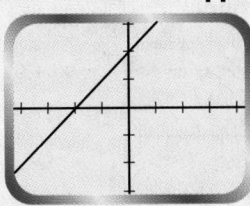

8.
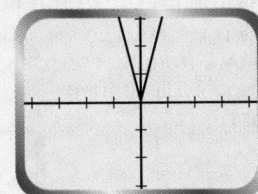

9.
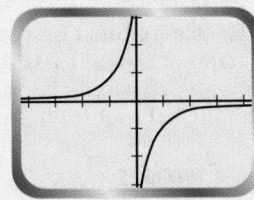

Lesson 9.3

➤ *Key Skills*

Describe the effect of a stretch on the graph of a function.

The parent function of $y = 3|x|$ is $y = |x|$. Each of the y-values is 3 times the corresponding value of the parent function.

x	3|x|	y
−2	3|−2|	6
−1	3|−1|	3
0	3|0|	0
1	3|1|	3
2	3|2|	6

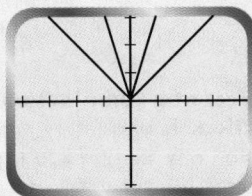

➤ **Exercises**

Graph each of the following.

10. $y = 4|x|$ **11.** $y = 3x^2$ **12.** $y = \dfrac{1}{4x}$ **13.** $y = 2\,\text{INT}(x)$

Lesson 9.4

➤ *Key Skills*

Describe the effect of a reflection on the graph of the parent function.

The parent function of $y = -|x|$ is $y = |x|$. The values of $-|x|$ are opposite of those for $|x|$. This is a reflection through the x-axis of the parent graph, $y = |x|$.

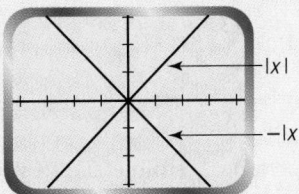

➤ **Exercises**

Sketch a graph of each of the following.

14. $y = -x$ **15.** $y = -\dfrac{1}{x}$ **16.** $y = -3^x$ **17.** $y = -3x^2$

The answers to Exercises 10–17 can be found in Additional Answers beginning on page 729.

7. $y = x$; vertical shift
8. $y = |x|$; stretch
9. $y = \dfrac{1}{x}$; reflection through the x-axis

Lesson 9.5

➤ *Key Skills*

Describe the effect of a translation on the graph of a function.

The parent function of $y = |x| + 2$ is $y = |x|$. The graph of $y = |x| + 2$ is shifted up by 2 from the graph of $y = |x|$. The graph of $y = |x + 1|$ is shifted to the left by 1 from the graph of $y = |x|$.

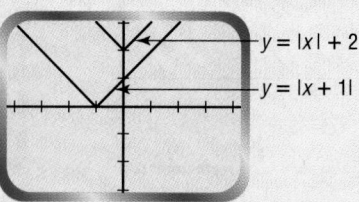

➤ *Exercises*

Identify the parent function. Tell what type of transformation is applied to each, and then draw a carefully labeled graph.

18. $y = x^2 + 1$ **19.** $y = (x + 1)^2$ **20.** $y = |x| - 4$ **21.** $y = |x - 4|$

Lesson 9.6

➤ *Key Skills*

Describe the effect of a combination of transformations on the graph of a function.

The parent function of $y = -2|x + 2|$ is $y = |x|$. Because of the factor -2, the y-values will be twice as great as the parent function and reflected through the x-axis. This is a reflection and a vertical stretch by 2. Because 2 is added to the variable, x-values are less than the same y-values for the parent function. This results in a horizontal translation to the left by 2.

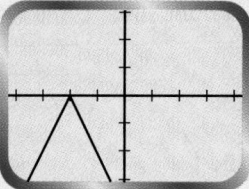

➤ *Exercises*

Graph each function.

22. $f(x) = 3x^2 + 1$ **23.** $g(x) = |x - 6| + 2$ **24.** $h(x) = -\dfrac{2}{x} + 1$

Applications

25. Are meters a function of centimeters? (If you know the number of meters, do you necessarily know the number of centimeters?)   yes

26. Are centimeters a function of meters? (If you know the number of centimeters, do you necessarily know the number of meters?)   yes

27. Temperature Kelvins are often used to measure the temperature of substances in science. To change degrees Celsius to kelvins, add 273.16 to degrees Celsius. What transformation relates degrees Celsius to kelvins?   shift up by 273.16

28. Business Tomas decides to have a sale at his art store. He decides to reduce the price of everything in the store by 20%. Find the scale factor that shows the amount by which the cost of the items were "stretched."   0.80

29. Geometry Belinda reflected the line segment with endpoints (2, 4) and (6, 6) through the x-axis and then shifted it to the left 5 units. What were the endpoints of the new line segment? $(-3, -4); (1, -6)$

The graphs for Exercises 18–21 and answers for Exercises 22–24 can be found in Additional Answers beginning on page 729.

18. $y = x^2$; Shifted up 1 unit
19. $y = x^2$; Shifted left 1 unit
20. $y = |x|$; Shifted down 4 units
21. $y = |x|$; Shifted right 4 units

Chapter 9 Assessment

1. When is a relation a function? No two points have the same first coordinate.

2. Explain how to use the vertical line test to identify the graph of a function. Draw a vertical line. If it intersects in more than one point, the relation is not a function.

Evaluate the following functions when $x = -3$.

3. $f(x) = \text{INT}(x)$ -3 4. $g(x) = -x^2$ -9 5. $h(x) = 2|x|$ 6

Graph each function. Then tell whether the graph of $y = |x|$ is stretched, reflected, of shifted.

6. $y = 3|x|$ 7. $y = |x| - 5$ 8. $y = -|x|$ 9. $y = |x + 4|$

Given the function $y = \dfrac{2}{x}$,

10. identify the parent function. $y = \dfrac{1}{x}$

11. sketch the graphs of the function and the parent function on the same axes.

12. tell how the graph of the parent function was changed by the 2.

Tell what happens to the point (3, 2) when each of the following is applied.

13. vertical translation by 5 (3, 7)   14. horizontal translation by -3 (0, 2)

15. vertical translation by -6 16. vertical stretch by 4 (3, 8)
(3, -4)

17. What happens to the average hourly wage of the workers at a restaurant if all hourly wages are increased by 5%? increases by 5%

18. Explain how you know when a parabola opens downward.
If $y = ax^2$ and $a < 0$.

Graph each function.

19. $f(x) = (x - 3)^2$ 20. $h(x) = 5|x + 3|$ 21. $t(x) = \dfrac{-2}{x + 2}$

22. **Business** A local department store is having a sale. They marked down their clearance items by 25%, and for one day only they are offering an additional 30% off each clearance item. Miki decided to leave the original prices on the price tag and make a sign that shows 55% off. Is Miki's sign correct? Why or why not? no; The reduction is off the sale price, not the original price.

23. Explain how to use a graphics calculator to approximate the minimum point of a quadratic function like $y = 2(x + 4)^2 - 1$.

Explain how to use a graphics calculator to solve each of the following for x.

24. $\dfrac{4}{x} = 3 + x$ 25. $|1 - 3x| = |x| + 5$

The graphs for Exercises 6–9 and answers for Exercises 11–12, 19-21 can be found in Additional Answers beginning on page 729.

6. stretched vertically by 3
7. shifted down 5 units
8. reflected vertically
9. shifted left 4 units
23. Graph the function and then use the trace function to find the vertex.

24. Graph $y = \dfrac{4}{x}$ and $y = 3 + x$; Find their points of intersection OR graph $y = \dfrac{4}{x} - (3 + x)$ and determine where it crosses the x-axis. $x = -4, 1$

25. Graph $y = |1 - 3x|$ and $y = |x| + 5$. Find their points of intersection OR graph $y = |1 - 3x| - (|x| + 5)$ and determine where it crosses the x-axis. $x = 3, -2$

10 Exponents

Meeting Individual Needs

10.1 Exploring Exponents

Core Resources

Practice Master 10.1
Enrichment, p. 458
Technology Master 10.1
Interdisciplinary
   Connection, p. 457

[1 day]

Core Two-Year Resources

Inclusion Strategies, p. 458
Reteaching the Lesson, p. 459
Practice Master 10.1
Enrichment Master 10.1
Technology Master 10.1
Lesson Activity Master 10.1

[3 days]

10.2 Multiplying and Dividing Monomials

Core Resources

Practice Master 10.2
Enrichment, p. 464
Technology Master 10.2

[1 day]

Core Two-Year Resources

Inclusion Strategies, p. 464
Reteaching the Lesson, p. 465
Practice Master 10.2
Enrichment Master 10.2
Technology Master 10.2
Lesson Activity Master 10.2

[2 days]

10.3 Negative and Zero Exponents

Core Resources

Practice Master 10.3
Enrichment, p. 470
Technology Master 10.3
Mid-Chapter Assessment
   Master

[1 day]

Core Two-Year Resources

Inclusion Strategies, p. 470
Reteaching the Lesson, p. 471
Practice Master 10.3
Enrichment Master 10.3
Technology Master 10.3
Lesson Activity Master 10.3
Mid-Chapter Assessment Master

[2 days]

10.4 Scientific Notation

Core Resources

Practice Master 10.4
Enrichment, p. 478
Technology Master 10.4
Interdisciplinary
   Connection, p. 477

[1 day]

Core Two-Year Resources

Inclusion Strategies, p. 478
Reteaching the Lesson, p. 479
Practice Master 10.4
Enrichment Master 10.4
Technology Master 10.4
Lesson Activity Master 10.4

[3 days]

10.5 Exponential Functions

Core Resources

Practice Master 10.5
Enrichment, p. 484
Technology Master 10.5
Interdisciplinary
   Connection, p. 484

[2 days]

Core Two-Year Resources

Inclusion Strategies, p. 485
Reteaching the Lesson, p. 486
Practice Master 10.5
Enrichment Master 10.5
Technology Master 10.5
Lesson Activity Master 10.5

[3 days]

10.6 Applications of Exponential Functions

Core Resources

Practice Master 10.6
Enrichment, p. 491
Technology Master 10.6
Interdisciplinary
   Connection, p. 490

[2 days]

Core Two-Year Resources

Inclusion Strategies, p. 491
Reteaching the Lesson, p. 492
Practice Master 10.6
Enrichment Master 10.6
Lesson Activity Master 10.6
Technology Master 10.6

[4 days]

Chapter Summary

Core Resources

Chapter 10 Project,
    pp. 496–497
Lab Activity
Long-Term Project
Chapter Review,
    pp. 498–500
Chapter Assessment,
    p. 501
Chapter Assessment, A/B
Alternative Assessment
Cumulative Assessment,
    pp. 502–503
      [2 days]

Core Two-Year Resources

Chapter 10 Project, pp. 496–497
Lab Activity
Long-Term Project
Chapter Review, pp. 498–500
Chapter Assessment, p. 501
Chapter Assessment A/B
Alternative Assessment
Cumulative Assessment,
    pp. 502–503
      [4 days]

Reading Strategies

Ask a volunteer to explain how interest on savings and loans is accrued. Point out that this type of interest typically grows exponentially. When students have some knowledge of what they are reading, they will find it easier to comprehend. Encourage students to write down concepts like the definitions of exponents and negative exponents in Lessons 1 and 3, the Product-of-a-Power Property in Lesson 2, how to write numbers in scientific notation in Lesson 4, the general growth formula in Lesson 5, and the various applications of exponential functions in Lesson 6. After reading, have students write about the problems of population growth. Have them relate the general growth formula to the exponential growth of world population.

Visual Strategies

Graphing exponential functions on a graphics calculator allows students to see what the function looks like and what happens to it when parameters are changed. This is a very important strategy for teaching this chapter.

Tree diagrams provide another visual tool that allows students to see all the possible outcomes of a situation.

Hands-on Strategies

Hands-on activities are found in the Teacher's Edition comments. For instance, in Lesson 10.5 on page 485, the Inclusion Strategies suggests that students try folding a piece of paper in half as many times as possible. The resulting paper shows 2^n divisions and usually a piece of paper can be folded no more than 10 times.

Cooperative Learning

GROUP ACTIVITIES	
exploring exponents	Lesson 10.1 Explorations 1, 2, and 3
property of sums of squares	Lesson 10.2 Exercises 65–67
defining x⁰	Lesson 10.3 Exploration
Please Don't Sneeze	Chapter Project

You may wish to have students work with partners or in small groups to complete any of these activities. Additional suggestions for cooperative group activities can be found in the teacher's notes in the lessons.

Multicultural

The cultural connections in this chapter include references to France, England, India, Spain, and Italy.

CULTURAL CONNECTIONS	
Europe: Fermat's Theories	Lesson 10.3, Exercises 47–48
Europe: Sir Isaac Newton	Lesson 10.4, Discussion
Asia: Population of India	Lesson 10.5, Example 2
Europe: Population	Lesson 10.5, Exercises 25–34

Portfolio Assessment

Below are portfolio activities for the chapter listed under seven activity domains that are appropriate for portfolio development.

1. **Investigation/Exploration** The methods used to explore exponents in Lesson 10.1 allow students to make and test mathematical conjectures.

2. **Applications** Recommended to be included are any of the following: Biology, Lesson 10.1, Exercises 49–52; Astronomy, Lesson 10.4, Opening discussion and Examples 1 and 2 and Lesson 10.4, Exercises 44–47; Physics, Lesson 10.4, Example 4; Social Studies, Lesson 10.4, Exercises 22–25; Finance, Lesson 10.6, Example 2; Demographics, Lesson 10.6, Example 5.

3. **Non-Routine Problems** The "All Mixed Up?" discussion about shuffling cards in the Eyewitness Math feature provides a mathematical application of a non-routine problem. The exploration of the sum of squares in Look Beyond in Lesson 10.2, Exercises 65–68, challenges students to use critical thinking.

4. **Project** "Please Don't Sneeze": see pages 496–497. Students study the spread of a contagious disease and use a mathematical simulation to predict the course of the disease.

5. **Interdisciplinary Topics** Students may choose from the following: Geometry, Lesson 10.2, Exercises 41–42; Probability, Lesson 10.3, Exercises 25–28 and Lesson 10.6, Example 4 and Exercises 21–23; Number Theory, Lesson 10.3, Exercise 48.

6. **Writing** *Communicate* exercises that ask students to describe and explain procedures involved in solving probability problems offer excellent writing selections for the portfolio.

7. **Tools** Chapter 4 uses both computer spreadsheets and graphics calculators to work with exponential functions, simulations, and random numbers.

Technology

The use of graphics calculators allows students to explore and study many topics that could only be explored with the use of a computer a few years ago. Exponential equations is one of these topics. Many of the concepts explored in Chapter 10 would require tedious, time-consuming effort without the availability of the graphics calculator or computer. All activities completed on a graphics calculator in this chapter can also be programmed and carried out on a computer. Detailed instructions for calculators and other technologies can be found in the *HRW Technology Handbook.*

Graphics Calculator

Student's learn to use the power key in Lessons 10.1 and 10.5. This key varies, depending upon the calculator being used.

x^y	$\wedge$	a^b
Casio	Texas Instruments	Sharp

With these calculators, to find 5^3, press: 5 x^y 3 ENTER, 5 $\wedge$ 3 ENTER, or 5 a^b 3 ENTER, and the display will show 125.

Students are also asked to evaluate 2^{2^3}. Have students determine whether parentheses are required on their calculators.

Another key students use in this chapter is the key used for scientific or exponential notation.

This key might be EE or EXP. Again, allow students to check exactly how their calculators work.

In Lesson 10.4, students study the similarities and differences between $y = 3^x$ and $y = 4^x$. The use of graphics calculator allows students to expand this study to $x = 5^x$ and $y = 6^x$. Example 6 in Lesson 10.6 explores the transformations of exponential functions. Be sure students understand the necessity of using parentheses when graphing 2^{x-3} on the calculator. Without a graphics calculator or a computer, studies of this type would be difficult and time-consuming for students.

Error Analysis

When evaluating or graphing negative powers, students may try to use the - key rather than the (-) key. If students consistently get an error message, suggest that they practice with the (-) key until they learn to use it correctly.

Integrated Software

f(g) Scholar™ is an integrated computer-based mathematics productivity tool that combines calculator, spreadsheet, and graphics capabilities to provide a dynamic and interactive environment for explorations in mathematics. It is appropriate to use *f(g) Scholar*™ for any lesson needing a spreadsheet, calculator, graphics calculator, or any combination of the three.

10 Transformations

Exponents

LESSONS

10.1 *Exploring* Exponents

10.2 Multiplying and Dividing Monomials

10.3 Negative and Zero Exponents

10.4 Scientific Notation

10.5 Exponential Functions

10.6 Applications of Exponential Functions

Chapter Project
Please Don't Sneeze

ABOUT THE CHAPTER

Background Information

Most students have heard of carbon-dating fossils, compound interest, and earthquakes. What they probably do not know is that these are all examples of exponential functions.

CHAPTER RESOURCES

- Practice Masters
- Enrichment Masters
- Technology Masters
- Lesson Activity Masters
- Lab Activity Masters
- Long-Term Project Masters
- Assessment Masters
    Chapter Assessments, A/B
    Mid-Chapter Assessment
    Alternative Assessments, A/B
- Teaching Transparencies
- Cumulative Assessment
- Spanish Resources

CHAPTER OBJECTIVES

- Understand the concept of exponents and powers.
- Understand the structure and concept of a monomial.
- Use the properties of exponents to simplify expressions.
- Understand the concept of negative and zero exponents.
- Simplify expressions containing negative and zero exponents.
- Recognize the need for a special notation to facilitate computation with very large and very small numbers.
- Perform computations using scientific notation, with and without a calculator.

Exponential functions model a variety of important real-world activities. The growth of living organisms, population growth and decline, radioactive dating of fossils, and the spread of certain diseases are all examples of exponential growth or decay. You have already investigated some examples briefly in Chapter 2. This Chapter will give you the opportunity to explore them in greater depth.

One of the areas in which exponential functions play an important role is finance. The portfolio activity gives you an example of interest on savings and loans that typically behaves exponentially.

ABOUT THE PHOTOS

Have students discuss the photographs on these pages and speculate as to how they might relate to exponents. Students may give answers like the following:

- Some bacteria double in number every so often.

- Carbon-dating helps scientists tell how old fossils are.
- Animal life that is allowed to grow without restriction.

Students should notice that the operation ^ is used for the computer instruction 2^1000. It is used to generate the printout of the number 2^{1000}.

Exponential
Growth Function

$y = 2^x$

Exponential
Decay Function

$y = 2^{-x}$

PORTFOLIO ACTIVITY

A credit card company charges 1.5% on the unpaid balance and requires a minimum payment of $10 or 3% of the unpaid balance, whichever is greater. Suppose you borrow $1000 and pay it back making the minimum payment each month. How long will it take to pay it back? What is the total amount that you will pay back?

ABOUT THE CHAPTER PROJECT

In the Chapter Project on pages 496–497, students investigate the spread of a contagious disease. Drawing a diagram and using the tools of technology may help students in the problem solving activities for the Chapter Project.

- Understand the exponential function and how it is used.
- Identify the graph of the exponential function for different bases.
- Use exponential functions to model applications.
- Use exponential functions to model growth and decay in different contexts.

PORTFOLIO ACTIVITY

Assign the Portfolio Activity as part of the assignment for Lesson 10.5 on page 488.

One of the most important skills students can learn is how to handle credit wisely. This activity helps students realize how very expensive it can be to use credit cards. Have students work in small groups to explore how credit is used. Have each group find out the interest rate for each card and the minimum payments for various cards. The Portfolio Activity can be extended by having students investigate the types of loans or credit cards available to people under 21. Include a discussion of the problems of credit card debt.

Clever use of technology is useful for completing the Portfolio Activity. It will allow students to avoid excessive computation and to experiment with "what if . . ." situations.

For students without access to a calculator or a computer, the task can be modified by asking for the balance after *one* year.

Exploring Exponents

PREPARE

Objectives

- Understand the concepts of exponents and powers.
- Use properties of exponents to simplify expressions.

RESOURCES

- Practice Master   **10.1**
- Enrichment Master   **10.1**
- Technology Master   **10.1**
- Lesson Activity Master   **10.1**
- Quiz   **10.1**
- Spanish Resources   **10.1**

Assessing Prior Knowledge

Evaluate.

1. a^2 when $a = 4$ **[16]**
2. $\frac{1}{b^2}$ when $b = 3$ $\left[\frac{1}{9}\right]$
3. $c^3 d$ when $c = 4$ and $d = 5$ **[320]**
4. $e^4 \div f^2$ when $e = 3$ and $f = 9$ **[1]**
5. $(hj)^2$ when $h = 3$ and $j = 8$ **[576]**

TEACH

why Have students discuss how they have used exponents in their past experiences in mathematics. Have volunteers make a list, including examples.

why *Numbers with many digits can be expressed in a simpler way by using exponents.*

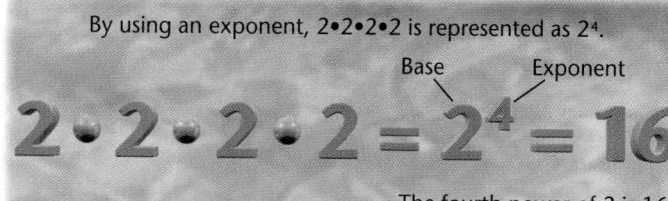

By using an exponent, 2•2•2•2 is represented as 2⁴.

Base   Exponent

$$2 \cdot 2 \cdot 2 \cdot 2 = 2^4 = 16$$

The fourth power of 2 is 16.

The expression 2^{1000} represents 2 to the thousandth power and is the product when a thousand 2s are multiplied.

Some powers have special names. The first power of 2 is 2^1. It is read *two to the first*. The second power of 2 is 2^2. It is read *two squared*. The third power of 2 is 2^3. It is read *two cubed*. The nth power of 2 is 2^n. It is read *two to the nth*.

PRINTOUT OF 2^1000
107150860718626732094842504906000181056140481170553360744375038830351051124961404811705533607443750388303510511249361224931983788156958585858127594672917553146825187145285692314043598457757469857481393456777482423098542107460506237114187795418215304647498358194126739876755916554394607706291457119647768654217576604298376526243868372056680693766766042983765262438683720566806937

TEACHING *tip*

Technology If a computer is available in the classroom, display the value of 2^{1000} on the screen. Otherwise, you may display the number on a poster.

ALTERNATIVE teaching strategy

Using Patterns Have students study the patterns below.

$10^1 \cdot 10^1 = 10 \cdot 10 = 100$
$10^1 \cdot 10^2 = 10 \cdot 100 = 1000$
$10^1 \cdot 10^3 = 10 \cdot 1000 = 10,000$
$10^1 \cdot 10^4 = 10 \cdot 10,000 = 100,000$
$10^2 \cdot 10^2 = 100 \cdot 100 = 10,000$
$10^2 \cdot 10^3 = 100 \cdot 1000 = 100,000$
$10^2 \cdot 10^4 = 100 \cdot 10,000 = 1,000,000$

Have students write each product using exponents. Then have students make and test a conjecture about the product of $a^m \cdot a^n$. [a^{m+n}]

EXPONENTS AND POWERS

If x is any number and a is an integer greater than 1, then

$$x^a = \underbrace{x \cdot x \cdot x \cdot \cdots \cdot x.}_{a \text{ factors}}$$

When $a = 1$, $x^a = x^1 = x$.

Our number system uses 10 as its base. A number written in the form **2315** is in *customary notation*. It can be expressed in *expanded notation* as

$$2 \cdot 1000 + 3 \cdot 100 + 1 \cdot 10 + 5$$

or in *exponential notation* as

$$2 \cdot 10^3 + 3 \cdot 10^2 + 1 \cdot 10^1 + 5.$$

How would you write 15,208 in exponential notation?

Exploration 1 *Multiplying Powers*

1 Examine the steps in the example.

Product Form ⟶ $10^3 \cdot 10^1$ ⟶ $1000 \cdot 10$ ⟶ $10{,}000$ ⟶ 10^4 ⟵ Simplified Form

Complete the steps as shown in Step 1.

a. $10^3 \cdot 10^2 = 1000 \cdot 100 = \underline{?} = \underline{?}$

b. $10^3 \cdot 10^3 = \underline{(?)(?)} = \underline{?} = \underline{?}$

2 Complete the table for $10^4 \cdot 10^2$. Write the missing product as a power of ten. This product is now in simplest form.

Decimal form	$10{,}000 \cdot 100 = 1{,}000{,}000$
Exponential form	$10^4 \cdot 10^2 = \underline{?}$

3 Make a similar table for each of the following.

a. $10^3 \cdot 10^6$ **b.** $10^1 \cdot 10^3$ **c.** $2^2 \cdot 2^4$ **d.** $3^2 \cdot 3^2$

4 How would you simplify $a^m \cdot a^n$ using exponents? Make a table to check your guess.

5 For any positive number a and any positive integers m and n, what is the simplified form for $a^m \cdot a^n$? ❖

The statement in Step 5 suggests that multiplying powers *with the same base* involves the addition of exponents. Guess what happens to the exponents when you divide powers.

---

interdisciplinary CONNECTION

Science Start the application on page 459 with this puzzle:

The bacteria colony in a culture double in number every hour. One scientist estimated that at 10:30 A.M. there were about 1,500,000 bacteria in the colony. At what time were there half that many?   **[9:30 A.M.]**

Cooperative Learning

Allow students to work through the explorations in cooperative-learning groups. Remind students that all members are to contribute to the discussion and findings of the group. You could encourage students to take turns answering the questions so that all members of the group have an opportunity to participate.

ongoing ASSESSMENT

$$1 \cdot 10^4 + 5 \cdot 10^3 + 2 \cdot 10^2 + 8$$

Exploration 1 Notes

The purpose of this exploration is to allow students to discover the algorithm for multiplying powers by adding exponents rather than just giving them the rule. In Part 1, you could point out the number of zeros in the product and then ask students to compare this with the exponent.

ongoing ASSESSMENT

5. $a^m \cdot a^n = a^{m+n}$

TEACHING tip

Have students make and record their conjecture about dividing powers before continuing the lesson.

Use Transparency 44

Students should be reminded that division and multiplication are inverse operations. Since multiplication resulted in adding exponents, ask students to guess what will happen when division is involved. After students have completed the exploration, they can compare their guesses with the actual results.

Aongoing SSESSMENT

4. a^{m-n}

6. **No, the bases of numbers expressed in exponential form must be the same in order to combine the exponents when dividing powers.**

CRITICAL Thinking

$(ab)^n$ means that ab is a factor n times. By rearranging the factors, a is a factor n times and b is a factor n times. Hence $(ab)^n = a^n b^n$. Similarly $\left(\dfrac{a}{b}\right)^n$ means $\dfrac{a}{b}$ is a factor n times. So, a is a factor in the numerator n times and b is a factor in the denominator n times. Hence $\left(\dfrac{a}{b}\right)^n = \dfrac{a^n}{b^n}$.

Exploration 3 Notes

You might start this exploration by asking students to write a^5 in an expanded form. Then point out to students that you could substitute a value such as (x^2) for a in a^5 (both in the original and expanded forms).

Aongoing SSESSMENT

5. $(a^m)^n = a^{mn}$

•Exploration 2 Dividing Powers

1 Use the table to write the quotient for $\dfrac{10^6}{10^2}$ as a power of 10.

FORM	NUMERATOR	DENOMINATOR	QUOTIENT
Decimal	1,000,000	100	10,000
Exponential	10^6	10^2	?

2 Make a similar table for each of the following.

a. $\dfrac{10^5}{10^2}$ b. $\dfrac{2^6}{2^2}$ c. $\dfrac{3^4}{3^3}$

3 Guess the simplified form in exponent notation for $\dfrac{10^5}{10^1}$. Make a table to check your guess.

4 For any positive number a and any positive integers m and n with $m > n$, what is the simplified form for $\dfrac{a^m}{a^n}$?

5 Explain how you would simplify $\dfrac{2^5}{2^3}$.

6 Can you use this procedure to simplify $\dfrac{5^2}{3^2}$? Make a conjecture to identify the restrictions on combining the exponents when dividing powers. ❖

CRITICAL Thinking Explain why $(ab)^n = a^n b^n$ and $\left(\dfrac{a}{b}\right)^n = \dfrac{a^n}{b^n}$ when n is any positive integer.

You can multiply and divide powers. What is the result when you find a power of a power?

•Exploration 3 The Power of a Power

1 Write both 10^5 and 10^6 in customary notation.

2 Since $(10^3)^2$ means $10^3 \cdot 10^3$, change 10^3 to customary notation and square the number.

3 Change the resulting number back from customary notation to exponential notation. Does $(10^3)^2$ equal 10^5 or 10^6?

4 What was done with the exponents? Does $(5^2)^4$ equal 5^6 or 5^8?

5 For any positive number a and any positive integers m and n, what is the equivalent expression for $(a^m)^n$? ❖

ENRICHMENT Have students research the meaning of the word *googol* and make a display to show its meaning. [A googol is 10^{100} or a 1 followed by 100 zeros.]

INCLUSION strategies **Using Visual Models** Allow students to use the definition on page 457 to help them develop understanding of these rules for exponents. For example, to find $2^3 \cdot 2^4$, have students write $(2 \cdot 2 \cdot 2) \cdot (2 \cdot 2 \cdot 2 \cdot 2) = 2 \cdot 2 \cdot 2 \cdot 2 \cdot 2 \cdot 2 \cdot 2$. For division, show how to eliminate matching factors in the numerator and denominator (they equal the identity) to simplify the expression.

$$\dfrac{2^5}{2^3} = \dfrac{2 \cdot 2 \cdot 2 \cdot 2 \cdot 2}{2 \cdot 2 \cdot 2}$$

$$\left(\dfrac{2}{2}\right)\left(\dfrac{2}{2}\right)\left(\dfrac{2}{2}\right) \cdot 2 \cdot 2 = 2 \cdot 2, \text{ or } 2^2$$

APPLICATION

Suppose a certain bacteria colony doubles every hour. If a colony contains 1000 bacteria at noon, the properties of exponents can be used to find how many bacteria the culture contains at 3 P.M. and 2 hours later.

At 3 P.M. there will be $1000 \cdot 2^3$ or 8000 bacteria. Two hours later the bacteria will double two more times. There will be $(1000 \cdot 2^3) \cdot 2^2 = 1000 \cdot 2^{3+2} = 1000 \cdot 2^5$ or 32,000 bacteria in the culture. ❖

Once you are familiar with the properties of exponents, you can simplify many exponential expressions.

a. $3^2 \cdot 3^3 = 3^5$ **b.** $5^4 \div 5^1 = 5^3$ **c.** $(2^2)^3 \div (2^3)^2 = 2^6 \div 2^6$
     $9 \cdot 27 = 243$ $625 \div 5 = 125$ $64 \div 64 = 1$

EXTENSION

Calculator

Most computations with exponents can be performed easily on a calculator. Check the calculator you use to find how it performs these computations.

Most calculators will use the $\boxed{y^x}$ or $\boxed{\wedge}$ key to indicate an exponent. Parentheses are used for more complicated numerators and denominators.

For example, to simplify $\dfrac{10^3 + 10^4}{10^2}$, the numerator contains addition of the powers, so it should be included in parentheses. On many calculators, you can use

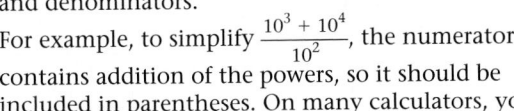

$\boxed{(}$ 10 $\boxed{y^x}$ 3 $\boxed{+}$ 10 $\boxed{y^x}$ 4 $\boxed{)}$ $\boxed{\div}$ 10 $\boxed{y^x}$ 2 $\boxed{=}$ or

$\boxed{(}$ 10 $\boxed{\wedge}$ 3 $\boxed{+}$ 10 $\boxed{\wedge}$ 4 $\boxed{)}$ $\boxed{\div}$ 10 $\boxed{\wedge}$ 2 $\boxed{ENTER}$.

Notice the difference in each computation.

a. $\dfrac{10^3 + 10^4}{10^2} = \dfrac{10^3(1 + 10)}{10^2} = 10^1(11) = 110$

b. $\dfrac{10^3 \times 10^4}{10^2} = \dfrac{10^7}{10^2} = 10^5 = 100,000$ ❖

CRITICAL *Thinking*

Compare the way the exponents are treated when the powers are added with the way they are treated when they are multiplied. Make a conjecture about how to treat exponents when powers are added.

Assignment Guide

Core 1–9, 10–48 even, 49–70

Core Two-Year 1–63

Technology

Students should solve Exercises 26–29 without a calculator and then use a calculator to check their work. Students should use their calculators to solve Exercises 40–43 and 66–69.

Error Analysis

When evaluating expressions containing exponents, some students may write $5^3 = 5 \cdot 3 = 15$. Remind students of the definition found on page 457 and show students that $5^3 = 5 \cdot 5 \cdot 5 = 125$.

Practice Master

Practice & Apply
10.1 Exploring Exponents

1. Make a table showing decimal form and exponential form for $10^3 \cdot 10^5$. Write the missing product in simplified form.

2. Make a table showing decimal and exponential form for the numerator, denominator, and quotient for $\frac{10^5}{10^1}$. Write the quotient as a power of ten.

Evaluate each of the following.

3. $(3^4)^2$ _____ 4. $\frac{5^7}{5^3}$ _____ 5. $2^4 \cdot \frac{2^2}{2^3}$ _____ 6. $3^2 + \frac{3}{3}$ _____

7. $(a^4)^2$ _____ 8. $\frac{3^7}{3^3}$ _____ 9. $m^4 \cdot \frac{m^2}{m^3}$ _____ 10. $y^2 + \frac{y}{y}$ _____

Match each with the letter that indicates the equivalent value.

A. 10,000 B. 1,000 C. 110 D. 200 E. 100

11. $10 \cdot 10^2$ _____ 12. $10 + 10^2$ _____ 13. $(10)^2$ _____ 14. 10^{2+2} _____

How old is the universe? Scientists measure the rate at which distant stars have cooled down since the universe began.

10–35 seconds	big bang ends
100 seconds	helium forms
10,000 years	matter begins
100,000 years	stars form
one billion years	Earth forms
ten billion years	life on Earth begins

Express each of the following as a power of ten.

15. The number of years after the big bang that stars form. _____

16. The number of years after the big bang that the Earth formed. _____

17. The number of years after the big bang that life on Earth began. _____

E**XERCISES** & P**ROBLEMS**

Communicate

1. Explain the relationship between the exponent in a power of ten and the number of zeros in a number written in customary notation. Use 10^{12} as an example.

2. How can you tell how many times the base is used as a factor in 3^6?

3. When you multiply powers, what operation is applied to the exponents of the common base? Explain why.

4. When you divide powers, what operation is applied to the exponents of the common base? Explain why.

5. When you raise a power to a power, what operation is applied to the exponents of the common base? Explain why.

6. Explain how you can determine, without computing anything, which value is greatest: 5^{17}, 8^{16}, or 8^{17}.

Explain why the expressions in each of the following equations are equivalent.

7. $a^r \cdot a^s = a^{r+s}$ 8. $\dfrac{a^p}{a^q} = a^{p-q}$ 9. $(a^s)^t = a^{st}$

Practice & Apply

Make a table like the one in Exploration 1 for each product.

10. $10^6 \cdot 10^4$ 11. $10^3 \cdot 10^5$ 12. $2^3 \cdot 2^4$ 13. $3^4 \cdot 3^2$

Make a table like the one in Exploration 2 for each quotient.

14. $\dfrac{10^3}{10^2}$ 15. $\dfrac{10^6}{10^3}$ 16. $\dfrac{2^6}{2^3}$ 17. $\dfrac{3^5}{3^4}$

Change each of the following to customary notation.

18. 2^5 32   19. 2^2 4

20. Use the evaluations of 2^5 and 2^2 to find $2^5 \cdot 2^2$. 128

21. Express the product of $2^5 \cdot 2^2$ as a single power of 2. 2^7

22. Use the evaluations of 2^5 and 2^2 to find $\dfrac{2^5}{2^2}$. 8

23. Express the quotient $\dfrac{2^5}{2^2}$ as a single power of 2. 2^3

24. Use the evaluation of 2^5 to find $(2^5)^2$. 1024

25. Express the result in Exercise 24 as a single power of 2. 2^{10}

Evaluate each of the following using the rules for exponents when they apply.

26. $(6^2)^3$ 46,656   27. $\dfrac{4^3}{4^2}$ 4   28. $\dfrac{10^2 \cdot 10^5}{10^3}$ 10,000   29. $\dfrac{10^2 + 10^5}{10^3}$ 100.1

10. Decimal form
$$1,000,000 \cdot 10,000 = 10,000,000,000$$
Exponent form $10^6 \cdot 10^4 = 10^{10}$

11. Decimal form
$$1000 \cdot 100,000 = 100,000,000$$
Exponent form $10^3 \cdot 10^5 = 10^8$

12. Decimal form $8 \cdot 16 = 128$
Exponent form $2^3 \cdot 2^4 = 2^7$

13. Decimal form $81 \cdot 9 = 729$
Exponent form $3^4 \cdot 3^2 = 3^6$

	Form	Numer	Denom	Quot
14.	Decimal	1000	100	10
	Exponent	10^3	10^2	10^1
15.	Decimal	1,000,000	1000	1000
	Exponent	10^6	10^3	10^3
16.	Decimal	64	8	8
	Exponent	2^6	2^3	2^3
17.	Decimal	243	81	3
	Exponent	3^5	3^4	3^1

Evaluate each of the following.

30. $(x^2)^3$ x^6 **31.** $\dfrac{c^4}{c^1}$ c^3 **32.** $\dfrac{r^3(r^4)}{r^2}$ r^5 **33.** $\dfrac{2^3 + 2^4}{2^2}$ 6

Some of the following can be expressed as a single power of 10; others cannot. If you can, write the given expression as a power of ten. Then evaluate each expression.

34. $10^4 \cdot 10^2$ **35.** $10^4 + 10^2$ **36.** $10^4 \div 10^2$

37. $10^4 - 10^2$ **38.** $(10^2)^3$ **39.** 10^{2+3}

Technology For exponents, 2^3 is entered as 2 $\boxed{\wedge}$ 3 on some calculators and as 2 $\boxed{y^x}$ 3 on others. Evaluate 2^3. Familiarize yourself with the way the calculator you are using handles exponents. Then use the calculator to evaluate the following.

40. 2^{20} **41.** 3^{10} **42.** 5^{10} **43.** 0.5^4
1,048,576   59,049   9,765,625   0.0625

The table shows how to read certain large numbers.

Customary	Exponential	Word
1,000,000	10^6	million
1,000,000,000	10^9	billion
1,000,000,000,000	10^{12}	trillion

For example, the number 2,450,000,000 is read "two billion, four hundred fifty million," and 3×10^6 is read "three million." Write the following in words.

44. 7,000,000,000,000   **45.** 2,030,400,000,000   **46.** 5×10^9 **47.** 3×10^{12}

48. A newspaper article refers to a debt of 4.2 trillion dollars. Write this in customary notation. 4,200,000,000,000

Algae

Yellowtail Fish

Krill (zooplankton)

Biology The food chain explains why even small organisms affect much larger ones. It works in the following way.

- A killer whale eats about 10 pounds of seal per day.
- Each pound of seal requires about 10 pounds of fish or squid per day.
- Each pound of fish or squid requires about 10 pounds of zooplankton per day.
- Each pound of zooplankton requires about 10 pounds of algae per day.

Express each number of pounds per day as a power of 10.

49. seals needed to sustain a killer whale 10^1 **50.** fish or squid needed to sustain a killer whale 10^2

51. zooplankton needed to sustain a killer whale 10^3 **52.** algae needed to sustain a killer whale 10^4

34. $10^6 = 1,000,000$
35. 10,100
36. $10^2 = 100$
37. 9900
38. $10^6 = 1,000,000$
39. $10^5 = 100,000$

44. seven trillion
45. two trillion, thirty billion, four hundred million
46. five billion
47. three trillion

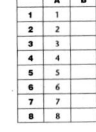

53. Bar graph should display
the following frequencies.

Digit	Frequency
0	28
1	34
2	23
3	25
4	35
5	35
6	34
7	35
8	30
9	23

56. $-4 \leq x \leq 10$

57. $x < -6$ or $x > 2$

58. $y = \dfrac{19}{30}x - \dfrac{178}{15}$

59. $A^{-1} = \begin{bmatrix} 1 & 1 \\ 2 & 3 \end{bmatrix}$

$A \cdot A^{-1} = \begin{bmatrix} 1 & 0 \\ 0 & 1 \end{bmatrix}$

Look Beyond

Exercises 64 and 65 foreshadow the concept of zero as an exponent, which will be covered in Lesson 10.3. Exercises 66–70 foreshadow the study of radicals, which will be covered in later chapters.

53. **Statistics** Make a bar graph from a frequency distribution of the digits of 2^{1000} given at the beginning of this lesson. Count how many 0s, 1s, 2s, . . . , 9s there are. Would you say the distribution is about the same for each digit, or does it favor some particular digits?

54. Write 2^{10} in customary notation.   1024

55. **Technology** A calculator shows the value 20^{10}. Compare this answer to the answer in Exercise 54. What do you think this notation means? Write out the answer in customary notation.
10,240,000,000,000

20^10

1.024ᴇ13

Look Back

Solve for x. [Lesson 4.8]

56. $|x - 3| \leq 7$ **57.** $|2x + 4| > 8$

58. **Coordinate Geometry** A line passes through the points $P(14, -3)$ and $Q(-5, 27)$. Write an equation for the line through point P perpendicular to line PQ. **[Lessons 5.4, 5.6]**

59. Find the inverse of $A = \begin{bmatrix} 3 & -1 \\ -2 & 1 \end{bmatrix}$. What is the product of $A \cdot A^{-1}$?
[Lesson 7.4]

60. **Probability** A study of more than 50,000 nurses was conducted for over eight years. Those whose diets were rich in vitamin A were 40% less likely to have cataracts that impaired their vision than those whose diets were low in the vitamin. Is this an example of experimental probability or theoretical probability? Explain. **[Lesson 8.1]**

61. How many different ways can you order an ice cream cone with a topping? **[Lesson 8.5]** 252

62. **Geometry** The area in square centimeters of a square piece of cardboard is determined from the length of its sides. Write the function for the area if 3 centimeters is added to each side of the original square and that area is doubled. **[Lesson 9.6]**
$A(x) = 2(x + 3)^2$

63. Describe the graph of the function in Exercise 62 without drawing it. Use your understanding of transformations. **[Lesson 9.6]**

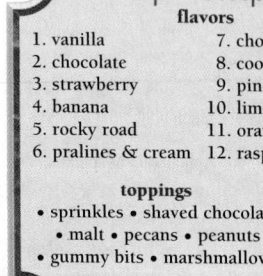

Super Scooper

flavors

1. vanilla
2. chocolate
3. strawberry
4. banana
5. rocky road
6. pralines & cream

7. chocolate chip
8. cookies & cream
9. pineapple
10. lime sherbet
11. orange sherbet
12. raspberry sherbet

toppings
• sprinkles • shaved chocolate
• malt • pecans • peanuts
• gummy bits • marshmallows

cones
plain
sugar
waffle

Look Beyond

64. Look at the pattern in the table for Exploration 1. Based on the table, 10^n can be written as a 1 followed by how many zeros? n

65. What does Exercise 64 suggest for the value of 10^0? $10^0 = 1$

Use a calculator to evaluate each of the following.

66. $4^{0.5}$ 2   **67.** $9^{0.5}$ 3   **68.** $81^{0.5}$ 9   **69.** $100^{0.5}$ 10

70. What do you think $b^{0.5}$, for any positive number b, means? $\sqrt{b}$

60. Experimental; because the data was obtained from a sample.

63. $y = 2(x + 3)^2$ is a parabola shifted 3 units to the left and stretched vertically by a factor of 2.

LESSON 10.2 Multiplying and Dividing Monomials

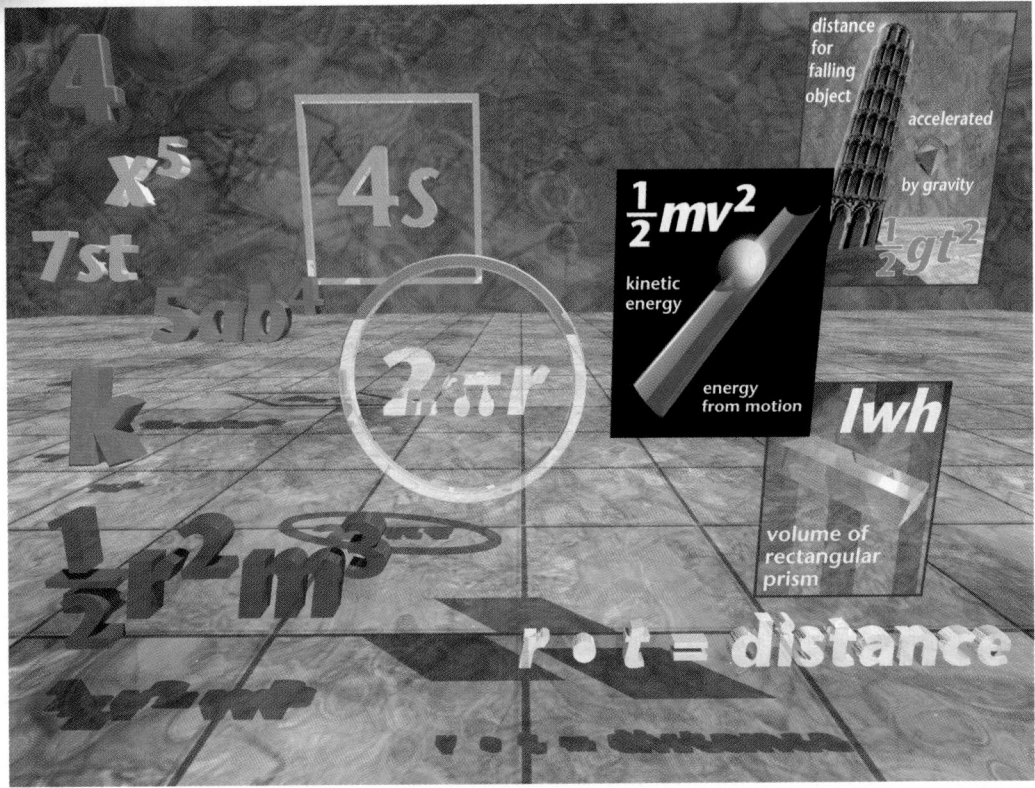

PREPARE

Objectives

- Understand the structure and concept of a monomial.
- Use properties of exponents to simplify expressions.

RESOURCES

- Practice Master   10.2
- Enrichment Master   10.2
- Technology Master   10.2
- Lesson Activity Master   10.2
- Quiz   10.2
- Spanish Resources   10.2

Assessing Prior Knowledge

Evaluate.

1. $(-2)^2$ **[4]**
2. $(-2)^3$ **[-8]**
3. $(2)(3)(4)$ **[24]**
4. $(2)(-3)(4)$ **[-24]**
5. $(-2)(-3)(-4)$ **[-24]**
6. $(-2)(3)(-4)$ **[24]**

TEACH

Computation with powers of variables can be simplified by the same properties you used for numerical powers.

In the previous lesson you discovered a property that lets you simplify the product of two powers with the same base. Just keep the base and add the exponents. For example,

$$x^3 \cdot x^5 = x^{3+5} = x^8.$$

PRODUCT-OF-POWERS PROPERTY

If x is any number and a and b are any positive integers, then

$$x^a \cdot x^b = x^{a+b}.$$

The Product-of-Powers Property can be used to find the product of more complex expressions such as $5a^2b$ and $-2ab^3$. Expressions like $5a^2b$ and $-2ab^3$ are *monomials*.

Using the properties of exponents often simplifies computation and may allow the student to complete the computation mentally. Explain to students that in this lesson they will formalize what they explored in Lesson 10.1.

Using Properties Have students discuss how the Commutative and Associative Properties are used when multiplying and dividing

monomials. For instance, $(2m)^2(-3m^3n) = 2m \cdot 2m \cdot (-3m^3n) = (2)(2)(-3) \cdot (m^2 \cdot m^3) \cdot n = -12m^5n$. This also applies for division:

$$\frac{-12(x^2y)^3}{3xy^2} = \frac{-12}{3} \cdot \frac{x^6}{x} \cdot \frac{y^3}{y^2} = -4x^5y.$$

TEACHING *tip*

To find $(5a^2b^2)(-2ab^3)$, ask students to write out the step-by-step method of finding the product so that they notice that shortcuts actually involve the Commutative and Associative Properties.

Alternate Example 1

Find each product.

a. $(-5a^2)(2ab^3)$ $[-10a^3b^3]$

b. $(-m^2n^3)(-3m)(-8m^4n)$
  $[-24m^7n^4]$

c. $(rst^2)(-st^3)(-r^3t)$
  $[r^4s^2t^6]$

CRITICAL *Thinking*

a. $x^{3c} \cdot x^{4c} = x^{(3c+4c)} = x^{7c}$

b. $(x+c)^d \cdot (x+c)^{2d} =$
  $(x+c)^{(d+2d)} = (x+c)^{3d}$

Alternate Example 2

Simplify.

a. $(-2^3)^2$ [64]

b. $(z^5)^4$ $[z^{20}]$

A **monomial** is an algebraic expression that is either a constant, a variable, or a product of a constant and one or more variables. A monomial can also contain powers of variables with positive integer exponents.

$(5a^2b^2)(-2ab^3) = -10(a^2 \cdot a^1)(b^2 \cdot b^3)$ Multiply the coefficients.
  Then group powers of like bases.

$= -10a^3b^5$ Product-of-Powers Property

EXAMPLE 1

Find the products.

Ⓐ $(6st)(-2s^2t)$

Ⓑ $(-4a^2b)(-ac^2)(3b^2c^2)$

Solution ➤

Multiply the coefficients and group powers of the same base, then use the Product-of-Powers Property.

Ⓐ $(6st)(-2s^2t) = -12s^3t^2$

Ⓑ $(-4a^2b)(-ac^2)(3b^2c^2) = 12a^3b^3c^4$ ✤

CRITICAL *Thinking*

Explain how to use the Product-of-Powers Property to simplify each of the following expressions.

a. $x^{3c} \cdot x^{4c}$ **b.** $(x+c)^d \cdot (x+c)^{2d}$

The Product-of-Powers Property can be used to simplify $(a^2)^3$.

$$(a^2)^3 = a^2 \cdot a^2 \cdot a^2 = a^{2+2+2} = a^6$$

Notice that $(a^2)^3 = a^{2 \cdot 3}$. To raise a power to a power, multiply the exponents.

POWER-OF-A-POWER PROPERTY

If x is any number and a and b are any positive integers, then

$$(x^a)^b = x^{ab}.$$

EXAMPLE 2

Simplify.

Ⓐ $(3^2)^4$

Ⓑ $(p^2)^5$

Solution ➤

Use the Power-of-a-Power Property to multiply the exponents.

Ⓐ $(3^2)^4 = 3^{2 \cdot 4} = 3^8$

Ⓑ $(p^2)^5 = p^{2 \cdot 5} = p^{10}$ ✤

ENRICHMENT

The volume of a rectangular prism is found by using the formula $V = lwh$. Have students find the volume of rectangular prisms with the following measurements.

Length	Width	Height
$2z$	$3y$	yz
$4a^2$	$2ab^3$	$3b$
$2m^2n$	$5mn^3$	$3m^3n^3$

INCLUSION strategies

Hands-on Strategies

Have students work at the board in groups of three. Write a problem on the board. One person in the group will be the "sign" person and determine whether the answer will need a negative sign. Another person will determine the numerical coefficient. A third person will select the monomials to be used. Check the answer. Then have students change roles, and present another problem.

Sometimes an exponential expression has a monomial for a base.

EXAMPLE 3

Simplify $(xy^2)^3$.

Solution ➤

The exponent, 3, outside the parentheses indicates that the monomial xy^2 is used as a factor 3 times. Simplify by regrouping and multiplying.

$$(xy^2)^3 = (xy^2)(xy^2)(xy^2)$$
$$= (x \cdot x \cdot x)(y^2 \cdot y^2 \cdot y^2)$$
$$= x^3y^6 \; ❖$$

POWER-OF-A-PRODUCT PROPERTY

If x and y are any numbers and n is a positive integer, then

$$(xy)^n = x^ny^n.$$

If a is the numerical coefficient, then $(axy)^n = a^nx^ny^n$. For example,

$$(3xy)^2 = (3xy)(3xy) = (3^2)(x^2)(y^2) = 9x^2y^2.$$

Remember to apply the exponent outside of the parentheses to *each* factor of the monomial inside the parentheses. This includes the coefficient, 3. Thus, $(3xy)^2 = 9x^2y^2$.

EXAMPLE 4

Simplify each expression.

Ⓐ $(3x^2y^3)^3$ Ⓑ $(-t)^5$ Ⓒ $(-t)^6$ Ⓓ $-t^4$ Ⓔ $(-5x)^3$

Solution ➤

Ⓐ $(3x^2y^3)^3 = (3)^3(x^2)^3(y^3)^3 = (3^3)(x^{2 \cdot 3})(y^{3 \cdot 3}) = 27x^6y^9$

Ⓑ $(-t)^5 = (-1 \cdot t)^5 = (-1)^5 \cdot t^5 = -1 \cdot t^5 = -t^5$

Ⓒ $(-t)^6 = (-1 \cdot t)^6 = (-1)^6 \cdot t^6 = 1 \cdot t^6 = t^6$

Ⓓ $-t^4$ is in simplest form. $(-1) \cdot t^4 = -t^4$
The exponent applies only to t and not to the -1.

Ⓔ $(-5x)^3 = (-5 \cdot x)^3 = (-5)^3 \cdot x^3 = -125x^3 \; ❖$

Explain whether $(-x)^m$ is positive or negative when x is positive and m is even. Do the same when m is odd. When are the exponents of a power added, and when are they multiplied?

RETEACHING the lesson

Cooperative Learning
Allow students to work together to make up study cards for each of the four properties studied in this lesson. Each study card should contain the following:

• Name of property
• Property written in correct mathematical form

• An example that uses only numbers
• An example that uses only variables and exponents (no coefficients)
• An example that has both numbers and variables

Students can then refer to these cards as needed when completing the homework in this lesson or in other lessons throughout the textbook.

Alternate Example 3

Simplify $(v^2w)^3$. $[v^6w^3]$

Alternate Example 4

Simplify each expression.

a. $(-2xy^2)^3$ $[-8x^3y^6]$
b. $-x^3$ **[is already in simplest form]**
c. $-(3m^2n^7)^4$ $[-81m^8n^{28}]$
d. $(-5)^4$ **[625]**
e. $(-5m^3)^3$ $[-125m^9]$

CRITICAL *Thinking*

When x is positive, $(-x)^m$ is negative if m is odd, and positive if m is even.

Exponents of a power are added if powers with the same base are multiplied. The exponents of a power are multiplied to find a power of a power.

Alternate Example 5

Simplify each expression.

a. $\dfrac{-30x^3y^4}{-5xy^3}$ $[6x^2y]$

b. $\dfrac{2m^5n^4}{-m^3n^7}$ $\left[\dfrac{-2m^2}{n^3}\right]$

ASSESS

Selected Answers

Odd-numbered Exercises 7–63

Assignment Guide

Core 1–5, 6–38 even, 39–68

Core Two-Year 1–64

Earlier you found that if you simplify quotients like $\dfrac{10^6}{10^2}$, the quotient can be written without a denominator. A similar property holds for variables and monomials.

EXAMPLE 5

Simplify. **Ⓐ** $\dfrac{-4x^2y^5}{2xy^3}$ **Ⓑ** $\dfrac{c^4b}{c^2a}$

Solution ➤

Simplify any numerical coefficients. Then subtract the exponent in the denominator from the exponent in the numerator with the same base.

Ⓐ $\dfrac{-4x^2y^5}{2xy^3} = \left(\dfrac{-4}{2}\right)(x^{2-1})(y^{5-3}) = -2xy^2$

Ⓑ $\dfrac{c^4b}{c^2a} = (c^{4-2})\left(\dfrac{b}{a}\right) = \dfrac{c^2b}{a}$ ❖

Try This   Simplify the expressions.

a. $\dfrac{6ab^2}{-2b}$ b. $\dfrac{(25^2)(-81t^3)}{45^3t}$

EXERCISES & PROBLEMS

Communicate

1. Explain why $3x^2$ is a monomial but $3x^{-2}$ is not.

2. Compare the Product-of-Powers Property with the Quotient-of-Powers Property.

3. Explain why the exponent of each factor of a monomial is multiplied by the outside exponent to calculate the power of a power.

4. Explain why $(x^a)^b = (x^b)^a$.

5. Explain why $y^2 \cdot y^4 \neq y^8$.

17. $2r$

18. $10x$

19. $\dfrac{-a^3}{2}$

20. $-\dfrac{p}{10}$

21. $\dfrac{2a^2b^6}{5}$

22. xy^3

23. $\dfrac{-40a^3b^5}{c^3}$

24. $200r^3$

25. $40u^2v^7w^2$

26. $8x^{12}$

27. $243b^{10}$

28. $4r^6$

29. $-1000m^{12}$

30. $625j^8k^{12}$

31. $9x^2y^8z^{10}$

32. $54a^6$

33. $250b^{10}$

34. $512n^6p^3$

35. $343j^6$

36. a^7b^5

Practice & Apply

Evaluate each of the following.

6. $(3x)^2$
$9x^2$

7. $\left(\dfrac{a}{b}\right)^3$ $\dfrac{a^3}{b^3}$

8. $\left(\dfrac{10x}{y^3}\right)^2$ $\dfrac{100x^2}{y^6}$

Find each product.

9. $(8r^2)(4r^3)$ $32r^5$

10. $(70x^4)(7x^3)$ $490x^7$

11. $(-2a^5)(4a^2)$ $-8a^7$

12. $(-p^2)(10p^3)$ $-10p^5$

13. $(-2a^2)(-5a^4)$ $10a^6$

14. $(-x^2y^5)(-x^3y^2)$ x^5y^7

15. $(48a^4b^2)(-0.2ac^5)$
$-9.6a^5b^2c^5$

16. $(4.3d^2n^{10}k)(0.1n^2k^3)$
$0.43d^2n^{12}k^4$

Find each quotient.

17. $\dfrac{8r^3}{4r^2}$

18. $\dfrac{70x^4}{7x^3}$

19. $\dfrac{-2a^5}{4a^2}$

20. $\dfrac{-p^4}{10p^3}$

21. $\dfrac{-2a^6b^7}{-5a^4b}$

22. $\dfrac{-x^4y^5}{-x^3y^2}$

23. $\dfrac{48a^4b^5}{-1.2ac^3}$

24. $\dfrac{0.8r^6}{0.004r^3}$

25. $\dfrac{4u^2v^{10}w^4}{0.1w^2v^3}$

Simplify each of the following.

26. $(2x^4)^3$

27. $(3b^2)^5$

28. $(-2r^3)^2$

29. $(-10m^4)^3$

30. $(5j^2k^3)^4$

31. $(3xy^4z^5)^2$

32. $2(3a^2)^3$

33. $10(-5b^5)^2$

34. $(8n^2p)^3$

35. $(7j^2)^3$

36. $(ab^5)(a^3)^2$

37. $(v^4w^3)^2(v^3)^4$

38. Evaluate the two monomials $2A^3$ and $(2A)^3$ for $A = 10$. 2000; 8000

39. Evaluate each of the following. $(-1)^1, (-1)^2, (-1)^3, (-1)^4, (-1)^5$

40. Use the pattern in Exercise 39 to find the value $(-1)^{100}$. Explain what you notice about $(-1)^n$ for various values of n.

41. 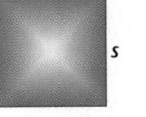 **Geometry** Find the area of a square if each side is twice as long as the side of the square shown. $4s^2$

s

42. 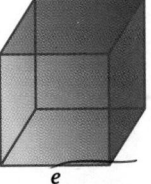 **Geometry** Find the volume of a cube if each edge is double the edge of the cube shown. $8e^3$

e

Technology Calculators can be used to evaluate expressions involving monomials. Store 10 for A in memory and 2 for B in memory on a graphics calculator. Use a calculator to evaluate each of the following.

43. A^2
100

44. B^3
8

45. $A^2 + B^3$
108

46. A^2B^3
800

47. $\dfrac{A^2}{B^3}$
12.5

48. $\dfrac{1}{A^2 + B^3}$
0.0093

Evaluate the following for $A = 9.8$ and $B = 2.1$.

49. A^2
96.04

50. B^3
9.261

51. $A^2 + B^3$
105.301

52. A^2B^3
889.42644

53. $\dfrac{A^2}{B^3}$

54. $\dfrac{1}{A^2 + B^3}$

55. Does $(10^2)^3$ equal 10^8 or 100^3? Explain your reasoning.

37. $v^{20}w^6$

39. $-1; 1; -1; 1; -1$

40. $(-1)^{100} = 1;$ $(-1)^n = 1$ if n is even, $(-1)^n = -1$ if n is odd.

53. 10.37

54. 0.0095

55. $(10^2)^3 = 100^3$ or $(10^2)^3 = 10^{2 \cdot 3} = 10^6$, not 10^8

Technology

The use of a calculator is appropriate for Exercises 43–54.

Error Analysis

Students may think that $\dfrac{(-x)^2}{-x^2}$ $= 1$. Point out that $(-x)^2 = x^2$, and $-x^2$ is just $-x^2$, so the quotient should be -1.

Technology Master

Technology
10.2 Visualizing Products

In this chapter, you have learned that $x^a \cdot x^b = x^{a+b}$. As you know, you apply this law of exponents when you multiply monomials. With a graphics calculator, you can visualize the factors and draw some conclusions about the products.

Use a graphics calculator to graph each monomial and the product $Y_1 \cdot Y_2$.

1. $Y_1 = 2x^2$ and $Y_2 = 3x^2$

2. $Y_1 = -2x^2$ and $Y_2 = -3x^4$

3. $Y_1 = 2x^3$ and $Y_2 = 3x^3$

4. $Y_1 = -2x^3$ and $Y_2 = -3x^3$

5. Suppose that you have two monomials rx^a and tx^b, where r and t are fixed numbers and a and b are positive integers. From your graphs in Exercises 1–4, what can you say about the sign of the product, rtx^{a+b}, if r and t have the same sign and $a + b$ is even?

Use a graphics calculator to graph each monomial and the product $Y_1 \cdot Y_2$.

6. $Y_1 = 2x^3$ and $Y_2 = -3x^2$

7. $Y_1 = 2x^3$ and $Y_2 = -3x^5$

8. $Y_1 = -2x^2$ and $Y_2 = 3x^3$

9. $Y_1 = 2x^2$ and $Y_2 = -3x^3$

10. When you multiply the monomials rx^a and tx^b, you get $(rt)x^{a+b}$. Based on the product graphs from Exercises 6–9, describe the basic shape(s) of the graph of $y = (rt)x^{a+b}$ when $a + b$ is even and when $a + b$ is odd.

Performance Assessment

Have students choose an example that illustrates each of the properties of powers. For each property and example, have students explain in their own words what the property mean and how it is demonstrated in the example.

Use Transparency 45

Look Beyond

Exercises 64–68 introduce some interesting and unexpected properties of the sum of two squares.

Look Back

Solve the equations for *n*. **[Lesson 4.6]**

56. $(4n - 20) - n = n + 12$ 16

57. $9(n + k) = 5n + 17k + 12$ $2k + 3$

58. Travel   Distance, *d*, varies directly as time, *t*. If a car travels 195 miles in 3 hours, determine the constant, *k*, and write the equation for the direct variation. What is the distance the car will travel in 4 hours?   **[Lesson 5.3]**
$k = 65$; $d = 65t$, 260 miles

Solve the following systems.
[Lessons 6.2, 6.3]

59. $\begin{cases} 5x - 3y = 32 \\ 2x = 12.8 \end{cases}$
(6.4, 0)

60. $\begin{cases} 7x - 3y = 2 \\ 2x - y = -5 \end{cases}$
(17, 39)

When traveling in foreign countries, the distances are usually given in kilometers.

Statistics
Use the data to find

61. the median.   3

62. the mean.   4

63. the mode.   **[Lesson 8.3]**
3

64. A math class has 28 students. Of those students, 16 are girls. What is the probability that a random selection of one student will be a boy?   **[Lesson 8.6]** $\frac{3}{7}$

Look Beyond

Work all four exercises to discover a property of sums of squares.

65. Show that $1^2 + 8^2 = 4^2 + 7^2$. Show that $14^2 + 87^2 = 41^2 + 78^2$. How were the numbers 14, 87, 41, and 78 chosen?

66. Does $17^2 + 84^2 = 71^2 + 48^2$? How were the numbers 17, 84, 71, and 48 chosen?

67. Show that $0^2 + 5^2 = 3^2 + 4^2$. Find new numbers *x* and *y* so that $(03)^2 + 54^2 = x^2 + y^2$.

68. Find new numbers *x* and *y* so that $(04)^2 + 53^2 = x^2 + y^2$. Show that $4^2 + 6^2 = 3^2 + 7^2$ is *false*. Does $43^2 + 67^2 = 34^2 + 76^2$?

---

65. $1^2 + 8^2 = 65$; $4^2 + 7^2 = 65$
$14^2 + 87^2 = 7765$; $41^2 + 78^2 = 7765$
14 and 87 are each composed of two digits, one from the base numbers in the two members (sides) of the equation $1^2 + 8^2 = 4^2 + 7^2$; the same is true for 41 and 78.

66. $17^2 + 84^2 = 7345$; $71^2 + 48^2 = 7345$
Yes; 17 and 71 are composed of the same two digits reversed, as are 84 and 48.

67. $0^2 + 5^2 = 25$; $3^2 + 4^2 = 25$
$03^2 + 54^2 = 30^2 + 45^2 = 2925$: $x = 30$, $y = 45$

68. $x = 40$, $y = 35$
$4^2 + 6^2 = 52$; $3^2 + 7^2 = 58$: so $4^2 + 6^2 \neq 3^2 + 7^2$; no

LESSON 10.3 Negative and Zero Exponents

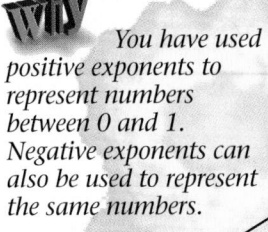

 *You have used positive exponents to represent numbers between 0 and 1. Negative exponents can also be used to represent the same numbers.*

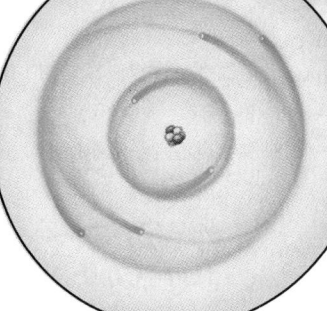

In chemistry, the approximate mass of a neutron or proton of an atom is 1.660×10^{-24} grams.

The expression 10^1 is defined to be 10. How should the expression 10^{-1} be defined?

Calculator

When the Quotient-of-Powers Property was defined, the exponent in the numerator was greater than the exponent in the denominator. For example, $\frac{10^4}{10^3} = 10^{4-3} = 10^1$. If the Quotient-of-Powers Property is applied when the exponent in the numerator is less than the exponent in the denominator, the result is $\frac{10^3}{10^4} = 10^{3-4} = \mathbf{10^{-1}}$.

If you write the powers as factors the result is

$$\frac{10^3}{10^4} = \frac{10 \cdot 10 \cdot 10}{10 \cdot 10 \cdot 10 \cdot 10} = \frac{1}{10}.$$

The results above indicate that 10^{-1} should be defined as $\frac{1}{10}$. How can you find 10^{-1} on your calculator?

ALTERNATIVE teaching strategy

Using Patterns Have students discuss the patterns of zeros and of exponents shown in this display. Help them find a simple pattern.

$10^4 = 10 \cdot 10 \cdot 10 \cdot 10 = 10,000$
$10^3 = \ 10 \cdot 10 \cdot 10 \ = 1000$

$10^2 =$	$10 \cdot 10$	$= 100$
$10^1 =$	10	$= 10$
$10^0 =$	1	$= 1$
$10^{-1} =$	$\frac{1}{10}$	$= 0.1$
$10^{-2} =$	$\frac{1}{100}$	$= 0.01$
$10^{-3} =$	$\frac{1}{1000}$	$= 0.001$

PREPARE

Objectives

- Understand the concept of negative and zero exponents.
- Simplify expressions containing negative and zero exponents.

RESOURCES

- Practice Master   10.3
- Enrichment Master   10.3
- Technology Master   10.3
- Lesson Activity Master   10.3
- Quiz   10.3
- Spanish Resources   10.3

Assessing Prior Knowledge

Evaluate.
1. $5 - 9$ **[−4]**
2. $-2 - 4$ **[−6]**
3. $-9 - (-9)$ **[0]**
4. $2 - (-4)$ **[6]**

TEACH

 In early 1995, another "quark" was discovered by scientists. Quarks are minute particles that protons and neutrons can be split into. Such developments in science show the necessity of being able to deal with extremely small numbers.

ongoing ASSESSMENT

Technology

On most calculators 10^{-1} can be evaluated in two ways.

a. 10 [x^{-1}] [ENTER].
b. 10 [y^x] $- 1$ [$=$].

Allow students to become familiar with how to use these keys on their calculators.

Lesson 10.3 **469**

NEGATIVE EXPONENT

If x is any number except zero and n is any integer, then

$$x^{-n} = \frac{1}{x^n}.$$

EXAMPLE 1

Simplify each expression.

Ⓐ $2^{-3} \cdot 2^2$ Ⓑ $\dfrac{10^3}{10^{-1}}$ Ⓒ $10^2 \cdot 3^2$

Solution ▸

Ⓐ Use the Product-of-Powers Property.

$$2^{-3} \cdot 2^2 = 2^{-3+2} = 2^{-1} = \frac{1}{2^1} = \frac{1}{2}$$

Ⓑ Use the Quotient-of-Powers Property. Subtract the exponents, and simplify the expression.

$$\frac{10^3}{10^{-1}} = 10^{3-(-1)} = 10^4 = 10,000$$

Ⓒ Since 10^2 and 3^2 do not have the same base, $10^2 \cdot 3^2$ cannot be simplified using the properties of exponents.

$$10^2 \cdot 3^2 = 100 \cdot 9 = 900 \; ❖$$

CRITICAL
Thinking

Compare the expressions $(-2)^3$ and $(-2)^{-3}$. Simplify each expression. How are the results related?

You have simplified expressions containing positive and negative exponents. What is the significance of an expression that contains a zero exponent?

·Exploration *Defining x^0*

1. Simplify each expression.

   a. $\dfrac{10^5}{10^5}$ b. $\dfrac{10^3}{10^3}$ c. $\dfrac{10^1}{10^1}$

2. Now use the Quotient-of-Powers Property to simplify each expression in Step 1.

3. What is the simplified numerical value of 10^0?

4. How would you define x^0 for any positive number x? ❖

Expressions containing variables with negative and zero exponents can be simplified if you use the properties of powers.

EXAMPLE 2

Simplify the following expressions.

A $c^{-4} \cdot c^4$ **B** $-3y^{-2}$ **C** $\dfrac{m^2}{n^{-3}}$

Solution ➤

A $c^{-4} \cdot c^4 = c^{-4+4} = c^0 = 1$
Does this agree with the definition you wrote in Step 4 of the Exploration?

B $-3y^{-2} = (-3)(y^{-2}) = \dfrac{-3}{y^2}$

C $\dfrac{m^2}{n^{-3}} = m^2 \cdot n^{-(-3)} = m^2 n^3$

Notice that $\dfrac{m^2}{n^{-3}} = \dfrac{m^2}{\frac{1}{n^3}} = m^2 n^3$. ❖

SUMMARY OF POWER PROPERTIES
Let a and b be any numbers with integer exponents m and n.

Product of Powers	Quotient of Powers ($b \neq 0$)	Power of a Power	Power of a Product
$b^m \cdot b^n = b^{m+n}$	$\dfrac{b^m}{b^n} = b^{m-n}$	$(b^m)^n = b^{mn}$	$(ab)^m = a^m b^m$

CRITICAL Thinking

If a and b are non-zero numbers and n is any integer, explain why the following is true.

$$\left(\dfrac{a}{b}\right)^{-n} = \left(\dfrac{b}{a}\right)^n$$

EXERCISES & PROBLEMS

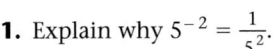
Communicate

1. Explain why $5^{-2} = \dfrac{1}{5^2}$.

2. Use a pattern to show that $2^0 = 1$. For what other bases does this apply?

3. Can $5^2 \cdot 4^{-3}$ be simplified using the properties of exponents? Why or why not?

4. Explain why $(3a)^{-2}$ does not equal $\dfrac{3}{a^2}$.

5. What is the meaning of a negative exponent?

6. What is the meaning of a zero exponent?

RETEACHING the lesson

Have students work in small groups. Each group will need two dice and a piece of paper to play this game. A line is drawn on the paper to separate it into equal parts. A "+" is written on one part and a "−" on the other. Each person takes a turn. A turn consists of rolling one die onto the paper to find the base number and determine whether it is positive or negative. Then the second die is rolled to determine the exponent. The player then must give the simplified value. If the value is correct, the player scores 1 point. The player with the most points after five turns is the winner. Optional scoring: Each player scores the value of his or her roll, and the scores are multiplied to find the game score after five turns.

Alternate Example 2
Simplify.

a. $n^{-3} \cdot n^4$ $[n]$

b. $4^{-2}a^2 b^{-1}$ $\left[\dfrac{a^2}{16b}\right]$

c. $\dfrac{x^{-5}}{y^{-5}}$ $\left[\dfrac{y^5}{x^5}\right]$

TEACHING tip

When completing exercises such as (c) in Alternate Example 2 above, students may try to reduce the answer because the exponents are the same. Point out that the **bases**, not the exponents, must be the same.

CRITICAL Thinking

$$\left(\dfrac{a}{b}\right)^{-n} = \dfrac{a^{-n}}{b^{-n}} = \dfrac{\frac{1}{a^n}}{\frac{1}{b^n}} = \dfrac{b^n}{a^n} = \left(\dfrac{b}{a}\right)^n$$

ASSESS

Selected Answers
Odd-numbered Exercises 7–45

Assignment Guide
Core 1–48

Core Two-Year 1–46

Use Transparency ▶ 46

Practice & Apply

Evaluate each expression.

7. -3^{-2} $-\frac{1}{9}$ **8.** $(-3)^2$ 9 **9.** 3^{-2} $\frac{1}{9}$ **10.** -3^2 -9

11. Copy the table and fill in the remaining entries by continuing the pattern.

Decimal form	10,000	1000	100	10	?	?	?
Exponential form	10^4	10^3	10^2	10^1	?	?	?

12. The corresponding numbers 10,000 and 10^4 are equal. Are the corresponding numbers you wrote in the table equal? yes

Write each of the following without negative or zero exponents.

13. 2^{-3} **14.** 10^{-5} **15.** a^3b^{-2} **16.** $c^{-4}d^3$

17. $v^0w^2y^{-1}$ **18.** $(a^2b^{-7})^0$ **19.** r^6r^{-2} **20.** $-t^{-1}t^{-2}$

21. $\frac{m^2}{m^{-3}}$ **22.** $\frac{2a^{-5}}{a^{-6}}$ **23.** $\frac{(2a^3)(10a^5)}{4a^{-1}}$ **24.** $\frac{b^{-2}b^4}{b^{-3}b^4}$

Probability The probability p of getting a multiple-choice question with four choices for each item correct by guessing is $\frac{1}{4}$, and the probability q of getting it wrong is $\frac{3}{4}$. On a test with 5 items, a formula for finding the probability of getting a certain number of items correct is shown in the chart.

25. Find the remaining values in the chart to the nearest thousandth.

No. Correct	Probability	Value
0	p^0q^5	$\left(\frac{1}{4}\right)^0\left(\frac{3}{4}\right)^5 \approx 0.237$
1	$5p^1q^4$	?
2	$10p^2q^3$	?
3	$10p^3q^2$	?
4	$5p^4q^1$	?
5	p^5q^0	?

26. Which number of correct items is most likely? 1

27. Which number of correct items is least likely? 5

28. Find the sum of all six probabilities in Exercise 25. 1

Technology Tell whether each of the following is easier to do on a calculator or easier to do mentally. Explain why you think so. Then simplify.

29. $\frac{2.56^7}{2.56^6}$ **30.** $\frac{2.56^6}{2.56^7}$ **31.** 0^7

32. 7^0 **33.** $(2.992 \times 9.554)^0$ **34.** $(19.43 \times 0)^{18}$

The answer to Exercise 11 can be found in Additional Answers beginning on page 729.

13. $\frac{1}{2^3}$ **17.** $\frac{w^2}{y}$ **21.** m^5

14. $\frac{1}{10^5}$ **18.** 1 **22.** $2a$

15. $\frac{a^3}{b^2}$ **19.** r^4 **23.** $5a^9$

16. $\frac{d^3}{c^4}$ **20.** $\frac{-1}{t^3}$ **24.** b

25. 0.396; 0.264; 0.088; 0.015; 0.001

29. mentally: subtract exponents to obtain 2.56^1 or 2.56.

30. mentally: except for the final division. $2.56^{-1} = \frac{1}{2.56} = 0.391$

31. mentally: any nonzero power of 0 is 0, so $0^7 = 0$.

32. mentally: the exponent 0 gives the value 1, so $7^0 = 1$.

33. mentally: the 0 exponent makes the value of the product 1.

35. Technology Try to compute 2^{236} on a calculator. What happens?

36. Try to compute 0^0 on a calculator. What happens?

37. Draw a graph of $Y_1 = 0^x$ for $x > 0$. What does this suggest the value of 0^0 should be?

38. Draw a graph of $Y_2 = x^0$ for $x > 0$. What does this suggest the value of 0^0 should be?

Look Back

39. What are the next three terms of the sequence 8, 11, 16, 23, 32 . . . ? **[Lesson 1.2]** 43, 56, 71

Solve each equation for y. How are the equations related? [Lesson 4.6]

40. $5(3y) = 3y - 24$ $y = -2$ **41.** $aby = by - c$ $y = \dfrac{-c}{ab - b}$

42. Draw the number line graph for the values of x that solve the inequality $3x - 6 > -4x + 12$. Describe the solution set of this inequality in words. **[Lesson 4.8]**

43. How do the slopes of vertical and horizontal lines differ?   **[Lesson 5.5]**

44. Transformations The graph of the quadratic parent function is stretched vertically by a scale factor of 2. It is then translated -3 units horizontally and 4 units vertically. Finally, it is reflected through the x-axis. Draw a graph and write a formula for this transformation. **[Lesson 9.6]**

45. Let $A = 3 \cdot 3 \cdot 3 \cdot 3 \cdot 3 \cdot x \cdot x \cdot x \cdot x \cdot x \cdot x \cdot y \cdot y \cdot y \cdot y$ and $B = 2 \cdot 2 \cdot 2 \cdot 2 \cdot 3 \cdot 3 \cdot 3 \cdot x \cdot x \cdot x \cdot x \cdot y$. Simplify each expression. **[Lesson 10.2]** $A = 243x^6y^4$; $B = 432x^4y$

46. Write the product $A \cdot B$ from Exercise 45 using powers.   **[Lesson 10.2]** $2^4 \cdot 3^8 x^{10} y^5$ or $104{,}976 x^{10} y^5$

Look Beyond

Cultural Connection: Europe   Pierre de Fermat (1601–1665) was a French mathematician who helped develop early number theory and probability. He studied numbers of the form $2^{2^n} + 1$, now called Fermat numbers.

47. Copy the table and complete the Fermat numbers.

48. Number Theory Fermat observed that each of the numbers was prime. He made the conjecture that all such numbers were prime. Though an unusually gifted mathematician, he was wrong in thinking the sixth number, $2^{2^5} + 1$, was prime, since it has a factor of 641. Use a calculator to find the other factor. Later research suggested that *no* more Fermat numbers are prime, so even great mathematicians can be wrong.

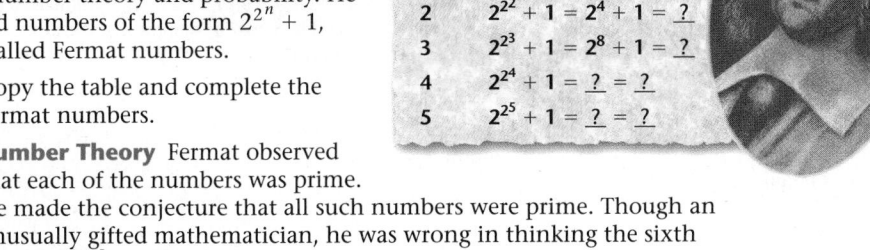

n	Fermat number
0	$2^{2^0} + 1 = 2^1 + 1 = 3$
1	$2^{2^1} + 1 = 2^2 + 1 = 5$
2	$2^{2^2} + 1 = 2^4 + 1 = \underline{?}$
3	$2^{2^3} + 1 = 2^8 + 1 = \underline{?}$
4	$2^{2^4} + 1 = \underline{?} = \underline{?}$
5	$2^{2^5} + 1 = \underline{?} = \underline{?}$

---

34. mentally: anything multiplied by 0 is zero, any nonzero power of 0 is 0.

35. On some calculators an error code appears because the value is too large. On other calculators the answer is given in scientific notation.

36. Error code appears because the value is undefined.

37. A horizontal line, lying on the x-axis, beginning just to the right of the origin. It suggests that the value of 0^0 should be 0.

38. a horizontal line at $y = 1$, beginning just to the right of the y-axis; it suggests that the value of 0^0 should be 1.

43. vertical: undefined slope; horizontal: slope 0

48. 6,700,417

MOTIVATE

Before students read the news excerpt, discuss the shuffling-card questions at the top of the page.

During the reading, consider pausing after the first seven paragraphs to make sure students are following the story. Ask questions like:
• *What did the mathematicians discover?*
• *Why was the problem complicated?*
• *Why does the article mention "analyzing speech patterns," which seems to have nothing to do with shuffling cards?*
• *What is the "dovetail" or "riffle" shuffle?*

After reading the body of the article, go over the article on page 475 titled *Getting Lost in the Shuffle*.
• Discuss the diagram, which illustrates one imperfect shuffle for a "deck" of 13 cards.
• Discuss the meaning of the graph. Talk about how the curve labeled *first shuffle* shows that after only one shuffle, the first card is much more likely to still be near the front of the deck than farther back in the deck. In looking at the curves for successive shuffles, help students see how a flatter curve means the card is <u>not</u> much more likely to be in one place than another.

All Mixed Up?

You are playing a card game and it's your turn to deal. How many times do you shuffle the deck? Is that enough to be sure the cards are thoroughly mixed? Is there such a thing as shuffling too much? Read the article below to learn how two mathematicians, one of them a magician as well, found some answers that surprised shufflers around the world.

In Shuffling Cards, 7 Is Winning Number

By Gina Kolata

It takes just seven ordinary, imperfect shuffles to mix a deck of cards thoroughly, researchers have found. Fewer are not enough and more do not significantly improve the mixing.

The mathematical proof, discovered after studies of results from elaborate computer calculations and careful observation of card games, confirms the intuition of many players that most shuffling is inadequate.

The finding has implications for everyone who plays cards and everyone who has a stake in knowing whether a shuffle is random. . . .

No one expected that the shuffling problem would have a simple answer, said Dr. Dave Bayer, a mathematician and computer scientist at Columbia who is coauthor of the recent discovery. Other problems in statistics, like analyzing speech patterns to identify speakers, might be amenable to similar approaches, he said. . . .

Dr. Persi Diaconis, a mathematician and statistician at Harvard University who is another author of the discovery, said the methods

used are already helping mathematicians analyze problems in abstract mathematics that have nothing to do with shuffling or with any known real-world phenomena.

Dr. Diaconis has been carefully watching card players shuffle for the past 20 years. The researchers studied the dovetail or riffle shuffle, in which the deck of cards is cut in half and the two halves are riffled together. They said this was the most commonly used method of shuffling cards. But, Dr. Diaconis said it produces a card order that is "far from random." . . .

Dr. Diaconis began working with Dr. Jim Reeds at Bell Laboratories and showed that a deck is perfectly mixed if it is shuffled between 5 and 20 times.

Next, Dr. Diaconis worked with Dr. Aldous and showed that it takes 5 to 12 shuffles to perfectly mix a deck.

In the meantime, he also worked on "perfect shuffles," those that exactly interlace the cards. He derived a mathematical proof showing that if a deck is perfectly shuffled eight times, the cards will be in the same order as they were before the shuffling.

To find out how many ordinary shuffles were necessary to mix a deck, Dr. Diaconis and Dr. Bayer watched players shuffle. He also watched Las Vegas dealers to see how perfectly they would interlace the cards they shuffled . . .

The researchers did extensive simulations of shuffling on a computer. To get the proof, the researchers looked at a lot of shuffles, guessed that the answer was seven, and finally proved it by finding an abstract way to describe what happens when cards are shuffled.

"When you take an honest description of something realistic and try to write it out in mathematics, usually it's a mess," Dr. Diaconis said. "We were lucky that the formula fit the real problem. That is just miraculous, somehow."

1. About 6 or 7 times greater (although other estimates may be reasonable, as well). The probability of the card still being first is about 0.07 and the probability of it being thirty-fifth is about 0.01.

2a. In a perfect shuffle, the cards from the two halves of the deck alternate: you get one card from one half, one from the other half, then one from the first half, and so on. In an ordinary shuffle, you may get two cards from one half, then three from the other, then one from the first, and so on.

Getting Lost in the Shuffle

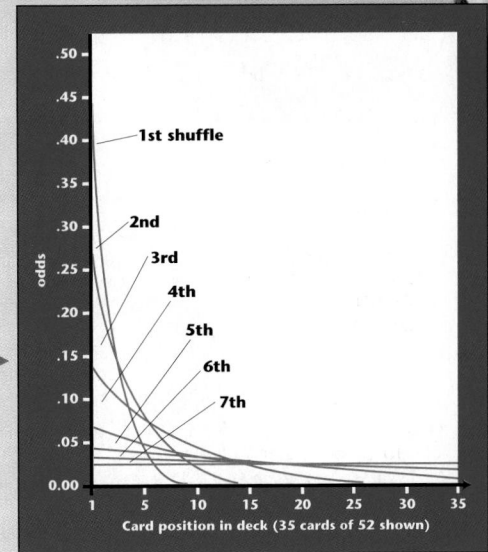

Even after a deck of cards is cut and shuffled, fragments of the original arrangement remain. In this example, the hearts are cut into two sequences: ace through six, and seven through king. Then the sequences are shuffled.

The cards are no longer ordered, but the sequence ace through six remains, with cards from the other fragment interspersed, also in sequence. The larger the number of cards in play, the more shuffles are required to mix them.

Curved lines show the odds that the first card in a deck will occupy any other position in the deck after one to seven shuffles. After one shuffle, for example, the first card is very likely to be one of the first few cards in the deck and very unlikely to be even five or six cards back. After four shuffles, it is still far more likely to be at the beginning of the deck than at the end. Only after seven shuffles does the card have about the same odds of being in any given position.

Graph labels: 1st shuffle, 2nd, 3rd, 4th, 5th, 6th, 7th

odds axis: .50, .45, .40, .35, .30, .25, .20, .15, .10, .05, 0.00

Card position in deck (35 cards of 52 shown): 1, 5, 10, 15, 20, 25, 30, 35

Cooperative Learning

1. Based on the graph in the article, after 4 shuffles, how many times more likely is the first card to be in the first position than in the 35th position? Explain.

2. a. What is the difference between a perfect shuffle and an ordinary shuffle?
   b. Is the shuffle in the diagram above perfect? Explain.

3. a. Use a diagram or model to find out how many perfect shuffles it takes for a deck to return to its original order if the deck consists of
   a. 2^2 cards.   b. 2^3 cards.   c. 2^4 cards.

4. a. How many perfect shuffles do you think it would take for a deck of 32 cards to return to its original order? Why?
   b. Use a diagram or model to check your prediction. What did you find?

5. Would a machine that always made perfect shuffles be useful when you play a card game? Why or why not?

2b. No. The first four cards alternate, but then the 3 of hearts should come before the 9 of hearts. Also, the 4 and 5 of hearts are together; they should be separated by a card from the right half of the deck.

3a. 2   **3b.** 3   **3c.** 4

4a. 5, because 32 is 2^5, which continues the pattern in activity 3.

4b. It does take 5 perfect shuffles for the deck of 32 cards to return to its original order.

5. No, because the machine wouldn't mix the cards randomly, and, with a certain number of shuffles, it might not mix them at all.

Objectives

- Recognize the need for a special notation to facilitate computation with very large and very small numbers.
- Perform computations using scientific notation, with and without a calculator.

RESOURCES

- Practice Master   **10.4**
- Enrichment Master   **10.4**
- Technology Master   **10.4**
- Lesson Activity Master **10.4**
- Quiz   **10.4**
- Spanish Resources   **10.4**

Assessing Prior Knowledge

Simplify.

1. $20 \cdot 100$ **[2000]**
2. $2.64 \cdot 10^2$ **[264]**
3. $10^3 \cdot 10^{-3}$ **[1]**
4. $200 \cdot 10^{-3}$ **[0.2]**
5. $1.36 \cdot 10^{-4}$ **[0.000136]**

TEACH

why Scientific notation is very helpful when working with very large and very small numbers. For example, when students study the planets, very large distances are involved. X-rays, on the other hand, involve the use of very small numbers.

LESSON 10.4 Scientific Notation

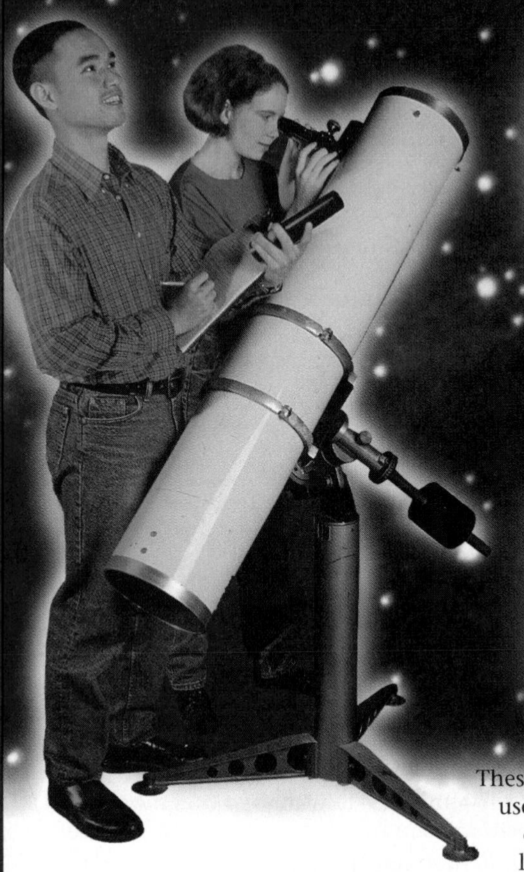

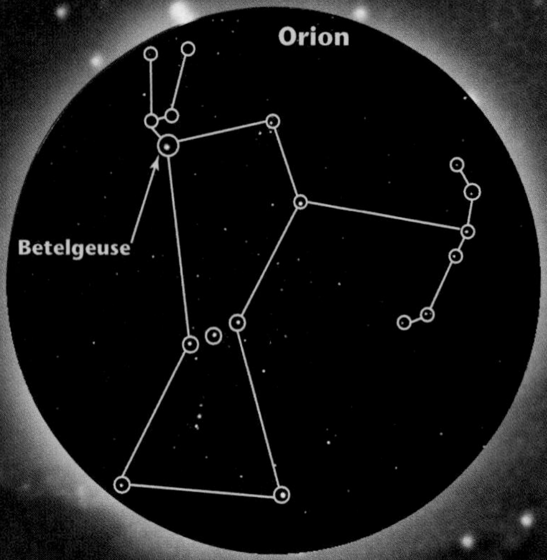

Orion

Betelgeuse

why *The study of the universe requires measurements of very large and very small quantities. The distance from the Earth to one of the stars, for instance, is about 6,000,000,000,000,000 miles. The wavelength of an X-ray is approximately 0.0000000001 meters.*

These numbers with all their zeros are hard to read and to use for calculations. They do not even fit on a calculator display. A special way to express these large and small quantities in a compact number form is to use *scientific notation*.

Astronomy   Betelgeuse is one of the stars in the constellation Orion, named for the great hunter of Greek and Roman mythology. The approximate distance in miles from Earth to Betelgeuse is written as a 6 followed by 15 zeros. In scientific notation this distance is written 6×10^{15}.

A number written in **scientific notation** is written with two factors, a number from 1 to 10, but not including 10, and a power of ten.

$$6 \times 10^{15}$$

First factor
Number from 1 to 10

Second factor
Power of 10

ALTERNATIVE teaching strategy

Using Discussion Have students write these numbers as you say them.

a. 4 million
b. 3.23 trillion
c. 16 ten-thousandths

Have students discuss how each number is written in standard form and in scientific notation. Then have students discuss how to multiply or divide two numbers that are written in scientific notation.

EXAMPLE 1

The distance from earth to Proxima Centauri, the nearest star other than the sun, is about 24,000,000,000,000 miles. It is read as 24 trillion miles. Write this number using scientific notation.

Solution ➤

To express the first factor, place the decimal point after the 2. This moves the decimal point 13 places to the left. To compensate for moving the decimal point, multiply the expression 2.4 by 10^{13}. This moves the decimal point 13 places to the right of 2. The two representations are now equivalent.

Customary Notation	Scientific Notation
24,000,000,000,000	2.4×10^{13} ❖

Try This Write 875,000 in scientific notation.

EXAMPLE 2

Space Travel If you travel at a speed of 1000 miles per hour, how long will it take to get to the star Betelgeuse?

Solution A ➤

Find $6 \times 10^{15} \div 1000$. If you use long division,

$$1000 \overline{)6,000,000,000,000,000} = 6 \times 10^{12}.$$
$$6,000,000,000,000$$

The flight will take about 6 trillion hours!

Solution B ➤

You can also use the properties of exponents.

$$6 \times 10^{15} \div 1000 = \frac{6 \times 10^{15}}{10^3} = 6 \times 10^{15-3} = 6 \times 10^{12} ❖$$

Try This How many years are there in 6 trillion hours?

Numbers for small quantities can also be written in scientific notation. Recall that a decimal such as 0.01 can be written with a negative exponent, 10^{-2}.

interdisciplinary
CONNECTION

Astronomy Have students find the distance of each planet in our solar system from the Sun. Then have them find how long it takes light from the Sun to reach each planet.

Alternate Example 1

Light travels at about 300,000,000 meters per second. Write that speed in scientific notation. [3×10^8]

ongoing
ASSESSMENT

Try This

875,000 [8.75×10^5]

Alternate Example 2

The distance from the Moon to Earth is about 200,000 miles. The distance from the Sun to Earth is about 93 million miles. About how many times farther is it from Earth to the Sun than from the Sun to the Moon? [465]

ongoing
ASSESSMENT

Try This

684,931,506.86 at 365 days per year; 684,462,696.8 at 365.25 days per year

Alternate Example 3

UV (ultraviolet) rays cause sunburn. The wavelength of a UV ray is 0.00000028 meter. Write this number in scientific notation. [2.8×10^{-7}]

Try This

4.02×10^{-8}

Alternate Example 4

How many times longer is a radio wave of 10^0 meters than a gamma ray of 10^{-14} meters. [10^{14}]

CRITICAL *Thinking*

$12 \times 10^{-2} =$
$1.2 \times 10^1 \times 10^{-2} =$
1.2×10^{-1}

$0.12 \times 10^{-2} =$
$1.2 \times 10^{-1} \times 10^{-2} =$
1.2×10^{-3}

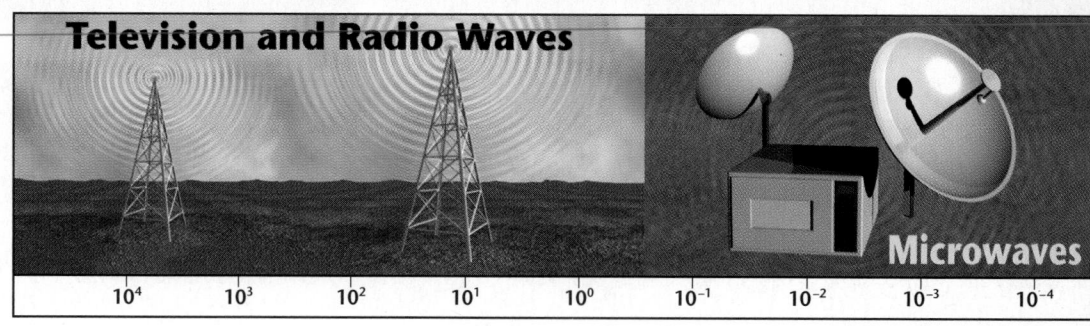

Television and Radio Waves

Microwaves

| 10^4 | 10^3 | 10^2 | 10^1 | 10^0 | 10^{-1} | 10^{-2} | 10^{-3} | 10^{-4} |

EXAMPLE 3

A typical X-ray has a wavelength of about 0.0000000001 meter. Write this number in scientific notation.

Solution ➤

To express the first factor, place the decimal after the 1. This, in effect, moves the decimal point 10 places to the right.

To compensate for moving the decimal point, multiply 1 by 10^{-10}. This is equivalent to dividing by 10^{10}, which moves the decimal point 10 places to the left, where it belongs.

Customary Notation	**Scientific Notation**
Thus, 0.0000000001	= 1×10^{-10} ❖

Try This Write 0.0000000402 in scientific notation.

Cultural Connection: Europe About 300 years ago, the English mathematician, Sir Isaac Newton, passed a ray of sunlight through a prism. Newton observed the colors of the rainbow. Today we know that visible light is only part of a larger set of waves that makes up the electromagnetic spectrum.

EXAMPLE 4

Physics How many times longer is a microwave of 10^{-2} meters than an X-ray of 10^{-10} meters?

Solution ➤

Divide the length of the microwave by the length of the X-ray.

$$\frac{10^{-2}}{10^{-10}} = 10^{-2-(-10)} = 10^{-2+10} = 10^8$$

The microwave is 10^8, or 100 million, times longer than the X-ray. ❖

CRITICAL *Thinking* How can you change 12×10^{-2} and 0.12×10^{-2} to scientific notation using mental arithmetic?

ENRICHMENT Have students read about ultra violet (UV) radiation and find out the difference between UVB and UVA rays. Which are shorter? [**UVB rays are shorter and cause sunburn. UVA rays are longer and cause wrinkling and aging of skin.**]

INCLUSION strategies **Inviting Participation** Students with vision problems may have difficulty reading numbers that have many zeros. Point out that using scientific notation can ease this problem. However, for Exercises 5–12 on page 480, you may wish to furnish visually challenged students with enlargements or with copies that are written with extra space between digits.

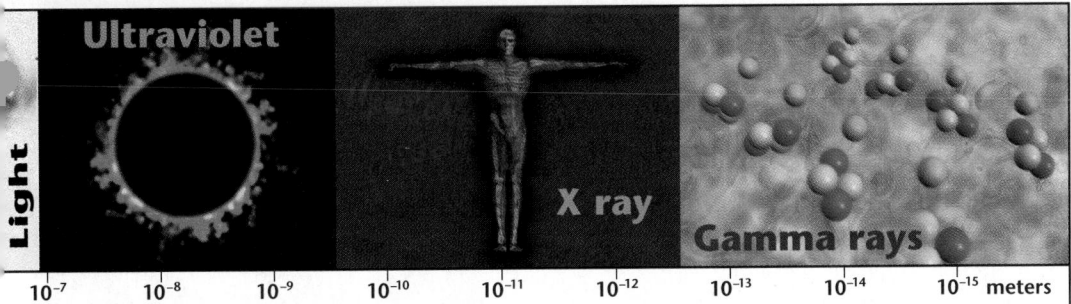

When you multiply or divide two numbers written in scientific notation, you are working with pairs of numbers that are each represented by two factors. Rearrange the factors to simplify the expression.

EXAMPLE 5

Express the product and quotient in scientific notation.

A $(3 \times 10^3)(4 \times 10^{-5})$ **B** $\dfrac{2 \times 10^6}{5 \times 10^2}$

Solution ➤

A
$$\begin{aligned}(3 \times 10^3)(4 \times 10^{-5}) &= (3 \cdot 4)(10^3 \times 10^{-5}) \\ &= 12 \times 10^{-2} \\ &= 1.2 \times 10^{-1}\end{aligned}$$

B
$$\begin{aligned}\frac{2 \times 10^6}{5 \times 10^2} &= \left(\frac{2}{5}\right)(10^6 \times 10^{-2}) \\ &= 0.4 \times 10^4 \\ &= 4 \times 10^3 ❖\end{aligned}$$

Most computations with scientific notation can be done on a calculator. Check to see if the calculator you use has an [EE] key or [EXP] key.

EXAMPLE 6

Calculator

The speed of light is about 2.98×10^5 kilometers per second. If the sun is about 1.49×10^8 kilometers from the Earth, how long does it take sunlight to reach Earth?

Solution ➤

Depending on your calculator use the [EE] or [EXP] key.

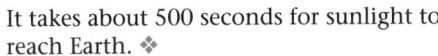

1.49E8/2.98E5
500

It takes about 500 seconds for sunlight to reach Earth. ❖

Alternate Example 5

Express the product or quotient in scientific notation.

a. $(5 \times 10^{-4})(3 \times 10^{-2})$
[1.5×10^{-5}]

b. $\dfrac{3 \times 10^5}{8 \times 10^7}$ [3.75×10^{-3}]

TEACHING *tip*

Technology Be sure to allow students time to find out how their calculators handle scientific notation. For example, ask students to record the number 489,000,000,000 in scientific notation.

Alternate Example 6

In 1990, the United States population was about 250 million people and the federal budget deficit was $220 billion. How much would each American have had to pay to erase the deficit in 1990? $\left[\dfrac{2.2 \times 10^{11}}{2.5 \times 10^8}\right. =$

8.8×10^2 or about 880]

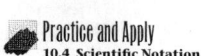

EXERCISES & PROBLEMS

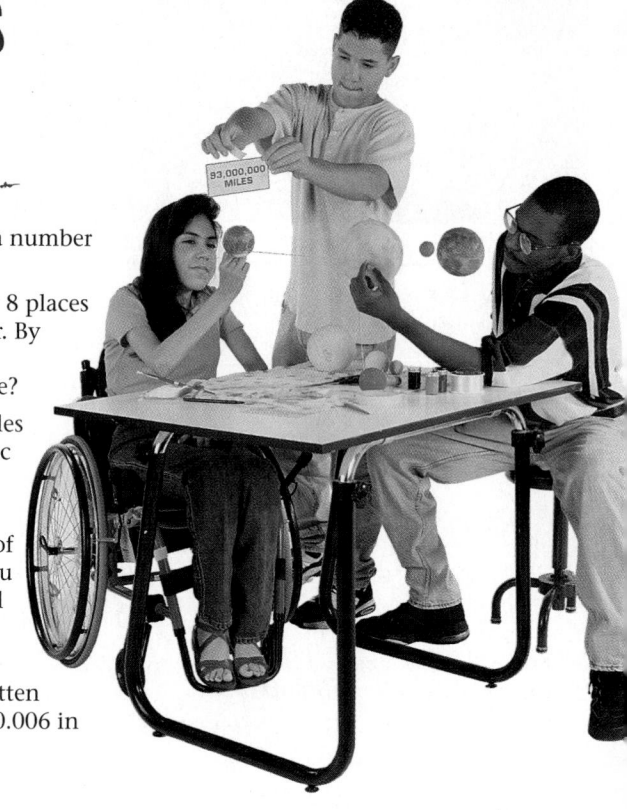

Communicate

1. Why is it important to be able to write a number with many zeros in scientific notation?

2. Explain what moving the decimal point 8 places to the left does to the value of a number. By what power of 10 must you multiply to compensate for that decimal point move?

3. Explain how to write the distance in miles from the Earth to the sun using scientific notation.

4. Explain what moving the decimal point 12 places to the right does to the value of a number. By what power of 10 must you multiply to compensate for that decimal point move?

5. Explain how to use the properties of exponents to multiply two numbers written in scientific notation. Use 240,000 and 0.006 in your discussion.

Practice & Apply

Write each number in scientific notation.

6. 2,000,000   7. 8,000,000,000   8. 340,000   9. 58,000

10. 0.00008   11. 0.0000005   12. 0.000234   13. 0.000000082

Write each number in customary notation.

14. 3×10^4 15. 4×10^8 16. 6.7×10^{10} 17. 9.01×10^5

18. 4×10^{-7} 19. 5×10^{-9} 20. 8.8×10^{-12} 21. 7.2×10^{-10}

Social Studies The following information shows how large numbers appear when studying different aspects of American culture. Write each in scientific notation.

22. 125,000: the number of passengers bumped on United States airlines each year due to overbooking 1.25×10^5

23. 5 trillion: the total value of outstanding stock in the stock market 5×10^{12}

24. 5.4 million: the number of American businesses owned by women 5.4×10^6

25. 3.7 million: the number of square feet in the Pentagon, the world's largest office building 3.7×10^6

Perform the following computations. Answer using scientific notation.

26. $(2 \times 10^4)(3 \times 10^5)$ **27.** $(6 \times 10^8)(1 \times 10^6)$ **28.** $(8 \times 10^6)(2 \times 10^{10})$

29. $(9 \times 10^6)(7 \times 10^6)$ **30.** $(9 \times 10^6) + (7 \times 10^6)$ **31.** $(9 \times 10^6) - (7 \times 10^6)$

32. $(8.2 \times 10^6)(3.1 \times 10^6)$ **33.** $(1.9 \times 10^8)(2 \times 10^{10})$ **34.** $\dfrac{8 \times 10^6}{2 \times 10^2}$

35. $\dfrac{9 \times 10^8}{2 \times 10^4}$ **36.** $\dfrac{3 \times 10^{10}}{6 \times 10^4}$ **37.** $\dfrac{2 \times 10^6}{5 \times 10^5}$

38. $\dfrac{(2 \times 10^6)(9 \times 10^4)}{6 \times 10^4}$ **39.** $(2 \times 10^{10})(6 \times 10^7)(4 \times 10^{-8})$

Economics The national debt for the United States in 1992 was listed as 4,064.6 billion dollars.

40. Write 4,064.6 billion in scientific notation. 4.0646×10^{12}

41. This can be expressed as about 4 ? dollars. trillion

42. Assuming the population of the United States was about 250 million in 1992, how much debt was this per person? about $16,000

43. Write your age at your last birthday in *seconds*. Use scientific notation.
1 year = 3.1536×10^7 seconds

44. Astronomy Write the distance in miles from Earth to the moon in scientific notation. 2.48×10^5

The distance from Earth to the moon is about 248,000 miles.

Astronomy The distance light travels in one year, 5.87×10^{12} miles, is called a light-year.

45. Write this number in customary notation. 5,870,000,000,000

46. If the distance from Betelgeuse to Earth is 6×10^{15} miles, how long does it take light to travel to Earth from that star? $\approx 1.022 \times 10^3$

47. If the distance from Proxima Centauri to Earth is 2.4×10^{13} miles, how long does it take light to travel to Earth from Proxima Centauri? ≈ 4.089

48. Science According to Nobel Prize-winning physicist Leon Lederman, the universe is 10^{18} seconds old. How many years is this? Write your answer in words. 31.7 billion

Sometimes scientific notation is used to indicate a level of accuracy. For example, an attendance figure for a rock concert, given as 1.2×10^4, (or 12,000), would be assumed to be given to the nearest *thousand*, while a figure of 1.20×10^4 (also 12,000) would be assumed to be given to the nearest hundred. This is because the appearance of the 0 in 1.20 would indicate that it is significant. Similarly, 1.2000×10^4 would indicate that all the zeros are significant, so it is accurate to the nearest *one*.

Write the following in scientific notation to indicate which digits are significant.

49. 76,000 to the nearest thousand **50.** 76,000 to the nearest hundred

51. 8,000,000 to the nearest million **52.** 8,000,000 to the nearest thousand

53. 8,000,000 to the nearest hundred **54.** 8,000,000 to the nearest ten

55. 0.002 to the nearest thousandth **56.** 0.0020 to the nearest ten-thousandth

26. 6×10^9 **34.** 4×10^4 **51.** 8×10^6

27. 6×10^{14} **35.** 4.5×10^4 **52.** 8.000×10^6

28. 1.6×10^{17} **36.** 5×10^5 **53.** 8.0000×10^6

29. 6.3×10^{13} **37.** 4×10^0 **54.** 8.00000×10^6

30. 1.6×10^7 **38.** 3×10^6 **55.** 2×10^{-3}

31. 2×10^6 **39.** 4.8×10^{10} **56.** 2.0×10^{-3}

32. 2.542×10^{13} **49.** 7.6×10^4

33. 3.8×10^{18} **50.** 7.60×10^4

57. A **googol** is the number formed by writing 1 followed by 100 zeros. Write this number in scientific notation. 1×10^{100}

Technology Write the following numbers in customary notation.

58. 2.3 E 04   **59.** 5.6 E 03   **60.** 7.22 E −03

61. 1.01 E −04   **62.** −2.8 E 02   **63.** −9.303 E −04

64. Technology What is the largest number that can be shown on your calculator without using scientific notation?

65. What is the largest number that can be shown on your calculator using scientific notation?

66. What is the smallest number that can be shown on your calculator without using scientific notation?

67. What is the smallest number that can be shown on your calculator using scientific notation?

Look Back

68. 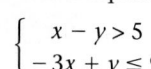 **Statistics** The scores on a test and a measure of student behavior have a correlation coefficient of $r = -0.90$. If there are 10 subjects, draw the scatter plot as you think it might look. **[Lesson 1.6]**

69. Find the value of the expression. $3 + (-4) - (-9) - [-8 + 2(3 - 5)]$ **[Lessons 3.2, 4.1]** 20

70. Find the product. $(5)(-3)[-2(4)](-12)$ **[Lesson 4.1]** −1440

71. Simplify. $(6x + 4y - 8) - 5(-7x + y)$ **[Lesson 4.2]** $41x - y - 8$

72. **Coordinate Geometry** Given points on the coordinate graph $A(3, 7)$ and $B(-3, 12)$, find the slope of the line joining the points. **[Lesson 5.1]** $-\frac{5}{6}$

73. Graph the following inequalities, and describe the region that solves both inequalities. **[Lesson 6.5]**

$$\begin{cases} x - y > 5 \\ -3x + y \le 9 \end{cases}$$

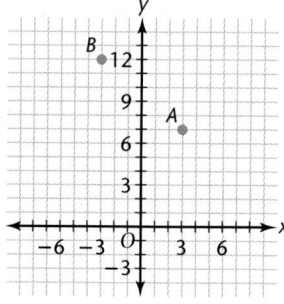

Look Beyond

Technology Use the constant feature of your calculator to write the following sequences in customary notation.

74. $0.9^1, 0.9^2, 0.9^3, 0.9^4, 0.9^5, \ldots$

75. $1.1^1, 1.1^2, 1.1^3, 1.1^4, 1.1^5, \ldots$

76. Compare what happens when these numbers are raised to higher and higher powers.

The answers to Exercises 68 and 73 can be found in Additional Answers on page 729.

58. 23,000   **61.** 0.000101

59. 5600   **62.** −280

60. 0.00722   **63.** −0.0009303

64–67. Answers may vary depending on the brand of calculator: check the manual.

74. $0.9, 0.81, 0.729, 0.6561, 0.59049, \ldots$

75. $1.1, 1.21, 1.331, 1.4641, 1.61051, \ldots$

76. 1.1^n keeps getting larger; 0.9^n gets smaller, approaching 0.

LESSON 10.5 Exponential Functions

The News
Weather Cold & Snow

World Population Hits 5.5 Billion—Growing at a Rate of 1.7% per year

Why *An exponential function models problems of growth and decline in population, finance, and science.*

Social Studies To better understand the effects of population growth it is helpful to understand the underlying mathematics. This involves a look back to the exponential function introduced in the second chapter.

According to the headline above, how many people are added to the population in one year? How many are added in one minute?

$$1.7\% \times 5.5 \text{ billion} = 0.017 \times 5,500,000,000$$
$$= 93,500,000 \text{ (in one year)}$$
$$\approx 256,000 \text{ (in one day)}$$
$$\approx 10,700 \text{ (in one hour)}$$
$$\approx 178 \text{ (in one minute)}$$

In the time it took you to read this far, the world population probably increased by several dozen people.

CRITICAL Thinking Explain how the increase for one day was obtained from the increase for one year.

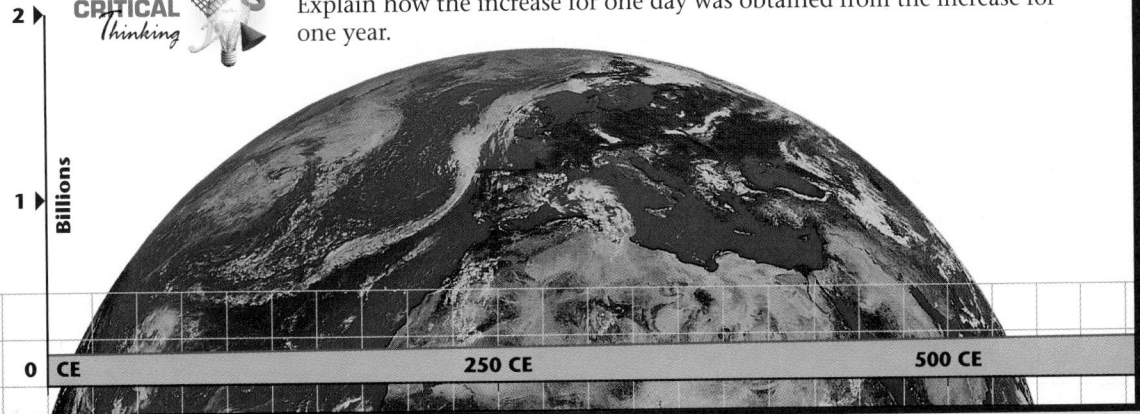

(vertical axis: Billions, marked 0, 1, 2; horizontal axis: CE, 250 CE, 500 CE)

ALTERNATIVE teaching strategy

Cooperative Learning Pose this situation: A class of 20 students enters kindergarten. The average increase per year in the number of students is 5% (that means that there will be one more student in first grade). Have students use their calculator and the general growth formula to find out how many students will be in the class when they graduate from high school. **[36 students]**
Have students find the number of graduates if the growth rate were
10%   **[63 students]**,
15%   **[107 students]**,
20%   **[178 students]**,
25%   **[271 students]**.

PREPARE
Objectives
- Understand the exponential function and how it is used.
- Identify the graph of the exponential function for different bases.

RESOURCES

- Practice Master   10.5
- Enrichment Master   10.5
- Technology Master   10.5
- Lesson Activity Master   10.5
- Quiz   10.5
- Spanish Resources   10.5

Assessing Prior Knowledge

Use a calculator to find each of the following.

1. 3^3 **[27]**
2. 3^5 **[243]**
3. 3^7 **[2187]**
4. 3^9 **[19,683]**

TEACH

Why  Ask students how many natural grandparents each student has. **[4]** How many great-grandparents? **[8]** How many great-great-grandparents? **[16]** Tell students that the number of grandparents increases exponentially and ask what they think that means. Explain by drawing a chart.

Self	$2^0 = 1$
Parents	$2^1 = 2$
Grandparents	$2^2 = 4$
Great-grandparents	$2^3 = 8$

CRITICAL Thinking

The figure for one year was divided by 365.25 and rounded to the nearest thousandth.

Alternate Example 1

A town of 3.5 thousand people expects to grow at the rate of 10% per year. Find the town's expected population in

5 years   [**5.6 thousand**],
10 years   [**9 thousand**],
20 years. [**23.6 thousand**]

Cooperative Learning

Have students work in small groups. Each group will choose a town, a group, or an organization, and find its current growth rate. Then they will find its anticipated population in 5 years and in 10 years. Each group can also decide whether the results are practical.

EXAMPLE 1

(A) Estimate the world's population in the year 2000.

(B) If population growth continues at the same rate, how many people will be added to the population each minute in the year 2000?

Solution ➤

B3	=B2*1.017	
	A	**B**
1	Year	Population
2	1990	5.50
3	1991	**5.59**
4	1992	5.69
5	1993	5.79
6	1994	5.88
7	1995	5.98
8	1996	6.09
9	1997	6.19
10	1998	6.29
11	1999	6.40
12	2000	6.50

(A) Start with the 1990 population of 5.5 billion. Since the population increases by 1.7% each year, the new population will be 101.7%, or 1.017, times the old population. Use 1.017 as a multiplier to get the population for 1991, 1992, 1993, and so on.

If you round to the nearest 0.1 billion, the population in the year 2000 will be about 6.5 billion.

(B) If this new population of 6.5 billion increases by 1.7%, the number of people added is calculated as follows.

$$1.7\% \times 6.5 \text{ billion} = 0.017 \times 6,500,000,000$$
$$= 110,500,000 \text{ (in one year)}$$
$$\approx 303,000 \text{ (in one day)}$$
$$\approx 12,600 \text{ (in one hour)}$$
$$\approx 210 \text{ (in one minute)}$$

This means that while the population was increasing by 178 people per minute in 1990, it will be increasing by 210 people per minute in 2000. ❖

What is the formula for finding the population if you know its rate of growth? Examine this population pattern.

Years	Population (billions)	Exponential notation
0	5.5	$5.5(1.017)^0$
1	5.5(1.017)	$5.5(1.017)^1$
2	5.5(1.017)(1.017)	$5.5(1.017)^2$
3	5.5(1.017)(1.017)(1.017)	$5.5(1.017)^3$

750 CE   1000 CE   1250 CE

interdisciplinary CONNECTION **Social Studies** Have students use a social studies textbook as a source of data for writing word problems that involve exponential growth.

ENRICHMENT Have students graph $f(x) = 0.5^x$ and $f(x) = 2^x$ on the same axes. Have students compare and contrast the graphs. [**If the base > 1, the value of $f(x)$ increases. If the base < 1, the value of $f(x)$ decreases.**]

Each year the world population increases by another factor of 1.017. The population, *P* in billions, after *t* years can be represented by

$$P = 5.5(1.017)^t.$$

The formula $P = 5.5(1.017)^t$ shows the population, *P*, after *t* years, if the original population is 5.5 (billion) and the rate of growth is 1.7% or 0.017. This leads to a general formula for yearly growth.

GENERAL GROWTH FORMULA

Let *P* be the amount after *t* years at a yearly growth rate of *r*, expressed as a decimal. If the original amount is *A*, then

$$P = A(1 + r)^t.$$

This formula works for population, money, and many other situations where growth takes place exponentially.

EXAMPLE 2

Social Studies   India had a 1992 estimated population of about 886 million and was growing at a yearly rate of 1.9%. How much will the population increase in 10 years?

Solution ➤

Use the population growth formula. If you have a calculator use the [y^x] or [^] key. You may also use a constant multiplier of 1.019 in your computation.

$$P = A(1 + r)^t$$
$$= 886(1.019)^{10}$$
$$\approx 1069$$

The increase is $1069 - 886 = 183$ or about 183 million people. ❖

Try This   Suppose the population of a country was 200,000,000 inhabitants at the last census. If the population continues to grow at an annual rate of 1.5%, how much will the population increase in the next 6 years?

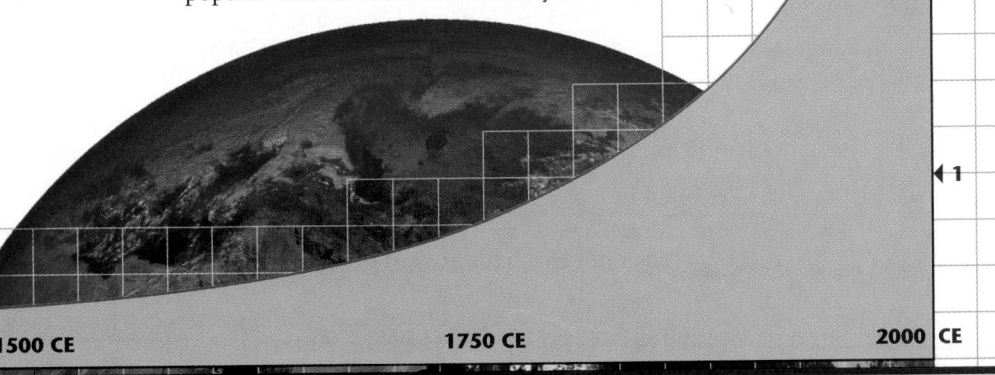

Billons

8
7
6
5
4
3
2
1

1500 CE   1750 CE   2000 CE

Alternate Example 2

Suppose your town is growing at a rate of 0.5% per year. Find the expected population of your town in 10 years. [**Answers will vary depending on the town's current population.**]

TEACHING *tip*

Technology   Be sure students understand how to use the necessary keys for exponential equations.

ongoing ASSESSMENT

Try This

Approximately 19 million people

INCLUSION strategies

Hands-On Strategies   Have students fold a sheet of typing paper in half, and then fold it in half again. Ask them how many times they think they can fold that piece of paper in half in that way. Then have students write an exponential expression to show the maximum number of times they actually could fold the paper in half. Then have them evaluate that expression to find how many sections the typing paper should have if the folding were properly done.   [**Most students will only be able to fold the paper 6 times, $2^6 = 64$.**]

Ask students whether a larger sheet of paper would make a difference in the number of folds. [**Maybe, but probably only one or two more folds can be made.**]   Ask students whether having a thinner sheet of paper would make a difference. [**Probably, but the maximum number of folds still would be about 10.**]

Alternate Example 3

Make a table of values. Graph and describe the function $f(x) = 3^x$. Show that $f(x) = 3^x$ is a function. [Check students' graphs.

x	0	1	2	3	4
3^x	1	3	9	27	81

Alternate Example 4

Graph $f(x) = 2.5^x$ and $f(x) = 3.5^x$. Compare and contrast the graphs. [Check students' graphs. They both cross the *y*-axis at $y = 1$. Each gets closer and closer to the *x*-axis as *x* decreases. The graph of $f(x) = 3.5^x$ rises faster than the graph of $f(x) = 2.5^x$ for values of *x* greater than 0.]

Aongoing ASSESSMENT

For the graphs of $f(x) = a^x$, for $a > 0$, as *a* increases the graph rises faster.

Practice Master

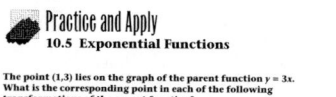

Practice and Apply
10.5 Exponential Functions

The point (1,3) lies on the graph of the parent function $y = 3x$. What is the corresponding point in each of the following transformations of the parent function?

1. $y = 3^x + 2$ _____
2. $y = 3^{2x}$ _____
3. $y = 2 \cdot 3^x$ _____
4. $y = 3^x - 2$ _____

According to the founders of a leading manufacturer of computer chips, the growth of the number of transistors on a chip since 1970 is modeled by the equation $y = 3^{x-1}$.

5. Use a spreadsheet or calculator to complete the table.

Year	1975	1980	1985	1990	1995
Population					

6. Use this information to estimate the growth of the number of transistors on a chip in 1988. _____

7. If the number of transistors on a chip continues to increase at the same rate, how many times greater will the number of transistors on a chip be in the year 2000 than in the year 1970? _____

Match each graph with the letter that indicates the function.

A. $y = 1^x$ B. $y = 0.1^x$ C. $y = -(1^x)$ D. $y = (10)^{x-1}$

8.    9.   10.   11.

If $f(x) = \left(\frac{1}{2}\right)^x$, find the following values for the function.

12. $f(2)$ _____
13. $f(0)$ _____
14. $f(-1)$ _____
15. $f(-2)$ _____

EXAMPLE 3

Make a table of values. Graph and describe the function $f(x) = 2^x$. Show why it is a function.

Solution ➤

x	−3	−2	−1	0	1	2	3
2^x	0.125	0.25	0.5	1	2	4	8

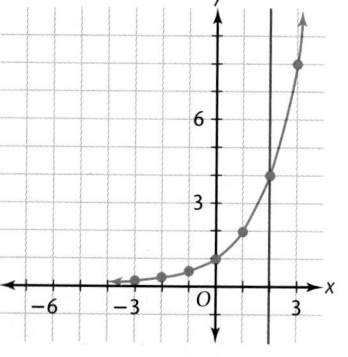

The function $f(x) = 2^x$ shows the characteristics of an exponential function. The graph remains above the *x*-axis for all values of *x*. When *x* is 0 the function equals 1. After crossing the *y*-axis the function increases rapidly.

The vertical line test shows that $f(x) = 2^x$ is a function.

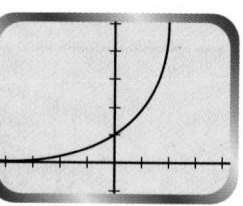

A calculator produces the same distinctive curve. ❖

EXAMPLE 4

Graph $y = 3^x$ and $y = 4^x$ on the same axes. Explain how they are similar and how are they different.

Solution ➤

You can graph each function using a graphics calculator, or you can make a table of values and graph them by hand. Include positive and negative values of *x*.

Graphics Calculator

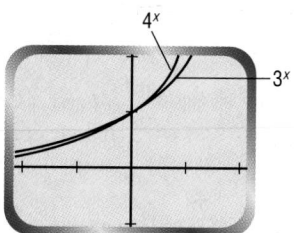

x	−3	−2	−1	0	1	2	3
3^x	0.04	0.11	0.33	1	3	9	27
4^x	0.0156	0.0625	0.25	1	4	16	64

The graphs of the functions are similar in that each of them rises from left to right. Each crosses the *y*-axis at $y = 1$. Each gets closer and closer to the *x-axis* as *x* decreases.

They are different in that the graph of $y = 4^x$ rises faster than the graph of $y = 3^x$ for values of *x* greater than 0. ❖

Continue graphing powers on the same graph using other bases, such as 5^x, 6^x, and so on. What do you notice about the pattern?

RETEACHING **the lesson**

Have students graph the equation from Example 1 on a graphics calculator with ranges set as follows:

Xmin $= -10$, Xmax $= 300$, Xscl $= 1$
Ymin$= -10$, Ymax $= 300$, Yscl $= 1$

Then have students use the ZOOM and TRACE features to find the value of the function after 10 years.

EXERCISES & PROBLEMS

Communicate

1. The population of Zimbabwe in 1995 was about 11 million and was growing at a rate of about 1.9% per year. Explain how to find the constant multiplier.

2. Explain how mathematics can give you a clearer understanding of the effects of population growth.

3. Identify and explain each of the variables in the growth formula, $P = A(1 + r)^t$.

4. Describe how the exponential function changes as the base increases.

5. Describe the relationship between the constant multiplier on a calculator and the exponent in mathematics.

6. Explain why population growth and invested money can be represented by the same formula.

Practice & Apply

Graph each of the following.

7. $y = 2^x$ **8.** $y = 4^x$ **9.** $y = 5^x$ **10.** $y = 10^x$ **11.** $y = 0.5^x$ **12.** $y = 0.1^x$

Generalize the results in Exercises 7–12.

13. For what values of b does the graph of $y = b^x$ rise from left to right? $b > 1$

14. For what values of b does the graph of $y = b^x$ fall from left to right? $0 < b < 1$

Graph each of the following. In each case describe the effect of the 3 as a transformation of the parent function $y = 2^x$.

15. $y = 2^x + 3$ **16.** $y = 2^{x+3}$ **17.** $y = 3 \cdot 2^x$ **18.** $y = 2^{3x}$

19. Is the graph of $y = x^4$ an exponential function? Explain.

20. Is the graph of $y = 4^x$ an exponential function? Explain.

If $f(x) = 2^x$, find the following values for the function.

21. $f(3)$ 8   **22.** $f(0)$ 1   **23.** $f(-1)$ $\frac{1}{2}$ **24.** $f(-4)$ $\frac{1}{16}$

The population of Spain was about 39 million in 1992 and was growing at a rate of about 0.3% per year.

25. What multiplier is used to find the new population each year? 1.003

26. Use this information to estimate the population for 1993. 39.117 million

27. Estimate the population for the year 2000. about 39.95 million

Answers to Exercises 7–12 and 15–18 can be found in Additional Answers beginning on page 729.

19. no; variable is not in the exponent

20. yes; variable is in the exponent

Assess

Selected Answers
Odd-numbered Exercises 7–33, 37–43

Assignment Guide
Core 1–27, 35–46

Core Two-Year 1–43

Technology
Students will need to use a graphics calculator for Exercises 7–12, 15–18, and 44–46.

Error Analysis
When using a graphics calculator, students may neglect to set the range or window for the graph of a function. If a range is set too small, the graph will not show at all or it might appear as a straight line rather than a curve.

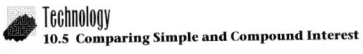

Technology Master

Technology
10.5 Comparing Simple and Compound Interest

When you deposit money into a savings account, your money earns interest. Your money may earn simple interest, interest on the deposit only, or compound interest, interest on all money in the account at the time interest is paid. If you deposit P dollars at an annual percentage rate r, then the amount of money A you will have after t years is shown.

Simple Amount: $A = P(1 + rt)$ Compound Amount: $A = P(1 + r)^t$

Suppose you deposit $1000 into an account that pays 5% interest annually. With a spreadsheet, you can compare the growth in the account if interest is simple or compound. In the spreadsheet, cell D2 contains A2*B$2*C2 and cell E2 contains 1000+D2. Cell F2 contains A$2*(1+B$2)^C2 − 1000 and cell G2 contains A$2*(1+B$2)^C2. FILL DOWN columns D, E, F, and G.

	A	B	C	D	E	F	G
	P	R	T	SIMP INT	SIMP AMT	COMP INT	COMP AMT
1							
2	1000	0.05	0				
3			1				
4			2				
5			3				

Use the spreadsheet to answer the questions.

1. By how much does the simple amount differ from the compound amount after 1 year? after 10 years?

2. By how much does the simple interest differ from the compound interest after 1 year? after 10 years?

3. How many years will it take your deposit to double if interest is simple? compound?

4. How many years will it take your deposit to triple if interest is simple? compound?

5. How would you modify the spreadsheet so that amounts are reported to the nearest cent?

Demographics The population of Italy was about 58 million in 1995 and was growing at a rate of about 0.1% per year. The population of France was about 57.8 million in 1995 and was growing at a rate of about 0.4% per year. Use this information to estimate the population of

28. Italy in 1996. about 58.058 million

29. Italy in 1998. about 58.174 million

30. Italy in 2000. about 58.291 million

31. France in 1996. about 58.031 million

32. France in 1998. about 58.496 million

33. France in 2000. about 58.965 million

34. In what year does the population of France surpass that of Italy? 1997

35. **Portfolio Activity** Complete the problem in the portfolio activity on page 455.

Look Back

Refer to matrix *A*. $A = \begin{bmatrix} 1 & -3 \\ 5 & 2 \end{bmatrix}$ **[Lesson 7.3]**

36. What is A^2?   **37.** What is A^3?

38. Explain the characteristic that a relation must have to be a function. **[Lesson 9.1]**

39. What must happen to the parent function for it to be reflected through the *x*-axis? **[Lesson 9.4]**

Evaluate each of the following. **[Lesson 10.3]**

40. 10^{-2} **41.** -10^2 **42.** $(-10)^2$ **43.** $(-10)^{-2}$
  $\frac{1}{100}$ -100 100 $\frac{1}{100}$

Look Beyond

Cultural Connection: Europe Maria Gaetana Agnesi was born in Italy in 1718. She is considered to be the first woman in modern times to achieve a reputation for mathematical works. She wrote her best known work in 1748 in which she first discussed a curve that is now known as the Agnesi curve.

This curve is represented by the function $f(x) = \frac{a^3}{x^2 + a^2}$. Let the domain of *f* be {−3, −2, −1, 0, 1, 2, 3}.

44. Graph the equation for $a = 1$.

45. Graph the equation for $a = 5$.

46. How do the graphs compare?

The answers to Exercises 35, and 44–46 can be found in Additional Answers beginning on page 729.

36. $\begin{bmatrix} -14 & -9 \\ 15 & -11 \end{bmatrix}$

37. $\begin{bmatrix} -59 & 24 \\ -40 & -67 \end{bmatrix}$

38. For each *x*-value there is only one *y*-value.

39. $-f(x)$ is a reflection of $f(x)$ through the *x*-axis.

Applications of Exponential Functions

How old is a dinosaur bone? How much money should be invested to reach a savings target? These are a few of the many questions that can be answered with the use of exponential functions.

Exponential functions can be used to determine the age of fossils and archeological finds. This method is based on the fact that the amount of radioactivity in an object decreases, or decays, over time.

Science

Carbon-14 dating is a reliable method of determining the age of objects up to 40,000 years old. Carbon-14 is a radioactive form of carbon that has a half-life of 5700 years. This means that half of the substance is converted to nonradioactive carbon every 5700 years. For example, it can be concluded that an object with half as much carbon-14 as its living counterpart died 5700 years ago. Then, each 5700 years, half of the remaining amount of carbon-14 decays.

Years	Fraction of Carbon-14 Remaining
0	$1 = \left(\frac{1}{2}\right)^0$
5,700	$\frac{1}{2} = \left(\frac{1}{2}\right)^1$
11,400	$\frac{1}{4} = \left(\frac{1}{2}\right)^2$
17,100	$\frac{1}{8} = \left(\frac{1}{2}\right)^3$
22,800	$\frac{1}{16} = \left(\frac{1}{2}\right)^4$
28,500	$\frac{1}{32} = \left(\frac{1}{2}\right)^5$
34,200	$\frac{1}{64} = \left(\frac{1}{2}\right)^6$

ALTERNATIVE teaching strategy

Using Technology

Have students use a computer by programming the necessary functions into a spreadsheet program. Then students can create a graph and print both the equation and the graph for analysis and class discussion.

PREPARE

Objectives

• Use exponential functions to model applications.

• Use exponential functions to model growth and decay in different contexts.

RESOURCES

• Practice Master	**10.6**
• Enrichment Master	**10.6**
• Technology Master	**10.6**
• Lesson Activity Master	**10.6**
• Quiz	**10.6**
• Spanish Resources	**10.6**

Assessing Prior Knowledge

Use a calculator to find each of the following:

1. $3 \cdot 0.2^2$ **[0.12]**
2. $3 \cdot 0.2^3$ **[0.024]**
3. $3 \cdot 0.2^4$ **[0.0048]**

TEACH

Exponential functions were used in Lesson 10.5 to find future values and positive trends where values increased. Such functions can also be used to look at past values and trends where values decrease.

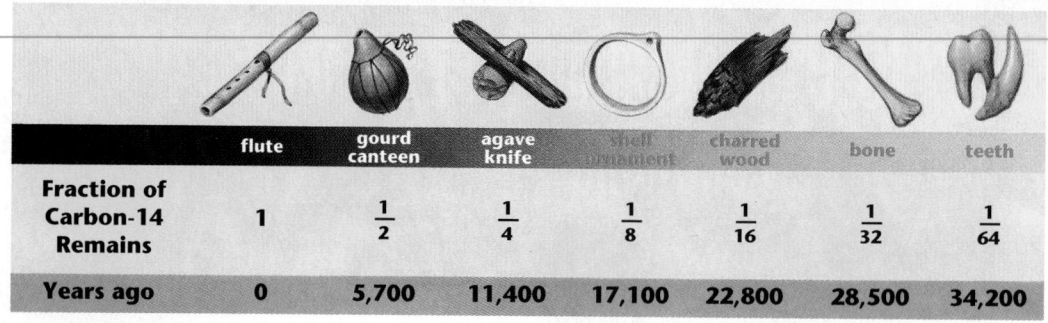

	flute	gourd canteen	agave knife	shell ornament	charred wood	bone	teeth
Fraction of Carbon-14 Remains	1	$\frac{1}{2}$	$\frac{1}{4}$	$\frac{1}{8}$	$\frac{1}{16}$	$\frac{1}{32}$	$\frac{1}{64}$
Years ago	0	5,700	11,400	17,100	22,800	28,500	34,200

Alternate Example 1

Use the table on page 489 to estimate the age of an object that has 5% of its original carbon-14. [approximately 25,000 years]

Alternate Example 2

You would like to buy a $2000 computer system 4 years from now. Find the amount of money you will need to invest in a savings account that pays 6.5% compounded annually. [$1554.65]

Alternate Example 3

Travis's aunt opened a savings account for him when he was born. The account pays 5% interest annually and now contains $1091.43. If Travis is 16, how much did his aunt put in the savings account when he was born? [$500]

Use Transparency ▶ 48

EXAMPLE 1

Use the table to estimate the age of an object which has 10% of its original carbon-14.

Solution ➤

Since 10% or $\frac{1}{10}$ is between $\frac{1}{8}$ and $\frac{1}{16}$, the age is between 17,100 and 22,800 years ago. A reasonable estimate is about 20,000 years. ❖

EXAMPLE 2

Finance Suppose that you wish to save $1000 for a down payment on a car and that you want to buy the car when you graduate in 3 years. The current interest rate for savings is 5% compounded annually. How much will you need to invest in a savings account now to buy the car in 3 years?

Solution ➤

If A represents the original amount, then $P = A(1 + r)^t$ is the formula for compound interest. Let P represent the amount of money after t years. Let r represent the yearly rate of growth (expressed as a decimal) compounded once each year. Substitute the values in the formula, and solve for A.

$$P = A(1 + r)^t$$
$$1000 = A(1.05)^3$$
$$1000 = A(1.157625)$$
$$\frac{1000}{1.157625} = A$$

Since $863.84 \approx A$, you will need to invest $863.84. ❖

The example can also be solved by working backward. Think of the $1000 as being invested 3 years ago, and use -3 as the number of years. Substitute the values into the formula, then solve for P.

$$P = A(1 + r)^t$$
$$P = 1000(1.05)^{-3}$$
$$P \approx 863.84$$

interdisciplinary

Archeology Archaeologists use carbon-14 dating to help them determine the age of artifacts and remains. Until recently there was no way to use this method for dating pictographs or cave drawings. These were dated by carbon-dating items found near the drawings. Now scientists have learned how to separate the paint used for the drawing from the rock that the drawing is on. Thus, cave drawings can be more accurately dated.

EXAMPLE 3

Investment

If the sculpture is now worth $14,586.08, what was the value of the investment 4 years ago?

The value of a sculpture has been growing at a rate of 5% per year for 4 years.

Solution ➤

Use the formula. Since you are trying to find the value 4 years ago, substitute -4 for t.

$$P = A(1 + r)^t$$
$$= 14,586.08(1.05)^{-4}$$
$$\approx 14,586.08(0.8227)$$
$$\approx 12,000$$

The value was $12,000. ❖

Try This

An art museum recently sold a painting for $4,500,000. The directors claimed that its value had been growing at a rate of 7% for the past 50 years. What did the museum pay for the painting 50 years ago?

EXAMPLE 4

PROBABILITY
Connection

Ginny has a 90% probability of getting a given step correct in an algebra problem. She does 100 problems each with two steps.

A In how many problems can you expect her to get both steps correct?

B Calculate the probability if the problems have three steps.

C Calculate the probability if the problems have ten steps.

Solution ➤

A In 100 problems you can expect her to get the first step correct 90 times. Of these 90 times, she would get the second step correct 90% of the time. Since 90% of 90 is 81, she would get both steps correct 81 times out of 100. Notice that the answer is $100(0.90)^2$.

B From the solution for A, you know that in 100 problems she would have the first two steps correct 81 times. She would also get the third step correct 90% of these times or about 73 times. Notice that the answer is $100(0.90)^3$.

C By the same reasoning, the answer is $100(0.90)^{10}$. If you use the power key on a calculator, you get approximately 35. Thus, she gets only about 35% correct when doing 10-step problems, even though she has a 90% chance of getting any particular step correct. ❖

CRITICAL
Thinking

How many steps would a problem have to have for Ginny to have less than a 20% probability of getting the problem correct?

ENRICHMENT The formula $S = P\left(\dfrac{(1 + r)^n - 1}{r}\right)$ gives the amount of money accumulated immediately after the final payment (S), where P is the amount of each payment, r is the annual interest rate, and n is the number of payments. If you are making 36 payments of $200 per month on an auto loan at a 12% annual interest rate, how much will you have paid at the end of 3 years? [$4826.63]

INCLUSION strategies **Technology** Provide opportunities for students to print out graphs of exponential equations by using *f(g) Scholar*™. Also allow students to discuss and describe the graphs and their relation to the exponential function.

In this lesson you have seen several examples of exponential growth and decay. The general formula for exponential growth gives the amount, P, when A is the initial amount, r is the percent of increase expressed in decimal form, and t is the time expressed in years, or

$$P = A(1 + r)^t.$$

For exponential growth, r is positive, but when decay or decrease takes place, the value of r is negative.

EXAMPLE 5

Demographics

The population of a city in the United States was 152,494 in 1990 and was *decreasing* at a rate of about 1% per year. When will the city's population fall below 140,000? Solve the problem

A graphically. **B** using a table.

Solution ➤

Each year the population will be 99% of what it was the previous year, so the multiplier is $1.00 - 0.01$ or 0.99. The population x years after 1990 is given by $Y = 152,494\,(0.99)^x$.

Graphics Calculator

A Enter the formula $Y = 152,494\,(0.99)^x$ into a graphics calculator and graph it. Use X-values between 0 and 10 and Y-values between 130,000, and 160,000. Trace the function to find when it falls below 140,000.

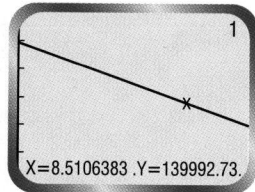

X=8.5106383 Y=139992.73

It falls below 140,000 in about $8\frac{1}{2}$ years.

B The population decreases by a factor of 0.99 each year. Use the constant feature on a calculator to generate a table of values for the population. Continue the table until you reach the year that the value passes below 140,000.

x	$Y = 152,494(0.99)^x$ Population	x	$Y = 152,494(0.99)^x$ Population
0	152,494	5	145,020
1	150,969	6	143,570
2	149,459	7	142,134
3	147,965	8	140,713
4	146,485	9	139,306

According to the table, the population falls below 140,000 between 8 and 9 years. ❖

In Chapter 9 you learned how to use transformations to translate the graph of a parent function. Graphs that represent the exponential function $y = a^x$ can be translated in the same way.

EXAMPLE 6

Compare the graphs of the transformations with the graph of the parent function $y = 2^x$.

A $y = 2^x - 3$ **B** $y = 2^{x-3}$

Solution ➤

Graph each function.

Parent function
$y = 2^x$

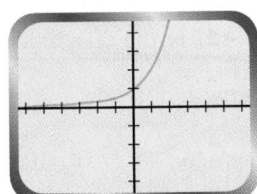   **A** $y = 2^x - 3$ 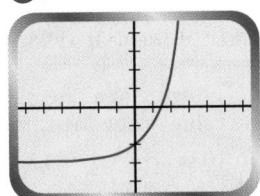   **B** $y = 2^{x-3}$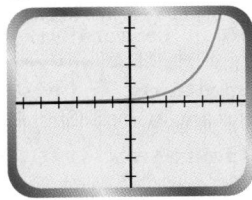

Note that all of the graphs have the same shape.

A The graph of $y = 2^x - 3$ is shifted *down* 3 units from the parent graph $y = 2^x$. Each y-value will be 3 units less than the corresponding point in the parent graph.

B The graph of $y = 2^{x-3}$ is shifted 3 units *to the right* of the parent graph $y = 2^x$. Subtracting 3 before you raise 2 to the power means that to get the same y-value as in the parent graph, you have to substitute an x-value that is 3 units greater. ❖

EXERCISES & PROBLEMS

Communicate

1. List several examples of situations that are described by exponential functions.

2. Describe the formula for exponential decay.

3. Explain why r is negative in the formula for exponential decay.

4. Tell how a horizontal translation is shown in the function rule for exponential functions.

5. Tell how a vertical translation is shown in the rule for exponential functions.

6. Choose a base and form an exponential sequence of 5 or 6 terms. Describe what you find when you examine the differences.

Technology

Use of a graphics calculator is appropriate for Exercises 7–34 and 42 and 43.

Error Analysis

When using a graphics calculator, students may neglect to set or reset the range for the graph of a function. Remind them to consider the range or window settings.

Students may fail to realize when an amount is decreasing and use the wrong formula. Encourage students to estimate whether the answer should be greater than or less than the starting amount or value. Then students can decide whether the increase or decrease is reasonable.

Practice & Apply

Carbon Dating Use the method of carbon-14 dating to estimate the age of an object which has

7. 25% of its original carbon-14 remaining. about 11,400 years

8. 1% of its original carbon-14 remaining. about 37,000 years

9. Technology The population of the metropolitan area of Corpus Christi, Texas, was 361,280 in 1993 and was *decreasing* at a rate of about 0.1% per year. Draw a graph of the population after x years using a "friendly" window.

Demography Trace the graph to estimate the population of Corpus Christi, Texas,

10. in the year 1995.
360,558

11. in the year 1998.
359,477

12. in the year 2000.
358,759

Investment An investment is growing at a rate of 8% per year and now has a value of $8200. Find the value of the investment

13. in 5 years.
$12,048

14. in 10 years.
$17,703

15. 5 years ago.
$5581

16. 10 years ago.
$3798

Investment An investment is losing money at a rate of 2% per year and now has a value of $94,000. Find the value of the investment

17. in 5 years.
$84,969

18. in 10 years.
$76,805

19. 5 years ago.
$103,991

20. 10 years ago.
$115,045

21. **Probability** If Ginny (Example 4) does 100 four-step problems, in how many problems will she get *all* of the steps correct? 66

22. **Probability** If x is chosen randomly from the list 1, 2, 3, 4, 5, find the probability that 2^x is greater than 10. $\frac{2}{5}$

23. Repeat Exercise 22 if x is chosen from the list 1, 2, 3, . . . , 100. $\frac{97}{100}$

The digits of a repeating decimal like $\frac{1}{3} = 0.33333$ . . . can be described in terms of an exponential function. Evaluate each of the following.

24. $3\left(\frac{1}{10}\right)^1$
0.3

25. $3\left(\frac{1}{10}\right)^2$
0.03

26. $3\left(\frac{1}{10}\right)^3$
0.003

27. $3\left(\frac{1}{10}\right)^4$
0.0003

9.
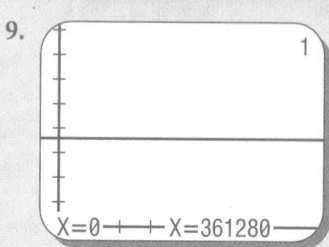

X=0   X=361280

31. shifts parent graph up 5 units

32. shifts parent graph 5 units to the left

33. stretched vertically by a factor of 5

34. stretched vertically by a power of 5

Write the digit in the sixth decimal place of

28. $\frac{1}{3}$. 3　　　　　**29.** $\frac{2}{3}$. 6　　　　　**30.** $\frac{4}{9}$. 4

Technology Graph each of the following, and describe the effect of the 5 as a transformation of the parent function $y = 10^x$.

31. $y = 10^x + 5$　　**32.** $y = 10^{x+5}$　　**33.** $y = 5 \cdot 10^x$　　**34.** $y = 10^{5x}$

Look Back

35. Calculate.　$5(3 - 6) \div 3 \times 2$　**[Lesson 1.4]**　-10

36. Jamie bought tapes and cassettes and paid the cashier $42.75. If the cost of the items before tax was $39.33, what was the percent of the tax? **[Lesson 4.5]**　about 8.7%

37. Technology　What formula would you use in a calculator or spreadsheet to produce random 0s and 1s?　**[Lesson 8.1]**

38. What formula would produce random numbers from 0 to 9 in a calculator or spreadsheet?　**[Lesson 8.1]**

39. There are 44 students that signed up for band or art. Of those students, 21 chose band, and 29 chose art. How many students are taking both band and art?　**[Lesson 8.4]**　6

40. Tamitra flips a coin, records the result, and then flips it again. She wants to know the probability of heads on both tosses. Are the events independent? How do you know?　**[Lesson 8.7]**

41. Describe the differences between the graph of $y = 2^x$ and the graph of $y = 2^{-x}$.　**[Lesson 10.3]**

Look Beyond

42. Technology　A number that is often used to describe growth of natural phenomena and money is e, which is about 2.718. Find e^x on your calculator. Write the first 8 digits of the decimal expansion. Use the $\boxed{e^x}$ key and 1 for x. Then draw the graph of $y = e^x$.

43. Compare.

　a. The value of $1000 invested at 8% interest compounded *once a year* is given by $Y = 1000(1.08)^x$, where x is the number of years. Find the value after 10 years.

　b. The value of $1000 invested at 8% interest compounded *continuously* is given by $y = 1000e^{0.08x}$. Find the value after 10 years.

37. INT(RAND*2) or INT(RAND()*2)

38. INT(RAND*10) or INT(RAND()*10)

40. Yes, because the result of the first toss does not affect the outcome of the next toss.

41. $y = 2^x$ increases from left to right; $y = 2^{-x}$ decreases from left to right.

42. 2.7182818

43. a. $2159

　b. $2226; this method earns more interest.

Focus

This project tracks the spread of a contagious disease. The use of technology and drawing a diagram are relevant tools for problem-solving activities associated with the project.

Motivate

Ask students in class how many of them have had a cold or the flu during the school year. Tell students that they can build a mathematical model to show how a contagious disease like colds can spreads.

Please Don't Sneeze

Suppose that one person in your class gets the flu. While contagious, this person makes contact with two other people so that two more people are exposed to the virus. If any have not yet had the flu, assume that each of them will now get it and make contact with two more people. If these people already have had contact, they are immune and will not pass it on. Use a simulation to find out how many people in the class will get the flu before the epidemic is over.

To do this, assign each person in your class a number. Then use random numbers from a calculator or computer spreadsheet or use numbers drawn from a hat to see who is exposed to the flu virus according to the conditions given above.

The formula INT(RAND*30)+1 will generate a random number from 1 to 30, where RAND is a random number from 0 to 1 including 0, but not 1. (If your class has 28 students, replace the 30 in the formula by 28, etc.) Summarize the results in a tree diagram.

In the example, the following thirteen random numbers were initially generated. 10, 23, 12, 25, 29, 25, 4, 5, 29, 23, 9, 3, 5 . . .

The first random number was 10, so student 10 was the first to get the flu.

The next two random numbers were 23 and 12. This means that while contagious, student 10 had contact with students 23 and 12 who also got the flu.

The next two random numbers were 25 and 29, indicating that student 23 passed it on to students 25 and 29.

The next two random numbers were 25 and 4, indicating that student 12 had contact with 25 and 4. Since student 25 already had the virus, this branch ends.

The next random numbers generated were 5, 29, 23, 9, 3, and 5, so the tree diagram so far looks like this. Remember that a circle around a number indicates that the branch ends.

When the above simulation was continued, all students except 1, 16, 17, 18, 22, 27, 28, and 30 got the virus before all the branches ended. Thus 22 of the 30 students got the flu.

Cooperative Learning

Have students work in pairs to solve the given situation by using a simulation. Remind students to draw a tree diagram and to circle any number that occurs a second time. This indicates that branch has stopped, because the person has immunity.

Discuss

You may wish to discuss with students why there is a government agency that tracks infectious diseases.

Activity

Conduct a simulation of an epidemic. Then pool your data for the class, and summarize the results. What is the average number of people who are infected before the simulated epidemic comes to an end?

Chapter 10 Review

Vocabulary

base	456	monomial	464	Product-of-Powers Property	463
expanded notation	457	negative exponent	470	Quotient-of-Powers Property	466
exponential decay	492	Power-of-a-Power	464	scientific notation	476
exponential form	457	Power-of-a-Power Property	464	zero exponent	470
growth formula	485	Power-of-a-Product Property	463		

Key Skills & Exercises

Lesson 10.1

➤ **Key Skills**

Use the definition of exponent to simplify expressions.

$$3^5 = 3 \cdot 3 \cdot 3 \cdot 3 \cdot 3 = 243 \qquad 4^3 = 4 \cdot 4 \cdot 4 = 64$$

The exponent in a power of 10 tells you the number of zeros to use when writing the number in decimal notation.

$$10^4 = 10,000$$

➤ **Exercises**

Simplify each of the following.

1. 6^3 216   **2.** 5^5 3125   **3.** 10^8 100,000,000   **4.** 10^6 1,000,000

Lesson 10.2

➤ **Key Skills**

Use Product-of-Powers Property or Quotient-of-Powers Property to simplify expressions.

Product of Powers
$$(6x^2y^3)(-3x^4y) = (6 \cdot -3)(x^{2+4})(y^{3+1})$$
$$= -18x^6y^4$$

Quotient of Powers
$$\frac{15x^4y^3z}{3xy^2} = (5)(x^{4-1})(y^{3-2})(z) = 5x^3yz$$

Use Power-of-a-Power Property or Power-of-a-Product Property to simplify expressions.

Power of a Power
$$(x^3)^5 = x^{3 \cdot 5} = x^{15}$$

Power of a Product
$$(4x^4y^3)^2 = (4^2)(x^{4 \cdot 2})(y^{3 \cdot 2}) = 16x^8y^6$$

➤ **Exercises**

Simplify each of the following.

5. $(2t^4)(3t^2)$ $6t^6$ **6.** $(-a^2b^3)(4ab^2)$ $-4a^3b^5$ **7.** $(-0.1w^2x^2y)(-0.5w^5)$ $0.05w^7x^2y$

8. $\frac{10w^5}{2w^2}$ $5w^3$ **9.** $\frac{-21wx^2y^4}{-3xy^3}$ $7wxy$ **10.** $(r^3)^4$ r^{12}

11. $(-3c^2)^3$ $-27c^6$ **12.** $(2p^2q^3)^2$ $4p^4q^6$ **13.** $(m^4n^2)^2(m^3)^4$ $m^{20}n^4$

Lesson 10.3

➤ Key Skills

Simplify expressions containing negative exponents.

$$5x^{-3}y^2 = \frac{5y^2}{x^3}$$

$$\frac{m^{-3}p^2}{n^{-2}p^2} = \frac{n^2}{m^3}$$

Simplify expressions containing zero exponents.

$$(3x)^0 = 1$$

$$7a^0b^2 = 7b^2$$

➤ Exercises

Write each of the following without negative or zero exponents.

14. 3^{-2} $\frac{1}{9}$

15. a^2b^{-3} $\frac{a^2}{b^3}$

16. $(-3)^0$ 1

17. $a^0b^{-1}c^2$ $\frac{c^2}{b}$

18. $(2c^2d^{-1})^0$ 1

19. $-q^{-3}q^2$ $-\frac{1}{q}$

20. $\frac{b^{-2}}{b^3}$ $\frac{1}{b^5}$

21. $\frac{t^{-2}u}{t^{-4}u^2}$ $\frac{t^2}{u}$

Lesson 10.4

➤ Key Skills

Write numbers in scientific notation.

Customary Notation	Scientific Notation
32,000,000,000	3.2×10^{10}
0.000000784	7.84×10^{-7}

Perform computations with numbers written in scientific notation.

$$(3 \times 10^{-3})(4 \times 10^8) = (3 \cdot 4)(10^{-3} \times 10^8)$$
$$= 12 \times 10^5$$
$$= 1.2 \times 10^6$$

$$\frac{4 \times 10^6}{2 \times 10^3} = \left(\frac{4}{2}\right)(10^{6-3}) = 2 \times 10^3$$

➤ Exercises

Write each number in scientific notation.

22. 5,900,000

23. 368,000,000,000

24. 0.0000075

Write each number in decimal notation.

25. 2×10^3

26. 7.9×10^{-5}

27. 8.34×10^9

Perform the following computations. Answer using scientific notation.

28. $(3 \times 10^2)(5 \times 10^5)$

29. $(2.1 \times 10^5)(4 \times 10^{-3})$

30. $(8 \times 10^2) + (2 \times 10^2)$

31. $(9 \times 10^5) - (3 \times 10^5)$

32. $\frac{9 \times 10^7}{3 \times 10^2}$

33. $\frac{8 \times 10^4}{2 \times 10^{-2}}$

---

22. 5.9×10^6

23. 3.68×10^{11}

24. 7.5×10^{-6}

25. 2000

26. 0.000079

27. 8,340,000,000

28. 1.5×10^8

29. 8.4×10^2

30. 1×10^3

31. 6×10^5

32. 3×10^5

33. 4×10^6

Lesson 10.5

➤ Key Skills

Use exponential functions to estimate population growth.

The population of a country was estimated to be about 750 million in 1990 and was growing at a yearly rate of about 1.7% At this rate of growth, the population of this country in 10 years will be $P = 750(1 + 0.017)^{10} \approx 888$, or about 888 million people.

➤ Exercises

The population of New Zealand was about 3.3 million in 1991 and was growing at a rate of about 0.8% per year. Estimate the population

34. in the year 1998.
3.49 million

35. in the year 2000.
3.55 million

Lesson 10.6

➤ Key Skills

Use exponential functions to estimate growth or decay.

An investment is losing money at a rate of 3% per year and now has a value of $6500. The value of the investment in 5 years will be $y = 6500(1 - 0.03)^5 \approx \$5581.77$. The value of the investment 5 years ago was $y = 6500(1 - 0.03)^{-5} \approx \$7569.28$.

➤ Exercises

An investment is growing at a rate of 5.5% per year and now has a value of $6300. Find the value of the investment

36. in 5 years. $8234

37. in 10 years. $10,761

38. 5 years ago. $4820

39. 10 years ago. $3688

Applications

40. Geometry Suppose the radius of a circle has length r. Then the area of the circle is πr^2. Find the area of a circle if the radius is 5 times as long as the radius of the first circle. $25\pi r^2$

41. Astronomy The sun has a diameter of about 864,000 miles. Write this number in scientific notation. 8.64×10^5

42. The interior temperature of the sun is about 3.5×10^7 °F. Write this number in decimal notation. 35,000,000

43. Time The calendar year is increasing at a rate of about 1.7×10^{-6} minutes each year. How many seconds is this? Write your answer in scientific notation. about 1.02×10^{-4}

Probability Suppose x is chosen randomly from the list $-1, -2, -3, -4, -5$. Find the probability that

44. 2^x is less than 0.5. 80%

45. 3^x is greater than $\frac{1}{27}$. 40%

46. 10^x is greater than 0.001. 40%

47. $\frac{1}{2^x}$ is less than 16. 60%

Chapter 10 Assessment

1. What is a monomial?

2. The product of two powers can be found by adding the exponents. What must be true about these powers?

Simplify each of the following.

3. $(2y)^3$

4. $(2f^3)(3f^4)$

5. $(-x^3y)(5x^2y^2z)$

6. $\dfrac{36t^5}{6t^2}$

7. $\dfrac{-2w^3z^4}{-3wz^5}$

8. $\dfrac{ab^4}{-2b^3}$

9. $(g^3)^4$

10. $(4c^2)^5$

11. $6(2cd^3)^2$

12. What is the difference between an exponent of 0 and an exponent of 1?

Write each of the following without negative or zero exponents.

13. 2^{-4}

14. $a^{-3}b^2$

15. $m^{-4}n^0p^4$

16. $-2t^{-2}$

17. y^5y^{-3}

18. $\dfrac{3h^{-2}}{h^{-1}}$

19. What is the meaning of a negative exponent in a number written in scientific notation?

Taxes In 1970, the IRS collected about 196 million dollars in taxes. In 1992, about 1.1 billion dollars was collected.

20. Write 196 million in scientific notation. 1.96×10^8

21. Write 1.1 billion in scientific notation. 1.1×10^9

22. How much more did the IRS collect in 1992 than in 1970? Answer using scientific notation and decimal notation. 9.04×10^8; $904,000,000

Perform the following computations. Answer using scientific notation.

23. $(3 \times 10^{-2})(5 \times 10^5)$ 1.5×10^4

24. $\dfrac{3 \times 10^6}{5 \times 10^4}$ 6×10^1

Demographics The population of Peru in 1992 was about 23 million and growing at a rate of 2% per year.

25. What is the multiplier to get the new population each year? 1.02

26. Estimate the population for the year 1997. 25.4 million

27. Estimate the population for the year 2000. 26.9 million

Investment An investment is losing money at a rate of 0.5% per year and had a value of $8500 five years ago. Find the value of the investment

28. 10 years ago. $8716

29. now. $8290

30. in 5 years. $8084

31. in 10 years. $7884

---

1. A monomial is a constant, a variable, or the product of a constant and one or more variables.

2. The powers must have the same base.

3. $8y^3$

4. $6f^7$

5. $-5x^5y^3z$

6. $6t^3$

7. $\dfrac{2w^2}{3z}$

8. $\dfrac{ab}{-2}$

9. g^{12}

10. $1024c^{10}$

11. $24c^2d^6$

12. exponent 0: the value of the power is 1; exponent 1: the value of the power is the base

13. $\dfrac{1}{2^4} = \dfrac{1}{16}$

14. $\dfrac{b^2}{a^3}$

15. $\dfrac{p^4}{m^4}$

16. $\dfrac{-2}{t^2}$

17. y^2

18. $\dfrac{3}{h}$

19. The value of the number is less than 1.

Chapters 1 – 10
Cumulative Assessment

College Entrance Exam Practice

COLLEGE ENTRANCE-EXAM PRACTICE

Multiple-Choice and Quantitative-Comparison Samples

The first half of the Cumulative Assessment contains two types of items found on standardized tests—multiple-choice questions and quantitative-comparison questions. Quantitative-comparison items emphasize the concepts of equality, inequality, and estimation.

Free-Response Grid Samples

The second half of the Cumulative Assessment is a free-response section. A portion of this part of the Cumulative Assessment consists of student-produced response items commonly found on college entrance exams. These questions require the use of machine-scored answer grids. You may wish to have students practice answering these items in preparation for standardized tests.

Sample answer-grid masters are available in the *Chapter Teaching Resources Booklets*.

Quantitative Comparison   For Questions 1–4, write

A if the quantity in Column A is greater than the quantity in Column B;
B if the quantity in Column B is greater than the quantity in Column A;
C if the two quantities are equal; or
D if the relationship cannot be determined from the information given.

	Column A	Column B	Answers
1.	INT(4.5)	ABS(-4.5)	Ⓐ Ⓑ Ⓒ Ⓓ [Lesson 2.4] B
2.	5×10^{-3}	0.005	Ⓐ Ⓑ Ⓒ Ⓓ [Lesson 10.4] C
3.	slope of $y = -2$	slope of $y = x - 2$	Ⓐ Ⓑ Ⓒ Ⓓ [Lesson 5.1] B
4.	7^0	$56[9 + (-9)]$	Ⓐ Ⓑ Ⓒ Ⓓ [Lesson 10.3] A

5. What is the sum of $-14 + (-6)$?   **[Lesson 3.1]** a
  a. -20 **b.** -8 **c.** 20   **d.** 8

6. What is 65% of 20?   **[Lesson 4.5]** c
  a. 1300   **b.** 0.0325   **c.** 13   **d.** 3.25

7. Which command generates a random number from the list 3, 4, 5?   **[Lesson 8.2]** d
  a. INT(RAND() * 5) + 5   **b.** INT(RAND() * 3) + 5
  c. INT(RAND() * 5) + 3   **d.** INT(RAND() * 3) + 3

8. There are 3 blue chips, 7 red chips, and 2 yellow chips in a bag. If you select a chip, replace it, and then select another chip, what is the probability of selecting a red chip and a yellow chip?   **[Lesson 8.6]** b
  a. $\frac{3}{4}$ **b.** $\frac{7}{72}$ **c.** $\frac{7}{66}$ **d.** $\frac{1}{14}$

9. What is the parent function for the graph?   **[Lesson 2.5]**
  a. $y = x^2$ **b.** $y = x$ c
  c. $y = |x|$ **d.** $y = \text{INT}(x)$

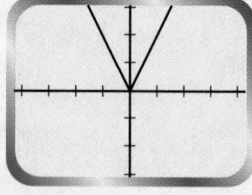

13. 1.0924×10^7

14. $y = -\dfrac{4}{3}x$

15. $\begin{bmatrix} 0.1 & 0.3 \\ 0.4 & 0.2 \end{bmatrix}$

10. Cathi has $17.40 in dimes and quarters. If she has a total of 81 coins, how many of each type of coin does she have? **[Lesson 6.3]** 19*d*; 62*q*

11. Solve $\frac{x}{-10} = \frac{-3}{5}$. **[Lesson 4.3]** 6

12. The first three terms of a sequence are 2, 4, and 9. The second differences are a constant 3. What are the next three terms? **[Lesson 1.2]** 17, 28, 42

13. In 1992, a total of 10,924,000 individual tax returns were filed electronically. Write this number in scientific notation. **[Lesson 10.4]**

14. Find the equation of the line passing through the origin and the point (3, −4). **[Lesson 5.3]**

15. Find the inverse for $\begin{bmatrix} -2 & 3 \\ 4 & -1 \end{bmatrix}$ if it exists. **[Lesson 7.4]**

16. Identify the parent function for $y = \frac{-2}{x}$. Tell what type of transformation is applied and then draw a carefully labeled graph. **[Lesson 9.3]**

17. There are 48 students in the choir and 50 students in the orchestra. If 15 students are in both the choir and the orchestra, what is the total number of students in the choir and the orchestra? **[Lesson 8.4]** 83

Free-Response Grid The following questions may be answered using a free-response grid commonly used by standardized test services.

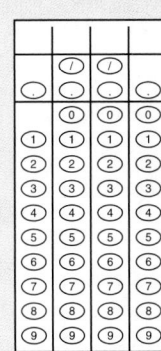

18. Simplify $3(4 + 5) \cdot (4 - 3)2 \div 5 \cdot 3$. **[Lesson 1.4]** 32.4

19. The perimeter of a rectangle is 30 centimeters and its length is twice its width. Find the length of this rectangle. **[Lesson 4.6]** 10 cm

20. Suppose you roll a number cube. What is the probability that you will roll a number greater than 4? **[Lesson 8.6]**

21. What is the slope of a line parallel to the line passing through (−7, −5) and (1, −3)? **[Lesson 5.6]**

16.

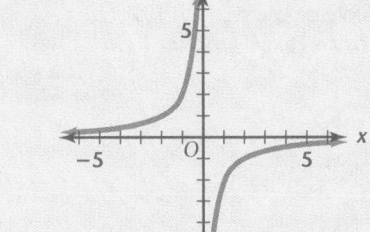

The parent function is $y = \frac{1}{x}$;

reflection through the *x*-axis and a vertical stretch of factor 2

20. $\frac{1}{3}$

21. $\frac{1}{4}$

11 Polynomials

CHAPTER 11

Meeting Individual Needs

11.1 Adding and Subtracting Polynomials

Core Resources

Practice Master 11.1
Enrichment, p. 508
Technology Master 11.1

[1 day]

Core Two-Year Resources

Inclusion Strategies, p. 508
Reteaching the Lesson, p. 509
Practice Master 11.1
Enrichment Master 11.1
Technology Master 11.1
Lesson Activity Master 11.1

[2 days]

11.2 Exploring Products and Factors

Core Resources

Practice Master 11.2
Enrichment, p. 515
Technology Master 11.2

[1 day]

Core Two-Year Resources

Inclusion Strategies, p. 515
Reteaching the Lesson, p. 516
Practice Master 11.2
Enrichment Master 11.2
Technology Master 11.2
Lesson Activity Master 11.2

[3 days]

11.3 Multiplying Binomials

Core Resources

Practice Master 11.3
Enrichment, p. 520
Technology Master 11.3
Mid-Chapter Assessment
    Master

[1 day]

Core Two-Year Resources

Inclusion Strategies, p. 520
Reteaching the Lesson, p. 521
Practice Master 11.3
Enrichment Master 11.3
Technology Master 11.3
Lesson Activity Master 11.3
Mid-Chapter Assessment Master

[2 days]

11.4 Common Factors

Core Resources

Practice Master 11.4
Enrichment, p. 525
Technology Master 11.4

[1 day]

Core Two-Year Resources

Inclusion Strategies, p. 525
Reteaching the Lesson, p. 526
Practice Master 11.4
Enrichment Master 11.4
Technology Master 11.4
Lesson Activity Master 11.4

[3 days]

11.5 Factoring Special Polynomials

Core Resources

Practice Master 11.5
Enrichment, p. 531
Technology Master 11.5
Interdisciplinary
    Connection, p. 530

[2 days]

Core Two-Year Resources

Inclusion Strategies, p. 531
Reteaching the Lesson, p. 532
Practice Master 11.5
Enrichment Master 11.5
Technology Master 11.5
Lesson Activity Master 11.5

[3 days]

11.6 Factoring Trinomials

Core Resources

Practice Master 11.6
Enrichment, p. 536
Technology Master 11.6

[1 day]

Core Two-Year Resources

Inclusion Strategies, p. 536
Reteaching the Lesson, p. 537
Practice Master 11.6
Enrichment Master 11.6
Technology Master 11.6
Lesson Activity Master 11.6

[4 days]

Chapter Summary

Core Resources	Core Two-Year Resources
Chapter 11 Project, pp. 540–541	Chapter 11 Project, pp. 540–541
Lab Activity	Lab Activity
Long-Term Project	Long-Term Project
Chapter Review, pp. 542–544	Chapter Review, pp. 542–544
Chapter Assessment, p. 545	Chapter Assessment, p. 545
Chapter Assessment, A/B	Chapter Assessment, A/B
Alternative Assessment	Alternative Assessment
[3 days]	**[5 days]**

Visual Strategies

Polynomials can be used to express the area of a square or rectangle or to find the length of a side from the area . Students can sketch rectangles with given dimensions in the form $nx + c$ and find the area in the form of the polynomial $ax^2 + bx + c$. When trying to find factors of a general trinomial, the use of a pattern board and index cards provides a visual layout where students can experiment with various pairs of factors. This provides a systematic, yet visual, way to factor general trinomials. (See Reteaching, pages 532, for the pattern board.)

Hands-on Strategies

Chapter 11 requires the use of algebra tiles. You may want to start by having students use algebra tiles to show various polynomials, such as $x^2 + x, 2x^2, 4x + 3$, and so on. Lesson 11.1 on adding and subtracting polynomials is greatly enhanced by having students use algebra tiles to complete the addition or subtraction. Lesson 11.2 provides for active learning and requires hands-on participation as students explore factors and products. The remaining lessons in the chapter also are enhanced through the use of algebra tiles.

Cooperative Learning

GROUP ACTIVITIES	
adding/subtracting polynomials	Lesson 11.1, entire lesson
exploring products	Lesson 11.2, Explorations 1, 2, and 3
finding factors	Lesson 11.2, Exploration 4; Lesson 11.6, Factoring With Files
multiplying binomials	Lesson 11.3, Explorations 1 and 2

You may wish to have students work with partners or in small groups to complete any of these activities. Additional suggestions for cooperative group activities can be found in the teacher's notes in the lessons.

Multicultural

The cultural connections in this chapter include references to Egypt, Guatemala, France, and China.

CULTURAL CONNECTIONS	
Africa: Geometric Models	Lesson 11.5, Exercise 41
Americas: Number Problems	Lesson 11.6, Exercise 47
Asia: Pascal's Triangle	Chapter Project, pp. 540–541

Portfolio Assessment

Below are portfolio activities for the chapter, listed under seven activity domains that are appropriate for portfolio development.

1. **Investigation/Exploration** In the first two lessons, students investigate how algebra tiles can be used to represent polynomials and how to add and subtract polynomials. The remaining lessons consider products and factors of polynomials.

2. **Applications** Advertising, Lesson 11.2, Exercise 44; Small Business, Lesson 11.3, Example 2; Physics, Lesson 11.4, Exercise 57; and Sales Tax, Lesson 11.6, Exercise 42.

3. **Non-Routine Problems** Punnett Squares in the Portfolio Activity and in Example 3 in Lesson 11.3 provide an application of probability to a non-routine problem. Another application of probability is explored in the chapter project.

4. **Project** Powers, Pascal, and Probability: see pages 540–541. Students study Pascal's triangle and relate the patterns involved to the coefficients formed by finding powers of a binomial. This is then applied to probability.

5. **Interdisciplinary Topics** Students may choose from the following: Probability, Lesson 11.3, Example 3; Lesson 11.4, Exercise 53; Lesson 11.5, Exercise 59; and Chapter Project on pages 540–541; Physics, Lesson 11.4, Exercise 57; Geometry, Lesson 11.1, Example 2 and Exercises 40–43; Lesson 11.2, Exercises 37–40, and 43; Lesson 11.3, Exercises 32–36; Lesson 11.4, Example 2 and Exercises 44–45; and Lesson 11.5, Exercises 41–50.

6. **Writing** *Communicate* exercises offer excellent writing selections for the portfolio. Suggested selections include: Lesson 11.1, Exercises 1–6; Lesson 11.2, Exercises 1–4; Lesson 11.5, Exercises 1–3; and Lesson 11.6, Exercises 1–8.

7. **Tools** Chapter 11 often makes use of algebra tiles. Calculators and computer spreadsheets can be used to evaluate factors of a trinomial. The *IBM® Mathematics Exploration Toolkit* can also be used with polynomials. See next page.

Technology

Computer software has been developed that can perform basic and repetitive computations very efficiently. Students should be aware of the tools that are available for simplifying mathematical tasks. If students continue in mathematics, science, or engineering they will eventually need to gain experience using this software.

Computer spreadsheets can be used to check whether pairs of numbers work when trying to factor a trinomial. Students enter the coefficients of a given trinomial, written in $ax^2 + bx + c$ form. Then pairs of factors, (d, f) and (e, g), are entered. The computer checks to see if $(dx + e)(fx + g)$ are factors by checking whether $a = df$; $b = dg + ef$; and $c = eg$.

The *Maple®* software for mathematics is an example of a computer program that can be used to explore polynomials. To do this, you must know some fundamental instructions. The commands needed to work with polynomials include:

expr	identifies a symbolic expression
expand	multiplies over additions
normal	is one command used for dividing expressions
factor	factors polynomials
sort	arranges the order of a polynomial
+, −, *, /	are operation commands
;	denotes the end of a command and must appear before executing the command
Enter	executes the command

When you add $x^2 + 3x$ and $-x + 3x^2 - 6$, you can enter the expressions directly after the prompt symbol, $>$. You can also assign a variable to an expression and add the variables.

> x ^ 2 + 3*x + (−x + 3 * x ^ 2 − 6; *[Enter]*

$$4x^2 + 2x - 6$$

The result of $4x^2 + 2x - 6$ will be shown quickly on the screen. Subtraction multiplication, and division can be done in the same manner. For multiplication, use the * key. For example, you can enter

> expand((3 * x ^ 2 + 5)* x); *[Enter]*

$$3x^3 + 5x$$

For division, you can use the / and parentheses together. (You must use parentheses around the divisor.) To divide $12x^2$ by $4x$, you would enter

> (12 * x ^ 2) / (4 * x); *[Enter]*

$$3x$$

The *Maple®* can also be used to factor polynomials. To find the factors of $28x^3 + 35x^2 - 21x$, enter the following:

> factor(28 * x ^ 3 + 35 * x ^ 2 − 21 * x); *[Enter]*

$$7x(4x^2 + 5x - 3)$$

The result $7x(4x^2 + 5x - 3)$ will be displayed on the screen.

Integrated Software

f(g) Scholar™ is an integrated computer-based mathematics productivity tool that combines calculator, spreadsheet, and graphics capabilities to provide a dynamic and interactive environment for explorations in mathematics. It is appropriate to use *f(g) Scholar*™ for any lesson needing a spreadsheet, calculator, or any combination of the three.

11 Polynomials and Factoring

CHAPTER 11

LESSONS

11.1 Adding and Subtracting Polynomials

11.2 *Exploring* Products and Factors

11.3 Multiplying Binomials

11.4 Common Factors

11.5 Factoring Special Polynomials

11.6 Factoring Trinomials

Chapter Project
Powers, Pascal, and Probability

Polynomials and Factoring

$$(a+b)^3$$

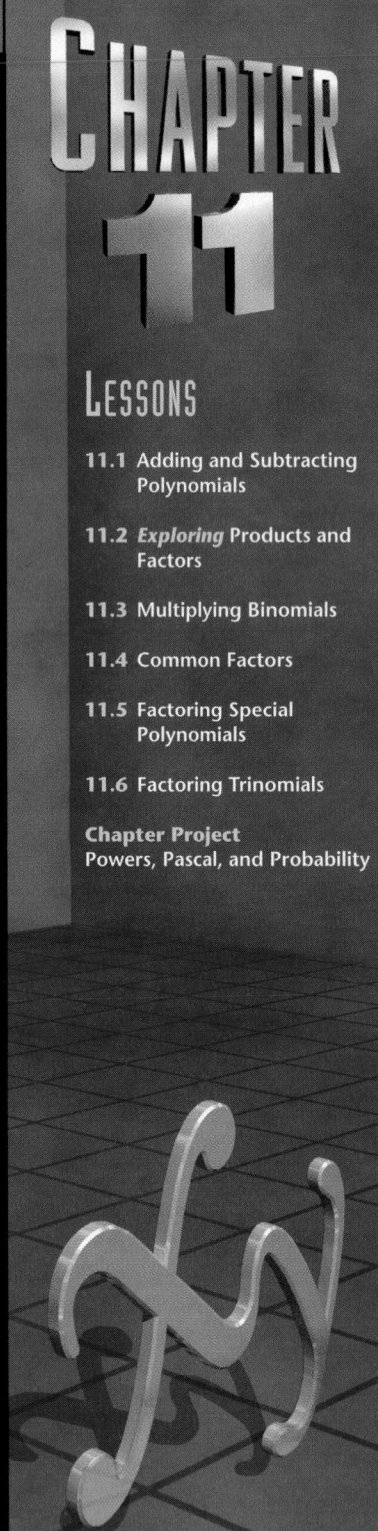

ABOUT THE CHAPTER

Background Information

Students should be familiar with terms such as *factor* and *product*. This chapter deals with multiplying binomials and factoring polynomials. Students will learn to use patterns to simplify operations with or find the factors of polynomials.

CHAPTER RESOURCES

- Practice Masters
- Enrichment Masters
- Technology Masters
- Lesson Activity Masters
- Lab Activity Masters
- Long-Term Project Masters
- Assessment Masters
  Chapter Assessments, A/B
  Mid-Chapter Assessments
  Alternative Assessments, A/B
- Teaching Transparencies
- Spanish Resources

CHAPTER OBJECTIVES

- Add and subtract polynomials.
- Use tiles and the Distributive Property to show the product of a monomial and a binomial.
- Show the greatest common factor as part of the factored form of a polynomial.
- Use the Distributive Property to multiply binomials.
- Use the FOIL method to multiply binomials.
- Factor the greatest common factor from a polynomial.
- Factor a binomial as the greatest common factor.

ABOUT THE PHOTOS

Have students explain how the components of the cube can be transformed into a polynomial. Be sure students recognize that the volume of each of the components is expressed as a the product of length, width, and depth. The mathematics involved in trying to fit together the different parts of an exploded cube illustrates how monomial components form a polynomial.

$$a^3+3a^2b+3ab^2+b^3$$

PORTFOLIO ACTIVITY

Punnett Squares are used in genetics to determine the possible characteristic of offspring. Assume that each parent has both the recessive and the dominant form of the gene for height. Show how to determine the probability that the offspring from a cross between these two parents will be tall if tall is dominant and short is recessive. How is this related to polynomials?

ABOUT THE CHAPTER PROJECT

In the Chapter Project on pages 540–541, students investigate Pascal's Triangle and relate it to the coefficients of the terms in the product of two binomials. This investigation is then extended to examine the concepts of combinations and probability.

- Factor the difference of two squares.
- Factor perfect-square trinomials.
- Use algebra tiles and guess-and-check to determine factors of a trinomial.
- Use sign and number patterns to determine factors of a trinomial.

PORTFOLIO ACTIVITY

Assign the Portfolio Activity as part of the assignment for Lesson 11.4 on page 528.

Have students work in small groups to develop a Punnett Square to determine the probability that a garden pea plant will have tall offspring.

In the Punnett Square below, dominant genes are shown with a capital T and recessive genes are shown with a lowercase t (T Tall; t short). Parents that have one of each type of gene are called hybrids. Tt and tT are both hybrid tall. The pure form TT indicates tall, while tt indicates short.

	T	t
T	TT	Tt
t	Tt	tt

Thus, the probability is 1 out of 4: pure tall, 2 out of 4: hybrid tall, and 1 out of 4: pure short. In all, 3 out of 4 of the offspring will be tall.

This can be related to the coefficients of the expanded binomial square.
$$(T + t)^2 = T^2 + 2Tt + t^2.$$

- Add and subtract polynomials.

RESOURCES

- Practice Master   **11.1**
- Enrichment Master   **11.1**
- Technology Master   **11.1**
- Lesson Activity Master   **11.1**
- Quiz   **11.1**
- Spanish Resources   **11.1**

Assessing Prior Knowledge

Simplify.

1. $a + 2a$ [**3a**]
2. $8b - 3b$ [**5b**]
3. $3cd + 7cd$ [**10cd**]
4. $2e - (3f + 5e) + 3f$
                  [**−3e**]

TEACH

 Have students discuss how they find the volume of a cube [$V = s^3$]. Tell students that if they had to pack different sized cubes in boxes, they could use polynomials to express the volume of the box.

LESSON 11.1   # Adding and Subtracting Polynomials

The volume of a rectangular prism can be found by using the formula V = lwh. If the side of a cube is x + 2, the volume is $(x + 2)^3$ or $x^3 + 6x^2 + 12x + 8$.

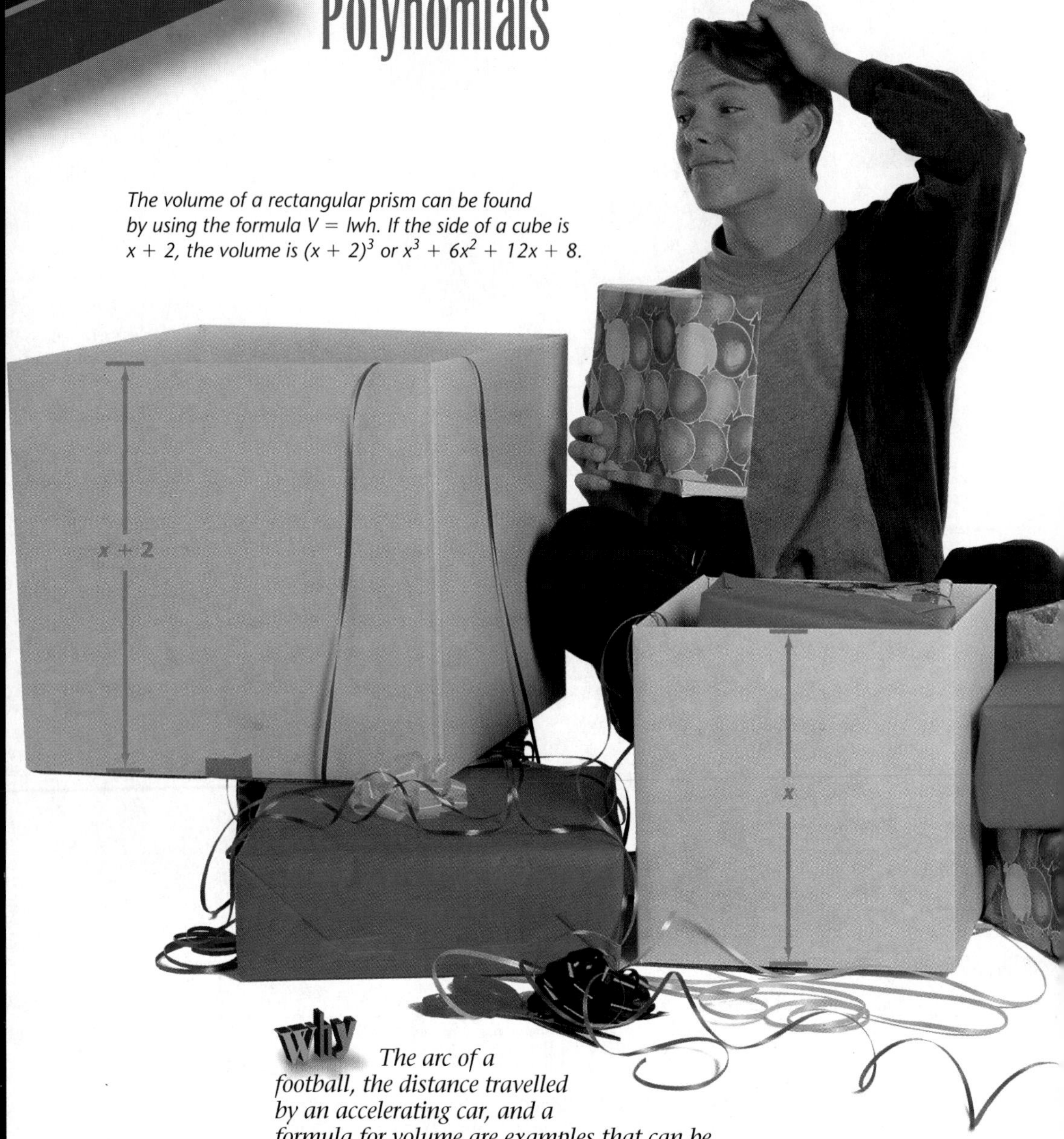

x + 2

x

 *The arc of a football, the distance travelled by an accelerating car, and a formula for volume are examples that can be modeled by a polynomial.*

A **polynomial** is a monomial or the sum or difference of two or more monomials. An example of a polynomial in one variable is
$x^3 + 6x^2 + 12x + 8$.

ALTERNATIVE teaching strategy

Using Manipulatives Tell students that the length of their pencil or pen is x units and the width of their thumb is one unit. Have students find the length and width of their desk or table in terms of x and units. Then have students discuss how to use these measurements to find the perimeter of the table or desk in terms of x and units.

The **degree** of a polynomial in **one variable** is the *exponent with the greatest value of any of the polynomial's terms.* The degree of $9 - 4x^2$ is 2. What is the degree of $2a^2 + 5a^4 - 3a^3 + 1$?

The terms of a polynomial may appear in any order. However, in **standard form**, the terms of a polynomial are ordered from left to right, from the greatest to the least degree of the variable. This is called **descending order**.

The polynomial $3x^2 - 4x + 6$ is in standard form. The degrees of the x-terms are in order from greatest to least. The polynomial $9 - 4x^2$ written in standard form is $-4x^2 + 9$. Sometimes it is convenient to write $-4x^2 + 9$ as $(-1)(4x^2 - 9)$.

Write $5x^2 + x^3$ in standard form.

Some polynomial expressions have special names that are dependent on either their *degree* or their *number of terms*, as illustrated in the table.

Polynomial	# of terms	Name by # of terms	Degree	Name by degree
12	1	monomial	0	constant
$8x$	1	monomial	1	linear
$9 - 4x^2$	2	binomial	2	quadratic
$5x^2 + x^3$	2	binomial	3	cubic
$3x^2 - 4x + 6$	3	trinomial	2	quadratic
$3x^4 - 4x^3 + 6x^2 - 7$	4	polynomial	4	quartic

Second degree polynomials can be modeled by algebra tiles.

x^2	$-x^2$	x	$-x$	1	-1
	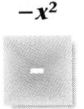				

1.  represents $3x^2 - 4x + 6$.

2.  represents $5x^2 + x - 3$.

3. 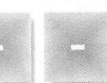 represents $9 - 4x^2$.

Alternate Example 1

Add $4x^3 + 5x^2 - x + 6$ and $6x^3 - 2x^2 - x - 8$.
$[10x^3 + 3x^2 - 2x - 2]$

Adding Polynomials

Algebra tiles can be used to model polynomial addition. To add $x^2 + x + 1$ and $2x^2 + 3x + 2$, model each polynomial and group like tiles.

Polynomials can also be added in vertical or horizontal form. In vertical form align the like terms and add.

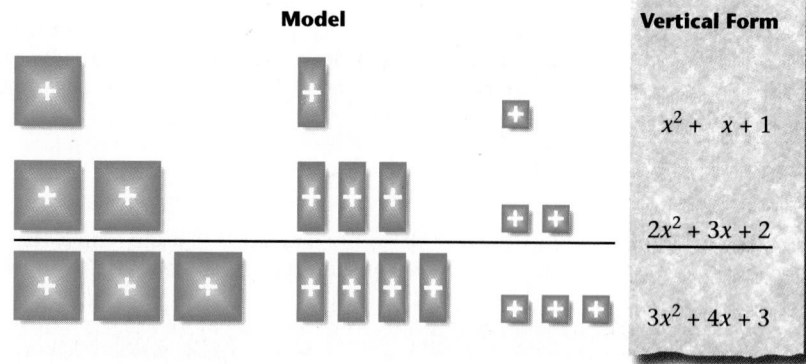

In horizontal form, write $(x^2 + x + 1) + (2x^2 + 3x + 2)$ and add like terms.

EXAMPLE 1

Use algebra tiles to model $(2x^2 - 3x + 5) + (4x^2 + 7x - 2)$. Find the sum.

Solution A ➤

Model each polynomial, and group like tiles. Notice that when an equal number of negative and positive tiles of the same type are put together, the combination represents zero.

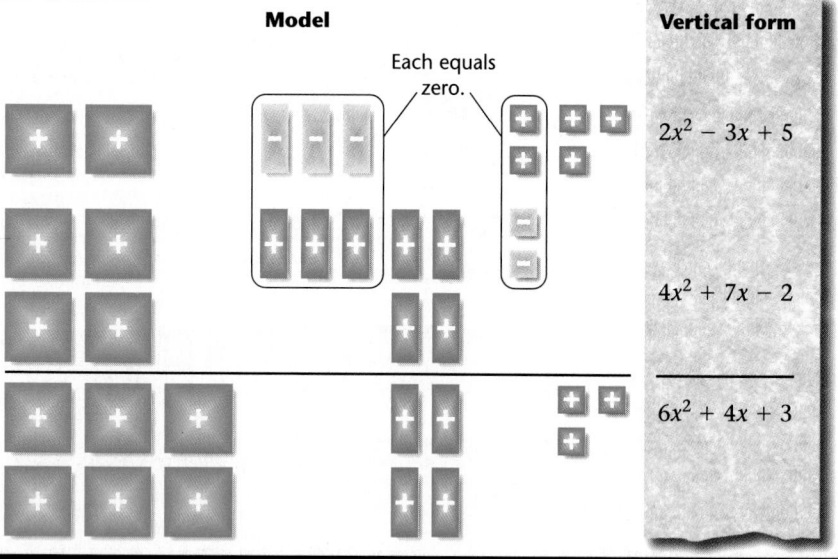

RETEACHING **the lesson**

Using Tables When adding and subtracting polynomials in vertical form, you may wish students to use a table like the one below to assure the proper alignment of the monomial terms.

This encourages students to watch for like terms and to write the answer in standard form.

Solution B ➤

In horizontal form, add like terms.

$$(2x^2 - 3x + 5) + (4x^2 + 7x - 2)$$
$$= (2x^2 + 4x^2) + (-3x + 7x) + (5 - 2)$$
$$= 6x^2 + 4x + 3 \; ❖$$

Try This Use algebra tiles to model $(3x^2 + 5x) + (4 - 6x - 2x^2)$. Find the sum using the horizontal form.

EXAMPLE 2

Find the perimeter of each polygon. Check the addition by substituting a value for each variable.

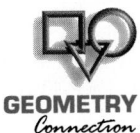

GEOMETRY
Connection

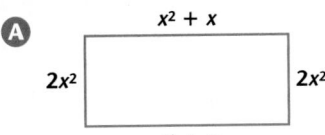

Ⓐ $x^2 + x$ ... $2x^2$ ... $2x^2$ ... $x^2 + x$

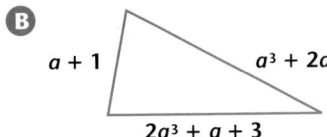

Ⓑ $a + 1$ ... $a^3 + 2a$ ... $2a^3 + a + 3$

Solution ➤

Use vertical form to add like terms. Substitute 10 for each variable to check.

Ⓐ
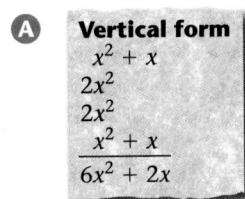

Vertical form
$$\begin{array}{l} x^2 + x \\ 2x^2 \\ 2x^2 \\ \underline{x^2 + x} \\ 6x^2 + 2x \end{array}$$

Check: $x = 10$
$$\begin{array}{r} 10^2 + \quad 10 = 110 \\ 2 \cdot 10^2 \qquad = 200 \\ 2 \cdot 10^2 \qquad = 200 \\ \underline{10^2 + \quad 10 = 110} \\ 6 \cdot 10^2 + 2 \cdot 10 = 620 \end{array}$$

Ⓑ
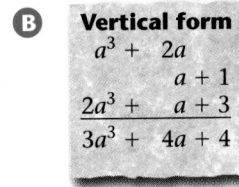

Vertical form
$$\begin{array}{l} a^3 + \; 2a \\ \qquad a + 1 \\ \underline{2a^3 + \; a + 3} \\ 3a^3 + \; 4a + 4 \end{array}$$

Check: $x = 10$
$$\begin{array}{r} 10^3 + 2 \cdot 10 \qquad = 1020 \\ 10 + 1 = \quad 11 \\ \underline{2 \cdot 10^3 + \quad 10 + 3 = 2013} \\ 3 \cdot 10^3 + 4 \cdot 10 + 4 = 3044 \; ❖ \end{array}$$

The perimeter can also be found using horizontal form.

Ⓐ $(x^2 + x) + 2x^2 + 2x^2 + (x^2 + x) = 6x^2 + 2x$

Ⓑ $(a^3 + 2a) + (a + 1) + (2a^3 + a + 3) = 3a^3 + 4a + 4 \; ❖$

CRITICAL
Thinking

Explain how using the horizontal form allows you to add polynomials mentally.

Subtracting Polynomials

Recall that the definition of subtraction is to add the opposite.

EXAMPLE 3

Subtract $x^2 - 4$ from $3x^2 - 2x + 8$.

Solution ►

The binomial $x^2 - 4$ can be written as $x^2 + 0x - 4$, where $0x$ is a "placeholder" for the missing x term. Remember to change signs when subtracting.

Vertical form
$$3x^2 - 2x + 8$$
$$- (x^2 + 0x - 4)$$
$$\overline{2x^2 - 2x + 12}$$

Horizontal form
$$3x^2 - 2x + 8 - (x^2 - 4) = 3x^2 - 2x + 8 - x^2 + 4$$
$$= 2x^2 - 2x + 12$$

CRITICAL
Thinking

Study the two examples. How is the addition of whole numbers related to the addition of polynomials?

1. $485 = 4 \cdot 10^2 + 8 \cdot 10 + 5$
   $\underline{113 = 1 \cdot 10^2 + 1 \cdot 10 + 3}$
   $598 = 5 \cdot 10^2 + 9 \cdot 10 + 8$

2. $4x^2 + 8x + 5$
   $\underline{x^2 + x + 3}$
   $5x^2 + 9x + 8$

EXERCISES & PROBLEMS

Communicate

1. Explain how to use algebra tiles to represent $5x^2 - 2x + 3$.

2. How do you determine the degree of a polynomial?

3. Give two different names for the polynomial $3x^3 + 4x^2 - 7$.

4. Draw an algebra-tile model of $(2x^2 + 3) + (x^2 - 1)$. Find the sum.

5. Explain how to find $(3b^3 - 2b + 1) - (b^3 + b - 3)$.

6. Explain how to write a polynomial in standard form. Use $5x^2 - 2x + 3x^4 - 6$ as a model.

7.    $x^2 + 4x + 4$

Practice & Apply

7. Use algebra tiles to represent $x^2 + 4x + 4$.

8. Draw an algebra-tile model of $(3y^2 + 6) + (2y^2 - 1)$.

Rewrite each polynomial in standard form.

9. $6 + c + c^3$
$c^3 + c + 6$

10. $5x^3 - 1 + 5x^4 + 5x^2$
$5x^4 + 5x^3 + 5x^2 - 1$

11. $10 + p^7$
$p^7 + 10$

Write the degree of each polynomial.

12. $4r + 1$
1

13. $x^3 + x^4 + x - 1$
4

14. $y + y^3$
3

Identify each polynomial by name from the number of terms and degree.

15. $3x + 1$
binomial; linear

16. $8x^2 - 1$
binomial; quadratic

17. $8x^2 - 2x + 3$
trinomial; quadratic

Give an example of a polynomial that is a

18. quadratic trinomial.
example: $6x^2 - 5x + 9$

19. linear binomial.
example: $10n + 3$

20. cubic monomial.
example: $8d^3$

Use vertical form to add. $7x^4 + 4x^2 - x - 5$

21. $3x^2 + 4x^4 - x + 1$ and $3x^4 + x^2 - 6$

22. $2y^3 + y^2 + 1$ and $3y^3 - y^2 + 2$ $\quad 5y^3 + 3$

23. $4r^4 + r^3 - 6$ and $r^3 + r^2$
$4r^4 + 2r^3 + r^2 - 6$

24. $2c - 3$ and $c^2 + c + 4$
$c^2 + 3c + 1$

Use horizontal form to add.

25. $y^3 - 4$ and $y^2 - 2$ $\quad y^3 + y^2 - 6$

26. $x^3 + 2x - 1$ and $3x^2 + 4$ $\quad x^3 + 3x^2 + 2x + 3$

27. $3s^2 + 7s - 6$ and $s^3 + s^2 - s - 1$
$s^3 + 4s^2 + 6s - 7$

28. $w^3 + w - 2$ and $4w^3 - 7w + 2$
$5w^3 - 6w$

Use vertical form to subtract.

29. $x^2 + x$ from $x^3 + x^2 + 7$ $\quad x^3 - x + 7$

30. $3y^2 - 4$ from $4y^2 - y + 6$ $\quad y^2 - y + 10$

31. $4c^3 - c^2 - 1$ from $5c^3 + 10c + 5$
$c^3 + c^2 + 10c + 6$

32. $8x^3$ from $x^3 - x + 4$
$-7x^3 - x + 4$

Use horizontal form to subtract.

33. $y^2 + 3y + 2 - (3y - 2)$ $\quad y^2 + 4$

34. $3x^2 - 2x + 10 - (2x^2 + 4x - 6)$ $\quad x^2 - 6x + 16$

35. $3x^2 - 5x + 3 - (2x^2 - x - 4)$
$x^2 - 4x + 7$

36. $2x^2 + 5x - (x^2 - 3)$ $\quad x^2 + 5x + 3$

Simplify. Express all answers in standard form.

37. $(1 - 4x - x^4) + (x - 3x^2 + 9)$ $\quad -x^4 - 3x^2 - 3x + 10$

38. $5 - 3x - 1.4x^2 - (13.7x - 62 + 5.6x^2)$ $\quad -7x^2 - 16.7x + 67$

39. Subtract $x^5 - 2x^2 + 3x + 5$ from $4x^5 - 3x^3 - 3x - 5$. $\quad 3x^5 - 3x^3 + 2x^2 - 6x - 10$

$$34{,}276 = 3(10^4) + 4(10^3) + 2(10^2) + 7(10^1) + 6$$

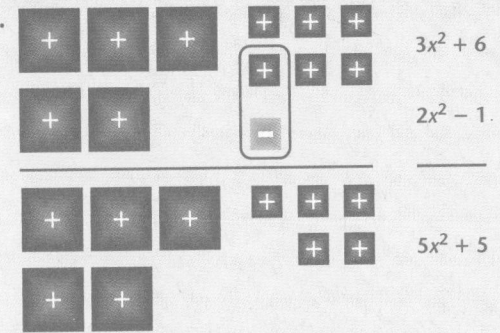

8.

$3x^2 + 6$

$2x^2 - 1$

$5x^2 + 5$

Technology

For Exercise 40, use a scientific calculator with programming capabilities to find the perimeter of the triangle. Change the variable to x for the calculator, but use y as the variable for writing the answer. Perimeter = $(3y^2 + 1) + (3y^2 + 1) + (2y^2 + y - 3) = 8y^2 + y - 1$. Then have students find a numerical value for the perimeter if $y = 12$. **[1163 units]**

Aalternative
SSESSMENT

Portfolio Assessment

Have students choose exercises that demonstrate that they can add and subtract polynomials and include these in their portfolio. Students should also explain why writing answers in standard form is important.

Look Beyond

Exercise 54 foreshadows factoring perfect-square trinomials, which is studied later in this chapter.

---

40. 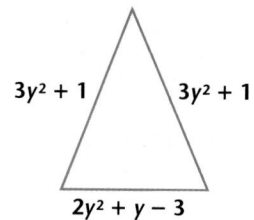 **Geometry** Find the perimeter of the triangle. $8y^2 + y - 1$

Triangle with sides labeled $3y^2 + 1$ (left), $3y^2 + 1$ (right), and $2y^2 + y - 3$ (bottom).

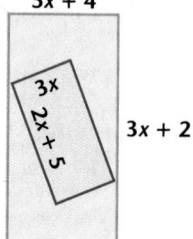 **Geometry** Exercises 41–43 refer to the dimensions of the figure at the right.

41. Find the polynomial expression that represents the perimeter of the large rectangle. $16x + 12$

42. Find the polynomial expression that represents the perimeter of the smaller rectangle. $10x + 10$

43. Find the difference between the perimeters of the two rectangles. $6x + 2$

Figure labels: $5x + 4$, $3x$, $2x + 5$, $3x + 2$

Look Back

44. State the pattern, and find the next three terms of the sequence.
1, 3, 9, 27, 81, ___, ___, ___ **[Lesson 1.1]**   multiply by 3; 243, 729, 2187

45. Divide $\dfrac{6x^2 - 12x + 18}{3}$.   **[Lesson 4.2]** $2x^2 - 4x + 6$

46. Solve $-2(a + 3) = 5 - 6(2a - 7)$.   **[Lesson 4.6]**   5.3

47. Solve. $\begin{cases} 6x = 4 - 2y \\ 12x - 4y = 16 \end{cases}$ **[Lesson 6.3]** $(1, -1)$

48. Without repeating any digits, how many different 3-digit codes can be made with the digits 1–9?   **[Lesson 8.5]**   504

Multiply.
49. $6x(x^2 - 4x)$ $6x^3 - 24x^2$ **50.** $8m^3(m^2 + 6m)$ **[Lesson 10.2]** $8m^5 + 48m^4$
51. Simplify $(3x^5)(-2x^3)(-x)$. $6x^9$

Write each number in scientific form.
52. 7,100,000   **53.** 8,900,000,000   **[Lesson 10.4]**
      7.1×10^6 8.9×10^9

Look Beyond

54. Use algebra tiles to model a square whose area can be represented as $x^2 + 6x + 9$. What is the length of a side of the square? $x + 3$

LESSON 11.2

Exploring Products and Factors

Factor
tiles

$x \cdot x = x^2$

Product
tile

*Multiplication of polynomials can be
represented by a geometric model.*

why *Algebra tiles can be used to model polynomials,
their sums, and differences. These tiles can also be used to
explore products and factors.*

To model $x \cdot x$, the factor tiles are placed
above and next to the crossed lines. A
product tile is placed within the crossed
lines. The rules for the signs of the factor
tiles are the same as those developed for
integers. If the signs of the factor tiles are
the same, the product tile is positive. If the
signs are opposite, the product tile is
negative.

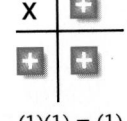 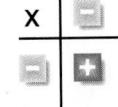 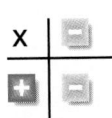 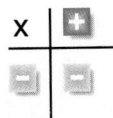

$(1)(1) = (1)$ $(-1)(-1) = (1)$ $(1)(-1) = (-1)$ $(-1)(1) = (-1)$

PREPARE

Objectives

- Use tiles and the Distributive Property to show the product of a monomial and a binomial.
- Show the greatest common factor as part of the factored form of a polynomial.

RESOURCES

- Practice Master   **11.2**
- Enrichment Master   **11.2**
- Technology Master   **11.2**
- Lesson Activity Master **11.2**
- Quiz   **11.2**
- Spanish Resources   **11.2**

Assessing Prior Knowledge

Evaluate.

1. $(-2)(-3)$ **[6]**
2. $(-2)(3)$ **[-6]**
3. $(2)(3-4)$ **[-2]**

Find the greatest common factor of each pair of numbers.

4. 4 and 14   **[2]**
5. 9 and 27   **[9]**
6. 17 and 39   **[1]**

TEACH

why Using the Distributive Property often simplifies computation and may even allow the computation to be completed mentally. Explain to students that in this lesson they will use the Distributive Property when working with polynomials.

Using Patterns Have
students explore an area
model of quadratic poly-
nomials. Students should
begin with a piece of graph paper. Specify an x
value such as a length of 5 squares. Assume each
square has a length of 1. With this condition,
have students mark the horizontal margins with
the factor $(2x + 3)$ and the vertical with $(x +
4)$. They should form the rectangular area for
the product of the factors, $2x^2 + 11x + 12$.
Have the students repeat the process with different values of x. Then have them repeat the
process with different combinations of monomial and binomial factors. Students should discuss their observations.

Provide students with tiles, and have them work along with you or in small groups to complete this exploration.

Cooperative Learning

Allow students to work in small groups and discuss the examples shown in the center of page 514. Groups should discuss how each answer is found and how the tiles relate to using the Distributive Property. Then have each group act out the Application at the bottom of the page.

Use Transparency 50

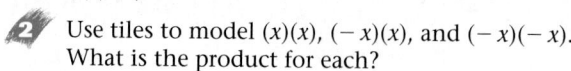

Exploration 1 — Modeling With Tiles

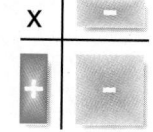

1. What product is shown in the diagram that models $(x)(-x)$?

2. Use tiles to model $(x)(x)$, $(-x)(x)$, and $(-x)(-x)$. What is the product for each?

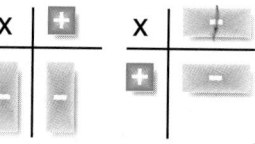

3. The diagram models $(-x)(1)$ and $(1)(-x)$. What product is shown for each?

4. Use tiles to model $(x)(-1)$, $(x)(1)$, $(-x)(-1)$, $(1)(x)$, $(1)(-x)$, and $(-1)(-x)$. What is the product for each? ❖

Either tiles or the Distributive Property can be used to find the product of 3 and $2x + 1$.

Algebra-Tile Model

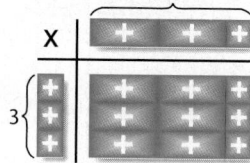

Count the tiles for the product.
$6x + 3$

Distributive Property

$$3(2x + 1) = 3(2x) + 3(1)$$
$$= 6x + 3$$

The model shows that the product of 3 and $2x + 1$ is $6x + 3$. The tiles modeling 3 and $2x + 1$ represent two **factors** of $6x + 3$. Notice that applying the Distributive Property results in the same algebraic product.

APPLICATION

Suppose each of three students has 2 boxes of pencils and 1 loose pencil. If b represents the number of pencils in a box, $2b + 1$ represents the number of pencils each student has. Since there are 3 students, $3(2b + 1)$ represents the total number of pencils. This product can be rewritten $6b + 3$, that is, $3(2b + 1) = 6b + 3$. ❖

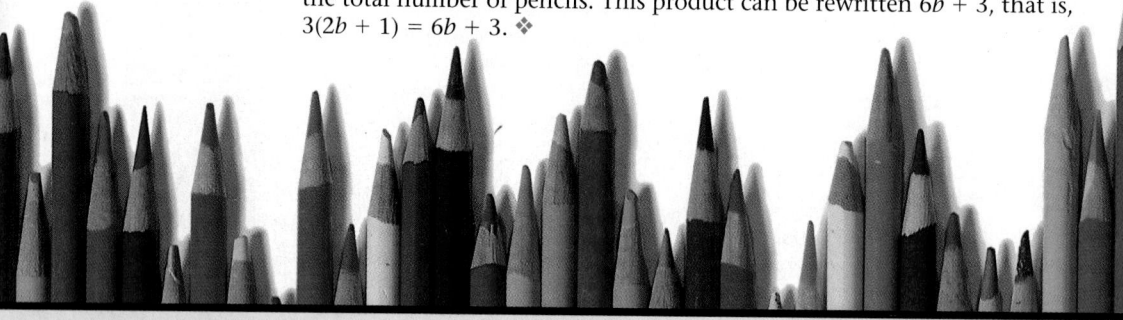

ENRICHMENT The perimeter of a square is four times the length of a side. Have students find the perimeter of squares that have sides with these measurements.

1. $z + 2$ [$4z + 8$]
2. $2a^2$ [$8a^2$]
3. $3m^2 - 1$ [$12m^2 - 4$]

INCLUSION strategies **Cooperative Learning** Have students work in small groups and explain the relationship between the layout of the algebra tiles and the product of a monomial and a binomial.

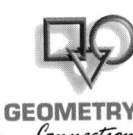 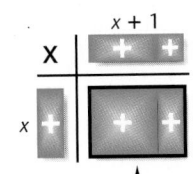

Exploration 2 Finding Products

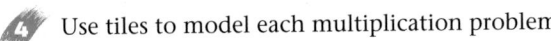

GEOMETRY
Connection

X $\qquad$ $x + 1$

x

↑
Product rectangle

Examine the product rectangle.

1 What are the two factors?

2 What is the product?

3 Use the Distributive Property to find the product $x(x + 1) = \underline{\ ?\ }$.

4 Use tiles to model each multiplication problem.
  a. $2x(x + 3)$ **b.** $3(3x + 2)$

5 Use the Distributive Property to find each product.
  a. $2x(x + 3)$ **b.** $3(3x + 2)$

6 Explain how to use the Distributive Property to find the product $2x(x + 1)$. ❖

Rectangles that model products can be built using positive and negative tiles. When you build *product rectangles* be careful to obey the laws of signs for multiplying integers.

Exploration 3 Using Negative Tiles

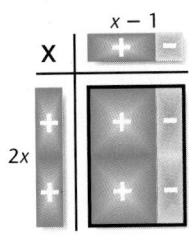

X $\qquad$ $x - 1$

2x

Examine the product rectangle.

1 What are the two factors?

2 What is the product?

3 Use the Distributive Property to find the product $2x(x - 1) = \underline{\ ?\ }$.

4 Use tiles to model each multiplication.
  a. $4(-x + 2)$ **b.** $3(2x - 1)$

5 Use the Distributive Property to find each product.
  a. $4(-x + 2)$ **b.** $3(2x - 1)$

6 Explain how to use the Distributive Property to find $-3(-x + 4)$. ❖

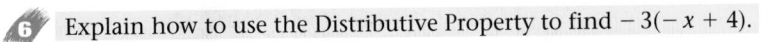

Have students arrange the tiles for $4x^2$ and $6x$ and complete the model rectangle.

6. $5x^2 + 15x = 5x(x + 3)$

This is equivalent to a rectangle with one side having 5 x-tiles and the other having 1 x-tile and 3 unit tiles.

TEACHING *tip*

You may wish to review the concept of greatest common factor for two whole numbers.

CRITICAL *Thinking*

No. Students should cite a counterexample. For instance, $x^2 + x + 1$ cannot be arranged in a rectangle.

Practice Master

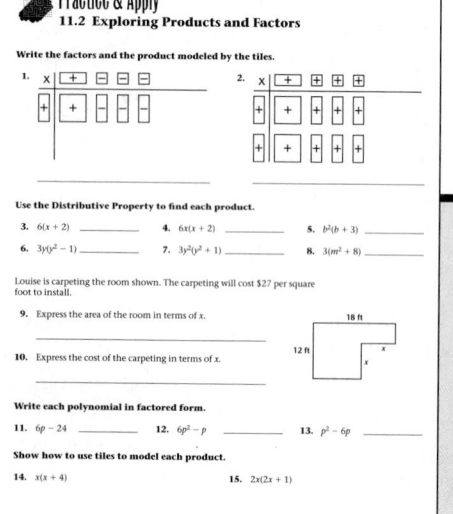

Practice & Apply
11.2 **Exploring Products and Factors**

Write the factors and the product modeled by the tiles.

1. 2.

Use the Distributive Property to find each product.

3. $6(x + 2)$ _____ 4. $6x(x + 2)$ _____ 5. $b^2(b + 3)$ _____

6. $3y(y^2 - 1)$ _____ 7. $3y^2(y^2 + 1)$ _____ 8. $3(m^2 + 8)$ _____

Louise is carpeting the room shown. The carpeting will cost $27 per square foot to install.

9. Express the area of the room in terms of x.
_____

10. Express the cost of the carpeting in terms of x.
_____

Write each polynomial in factored form.

11. $6p - 24$ _____ 12. $6p^2 - p$ _____ 13. $p^2 - 6p$ _____

Show how to use tiles to model each product.

14. $x(x + 4)$ 15. $2x(2x + 1)$

Once a product rectangle is formed, you can tell what two factors have been multiplied to form the product.

Exploration 4 *From Product Rectangles to Factors*

Begin with the tiles for $4x^2$ and $6x$.

1. Form a product rectangle with $2x$ as one factor.

2. What is the second factor?

3. Start with the tiles for $3x^2$ and $6x$. Form a product rectangle with $3x$ as one factor.

4. What is the second factor?

5. Explain how to use the Distributive Property to complete $3x^2 + 6x = 3x(\underline{?} + \underline{?})$.

6. Explain how to use a product rectangle and the Distributive Property to complete $5x^2 + 15x = 5x(\underline{?} + \underline{?})$. ❖

The **greatest common factor**, or **GCF**, of $4x^2$ and $6x$ is $2x$. For the polynomial $4x^2 + 6x$, the factored form is $2x(2x + 3)$. Here are some examples.

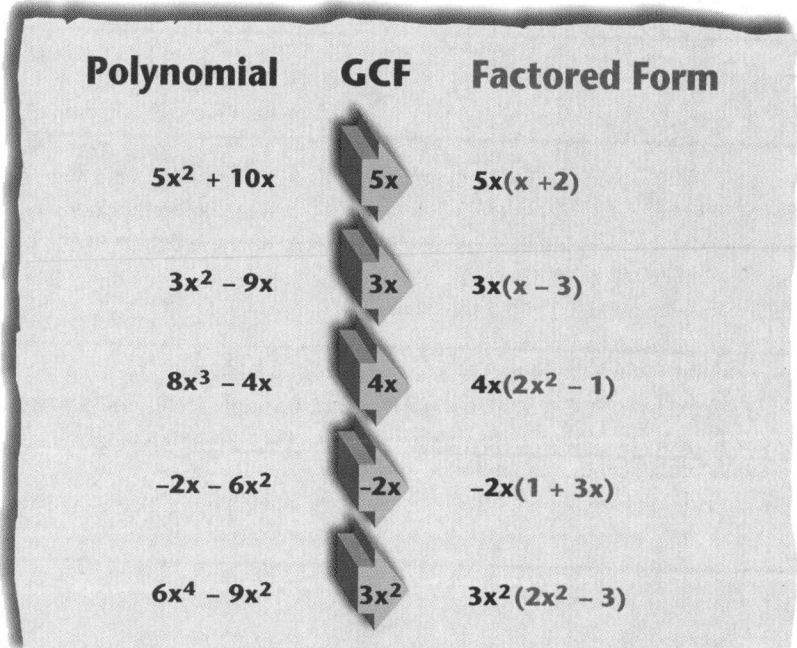

Polynomial	GCF	Factored Form
$5x^2 + 10x$	$5x$	$5x(x + 2)$
$3x^2 - 9x$	$3x$	$3x(x - 3)$
$8x^3 - 4x$	$4x$	$4x(2x^2 - 1)$
$-2x - 6x^2$	$-2x$	$-2x(1 + 3x)$
$6x^4 - 9x^2$	$3x^2$	$3x^2(2x^2 - 3)$

CRITICAL *Thinking*

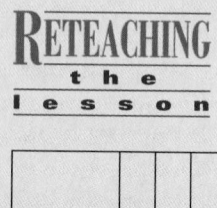

Given the tiles that model a second-degree polynomial, can you always form a rectangle from the tiles? Explain.

RETEACHING *the lesson*

Using Models Display the diagram and ask the questions that follow.

1. What is the length? $[x + 3]$
2. What is the width? $[2x]$
3. What is the area? $[2x^2 + 6x]$
4. Inside each small square or rectangle, write its area. Add those areas. $[2x^2 + 6x]$
5. Compare your answers for 3 and 4. [They are the same.]

EXERCISES & PROBLEMS

Communicate

1. Explain how to use algebra tiles to show the area of a rectangle with sides $2x$ and $x + 5$.

2. Explain how to use the Distributive Property to find the area of a rectangle with sides $2x$ and $x + 5$.

3. Draw a tile model to find the sides of a rectangle that has a product of $4x^2 + 2x$.

4. Tell what it means to write the polynomial $x^3 + x$ in factored form.

Practice & Apply

5. Draw a tile model to show the area of a rectangle that has factors of $x + 7$ and x.

6. Draw a tile model to find the sides of a rectangle that has a product of $2x^2 + 2x$.

Geometry Find the factors for the rectangle modeled with positive tiles and write the product shown by the tiles.

7.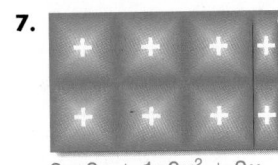
$2x,\ 3x + 1;\ 6x^2 + 2x$

8.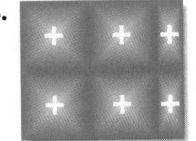
$2x,\ 2x + 1;\ 4x^2 + 2x$

9.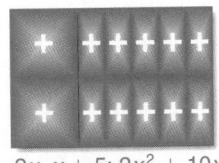
$2x,\ x + 5;\ 2x^2 + 10x$

10.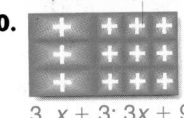
$3,\ x + 3;\ 3x + 9$

11.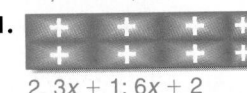
$2,\ 3x + 1;\ 6x + 2$

12.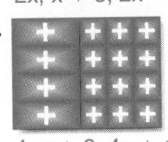
$4,\ x + 3;\ 4x + 12$

Use the Distributive Property to find each product.

13. $4(x + 2)$
14. $6(2x + 7)$
15. $5(y + 10)$
16. $3(m + 8)$

17. $x(x + 2)$
18. $2y(y - 4)$
19. $3r(r^2 - 3)$
20. $p^2(p + 7)$

21. $8(3x - 4)$
22. $4y(y - 4)$
23. $5x(4x + 9)$
24. $3w(w^2 - w)$

25. $2y^2(y^2 + y)$
26. $2t^2(2t^2 + 8t)$
27. $3y^2(3y - 6)$
28. $z(11z^2 + 22z)$

Write each polynomial in factored form.

29. $3x^2 + 6$
30. $5x^2 - 20$
31. $y^2 - y^3$
32. $4p + 12p^3$

33. $7y^4 + 49y$
34. $m^4 - 6m^2$
35. $-4r^2 - 6r$
36. $9n^2 - 27n^4$

The answers to Exercises 5 and 6 can be found in Additional Answers beginning on page 729.

13. $4x + 8$
14. $12x + 42$
15. $5y + 50$
16. $3m + 24$
17. $x^2 + 2x$
18. $2y^2 - 8y$
19. $3r^3 - 9r$
20. $p^3 + 7p^2$
21. $24x - 32$
22. $4y^2 - 16y$
23. $20x^2 + 45x$
24. $3w^3 - 3w^2$
25. $2y^4 + 2y^3$
26. $4t^4 + 16t^3$
27. $9y^3 - 18y^2$
28. $11z^3 + 22z^2$
29. $3(x^2 + 2)$
30. $5(x^2 - 4)$
31. $y^2(1 - y)$
32. $4p(1 + 3p^2)$
33. $7y(y^3 + 7)$
34. $m^2(m^2 - 6)$
35. $-2r(2r + 3)$
36. $9n^2(1 - 3n^2)$

ASSESS

Selected Answers
Odd-numbered Exercises 5–51

Assignment Guide
Core 1 – 4, 8 – 38 even, 37 – 52

Core Two-Year 1 – 51

Technology
If available, have students use interactive software that models algebra tiles while completing this assignment.

Error Analysis
Students may fail to complete the multiplication due to carelessness or due to lack of understanding of the Distributive Property. For example, they may erroneously write $6(2x + 7) = 12x + 7$. Use algebra tiles, drawings, or other models to reinforce the need to distribute the 6 over all terms within the parentheses to get the correct answer of $12x + 42$.

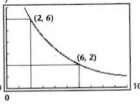

Geometry Mary wants a rectangular flower bed and lawn in her backyard.

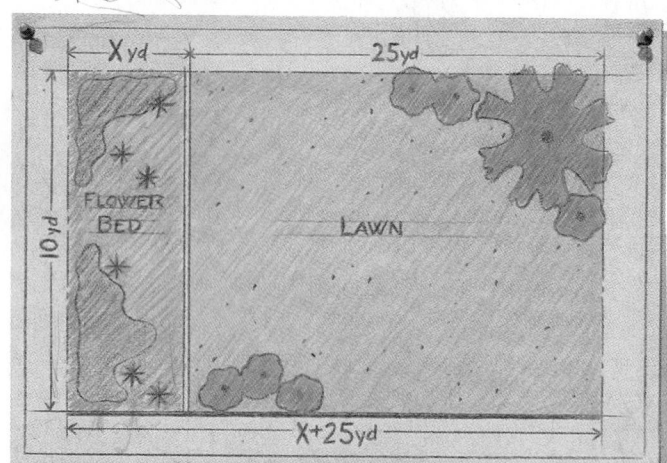

A landscape artist is designing Mary's backyard.

37. Write an expression for the area of the flower bed. $10x$ yd^2

38. Write an expression for the area of the lawn. $(25)(10)$ yd^2

39. Write an expression for the area of her backyard. $(10x + 250)$ yd^2

40. What conjecture can you make about the areas of the flower bed and the lawn compared with the area of her backyard?
flower bed area + lawn area = backyard area

41. Discount Sales Wholesale Grocery has pieces of dinnerware for sale in the houseware department. Glasses are $1.50, plates are $2.50, cups are $1.25, and saucers are $1.00. A 25% discount is given for orders over $10.00. If a customer buys x glasses and y plates and receives a discount, write an expression for how much discount the customer receives.
$0.375x + 0.625y$

42. Geometry Find the perimeter of a rectangle with a width of $2x + 3$ and a length of $3x$. $10x + 6$

Look Back

43. Coordinate Geometry Find the equation of the line between $A(2, -7)$ and $B(2, 11)$. **[Lesson 5.2]** $x = 2$

44. Advertising In how many different orders can four different models of bicycles be displayed in a store's window? **[Lesson 8.5]** 24

Let $f(x) = 2^x$. **45.** Find $f(7)$. 128 **46.** Find $f(4)$. **[Lesson 9.1]** 16

Use the Distributive Property to find each polynomial. **[Lesson 10.1]**

47. $2a(a - 6b)$
$2a^2 - 12ab$

48. $x(z - x^2)$
$xz - x^3$

49. $9y^2(y - 3)$
$9y^3 - 27y^2$

Write each number in scientific notation. **[Lesson 10.4]**

50. 1,900,000
1.9×10^6

51. 0.0000000001
1×10^{-10}

Look Beyond

52. Explain how factoring $x^2 - x - 12$ will help simplify $(x^2 - x - 12) \div (x + 3)$.

52. $(x^2 - x - 12) \div (x + 3)$
$$\frac{(x + 3)(x - 4)}{(x + 3)}; x \neq -3$$
$1(x - 4)$
$x - 4$

The Distributive Property can be used to multiply a monomial and a binomial. It can also be used to multiply two binomials.

Exploration 1 Multiplying With Tiles

1 Complete the algebra-tile model.

2 What two factors are multiplied?

3 What is the product represented by the completed model?

4 If a rectangle has sides of $x + 3$ and $x + 2$, what is the area?

5 Explain how to use an algebra-tile model to show that $(x + 4)(x + 5) = x^2 + 9x + 20$. ❖

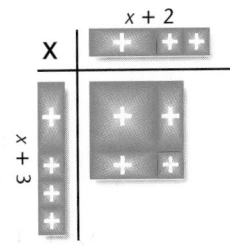

The Distributive Property can be used to show that $(x + 3)(x + 2) = x^2 + 5x + 6$.

$(x + 3)(x + 2)$	Given
$= x(x + 3) + 2(x + 3)$	Distribute the first binomial to *each term of the second binomial.*
$= x^2 + 3x + 2x + 6$	Distribute *each term* of the second binomial *to each term* of the first binomial.
$= x^2 + \quad 5x \quad + 6$	Add like terms.

Use the Distributive Property to show that $(x + 4)(x + 5) = x^2 + 9x + 20$.

Using Algorithms

Show students how to use the traditional multiplication algorithm for whole numbers when multiplying binomials.

$$
\begin{array}{r}
x + 3 \\
\times\, x - 2 \\
\hline
-2x - 6 \\
x^2 + 3x \\
\hline
x^2 + x - 6
\end{array}
$$

$(x + 3)(-2)$
$(x + 3)(x)$
Add.

PREPARE

Objectives

- Use the Distributive Property to multiply binomials.
- Use the FOIL method to multiply binomials.

RESOURCES

- Practice Master   **11.3**
- Enrichment Master   **11.3**
- Technology Master   **11.3**
- Lesson Activity Master   **11.3**
- Quiz   **11.3**
- Spanish Resources   **11.3**

Assessing Prior Knowledge

Multiply. Write each answer in standard form.

1. $x(x^2)$ $[x^3]$
2. $x(x - 3)$ $[x^2 - 3x]$
3. $-2(x + 5)$ $[-2x - 10]$

TEACH

Ask students to find the product of 8 and 98 within 15 seconds. If any students can do this, ask them for their method. The Distributive Property makes this possible.

ongoing ASSESSMENT

5. Use algebra tiles to build a rectangle from $x + 4$ and $x + 5$. The rectangle will consist of 1 x^2-tile, 9 x-tiles, and 20 unit tiles. $x^2 + 9x + 20$.

ongoing ASSESSMENT

$(x + 4)(x + 5)$
$= x(x + 4) + 5(x + 4)$
$= x^2 + 4x + 5x + 20$
$= x^2 + 9x + 20$

Exploration 2 Notes

Explain that when multiplying a binomial by another binomial, *all* terms must be multiplied by *all other* terms. In the case of the middle term, the coefficients are added (watch out for minus signs). This introduces the FOIL method of the next section.

A ongoing SSESSMENT

6. $x^2 + (a + b)x + ab$ OR
   $x^2 + ax + bx + ab$

Exploration 2 *Multiplying Binomials*

$$(x + 3)(x + 5) = x^2 + 8x + 15$$

$$(x + 3)(x + 5) = x^2 + 5x + 3x + 15$$

$$(x + 3)(x + 5) = x^2 + \quad 8x \quad + 15$$

$$(x + 3)(x + 5) = x^2 + 8x + 15$$

1 What is the relationship of the first term of each binomial factor to the first term of the trinomial?

2 What is the relationship of the outside terms and the inside terms of the binomial factors to the middle term of the trinomial?

3 What is the relationship of the last term of each binomial factor to the last term of the trinomial?

4 Explain how to find $(x + 3)(x + 5)$ using what you learned in the first 3 steps.

5 Use this shortcut to write each product. Check the result by using a tile model or the Distributive Property.

   a. $(x + 3)(x - 5)$ **b.** $(x - 3)(x + 5)$ **c.** $(x - 3)(x - 5)$

6 Show how you would use this shortcut to complete $(x + a)(x + b) = \underline{\ ?\ }$ ❖

FOIL Method

In Exploration 2 you used a method for multiplying two binomials that is usually referred to as **FOIL**.

• Multiply **F**irst terms.

• Multiply **O**utside terms. Multiply **I**nside terms. Add outside and inside products.

• Multiply **L**ast terms.

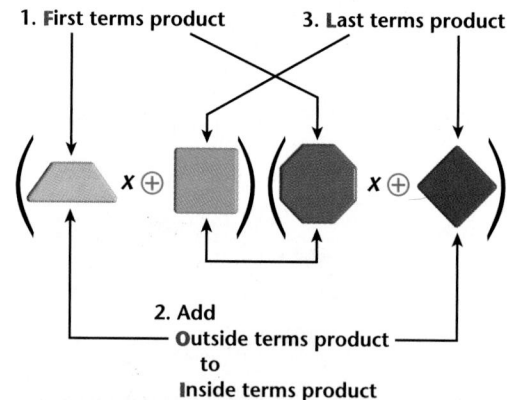

1. **First terms product**   3. **Last terms product**

2. **Add** Outside terms product to Inside terms product

EXAMPLE 1

Use the FOIL method to find each product.

Ⓐ $(3x + 1)(3x - 2)$ **Ⓑ** $(2a - 1)(5a + 3)$ **Ⓒ** $(5m - 2)(5m + 2)$

Solution ➤

Multiply to find each term, then simplify.

Ⓐ $(3x + 1)(3x - 2) = 9x^2 - 6x + 3x - 2 = 9x^2 - 3x - 2$

Ⓑ $(2a - 1)(5a + 3) = 10a^2 + 6a - 5a - 3 = 10a^2 + a - 3$

Ⓒ $(5m - 2)(5m + 2) = 25m^2 + 10m - 10m - 4 = 25m^2 - 4$ ✦

CRITICAL Thinking

Draw an algebra-tile diagram to show that $(2x + 3)(x + 1) = 2x^2 + 5x + 3$. Explain how you can check your diagram by the Distributive Property and the FOIL method.

EXAMPLE 2

Small Business Cruz's Frame Shop makes a mat by cutting out the inside of a rectangular mat board.

Find the length and width of the original mat board if the area of the mat is 148 square inches.

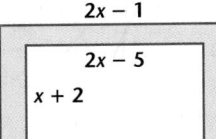

Solution ➤

Find the area of each rectangle.

Area of the outside rectangle	Area of the inside rectangle
$(2x - 1)(x + 6)$	$(x + 2)(2x - 5)$
$2x^2 + 12x - x - 6$	$2x^2 - 5x + 4x - 10$
$2x^2 + 11x - 6$	$2x^2 - x - 10$

Substitute each area into the formula, and then solve for x.

Area outside rectangle	−	Area inside rectangle	=	Area of the mat
$2x^2 + 11x - 6$	−	$(2x^2 - x - 10)$	=	148
$2x^2 + 11x - 6$	−	$2x^2 + x + 10$	=	148
		$12x + 4$	=	148
		$12x$	=	144
		x	=	12

The length of the original rectangle is $2x - 1$. Substitute.

$$2x - 1 = 2(12) - 1 = 23$$

The width is $x + 6$. Substitute. $x + 6 = 12 + 6 = 18$

The original mat board was 23 inches by 18 inches. ✦

RETEACHING the lesson

Using Manipulatives

Allow students to use algebra tiles to complete Exercises 5, 9, 11, 12, and 15. Have students explain the relationship between the algebra tiles, the Distributive Property, and the FOIL method of multiplication.

Alternate Example 1

Use the FOIL method to find each product.

a. $(2n - 1)(3n + 5)$
$[6n^2 + 7n - 5]$

b. $(5n + 4)(4n - 3)$
$[20n^2 + n - 12]$

c. $(3n - 7)(3n + 7)$
$[9n^2 - 49]$

CRITICAL Thinking

Check students' diagrams and explanations. Students should show that $(2x)(x) = 2x^2$; $(2x)(1) + (3)(x) = 5x$; and $(3)(1) = 3$.

Alternate Example 2

Find the length of the larger rectangle if the difference in the areas of the two rectangles shown in Example 2 is 208 square centimeters.
[33 cm]

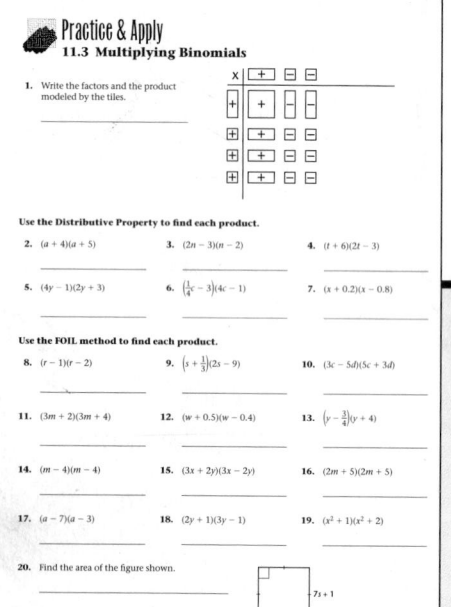

The Punnett Square shows the possible results of crossing two flowers, each containing a purple gene and a white gene. Since the purple gene dominates, only flowers with a ww genetic makeup are white.

EXAMPLE 3

Show how squaring a binomial models the probability that crossing two hybrid purple (Pw) flowers results in a flower that has a 25% probability of being white.

Solution ▶

Square the binomial $P + w$.

$$(P + w)^2 = P^2 + 2Pw + w^2 = PP + Pw + Pw + ww$$

Since there are four possibilities and one is w^2 or ww, the probability that the offspring will be white is $\frac{1}{4}$ or 25%. ❖

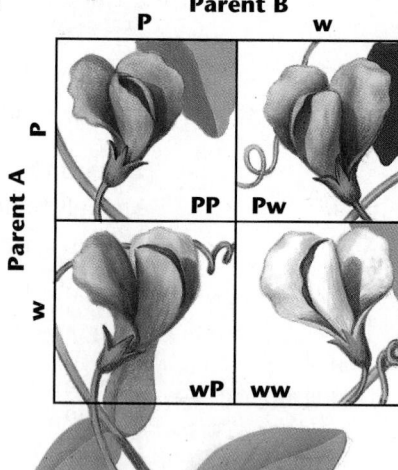

Parent B

Parent A

EXERCISES & PROBLEMS

Communicate

1. Show how to model each product.
   a. $(x + 1)(x + 2)$
   b. $(x - 1)(x - 2)$

2. Explain how to use the Distributive Property to find each product.
   a. $(x + 1)(x + 2)$
   b. $(x - 1)(x - 2)$

3. Describe how to find $(2x + 3)(x - 4)$ using the FOIL method.

4. Use algebra tiles to model the areas of $(x + 2)^2$ and $x^2 + 2^2$. How are the two models different?

Practice & Apply

Use the Distributive Property to find each product.

5. $(x + 2)(x + 5)$
6. $(a - 3)(a + 4)$
7. $(b - 1)(b - 3)$
8. $(y + 3)(y - 2)$
9. $(c + 5)(c + 5)$
10. $(d + 5)(d - 5)$

5. $x^2 + 7x + 10$
6. $a^2 + a - 12$
7. $b^2 - 4b + 3$
8. $y^2 + y - 6$
9. $c^2 + 10c + 25$
10. $d^2 - 25$

Use the FOIL method to find each product.

11. $(y + 5)(y + 3)$ **12.** $(w + 9)(w + 1)$ **13.** $(b - 7)(b + 3)$

14. $(3y - 2)(y - 1)$ **15.** $(5p + 3)(p + 1)$ **16.** $(2q - 1)(2q + 1)$

17. $(2x + 5)(2x - 3)$ **18.** $(4m + 1)(5m - 3)$ **19.** $(2w - 9)(3w - 8)$

20. $(3x - 5)(3x - 5)$ **21.** $(7s + 2)(2s - 3)$ **22.** $\left(y - \frac{1}{2}\right)\left(y + \frac{1}{2}\right)$

23. $\left(y - \frac{1}{3}\right)\left(y + \frac{5}{9}\right)$ **24.** $(c^2 + 1)(c^2 + 2)$ **25.** $(2a^2 + 3)(2a^2 + 3)$

26. $(a + c)(a + 2c)$ **27.** $(p + q)(p + q)$ **28.** $(a^2 + b)(a^2 - b)$

29. $(x^2 + y)(x + y)$ **30.** $(c + d)(2c + d)$ **31.** $(1.2m + 5)(0.8m - 4)$

Geometry A rectangular garden has a length of $x + 8$ units and a width of $x - 4$ units.

32. Draw a diagram, and label the dimensions. $(x + 8)$ by $(x - 4)$

33. Find the area. $x^2 + 4x - 32$

34. **Geometry** Find the area of a square rug that is $y + 6$ units on a side. $y^2 + 12y + 36$

35. **Geometry** Which has the greater area, a square with sides $(x + 1)$ units long or a rectangle with length $(x + 2)$ units and width x units? How much greater is it? square; 1 unit

36. **Geometry** Which has a greater area, a square with sides $(x + 1)$ units long or a rectangle with length x units and width $(x - 2)$ units? How much greater is it? square; $4x + 1$ units

Look Back

Evaluate. **[Lesson 2.4]**

37. INT(6.5)   6   **38.** INT$\left(2\frac{3}{4}\right)$ 2   **39.** INT(7.91)   7

Find the slope of the following lines. **[Lesson 5.3]**

40. $4x + 3y = 12$ $-\frac{4}{3}$ **41.** $y = 4x$ 4

42. Write the equation of the line through $(-2, -1)$ and parallel to $y = -2x - 4$. **[Lesson 5.6]** $y = -2x - 5$

Graph each system of equations. Check by the substitution method. **[Lesson 6.2]**

43. $\begin{cases} 2x + y = 9 \\ x - 2y = 2 \end{cases}$ **44.** $\begin{cases} x + y = 4 \\ x + y = 10 \end{cases}$

Look Beyond

Where does the graph of each function cross the x-axis?

45. $f(x) = (x + 3)(x - 2)$ $-3, 2$ **46.** $g(x) = (x + 2)^2$ -2 **47.** $h(x) = x^2 - 4$ $-2, 2$

---

11. $y^2 + 8y + 15$
12. $w^2 + 10w + 9$
13. $b^2 - 4b - 21$
14. $3y^2 - 5y + 2$
15. $5p^2 + 8p + 3$
16. $4q^2 - 1$
17. $4x^2 + 4x - 15$
18. $20m^2 - 7m - 3$
19. $6w^2 - 43w + 72$

20. $9x^2 - 30x + 25$
21. $14s^2 - 17s - 6$
22. $y^2 - \frac{1}{4}$
23. $y^2 + \frac{2}{9}y - \frac{5}{27}$
24. $c^4 + 3c^2 + 2$
25. $4a^4 + 12a^2 + 9$
26. $a^2 + 3ac + 2c^2$

27. $p^2 + 2pq + q^2$
28. $a^4 - b^2$
29. $x^3 + x^2y + xy + y^2$
30. $2c^2 + 3cd + d^2$
31. $0.96m^2 - 0.8m - 20$

The answers to Exercises 43 and 44 can be found in Additional Answers beginning on page 729.

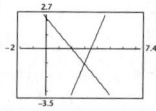

Objectives

- Factor the greatest common factor from a polynomial.
- Factor a binomial as the greatest common factor.

RESOURCES

- Practice Master   **11.4**
- Enrichment Master   **11.4**
- Technology Master   **11.4**
- Lesson Activity Master   **11.4**
- Quiz   **11.4**
- Spanish Resources   **11.4**

Assessing Prior Knowledge

Write the prime factorization of each number.

1. 18 $[2 \cdot 3^2]$
2. 20 $[2^2 \cdot 5]$
3. 15 $[3 \cdot 5]$
4. 42 $[2 \cdot 3 \cdot 7]$
5. 400 $[2^4 \cdot 5^2]$

TEACH

 Ask students to relate a real-world experience that could be solved by using the equation $12x = 48$. Explain that 12 and x are factors and 48 is the product. Ask students to write a division equation that shows the value of x. $\left[x = \dfrac{48}{12}\right]$ Tell students that they will learn about common factors of a polynomial in this lesson.

LESSON 11.4 Common Factors

The product of two factors is found by multiplication. Sometimes the product and one of the two factors are known and it is necessary to find the second factor.

Two unit squares can be used to form a rectangle of height 1 and length 2. A rectangle of height 2 and length 1 can also be formed.

However, the two rectangles will be considered the same, because the second rectangle is a rotation of the first.

Four unit squares form two different rectangles, a 2-by-2 and a 1-by-4. A 4-by-1 rectangle is the same as a 1-by-4 rectangle.

 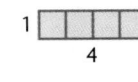

Exploration — Prime Numbers

1 Use unit squares to form rectangles. Record the number of different rectangles you can make.

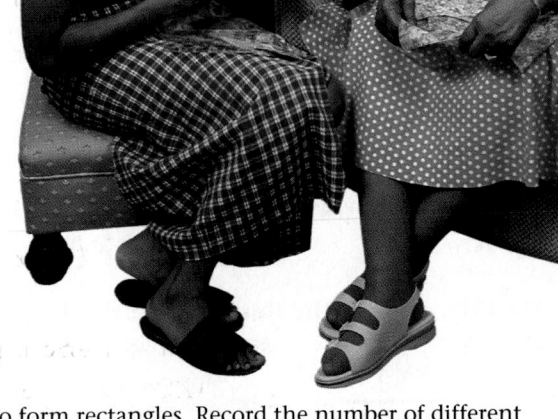

	Number of Unit Squares	Number of Rectangles		Number of Unit Squares	Number of Rectangles
a.	2	1	g.	8	?
b.	3	1	h.	9	?
c.	4	2	i.	10	?
d.	5	?	j.	11	?
e.	6	?	k.	12	?
f.	7	?	l.	13	?

ALTERNATIVE teaching strategy

Using Algorithms

Show students how to use the Distributive Property to factor polynomials. For instance, in $3x^2 - 12x$, each term contains a 3 and an x. By dividing the binomial by $3x$ you get $\dfrac{3x^2 - 12x}{3x} = \dfrac{3x^2}{3x} - \dfrac{12x}{3x} = x - 4$. This provides the factors $3x$ and $x - 4$ that fit the pattern for the Distributive Property.

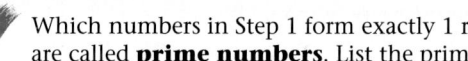 Which numbers in Step 1 form exactly 1 rectangle? These numbers are called **prime numbers**. List the prime numbers less than 30.

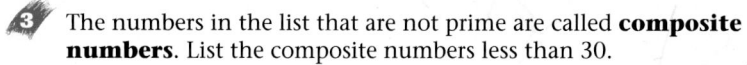 The numbers in the list that are not prime are called **composite numbers**. List the composite numbers less than 30.

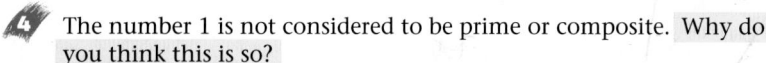 The number 1 is not considered to be prime or composite. Why do you think this is so?

5 Give a definition of a prime number using the word factor. ❖

When two numbers are multiplied they form a product. Each number is called a **factor** of that product. Thus, 3 and 8 are factors of 24 because $3 \cdot 8 = 24$, and one way to express 24 in factored form is $3 \cdot 8$. Sometimes individual factors can be factored further. The process ends when all the factors are prime. Name the prime factors of 24.

When you are asked to factor $3x^2 + 12x$, the first step is to examine the terms for a common monomial factor. Since $3x$ is common to both terms, write

$$3x^2 + 12x = 3x(x + 4).$$
$$\uparrow$$
common monomial factor

The polynomial $3x^2 + 12x$ has been factored over the integers because each term in $3x(x + 4)$ has an integer coefficient. When a polynomial has no polynomial factors with integral coefficients except itself and 1, it is a **prime polynomial** with respect to the integers. In this text, to factor means to factor over the integers.

EXAMPLE 1

Factor each polynomial.

Ⓐ $5am - 5an$ Ⓑ $5x^3 - 3y^2$ Ⓒ $2c^4 - 4c^3 + 6c^2$

Solution ➤

Factor out the greatest common factor, or GCF, from each term.

Ⓐ The GCF is **5a**. $5am - 5an = 5a(m - n)$

Ⓑ The GCF is 1. $5x^3 - 3y^2$ is prime.

Ⓒ The GCF is **2c²**. $2c^4 - 4c^3 + 6c^2 = 2c^2(c^2 - 2c + 3)$

To check each factorization, multiply the two factors to find the original expression. ❖

Substitute numbers for the variables in the previous example and simplify. Compare the results. What do you find?

 **CRITICAL** *Thinking*

Even though the expression $2x + 3y$ can be written as the product of a monomial and a binomial, $6\left(\frac{1}{3}x + \frac{1}{2}y\right)$, the binomial $2x + 3y$ is considered prime. Explain why.

Sometimes a polynomial has a common factor that contains two terms. This expression is called a **common binomial factor**.

EXAMPLE 2

Factor $r(t + 1) + s(t + 1)$.

Solution ➤

Notice that $r(t + 1) + s(t + 1)$ contains the common binomial factor $t + 1$. Think of $t + 1$ as c.

$$r(t + 1) + s(t + 1) = rc + sc$$
$$= (r + s)c$$
$$= (r + s)(t + 1)$$

The expression can also be factored mentally if you apply the Distributive Property.

$$r(t + 1) + s(t + 1) = (r + s)(t + 1) ❖$$

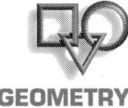

GEOMETRY
Connection

You can visualize this factoring procedure with an area model. It shows the same total area in different arrangements.

$$r(t + 1) \quad + \quad s(t + 1) \quad = \quad (r + s)(t + 1)$$

Try This   Factor $a(x - 3) + 6(x - 3)$.

When a polynomial has terms with more than one common factor, the expression can sometimes be factored by grouping terms.

EXAMPLE 3

Factor $x^2 + x + 2x + 2$ by grouping.

Solution ➤

There are several ways to factor this expression by grouping. One way is to group the terms that have a common coefficient or variable. Treat $x^2 + x$ as one expression, and treat $2x + 2$ as another expression.

$x^2 + x + 2x + 2 = (x^2 + x) + (2x + 2)$ Group terms.
$\qquad\qquad\qquad = x(x + 1) + 2(x + 1)$ Factor the GCF from each group.
$\qquad\qquad\qquad = (x + 2)(x + 1)$ Write the expression as the product of two binomials.

To check, multiply the factors. ❖

Try This   Factor $ax + bx + ay + by$ by grouping.

RETEACHING the lesson

Cooperative Learning Have students work together in groups. Each group is to discuss the exploration and examples. Have them explain their reasoning process for factoring various polynomials. Have one student in each group find a common factor of a polynomial and challenge others to find more. For example, in Example 1, $5am - 5an$ may initially be factored as $5(am - an)$, but factoring a out also yields $5a(m - n)$, the correct answer. Have students check that all the factoring is done, and then have them multiply the factors to find the original polynomial as a final check.

EXERCISES & PROBLEMS

Communicate

1. Define a *prime* polynomial.

2. What property of mathematics is being used when you factor by removing a GCF?

3. Once you have found the GCF, how do you find the remaining factor of the polynomial?

Each of the given pairs contains a GCF. Find the GCF, and explain how you find it.

4. 60 and 150

5. x^3y^5 and x^5y^2

6. $25(x + y)$ and $39(x + y)$

7. Explain how to group the terms to factor $y^2 + 2y + 3y + 6$.

Practice & Apply

Identify each polynomial as prime or composite.

8. $4x^2 - 16$ composite

9. $r^2 + 10$ prime

10. $n^2 + 4$ prime

Factor each polynomial by removing the GCF.

11. $2x^2 - 4$

12. $5n^2 - 10$

13. $3x^2 + 6x$

14. $x^9 - x^2$

15. $k^5 + k^2$

16. $4a^8 - 20a^6 + 8a^4$

17. $4x^2 + 2x - 6$

18. $7x^2 - 28x - 14$

19. $27y^3 + 18y^2 - 81y$

20. $3m^3 - 9m^2 + 3m$

21. $90 + 15a^5 - 45a$

22. $2x^3y - 18x^2y^2 + 17xy^3$

Write each as the product of two binomials.

23. $x(x + 1) + 2(x + 1)$

24. $5(y + 3) - x(y + 3)$

25. $a(x + y) + b(x + y)$

26. $(4 + p)3q - 4(4 + p)$

27. $x(x - 1) + 2(x - 1)$

28. $r(x - 4) + t(x - 4)$

29. $5a(a - 3) + 4(a - 3)$

30. $2w(w + 4) - 3(w + 4)$

31. $2(x - 2) + x(2 - x)$

32. $8(y - 1) - x(y - 1)$

33. $2r(r - s)^2 - 3(r - s)^2$

34. $ax(u - v)^n + bz(u - v)^n$

Factor.

35. $2x + 2y + ax + ay$

36. $nu + nv + 3u + 3v$

37. $12ab - 15a - 8b + 10$

38. $ax + ay + 12x + 12y$

39. $3(x + y) + 12(x + y)$

40. $x^2 + 3x + 4x + 12$

41. $2n^2 - 6n + 14n - 42$

42. $6pq + 12p^2 - 8qp + 2p^2$

43. $5x(2d + 3)^3 - 10(2d + 3)^3$

---

11. $2(x^2 - 2)$

12. $5(n^2 - 2)$

13. $3x(x + 2)$

14. $x^2(x^7 - 1)$

15. $k^2(k^3 + 1)$

16. $4a^4(a^4 - 5a^2 + 2)$

17. $2(2x^2 + x - 3)$

18. $7(x^2 - 4x - 2)$

19. $9y(3y^2 + 2y - 9)$

20. $3m(m^2 - 3m + 1)$

21. $15(6 + a^5 - 3a)$

22. $xy(2x^2 - 18xy + 17y^2)$

23. $(x + 2)(x + 1)$

24. $(5 - x)(y + 3)$

25. $(a + b)(x + y)$

26. $(3q - 4)(4 + p)$

27. $(x + 2)(x - 1)$

28. $(r + t)(x - 4)$

29. $(5a + 4)(a - 3)$

30. $(2w - 3)(w + 4)$

31. $(x - 2)(2 - x)$
   or $-(x - 2)^2$

32. $(8 - x)(y - 1)$

33. $(2r - 3)(r - s)^2$

34. $(ax + bz)(u - v)^n$

35. $(2 + a)(x + y)$

36. $(n + 3)(u + v)$

37. $(3a - 2)(4b - 5)$

38. $(a + 12)(x + y)$

39. $15(x + y)$

ASSESS

Selected Answers
Odd-numbered Exercises 9–55

Assignment Guide
Core 1–7, 8–42 even, 44–57

Core Two-Year 1–56

Technology
Students may use a calculator to check Exercises 44–45 and to complete Exercise 57.

Error Analysis
When factoring, students may fail to find the GCF. For example, for $12x^2 + 6x + 18$, students may factor it as $2(6x^2 + 3x + 9)$ instead of $6(2x^2 + x + 3)$.

40. $(x + 4)(x + 3)$

41. $2(n + 7)(n - 3)$

42. $2p(7p - q)$ or $-2p(q - 7p)$

43. $5(x - 2)(2d + 3)^3$

Technology Master

Technology
11.4 An Algorithm for Finding the GCF

When you factor an expression, you must look for the GCF of the numerical coefficients in the expression. To factor $91x^2 + 182x + 117$, you must find the GCF of 91, 182, and 117. This is not so easy. Notice that you can make the following observations.

① The GCF must be an integer that evenly divides 91, 182, and 117.

② The GCF must be the largest integer that divides 91, 182, and 117.

③ The GCF of 91, 117, and 182 cannot be more than 91, the smallest of the three numbers.

Since finding the GCF of a set of numbers is simply a matter of division, you can use a spreadsheet like the one shown to find the GCF.

	A	B	C	D
1	1	·		
2	2		·	
3	3			
4	4			
5	5			
6	6			

Cell A1 contains 1.
Cell A2 contains 1+A1.
Cell B1 contains 91/A1.
Cell C1 contains 117/A1.
Cell D1 contains 182/A1.

Fill rows 1 through 91 of the spreadsheet.

1. Find the GCF of 91, 182, and 117. How can you tell from the spreadsheet what the GCF is?

Use the spreadsheet above or a modification of it. Find the GCF of each set of numbers.

2. 21, 140, 105

3. 15, 21, 729

4. 5, 10, 15

5. 256, 1024, 6144

6. 7, 243, 49

7. 24, 36, 96, 1296

8. Explain how you know that the GCF of positive integers a, b, and c cannot be greater than the smallest of a, b, and c.

Performance Assessment

Have students use algebra tiles and then make sketches to show how to factor each of the following:

a. $6x^2 + 2x$ $[(2x)(3x + 1)]$

b. $2x^2 + 4x + 3x + 6$
   $[(2x + 3)(x + 2)]$

c. $np - 2nr + 3pt - 6rt$
   $[(n + 3t)(p - 2r)]$

An annulus is an object shaped like a washer.

Geometry The shaded area between the two concentric circles is called an **annulus**. The formula for the area of the *annulus* is $\pi R^2 - \pi r^2$ where the radius of the larger circle is R and the radius of the smaller is r.

44. Write the formula for the area of an annulus in factored form. $\pi(R^2 - r^2)$

45. Use the factored form to find the area of an annulus formed by concentric circles of radii 8 and 5.   122.522

46.  **Portfolio Activity**
Complete the problem in the portfolio activity on page 505. $\frac{3}{4}$

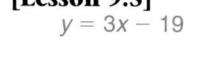 **Look Back**

47. Is the given sequence 3, 6, 12, 24, 48 representative of a linear relationship? How do you know?
[Lesson 2.1]

48. Evaluate $8 - 5(2 - 6) - 3^2$. **[Lesson 3.1]**   19

49. Solve the formula $A = \frac{1}{2}h(b_1 + b_2)$ for h.   **[Lesson 4.6]** $\frac{2A}{b_1 + b_2}$

50. Solve $3(x - 5) - 2(x + 8) = 7x + 3$. **[Lesson 4.6]** $-\frac{17}{3}$

51. Graph $x - 4y = 8$. **[Lesson 5.2]**

52. Solve. $\begin{cases} 4x = 11 + 15y \\ 6x + 5y = 0 \end{cases}$ **[Lesson 6.3]** $\left(\frac{1}{2}, -\frac{3}{5}\right)$

53. **Probability** Find the probability of drawing a heart or a queen from a deck of cards. **[Lesson 8.5]** $\frac{4}{13}$

Given $y = 3x - 5$ as the original function, write the equation that would represent each of the following transformations.

54. A reflection across the x-axis   **[Lesson 9.4]** $y = -3x + 5$

55. A translation of 4 units to the right and 2 units down   **[Lesson 9.5]**
           $y = 3x - 19$

56. Evaluate $\dfrac{2^2 \cdot 2^4}{2^5} - 2^0$. **[Lesson 10.3]**   1

Look Beyond

57. **Physics** When a projectile is fired vertically into the air, its motion can be modeled by the equation $h = -16t^2 + 320t$, where h is the height in feet at time t in seconds. From the graph of the equation, find the time when $h = 1200$.

Look Beyond

Exercise 57 has students use a graph to solve a quadratic equation. This foreshadows the use of factoring to find the solutions to a quadratic equation.

---

47. No; the first difference is not constant.

51.

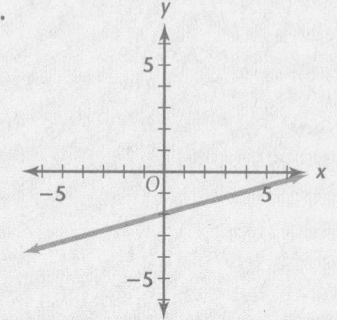

57. 5 and 15

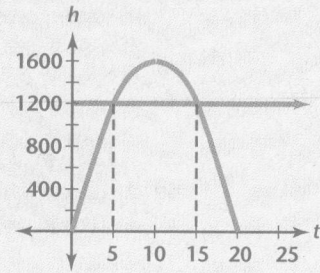

Factoring Special Polynomials

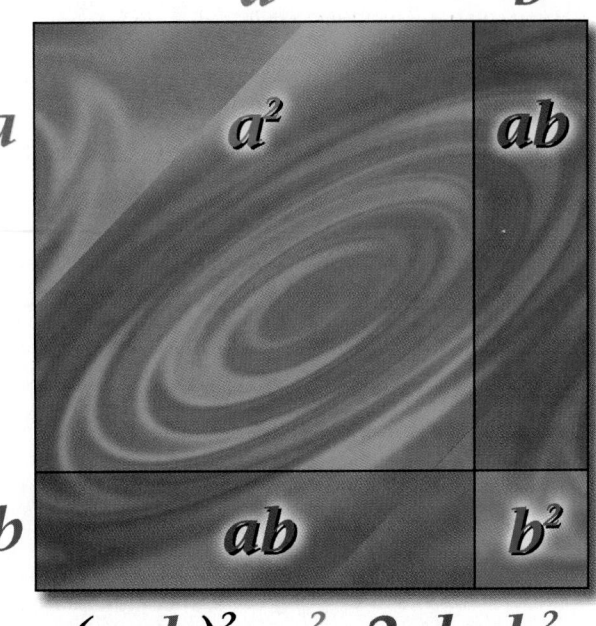

$$(a+b)^2 = a^2 + 2ab + b^2$$

 *If the length of one side of a square is a, the expression for its area is a^2. If the length of the side is increased, then the new expression for the area exhibits a simple pattern.*

Perfect-Square Trinomial

GEOMETRY
Connection

You can see from the geometry of the figure that the area of the large square is $(a + b)^2$, which is $a^2 + 2ab + b^2$. This shows why the product $a^2 + 2ab + b^2$ is called a **perfect-square trinomial**.

The same pattern is used to square the expression $a - b$. Square the first term, subtract twice the product of the binomial terms, and add the square of the last term.

$$(a - b)^2 = a^2 - 2ab + b^2$$

Explain how to find the perfect-square trinomials for $(2x + 6)^2$ and $(2x - 6)^2$. How do their products compare?

To see if a trinomial such as $4x^2 + 12x + 9$ is a perfect square, use this pattern.

- The first term $4x^2$ is a perfect square $(2x)^2$.
- The last term 9 is a perfect square $(3)^2$.
- The middle term is $12x$ or $2(2x)(3)$.

Thus, $4x^2 + 12x + 9$ is a perfect square trinomial.

ALTERNATIVE
teaching
strategy

Using Patterns Allow students to work in small groups. Remind them to look for patterns involving perfect squares ($a^2 \pm 2ab + b^2$) and difference between two squares ($a^2 - b^2$).

Present them with several examples for practice. Then have each group make up 5 factoring problems. Use all the made-up problems in a "factor-down" tournament like an old-fashioned spell-down.

PREPARE

Objectives
- Factor the difference of two squares.
- Factor perfect-square trinomials.

RESOURCES

- Practice Master   11.5
- Enrichment Master   11.5
- Technology Master   11.5
- Lesson Activity Master   11.5
- Quiz   11.5
- Spanish Resources   11.5

Assessing Prior Knowledge

Tell whether each sentence is true or false. If it is false, change the sentence to make it true.

1. The product of $(a + b)(a + b)$ is $a^2 + b^2$. [**False, the product is $a^2 + 2ab + b^2$.**]

2. The product of $(a + b)(a - b)$ is $a^2 - b^2$. [**True**]

TEACH

Ask students to draw a square on a sheet of graph paper. Have them find the area of that square in square units. Tell the students to extend two sides of the square and make a larger square. Have them find the area of the larger square.

ongoing ASSESSMENT

$(2x + 6)^2 = (2x)^2 +$
$\qquad (2)(2x)(6) + 6^2$
$\qquad = 4x^2 + 24x + 36$
$(2x - 6)^2 = (2x)^2 +$
$\qquad (2)(2x)(-6) + 6^2$
$\qquad = 4x^2 - 24x + 36$
Middle terms have opposite signs.

Examine each expression. Is the expression a perfect-square trinomial? If not, why not?

a. $a^2 + b^2$ [**No, there is no middle term.**]

b. $4a^2 + 4ab + b^2$ [**Yes**]

c. $25a^2 + 50ab + 4b^2$ [**No, 50ab is not equal to 2(5a)(2b).**]

d. $100a^2 - 20ab - b^2$ [**No, the sign in front of b^2 is negative.**]

Use Transparency ▸ 54

Alternate Example 2

Factor each expression.

a. $n^2 - 2np + p^2$
$[(n - p)^2]$

b. $64n^2 - 48np + 9p^2$
$[(8n - 3p)^2]$

c. $25n^2 - 10np + p^2$
$[(5n - p)^2]$

d. $n^2 - 18np + 81p^2$
$[(n - 9p)^2]$

CRITICAL
Thinking

$x^2 + 6x + 9 = (x + 3)^2$,
$x + 3 = 5, x = 2$

EXAMPLE 1

Examine each expression. Is the expression a perfect-square trinomial? If not, why not?

A $x^2 + 8x + 16$ **B** $3x^2 + 16x + 16$ **C** $4x^2 - 3xy - y^2$

D $4a^2 + 12ab + 9b^2$ **E** $9y^2 + 6xy + x^2$ **F** $16x^2 + 12xy + y^2$

Solution ▸

A Yes.   **B** No, $3x^2$ is not a perfect square.

C No, $4x^2$ and y^2 are perfect squares, but the sign in front of y^2 is negative, not positive.

D Yes.   **E** Yes.

F No, $16x^2$ and y^2 are perfect squares, but the middle term is not $2(4x)(y)$. ❖

The perfect-square trinomial pattern can be used to factor expressions in the form of $a^2 + 2ab + b^2$ or $a^2 - 2ab + b^2$.

For example, $9m^2 + 12mn + 4n^2$ fits the pattern.
The first term is a perfect square, $(3m)^2$.
The last term is a perfect square, $(2n)^2$.
The middle term is twice the product of ($3m$ and $2n$), $(12mn)$.
Thus, $9m^2 + 12mn + 4n^2 = (3m + 2n)^2$.

FACTORING A PERFECT-SQUARE TRINOMIAL
For all numbers a and b,

1. $a^2 + 2ab + b^2 = (a + b)(a + b) = (a + b)^2$.
2. $a^2 - 2ab + b^2 = (a - b)(a - b) = (a - b)^2$.

EXAMPLE 2

Factor each expression.

A $x^2 - 10x + 25$ **B** $9s^2 + 24s + 16$

C $64a^2 - 16ab + b^2$ **D** $49y^4 + 14y^2 + 1$

Solution ▸

A $x^2 - 10x + 25 = (x - 5)^2$

B $9s^2 + 24s + 16 = (3s + 4)^2$

C $64a^2 - 16ab + b^2 = (8a - b)^2$

D $49y^4 + 14y^2 + 1 = (7y^2 + 1)^2$

To check, multiply the factors. ❖

CRITICAL
Thinking

The area of a square is $x^2 + 6x + 9$ square units. If the length of one side is 5 units, find the value of x by factoring.

interdisciplinary
CONNECTION

Creative Arts After a beautiful painting is completed or a great action photograph is developed, the picture needs to be displayed. Often the display involves mounting the picture in a frame with a mat around the picture to create a border. A prize-winning square photograph is to be displayed with an even border of 3 inches. If the mat used for the border has an area of 180 square inches, how large is the photograph?

Difference of Two Squares

When you multiply $(a + b)(a - b)$, the product is $a^2 - b^2$. This product is called the **difference of two squares**.

$$(a + b)(a - b) = a(a - b) + b(a - b)$$
$$= a^2 - ab + ab - b^2$$
$$= a^2 \quad - \quad b^2$$

EXAMPLE 3

Examine each expression. Is the expression a difference of two squares? If not, why not?

A $4a^2 - 25$ **B** $9x^2 - 15$ **C** $b^2 + 49$

D $c^2 - 4d^2$ **E** $a^3 - 9$

Solution ➤

A Yes.

B No, 15 is not a perfect square.

C No, $b^2 + 49$ is not a difference; it is a sum.

D Yes.

E No, a^3 is not a perfect square. ❖

The difference-of-two-squares pattern can be used to factor expressions in the form $a^2 - b^2$.

For example, $4c^2 - 81d^2$ fits the pattern.
The first term is a perfect square, $(2c)^2$.
The second term is a perfect square, $(9d)^2$.
The terms are subtracted.

Thus, $4c^2 - 81d^2 = (2c + 9d)(2c - 9d)$.

FACTORING A DIFFERENCE OF TWO SQUARES
For all numbers a and b, $a^2 - b^2 = (a + b)(a - b)$.

EXAMPLE 4

Factor each expression.

A $x^2 - 4$ **B** $36a^2 - 49b^2$ **C** $16x^2 - 25$ **D** $m^4 - n^4$

Solution ➤

A $x^2 - 4 = (x + 2)(x - 2)$ **B** $36a^2 - 49b^2 = (6a + 7b)(6a - 7b)$

C $16x^2 - 25 = (4x + 5)(4x - 5)$ **D** $m^4 - n^4 = (m^2 + n^2)(m^2 - n^2)$
$$= (m^2 + n^2)(m + n)(m - n) \text{ ❖}$$

Try This   Factor $25w^2 - 81$.

CRITICAL
Thinking

When $a^8 - b^8$ is factored, you get $(a^4 + b^4)(a^4 - b^4)$. Notice that $a^4 - b^4$ can also be factored, $(a^4 + b^4)(a^2 + b^2)(a^2 - b^2)$. Finally, the difference of two squares can be factored, $(a^2 - b^2) = (a + b)(a - b)$. So $a^8 - b^8 = (a^4 + b^4)(a^2 + b^2)(a + b)(a - b)$.

Alternate Example 5

Find each product.

a. $51 \cdot 49$ **[2499]**
b. $103 \cdot 97$ **[9991]**
c. $75 \cdot 65$ **[4875]**

ASSESS

Selected Answers

Odd-numbered Exercises 5–63

Assignment Guide

Core 1–3, 4–40 even, 41–64

Core Two-Year 1–63

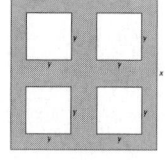

Practice & Apply
11.5 Factoring Special Polynomials

Factor each of these polynomials if possible.

1. $16y^2 - 25$
2. $121 - x^2$
3. $16y^2 + 9$
4. $64x^2 - 48xy + 9y^2$
5. $25 - 10a + a^2$
6. $4m^2 + 4m + 1$
7. $81r^2s^2 - 100t^4q^4$
8. $16x^2 + 24x + 9$
9. $y^2 - x^2$
10. $25a^2 - 1$
11. $1 - x^2y^4$
12. $144c^2 - 120cd + 25d^2$

Use any method to factor each polynomial completely.

13. $6a^2 - 216b^2$
14. $x^4 - 18x^2 + 81$
15. $49x - xy^2$
16. $2x^4 - 8x^3 + 8x^2$
17. $12x^4 - 12$
18. $b^4 - 2b^2 + b^2 - 2$
19. $4c^2 - 24c + 36$
20. $4m^4n^2 + 4m^3n^2 + m^2n^2$

21. Write the formula for the area of the shaded region in factored form.

CRITICAL
Thinking

Explain why $a^8 - b^8$ has more than two factors. What are they?

EXAMPLE 5

Find each product by using the difference of two squares.

A $31 \cdot 29$ **B** $17 \cdot 13$ **C** $34 \cdot 26$

Solution ➤

A Think of $31 \cdot 29$ as $(30 + 1)(30 - 1)$.
The product is $30^2 - 1^2 = 900 - 1 = 899$.

B Think of $17 \cdot 13$ as $(15 + 2)(15 - 2)$.
The product is $15^2 - 2^2 = 225 - 4 = 221$.

C Think of $34 \cdot 26$ as $(30 + 4)(30 - 4)$.
The product is $30^2 - 4^2 = 900 - 16 = 884$. ❖

Graphics Calculator

Graph $y = x^2 - 10x + 25$. Explain how the graph of a perfect-square trinomial can give you the factors of the expression. Use your graphics calculator to factor $x^2 - 16x + 64$.

EXERCISES & PROBLEMS

Communicate

1. Describe the process for determining the factors for the perfect-square trinomial $x^2 + 20x + 100$.

2. Explain how to factor $4x^2 - 12x + 9$.

3. Explain how to factor the difference of two squares, $p^2 - 121$.

Practice & Apply

Use the generalization of a perfect-square trinomial or the difference of two squares to find each product.

4. $(p + 3)^2$
5. $(2x - 1)^2$
6. $(a - 4)(a + 4)$
7. $(7y - 9)(7y + 9)$
8. $(8x - 3y)(8x + 3y)$
9. $(5z - 12)^2$

Find the missing terms in each perfect-square trinomial.

10. $x^2 - 14x + \underline{\ ?\ }$ 49
11. $16y^2 + \underline{\ ?\ } + 9$ 24y
12. $25a^2 + 60a + \underline{\ ?\ }$ 36
13. $9x^2 + \underline{\ ?\ } + 25$ 30x
14. $x^2 - 12x + \underline{\ ?\ }$ 36
15. $\underline{\ ?\ } - 36y + 81$ $4y^2$

RETEACHING the lesson

Using Patterns Have students create a pattern board to help factor trinomials. On a large sheet of paper, have students create the following:

$$\boxed{}\ x^2 + \boxed{}\ x + \boxed{}$$

$$(\boxed{}\ x + \boxed{}\ )(\boxed{}\ x + \boxed{}\ )$$

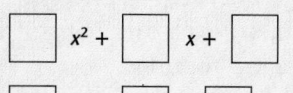

Then have students use index cards with numbers written on them to complete various problems.

4. $p^2 + 6p + 9$
5. $4x^2 - 4x + 1$
6. $a^2 - 16$
7. $49y^2 - 81$
8. $64x^2 - 9y^2$
9. $25z^2 - 120z + 144$

Factor each polynomial completely.

16. $x^2 - 4$

17. $x^2 + 4x + 4$

18. $y^2 - 100$

19. $y^2 + 8y + 16$

20. $16c^2 - 25$

21. $4t^2 - 1$

22. $81 - 4m^2$

23. $25x^2 - 9$

24. $r^2 - 18r + 81$

25. $4x^2 - 20x + 25$

26. $100 - 36q^2$

27. $36d^2 + 12d + 1$

28. $p^2 - q^2$

29. $9c^2 - 4d^2$

30. $16x^2 + 72xy + 81y^2$

31. $9a^2 - 12a + 4$

32. $49x^2 - 42xy + 9y^2$

33. $a^2x^2 + 2axb + b^2$

34. $4m^2 + 4mn + n^2$

35. $81a^4 - 9b^2$

36. $x^4 - y^4$

37. $x^2(25 - x^2) - 4(25 - x^2)$

38. $(x - 1)x^2 - 2x(x - 1) + x - 1$

39. $(3x + 5)(x^2 - 3) - (3x + 5)$

40. $(x^2 - y^2)(x^2 + 2xy) + (x^2 - y^2)(y^2)$

41. Cultural Connection: Africa Abu Kamil, known as "The Egyptian Calculator," used geometric models in 900 C.E. to solve problems. You can make his models with algebra tiles or paper rectangles that you can cut.

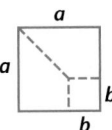

Use a sheet of paper to create a model of $a^2 - b^2$ as shown. Remove the b^2 from the corner. Next cut along the dotted line, and move the pieces to form a rectangle. Explain how this shows the factorization of $a^2 - b^2$.

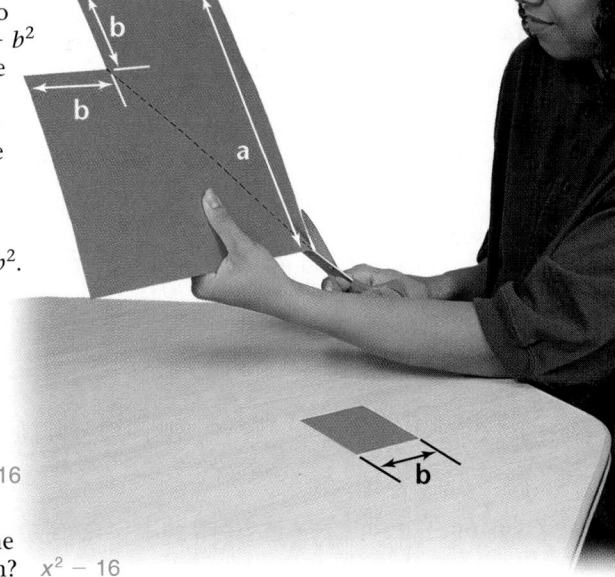

Geometry Exercises 42–47 refer to the square figure shown.

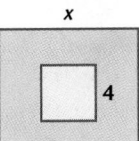

42. What is the area of the large square? x^2

43. What is the area of the smaller square? 16

44. If the smaller region is removed, what is the area of the blue region? $x^2 - 16$

45. Factor the polynomial that represents the area of the blue region. $(x + 4)(x - 4)$

46. Draw a rectangle whose dimensions are the factors you just found.

47. Show that the area of this rectangle equals the original blue area.

Geometry The area of a square is represented by $n^2 - 12n + 36$.

48. Find the length of each side. $n - 6$

49. Find the perimeter of the square. $4n - 24$

a^2; 9; $a^2 - 9$; $(a + 3)(a - 3)$

50. Geometry Within a large square whose side is a units is a smaller square whose side is 3 units. What is the area of the large square? What is the area of the smaller square? What is the area of the yellow surface between the two squares? Factor the expression for the yellow area.

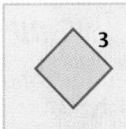

16. $(x + 2)(x - 2)$

17. $(x + 2)^2$

18. $(y + 10)(y - 10)$

19. $(y + 4)^2$

20. $(4c + 5)(4c - 5)$

21. $(2t + 1)(2t - 1)$

22. $(9 + 2m)(9 - 2m)$

23. $(5x + 3)(5x - 3)$

24. $(r - 9)^2$

25. $(2x - 5)^2$

26. $4(5 + 3q)(5 - 3q)$

27. $(6d + 1)^2$

28. $(p + q)(p - q)$

29. $(3c + 2d)(3c - 2d)$

30. $(4x + 9y)^2$

31. $(3a - 2)^2$

32. $(7x - 3y)^2$

33. $(ax + b)^2$

34. $(2m + n)^2$

35. $9(3a^2 + b)(3a^2 - b)$

36. $(x^2 + y^2)(x + y)(x - y)$

37. $(x + 2)(x - 2)(5 + x)(5 - x)$

38. $(x - 1)^3$

39. $(3x + 5)(x + 2)(x - 2)$

40. $(x - y)(x + y)^3$

The answers to Exercises 41, 46, 47 can be found in Additional Answers beginning on page 729.

Technology

Students can use a calculator to check their work by evaluating the expressions in Exercises 48–50 for some appropriate value of the unknown.

Error Analysis

Some students may believe that $x^2 - 6x - 9$ is a perfect square. This is an error. For a perfect square, the middle term can be positive or negative, but the first and last terms must be positive. Remind students that a negative times a negative is positive.

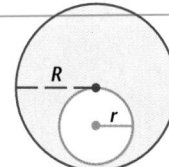

51. **Geometry** A smaller circle whose radius is r units is drawn within a large circle whose radius is R units. What is the area of the large circle? What is the area of the shaded part? Factor the expression for the shaded area. HINT: First, factor out the common monomial factor, π. Then factor the binomial.
πR^2; $\pi R^2 - \pi r^2$; $\pi(R + r)(R - r)$

Look Back

52. There is exactly one line that connects any two points. Use the table and the number pattern to find how many lines can be drawn that connect two points at a time for 10 points. **[Lesson 1.1]**
45

points	2	3	4	5	6	. . .	10
lines	1	3	6	?	?	. . .	?

53. 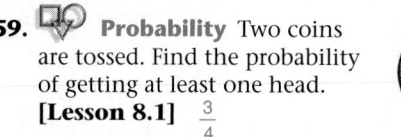 **Statistics** Mark has scores of 78, 83, and 92 on his first 3 tests. What must his score be on the next test to have an average of 87? **[Lesson 4.1]** 95

54. Graph the solution of $|x| < 4$ on a number line. **[Lesson 4.8]**

A line is represented by the equation $3y = 5 - 4x$.

55. Find the slope of a line parallel to the given line. **[Lesson 5.6]** $-\dfrac{4}{3}$

56. Find the slope of a line perpendicular to the given line. **[Lesson 5.6]** $\dfrac{3}{4}$

57. Graph the solution common to the following system. Label the solution clearly. $y < x - 5$ and $y \geq 3$ **[Lesson 6.5]**

58. Find the inverse matrix of $\begin{bmatrix} 1 & 2 \\ -2 & 1 \end{bmatrix}$. **[Lesson 7.4]** $\begin{bmatrix} 0.2 & -0.4 \\ 0.4 & 0.2 \end{bmatrix}$

59.  **Probability** Two coins are tossed. Find the probability of getting at least one head. **[Lesson 8.1]** $\dfrac{3}{4}$

Determine whether the ordered pairs represent a function. [Lesson 9.1]

60. {(1, 2), (2, 2), (3, 2), (4, 2)} yes   **61.** {(2, 7), (3, 5), (−2, 4), (3, −2)} no

62. What is the graph of the equation $y = (x - 5)^2$ called? **[Lesson 9.5]** parabola

63. What is the name for the point where the graph of a quadratic function changes direction? **[Lesson 9.4]** vertex

Look Beyond

64. Given the trinomial $x^2 + x - 42$, what are possible factor pairs for 42? Which factors result in the sum of 1? What does the sign before the last term determine?

54.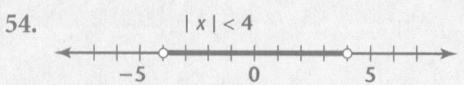
$|x| < 4$

57.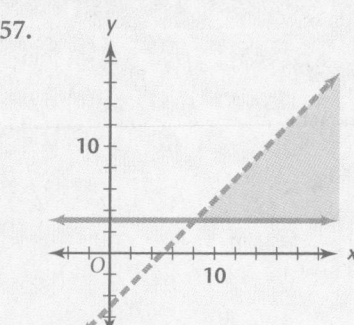

64. The factor pairs are 1 and 42, 2 and 21, 3 and 14, 6 and 7; 6 and 7; One of the factors must be negative.

LESSON 11.6 Factoring Trinomials

The area of a rectangular garden can be found by using the formula A = lw.

Factoring can be thought of as undoing multiplication. If you examine certain products carefully, you will discover patterns that enable you to recognize the factors.

Factoring With Tiles

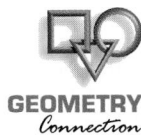

GEOMETRY
Connection

Suppose you know that the area of a rectangle is represented by $x^2 + 6x + 8$. How can you find the representation of its length and width? One way to find the length and width is to use algebra tiles. If the tiles can be arranged to form a rectangle, the length and width can be determined.

Start with the tiles that model $x^2 + 6x + 8$.

$x^2 + 6x + 8$

Arrange the tiles in a rectangle.

The length is represented by $x + 4$, and the width is represented by $x + 2$. To check, multiply the factors.

$$(x + 2)(x + 4) = x^2 + 2x + 4x + 8$$
$$= x^2 + 6x + 8$$

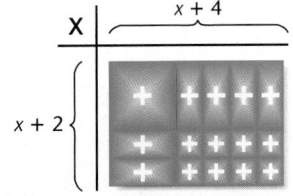

ALTERNATIVE teaching strategy

Using Technology
Have students use a spreadsheet program to check whether pairs of numbers work when trying to factor a trinomial. A simple way to do this is to have students enter the coefficients of the given trinomial and of the pairs of factors they want to try. For instance:

$$ax^2 + bx + c^2 = (dx + e)(fx + g)$$

Then have the spreadsheet check each of the following relationships:

$$a = df; b = dg + ef; c = eg.$$

If all are true, then the selected number pairs form factors of the equation. If any are not true, the factors are not correct.

PREPARE

Objectives

- Use algebra tiles and guess-and-check to determine factors of a trinomial.
- Use sign and number patterns to determine factors of a trinomial.

RESOURCES

- Practice Master   11.6
- Enrichment Master   11.6
- Technology Master   11.6
- Lesson Activity Master 11.6
- Quiz   11.6
- Spanish Resources   11.6

Assessing Prior Knowledge

Factor each polynomial.

1. $x^2 - 6x + 9$ $\quad[(x - 3)^2]$
2. $6x^2 - 18x$ $\quad[6x(x - 3)]$
3. $16x^2 + 40x + 25$
   $\quad[(4x + 5)^2]$
4. $x^2 - 144$
   $\quad[(x + 12)(x - 12)]$
5. $36x^2 + 12x + 1$
   $\quad[(6x + 1)^2]$
6. $x^4 - 81$
   $\quad[(x^2 + 9)(x + 3)(x - 3)]$

TEACH

Have students discuss how they find the area of a rectangle if they know its length and width. Then ask how they could find the possible dimensions of the rectangle if they know the area. In real life, there are many such related activities—open a door, shut a door; make money, spend money; gain 5 yards, lose 5 yards. Tell students that multiplying binomials and factoring trinomials are also related activities.

TEACHING tip

Have students use algebra tiles and work along with the lesson development on the page.

TEACHING tip

Be sure students understand that a "neutral pair" of tiles adds to zero. That is, $x + (-x) = 0$.

Sometimes using tiles may take more than one try. Examine the tiles for $x^2 + 2x - 8$.

Place the x^2-tile. Try a 1-by-8 arrangement of the negative 1-tiles. It is impossible to complete a rectangle.

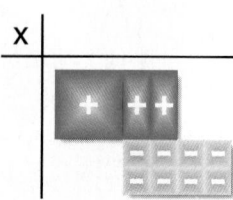

Try a 2-by-4 arrangement of the negative 1-tiles.

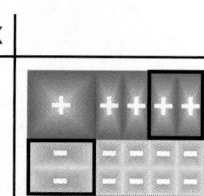

Complete a rectangle by adding 2 positive and 2 negative x-tiles.

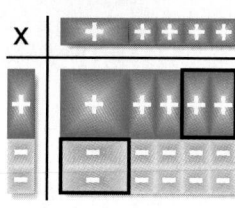

Finally, decide on the factor tiles.

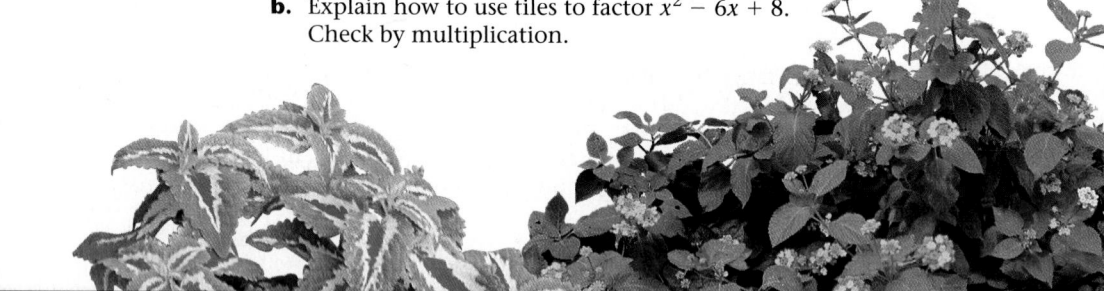

Thus, $x^2 + 2x - 8 = (x - 2)(x + 4)$. To check, multiply the factors.

$$(x - 2)(x + 4) = x^2 - 2x + 4x - 8$$
$$= x^2 + 2x - 8$$

Try This   **a.** Explain how to use tiles to factor $x^2 - 2x - 8$. Check by multiplication.

b. Explain how to use tiles to factor $x^2 - 6x + 8$. Check by multiplication.

ongoing ASSESSMENT

Try This

a. To make a rectangle, you have to add two pairs of neutral x-tiles. Then the rectangle has dimensions of $x - 4$ and $x + 2$. Check: $(x - 4)(x + 2) = x^2 - 2x - 8$

b. To make a rectangle, arrange the tiles so the dimensions are $x - 2$ and $x - 4$. Since $(-)(-) = (+)$, the units are positive. Check: $(x - 2)(x - 4) = x^2 - 6x + 8$

ENRICHMENT

What values of n make each trinomial factorable? Consider both positive and negative values.

1. $x^2 + nx + 6$ $[\pm 5, \pm 7]$
2. $x^2 + nx - 9$ $[0, \pm 8]$
3. $2x^2 + nx + 3$ $[\pm 5, \pm 7]$
4. $5x^2 + nx - 6$ $[\pm 1, \pm 7, \pm 13, \pm 29]$

INCLUSION strategies

Using Cognitive Strategies Allow students to work together in pairs or small groups. Have students develop a summary table for factoring. For instance, if a polynomial has only two terms, the only factoring that can be done is finding a common monomial factor or factoring the difference of two squares.

Factoring by Guess-and-Check

Examine the following trinomial patterns and their factors.

a. $x^2 + 6x + 8 = (x + 2)(x + 4)$ **c.** $x^2 + 2x - 8 = (x - 2)(x + 4)$

b. $x^2 - 6x + 8 = (x - 2)(x - 4)$ **d.** $x^2 - 2x - 8 = (x + 2)(x - 4)$

In each trinomial, the coefficient of x^2 is 1, and the last term is either 8 or -8. Since factoring is related to multiplication, a trinomial can be factored by working backward using the FOIL method. You can use guess-and-check to write the correct factors and signs in the FOIL model.

Follow the steps to factor $x^2 + x - 12$.

1. Examine the last term of the trinomial. Since there is a negative sign before the 12 and the coefficient of x is positive, $x^2 + x - 12$ is an example like trinomial **c** above. Write the appropriate signs in the FOIL model.

2. To find the values in ■ and ◆, remember that their product must be 12. Possible factor pairs of 12 are $1 \cdot 12$, $2 \cdot 6$, or $3 \cdot 4$. Choose the factors 3 and 4, since the sign for the middle term of the trinomial is positive and the coefficient is 1. Let 3 replace ■ and 4 replace ◆, so that $-3x + 4x$ will equal x.

Thus, the result of factoring $x^2 + x - 12$ is $(x - 3)(x + 4)$.

3. Explain why the other factor pairs do not work.

4. Check the factors. $(x - 3)(x + 4) = x^2 + 4x - 3x - 12$
$$= x^2 + x - 12$$

EXAMPLE

Factor each trinomial.

Ⓐ $x^2 - x - 20$ Ⓑ $x^2 - 10x + 16$

Ⓒ $x^2 + 4x - 21$ Ⓓ $x^2 + 9x + 18$

Solution ▸

Ⓐ Use pattern **d**.
$x^2 - x - 20 = (x + 4)(x - 5)$

Ⓑ Use pattern **b**.
$x^2 - 10x + 16 = (x - 2)(x - 8)$

Ⓒ Use pattern **c**.
$x^2 + 4x - 21 = (x - 3)(x + 7)$

Ⓓ Use pattern **a**.
$x^2 + 9x + 18 = (x + 6)(x + 3)$ ❖

Common monomial factors should always be factored out first.

A<small>SSESS</small>

Selected Answers

Odd-numbered Exercises 9–47

Assignment Guide

Core 1–8, 10–36 even, 38–48

Core Two-Year 1–46

Error Analysis

Students often fail to factor polynomials *completely*.

a. Students fail to factor common monomial factors.

b. Students fail to recognize the difference of two squares after completing one factorization.

Encourage students to always check whether the polynomial is completely factored.

Practice Master

Practice & Apply
11.6 Factoring Trinomials

1. Complete the model. Then write the factors and their product.

Write each trinomial in factored form.

2. $x^3 + x - 30$ _____   3. $m^2 + 9m + 20$ _____   4. $c^2 - c - 72$ _____

5. $d^2 - 7d + 12$ _____   6. $y^2 + y - 156$ _____   7. $f^2 - 2f - 48$ _____

For each polynomial, write all the factor pairs of the third term, then circle the pair that would successfully factor the polynomial.

8. $n^2 - 8n + 15$ _____   9. $t^2 - 121$ _____

10. $s^2 + 5s + 4$ _____   11. $q^2 - 2q - 35$ _____

Write each trinomial as a product of its factors. Use factoring patterns, graphing, or algebra tiles to assist you in your work.

12. $g^2 - 3g - 40$ _____   13. $h^2 + 6h - 40$ _____

14. $j^2 + 22j + 40$ _____   15. $k^2 - 39k - 40$ _____

16. $x^2 - x - 12$ _____   17. $y^2 - 7y - 18$ _____

18. $a^2 - 9a + 14$ _____   19. $x^2 - 5x - 6$ _____

20. $x^2 - 8x + 15$ _____   21. $p^2 + 18p + 45$ _____

Sometimes a trinomial has a common monomial factor. Explain why $2x^3 + 16x^2 + 24x$ written in factored form is $2x(x + 2)(x + 6)$.

There are some trinomials such as $x^2 - 6x - 8$ that cannot be factored by any method. For the middle term to equal $-6x$, the factors for the constant term, 8, would have to be -2 and -4. The constant term would then equal $+8$. There are no signs for the factor pair that fit the pattern of signs for the trinomial. The polynomial $x^2 - 6x - 8$ is prime or *irreducible* over the integers.

E<small>XERCISES</small> & P<small>ROBLEMS</small>

Communicate

Explain how to use algebra tiles to factor these polynomials.

1. $x^2 - 5x + 4$ **2.** $x^2 - 4x - 12$ **3.** $x^2 + 6x + 9$

Write each trinomial in factored form. Tell why the signs of the factors are the same or opposite.

4. $x^2 + x - 6$ **5.** $x^2 - 7x + 10$ **6.** $x^2 + 2x - 15$

7. Explain how to use the guess-and-check method to factor $x^2 - 5x - 24$.

8. If the third term of a trinomial is 36, write the possible factor pairs.

Practice & Apply

9. Use algebra tiles to factor $x^2 + 5x + 6$. $(x + 2)(x + 3)$

Write each trinomial in factored form. Are the signs of the factors the same or opposite?

10. $x^2 - 13x + 36$ **11.** $x^2 + 11x + 24$

12. $x^2 + 10x - 24$ **13.** $x^2 - 35x - 36$

For each polynomial, write all the factor pairs of the third term, then circle the pair that would successfully be used to factor the polynomial.

Example: $x^2 - 3x - 4$ has factor pairs of $(1 \cdot 4)$ and $2 \cdot 2$.

14. $y^2 - 9y - 36$ **15.** $x^2 - 21x + 54$

16. $x^2 + 19x + 48$ **17.** $z^2 + 10z - 144$

10. $(x - 4)(x - 9)$; same

11. $(x + 3)(x + 8)$; same

12. $(x + 12)(x - 2)$; opposite

13. $(x + 1)(x - 36)$; opposite

14. 1, 36; 2, 18; 3, 12; 4, 9; 6, 6;   [circle around 3, 12]

15. 1, 54; 2, 27; 3, 18; 6, 9;   [circle 3, 18]

16. 1, 48; 2, 24; 3, 16; 4, 12; 6, 8   [circle 3, 16]

17. 1, 144; 2, 72; 3, 48; 4, 36; 6, 24; 8, 18, 9, 16; 12, 12;   [circle 8, 18]

Write each trinomial as a product of its factors. Use factoring patterns or algebra tiles to assist you in your work.

18. $a^2 - 2a - 35$ **19.** $p^2 + 4p - 12$ **20.** $y^2 - 5y + 6$ **21.** $b^2 - 5b - 24$

22. $n^2 - 11n + 18$ **23.** $z^2 + z - 20$ **24.** $x^2 - 3x - 28$ **25.** $s^2 - 24s + 63$

This set of exercises includes all factoring patterns used throughout the chapter. Write each polynomial as a product of its factors.

26. $4x^3y - 20x^2y + 16xy$ **27.** $6y^3 - 18y^2 + 12y$ **28.** $x^2 - 18x + 81$

29. $(a + 3)(a^2 + 5a) - 6(a + 3)$ **30.** $5x^3 - 50x^2 + 45x$ **31.** $x^3 + 2x^2 - 36x - 72$

32. $-x^4 + 2x^2 + 8$ **33.** $64p^4 - 16$ **34.** $z^2 - 5z - 36$

35. $x^2 - 2x + 1$ **36.** $125x^2y - 5x^4y$ **37.** $2ax + ay + 2bx + by$

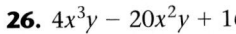 **Look Back**

38. Write the next three terms in the sequence
3, 9, 19, 33, . . . **[Lesson 1.3]**   51, 73, 99

Solve. [Lesson 4.4]

39. $\frac{3}{4}a = 363$ **40.** $\frac{z}{-8} = \frac{9}{12} - 6$ **41.** $w - \frac{7}{9} = 93$ $93\frac{7}{9}$
   484

42. Sales Tax If sales tax is 5.5%, what is the total cost of the tree? **[Lesson 4.5]** $146.65

 $139 ^{00}

43. A line with slope -2 passes through point $A(4, -1)$. Write the equation of the line that passes through the point and is perpendicular to the original line. **[Lesson 5.6]** $y = \frac{1}{2}x - 3$

A freshman class donates a tree to the childrens' hospital.

44. Represent the inequality $-4 < x < 5$ on the number line. **[Lesson 6.5]**

45. Write the system of equations as a matrix, and solve using the matrix method. **[Lesson 7.5]** $\begin{cases} 3x + y = 7 \\ 2x - y = 3 \end{cases}$

46. List the integers from 1 to 25, inclusive, that are multiples of 2. **[Lesson 8.4]**

Look Beyond

Cultural Connection: Americas De Padilla, a Guatemalan mathematician, wrote this problem 250 years ago.

47. Find two numbers, given that the second is triple the first. If you multiply the first number by the second number and the second number by 4, the sum of the products is 420.   10, 30; -14, -42

48. Graph $y = |x|$ and $y = x^2$. If these graphs are "folded" together along a vertical line, what is the equation of the line that divides them both equally?

18. $(a - 7)(a + 5)$
19. $(p + 6)(p - 2)$
20. $(y - 3)(y - 2)$
21. $(b - 8)(b + 3)$
22. $(n - 2)(n - 9)$
23. $(z - 4)(z + 5)$
24. $(x - 7)(x + 4)$
25. $(s - 21)(s - 3)$
26. $4xy(x - 1)(x - 4)$
27. $6y(y - 1)(y - 2)$

28. $(x - 9)^2$
29. $(a + 6)(a - 1)(a + 3)$
30. $5x(x - 1)(x - 9)$
31. $(x + 6)(x - 6)(x + 2)$
32. $-(x - 2)(x + 2)(x^2 + 2)$
33. $16(2p^2 + 1)(2p^2 - 1)$
34. $(z - 9)(z + 4)$
35. $(x - 1)^2$
36. $5x^2y(5 - x)(5 + x)$
37. $(a + b)(2x + y)$

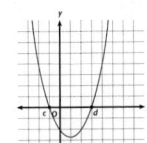

FOCUS

This project involves Pascal's Triangle and relates it to the co-efficients of the terms in the product of two binomials. The project is then extended to the study of probability.

MOTIVATE

Draw this triangle of numbers on the board and ask students to guess what the next line will be. Have them explain their reasoning for their suggested answers.

$$
\begin{array}{ccccccccc}
 & & & & 1 & & & & \\
 & & & 1 & & 1 & & & \\
 & & 1 & & 2 & & 1 & & \\
 & 1 & & 3 & & 3 & & 1 & \\
[1 & & 4 & & 6 & & 4 & & 1]
\end{array}
$$

Tell students that they will learn more about this pattern as they complete the project.

PROJECT

Powers, PASCAL, & PROBABILITY

cultural connection: ASIA If a great mathematical discovery were made today, the discoverer and his work would be on the evening news. This has not been the case throughout history. Pascal's Triangle, named after the French mathematician Blaise Pascal, appeared in work published in Europe in 1665. However, historical writings show that mathematicians from the ancient Islamic civilization were already using the remarkable properties of Pascal's Triangle in the tenth century.

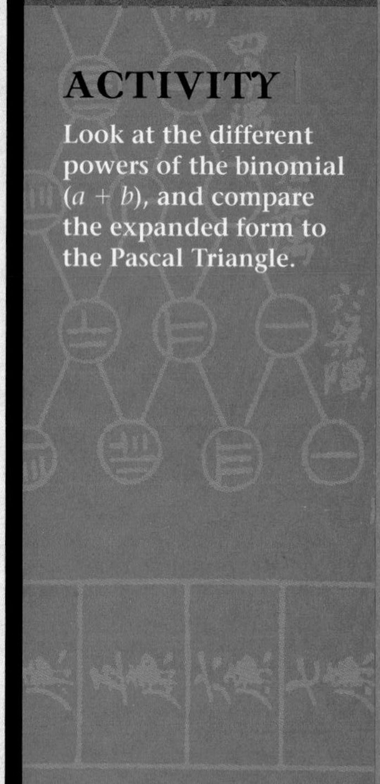

ACTIVITY

Look at the different powers of the binomial $(a + b)$, and compare the expanded form to the Pascal Triangle.

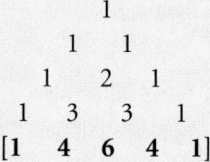

A Chinese version of the triangle appeared in 1303 C.E.

POWER	EXPANDED FORM	PASCAL'S TRIANGLE
$(a+b)^0 =$	1	1
$(a+b)^1 =$	$1a+1b$	$1\ 1$
$(a+b)^2 =$	$1a^2+2ab+1b^2$	$1\ 2\ 1$
$(a+b)^3 =$	$1a^3 +3\ a^2b+3ab^2+1b^3$	$1\ 3\ 3\ 1$
$(a+b)^4 =$	$1a^4+4a^3b+6a^2b^2+4ab^3+1b^4$	$1\ 4\ 6\ 4\ 1$
$(a+b)^5 =$	$1a^5+5a^4b+10a^3b^2+10a^2b^3+5ab^4+1b^5$	$1\ 5\ 10\ 10\ 5\ 1$

1. Start with the expanded form of $(a + b)^2$, and multiply it by $a + b$. What do you get?
2. Do the same with the expanded forms of $(a + b)^3$ and $(a + b)^4$. What do you get?
3. How does the number of terms in each row of the expanded form compare with the exponent of the binomial?
4. How many terms would you expect in the expansion of $(a + b)^7$?
5. Recalling that the triangle's elements are created by adding pairs of elements from a previous row, extend the rows of the triangle through row 10.
6. Write the coefficients for each term in the expansion of $(a + b)^7$.
7. Look at the algebraic terms of the expanded form. Describe the pattern in the exponents as you read them from left to right. What happens to the exponents of a? What happens to the exponents of b?
8. Write the complete algebraic expansion of $(a + b)^7$ without multiplying.

The answers for Activity 1 can be found in Additional Answers beginning on page 729.

Activity 2

1–3. Answers may vary but should reflect the theoretical probabilities of $\frac{1}{8}, \frac{3}{8}, \frac{3}{8}$, and $\frac{1}{2}$.

	3 heads	2 heads 1 tail	1 head 2 tails	3 tails
Tally marks				
Totals				
Totals ÷ 4				

ACTIVITY 2

1. Copy the table.
2. Flip three coins a total of 32 times. Record the results in the table.
3. Divide each total by 4, and round the answer to the nearest whole number. What are the results for each column?

ACTIVITY 3

1. Look at the tree diagram for the coin-tossing activity.
2. How many ways are there to toss 3 heads?
3. How many ways are there to toss 3 tails?
4. How many ways are there to toss 2 heads and 1 tail?
5. How many ways are there to toss 1 head and 2 tails?
6. What is the probability of each outcome?

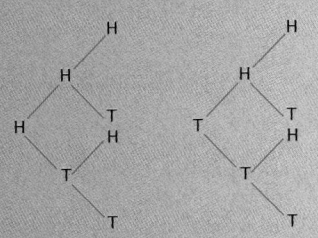

ACTIVITY 4

1. Write the expanded form of $(h + t)^3$.
2. The probability of getting a head when tossing a coin is $\frac{1}{2}$. The probability of getting a tail when tossing a coin is $\frac{1}{2}$. Let h and t each equal $\frac{1}{2}$, and find the value of *each term* of the expanded form of $(h + t)^3$.

3. How are the values for each term related to
   a. the probabilities you calculated in Activity 3?
   b. the results you obtained in Activity 2?
4. If you tossed 5 coins at a time, what would be the probability of getting 5 heads? 4 heads and 1 tail?

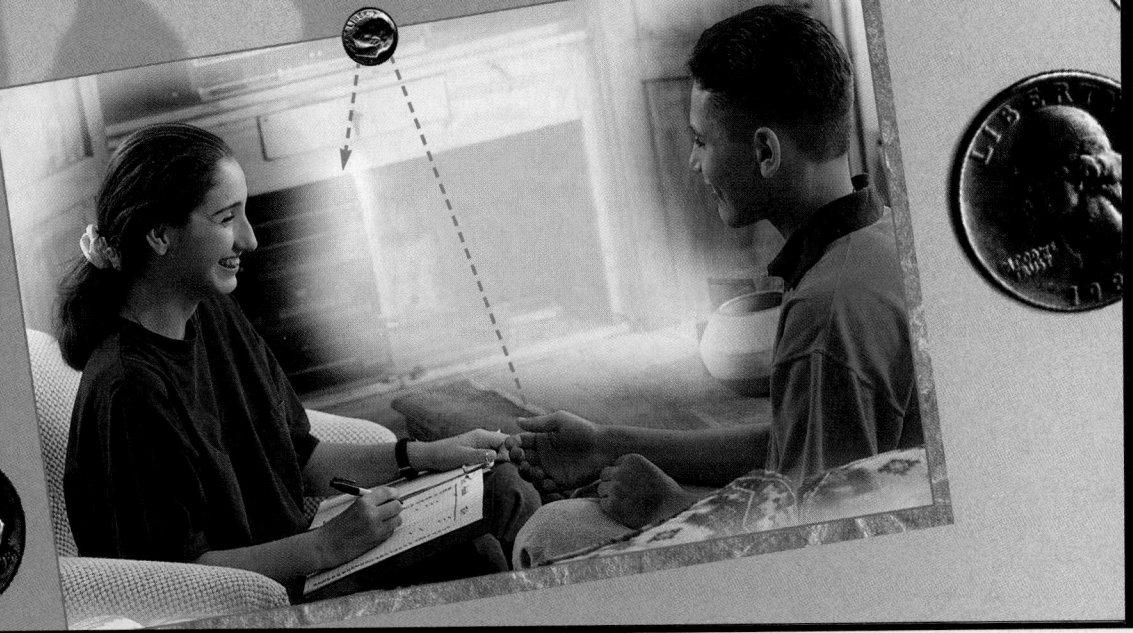

Activity 3
2. one way
3. one way
4. three ways
5. three ways
6. 3 heads: $\frac{1}{8}$; 3 tails: $\frac{1}{8}$; 2 heads and one tail: $\frac{3}{8}$;

   1 head and two tails: $\frac{3}{8}$;

Activity 4
1. $h^3 + 3h^2t + 3ht^2 + t^3$
2. $\frac{1}{8} + \frac{3}{8} + \frac{3}{8} + \frac{1}{8}$
3a. The values of each term are equal to the values calculated in Activity 3
3b. The values of each term are in the same proportion as those calculated in Activity 2
4. 5 heads: $\frac{1}{32}$, 4 heads and 1 tail: $\frac{5}{32}$

Chapter 11 Review

Vocabulary

common binomial factor	526	factor	525	polynomial	506
common monomial factor	525	FOIL method	520	prime numbers	525
composite numbers	525	greatest common factor	516	prime polynomial	525
degree	507	perfect-square trinomial	529	standard form	507
difference of two squares	531				

Key Skills & Exercises

Lesson 11.1

➤ **Key Skills**

Use vertical form to add polynomials.

Add $4x^2 + 6x - 2$ and $x^3 - 2x^2 + 7$.

$$\begin{array}{r} 4x^2 + 6x - 2 \\ + (x^3 - 2x^2 \quad + 7) \\ \hline x^3 + 2x^2 + 6x + 5 \end{array}$$

Use vertical form to subtract polynomials.

Subtract $5m^2 - 4$ from $m^2 - 5m - 10$.

$$\begin{array}{r} m^2 - 5m - 10 \\ -(5m^2 \qquad - 4) \\ \hline -4m^2 - 5m - 6 \end{array}$$

Use horizontal form to add polynomials.

Add $5y^2 - 3y + 8$ and $y^2 + y - 9$.

$(5y^2 - 3y + 8) + (y^2 + y - 9) = 6y^2 - 2y - 1$

Use horizontal form to subtract polynomials.

Subtract $6b^2 + 4b - 8$ from $10b^2 - 13$.

$$\begin{aligned} (10b^2 &- 13) - (6b^2 + 4b - 8) \\ &= 10b^2 - 13 - 6b^2 - 4b + 8 \\ &= 4b^2 - 4b - 5 \end{aligned}$$

➤ **Exercises**

Simplify. Express all answers in standard form.

1. $(3x^2 - 4x + 2) + (2x^2 + 3x - 2)$ **2.** $(c^3 + 4c^2 + 6) + (c^2 + 3c - 5)$

3. $(8d^2 - d) - (2d^2 + 4d - 5)$ **4.** $(w^3 - 3w + 9) - (8w^3)$

5. $(10m^2 - m + 4) - (2m^2 + m)$ **6.** $(7c + 3) + (3c^2 - 7c - 2)$

Lesson 11.2

➤ **Key Skills**

Use the Distributive Property to find products.

$3x(2x - 3) = 3x(2x) - 3x(3) = 6x^2 - 9x$

Write a polynomial in factored form.

$-15z^2 - 20z^3 = -5z^2(3 + 4z)$

1. $5x^2 - x$
2. $c^3 + 5c^2 + 3c + 1$
3. $6d^2 - 5d + 5$
4. $-7w^3 - 3w + 9$
5. $8m^2 - 2m + 4$
6. $3c^2 + 1$

➤ **Exercises**

Use the Distributive Property to find each product.

7. $5(x - 5)$　　　　　**8.** $y(y + 4)$　　　　　**9.** $4t(t^2 + 7)$　　　　　**10.** $2r^2(r^2 - 3r)$

11. $b(12b^2 + 11b)$　　**12.** $4y(y + 5)$　　　　**13.** $5x^2(2x^2 - x)$　　　**14.** $6d^2(d^2 - 1)$

Write each polynomial in factored form.

15. $6x^2 + 8$　　　　　**16.** $5c^3 - 25c$　　　　**17.** $n^4 + 2n^3$　　　　　**18.** $-9w^2 - 21w^4$

19. $8y^2 - 3y^2$　　　　**20.** $-8p - 14p^2$　　　**21.** $z^4 + 5z^2$　　　　　**22.** $16y^5 - 4y^3$

Lesson 11.3

➤ **Key Skills**

Use the Distributive Property to find a product of two binomials.

$$(x + 6)(x + 2)$$
$$x(x + 6) + 2(x + 6)$$
$$x^2 + 6x + 2x + 12$$
$$x^2 + 8x + 12$$

Use the FOIL method to multiply binomials.

$$(2x + 3)(x - 1)$$

$(2x + 3)\ (x - 1) = 2x^2 - 2x + 3x - 3$
$$2x^2 + x - 3$$

➤ **Exercises**

Use the Distributive Property to find a product of two binomials.

23. $(x + 6)(x - 1)$　　　**24.** $(y + 9)(y - 2)$　　　**25.** $(z - 3)(z - 6)$

26. $(3m + 5)(m + 5)$　　**27.** $(2p - 9)(p + 5)$　　**28.** $(2d + 7)(2d - 6)$

Use the FOIL method to find each product.

29. $(x + 3)(x - 4)$　　　**30.** $(5d - 8)(d - 1)$　　**31.** $(4w + 3z)(w + z)$

32. $(y + 4)(y + 5)$　　　**33.** $(x + 2)(x - 4)$　　　**34.** $(3z + 1)(4z - 1)$

Lesson 11.4

➤ **Key Skills**

Factor a polynomial by removing the GCF.

$$8m^6 + 4m^4 - 2m^2$$
$$= 2m^2(4m^4 + 2m^2 - 1)$$

Find a binomial as the greatest common factor.

$$2x^2 + 2 + x^3 + x$$
$$= (2x^2 + 2) + (x^3 + x)$$
$$= 2(x^2 + 1) + x(x^2 + 1)$$
$$= (2 + x)(x^2 + 1)$$

➤ **Exercises**

Factor each polynomial by removing the GCF.

35. $16x^3 + 8x^2$　　　　　**36.** $9y^7 + 6y^3 + 3y$　　　　**37.** $b^6 + 15b^3 - 30b^2$

38. $24m^9 - 16m^4 + 8m^3$　　**39.** $60a^4 + 20a^3 + 10a^2$　　**40.** $100p^8 - 50p^6 - 25p$

Factor the common binomial factor from each polynomial.

41. $d(f + 1) + h(f + 1)$　　**42.** $3y(y - 3) - 4(y - 3)$　　**43.** $5x - 5y + x^2 - xy$

44. $x(z - 4) + y(z - 4)$　　**45.** $10 - 5t - 2t + t^2$　　　**46.** $6c^2 + 6 + c^3 + c$

---

7. $5x - 25$

8. $y^2 + 4y$

9. $4t^3 + 28t$

10. $2r^4 - 6r^3$

11. $12b^3 + 11b^2$

12. $4y^2 + 20y$

13. $10x^4 - 5x^3$

14. $6d^4 - 6d^2$

15. $2(3x^2 + 4)$

16. $5c(c^2 - 5)$

17. $n^3(n + 2)$

18. $-3w^2(3 + 7w^2)$

19. $y^2(8 - 3)$ or $5y^2$

20. $-2p(4 + 7p)$

21. $z^2(z^2 + 5)$

22. $4y^3(4y^2 - 1)$ or $4y^3(2y - 1)$ $\times(2y + 1)$

23. $x^2 + 5x - 6$

24. $y^2 + 7y - 18$

25. $z^2 - 9z + 18$

26. $3m^2 + 20m + 25$

27. $2p^2 + p - 45$

28. $4d^2 + 2d - 42$

29. $x^2 - x - 12$

30. $5d^2 - 13d + 8$

31. $4w^2 + 7wz + 3z^2$

32. $y^2 + 9y + 20$

33. $x^2 - 2x - 8$

34. $12z^2 + z - 1$

35. $8x^2(2x + 1)$

36. $3y(3y^6 + 2y^2 + 1)$

37. $b^2(b^4 + 15b - 30)$

38. $8m^3(3m^6 - 2m + 1)$

39. $10a^2(6a^2 + 2a + 1)$

40. $25p(4p^7 - 2p^5 - 1)$

41. $(d + h)(f + 1)$

42. $(3y - 4)(y - 3)$

43. $(5 + x)(x - y)$

44. $(x + y)(z - 4)$

45. $(2 - t)(5 - t)$

46. $(6 + c)(c^2 + 1)$

Lesson 11.5

➤ *Key Skills*

Use a generalization of special polynomials to find each product.

$(x + 7)^2 = (x + 7)(x + 7) = x^2 + 14x + 49$

$(p + 8)(p - 8) = p^2 - 64$

Factor perfect-square trinomials.

$x^2 - 18x + 81 = (x - 9)^2$

Factor the difference of two squares.

$4x^2 - 9 = (2x + 3)(2x - 3)$

➤ *Exercises*

Use a generalization of special polynomials to find each product.

47. $(c - 9)^2$ **48.** $(b - 10)(b + 10)$ **49.** $(5a - 2)(5a + 2)$

Factor each polynomial.

50. $a^2 + 6a + 9$ **51.** $w^2 - 16w + 64$ **52.** $36p^2 + 12p + 1$

53. $y^2 - 81$ **54.** $16r^2 - 25$ **55.** $9z^2 - 1$

Lesson 11.6

➤ *Key Skills*

Use sign and number patterns to determine the factors of a trinomial.

$x^2 - x - 12 = (x + \underline{?})(x - \underline{?})$ Factor pairs: $1 \cdot 12,\ 2 \cdot 6,\ 3 \cdot 4$

The factors are $(x + 3)(x - 4)$.

➤ *Exercises*

Write each polynomial as a product of its factors.

56. $n^2 - 2n - 24$ **57.** $h^2 + 14h + 40$ **58.** $y^2 + 3y - 18$

59. $a^2 - 7a + 12$ **60.** $x^2 - 5x + 4$ **61.** $2a^2 + 8a + 6$

Applications

Geometry Given the rectangle with sides as shown, solve the following.

62. What is the area of the large rectangle? $x^2 + 6x$

63. What is the area of the small rectangle? $4x + 8$

64. If the smaller region is removed, what is the area of the shaded region? $x^2 + 2x - 8$

65. Factor the polynomial that represents the area of the shaded region.
$(x + 4)(x - 2)$

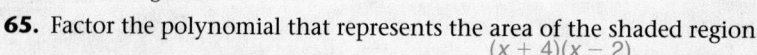

Crafts Marcus works at a local craft store mounting pictures and posters on mats for framing. He is mounting a picture on a mat that has an area of 99 square inches.

66. The length of the mat is 2 inches greater than the width of the mat. The equation representing the area of the mat is $x(x + 2) = 99$, or $x^2 + 2x - 99 = 0$. What are the dimensions of the mat? 9 by 11

67. An equal amount of mat will be showing on each side of a picture. Let x represent the amount of mat showing on each side. What are the dimensions of the picture? What is the area of the picture?

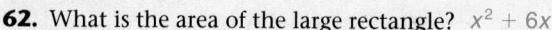

47. $c^2 - 18c + 81$

48. $b^2 - 100$

49. $25a^2 - 4$

50. $(a + 3)^2$

51. $(w - 8)^2$

52. $(6p + 1)^2$

53. $(y - 9)(y + 9)$

54. $(4r - 5)(4r + 5)$

55. $(3z + 1)(3z - 1)$

56. $(n - 6)(n + 4)$

57. $(h + 10)(h + 4)$

58. $(y + 6)(y - 3)$

59. $(a - 4)(a - 3)$

60. $(x - 1)(x - 4)$

61. $2(a + 1)(a + 3)$

67. $11 - 2x$ and $9 - 2x$; $4x^2 - 40x + 99$

Chapter 11 Assessment

Simplify. Express all answers in standard form.

1. $(2 + 4x + 2x^2) + (4x^2 - 6)$
$6x^2 + 4x - 4$

2. $9 - 5v^2 + 5v^4 - (10v^2 + 7v - 11)$
$5v^4 - 15v^2 - 7v + 20$

Use the Distributive Property to find each product.

3. $8(x - 7)$
$8x - 56$

4. $r(r - 6)$
$r^2 - 6r$

5. $3a^2(a - 1)$
$3a^3 - 3a^2$

6. $q(2q^2 + q)$
$2q^3 + q^2$

Write each polynomial in factored form.

7. $6x^6 + 3x^4$
$3x^4(2x^2 + 1)$

8. $25a^8 - 15a^4$
$5a^4(5a^4 - 3)$

9. $8x^5 - 4x^4 + x^3$
$x^3(8x^2 - 4x + 1)$

10. $25y^9 + 15y^7 - 5y^2$
$5y^2(5y^7 + 3y^5 - 1)$

11. Describe how to find $(x + 4)(x - 8)$ using the FOIL method.

12. Write the degree of the polynomial $6y^2 + 3y^3 - 1$. 3

Factor the common binomial factor from each polynomial.

13. $5(r + 1) - t(r + 1)$
$(5 - t)(r + 1)$

14. $p^2 + 2p + 3p + 6$
$(p + 2)(p + 3)$

Use the Distributive Property or the FOIL method to find each product.

15. $(y + 3)(y + 2)$

16. $(c + 3)(c - 4)$

17. $\left(\frac{1}{2} a + 1\right)(a - 1)$

18. $(4x - y)(3x + 2y)$

19. $(w - 9)(w - 6)$

20. $(2n + 1)(n + 1)$

Use a generalization of special polynomials to find each product.

21. $(x + 11)^2$

22. $(p + 12)(p - 12)$

23. $(m - 9)^2$

24. Find the area of the floor of a square garage that is $t + 5$ units on a side.

25. Can $7c^2 + 45c - 28$ be factored? Explain why or why not.

Use any method to factor each polynomial completely.

26. $4r + 4s + rt + st$

27. $14x^3 + 7x - 8x^2 - 4$

28. $4v^2 - 144$

29. $a^4 - 81$

30. $n^2 - 12n + 36$

31. $15z^2 - z - 2$

32. $w^4 - 7w^3 - 18w^2$

33. $d^4 + 5d^2 + 6$

Find the greatest common factor.

34. $y^3 - y^2 + 4y$
y

35. $6w^3 + 3w^2 + 3$
3

36. $7m^3 - m^2$
m^2

Within a large rectangle there is a smaller rectangle.

37. What is the area in square units of the large rectangle?

38. What is the area in square units of the smaller rectangle?

39. What is the area in square units of the shaded area between the two figures?

$x + 12$

$x + 6$

x

$x + 8$

26. $(4 + t)(r + s)$

27. $(2x^2 + 1)(7x - 4)$

28. $4(v - 6)(v + 6)$

29. $(a^2 + 9)(a - 3)(a + 3)$

30. $(n - 6)^2$

31. $(5z - 2)(3z + 1)$

32. $w^2(w - 9)(w + 2)$

33. $(d^2 + 3)(d^2 + 2)$

37. $x^2 + 20x + 96$ units^2

38. $x^2 + 6x$ units^2

39. $14x + 96$ units^2

11. Multiply x by x for x^2, x by -8 for $-8x$, 4 by x for $4x$, and 4 by -8 for -32. Collect like terms for $x^2 - 4x - 32$.

15. $y^2 + 5y + 6$

16. $c^2 - c - 12$

17. $\frac{1}{2} a^2 + \frac{1}{2} a - 1$

18. $12x^2 + 5xy - 2y^2$

19. $w^2 - 15w + 54$

20. $2n^2 + 3n + 1$

21. $x^2 + 22x + 121$

22. $p^2 - 144$

23. $m^2 - 18m + 81$

24. $t^2 + 10t + 25$ units^2

25. Yes; there is no GCF, but by factoring with guess-and-check, it can be factored into $(c + 7)(7c - 4)$.

12 CHAPTER Quadratic Functions

Meeting Individual Needs

12.1 Exploring the Parabola

Core Resources

Practice Master 12.1
Enrichment, p. 549
Technology Master 12.1

[1 day]

Core Two-Year Resources

Inclusion Strategies, p. 549
Reteaching the Lesson, p. 550
Practice Master 12.1
Enrichment Master 12.1
Technology Master 12.1
Lesson Activity Master 12.1

[2 days]

12.2 Solving Equations of the Form $x^2 + k$

Core Resources

Practice Master 12.2
Enrichment, p. 554
Technology Master 12.2
Interdisciplinary
    Connection, p. 553

[1 day]

Core Two-Year Resources

Inclusion Strategies, p. 554
Reteaching the Lesson, p. 555
Practice Master 12.2
Enrichment Master 12.2
Technology Master 12.2
Lesson Activity Master 12.2

[2 days]

12.3 Completing the Square

Core Resources

Practice Master 12.3
Enrichment, p. 559
Technology Master 12.3
Mid-Chapter Assessment
    Master

[2 days]

Core Two-Year Resources

Inclusion Strategies, p. 559
Reteaching the Lesson, p. 560
Lesson Activity Master 12.3
Enrichment Master 12.3
Practice Master 12.3
Technology Master 12.3
Mid-Chapter Assessment Master

[3 days]

12.4 Solving Equations of the Form $x^2 + bx + c = 0$

Core Resources

Practice Master 12.4
Enrichment, p. 568
Technology Master 12.4
Interdisciplinary
    Connection, p.567

[2 days]

Core Two-Year Resources

Inclusion Strategies, p. 568
Reteaching the Lesson, p. 569
Practice Master 12.4
Enrichment Master 12.4
Technology Master 12.4
Lesson Activity Master 12.4

[3 days]

12.5 The Quadratic Formula

Core Resources

Practice Master 12.5
Enrichment, p. 574
Technology Master 12.5

[2 days]

Core Two-Year Resources

Inclusion Strategies, p. 575
Reteaching the Lesson, p. 578
Practice Master 12.5
Enrichment Master 12.5
Technology Master 12.5
Lesson Activity Master 12.5

[4 days]

12.6 Graphing Quadratic Inequalities

Core Resources

Practice Master 12.6
Enrichment, p. 583
Technology Master 12.6
Interdisciplinary
    Connection, p. 582

[2 days]

Core Two-Year Resources

Inclusion Strategies, p. 583
Reteaching the Lesson, p. 584
Practice Master 12.6
Enrichment Master 12.6
Technology Master 12.6
Lesson Activity Master 12.6

[4 days]

Chapter Summary

Core Resources	Core Two-Year Resources
Eyewitness Math, pp.564–565	Eyewitness Math, pp. 564–565
Chapter 12 Project, pp. 586–587	Chapter 12 Project, pp. 586–587
Lab Activity	Lab Activity
Long-Term Project	Long-Term Project
Chapter Review, pp. 588–590	Chapter Review, pp. 588–590
Chapter Assessment, p. 591	Chapter Assessment, p. 591
Chapter Assessment, A/B	Chapter Assessment, A/B
Alternative Assessment	Alternative Assessment
Cumulative Assessment, pp. 592–593	Cumulative Assessment, pp. 592–593
[3 days]	**[5 days]**

Reading Strategies

During reading, have students use their journals to keep track of key concepts from the chapter, such as the relationship between parabolas and constant differences in Lesson 1, the $x^2 = k$ *Generalization* in Lesson 2, how to complete the square in Lesson 3, the Zero Product Property in Lesson 4, the Quadratic Formula in Lesson 5, and how to graph quadratic inequalities in Lesson 6.

Visual Strategies

The use of algebra tiles provides a visual interpretation of a quadratic expression. Further, algebra tiles can be used to help students understand the relationship between building a square and the process of completing the square. Graphical depictions of quadratic equations are central to students' development and understanding of such equations. Students can visually identify the axis of symmetry, vertex, and zeros of the equation, and then relate those concepts to their algebraic interpretations.

Hands-on Strategies

Chapter 12 uses algebra tiles to develop the concept of solving quadratic equations. By developing the concept of area with algebra tiles, students learn that the equation for the area of a rectangle can be written as a quadratic equation.

Cooperative Learning

GROUP ACTIVITIES	
exploring parabolas	Lesson 12.1, Explorations 1, 2, and 3
completing the square	Lesson 12.3, Opening Discussion and Exploration
the quadratic formula	Lesson 12.5, Development on p. 575
graphing quadratic inequalities	Lesson 12.6, Exploration

You may wish to have students work with partners or in small groups to complete any of these activities. Additional suggestions for cooperative group activities can be found in the teacher's notes in the lessons.

Multicultural

The cultural connections in this chapter include references to ancient Babylon and the United States.

CULTURAL CONNECTIONS	
Asia: Ancient Babylonian Scholars	Lesson 12.5, Opening Discussion
Europe: Galileo Galilei	Lesson 12.2, Opening Discussion

Portfolio Assessment

Below are portfolio activities for the chapter, listed under seven activity domains that are appropriate for portfolio development.

1. **Investigation/Exploration** Students start by exploring parabolas. Students then investigate solving quadratic equations of the form $x^2 + k$, use algebra tiles and graphics calculators to explore the process of completing the square, derive the quadratic formula, and graph quadratic inequalities.

2. **Applications** Physics: Lesson 12.1, Applications and Exercises 32–37; Lesson 12.2, Opening Discussion and Exercises 56–57; Lesson 12.3, Exercises 44–51; Lesson 12.4, Exercise 51; and Lesson 12.6, Example 3 and Exercises 36–38; Photography: Lesson 12.4, Exercise 48; Accounting: Lesson 12.5, Exercise 46; Production and Profit: Lesson 12.6, Exploration.

3. **Non-Routine Problems** Patterns are used in Lesson 12.1 in the Opening Discussion, in the Chapter Project, and in Exercise 63 in Lesson 12.4.

4. **Project** "What's the Difference?" pages 586–587. Students find a rule that will generate a numerical sequence.

5. **Interdisciplinary Topics** Maximum/Minimum: Lesson 12.1, Exploration 3; Lesson 12.3, Discussion; Coordinate Geometry: Lesson 12.2 Discussion; Geometry: Lesson 12.2, Exercises 54–55; Lesson 12.4, Example 4 and Exercise 47; Probability: Lesson 12.3, Exercise 55.

6. **Writing** *Communicate* exercises offer excellent writing selections for the portfolio.

7. **Tools** Chapter 12 uses algebra tiles and the graphics calculator.

Technology

Graphics Calculator

The graphics calculator plays an important part in Chapter 12. To use the calculator successfully to graph and interpret quadratic equations, students must understand how to use such features as the ZOOM and TRACE. They must also understand how to set the RANGE or WINDOW for the equation being graphed. Try this with students: Graph $y = -15x^2 - 6x + 3$.

The keystrokes on the Casio fx-7000G, on the Sharp EI-9300C, and on the TI-82 are similar. *Clear the graphics window first.

Casio:

GRAPH – 15 ALPHA x x^2 – 6 ALPHA x + 3 exe

Sharp:

eqtn (–) 15 ALPHA x x^2 – 6 ALPHA x + 3 (Graphing key)

TI-82:

Y= (–) 15 ALPHA x x^2 – 6 ALPHA x + 3 GRAPH

The variable can also be entered easily using the X,T,θ key.

If the graph is small and hard to interpret. Have students use the RANGE and ZOOM keys to enlarge the graph and find the approximate solutions. $[x = -0.7, x = 0.3]$

When students learn to use the quadratic formula, the calculator can be used as an aid in the more complicated computations.

Integrated Software

f(g) Scholar™ is an integrated computer-based mathematics productivity tool that combines calculator, spreadsheet, and graphics capabilities to provide a dynamic and interactive environment for explorations in mathematics. It is appropriate to use *f(g) Scholar*™ for any lesson needing a spreadsheet, calculator, graphics calculator, or any combination of the three.

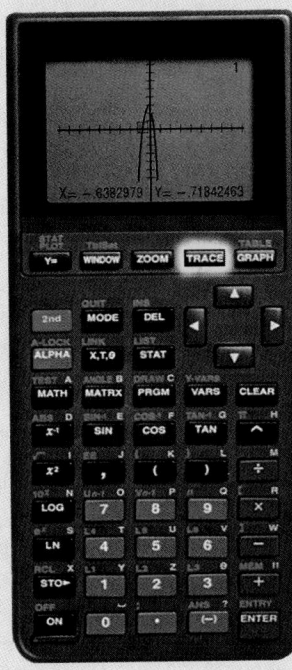
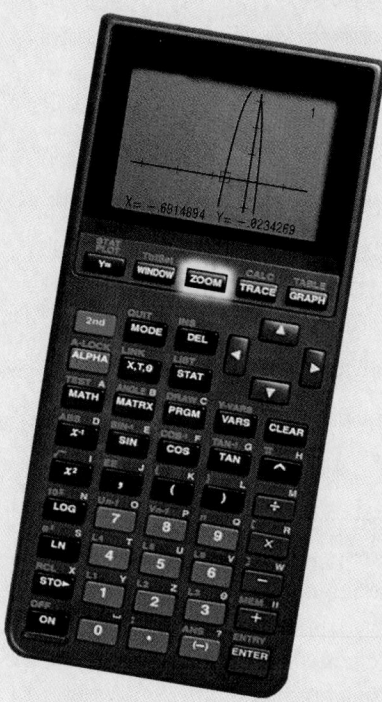

Quadratic Functions

ABOUT THE CHAPTER

Background Information

Students should be familiar with terms such as *factor* and *product*. This chapter deals with multiplying binomials and factoring polynomials. Students will learn to use patterns to simplify operations on polynomials or to factor polynomials.

CHAPTER RESOURCES

- Practice Masters
- Enrichment Masters
- Technology Masters
- Lesson Activity Masters
- Lab Activity Masters
- Long-Term Project Masters
- Assessment Masters
    Chapter Assessments, A/B
    Mid-Chapter Assessments
    Alternative Assessments, A/B
- Teaching Transparencies
- Cumulative Assessment
- Spanish Resources

CHAPTER OBJECTIVES

- Explore the parabola as the graph of a quadratic function.
- Examine graphs of quadratic functions to determine the vertex, the axis of symmetry, and the zeros.
- Use different methods to find the value of a function.
- Use square roots to solve quadratic equations.
- Complete the square.
- Rewrite a quadratic in the form $y = (x - h)^2 + k$.
- Identify the vertex of a parabola.
- Find the minimum value of a quadratic from the vertex.

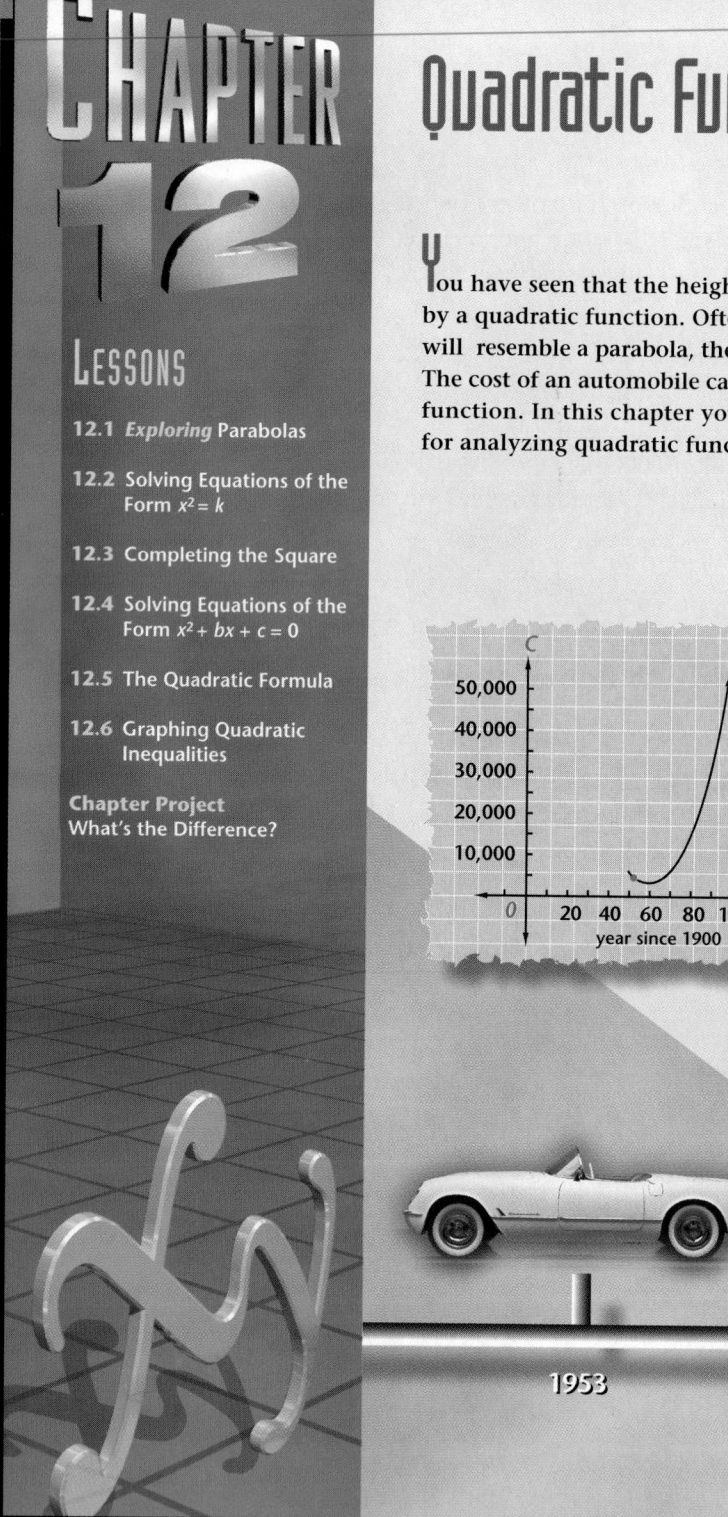

CHAPTER 12

LESSONS

12.1 *Exploring* Parabolas

12.2 Solving Equations of the Form $x^2 = k$

12.3 Completing the Square

12.4 Solving Equations of the Form $x^2 + bx + c = 0$

12.5 The Quadratic Formula

12.6 Graphing Quadratic Inequalities

Chapter Project
What's the Difference?

Quadratic Functions

You have seen that the height of a projectile can be modeled by a quadratic function. Often the trajectory of a projectile will resemble a parabola, the graph of a quadratic function. The cost of an automobile can be modeled using a quadratic function. In this chapter you will learn different methods for analyzing quadratic functions and quadratic equations.

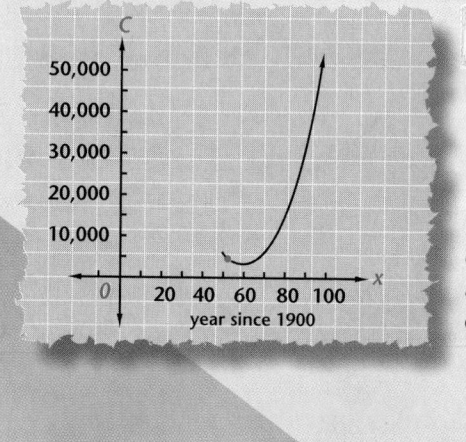

Sometimes the cost of a product can be approximated by a quadratic function.

1953   1962

ABOUT THE PHOTOS

The photographs on these pages show an example of how a quadratic function can model the increase in the cost of a sportscar over several years. Allow students to discuss briefly the different cars shown on the page. Then have students read and discuss the time line shown. The Portfolio Activity is based on the cost function.

PORTFOLIO ACTIVITY

This quadratic function approximates the annual cost of a popular car from 1953-1995.

$$C = 34x^2 - 4206x + 132,539$$

The variable C represents the cost. The variable x represents the number of years since 1900. For example, x is 80 for the year 1980.

To find the cost of the car for 1980, substitute 80 for x and compute the value of C.

$$C = 34x^2 - 4206x + 132,539$$
$$C = 34(80)^2 - 4206(80) + 132,539$$
$$= 13,659$$

The cost of this sports car for 1980 was approximately $13,659.

Use the formula to project the cost of this sports car for the year 2000. In what year did the car cost $25,000?

1989   1990   1995

ABOUT THE CHAPTER PROJECT

In the Chapter Project on pages 586–587, students investigate number patterns generated by using a quadratic equation and learn how to identify the rule that will produce the sequence. This project is an introduction to finite differences.

- Solve quadratic equations by completing the square or by factoring.
- Use the quadratic formula to find solutions to quadratic equations.
- Use the quadratic formula to find the zeros of a quadratic function.
- Evaluate the discriminant to determine how many real roots the quadratic has and whether the quadratic can be factored over the integers.
- Solve and graph quadratic inequalities, and test solution regions.

PORTFOLIO ACTIVITY

Assign the Portfolio Activity as part of the assignment for Lesson 12.5 on page 580.

Ask students what they know about inflation. Students should understand that when the costs of goods and services increase, inflation is present. By studying the changes in costs over the years, economists can relate the rate of inflation to a mathematical equation. Frequently, these equations are quadratic functions, as illustrated in this Portfolio Activity.

The Portfolio Activity can be extended by having students extend their study of the given function. For instance, have them find the value of the sports car in the year they were born. Then have students try to find the value of the car in a year prior to 1962. What did they notice? [**The costs were higher than in 1962.**] Have students decide whether this is realistic. Have them consider what limitations must be placed on the functions.

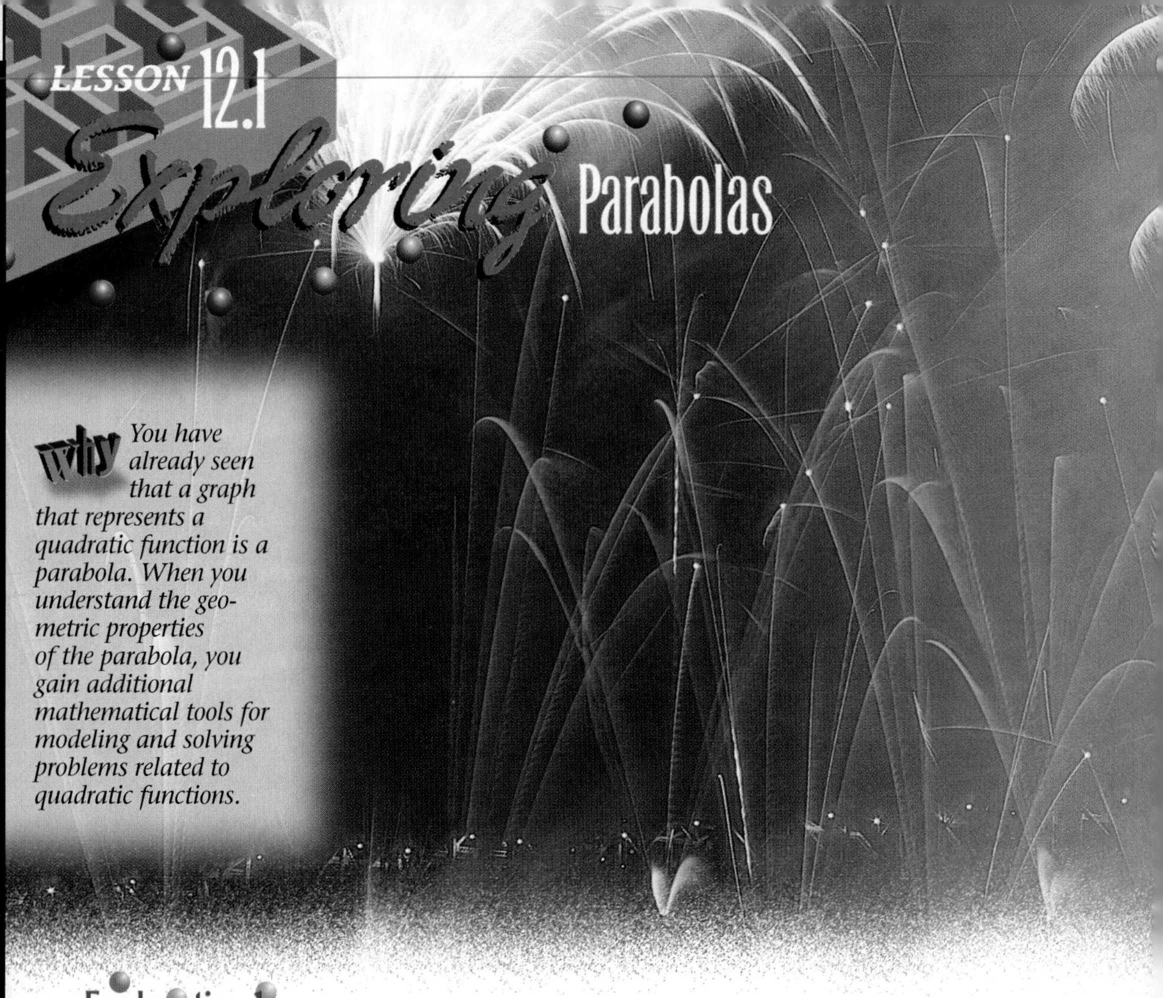

LESSON 12.1 Exploring Parabolas

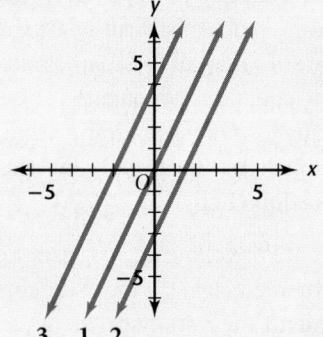

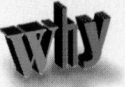

You have already seen that a graph that represents a quadratic function is a parabola. When you understand the geometric properties of the parabola, you gain additional mathematical tools for modeling and solving problems related to quadratic functions.

Exploration 1 *Parabolas and Constant Differences*

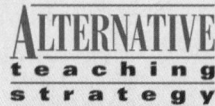

The table shows some points on the graph of the function $y = f(x)$.

x	0	1	2	3	4	5	6
$f(x)$	10	0	−6	−8	−6	?	?

1. Calculate the first and second differences for $f(x)$.
2. What do you notice about the second differences?
3. Work backward using the pattern, and predict $f(5)$ and $f(6)$.
4. Plot $y = f(x)$. Connect the points with a smooth curve.
5. What kind of curve did you draw? Explain the relationship between second differences and the curve that you drew. ❖

In chapter 9, you learned that the function $g(x) = a(x - h)^2 + k$ transforms the parent quadratic function $f(x) = x^2$ by stretching the parent function by a factor of a and moving its vertex from (0, 0) to (h, k). The axis of symmetry is $x = h$. If $a > 0$ the minimum value of g is k.

PREPARE

Objectives

- Explore the parabola as the graph of a quadratic function.
- Examine graphs of quadratic functions to determine the vertex, the axis of symmetry, and the zeros.

RESOURCES

• Practice Master	12.1
• Enrichment Master	12.1
• Technology Master	12.1
• Lesson Activity Master	12.1
• Quiz	12.1
• Spanish Resources	12.1

Assessing Prior Knowledge

Graph each equation.

1. $y = 2x$
2. $y = 2x - 3$
3. $y = 2x + 4$

TEACH

Why Have students draw what the flight of the arrow would look like on the board. The student should draw a parabola-type arc.

ALTERNATIVE teaching strategy

Technology Have students use their graphics calculators to graph several quadratic equations in the form $y = ax^2 + bx + c$. Explain that the *vertex* of the graph is the maximum or minimum of the graph. Have students identify the vertex of each graph. Ask students to relate the x-coordinate of the vertex to the values of a and b in the general form of the equation. They should find that the x-coordinate is at $-\frac{b}{2a}$. The y-coordinate is at $f\left(-\frac{b}{2a}\right)$. Repeat the discovery and discussion process for the *axis of symmetry* for the equation if $x = -\frac{b}{2a}$.

ongoing ASSESSMENT

5. A parabola. The second differences are constant.

Exploration 2 Parabolas and Transformations

1 Consider the graph of the function $g(x) = 2(x - 3)^2 - 8$. What is its vertex?

2 Find 4 more points on the graph.

3 Plot the points and sketch the graph.

4 What kind of curve did you draw? How does your graph compare with the graph you drew in Exploration 1? ❖

The graph of a quadratic function is a parabola. If the parabola crosses the x-axis, its x-value at the intersection is called a **zero of the function**. How many zeros can a quadratic function possibly have? Explain your answer.

Exploration 3 Parabolas and Polynomials

MAXIMUM
MINIMUM
Connection

1 Graph the function $t(x) = 2x^2 - 12x + 10$.

2 Find the zeros of the function.

3 Find the average, h, of the two zeros.

4 Find $t(h)$.

5 What is the vertex of the graph?

6 What is the axis of symmetry of the graph?

7 What is the minimum value of the function?

8 Compare the graphs for Exploration 2 and 3.

9 Simplify $2(x - 3)^2 - 8$. Compare $g(x)$ with $t(x)$. ❖

APPLICATION

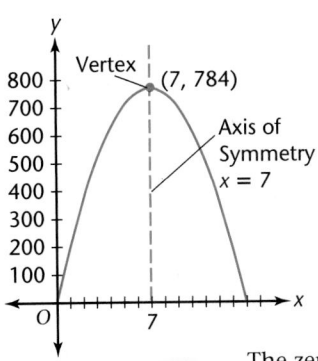

Vertex (7, 784)

Axis of Symmetry
$x = 7$

In Lesson 1.2, you used a table and differences to find the maximum height of the flight of a small rocket. The flight can be modeled with the function $f(x) = -16(x - 7)^2 + 784$. In this form, the vertex, the axis of symmetry and the maximum value can be read directly from the equation. The zeros can be read from the graph.

function	$f(x) = -16(x - 7)^2 + 784$
vertex	(7, 784)
axis of symmetry	$x = 7$
maximum value	784
zeros	0, 14

The zeros of f are 0 and 14. Check by substitution. ❖

CRITICAL
Thinking

How can the zeros of $f(x) = -16(x - 7)^2 + 784$ be interpreted in terms of the flight of the rocket?

ENRICHMENT Have students draw a line, *d*, near the left edge of a sheet of wax paper. Then have students locate point *P* near the center of the paper. Holding the paper to the light, fold the paper so *P* is on *d*. When the point is on the line, crease the fold of the paper. Reposition the point so that it touches the line in a different position, and crease the paper again. Repeat this procedure many times. When finished, the students should see that the folds form a parabola.

INCLUSION
strategies

Hands-On Strategies
Some students may have trouble graphing on small grid paper. Suggest that those students use graph paper with a larger grid so they can better see and accurately graph quadratic equations.

Assignment Guide

Core 1–13, 14–30 even, 32–45,

Core Two-Year 1–44

Technology

For Exercises 8–28, have students first complete them *without* using a calculator and then check their answers by using a calculator. The use of a calculator is also appropriate for Exercise 37.

Error Analysis

Students may analyze ax^2 as $(ax)^2$. Emphasize that the term is $(a)(x^2)$.

Practice Master

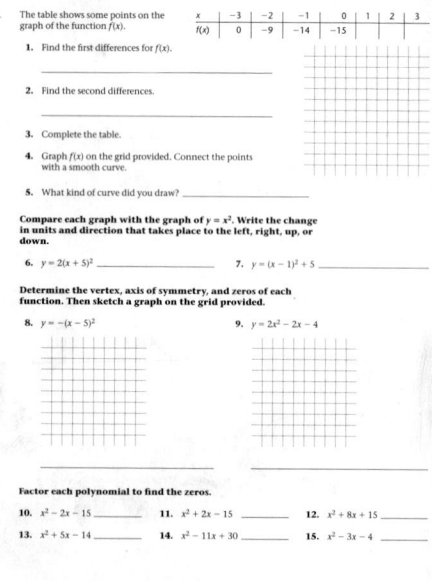

Practice & Apply
12.1 Exploring Parabolas

The table shows some points on the graph of the function $f(x)$.

x	–3	–2	–1	0	1	2	3
f(x)	0	–9	–14	–15			

1. Find the first differences for $f(x)$.

2. Find the second differences.

3. Complete the table.

4. Graph $f(x)$ on the grid provided. Connect the points with a smooth curve.

5. What kind of curve did you draw? _____

Compare each graph with the graph of $y = x^2$. Write the change in units and direction that takes place to the left, right, up, or down.

6. $y = 2(x + 5)^2$ _____   7. $y = (x – 1)^2 + 5$ _____

Determine the vertex, axis of symmetry, and zeros of each function. Then sketch a graph on the grid provided.

8. $y = –(x – 5)^2$ 9. $y = 2x^2 – 2x – 4$

Factor each polynomial to find the zeros.

10. $x^2 – 2x – 15$ _____   11. $x^2 + 2x – 15$ _____   12. $x^2 + 8x + 15$ _____
13. $x^2 + 5x – 14$ _____   14. $x^2 – 11x + 30$ _____   15. $x^2 – 3x – 4$ _____

EXERCISES & PROBLEMS

Communicate

1. Explain how you can tell from second differences that the values of a function represent a quadratic.

x	–4	–3	–2	–1	0
f(x)	0	–1	0	3	8

2. Discuss how the graph of the quadratic function $g(x) = 2(x – 3)^2 – 8$ differs from the graph of the parent function $y = x^2$.

Refer to the function $g(x) = a(x – h)^2 + k$ to complete Exercises 3-4.

3. Explain how to determine the vertex of $g(x) = (x – 3)^2 – 4$.

4. Explain how to determine the axis of symmetry.

5. Discuss how to determine the zeros of $g(x) = x^2 – 8x + 16$.

6. Describe how to determine the axis of symmetry for $g(x) = x^2 – 8x + 16$ from the zeros of the polynomial.

Practice & Apply

7. Do the values represent a quadratic function? Graph to check. yes; the second differences are constant.

x	0	1	2	3	4	5	6
f(x)	–5	0	3	4	3	0	–5

Compare each graph with the graph of $y = x^2$. Write the change in units and direction that takes place to the left, right, up, or down.

8. $y = (x – 2)^2 + 3$ 9. $y = 3(x – 5)^2 – 2$ 10. $y = –(x – 2)^2 + 1$
11. $y = (x + 1)^2$ 12. $y = –3(x – 2)^2 + 1$ 13. $y = \frac{1}{2}(x – 2)^2 + 3$

Determine the vertex and axis of symmetry for each function, and then sketch a graph from the information.

14. $y = –2(x + 4)^2 – 3$ 15. $y = \frac{1}{2}(x – 2)^2 + 3$ 16. $y = (x – 3)^2 – 7$
17. $y = –3(x – 5)^2 + 2$ 18. $y = –(x + 3)^2 – 2$ 19. $y = 2(x + 5)^2 + 7$

RETEACHING the lesson

Using Visual Models
Have students graph $h(x) = 2x^2 + 5x – 3$ on graph paper, with axes marked so that 4 squares = 1 unit. Ask students where the graph crosses the x-axis.
$\left[(-3, 0) \text{ and } \left(\frac{1}{2}, 0\right)\right]$

Then have students fold the paper so that the two branches of the parabola match. Ask for the equation of that fold. $\left[x = -1\frac{1}{4}\right]$

Tell students that the fold is the axis of symmetry of the parabola. Have students identify where the fold intersects the parabola.
$\left[-1\frac{1}{4}, -6\frac{1}{8}\right]$

This is the vertex of the parabola.

Point out that algebraic methods allow for more precise answers in some cases.

The answers to Exercises 7–19 can be found in Additional Answers beginning on page 729.

Find the zeros of each function by graphing.

20. $y = x^2 + 8x - 9$ $1, -9$ **21.** $y = x^2 - 20x + 100$ 10 **22.** $y = x^2 - x - 72$ $9, -8$

23. $y = x^2 + 6x - 7$ $1, -7$ **24.** $y = x^2 + 4x - 5$ $1, -5$ **25.** $y = x^2 + 2x - 24$ $4, -6$

26. $y = 2x^2 - 2x - 144$ $9, -8$ **27.** $y = 3x^2 + 9x - 12$ $1, -4$ **28.** $y = 5x^2 + 15x - 20$ $1, -4$

First graph the function. Then find the zeros, the axis of symmetry, and the minimum value of the functions.

29. $f(x) = x^2 + 18x + 81$ **30.** $f(x) = x^2 + 6x - 7$ **31.** $f(x) = x^2 - 5x + 6$

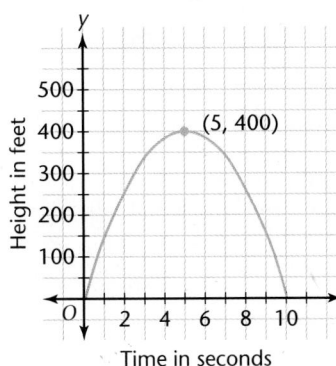

Physics The graph represents a relationship between the time when a projectile is propelled vertically into the air and the height that it reaches.

32. What is the maximum height reached by the projectile? 400 feet

33. How long does it take for the projectile to reach its maximum height? 5 seconds

34. How long does it take for the projectile to return to Earth? 10 seconds

35. What is the equation of the axis of symmetry of the graph? $x = 5$

36. Which equation was used to graph the parabola? c

  a. $y = -16(x + 5)^2 + 400$ **b.** $y = -16(x + 5)^2 - 400$
  c. $y = -16(x - 5)^2 + 400$ **d.** $y = -16(x - 5)^2 - 400$

37. Physics Use the formula in the caption to find the height in meters after 5 seconds. 75 m

The formula $h = 40t - 5t^2$ can be used to find the height in meters of the arrow after t seconds.

Look Back

The expression $2(x - 3)^2 + 1$ involves multiplication, an operation inside parentheses, an exponent, and addition. **[Lesson 1.4]**

38. Which of these operations should be done first? second? third?

Perform each matrix operation. [Lesson 7.2]

39. $\begin{bmatrix} 6 & -2 \\ 8 & 3 \end{bmatrix} + \begin{bmatrix} -2 & 4 \\ 3 & 6 \end{bmatrix}$ **40.** $\begin{bmatrix} -4 & 5 \\ 6 & 2 \end{bmatrix} - \begin{bmatrix} 3 & -1 \\ -5 & 6 \end{bmatrix}$ **41.** $\begin{bmatrix} -3 & 7 \\ 4 & -5 \end{bmatrix} + \begin{bmatrix} 6 & 8 \\ -1 & -3 \end{bmatrix}$

Factor. [Lesson 11.6]

42. $x^2 + 10x + 25$ $(x + 5)^2$ **43.** $x^2 + 14x + 49$ $(x + 7)^2$ **44.** $x^2 - 18x + 81$ $(x - 9)^2$

Look Beyond

45. Several students are able to collect 100,000 cans to recycle. What is the minimum space the cans will take? Assume that a can has a circumference of 21 centimeters and a height of 12.5 centimeters.

29. zero at $x = -9$; axis of sym. $x = -9$; min $= 0$

30. zeros at 1 and -7; axis of sym. $x = -3$; min $= -16$

31. zeros at 2 and 3; axis of sym. $x = 2.5$; min $= -0.25$

38. exponent, $(x - 3)^2$, multiplication, $\times 2$, addition, $+1$

39. $\begin{bmatrix} 4 & 2 \\ 11 & 9 \end{bmatrix}$ **40.** $\begin{bmatrix} -7 & 6 \\ 11 & -4 \end{bmatrix}$

41. $\begin{bmatrix} 3 & 15 \\ 3 & -8 \end{bmatrix}$

45. ≈ 43.9 meters^3, the minimum volume is the area of the base times the height of the 100,000 cans. $V = \pi r^2 h$, r is the radius of the base of the can and h is the height.

A **alternative**
ASSESSMENT

Authentic Assessment
Have students describe in their own words how to find the axis of symmetry, the vertex, and the zeros of a quadratic function.

Look Beyond
Exercise 45 is a challenge problem to encourage student's creativity in problem solving. Extensions of this problem can provide open ended exploration. Students might wonder about the total volume necessary to contain cylindrical cans in a rectangular box.

Technology
12.1 Patterns Among Second Differences

When you evaluate a function such as $y = x^2$ for a set of consecutive numbers such as 1, 2, 3, 4, 5, 6, you will see a pattern. In the table shown, the second differences are all equal to 2.

x	1	2	3	4	5	6
x^2	1	4	9	16	25	36
First Differences		3	5	7	9	11
Second Differences			2	2	2	2

Using a spreadsheet, you can explore many functions of the form $y = ax^2$ and see even more patterns.

1. Create a spreadsheet in which column A contains 1, 2, 3, 4, 5, and 6, and column B contains the value of a given function for each value in column A. Column C contains the differences of consecutive entries in column B, and column D contains the differences of consecutive entries in column C.

In Exercises 2–7, use the spreadsheet created in Exercise 1 to make a table of differences for each function.

2. $y = 2x^2$ **3.** $y = 3x^2$ **4.** $y = 7x^2$

5. $y = -2x^2$ **6.** $y = 12x^2$ **7.** $y = -4.2x^2$

8. Write a statement that relates the list of second differences to the value of a in $y = ax^2$.

9. Use the result of Exercise 8 to predict the number in the list of second differences for $y = 112.56x^2$.

10. Let $y = ax^2$. Use algebra to find the first and second differences. Simplify as you go. Does this prove your statement in Exercise 8?

Value of ax^2	an^2	$a(n + 1)^2$	$a(n + 2)^2$
First Differences			
Second Differences			

Objectives

RESOURCES

• Practice Master	**12.2**
• Enrichment Master	**12.2**
• Technology Master	**12.2**
• Lesson Activity Master	**12.2**
• Quiz	**12.2**
• Spanish Resources	**12.2**

Assessing Prior Knowledge

Find each product.

1. $8 \cdot 8$ **[64]**
2. $17 \cdot 17$ **[289]**

Factor the polynomial.

3. $x^2 - 4$ **[$(x - 2)(x + 2)$]**

TEACH

Suggest that a pound of feathers and a pound of gold were dropped from the top of a building at the same time. If the contents are identical in shape and size, which will hit the ground first? Students should conclude that both will hit the ground at the same time.

ongoing ASSESSMENT

Make another table that uses 3.40 to 3.50 for values of t.

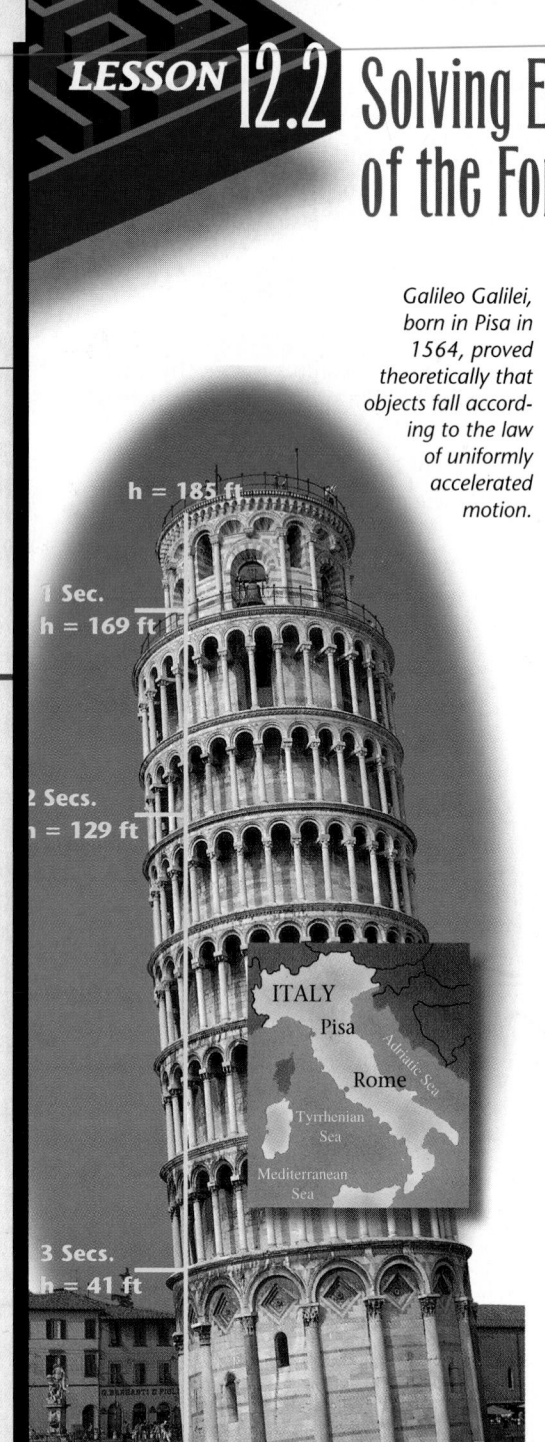

h = 185 ft

1 Sec.
h = 169 ft

2 Secs.
h = 129 ft

ITALY
Pisa
Adriatic Sea
Rome
Tyrrhenian Sea
Mediterranean Sea

3 Secs.
h = 41 ft

Ground Level 0

Solving Equations of the Form $x^2 = k$

Galileo Galilei, born in Pisa in 1564, proved theoretically that objects fall according to the law of uniformly accelerated motion.

Why *You already know how to find approximate solutions to a quadratic equation by using technology or graphing. Algebraic techniques can be used to find exact solutions.*

An object falls from the top of the Leaning Tower of Pisa, 185 feet above ground level. Its height, h, after t seconds is given by the function $h = -16t^2 + 185$. How long will it take the object to reach the ground?

When the object reaches the ground, the height is 0 feet. Thus, we substitute 0 for h in the given quadratic function to obtain the quadratic equation $0 = -16t^2 + 185$.

Tabular Method

Begin by making a table of values for the function $h = -16t^2 + 185$. Substitute integer values of t into $h = -16t^2 + 185$ to find h. According to the table, h will be 0 somewhere between $t = 3$ and $t = 4$ seconds. A new table can be constructed where t changes by tenths.

time (t)	height (h)	
0	185	
1	169	
2	121	
3	41	
4	−71	← Stop here.

time (t)	height (h)	
3.0	41.00	
3.1	31.24	
3.2	21.16	
3.3	10.76	
3.4	0.04	
3.5	−11.00	← Stop here.

Closer investigation using the table shows that 3.4 is an approximate solution to the nearest tenth for the equation $0 = -16t^2 + 185$.

How can you get a solution to the nearest hundredth for the equation $0 = -16t^2 + 185$?

ALTERNATIVE teaching strategy

Using Manipulatives

Have students use algebra tiles to show that $2^2 = 4$, $3^2 = 9$, and $4^2 = 16$. Then have students use grid paper to show the squares for numbers 5 through 12. Finally, have them use their calculators to find the squares of 13 through 20.

Graphing Method

Graphics Calculator

Graph the function $h = -16t^2 + 185$ with t on the x-axis and h on the y-axis. Use the trace function to find a point where $h \approx 0$. According to the graph, $t \approx 3.4$.

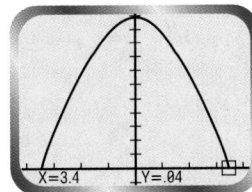

X=3.4   Y=.04

In fact, the quadratic equation $0 = -16t^2 + 185$ has 2 solutions. A more complete picture of the graph of $0 = -16t^2 + 185$ suggests the solutions are approximately 3.4 or -3.4. Since the falling-object problem is about height and time, only the positive time values make sense. In this case, ignore the negative time values.

Finding Square Roots

Finding the exact solutions for $0 = -16t^2 + 185$ involves finding square roots. Every positive number has two square roots. For example,

The square root of 9 is 3. $\sqrt{9} = 3.$
The negative square root of 9 is -3. $-\sqrt{9} = -3.$

If $x^2 = 9$, $x = 3$ or $x = -3$. The two solutions can be combined using the symbol $\pm$, read plus or minus.

$$\text{If } x^2 = 9, x = \pm\sqrt{9} = \pm 3.$$

EXAMPLE 1

Show that $\sqrt{\frac{4}{9}} = \frac{\sqrt{4}}{\sqrt{9}}$.

Solution ➤

Simplify the square root under the radical signs.

$\sqrt{\frac{4}{9}} = \frac{2}{3}$ because $\frac{2}{3} \cdot \frac{2}{3} = \frac{4}{9}$.

$\frac{\sqrt{4}}{\sqrt{9}} = \frac{2}{3}$ because $\sqrt{4} = 2$ and $\sqrt{9} = 3$.

Since $\sqrt{\frac{4}{9}} = \frac{2}{3}$ and $\frac{\sqrt{4}}{\sqrt{9}} = \frac{2}{3}$, $\sqrt{\frac{4}{9}} = \frac{\sqrt{4}}{\sqrt{9}}$ by substitution. ❖

Example 1 shows two ways to find the square root of $\frac{4}{9}$. But, it does not show how you find the solution or solutions to $x^2 = \frac{4}{9}$.

---

interdisciplinary

CONNECTION

Physics The equation in the opening of the lesson is derived from the equation $d = v_1 t - \frac{1}{2}gt^2$, where d is the height, v_1 is the initial velocity, t is time, and g is the acceleration of an object as the result of the force of gravity. In the given example, the initial velocity is zero. The force of gravity is 32 feet per second^2. Thus, the equation becomes $d = 0 - \frac{1}{2}(32)t^2 = -16t^2$. Then, if $d = (-185)$, we have $0 = -16t^2 + 185$.

Show that $\sqrt{\dfrac{81}{16}} = \dfrac{\sqrt{81}}{\sqrt{16}}$.

$$\left[\sqrt{\frac{9}{4} \cdot \frac{9}{4}} = \frac{9}{4} \text{ and } \frac{\sqrt{9 \cdot 9}}{\sqrt{4 \cdot 4}} = \frac{9}{4}; \right.$$
$$\left. \frac{9}{4} = \frac{9}{4} \right]$$

EXAMPLE 2

Solve the following equations.

A $x^2 = \dfrac{4}{9}$ **B** $x^2 = 1.44$ **C** $x^2 = 10$

Solution ➤

A There are two solutions: $\sqrt{\dfrac{4}{9}} = \dfrac{2}{3}$ and $-\sqrt{\dfrac{4}{9}} = -\dfrac{2}{3}$.

B There are two solutions: $\sqrt{1.44} = 1.2$ and $-\sqrt{1.44} = -1.2$.

C The solutions are $\sqrt{10}$ and $-\sqrt{10}$. There is no rational number answer. An approximation can be found by using [√] on your calculator.

$$\boxed{\sqrt{}} \; 10 \; \boxed{=} \; 3.16227766$$

The approximate solutions are 3.16 and −3.16.

The results of Example 2 lead to a generalization for solving a quadratic equation of the form $x^2 = k$. ❖

SOLVING $x^2 = k$ WHEN $k \geq 0$

If $x^2 = k$, $k \geq 0$, then
1. $x = \pm \sqrt{k}$ and
2. the solutions are $\sqrt{k}$ and $-\sqrt{k}$.

For example, if $x^2 = 16$, Then $x = \pm\sqrt{16} = \pm 4$. The solutions are 4 and −4.

Algebraic Method

Now the time can be found for the object falling from the Tower of Pisa.

$$-16t^2 + 185 = 0$$
$$-16t^2 = -185$$
$$t^2 \approx 11.56$$
$$t \approx \pm\sqrt{11.56}$$

The time is approximately $\sqrt{11.56}$ seconds.

Why is $-\sqrt{11.56}$ not considered as a solution?

A calculator gives the solution 3.4, which agrees with the tabular and graphing methods.

CRITICAL *Thinking*

Explain why the algebraic method for solving the quadratic equation $-16t^2 + 185 = 0$ gives another way to find the zeros for the quadratic function $h = -16t^2 + 185$.

The negative value is not a solution because time cannot be negative.

CRITICAL *Thinking*

The zeros of a function are the values for x which make $f(x) = 0$, so solving $f(x) = 0$ algebraically finds the zeros.

ENRICHMENT Have students graph $y = \sqrt{x}$ and $y = -\sqrt{x}$ on the same set of axes. Ask students what the combined graphs form. [**a parabola**] Ask whether the graph is a function. [**No**] Ask what equation would generate the entire graph. [$y = \pm\sqrt{x}$ or $y^2 = x$]

INCLUSION **strategies** **Using Models** Have students use a ball and stopwatch to repeat this experiment several times. Under careful supervision, the ball is dropped from the top of a building, and the time it takes to reach the ground is recorded. Have students estimate the height of the building by using the formula $h = 16t^2$.

EXAMPLE 3

Solve the following equations.

A $(a - 2)^2 - 9 = 0$ **B** $(x - 2)^2 = 11$

Solution ➤

A In this equation, the expression $a - 2$ plays the role of x in the statement $x^2 = k$.

$$(a - 2)^2 - 9 = 0$$
$$(a - 2)^2 = 9$$
$$a - 2 = \pm 3$$
$$a = 2 \pm 3$$

The solutions are 5 and -1.

Check each solution in the original equation.

B $(x - 2)^2 = 11$
$$x - 2 = \pm\sqrt{11}$$

The approximate solutions are 5.32 and -1.32.

Check each solution in the original equation. ❖

The solution to Example 3a will help you sketch the graph of the function $f(x) = (x - 2)^2 - 9$.

COORDINATE GEOMETRY
Connection

In the form $y = (x - 2)^2 - 9$, the vertex is $(2, -9)$ and the axis of symmetry is $x = 2$. Since the coefficient of the quadratic term is positive, the parabola opens upward and has a minimum value. The values where $y = 0$ are the zeros of the function. When $(x - 2)^2 - 9 = 0$, x is 5 or x is -1. Thus, the graph is a parabola that crosses the x-axis at 5 and -1.

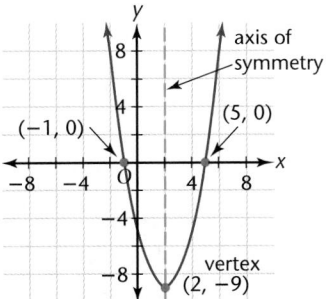

Alternate Example 3

Solve the following equations.

a. $(y + 4)2 = 25$
   [$y = 1$ or $y = -9$]
b. $(r - 3)2 + 5 = 19$
   [$r < 6.74$ or < -0.74]

TEACHING *tip*

Students can use mental estimation to check that the answers are within a reasonable range, and then substitute to check the exact answer.

Math Connection
Coordinate Geometry

Students use coordinate geometry to relate the function $y = (x - h)^2 + k$ to the vertex and axis of symmetry of a parabola.

Use Transparency ▶ **56**

EXERCISES & PROBLEMS

Communicate

1. Explain how to make a table of values for the function $y = -12x^2 + 300$ to determine where the function crosses the x-axis. Why would it be important to know when the values of the function are no longer positive?

2. Discuss how to find the square roots of 100. What is meant by the sign $\pm$?

Explain how to solve the following equations.

3. $x^2 = 64$

4. $x^2 = 8$

5. $x^2 = \dfrac{16}{100}$

Discuss how to solve each equation using square roots.

6. $(x + 3)^2 - 25 = 0$

7. $(x - 8)^2 = 2$

8. Explain how to sketch a graph from the vertex, axis of symmetry, and zeros for $f(p) = (p + 4)^2 - 5$.

Practice & Apply

9. Make a table of values for the function $f(t) = -14t^2 + 300$. Where does the function cross the x-axis?

Find each square root. Round answers to the nearest hundredth when necessary.

10. $\sqrt{121}$ 11

11. $\sqrt{144}$ 12

12. $\sqrt{625}$ 25

13. $\sqrt{36}$ 6

14. $\sqrt{44}$ 6.63

15. $\sqrt{90}$ 9.49

16. $\sqrt{88}$ 9.38

17. $\sqrt{19}$ 4.36

Solve each equation. Round answers to the nearest hundredth when necessary.

18. $x^2 = 25$ ± 5

19. $x^2 = 169$ ± 13

20. $x^2 = 81$ ± 9

21. $x^2 = 625$ ± 25

22. $x^2 = 12$ ± 3.46

23. $x^2 = 24$ ± 4.90

24. $x^2 = 18$ ± 4.24

25. $x^2 = 54$ ± 7.35

26. $x^2 = \dfrac{25}{81}$ $\pm\dfrac{5}{9}$

27. $x^2 = \dfrac{49}{121}$ $\pm\dfrac{7}{11}$

28. $x^2 = \dfrac{36}{49}$ $\pm\dfrac{6}{7}$

29. $x^2 = \dfrac{4}{100}$ ± 0.2

30. $x^2 = 32$ ± 5.66

31. $x^2 = 45$ ± 6.71

32. $x^2 = 28$ ± 5.29

33. $x^2 = 63$ ± 7.94

34. $(x + 4)^2 - 25 = 0$

35. $(x - 5)^2 - 9 = 0$

36. $(x + 1)^2 - 1 = 0$

37. $(x - 2)^2 - 6 = 0$

38. $(x + 7)^2 - 5 = 0$

39. $(x - 3)^2 - 2 = 0$

40. $(x + 3)^2 = 36$

41. $(x - 2)^2 = 144$

42. $(x - 8)^2 = 81$

43. $(x - 1)^2 = 11$

44. $(x + 5)^2 = 10$

45. $(x + 6)^2 = 15$

Find the vertex, axis of symmetry, and zeros of each function. Sketch a graph.

46. $f(x) = (x - 4)^2 - 9$

47. $g(x) = (x + 2)^2 - 1$

48. $h(x) = (x - 4)^2 - 3$

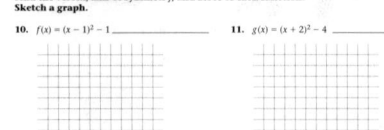

9.

t	y
5	-50
4	76
3	174
2	244
1	286
0	300

t	y
-1	286
-2	244
-3	174
-4	76
-5	-50

The graph crosses the x-axis about 4.5 and -4.5.

34. $1, -9$

35. $2, 8$

36. $0, -2$

37. $4.45, -0.45$

38. $-4.76, -9.24$

39. $4.41, 1.59$

40. $3, -9$

41. $14, -10$

42. $17, -1$

43. $4.32, -2.32$

44. $-1.84, -8.16$

45. $-2.13, -9.87$

The answers to Exercises 46–48 can be found in Additional Answers beginning on page 729.

Refer to the function $f(x) = (x + 4)^2 - 4$.

49. Find the vertex. $(-4, -4)$ **50.** Find the axis of symmetry. $x = -4$

51. Find the zeros. $-2, -6$ **52.** Sketch a graph.

53. Technology Check Exercises 49–52 by using a graphics calculator.   13 feet

54. 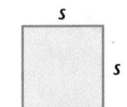 **Geometry** The area of a square garden is 169 square feet. Find the length of each side. ($A = s^2$)

s

s

55. 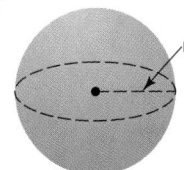 **Geometry** Use the formula $S = 4\pi r^2$ to find the radius of a sphere with a surface area of 90 square meters. (Use $\pi = 3.14$.)
≈ 2.68 m

radius

700 ft

The height of the Met Life Building in New York City is 700 feet.

Physics If an object falls from the Met Life Building, and its height, h, after t seconds is given by the function $h = -16t^2 + 700$, find

56. how long it will take for the object to reach the ground.
≈ 6.61 seconds

57. how long it will take for the object to reach a height of 100 feet. ≈ 6.12 seconds

Look Back

Find the products. [Lesson 7.3]

58. $\begin{bmatrix} 4 & -1 \\ 6 & 0 \end{bmatrix}\begin{bmatrix} 4 & -3 \\ -7 & 2 \end{bmatrix}$ **59.** $\begin{bmatrix} 9 & -4 \\ 2 & -4 \end{bmatrix}\begin{bmatrix} 6 & 6 \\ -2 & -4 \end{bmatrix}$

Simplify each expression. [Lesson 10.2]

60. $\dfrac{-6x^2y^2}{2x}$ **61.** $\dfrac{b^3c^4}{bc}$ **62.** $(p^2)^4(2a^2b^3)^2$

Add or subtract each polynomial. [Lesson 11.1]

63. $(3x^2 - 2x + 1) + (2x^2 + 4x + 6)$ $5x^2 + 2x + 7$

64. $x^2 + 2x - 1 - (2x^2 - 5x + 7)$ $-x^2 + 7x - 8$

65. Find the factors of $a^2 - b^2$, $a^4 - b^4$, and $a^8 - b^8$. How many factors does $a^{64} - b^{64}$ have? Explain.

Look Beyond

66. Find $\sqrt{5^2 - 4 \cdot 1 \cdot 2}$.   **67.** Find $\sqrt{6^2 - 4(1)(9)}$.

The answers to Exercises 52 and 53 can be found in Additional Answers beginning on page 729.

58. $\begin{bmatrix} 23 & -14 \\ 24 & -18 \end{bmatrix}$ **59.** $\begin{bmatrix} 62 & 70 \\ 20 & 28 \end{bmatrix}$

60. $-3xy^2$ **61.** b^2c^3 **62.** $4a^4b^6p^8$

65. $(a + b)(a - b)$; $(a^2 + b^2)(a + b)(a - b)$; $(a^4 + b^4)(a^2 + b^2)(a + b)(a - b)$; 7; 64 is 2^6, so $(a^{64} - b^{64})$ has $6 + 1$ factors.

66. ≈ 4.12 **67.** 0

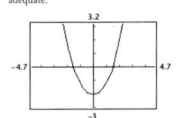
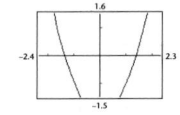

PREPARE

Objectives

- Complete the square.
- Rewrite a quadratic in the form $y = (x - h)^2 + k$.
- Identify the vertex of a parabola.
- Find the minimum value of a quadratic from the vertex.

RESOURCES

- Practice Master   **12.3**
- Enrichment Master   **12.3**
- Technology Master   **12.3**
- Lesson Activity Master   **12.3**
- Quiz   **12.3**
- Spanish Resources   **12.3**

Assessing Prior Knowledge

Factor each of the following:

1. $x^2 - 4x + 4$ $[(x - 2)^2]$
2. $x^2 + 2x + 1$ $[(x + 1)^2]$
3. $x^2 - 12x + 36$
    $[(x - 6)^2]$
4. $x^2 + 20x + 100$
    $[(x + 10)^2]$

TEACH

 Ask students if there is always a single way to solve a word problem. Students should agree that one word problem can be solved in different ways. Quadratic equations can also be solved by using many techniques. One such technique is called *completing the square.*

ongoing ASSESSMENT

When the nine 1-tiles are added, the figure is a square.

LESSON 12.3 Completing the Square

If you rewrite the quadratic function $y = x^2 + bx + c$ in the form $y = (x - h)^2 + k$, you can find the vertex and axis of symmetry of the parabola. A technique called completing the square will enable you to do this.

In the previous chapter you used algebra tiles to factor perfect-square trinomials such as $x^2 + 6x + 9$. How can a square be constructed when you begin with the following tiles?

To make a square, start with the x^2-tile and arrange the six x-tiles into 2 groups of 3.

How many 1-tiles will it take to fill in the corner?

$x^2 + 6x$

$x^2 + 6x + 9$

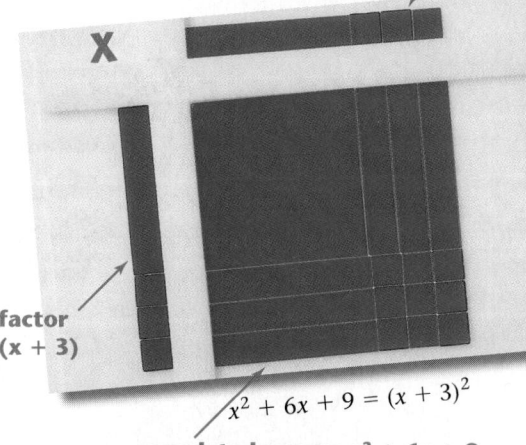

factor $(x + 3)$

factor $(x + 3)$

$x^2 + 6x + 9 = (x + 3)^2$

completed square $x^2 + 6x + 9$

Why do you think the process of adding the nine 1-tiles is called completing the square?

ALTERNATIVE teaching strategy

Using Analytical Thinking Have students expand $(x - h)^2$ and write $y = (x - h)^2 + k$ without parentheses as $y = x - 2hx + h^2 + k$. Point out that $h^2 + k$ has a numerical value. Have students analyze $y = x^2 - 8x + 7$ in terms of the expanded equation.

Exploration *How Many Ones?*

Use algebra tiles to complete the square for $x^2 + 10x$.

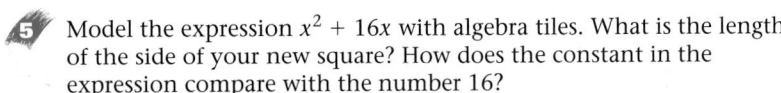

1   Into how many groups should the x-tiles be divided?

2   How many 1-tiles must you add?

3   Write the algebraic expression for the area of the square.

4   Fill in the blanks to complete the square for $x^2 + 10x$.

    a. Divide the x-tiles into 2 sets of _?_ x-tiles.

    b. Add _?_ 1-tiles.

    c. Write $x^2 + 10x + \underline{?} = (x + 5)^2$.

5   Model the expression $x^2 + 16x$ with algebra tiles. What is the length of the side of your new square? How does the constant in the expression compare with the number 16?

    a. Write the area of your new square in the form $x^2 + bx + c$.

    b. Write the area of your new square in the form $(x + \underline{?})^2$.

6   Suppose you have 1 x-square tile and 74 x-tiles. How many 1-tiles will you need to complete the square? Explain how you get your answer. ❖

In the Exploration, the coefficient of x is an even number. Thus, it is easy to divide the x-tiles into 2 sets. What happens if the coefficient of x is odd? What happens if the coefficient of x is negative?

EXAMPLE 1

Complete the square for $x^2 + 5x$.

Solution ➤

It is not possible to use actual tiles to complete the square. However, a picture will serve as a good model in this case.

Split the 5 x-tiles into 2 groups, each with $2\frac{1}{2}$ x-tiles. Then add 1-tiles to complete the square.

You need $\left(\frac{5}{2}\right)^2$, or $\frac{25}{4}$, 1-tiles.

By adding $6\frac{1}{4}$ 1-tiles, the result is a square whose area is $x^2 + 5x + 6\frac{1}{4}$, or $\left(x + \frac{5}{2}\right)^2$. ❖

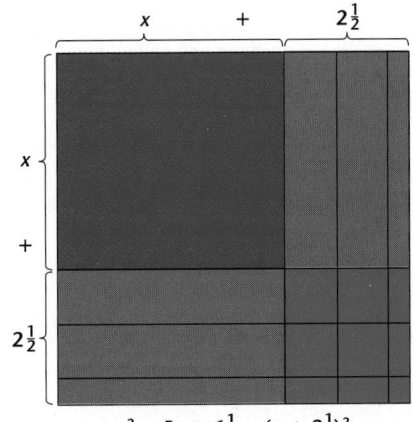

$$x^2 + 5x + 6\tfrac{1}{4} = (x + 2\tfrac{1}{2})^2$$

Exploration Notes

Have students use tiles and work in small groups to complete this exploration. Remind students that the goal is to create a *square*—that is, each side must have the same dimensions.

ongoing ASSESSMENT

6. **You will need 1369 1-tiles. Divide 74 by 2, and square the result.**

Alternate Example 1

Complete the square for
$x^2 + 7x$. $\left[\left(x + \frac{7}{2}\right)^2\right]$

ENRICHMENT   Challenge students to find the value of n that makes each trinomial a perfect square.

   1. $x^2 + nx + 81$ $[n = 18]$
   2. $x^2 - nx + 121$ $[n = 22]$
   3. $x^2 + nx + c$ $[n = 2\sqrt{c}]$

INCLUSION strategies   **Using Visual Models** Some students may need additional help when completing the square. Suggest they draw a flowchart that shows each step involved in the process.

EXAMPLE 2

Complete the square for $x^2 - 6x$.

Solution ➤

One-half the coefficient of x is -3.
Add $(-3)^2$ or 9 to form a perfect square.
The perfect square is $x^2 - 6x + 9 = x^2 - 6x + (-3)^2 = (x - 3)^2$. ❖

How do you decide what to add to $x^2 + 6x$ and $x^2 - 6x$
to form a perfect square?

Completing the square gives a technique for rewriting the quadratic
$y = x^2 + 6x$ in the form $y = (x - h)^2 + k$.

Start with the quadratic.	$y = x^2 + 6x$
Find what you need to add to complete the square.	$\dfrac{1}{2} \cdot 6 = 3;\quad 3^2 = 9$
Add and subtract the square from the quadratic.	$y = x^2 + 6x + 9 - 9$
Group the terms.	$y = (x^2 + 6x + 9) - 9$
Write in the form $y = (x - h)^2 + k$.	$y = (x + 3)^2 - 9$

Why can you add $9 - 9$ to $x^2 + 6x$ without changing the value of the
expression?

**MAXIMUM
MINIMUM**
Connection

*The **minimum** value
for a quadratic function
$y = (x - h)^2 + k$
is the y-value of the
vertex.*

Minimum

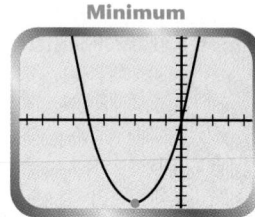

The vertex of the graph of
$y = (x + 3)^2 - 9$ is $(-3, -9)$.
Thus, the minimum value
of $y = x^2 + 6x$ is -9.

EXAMPLE 3

Find the minimum value for the function $f(x) = x^2 - 3x$.

Solution ➤

The vertex of the parabola can be found by completing the square. The
coefficient of x is -3. Add and subtract the square of $-\dfrac{3}{2}$ to the equation.

$$f(x) = x^2 - 3x + \left(-\dfrac{3}{2}\right)^2 - \left(-\dfrac{3}{2}\right)^2$$

$$= \left(x - \dfrac{3}{2}\right)^2 - \dfrac{9}{4}$$

The graph of $f(x) = \left(x - \dfrac{3}{2}\right)^2 - \dfrac{9}{4}$ is a parabola with vertex $\left(\dfrac{3}{2}, -\dfrac{9}{4}\right)$ and axis

of symmetry $x = \dfrac{3}{2}$. Thus, the minimum value is $-\dfrac{9}{4}$. ❖

RETEACHING
**the
lesson**

Cooperative Learning
Have students work
in cooperative-learning
groups as they complete
the square for several quadratic expressions of
the form $x^2 + nx$. Have each group write a
summary of the steps used to complete the
square.

You can now complete the square for a quadratic of the form $y = x^2 + bx$. The next step is to complete the square for a quadratic of the form $y = x^2 + bx + c$.

Alternate Example 4

a. Rewrite $y = x^2 - 6x + 4$ in the form $y = (x - h)^2 + k$. $[y = [(x - 3)^2 - 5]$

b. Find the vertex of the parabola. $[(3, -5)]$

EXAMPLE 4

A Rewrite $y = x^2 - 8x + 7$ in the form $y = (x - h)^2 + k$.

B Find the vertex of the parabola.

Solution ➤

A The new feature in this problem is the constant 7.

Original function	$y = x^2 - 8x + 7$
Group the x^2 and x terms.	$y = (x^2 - 8x) + 7$

First complete the square using the coefficient of the x-term, -8.

$$\frac{1}{2}(-8) = -4 \quad \text{and} \quad (-4)^2 = \mathbf{16}$$

Now add and subtract **16**.	$y = (x^2 - 8x + \mathbf{16}) + 7 - \mathbf{16}$
Write in the form $y = (x - h)^2 + k$.	$y = (x - 4)^2 - 9$

B The vertex is $(4, -9)$. Check by graphing. ❖

Try This Find the minimum value for $y = x^2 - 6x + 11$.

ongoing
ASSESSMENT

Try This

minimum value: $(3, 2)$

ASSESS

Selected Answers
Odd-numbered Exercises 7–59

Assignment Guide
Core 1–10, 12–28 even, 29–47, 52–61

Core Two-Year 1–59

EXERCISES & PROBLEMS

Communicate

1. Describe how to use algebra tiles to complete the square for $x^2 + 4x$.

2. Discuss how to complete the square algebraically for $x^2 - 7x$.

3. How do you rewrite $y = x^2 + 10x + 25 - 25$ in the form $y = (x - h)^2 + k$?

Describe how to find the minimum value for each quadratic function.

4. $y = x^2 + 2$ 5. $y = x^2 - x$

6. Explain how to rewrite $y = x^2 - 10x + 11$ in the form $y = (x - h)^2 + k$. How do you find the vertex from this form of the equation?

Technology

Students can use a calculator to check Exercises 32–43 and to solve Exercises 44–51.

Error Analysis

When working with tiles arranged in a square, students may incorrectly name the factors. For example, the layout on the bottom of page 559 might be described as having factors of $x^2 + 2\frac{1}{2}x$. Remind students how to read the factors of such a display.

Practice & Apply

Use algebra tiles to complete the square for each of the following. Write the area of each new square in the form $x^2 + bx + c$.

7. $x^2 + 14x + 49$ **8.** $x^2 - 14x + 49$ **9.** $x^2 + 8x + 16$ **10.** $x^2 - 8x$ $+ 16$

Complete the square.

11. $x^2 + 6x + 9$ **12.** $x^2 - 2x + 1$ **13.** $x^2 + 12x + 36$ **14.** $x^2 - 12x$ $+ 36$

15. $x^2 + 7x$ **16.** $x^2 - 10x$ **17.** $x^2 + 15x$ **18.** $x^2 - 5x$

19. $x^2 + 16x$ **20.** $x^2 + 20x$ **21.** $x^2 - 9x$ **22.** $x^2 + 40x$

Rewrite each function in the form $y = (x - h)^2 + k$.

23. $y = x^2 + 8x + 16 - 16$ **24.** $y = x^2 - 4x + 4 - 4$

25. $y = x^2 - 10x + 25 - 25$ **26.** $y = x^2 + 14x + 49 - 49$

27. $y = x^2 - 16x + 64 - 64$ **28.** $y = x^2 + 20x + 100 - 100$

Refer to the equation $y = x^2 + 10x$ to complete Exercises 29-31.

29. Complete the square and rewrite it in the form $y = (x - h)^2 + k$. $y = (x + 5)^2 - 25$

30. Find the vertex. $(-5, -25)$

31. Find the maximum or minimum value.

Find the minimum value for each quadratic function.

32. $f(x) = x^2 + 33$ **33.** $f(x) = x^2 - 4$ -4 **34.** $f(x) = x^2 + 2$ 2

35. $f(x) = x^2 - 1$ -1 **36.** $f(x) = x^2 + 6$ 6 **37.** $f(x) = x^2 - 5$ -5

Rewrite each function in the form $y = (x - h)^2 + k$. Find each vertex.

38. $y = x^2 + 4x - 1$ **39.** $y = x^2 - 2x - 3$ **40.** $y = x^2 + 10x - 12$

41. $y = x^2 - x + 2$ **42.** $y = x^2 - 6x - 5$ **43.** $y = x^2 + \frac{1}{3}x - 3$

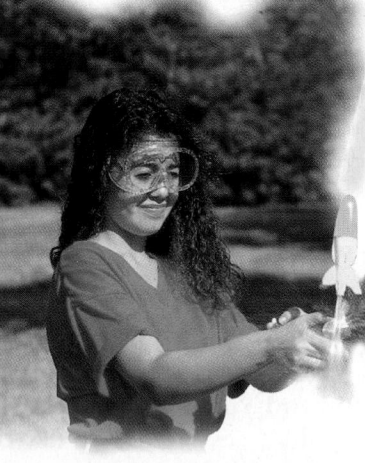

A water rocket is shot vertically into the air with an initial velocity of 192 feet per second.

Physics   The relationship between time, t, and height, h, is given by the formula $h = -16t^2 + 192t$.

44. Graph the function. (If you use graphing technology you will need to use x in place of t and y in place of h.)

45. Complete the square. Find the vertex of the parabola. $(6, 576)$

46. What is the maximum height reached by the projectile? 576 feet

47. How long does it take for the projectile to reach its maximum height? 6 seconds

15. $x^2 + 7x + \dfrac{49}{4}$

16. $x^2 - 10x + 25$

17. $x^2 + 15x + \dfrac{225}{4}$

18. $x^2 - 5x + \dfrac{25}{4}$

19. $x^2 + 16x + 64$

20. $x^2 + 20x + 100$

21. $x^2 - 9x + \dfrac{81}{4}$

22. $x^2 + 40x + 400$

23. $y = (x + 4)^2 - 16$

24. $y = (x - 2)^2 - 4$

25. $y = (x - 5)^2 - 25$

26. $y = (x + 7)^2 - 49$

27. $y = (x - 8)^2 - 64$

28. $y = (x + 10)^2 - 100$

38. $y = (x + 2)^2 - 5; V(-2, -5)$

39. $y = (x - 1)^2 - 4; V(1, -4)$

40. $y = (x + 5)^2 - 37; V(-5, -37)$

41. $y = \left(x - \dfrac{1}{2}\right)^2 + 1\dfrac{3}{4}; V(0.5, 1.75)$

42. $y = (x - 3)^2 - 14; V(3, -14)$

43. $y = \left(x + \dfrac{1}{6}\right)^2 - 3\dfrac{1}{36}; V\left(-\dfrac{1}{6}, -3\dfrac{1}{36}\right)$

The answer to Exercises 31 and 44 can be found in Additional Answers beginning on page 729.

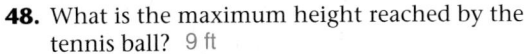
A tennis player throws a ball vertically into the air and watches as it begins to descend. He extends his racket to make contact with the ball.

Physics The motion of the ball can be modeled by the function $h = -(t - 1)^2 + 9$, where h is the height of the ball in feet after t seconds have elapsed.

48. What is the maximum height reached by the tennis ball? 9 ft

49. How long did it take for the tennis ball to reach its maximum height? 1 second

50. The player can make contact with the ball at a height of 9 feet. How much time will elapse before he hits the tennis ball? 1 second

51. If the tennis ball is allowed to fall to the ground, how much time will elapse? 4 seconds

Look Back

Solve each system by graphing. [Lesson 6.1]

52. $\begin{cases} x - 3y = 3 \\ 2x - y = -4 \end{cases}$ **53.** $\begin{cases} x - 2y = 0 \\ x + y = 3 \end{cases}$

54. Can you find the inverse of $\begin{bmatrix} 4 & 6 \\ 2 & 3 \end{bmatrix}$? Explain.   [Lesson 7.4]

55. **Probability** Find the probability that an even number will be drawn from a bag containing the whole numbers from 1 to 20 inclusive. [Lesson 8.6] $\frac{1}{2}$

Write each of the following without negative exponents. [Lesson 10.3]

56. a^2b^{-3} $\frac{a^2}{b^3}$ **57.** $\frac{m^5}{m^{-2}}$ m^7 **58.** $\frac{2n^{-6}}{n^{-4}}$ $\frac{2}{n^2}$

59. **Geometry** Find the area of the square rug. [Lesson 11.3] $x^2 + 8x + 16$

Look Beyond

The equation for the height in feet of a model rocket after t seconds is given by $h(t) = -16t^2 + 200t$. Find t when the rocket has a height of 400 feet

60. using the graphing method.

61. using the algebraic method.

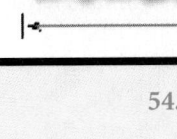
$|\!\leftarrow\!\!\!\!-\!-\!-\!-\!-\!-\!- x+4 -\!-\!-\!-\!-\!-\!-\!\!\!\!\rightarrow\!|$

---

52. $(-3, -2)$

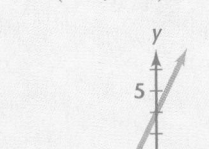

53. $(2, 1)$

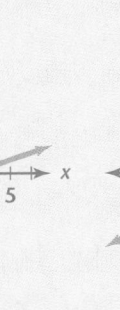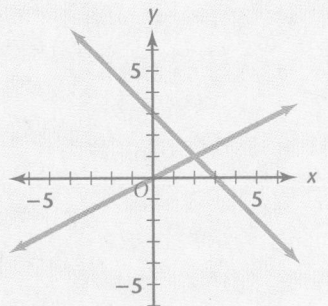

54. The matrix has no inverse because the determinant is 0.

The answers to Exercises 60 and 61 can be found in Additional Answers beginning on page 729.

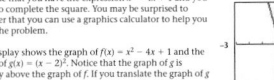
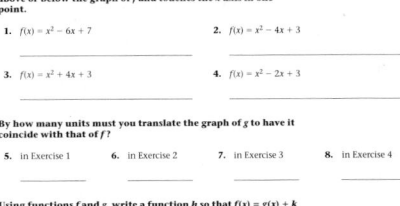

MOTIVATE

After students read the article, have volunteers describe the rescue in their own words, including their hunches about the times and speeds involved.

Ask students how they think the data, specifically the altitudes at which different events took place, were arrived at. (Although skydivers wear altimeters, it's doubtful Robertson or Williams would bother to look at them during this life-and-death episode. Altitude data in the story were probably based on estimates made by the divers and ground observers.

Review the relationship:
speed = distance/time

Go over the diagrams at the bottom of the page. Make these points:

· Before reaching terminal velocity, the motion of a skydiver is not as simple. As the force of air resistance keeps changing, the acceleration keeps changing.

· Once terminal velocity is reached, the motion is at a constant speed and therefore much easier to calculate.

· The arrows do not represent velocity. In fact, when a skydiver first exits the plane, most of the divers motion is horizontal. As

EYEWITNESS MATH

Rescue at 2000 Feet

A Miraculous Sky Rescue

The jump began as a routine skydiving exercise, part of a convention of 420 parachutists sponsored by Skydive, Arizona, but it quickly turned into a test of nerve, instinct and courage...

Moments after he went out the open hatch of a four-engine DC-4 airplane near Coolidge, Ariz., Sky Diver Gregory Robertson, 35, could see that Debbie Williams, 31, a fellow parachutist with a modest 50 jumps to her credit, was in big trouble. Instead of "floating" in the proper stretched-out position parallel to the earth, Williams was tumbling like a rag doll. In attempting to join three other divers in a hand-holding ring formation, she had slammed into the backpack of another chutist, and was knocked unconscious.

From his instructor's position above the other divers, Robertson reacted with instincts that had been honed by 1,700 jumps during time away from his job as an AT&T engineer in Phoenix. He straightened into a vertical dart, arms pinned to his body, ankles crossed, head aimed at the ground in what chutists call a "no-lift" dive, and plummeted toward Williams... <u>At 3,500 ft., about ten seconds before impact, Robertson caught up with Williams,</u> almost hitting her but slowing his own descent by assuming the open-body frog-like position. He angled the unconscious sky diver so her chute could open readily and <u>at 2,000 ft., with some six seconds left, yanked the ripcord</u> on her emergency chute, then pulled his own ripcord. The two sky divers floated to the ground. Williams, a fifth-grade teacher from Post, Texas, landed on her back, suffering a skull fracture and a perforated kidney— but alive. In the history of recreational skydiving, there has never been such a daring rescue in anyone's recollection.

Does the article give you a complete picture of the rescue? How can you tell if the numbers in the article are accurate?

You can use what you know about distance, time, and speed to check the facts in the article and to get a fuller sense of what took place during this amazing feat.

First, study these diagrams to get an idea of some of the forces that affect sky divers in free fall.

Robertson explains how he saved a life at 2,000 ft.

Force of Gravity

When a parachutist jumps out of a plane, the force of gravity causes the diver to accelerate downward.

Drag

Force of Gravity

As the sky diver's speed increases, the force of air resistance (or drag) plays a larger and larger role.

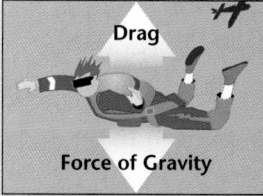

Drag

Force of Gravity

At some point, the two forces balance and the sky diver no longer accelerates, but falls at a steady rate, called the terminal velocity.

air resistance slows the horizontal velocity and gravity increases the vertical velocity, the divers direction of motion becomes more and more vertical.

1a. 4 sec.
1b. 1500 ft.
1c. No; 1500 ft/4s = 375 ft/s or 256 mph

2a. 135 mph = 198 ft/s
time to impact at 3500 ft:
(3500 ft)/(198 ft/s) or 17.7s
time to impact at 2000 ft:
(2000 ft)/(198 ft/s) or 10.1s
2b. 17.7s − 10.1s = 7.5s
(or about 7 or 8 seconds)

Cooperative Learning

1. Use the fact that 88 feet per second equals 60 miles per hour to convert. Answer Part 1 to see if the underlined part of the article makes sense.

   Based on the data in the article, how long did it take Robertson to pull Williams's ripcord after he caught up to her? During that time, how far did the two divers fall according to the article? Given that the maximum speed of a person in free fall is about 190–200 miles per hour, is the data in the article accurate? Explain.

2. After Robertson caught up with Williams, the two divers were probably falling at a speed between 125 and 150 miles per hour. Suppose they fell at 135 miles per hour. You can use that value to find more plausible time estimates for the rescue.

   About how many seconds from impact would the two sky divers be at 3500 feet? at 2000 feet? Suppose Robertson did catch up to Williams at 3500 feet and did pull her ripcord at 2000 feet. About how much time would have elapsed between those two events if the divers were falling at 135 miles per hour?

3. While Robertson was catching up to Williams, he changed his body position to speed up and then slow down. Such motion is difficult to calculate exactly, but you can make estimates for this part of the rescue by using a single average velocity for Robertson. Assume that Robertson was at 8500 feet and Williams was at 8300 feet when Robertson started trying to catch up.

   Write an equation for t in terms of v_r and v_w, where

   v_r is Robertson's average velocity in feet per second while catching up to Williams.

   v_w is Williams' velocity in feet per second.

   t is the time in seconds from when Robertson starts to go after Williams to when he reaches her.

4. Use your equation to find t if $v_r = 206$ (about 140 miles per hour) and $v_w = 198$ (about 135 miles per hour). Based on your results, at what altitude would Robertson catch up to Williams?

5. Imagine you are a journalist writing about the rescue. Use your data from Parts 2, 3, and 4 to create a paragraph that will give your readers a reasonably accurate sense of the event.

Cooperative Learning

Have students work in pairs or small groups. At the outset you may wish to suggest some ways students might convert from feet per second to miles per hour. Then have students work through the five activities.

Discuss

Have students share their paragraphs from Activity 5 with the class. Have students evaluate each others work. Ask them to suggest ways to improve each story.

3. $t = (200)/v_r - v_w)$

4a. $t = (200)/v_r - v_w)$
   $t = 200/(206-198) = 200/8 = 25$
   25 seconds

4b. 3350 feet

5. Answers will vary, but numbers used should be consistent with figures given in the lesson and with each other. For example, speeds, times, and distances should satisfy basic equations of motion.

Objectives

- Solve quadratic equations by completing the square or by factoring.

- Practice Master   **12.4**
- Enrichment Master   **12.4**
- Technology Master   **12.4**
- Lesson Activity Master   **12.4**
- Quiz   **12.4**
- Spanish Resources   **12.4**

Assessing Prior Knowledge

Factor.

1. $x^2 + 5x + 6$
   $[(x + 3)(x + 2)]$
2. $x^2 + x - 6$
   $[(x + 3)(x - 2)]$
3. $x^2 - x - 6$
   $[(x - 3)(x + 2)]$
4. $x^2 - 4x + 4$ $[(x - 2)^2]$
5. $x^2 + 2x + 1$ $[(x + 1)^2]$

TEACH

 Students have worked with quadratic expressions and equations at this point. They can now use quadratic equations for solving problems.

LESSON 12.4 Solving Equations of the Form $x^2 + bx + c = 0$

Why *You have already solved equations of the form $x^2 = k$. Equations of the form $x^2 + bx + c = 0$ can be solved by completing the square.*

In Lesson 12.3 you found the vertex of $y = x^2 - 8x + 7$ by completing the square. When the function is rewritten in the form $y = (x - 4)^2 - 9$, the vertex $(4, -9)$ can be noted by inspection.

$$(x - 4)^2 - 9 = 0$$
$$(x - 4)^2 = 9$$
$$x - 4 = \pm 3$$
$$x - 4 = 3 \quad \text{or} \quad x - 4 = -3$$

The solutions are 7 and 1. These two solutions are the x-intercepts of $y = x^2 - 8x + 7$ and the zeros of $f(x) = x^2 - 8x + 7$.

$$f(7) = 7^2 - 8(7) + 7 = 49 - 56 + 7 = 0$$
$$f(1) = 1^2 - 8(1) + 7 = 1 - 8 + 7 = 0$$

Thus, the parabola intersects the x-axis at $(7, 0)$ and $(1, 0)$.

Cooperative Learning Explain the Zero Product Property to the class. Then separate the class into four groups. Prepare 4 index cards, each containing a different, factorable quadratic equation. On one card write FACTOR, on the second write GRAPH, on the third write COMPLETE THE SQUARE, and on the fourth write ALGEBRA TILES. The fourth card must contain an equation with all positive terms. Each group is to work together to solve the equation and check their solutions by substituting the values into the original equation. After groups have solved their equations, have them exchange cards until they have used all four methods.

CRITICAL Thinking Why are the zeros of the function $f(x) = x^2 - 8x + 6$ and the solutions to the equation $x^2 - 8x + 6 = 0$ the same?

EXAMPLE 1

A Solve $x^2 - 6x + 8 = 0$.

B Find the zeros of $h(x) = x^2 - 6x + 7$.

Solution ➤

For each problem, write the expression in the form $(x - h)^2 + k$ by completing the square. Then use the $x^2 = k$ form to find the solution. Check by substitution.

A
$$x^2 - 6x + 8 = 0$$
$$x^2 - 6x = -8$$
$$x^2 - 6x + 9 = -8 + 9$$
$$(x - 3)^2 = 1$$
$$x - 3 = \sqrt{1} \quad \text{or} \quad x - 3 = -\sqrt{1}$$
$$x = 4 \quad \text{or} \quad x = 2$$
The solutions are 4 and 2.

B
$$x^2 - 6x + 7 = 0$$
$$x^2 - 6x = -7$$
$$x^2 - 6x + 9 = -7 + 9$$
$$(x - 3)^2 = 2$$
$$x - 3 = \sqrt{2} \quad \text{or} \quad x - 3 = -\sqrt{2}$$
$$x = 3 + \sqrt{2} \quad \text{or} \quad x = 3 - \sqrt{2}$$
The zeros are $3 + \sqrt{2}$ and $3 - \sqrt{2}$. ❖

Try This Solve $x^2 - 7x + 10 = 0$.

When a quadratic can be factored mentally you can quickly solve the quadratic equation. For example, factor $x^2 - 6x + 8$ into $(x - 4)(x - 2)$.

Thus, $x^2 - 6x + 8 = 0$ is the same as $(x - 4)(x - 2) = 0$.

If the product of two factors is equal to zero, then at least one of the factors must be zero. This generalization is called the **Zero Product Property.**

ZERO PRODUCT PROPERTY
If a and b are real numbers such that $ab = 0$, then $a = 0$ or $b = 0$.

The Zero Product Property can be used to solve $(x - 4)(x - 2) = 0$.

$$x - 4 = 0 \quad \text{or} \quad x - 2 = 0$$
$$x = 4 \quad \text{or} \quad x = 2$$

The solutions are 4 and 2.

Check

$$x^2 - 6x + 8 = (4)^2 - 6(4) + 8 = 16 - 24 + 8 = 0 \quad \text{True}$$

or

$$x^2 - 6x + 8 = (2)^2 - 6(2) + 8 = 4 - 12 + 8 = 0 \quad \text{True}$$

interdisciplinary CONNECTION

Marketing A company is currently advertising on a painted sign that is 8 meters by 12 meters on their building. They want to increase the length and the width of the sign by the same amount and double the area of the original. What should the new dimensions of the sign be? **[12 meters by 16 meters]**

You may want to help students set up the equation for solving this problem. $(8 + x)(12 + x) = 192$

EXAMPLE 2

Solve $x^2 + x = 2$ by the indicated method.

A Completing the square   **B** Factoring

Solution ➤

A
$$x^2 + x = 2$$
$$x^2 + x + \left(\frac{1}{2}\right)^2 = 2 + \frac{1}{4}$$
$$\left(x + \frac{1}{2}\right)^2 = \frac{9}{4}$$

$$x + \frac{1}{2} = \frac{3}{2} \quad \text{or} \quad x + \frac{1}{2} = -\frac{3}{2}$$
$$x = \frac{3}{2} - \frac{1}{2} \quad \text{or} \quad x = -\frac{3}{2} - \frac{1}{2}$$
$$x = 1 \quad \text{or} \quad x = -2$$

The solutions are 1 and -2. ❖

B
$$x^2 + x = 2$$
$$x^2 + x - 2 = 0$$
$$(x + 2)(x - 1) = 0$$

$$x + 2 = 0 \quad \text{or} \quad x - 1 = 0$$
$$x = -2 \quad \text{or} \quad x = 1$$

How would you decide which method to use?

EXAMPLE 3

Let $f(x) = x^2 + 3x - 15$. Find the value of x when $f(x)$ is -5.

Graphics Calculator

Solution A ➤

Graph $y = x^2 + 3x - 15$ and $y = -5$ on the same set of axes. Use a graphics calculator if you have one. Find the point or points of intersection. You see that when $x = 2$ and $x = -5$, then $f(x) = -5$.

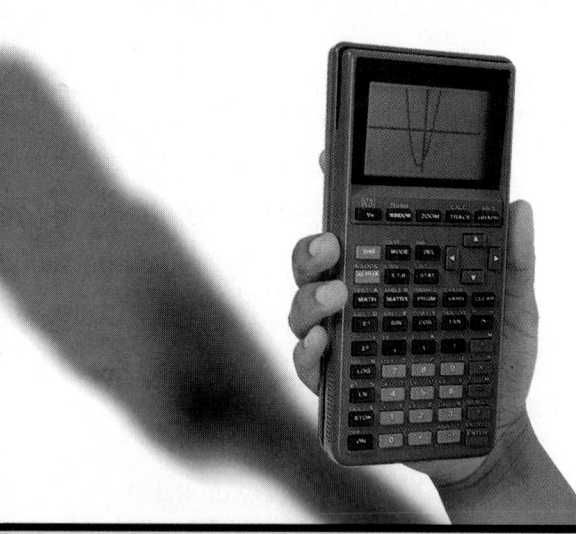

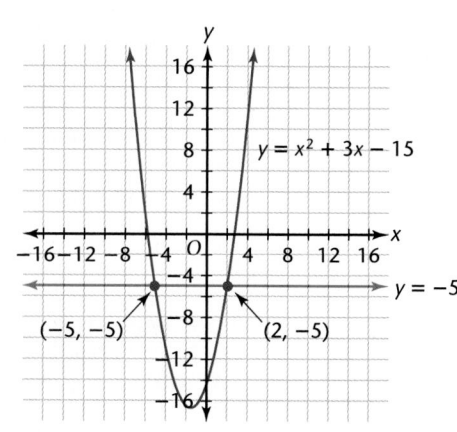

$y = x^2 + 3x - 15$

$y = -5$

$(-5, -5)$ $(2, -5)$

Solution B ➤

Since we want $x^2 + 3x - 15$ to equal -5, write $x^2 + 3x - 15 = -5$. Add 5 to both sides of the equation. Now, $x^2 + 3x - 10 = 0$. Factor the quadratic, and use the Zero Product Property to find the possible values of x.

$$(x - 2)(x + 5) = 0$$
$$x - 2 = 0 \quad \text{or} \quad x + 5 = 0$$
$$x = 2 \quad \text{or} \quad x = -5 \qquad \text{Check by substitution.} ❖$$

Try This   Solve $x^2 - 8x + 22 = 10$.

EXAMPLE 4

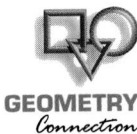

GEOMETRY
Connection

A box with a square base and no top is to be made by cutting out 2-inch squares from each corner of a square piece of cardboard and folding up the sides. The volume of the box is to be 8 cubic inches. What size should the piece of cardboard be?

Solution ➤

Let x be the length (and width) of the piece of cardboard.

$V = 8$ cubic inches

The volume of a rectangular box is $V = \text{length} \cdot \text{width} \cdot \text{height}$.

$8 = (x - 4)(x - 4)(2)$	$V = lwh$
$4 = (x - 4)(x - 4)$	Divide both sides by 2.
$4 = x^2 - 8x + 16$	Simplify.
$x^2 - 8x + 12 = 0$	Subtract 4 from both sides.
$(x - 6)(x - 2) = 0$	Factor.
$x - 6 = 0 \quad \text{or} \quad x - 2 = 0$	Zero Product Property
$x = 6 \quad \text{or} \quad x = 2$	Addition Property of Equality

Can you cut 2-inch squares from each corner of a 2-inch square?

The value of x cannot equal 2. The piece of cardboard should be 6 inches by 6 inches. ❖

RETEACHING
the
lesson

Hands-On Strategies
Separate the class into small groups. Half of the small groups are to use factoring or completing the square to solve equations. The other half are to use a graphics calculator. After the groups have solved Exercises 9–11, allow for a class discussion on problems or limitations of the methods the groups were told to use. Encourage students to use a variety of approaches to solve quadratic equations.

Try This

$x = 2$ or $x = 6$

Alternate Example 4

A box with a square base and no top is to be made by cutting out 4-inch squares from each corner of a square piece of cardboard and folding up the sides. The volume of the box is to be 36 cubic inches. What size should the piece of cardboard be? [**11 inches by 11 inches**]

TEACHING *tip*

Be sure students understand why $x - 4$ is the length of each side and why 2 is established as the height of the box.

Aongoing
SSESSMENT

No, you cannot cut 2-inch squares from each corner of a 2-inch square.

EXAMPLE 5

Alternate Example 5

Find the points where the graph of $y = x + 1$ intersects the graph of $y = x^2 - 4x + 5$ by factoring.   [$(1, 2)$ and $(4, 5)$]

Find the points where the graph of $y = x - 3$ intersects the graph of $y = x^2 - 10x + 21$.

Solution A ➤

Graph the equations, and find the points of intersection.

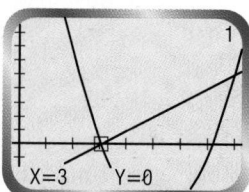

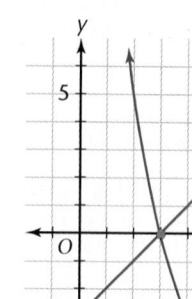

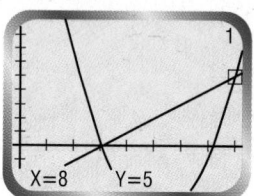

The graphs intersect at $(8, 5)$ and $(3, 0)$.

Solution B ➤

Solve the system. $\begin{cases} x^2 - 10x + 21 = y \\ y = x - 3 \end{cases}$

First, substitute $x - 3$ (from the second equation) for y (in the first equation) and simplify.

$x^2 - 10x + 21 = x - 3$
$x^2 - 11x + 24 = 0$

Then solve the new quadratic by factoring.

$(x - 8)(x - 3) = 0$
$x - 8 = 0 \quad \text{or} \quad x - 3 = 0$
$\qquad x = 8 \quad \text{or} \qquad x = 3$

Substitute each value for x into the equation $y = x - 3$.

$y = x - 3 \qquad\quad y = x - 3$
$y = 8 - 3 \qquad\quad y = 3 - 3$
$y = 5 \qquad\qquad\ y = 0$

The graphs intersect at $(8, 5)$ and $(3, 0)$. ❖

Aongoing SSESSMENT

Try This

$(3, -2)$ and $(-1, 6)$

Try This   Find the points where the graph of $y = -2x + 4$ intersects the graph of $y = x^2 - 4x + 1$.

EXERCISES & PROBLEMS

Communicate

1. How are *x*-intercepts related to the zeros of a function?

2. Explain how to find the zeros of $f(x) = x^2 - 6x + 8$ by completing the square.

3. Describe how to solve $x^2 - 2x = 15$ by completing the square.

4. Discuss how to solve $x^2 + 10x = 24$ by factoring.

Which method would you choose to solve the following problems? Explain why.

5. $x^2 + 12x + 36 = 0$ **6.** $x^2 - 8x - 9 = 0$

Graph $f(x) = x^2 - 7x + 12$.

7. Explain how to find the value of *x* when $f(x)$ is 2.

8. Describe how to find where the graph of $y = x - 1$ intersects the graph of $y = x^2 - 3x + 2$.

ASSESS

Selected Answers
Odd-numbered Exercises 9–61

Assignment Guide
Core 1–11, 12–28 even, 30–48, 52–64

Core Two-Year 1–61

Technology
Students can use graphics calculators to check Exercises 9–45 and 51.

Error Analysis
When factoring, students may not set the equation equal to zero. They let $x^2 + 18x = 19$ and attempt to solve $x(x + 18) = 19$. This gives the erroneous values $x = 19$ and $x = 1$. Explain the reason for setting the quadratic equal to zero to emphasize the proper procedure for solving these equations.

Practice & Apply

Find the *x*-intercepts or zeros of each function. Graph to check.

9. $y = x^2 - 2x - 8$ 4, −2   **10.** $y = x^2 + 6x + 5$ −1, −5   **11.** $y = x^2 - 4x + 4$ 2

Find the solution of each equation by factoring.

12. $x^2 - 2x - 3 = 0$ −1, 3   **13.** $x^2 + 4x - 5 = 0$ 1, −5   **14.** $x^2 + 7x + 12 = 0$ −3, −4

15. $x^2 - 10x + 24 = 0$ 4, 6   **16.** $x^2 - 3x - 10 = 0$ 5, −2   **17.** $x^2 - 8x + 15 = 0$ 3, 5

18. $x^2 - 6x + 9 = 0$ 3   **19.** $x^2 + 10x + 25 = 0$ −5   **20.** $x^2 - 2x + 1 = 0$ 1

Find the solution of each equation by completing the square.

21. $x^2 - 2x - 15 = 0$ 5, −3   **22.** $x^2 + 4x - 5 = 0$ 1, −5   **23.** $x^2 - x - 20 = 0$ 5, −4

24. $x^2 + 2x - 8 = 0$ 2, −4   **25.** $x^2 - 4x - 12 = 0$ 6, −2   **26.** $x^2 + x - 6 = 0$ 2, −3

27. $x^2 + 2x - 5 = 0$ **28.** $x^2 + 4x - 1 = 0$ **29.** $x^2 + 8x + 13 = 0$

27. $\approx 1.45, -3.45$

28. $\approx 0.24, -4.24$

29. $\approx -2.27, -5.73$

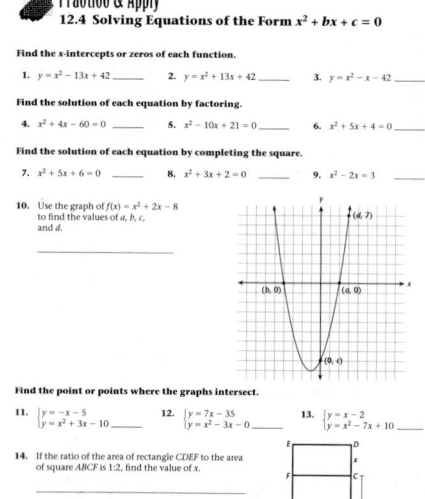

Alternative
ASSESSMENT

Portfolio Assessment

Have students include an example of each different method they have learned for solving quadratic equations. Tell students to include in their portfolio for future use any personal notes about the methods.

Solve each equation by factoring or completing the square.

30. $b^2 + 10b = 0$ **31.** $r^2 - 10r + 24 = 0$ **32.** $x^2 - x - 3 = 0$ **33.** $x^2 + 4x - 12 = 0$

34. $x^2 + 3x - 5 = 0$ **35.** $x^2 - 3x - 1 = 0$ **36.** $x^2 - 5x - 3 = 0$ **37.** $x^2 - x = 0$

Refer to the equation $f(x) = x^2 - 4x + 3$ to complete Exercises 38-41. Graph the equation to find a value for x when $f(x)$ is the following value. Check by substituting the given value of $f(x)$.

38. $f(x) = 3$ 0, 4 **39.** $f(x) = 8$ 5, -1 **40.** $f(x) = -1$ 2 **41.** $f(x) = 15$ 6, -2

Find the point or points where the graphs intersect. Graph to check.

42. $\begin{cases} y = 9 \\ y = x^2 \end{cases}$ **43.** $\begin{cases} y = 4 \\ y = x^2 - 2x + 1 \end{cases}$ **44.** $\begin{cases} y = x - 1 \\ y = x^2 - 3x + 3 \end{cases}$ **45.** $\begin{cases} y = x + 3 \\ y = x^2 - 4x + 3 \end{cases}$

$(-3, 9), (3, 9)$ $(3, 4), (-1, 4)$ $(2, 1)$ $(0, 3), (5, 8)$

46. Find two consecutive even integers whose product is 224.
14 and 16; -14 and -16

47. 🔷 **Geometry** The length of a rectangle is 4 yards longer than the width. Find the length and the width.
(10 by 6) yards

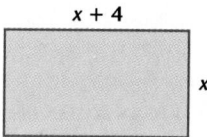

$x + 4$

x

$A = 60$ square yards

48. Photography A perimeter of a photograph is 80 centimeters. Find the dimensions of the photo for a frame if the area is 396 square centimeters.
(22 by 18) cm

A marching band uses a rectangular formation with 8 columns and 10 rows.

49. An equal number of columns and rows were added to the formation when 40 new members joined at the start of a new school year. How many rows and columns were added? 2 rows, 2 columns

30. $0, -10$

31. $4, 6$

32. $\approx 2.30, \approx -1.30$

33. $2, -6$

34. $\approx 1.19, \approx -4.19$

35. $\approx 3.30, \approx -0.30$

36. $\approx 5.54, \approx -0.54$

37. $0, 1$

50. The difference of two numbers is 3. What are the numbers if the sum of their squares is 117?

9 and 6; −9 and −6

51. Physics A boat fires an emergency flare that travels upward at an initial velocity of 25 meters per second. To find the height in meters, h, of the flare at t seconds, use the formula $h = 25t − 5t^2$. After how many seconds will the flare be at a height of 10 meters? Round answers to the nearest tenth.

0.4 seconds and 4.6 seconds

Look Back

Write each system in its matrix equation form.   [Lesson 7.5]

52. $\begin{cases} x − 2y = 1 \\ 4x + 2y = −1 \end{cases}$ **53.** $\begin{cases} 5x − 2y = 11 \\ 3x + 5y = 19 \end{cases}$

Statistics Find the mean and the mode for the following sets of numbers.   [Lesson 8.3]

54. 10, 15, 20, 36, 10, 25 ≈ 19.3; 10

55. 8, 10.5, 30, 15, 12, 10.5 ≈ 14.3; 10.5

Write each number in decimal form.   [Lesson 10.4]

56. $4 \cdot 10^4$ 40,000   **57.** $6.5 \cdot 10^7$ 65,000,000   **58.** $9.6 \cdot 10^{−5}$ 0.000096

Write each polynomial in factored form.   [Lesson 11.2]

59. $4b^2 + 6b − 36$ **60.** $6w(w^2 − w)$ **61.** $8p^4 − 16p$
$2(2b^2 + 3b − 18)$ $6w^2(w − 1)$ $8p(p^3 − 2)$

Look Beyond

62. Factor $2x^2 + x − 6$. $(2x − 3)(x + 2)$

63. Earlier you learned a pattern that simplifies the process of determining the powers of binomials. Can you find a pattern to determine the powers of a trinomial, $(a + b + c)^n$? What are the first 3 powers?

64. Imagine that someone invented a number for $\sqrt{−1}$ and called it i. Consider i^0 to be 1. If $i = \sqrt{−1}$, then what is i^2 or i^3? What is i^6? Can you find a pattern that will let you predict the value of i^{92}?

52. $\begin{bmatrix} 1 & −2 \\ 4 & 2 \end{bmatrix} \begin{bmatrix} x \\ y \end{bmatrix} = \begin{bmatrix} 1 \\ −1 \end{bmatrix}$

53. $\begin{bmatrix} 5 & −2 \\ 3 & 5 \end{bmatrix} \begin{bmatrix} x \\ y \end{bmatrix} = \begin{bmatrix} 11 \\ 19 \end{bmatrix}$

63. $(a + b + c)^1 = a + b + c$
$(a + b + c)^2 = a^2 + 2a(b + c) + (b + c)^2$
$(a + b + c)^3 = a^3 + 3a^2(b + c) + 3a(b + c)^2 + (b + c)^3$

$(a + b + c)^4 = a^4 + 4a^3(b + c) + 6a^2(b + c)^2 + 4a(b + c)^3 + (b + c)^4$
$(a + b + c)^n = a^n + na^{n−1}(b + c) + \dfrac{n(n − 1)}{2} a^{n−2}(b + c)^2 + \cdots + na(b + c)^{n−1} + (b + c)^n$

64. $i^2 = −1, i^3 = −i, i^6 = −1, i^{92} = 1$

Look Beyond

Exercise 62 involves factoring quadratic expressions when the coefficient of the x^2 term is not one. Exercise 63 encourages students to extend their study of patterns in mathematics. Exercise 64 foreshadows the study of imaginary numbers.

PREPARE

Objectives

- Use the quadratic formula to find solutions to quadratic equations.
- Use the quadratic formula to find the zeros of a quadratic function.
- Evaluate the discriminant to determine how many real roots the quadratic has and whether the quadratic can be factored over the integers.

RESOURCES

- Practice Master   **12.5**
- Enrichment Master   **12.5**
- Technology Master   **12.5**
- Lesson Activity Master   **12.5**
- Quiz   **12.5**
- Spanish Resources   **12.5**

Assessing Prior Knowledge

Solve by factoring. If not factorable, write NF.

1. $x^2 - 5x - 6$
   $[(x - 6)(x + 1)]$
2. $x^2 + 5x - 6$
   $[(x + 6)(x - 1)]$
3. $x^2 + 5x + 6$
   $[(x + 3)(x + 2)]$
4. $x^2 - 5x + 16$ **[NF]**

TEACH

 Ask the students whether all quadratic functions they have encountered can be easily solved by factoring. Tell students that there is a method that works for all quadratic equations.

In the Shot-put event in track and field, a shot is tossed into the air from a position about 6 feet above the ground.

why *An object thrown in the air can be modeled by a quadratic function. The zeros of the quadratic function tell when the object is at ground level. When you derive the quadratic formula by completing the square, the formula will provide a way to find those zeros.*

If the shot is tossed with an initial horizontal and vertical velocity of 31 feet per second, the time that the shot travels can be found by solving the quadratic equation $-16t^2 + 31t + 6.375 = 0$. You can then use t to solve for the horizontal distance, d, in the formula $d = 31t$.

Cultural Connection: Asia Since early Babylonian times scholars have known how to solve the standard quadratic equation $ax^2 + bx + c = 0$ by using a formula.

To see how this quadratic formula can be derived, follow the steps as the equations $2x^2 + 5x + 1 = 0$ and $ax^2 + bx + c = 0$ are solved.

 **ALTERNATIVE teaching strategy**

Technology Have students use their graphics calculators to solve each of the equations in Examples 1 and 2. For the equation in Example 1, part b, tell students to round their answer to the nearest tenth. [**1.6, 4.4**]

ENRICHMENT The golden ratio produces interesting numbers. To find the golden ratio, write a proportion. Begin with the ratio of 1 to n. Set that ratio equal to the ratio of n to 1 minus n. Solve this proportion for n. The result is the value of the golden ratio. Since the value is an irrational number, round to the nearest hundredth for convenience.
$$\left[\frac{1}{n} = \frac{n}{1 - n}; n^2 = 1 - n; n \approx 0.62\right]$$

Divide both sides of each equation by the coefficient of x^2.

$$2x^2 + 5x + 1 = 0 \qquad\qquad ax^2 + bx + c = 0$$

$$x^2 + \frac{5}{2}x + \frac{1}{2} = 0 \qquad\qquad x^2 + \frac{b}{a}x + \frac{c}{a} = 0$$

Subtract the constant term from both sides of each equation.

$$x^2 + \frac{5}{2}x = -\frac{1}{2} \qquad\qquad x^2 + \frac{b}{a}x = -\frac{c}{a}$$

Complete the square.

$$x^2 + \frac{5}{2}x + \left(\frac{5}{4}\right)^2 = -\frac{1}{2} + \left(\frac{5}{4}\right)^2 \qquad x^2 + \frac{b}{a}x + \left(\frac{b}{2a}\right)^2 = -\frac{c}{a} + \left(\frac{b}{2a}\right)^2$$

$$\left(x + \frac{5}{4}\right)^2 = -\frac{1}{2} + \frac{25}{16} \qquad\qquad \left(x + \frac{b}{2a}\right)^2 = -\frac{c}{a} + \frac{b^2}{4a^2}$$

Work with the right side of each equation to simplify.

$$\left(x + \frac{5}{4}\right)^2 = -\frac{8 \cdot 1}{8 \cdot 2} + \frac{25}{16} \qquad\qquad \left(x + \frac{b}{2a}\right)^2 = -\frac{4a \cdot c}{4a \cdot a} + \frac{b^2}{4a^2}$$

$$\left(x + \frac{5}{4}\right)^2 = \frac{25 - 8}{16} \text{ or } \frac{17}{16} \qquad\qquad \left(x + \frac{b}{2a}\right)^2 = \frac{b^2 - 4ac}{4a^2}$$

Use $x^2 = k$ to solve each equation.

$$x + \frac{5}{4} = \pm\sqrt{\frac{17}{16}} \qquad\qquad x + \frac{b}{2a} = \pm\sqrt{\frac{b^2 - 4ac}{4a^2}}$$

$$x + \frac{5}{4} = \pm\frac{\sqrt{17}}{\sqrt{16}} \qquad\qquad x + \frac{b}{2a} = \pm\frac{\sqrt{b^2 - 4ac}}{\sqrt{4a^2}}$$

$$x = -\frac{5}{4} \pm \frac{\sqrt{17}}{4} \qquad\qquad x = -\frac{b}{2a} \pm \frac{\sqrt{b^2 - 4ac}}{2a}$$

$$x = \frac{-5 \pm \sqrt{17}}{4} \qquad\qquad x = \frac{-b \pm \sqrt{b^2 - 4ac}}{2a}$$

You have derived the quadratic formula.

THE QUADRATIC FORMULA

The solutions of the quadratic equation $ax^2 + bx + c = 0$ are

$$\frac{-b \pm \sqrt{b^2 - 4ac}}{2a}.$$

Notice that if you substitute the values $a = 2$, $b = 5$, and $c = 1$ from $2x^2 + 5x + 1 = 0$ into the formula $x = \dfrac{-b \pm \sqrt{b^2 - 4ac}}{2a}$, you get the following result.

$$x = \frac{-5 \pm \sqrt{25 - (4 \cdot 2 \cdot 1)}}{2 \cdot 2} = \frac{-5 \pm \sqrt{17}}{4}$$

EXAMPLE 1

Alternate Example 1

Use the quadratic formula to find the solutions to
$x^2 - 3x - 28 = 0$.

$$\left[x = \frac{3 \pm \sqrt{121}}{2} ; \right.$$

$$\left. x = 7 \text{ or } x = -4 \right]$$

Aongoing ASSESSMENT

Try This

$x = -4 \text{ or } 3$

CRITICAL Thinking

When solving quadratic equations, it is best to use a calculator and the quadratic formula when working with decimals and fractions. Such equations are not easily factored.

Alternate Example 2

Find the zeros of
$h(x) = x^2 - 4 - 5x$.

$$\left[x = \frac{2 \pm \sqrt{41}}{2} ; \right.$$

$$\left. x \approx 5.7 \text{ or } -0.7 \right]$$

Aongoing ASSESSMENT

Write the equation as $3n^2 - 2n - 2 = 0$, and substitute in the quadratic formula.

EXAMPLE 1

Use the quadratic formula to solve $x^2 - 10 + 3x = 0$.

Solution ➤

Rewrite $x^2 - 10 + 3x = 0$ in the form $x^2 + 3x - 10 = 0$. Notice that $a = 1$, $b = 3$, and $c = -10$. Substitute and simplify.

$$x = \frac{-3 \pm \sqrt{(3)^2 - 4(1)(-10)}}{2(1)}$$

$$x = \frac{-3 \pm \sqrt{9 + 40}}{2}$$

$$x = \frac{-3 + 7}{2} \quad \text{or} \quad x = \frac{-3 - 7}{2}$$

$$x = \frac{4}{2} = 2 \quad \text{or} \quad x = \frac{-10}{2} = -5$$

The solutions are 2 and -5. ❖

Try This Use the quadratic formula to solve $x^2 + x - 12 = 0$.

You now have three different methods for solving a quadratic equation.

1. Factoring

2. Completing the Square

3. The Quadratic Formula

CRITICAL Thinking Use all three methods to solve $2x^2 - 7x - 4 = 0$. Which method do you think is best for solving real applications? When is factoring the best way to solve a quadratic equation?

EXAMPLE 2

Find the zeros of $h(x) = 3x^2 + 2x - 4$.

Solution ➤

Use the quadratic formula, where $a = 3$, $b = 2$, and $c = -4$.

$$x = \frac{-2 \pm \sqrt{4 - (-48)}}{6}$$

$$x = \frac{-2 \pm \sqrt{52}}{6}$$

A calculator shows that $x \approx 0.87$ or $x \approx -1.54$. Check by substitution. ❖

How would you use the quadratic formula to solve $3n^2 - 2n = 2$?

You can now solve the quadratic equation $-16t^2 + 31t + 6.375 = 0$ that models the shot-put problem.

Let $a = -16$, $b = 31$ and $c = 6.375$.

$$t = \frac{-31 \pm \sqrt{(31)^2 - (4)(-16)(6.375)}}{2(-16)} = \frac{-31 \pm \sqrt{961 + 408}}{-32}$$

$t = -0.1875$ or $t = 2.125$ and $d = 31t$, so
$d = 31(2.125) = 65.875$

The shot travels a horizontal distance of about 66 feet.

Exploration *The Discriminant*

Examine each graph.

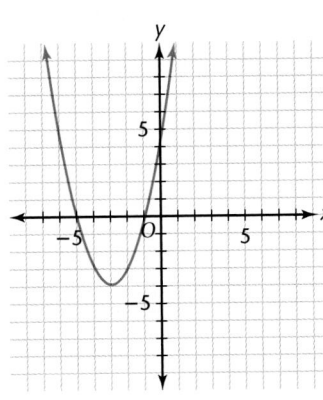

$f(x) = x^2 + 6x + 5$

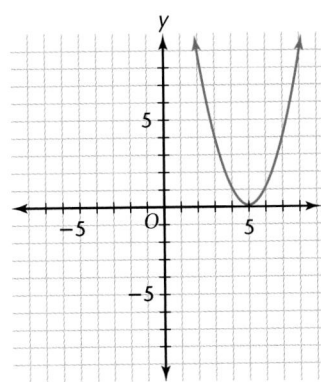

$g(x) = x^2 - 10x + 25$

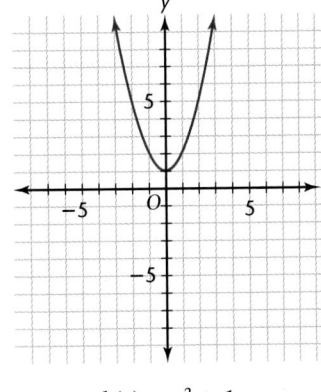

$h(x) = x^2 + 1$

Each expression is written in the form $ax^2 + bx + c$.

 Find the values of a, b, and c, and complete the chart. The expression $b^2 - 4ac$ is called the **discriminant** of the quadratic formula.

Function	a	b	c	b² – 4ac	Number of x-intercepts
f	?	?	?	?	?
g	?	?	?	?	?
h	?	?	?	?	?

 Explain the role of the discriminant in determining the number of solutions to a quadratic equation. ❖

Exploration Notes

Ask students if they have ever tried to solve a puzzle only to have someone tell them that it is not possible. Ask why it would be nice to know that ahead of time. [**Usually to avoid wasting time**] Tell students that in this exploration they will learn to tell whether there are real solutions to quadratic equations.

Use Transparency ▶ **59**

TEACHING *tip*

Technology Allow students to use their graphics calculators to test other equations to see if the results are the same.

Aongoing **SSESSMENT**

2. If the discriminant is greater than zero, there are two solutions. If the discriminant is zero, there is one solution (or two solutions that are the same). If the discriminant is negative, there are no real roots because there are no zeros of the equation—the graph does not cross the x-axis.

Alternate Example 3

What does the discriminant tell you about the solution to each of the following equations?

a. $6x^2 - 4x - 5 = 0$ [**The equation has two distinct solutions.**]

b. $-2x^2 = -8x + 8$ [**The equation has exactly one real solution.**]

c. $6x^2 - 9x = -4$ [**The discriminant is negative; there are no real solutions.**]

Alternate Example 4

Factor $12x^2 + 8x - 15$.
[**$(2x + 3)(6x - 5)$**]

Practice Master

Practice & Apply
12.5 The Quadratic Formula

Examine each graph. Then complete the chart.

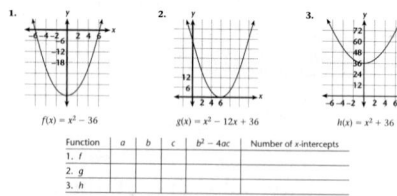

$f(x) = x^2 - 36$ $g(x) = x^2 - 12x + 36$ $h(x) = x^2 + 36$

Function	a	b	c	$b^2 - 4ac$	Number of x-intercepts
1. f					
2. g					
3. h					

Find the value of the discriminant for each quadratic equation. Describe the number of solutions.

4. $4x^2 + 3x + 1 = 0$ _____
5. $x^2 + 5x + 1 = 0$ _____
6. $x^2 - 4x + 4 = 0$ _____
7. $3 - 4x + 2x^2 = 0$ _____

Use the quadratic formula to solve each equation. Give answers to the nearest hundredth when necessary.

8. $3x^2 - 4x - 2 = 0$ _____
9. $2x^2 - 5x - 4 = 0$ _____
10. $3x^2 - 8 = 0$ _____
11. $6x^2 - 5x + 1 = 0$ _____
12. $x^2 - 8x + 8 = 0$ _____
13. $x^2 - 5 = 0$ _____

Choose any method to solve the following quadratic equations. Give answers to the nearest hundredth when necessary.

14. $3x^2 + 2x - 5 = 0$ _____
15. $x^2 + 3x + 1 = 0$ _____
16. $2x^2 - 10x + 8 = 0$ _____
17. $x^2 + 5x + 4 = 0$ _____

EXAMPLE 3

What does the discriminant tell you about the solution to each of the following equations?

Ⓐ $3x^2 - 2x + 1 = 0$ Ⓑ $4x = 4x^2 + 1$ Ⓒ $2x^3 + 3x + 1 = 0$

Solution ➤

Ⓐ Find $b^2 - 4ac$ where $a = 3$, $b = -2$, and $c = 1$. The discriminant is $(-2)^2 - 4 \cdot 3 \cdot 1 = 4 - 12 = -8$. If the discriminant is negative, there are no x-intercepts, and $3x^2 - 2x + 1 = 0$ has no real solutions.

Ⓑ Rewrite the equation in standard form to find a, b, and c. If $4x^2 - 4x + 1 = 0$, then $a = 4$, $b = -4$, and $c = 1$. The discriminant is $(-4)^2 - 4 \cdot 4 \cdot 1 = 16 - 16 = 0$. If the discriminant equals 0, there is exactly one real solution, and $4x = 4x^2 + 1$ has one real solution.

Ⓒ The equation is not a quadratic. There is no discriminant. ❖

When the discriminant is a perfect square, the quadratic equation can be factored over the integers. You can use the quadratic formula to find the factors.

EXAMPLE 4

Factor $6x^2 + 23x + 20$.

Solution ➤

Evaluate the discriminant.

$$b^2 - 4ac = (23)^2 - 4(6)(20) = 49$$ which is a perfect square.

Find the solution to $6x^2 + 23x + 20 = 0$.

$$x = \frac{-b \pm \sqrt{b^2 - 4ac}}{2a} = \frac{-23 \pm \sqrt{49}}{12} = \frac{-23 \pm 7}{12}$$

Simplify.

$$x = -\frac{30}{12} = -\frac{5}{2} \text{ or } x = -\frac{16}{12} = -\frac{4}{3}$$

Work backward to find the factors.

$x = -\frac{5}{2}$ or	$x = -\frac{4}{3}$	Given
$2x = -5$ or	$3x = -4$	Multiplication Property of Equality
$2x + 5 = 0$ or $3x + 4 = 0$		Addition Property of Equality

The factors of $6x^2 + 23x + 20$ are $2x + 5$ and $3x + 4$. Check by multiplying the factors. ❖

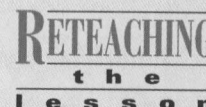

RETEACHING the lesson

Using Algorithms Display the quadratic formula on the chalkboard or overhead. Help students understand how to complete each step:

1. Identify the values of a, b, and c. Be careful to identify the signs of each value.

2. Substitute values in the quadratic formula.

3. Simplify the numerical expression, being careful to use the signs of the variables correctly.

4. Identify both solutions, being sure to check each in the original equation.

EXERCISES & PROBLEMS

Communicate

1. Discuss how to identify the coefficients a, b, and c in the equation $2y^2 + 3y = 7$ in order to use the quadratic formula. Why is the standard form of the equation important?

2. How do you find the discriminant for $2y^2 + 3y - 7 = 0$?

Discuss how to use the discriminant to determine the number of solutions to each equation.

3. $2x^2 - x - 2 = 0$ 4. $x^2 - 2x = -7$

5. Describe how to use the quadratic formula to find the solution to $4y^2 + 12y = 7$.

6. If the solutions to $4x^2 + 12x + 5 = 0$ are $-\frac{1}{2}$ and $-\frac{5}{2}$, discuss how to work backward to find the factors of $4x^2 + 12x + 5$.

ASSESS

Selected Answers
Odd-numbered Exercises 7–57

Assignment Guide
Core 1–6, 8–44 even, 46–59

Core Two-Year 1–58

Technology
A graphics calculator can be used to check Exercises 22–36 and to solve or check Exercises 37–47.

Error Analysis
Students may not correctly substitute in the quadratic formula because the given equation is not written in the form $ax^2 + bx + c$. Caution students to substitute correct values in the quadratic formula and to simplify the expression correctly.

Practice & Apply

Identify a, b, and c for each quadratic equation.

7. $x^2 - 4x - 5 = 0$ 8. $r^2 - 5r - 4 = 0$ 9. $5n^2 - 2n - 1 = 0$

10. $-p^2 - 3p + 7 = 0$ 11. $3t^2 - 45 = 0$ 12. $-3m + 7 - m^2 = 0$

Find the value of the discriminant for each quadratic equation.

13. $x^2 - x - 3 = 0$ 13   14. $x^2 + 2x - 8 = 0$ 36   15. $x^2 + 8x + 13 = 0$ 12

16. $z^2 + 4z - 21 = 0$ 100   17. $2y^2 - 8y - 8 = 0$ 128   18. $8y^2 - 2 = 0$ 64

19. $4k^2 - 3k = 5$ 89   20. $3x^2 - x + 4 = 0$ -47 21. $4x^2 - 12x + 9 = 0$ 0

Use the quadratic formula to solve each equation. Give answers to the nearest hundredth when necessary. Check by substitution.

22. $a^2 - 4a - 21 = 0$ 7, -3 23. $t^2 + 6t - 16 = 0$ 2, -8 24. $m^2 + 4m - 5 = 0$ 1, -5

25. $w^2 - 4w = 0$ 0, 4   26. $x^2 + 9x = 0$ 0, -9 27. $x^2 - 9 = 0$ 3, -3

28. $x^2 + 2x = 0$ 0, -2 29. $3m^2 = 2m + 1$ 1, -0.33 30. $-x^2 + 6x - 9 = 0$ 3

31. $2r^2 - r - 3 = 0$ 1.5, -1 32. $2w^2 - 5w + 3 = 0$ 1, 1.5   33. $x^2 + 6x + 3 = 0$

34. $10y^2 + 7y = 12$ 35. $3x^2 - 2x - 7 = 0$ 36. $8y^2 + 7y - 2 = 0$

Technology Master

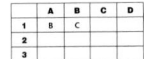

Technology
12.5 Seeing the Power of the Quadratic Formula

In Lesson 12.5, you learned about the quadratic formula. It allows you to find x when you know the values of a, b, and c in the equation $ax^2 + bx + c = 0$.

If $ax^2 + bx + c = 0$, $x = \frac{-b \pm \sqrt{b^2 - 4ac}}{2a}$.

Suppose that you need to solve a large number of quadratic equations. A calculator is helpful. However, the power of the quadratic formula is evident when you use it in a spreadsheet.

Consider $x^2 + bx + c = 0$. The spreadsheet shown will give the real roots of the equation when b and c are specified. In the spreadsheet, enter the values of b and c in cells A2 and B2. Cell C2 contains the formula $(-A2 + SQRT(A2^2 - 4*B2))/2$ and cell D2 contains the formula $(-A2 - SQRT(A2^2 - 4*B2))/2$.

	A	B	C	D
1	B	C		
2				
3				

Use the spreadsheet shown to solve $x^2 + bx + c = 0$ for each pair of values of b and c.

1. $b = 2$ and $c = -2$ 2. $b = 2$ and $c = -1$ 3. $b = 2$ and $c = 1$

4. $b = 2$ and $c = 2$ 5. $b = 2$ and $c = 3$ 6. $b = 2$ and $c = 4$

7. Suppose that $b = 2$. At what value of c from Exercises 1–6 does the spreadsheet stop reporting real roots?

8. Suppose that $x^2 + bx + c = 0$. Based on the results of Exercises 1–6, write a condition placed on b and c that will tell when the roots are real numbers.

---

7. $a = 1, b = -4, c = -5$
8. $a = 1, b = -5, c = -4$
9. $a = 5, b = -2, c = -1$
10. $a = -1, b = -3, c = 7$
11. $a = 3, b = 0, c = -45$
12. $a = -1, b = -3, c = 7$

33. $-0.55, -5.45$
34. $0.8, -1.5$
35. $1.90, -1.23$
36. $0.23, -1.10$

Authentic Assessment

Have students fold a sheet of notebook paper in half vertically. Give them a quadratic equation and tell students to solve it by using the quadratic formula. Students are to solve the equation on the left half of the paper and explain their steps and reasoning on the right half.

37. $2, -6$
38. $3.61, -3.61$
39. $0.5, -3$
40. $-2, -5$
41. $4, -1.5$
42. $1.22, -0.55$
43. $2.70, -0.37$
44. $-0.5, -0.75$
45. $0.44, -3.77$
48. Cost is $51,939 in the year 2000; cost was $25,000 in 1987. Use 87.6 for x for a cost of $25,001.24.

Look Beyond

Exercise 59 foreshadows the study of quadratic inequalities.

Choose any method to solve the following quadratic equations. Report answers to the nearest hundredth when necessary.

37. $x^2 + 4x - 12 = 0$
38. $x^2 - 13 = 0$
39. $2x^2 + 5x - 3 = 0$
40. $x^2 + 7x + 10 = 0$
41. $2a^2 - 5a - 12 = 0$
42. $3x^2 - 2x - 2 = 0$
43. $3x^2 - 7x - 3 = 0$
44. $8x^2 + 10x + 3 = 0$
45. $3x^2 + 10x - 5 = 0$

46. **Accounting** To approximate profits per day for her business, Mrs. Haag uses the following formula.

$$p = -x^2 + 50x - 350$$

The profit, p, is derived from selling x cases of decorator napkins. How many cases of napkins must she sell to make a maximum profit? Find the maximum profit. 25 cases, $275

47. The number of seats in each row of a theater is 16 fewer than the number of rows. How many seats are in each row of a 1161-seat theater?
27 seats

48. **Portfolio Activity** Complete the problem in the portfolio activity on page 547.

Look Back

49. **Probability** Find the probability of guessing at least 60% of the correct answers on a 5-question multiple-choice quiz in which each item has 4 possible responses. **[Lesson 8.2]** 0.1035

Write each of the following numbers in scientific notation. [Lesson 10.4]

50. 3,000,000
3×10^6
51. 67,000
6.7×10^4
52. 0.006
6×10^{-3}
53. 0.0000009
9×10^{-7}

Graph. [Lesson 10.5]

54. $y = 2^x + 4$
55. $y = 2 \cdot 3^x$

56. **Geometry** Write the expression for the perimeter of the rectangle. **[Lesson 11.2]**
$4x + 28$

$x + 9$

$x + 5$

Multiply. [Lesson 11.3]

57. $(x - 5)(x + 3)$
$x^2 - 2x - 15$
58. $(2x + 4)(2x - 2)$
$4x^2 + 4x - 8$

Look Beyond

59. From what you know about graphing linear inequalities, make a conjecture about how you might graph $x^2 + 3x + 2 < 0$.

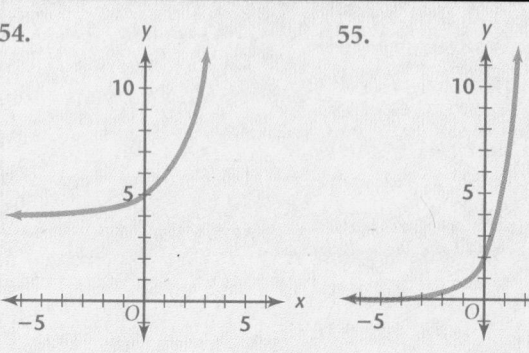

54.

55.

59. First graph the parabola $f(x) = x^2 + 3x + 2$. Determine where $f(x) < 0$.

LESSON 12.6 Graphing Quadratic Inequalities

 *Not all problems involving quadratics use equations. Sometimes you will need to solve an inequality to get an answer. Solving a quadratic inequality includes techniques you already know with one more important step.*

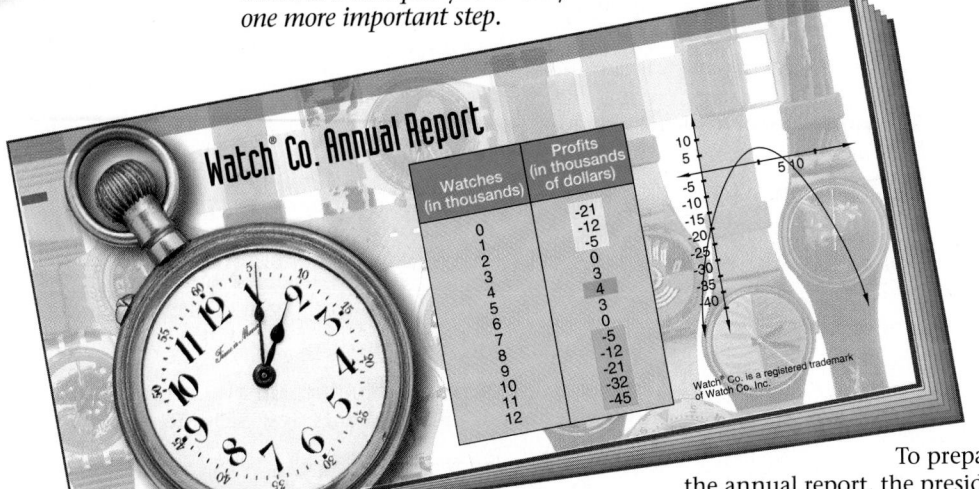

Watch® Co. Annual Report

Watches (in thousands)	Profits (in thousands of dollars)
0	-21
1	-12
2	-5
3	0
4	3
5	4
6	3
7	0
8	-5
9	-12
10	-21
11	-32
12	-45

Watch® Co. is a registered trademark of Watch Co. Inc.

To prepare the annual report, the president of a watch company gathers data about the number of watches produced per week and the corresponding profit. He uses the model $p = -x^2 + 10x - 21$ in which p represents profit and x represents the number of watches produced per week.

Exploration — Production and Profit

MAXIMUM MINIMUM
Connection

Use the graph of the profit function and the table of values to answer each question.

1 At what levels of production is the profit 0?

2 At what levels of production is the profit greater than 0?

3 At what levels of production is the profit less than 0?

4 How many watches should be produced to maximize profit?

5 According to the consultant's model, what is the maximum profit that the company can make? ❖

In the Exploration, you solved a quadratic equation and two quadratic inequalities.

$-x^2 + 10x - 21 = 0$	$-x^2 + 10x - 21 > 0$	$-x^2 + 10x - 21 < 0$
quadratic equation	**quadratic inequality**	**quadratic inequality**

ALTERNATIVE teaching strategy

Using Cognitive Strategies Students can use the signs of factors to solve a quadratic inequality.

1. Solve the related equation by factoring, and locate the solutions on a number line.
2. Choose a number from each interval. Substitute that number in each factor, and record the sign of the result in a table.
3. Use the results of Step 2 to find the sign of the product of the factors. Examine the table for $x^2 - 2x - 15 < 0$.

	$x < -3$	$-3 < x < 5$	$x > 5$
$(x - 5)$	−	−	+
$(x + 3)$	−	+	+
$(x - 5)(x + 3)$	+	−	+

Since the polynomial is less than 0, the interval between -3 and 5 will be shaded.

Alternate Example 1

Solve $x^2 - x - 6 > 0$.
[Check students' graphs. All numbers less than -2 and all numbers greater than 3 should be shaded.]

TEACHING *tip*

Technology To duplicate the graph in the book, students should set the range for x at min: -18, max: 18, scale: 2.

Aongoing
SSESSMENT

The points -3 and 5 should be darkened only if the inequality includes the equality.

Aongoing
SSESSMENT

Try This

$x \leq -4$ or $x \geq 3$ Check students' graphs.

CRITICAL
Thinking

Test the interval between the two solutions. If it is a solution, shade that area. If not, shade the areas outside the solutions.

EXAMPLE 1

Solve $x^2 - 2x - 15 > 0$.

Graphics Calculator

Solution A ➤

A graphics calculator can be used to graph $Y = x^2 - 2x - 15$. Trace the graph to see where the values are greater than 0.

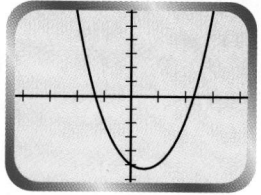

Solution B ➤

Use the x-axis as a number line.

1. Solve the related quadratic equation using the Zero Product Property.

$$x^2 - 2x - 15 = 0$$
$$(x - 5)(x + 3) = 0$$
$$x - 5 = 0 \quad \text{or} \quad x + 3 = 0$$
$$x = 5 \quad \text{or} \quad x = -3$$

2. Plot the solutions on an x-axis. When would you darken the points at -3 and 5?

3. The two solutions divide the x-axis into three intervals, to the left, between, and to the right of the two points. Test each interval.

$$\xleftarrow{\hspace{1cm}} \circ \hspace{1cm} \circ \xrightarrow{\hspace{1cm}}$$
$$-6 \quad -4 \quad -2 \quad 0 \quad 2 \quad 4 \quad 6$$

Choose a number from each interval. Substitute those numbers for x in the inequality.

$x^2 - 2x - 15 > 0$	$x^2 - 2x - 15 > 0$	$x^2 - 2x - 15 > 0$
$(-5)^2 - 2(-5) - 15 \overset{?}{=} 0$	$(0)^2 - 2(0) - 15 \overset{?}{=} 0$	$(6)^2 - 2(6) - 15 \overset{?}{=} 0$
$25 + 10 - 15 \overset{?}{=} 0$	$0 - 0 - 15 \overset{?}{=} 0$	$36 - 12 - 15 \overset{?}{=} 0$
$20 > 0$	$-15 > 0$	$9 > 0$
Solution interval	**No**	**Solution interval**

4. Graph the solutions.

$$\xleftarrow{\hspace{1cm}} \circ \hspace{1cm} \circ \xrightarrow{\hspace{1cm}}$$
$$-3 \quad 0 \quad 5$$
$$x < -3 \quad \text{or} \quad x > 5 \quad ❖$$

Try This Solve $x^2 + x - 12 \geq 0$.

CRITICAL
Thinking

Explain how you could tell which intervals should be shaded without testing numbers from each interval.

interdisciplinary
CONNECTION

Communications Have students investigate parabolic satellite dish antennas. In such antennas, all of the incoming communication waves are directed to a feed horn. The feed horn is located at the focus of the parabola.

EXAMPLE 2

Graph $y > -x^2 + 1$.

Solution ➤

1. Graph $y = -x^2 + 1$. Draw the parabola with a dashed line to show the solutions of $y = x^2 + 1$ are *not* on the graph.

2. To graph $y > -x^2 + 1$, shade the points above the parabola.

3. As a check, test points inside and outside the parabola.

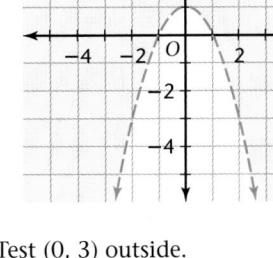

Test (0, 0) inside.	Test (0, 3) outside.
$y > -x^2 + 1$	$y > -x^2 + 1$
$0 > 0 + 1$?	$3 > 0 + 1$?
$0 > 1$?　No	$3 > 1$?　Yes

4. Shade the region outside the parabola.

The graph of $y > -x^2 + 1$ is the **set of points outside the parabola**. ❖

You can test points inside and outside the graph of $y = -x^2 + 1$ to show that the shaded region on each graph below represents the given set of points.

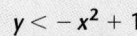

$y < -x^2 + 1$

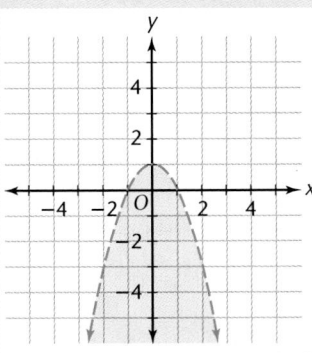

$y \geq -x^2 + 1$

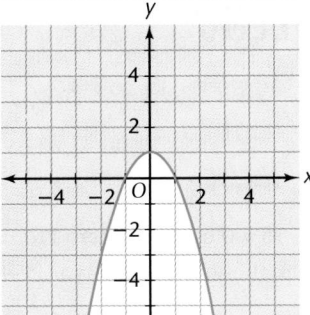

$y \leq -x^2 + 1$

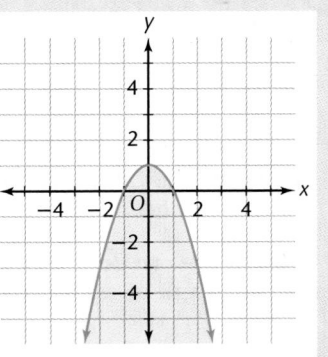

TEACHING *tip*

Remind students that when $\leq$ or $\geq$ is used in an equation, the equation itself is part of the solution and the line for the equation should be a solid line. If $<$ or $>$ is used, the equation is not part of the solution and the line should be dashed.

Alternate Example 2

Graph $y \leq x^2 - 5$. [**Check students' graphs. The section below the parabola (not including the origin) should be shaded.**]

TEACHING *tip*

Ask students why using the origin (0, 0) as a test point simplifies the process. Students should realize that any term containing an x then becomes zero.

ENRICHMENT Tell students that a quick way to check whether they have correctly found the roots of a quadratic equation is to use the sum and product of the roots. For $ax^2 - bx + c = 0$, where $a \neq 0$, the sum of the roots is $\dfrac{-b}{a}$ and the product of the roots is $\dfrac{c}{a}$.

INCLUSION strategies **Using Discussion** Have students orally discuss the steps involved in solving a quadratic inequality. Be sure students can explain in their own words how to decide which intervals of the graph, whether on a number line or on a coordinate grid, need to be shaded.

EXAMPLE 3

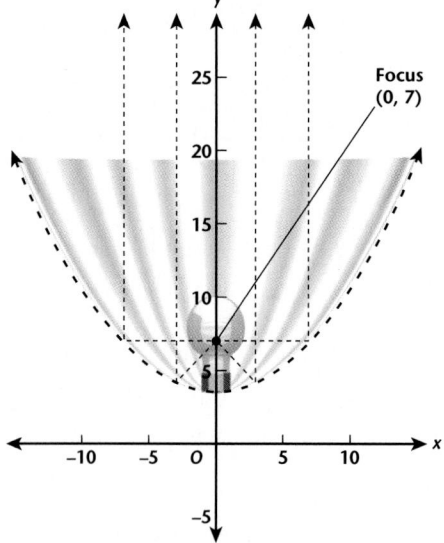

Focus
(0, 7)

A cross section of the parabolic light reflector at the right is described by the equation $y = \frac{1}{14} x^2 + \frac{7}{2}$. The bulb inside the reflector is located at the point (0, 7). Light bounces off the reflector in parallel rays. This quality allows flashlights to direct light in a narrow beam. Determine which of the inequalities indicate the region in which the light bulb is located.

Ⓐ $y > \frac{1}{14} x^2 + \frac{7}{2}$ **Ⓑ** $y < \frac{1}{14} x^2 + \frac{7}{2}$

Solution ➤

Ⓐ $y > \frac{1}{14} x^2 + \frac{7}{2}$ **Ⓑ** $y < \frac{1}{14} x^2 + \frac{7}{2}$

Test (0, 7).   Test (0, 7).

$7 > \left(\frac{1}{14}\right)(0) + \frac{7}{2}$ $7 < \left(\frac{1}{14}\right)(0) + \frac{7}{2}$

$7 > \frac{7}{2}$ Yes $7 < \frac{7}{2}$ No

The inequality $y > \frac{1}{14} x^2 + \frac{7}{2}$ contains the point of focus inside the parabola. ❖

EXERCISES & PROBLEMS

Communicate

1. Explain how to solve $x^2 - 9x + 18 > 0$ using the Zero Product Property.
2. Describe how to graph and check the solution intervals to Exercise 1 on a number line.

Explain how to determine if the boundary line is solid or dashed for the graph of each inequality.

3. $y < x^2 - 3$ 4. $y \leq x^2 - 3$ 5. $y \geq x^2 - 3$

Discuss how to graph each inequality. Explain how to decide whether to shade inside or outside the parabola for each solution.

6. $y < x^2 - 3$ 7. $y \leq x^2 - 3$ 8. $y \geq x^2 - 3$

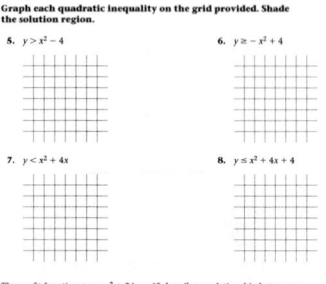

Practice & Apply

Solve each quadratic inequality using the Zero Product Property. Graph the solution on a number line.

9. $x^2 - 6x + 8 > 0$

10. $x^2 + 3x + 2 < 0$

11. $x^2 - 9x + 14 \leq 0$

12. $x^2 + 2x - 15 > 0$

13. $x^2 - 7x + 6 < 0$

14. $x^2 - 8x - 20 < 0$

15. $x^2 - 8x + 15 \leq 0$

16. $x^2 - 3x - 18 \leq 0$

17. $x^2 - 11x + 30 \geq 0$

18. $x^2 - 4x - 12 \geq 0$

19. $x^2 + 5x - 14 > 0$

20. $x^2 + 10x + 24 \geq 0$

Graph each quadratic inequality. Shade the solution region.

21. $y \geq x^2$

22. $y \geq x^2 + 4$

23. $y < -x^2$

24. $y > x^2$

25. $y > x^2 - 2x$

26. $y \geq 1 - x^2$

27. $y > -2x^2$

28. $y \leq x^2 - 3x$

29. $y \leq x^2 + 2x - 8$

30. $y \geq x^2 - 2x + 1$

31. $y \leq x^2 - 4x + 4$

32. $y \leq x^2 - 5x - 6$

33. $y > x^2 + 5x + 6$

34. $y < x^2 - 3x$

35. $y < x^2 - x - 6$

A projectile is fired vertically into the air. Its motion is described by $h = -16t^2 + 320t$, where h is its height (in feet) after t seconds.

Physics Use a quadratic inequality to answer each of the following questions.

36. During what time intervals will the height of the projectile be below 1024 feet?

37. During what time intervals will the height of the projectile be above 1024 feet?

38. Technology Use graphing technology to graph $h = -16t^2 + 320t$. Trace the curve to check your answers to Exercises 36 and 37.

Look Back

Find the products. [Lesson 10.2]

39. $(-ab^2c)(a^2bc^3)$
$-a^3b^3c^4$

40. $(3p^2q^3r^4)(-2pqr)$
$-6p^3q^4r^5$

Write each number in scientific notation. [Lesson 10.4]

41. 0.8
8×10^{-1}

42. 0.000001
1×10^{-6}

43. 0.0000074
7.4×10^{-6}

Factor each polynomial by first removing a greatest common factor. [Lesson 11.4]

44. $12y^2 - 2y$
$2y(6y - 1)$

45. $2ax + 6x + ab + 3b$
$(a + 3)(2x + b)$

46. $4x^2 - 24x + 32$
$4(x - 4)(x - 2)$

Look Beyond

47. Three identical rectangles with sides of positive integer length, each have a perimeter of 24 inches. Find the area of the square formed when the rectangles are joined. 81 square inches

9. $x < 2 \text{ or } x > 4$

10. $-2 < x < -1$

11. $2 \leq x \leq 7$

12. $x < -5 \text{ or } x > 3$

13. $1 < x < 6$

14. $-2 < x < 10$

15. $3 \leq x \leq 5$

16. $-3 \leq x \leq 6$

17. $x \leq 5 \text{ or } x \geq 6$

18. $x \geq 6 \text{ or } x \leq -2$

19. $x < -7 \text{ or } x > 2$

20. $x \leq -6 \text{ or } x \geq -4$

The graphs for Exercises 9–20 and answers to Exercises 21–35 and 38 can be found in Additional Answers beginning on page 729.

36. less than 4 seconds and greater than 16 seconds

37. between 4 and 16 seconds

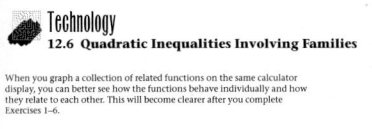

FOCUS

This project involves number patterns that are generated by using a quadratic equation.

MOTIVATE

Remind students of their study of Pascal's triangle in the last chapter.

```
          1
        1   2   1
      1   3   3   1
    1   4   6   4   1
  1   5  10  10   5   1
```

Tell students that they will explore different patterns as they complete this project.

WHAT'S THE DIFFERENCE?

If n represents the dots on a side, the sequence shows the total number of dots in a square pattern.

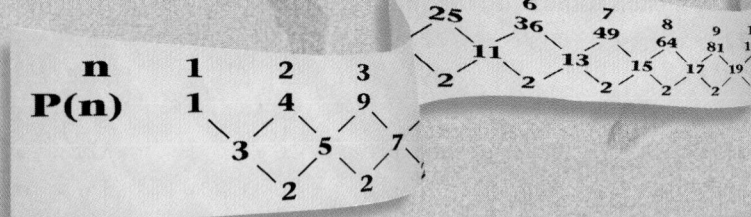

A function whose domain is consecutive positive integers can produce a sequence. Many sequences have a function rule that you can discover.

A function that will produce this sequence is $A(n) = n^2$. The variable n is usually used when working with sequences, and it represents the number of the term in the sequence.

How do you determine the function rule? Sometimes the pattern in the sequence is clear enough to guess and check to find the rule. When the sequence is more complicated, other methods are needed. If a sequence eventually produces a constant difference, one way to find the general rule for that sequence is to use **finite differences**.

Examine the sequence 0, 5, 12, 21, 32, 45, 60. You want to find a general rule that will generate any term of this sequence. Begin by finding differences until they are constant.

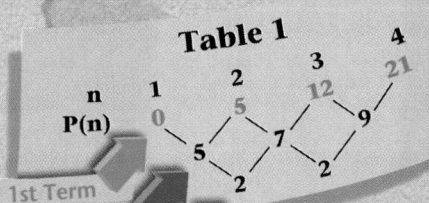

Table 1

$$P(n) = an^2 + bn + c$$
$$P(1) = a(1)^2 + b(1) + c = a + b + c$$
$$P(2) = a(2)^2 + b(2) + c = 4a + 2b + c$$
$$P(3) = a(3)^2 + b(3) + c = 9a + 3b + c$$
$$P(4) = a(4)^2 + b(4) + c = 16a + 4b + c$$

If the sequence produces constant second differences, you expect a quadratic equation. Write the quadratic in the form of a polynomial function, $P(n) = an^2 + bn + c$. Evaluate the function for n from 1 to at least 3. Next, make a table of the terms and differences using the expressions you get from evaluating the general quadratic polynomial.

The sequence in Table 1 is equal to the corresponding general expressions in Table 2. Match the first expression in each row of Table 2 with the corresponding expression in Table 1.

Table 2		Table 1
$a + b + c$	$=$	0
$3a + b$	$=$	5
$2a$	$=$	2

Use *substitution* to find the values of a, b, and c.

Use $2a = 2$ to solve for a.
$a = 1$

Use $3a + b = 5$ and the value for a to solve for b.
$3 + b = 5$, so $b = 2$

Use $a + b + c = 0$ and the values for a and b to solve for c.
$1 + 2 + c = 0$, so $c = -3$

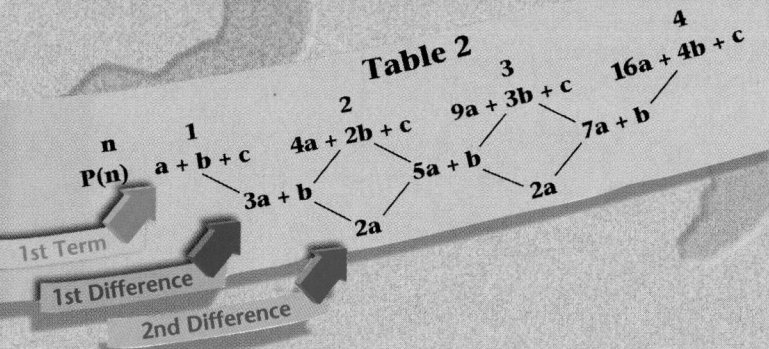

Finally, replace the letters in the quadratic polynomial, $P(n) = an^2 + bn + c$ with the values $a = 1$, $b = 2$, and $c = -3$.

$$P(n) = n^2 + 2n - 3$$

Check your rule by substitution to see if it generates the appropriate sequence. $0, 5, 12, 21, 32, 45, 60, \ldots$

Thus, $P(n) = n^2 + 2n - 3$ is a rule that generates the sequence.

Activity

Find a Rule

In groups of 2 to 5 people, the leader begins by choosing a quadratic equation with integer coefficients and generating a sequence of 5 numbers. The leader then shows only the sequence to the other people in the group. The group then tries to find the function rule that generated the sequence. Finally, the group uses the rule they found to generate any 2 terms between 6 and 20. The leader confirms the results by showing the original function rule and sequence. Let the group members take turns being the leader.

Discuss

Discuss the process used in determining the pattern. Be sure that students understand how the pattern is generated before they try to develop their own patterns.

Chapter Review

1. $V(-1, -4)$; axis of symmetry: $x = -1$; zeros 1, -3

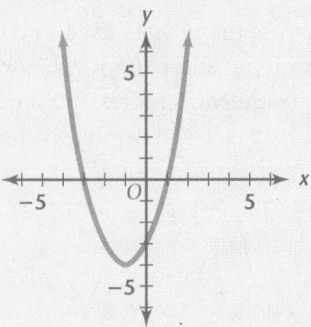

2. $V(5, -1)$; axis of symmetry: $x = 5$; zeros 4, 6

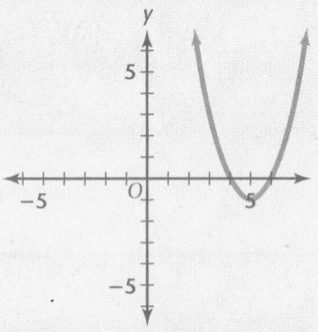

3. $V(3, -2)$; axis of symmetry: $x = 3$; zeros 2, 4

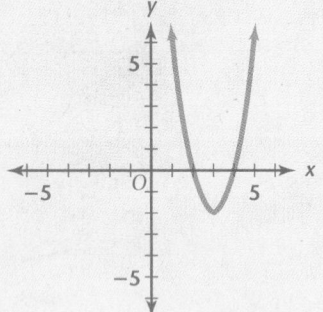

Chapter 12 Review

Vocabulary

axis of symmetry	548	minimum value	560	square root	553
discriminant	577	quadratic formula	575	vertex	548
maximum value	560	quadratic inequality	581	Zero Product Property	567

Key Skills & Exercises

Lesson 12.1

➤ **Key Skills**

Examine quadratic functions to find the vertex, axis of symmetry, and zeros of the quadratic function to sketch a graph.

$$g(x) = (x - 1)^2 - 4$$

The vertex is $(1, -4)$ and the axis of symmetry is $x = 1$.

$$g(x) = (x - 1)^2 - 4$$
$$= x^2 - 2x - 3$$
$$= (x - 3)(x + 1)$$

The zeros of $g(x)$ are 3 and -1.

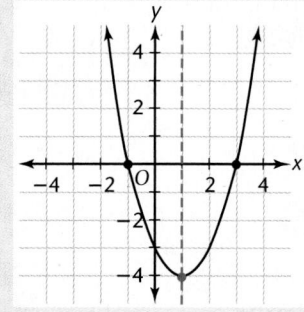

➤ **Exercises**

Determine the vertex, axis of symmetry, and zeros for each function. Then sketch a graph from the information.

1. $y = (x + 1)^2 - 4$ **2.** $y = (x - 5)^2 - 1$ **3.** $y = 2(x - 3)^2 - 2$

Lesson 12.2

➤ **Key Skills**

Use square roots to solve quadratic equations.

$x^2 = 36 \to x = \sqrt{36}$ or $-\sqrt{36} \to x = 6$ or -6

$(x - 3)^2 = 15 \to x - 3 = \pm\sqrt{15} \to x = 3 + \sqrt{15} \approx 6.87$ or $3 - \sqrt{15} \approx -0.87$

➤ **Exercises**

Solve each equation. Give answers to the nearest hundredth when necessary.

4. $x^2 = 225$ ± 15 **5.** $x^2 = \frac{9}{144}$ $\pm\frac{1}{4}$ **6.** $x^2 = 50$ ± 7.07

7. $(x + 1)^2 - 4 = 0$ **8.** $(x + 7)^2 - 81 = 0$ **9.** $(x - 3)^2 - 3 = 0$

7. $1, -3$
8. $2, -16$
9. $x \approx 4.73, 1.27$

Lesson 12.3

➤ Key Skills

Complete the square.

To complete the square, find one half of the coefficient of $-6x$, which is -3. Square -3, and add it to $x^2 - 6x$.

$$x^2 - 6x + (-3)^2 = x^2 - 6x + 9$$

Write quadratic equations in the form $y = (x - h)^2 + k$. Find the vertex.

To find the vertex of $y = x^2 - 6x + 2$, complete the square. Add and subtract 9.

$$y = (x^2 - 6x + 9) + 2 - 9 = (x - 3)^2 - 7$$

The vertex is $(3, -7)$.

➤ Exercises

Complete the square.

10. $x^2 + 4x \; +4$ **11.** $x^2 - 16x \; +64$ **12.** $x^2 - x \; +\frac{1}{4}$ **13.** $x^2 + 3x \; +\frac{9}{4}$

Rewrite each function in the form $y = (x - h)^2 + k$. Find each vertex.

14. $y = x^2 + 4x + 3$ **15.** $y = x^2 - 8x + 18$ **16.** $y = x^2 - 12x + 32$
   $y = (x + 2)^2 - 1; (-2, -1)$ $y = (x - 4)^2 + 2; (4, 2)$ $y = (x - 6)^2 - 4; (6, -4)$

Lesson 12.4

➤ Key Skills

Solve quadratic equations by completing the square.

$$x^2 + 2x = 24$$
$$x^2 + 2x + 1 = 24 + 1$$
$$(x + 1)^2 = 25$$
$$x + 1 = 5 \quad \text{or} \quad x + 1 = -5$$
$$x = 4 \quad \text{or} \qquad x = -6$$

Solve quadratic equations by factoring.

$$x^2 + 2x = 24$$
$$x^2 + 2x - 24 = 0$$
$$(x + 6)(x - 4) = 0$$
$$x + 6 = 0 \quad \text{or} \quad x - 4 = 0$$
$$x = -6 \quad \text{or} \qquad x = 4$$

➤ Exercises

Solve each equation by factoring or completing the square.

17. $x^2 + 3x - 10 = 0$ **18.** $x^2 + 5x + 6 = 0$ **19.** $x^2 - 8x + 16 = 0$
    $2, -5$ $-2, -3$ 4

Lesson 12.5

➤ Key Skills

Evaluate the discriminant. Determine the number of solutions.

$$3x^2 + 4x + 7 = 0$$

Substitute $a = 3$, $b = 4$, and $c = 7$ in $b^2 - 4ac$. Simplify. If the discriminant is negative, the equation has no real solutions. If the discriminant is 0, there is exactly one real solution. If the discriminant is positive, the equation has 2 real solutions.

$$b^2 - 4ac = (4)^2 - (4 \cdot 3 \cdot 7) = -68$$

The equation has no real solutions.

Use the quadratic formula to find solutions to quadratic equations.

For the equation $x^2 + 3x - 14 = 0$, $a = 1$, $b = 3$, and $c = -14$. Substitute these values in the quadratic formula and simplify.

$$x = \frac{-3 \pm \sqrt{9 - (4 \cdot 1 \cdot -14)}}{2 \cdot 1}$$

$$x = \frac{-3 \pm \sqrt{65}}{2}$$

$$x \approx 2.53 \text{ or } x \approx -5.53$$

29.

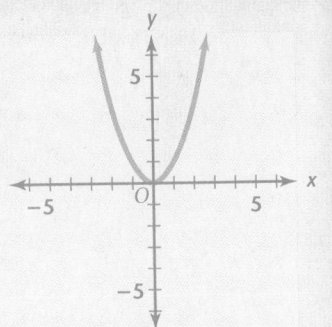

30.

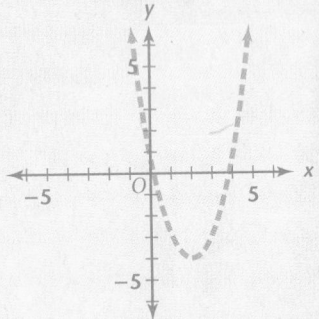

31.

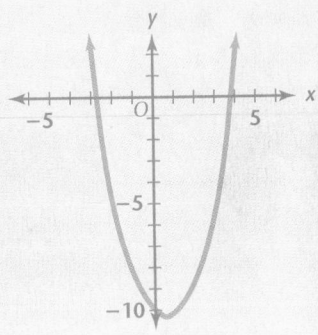

> **Exercises**

Evaluate the discriminant to determine the number of solutions. Solve. Give answers to the nearest hundredth.

20. $x^2 - 2x + 9 = 0$ **21.** $4x^2 - 5x - 4 = 0$ **22.** $4x^2 - 4x + 1 = 0$

23. $x^2 - 9x + 18 = 0$ **24.** $6x^2 - 7x - 3 = 0$ **25.** $8x^2 + 2x - 1 = 0$

Lesson 12.6

> *Key Skills*

Solve and graph quadratic inequalities.

Use factoring and the Zero Product Property to solve $x^2 + 3x - 4 > 0$.

$$x^2 + 3x - 4 = 0$$
$$(x + 4)(x - 1) = 0$$
$$x = -4 \text{ or } x = 1$$

Test numbers in the inequality to determine the solutions. Since numbers less than -4 or numbers greater than 1 produce a true inequality, the solutions are $x < -4$ or $x > 1$.

To graph $y \leq x^2 + 2$, graph $y = x^2 + 2$. Draw a solid parabola to show that the solutions are on the graph of $y = x^2 + 2$. Since $y \leq x^2 + 2$, shade the points below the parabola.

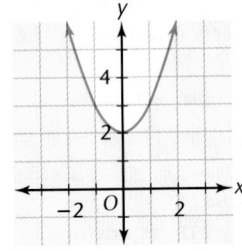

> **Exercises**

Solve each quadratic inequality using the Zero Product Property. Graph the solution on a number line.

26. $x^2 + 6x + 8 \geq 0$ **27.** $x^2 - 3x - 10 < 0$ **28.** $x^2 + 9x + 18 > 0$

Graph each inequality. Shade the solution region.

29. $y \leq x^2$ **30.** $y > x^2 - 4x$ **31.** $y \geq x^2 - x - 10$

Applications

Sports The graph represents a relationship between the number of yards a football is thrown and its height above the ground.

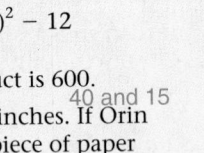

32. What is the maximum height reached by the football? ≈ 12 ft

33. How far does the football travel before it reaches its maximum height? 15 yd

34. If a receiver catches the ball 30 yards from where the ball was thrown, what was the height of the ball? ≈ 5 ft

35. Which equation was used to graph the parabola? a

   a. $y = -\frac{1}{30}(x - 15)^2 + 12$ **b.** $y = -\frac{1}{30}(x + 15)^2 + 12$

   c. $y = -\frac{1}{30}(x - 15)^2 - 12$ **d.** $y = -\frac{1}{30}(x + 15)^2 - 12$

36. Find two numbers whose sum is 55 and whose product is 600.
   40 and 15

37. A piece of wrapping paper has an area of 480 square inches. If Orin cuts out a square piece from the paper and the new piece of paper has an area of 455 square inches, what is the length of each side of the square?
   5 inches

20. -32; no solution

21. 89; two solutions; $x \approx 1.80, -0.55$

22. 0; one solution; 0.5

23. 9; two solutions; 3, 6

24. 121; two solutions; 1.5, ≈ -0.33

25. 36; two solutions; 0.25, -0.5

26. $x \leq -4, x \geq -2$

27. $-2 < x < 5$

28. $x < -6 \text{ or } x > -3$

Chapter 12 Assessment

1. Compare the graph of $y = (x + 2)^2 - 4$ to the graph of $y = x^2$. Write the change in units and direction that takes place to the left, right, up, or down. left 2 units; down 4 units

Determine the vertex, axis of symmetry, and zeros for each function. Then sketch a graph from the information.

2. $y = (x - 6)^2 - 4$ **3.** $y = (x + 3)^2 - 1$

4. Why does an equation of the form $x^2 = k$ have two solutions?
$\sqrt{k} \cdot \sqrt{k} = k$ and $-\sqrt{k} \cdot -\sqrt{k} = k$

Solve each equation. Give answers to the nearest hundredth when necessary.

5. $x^2 = 900$ ± 30 **6.** $x^2 = 12$ $\approx \pm 3.46$ **7.** $(x + 5)^2 - 16 = 0$ $-1, -9$

8. Use the formula $A = \pi r^2$ to find the radius of a circle with an area of 100 square feet. (Use $\pi = 3.14$.) 5.64 feet

9. An object is dropped and its height, h, after t seconds is given by the function $h = -16t^2 + 550$. About how long will it take for the object to reach the ground? 5.86 seconds

Complete the square.

10. $x^2 + 8x$ $+ 16$ **11.** $x^2 - 30x$ $+ 225$ **12.** $x^2 + 11x$ $+ \frac{121}{4}$

Rewrite each function in the form $y = (x - h)^2 + k$. Find each vertex.

13. $y = x^2 + 4x + 1$
$y = (x + 2)^2 - 3; (-2, -3)$

14. $y = x^2 - 14x + 50$
$y = (x - 7)^2 + 1; (7, 1)$

Solve each equation by factoring or completing the square.

15. $x^2 - 35 - 2x = 0$
$7, -5$

16. $x^2 + 6x - 12 = 0$
$\approx 1.58, \approx -7.58$

17. Why does a quadratic equation need to be written in standard form in order to use the quadratic formula to solve it?
The formula was derived from the standard form $ax^2 + bx + c = 0$.

Find the value of the discriminant for each quadratic equation. Describe the number of solutions.

18. $x^2 + 10x - 13 = 0$
$152; 2$

19. $5x^2 - 4x + 2 = 0$
$-24;$ none

Use the quadratic formula to solve each equation. Give answers to the nearest hundredth when necessary.

20. $x^2 - 4x - 60 = 0$
$10, -6$

21. $2x^2 + 3x - 4 = 0$
$0.85, -2.35$

Solve each quadratic inequality using the Zero Product Property. Graph the solution on a number line.

22. $x^2 + 4x - 5 > 0$ **23.** $x^2 + 10x + 21 \leq 0$

Graph each quadratic inequality. Shade the solution region.

24. $y \leq -x^2 + 2$ **25.** $y > x^2 + 5x + 6$

2. V(6, −4); axis of sym. $x = 6$; zeros 4, 8

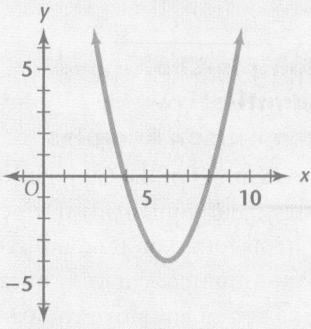

3. V(−3, −1); axis of sym. $x = -3$; zeros −2, −4

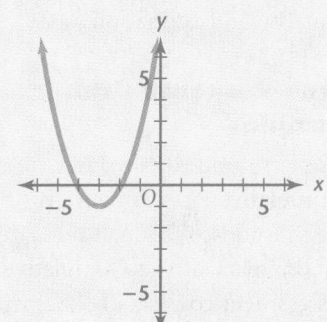

22. $x < -5, x > 1$

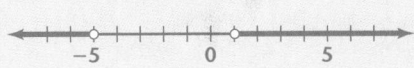

23. $-7 \leq x \leq -3$

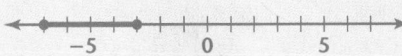

24.

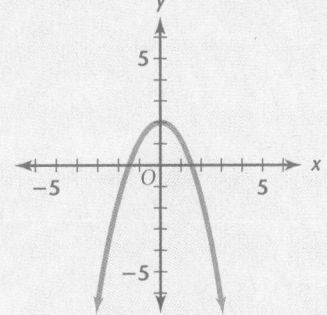

25.

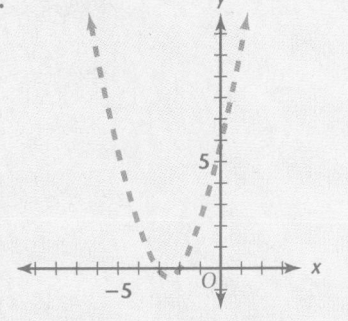

Chapters 1 – 12
Cumulative Assessment

COLLEGE ENTRANCE-
EXAM PRACTICE

Multiple-Choice and Quantitative-Comparison Samples

The first half of the Cumulative Assessment contains two types of items found on standardized tests—multiple-choice questions and quantitative-comparison questions. Quantitative-comparison items emphasize the concepts of equality, inequality, and estimation.

Free-Response Grid Samples

The second half of the Cumulative Assessment is a free-response section. A portion of this part of the Cumulative Assessment consists of student-produced response items commonly found on college entrance exams. These questions require the use of machine-scored answer grids. You may wish to have students practice answering these items in preparation for standardized tests.

Sample answer grid masters are available in the *Chapter Teaching Resources Booklets.*

College Entrance Exam Practice

Quantitative Comparison   For Questions 1–4, write

A if the quantity in Column A is greater than the quantity in Column B;
B if the quantity in Column B is greater than the quantity in Column A;
C if the two quantities are equal; or
D if the relationship cannot be determined from the information given.

	Column A	Column B	Answers	
1.	$-6 \cdot 9$	$6 \cdot -9$	(A) (B) (C) (D) **[Lesson 4.1]**	C
2.	correlation of a line of best fit that rises	correlation of a line of best fit that falls	(A) (B) (C) (D) **[Lesson 1.6]**	A
3.	70% of a number	90% of a number	(A) (B) (C) (D) **[Lesson 4.5]**	D
4.	$\dfrac{5 \times 10^{-2}}{7 \times 10^{-5}}$	$\dfrac{5 \times 10^{2}}{7 \times 10^{5}}$	(A) (B) (C) (D) **[Lesson 10.3]**	A

5. What is the reciprocal of 4?   **[Lesson 2.3]**   b
   a. -4 **b.** $\frac{1}{4}$ **c.** 4   **d.** $-\frac{1}{4}$

6. What is the simplest form of the expression $5r - (3r + 2)$?   **[Lesson 4.2]**   c
   a. $8r + 2$ **b.** $2r + 2$ **c.** $2r - 2$ **d.** $8r - 2$

7. What is the solution to $6 - 4t > 18$?   **[Lesson 4.7]**   a
   a. $t < -3$ **b.** $t > -3$ **c.** $t > -6$ **d.** $t < -6$

8. What is the slope of a line passing through $(-1, 2)$ and $(4, -3)$?   **[Lesson 5.3]**   d
   a. -2 **b.** 2   **c.** 1   **d.** -1

9. Which equation is graphed at the right?
   a. $y = 2x^2 - 1$
   b. $y = -2x^2 - 1$
   c. $y = \frac{1}{2}x^2 - 1$
   d. $y = -\frac{1}{2}x^2 - 1$
   [Lesson 12.1]   a

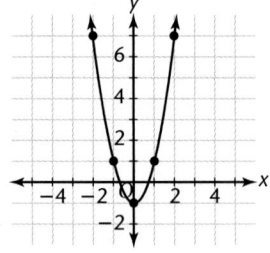

10. What is the factored form of $4x^2 + 25$? **[Lesson 11.4]** d
    a. $(2x - 5)^2$ **b.** $(2x + 5)(2x - 5)$
    c. $(2x + 5)^2$ **d.** Cannot be factored.

11. If tickets to a concert cost $24 per person, plus a $7 handling charge per order, regardless of how many tickets are ordered, how many tickets could you order for $199? **[Lesson 4.2]** 8

12. Write an inequality that describes the points graphed on the number line. **[Lesson 3.6]** $x < -3$ or $x \geq 2$

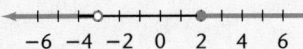

$$-6 \quad -4 \quad -2 \quad 0 \quad 2 \quad 4 \quad 6$$

13. Simplify $\dfrac{-100a^8 b^{10}}{-20a^5 b^9}$. **[Lesson 10.2]** $5a^3 b$

14. Vernon scored 86 on his first two tests, 90 on his third test, and 78 on his fourth test. Find the median test score. **[Lesson 8.3]** 86

15. Write the equation with slope -2 and passing through $(-1, 0)$. **[Lesson 5.3]** $y = -2x - 2$

16. Write an equation for a line that is perpendicular to the line in Item 15. **[Lesson 5.6]** $y = \dfrac{1}{2}x + b$

17. Factor $3a(2c + d)^2 - 5(2c + d)^2$. **[Lesson 11.4]** $(3a - 5)(2c + d)^2$

18. Solve the system. $\begin{cases} 3x + y = -2 \\ 2x + 3y = 8 \end{cases}$ **[Lesson 6.2]** $(-2, 4)$

19. Does the matrix $\begin{bmatrix} -2 & 2 \\ 1 & -1 \end{bmatrix}$ have an inverse? **[Lesson 7.4]** no

20. Simplify $(3x^2 y^3)^0$. **[Lesson 10.3]** 1

21. Suppose you choose a card from an ordinary deck of 52 cards. What is the probability that you will choose an ace or an eight? **[Lesson 8.7]**

Free-Response Grid The following questions may be answered using a $\dfrac{2}{13}$ free-response grid commonly used by standardized test services.

22. How many ways are there to arrange 6 different letters? **[Lesson 8.5]** 720

23. Solve $\dfrac{2x}{3} = 12$. **[Lesson 4.6]** 18

24. Evaluate $h(x) = 2x^2 + 7x - 20$ for $x = 5$. **[Lesson 12.4]** 65

25. Find the slope of the line for the equation $y = 2x + 5$. **[Lesson 5.1]** 2

Radicals and Coordinate Geometry

Meeting Individual Needs

13.1 Exploring Square Root Functions

Core Resources

Practice Master 13.1
Enrichment, p. 597
Technology Master 13.1

[1 day]

Core Two-Year Resources

Inclusion Strategies, p. 597
Reteaching the Lesson, p. 598
Practice Master 13.1
Enrichment Master 13.1
Technology Master 13.1
Lesson Activity Master 13.1

[2 days]

13.2 Operations With Radicals

Core Resources

Practice Master 13.2
Enrichment, p. 605
Technology Master 13.2

[1 day]

Core Two-Year Resources

Inclusion Strategies, p. 605
Reteaching the Lesson, p. 606
Practice Master 13.2
Enrichment Master 13.2
Technology Master 13.2
Lesson Activity Master 13.2

[2 days]

13.3 Solving Radical Equations

Core Resources

Practice Master 13.3
Enrichment, p. 613
Technology Master 13.3
Interdisciplinary
  Connection, p. 612
Mid-Chapter Assessment
  Master

[2 days]

Core Two-Year Resources

Inclusion Strategies, p. 613
Reteaching the Lesson, p. 615
Lesson Activity Master 13.3
Enrichment Master 13.3
Practice Master 13.3
Technology Master 13.3
Mid-Chapter Assessment Master

[3 days]

13.4 The "Pythagorean" Right-Triangle Theorem

Core Resources

Practice Master 13.4
Enrichment, p. 620
Technology Master 13.4

[2 days]

Core Two-Year Resources

Inclusion Strategies, p. 620
Reteaching the Lesson, p. 621
Practice Master 13.4
Enrichment Master 13.4
Technology Master 13.4
Lesson Activity Master 13.4

[3 days]

13.5 The Distance Formula

Core Resources

Practice Master 13.5
Enrichment, p. 627
Technology Master 13.5
Interdisciplinary
  Connection, p. 626

[2 days]

Core Two-Year Resources

Inclusion Strategies, p. 627
Reteaching the Lesson, p. 628
Practice Master 13.5
Enrichment Master 13.5
Technology Master 13.5
Lesson Activity Master 13.5

[4 days]

13.6 Exploring Geometric Properties

Core Resources

Practice Master 13.6
Enrichment, p. 634
Technology Master 13.6
Interdisciplinary
  Connection, p. 633

[2 days]

Core Two-Year Resources

Inclusion Strategies, p. 634
Reteaching the Lesson, p. 635
Practice Master 13.6
Enrichment Master 13.6
Technology Master 13.6
Lesson Activity Master 13.6

[4 days]

13.7 Trigonometric Functions

Core Resources

Practice Master 13.7
Enrichment, p. 639
Technology Master 13.7

[2 days]

Core Two-Year Resources

Inclusion Strategies, p. 639
Reteaching the Lesson, p. 640
Practice Master 13.7
Enrichment Master 13.7
Technology Master 13.7
Lesson Activity Master 13.7

[4 days]

Chapter Summary

Core Resources

Chapter 13 Project,
   pp. 644–645
Lab Activity
Long-Term Project
Chapter Review,
   pp. 646–648
Chapter Assessment,
   p. 649
Chapter Assessment,
   A/B
Alternative Assessment

[3 days]

Core Two-Year Resources

Chapter 13 Project, pp. 644–645
Lab Activity
Long-Term Project
Chapter Review, pp. 646–648
Chapter Assessment,
   p. 649
Chapter Assessment, A/B
Alternative Assessment

Visual Strategies

When solving problems that involve right triangles, making a diagram or sketch is a very useful problem-solving technique. Showing an oblique line segment on a coordinate grid and developing the distance formula adds to students' understanding because they can see the triangle created and integrate that with previously learned concepts about right triangles. Using visual strategies helps students understand concepts like the distance on a coordinate plane rather than just having them learn a formula without comprehending its meaning.

Hands-on Strategies

The chief hands-on strategy employed in Chapter 13 is the use of the coordinate grid. Much of the graphing can be completed on a graphics calculator, but it is also beneficial for students to use paper and pencil to see relationships between quantities being graphed. When working with the "Pythagorean" Right-Triangle Theorem (Lesson 13.4) and trigonometric functions (Lesson 13.7), you may wish to have some physical models of right triangles available for students to measure.

Cooperative Learning

GROUP ACTIVITIES	
Exploring the square root function	Lesson 13.1, Explorations 1 and 2
Exploring pendulums	Lesson 13.3, Opening Discussion on pp. 611–612
Distance and midpoint	Lesson 13.5, Opening Development and Explorations 1 and 2
Angles and tangents	Lesson 13.6, Exploration

You may wish to have students work with partners or in small groups to complete any of these activities. Additional suggestions for cooperative group activities can be found in the teacher's notes in the lessons.

Multicultural

The cultural connections in this chapter include references to China, ancient Babylon, India, and Europe.

CULTURAL CONNECTIONS	
Asia: Ancient Chinese, Babylonians, and Hindus	Lesson 13.4, Opening Discussion
Europe: "Pythagorean" Right-Triangle Theorem	Lesson 13.4, Discussion, p. 620

Portfolio Assessment

Below are portfolio activities for the chapter listed under seven activity domains that are appropriate for portfolio development.

1. **Investigation/Exploration** Students start by exploring the square root function in Lesson 13.1, Explorations 1 and 2. An extended application about pendulums is found in Lesson 13.3. The distance between two points is covered in Lesson 13.5, Opening Development and Exploration 1. The midpoint of a segment is explored and investigated in Lesson 13.5, Exploration 2. An investigation of the relationship of angles and tangents is explored in Lesson 13.6. Finally, golden rectangles are investigated in the Chapter Project.

2. **Applications** Recommended for inclusion are any of the following: Physics: Lesson 13.3, Opening Discussion Exercises 25–28; Sports: Lesson 13.3, Exercises 29–30; Boating: Lesson 13.4, Example 2; Rescue Services: Lesson 13.5, Exercise 14; Surveying: Lesson 13.6, Application on p. 634.

3. **Non-Routine Problems** In Lesson 13.3, Exercise 56 asks students to find the center of a circle if they are given only the circle itself.

4. **Project** Radical Rectangles on pages 644–645 provides an opportunity for students to explore the golden ratio.

5. **Interdisciplinary Topics** Students may choose from the following: Geometry: Lesson 13.1, Exploration 1; Lesson 13.3, Example 5; Lesson 13.4, entire lesson; Lesson 13.6, Opening Discussion and Exercises 44–45.

6. **Writing** Communicate exercises that ask students to describe and explain procedures involved in solving probability problems offer excellent writing selections.

7. **Tools** Chapter 13 makes extensive use of the graphics calculator. The graphics calculator is used to explore the graphs of quadratic equations that contain radicals.

Technology

Chapter 13 provides an opportunity for students to use calculators to simplify arithmetic calculations, thus allowing students to concentrate on mathematical concepts rather than on mechanical manipulations.

Coordinate grids with different axis scales can show the same graph quite differently. When dealing with graphs on the graphics calculators, students must learn to properly set the range window in order to

- see a graph realistically by setting the ranges so that the graphs are not distorted. (For example, the graph of $x = y$ should make a 45° angle with the axes.)
- display the important parts of the graph, such as the maximum, minimum, or intercepts.

Students can expand their understanding of exponents by using a calculator to explore fractional exponents.

1. Graph $x^{\frac{1}{2}}$. (Depending on the calculator, students may have to use decimal equivalents and graph $x^{0.5}$.)

2. Use the ZOOM and TRACE features to find the value of $x^{\frac{1}{2}}$ when $x = 3$ ($\approx$**1.7**), $x = 4$ (**2**), $x = 5$ ($\approx$**2.2**). Have students make a conjecture about the meaning of a fractional exponent.

3. Have students test their conjecture by finding $x^{\frac{1}{4}}$ for $x = 81$ and $x^{\frac{1}{3}}$ for $x = 8$.

4. Encourage students to make conjectures about properties of and operations on fractional exponents and then use the calculator to check their conjectures.

Students can use a computer or programmable calculator to find the distance between two points by using this BASIC program.

```
10   PRINT "ENTER COORDINATES OF POINT (X1,Y1)."
20   INPUT X1,Y1
30   PRINT "ENTER COORDINATES OF POINT (X2,Y2)."
40   INPUT X2,Y2
50   D=SQR((X2-X1)*(X2-X1)+(Y2-Y1)*(Y2-Y1))
60   PRINT "THE DISTANCE BETWEEN THE POINTS
     IS";D;"UNITS."
```

Challenge students to create a program for finding the midpoint of a line segment.

Integrated Software

f(g) Scholar[TM] is an integrated computer-based mathematics productivity tool that combines calculator, spreadsheet, and graphics capabilities to provide a dynamic and interactive environment for explorations in mathematics. It is appropriate to use **f(g) Scholar**[TM] for any lesson needing a spreadsheet, calculator, graphics calculator, or any combination of the three.

13 Radicals and Coordinate Geometry

ABOUT THE CHAPTER

Background Information

In this chapter students will extend their understanding of square roots. They will learn about operating with radicals and solving radical equations. The use of radicals will be applied as the "Pythagorean" Right-Triangle Theorem, distance formula, and trigonometric functions.

CHAPTER RESOURCES

- Practice Masters
- Enrichment Masters
- Technology Masters
- Lesson Activity Masters
- Lab Activity Masters
- Long-Term Project Masters
- Assessment Masters:
    Chapter Assessments, A/B
    Mid-Chapter Assessments
    Alternative Assessments,
    A/B
- Teaching Transparencies
- Spanish Resources

CHAPTER OBJECTIVES

- Find or estimate square roots.
- Explore transformations of the square root function using graphs.
- Identify positive and negative square roots.
- Define square root and write square roots in simplest radical form.
- Examine mathematical operations with radicals.
- Solve equations containing radicals.
- Solve equations by using radicals.

CHAPTER 13

LESSONS

13.1 *Exploring* Square Root Functions

13.2 Operations With Radicals

13.3 Solving Radical Equations

13.4 The "Pythagorean" Right-Triangle Theorem

13.5 The Distance Formula

13.6 *Exploring* Geometric Properties

13.7 Trigonometric Functions

Chapter Project
Radical Rectangles

Radicals and Coordinate Geometry

In this chapter you can examine square roots and their relationship to the process of squaring. Square roots are used whenever the Quadratic formula or the distance formula are used. For this reason, they play a central and useful role in both algebra and geometry.

Square roots appear in many applications. For example,

- You know the length of a pendulum and want to know how long it will take for one complete swing.

- You know the braking distance of a car and want to know the speed of the car.

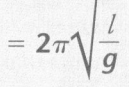

$$t = 2\pi\sqrt{\frac{l}{g}}$$

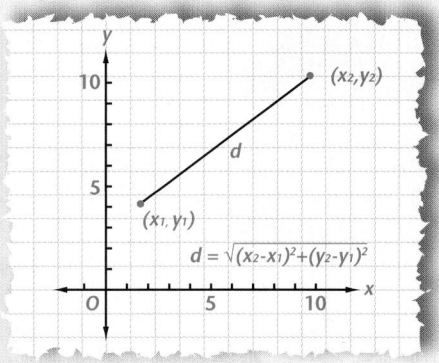

$$d = \sqrt{(x_2-x_1)^2+(y_2-y_1)^2}$$

ABOUT THE PHOTOS

Ask students what a pendulum, a car trying to stop, and a right triangle have in common.

When the mathematical properties of all three are studied, students discover that they all involve radicals, or square roots. Have students read and analyze the graph of braking distances.

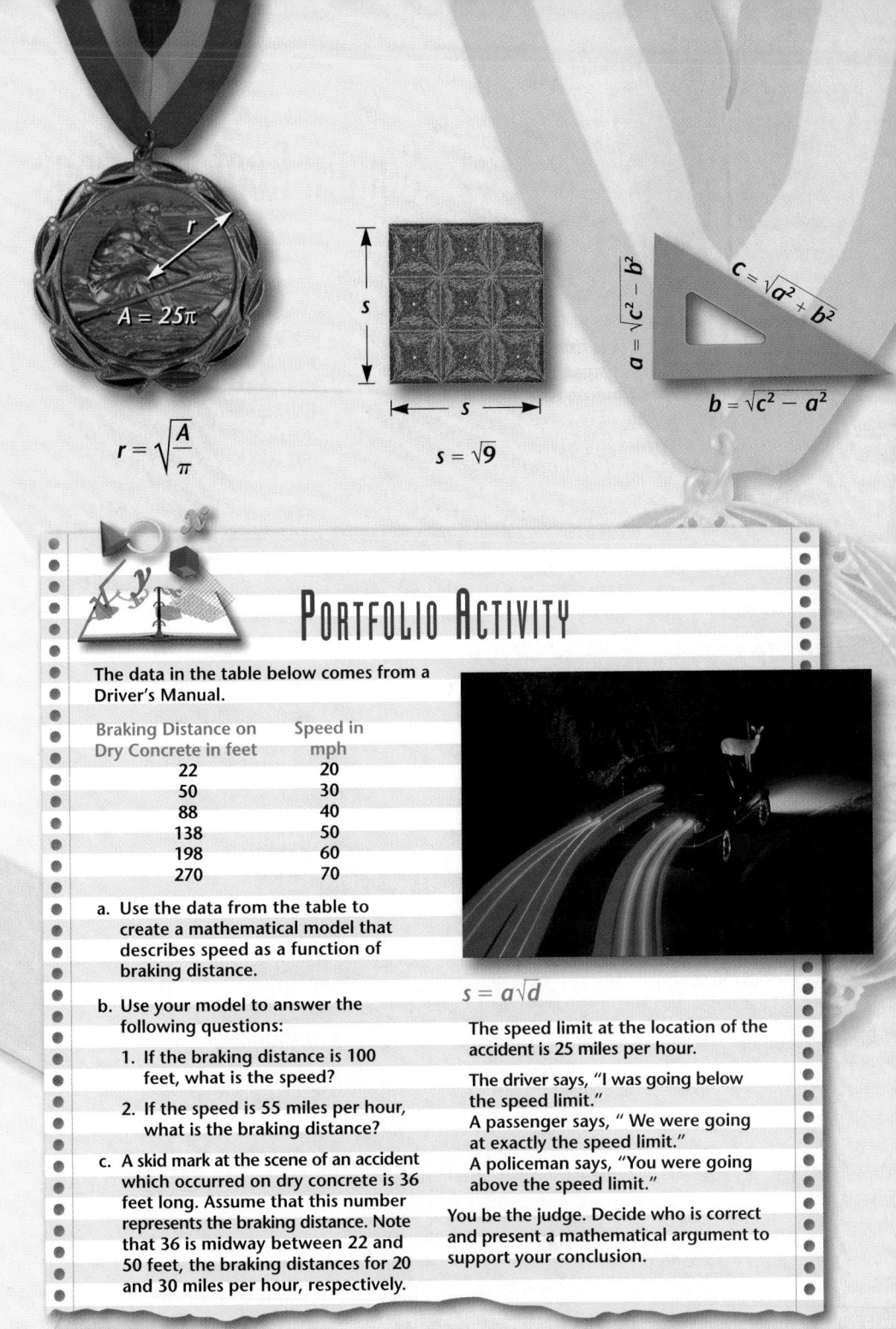

$A = 25\pi$

$r = \sqrt{\dfrac{A}{\pi}}$

$s = \sqrt{9}$

$a = \sqrt{c^2 - b^2}$

$c = \sqrt{a^2 + b^2}$

$b = \sqrt{c^2 - a^2}$

- Examine a right triangle to determine a relationship among the sides.
- Use the "Pythagorean" Right Triangle Theorem to solve problems involving right triangles.
- Use the distance formula to find the distance between two points on the coordinate plane.
- Determine whether a triangle is a right triangle.
- Use the midpoint formula to find the midpoint of a segment.
- Write the equation of a circle, given its center and radius.
- Use the distance formula to find the diagonals of a rectangle.
- Define the tangent function for right triangles.
- Use the tangent of an acute angle to solve problems.

PORTFOLIO ACTIVITY

The data in the table below comes from a Driver's Manual.

Braking Distance on Dry Concrete in feet	Speed in mph
22	20
50	30
88	40
138	50
198	60
270	70

a. Use the data from the table to create a mathematical model that describes speed as a function of braking distance.

b. Use your model to answer the following questions:

   1. If the braking distance is 100 feet, what is the speed?

   2. If the speed is 55 miles per hour, what is the braking distance?

c. A skid mark at the scene of an accident which occurred on dry concrete is 36 feet long. Assume that this number represents the braking distance. Note that 36 is midway between 22 and 50 feet, the braking distances for 20 and 30 miles per hour, respectively.

$s = a\sqrt{d}$

The speed limit at the location of the accident is 25 miles per hour.

The driver says, "I was going below the speed limit."
A passenger says, "We were going at exactly the speed limit."
A policeman says, "You were going above the speed limit."

You be the judge. Decide who is correct and present a mathematical argument to support your conclusion.

ABOUT THE CHAPTER PROJECT

In the Chapter Project on pages 644–645, students investigate golden rectangles. This interesting application provides a dramatic link between algebra, irrational numbers, geometry, and art.

PORTFOLIO ACTIVITY

Assign the Portfolio Activity as part of the assignment for Lesson 13.5 on page 631.

To introduce the activity, ask students whether they think a car traveling 40 miles per hour would take twice the distance to stop once the brakes were applied as a car traveling 20 miles per hour. After allowing discussion, tell students that it takes 4 times the distance to stop at 40 miles per hour. This activity will allow students to investigate these relationships.

Allow students to work in small groups to develop a mathematical model to describe speed as a function of breaking distance.

Exploring Square Root Functions

why
How can you find the length of one side of a square if you know its area? Finding the length of the side of a square is the same as finding the positive square root of a number.

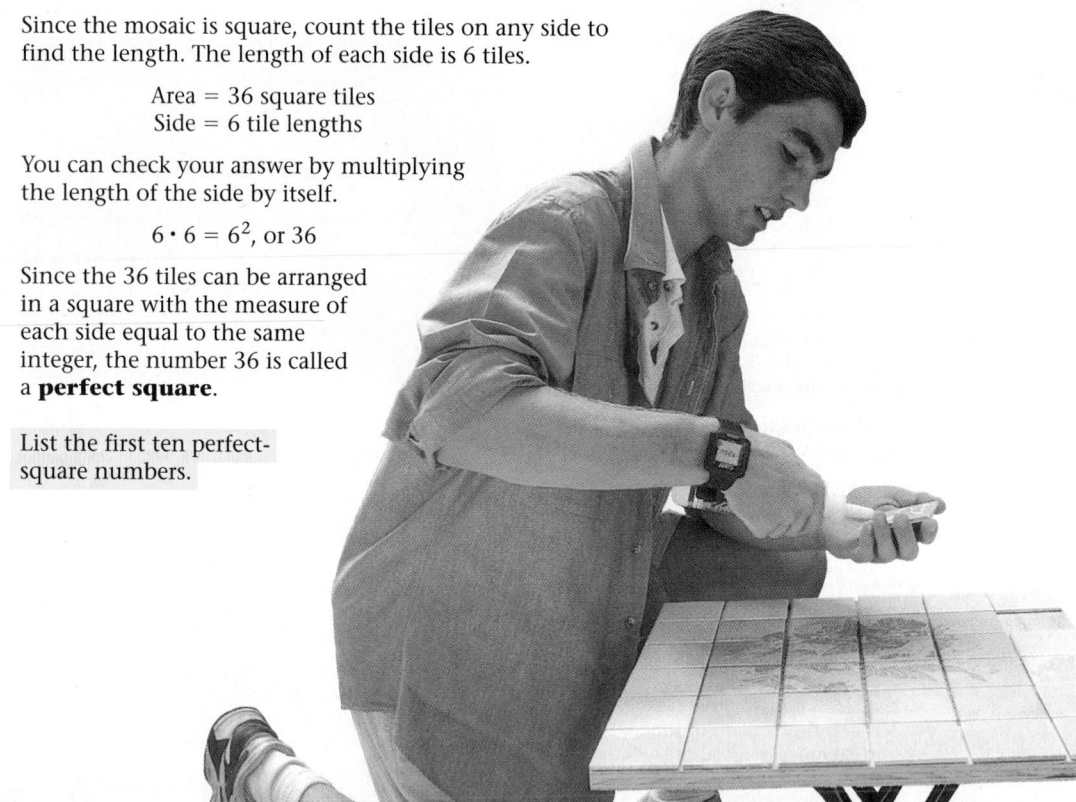

An artist is working on a square table-top mosaic with an area covered by 36 square tiles. What is the length of one side of that table?

Since the mosaic is square, count the tiles on any side to find the length. The length of each side is 6 tiles.

> Area = 36 square tiles
> Side = 6 tile lengths

You can check your answer by multiplying the length of the side by itself.

> $6 \cdot 6 = 6^2$, or 36

Since the 36 tiles can be arranged in a square with the measure of each side equal to the same integer, the number 36 is called a **perfect square**.

List the first ten perfect-square numbers.

Exploration 1 · Estimating Square Roots

How can you determine the length of the side of the blue square if its area is 12 square centimeters?

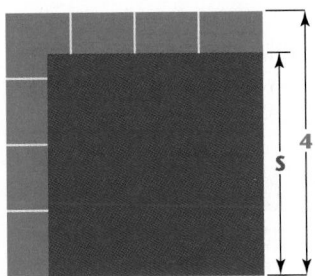

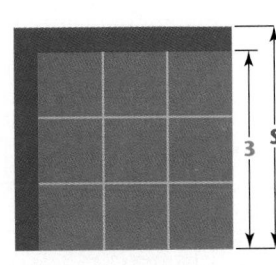

Side s is between 3 and 4.

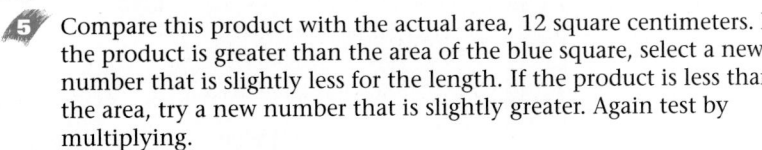

1. What is the area of the large green square?

2. What is the area of the large red square?

3. Estimate the length of a side of the blue square.

4. Multiply the length you find in Step 3 by itself.

5. Compare this product with the actual area, 12 square centimeters. If the product is greater than the area of the blue square, select a new number that is slightly less for the length. If the product is less than the area, try a new number that is slightly greater. Again test by multiplying.

6. Continue to guess and check until you have an estimate with 2 decimal places. You have estimated $\sqrt{12}$.

7. Use this procedure to estimate the decimal value for the length of a side of a square with area 20 square centimeters. Describe how to find $\sqrt{20}$ to the nearest hundredth. ❖

Irrational Numbers

Calculator

Every *rational number* can be represented by a terminating or repeating decimal. A repeating decimal is written with a bar over the numbers that repeat. On the other hand, when you calculate $\sqrt{12}$, the decimal part will never terminate or repeat. Numbers such as $\sqrt{12}$ or π are called *irrational numbers*.

Rational Numbers	Irrational Numbers
$\frac{5}{8} = 0.625$	$\sqrt{2} = 1.4142 \ldots$ $\pi = 3.1415926 \ldots$
$\frac{2}{3} = 0.\overline{6}$ (6 repeats forever.)	$\sqrt{12} = 3.4641 \ldots$ $a = 0.121221222 \ldots$

To save time and effort, many people use a calculator to find the square root. The square root key on most calculators is identified by the **radical sign**, $\sqrt{}$. When the $\boxed{\sqrt{}}$ is pressed, the square root is given as an exact number or as an approximation. On some calculators, $\sqrt{2}$ is represented by 1.414213562. This, however, is only an approximation. When a calculator displays an irrational number, it cannot keep producing digits indefinitely, so it rounds the value.

The *real numbers* include all the rational and irrational numbers. The real numbers form a **dense set**. That is, between any two real numbers there is another real number. Explain why the rationals are dense but the integers are not.

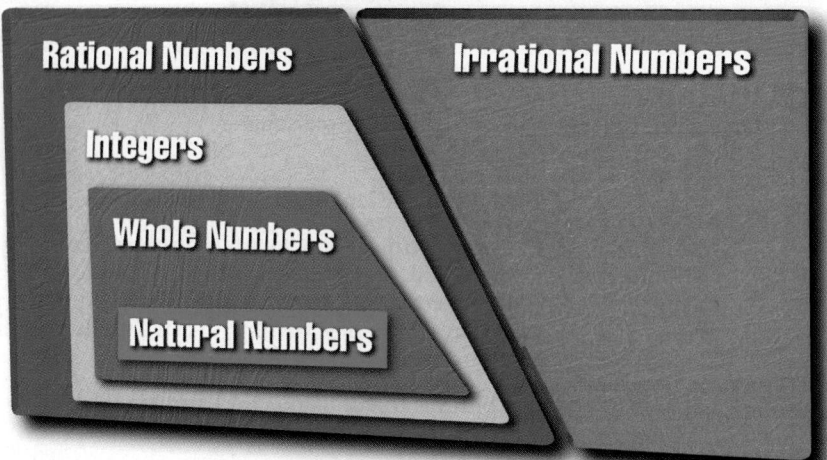

The Principal Square Root Function

In Lesson 12.3 you found the positive and negative square roots of a positive number. For example, 2 and -2 are square roots of 4 because $(-2)^2 = 4$ and $2^2 = 4$. The positive square root, sometimes called the **principal square root**, is indicated by the radical sign $\sqrt{}$.

Thus, $\sqrt{4} = 2$.

The function $f(x) = \sqrt{x}$ represents the parent function of the principal square root.

What are the domain and range of $f(x) = \sqrt{x}$?

The squaring function and the principal square root function are related. Plot $f(x) = \sqrt{x}$ and $g(x) = x^2$ on the same axes. Compare the two graphs. What is the relationship? Give as many similarities and differences for the two functions as you can.

•Exploration 2 *Transforming the Square Root Function*

1. Graph each function on the same axes.
   a. $g(x) = \sqrt{x} + 2$ **b.** $h(x) = \sqrt{x} + 5$ **c.** $p(x) = \sqrt{x} - 3$

2. Compare each function with $f(x) = \sqrt{x}$. What transformation results when $f(x) = \sqrt{x}$ is changed to $q(x) = \sqrt{x} + a$?

3. Graph each function on the same axes.
   a. $g(x) = \sqrt{x + 2}$ **b.** $h(x) = \sqrt{x + 5}$ **c.** $p(x) = \sqrt{x - 3}$

4. Compare each function with $f(x) = \sqrt{x}$. What transformation results when $f(x) = \sqrt{x}$ is changed to $q(x) = \sqrt{x - a}$?

5. Graph each function on the same axes.
   a. $g(x) = 2\sqrt{x}$ **b.** $h(x) = 3\sqrt{x}$ **c.** $p(x) = 5\sqrt{x}$

6. Compare each function with $f(x) = \sqrt{x}$. What transformation results when $f(x) = \sqrt{x}$ is changed to $g(x) = a\sqrt{x}$, when $a > 0$?

7. Graph each function on the same axes.
   a. $g(x) = -2\sqrt{x}$ **b.** $h(x) = -3\sqrt{x}$ **c.** $p(x) = -5\sqrt{x}$

8. Compare each function to $f(x) = \sqrt{x}$. What transformation results when $f(x) = \sqrt{x}$ is changed to $g(x) = a\sqrt{x}$, when $a < 0$?

9. Summarize your findings by explaining the results of statements 2, 4, 6, and 8. ❖

Compare the graph of $y = \sqrt{x}$ with the graph of $y = -\sqrt{x}$. The graph of $y = -\sqrt{x}$ shows that it is also a function. However, the relation $y = \pm\sqrt{x}$ is not a function.

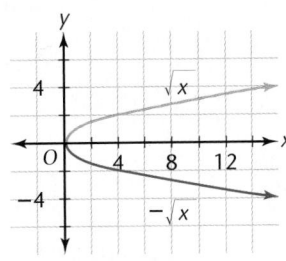

Explain the transformations that were applied to $f(x) = \sqrt{x}$ to produce g and h graphed below.

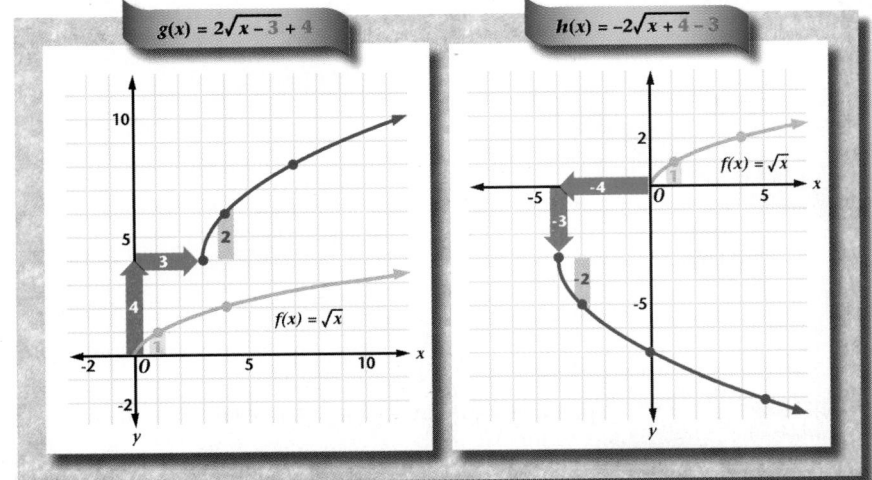

$g(x) = 2\sqrt{x - 3} + 4$

$h(x) = -2\sqrt{x + 4} - 3$

Exploration 2 Notes

Remind students that these exercises will allow them to come to a generalization about how a function that contains a radical can be transformed.

As students complete Exploration 2, suggest that they use black to graph the function $f(x) = \sqrt{x}$ and use a different color to write the function and to graph each part of the odd-numbered exercises; for instance, part a: red, part b: blue, part c: green.

Aongoing
SSESSMENT

9. $(\sqrt{x}) + a$ moves the graph vertically a units; $\sqrt{x} - a$ moves the graph horizontally to the right a units; $2\sqrt{x}$ stretches the values of y by a scale factor of 2; $-2\sqrt{x}$ reflects $2\sqrt{x}$ through the x-axis.

Aongoing
SSESSMENT

g: Stretch of 2, vertical shift of 4, horizontal shift of 3 to the right
h: Stretch of -2, vertical shift of -3, horizontal shift of 4 to the left

Have students work in small groups. On an index card, one person in the group writes a simple expression involving a radical in the form $g(x) = \pm\sqrt{x + a} + b$. Other members of the group graph the equation. When all members agree on the graph, draw it on the back of the card. After each group has completed 4 cards, have groups exchange cards. Then have each group look at the graph and write the equation.

TEACHING tip

Technology When using the radical key, be sure students determine whether they need to use parentheses.

CRITICAL
Thinking

There is no number multiplied by itself that yields a negative number as the product.

EXTENSION

The transformations in Exploration 2 can be combined to graph functions such as $f(x) = 3 - \sqrt{x + 2}$. First, write the function in the form $f(x) = -\sqrt{x + 2} + 3$.

The transformation $\sqrt{x + 2}$ shifts the graph of $\sqrt{x}$ left 2 units. Then $-\sqrt{x + 2}$ reflects the graph through the x-axis. Finally, $-\sqrt{x + 2} + 3$ shifts the graph up 3 units.

Thus, the transformation $f(x) = -\sqrt{x + 2} + 3$ shifts the graph to the left 2 units, reflects it through the x-axis, then shifts it up 3 units. ❖

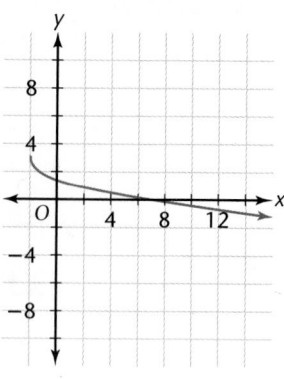

Janice Brown made the first solar-powered flight in the Solar Challenger on December 3, 1980.

APPLICATION

The approximate distance to the horizon (in kilometers) is given by the formula $D = 3.56\sqrt{x}$. D represents the distance to the horizon in kilometers and x is the altitude in meters. Find the approximate distance in kilometers to the horizon. Use the formula $3.56\sqrt{x}$ with $x = 20$.

$$D = 3.56\sqrt{20}$$
$$D \approx 3.56 \cdot 4.472$$
$$D \approx 15.92$$

The distance to the horizon is approximately 15.92 kilometers. ❖

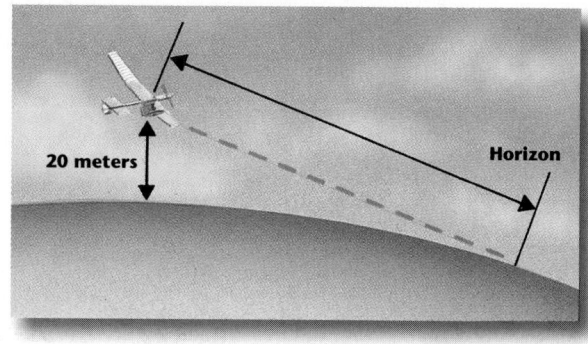

20 meters

Horizon

CRITICAL
Thinking

Explain why the square root of a negative number does not make sense for real numbers.

EXERCISES & PROBLEMS

Communicate

1. Describe how you can determine the square root of a perfect square with graph paper.

2. How can you estimate the square root of a number that is not a perfect square with graph paper?

3. What are the characteristics of an irrational number? Give three examples.

4. Explain how to estimate $\sqrt{7}$.

5. Explain how to write the transformation that moves the graph of the square root function up 3 and right 2.

6. Explain why every positive number has two square roots.

Practice & Apply

Estimate each square root. If the square root is irrational, find the value to the nearest hundredth.

7. $\sqrt{225}$ 15   **8.** $-\sqrt{169}$ -13 **9.** $\sqrt{11}$ 3.32   **10.** $\sqrt{\dfrac{4}{9}}$ $\dfrac{2}{3}$ **11.** $-\sqrt{40}$ -6.32

12. $-\sqrt{27}$ -5.2 **13.** $\sqrt{1000}$ 31.62   **14.** $\sqrt{10,000}$ 100   **15.** $-\sqrt{0.04}$ -0.2 **16.** $\sqrt{0.059}$ 0.24

Geometry Find the length of the side of each square that has the given area.

17. 250 square meters 15.81 m

18. 144 square centimeters 12 cm

19. 28 square miles 5.29 miles

20. The Smiths have a yard in the shape of a square. If the area of the yard is 676 square feet, what is the length of each side? 26 feet

21. Use the formula, $D = 1.22\sqrt{A}$ to find the distance in miles to the horizon if an airplane is flying at an altitude of 30,000 feet. ≈ 211 miles

22. Technology Use a calculator for the following activities. Choose five positive numbers greater than 50. Find the square of each positive number. Then find the square root of the result. Record your results in a table with the following headings. One example is started for you.

Positive Number	(Positive Number)^2	$\sqrt{(\text{Positive Number})^2}$
52	2704	?

Formulate a conjecture based on the data in your table.

22. Answers may vary. For example:

+ number	(+ number)^2	$\sqrt{(+\text{ number})^2}$
52	2704	52
54	2916	54
60	3600	60
75	5625	75
90	8100	90

The square root of a positive number squared is the positive number.

ASSESS

Selected Answers
Odd-Numbered Exercises 7–55

Assignment Guide
Core 1–6, 8–20 even, 21–35, 36–42 even, 43–60

Core Two-Year 1–56

Technology
Students will need to use their calculators to complete Exercises 7–28 and Exercises 41 and 46.

Error Analysis
Students may make errors when using their calculators with radicals. Remind students to check whether parentheses are required for expressions entered under the radical sign.

Performance Assessment

Provide each student with grid paper. Have students fold the grid paper into fourths. Tell students to graph one expression in each quarter. The following expressions are to be graphed.

a. $\sqrt{x}$

b. $\sqrt{x + 3}$

c. $\sqrt{x} + 2$

d. $\sqrt{x + 3} + 2$

23. Technology Choose five negative numbers. Find the square of each negative number, and then find the square root of the result. Record your results in a table with the following headings.

Negative Number	(Negative Number)^2	$\sqrt{\text{(Negative Number)}^2}$
?	?	?

Formulate a conjecture based on the data in your table.

Visualize the transformations of the principal square root function. Use a calculator or graph paper to graph the following.

24. $y = \sqrt{x} + 6$ **25.** $y = 4\sqrt{x}$ **26.** $y = \sqrt{x - 2}$

27. $y = -\sqrt{x} + 1$ **28.** $y = 2\sqrt{x + 1} - 3$ **29.** $y = -2\sqrt{x - 1} + 6$

30. Write equations for two radical functions whose graphs create a parabola with the vertex at (0, 3).

31. Find the vertex and three additional ordered pairs for each pair of functions. Plot the ordered pairs on the same coordinate axes. State the domain and range.

a. $y = \sqrt{4 - x}$ and $y = -\sqrt{4 - x}$ b. $y = \sqrt{-x} + 3$ and $y = -\sqrt{-x} + 3$
V: (4, 0); D: $x \le 4$; R: all numbers   V: (0, 3); D: $x \le 0$; R: all numbers

32. Test each statement with numbers. Then determine which statements are always true, and explain your reasoning. Give a counterexample if the statement is false.

a. $\sqrt{k^2} = k$, for any k. F   b. $\sqrt{k^2} = |k|$, for any k. T

c. $\sqrt{k^2}$ must be written two different ways, depending on the value of k. T

33. For what values of x is $\sqrt{x}$ defined? $x \ge 0$

34. For what values of x is $\sqrt{x + 5}$ defined? $x \ge -5$

35. What is the domain of the function $y = \sqrt{x + 5}$? $x \ge -5$

Identify the transformation(s), and describe the effect on the parent function.

36. $y = -\sqrt{x - 4}$ **37.** $y = 5\sqrt{x + 3} - 4$ **38.** $y = 2.00 - 3.50\sqrt{x + 3.75}$

Write the equation of the function graphed.

39.
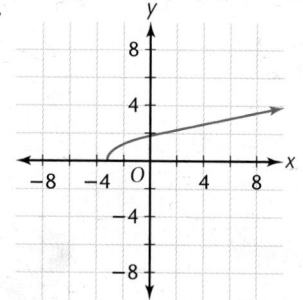

40.

41.
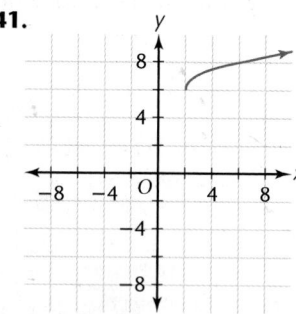

23. Answers may vary. For example:

$-$ number	$(-\text{number})^2$	$\sqrt{(-\text{number})^2}$
-52	2704	52
-54	2916	54
-60	3600	60
-75	5625	75
-90	8100	90

The square root of a negative number squared is the positive value of the number.

The answers to Exercises 24–29, 31, and 37–41 can be found in Additional Answers beginning on page 729.

30. Answers may vary, for example, $y = \sqrt{x} + 3$ and $y = -\sqrt{x} + 3$.

36. The graph of the parent function is shifted horizontally to the right by 4 units, and then reflected through the x-axis.

42. Technology Copy and complete the table below. Use a calculator to obtain decimal approximations for the square roots to the nearest tenth.

x	0	1	2	3	4	5	6	7	8	9	10
$\sqrt{x}$											

43. Plot the data in your table on graph paper with x on the horizontal axis and $\sqrt{x}$ on the vertical axis.

Is it possible to have a point on the graph in Exercise 43 in the

44. second quadrant?   **45.** third quadrant?   **46.** fourth quadrant?
no   no   no

47. Technology Use graphing technology to graph $y = \sqrt{x}$. Trace the function. Do the values in your table agree with trace values on the calculator?

Look Back

Without using graphing technology, sketch the graph from what you know about transformations. [Lesson 9.2]

48. $y = 3x + 2$ **49.** $y = x^2 - 4$ **50.** $y = 3 - 2^x$ **51.** $y = -x^2 + 6$

Find the vertex of each absolute value function. Use graphing technology to check your answer. [Lesson 9.6]

52. $y = |x - 2|$ **53.** $y = |x| + 5$
54. $y = 2|x + 3| - 4$ **55.** $y = -|x - 1| + 3$

56. Solve the equation $x^2 - 5x - 7 = 0$. [Lesson 12.3]

$\dfrac{5 \pm \sqrt{53}}{2}$

Look Beyond

Decide whether the following statements are true or false. If you believe that a statement is false, give a counterexample.

57. If $a^2 = b^2$, then $a = b$. F

58. If $a = b$, then $a^2 = b^2$. T

59. If $a^2 = b^2$ and both a and b have the same sign, then $a = b$. T

60. In a long-distance race of 4 runners, Caitlin finished ahead of Andrew. David finished with a time 0.1 second more than Barbara, who finished 1 second after the winner. David was faster than only 1 other runner. Who won the race, and what was the order of finish? Caitlin; Barbara; David; Andrew

Look Beyond

Exercises 57–59 extend the study of the square root function to solving equations by using radicals or by finding the square.

Technology Master

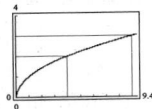

Technology
13.1 Almost a Square Root

From the diagram of the square root function, you can see that the curve appears continuous, unbroken, and rises to the right. This observation allows you to explore in detail square roots of numbers. Notice from the inequality shown that 2 and 3 are not square roots of 5, but they are somewhat close to it.

$2 < \sqrt{5} < 3$

Since $\sqrt{5}$ is an irrational number, you can never represent it exactly. You can, however, find numbers that are almost equal to $\sqrt{5}$. A spreadsheet is a good tool to use to understand how this can be done.

Use a spreadsheet in the exercises.

1. Create a spreadsheet in which A1 contains 2, cell A2 contains A1+0.1, and the first 11 rows are filled. In column B, enter the formula for squaring each entry in column A. What do the numbers in column B tell you about numbers that are almost equal to $\sqrt{5}$?

2. Explain how to modify the spreadsheet in Exercise 1 to obtain two rational numbers between which $\sqrt{5}$ is located and that differ by 0.01. Use the spreadsheet to find two numbers that are almost equal to $\sqrt{5}$ and closer to $\sqrt{5}$ than the numbers in column A from Exercise 1.

Use a spreadsheet to find two rational numbers between which the given square root is located and that differ by 0.001. Show your results from each pass through the approximation process.

3. $\sqrt{5}$ 4. $\sqrt{17}$ 5. $\sqrt{63}$ 6. $\sqrt{630}$

7. Suppose that your spreadsheet had the capability of calculating numbers to any degree of accuracy and display them. Suppose also that you found two rational numbers a and b between which $\sqrt{5}$ is located and that differ by 10^{-n}. Explain how to modify your spreadsheet from Exercise 1 to find two rational numbers a' and b' between which $\sqrt{5}$ is located and that differ by even less.

42.

x	0	1	2	3	4	5	6	7	8	9	10
$\sqrt{x}$	0	1	1.4	1.7	2	2.2	2.4	2.6	2.8	3	3.2

The answers to Exercises 43 and 47–51 can be found in Additional Answers beginning on page 729.

52. $(2, 0)$ **54.** $(-3, -4)$
53. $(0, 5)$ **55.** $(1, 3)$

- Define square root and write square roots in simplest radical form.
- Examine mathematical operations with radicals.

RESOURCES

- Practice Master   13.2
- Enrichment Master   13.2
- Technology Master   13.2
- Lesson Activity Master 13.2
- Quiz   13.2
- Spanish Resources   13.2

Assessing Prior Knowledge

Simplify.

1. $\dfrac{a^4b^3}{a^3b^5}$ $\left[\dfrac{a}{b^2} \text{ or } ab^{-2}\right]$
2. $(-5c^2)^2$ $[25c^4]$
3. $-5(c^2)^2$ $[-5c^4]$
4. $m^3n^6 \cdot m^4n^2$ $[m^7n^8]$

TEACH

why Ask students the difference between adding fractions and adding whole numbers. Answers will vary, but students should understand that there are different procedures involved. Tell students that this lesson will present new procedures that will allow them to work with radicals.

ongoing ASSESSMENT

If x is not restricted, x can be positive or negative. However, x^2 will still be positive.

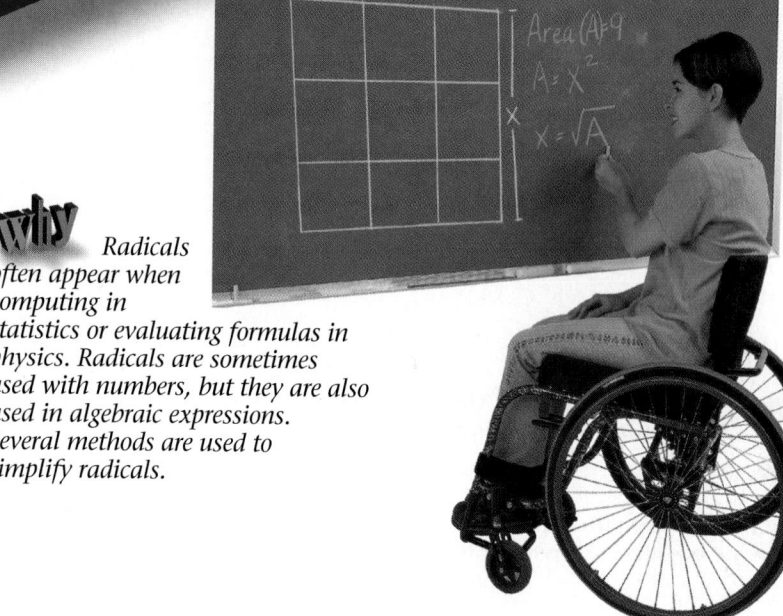

LESSON 13.2 Operations With Radicals

why *Radicals often appear when computing in statistics or evaluating formulas in physics. Radicals are sometimes used with numbers, but they are also used in algebraic expressions. Several methods are used to simplify radicals.*

The area model is restricted to positive values. What happens when $A = x^2$ and x is not restricted? Recall that 9 has two square roots, 3 and -3.

DEFINITION OF SQUARE ROOT

If a is a number greater than or equal to zero, $\sqrt{a}$ and $-\sqrt{a}$ represent the square roots of a. Each square root of a has the following property.

$$\sqrt{a} \cdot \sqrt{a} = a \qquad (-\sqrt{a})(-\sqrt{a}) = a$$

When you work with radicals, you should know the name for each part.

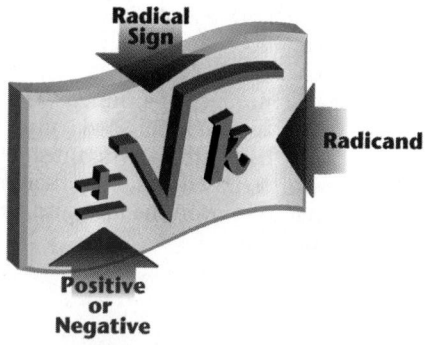

Radical Sign

Radicand

Positive or Negative

ALTERNATIVE teaching strategy

Using Discussion Ask students how they simplify fractions and how they know when the simplification is correct. [**Divide numerator and denominator by common factors; when the numerator and denominator only have 1 as a common factor.**] Tell students that a radical expression is in simplest form if

1. the radical sign contains no perfect squares greater than 1,
2. the radical sign contains no fractions, and
3. the radical sign is not in the denominator of a fraction.

Make up radical expressions that are not in simplest form, and have the class discuss how to simplify each expression.

EXAMPLE 1

Find each square root. $\textbf{A}$ $\sqrt{25}$ $\textbf{B}$ $-\sqrt{144}$ $\textbf{C}$ $\sqrt{7}$

Solution ➤

$\textbf{A}$ $\sqrt{25} = 5$ because $5 \cdot 5 = 25$.

$\textbf{B}$ $-\sqrt{144} = -12$ because $-(12 \cdot 12) = -144$.

$\textbf{C}$ $\sqrt{7}$ is not a perfect square. It is approximately 2.646. ❖

Try This Find each square root. **a.** $\sqrt{196}$ **b.** $-\sqrt{225}$

Simplifying Radicals and Radical Expressions

Radical expressions that are simplified allow you to manipulate algebraic expressions more easily. In some cases the calculations can be done without technology. A radical expression is in **simplest radical form** if the expression within the radical sign

1. contains no perfect squares greater than 1,

2. contains no fractions, and

3. is not in the denominator of a fraction.

EXAMPLE 2

Simplify.

$\textbf{A}$ $\sqrt{20}$ $\textbf{B}$ $\dfrac{\sqrt{12}}{4}$ $\textbf{C}$ $\dfrac{\sqrt{18}}{\sqrt{2}}$

Solution ➤

$\textbf{A}$ $\sqrt{20}$ is not a perfect square, but it can be factored so the radicand contains a perfect square. Simplify further.

$$\sqrt{20} = \sqrt{2^2 \cdot 5} = \sqrt{2^2}\sqrt{5} = 2\sqrt{5}$$

$\textbf{B}$ $\dfrac{\sqrt{12}}{4} \cdot \dfrac{\sqrt{2^2 \cdot 3}}{4} = \dfrac{2\sqrt{3}}{4} = \dfrac{\sqrt{3}}{2}$

$\textbf{C}$ $\dfrac{\sqrt{18}}{\sqrt{2}} = \sqrt{\dfrac{18}{2}} = \sqrt{9} = 3$ ❖

When the radicand contains variables such as a^4 or b^5, rewrite the expression as a product of even powers. Then simplify.

$$\sqrt{a^4 \cdot b^5} = \sqrt{a^2 \cdot a^2 \cdot b^2 \cdot b^2 \cdot b}$$
$$= \sqrt{(a^2)^2 \cdot (b^2)^2 \cdot b}$$
$$= a^2 b^2 \sqrt{b}$$

Alternate Example 3

Simplify.

a. $\sqrt{147}$ $[7\sqrt{3}]$
b. $\sqrt{6400}$ $[80]$
c. $\sqrt{a^3b^9}$ $[ab^4\sqrt{ab}]$

Aongoing
ASSESSMENT

Try This

$6mn^2\sqrt{2n}$

Alternate Example 4

Simplify.

a. $\sqrt{\dfrac{9}{121}}$ $\left[\dfrac{3}{11}\right]$

b. $\sqrt{\dfrac{11}{196}}$ $\left[\dfrac{\sqrt{11}}{14}\right]$

c. $\sqrt{\dfrac{100}{27}}$ $\left[10\dfrac{\sqrt{3}}{9}\right]$

d. For $x \geq 0$, and $z, y > 0$,

$\sqrt{\dfrac{(xy^5)}{y^3z^2}}$ $\left[\dfrac{y}{z}\sqrt{x}\right]$

> **MULTIPLICATION PROPERTY OF SQUARE ROOTS**
> For all numbers a and b, such that a and $b \geq 0$,
> $$\sqrt{ab} = \sqrt{a}\sqrt{b}.$$

EXAMPLE 3

Simplify.

Ⓐ $\sqrt{12}$ Ⓑ $\sqrt{400}$ Ⓒ $\sqrt{a^5b^{12}}$

Solution ➤

Look for perfect square factors, and apply the Multiplication Property of Square Roots. Then find the square roots of the perfect squares. Leave the factor that is not a perfect square in radical form.

Ⓐ $\sqrt{12} = \sqrt{4 \cdot 3} = \sqrt{4} \cdot \sqrt{3} = 2\sqrt{3}$

Ⓑ $\sqrt{400} = \sqrt{4 \cdot 100} = \sqrt{4} \cdot \sqrt{100} = 2 \cdot 10 = 20$

Ⓒ $\sqrt{a^5b^{12}} = \sqrt{(a^2)^2 \cdot a \cdot (b^6)^2} = a^2b^6\sqrt{a}$ ❖

Try This   Simplify $\sqrt{72m^2n^5}$.

> **DIVISION PROPERTY OF SQUARE ROOTS**
> For all numbers $a \geq 0$ and $b > 0$,
> $$\sqrt{\dfrac{a}{b}} = \dfrac{\sqrt{a}}{\sqrt{b}}.$$

EXAMPLE 4

Simplify.

Ⓐ $\sqrt{\dfrac{16}{25}}$ Ⓑ $\sqrt{\dfrac{7}{16}}$ Ⓒ $\sqrt{\dfrac{9}{5}}$ Ⓓ $\sqrt{\dfrac{a^2b^3}{c^2}}$

Solution ➤

Rewrite the square root using the Division Property of Square Roots. Then simplify the numerator and denominator separately.

Ⓐ $\sqrt{\dfrac{16}{25}} = \dfrac{\sqrt{16}}{\sqrt{25}} = \dfrac{4}{5}$

Ⓑ $\sqrt{\dfrac{7}{16}} = \dfrac{\sqrt{7}}{\sqrt{16}} = \dfrac{\sqrt{7}}{4}$

RETEACHING
the
lesson

When simplifying radical expressions, have students completely factor expressions and then identify pairs of identical factors under the radical symbol. Simplify $\sqrt{m^4n^3}$.

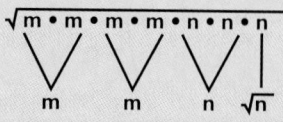

C $\sqrt{\dfrac{9}{5}} = \dfrac{\sqrt{9}}{\sqrt{5}} = \dfrac{3}{\sqrt{5}}$ You can can simplify this expression by multiplying

the fraction by $\dfrac{\sqrt{5}}{\sqrt{5}}$ or 1. $\qquad \dfrac{3}{\sqrt{5}} \cdot \dfrac{\sqrt{5}}{\sqrt{5}} = \dfrac{3\sqrt{5}}{\sqrt{5}\sqrt{5}} = \dfrac{3\sqrt{5}}{\sqrt{25}} = \dfrac{3\sqrt{5}}{5}$

This is known as **rationalizing the denominator**.

D For a, $b \geq 0$, and $c > 0$, $\sqrt{\dfrac{a^2 b^3}{c^2}} = \dfrac{\sqrt{a^2 b^2 b}}{\sqrt{c^2}} = \dfrac{ab\sqrt{b}}{c}$. ❖

Operations and Radical Expressions

You know that $\sqrt{16 \cdot 9} = \sqrt{16} \cdot \sqrt{9}$.
What is the relationship between
$\sqrt{16 + 9}$ and $\sqrt{16} + \sqrt{9}$? Complete
the computations to find the answer.

$$\sqrt{16 + 9} = \sqrt{25} \text{ or } 5$$
$$\sqrt{16} + \sqrt{9} = 4 + 3 \text{ or } 7$$

Since 5 does not equal 7, $\sqrt{16 + 9}$
does not equal $\sqrt{16} + \sqrt{9}$. In general,
$\sqrt{a + b} \neq \sqrt{a} + \sqrt{b}$.

Radicands must be the same to add
expressions that contain radicals.
Like terms that contain $\sqrt{5}$, for
example, can be added by the
Distributive Property.

$$2\sqrt{5} + 4\sqrt{5} = (2 + 4)\sqrt{5} = 6\sqrt{5}$$

EXAMPLE 5

Simplify.

A $5\sqrt{6} - 2\sqrt{6}$

B $5 + 6\sqrt{7} - 2\sqrt{7} - 3$

C $8\sqrt{3} + 6\sqrt{2} - \sqrt{3} + 2\sqrt{2}$

D $a\sqrt{x} + b\sqrt{x}$

Solution ➤

A Use the Distributive Property to combine like terms. In this case, $\sqrt{6}$ is
the common factor. Then simplify.
$$5\sqrt{6} - 2\sqrt{6} = (5 - 2)\sqrt{6} = 3\sqrt{6}$$

B Rearrange and combine like terms.
$$5 + 6\sqrt{7} - 2\sqrt{7} - 3 = (5 - 3) + (6 - 2)\sqrt{7} = 2 + 4\sqrt{7}$$

C Rearrange and combine like terms.
$$8\sqrt{3} + 6\sqrt{2} - \sqrt{3} + 2\sqrt{2} = (8 - 1)\sqrt{3} + (6 + 2)\sqrt{2} = 7\sqrt{3} + 8\sqrt{2}$$

D Literal expressions are treated in a similar way.
$$a\sqrt{x} + b\sqrt{x} = (a + b)\sqrt{x} ❖$$

Alternate Example 5

a. $7\sqrt{3} - 9\sqrt{3}$ $\quad [-2\sqrt{3}]$
b. $6 - 4\sqrt{2} - 9 + 3\sqrt{2}$
$\qquad [-3 - \sqrt{2}]$
c. $4\sqrt{5} + \sqrt{11} - 2\sqrt{11} +$
$\quad 9\sqrt{5}$ $\;[13\sqrt{5} - \sqrt{11}]$
d. $m\sqrt{x} - n\sqrt{x}$
$\qquad\qquad [(m - n)\sqrt{x}]$

No. Explanations may vary. The only condition that allows the statement to be true is if b is 0 or if both a and b are 0.

Alternate Example 6

Simplify.

a. $(2\sqrt{3})^2$ [12]

b. $\sqrt{8}\sqrt{25}$ [$10\sqrt{2}$]

c. $\sqrt{3}(\sqrt{6} + \sqrt{25})$

    [$3\sqrt{2} + 5\sqrt{3}$]

d. $(8 - \sqrt{7})(2 + 3\sqrt{7})$

    [$-5 + 22\sqrt{7}$]

Aongoing
ASSESSMENT

19

Practice Master

Does $\sqrt{a} - \sqrt{b}$ equal $\sqrt{a - b}$ for any a and b? Substitute values to test the idea. Explain the results of the test. Are there any values of a and b that allow the statement to be true?

There are several ways that radicals may appear in multiplication.

EXAMPLE 6

Simplify.

Ⓐ $(5\sqrt{3})^2$

Ⓑ $\sqrt{3}\sqrt{6}$

Ⓒ $\sqrt{2}(6 + \sqrt{12})$

Ⓓ $(3 - \sqrt{2})(4 + \sqrt{2})$

Solution ➤

Ⓐ Recall from the properties of exponents that the second power means the product of two identical factors. Rearrange the factors and multiply.

$$(5\sqrt{3})^2 = (5\sqrt{3})(5\sqrt{3}) = (5 \cdot 5)(\sqrt{3}\sqrt{3}) = 25 \cdot 3 = 75$$

Ⓑ The Multiplication Property of Square Roots allows you to multiply separate radicals. You can then factor the new product in a different way and simplify.

$$\sqrt{3}\sqrt{6} = \sqrt{3 \cdot 6} = \sqrt{18} = \sqrt{9 \cdot 2} = \sqrt{9}\sqrt{2} = 3\sqrt{2}$$

Ⓒ Use the Distributive Property to multiply, then factor again and simplify the result. Notice that the radical is usually written last in each term.

$$\sqrt{2}(6 + \sqrt{12}) = \sqrt{2} \cdot 6 + \sqrt{2} \cdot \sqrt{12}$$
$$= 6\sqrt{2} + \sqrt{2 \cdot 12}$$
$$= 6\sqrt{2} + \sqrt{24}$$
$$= 6\sqrt{2} + \sqrt{4 \cdot 6}$$
$$= 6\sqrt{2} + 2\sqrt{6}$$

Ⓓ To multiply differences and sums, multiply the two binomials.

$$(3 - \sqrt{2})(4 + \sqrt{2}) = 12 + 3\sqrt{2} - 4\sqrt{2} - 2 = 10 - \sqrt{2} ❖$$

Calculator

When you enter more complicated expressions into a calculator, you may need to use parentheses. To find the square root of $3^2 + 4^2$, enter $\sqrt{(3^2 + 4^2)}$ into the calculator.

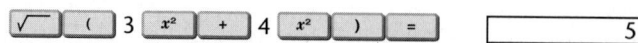

What do you think you will get on a calculator if the parentheses are not used?

EXERCISES & PROBLEMS

Communicate

1. Explain when $\sqrt{a^2} = a$.

2. How is factoring used to simplify expressions that contain radicals? What is simplest radical form?

3. For what values of a can you find $\sqrt{-a + 3}$?

4. Explain in your own words what the Multiplication and Division Properties of Square Roots allow you to do.

5. When is it possible to add two radical expressions?

6. Describe a procedure for multiplying $(\sqrt{3} + 2)(\sqrt{2} + 3)$.

Practice & Apply

Simplify the radicals by factoring.

7. $\sqrt{49}$ 7
8. $\sqrt{196}$ 14
9. $\sqrt{576}$ 24
10. $\sqrt{3600}$ 60
11. $\sqrt{75}$ $5\sqrt{3}$
12. $\sqrt{98}$ $7\sqrt{2}$
13. $\sqrt{1620}$ $18\sqrt{5}$
14. $\sqrt{264}$ $2\sqrt{66}$

Decide whether the given statement is true or false. Assume that $a, b \geq 0$.

15. $\sqrt{a + b} = \sqrt{a} + \sqrt{b}$ F
16. $\sqrt{a - b} = \sqrt{a} - \sqrt{b}$ F
17. $\sqrt{ab} = \sqrt{a}\sqrt{b}$ T
18. $\sqrt{\dfrac{a}{b}} = \dfrac{\sqrt{a}}{\sqrt{b}}, \, b \neq 0$ T

Express in simplest radical form.

19. $\sqrt{3}\sqrt{12}$ 6
20. $\sqrt{8}\sqrt{18}$ 12
21. $\sqrt{48}\sqrt{3}$ 12
22. $\sqrt{54}\sqrt{6}$ 18
23. $\sqrt{\dfrac{64}{16}}$ 2
24. $\sqrt{\dfrac{96}{2}}$ $4\sqrt{3}$
25. $\dfrac{\sqrt{50}}{\sqrt{8}}$ $\dfrac{5}{2}$
26. $\dfrac{\sqrt{150}}{\sqrt{6}}$ 5
27. $\sqrt{5}\sqrt{15}$ $5\sqrt{3}$
28. $\sqrt{98}\sqrt{14}$ $14\sqrt{7}$
29. $\sqrt{\dfrac{56}{8}}$ $\sqrt{7}$
30. $\dfrac{\sqrt{96}}{\sqrt{8}}$ $2\sqrt{3}$

Simplify each of the following. Assume that all variables are non-negative and that all denominators are non-zero.

31. $\sqrt{a^4 b^6}$
32. $\sqrt{x^8 y^9}$
33. $\sqrt{\dfrac{p^9}{q^{10}}}$
34. $\sqrt{\dfrac{x^3}{y^6}}$

31. $a^2 b^3$

32. $x^4 y^4 \sqrt{y}$

33. $\dfrac{p^4 \sqrt{p}}{q^5}$

34. $\dfrac{x \sqrt{x}}{y^3}$

Portfolio Assessment

Have students choose examples and exercises that show their understanding of how to operate with radical expressions, including using the Distributive Property and finding products, quotients, and powers.

If possible, perform the indicated operation, and simplify your answer.

35. $3\sqrt{5} + 4\sqrt{5}$
$7\sqrt{5}$

36. $\sqrt{10} + \sqrt{15}$
$\sqrt{10} + \sqrt{15}$

37. $\sqrt{7} + \sqrt{29}$
$\sqrt{7} + \sqrt{29}$

38. $4\sqrt{5} + 2\sqrt{5} - 5\sqrt{5}$
$\sqrt{5}$

39. $\sqrt{6} + 2\sqrt{3} - \sqrt{6}$
$2\sqrt{3}$

40. $(4 + \sqrt{3}) + (1 - \sqrt{2})$
$5 - \sqrt{2} + \sqrt{3}$

41. $\frac{6 + \sqrt{18}}{3}$
$2 + \sqrt{2}$

42. $\frac{\sqrt{15} + \sqrt{10}}{\sqrt{5}}$
$\sqrt{3} + \sqrt{2}$

Simplify.

43. $(3\sqrt{5})^2$ 45

44. $(4\sqrt{25})^2$ 400

45. $\sqrt{12}\sqrt{6}$ $6\sqrt{2}$

46. $\sqrt{72}\sqrt{32}$ 48

47 $3(\sqrt{5} + 9)$ $3\sqrt{5} + 27$

48. $\sqrt{5}(6 - \sqrt{15})$ $6\sqrt{5} - 5\sqrt{3}$

49. $\sqrt{6}(6 + \sqrt{18})$ $6\sqrt{6} + 6\sqrt{3}$

50. $(\sqrt{5} - 2)(\sqrt{5} + 2)$ 1

51. $(\sqrt{3} - 4)(\sqrt{3} + 2)$ $-5 - 2\sqrt{3}$

52. $(\sqrt{5} + 7)(\sqrt{2} - 8)$
$\sqrt{10} - 8\sqrt{5} + 7\sqrt{2} - 56$

53. $(\sqrt{6} + 2)^2$
$10 + 4\sqrt{6}$

54. $\sqrt{3}(\sqrt{3} + 2)^2$
$12 + 7\sqrt{3}$

55. **Geometry** Copy and complete the table. Each entry should be simplified.

56. Add three more columns to your table according to the number pattern established in the Area of Square column.

Area of Square	6	12	24	48	96
Length of side	?	?	?	?	?

Look Back

57. Find the solution to the system of equations. **[Lesson 6.2]** (7, 13)
$\begin{cases} y = -x + 20 \\ y = 2x - 1 \end{cases}$

Simplify. [Lesson 10.2]

58. $(-a^2b^2)^3(a^4b)^2$
$-a^{14}b^8$

59. $\frac{x^5y^7}{x^2y^3}$ x^3y^4

60. $\left(\frac{20x^3}{-4x^2}\right)^3$ $-125x^3$

Find each product. [Lesson 11.3]

61. $(2x - 4)(2x - 4)$
$4x^2 - 16x + 16$

62. $(3a + 5)(2a - 6)$
$6a^2 - 8a - 30$

63. $(6b + 1)(3b - 1)$
$18b^2 - 3b - 1$

Solve each equation by completing the square. [Lesson 12.3]

64. $y^2 - 8y + 12 = 0$
2, 6

65. $x^2 + 14x - 15 = 0$
1, -15

66. $r^2 - 24r + 63 = 0$
21, 3

Look Beyond

Exercises 67–71 extend the study of radicals to include cube roots.

Look Beyond

67. Notice that $2 \cdot 2 \cdot 2 = 2^3 = 8$. Then $\sqrt[3]{8} = 2$ (read the cube root of 8 equals 2). Copy and complete the following table.

x	1	2	3	4	5	6	7	8	9	10
x^3	1	8	27	?	?	?	?	?	?	?

Based on the data in the table, answer Exercises 68–71.

68. $\sqrt[3]{125} = $? 5

69. $\sqrt[3]{1000} = $? 10

70. $\sqrt[3]{100}$ is between what two whole numbers? 4 and 5

71. **Technology** Use a calculator to approximate $\sqrt[3]{100}$ to the nearest tenth. 4.6

55–56.

Area of square	6	12	24	48	96	192	384	768
Length of side	$\sqrt{6}$	$2\sqrt{3}$	$2\sqrt{6}$	$4\sqrt{3}$	$4\sqrt{6}$	$8\sqrt{3}$	$8\sqrt{6}$	$16\sqrt{3}$

67.

x	1	2	3	4	5	6	7	8	9	10
x^3	1	8	27	64	125	216	343	512	729	1000

LESSON 13.3 Solving Radical Equations

PREPARE

Objectives

- Solve equations containing radicals.
- Solve equations by using radicals.

RESOURCES

- Practice Master   13.3
- Enrichment Master   13.3
- Technology Master   13.3
- Lesson Activity Master   13.3
- Quiz   13.3
- Spanish Resources   13.3

Equations that contain radicals are often used in science to model natural phenomena. Scientists also use radicals to solve equations that contain squares of numbers and variables. The mathematics of radicals provides a necessary tool to solve many problems that occur in science and other areas.

Assessing Prior Knowledge

Simplify.

1. $(3\sqrt{5})^2$ [**45**]

2. $\sqrt{9} \cdot \sqrt{8}$ [**$6\sqrt{2}$**]

3. $7\sqrt{5} - 9\sqrt{5}$ [**$-2\sqrt{5}$**]

4. $\sqrt{\dfrac{120}{16}}$ $\left[\dfrac{\sqrt{30}}{2}\right]$

5. $\dfrac{\sqrt{9}}{\sqrt{8}}$ $\left[\dfrac{3\sqrt{2}}{4}\right]$

The relationship of the length of a pendulum to the number of swings per minute is one example of an equation that contains a radical.

Physics In a physical science class, six groups are given six strings of different sizes, each tied to a washer. They are asked to time one complete swing of each pendulum. Then they are asked to put all their data together, make a table, and draw a graph of the data. The point (0, 0) represents the time for a string of zero length.

TEACH

Hold a long string with a weight attached to one end and let it swing like a pendulum. Then hold the string closer to the weight and swing the weight again. Ask: Which takes longer, for the weight to go from one side to the other when the string is short or when it is long? [**long**] Tell students that they will learn how to make a mathematical model of this situation in today's lesson.

ALTERNATIVE teaching strategy

Using Algorithms Show that the quadratic formula can also be used to solve quadratic equations, and show the need for checking all solutions.

Solve $x + 2 = x\sqrt{3}$ using the quadratic formula.

Square both sides. $x^2 + 4x + 4 = 3x^2$
Write in standard form. $2x^2 - 4x - 4 = 0$

Simplify by dividing both sides by 2.
$x^2 - 2x - 2 = 0$

Apply formula. $x = \dfrac{2 \pm \sqrt{4 + 8}}{2} = 1 \pm \sqrt{3}$

CHECK: $1 + \sqrt{3} + 2 \overset{?}{=} (1 + \sqrt{3})(\sqrt{3})$
$\qquad 3 + \sqrt{3} = \sqrt{3} + 3$ True

$\qquad 1 - \sqrt{3} + 2 \overset{?}{=} (1 - \sqrt{3})(\sqrt{3})$
$\qquad 3 - \sqrt{3} \neq \sqrt{3} - 3$
$\qquad 1 - \sqrt{3}$ DOES NOT CHECK

SOLUTION: $1 + \sqrt{3}$.

Have students work in cooperative learning groups. Provide each group with a meter stick, string, and a weight. Students will also need to be able to see a clock with a second hand. Have each group experiment to see whether their results agree with those in the text. After all groups have completed the experiment, have the class, as a whole, discuss the experiment.

Use Transparency ▶ **62**

A **ongoing**
SSESSMENT

3 seconds

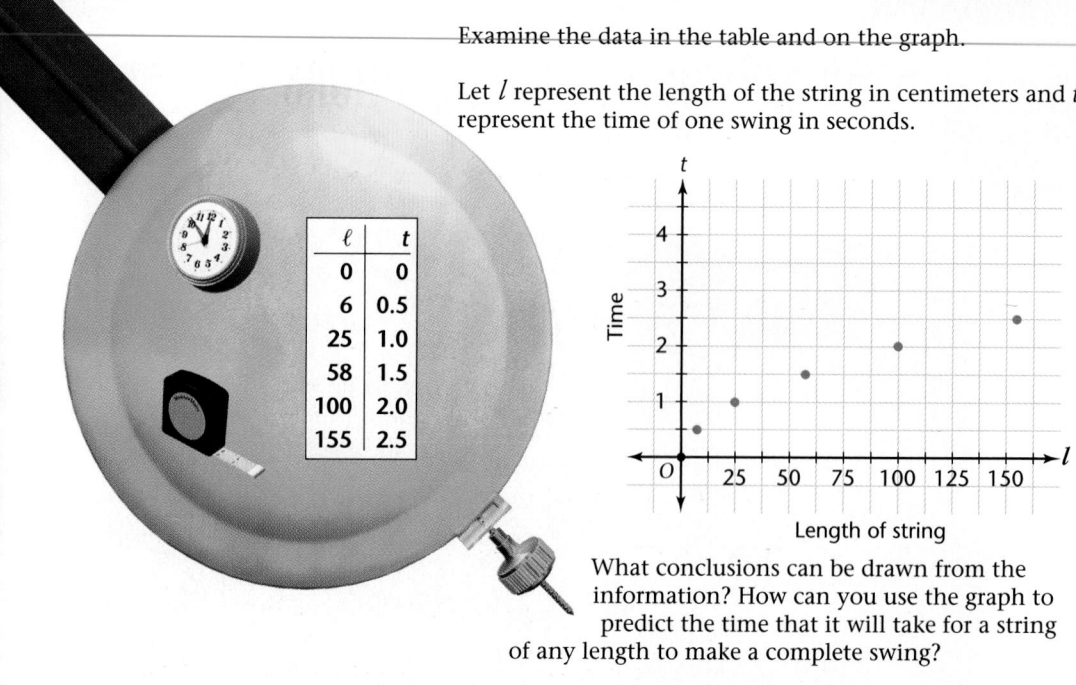

Examine the data in the table and on the graph.

Let l represent the length of the string in centimeters and t represent the time of one swing in seconds.

l	t
0	0
6	0.5
25	1.0
58	1.5
100	2.0
155	2.5

Length of string

What conclusions can be drawn from the information? How can you use the graph to predict the time that it will take for a string of any length to make a complete swing?

First, notice that the curve resembles a transformation of the parent function $y = \sqrt{x}$. In this case, the function is written $t = \sqrt{l}$. The transformation for a stretch of the function is $t = a\sqrt{l}$. The physical science class found that $a = \frac{1}{5}$ worked for their data. Check to see if the class was right. Use the function to predict the time it will take for a string 225 centimeters long to make a complete swing.

The equation that models the motion of a pendulum is

$t = 2\pi \sqrt{\dfrac{l}{g}}$. The variable l is the length in centimeters,

t is the time in seconds and g is acceleration due to gravity (980 centimeters per second per second). This formula can be used to see how close the physical science class's data is to the theoretical value determined by the formula.

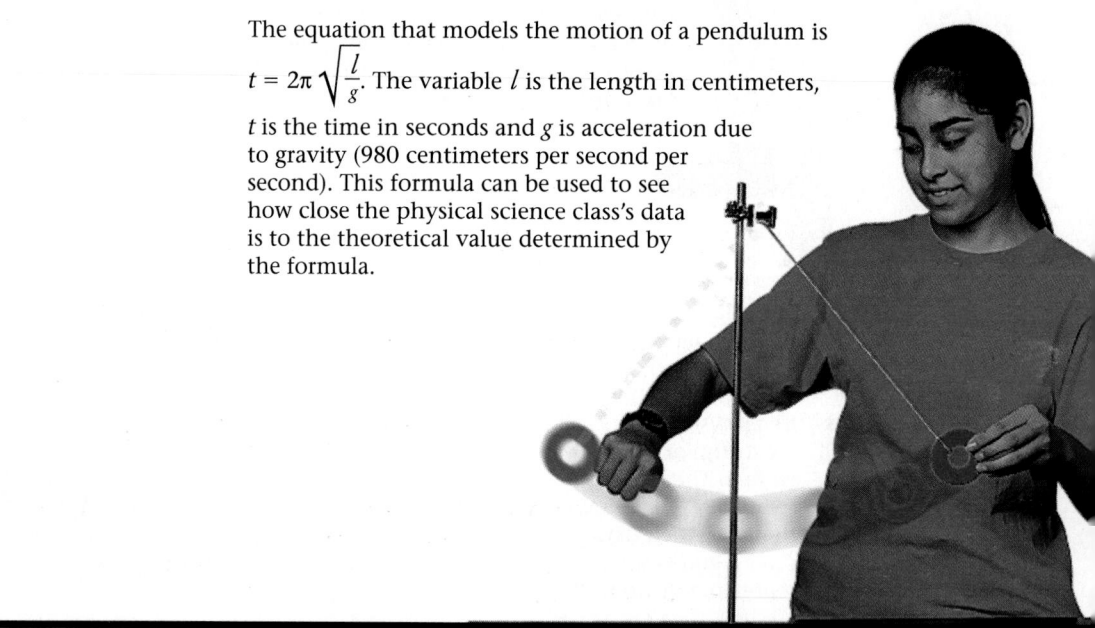

interdisciplinary
CONNECTION

Science: Solar Energy
Solar cells convert the energy of sunlight to electrical energy. For each square centimeter of a cell that is in direct sunlight, approximately 0.01 watt of electrical power is produced. For instance, a circular solar cell with a radius of 10 cm has an area of about 314 cm² and produces about 314(0.01) or 3.14 watts of electrical power. Ask students how large a square collector would be need to be to create 1000 watts of electrical power. [≈ **31.6 cm by 31.6 cm**]

EXAMPLE 1

Ⓐ Find the time it takes a 75-centimeter pendulum to make a complete swing.

Ⓑ Find the length of a pendulum that takes 3 seconds to make a complete swing.

Solution ➤

Scientific Calculator

Ⓐ Substitute into the pendulum formula.

$$t = 2\pi \sqrt{\frac{75}{980}}$$

Use a calculator to help in the calculation.

$$t \approx 1.74$$

One swing takes about 1.7 seconds.

Ⓑ Substitute the known information into the formula, and solve for l.

$$3 = 2\pi \sqrt{\frac{l}{980}} \qquad \text{Given}$$

$$3\sqrt{980} = 2\pi\sqrt{l} \qquad \text{Multiply both sides by } \sqrt{980}.$$

$$\frac{3\sqrt{980}}{2\pi} = \sqrt{l} \qquad \text{Divide both sides by } 2\pi.$$

$$\left(\frac{3\sqrt{980}}{2\pi}\right)^2 = l \qquad \text{Use the definition of square root.}$$

$$223.4 \approx l \qquad \text{Use a calculator.}$$

The length of the pendulum is about 223.4 centimeters. ❖

Solving Equations Containing Radicals

EXAMPLE 2

Solve $\sqrt{x + 2} = 3$. Ⓐ graphically Ⓑ algebraically

Solution ➤

Graphics Calculator

Ⓐ You can solve the equation on a graphics calculator. Set a *friendly window*. Enter the left side of the equation in the graphics calculator as $Y_1 = \sqrt{x + 2}$. Then enter the right side of the equation, $Y_2 = 3$.

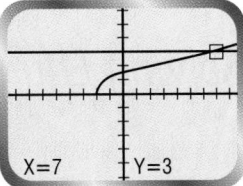
X=7   Y=3

The calculator will graph both equations on the same set of axes. You can then use TRACE to find the point(s) of intersection. Since the point of intersection is (7, 3), $x = 7$.

---

ongoing
ASSESSMENT

Try This

$x = 9\frac{1}{2}$

ongoing
ASSESSMENT

$x \geq -2$. **The radicand must be positive.**

Alternate Example 3
Solve $n = \sqrt{7n - 12}$.
[$n = 3$ or 4]

ongoing
ASSESSMENT

The graphs intersect at only one point.

B The square root and the squaring operations are inverses and can be used to solve equations.

$$\sqrt{x + 2} = 3 \qquad \text{Given}$$
$$(\sqrt{x + 2})^2 = 3^2 \qquad \text{Square both sides.}$$
$$x + 2 = 9 \qquad \text{Simplify.}$$
$$x = 7 \qquad \text{Subtraction Property of Equality}$$

Check by substituting. Since $\sqrt{7 + 2} = \sqrt{9} = 3$, the solution is 7. ❖

Try This Solve $\sqrt{2x - 3} = 4$.

What is the domain of the function $y = \sqrt{x + 2}$? Explain.

EXAMPLE 3

Solve $\sqrt{x + 6} = x$.

Solution ➤

Notice that the variable appears on both sides of the equation.

$$\sqrt{x + 6} = x \qquad\qquad \text{Given}$$
$$(\sqrt{x + 6})^2 = x^2 \qquad\qquad \text{Square both sides of the equation.}$$
$$x + 6 = x^2 \qquad\qquad \text{Simplify.}$$
$$0 = x^2 - x - 6 \qquad\qquad \text{Form a quadratic equal to 0.}$$
$$0 = (x - 3)(x + 2) \qquad\qquad \text{Factor.}$$
$$x - 3 = 0 \quad \text{or} \quad x + 2 = 0 \qquad \text{Set each factor equal to zero.}$$

The possible solutions are 3 and -2. You **must** check each possible solution by substituting it into the original equation to determine whether it is, in fact, a solution.

Check

For $x = 3$, $\sqrt{3 + 6} = \sqrt{9} = 3$,
so 3 is a solution.

For $x = -2$, $\sqrt{-2 + 6} = \sqrt{4} \neq -2$,
so -2 is *not* a solution. ❖

The graphs of $y = \sqrt{x + 6}$ and $y = x$ show that the equation $\sqrt{x + 6} = x$ has only one solution. How do you know this?

Solving Equations by Using Radicals

You can solve the equation $x^2 = 225$ by using the generalization on page 554. Recall that the square root may have two answers, since the values of x can be positive or negative.

$$x^2 = 225$$
$$x = \pm\sqrt{225}$$
$$x = \pm 15$$

EXAMPLE 4

Solve the following equations for x.

A $x^2 = 150$ **B** $x^2 = 4^2 + 3^2$ **C** $x^2 = y^2 + z^2$

Solution ➤

For each problem, use the generalization on page 554.

A $x^2 = 150$

$x = \pm\sqrt{150} = \pm\sqrt{25 \cdot 6} = \pm 5\sqrt{6}$

Thus, $x = 5\sqrt{6}$ or $x = -5\sqrt{6}$.

Check $\qquad 5\sqrt{6} \cdot 5\sqrt{6} = 25 \cdot 6 = 150$

$\qquad\qquad (-5\sqrt{6}) \cdot (-5\sqrt{6}) = 25 \cdot 6 = 150$

B $x^2 = 4^2 + 3^2$

$x = \pm\sqrt{4^2 + 3^2} = \pm\sqrt{25} = \pm 5$

C For a literal equation, follow the same procedure that you would with numbers. Remember to include everything on the right side of the equation under one radical. Account for positive and negative values as possible solutions.

$$x^2 = y^2 + z^2$$
$$x = \pm\sqrt{y^2 + z^2}$$
$$x = \sqrt{y^2 + z^2} \text{ or } -\sqrt{y^2 + z^2} \; ❖$$

Problems that have two possible answers must also be considered in context. For example, if a geometric problem relates to distance, the answer must be positive. In this case, the negative value should be disregarded.

EXAMPLE 5

$A = \pi r^2$

The area of the circular flower garden is 23 square yards. Find the radius.

Solution ➤

Substitute 23 for A, and solve for r.
Since $\pi r^2 = 23$ and $r^2 = \dfrac{23}{\pi}$,

$$r = \pm\sqrt{\dfrac{23}{\pi}} \approx \pm\sqrt{\dfrac{23}{3.14}} \approx \pm\sqrt{7.325} \approx \pm 2.706.$$

Only the positive square root can be used. The length of the radius is approximately 2.7 yards. ❖

Using Algorithms

When working with equations like the model of motion of a pendulum, it is sometimes easier to solve the equation for the needed variable before substituting in numerical values. For example, solve

$t = 2\pi\sqrt{\dfrac{l}{g}}$ for l. $\left[l = \dfrac{t^2 g}{4\pi^2}\right]$

Alternate Example 6

Solve the equation
$x^2 + 12x + 36 = 169$.
$[x = 7 \text{ or } x = -19]$

ongoing ASSESSMENT

$10^2 - 8(10) + 16 \underset{?}{=} 36;$

$\qquad\qquad 36 = 36 \quad$ True

$(-2)^2 - 8(-2) + 16 \underset{?}{=} 36;$

$\qquad\qquad 36 = 36 \quad$ True

CRITICAL Thinking

The graphs of $y = x^2 - 8x + 16$ and $y = 36$ intersect at $(-2, 36)$ and $(10, 36)$.

ASSESS

Selected Answers

Odd-Numbered Exercises 7–55

Assignment Guide

Core 1–6, 8–24 even, 25–30, 32–42 even, 43–56

Core Two-Year 1–55

Practice Master

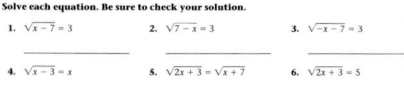

Practice & Apply
13.3 Solving Radical Equations

Solve each equation. Be sure to check your solution.

1. $\sqrt{x-7} = 3$ 2. $\sqrt{7-x} = 3$ 3. $\sqrt{-x-7} = 3$

4. $\sqrt{x-3} = x$ 5. $\sqrt{2x+3} = \sqrt{x+7}$ 6. $\sqrt{2x+3} = 5$

Make separate graphs from the left and right sides of the equation. Use graphing technology or sketch a graph on your own graph paper. Solve the equations that have solutions by finding the intercepts of the graphs.

7. $4\sqrt{x} = -x - 3$ 8. $\sqrt{x} + 0.1 = 1.0$

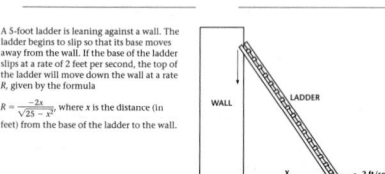

A 5-foot ladder is leaning against a wall. The ladder begins to slip so that its base moves away from the wall. If the base of the ladder slips at a rate of 2 feet per second, the top of the ladder will move down the wall at a rate R, given by the formula
$R = \dfrac{-2x}{\sqrt{25-x^2}}$, where x is the distance (in feet) from the base of the ladder to the wall.

9. How fast will the top of the ladder be moving down the wall when the base is 4 feet from the wall? _____

10. Explain why the answer is negative. _____

Solve each equation and simplify the solution.

11. $5x^2 = 125$ _____ 12. $x^2 - 1 = 120$ _____ 13. $4x^2 = 120$ _____

EXAMPLE 6

Solve the equation $x^2 - 8x + 16 = 36$ using radicals.

Solution ➤

Recall the methods from the previous two chapters. The trinomial $x^2 - 8x + 16$ is a perfect square. When it is factored it becomes $(x - 4)^2$. Write the equation in this form, and solve.

$$(x - 4)^2 = 36$$
$$x - 4 = \pm\sqrt{36}$$
$$x - 4 = \pm 6$$
$$x = 10 \text{ and } x = -2$$

Since x can be any number, you must account for both positive and negative values. ❖

Check by substitution to see if the values actually fit the original equation.

CRITICAL Thinking

Solve the equation $x^2 - 8x + 16 = 36$ by graphing it as a transformation. How does the graph justify 2 solutions for x?

EXERCISES & PROBLEMS

Communicate

1. Explain why you must consider both positive and negative values if you take the square root of a number.

2. Describe a step-by-step method for finding the solution to an equation of the form $\sqrt{x + 7} = 114$.

3. Does $\sqrt{-x} = -3$ have a solution? Explain.

4. Explain the difference between solving $a^2 = b^2 + c^2$ and solving $a^2 = (b + c)^2$, for a.

5. Give a step-by-step procedure for finding the solution to the equation $x^2 + 2bx + b^2 = c^2$ for x. Then test the procedure by substituting numbers for b and c.

6. Find the explanation of the quadratic formula in Lesson 12.5, and identify how radicals are used to develop the formula.

Practice & Apply

Solve each equation algebraically. Be sure to check your solution.

7. $\sqrt{x-5} = 2$ 9   **8.** $\sqrt{x+7} = 5$ 18   **9.** $\sqrt{2x} = 6$ 18

10. $\sqrt{10-x} = 3$ 1   **11.** $\sqrt{2x+9} = 7$ 20   **12.** $\sqrt{2x-1} = 4$ $\frac{17}{2}$

Solve each equation algebraically. Be sure to check your solution.

13. $\sqrt{x+2} = x$ 2   **14.** $\sqrt{6-x} = x$ 2   **15.** $\sqrt{5x-6} = x$ 2, 3

16. $\sqrt{x-1} = x-7$ 10   **17.** $\sqrt{x+3} = x+1$ 1   **18.** $\sqrt{2x+6} = x-1$ 5

19. $\sqrt{x^2+3x-6} = x$ 2   **20.** $\sqrt{x-1} = x-1$ 1, 2   **21.** $\sqrt{x^2+5x+11} = x+3$ 2

Make separate graphs from the left and right sides of the equation. Use graphing technology or sketch a graph. Solve the equations that have solutions by finding the intercepts of the graphs.

22. $\sqrt{x+12} = x$ **23.** $\sqrt{x-2} = x$ **24.** $\sqrt{x} = \frac{1}{3}x + \frac{2}{3}$

Physics The motion of a pendulum can be modeled by $t = 2\pi\sqrt{\dfrac{l}{32}}$, where l is the length of the pendulum in feet and t is the number of seconds required for one complete swing.

25. A grandfather clock is based on the motion of a pendulum. About how long should a pendulum be so that the time required for 1 complete swing is 1 second?   0.81 feet

26. If the time required for 1 complete swing is doubled, by what is the length multiplied?   4

27. If the time is multiplied by 3, by what is the length multiplied?   9

28. The time required for 1 complete swing of a pendulum is multiplied by c. By what will the corresponding length be multiplied? Generalize. c^2

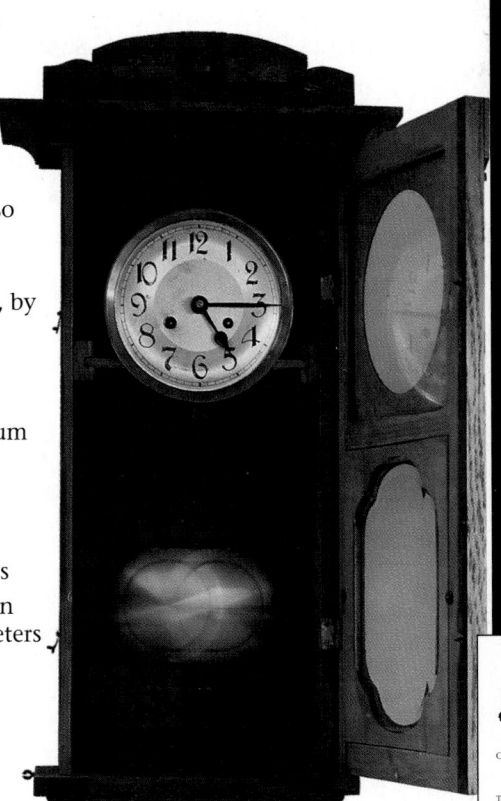

Sports The formula for kinetic energy is $E = \frac{1}{2}mv^2$. E is the kinetic energy in joules, m is the mass of the object in kilograms, and v is the velocity of a moving object in meters per second.

29. Solve the formula for v. $v = \dfrac{\sqrt{2Em}}{m}$

30. If a baseball is thrown with 50 joules of energy and has a mass of 0.14 kilograms, what is its velocity?   26.73 m/s

Solve each equation and simplify the solution.

31. $x^2 = 90$ **32.** $4x^2 = 7$ **33.** $3x^2 - 27 = 0$

34. $2x^2 = 48$ **35.** $x^2 - 8x + 16 = 0$ **36.** $x^2 - 12x + 36 = 0$

The answers to Exercises 22–24 can be found in Additional Answers beginning on page 729.

31. $3\sqrt{10}, -3\sqrt{10}$

32. $\dfrac{\sqrt{7}}{2}, \dfrac{-\sqrt{7}}{2}$

33. $3, -3$

34. $2\sqrt{6}, -2\sqrt{6}$

35. 4

36. 6

Technology

Exercises 22–30 can be solved with the aid of a graphics calculator.

Error Analysis

When squaring both sides of an equation, students may make any of several different errors. For example, when squaring both sides of $\sqrt{x+2} = 5$, students might write any of the following errors:

$x + 2 = 5$
$x^2 + 4 = 25$
$x + 2 = 10$
$x^2 + 4x + 4 = 25$

Remind students of the proper procedures for squaring both sides of an equation.

$[x + 2 = 25]$

Authentic Assessment

Have students solve each equation and explain their method of solution.

1. $\sqrt{x + 3} = 7$ [$x = 46$]

2. $\sqrt{x} + 3 = 7$ [$x = 16$]

3. $\sqrt{x + 6} = x$
   [$x = 3$ or $x = -2$]

Solve each equation if possible. If not possible, explain why.

37. $x^2 = 9$ $3, -3$

38. $\sqrt{x} = 9$ 81

39. $|x| = 9$ $9, -9$

40. $x^2 = -9$ no

41. $\sqrt{x} = -9$ no

42. $|x| = -9$ no

Physics   A beam is balanced when the product of the weight and the distance from the fulcrum on one side is equal to the product of the weight and the distance from the fulcrum on the right side.

$$W_l d_l = W_r d_r \text{ or } \frac{W_l}{W_r} = \frac{d_r}{d_l}$$

To find the weight of the metal cube, W, first place W on the right, and balance it with weight W_l. Then switch sides, and balance W with weight W_r.

43. The ratios $\dfrac{W_l}{W} = \dfrac{W}{W_r}$ are equal. So $W_l W_r = W^2$. Express the value of W in terms of the other weights. $W = \sqrt{W_l W_r}$

44. A *geometric mean* is the square root of the product of two factors. If 25 and 49 are two factors, what is the geometric mean?   35

45. If W_l is 9 pounds, and W_r is 16 pounds, what is the true weight of W?
   12

Look Back

46. Solve $-2x + 3 < 11$.   **[Lesson 4.7]** $x > -4$

47. Find the inverse of the matrix $A = \begin{bmatrix} 6 & -1 \\ 2 & -7 \end{bmatrix}$.   **[Lesson 7.4]**

Graph each pair of functions on the same set of axes. Describe the relationship between the two graphs.   [Lesson 9.6]

48. $y = x + 3$
   $y = (x - 2) + 3$

49. $y = x^2 + 3$
   $y = (x - 2)^2 + 3$

50. $y = |x| + 3$
   $y = |x - 2| + 3$

51. Factor $4x^4 - 16y^4$ completely.   **[Lesson 11.5]** $4(x^2 - 2y^2)(x^2 + 2y^2)$

52. What conditions for the discriminant indicates that a quadratic has no real solution? Give an example.   **[Lesson 12.5]**   less than zero

Solve and graph each inequality.   [Lesson 12.6]

53. $x^2 - x - 12 > 0$

54. $y^2 + 5y + 6 < 0$

55. $a^2 - 6a - 16 < 0$

Look Beyond

56. What is the center of the circle $x^2 + y^2 = 4$?   (0, 0)

Look Beyond

Exercise 56 provides an opportunity for students to solve non-routine problems and precedes the study of coordinate geometry in Lesson 13.6.

---

47. $\begin{bmatrix} 0.175 & -0.025 \\ 0.05 & -0.15 \end{bmatrix}$

The answers to Exercises 48–50 can be found in Additional Answers beginning on page 729.

53. $x < -3$ or $x > 4$

54. $-3 < y < -2$

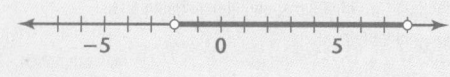

55. $-2 < a < 8$

56. (0, 0)

LESSON 13.4 The "Pythagorean" Right-Triangle Theorem

Right Triangle ABC

hypotenuse *c*

B

leg *a*

right angle *C*

leg *b*

A

The "Pythagorean" Right-Triangle theorem is one of the most familiar and significant theorems in mathematics. It makes extensive use of squares and square roots. You can use this theorem to find the length of one side of a right triangle if you know the lengths of the other two.

Copy and complete the table for the given sides of a right triangle.

Side *a*	Side *b*	Side *c*	a^2	b^2	c^2	$a^2 + b^2$
3	4	5				
5	12	13				
16	30	34				
13	84	85				

Compare the last two columns of the table. What do you notice about the relationship between c^2 and $a^2 + b^2$?

Cultural Connection: Asia The relationships of the sides of right triangles have been studied in many parts of the world throughout history. The ancient Chinese, Babylonians, and Hindus had all discovered the right-triangle relationship. Throughout history, many people have discovered clever proofs of the relationship.

ALTERNATIVE teaching strategy

Using Manipulatives
Provide students with typing paper. Have students mark point *A* on one edge of the paper near the center. Then have them draw a 60° angle at *A* so that one side of the angle is the edge of the paper. Have students cut out the triangle and measure each side. Have students find the following ratios. (*C* is the vertex at the right angle and *B* is the other vertex.)

1. $\dfrac{AB}{AC}$ $\left[\dfrac{2}{1}\right]$

2. $\dfrac{BC}{AC}$ $\left[\dfrac{\sqrt{3}}{1}\right]$

Have students repeat the experiment with an angle of 45°.

3. $\dfrac{AB}{AC}$ $\left[\dfrac{\sqrt{2}}{1}\right]$

4. $\dfrac{BC}{AC}$ $\left[\dfrac{1}{1}\right]$

PREPARE

Objectives
- Examine a right triangle to determine a relationship among the sides.
- Use the "Pythagorean" Right-Triangle Theorem to solve problems involving right triangles.

RESOURCES

• Practice Master	**13.4**
• Enrichment Master	**13.4**
• Technology Master	**13.4**
• Lesson Activity Master	**13.4**
• Quiz	**13.4**
• Spanish Resources	**13.4**

Assessing Prior Knowledge

1. Solve $a^2 + b^2 = c^2$ for *c*.
   $[c = \sqrt{a^2 + b^2}]$

2. Solve $a^2 + b^2 = c^2$ for *b*.
   $[b = \sqrt{c^2 - a^2}]$

3. Solve $a^2 + b^2 = c^2$ for *a*.
   $[a = \sqrt{c^2 - b^2}]$

TEACH

Have students measure the length, the width, and length of the diagonal of rectangular objects in the room (table, door, the room itself). Challenge students to find a relationship among the three measurements. Tell students they will learn about this relationship in this lesson.

European culture provided the name we currently associate with this right-triangle relationship. It is usually referred to as the "Pythagorean" Right-Triangle Theorem, although Babylonians knew of the relationship over 1000 years earlier.

> **"PYTHAGOREAN" RIGHT-TRIANGLE THEOREM**
> Given a right triangle with legs of length a and b and hypotenuse of length c, then $c^2 = a^2 + b^2$ or $a^2 + b^2 = c^2$.

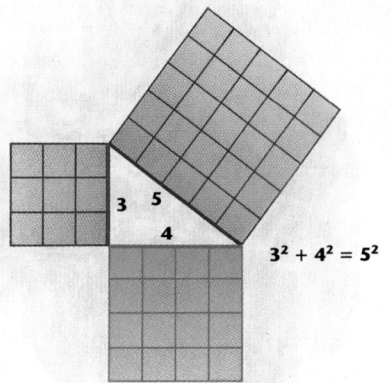

3 5
4
$3^2 + 4^2 = 5^2$

Over the years this illustration has been used to show that the sum of the squares of the legs of a right triangle is equal to the square of the hypotenuse.

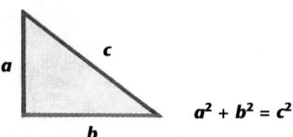

$a^2 + b^2 = c^2$

Alternate Example 1

Find the length of the missing side of each right triangle.

	side a	side b	hypotenuse
1.	9	15	?
2.	?	6	9
3.	8	?	16

[**side** $a \approx$; **6.7; side** $b \approx 13.9$; **hypotenuse** $c \approx 17.5$]

EXAMPLE 1

Find the length of x, y, and z.

A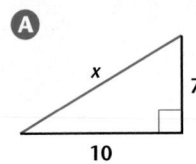

B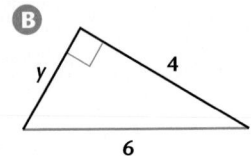

C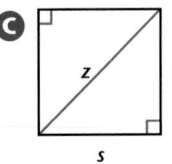

Solution ➤

A Since x is opposite the right angle, x represents the length of the hypotenuse.

$$x^2 = 7^2 + 10^2$$
$$x^2 = 149$$
$$x = \pm\sqrt{149} \approx \pm 12.21$$

Since x represents the length of a segment, it must be positive. Therefore, $x = \sqrt{149} \approx 12.21$.

B The length of the hypotenuse is 6.

$$y^2 + 4^2 = 6^2$$
$$y^2 + 16 = 36$$
$$y^2 = 20$$
$$y = \pm\sqrt{20} \text{ or } \pm 2\sqrt{5}$$
$$y \approx 4.47 \qquad \text{The negative root is not a solution.}$$

ENRICHMENT Provide a standard 7-piece tangram puzzle to students. Tell them that each leg of the smallest triangle has a measure of 1 unit. Have students find the length of the sides of the seven pieces of the puzzle. [**small triangles 1, 1,$\sqrt{2}$; medium triangle $\sqrt{2}$, $\sqrt{2}$, 2; large triangle 2, 2, 2$\sqrt{2}$; square 1, 1, 1, 1; parallelogram 1, $\sqrt{2}$, 1, $\sqrt{2}$**]

INCLUSION **strategies** **Using Cognitive Strategies** Have students use different colors to color-code the legs and hypotenuse of a right triangle. Then have students use the same color code as they write the "Pythagorean" Right-Triangle Theorem.

C The diagonal, z, of a square is the hypotenuse. Each leg is s.

$$z^2 = s^2 + s^2$$
$$z^2 = 2s^2$$
$$z = \sqrt{2s^2}$$
$$z = s\sqrt{2} \qquad \text{Only positive numbers represent a length.}$$

Thus, the formula for the diagonal of a square with sides s is $s\sqrt{2}$. ❖

EXAMPLE 2

Boating Navigation

Your boat is traveling due north at 20 miles per hour. Your friend's boat left at the same time from the same location headed due west at 15 miles per hour. After an hour you get a call from your friend that he has engine trouble. How far must you travel to reach your friend?

Solution ➤

The legs will be 20 miles and 15 miles. The distance you must travel is the length of the hypotenuse. Use the "Pythagorean" Right-Triangle Theorem to find the distance. Add and use the definition of square root to solve.

$$c^2 = a^2 + b^2$$
$$c^2 = 20^2 + 15^2$$
$$c^2 = 625$$

Since distance is involved, only positive values are used.

$$c = \sqrt{625} \rightarrow c = 25 \text{ miles} ❖$$

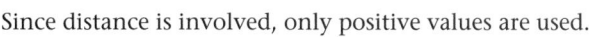

Try This Find the hypotenuse of a right triangle if the legs are 36 and 48.

What do you notice about multiples of the sides of a 3-by-4-by-5 right triangle? If you do not know, make a table of multiples of the legs and see if you can guess the hypotenuse. Then check your results.

Use algebra to solve the equation $a^2 + b^2 = c^2$ for a or b. Then you can use the new equation to determine the unknown side of the right triangle. All you need to know is the hypotenuse and the other side.

RETEACHING the lesson

Using Symbols For students who have difficulty remembering formulas, have them create reference cards that include the following information along with appropriate drawings.

• Pythagorean Theorem
• 30-60-90 right triangle
• isosceles right triangle

Optional information might include how to find the *slant height* of a square pyramid or the *apothem* of a hexagon.

A plane leaves an airport and flies 100 kilometers north and then 50 kilometers east to avoid a storm. At that point, how far is the plane from the airport? [≈ **111.8 kilometers**]

Aongoing ASSESSMENT

Try This

60

Aongoing ASSESSMENT

All triangles with sides that are multiples of 3, 4, and 5, respectively, are similar, thus all are right triangles.

EXAMPLE 3

GEOMETRY
Connection

The height of a pyramid with a square base can be determined from measurements of the base and the slant height. The measurement of the base is 40 meters, and the slant height is 52 meters. How high is the pyramid?

Solution ➤

Use half the base of the pyramid, 20 meters, as one side of the triangle. Use the slant height, 52 meters, as the hypotenuse.

If you solve for side a, the formula from the "Pythagorean" Right-Triangle Theorem will be $a = \sqrt{c^2 - b^2}$. Let the height, a, represent the leg whose length we do not know. Since distance is positive, you can disregard the negative value. Substitute the values for b and c to find the height of the pyramid.

$$a^2 = 52^2 - 20^2$$
$$a = \sqrt{52^2 - 20^2}$$
$$a = \sqrt{2304} = 48$$

The height of the pyramid is 48 meters. ❖

Height (*a*)
Slant height
Half the base

CRITICAL
Thinking

An important theorem in geometry states that for all positive numbers a and b, $\sqrt{a^2 + b^2} < a + b$. Use this theorem to explain why the path along the hypotenuse is always shorter than the path along the two legs.

EXERCISES & PROBLEMS

Communicate

1. Name the sides of a right triangle.
2. State the "Pythagorean" Right-Triangle Theorem in your own words.
3. Explain how to find the hypotenuse of a right triangle if you know the lengths of the legs.
4. Describe how to change the basic equation of the "Pythagorean" Right-Triangle Theorem to let you find the length of one of the legs.
5. How would you show someone that the "Pythagorean" Right-Triangle Theorem is true?
6. Explain how to use the "Pythagorean" Right-Triangle Theorem to find the length of a diagonal of a rectangle.

	Leg	Leg	Hypotenuse			Leg	Leg	Hypotenuse
7.	24	45	51	11.		12	21.9	25
8.	10	24	26	12.		24.3	6	25
9.	15	8	17	13.		30	40	50
10.	5	9	10.3	14.		0.75	1	1.25

Practice & Apply

Technology Copy and complete the table. Use a calculator and round the answer to the nearest tenth.

	Leg	Leg	Hypotenuse		Leg	Leg	Hypotenuse
7.	24	45	?	**8.**	10	?	26
9.	?	8	17	**10.**	5	9	?
11.	12	?	25	**12.**	?	6	25
13.	30	40	?	**14.**	0.75	?	1.25

Solve for x.

15. 10

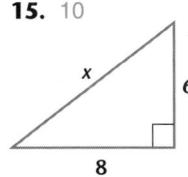

x
6
8

16. 13.93

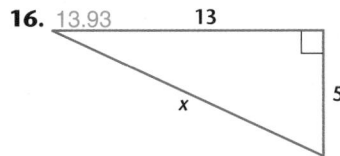

13
x
5

17.

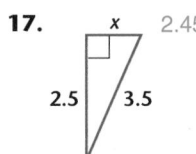

x   2.45
2.5   3.5

Landscaping A garden in the shape of a right triangle is shown in the figure.

18. How many feet of fencing must be bought to enclose the garden? Assume that fencing is sold by the foot.

19. If the cost of fencing is $4.98 per foot, how much will it cost to enclose the garden? $169.32

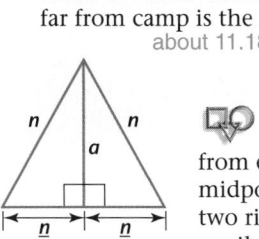

34 feet
14 feet
10 feet

20. Sports To the nearest foot, how long is a throw from third base to first base?
127 feet

21. Find the diagonal of a square that is 5 inches on each side. about 7.07

22. A hiker leaves camp and walks 5 miles east. He then walks 10 miles south. How far from camp is the hiker?
about 11.18 miles

A baseball diamond is a square with sides of 90 feet.

n   n
a
$\frac{n}{2}$ $\frac{n}{2}$

Geometry An equilateral triangle has an axis of symmetry from each vertex that is perpendicular to the opposite side at its midpoint. The axis of symmetry divides the equilateral triangle into two right triangles that have the same size and shape. A representative equilateral triangle is shown where a is the altitude of the triangle.

23. Fill in the table. Simplify each answer.

Equilateral Triangles

Side	4	6	8	10	12	20	n	?
Half of a Side	2	?	?	?	?	?	?	?
Altitude	?	?	?	?	?	?	?	$17\sqrt{3}$

23.

Equilateral Triangles

Side	4	6	8	10	12	20	n	34
Half of a Side	2	3	4	5	6	10	$\frac{n}{2}$	17
Altitude	$2\sqrt{3}$	$3\sqrt{3}$	$4\sqrt{3}$	$5\sqrt{3}$	$6\sqrt{3}$	$10\sqrt{3}$	$\frac{n}{2}\sqrt{3}$	$17\sqrt{3}$

Technology

Students can use a calculator to help them complete Exercises 7–22 and 27. Note that for Exercises 23–26, answers should be left in simplest radical form.

Error Analysis

When solving a right triangle, students may incorrectly use the "Pythagorean" Right-Triangle Theorem. Some students will always substitute the numbers for a and b, even though one of the lengths given is for the hypotenuse of the triangle.

Technology Master

Technology
13.4 Lines of Sight

Have you ever watched a plane high in the sky come closer to you and then disappear into the distance after it passes by? If you are at point O, the length of your *line of sight* $\overrightarrow{OP}$ changes as the position of the plane relative to you changes. The diagrams show three situations involving lines of sight $\overrightarrow{OP}$ that you can examine by using the Pythagorean theorem.

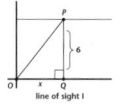

line of sight I

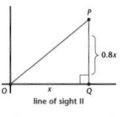

line of sight II

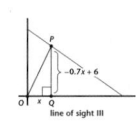

line of sight III

To study line of sight problems, you will need a graphics calculator as well.

Use the Pythagorean theorem to write an equation for the length d of the line of sight $\overrightarrow{OP}$ in each situation.

1. line of sight I   **2.** line of sight II   **3.** line of sight III

4. Graph the equation you wrote in Exercise 2. Describe the graph in your own words. Use the Distributive Property and the Multiplication Property of Square Roots to confirm your description.

5. Graph the equation you wrote in Exercise 1. Describe the graph in your own words. In what way are the equations for d in Exercises 1 and 2 alike? How are they different?

6. Graph the equation you wrote in Exercise 3. Describe the graph in your own words. For what value(s) of x will d be a minimum? Find the minimum value of d.

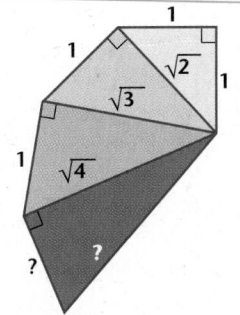

Alternative ASSESSMENT

Portfolio Assessment

Have students choose examples or exercises from this lesson or make up their own problems that show that they understand the "Pythagorean" Right-Triangle Theorem and associated concepts. Items that must be included are the following:

- "Pythagorean" Right-Triangle Theorem
- special triangles (either isosceles right triangle or 30-60-90 triangle)
- at least one application of the "Pythagorean" Right-Triangle Theorem.

Cultural Connection: Europe The given figure is known as the wheel of Theodorus, a Greek mathematician who lived in the mid-fifth century B.C.

24. Continue the spiral by extending the length of the hypotenuse of each new right triangle. Check the numbers with the "Pythagorean" Right-Triangle Theorem. Make a table and extend the number to 9 right triangles.

Right Triangle Number	1	2	3	4	5	6	7	8	9
Hypotenuse	$\sqrt{2}$	?	?	?	?	?	?	?	?

25. Predict the length of the hypotenuse for the tenth right triangle. $\sqrt{11}$

26. Describe the length of the hypotenuse for the nth right triangle in terms of n. $\sqrt{n+1}$

27. The hypotenuse of a right triangle is 8 centimeters. Find the lengths of the two legs if one leg is 1 centimeter longer than the other. Give both an exact and an approximate answer rounded to the nearest hundredth.

Look Back

Darnell is paid a weekly fee of $20 plus $0.50 for every T-shirt that he sells. Copy and complete the table. [Lesson 4.6]

T-Shirts Sold	0	1	2	3	4
Total	28. ___	29. ___	30. ___	31. ___	32. ___

33. Write an equation that describes w, weekly salary, in terms of t, the number of T-shirts sold. $w = 20.00 + 0.50t$

34. If Darnell sells 29 T-shirts in a given week, what will his salary be? $34.50

35. If Darnell needs to earn $49 in a given week in order to buy a particular sweater, how many T-shirts must he sell?
58

Decide whether the lines are parallel, perpendicular, or neither. [Lesson 5.6]

36. $y = \frac{1}{2}x + 3$
$y = 4x + 3$
neither

37. $y = 3x - 4$
$y = -\frac{1}{3}x + 2$
perpendicular

38. $-2x + y = 8$
$-6x + 3y = 15$
parallel

Look Beyond

Exercises 39–42 precede the development and use of the distance formula in Lesson 13.5.

Look Beyond

Plot each pair of points on graph paper, and find the distance between them.

39. $(3, 2), (10, 2)$ **40.** $(-7, -1), (5, -1)$ **41.** $(4, 1), (4, 8)$ **42.** $(-6, -4), (-6, -1)$

27. $\dfrac{-1 + \sqrt{127}}{2}, \dfrac{1 + \sqrt{127}}{2}$; 5.13, 6.13

The answers to Exercises 24 and 39–42 can be found in Additional Answers beginning on page 729.

	T-Shirts sold	Total (dollars)
28.	0	20.00
29.	1	20.50
30.	2	21.00
31.	3	21.50
32.	4	22.00

The Distance Formula

Why _The ability to find the distance between two points is of critical importance to sailors, surveyors, mathematicians, architects, astronomers, and draftsmen. Countless others, taxi drivers, carpenters, and emergency medical technicians, must possess the ability to measure indirect distance._

Emergency Service
A medical center learns of an auto accident located 7 miles east and 11 miles north of its location. If the medical center's helicopter is now located 1 mile east and 3 miles north of the medical center, what distance must the helicopter travel to get to the scene of the accident?

Visualize a segment drawn from *H* to *A*. This segment represents the distance the helicopter must travel. Now visualize a line through *A* perpendicular to the west-east axis and a line through *H*, perpendicular to the north-south axis. The intersection of these two lines meet at point *V* (7, 3). The 3 lines form the right triangle *AVH* with the right angle at *V*. How can you find the length of *HA*?

Using Manipulatives
Have students use a geoboard where the bottom row and left-hand-side row of pegs are the axes. Have students locate two points on the geoboard and use the distance formula to find the distance between those points. Then have students create a pentagon on the geoboard and find its perimeter.

PREPARE

Objectives

- Use the distance formula to find the distance between two points on the coordinate plane.
- Determine whether a triangle is a right triangle.
- Use the midpoint formula to find the midpoint of a segment.

RESOURCES

- Practice Master   13.5
- Enrichment Master   13.5
- Technology Master   13.5
- Lesson Activity Master   13.5
- Quiz   13.5
- Spanish Resources   13.5

Assessing Prior Knowledge

Find the length of the hypotenuse, given the lengths of the two sides. Round to the nearest tenth if needed.

1. side a = 6 cm
   side b = 8 cm
   [**hypotenuse = 10 cm**]

2. side a = 16 cm
   side b = 11 cm
   [**hypotenuse ≈ 19.4 cm**]

3. side a = 45 cm
   side b = 45 cm
   [**hypotenuse = $45\sqrt{2}$ or ≈ 63.6**]

TEACH

Ask students whether they ever needed to know a distance and had no way to measure it. Ask students how to find the distance of a plane from an airport if it flew directly west and then directly south. [Use the "Pythagorean" Right-Triangle Theorem.]

COORDINATE GEOMETRY
Connection

1. You can find the length of $\overline{HV}$ by subtracting the first coordinates of V and H.

   length of $\overline{HV} = 7 - 1 = 6$

2. You can find the length of $\overline{AV}$ by subtracting the second coordinates of A and V.

   length of $\overline{AV} = 11 - 3 = 8$

3. Now use the "Pythagorean" Right-Triangle Theorem.

   $$c^2 = a^2 + b^2$$
   $$(\text{length of } \overline{AH})^2 = 6^2 + 8^2 = 100$$
   $$\text{length of } \overline{AH} = \sqrt{100} = 10$$

The medical center's helicopter must travel 10 miles.

•Exploration 1 Finding the Distance Formula

COORDINATE GEOMETRY
Connection

1 Use graph paper and the method illustrated in the medical center problem to find the distance between each pair of given points.

   a. $K(11, 2)$, $L(14, 6)$ **b.** $M(2, 3)$, $N(7, 15)$

   c. $R(-4, 3)$, $S(5, 8)$ **d.** $T(-2, 3)$, $U(-6, -2)$

The **distance** between two points A and B is the **length** of $\overline{AB}$. This is the segment with A and B as endpoints.

2 Follow the same procedure as before to find the distance from A to B, but use the letters for the coordinates of the points $A(x_1, y_1)$ and $B(x_2, y_2)$ instead of numbers. Compare your results with the distance formula. ❖

DISTANCE FORMULA

The distance, d, between two points $A(x_1, y_1)$ and $B(x_2, y_2)$ is given by

$$d = \sqrt{(x_2 - x_1)^2 + (y_2 - y_1)^2}.$$

interdisciplinary
CONNECTION

Sports Many playing fields in sports are rectangles. For instance, in baseball the infield is a square with sides of 90 feet. If the pitcher's mound is halfway between home plate and second base, how far is it from home plate to the pitcher's mound? How far is it from first base to third base? Have students research other sports' playing fields and make up problems that require the use of the "Pythagorean" Right-Triangle Theorem, the distance formula, or the midpoint formula.

The Converse of the Theorem

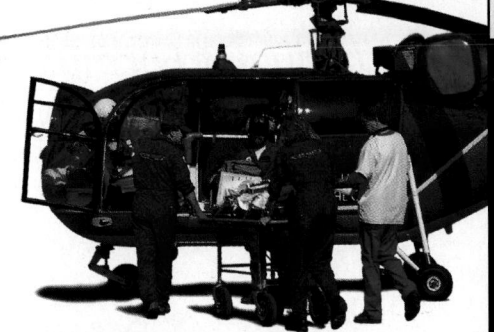

When the *if* and *then* portions of a theorem are interchanged, the new statement is called a **converse**.

TEACHING tip

You may wish to review the meaning of the *converse* of a statement before introducing this page.

> **CONVERSE OF THE "PYTHAGOREAN" RIGHT-TRIANGLE THEOREM**
> If a, b, and c are the lengths of the sides of a triangle with $a^2 + b^2 = c^2$, $c > a$ and $c > b$, then the triangle is a right triangle.

CRITICAL Thinking

Is the *converse* of an if-then statement always true? If not, give an example.

CRITICAL Thinking

No. Counterexamples will vary. For instance, "If you are in Chicago, then you are in Illinois." is true. The converse "If you are in Illinois, then you are in Chicago," is not true.

EXAMPLE 1

COORDINATE GEOMETRY
Connection

Given points $P(1, 2)$, $Q(3, -1)$, and $R(-5, -2)$, determine whether triangle PQR is a right triangle.

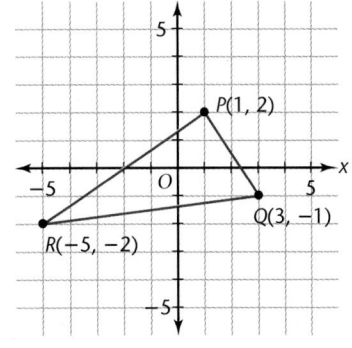

Solution ➤

Draw triangle PQR on graph paper.

Use the Distance Formula to find the lengths of each side.

$$\text{Length of } \overline{PR} = \sqrt{(-5 - 1)^2 + (-2 - 2)^2} = \sqrt{(-6)^2 + (-4)^2} = \sqrt{52}$$
$$\text{Length of } \overline{PQ} = \sqrt{(3 - 1)^2 + (-1 - 2)^2} = \sqrt{2^2 + (-3)^2} = \sqrt{13}$$
$$\text{Length of } \overline{RQ} = \sqrt{[3 - (-5)]^2 + [-1 - (-2)]^2} = \sqrt{8^2 + 1^2} = \sqrt{65}$$

Now use the converse of the "Pythagorean" Right-Triangle Theorem to test whether the three lengths can be the sides of a right triangle.

Does $(\sqrt{52})^2 + (\sqrt{13})^2 = (\sqrt{65})^2$?

Since $52 + 13 = 65$, the triangle is a right triangle. ❖

Alternate Example 1

Given points $D(-3, -4)$, $E(2, -2)$, $F(0, 3)$. Determine whether $\triangle DEF$ is a right triangle. **[Yes]**

Aongoing ASSESSMENT

Try This

No. They are close to right triangles, but are not.

Try This

The sides of a triangle are 60, 90, and 109 centimeters. Is it a right triangle? Is a triangle with sides 54, 71, 90 a right triangle?

CRITICAL Thinking

How can you use the idea of slope to show that the triangle PQR in Example 1 is a right triangle?

CRITICAL Thinking

If the slopes of the two legs are negative reciprocals of each other, then the lines are perpendicular. In other words, if $m_1 \cdot m_2 = -1$, the lines are perpendicular.

ENRICHMENT Have students prove that a quadrilateral with vertices at $(0, 0)$, $(5, 0)$, $(9, 3)$, and $(4, 3)$ is a rhombus.

INCLUSION strategies

Cooperative Learning Allow students to work in small groups while studying the distance and midpoint formulas. Encourage groups to discuss the formulas and procedures used in applying them. Finally, have the group make up a problem that requires the use of at least one of the formulas. Then have groups exchange problems and solve them.

Exploration 2 Finding the Midpoint Formula

COORDINATE
GEOMETRY
Connection

1 Locate the points on graph paper. Guess the midpoint of $\overline{PQ}$.

    a. $P(1, 6)$, $Q(7, 10)$ **b.** $P(-2, 3)$, $Q(6, 7)$ **c.** $P(-4, -1)$, $Q(8, 7)$

2 Use the distance formula to verify that your proposed midpoint is the midpoint of the given segment.

3 What rule can be applied to the x-coordinates of the endpoints of a segment to obtain the x-coordinate of the midpoint of the segment?

4 What rule can be applied to the y-coordinates of the endpoints of a segment to obtain the y-coordinate of the midpoint of the segment?

5 Apply your rule in Steps 3 and 4 to find the coordinates of the midpoint of $\overline{AB}$, with $A(12, 63)$ and $B(43, 20)$.

6 Use the distance formula to verify your answer. ❖

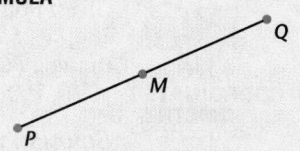

> **MIDPOINT FORMULA**
>
> Given: $\overline{PQ}$ with $P(x_1, y_1)$ and $Q(x_2, y_2)$.
>
> The coordinates of the midpoint, M,
>
> of $\overline{PQ}$, are $\left(\dfrac{x_1 + x_2}{2}, \dfrac{y_1 + y_2}{2}\right)$.

EXAMPLE 2

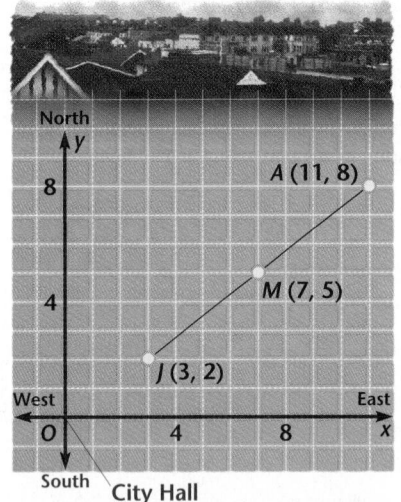

The streets of a city are laid out like a coordinate plane with the origin at City Hall. Jacques lives 3 blocks east and 2 blocks north of City Hall. His friend Alise lives 11 blocks east and 8 blocks north of City Hall. They want to meet at the point midway between their two locations. At what point should they meet?

Solution ➤

Plot $J(3, 2)$ and $A(11, 8)$ on a coordinate plane as shown in the figure. To find the point midway between J and A, use the midpoint formula, which gives the "average" coordinate, $(\overline{x}, \overline{y})$.

$$\overline{x} = \frac{x_1 + x_2}{2} = \frac{3 + 11}{2} = \frac{14}{2} = 7$$

$$\overline{y} = \frac{y_1 + y_2}{2} = \frac{2 + 8}{2} = \frac{10}{2} = 5$$

The midpoint $M = (7, 5)$ is where they should meet. ❖

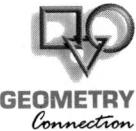

GEOMETRY
Connection

EXAMPLE 3

The midpoint of a diameter of a circle is $M(3, 4)$. If one endpoint of a diameter is $A(-3, 6)$, what are the coordinates of the other endpoint, $B(x_2, y_2)$?

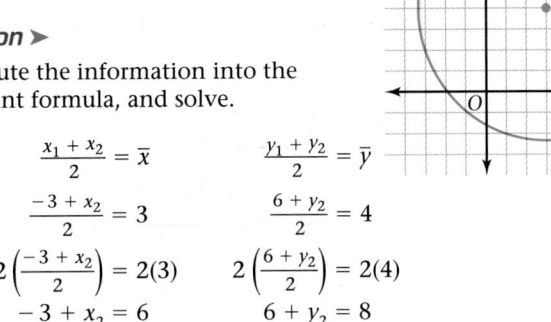

Solution ➤

Substitute the information into the midpoint formula, and solve.

$$\frac{x_1 + x_2}{2} = \bar{x} \qquad \frac{y_1 + y_2}{2} = \bar{y}$$

$$\frac{-3 + x_2}{2} = 3 \qquad \frac{6 + y_2}{2} = 4$$

$$2\left(\frac{-3 + x_2}{2}\right) = 2(3) \qquad 2\left(\frac{6 + y_2}{2}\right) = 2(4)$$

$$-3 + x_2 = 6 \qquad\qquad 6 + y_2 = 8$$

$$x_2 = 9 \qquad\qquad\qquad y_2 = 2$$

The other endpoint is $B(9, 2)$. ❖

EXERCISES & PROBLEMS

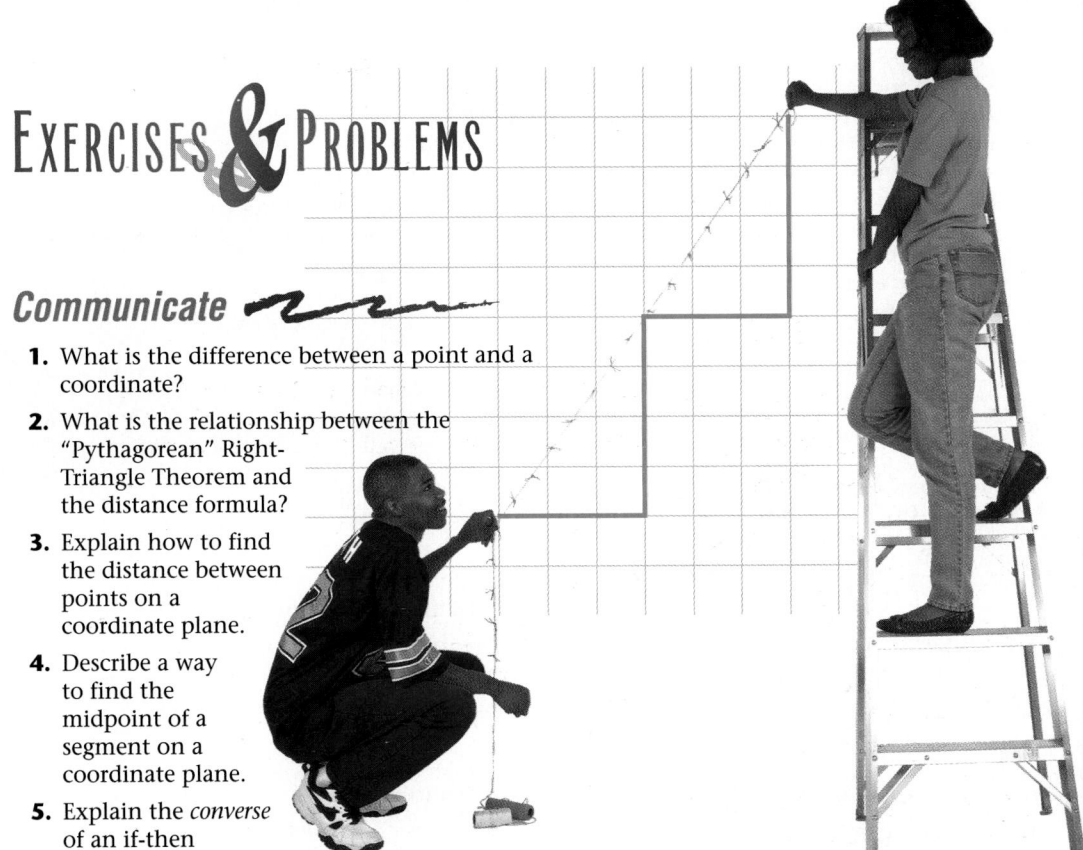

Communicate

1. What is the difference between a point and a coordinate?

2. What is the relationship between the "Pythagorean" Right-Triangle Theorem and the distance formula?

3. Explain how to find the distance between points on a coordinate plane.

4. Describe a way to find the midpoint of a segment on a coordinate plane.

5. Explain the *converse* of an if-then statement.

Practice & Apply

For each figure, find the coordinates of Q and the lengths of the three sides of the triangle.

6.

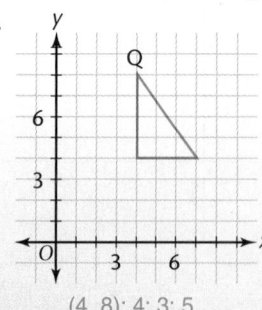

(4, 8); 4; 3; 5

7.

(7, 2); 5; 7; ≈ 8.60

8.

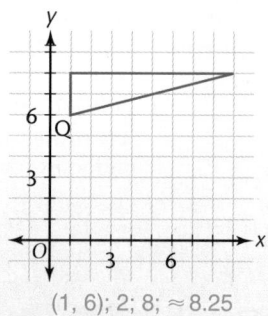

(1, 6); 2; 8; ≈ 8.25

Find the distance between the two given points.

9. $A(4, 7)$, $B(1, 3)$ 5

10. $P(5, 6)$, $Q(17, 11)$ 13

11. $R(-5, -2)$, $S(-9, 3)$
≈ 6.40

Plot △PQR on graph paper. Decide which description(s) apply to △PQR: scalene (no sides equal), isosceles (2 sides equal), equilateral (3 sides equal), or right triangle. Justify your answer.

12. $P(-1, 3)$, $Q(4, 6)$, $R(4, 0)$

13. $P(-2, 2)$, $Q(2, 5)$, $R(8, -3)$

14. Rescue Services A Coast Guard communications center is located at $(0, 0)$. A rescue helicopter located 3 miles west and 2 miles north of the center must respond to a ship located 5 miles east and 7 miles south of the center. What distance must the helicopter travel to reach the ship?
≈ 12.04 miles

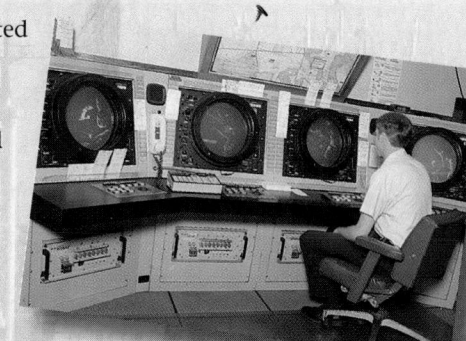

Plot each segment in a coordinate plane, and find the coordinates of the midpoint of the segment.

15. $\overline{AB}$, with $A(2, 5)$ and $B(9, 5)$

16. $\overline{EF}$, with $E(-4, -1)$ and $F(-2, -7)$

17. $\overline{GH}$, with $G(-5, 6)$ and $H(-1, -2)$

18. $\overline{IJ}$, with $I(4, 3)$ and $J(9, -4)$

The locations of two points are given relative to the origin, $(0, 0)$. Find the coordinates of the midpoint of the segment connecting the two points.
(−1, 4)

19. A is 2 units east, 3 units north of 0.
B is 4 units west, 5 units north of 0.

20. C is 2 units west, 5 units south of 0. (−4, −3)
D is 6 units west, 1 unit south of 0.

21. Use the method in Example 2 to find the coordinates of the midpoint of $\overline{PQ}$ with $P(10, 7)$ and $Q(2, 3)$. (6, 5)

22. If the coordinates of P are (a, b) and the coordinates of Q are (c, d), what are the coordinates of the midpoint of $\overline{PQ}$? $\left(\dfrac{a+c}{2}, \dfrac{b+d}{2}\right)$

The midpoint of $\overline{PQ}$ is M. Find the missing coordinates.

23. $P(2, -5)$, $Q(6, 7)$, $M(\underline{?}, \underline{?})$ (4, 1)

24. $P(4, 8)$, $Q(\underline{?}, \underline{?})$, $M(10, 7)$ (16, 6)

25. $P(\underline{?}, \underline{?})$, $Q(6, -2)$, $M(9, 4)$ (12, 10)

26. $P(3, \underline{?})$, $Q(\underline{?}, 5)$, $M(2, 8)$ (3, 11); (1, 5)

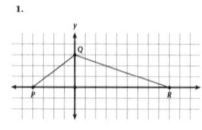
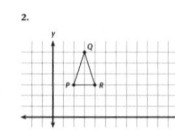
The graphs for Exercises 12, 13, 15–18 can be found in Additional Answers beginning on page 729.

12. isosceles; the length of both PQ and PR is $\sqrt{34}$ units.

13. right scalene; the length of PQ, QR, and RP is 5, 10, and $5\sqrt{5}$ units, respectively, and $PQ^2 + QR^2 = RP^2$.

15. (5.5, 5)

16. (−3, −4)

17. (−3, 2)

18. (6.5, −0.5)

Complete Exercises 27–32.

27. Plot $\overline{RS}$ with $R(2, 1)$ and $S(6, 3)$.

28. Find the coordinates of M, the midpoint of $\overline{RS}$, and plot M.

29. Find the slope of $\overline{RS}$.

30. What is the slope of any line perpendicular to $\overline{RS}$?

31. Find an equation of k, the line that is perpendicular to $\overline{RS}$ and contains M. Draw k on your graph. The line k is called the *perpendicular bisector* of $\overline{RS}$ because it is perpendicular to $\overline{RS}$ and contains the midpoint of $\overline{RS}$.

32. Choose any two points on k other than M. Name them A and B. Compute the distance from A to R and the distance from A to S. Compute the distance from B to R and the distance from B to S.

33. **Portfolio Activity** Complete the problem in the portfolio activity on page 595.

34. Use right triangles and the "Pythagorean" Right-Triangle Theorem to show that $(4, 7)$ is the midpoint between $(1, 3)$ and $(7, 11)$.

Look Back

Factor each trinomial. **[Lesson 11.5]**

35. $y^2 + 35y + 300$
$(y + 20)(y + 15)$

36. $x^2 + 30x + 216$
$(x + 12)(x + 18)$

Technology Find the vertex of each parabola. Use graphing technology to assist you. **[Lesson 12.1]**

37. $y = (x - 2)^2 + 3$ $(2, 3)$

38. $y = (x + 3)^2 - 1$ $(-3, -1)$

39. $y = -2(x - 1)^2 - 4$
$(1, -4)$

40. $y = -(x + 4)^2 + 1$
$(-4, 1)$

41. Use an algebraic technique to find the x-intercept(s) of the graph of the function $y = x^2 - 6x - 10$. Give an exact solution(s). **[Lesson 12.4]**
$3 \pm \sqrt{19}$

Use properties of radicals to simplify each expression. Use a calculator to check your answers. **[Lesson 13.2]**

42. $\sqrt{3}(\sqrt{12} - \sqrt{75})$
-9

43. $\dfrac{\sqrt{75}}{\sqrt{3}}$
5

44. $\dfrac{\sqrt{36} + \sqrt{81}}{\sqrt{9}}$
5

Look Beyond

Technology Use a scientific calculator to evaluate each expression.
NOTE: To evaluate 9^(1/2), use the key strokes of 9 y^x (1 ÷ 2) or 9 y^x .5. If your calculator has an exponent key, ∧ , you may use 9 ∧ (1 ÷ 2) or 9 ∧ .5.

45. 9^(1/2) 3

46. 16^(1/2) 4

47. 39^(1/2)
≈ 6.24

48. 2^(1/2)
≈ 1.41

49. Make a conjecture about the meaning of the exponent $\dfrac{1}{2}$.

27–28.

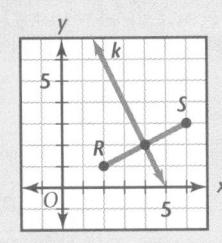

$M(4, 2)$

29. $\dfrac{1}{2}$

30. -2

31.

$y = -2x + 10$

32. Answers may vary. Any point on the perpendicular bisector of a segment is equidistant from the endpoints of the segment.

The answers to Exercises 33–34 can be found in Additional Answers beginning on page 729.

49. Raising a number to the power of $\dfrac{1}{2}$ is the same as taking the square root of the number.

Exploring Geometric Properties

PREPARE

Objectives

* Write the equation of a circle, given its center and radius.
* Use the distance formula to find the diagonals of a rectangle.

RESOURCES

* Practice Master   13.6
* Enrichment Master   13.6
* Technology Master   13.6
* Lesson Activity Master   13.6
* Quiz   13.6
* Spanish Resources   13.6

Assessing Prior Knowledge

Point *A* is at $(-3, 4)$ and point *B* is at $(3, -4)$.

1. Find the length of line segment *AB*. [**10 units**]
2. Find the coordinates of the midpoint of line segment *AB*. [**(0, 0)**]
3. If point *C* is at $(-4, -3)$, is $\triangle ABC$ a right triangle? [**Yes**]

TEACH

 An airline pilot wants to know how much fuel is in the plane. The amount of fuel affects the range that the plane can fly before refueling. Discuss how the pilot can plot the plane's range on a map.

In coordinate geometry, algebra is used to find relationships between the line segments that form geometric figures. The distance formula and the midpoint formula are useful in connecting geometry with algebra. This connection can then help you solve problems.

The sandstone blocks of Stonehenge, an ancient monument in England, once formed a circle with a 160-foot radius.

COORDINATE GEOMETRY
Connection

A *circle* is the set of all points in a plane that are the same distance from a given point called the *center*. You can use the distance formula to find the equation of the circle with *center* (0, 0) and *radius* 160.

If (x, y) is on the circle, then its distance from (0, 0) must be 160.

$$d = \sqrt{(x - 0)^2 + (y - 0)^2} = 160$$

$$\sqrt{x^2 + y^2} = 160$$

Square both sides and simplify.

$$x^2 + y^2 = 160^2 \text{ or}$$
$$x^2 + y^2 = 25{,}600$$

The equation $x^2 + y^2 = 160^2$ represents all possible points on the circle.

ALTERNATIVE teaching strategy

Using Manipulatives Before students start this lesson, have them construct several rectangles with a compass and straightedge. Then have students draw the diagonals in color. Have students measure the diagonals of each rectangle and record their results. Have students make a conjecture based on their results.

Exploration 1 — The Equation of a Circle

1. Construct a circle with center $(0, 0)$ and radius 3. Then repeat with a radius of 1.

2. Find the equation of each circle.

3. What is the equation of a circle with center $(0, 0)$ and radius r?

4. What is the equation of a circle with center $(1, 1)$ and radius 1? ❖

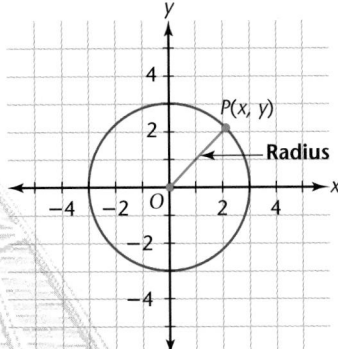

Exploration 2 — The Diagonals of a Rectangle

1. Use the distance formula to find the length of each diagonal of rectangle *ABCO*.

2. Construct two other rectangles and find the lengths of the diagonals for each rectangle.

3. Make a conjecture about the diagonals of a rectangle. ❖

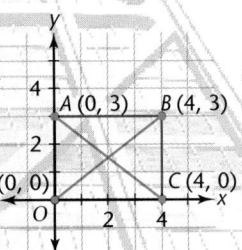

When you change from numbers to letters, the conjecture becomes more general. It can now fit whatever numbers are specified for the letters.

Exploration 3 — Finding the Formula

1. Use the 0, a, and b labels in the diagram to write a formula for finding the length of a diagonal of a rectangle.

2. Explain the general case from your discoveries and your conjecture. ❖

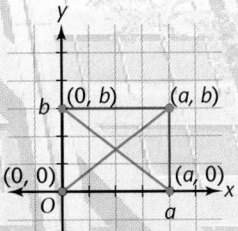

interdisciplinary **CONNECTION** — **Seismology** When earthquakes occur, the place where they begin is called the epicenter. Seismologists know that the distance a building is from the epicenter of a quake often determines how much damage will occur to that building. Have students research seismology and the various degrees of damage that can occur within various distance ranges. Have students prepare a coordinate map to show such potential damage.

Exploration 1 Notes

Students should understand from the opening development on page 632 that the general equation for a circle with its center at the origin is $x^2 + y^2 = r^2$, where (x, y) is a point on the circle and r is the length of the radius.

ongoing ASSESSMENT

4. $(x - 1)^2 + (y - 1)^2 = r^2$

Exploration 2 Notes

Tell students they will discover a relationship between the diagonals of a rectangle. Allow students to work together in cooperative learning groups to complete this exploration. Each group should construct many rectangles to test their conjectures.

ongoing ASSESSMENT

3. **The diagonals of a rectangle have equal length.**

Exploration 3 Notes

Students will discover a formula for the length of a diagonal of a rectangle. Check that students have written the correct equation for the formula in this exploration.

ongoing ASSESSMENT

2. **For any rectangle with one corner at (a, b) and the opposite corner at (c, d), the length of the diagonal is equal to $\sqrt{(a - c)^2 + (b - d)^2}$.**

Students will discover the midpoint formula by comparing values for several triangles.

After students have completed the exploration, have them draw a general triangle, but not on grid paper. Does their conjecture work for any triangle? [Yes]

4. Place $\triangle ABC$ so that A is at $(0, 0)$, C is at $(c, 0)$ and B is at (a, b). Let D be the midpoint of side AB, then the coordinates of D are $\left(\frac{a}{2}, \frac{b}{2}\right)$. Let E be the mid-point of side BC, then the coordinates are $\left(\frac{a+c}{2}, \frac{b}{2}\right)$. The length of DE and AC can be found by applying the distance formula.

length of $DE =$

$$\sqrt{\left(\frac{a}{2} - \frac{a+c}{2}\right)^2 + \left(\frac{b}{2} - \frac{b}{2}\right)^2} = \frac{c}{2}$$

length of $AC = \sqrt{(c - 0)^2} = c$

Therefore, the length of the line segment connecting the midpoints of the two sides of a triangle is half the length of the third side.

TEACHING *tip*

Have students try to create a drawing that shows the figure formed by the set of all points the same distance from a given line and a given point. Have successful students show and explain their method to the class.

The midpoint formula is also helpful for showing geometric relationships algebraically.

Exploration 4 *Segments and Midpoints*

1 Construct three triangles on a coordinate plane like the one shown. For each, find the midpoints of two sides, and draw a segment to connect those midpoints.

What do you notice about the length of the segment AB and the length of the side parallel to that segment?

2 Make a conjecture that describes this relationship.

3 Test the other triangles to see if your conjecture is true for them.

4 Use coordinate geometry with letters to represent the nonzero coordinates to show that your conjecture is true. ❖

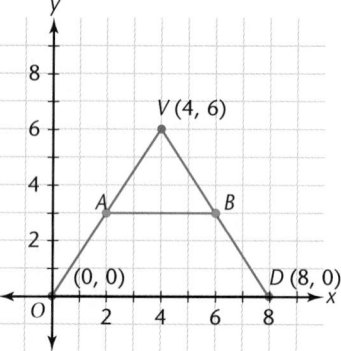

One of the interesting properties of a parabola is that it represents the set of all points the same distance from a given line and a given point. This is illustrated in the following application.

APPLICATION

Surveying Engineers plan to locate a maintenance facility at a point that is the same distance from a TV tower and a road. The TV tower is located 2 miles from the road. Point P is the same distance from the TV tower, T, and the road, X. Is there an equation that describes the possible placement of the maintenance facility?

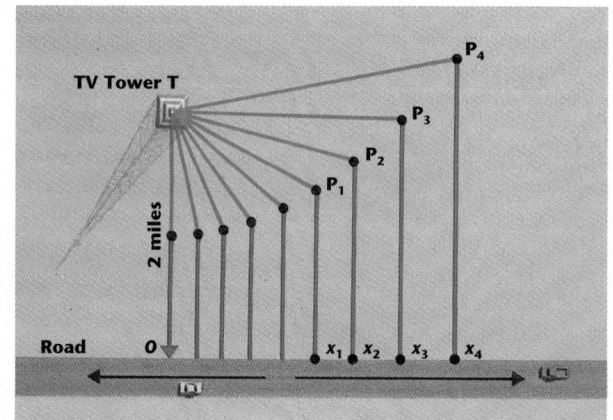

A coordinate grid can be used to model the problem. Let $T(0, 2)$ be the TV tower and $X(x, 0)$ be a point on the road x units away from the origin. Let any point $P(x, y)$ be any point equidistant from T and X. You can use coordinate geometry to find the equation of a graph that is equidistant from the x-axis $(y = 0)$ and the point $(0, 2)$.

ENRICHMENT Use coordinate geometry to prove that the hypotenuse of an isosceles right triangle is equal to the length of leg multiplied by $\sqrt{2}$.

INCLUSION **strategies** **Visual Strategies** For students who have visual problems, provide graph paper with large squares, such as 1-inch grid paper. Replicate explanations found in the textbook on the chalkboard or overhead projector to help these students in reading and interpreting the examples. Also, allow students to work on the chalkboard to complete exercises.

Let $P(x, y)$ be any point that is equidistant from $(x, 0)$ and $(0, 2)$.

$$d_1 = \sqrt{(x - x)^2 + (y - 0)^2} = \sqrt{y^2} = y$$
$$d_2 = \sqrt{(x - 0)^2 + (y - 2)^2} = \sqrt{x^2 + (y - 2)^2}$$

Since d_1 and d_2 are equal distances, the two expressions are equal. Substitute the values, square both sides of the equation, and simplify. This is the function that will produce the graph.

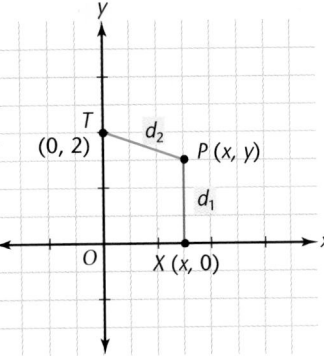

$d_1 = d_2$	Given
$y = \sqrt{x^2 + (y - 2)^2}$	Substitute the values.
$y^2 = x^2 + (y - 2)^2$	Square both sides.
$y^2 = x^2 + y^2 - 4y + 4$	Simplify.
$4y = x^2 + 4$	Addition Property of Equality
$y = \frac{1}{4}x^2 + 1$	Division Property of Equality

The equation that describes the possible placement of the maintenance facility is $y = \frac{1}{4}x^2 + 1$. ❖

CRITICAL Thinking

What is the name for the graph of $y = \frac{1}{4}x^2 + 1$? What is the vertex? What is the line of symmetry?

CRITICAL Thinking

The equation is for a parabola. The vertex is $(0, 1)$. The axis of symmetry is the y-axis.

Cultural Connection: Africa In ancient Egypt, African surveyors used geometry to solve algebraic problems. They noticed that by using the diagonal of a 1-cubit square as the side of a larger square, the area of the new square would be double the original square. Recall that $(\sqrt{2})^2 = 2$. They made the diagonal of a 1-cubit square part of their system of measurement. It was called a *double-remen* and measured $\sqrt{2}$ cubits.

1. Square B is 1 square unit in area. How can you determine that square C has an area of 2 square units?

2. Since the area of square C is 2, how would you explain why the side of square C has length $\sqrt{2}$?

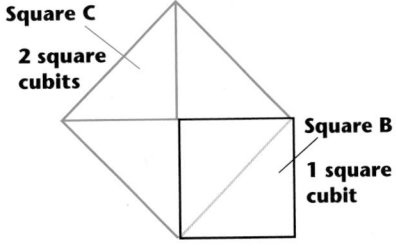

Square C
2 square cubits
Square B
1 square cubit

ongoing ASSESSMENT

It is made of four triangles, each with an area of $\frac{1}{2}$ square cubit.

RETEACHING the lesson

Using Visual Models Have students label a pair of axes from -10 to 10. Tell students to draw line segments from $A(-3, -4)$ to $B(3, 2)$ and from $C(2, -4)$ to $D(-2, 2)$. Ask students to find the midpoint of each segment. **[$(0, -1)$ is the midpoint for both]** Have students connect the endpoints to form a quadrilateral. Ask students what kind of quadrilateral is formed. **[parallelogram]**

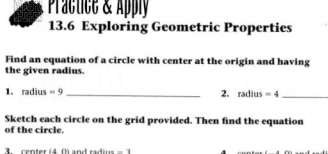
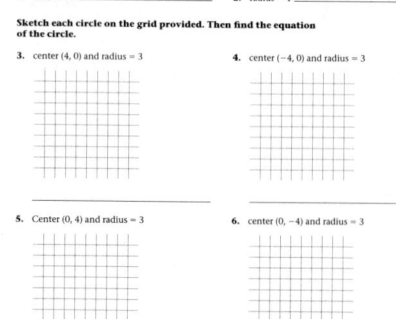
EXERCISES & PROBLEMS

Communicate

1. Discuss the ways that geometry can help solve problems in algebra.

2. What is meant by the set of all points equidistant from a given point?

3. Describe how to find the distance between two points using the distance formula.

4. Explain how the distance formula is used to express the set of all points equidistant from a given point.

5. How do you find the length of a diagonal of a rectangle by using the distance formula?

Practice & Apply

Find an equation of a circle with center at the origin and having the given radius.

6. radius = 3
$x^2 + y^2 = 9$

7. radius = 7
$x^2 + y^2 = 49$

8. radius = 15
$x^2 + y^2 = 225$

9. radius = r
$x^2 + y^2 = r^2$

10. Copy and complete the table of ordered pairs that satisfy the equation $x^2 + y^2 = 100$.

x	0	6	−6	8	−8	10	−10	?	?	?	?	?	?	?
y	?	?	?	?	?	?	?	0	6	−6	8	−8	10	−10

11. Plot the ordered pairs on graph paper.

12. What are the x-intercepts?

13. What are the y-intercepts?

14. What shape is the graph?

Use graph paper to sketch each circle below. Find an equation of a circle with

15. center (2, 3) and radius = 5.

16. center (−2, 3) and radius = 5.

17. center at (2, −3) and radius = 5.

18. center (−2, −3) and radius = 5.

19. Rewrite the equations for Exercises 15–18 with a radius of 7.

20. Generalize. Find an equation of a circle with center (h, k) and radius, r.
$(x − h)^2 + (y − k)^2 = r^2$

Use your generalization to identify the center and radius for each circle.

21. $(x − 1)^2 + (y − 5)^2 = 9$ (1, 5); 3

22. $(x + 3)^2 + (y − 4)^2 = 16$ (−3, 4); 4

23. $(x − 2)^2 + (y + 4)^2 = 50$ (2, −4); $5\sqrt{2}$

24. $(x + 5)^2 + (y + 10)^2 = 10$ (−5, −10); $\sqrt{10}$

10.

x	0	6	−6	8	−8	10	−10	±10	±8	±8	±6	±6	0	0
y	±10	±8	±8	±6	±6	0	0	0	6	−6	8	−8	10	−10

The graph for Exercise 11 can be found in Additional Answers beginning on page 729.

12. 10, −10

13. 10, −10

14. circle

15. $(x − 2)^2 + (y − 3)^2 = 25$

16. $(x + 2)^2 + (y − 3)^2 = 25$

17. $(x − 2)^2 + (y + 3)^2 = 25$

18. $(x + 2)^2 + (y + 3)^2 = 25$

19. $(x − 2)^2 + (y − 3)^2 = 49$;
$(x + 2)^2 + (y − 3)^2 = 49$;
$(x − 2)^2 + (y + 3)^2 = 49$;
$(x + 2)^2 + (y + 3)^2 = 49$

Geometry Plot triangle *PQR* on graph paper, with *P*(2, 1), *Q*(4, 7), and *R*(12, 3).

25. Find the coordinates of *M*, the midpoint of $\overline{PQ}$, and plot *M*.

26. Find the coordinates of *N*, the midpoint of $\overline{PR}$, and plot *N*.

27. Draw $\overline{MN}$ and find the length of $\overline{MN}$ (simplify your answer).

28. Find the length of $\overline{QR}$. $4\sqrt{5}$

29. Compare the lengths of $\overline{MN}$ and $\overline{QR}$. *QR* = 2*MN*

30. Find the slopes of $\overline{MN}$ and $\overline{QR}$. $-\frac{1}{2}, -\frac{1}{2}$

31. What does the slope information tell you about $\overline{MN}$ and $\overline{QR}$? parallel

32. Use the distance formula to find an equation for all points *P*(*x*, *y*) such that the sum of the distances from *P* to *A*(−8, 0) and from *P* to *B*(8, 0) is 20.

33. Find an equation for all points *P*(*x*, *y*) such that the sum of the distances from *P* to *A*(0, −4) and from *P* to *B*(0, 4) is 10.

Look Back

34. Identify which of the following correctly represents the distance between 2 and 5 on a number line. **[Lesson 4.8]** d, f, h

  a. 2 + 5   **b.** 2 − 5   **c.** 5 + 2   **d.** 5 − 2

  e. |2 + 5| **f.** |2 − 5| **g.** |5 + 2| **h.** |5 − 2|

35. Evaluate $\dfrac{3^0 \cdot 3^2}{3^2} - 3^2$. **[Lesson 10.3]** −8

Write each number in scientific form. **[Lesson 10.4]**

36. 2,300,000 2.3×10^6

37. 0.00000125 1.25×10^{-6}

Add or subtract each polynomial. **[Lesson 11.1]**

38. $(x^2 + 3x + 5) + (7x^2 - 5x - 10)$
  $8x^2 - 2x - 5$

39. $(8b^2 - 15) - (2b^2 + b + 1)$
  $6b^2 - b - 16$

Find the value of the discriminate for each equation.
[Lesson 12.5]

40. $3x^2 - 6x + 3 = 0$
  0

41. $2x^2 + 3x - 2 = 0$
  25

Graph each quadratic inequality. **[Lesson 12.6]**

42. $y < 2x^2 - 3x + 4$

43. $y > -4x^2 + 6x - 5$

Look Beyond

Geometry In geometry, you can prove several theorems by using coordinates. Prove each of the following.

44. The diagonals of a rectangle bisect each other.

45. The diagonals of an isosceles trapezoid have the same length.

25–27. (3, 4); (7, 2); $2\sqrt{5}$ or 4.47

32. $\sqrt{(x + 8)^2 + y^2} + \sqrt{(x - 8)^2 + y^2} = 20$

33. $\sqrt{x^2 + (y + 4)^2} + \sqrt{x^2 + (y - 4)^2} = 10$

The answers for Exercises 44 and 45 can be found in Additional Answers beginning on page 729.

42.

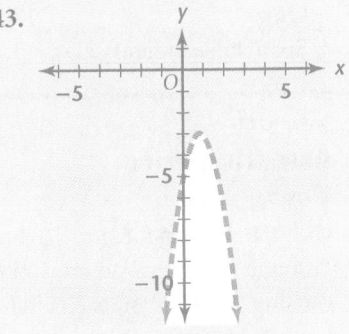

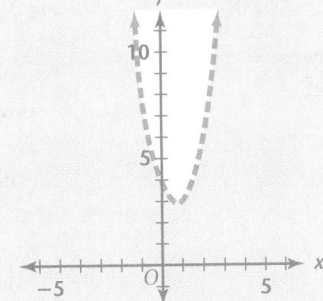

PREPARE

Objectives

- Define the tangent functions for right triangles.
- Use the tangent of an acute angle to solve problems.

RESOURCES

- Practice Master 13.7
- Enrichment Master 13.7
- Technology Master 13.7
- Lesson Activity Master 13.7
- Quiz 13.7
- Spanish Resources 13.7

Assessing Prior Knowledge

Use the "Pythagorean" Right-Triangle Theorem to find the missing lengths in each right triangle. Side c is the hypotenuse. Round answers to the nearest tenth if needed.

	side a	side b	side c
1.	4 cm	6 cm	? cm
2.	? ft	8 ft	9 ft
3.	9 m	9 m	? m
4.	$1\frac{1}{2}$ in.	? in.	2 in.

[1. 7.2; 2. 4.1; 3. 12.7; 4. 1.3]

TEACH

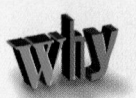

why Ask students what they remember about the sides of similar triangles. Then ask students to think of one application or real-world problem that they have solved using similar triangles or proportions. Tell students that they will learn more uses of similar triangles in this lesson.

LESSON 13.7 Trigonometric Functions

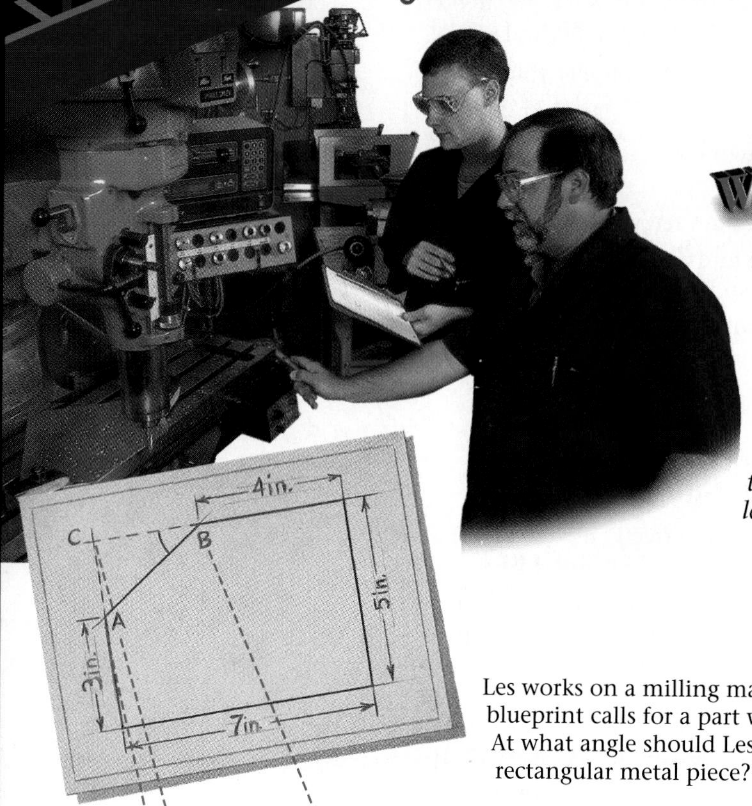

why *The tangent function is used for solving problems by indirect measure. If you know the measurement of an acute angle and one leg of a right triangle, you can determine the length of the other side without measuring it. In reverse, you have a way to determine the acute angle of the triangle if you know the lengths of the two legs.*

Les works on a milling machine in a machine shop. A blueprint calls for a part with the dimensions as shown. At what angle should Les make the cut from the rectangular metal piece?

Consider the triangle that is to be removed. Examine the following ratio with respect to angle B.

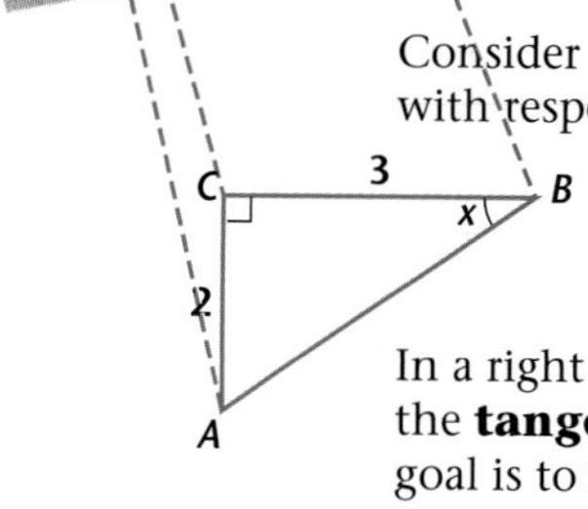

$$\frac{\textit{length of the leg opposite the angle}}{\textit{length of the leg adjacent to the angle}} = \frac{AC}{BC} = \frac{2}{3}$$

In a right triangle, this ratio is called the **tangent** of angle B or tan B. The goal is to find an angle whose tangent is $\frac{2}{3}$. "The angle whose tangent is $\frac{2}{3}$" is written $\tan^{-1}\left(\frac{2}{3}\right)$. $\text{Tan}^{-1}x$ is also read "the inverse tangent of x."

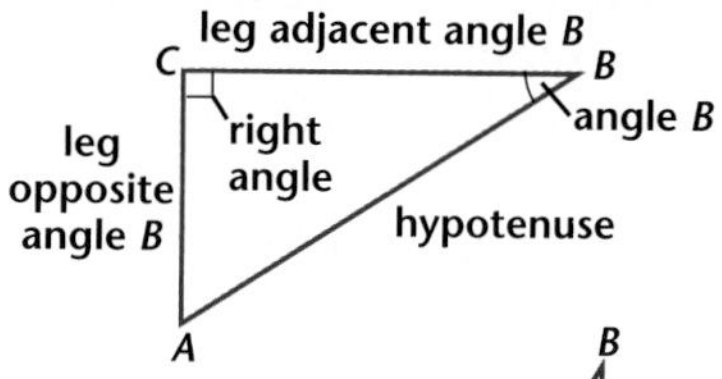

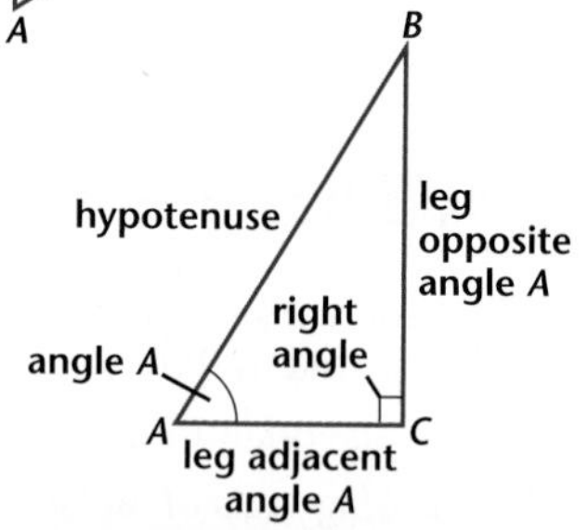

ALTERNATIVE teaching strategy

Using Visual Models Have students draw a large right triangle on grid paper so that the right angle is at the origin. Have students measure and record the length (to the nearest tenth) in a table with columns like the following

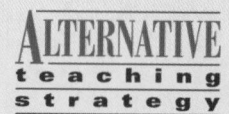

SIDES			RATIOS		
a	b	c	$\frac{a}{b}$	$\frac{a}{c}$	$\frac{b}{c}$

Have students find the appropriate ratios and use their calculators to find the value to the nearest tenth. Now tell students to draw the line for $x = 2$ to create a smaller right triangle. Again have students measure the sides and record the measurements in the table. Repeat the process by having students draw other vertical lines to create several right triangles of different sizes. Have students compare the ratios for the triangles. They should be roughly the same.

Exploration *Angles and Tangents*

$\overline{CA}$ is the base of all 7 triangles. For example, one triangle is CAB_1. As angle A increases in increments of $10°$, the length of the side opposite angle A for each triangle also increases.

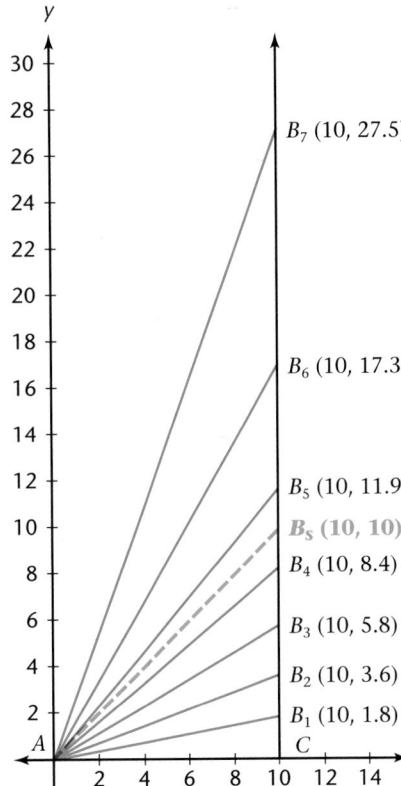

1 Find the tangent ratio for each given angle measure. Round the ratio to the nearest 10th.

a. $10°$ (Angle CAB_1)

b. $20°$ (Angle CAB_2)

c. $30°$ (Angle CAB_3)

d. $40°$ (Angle CAB_4)

e. $50°$ (Angle CAB_5)

f. $60°$ (Angle CAB_6)

g. $70°$ (Angle CAB_7)

Refer to your answers for **a–g** to answer the following questions.

2 How does the value of the tangent ratio increase as the measure of the angle doubles? (Compare, for example, tan 10° and tan 20°, and tan 30° and tan 60°.)

3 For which of the angles in Step 1 is the tangent ratio less than 1?

4 The measure of angle CAB_S is $45°$. What can you say about the length of the side opposite the angle and the length of the side adjacent to the angle?

5 How can you find the length of the side opposite angle A if you know that the measure of angle A is $45°$ and you know the length of the side adjacent to A? Explain. ❖

The function $f(x) = \tan x$ is an example of a trigonometric function. The domain of f is a set of angle measures. Thus $f(45°) = \tan 45° = 1$.

Calculator

You can use a calculator to find the tangent. Since you are using degrees to measure the angles, be sure the calculator mode is set for degrees. Enter the angle and use the TAN key. For example, for $f(x) = $ TAN x, press TAN 45 ENTER or = , and the calculator will return the value 1.

Exploration Notes

Students will discover that as the measurement of an angle in a triangle increases, the length of the side opposite that angle increases exponentially and thus so does the tangent of the angle. In other words, as the measure of the angle increases, the curve rises faster. Have students graph the ordered pairs for (number of degrees, y-value on diagram).

Aongoing ASSESSMENT

5. The opposite side will be the same length as the side adjacent to A.

Teaching tip

Technology Have students find $\tan^{-1} 0.67$ on their calculators. If they get an answer of $-0.62 \ldots$, the calculator is set in *radians*. Have students reset their calculators to *degrees*.

You can also find an acute angle if you know the tangent value. To find the angle x, where x represents the degree measure of the acute angle, use the [TAN⁻¹] key. For example, if tan x = 1, press [TAN⁻¹] 1 [ENTER] and the calculator will show 45.

3 units

7 units

Calculator

Now you can solve Les's problem. That is, find the angle whose tangent is $\frac{2}{3}$ or approximately 0.67.

$$\boxed{\text{TAN}^{-1}}\left(\frac{2}{3}\right) \approx \boxed{\text{TAN}^{-1}}\ 0.67 \approx 34$$

Les should cut the metal at a 34° angle.

EXAMPLE 1

Find the tangent of each acute angle of each triangle.

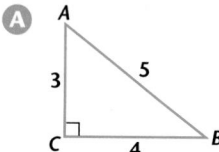

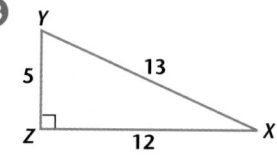

Solution ➤

Ⓐ $\tan A = \frac{4}{3}$, $\tan B = \frac{3}{4}$ Ⓑ $\tan Y = \frac{12}{5}$, $\tan X = \frac{5}{12}$ ❖

EXAMPLE 2

An airplane climbs to 30,000 feet at a steady rate. If the altitude is reached after the airplane has covered a distance of 35 miles on the ground, what is its angle of elevation?

Solution ➤

Draw a picture to model the problem. Change 35 miles to feet by multiplying by 5280.

Graphics Calculator

$$\tan A = \frac{30{,}000}{184{,}800}$$

In degree mode, use [TAN⁻¹] to find the angle whose tangent is $\frac{30{,}000}{184{,}800}$.

$$A = \tan^{-1}\frac{30{,}000}{184{,}800} \approx 9°$$

The airplane is climbing at an angle of 9°. The angle of elevation is 9°. ❖

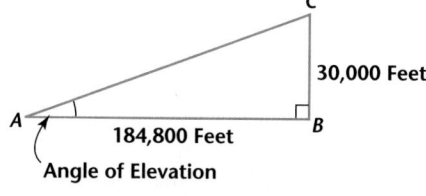

30,000 Feet

184,800 Feet

Angle of Elevation

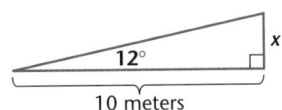

A group of skateboarders wants to build a ramp with an angle of 12°. How much should the ramp rise over each 10 meters of run to achieve a 12° angle?

Solution ➤

To find the rise, find the tangent of 12°.

$$\tan 12° = \frac{x}{10}$$

$$0.2126 = \frac{x}{10}$$

$$(10)(0.2126) = x$$

$$2.126 = x$$

The ramp should rise about 2.1 meters for each 10 meters of run. ❖

EXERCISES & PROBLEMS

Communicate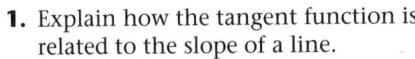

1. Explain how the tangent function is related to the slope of a line.

2. How is the length of the opposite leg of a right triangle determined from the acute angle *A* and the length of the other leg?

3. How is the angle determined from the ratio of the opposite over the adjacent legs of a right triangle?

4. Why is finding the tangent considered indirect measure?

Practice & Apply

Technology Use a calculator to find the tangents of the following angles to the nearest tenth.

5. 45° 1   **6.** 30° 0.6   **7.** 60° 1.7   **8.** 0° 0

9. 38° 0.8   **10.** 57° 1.5   **11.** 89° 57.3   **12.** 89.9° 573.0

Alternate Example 3

It is recommended that a wheelchair ramp have a slope no greater than 0.125. That means a ramp should rise only 1 foot for each 8 horizontal feet. To the nearest degree, what is the maximum recommended angle of elevation for a wheelchair ramp? [7°]

ASSESS

Selected Answers
Odd-Numbered Exercises 5–33

Assignment Guide
Core 1–4, 6–16 even, 17–38

Core Two-Year 1–36

Technology

A calculator is appropriate for Exercises 5–26. Caution students to be sure their calculators are set in the degree mode.

Error Analysis

When dealing with tangents, students often try to use the hypotenuse. Remind students that the tangent is the ratio of the opposite side to the adjacent side. Use the development on page 639 to reinforce the fact that as the angle gets larger, its tangent gets larger.

What are the corresponding angles for the following tangent approximations?

13. 1.732
   ≈ 60°

14. 1.192
   ≈ 50°

15. 0.268
   ≈ 15°

16. 3.732
   ≈ 75°

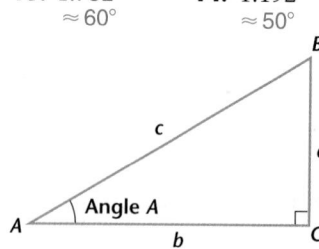

The small letters represent the sides, and the capital letters represent the angles of a general right triangle. Use the given information to find the indicated side to the nearest hundredth.

17. If angle A is 30° and side b is 12 meters, find a. 6.93 m

18. If angle B is 60° and side a is 10 feet, find b.
   17.32 feet

19. If side a is 3 centimeters and tan A is 0.75, what is the length of side b? 4 cm

20. If side b is 23 feet long and tan A is 1.0, what is the length of side a? 23 feet

21. If side a is 6 inches and angle A is approximately 31°, how long is side b? ≈ 10 in

22. If angle A is approximately 35° and side b is 20 millimeters, what is the length of side a? 14 mm

23. House Construction Mr. Fernandez wants the roof of his house to have an angle of 20°. The slope of the roof is the same on both sides and has the same length on either side of the center. How high does the roof rise if the whole length of the roof is 14 meters long? 2.55 m

24. A water tower is 600 feet from the observer. Sighting the top of the water reservoir forms an angle of 28°. A sighting at the bottom of the reservoir has an angle of 26°. How tall is the reservoir part of the water tower? 26.39 feet

25. Highways If a road has a 7% grade, its slope is $\frac{7}{100}$. What is the measure of the angle that the road inclines? ≈ 4°

26. Model Airplanes A model airplane and a car begin at the same place. After 15 seconds the car has driven $\frac{1}{4}$ mile, and the elevation angle from the starting point to the plane is 30°. What is the altitude of the plane?

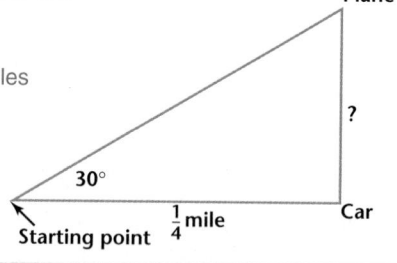

Plane

0.14 miles

?

30°

Starting point $\frac{1}{4}$ mile   Car

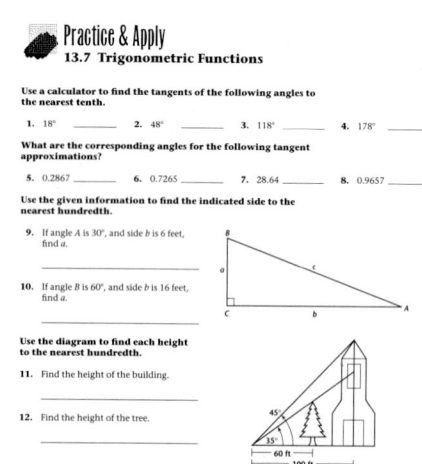

27. Solve the equation $4x - 5 = 7(x - 3) + 2$, for x. **[Lesson 4.6]** $\frac{14}{3}$

28. What is the probability that after 1 roll of 2 number cubes the sum of the dots equals 7? **[Lesson 8.6]** $\frac{1}{6}$

29. Determine whether the quadratic equation $2x^2 + x - 15 = 0$ has a real solution. If it does, use the quadratic formula to determine the solution.
[Lesson 12.5] yes; $2\frac{1}{2}, -3$

30. Solve $x^2 - 5x + 6 \geq 0$. **[Lesson 12.6]** **31.** Graph $x^2 - 16 < 0$. **[Lesson 12.6]**
   $x \geq 3$ or $x \leq 2$

32. **Geometry** The 45-45-90 degree triangle and the 30-60-90 degree triangle have sides with special relationships.
[Lesson 13.4]

Use the "Pythagorean" Right-Triangle Theorem to show that the sides actually have the relationship shown.

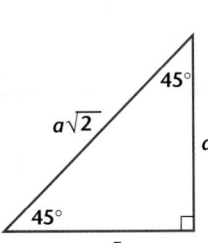
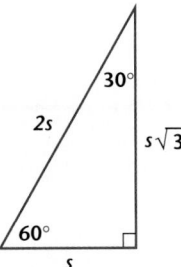

33. Points $P(4, 9)$ and $Q(10, 5)$ are given on a coordinate plane. Find the coordinates of M, the midpoint of segment PQ. **[Lesson 13.5]** $(7, 7)$

Find the length of each segment. **[Lesson 13.5]**
34. PQ. $2\sqrt{13}$ **35.** PM. $\sqrt{13}$ **36.** MQ. $\sqrt{13}$

Look Beyond

Besides the tangent (tan), the sine (sin) and cosine (cos) are two other basic trigonometric functions. They provide additional tools for solving problems by indirect measure.

$$f(x) = \tan (x) \text{ is } \frac{b}{a},$$

$$g(x) = \sin (x) \text{ is } \frac{b}{c},$$

$$h(x) = \cos (x) \text{ is } \frac{a}{c}$$

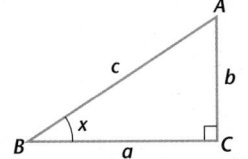

37. If the tangent of angle x is $\frac{3}{4}$, what is the sine and the cosine of x?

HINT: Use the "Pythagorean" Right-Triangle Theorem. $\frac{3}{5}, \frac{4}{5}$

38. Explain why $(\sin x)^2 + (\cos x)^2 = 1$ is true for any right triangle.

31.
   $x > -4$ and $x < 4$

32. 45-45-90 triangle:
   $(a\sqrt{2})^2 = a^2 + a^2$
   $2a^2 = 2a^2$
   30-60-90 triangle:
   $s^2 + (s\sqrt{3})^2 = (2s)^2$
   $s^2 + 3s^2 = 4s^2$
   $4s^2 = 4s^2$

38. $\sin (x) = \frac{b}{c}$ Solving for b, gives
   $b = c \sin (x)$.

   $\cos (x) = \frac{a}{c}$ Solving for c, gives
   $a = c \cos (x)$.
   $c^2 = c^2 (\sin (x))^2 + c^2 (\cos (x))^2$
   After dividing each term by c^2, $(\sin (x))^2 + (\cos (x))^2 = 1$. This is true for any right triangle because the hypotenuse value, c cancels and does not affect the equation.

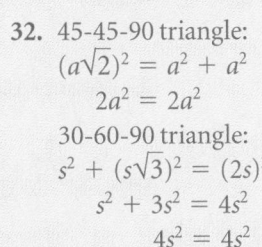

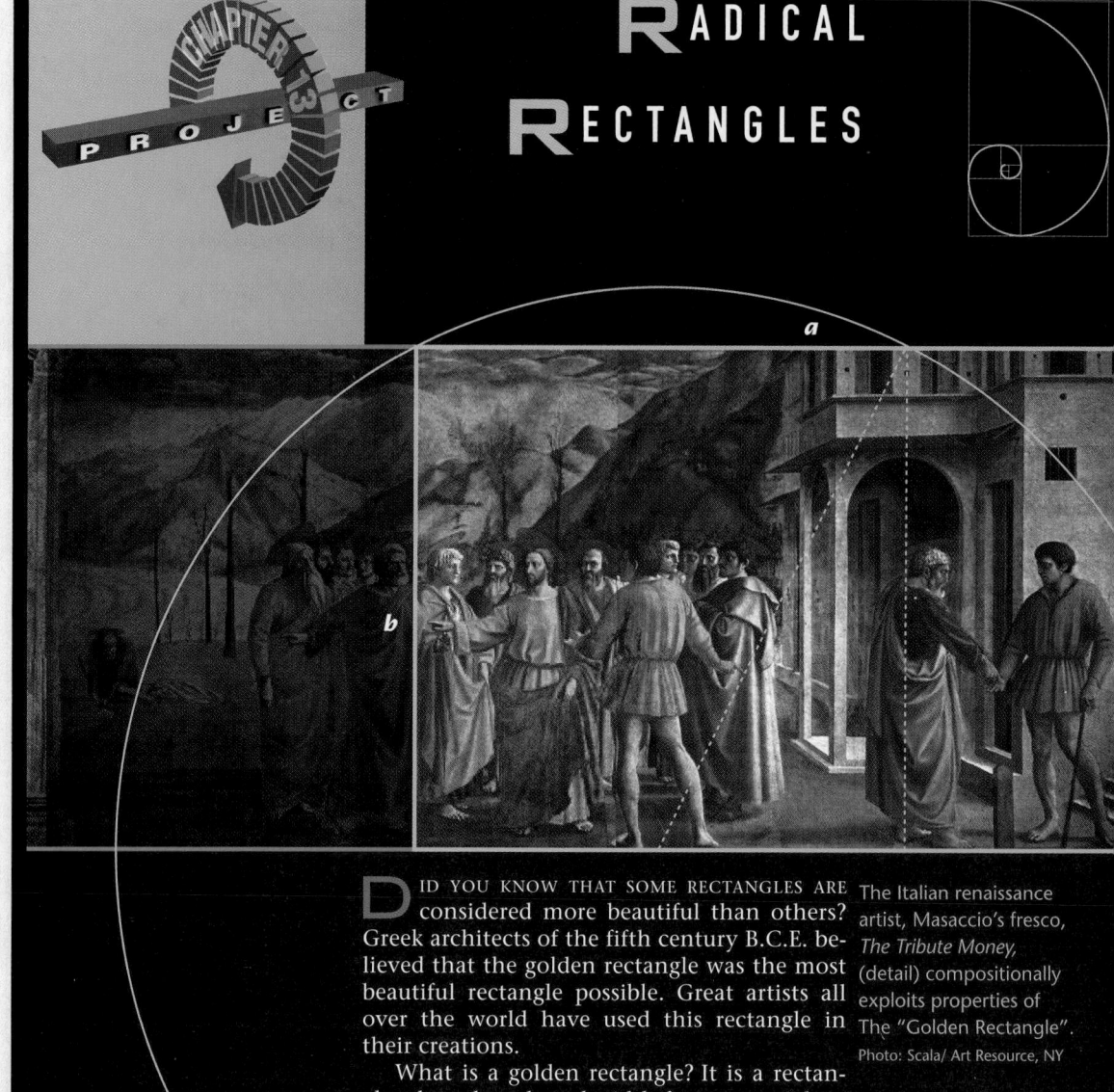

FOCUS

This project involves the study of the golden rectangle. It provides an opportunity for students to see the connections between algebra, geometry, sequences, and spirals.

MOTIVATE

Display rectangles with the following dimensions: 5 inches by 11 inches, 6 inches by 10 inches, 7 inches by 7 inches, and 8 inches by 11 inches. Take a survey of the class. Have each student choose the rectangle they like best, based solely on the shape. Then have students read the first paragraph of the Chapter Project and decide whether they agree with the Greek architects and artists.

D ID YOU KNOW THAT SOME RECTANGLES ARE considered more beautiful than others? Greek architects of the fifth century B.C.E. believed that the golden rectangle was the most beautiful rectangle possible. Great artists all over the world have used this rectangle in their creations.

The Italian renaissance artist, Masaccio's fresco, *The Tribute Money,* (detail) compositionally exploits properties of The "Golden Rectangle".
Photo: Scala/ Art Resource, NY

What is a golden rectangle? It is a rectangle whose length and width form a special ratio. If a is the longer side of the rectangle and b is the shorter side, then the golden ratio, m, is the proportion

$$\frac{b}{a} = \frac{a}{a+b}.$$

This proportion can be rewritten as $a^2 = b(a + b)$ and can be solved by the quadratic formula. The ratio is

$$m = \frac{\sqrt{5} - 1}{2} \approx 0.618.$$

A golden rectangle will produce a similar rectangle when a square is cut from the end of the original rectangle. A similar rectangle is one that has an identical shape, but has a different size.

ACTIVITY 1

Construct an approximate golden rectangle with a 3-by-5 inch note card. Since the golden ratio has the ratio of b to a or 0.61803, this is approximately $\frac{6}{10}$ or $\frac{3}{5}$.

You can see that the approximation is close. If you fold one side and draw a vertical line to form the square portion, the remaining rectangle will have a ratio of $\frac{2}{3} = 0.667$ which is close to 0.61803.

ACTIVITY 2

There is another way you can create a golden rectangle that is more accurate. Begin with a square. Find the midpoint of the base. Use a compass to measure the distance from, the midpoint, M, to the vertex, V, of the opposite side. Extend the base of the square, and draw an arc to the point of the intersection with the base, B. Use this to complete the rectangle.

ACTIVITY 3

Construct a large golden rectangle. Fold a corner of the rectangle and draw the line to create a smaller rectangle. Repeat the process with each of the resulting rectangles. Continue this process several times. You can produce a spiral by connecting the corresponding vertex of each square.

Examine the decreasing rectangles. Explore and describe the mathematical relationships that result. For example, consider the length of the side of the smallest square as 1. Use the "Pythagorean" Right-Triangle Theorem to determine the lengths of the diagonals.

The spiral pattern appears frequently in nature. It is shown here in the flower.

Cooperative Learning

Have students work in small groups to complete Activity 2. Have each member of the group start with a different size square. Have each member of the group compare his or her golden rectangle with those of other members. Students should notice that they are all similar in shape.

DISCUSS

Discuss the spiral pattern obtained by completing Activity 3. Relate that discussion to the photograph on page 644.

Chapter 13 Review

Vocabulary

converse	627	principal square root	598	rational numbers	597
Distance formula	626	"Pythagorean" Right-		real numbers	598
hypotenuse	619	Triangle Theorem	620	tangent function	638
irrational numbers	597				

Key Skills & Exercises

Lesson 13.1

➤ **Key Skills**

Estimate square roots. If the square root is irrational, find the value to the nearest hundredth.

a. $\sqrt{16} = 4$　　b. $\sqrt{\dfrac{9}{25}} = \dfrac{3}{5}$　　c. $\sqrt{18} = 4.24$　　d. $-\sqrt{0.012} = -0.11$

➤ **Exercises**

Estimate each square root. If the square root is irrational, find the value to the nearest hundredth.

1. $\sqrt{20}$ 4.47　　**2.** $\sqrt{115}$ 10.72　　**3.** $\sqrt{134}$ 11.58　　**4.** $-\sqrt{67}$ -8.19　　**5.** $\sqrt{\dfrac{9}{16}}$ $\dfrac{3}{4}$　　**6.** $-\sqrt{0.09}$ -0.3

Lesson 13.2

➤ **Key Skills**

Simplify radical expressions.

a. $\sqrt{32} = \sqrt{16 \cdot 2} = \sqrt{16} \cdot \sqrt{2} = 4\sqrt{2}$　　b. $\sqrt{\dfrac{4}{25}} = \dfrac{\sqrt{4}}{\sqrt{25}} = \dfrac{2}{5}$

c. $6\sqrt{3} + 5\sqrt{2} - 3\sqrt{3} + \sqrt{2} = (6 - 3)\sqrt{3} + (5 + 1)\sqrt{2} = 3\sqrt{3} + 6\sqrt{2}$

d. $(4\sqrt{5})^2 = (4\sqrt{5})(4\sqrt{5}) = (4 \cdot 4)(\sqrt{5} \cdot \sqrt{5}) = 16 \cdot 5 = 80$

➤ **Exercises**

Simplify.

7. $\sqrt{150}$ $5\sqrt{6}$　　**8.** $\sqrt{\dfrac{48}{27}}$ $\dfrac{4}{3}$　　**9.** $\sqrt{a^2 b^7}$ $ab^3\sqrt{b}$　　**10.** $\sqrt{2} + 3\sqrt{7} - 3\sqrt{2}$ $-2\sqrt{2} + 3\sqrt{7}$

11. $(2\sqrt{3})^2$ 12　　**12.** $\sqrt{3}(2 - \sqrt{12})$ $2\sqrt{3} - 6$　　**13.** $(\sqrt{2} + 2)^2$ $6 + 4\sqrt{2}$　　**14.** $(\sqrt{3} - 2)(\sqrt{5} + 6)$ $\sqrt{15} + 6\sqrt{3} - 2\sqrt{5} - 12$

Lesson 13.3

➤ **Key Skills**

Solve radical equations. Use radicals to solve equations.

$\sqrt{x + 2} = 5 \longrightarrow (\sqrt{x + 2})^2 = 5^2 \longrightarrow x + 2 = 25 \longrightarrow x = 23$

$2x^2 = 36 \longrightarrow x^2 = 18 \longrightarrow x = \pm\sqrt{18} \longrightarrow x = \pm\sqrt{9 \cdot 2} = \pm 3\sqrt{2}$

➤ **Exercises**

Solve each equation. Be sure to check your solution.

15. $\sqrt{x-7}=2$ 11

16. $\sqrt{3x+4}=1$ -1

17. $\sqrt{x+6}=x$ 3

18. $\sqrt{x^2-2x+1}=x-5$
     no solution

19. $x^2=40$
     $2\sqrt{10}, -2\sqrt{10}$

20. $2x^2-32=0$
     $4, -4$

Lesson 13.4

➤ **Key Skills**

Use the "Pythagorean" Right-Triangle Theorem to find the length of a missing side of a right triangle.

Find the length of the hypotenuse of the triangle.

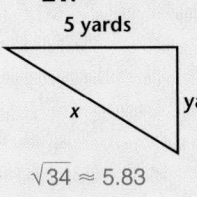

$$a^2+b^2=c^2$$
$$8^2+11^2=c^2$$
$$64+121=c^2$$
$$185=c^2$$
$$c=\pm\sqrt{185}\approx\pm13.6$$

Since c represents the length of a segment, it must be positive. Therefore, $c\approx13.6$ meters.

➤ **Exercises**

Use the "Pythagorean" Right-Triangle Theorem to find the length of the missing side of each right triangle.

21.

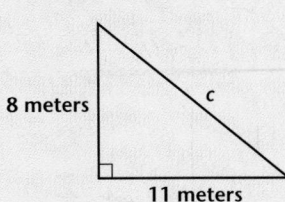

5 yards

x

3 yards

$\sqrt{34}\approx5.83$

22.

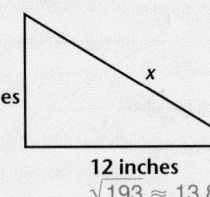

7 inches

x

12 inches

$\sqrt{193}\approx13.89$

23.

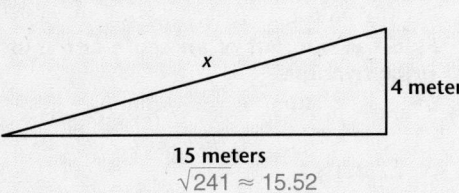

x

4 meters

15 meters

$\sqrt{241}\approx15.52$

Lesson 13.5

➤ **Key Skills**

Use the Distance formula to find the distance between two points.

Use the Distance formula to find the distance between $(2, 3)$ and $(-4, -1)$.

$$d=\sqrt{(x_2-x_1)^2+(y_2-y_1)^2}$$
$$d=\sqrt{(-4-2)^2+(-1-3)^2}$$
$$d=\sqrt{36+16}$$
$$d=\sqrt{52}=2\sqrt{13}\approx7.2$$

Use the Midpoint formula to find the midpoint of a segment.

Use the Midpoint formula to find the midpoint of the segment whose endpoints are $(-2, 6)$ and $(3, 4)$.

$$\left(\frac{x_1+x_2}{2},\frac{y_1+y_2}{2}\right)=\left(\frac{-2+3}{2},\frac{6+4}{2}\right)=\left(\frac{1}{2},5\right)$$

➤ **Exercises**

Find the distance between the two given points.

24. $A(0, 3)$, $B(2, 8)$

25. $X(-1, 5)$, $Y(3, -8)$

26. $G(7, 2)$, $F(-6, -1)$

Find the midpoint of each segment whose endpoints are given.

27. $M(5, 2)$, $N(3, 6)$

28. $P(-3, 4)$, $Q(-7, -1)$

29. $R(8, -2)$, $S(-1, 4)$

24. $\sqrt{29}\approx5.39$

25. $\sqrt{185}\approx13.60$

26. $\sqrt{178}\approx13.34$

27. $(4, 4)$

28. $\left(-5,\frac{3}{2}\right)$

29. $\left(\frac{7}{2},1\right)$

Lesson 13.6

➤ **Key Skills**

Write the equation of a circle given the center and the radius.

The equation of a circle with center at the origin and radius 4 is $x^2 + y^2 = 16$.

The equation of a circle with center $(2, -3)$ and radius 5 is $(x - 2)^2 + (y + 3)^2 = 25$.

➤ **Exercises**

Write the equation of each circle with the given center and radius.

30. center $(0, 0)$, radius 2   **31.** center $(0, 2)$, radius 6   **32.** center $(-2, 5)$, radius 3

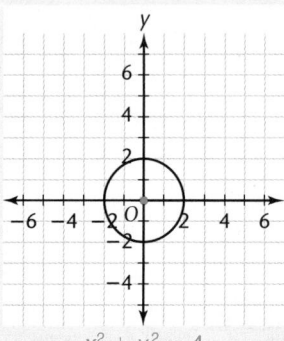
$x^2 + y^2 = 4$

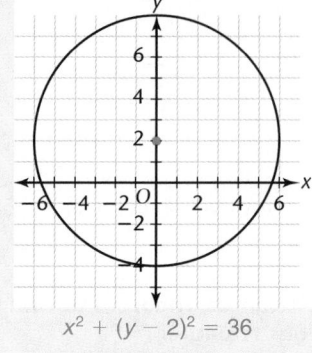
$x^2 + (y - 2)^2 = 36$

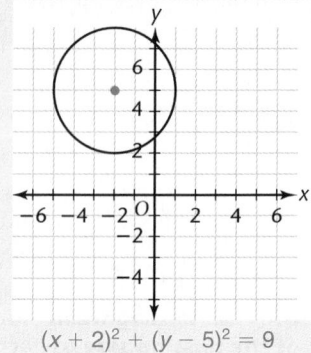
$(x + 2)^2 + (y - 5)^2 = 9$

Lesson 13.7

➤ **Key Skills**

Find the tangent of an acute angle of a right triangle.

$$\tan M = \frac{15}{8} \qquad \tan N = \frac{8}{15}$$

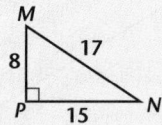

➤ **Exercises**

Find the tangent ratio for the given angle of the right triangle.

33. $\tan R$ $\frac{21}{20}$ **34.** $\tan S$ $\frac{20}{21}$

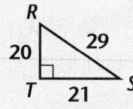

Applications

35. The bottom of a circular swimming pool has an area of 80 square feet. Find the radius to the nearest tenth. Use 3.14 for π. 5.0 feet

36. Physics The number of seconds it takes an object to fall d feet is given by the formula $s = \sqrt{\frac{d}{16}}$. About how long does it take an object dropped from the top of a 50-foot building to hit the ground? Give your answer to the nearest tenth. 1.8 seconds

Sports At the end of a ski lift, riders are 50 feet above the ground. The ski lift travels a horizontal distance of 120 feet.

37. How long is the cable? 130 feet

38. What angle does the cable make with the ground?

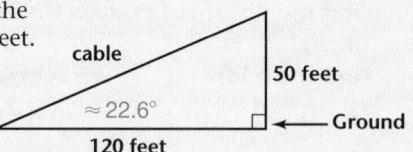

Chapter 13 Assessment

Simplify.

1. $\sqrt{50}$ $5\sqrt{2}$ **2.** $\sqrt{\dfrac{72}{2}}$ 6   **3.** $\sqrt{\dfrac{m^8}{n^{10}}}$ $\dfrac{m^4}{n^5}$

4. $6\sqrt{6} + 5\sqrt{6} - \sqrt{3}$ **5.** $(\sqrt{10} - 3)^2$

   $11\sqrt{6} - \sqrt{3}$ $19 - 6\sqrt{10}$

Solve each equation. Be sure to check your solution.

6. $\sqrt{4x + 2} = 8$ $15\frac{1}{2}$ **7.** $\sqrt{x + 2} = x + 2$ **8.** $4x^2 = 48$

                                    $-1, -2$ $2\sqrt{3}, -2\sqrt{3}$

Find the length of the missing side of each right triangle to the nearest hundredth.

9.

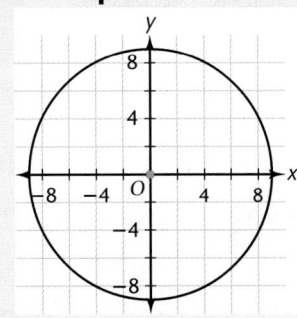

21 meters

7 meters

x

22.14 m

10.

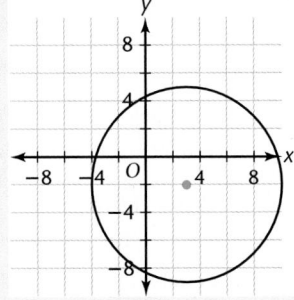

10 inches   8 inches

x

6 inches

11. One boat travels 9 miles south. Another boat travels 6 miles west and meets the first boat. How far apart were the two boats when they started? ≈ 10.82 miles

Find the indicated tangent ratio for the right triangles.

12. tan M
13. tan K
14. tan A
15. tan B
16. tan R
17. tan T

M, $15\sqrt{2}$, 15, K, 15, L

B, 13, C, 12, A

T, 40, S, 50, R

18. A ladder is leaning against a house. It forms an angle of 55° with the house. If the ladder touches the house at a point 12 feet high, how far is the bottom of the ladder from the bottom of the house? about 17.1 feet

Write the equation of each circle with the given center and radius.

19.

(circle graph)

20.

(circle graph)

12. 1

13. 1

14. $\frac{5}{12}$

15. $\frac{12}{5}$

16. $\frac{4}{5}$

17. $\frac{5}{4}$

19. $x^2 + y^2 = 81$

20. $(x - 3)^2 + (y + 2)^2 = 49$

CHAPTER 14 Rational Functions

Meeting Individual Needs

14.1 Rational Expressions

Core Resources

Practice Master 14.1
Enrichment, p. 652
Technology Master 14.1

[1 day]

Core Two-Year Resources

Inclusion Strategies, p. 653
Reteaching the Lesson, p. 654
Practice Master 14.1
Enrichment Master 14.1
Technology Master 14.1
Lesson Activity Master 14.1

[3 days]

14.2 Inverse Variation

Core Resources

Practice Master 14.2
Enrichment, p. 659
Technology Master 14.2

[1 day]

Core Two-Year Resources

Inclusion Strategies, p. 660
Reteaching the Lesson, p. 661
Practice Master 14.2
Enrichment Master 14.2
Technology Master 14.2
Lesson Activity Master 14.2

[2 days]

14.3 Simplifying Rational Expressions

Core Resources

Practice Master 14.3
Enrichment, p. 665
Technology Master 14.3

[2 days]

Core Two-Year Resources

Inclusion Strategies, p. 666
Reteaching the Lesson, p. 667
Practice Master 14.3
Technology Master 14.3
Enrichment Master 14.3
Lesson Activity Master 14.3

[3 days]

14.4 Operations With Rational Expressions

Core Resources

Practice Master 14.4
Enrichment, p. 671
Technology Master 14.4
Mid-Chapter Assessment Master

[2 days]

Core Two-Year Resources

Inclusion Strategies, p. 672
Reteaching the Lesson, p. 673
Practice Master 14.4
Enrichment Master 14.4
Technology Master 14.4
Lesson Activity Master 14.4
Mid-Chapter Assessment Master

[3 days]

14.5 Solving Rational Equations

Core Resources

Practice Master 14.5
Enrichment, p. 680
Technology Master 14.5
Interdisciplinary
   Connection, p. 679

[2 days]

Core Two-Year Resources

Inclusion Strategies, p. 680
Reteaching the Lesson, p. 681
Practice Master 14.5
Enrichment Master 14.5
Technology Master 14.5
Lesson Activity Master 14.5

[4 days]

14.6 Exploring Proportions

Core Resources

Practice Master 14.6
Enrichment, p. 684
Technology Master 14.6

[2 days]

Core Two-Year Resources

Inclusion Strategies, p. 685
Reteaching the Lesson, p. 686
Practice Master 14.6
Enrichment Master 14.6
Technology Master 14.6
Lesson Activity Master 14.6

[4 days]

14.7 Proof in Algebra

Core Resources

Practice Master 14.7
Enrichment, p. 690
Technology Master 14.7

[2 days]

Core Two-Year Resources

Inclusion Strategies, p. 690
Reteaching the Lesson, p. 691
Practice Master 14.7
Enrichment Master 14.7
Technology Master 14.7
Lesson Activity Master 14.7

[4 days]

Chapter Summary

Core Resources

Eyewitness Math,
  pp. 676–677
Chapter 14 Project,
  pp. 694–695
Lab Activity
Long-Term Project
Chapter Review,
  pp. 696–698
Chapter Assessment,
  p. 699
Chapter Assessment, A/B
Alternative Assessment
Cumulative Assessment,
  pp. 700–701

[3 days]

Core Two-Year Resources

Eyewitness Math, pp. 676–677
Chapter 14 Project, pp. 694–695
Lab Activity
Long-Term Project
Chapter Review, pp. 696–698
Chapter Assessment, p. 699
Chapter Assessment A/B
Alternative Assessment
Cumulative Assessment,
  pp. 700–701

[5 days]

Reading Strategies

Before students read the chapter, have them skim the material, paying particular attention to the lesson headings and graphs. You might also have the students read the Eyewitness Math article in Lesson 5 and predict how it will relate to the chapter. As they read, students should use journals to keep track of the key concepts from the chapter.

Visual Strategies

When working with rational equations, the ability to see the graph of the equation can help students understand the nature and approximate value of solutions to the equation. Having students clearly write each step in the solution to an equation or in a proof is also a type of visual strategy that is often overlooked. Seeing the steps and writing the words helps students remember concepts.

Hands-on Strategies

Lesson 14.2 is a study of inverse variation. Example 2 on page 660 has students analyze a physics experiment involving the placement of objects on a balance. They study the relationship of the distance from the fulcrum and the mass of the object required to achieve balance. Bring in a balance and objects so that students can complete this experiment in class. Then have students repeat the experiment with other objects to find the constant of variation.

Cooperative Learning

GROUP ACTIVITIES	
Operations with rational expressions	Lesson 14.4, opening discussion on pp. 670–671
How Worried Should You Be?	Eyewitness Math, pp. 676–677
Exploring proportions	Lesson 14.6, Explorations 1 and 2

You may wish to have students work with partners or in small groups to complete any of these activities. Additional suggestions for cooperative group activities can be found in the teacher's notes in the lessons.

Multicultural

The cultural connections in this chapter include references to the United States, the Blackfoot Indians, and Egypt.

CULTURAL CONNECTIONS	
North America: B. Banneker, Mathematical Puzzles	Lesson 14.1, Exercise 34
Africa: Commodity Exchange	Lesson 14.2, Discussion, p. 659
North America: Native American Games	Lesson 14.4, Exercises 40–41

Portfolio Assessment

Below are portfolio activities for the chapter listed under seven activity domains that are appropriate for portfolio development.

1. **Investigation/Exploration** Students explore and investigate operations with rational expressions in the opening discussion of Lesson 14.4. Medical testing is investigated in *How Worried Should You Be?* on pages 676–677. Students explore proportions in Explorations 1 and 2 and Applications in Lesson 14.6. Proof in algebra and logical reasoning are explored in Lesson 14.7.

2. **Applications** Recommended for inclusion are any of the following: Art, Lesson 14.1, opening discussion and Example 2; Ecology, Lesson 14.2, Example 1; Physics, Lesson 14.2, Example 2; Investments, Lesson 14.2, Example 3; Music, Lesson 14.2, Exercise 22.

3. **Non-Routine Problems** The Eyewitness Math involves data analysis on the topic *How Worried Should You Be?* Exercise 48 in Lesson 14.3 provides a geometric pattern problem for students.

4. **Project** In *A Different Dimension* on pages 694–695, students investigate dimensional analysis.

5. **Interdisciplinary Topics** Students may choose from the following: Physics, Lesson 14.2, Example 2 and Exercise 21; Geometry, Lesson 14.2, Exercises 20 and 27; Lesson 14.3, Example 4 and Exercises 34, 36 and 37; Lesson 14.6, Application on pages 686–687 and Exercise 34.

6. **Writing** Communicate exercises that ask students to describe and explain procedures involved in solving probability problems offer excellent writing selections.

7. **Tools** In Lesson 14.1, students use a spreadsheet to evaluate rational expressions for various values of the variable. Then students use a graphics calculator to explore the general rational expression $\frac{1}{x}$ and its transformations.

Technology

Lesson 14.1 includes the use of a spreadsheet to evaluate rational expressions for various values of the variable. Have students use a similar spreadsheet to solve other problems in the chapter. Students can use **Maple**^{TM} software to simplify rational expressions. To complete this activity, students need to be familiar with the following commands:

expr:-	Defines an expression
Simplify	Reduces an expression in the expression window to a simpler form
Factor	Factors an expression
;	Ends the command

Examples

Determine the factors of the numerator and denominator, then simplify $\dfrac{x^4 - 16}{x^2 - 4}$.

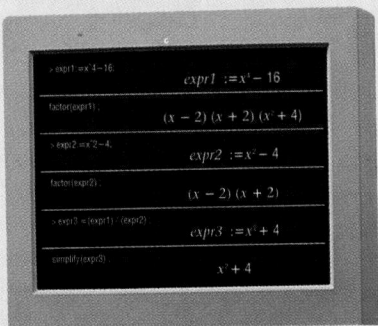

Simplify $\dfrac{(x-1)}{(x)} + \dfrac{(1)}{(x^2 - x)}$.

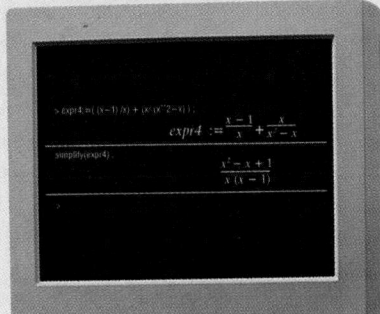

Integrated Software

f(g) **Scholar**^{TM} is an integrated computer-based mathematics productivity tool that combines calculator, spreadsheet, and graphics capabilities to provide a dynamic and interactive environment for explorations in mathematics. It is appropriate to use *f(g)* **Scholar**^{TM} for any lesson needing a spreadsheet, calculator, graphics calculator, or any combination of the three.

14 Rational Functions

ABOUT THE CHAPTER

Background Information

Students are familiar with fractions and know that they are called rational numbers. In this chapter, students will learn to use rational expressions. A **rational expression** is an algebraic fraction in which the numerator and denominator are polynomials. The chapter deals with simplifying and operating on rational expressions and then using those expressions to solve problems.

CHAPTER RESOURCES

- Practice Masters
- Enrichment Masters
- Technology Masters
- Lesson Activity Masters
- Lab Activity Masters
- Long-Term Project Masters
- Assessment Masters
   Chapter Assessments, A/B
   Mid-Chapter Assessment
   Alternative Assessments,
   A/B
- Teaching Transparencies
- Cumulative Assessment
- Spanish Resources

CHAPTER OBJECTIVES

- Define and evaluate rational functions.
- Determine the values for which a rational function is undefined.
- Describe transformations applied to the parent function, $f(x) = \frac{1}{x}$, and their effect on its graph.
- Define inverse variation.

CHAPTER 14

Rational Functions

LESSONS

14.1 Rational Expressions

14.2 Inverse Variation

14.3 Simplifying Rational Expressions

14.4 Operations With Rational Expressions

14.5 Solving Rational Equations

14.6 *Exploring* Proportions

14.7 Proof in Algebra

Chapter Project
A Different Dimension

Have you ever noticed how it takes you longer to row upstream than downstream? Have you ever used a lever to lift a heavy object? Do you know what a gear ratio is?

The rate of the current affects your time spent rowing. The distance from the balance point determines the amount of lift for a lever. The gear ratio affects how fast you ride a bicycle.

Rational expressions can be used to model these relationships. Look carefully at the words *rational* and *irrational*. At the heart of each is the word *ratio*. A **ratio** compares two quantities, and a **rational** expression is a fraction that compares two algebraic expressions.

ABOUT THE PHOTOS

Bicycle gears and gear ratios provide an opportunity for students to apply ratios and proportions. Allow students to briefly discuss what they know about bicycle gears. If possible, you may want students to make a chart showing when different gears are used—for instance, going uphill, racing, and riding leisurely.

- Use inverse variation to solve problems.
- Simplify rational expressions by factoring the numerator and denominator, and eliminating common factors by division.
- Multiply rational expressions.
- Add and subtract rational expressions with like and unlike denominators.
- Solve rational equations by graphing and by using the common denominator method.
- Use rational equations to solve problems.
- Identify the means and the extremes of a proportion.
- Determine whether an equation is a proportion.
- Use proportions to solve problems.
- Define *proof, hypothesis, conclusion,* and *converse.*
- Demonstrate proof by using properties and definitions.

PORTFOLIO ACTIVITY

A bicycle controls the speed and effort of pedaling by the relationship of the teeth on the pedal gears to the teeth on the wheel gears. For example, if the pedal gear has 54 teeth and the wheel gear has 16, the gear ratio is $\frac{16}{54} = 0.296$. The reciprocal is $\frac{54}{16} = 3.375$ or about 3.4. This represents the number of turns of the wheel for every full turn of the pedal. Assume the circumference of a bicycle tire is approximately 2.19 meters, how far will the rider travel during one full turn of the pedal?

a. The wheel gears on a bicycle might have 13, 14, 15, 16, 18, 21, and 24 teeth and the pedal gears, 49 or 54 teeth. Write a function for each pedal gear that will calculate the gear ratios. Then find the reciprocals.

b. Assume a bicycle rider turns the pedal 60 full turns per minute. Choose 2 gear combinations and calculate the miles per hour for each. Note: 1 Kilometer per hour is approximately 0.62 miles per hour.

PORTFOLIO ACTIVITY

Assign the Portfolio Activity as part of the assignment for Lesson 14.7 on page 688.

Have students complete the portfolio activity by determining the necessary ratios and by answering the questions.

Allow students to work in small groups to find out more about bicycle gears, chain wheels, cogs, number of teeth, and the different "speeds" of bicycles (for example, a 12-speed bicycle). Have students determine as many different ratios as they can and display their results in some manner, such as tables, charts, graphs, pictures, and so on.

ABOUT THE CHAPTER PROJECT

In the Chapter Project on pages 694–695, students investigate ratios and their reciprocals. Students should know how to form ratios but they may need direction at first to recall the correct placement of values in the numerator and denominator of a ratio. Students learn to form ratios and write functions to solve problems.

PREPARE

Objectives

- Define and evaluate rational functions.
- Determine the values for which a rational function is undefined.
- Describe transformations applied to the parent function, $f(x) = \frac{1}{x}$, and their effect on its graph.

RESOURCES

- Practice Master   14.1
- Enrichment Master   14.1
- Technology Master   14.1
- Lesson Activity Master 14.1
- Quiz   14.1
- Spanish Resources   14.1

Assessing Prior Knowledge

Factor each expression.

1. $m^2 - 2m + 1$
   $[(m - 1)^2]$

2. $5r^4s + 15r^3s^2 + 50r^2s$
   $[(5r^2s)(r^2 + 3rs + 10)]$

3. $w^2 - z^4$
   $[(w - z^2)(w + z^2)]$

4. $x^4 - 16$
   $[(x - 2)(x + 2)(x^2 + 4)]$

5. $a^2 - 7a - 18$
   $[(a - 9)(a + 2)]$

6. $6g^2 + 3g - 45$
   $[3(2g - 5)(g + 3)]$

TEACH

 Ask students whether an answer of $\frac{125}{875}$ mile would be sensible. Allow them to discuss why fractions are reduced to simplest form when reporting answers. Tell students that the same is true for rational expressions.

The art institute charges a $55 membership fee and $4.50 for each pottery lesson.

why *People are constantly comparing data. Baseball players compare their batting averages. Business owners compare profit to total income. Students figure their average expenses. Special expressions are needed to model these comparisons.*

When Saul's *total* expenses for institute membership and pottery lessons are considered, the following formula represents the average cost per lesson.

$$\text{Average cost} = \frac{\text{total expense}}{\text{number of lessons}}$$

If Saul takes only 1 lesson, his cost will be $55 + $4.50 or $59.50. If Saul takes 2 lessons and then stops, his average cost per lesson will be

$$\frac{\$55 + \$4.50(2)}{2} \text{ or } \$32.00.$$

If x is the number of lessons taken, then $55 + 4.50x$ is the total expense for all lessons. So,

$$\text{Average cost} = \frac{55 + 4.50x}{x}.$$

ALTERNATIVE teaching strategy

Using Discussion
Have students evaluate each of the following expressions when $a = -3$.

1. $a - 3 \ [-6]$
2. $a + 3 \ [0]$
3. $a^2 - 9 \ [0]$
4. $\frac{(a^2 - 9)}{(a + 3)}$ **[undefined]**

Remind students that division by zero is undefined.

ENRICHMENT

Rational expressions that contain fractions within the denominator and numerator are called *complex fractions*. Have students work in small groups to simplify these complex fractions.

1. $\dfrac{\dfrac{x^2 - 1}{6}}{\dfrac{x - 1}{3}}$ $\left[\dfrac{x - 1}{2}\right]$

2. $\dfrac{\dfrac{12x + 6}{x}}{\dfrac{12x^2 - 3}{x^2}}$ $\left[\dfrac{2x}{2x - 1}\right]$

How can Saul compare the average cost per lesson for taking 2 lessons per week with the average cost for taking 3 lessons per week? Remember, he has 10 weeks to take lessons.

To compare the average cost per lesson, evaluate the expression

$$\frac{55 + 4.50x}{x}$$

for 20 lessons and 30 lessons. A spreadsheet or graphics calculator can quickly evaluate the expression for any value of x.

	A	B	C
1	x	55 + 4.50x	(55 + 4.50x)/x
2	0	55.00	error
3	1	59.50	59.50
4	2	64.00	32.00
5	3	68.50	22.83
6	4	73.00	18.25
7	5	77.50	15.50
8	6	82.00	13.67
9	20	145.00	**7.25**
10	30	190.00	**6.33**

Saul's average cost per lesson for 20 lessons is $7.25. If Saul takes 30 lessons, his average cost is $6.33 per lesson. Why does the entry in row 2, column C, read "error"?

The expression $\frac{55 + 4.50x}{x}$ is an example of a *rational expression*.

RATIONAL EXPRESSION
If P and Q are polynomials and $Q \neq 0$, then an expression in the form $\frac{P}{Q}$ is a rational expression.

Since $55 + 4.50x$ and x are polynomials and $x \neq 0$, $\frac{55 + 4.50x}{x}$ is a rational expression.

A function of the form $y = \frac{P}{Q}$, or $f(x) = \frac{P(x)}{Q(x)}$, where $\frac{P}{Q}$ is a rational expression is called a **rational function**. Thus, $g(x) = \frac{55 + 4.50x}{x}$ is a rational function. Since you cannot divide by zero, g is *undefined* when $x = 0$. A rational function is undefined or restricted when its denominator is equal to zero.

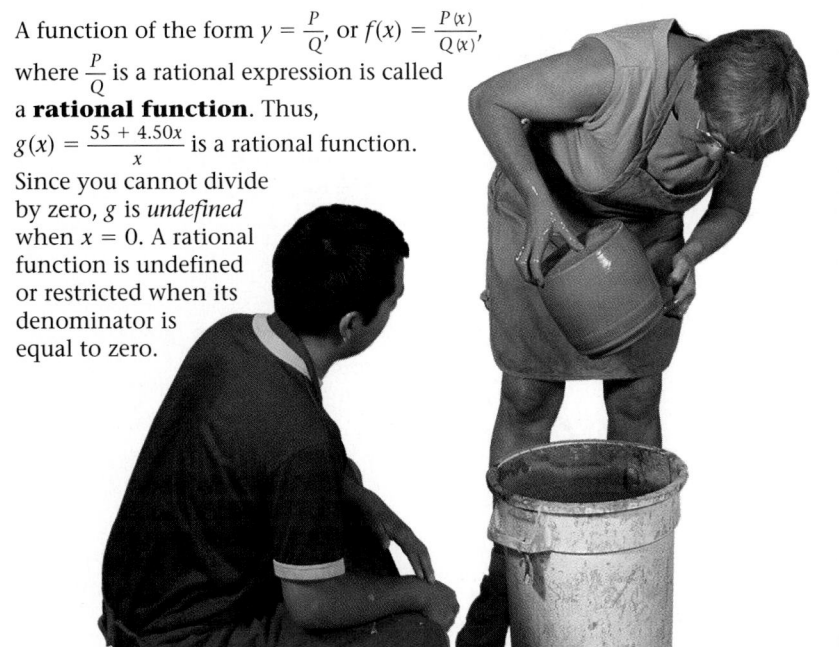

TEACHING *tip*

Technology Allow students to set up their own spreadsheets if possible. Be sure students correctly enter the formulas in cells B1 and C1. Have students save their spreadsheets for use with Example 2.

ongoing
ASSESSMENT

The error is caused by trying to divide by zero.

INCLUSION strategies

Using Algorithms Some students may have difficulty integrating the transformations when trying to solve a problem. Allow these students to experiment with each of the following functions. Tell students to use different numbers for b, such as $2, 4, -3,$ and so on.

1. $y = \frac{1}{x}$

2. $y = \frac{b}{x}$

3. $y = \frac{1}{x + b}$

4. $y = \frac{1}{x - b}$

5. $y = \frac{1}{x} + b$

6. $y = \frac{1}{x} - b$

Alternate Example 1

For what value or values is each rational function undefined?

a. $f = \dfrac{d}{t}$ $[t = 0]$

b. $f(x) = \dfrac{s^2}{(s-1)}$ $[s = 1]$

c. $f(x) = \dfrac{(a+b)}{(a^2-9)}$
$[a = 3 \text{ or } -3]$

ongoing ASSESSMENT

Try This

a. $y = 2$ or 3
b. $y = x$

Alternate Example 2

a. Describe the transformations applied to the parent
function $g(x) = \dfrac{(2+0.75)}{x}$.

[vertical stretch of 2 and translation up by 0.75]

b. A taxicab charges $2 base rate plus $0.75 per mile. Use the function to find how far you must travel to average $1.00 per mile.
[8 miles]

EXAMPLE 1

For what value or values is each rational function undefined?

Ⓐ $f(x) = \dfrac{1}{x}$ **Ⓑ** $y = \dfrac{a^2-b^2}{a-b}$ **Ⓒ** $t = \dfrac{m+n}{m^2-4m+3}$

Solution ➤

A rational function is undefined when its denominator is equal to zero.

Ⓐ $\dfrac{1}{x}$ is undefined when $x = 0$.

Ⓑ $\dfrac{a^2-b^2}{a-b}$ is undefined when $a - b = 0$. That is, when $a = b$.

Ⓒ $\dfrac{m+n}{m^2-4m+3}$ is undefined when $m^2 - 4m + 3 = 0$.

Since $m^2 - 4m + 3 = (m-1)(m-3)$, $\dfrac{m+n}{m^2-4m+3}$ is undefined when $m = 1$ or $m = 3$. ❖

Try This For what value or values is each rational function undefined?

a. $\dfrac{y-2}{y^2-5y+6}$ **b.** $\dfrac{3x+6x}{x-y}$

Recall that the parent function for a variety of rational functions is $f(x) = \dfrac{1}{x}$. In Chapter 9 you found its graph to be a hyperbola with vertical asymptote $x = 0$.

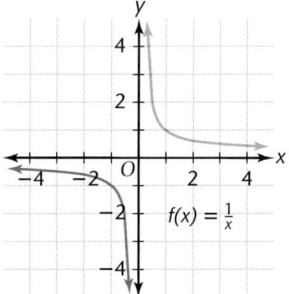

$f(x) = \frac{1}{x}$

EXAMPLE 2

Ⓐ Describe the transformations applied to the parent function $f(x) = \dfrac{1}{x}$ to produce the graph of $g(x) = \dfrac{55+4.50x}{x}$.

Ⓑ Use the graph to find the number of pottery lessons that can be taken for an average cost of $5.60 per lesson.

POTTERY LESSONS $5.60

RETEACHING  *the lesson*

Using Symbols Help students understand the various transformations of $f(x) = \dfrac{1}{x}$ by discussing each transformation. Emphasize that the coefficient of x must be 1 and that term must be the only term containing x. For instance, $\dfrac{1}{(2x)}$ should be rewritten as $\dfrac{1}{2}\left(\dfrac{1}{x}\right)$. Then

have students write a summary of what each transformation means in $f(x) = \dfrac{a}{(x+b)} + c$.

[a tells how much the graph is stretched vertically; b tells how far the graph shifts to the left or right; c tells how far the graph shifts up or down]

Solution ➤

Ⓐ The parent function is $f(x) = \frac{1}{x}$. To compare g with f, rewrite the expression using the Distributive Property and simplify.

$$\frac{55 + 4.50x}{x} = \frac{55}{x} + \frac{4.50x}{x}$$

$$= \frac{55}{x} + 4.5$$

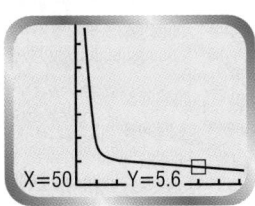

The graph of the function $g(x) = \frac{55}{x} + 4.5$ shows that the parent function has been transformed by a vertical stretch of 55 and a translation up by 4.5.

Ⓑ Examine the graph, or use the trace function to find the value of x so that $f(x) = 5.60$.

Graphics Calculator

The x value represents 50 lessons.

Therefore, in a 10-week period, the average cost for 50 lessons is $5.60. ❖

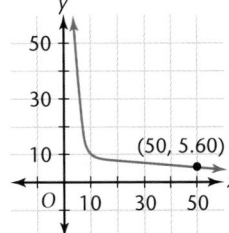

(50, 5.60)

X=50   Y=5.6

EXAMPLE 3

Describe the transformation applied to the parent function $f(x) = \frac{1}{x}$ to produce the graph of $g(x) = \frac{5}{x - 3} + 2$.

Solution ➤

The information from the function $g(x) = \frac{5}{x - 3} + 2$ shows that the parent function has been transformed by a vertical stretch of 5, a shift to the right by 3, and a shift up by 2. ❖

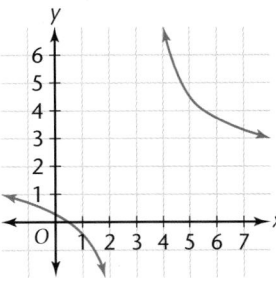

Try This   Describe the transformations applied to the parent function $f(x) = \frac{1}{x}$ to produce the graph of $g(x) = \frac{3}{2x + 3} - 5$.

CRITICAL Thinking

Begin with the function $f(x) = a\left(\frac{1}{x - h}\right) + k$. First vary the values of a, then h, then k, keeping the other 2 values constant, and graph the results. Describe the effects on the graph of varying a, h, and k.

TEACHING *tip*

Technology For part (a) of Example 2, be sure students set the correct range before graphing the equation. Xmin = −40, Xmax = 40, Xscl = 10, Ymin = −30, Ymax = 30, Yscl = 10. Have students use the ZOOM and TRACE functions to complete part (b).

Alternate Example 3

Describe the transformation applied to the parent function $f(x) = \frac{1}{x}$ to produce the graph of $g(x) = \frac{8}{(x + 3)} + 6$.

[vertical stretch of 8, shift to left of 3, shift up of 6]

TEACHING *tip*

Technology Be sure students correctly use parentheses when entering the function.

Aongoing **SSESSMENT**

Try This

Shift to the left 1.5 units, down 5 units, and stretch by a factor of $\frac{3}{2}$

CRITICAL *Thinking*

Changing a results in a vertical stretch; changing h results in a shift to the right or left; and changing k results in a shift up or down.

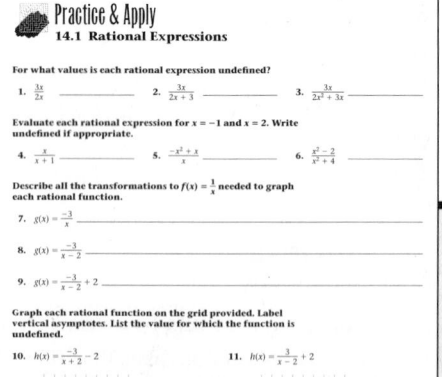
EXERCISES & PROBLEMS

Communicate

1. Explain how a rational expression is defined. Give two examples of a rational expression.

2. What condition causes a rational expression to be undefined?

Graph the function $h(x) = \frac{12x - 7}{x}$.

3. Discuss how to express $h(x)$ in a form that shows that it is a transformation of the parent function $g(x) = \frac{1}{x}$.

4. Describe the behavior of $h(x)$ when x is 0.

5. Describe all the transformations to $g(x) = \frac{1}{x}$ that produce the graph of $h(x)$.

Practice & Apply

For what values are these rational expressions undefined?

6. $\frac{3x + 9}{x}$ 0   7. $\frac{6}{y - 2}$ 2   8. $\frac{2m - 5}{6m^2 - 3m}$ 0, $\frac{1}{2}$ 9. $\frac{x^2 + 2x - 3}{x^2 + 4x - 5}$ 1, −5

Evaluate each rational expression for $x = 1$ and $x = -2$. Write "undefined" if appropriate.

10. $\frac{5x - 1}{x}$ 11. $\frac{4x}{x - 1}$ 12. $\frac{x^2 - 1}{x^2 - 4}$ 13. $\frac{x^2 + 2x}{x^2 + x + 2}$

Describe all the transformations to $f(x) = \frac{1}{x}$ needed to graph each of the following rational functions.

14. $g(x) = \frac{1}{x - 5}$ 15. $g(x) = \frac{2}{x}$ 16. $g(x) = \frac{-1}{x} + 3$

17. $g(x) = \frac{1}{2x + 3}$ 18. $g(x) = \frac{4}{x + 2} - 5$ 19. $g(x) = \frac{-2}{x - 1} + 4$

Graph each rational function. List the value for which the function is undefined.

20. $h(x) = \frac{1}{x} + 1$ 21. $h(x) = \frac{1}{x + 2}$ 22. $h(x) = \frac{-1}{x} + 3$

23. $h(x) = \frac{3}{x - 1}$ 24. $h(x) = \frac{2}{x - 3} + 4$ 25. $h(x) = \frac{-2}{x - 4} + 3$

Transportation   The air pressure, P, in pounds per square inch, in the tires of a car can be modeled by the rational equation $P = \frac{(1.1)10^5}{V}$, where V is volume in cubic inches.

26. Graph the rational function represented by P.

27. The owner's manual calls for a pressure of 32 pounds per square inch. What volume of air does a tire hold at this pressure? 3437.5 in³

Look Back

28. Your class orders 500 mugs with all the names of the class members. In the first week, 356 mugs were sold. What percent of the mugs sold the first week? **[Lesson 4.5]** 71.2%

29. Graph a line through the points (2, 1) and (−3, 5). **[Lesson 5.3]**

30. Find the inverse matrix to $\begin{bmatrix} 0.3 & 0.4 \\ 0.5 & 0.6 \end{bmatrix}$. **[Lesson 7.4]**

31. How many different 4-digit numbers are possible if the digits 1-9 can be used with repetition? **[Lesson 8.5]** 6561

32. What transformation is applied to $f(x) = x^2$ to obtain the graph of $y = (x - 5)^2$? **[Lesson 9.5]** shifts 5 units to the right

33. Solve $x^2 - 20 = -x$. **[Lesson 12.2]** 4, −5

34. **Cultural Connection: Americas** Benjamin Banneker, the first African American astronomer, is most famous for the almanacs he created from 1791 to 1797. Banneker also enjoyed solving mathematical problems, such as this one from his journal. **[Lesson 13.4]** 102.65 feet

37 ft.   **23 ft.**

If the length of each ladder is 60 feet, find the width of the street that runs between the buildings to the nearest hundredth.

Look Beyond

35. Describe the transformations that produce $y = \dfrac{5}{x - 1} + 2$ from $y = \dfrac{1}{x}$.

36. Sketch the graph.

29.

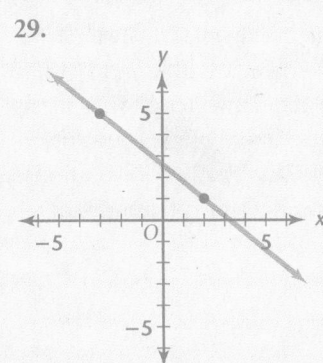

30. $\begin{bmatrix} -30 & 20 \\ 25 & -15 \end{bmatrix}$

35. vertical shift up by 2, horizontal shift right by 1 stretched by 5

36.

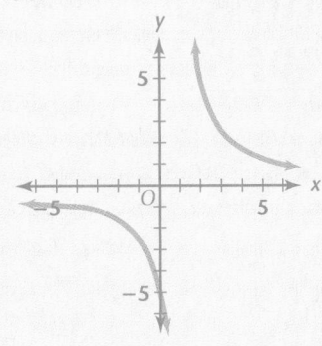

Objectives

- Define inverse variation.
- Use inverse variation to solve problems.

RESOURCES

• Practice Master	**14.2**
• Enrichment Master	**14.2**
• Technology Master	**14.2**
• Lesson Activity Master	**14.2**
• Quiz	**14.2**
• Spanish Resources	**14.2**

Assessing Prior Knowledge

Solve each equation.

1. $128 = 4r$ $[r = 32]$

2. $\dfrac{n}{9} = \dfrac{1326}{234}$ $[n = 51]$

3. $\dfrac{2}{x} = \dfrac{x}{32}$ $[x = \pm 8]$

TEACH

Tell students you are going on a 200-mile car trip. You are driving on an interstate highway and can average 60 miles per hour. How long will the trip take? $\left[3\frac{1}{3}\text{ hours}\right]$ Write the results in a chart with labels for distance, rate, and time. How long would it take if you drove on a local highway and averaged 50 miles per hour? [**4 hours**] What if you only averaged 32 miles per hour? $\left[6\frac{1}{4}\text{ hours}\right]$ Ask students to look at the results in the table. As the average rate of speed decreases, what happens to the time? [**It increases.**] Tell students there is an inverse relation between speed and time when distance is constant.

LESSON 14.2 Inverse Variation

It takes a person 6 minutes to plant each tree for this project.

why

Variation describes how a change in one quantity changes another quantity. You have seen how quantities vary directly. They can also vary inversely.

In Chapter 2 you solved a fund-raising problem about students raising $1000 to help a needy family. You found that the number of contributors needed decreased as the amount of money each person gave increased. This relationship is an example of inverse variation.

> **INVERSE VARIATION**
>
> If $x \cdot y = k$, or $y = \frac{k}{x}$, when $x \neq 0$ and k is a constant, then y varies inversely as x. The constant k is called the constant of variation.

EXAMPLE 1

Ecology   A volunteer organization has 5000 trees to plant to beautify an area. If 5 people need 100 hours to plant all the trees, how many hours are needed when 50 people help?

Using Manipulatives Bring in various gears, such as demonstration gears from a school science lab. Mark two teeth on two meshed gears. Have students turn the gears to find how many times each gear must turn until those two special teeth mesh together again. For instance, if one gear with 24 teeth and another gear with 8 teeth are meshed, the same teeth will mesh again when the large gear has made 1 revolution and the smaller gear has made 3 revolutions. Repeat the demonstration for different pairs of gears.

Solution ➤

Let n represent the number of people and t represent the number of hours to plant all the trees. You can determine the constant by the formula for inverse variation. Since $n \cdot t = k$, $k = 5 \cdot 100$ or $k = 500$. After you determine the constant of variation, you can use it and the remaining information in the problem to find the number of hours, t.

$$nt = 500$$
$$50t = 500$$
$$t = 10$$

Examine the table for various numbers of volunteers.

Persons n	1	2	4	5	40	50	125	250	500
Hours t	500	250	125	100	12.5	10	4	2	1
$nt = k$	500	500	500	500	500	500	500	500	500

To plant 500 trees, 50 volunteers must work 10 hours planting one tree every 6 minutes. ❖

Try This If y is 4 when x is 12, write an equation that shows how y is related to x. Assume that y varies inversely as x.

CRITICAL *Thinking* In Example 1, k is the constant of variation. What physical quantity does k represent?

Cultural Connection: Africa One of the earliest references to variation comes from an ancient Egyptian papyrus that describes the commodity exchange system used in some Egyptian cities. Bread was used as a form of currency, and its value was based on the amount of wheat that was used to make each loaf. A rating system was created to value the bread.

Loaves	1000	1000	1000	1000	1000
*Rating	10	20	25	40	50
**Value	100	50	40	25	20

* Rating is the number of loaves from 1 hekat.
** Value is the total number of hekats used.

The table shows the value of 1000 loaves of bread at various ratings.

Notice that as the rating for each 1000 loaves *increases* the *value* for each 1000 loaves *decreases*. This is an example of **inverse variation**.

Algebraically, the variables of rating and value are related by the following equation.

$$\text{Rating} \cdot \text{Value} = \text{Loaves, as a constant, } k$$
$$20 \ \tfrac{loaves}{hekat} \cdot 50 \ hekats = 1000 \text{ loaves}$$

The bread's rating is said to vary inversely as its value. ❖

$\mathbb{E}$NRICHMENT The volume of any gas varies inversely with its pressure as long as the temperature remains constant. If a balloon has a volume of 3400 cubic centimeters at a pressure of 120 kilopascals, what is its volume at 105 kilopascals? **[approximately 3885.7 cubic centimeters]**

Alternate Example 1
It takes Tom 45 hours to paint a long fence. How many friends must he get to help him paint so the job will be done in $7\frac{1}{2}$ hours? **[5 friends]**

$\mathbb{A}$ **ongoing** **SSESSMENT**

Try This

$xy = 48$

CRITICAL *Thinking*

The constant of variation, k, represents the total number of work hours needed to plant 500 trees.

Cooperative Learning
Tell students to note how changing one value (in this case, the rating) has an inverse effect on other values. Have students work in small groups and research other monetary systems, particularly ancient ones.

A 60-gram mass is balanced by placing an appropriate weight on the other end. This procedure is repeated for fulcrum distances of 10, 15, 20, and 25 centimeters from the end. The table gives the findings of a group of students who performed the experiment.

Distance	5	10	15	20	25
Mass	60	30	20	15	12

Examine the products of each distance and weight pair. If they vary inversely, what can you say about the products of these quantities?

Alternate Example 2

The current (I) in an electrical circuit varies inversely with the resistance (R) in a circuit.

I (in amperes)	2.0	2.5	3.0
R (in ohm)	3.0	2.4	2.0

a. Find the constant of variation. [6]

b. Write an equation that relates the variables.
   [$IR = 6$]

c. Find the resistance if the current is 5.0. [1.2]

EXAMPLE 2

Analyze the experiment just described.

Ⓐ Write a sentence that specifies the variation between the variables.

Ⓑ Write an equation that relates the variables.

Ⓒ Construct a graph of the equation, and analyze the type of function that the graph represents.

Solution ➤

This experiment illustrates inverse variation because the product of the distance from the balance point and the balance weight is constant.

Ⓐ The balance weight varies inversely as the distance to the balance point.

Ⓑ The inverse variation relationship means that

$$\text{weight} \cdot \text{fulcrum distance} = k.$$

For the values of 15 grams and 20 centimeters,

$$15 \cdot 20 = 300 = k.$$

The constant of variation is 300.

INCLUSION
strategies

Using Manipulatives Provide students with square tiles or squares cut from paper. Ask students to find what happens to the length of a rectangle as the width of the rectangle increases and the area of the rectangle remains the same. Have students use the square tiles to check their guesses. For example, start with an area of 12 square units. Make a 3-by-4 rectangle to show the area. Now shift some tiles so that the width is 6 instead of 4. What happens to the length? [It decreases to 2.] Tell students that there is an inverse relationship between the length and width of a rectangle with a fixed area.

C The equation for this inverse variation is $m \cdot d = 300$.

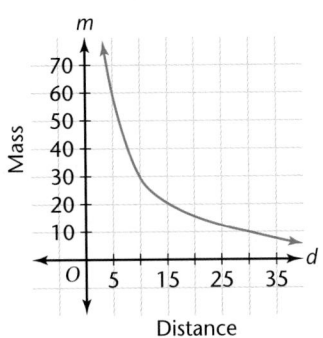

If $md = 300$, then $m = \frac{300}{d}$. The equation represents a rational function that is a transformation of the parent function $y = \frac{1}{x}$ by a vertical stretch of 300. ❖

Investment The amount of time it takes to double your money at *compound* interest can be approximated by the function $t = \frac{72}{r}$, where r is the interest rate written as a whole number. In business this function is usually referred to as the Rule of 72.

EXAMPLE 3

A How long does it take to double your money at 6% compound interest?

B Graph the Rule of 72 function.

Solution ➤

A To use the Rule of 72, substitute 6 for r so $t = \frac{72}{6} = 12$. It takes about 12 years to double your money at 6%.

B Choose interest rates for r. Make a table and sketch a graph from the points.

$r\%$	t
4	18
6	12
8	9
12	6

CRITICAL Thinking Describe the graph of the Rule of 72 as a transformation of its parent graph. What are the domain and range of the Rule of 72 function? What effect does the constant of variation have on its parent function?

Alternate Example 3

How long does it take to double your money at 8% interest?
[**9 years**]

CRITICAL Thinking

The graph of $t(r) = \frac{72}{r}$ has been stretched vertically by 72. The domain and range for the rule of 72 function are all positive real numbers. The constant of variation stretches the graph vertically.

Using Discussion Pose the following situations to students. First ask whether the situation represented presents an inverse variation. If it does, have students write and solve the associated equation. Then have students solve the problem.

1. If I read 20 pages per hour, it will take 5 hours to read this book. How long will it take to read if I only read 16 pages per hour? [$20 \cdot 5 = 16x, x = 6.25$; **It would take 6.25 hours.**]

2. The area of a rectangle is 36 square inches. Find possible lengths and widths of the rectangle. [$36 = lw$; **Answers will vary.**]

Practice Master

EXERCISES & PROBLEMS

Communicate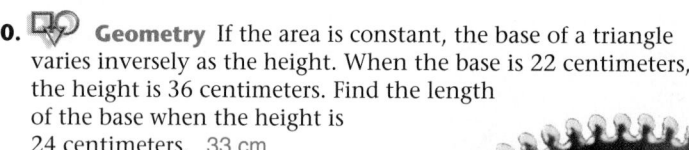

1. Give two equivalent equations expressing inverse variation if y varies inversely as x.
2. Explain what inverse variation between rate and time means. Give a real-life example of inverse variation.
3. If y is 3 when x is 8, and y varies inversely as x, explain how to find y when x is 2.
4. Explain how to use the Rule of 72 to find how long it would take to double your money at 4% interest compounded yearly.
5. Describe the parent function for inverse variation functions.

Practice & Apply

Determine if each of the following is an example of inverse variation.

6. $rt = 400$ yes   7. $y = \frac{-28}{x}$ yes   8. $x - y = 10$ no   9. $\frac{n}{5} = \frac{3}{m}$ yes

10. $\frac{x}{y} = \frac{1}{2}$ no   11. $x = 10y$ no   12. $a = \frac{42}{b}$ yes   13. $r = t$ no

For Exercise 14-19, y varies inversely as x. If y is

14. 8 when x is 6, find x when y is 12.   4

15. 9 when x is 12, find y when x is 36.   3

16. 3 when x is 32, find x when y is 4.   24

17. 3 when x is -8, find x when y is -4.   6

18. $\frac{3}{5}$ when x is -60, find y when x is 2. -18

19. $\frac{3}{4}$ when x is 12, find y when x is 27. $\frac{1}{3}$

20. **Geometry** If the area is constant, the base of a triangle varies inversely as the height. When the base is 22 centimeters, the height is 36 centimeters. Find the length of the base when the height is 24 centimeters.   33 cm

The speeds are inversely proportional to the number of teeth.

21. **Physics** If a gear with 12 teeth revolves at a speed of 500 revolutions per minute, at what speed should a gear with 16 teeth revolve?   375

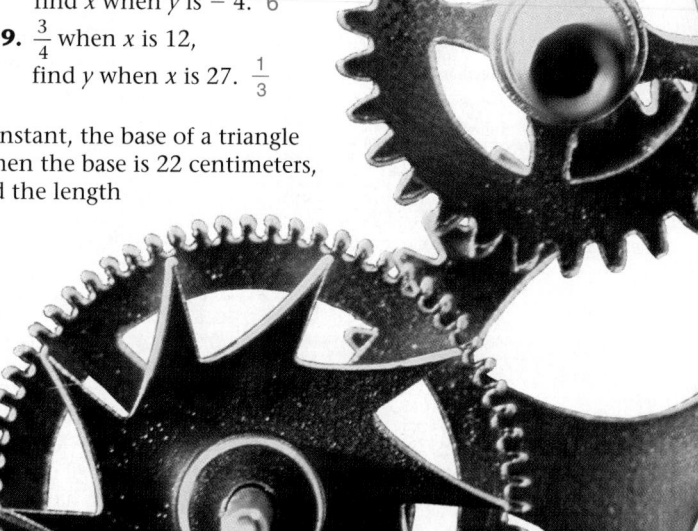

22. Music A harp string vibrates to produce sound. The number of these vibrations varies inversely to its length. If a string 28 centimeters long vibrates 510 times per second, how long is a string that vibrates 340 times per second?   42 cm

23. Time varies inversely as the average rate of speed over a given distance. When Mike travels 6 hours, his average rate is 80 kilometers per hour. Find Mike's time when his rate is 90 kilometers per hour.   5 hours 20 minutes

24. The frequency of a radio wave varies inversely as its wavelength. If a 200-meter wave has a frequency of 3000 kilocycles, what is the wavelength of a wave that has a frequency of 2000 kilocycles?   300 m

25. Transportation Time varies inversely as the average speed over a given distance. A jet takes about 2.7 hours to fly from Boston to Paris averaging 2200 kilometers per hour. How long would it take a plane traveling at an average speed of 1760 kilometers per hour to make the same trip?   3.375 hours

26. Investment If Harold invests $500 of his 4-H prize money at 5% interest compounded yearly, how long will it take him to double his money?   14 years 4.8 months

27. **Geometry** The area of a rectangle is 36 square centimeters. What is the length of a rectangle of the same area that has twice the original width? Explain why the formula for the area of a rectangle is a model for inverse variation.   half the original length

Look Back

28. The first three terms of a sequence are 1, 5, and 12. The second difference is a constant 3. Find the next three terms of the sequence.
[Lesson 1.1]   22, 35, 51

29. The 5-by-7 picture of your award winning cat, Kat, is surrounded by a frame of width 3x. Find the dimensions of the frame and picture unit.
[Lesson 3.3]   5 + 6x by 7 + 6x

30. The total cost for your order of computer discs, including the $1.95 handling charge, is $57.60. The discs are sold in boxes of 10 and were on sale for $7.95 a box. How many discs were in the order?
[Lesson 4.6]   70

31. What are the coordinates of the vertex of $y = -x^2 - 4x$?
[Lesson 12.1] $(-2, 4)$

Look Beyond

32. Subtract. $\dfrac{1}{x-3} - \dfrac{x}{x^2-9}$ $\dfrac{3}{x^2-9}$

33. The intensity, I, of light from a light source varies inversely as the square of the distance, d^2, from the light source. If I is 30 units when d is 6 meters, find I when d is 3 meters.   120 units

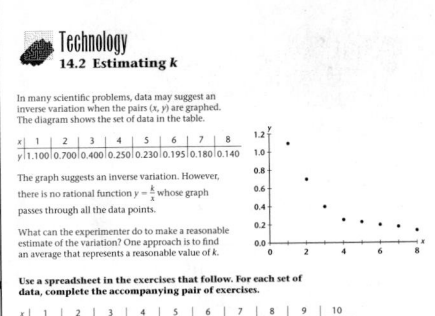

Objectives

• Simplify rational expressions by factoring the numerator and denominator, and eliminating common factors by division.

RESOURCES

• Practice Master   **14.3**
• Enrichment Master   **14.3**
• Technology Master   **14.3**
• Lesson Activity Master   **14.3**
• Quiz   **14.3**
• Spanish Resources   **14.3**

Assessing Prior Knowledge

Simplify by using the Distributive Property to factor the expression, and then divide. Check by performing the arithmetic.

a. $\dfrac{30 + 18}{6}$ **[8]**

b. $\dfrac{8(3^2 - 4^2)}{(3 + 4)(6 - 8)}$ **[4]**

TEACH

 Have students evaluate the following expression when $y = 2.53$.

$$\frac{6x^3 - 6}{2(3x + 3)}$$

Tell students that the answer is 1.53, or $x - 1$. In this lesson students will learn how to simplify rational expressions to avoid complicated expressions and tedious computations.

LESSON 14.3 Simplifying Rational Expressions

why *In arithmetic, a fraction is simplified when it is expressed in lowest terms. In algebra, rational expressions are simplified in much the same way. Algebraic operations on rational expressions follow the same mathematical steps that you use to simplify fractions.*

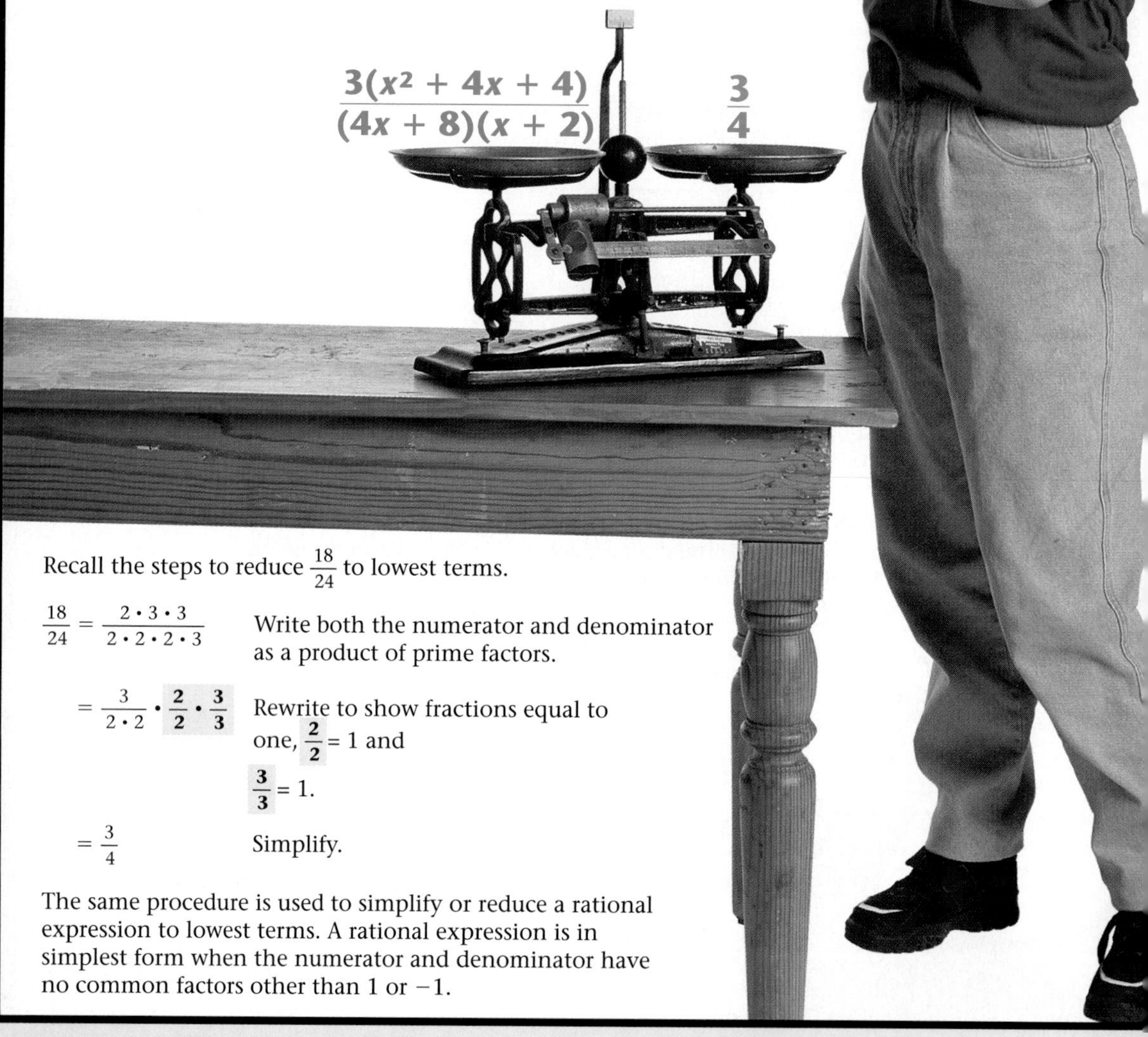

$$\frac{3(x^2 + 4x + 4)}{(4x + 8)(x + 2)} \qquad \frac{3}{4}$$

Recall the steps to reduce $\dfrac{18}{24}$ to lowest terms.

$\dfrac{18}{24} = \dfrac{2 \cdot 3 \cdot 3}{2 \cdot 2 \cdot 2 \cdot 3}$ Write both the numerator and denominator as a product of prime factors.

$= \dfrac{3}{2 \cdot 2} \cdot \dfrac{2}{2} \cdot \dfrac{3}{3}$ Rewrite to show fractions equal to one, $\dfrac{2}{2} = 1$ and

$\qquad\qquad\qquad\qquad \dfrac{3}{3} = 1$.

$= \dfrac{3}{4}$ Simplify.

The same procedure is used to simplify or reduce a rational expression to lowest terms. A rational expression is in simplest form when the numerator and denominator have no common factors other than 1 or −1.

Cooperative Learning Assign each person in the group a factoring task, such as checking for any common monomial factors, checking whether there is a difference of two squares, checking whether there is a perfect square, and finding general factors. The group then decides how to factor first the numerator and then the denominator. At this point any necessary restrictions on the value of x are noted. Finally, the group completes simplifying the expression.

EXAMPLE 1

Simplify $\dfrac{5a + 10}{5a}$.

Simplify $\dfrac{(27a^2 + 15a)}{(9a)}$. List any restricted values.

$\left[\dfrac{9a + 5}{3}, a \neq 0\right]$

Solution ➤

$\dfrac{5a + 10}{5a} = \dfrac{5(a + 2)}{5a}$, $a \neq 0$ Factor the numerator and denominator.

$= \dfrac{\boxed{5}}{\boxed{5}} \cdot \dfrac{a + 2}{a}$, $a \neq 0$ Rewrite to show fractions equal to 1, $\dfrac{5}{5} = 1$.

$= \dfrac{a + 2}{a}$, $a \neq 0$ Simplify.

One way to check your work is to replace the variable by a convenient value. For example, let $a = 10$.

Let $a = 10$.

Original form $\dfrac{5a + 10}{5a} = \dfrac{50 + 10}{50} = \dfrac{60}{50} = \dfrac{6}{5}$

Reduced form $\dfrac{a + 2}{a} = \dfrac{12}{10} = \dfrac{6}{5}$

The check shows that the value of the original form is the same as that of the reduced form. ❖

Why is it important that the check for the original fraction and the check for the reduced fraction be the same?

STEPS TO REDUCE RATIONAL EXPRESSIONS
1. Factor the numerator and denominator.
2. Express, or think of, each common factor pair as 1.
3. Simplify.

Be sure that only common factors are eliminated when you simplify. Notice the common factors in the examples below.

Expression	Common factor	Reduced form
$\dfrac{x(a + 3)}{2x}$	x	$\dfrac{a + 3}{2}$, $x \neq 0$
$\dfrac{x + 4}{x}$	none	Cannot be simplified further.
$\dfrac{2x(x + 1)}{4x^2}$	$2x$	$\dfrac{x + 1}{2x}$, $x \neq 0$
$\dfrac{2x^2 + 3}{x}$	none	Cannot be simplified further.

CRITICAL Thinking

Explain why $\dfrac{x}{x^2 + x} = \dfrac{1}{x + 1}$, $x \neq 0, -1$, is a true statement.

Ongoing ASSESSMENT

If they are different, an error has been made. It is important that the check be the same for the original fraction and its reduced form because you are trying to check that the two expressions are equivalent.

CRITICAL Thinking

The expression $\dfrac{x}{x^2 + x}$ can be reduced by dividing the numerator and denominator by x.

$\dfrac{\dfrac{x}{x}}{\dfrac{x^2 + x}{x}}$ which is equal to $\dfrac{1}{x + 1}$.

ENRICHMENT Show students the following "proof" that

shows that $1 = 2$.

$a = b$ Given

$a^2 = ab$ Multiply by a.

$a^2 - b^2 = ab - b^2$ Subtract b^2.

$(a - b)(a + b) = b(a - b)$ Factor.

$(a + b) = b$ Divide by $(a - b)$.

$b + b = b$ Substitute from given.

$2b = b$ Simplify.

$2 = 1$ Divide by b.

Have students find the flaw in the "proof". **[Dividing by $a - b$ is dividing by zero and is not allowed.]** Relate this to the necessity for identifying restricted values for the variables.

Alternate Example 2

Simplify $\dfrac{x^2 + 6x + 8}{x^3 + x^2 - 2x}$ and list the restrictions for the variable.

$$\left[\dfrac{x + 4}{x(x - 1)}, x \neq 0, 1, -2\right]$$

TEACHING tip

Remind students that the values of x must be determined from the factored form of the original expression, not from the answer.

---

ongoing ASSESSMENT

Try This

$x + 4; x \neq 1$

Alternate Example 3

Simplify $\dfrac{6x^4 - 54x^2}{45 - 5x^2}$ and list the restrictions.

$$\left[-\dfrac{6x^2}{5}, x \neq 3, -3\right]$$

---

ongoing ASSESSMENT

The opposite of $(x - 4)$ is $-(x - 4)$, or $-x + 4$, or $4 - x$

---

ongoing ASSESSMENT

Try This

$\dfrac{-5x}{3}, x \neq 2, -2$

CRITICAL Thinking

You must add the restriction $x \neq 1$ because if $x = 1, \dfrac{x - 1}{1 - x} = \dfrac{0}{0}$ which is undefined.

Sometimes a rational expression can be simplified by first factoring the numerator and denominator. Remember to check the restrictions on the denominator.

EXAMPLE 2

Simplify $\dfrac{3x - 6}{x^2 + x - 6}$.

Solution ➤

$$\dfrac{3x - 6}{x^2 + x - 6} = \dfrac{3(x - 2)}{(x + 3)(x - 2)} \qquad \text{Factor numerator and denominator.}$$

$$= \dfrac{3}{x + 3} \cdot \dfrac{x - 2}{x - 2}; \ x \neq 2, -3 \qquad \text{Rewrite to show fractions equal to 1.}$$

$$= \dfrac{3}{x + 3}; \ x \neq 2, -3 \qquad \text{Simplify. Write restrictions from original expression.} \ ❖$$

Try This Simplify $\dfrac{x^2 + 3x - 4}{x - 1}$.

Simplifying rational expressions can be tricky. Many steps may be needed. Examine each expression carefully to see if it can be factored.

EXAMPLE 3

Simplify $\dfrac{x^2 + x - 20}{16 - x^2}$.

Solution ➤

Rewrite the denominator with the variable first. In this case, factor -1 from the denominator.

$$\dfrac{x^2 + x - 20}{16 - x^2} = \dfrac{x^2 + x - 20}{(-1)(x^2 - 16)} \qquad \text{Factor } -1 \text{ from the denominator.}$$

$$= \dfrac{(x + 5)(x - 4)}{(-1)(x + 4)(x - 4)}, \ x \neq 4, -4 \qquad \text{Factor the numerator and denominator.}$$

$$= \dfrac{x + 5}{(-1)(x + 4)}, \ x \neq 4, -4 \qquad \text{The common factor is } x - 4. \text{ Simplify.}$$

$$= \dfrac{x + 5}{-x - 4}, \ x \neq 4, -4 \qquad \text{Multiply } x + 4 \text{ in the denominator by } -1. \ ❖$$

Explain why the opposite of $(x - 4)$ can be written $(4 - x)$.

Try This Simplify $\dfrac{5x^3 - 20x}{12 - 3x^2}$.

 **CRITICAL Thinking** Explain why you cannot write $\dfrac{x - 1}{1 - x} = -1$ without adding restrictions. What restrictions must be added?

INCLUSION strategies

Using Cognitive Strategies For students who have difficulty with the final step of simplifying a rational expression, have them try solving a simpler problem. For instance, in Example 3 have students follow these procedures:

• Write the numerator and denominator and factor.

• Let a different variable represent each factor. [$a = (x + 5)$, $b = (x - 4)$, $c = (x + 4)$, and -1 remains -1.]

• Write the division, using the new variables and simplify.

$$\left[\dfrac{ab}{(-1)cb} = \dfrac{a}{-c}\right]$$

EXAMPLE 4

Find the ratio of the volume of a right circular cylinder with radius r and height h to its surface area.

GEOMETRY
Connection

Solution ➤

$$\frac{\text{Volume}}{\text{Surface area}} = \frac{\pi r^2 h}{2\pi r^2 + 2\pi rh} \qquad \text{Given}$$

$$= \frac{\pi \mathbf{r} rh}{2\pi \mathbf{r}(r + h)} \qquad \text{Factor.}$$

$$= \frac{rh}{2(r + h)} \qquad \text{The common factors are } \pi \text{ and } r. \text{ Simplify.}$$

Therefore, the ratio $\frac{\text{Volume}}{\text{Surface area}}$ is $\frac{rh}{2(r + h)}$. ❖

Try This Find the ratio of the surface area of a cube to its volume.

EXERCISES & PROBLEMS

Communicate

1. Explain what makes a rational expression undefined.

2. Discuss how to find the restrictions of the denominator of a rational expression.

3. Explain what is meant by a common factor.

4. Describe the process that reduces $\frac{x + 1}{x^2 + 2x + 1}$ to lowest terms.

5. Explain how the expression $x - 7$ is related to $7 - x$.

Alternate Example 4

Find the ratio of the volume of a cone with radius r, height h, and slant height s to its surface area. You may wish to tell students that the total surface area of a cone is $\pi r^2 + \pi rs$.

Aongoing
ASSESSMENT

Try This

$$\frac{6}{s}$$

Assess

Selected Answers

Odd-Numbered Exercises 7–47

Assignment Guide

Core 1–17, 18–32 even, 34–48

Core Two-Year 1–47

- Substitute back for the old variables. Simplify.
$$\left[\frac{(x + 5)}{-(x + 4)} = \frac{(x + 5)}{(-x - 4)}\right]$$

RETEACHING
the
lesson

Using Visual Models
Review the steps involved in factoring polynomials. Have students create a flowchart for the steps in simplifying rational expressions. Part of that flowchart will be a sub-program for factoring the numerator and denominator. Included in the flowchart should also be a branch where the restrictions on the value of the variable are listed.

Error Analysis

When simplifying rational expressions, students often get carried away with canceling. For example, many students simplify $\dfrac{5a + 10}{5a}$ incorrectly. Typical errors include

1. the cancellation of $5a$ in the numerator and $5a$ in the denominator to result in $1 + 10 = 11$

2. the cancellation of a in the numerator and a in the denominator to result in $\dfrac{5 + 10}{5} = \dfrac{15}{5} = 3$

$$\left[\text{The answer is } \dfrac{a + 2}{a}. \right]$$

Practice & Apply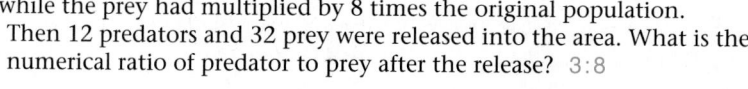

For what values of the variable is each rational expression undefined?

6. $\dfrac{10x}{x - 3}$ 3

7. $\dfrac{10}{5 - y}$ 5

8. $\dfrac{r - 6}{r}$ 0

9. $\dfrac{7p}{p - 4}$ 4

10. $\dfrac{k - 3}{3 - k}$ 3

11. $\dfrac{(a + 3)(a + 4)}{(a - 3)(a + 4)}$ 3, −4

12. $\dfrac{c - 4}{2c - 10}$ 5

13. $\dfrac{3}{y(y^2 - 5y + 6)}$ 0, 2, 3

Write the common factors.

14. $\dfrac{9}{12}$ 3

15. $\dfrac{3(x + 4)}{6x}$ 3

16. $\dfrac{x - y}{(x + y)(x - y)}$ $x - y$

17. $\dfrac{r + 3}{r^2 + 5r + 6}$ $r + 3$

Simplify.

18. $\dfrac{16(x + 1)}{30(x + 2)}$

19. $\dfrac{3(a + b)}{6(a - b)}$

20. $\dfrac{4(c + 2)}{10(2 + c)}$

21. $\dfrac{3(x + y)(x - y)}{6(x + y)}$

22. $\dfrac{6m + 9}{6}$

23. $\dfrac{7t + 21}{t + 3}$

24. $\dfrac{12 + 8x}{4x}$

25. $\dfrac{3d^2 + 2d}{3d + 1}$

26. $\dfrac{b + 2}{b^2 - 4}$

27. $\dfrac{x - 2}{x^2 + 2x - 8}$

28. $\dfrac{-(a + 1)}{a^2 + 8a + 7}$

29. $\dfrac{4 - k}{k^2 - k - 12}$

30. $\dfrac{c^2 - 9}{3c + 9}$

31. $\dfrac{3n - 12}{n^2 - 7n + 12}$

32. $\dfrac{y^2 + 2y - 3}{y^2 + 7y + 12}$

33. $\dfrac{a^2 - b^2}{(a + b)^2}$

34. **Geometry** The area of a certain rectangle is represented by $x^2 + 6x + 9$ and its length by $x + 3$. Find the width. $x + 3$

35. Biology Two equal populations of animals, one a predator, the other a prey were released into the wild. Five years later a survey showed the predator population had multiplied 3 times, while the prey had multiplied by 8 times the original population. Then 12 predators and 32 prey were released into the area. What is the numerical ratio of predator to prey after the release? 3:8

Geometry Exercises 36-37 refer to the cube below.

36. To find the volume of a cube, multiply the length by the width by the height. Write an expression to represent the volume of this cube. x^3

37. To find the total surface area of a cube, multiply the area of one face by 6. Write an expression to represent the total surface area of this cube. $6x^2$

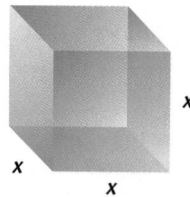

18. $\dfrac{8(x + 1)}{15(x + 2)}, x \neq -2$

19. $\dfrac{a + b}{2(a - b)}, a \neq b$

20. $\dfrac{2}{5}, c \neq -2$

21. $\dfrac{x - y}{2}, x \neq -y$

22. $\dfrac{2m + 3}{2}$

23. $7, t \neq -3$

24. $\dfrac{3 + 2x}{x}, x \neq 0$

25. $\dfrac{d(3d + 2)}{3d + 1}, d \neq -\dfrac{1}{3}$

26. $\dfrac{1}{b - 2}, b \neq 2, -2$

27. $\dfrac{1}{x + 4}, x \neq 2, -4$

28. $\dfrac{-1}{a + 7}, a \neq -1, -7$

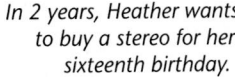 *Look Back*

38. Savings How much does Heather have to deposit in the bank at a 4.5% interest rate, compounded yearly, to have $1500 for the stereo on her sixteenth birthday? **[Lesson 2.1]** $1373.60

In 2 years, Heather wants to buy a stereo for her sixteenth birthday.

39. Chin-Lyn buys a new sweater at a 20% discount and pays $23.60, before tax. What is the sweater's original price? **[Lesson 4.5]** $29.50

40. Find the equation of the line segment connecting the points $\left(\frac{2}{3}, -2\right)$ and $\left(-\frac{5}{6}, 3\frac{1}{3}\right)$. **[Lesson 5.4]** $96x + 27y = 10$

41. Multiply $\begin{bmatrix} -3 & 2 \\ -1 & 4 \end{bmatrix} \cdot \begin{bmatrix} 5 & -3 \\ 3 & -6 \end{bmatrix}$. **[Lesson 7.3]** $\begin{bmatrix} -9 & -3 \\ 7 & -21 \end{bmatrix}$

42. **Probability** How many different 3-letter arrangements can be formed from the letters in the word PHONE? **[Lesson 8.5]** 125

43. If $f(x) = 3^x$, then explain how $g(x) = 3^{x+2} - 5$ compares with $f(x)$. **[Lesson 10.6]**

44. Multiply $(x + 2)(x - 2)$. **[Lesson 11.3]** $x^2 - 4$

45. Write the equation of the parabola whose vertex is (2, 3) and passes through the point (0, −1). **[Lesson 12.1]** $y = -(x - 2)^2 + 3$

46. Which of the parabolas open downward? **[Lesson 12.2]** b, c
   a. $y = 3x^2 - 8$ **b.** $y = 9 - x^2$ **c.** $y = -x^2 + 4$ **d.** $y = \frac{3}{4}x^2 - 6$

47. Simplify $\sqrt{\frac{2}{3}} - \sqrt{96}$. **[Lesson 13.2]** $\frac{-11\sqrt{6}}{3}$

Look Beyond

48. Eight squares are the same size, but each is a different color. The squares are stacked on top of each other as shown. Identify the order of the squares numerically from top to bottom. The top square is number 1.

29. $\frac{-1}{k + 3}$, $k \neq 4, -3$

30. $\frac{c - 3}{3}$, $c \neq -3$

31. $\frac{3}{n - 3}$, $n \neq 3, 4$

32. $\frac{y - 1}{y + 4}$, $y \neq -4, -3$

33. $\frac{a - b}{a + b}$, $a \neq -b$

43. $g(x)$ is obtained from $f(x)$ by a shift of 2 units to the left and 5 units down.

48. The top left square is number 2. The squares number counterclockwise ending at number 8 at the top right.

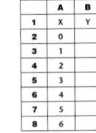

 PREPARE

Objectives

- Multiply rational expressions.
- Add and subtract rational expressions with like and unlike denominators.

RESOURCES

- Practice Master   14.4
- Enrichment Master   14.4
- Technology Master   14.4
- Lesson Activity Master   14.4
- Quiz   14.4
- Spanish Resources   14.4

Assessing Prior Knowledge

Compute:

1. $\frac{1}{2} + \frac{2}{3}$ $\left[1\frac{1}{6}\right]$

2. $1\frac{3}{4} - \frac{9}{10}$ $\left[\frac{17}{20}\right]$

3. $\frac{2}{3} \cdot \frac{3}{4} \cdot \frac{4}{5}$ $\left[\frac{2}{5}\right]$

4. $2\frac{2}{3} \cdot 2\frac{2}{5} \cdot 1\frac{1}{4}$ $[8]$

5. $\frac{1}{5} \div 5$ $\left[\frac{1}{25}\right]$

6. $\frac{1}{3} \div \frac{1}{3}$ $[1]$

TEACH

 An airplane flies to and from a city 400 miles away. Going to the city the plane is flying at a speed of 450 miles per hour and into a head wind of 50 miles per hour. Coming from the city the plane is flying at a speed of 450 miles per hour and has a tailwind of 50 miles per hour. Find the average speed of the plane for the round trip. [**444.4 miles per hour**]

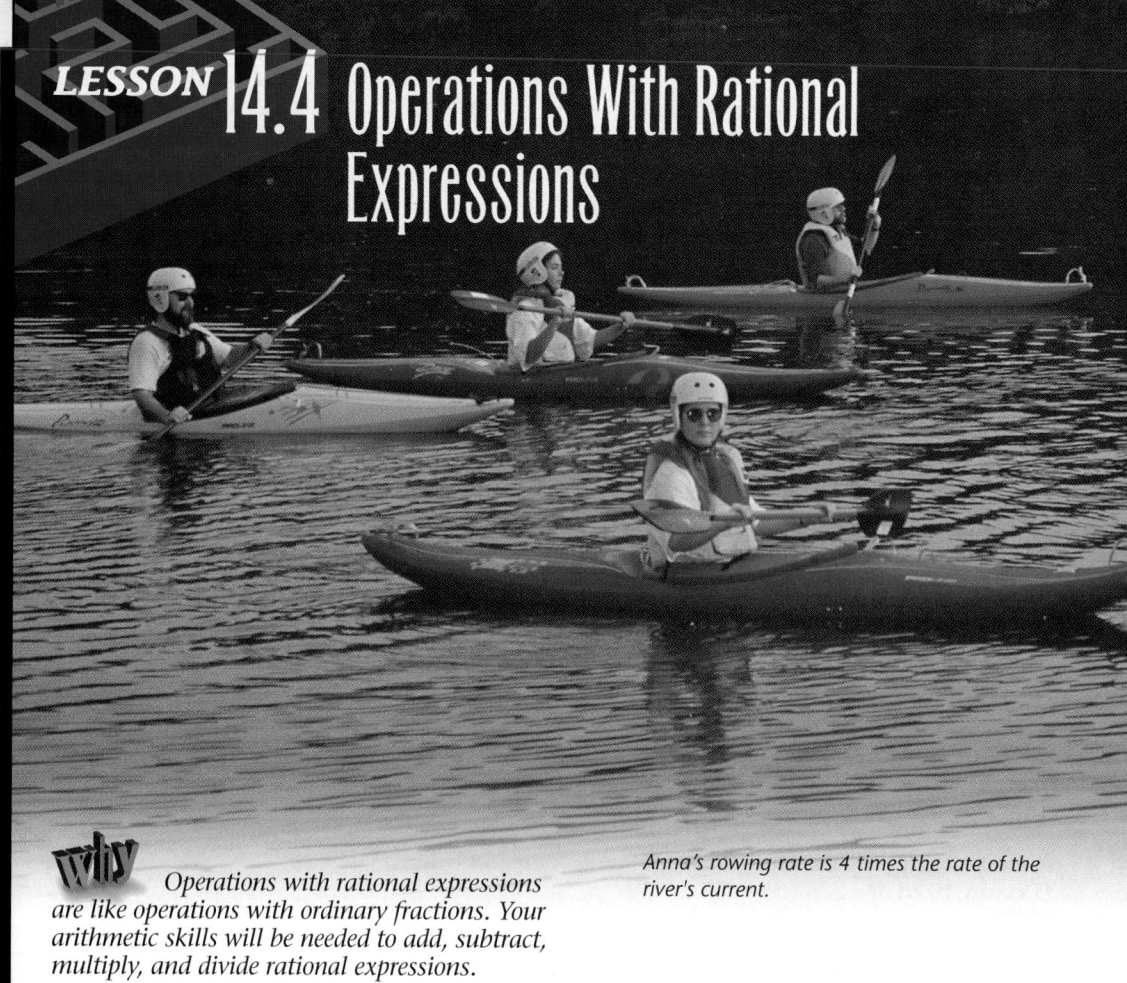

LESSON 14.4 Operations With Rational Expressions

Anna's rowing rate is 4 times the rate of the river's current.

Why *Operations with rational expressions are like operations with ordinary fractions. Your arithmetic skills will be needed to add, subtract, multiply, and divide rational expressions.*

Each year a local club sponsors River Challenge Day at the riverfront to feature safety and recreational activities. One of the most popular activities is the Current Challenge. The challenge requires a participant to paddle a kayak up and down a 1-mile portion of the river. Is there a simple rational expression that models Anna's round-trip time?

Write a formula for the total time it takes to paddle upstream and downstream.

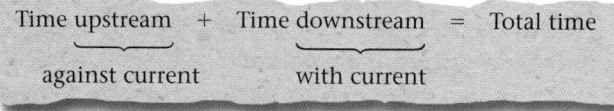

Time upstream + Time downstream = Total time
against current   with current

Write the formula for time in terms of distance and rate.

If distance = rate · time, then time = $\frac{\text{distance}}{\text{rate}}$.

Thus, $\frac{\text{distance}}{\text{rate against current}} + \frac{\text{distance}}{\text{rate with current}}$ = total time.

 **ALTERNATIVE teaching strategy**

Using Cognitive Strategies Begin by reviewing the addition and subtraction of fractions with like denominators. Then use addition and subtraction of rational expressions with like denominators only. After students are comfortable with these exercises, move on to those in the text.

The distance downstream, is 1 mile. Let x represent the rate of the current. Then Anna's rate is $4x$.

So, $\dfrac{1 \text{ mile}}{4x \text{ mph} - x \text{ mph}} + \dfrac{1 \text{ mile}}{4x \text{ mph} + x \text{ mph}} = \text{total time.}$

The equation that models the relationship is

$$\frac{1}{3x} + \frac{1}{5x} = \text{total time.}$$

To find a simple rational form for $\dfrac{1}{3x} + \dfrac{1}{5x}$, add the rational expressions in the same way you add rational numbers with unlike denominators. Compare the methods for adding $\dfrac{1}{30} + \dfrac{1}{50}$ and $\dfrac{1}{3x} + \dfrac{1}{5x}$.

Rational Numbers	Rational Expressions	
$\dfrac{1}{30} + \dfrac{1}{50}$	$\dfrac{1}{3x} + \dfrac{1}{5x}$	Given
$\dfrac{5}{5} \cdot \dfrac{1}{30} + \dfrac{3}{3} \cdot \dfrac{1}{50}$	$\dfrac{5}{5} \cdot \dfrac{1}{3x} + \dfrac{3}{3} \cdot \dfrac{1}{5x}$	Find a common denominator.
$\dfrac{5}{150} + \dfrac{3}{150}$	$\dfrac{5}{15x} + \dfrac{3}{15x}$	Change to common denominators.
$\dfrac{8}{150}$	$\dfrac{8}{15x}$	Simplify.

Anna's time in terms of the rate of the current is $\dfrac{8}{15x}$ where $x \neq 0$.

In working with rational expressions, why is it important to remember to check for the restrictions on the denominator in the original expression?

ENRICHMENT Simplify the complex fractions.

1. $\dfrac{1 - \dfrac{4}{x^2}}{1 + \dfrac{2}{x}}$ $\left[\dfrac{x - 2}{x} \right]$

2. $\dfrac{\dfrac{1}{x} - \dfrac{2}{x^2} - \dfrac{3}{x^2}}{\dfrac{9}{x} \div -x}$ $\left[\dfrac{5 - x}{9} \right]$

3. $\dfrac{\dfrac{1}{(1 + x)}}{1 - \dfrac{1}{(1 + x)}}$ $\left[\dfrac{1}{x} \right]$

Alternate Example 1

Simplify $\dfrac{x + 7}{5x + 10} - \dfrac{x + 5}{3x + 6}$.

$\left[\dfrac{-2}{15}; \; x \neq -2\right]$

A ongoing
SSESSMENT

Try This

$\dfrac{5}{2}, \; x \neq 3, -3$

A ongoing
SSESSMENT

Decide what factor you have to multiply the denominator of the fraction by to get the least common denominator. Then multiply the fraction by that factor divided by itself.

Alternate Example 2

Multiply

$\dfrac{x - 4}{3 - x} \cdot \dfrac{2x}{x - 4} \cdot \dfrac{x - 3}{8}$. $\left[\dfrac{-x}{4}\right]$

In adding the rational expressions on page 671, it is necessary to find a common denominator by multiplying $\dfrac{5}{5} \cdot \dfrac{1}{3x} = \dfrac{5}{15x}$. This illustrates the method used for multiplying any two rational expressions.

EXAMPLE 1

Multiply $\dfrac{x - 2}{x + 3} \cdot \dfrac{x + 3}{x - 5}$.

Solution ➤

$\dfrac{x - 2}{x + 3} \cdot \dfrac{x + 3}{x - 5} = \dfrac{(x - 2)(x + 3)}{(x + 3)(x - 5)}$ Multiply numerators and denominators.

$= \dfrac{x - 2}{x - 5} \cdot \dfrac{x + 3}{x + 3}$ Find fractions that equal 1.

$= \dfrac{x - 2}{x - 5}, \; x \neq 5, -3$ Simplify and write restrictions to original problem. ❖

Try This   Multiply $\dfrac{7}{x} \cdot \dfrac{x^2}{14} \cdot \dfrac{5}{x}$.

One way to add or subtract two rational expressions such as $\dfrac{1}{x + 2}$ and $\dfrac{1}{x - 1}$ is to rewrite the expressions to have a common denominator. This can be done by multiplying each expression by the appropriate name for 1.

$$\dfrac{x}{x + 2} + \dfrac{x}{x - 1} = \dfrac{x - 1}{x - 1} \cdot \dfrac{x}{x + 2} + \dfrac{x + 2}{x + 2} \cdot \dfrac{x}{x - 1}$$

Since the expressions now have common denominators, the numerators can be added.

$$\dfrac{(x - 1)\,x}{(x - 1)(x + 2)} + \dfrac{(x + 2)\,x}{(x + 2)(x - 1)} = \dfrac{(x^2 - x) + (x^2 + 2x)}{(x - 1)(x + 2)}$$

$$= \dfrac{2x^2 + x}{(x - 1)(x + 2)} \quad \text{or} \quad \dfrac{2x^2 + x}{x^2 + x - 2}, \; x \neq 1, -2$$

How do you choose the appropriate names for 1?

EXAMPLE 2

Simplify $\dfrac{a}{a^2 - 4} + \dfrac{2}{a + 2}$.

Solution ➤

Since $a^2 - 4 = (a + 2)(a - 2)$, multiply the second expression by $\dfrac{a - 2}{a - 2}$ and add.

$$\dfrac{a}{a^2 - 4} + \dfrac{2}{a + 2} = \dfrac{a}{(a + 2)(a - 2)} + \dfrac{2(a - 2)}{(a + 2)(a - 2)}$$

$$= \dfrac{a + 2a - 4}{(a + 2)(a - 2)}$$

$$= \dfrac{3a - 4}{a^2 - 4}, \; a \neq 2, -2 \; ❖$$

INCLUSION
strategies

Using Patterns For students who have difficulty finding the least common denominator, allow them to try using the common denominator that is found by finding the product of the denominators. For Example 3, students would find

$$\dfrac{(x - 4)(2x - 2) - (x - 3)(3x - 3)}{(2x - 2)(3x - 3)} =$$

$$\dfrac{(2x^2 - 10x + 8) - (3x^2 - 12x + 9)}{(2x - 2)(3x - 3)}.$$

Simplifying and factoring results in

$$\dfrac{-x^2 + 2x - 1}{(2x - 2)(3x - 3)} = \dfrac{(-1)(x - 1)(x - 1)}{(2)(x - 1)(3)(x - 1)}.$$

By dividing out like terms, the result is $\dfrac{-1}{6}$.

EXAMPLE 3

Simplify. $\dfrac{x-4}{3x-3} - \dfrac{x-3}{2x-2}$

Solution ➤

To find the common denominator, factor the denominators.

$$\frac{x-4}{3(x-1)} - \frac{x-3}{2(x-1)}$$

A common denominator is $3 \cdot 2(x-1) = 6(x-1)$, $x \neq 1$.

$\dfrac{x-4}{3(x-1)} - \dfrac{x-3}{2(x-1)} = \dfrac{2(x-4)}{6(x-1)} - \dfrac{3(x-3)}{6(x-1)}$, $x \neq 1$.   Change to common denominators.

$\qquad = \dfrac{2x-8-3x+9}{6(x-1)}$, $x \neq 1$ Subtract the numerators.

$\qquad = \dfrac{-x+1}{6(x-1)}$, $x \neq 1$ Simplify.

$\qquad = \dfrac{-1(x-1)}{6(x-1)}$, $x \neq 1$ Factor -1 from the numerator.

$\qquad = \dfrac{-1}{6}$, $x \neq 1$ Simplify. ❖

CRITICAL Thinking   Explain when to factor -1 from an expression. How does factoring -1 simplify a rational expression?

EXERCISES & PROBLEMS

Communicate

Refer to the expressions $\dfrac{x}{x+1}$ and $\dfrac{3}{x}$ for Exercises 1–5.

1. Explain how a common denominator may be found for the two rational expressions.

2. Describe how to find the sum of these rational expressions.

3. Describe how to find the difference of these rational expressions.

4. Describe how to find the product of these rational expressions.

5. How do you think you can find the quotient of the first expression divided by the second expression?

RETEACHING the lesson

Using Symbols When adding and subtracting rational expressions, review how to add or subtract each of the following when $a, b, c,$ and d are whole numbers and then when they are polynomials.

$\dfrac{1}{b} + \dfrac{1}{d}$ $\left[\dfrac{d+b}{bd}\right]$

$\dfrac{a}{b} + \dfrac{c}{b}$ $\left[\dfrac{a+c}{b}\right]$

$\dfrac{a}{b} + \dfrac{c}{bd}$ $\left[\dfrac{ad+c}{bd}\right]$

$\dfrac{a}{b} + \dfrac{c}{d}$ $\left[\dfrac{ad+bc}{bd}\right]$

Remind students that they can use the discriminant of the quadratic formula to check whether a quadratic expression is factorable. If $b^2 - 4ac$ is not a perfect square, then the expression is not factorable.

Alternate Example 3

Solve $\dfrac{3}{4x-8} - \dfrac{x+3}{2x^2-8}$.

$\left[\dfrac{x}{4(x+2)(x-2)}, x \neq 2, -2\right]$

CRITICAL Thinking

Factoring out -1 is sometimes helpful because it allows opposites to be expressed in a form in which a common factor can be identified, and then the expression can be reduced by that factor.

ASSESS

Selected Answers

Odd-Numbered Exercises 7–43

Assignment Guide

Core 1–5, 6–32 even, 33–47

Core Two-Year 1–44

The use of a calculator is appropriate for Exercises 34 and 44.

Error Analysis

When subtracting rational expressions, some students forget to distribute the subtraction sign through all terms of the expression being subtracted. When multiplying rational expressions, some students may divide the numerator and the denominator by like terms rather than by like factors. For instance, in $\dfrac{x-4}{x+2}$, students may try to divide the numerator and denominator by x. In each case, alert students to these common errors and encourage them to always check their solutions.

Practice & Apply

Perform the indicated operations. Simplify and state restrictions for variables.

6. $\dfrac{5}{3x} + \dfrac{2}{3x}$

7. $\dfrac{8}{x+1} - \dfrac{5}{x+1}$

8. $\dfrac{2x}{y+4} - \dfrac{5x}{y+4}$

9. $\dfrac{5x+4}{a-c} - \dfrac{7+3x}{a-c}$

10. $\dfrac{2}{a} + \dfrac{3}{b}$

11. $\dfrac{7}{3t} - \dfrac{8}{2t}$

12. $\dfrac{t}{2rs} + \dfrac{3s}{rt}$

13. $\dfrac{5}{pq} - \dfrac{m}{p^2q}$

14. $\dfrac{-2}{x+1} + \dfrac{3}{2(x+1)}$

15. $x - \dfrac{x-4}{x+4}$

16. $\dfrac{-3-d}{d-1} + 2$

17. $\dfrac{a}{b} + \dfrac{c}{d}$

18. $\dfrac{5+m}{3+n} \cdot \dfrac{3+n}{a+b}$

19. $\dfrac{x+2}{x(x+1)} \cdot \dfrac{x^2}{(x+2)(x+3)}$

20. $\dfrac{q^2-1}{q^2} \cdot \dfrac{q}{q+1}$

21. $\dfrac{1}{x+1} \cdot \dfrac{2}{x}$

22. $\dfrac{3}{x-1} \cdot \dfrac{5}{x}$

23. $\dfrac{y}{y-4} - \dfrac{1}{y}$

24. $\dfrac{-2}{x+1} + \dfrac{3}{x}$

25. $\dfrac{a-2}{a+1} + \dfrac{5}{a+3}$

26. $\dfrac{m+5}{m+2} \cdot \dfrac{m-3}{m-1}$

27. $\dfrac{2r}{(r+5)(r+1)} - \dfrac{r}{r+5}$

28. $\dfrac{3}{b^2+b-6} - \dfrac{5}{b-2}$

29. $\dfrac{4}{y-2} + \dfrac{5y}{y^2-4y+4}$

30. $\dfrac{2}{y-5} - \dfrac{5y-3}{y^2+y-30}$

31. $\dfrac{x}{1-x} + \dfrac{1}{x-1}$

32. $\dfrac{2}{x^2-9} - \dfrac{1}{2x+6}$

Ecology Students at an aquarium are creating an exhibit of a bay ecosystem. They determine that the water in the bay contains 4.0% salt-by-mass concentration. The water in the aquarium tanks for the sea exhibits is 6.0% salt-by-mass concentration. The students plan to dilute 750 kilograms of the water in the aquarium by adding enough pure water to get the 4.0% salt-by-mass concentration. The concentration can be determined by

$$\text{Concentration} = \frac{\text{sea-water salt concentration} \cdot \text{mass of sea water}}{\text{mass of sea water} + \text{mass of pure water}}.$$

33. Write a rational function for this problem to represent the resulting concentration when a mass of pure water, w, is added.

34. Graph the function.

35. How much pure water needs to be added to reach a concentration of 4.0% salt-by-mass?

The answers to Exercises 6–32 can be found in Additional Answers beginning on page 729.

33. $C = \dfrac{45}{750 + w}$, where w represents the number of kilograms of pure water added.

34.

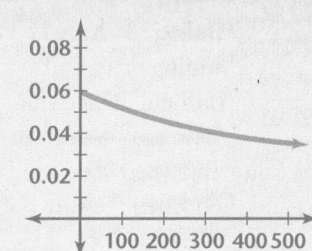

35. 375 kilograms

Look Back

36. State the pattern and find the next three terms of the following
sequence. $-1, 6, 15, 26, \underline{\quad}, \underline{\quad}, \underline{\quad}$ **[Lesson 1.1]**

37. Solve $0.3(x - 90) = 0.7(2x - 70)$.   **[Lesson 4.6]**   20

38. Graph the solution of $y < -2x + 6$ and $y \leq 3x - 4$.   **[Lesson 6.5]**

39. Add $\begin{bmatrix} -3 & 7 \\ 5 & 11 \end{bmatrix} + \begin{bmatrix} 7 & -13 \\ 22 & 2 \end{bmatrix}$.   **[Lesson 7.2]** $\begin{bmatrix} 4 & -6 \\ 27 & 13 \end{bmatrix}$

Cultural Connection: Americas The Blackfoot people of Montana played a game of chance in which they threw 4 decorated bones made from buffalo ribs. Three were blank on one side, but the fourth, called the chief, was engraved on both sides like a coin with heads and tails. The highest score was 6 points, for a throw of 3 blanks and the chief landing heads up. For 3 blanks and the chief reversed, the score was 3 points.

40. What is the probability
of throwing 3 blanks and the chief landing heads up?   **[Lesson 8.1]** $\frac{1}{16}$

41. What is the probability of throwing 3 blanks and the chief reversed?
[Lesson 8.1] $\frac{1}{16}$

42. **Probability** If a bag contains 4 red marbles, 3 black marbles,
and 1 white marble, what is the probability that you will select a
white marble?   **[Lesson 8.1]** $\frac{1}{8}$

43. Express 25,000,000,000 in scientific notation.   **[Lesson 10.5]** 2.5×10^{10}

44. Graph $y = (x - 1)^2 - 4$.   **[Lesson 12.1]**

Look Beyond

A proportion is the equality of two ratios such as $\frac{5}{10} = \frac{2}{4}$, since both ratios
equal $\frac{1}{2}$. This reduced fraction is called the constant of the proportion. Find
the missing term in each proportion and state the constant.

45. $\frac{26}{39} = \frac{x}{9}$ **46.** $\frac{14}{x} = \frac{49}{28}$ **47.** $\frac{15}{40} = \frac{27}{x}$

---

36. The second differences are $+2$; terms: 39,
54, 71

38.

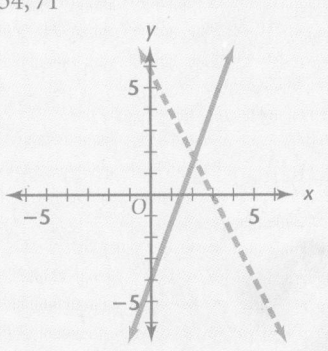

44.

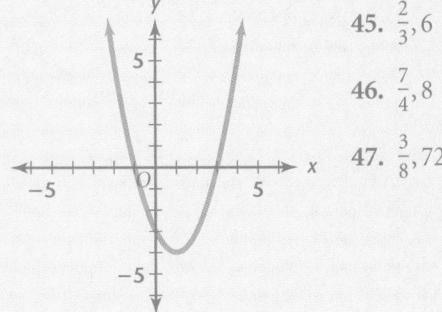

45. $\frac{2}{3}, 6$

46. $\frac{7}{4}, 8$

47. $\frac{3}{8}, 72$

Alternative Assessment

Performance Assessment

Have students perform the indicated operation on each of the
following. Remind students to include any restrictions on the variables.

$$a = \frac{x^2 - 4}{x^2 - 4x + 4}$$

$$b = \frac{2x - 4}{x + 2}$$

$$c = \frac{x^2 + x - 6}{4 - x^2}$$

1. $b + c$ $\left[\frac{(x - 7)}{(x + 2)}, \right.$

   $\left. x \neq 2, -2 \right]$

2. $c - a$ $\left[\frac{-(2x^2 + 5x - 2)}{(x + 2)(x - 2)}, \right.$

   $\left. x \neq 2, -2 \right]$

Look Beyond

Exercises 45–47 foreshadow the
use of proportions when solving problems, which will be
studied in Lesson 14.6.

Technology Master

Technology
14.4 Writing One Rational Expression as a Sum of Two

From Lesson 14.4, you know how to add two rational expressions to obtain
a third rational expression. You can reverse the process as well.

That is, given one rational expression, such as $\frac{2x - 1}{(x + 2)(x + 3)}$, you can write
it as the sum of two rational expressions $\frac{A}{x + 2}$ and $\frac{B}{x + 3}$ for some numbers
A and B. Here is how to begin.

$$\frac{2x - 1}{(x + 2)(x + 3)} = \frac{A}{x + 2} + \frac{B}{x + 3}$$

$$= \frac{A(x + 3) + B(x + 2)}{(x + 2)(x + 3)}$$

$$= \frac{(A + B)x + (3A + 2B)}{(x + 2)(x + 3)}$$

So, $2x - 1 = (A + B)x + (3A + 2B)$.
To find A and B, you must solve the system $\begin{bmatrix} A + B = 2 \\ 3A + 2B = -1 \end{bmatrix}$.
To solve the system for A and B, use a graphics calculator and matrix
inverses.

Write each rational expression as the sum of two simpler
rational expressions. Begin by writing a system of equations
involving A and B. Then use matrix inverses to find A and B.

1. $\frac{2x - 1}{(x + 2)(x + 3)} = \frac{A}{x + 2} + \frac{B}{x + 3}$

2. $\frac{2x}{(x + 2)(x + 3)} = \frac{A}{x + 2} + \frac{B}{x + 3}$

3. $\frac{x - 1}{(x + 2)(x + 3)} = \frac{A}{x + 2} + \frac{B}{x + 3}$

4. $\frac{-2x + 1}{(x + 2)(x + 3)} = \frac{A}{x + 2} + \frac{B}{x + 3}$

5. Graph the left side of the equation in Exercise 1 and your sum from
Exercise 1 on the same graphics calculator screen. Does the graphic
that results confirm that your solutions for A and B are correct?

6. Explain why you need to compute only one matrix inverse in
order to solve all of Exercises 1–4.

How Worried Should You Be?

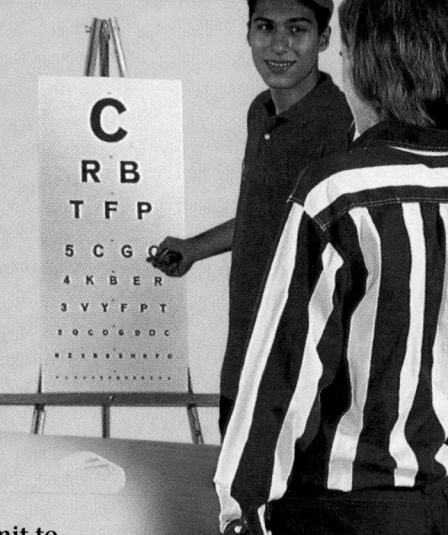

CENTRAL HIGH HEALTH FAIR

FOCUS

A hypothetical case involving a cancer test shows that there's more to interpreting test results than meets the eye. Students develop a formula to explore the conditions under which negative test results may be less worrisome than they seem at first glance.

MOTIVATE

After reading the article, have students write their responses so they can later look back at them.

Differences of opinion need not be resolved at this point because students will be calculating the relevant probability later in the lesson in Activity 1.

Now, the testing question

Concerns about iron's role in heart disease, if ultimately confirmed, surely will touch off a debate over iron tests. Who should be tested? and should testing be a matter of public-health policy? . . .

The virtue of such broad screening would be to help ferret out many of the estimated 32 million Americans who carry a faulty gene that prompts their body to harbor too much iron. There is no test for the gene itself, but anyone with a high blood level of iron is a suspect. Siblings and children would be candidates for testing as well, since the condition is inherited . . .

Two widely used blood tests for iron generally cost from $25 to $75 each. The first measures ferritin, a key iron-storing protein . . .

The results should always be discussed with a doctor, since some physicians take recent studies seriously enough to worry about even moderately elevated ferritin levels. Moreover, the test might bear repeating. Erroneously high readings are relatively common.

Although people submit to medical tests because they want to find out if something is wrong, the results may cause them to worry. Read the following excerpt about an imaginary case.

Assume that there is a test for cancer which is 98 percent accurate; i.e., if someone has cancer, the test will be positive 98 percent of the time, and if one doesn't have it, the test will be negative 98 percent of the time. Assume further that 0.5 percent—one out of two hundred people—actually have cancer. Now imagine that you've taken the test and that you've tested positive. The question is: How concerned should you be?

To find out whether your reaction would be justified, you need to look at the numbers more closely.

1a. 0.5% of $10,000 = (0.005)(10,000) = 50$

1b. 98% of $50 = (0.98)(50) = 49$

1c. 9950 Since 0.5% of the 10,000 are assumed to have cancer, then the remaining 99.5% can be assumed not to: $(0.995)(10,000) = 9950$. Or, since 50 out of 10,000 have it, $10,000 - 50$, or 9950, do not.

1d. 199 The test gives correct results 98% of the time and therefore gives incorrect results 2% of the time. So, $(9950)(0.02) = 199$

1e. 248 49 true positives plus 199 false positives

1f. $P = 49/248$, or about 0.2. You would have about a 20% chance of actually having cancer.

Cooperative Learning

Activity 1. Use the data in the excerpt to complete a–g. Explain each answer.

    a. Of the 10,000 people tested, how many would you expect to have cancer?

    b. How many of those people that have cancer would you expect to test positive? (We'll call such cases *true positives*.)

    c. Of the 10,000 people tested, how many would you expect to not have cancer?

    d. How many of those people that do not have cancer would you expect to test positive? (We'll call such cases *false positives*.)

    e. Add your results in b and d to find the total number of people expected to test positive.

    f. What is the probability (**P**) that if you tested positive, you actually have cancer?

    g. How does your answer in **f** compare with your response to the question, "How concerned should you be?"

> **Remember:**
>
> $$\text{Probability} = \frac{\text{considered outcome}}{\text{total number of outcomes}}$$
>
> $$P = \frac{\text{number of true positives}}{\text{total number expected to test positive}}$$

Activity 2. The data you used in Activity 1 are made up. To explore what happens with different sets of data, you can find a general formula for **P**, the probability that if you tested positive, you actually have cancer.

    Let **N** = the number of people tested
        a = the accuracy of the test
            (if test is 98% accurate, then **a** = 0.98)
        r = the portion of people tested that actually have what's being tested for (in the example, **r** = 0.005)

    a. Write a formula for the number of true positives. HINT: Look back at a and b of Activity 1.

    b. Write a formula for the number of false positives. HINT: Look back at c and d of Activity 1.

    c. Write a formula for **P**. HINT: Look back at e and f of Activity 1.

3. Look at your formula from 2c.

    a. Find **P** when **r** = 1. Explain why that result makes sense.

    b. In the imaginary case, suppose **r** = 0.5 instead of 0.005. Would the results in f of Activity 1 have been as surprising? Explain.

1g. Answers may vary. For those who originally figured there would be a 98% chance of having cancer, the 20% figure should be surprising. Though worried still, they should be much less worried with a 20% chance than with a 98% chance.

2a. number of true positives are **arN**.

2b. number of false positives are
$(1 - a)(N - rN)$ or $(1 - a)(1 - r)N$

2c. $P = \dfrac{arN}{(1 - a)(1 - r)N + arN}$

$\qquad = \dfrac{ar}{(1 - a)(1 - r) + ar}$

PREPARE

Objectives

• Solve rational equations by graphing and by using the common denominator method.

• Use rational equations to solve problems.

RESOURCES

• Practice Master   14.5
• Enrichment Master   14.5
• Technology Master   14.5
• Lesson Activity Master 14.5
• Quiz   14.5
• Spanish Resources   14.5

Assessing Prior Knowledge

Solve:

1. $\dfrac{x+1}{x} + \dfrac{2}{3x} = \dfrac{3}{2x}$ $\left[\dfrac{-1}{6}\right]$

2. $\dfrac{4}{x+2} = \dfrac{3}{x+1} - \dfrac{1}{x+2}$ $\left[\dfrac{1}{2}\right]$

TEACH

Tell students that you bought a book and a magazine for a total of $8. The book cost a dollar more than the magazine. How much did each cost? If the book is $4 and the magazine is $3, then the total would be $7 — too small. Try another guess. If the book is $4.50, the magazine is $3.50 and the total is $8 — correct. Tell students that in this lesson they will learn another method for solving these types of problems.

LESSON 14.5 Solving Rational Equations

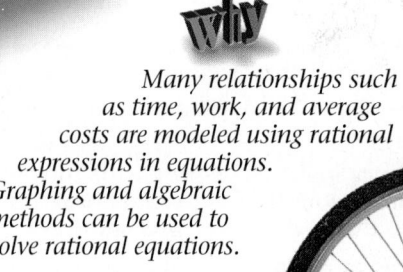

why

Many relationships such as time, work, and average costs are modeled using rational expressions in equations. Graphing and algebraic methods can be used to solve rational equations.

Sam rode his bike 6 miles before his chain broke, and he had to walk home.

Sports During the walk home, Sam estimated that he bikes 5 times as fast as he walks. If the entire trip took him 2 hours and 24 minutes, find his biking rate.

The time for each part of the trip can be represented using the formula $\dfrac{\text{Distance}}{\text{Rate}} = \text{Time}$. The total time can be represented by the formula

$$\frac{\text{Distance biked}}{\text{Rate}} + \frac{\text{Distance walked}}{\text{Rate}} = \text{Total time.}$$

If x is Sam's walking rate, then $5x$ is his biking rate.

$$\frac{6 \text{ miles}}{5x \text{ mph}} + \frac{6 \text{ miles}}{x \text{ mph}} = 2\frac{2}{5} \text{ hours}$$

The equation that models the problem is $\dfrac{6}{5x} + \dfrac{6}{x} = \dfrac{12}{5}$. It can be solved by either of two methods.

Common Denominator Method Multiply each of the terms of both sides of the original equation by a common denominator. This will clear the equation of fractions. Using the lowest common denominator saves extra steps in the solution.

$\dfrac{6}{5x} + \dfrac{6}{x} = \dfrac{12}{5}$	Given
$5x\left(\dfrac{6}{5x} + \dfrac{6}{x}\right) = 5x\left(\dfrac{12}{5}\right)$	Multiply both sides by $5x$, the lowest common denominator.
$6 + 30 = 12x$	Simplify.
$12x = 36$	Simplify.
$x = 3$	Division Property of Equality

ALTERNATIVE teaching strategy

Using Tables Show students how to use a table for problems like Example 3.

So, $\dfrac{3}{4}x + \dfrac{1}{2}x = 1$; $\dfrac{5}{4}x = 1$; $x = \dfrac{4}{5}$. Together, it will take $\dfrac{4}{5}$ hour, or 48 minutes.

	part done in 1 hour	no. of hours	=	part done in x hours
Max	$\dfrac{60}{80} = \dfrac{3}{4}$	x	=	$\dfrac{3}{4}x$
Carl	$\dfrac{60}{120} = \dfrac{1}{2}$	x	=	$\dfrac{1}{2}x$

Graphing Method Graph
$Y_1 = \dfrac{6}{5x} + \dfrac{6}{x}$ and $Y_2 = \dfrac{12}{5}$. You find that their intersection is at $x = 3$.

Graphics Calculator

Sam walks at 3 miles per hour. His biking rate is 5 times that, or 15 miles per hour.

CRITICAL Thinking

What is the "real-life" domain for the function in the previous example? What are the undefined value(s) for the rational expressions in the equations?

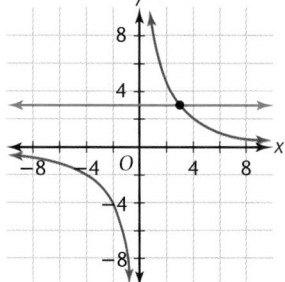

EXAMPLE 1

Solve $\dfrac{6}{x - 1} + 2 = \dfrac{12}{x^2 - 1}$ using the common denominator method and the graphing method.

Solution A ➤

Common Denominator Method
Factor. Notice that this equation is undefined for $x = 1$ and $x = -1$.

$$\frac{6}{x - 1} + 2 = \frac{12}{(x + 1)(x - 1)}$$

Multiply both sides by the lowest common denominator $(x + 1)(x - 1)$.

$$(x + 1)(x - 1)\left[\frac{6}{x - 1} + 2\right] = (x + 1)(x - 1)\left[\frac{12}{(x + 1)(x - 1)}\right], x \neq 1, -1$$

$6(x + 1) + 2(x + 1)(x - 1) = 12$	Divide out common factors.
$6x + 6 + 2x^2 - 2 = 12$	Simplify.
$2x^2 + 6x - 8 = 0$	Standard quadratic form
$2(x + 4)(x - 1) = 0$	Factor.
$x + 4 = 0$ or $x - 1 = 0$	Zero Product Property
$x = -4$ or $x = 1$	Solve.

Reject 1 as a solution since it makes the *original* equation undefined. Thus, -4 is the only solution. It is very important to check all possible solutions to be sure that the original equation is defined for them.

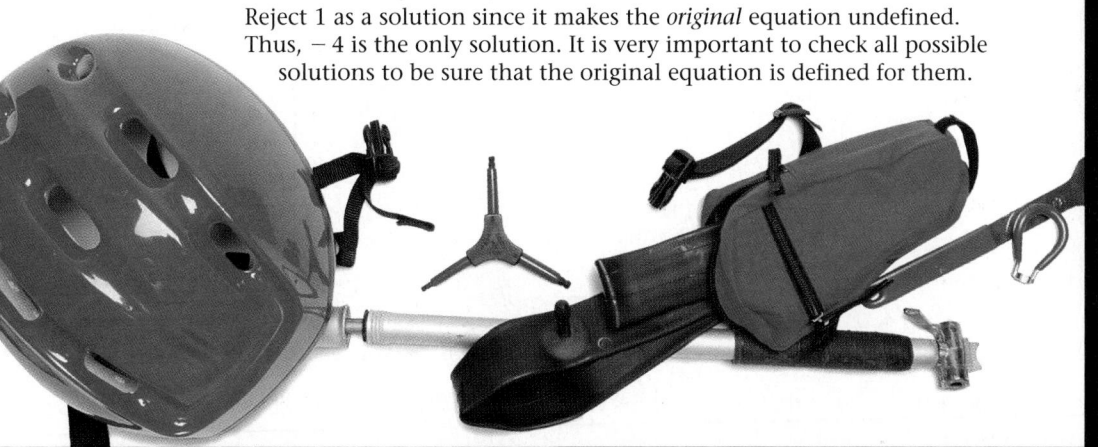

interdisciplinary

CONNECTION

Physical Education The track team used rowing as part of its preseason conditioning program. Suppose Pat and Jerry can row at 3 miles per hour in still water. On one outing, a 9-mile trip downstream with the current took as long as a 3-mile trip upstream against the current. What was the rate of the current? The resulting equation is $\dfrac{9}{(3 + c)} = \dfrac{3}{3 - c}$. The solution is $1\frac{1}{2}$ miles per hour.

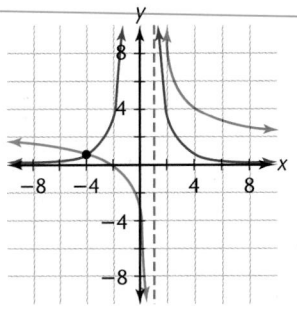

ASSESSMENT
ongoing

The undefined values become asymptotes of the graph.

Alternate Example 2

Diane told Alex that she earned $13 an hour last week. She told him that she earns $12 per hour and time-and-a-half for overtime (over 40 hours). How many hours did Diane work last week? $[12(40) + 18(x - 40) = 13x; x = 48]$

> **Use Transparency** 68

Solution B ▶

Graphing Method

Graph $Y_1 = \dfrac{6}{x-1} + 2$ and $Y_2 = \dfrac{12}{x^2-1}$. The two functions have a common asymptote at $x = 1$ and a clear point of intersection at $x = -4$.

The solution is $x = -4$. ❖

Explain the effect of an undefined value on the graph of a rational expression in simplest form.

EXAMPLE 2

Juanita is starting a lawn-mowing business in her neighborhood. The startup cost for the business is $400.

There are expenses of $6 per lawn that include a helper and the cost of gasoline. Juanita has the $6 expense each time she mows a lawn. How many lawns must she mow before the average cost per lawn is $20?

Solution ▶

Let x = the number of lawns mowed. Substitute the known values.

$$\begin{array}{c}\text{average} \\ \text{cost per} \\ \text{lawn}\end{array} = \dfrac{\begin{array}{c}\text{Start-up} \\ \text{costs}\end{array} + \left(\begin{array}{c}\text{Costs per} \\ \text{lawn}\end{array}\right) \times \left(\begin{array}{c}\text{Number} \\ \text{of lawns}\end{array}\right)}{\text{Number of lawns}}$$

$$20 = \dfrac{400 + 6x}{x}$$

Common Denominator Method

$$\dfrac{20}{1} = \dfrac{400 + 6x}{x}$$

$$20x = 400 + 6x$$

$$14x = 400$$

$$x = 28\dfrac{4}{7}$$

Graphing Method

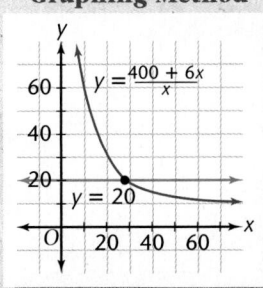

Juanita would have to mow 29 lawns before the average cost per lawn is $20. ❖

ENRICHMENT An experienced builder and an apprentice took 6 hours to complete a job. The experienced builder can work 3 times as fast as the apprentice. Working alone, how long would it take the experienced builder to complete the job? **[8 hours]**

INCLUSION strategies **Cooperative Learning** Students who need visual or auditory reinforcement should be allowed to work in small groups where each step of the solution of a rational equation is discussed and illustrated by members of the group. If extra practice is needed for a group, have them solve Exercises 16–24 on page 682 algebraically.

Try This Solve the rational equation by finding the lowest common denominator. Verify the solution by graphing. $\dfrac{20}{x^2-4}+6=\dfrac{10}{x-2}$

EXAMPLE 3

Max and his younger brother Carl are raising extra money by delivering papers. If the boys work together, how long will it take them to finish this job?

Max can deliver papers in 80 minutes, but it takes Carl 2 hours to deliver the same number of papers.

Solution ➤

Let t represent the number of minutes the boys work together. The completed job is equal to the sum of all the fractional parts and is represented by 1 complete job.

Fractional part of the job done by Max	+	Fractional part of the job done by Carl	=	The complete job, or 1 whole unit
$\left(\begin{array}{c}\text{Rate}\\ \text{of Max's}\\ \text{work}\end{array}\right) \cdot \left(\begin{array}{c}\text{Time}\\ \text{worked}\end{array}\right)$	+	$\left(\begin{array}{c}\text{Rate}\\ \text{of Carl's}\\ \text{work}\end{array}\right) \cdot \left(\begin{array}{c}\text{Time}\\ \text{worked}\end{array}\right)$	=	1
$\dfrac{1}{80} \cdot t$	+	$\dfrac{1}{120} \cdot t$	=	1
$\dfrac{t}{80}$	+	$\dfrac{t}{120}$	=	1

Continue to solve $\dfrac{t}{80}+\dfrac{t}{120}=1$. Multiply each term by the lowest common denominator of 80 and 120.

$$240 \cdot \dfrac{t}{80} + 240 \cdot \dfrac{t}{120} = 240 \cdot 1$$
$$3t + 2t = 240$$
$$5t = 240$$
$$t = 48$$

Working together it would take Carl and Max 48 minutes to deliver all the papers.

In 48 minutes, Max does $\dfrac{48}{80}$ of the job and Carl does $\dfrac{48}{120}$ of the job.

$\dfrac{48}{80}+\dfrac{48}{120}=\dfrac{6}{10}+\dfrac{4}{10}=\dfrac{10}{10}=1$ ❖

RETEACHING
the
l e s s o n

Cooperative Learning
Have students work in small groups as they re-examine and discuss the examples in the text. Some students may find that guess-and-check is a more efficient way to solve Example 1. However, after trying to use that strategy to solve Example 2, students will find that another strategy is needed. Encourage students to develop proficiency in both methods shown in the text.

EXERCISES & PROBLEMS

Communicate

1. Discuss what "undefined variable values of rational equations" means.

2. How is the "realistic domain" of a rational equation determined in applications?

Refer to the equation $\dfrac{3}{c} - \dfrac{2}{c-1} = \dfrac{6}{c}$ for Exercises 3–5.

3. Describe the steps needed to find the lowest common denominator.

4. Explain how to solve the rational equation by graphing.

5. How are the undefined values for the equation found? What are they?

6. Explain how to write the rate of work on a particular job as a fraction.

Practice & Apply

Solve the following rational equations by finding the lowest common denominator.

7. $\dfrac{x-1}{x} + \dfrac{7}{3x} = \dfrac{9}{4x}$ $\dfrac{11}{12}$

8. $\dfrac{10}{2x} + \dfrac{4}{x-5} = 4$ $6\dfrac{1}{4}, 1$

9. $\dfrac{x-3}{x-4} = \dfrac{x-5}{4+x}$ $3\dfrac{1}{5}$

10. $\dfrac{x+3}{x} + \dfrac{6}{5x} = \dfrac{7}{2x}$ $-\dfrac{7}{10}$

11. $\dfrac{x}{x-3} = 5 + \dfrac{x}{x-3}$ none

12. $\dfrac{x+3}{x^2-9} - \dfrac{6}{x-3} = 5$ 2

13. $\dfrac{4}{y-2} + \dfrac{5}{y+1} = \dfrac{1}{y+1}$ $\dfrac{1}{2}$

14. $\dfrac{10}{x+3} - \dfrac{3}{5} = \dfrac{10x+1}{3x+9}$ 2

15. $\dfrac{3}{x-2} - \dfrac{6}{x^2-2x} = 1$ 3

Solve the following rational equations by the graphing method.

16. $\dfrac{12x-7}{x} = \dfrac{17}{x}$ 2

17. $\dfrac{3}{7x+x^2} = \dfrac{9}{x^2+7x}$ none

18. $\dfrac{20-x}{x} = x$ $-5, 4$

19. $\dfrac{2}{x} + \dfrac{1}{3} = \dfrac{4}{x}$ 6

20. $\dfrac{1}{2} + \dfrac{1}{x} = \dfrac{1}{2x}$ -1

21. $\dfrac{x-3}{x-4} = \dfrac{x-5}{x+4}$ 3.2

22. $\dfrac{3}{x-2} - \dfrac{6}{x^2-2x} = 1$ 3

23. $\dfrac{1}{x-2} + \dfrac{16}{x^2+x-6} = -3$ $-\dfrac{1}{3}, -1$

24. $\dfrac{1}{x-2} + 3 = \dfrac{-16}{x^2+x-6}$ $-\dfrac{1}{3}, -1$

25. The sum of two numbers is 56. If the larger number is divided by the smaller, the quotient is 1 with a remainder of 16. Find the two numbers.   36, 20

Marisa pedals downstream at a still-water rate of 20 meters per minute for 500 meters before turning around to pedal upstream to where she started.

Recreation If Marisa maintains the same still-water rate, she will only pedal 300 meters upstream in the same time it takes her to pedal 500 meters downstream.

26. Write rational expressions that represent the time it takes Marisa to pedal downstream and upstream in the current.

27. What is the rate of the current?

28. Politics Jeff, Phil, and Allison decide to support their favorite candidate by stuffing envelopes for a candidate's political campaign. They find that they all work at different rates. Phil takes 2 hours longer than Jeff to stuff 1000 envelopes, and Allison takes 3 hours less than Phil to do the same. One day they work together for 4 hours. Write rational expressions that represent the rate at which each works.

Auto Painting In their shop, it takes Tom about 5 hours to paint a car working alone. It takes Shelley about 7 hours to paint a car.

29. Write rational expressions that represent the fractional parts that Tom and Shelley complete when painting a car together.

30. Write and solve a rational equation to find the time it would take to paint a car if they work together.

Look Back

31. Solve $x - 4 = \frac{-3}{4}(x + 2)$. **[Lesson 4.7]** $\frac{10}{7}$

32. Graph $|2x - 7| \le 7$. **[Lesson 4.8]**

33. Solve graphically $x + 5y = 9$ and $3x - 2y = 10$. **[Lesson 6.1]**

34. A club has 20 members. In how many ways can 3 officers be selected if all members are eligible for these positions? **[Lesson 8.1]** 6840

35. Multiply $(3x - 8)(5x + 2)$. **[Lesson 11.3]** $15x^2 - 34x - 16$

36. If $f(x) = \dfrac{x^2 - 3x - 10}{(x - 4)^2}$, find $f(-2)$. **[Lesson 14.1]** 0

Look Beyond

37. As a game, Antonio climbs up 6 steps in 3 seconds and hops back 4 steps in 2 seconds. How long will it take him to reach step 100? 250 seconds

---

26. upstream: $\dfrac{300}{20 - x}$, downstream: $\dfrac{500}{20 + x}$, where x represents the rate of the current

27. 5 meters per minute

28. Jeff $\dfrac{1}{x}$, Phil $\dfrac{1}{x + 2}$, Allison $\dfrac{1}{x - 1}$, where x represents the time for Jeff to complete the job alone.

29. Tom $\dfrac{x}{5}$, Shelley $\dfrac{x}{7}$, where x represents the time to paint one car together

30. $\dfrac{x}{5} + \dfrac{x}{7} = 1$; $x = 2\dfrac{11}{12}$ hrs or 2 hours 55 minutes

The answers to Exercises 32 and 33 can be found in Additional Answers beginning on page 729.

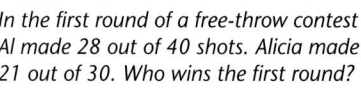
Exploring Proportions

PREPARE

Objectives

- Identify the means and the extremes of a proportion.
- Determine whether an equation is a proportion.
- Use proportions to solve problems.

RESOURCES

- Practice Master   14.6
- Enrichment Master   14.6
- Technology Master   14.6
- Lesson Activity Master   14.6
- Quiz   14.6
- Spanish Resources   14.6

Assessing Prior Knowledge

Solve:

1. $\frac{4}{x} = 8$ $\left[\frac{1}{2}\right]$

2. $\frac{4}{x} = \frac{8}{40}$ $[20]$

3. $\frac{4}{x} = \frac{8}{3}$ $\left[1\frac{1}{2}\right]$

4. $\frac{x}{4} = \frac{8}{3}$ $\left[10\frac{2}{3}\right]$

5. $\frac{x}{4} = \frac{3}{8}$ $\left[1\frac{1}{2}\right]$

TEACH

Why Ask students how much they would earn for 8 hours of work if they earn $9.75 for 2 hours of work. [$39] Have students discuss various methods they used to solve the problem. Suggest that one way would be to set up a proportion. Tell students they will learn more about using proportions in this lesson.

Why *You have used ratios and proportions to solve percent problems. Since a ratio containing a variable is a rational expression, proportions can be solved using the methods in the previous lesson.*

> In the first round of a free-throw contest Al made 28 out of 40 shots. Alicia made 21 out of 30. Who wins the first round?

There are several methods to determine the winner. One way is to use fractions.

Al	Alicia
$\frac{28}{40} = \frac{7}{10}$	$\frac{21}{30} = \frac{7}{10}$

Since both fractions reduce to $\frac{7}{10}$, $\frac{28}{40}$ is equal to $\frac{21}{30}$, and the contest ends in a tie. The equation $\frac{28}{40} = \frac{21}{30}$ is called a proportion.

A **proportion** is an equation that states that 2 ratios are equal.

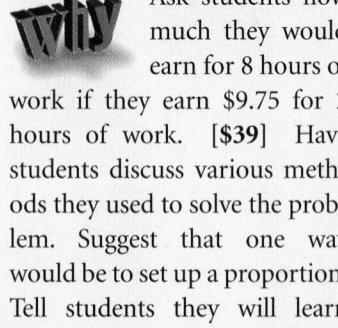

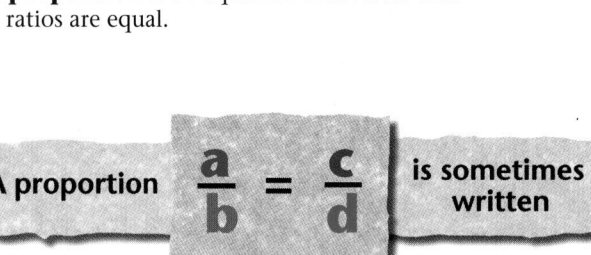 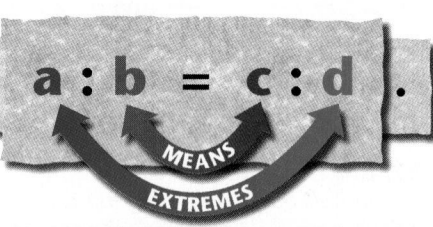 

A proportion $\dfrac{a}{b} = \dfrac{c}{d}$ is sometimes written $a : b = c : d$.

MEANS EXTREMES

ALTERNATIVE teaching strategy

Using Algorithms
Show the proportion $\dfrac{(2 + x)}{(x^2 - 1)} = \dfrac{(3 + x)}{(x + 1)}$.
Point out that an alternate method of solution is to subtract one term from both sides to get $\dfrac{2 + x}{(x^2 - 1)} - \dfrac{(3 + x)}{(x + 1)} = 0$. Now find a common denominator $(x^2 - 1)$ and subtract $\dfrac{(2 + x) - (3 + x)(x - 1)}{(x^2 - 1)} = \dfrac{-x^2 - x + 5}{x^2 - 1}$.

ENRICHMENT

The decoration committee started with some teal and maroon balloons. After they added 2 more maroon balloons, exactly $\frac{1}{5}$ of the balloons were teal. If they had added two teal balloons instead of maroon, exactly $\frac{1}{3}$ would have been teal. How many of each color did they start with? [3 teal and 10 maroon]

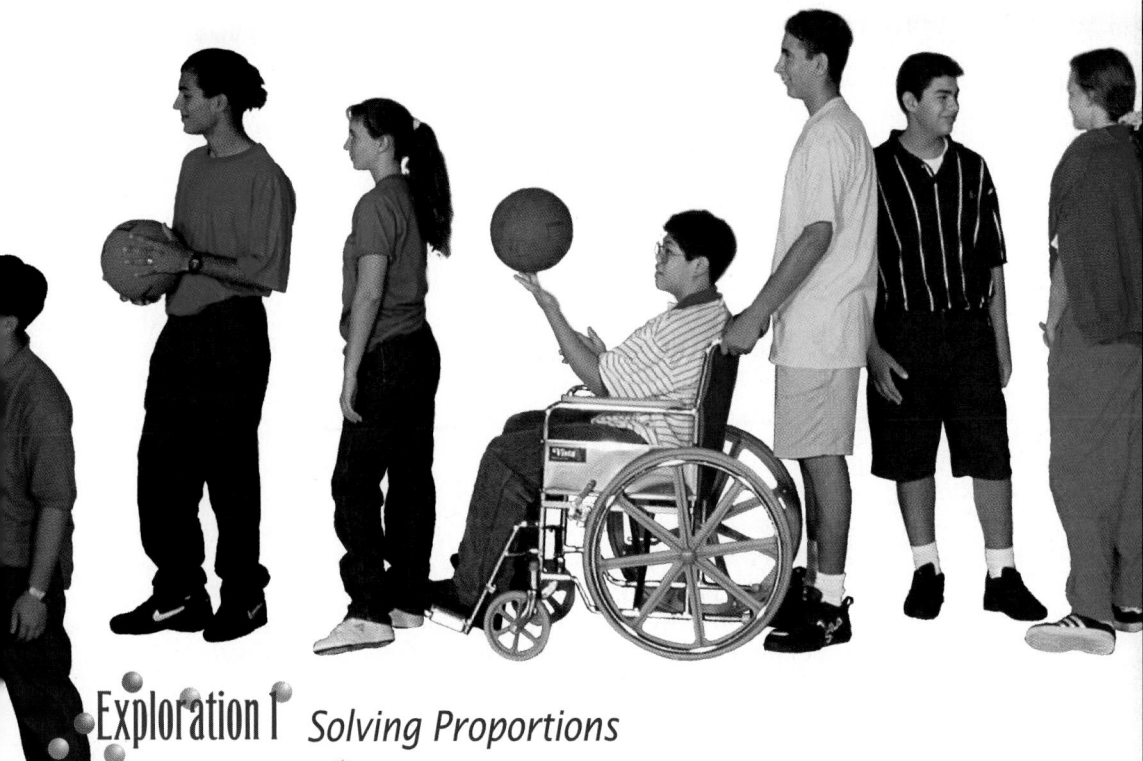

Exploration 1 Solving Proportions

1 For each statement, identify the extremes and the means.

a. $\dfrac{1}{2} = \dfrac{5}{10}$ **b.** $\dfrac{3}{4} = \dfrac{9}{12}$ **c.** $\dfrac{4}{6} = \dfrac{6}{9}$ **d.** $\dfrac{6}{2\sqrt{2}} = \dfrac{3\sqrt{2}}{2}$

2 For each statement find the product of the extremes and the product of the means.

a. $\dfrac{1}{2} = \dfrac{5}{10}$ **b.** $\dfrac{3}{4} = \dfrac{9}{12}$ **c.** $\dfrac{4}{6} = \dfrac{6}{9}$ **d.** $\dfrac{6}{2\sqrt{2}} = \dfrac{3\sqrt{2}}{2}$

3 Make a conjecture about the relationship between the product of the extremes and the product of the means in a proportion.

4 In $\dfrac{a}{b} = \dfrac{c}{d}$, ad and bc are often called the *cross products*. Restate your conjecture from Step 3 using cross products. Use your conjecture to tell which of these equations are proportions.

a. $\dfrac{6}{10} = \dfrac{20}{25}$ **b.** $\dfrac{2 + 4\sqrt{3}}{-4} = \dfrac{11}{2 - 4\sqrt{3}}$ **c.** $\dfrac{24}{15} = \dfrac{160}{100}$

5 Explain how to use your conjecture to solve the proportion $\dfrac{x}{4} = \dfrac{42}{24}$.

6 Use your conjecture to solve each of the following proportions.

a. $\dfrac{4}{b} = \dfrac{6}{b + 3}$ **b.** $\dfrac{c - 6}{7} = \dfrac{1}{c}$

7 How are the proportions in Step 6 the same? different? ❖

Exploration 1 Notes

Exploration 1 is designed to allow students to develop proficiency in the use of proper vocabulary and in working with proportions. Encourage correct use of vocabulary as well as correct mathematical procedures.

Aongoing ASSESSMENT

7. Each proportion contains a binomial term. In part (b), the variable c occurs in each of the means, thus yielding a squared term in the cross product. In part (a), the variable b occurs in one mean and one extreme, thus yielding a linear equation in the cross product.

This should be review for most students. However, be sure that students understand these exercises before continuing with the applications.

TEACHING *tip*

Geometry The length of a rectangle is twice the width. The ratio of the perimeter to the area is $\frac{1}{4}$. Find the dimensions of the rectangle.

[12 units by 24 units]

• Exploration 2 *True or False?*

1 Suppose $\frac{a}{b} = \frac{c}{d}$. Try various nonzero values for a, b, c, and d. Find the cross products in each proportion.

a. $\frac{b}{a} = \frac{d}{c}$ **b.** $\frac{a+b}{c} = \frac{a-b}{c}$ **c.** $\frac{a+b}{b} = \frac{c+d}{d}$

d. $\frac{a}{c} = \frac{b}{d}$ **e.** $\frac{a-b}{b} = \frac{c-d}{d}$ **f.** $\frac{a}{b} = \frac{a+c}{b+d}$

2 Which of the proportions a-f seem to be true for any replacements you make?

3 State a rule that gives the relationship between the original proportion and each true proportion. ❖

APPLICATION

Sports In competing for the best free-throw percentage, Al made 28 out of 40 free throws for 70%. To win the contest, Al needs to shoot 76%. If Al makes the rest of his free throws, how many free throws must he make?

Let x represent how many more successful free throws Al must make. Write a proportion using Al's new percentage.

$$70\% = \frac{28}{40} = \frac{\text{successes}}{\text{number attempted}} \qquad 76\% = \frac{\text{successes} + x}{\text{number made} + x}$$

$$\frac{76}{100} = \frac{28 + x}{40 + x} \qquad \text{Given, } 76\% = \frac{76}{100}$$

$$76(40 + x) = 100(28 + x) \qquad \text{Cross products are equal.}$$

$$3040 + 76x = 2800 + 100x \qquad \text{Distributive Property}$$

$$240 = 24x \qquad \text{Simplify.}$$

$$10 = x \qquad \text{Division Property of Equality}$$

Al must make 10 more consecutive free throws to shoot 76%. ❖

APPLICATION

GEOMETRY
Connection

The measures of the three angles of a triangle are in the ratio 1:2:3. Find the measure of each angle.

Draw a picture. Let the angle measures be x, $2x$, and $3x$. Recall that the sum of the degree measures of the angles of a triangle is 180°.

$$x + 2x + 3x = 180$$
$$6x = 180$$
$$x = 30 \qquad 2x = 60 \qquad 3x = 90$$

The angle measures are 30°, 60°, and 90°. ❖

RETEACHING the lesson

Cooperative Learning Have students work in cooperative learning groups to solve each problem.

1. $\frac{a}{2} = \frac{10}{4}$ $[a = 5]$ 2. $\frac{a-3}{2} = \frac{10}{4}$ $[a = 8]$

3. $\frac{a}{2} = \frac{10-a}{4}$ $\left[a = 3\frac{1}{3}\right]$ 4. $\frac{2}{a} = \frac{a}{32}$ $[a = \pm 8]$

Practice & Apply
14.6 Exploring Proportions

Identify the means and extremes of each proportion.

1. $\frac{3}{n} = \frac{4}{5}$ _____   2. $\frac{x+2}{5} = \frac{4}{x-2}$ _____   3. $\frac{2d-1}{4} = \frac{d}{5}$ _____

Solve for the variable indicated.

4. $\frac{3f-2}{6} = \frac{f}{3}$ _____   5. $\frac{3x}{8} = \frac{6}{x}$ _____   6. $\frac{3(x+4)}{5} = \frac{x+4}{4}$ _____

7. $\frac{b}{3} = \frac{2}{2b-1}$ _____   8. $\frac{x+3}{4} = \frac{1}{2}$ _____   9. $\frac{1}{4y-3} = \frac{-1}{y-2}$ _____

10. The dosage of cough syrup is 2 teaspoons for every 50 lb of body weight. How many teaspoons of cough syrup are required for a person who weighs 170 lb? _____

11. A rocket uses 534,000 gallons of fuel in 2.5 minutes. How much fuel does it use in 10 minutes? _____

12. When engineers design a new car, they first build a model of the car. The ratio of a part on the model to the actual size is 1:5. If the trunk of the car is 1.2 ft long on the model, what is the length of the trunk on the car? _____

The label shows nutrition facts from a producer of pasta.

13. If 1 oz is approximately 28.3495 g, about how many ounces are in the 1-cup serving size? _____

14. If the servings per container is 8, about how many ounces are in the pasta package? _____

Nutrition Facts
Serving Size 1 cup (56g)
Servings per Container 8

Amount per Serving	
Calories 200 Calories from Fat 10	
	%Daily Value*
Total Fat 1g	**2%**
Saturated Fat 0g	**0%**
Cholesterol 0mg	**0%**
Sodium 0mg	**0%**
Total Carbohydrate 39g	**13%**
Dietary Fiber 2g	**8%**
Sugars 1g	
Protein 7g	

Vitamin A 0% • Vitamin C	0%
Calcium 0% • Iron	10%
Thiamin 35% • Riboflavin	15%
Niacin 20%	

A theorem in geometry states that if a line is parallel to one side of a triangle and intersects the other two sides, then it divides the two sides proportionately. Illustrate the theorem with a proportion.

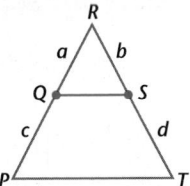

EXERCISES & PROBLEMS

Communicate

1. What is a proportion? Give an example.

2. Explain how to find the means and the extremes of a proportion.

3. Make a statement relating the cross products of the means and the extremes of a proportion.

4. Discuss how to solve for x in the proportion $\frac{1}{2} = \frac{2x + 1}{4}$.

5. Explain how to write a proportion to solve the following. A car travels 320 kilometers on 40 liters of gas. How far can the car travel on 75 liters?

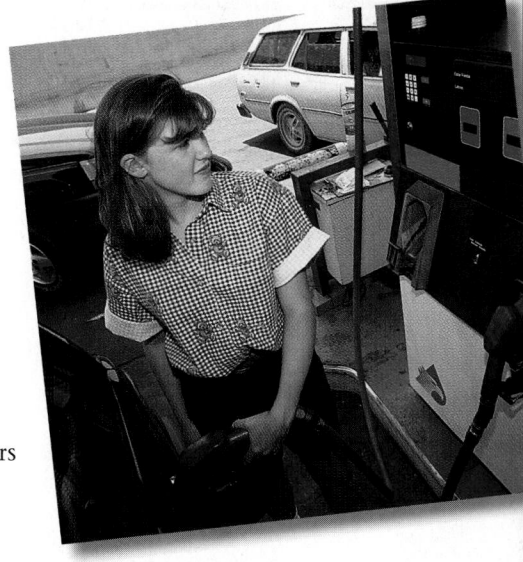

Practice & Apply

Identify the means and extremes of each proportion.

6. $\frac{x}{3} = \frac{7}{4}$

7. $\frac{x}{10} = \frac{2}{5}$

8. $\frac{3}{4} = \frac{y}{5}$

9. $\frac{17}{2} = \frac{8.5}{x}$

10. $\frac{10}{m - 2} = \frac{m}{7}$

11. $\frac{2w - 3}{5} = \frac{w}{3}$

12. $\frac{3}{n + 6} = \frac{2}{n}$

13. $\frac{c + 3}{5} = \frac{c - 2}{4}$

Solve for the variable indicated.

14. $\frac{8}{14} = \frac{4}{x}$ 7

15. $\frac{33}{n} = \frac{3}{1}$ 11

16. $\frac{p}{4} = \frac{4}{16}$ 1

17. $\frac{65}{13} = \frac{x}{5}$ 25

18. $\frac{2}{m - 3} = \frac{5}{m}$ 5

19. $\frac{w}{3} = \frac{w + 4}{7}$ 3

20. $\frac{x - 1}{3} = \frac{x + 1}{5}$ 4

21. $\frac{1}{y - 3} = \frac{3}{y - 5}$ 2

22. $\frac{p}{4} = \frac{10}{p - 3}$ 8, −5

23. $\frac{12}{c} = \frac{c + 4}{1}$ 2, −6

24. $\frac{x}{2} = \frac{30}{x - 4}$ −6, 10

25. $\frac{r}{3} = \frac{56}{r + 2}$ 12, −14

26. $\frac{2}{x} = \frac{x}{9}$ ±3√2

27. $\frac{m - 5}{4} = \frac{m + 3}{3}$ −27

28. $\frac{x}{3} = \frac{6}{2x}$ 3, −3

29. $\frac{2n}{3} = \frac{16}{n + 2}$ −6, 4

6. means: 3, 7; extremes: x, 4
7. means: 10, 2; extremes: 5, x
8. means: 4, y; extremes: 3, 5
9. means: 8.5, 2; extremes: 17, x
10. means: $m - 2$, m; extremes: 10, 7
11. means: 5, w; extremes: 3, $2w - 3$
12. means: 2, $n + 6$; extremes: 3, n
13. means: 5, $c - 2$; extremes: 4, $c + 3$

Performance Assessment

Have students solve the following problems, showing all work and checking all solutions.

1. A plane traveled 200 miles in 25 minutes. At that rate, how long will it take the plane to travel 760 miles? **[95 minutes]**

2. Find the measure of two complementary angles if their measures are in the ratio of $2:3$. **[36°, 54°]**

3. A recent poll showed that 12 out of 18 eligible voters had voted in local elections last year. If there were 2820 people who voted, how many eligible voters were there? **[4230 eligible voters]**

39. 7.45 m

39a. gear ratio $= \dfrac{\text{wheel teeth}}{\text{pedal teeth}}$

$\dfrac{13}{49}, \dfrac{14}{49}, \dfrac{15}{49}, \dfrac{16}{49}, \dfrac{18}{49}, \dfrac{21}{49}, \dfrac{24}{49}$;

$\dfrac{13}{54}, \dfrac{14}{54}, \dfrac{15}{54}, \dfrac{16}{54}, \dfrac{18}{54}, \dfrac{21}{54}, \dfrac{24}{54}$;

Reciprocal $= \dfrac{\text{pedal teeth}}{\text{wheel teeth}}$;

3.77, 3.5, 3.27, 3.06, 2.72, 2.33, 2.04; 4.15, 3.86, 3.6, 3.38, 3, 2.57, 2.25

39b. Answers may vary.

For Exercises 30–32, use a variable to represent the unknown. Write a proportion to solve each problem.

30. Twenty-five feet of copper wire weighs 1 pound. How much does 325 feet of copper wire weigh? 13 pounds

31. The distance between two cities is 750 kilometers. Find how far apart the cities are on a map with a scale that reads "1 centimeter = 250 kilometers." 3 cm

32. In a mixture of concrete, the ratio of sand to cement is $1:4$. How many bags of cement are needed for 100 bags of sand? 400

Solve each problem.

33. **Baking** A recipe for making blueberry muffins requires 1600 grams of flour to make 40 muffins. How many muffins can be made using 1000 grams of flour? 25

34.  **Geometry** Find the measures of two supplementary angles if their measures are in a ratio of $3:6$. The sum of the measures of supplementary angles is 180°. 60°, 120°

35. **Transportation** Mrs. Sanchez used 15 gallons of gasoline to drive on a trip of 450 miles. How far can she drive on a full tank of 20 gallons? 600 miles

It costs $8 to cover a 25-square foot area with wallpaper.

36. **Decorating** How much will it cost to cover a wall with an area of 165 square feet? $52.80

37. A recent poll found that 4 out of 6 people nationwide use Klean toothpaste. How many people can be expected to use this toothpaste in a city of 30,000? 20,000

38. **Travel** A plane has a cruising speed of 650 miles per hour. At this rate, how long does it take the plane to travel 3250 miles on a cross-country flight? 5 hours

39. 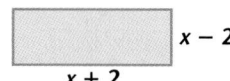 **Portfolio Activity** Complete the problem in the portfolio activity on page 651.

Look Back

40. Solve. $2(x - 3) = 6x$ **[Lesson 4.6]** $-1\frac{1}{2}$

41. Find the slope of the line parallel to $6x - 3y = 24$. **[Lesson 5.6]** 2

42. Simplify $\begin{bmatrix} 2 & -1 \\ 3 & 0 \end{bmatrix}\begin{bmatrix} 1 & 4 \\ -3 & 2 \end{bmatrix}$. **[Lesson 7.3]** $\begin{bmatrix} 5 & 6 \\ 3 & 12 \end{bmatrix}$

43. If x is chosen at random from {3, 4, 5, 7, 9}, what is the probability that the area of this rectangle will exceed 40 square units? $\frac{2}{5}$ **[Lesson 8.1]**

$x - 2$

$x + 2$

Complete the square.

44. $x^2 + 20x$ 45. $y^2 - 10y$ 46. $a^2 - 14a$ 47. $c^2 + c$ **[Lesson 12.3]**

44. $x^2 + 20x + 100$

45. $y^2 - 10y + 25$

46. $a^2 - 14a + 49$

47. $c^2 + c + \dfrac{1}{4}$

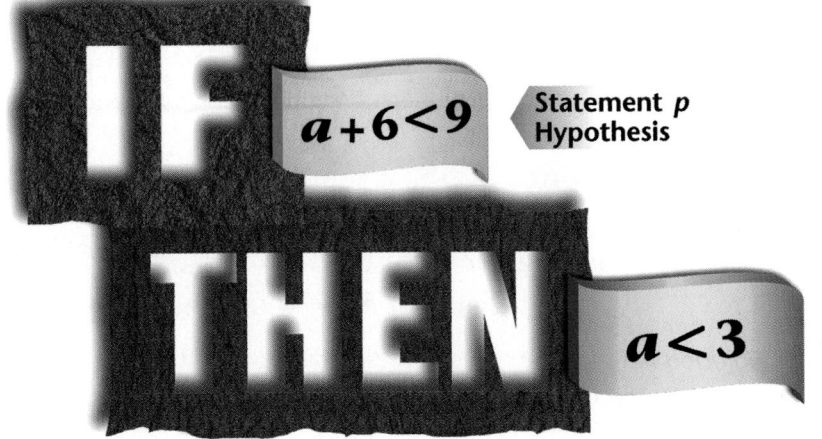

Statement *p*
Hypothesis

$a + 6 < 9$

$a < 3$

Statement *q*
Conclusion

PREPARE

Objectives

- Define *proof, hypothesis, conclusion*, and *converse*.
- Demonstrate proof by using properties and definitions.

RESOURCES

- Practice Master   14.7
- Enrichment Master   14.7
- Technology Master   14.7
- Lesson Activity Master   14.7
- Quiz   14.7
- Spanish Resources   14.7

Throughout this book statements have been made that are assumed to be true. Many of these statements can be proven by using a logical argument.

A statement that has been proven is called a **theorem**. In the *if p-then q* form of a theorem, the **hypothesis**, *p*, is the *if* part, and is assumed to be true. A **proof** provides a logical justification that shows that the **conclusion**, *q*, the *then* part, is also true.

Each step in a proof must be based on established definitions, on axioms or postulates (statements assumed to be true), or on previously proven theorems. Study the proof of Theorem A.

Theorem A	*If a + 6 < 9, then a < 3.*

$a + 6 < 9$	Hypothesis or Given
$a + 6 - 6 < 9 - 6$	Subtraction Property of Inequality
$a < 3$	Simplify.

The conclusion follows logically from the hypothesis, and the theorem is proven.

If you interchange the hypothesis and the conclusion, the new statement is called the **converse** of the original statement. "If *q*, then *p*" is the converse of "if *p*, then *q*." What is the converse of Theorem A? Is it true?

In Lesson 14.6 on proportions, one conjecture made is

$$\text{if } \frac{a}{b} = \frac{c}{d}, \text{ then } \frac{a+b}{b} = \frac{c+d}{d}.$$

Assessing Prior Knowledge

Tell whether each of these statements is true or false.

1. If two angles are complementary, then their sum is 90°. **[true]**
2. If you live in Texas, then you live in Dallas. **[false]**
3. If triangle *ABC* is a right triangle, then it has two acute angles and one right angle. **[true]**

TEACH

 If-then statements are used often in mathematics.

Have students discuss how they have used such statements. Tell students they will learn more about if-then statements in this lesson.

ONGOING ASSESSMENT

If *a* < 3, then *a* + 6 < 9. True

ALTERNATIVE teaching strategy

Cooperative Learning Have students think up if-then (*p-q*)-statements in which *p* is true and *q* is true and the converse is also true. [Example: If *a* = *b*, then *b* = *a*.] Then have stu- dents propose if-then statements in which the converse is not true. [Example: If *a* = *b* and *b* > *c*, then *a* > *c*. Converse: If *a* > *c*, then *a* = *b* and *b* > c; not true because *a* does not have to equal *b*.]

$$\textbf{Hypothesis: } \frac{a}{b} = \frac{c}{d} \qquad \textbf{Conclusion: } \frac{a+b}{b} = \frac{c+d}{d}$$

If $\frac{a}{b} = \frac{c}{d}$, **then** $\frac{a+b}{b} = \frac{c+d}{d}$.

The conjecture can be proven by adding 1 to each side of the original proportion.

$\frac{a}{b} = \frac{c}{d}$ Hypothesis

$\frac{a}{b} + 1 = \frac{c}{d} + 1$ Addition Property of Equality

$\frac{a}{b} + \frac{b}{b} = \frac{c}{d} + \frac{d}{d}$ $\frac{b}{b} = 1$ and $\frac{d}{d} = 1$, Substitution

$\frac{a+b}{b} = \frac{c+d}{d}$ Definition of Addition for Rational Numbers

EXAMPLE

Assume that you know all of the rules for operating with integers except the rule for multiplying two negatives. Prove the following statement.

$$\text{If} -3(2) = -6, \text{ then } -3(-2) = 6.$$

Solution ➤

Use the Definition of Opposites. Recall that if the sum of two numbers is zero, the numbers are opposites.

$-3(0) = 0$	Zero Property
$-3(-2 + 2) = 0$	$-2 + 2 = 0$, Substitution, Definition of Opposites
$-3(-2) + (-3)(2) = 0$	Distributive Property
$-3(-2) + (-6) = 0$	Hypothesis $-3(2) = -6$, Substitution
$-3(-2) + (-6) + 6 = 0 + 6$	Simplify. Addition Property of Equality
$-3(-2) + 0 = 0 + 6$	$-6 + 6 = 0$, Substitution, Definition of Opposites
$-3(-2) = 6$	Simplify. Addition Property of Zero ❖

DEFINITION OF EVEN NUMBERS

An even number is of the form $2n$, where n is an integer. For example, 16 is even because $16 = 2(8)$.

DEFINITION OF ODD NUMBERS

An odd number is of the form $2n + 1$ where n is an integer. For example, 17 is odd because $17 = 2(8) + 1$.

From the definition of even numbers, you can show that $2k + 2$ is even if k is an integer.

By the Distributive Property, $2k + 2 = 2(k + 1)$. Since k is an integer, $k + 1$ is also an integer. If $n = k + 1$, then $2k + 2$ can be written in the form $2n$, which is an even integer.

From the definition of odd numbers, you can show that $2k + 3$ is odd if k is an integer.

Notice that $2k + 3 = (2k + 2) + 1$. We just showed that $2k + 2$ is even if k is an integer. So, $(2k + 2) + 1$ can be written as $2n + 1$ and is an odd integer. Thus, $2k + 3$ is odd.

Now you can prove a theorem about odd numbers.

THEOREM B
If two numbers are odd, then their sum is even.

Hypothesis: **a and b are odd.**   **Conclusion:** **a + b is even.**

If $\begin{array}{c} \text{2 numbers} \\ \text{are odd,} \end{array}$ then $\begin{array}{c} \text{their sum} \\ \text{is even.} \end{array}$

Proof:

a and b are two odd numbers.	Hypothesis or Given
Let $a = 2k + 1$ and $b = 2p + 1$.	Definition of Odd Numbers (k and p are integers.)
$a + b = (2k + 1) + (2p + 1)$	Addition of Real Numbers
$= 2k + 2p + 2$	Simplify.
$= 2(k + p + 1)$	Distributive Property

$2(k + p + 1)$ is in the form $2n$, where n is equal to $k + p + 1$ and $k + p + 1$ is an integer. Thus, $2(k + p + 1)$ is an even number and Theorem B is proved.

Try This   Prove: If $a + c = b + c$, then $a = b$. Use the Addition Property of Equality and the Definition of Opposites, $-c + c = 0$.

CRITICAL *Thinking*   Suppose you are given the following statement: If I live in Texas, then I live in the United States. Is the converse true? Make up your own *if, then*-statement. Is the converse true? Explain your answer.

Cultural Connection: Africa The use of logic and proof was first formalized in the study of geometry. Sometime around 300 B.C.E., Euclid, a professor of mathematics at the University of Alexandria, wrote *The Elements*. This text used axioms and postulates to prove theorems. His method of proof incorporated what we know today as the rules of formal logic.

Try This

$a + c = b + c$
Given
$a + c - c = b + c - c$
Addition Property of Equality
$a + 0 = b + 0$
Definition of Opposites
$a = b$
Addition Property of Zero

CRITICAL *Thinking*

The converse statement would be "If I live in the United States, then I live in Texas." The converse statement is not always true. Some Americans live in Texas but many live in other states.

Use Transparency 69

ASSESS

Selected Answers
Odd-Numbered Answers 7–41

Assignment Guide
Core 1–16, 18–32 even, 33–42

Core Two-Year 1–42

Technology
Students can use calculators to check their conjectures for Exercises 28–29.

Error Analysis
Students assume the converse is always true. Provide several examples that are obviously false. For instance, if you live in Rhode Island, then you live in New England, and if you live in New England, then you live in Rhode Island.

Practice Master

EXERCISES & PROBLEMS

Communicate

1. What does a proof in algebra provide?
2. Name 3 types of reasons that can be used in proving theorems.
3. Identify the part of an *if, then*-statement that is assumed to be true. What part has to be proven?
4. Explain what is meant by the converse of a statement.
5. Explain how to prove that if $7x + 9 = -5$, then $x = -2$. Give a reason for each step.

Practice & Apply

Give a reason for each step.

Proof	
$3m - 4 = -19$	**Hypothesis or Given**
6. $3m - 4 + 4 = -19 + 4$	_____
7. $3m + 0 = -15$	_____
8. $3m = -15$	_____
9. $\dfrac{3m}{3} = \dfrac{-15}{3}$	_____
10. $1 \cdot m = -5$	_____
11. $m = -5$	_____

Give a reason for each step in the proof that $(a + b) + (-a) = b$.

Proof	
12. $(a + b) + (-a)$	_____
13. $= (b + a) + (-a)$	_____
14. $= b + [a + (-a)]$	_____
15. $= b + 0$	_____
16. $= b$	_____

Prove each of the following. Give a reason for each step.

17. For all y, $2y + 3y = 5y$.
18. For all x, $5x - 3x = 2x$.
19. For all a, $(-3a)(-2a) = 6a^2$.
20. If $7n - 3n = 32$, then $n = 8$.
21. If $3(x + 2) = -15$, then $x = -7$.
22. If $\dfrac{2}{3}x + 5 = -9$, then $x = -21$.
23. If $\dfrac{3}{4}m + 8 = -1$, then $m = -12$.

6. Addition Property of Equality
7. Definition of Opposites
8. Identity Property of Addition
9. Division Property of Equality
10. Reciprocal Property
11. Identity Property of Multiplication
12. Given

13. Commutative Property of Addition
14. Associative Property of Addition
15. Definition of Opposites
16. Identity for Addition

The answers to Exercises 17–23 can be found in Additional Answers beginning on page 729.

Write a proof for each statement. Let all variables represent real numbers. Complete Exercises 24–32.

24. If $a + (b + c)$, then $(a + b) + c$.

25. If $a = b$, then $a + c = b + c$.

26. $a(b - c) = ab - ac$

27. $(ax + b) + ay = a(x + y) + b$

28. The square of an even number is even.

29. The square of an odd number is odd.

30. For all real numbers a and b, $a - b$ and $b - a$ are opposites. HINT: Prove that $(a - b) + (b - a) = 0$ is true. Then refer to the Definition of Opposites.

31. If $x \neq 0$, then $(xy)\dfrac{1}{x} = y$.

32. Let a, b, c, and d be integers with $b, d \neq 0$. Prove that the arithmetic sum $\dfrac{a}{b} + \dfrac{c}{d} = \dfrac{da + bc}{bd}$.

 **Look Back**

33.  **Statistics** What is the average length in centimeters of the hand span from the tip of the right thumb to the tip of the little finger from this sample? 16.2, 18.5, 23.0, 21.2, 22.9, 21.1, 20.6, 19.2 **[Lesson 4.1]** 20.3 cm

34. Solve $x = 16x + 45$. **[Lesson 4.6]** -3

35. Write the equation of the line perpendicular to $2x - 3y = 9$ and passing through the point $(-6, 12)$. **[Lesson 5.6]** $y = -\dfrac{3}{2}x + 3$

36. Solve for x and y. $\begin{bmatrix} 1 & 3 \\ 2 & 1 \end{bmatrix} \cdot \begin{bmatrix} x \\ y \end{bmatrix} = \begin{bmatrix} 5 \\ 0 \end{bmatrix}$ **[Lesson 7.5]** $-1, 2$

37. Describe the transformations to $f(x) = x^2$ for $g(x) = -2(x - 1)^2 + 3$. **[Lesson 9.2]**

38. **Geometry** Find the area of a triangle whose dimensions are represented as base with length $2(x - 4)$ and height $x - 6$. **[Lesson 11.3]**

39. Solve $x^2 - 8 = 2x$. **[Lesson 12.5]** $-2, 4$

40. Solve $\sqrt{x - 5} = -3$. **[Lesson 13.3]** none

41. 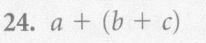 **Geometry** Find the midpoint of the segment joining the points $(-2.5, 4)$ and $(2.5, 6)$. **[Lesson 13.5]** $(0, 5)$

42. Simplify $\dfrac{x^2 - 9}{x + 3}$. **[Lesson 14.3]**

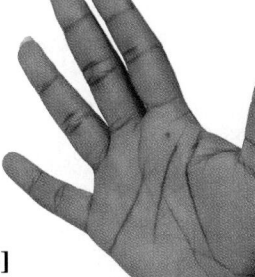

21 cm span

---

24. $a + (b + c)$ Given
$= (a + b) + c$ Associative Property of Addition

25. $a = b$ Given
$a + c = b + c$ Addition Property of Equality

26. $a(b - c)$ Given
$= ab - ac$ Distributive Property

27. $(ax + b) + ay$ Given
$= (ax + ay) + b$ Commutative Property for Addition
$= a(x + y) + b$ Distributive Property

The answers to Exercises 28–32 can be found in Additional Answers beginning on page 729.

A Different Dimension

When you apply mathematics to solve problems, the numbers are only one part of the problem to consider. Basic dimensions such as mass, length, force, time, and temperature contribute essential information about the problem. Some dimensions are derived from powers or combinations of dimensions. The measurement of these dimensions uses various units such as feet or centimeters to identify relative size.

Part of dimensional analysis is knowing how to convert measurements from one unit to another.

1 mile	= 5280 feet
1 hour	= 60 minutes
1 minute	= 60 seconds

A test car travels on the track at 200 *miles per hour*. What is the equivalent speed in feet per second? Express the original speed in terms of its dimensions.

$$\frac{200 \text{ miles}}{1 \text{ hour}}$$ The *per* indicates that hour is in the denominator.

Express the equivalent dimensions as a fraction, and multiply.

$$\frac{1 \text{ hour}}{60 \text{ minutes}} \cdot \frac{200 \text{ miles}}{1 \text{ hour}}$$

Since units can be treated as factors, the expression can be simplified by dividing identical expressions.

$$\frac{1 \cancel{\text{ hour}}}{60 \text{ minutes}} \cdot \frac{200 \text{ miles}}{1 \cancel{\text{ hour}}}$$

Now simplify, and multiply by the next equivalent expression.

$$\frac{200 \text{ miles}}{60 \cancel{\text{ minutes}}} \cdot \frac{1 \cancel{\text{ minute}}}{60 \text{ seconds}} = \frac{200 \text{ miles}}{3600 \text{ seconds}} = \frac{1 \text{ mile}}{18 \text{ seconds}}$$

Finally, multiply the result by the equivalent fraction to change from miles to feet.

$$\frac{1 \cancel{\text{ mile}}}{18 \text{ seconds}} \cdot \frac{5280 \text{ feet}}{1 \cancel{\text{ mile}}} = \frac{5280 \text{ feet}}{18 \text{ seconds}} = \frac{293.333 \text{ feet}}{1 \text{ second}}$$

Thus, 200 miles per hour is approximately equivalent to 293.333 feet per second.

Dimensional Analysis

Some derived dimensions are powers of a basic dimension. The centimeter is a unit for measuring length. Length squared is area. The units are square centimeters. Length cubed is volume. Its units are cubic centimeters.

You can use dimensional analysis to find the number of square feet in 16 square yards. Begin with the basic conversion 3 feet = 1 yard. This can be written $\frac{3 \text{ feet}}{1 \text{ yard}} = 1$.

$$\frac{16 \text{ yards}^2}{1} \cdot \left(\frac{3 \text{ feet}}{1 \text{ yard}}\right)^2 = \frac{16 \text{ yd}^2}{1} \cdot \frac{9 \text{ ft}^2}{1 \text{ yd}^2} = 144 \text{ square feet}$$

Since 1 inch is approximately 2.54 centimeters, you can use dimensional analysis to relate cubic centimeters and liters to cubic inches.

ACTIVITY 1

Use dimensional analysis to convert 220 kilometers per hour into meters per second. Find at least 3 other ways to express 220 kilometers using different units. Show the steps for converting to those units.

ACTIVITY 2

A racing automobile has a 7.1 liter engine. Use dimensional analysis to determine the cubic inch displacement of the engine. Consider a liter as 1000 cubic centimeters. Find the engine size of 3 different kinds of car and compare the displacement of these engines with the size of a racing engine.

ACTIVITY 3

Find two change-of-dimension problems that you might encounter in your activities, and show how to change the dimensions.

Activity 1: 61.1 m/sec
Activity 2: 433.3 in.³
Activity 3: Answers may vary.

Chapter 14 Review

Vocabulary

conclusion	689	inverse variation	651	proof	689
extremes	684	means	684	rational expression	653
hypothesis	689	proportion	684	rational function	653

Key Skills & Exercises

Lesson 14.1

➤ **Key Skills**

Identify transformations to $f(x) = \dfrac{1}{x}$ needed to graph a rational function. List the value(s) for which the function is undefined.

The transformation of $g(x) = \dfrac{-3}{x-1} - 6$ is a reflection over the x-axis, a vertical stretch by 3, a shift to the right by 1, and a shift down by 6. The vertical asymptote is $x = 1$. The rational expression is undefined when x is 1.

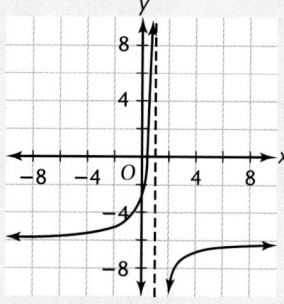

➤ **Exercises**

Describe the transformations to $f(x) = \dfrac{1}{x}$ needed to graph each of the following rational functions. List any values for which the function is undefined.

1. $g(x) = \dfrac{-2}{x} + 4$ **2.** $h(x) = \dfrac{1}{x-3} + 2$ **3.** $g(x) = \dfrac{-3}{x+4} - 1$

Lesson 14.2

➤ **Key Skills**

Identify and use inverse variation.

If y varies inversely as x, and y is 10 when x is 3, solve the equation $10 = \dfrac{k}{3}$ to find the constant of variation, k.

$$10 = \frac{k}{3} \qquad 30 = k$$

Use the constant of variation, 30, to find x when $y = 5$.

$$5 = \frac{30}{x} \qquad 5x = 30 \qquad x = 6$$

1. Stretch by 2, reflect through the x-axis and shift up 4 units; $x \neq 0$.

2. Shift to the right 3 units and up 2 units; $x \neq 3$.

3. Stretch by 3, reflect through the x-axis, shift 4 units left and 1 unit down; $x \neq -4$.

➤ *Exercises*

4. If x varies inversely as y, and $x = 75$ when $y = 20$, find y when $x = 25$.

5. If p varies inversely as q, and $q = 36$ when $p = 15$, find p when $q = 24$.

6. If m varies inversely as n, and $m = 0.5$ when $n = 5$, find m when $n = 50$.

Lesson 14.3

➤ *Key Skills*

Simplify rational expressions.

$$\frac{-x^2 + x + 2}{x^2 + 3x + 2} = \frac{(-1)(x + 1)(x - 2)}{(x + 1)(x + 2)} = \frac{(-1)(x - 2)}{x + 2} = \frac{-x + 2}{x + 2}, x \neq -1, -2$$

➤ *Exercises*

7. $\dfrac{8(x + 1)}{10(x + 1)^2}$ 　　　 **8.** $\dfrac{2x^2 + 6x}{6x - 30}$ 　　　 **9.** $\dfrac{2x^2 - x - 1}{1 - x^2}$

Lesson 14.4

➤ *Key Skills*

Add, subtract, and multiply rational expressions.

a. $\dfrac{5}{x - 3} + \dfrac{2}{x - 5} = \dfrac{5(x - 5)}{(x - 3)(x - 5)} + \dfrac{2(x - 3)}{(x - 5)(x - 3)}$

$$= \frac{5x - 25 + 2x - 6}{x^2 - 8x + 15} = \frac{7x - 31}{x^2 - 8x + 15}$$

b. $\dfrac{3x + 6}{x + 5} \cdot \dfrac{x^2 - 25}{x + 2} = \dfrac{3(x + 2) \cdot (x + 5)(x - 5)}{(x + 5)(x + 2)}$

$$= 3 \cdot \frac{x + 2}{x + 2} \cdot \frac{x + 5}{x + 5} \cdot (x - 5) = 3(x - 5)$$

➤ *Exercises*

Perform the indicated operations, and simplify.

10. $\dfrac{2}{5x} + \dfrac{5}{10y}$ 　　 **11.** $\dfrac{16}{y^2 - 16} - \dfrac{2}{y + 4}$ 　　 **12.** $\dfrac{t^2 - 9}{6} \cdot \dfrac{9}{3 - t}$

Lesson 14.5

➤ *Key Skills*

Solve rational equations. 　　　 $\dfrac{1}{x + 3} + \dfrac{3}{4x} = \dfrac{2}{x}$

$$4x(x + 3)\left(\frac{1}{x + 3} + \frac{3}{4x}\right) = 2 \cdot 4\left(\frac{x}{x}\right)(x + 3)$$

$$\frac{4x(x + 3)}{x + 3} + \frac{3(4x)(x + 3)}{4x} = \frac{x}{x} \cdot 4(x + 3)(2)$$

$$4x + 3(x + 3) = 8(x + 3)$$
$$7x + 9 = 8x + 24$$
$$x = -15$$

➤ *Exercises*

Solve.

13. $\dfrac{z + 2}{2z} + \dfrac{z + 3}{z} = 5$ 　　 **14.** $\dfrac{6 - x}{6x} = \dfrac{1}{x + 1}$ 　　 **15.** $\dfrac{5}{w + 6} - \dfrac{2}{w} = \dfrac{9w + 6}{w^2 + 6w}$

4. 60

5. 22.5

6. 0.05

7. $\dfrac{4}{5(x + 1)}$, $x \neq -1$

8. $\dfrac{x^2 + 3x}{3x - 15}$, $x \neq 5$

9. $\dfrac{2x + 1}{-x - 1}$, $x \neq 1, -1$

10. $\dfrac{4y + 5x}{10xy}$, $x \neq 0, y \neq 0$

11. $\dfrac{24 - 2y}{y^2 - 16}$, $y \neq 4, -4$

12. $\dfrac{-3t - 9}{2}$, $t \neq 3$

13. $\dfrac{8}{7}$

14. $2, -3$

15. -3

Lesson 14.6

➤ *Key Skills*

Solve proportions.

$$\frac{x}{8} = \frac{-2}{x + 10}$$
$$x(x + 10) = 8(-2)$$
$$x^2 + 10x = -16$$
$$x^2 + 10x + 16 = 0$$
$$(x + 8)(x + 2) = 0$$
$$x + 8 = 0 \quad \text{or} \quad x + 2 = 0$$
$$x = -8 \qquad x = -2$$

➤ *Exercises*

Solve.

16. $\frac{x}{3} = \frac{3}{9}$ **17.** $\frac{x+1}{2} = \frac{x-1}{3}$ **18.** $\frac{7}{x+3} = \frac{x}{4}$

Lesson 14.7

➤ *Key Skills*

Prove algebraic statements.

Prove that $a + 2(b + a) = 3a + 2b$.

  Proof

1. $a + 2(b + a)$ Hypothesis
2. $a + 2b + 2a$ Distributive Property
3. $3a + 2b$ Simplify.

➤ *Exercises*

Prove each of the following. Give a reason for each step.

19. If $2a + 3 = 5$, then $a = 1$.   **20.** For all a and b, $(a - b) - (b - a) = 2a - 2b$.

Applications

21. Fitness Millie joins a fitness club. There is a $300 fee to join and then a monthly charge of $20. Find a rational expression to represent Millie's average cost per month for belonging to the fitness club. $\frac{300 + 20x}{x}$, $x \neq 0$

22. Geometry The area of a certain triangle is represented by $x^2 - 2x - 15$ and its height is $x - 5$. Find the expression for the length of the base. $2(x + 3)$

23. It takes Andrea 3 days to hoe her garden. Helen can hoe the same garden in 4 days. Find the time it would take them to hoe the garden working together. $1\frac{5}{7}$

24. Travel A trip that is 189 miles long can take one hour less if Gerald increases the average speed of his car by 12 miles per hour. Find the original average speed of his car. 42 miles per hour

25. A tree that is 75 feet tall casts a shadow that is 40 feet long. Find the height of a tree that casts a shadow 15 feet long at the same time of day. 28.125 feet

16. 1
17. -5
18. $4, -7$
19. Proofs may vary.
20. Proofs may vary.

Chapter 14 Assessment

1. For what value is $f(x) = \frac{1}{x-2}$ undefined? $x = 2$

2. Describe all the transformations to $f(x) = \frac{1}{x}$ needed to graph $g(x) = \frac{-4}{x-5} + 1$. Write the restriction for the variable.

If x varies inversely as y, and $x = 0.2$ when $y = 7$, find

3. y when $x = 4$. 0.35

4. x when $y = 10$. 0.14

5. If 6 feet of wire weigh 0.7 kilograms, how much does 50 feet of wire weigh?   5.83 kilograms

6. Write the common denominator for $\frac{3}{x-2}$ and $\frac{2}{x+4}$. $(x-2)(x+4)$

Simplify.

7. $\frac{4x^2(x-1)}{6x}$

8. $\frac{x-y}{x^2-y^2}$

9. $\frac{5}{6y} + \frac{7}{4y}$

10. $\frac{6}{x-8} - \frac{7}{8-x}$

11. $\frac{4-x^2}{6x} \cdot \frac{3x^2-x}{5x+10}$

12. $\frac{a}{a-3} - \frac{2a}{a+4}$

13. The area of a certain rectangle is represented by $6x^2 + x - 35$ and its length is $2x + 5$. Find the width. $3x - 7$

Solve.

14. $\frac{5}{2y} - \frac{3}{10} = \frac{1}{y}$ 5

15. $\frac{x^2+5}{3} - \frac{6x}{3} = 0$ 1, 5

16. $\frac{2(y-1)}{y+2} + \frac{1}{y+2} = 1$ 3

17. $\frac{4}{7} = \frac{x}{49}$ 28

18. $\frac{x+4}{2} = \frac{7}{4}$ $-\frac{1}{2}$

19. $\frac{x}{3} = \frac{4}{x-1}$ 4, -3

20. A large truck has a load capacity of 3 tons more than a smaller truck. If the ratio of their load capacities is 5 to 2, find the capacity of both trucks.

21. The ratio of a number to 5 more than the same number is 2 to 5. Find the number.

22. A blueprint for a house shows a room as a square that has sides of 3 centimeters. The scale indicates that the actual length of a side of the square room is 12 meters. The actual length of another square room is 16 meters. What is the length of this room in centimeters on the blueprint?

23. Sean and Sam are bricklayers. Sean can build a wall with bricks in 5 hours, and Sam can build a wall of the same size in 4 hours. How long will it take them to build the wall if they work together?

24. A boat can travel 90 kilometers downstream in the same time that it can travel 60 kilometers upstream. If the boat travels 15 kilometers per hour in still water, find the rate of the current.

25. Prove that $3b - (b - a) = 2b + a$. Give a reason for each step.

22. 4 cm

23. $2\frac{2}{9}$ hours

24. 3 km/h

25. Proofs may vary.

---

2. Stretch by a factor 4, reflect in the x-axis, shift 5 units to the right and 1 unit up; $x \neq 5$.

7. $\frac{2x(x-1)}{3}$, $x \neq 0$

8. $\frac{1}{x+y}$, $x \neq \pm y$

9. $\frac{31}{12y}$, $y \neq 0$

10. $\frac{13}{x-8}$, $x \neq 8$

11. $\frac{-3x^2 + 7x - 2}{30}$, $x \neq 0, -2$

12. $\frac{10a - a^2}{a^2 + a - 12}$, $a \neq 3, -4$

20. 2

21. $\frac{10}{3}$

Chapters 1 – 14
Cumulative Assessment

College Entrance Exam Practice

Quantitative Comparison   For Questions 1–4 write

A if the quantity in Column A is greater than the quantity in Column B;

B if the quantity in Column B is greater than the quantity in Column A;

C if the two quantities are equal; or

D if the relationship cannot be determined from the information given.

	Column A	Column B	Answers
1.	the number of $15 tickets you can buy for $90	the number of $12 tickets you can buy for $84	Ⓐ Ⓑ Ⓒ Ⓓ [Lesson 4.3] B
2.	$-18 \cdot -4$	$18 \cdot -2$	Ⓐ Ⓑ Ⓒ Ⓓ [Lesson 4.1] A
3.	$\dfrac{2\sqrt{2}}{\sqrt{5}}$	$\dfrac{2\sqrt{10}}{5}$	Ⓐ Ⓑ Ⓒ Ⓓ [Lesson 13.2] C
4.	$\cos Y$	$\sin Y$	Ⓐ Ⓑ Ⓒ Ⓓ [Lesson 13.7] B

(Figure for 4: right triangle with vertices Y, X, Z; YX = 12, YZ = 20, XZ = 16, right angle at X.)

5. Find $|-3.6|$.   **[Lesson 2.4]** a
   a. 3.6   **b.** -3.6 **c.** 4   **d.** -4

6. Simplify $3\sqrt{2} - \sqrt{5} + 6\sqrt{2} + 9\sqrt{5}$.   **[Lesson 13.2]** d
   a. $9\sqrt{2} + 9$ **b.** $-3\sqrt{2} + 8\sqrt{5}$ **c.** $-3\sqrt{2} + 9$ **d.** $9\sqrt{2} + 8\sqrt{5}$

7. What is the solution to $x^2 = 144$?   **[Lesson 13.3]** c
   a. 12   **b.** 72   **c.** ± 12 **d.** ± 72

8. What is the slope of a line parallel to $2x + 3y = -5$?   **[Lesson 5.6]** b
   a. 2   **b.** $-\dfrac{2}{3}$ **c.** $\dfrac{3}{2}$ **d.** 3

9. What is the probability of choosing a blue chip from a bag containing 4 blue chips and 3 red chips?   **[Lesson 8.5]** c
   a. $\dfrac{3}{4}$ **b.** $\dfrac{4}{3}$ **c.** $\dfrac{4}{7}$ **d.** $\dfrac{3}{7}$

10. What is the simplified form of $(-2m^2n^4)^2$? **[Lesson 10.2]** d
   a. $-4m^4n^6$ **b.** $4m^4n^6$ **c.** $-4m^4n^8$ **d.** $4m^4n^8$

11. Name the type of function whose graph is shown at the right.
[Lesson 2.4]   quadratic

12. Explain how to use the FOIL method to multiply $(x + 4)(2x - 5)$.
Find the product. **[Lesson 11.3]**

13. Factor $4x^2 - 16$. **[Lesson 11.2]** $4(x - 2)(x + 2)$

14. If $A = \begin{bmatrix} 4 & -2 \\ 5 & -3 \end{bmatrix}$ and $B = \begin{bmatrix} -2 \\ 6 \end{bmatrix}$, find AB. **[Lesson 7.3]** $\begin{bmatrix} -20 \\ -28 \end{bmatrix}$

15. Simplify $3w(4w + 5) - (w + 10)$. **[Lesson 4.2]** $12w^2 + 14w - 10$

16. Solve $(x + 1)^2 - 4 = 0$. **[Lesson 12.3]** $1, -3$

17. The first three terms of a sequence are 4, 7, and 12. The second
differences are a constant 2. What are the next three terms?
[Lesson 1.2]   19, 28, 39

18. Find two numbers whose sum is 32 and whose product is 255.
[Lesson 6.6]   15, 17

19. A right triangle has legs with lengths 16 inches and 18 inches. Find
the length of the hypotenuse to the nearest tenth of an inch.
[Lesson 13.4]   24.1 inches

20. Find the distance between $(-2, 3)$ and $(4, 7)$. Give your answer as a
radical in simplest form. **[Lesson 13.5]** $2\sqrt{13}$

Free-Response Grid   The following questions may be answered using a
free-response grid commonly used by standardized test services.

21. Solve $\frac{5}{2} = \frac{2x}{5}$. **[Lesson 14.6]**   6.25

22. Evaluate $g(x) = x^3 - 4x + 7$ for $x = 3$. **[Lesson 12.4]**   22

23. 70% of 45 is ? . **[Lesson 4.5]**   31.5

24. Find the slope of the line through $(-1, -2)$ and $(3, 4)$.
[Lesson 5.3] $\frac{3}{2}$

25. Find INT(5.6). **[Lesson 2.4]**   5

---

12. Multiply the First terms, the Outside terms,
the Inside terms and the Last terms. Then
collect any like terms.
$2x^2 + 3x - 20$

Info Bank

Functions and Their Graphs . . . 704

Tables of Squares and Cubes . . 708

Tables of Square and Cube Roots . 708

Table of Random Digits 709

Glossary 710

Index 718

Credits 727

Additional Answers 729

Functions and Their Graphs

Throughout the text a variety of functions were studied—linear, exponential, absolute value, quadratic, and reciprocal. The simplest form of any function is called the parent function. Each parent function has a distinctive graph. Changes made to the parent function will alter the graph but retain the distinctive features of the parent function. Some changes produce a horizontal or vertical shift, others produce a stretch, and still others a reflection. Pages 704–707 summarize some parent functions and the changes that transform them.

Linear Function

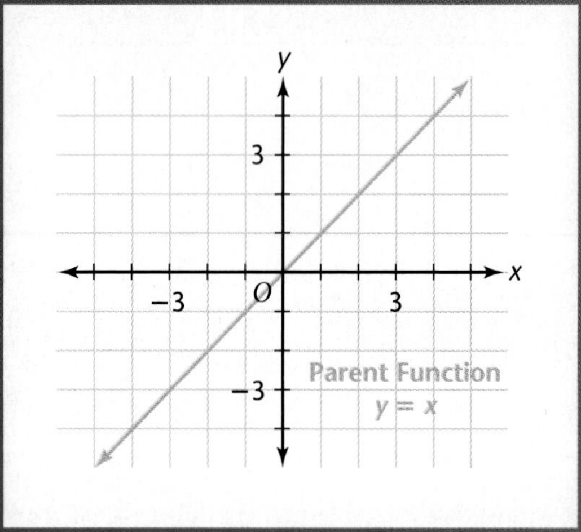

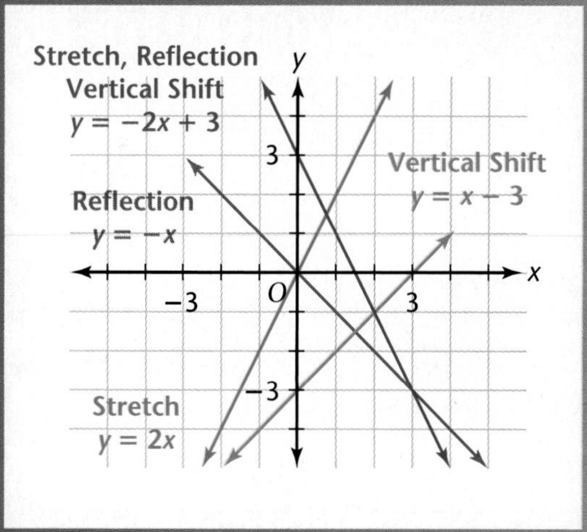

Parent Function
$y = x$

Stretch, Reflection
Vertical Shift
$y = -2x + 3$

Reflection
$y = -x$

Vertical Shift
$y = x - 3$

Stretch
$y = 2x$

Exponential Function

Parent Function
base greater
than 1
$y = 2^x$

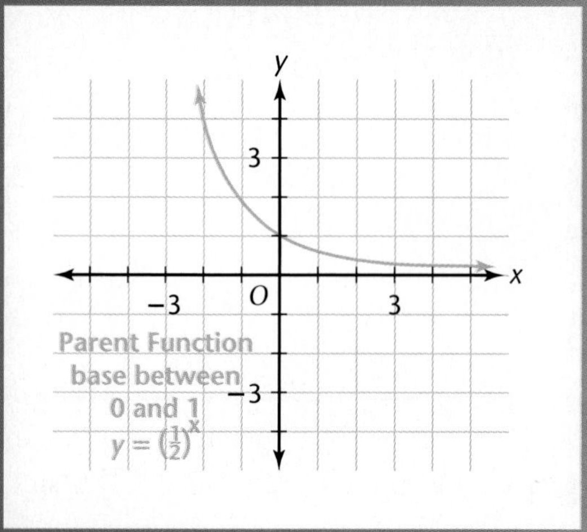

Parent Function
base between
0 and 1
$y = \left(\frac{1}{2}\right)^x$

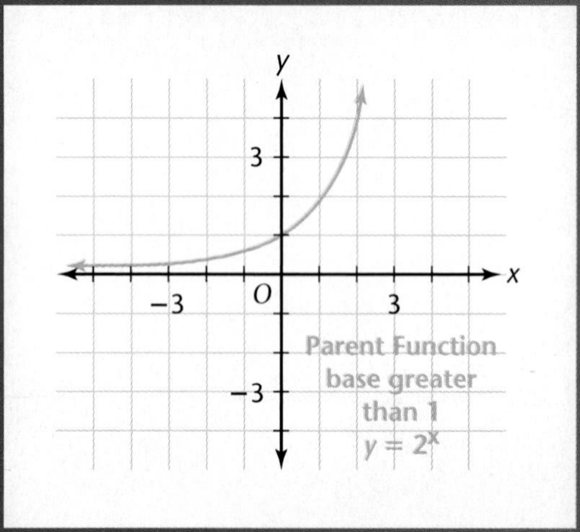

Absolute Value Function

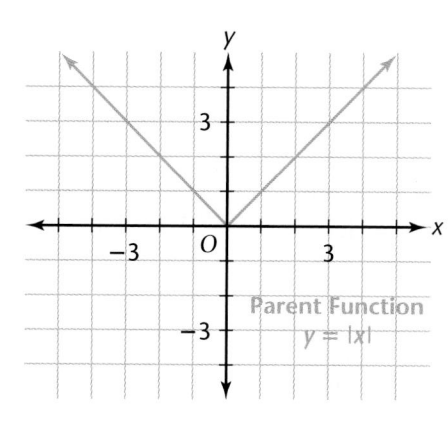

Parent Function
$y = |x|$

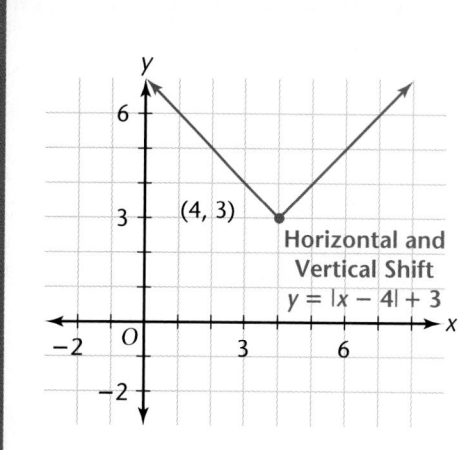

(4, 3)

Horizontal and
Vertical Shift
$y = |x - 4| + 3$

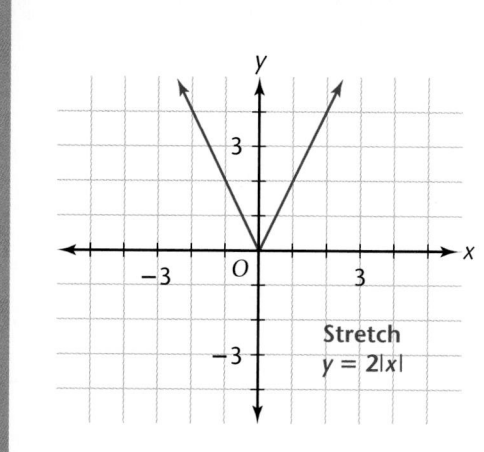

Stretch
$y = 2|x|$

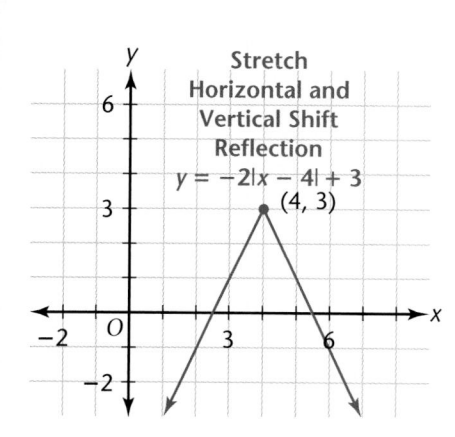

Stretch
Horizontal and
Vertical Shift
Reflection
$y = -2|x - 4| + 3$
(4, 3)

Quadratic Function

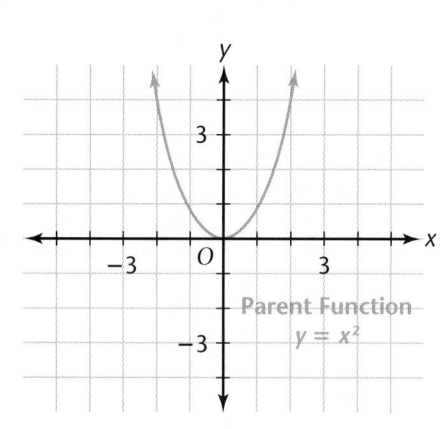

Parent Function
$y = x^2$

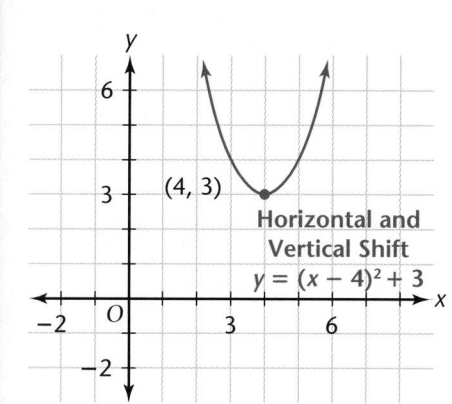

(4, 3)

Horizontal and
Vertical Shift
$y = (x - 4)^2 + 3$

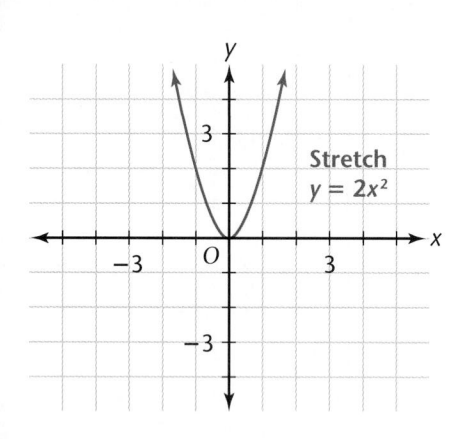

Stretch
$y = 2x^2$

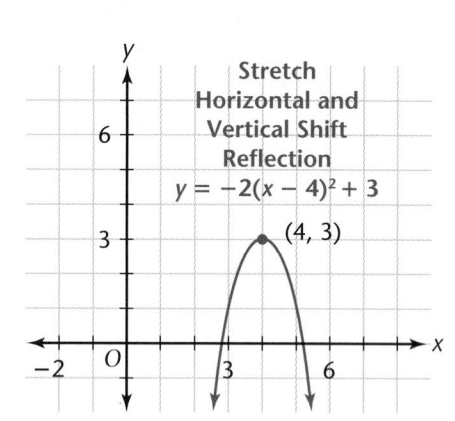

Stretch
Horizontal and
Vertical Shift
Reflection
$y = -2(x - 4)^2 + 3$

(4, 3)

Reciprocal Function

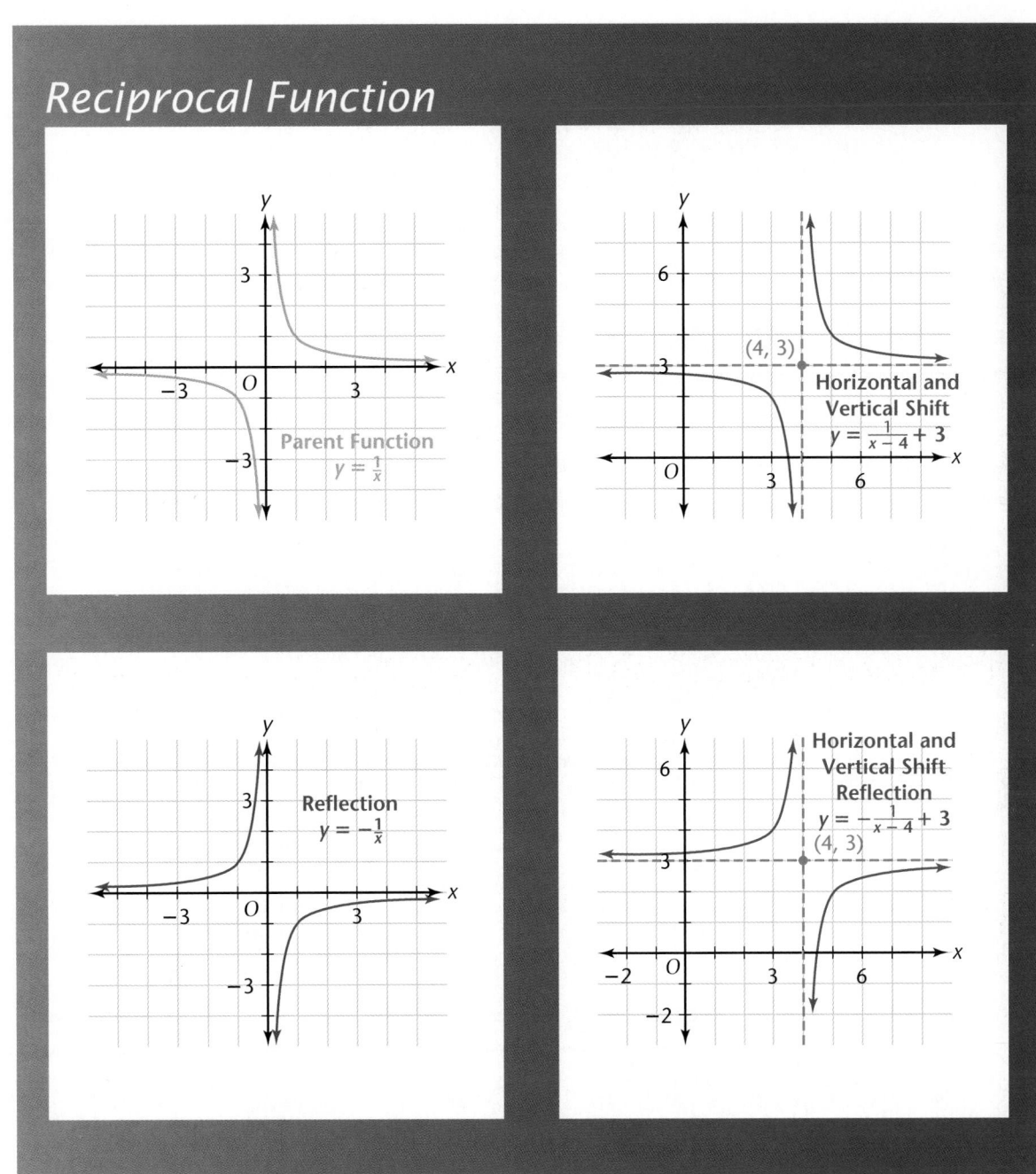

For additional transformations of the reciprocal function, see pages 654–655.

Table of Squares, Cubes, and Roots

No.	Squares	Cubes	Square Roots	Cube Roots	No.	Squares	Cubes	Square Roots	Cube Roots
1	1	1	1.000	1.000	51	2,601	132,651	7.141	3.708
2	4	8	1.414	1.260	52	2,704	140,608	7.211	3.733
3	9	27	1.732	1.442	53	2,809	148,877	7.280	3.756
4	16	64	2.000	1.587	54	2,916	157,464	7.348	3.780
5	25	125	2.236	1.710	55	3,025	166,375	7.416	3.803
6	36	216	2.449	1.817	56	3,136	175,616	7.483	3.826
7	49	343	2.646	1.913	57	3,249	185,193	7.550	3.849
8	64	512	2.828	2.000	58	3,364	195,112	7.616	3.871
9	81	729	3.000	2.080	59	3,481	205,379	7.681	3.893
10	100	1,000	3.162	2.154	60	3,600	216,000	7.746	3.915
11	121	1,331	3.317	2.224	61	3,721	226,981	7.810	3.936
12	144	1,728	3.464	2.289	62	3,844	238,328	7.874	3.958
13	169	2,197	3.606	2.351	63	3,969	250,047	7.937	3.979
14	196	2,744	3.742	2.410	64	4,096	262,144	8.000	4.000
15	225	3,375	3.873	2.466	65	4,225	274,625	8.062	4.021
16	256	4,096	4.000	2.520	66	4,356	287,496	8.124	4.041
17	289	4,913	4.123	2.571	67	4,489	300,763	8.185	4.062
18	324	5,832	4.243	2.621	68	4,624	314,432	8.246	4.082
19	361	6,859	4.359	2.668	69	4,761	328,509	8.307	4.102
20	400	8,000	4.472	2.714	70	4,900	343,000	8.367	4.121
21	441	9,261	4.583	2.759	71	5,041	357,911	8.426	4.141
22	484	10,648	4.690	2.802	72	5,184	373,248	8.485	4.160
23	529	12,167	4.796	2.844	73	5,329	389,017	8.544	4.179
24	576	13,824	4.899	2.884	74	5,476	405,224	8.602	4.198
25	625	15,625	5.000	2.924	75	5,625	421,875	8.660	4.217
26	676	17,576	5.099	2.962	76	5,776	438,976	8.718	4.236
27	729	19,683	5.196	3.000	77	5,929	456,533	8.775	4.254
28	784	21,952	5.292	3.037	78	6,084	474,552	8.832	4.273
29	841	24,389	5.385	3.072	79	6,241	493,039	8.888	4.291
30	900	27,000	5.477	3.107	80	6,400	512,000	8.944	4.309
31	961	29,791	5.568	3.141	81	6,561	531,441	9.000	4.327
32	1,024	32,768	5.657	3.175	82	6,724	551,368	9.055	4.344
33	1,089	35,937	5.745	3.208	83	6,889	571,787	9.110	4.362
34	1,156	39,304	5.831	3.240	84	7,056	592,704	9.165	4.380
35	1,225	42,875	5.916	3.271	85	7,225	614,125	9.220	4.397
36	1,296	46,656	6.000	3.302	86	7,396	636,056	9.274	4.414
37	1,369	50,653	6.083	3.332	87	7,569	658,503	9.327	4.431
38	1,444	54,872	6.164	3.362	88	7,744	681,472	9.381	4.448
39	1,521	59,319	6.245	3.391	89	7,921	704,969	9.434	4.465
40	1,600	64,000	6.325	3.420	90	8,100	729,000	9.487	4.481
41	1,681	68,921	6.403	3.448	91	8,281	753,571	9.539	4.498
42	1,764	74,088	6.481	3.476	92	8,464	778,688	9.592	4.514
43	1,849	79,507	6.557	3.503	93	8,649	804,357	9.644	4.531
44	1,936	85,184	6.633	3.530	94	8,836	830,584	9.695	4.547
45	2,025	91,125	6.708	3.557	95	9,025	857,375	9.747	4.563
46	2,116	97,336	6.782	3.583	96	9,216	884,736	9.798	4.579
47	2,209	103,823	6.856	3.609	97	9,409	912,673	9.849	4.595
48	2,304	110,592	6.928	3.634	98	9,604	941,192	9.899	4.610
49	2,401	117,649	7.000	3.659	99	9,801	970,299	9.950	4.626
50	2,500	125,000	7.071	3.684	100	10,000	1,000,000	10.000	4.642

Table of Random Digits

Line\Col	(1)	(2)	(3)	(4)	(5)	(6)	(7)	(8)	(9)	(10)	(11)	(12)	(13)	(14)
1	10480	15011	01536	02011	81647	91646	69179	14194	62590	36207	20969	99570	91291	90700
2	22368	46573	25595	85393	30995	89198	27982	53402	93965	34095	52666	19174	39615	99505
3	24130	48360	22527	97265	76393	64809	15179	24830	49340	32081	30680	19655	63348	58629
4	42167	93093	06243	61680	07856	16376	39440	53537	71341	57004	00849	74917	97758	16379
5	31570	39975	81837	16656	06121	91782	60468	81305	49684	60672	14110	06927	01263	54613
6	77921	06907	11008	42751	27756	53498	18602	70659	90655	15053	21916	81825	44394	42880
7	99562	72905	56420	69994	98872	31016	71194	18738	44013	48840	63213	21069	10634	12952
8	96301	91977	05463	07972	18876	20922	94595	56869	69014	60045	18425	84903	42508	32307
9	89579	14342	63661	10281	17453	18103	57740	84378	25331	12566	58678	44947	05585	56941
10	85475	36857	53342	53988	53060	59533	38867	62300	08158	17983	16439	11458	18593	64952
11	28918	69578	88231	33276	70997	79936	56865	05859	90106	31595	01547	85590	91610	78188
12	63553	40961	48235	03427	49626	69445	18663	72695	52180	20847	12234	90511	33703	90322
13	09429	93969	52636	92737	88974	33488	36320	17617	30015	08272	84115	27156	30613	74952
14	10365	61129	87529	85689	48237	52267	67689	93394	01511	26358	85104	20285	29975	89868
15	07119	97336	71048	08178	77233	13916	47564	81056	97735	85977	29372	74461	28551	90707
16	51085	12765	51821	51259	77452	16308	60756	92144	49442	53900	70960	63990	75601	40719
17	02368	21382	52404	60268	89368	19885	55322	44819	01188	65225	64835	44919	05944	55157
18	01011	54092	33362	94904	31273	04146	18594	29852	71585	85030	51132	01915	92747	64951
19	52162	53916	46369	58586	23216	14513	83149	98736	23495	64350	94738	17752	35156	35749
20	07056	97628	33787	09998	42698	06691	76988	13602	51851	46104	88916	19509	25625	58104
21	48663	91245	85828	14346	09172	30168	90229	04734	59193	22178	30421	61666	99904	32812
22	54164	58492	22421	74103	47070	25306	76468	26384	58151	06646	21524	15227	96909	44592
23	32639	32363	05597	24200	13363	38005	94342	28728	35806	06912	17012	64161	18296	22851
24	29334	27001	87637	87308	58731	00256	45834	15398	46557	41135	10367	07684	36188	18510
25	02488	33062	28834	07351	19731	92420	60952	61280	50001	67658	32586	86679	50720	94953
26	81525	72295	04839	96423	24878	82651	66566	14778	76797	14780	13300	87074	79666	95725
27	29676	20591	68086	26432	46901	20849	89768	81536	86645	12659	92259	57102	80428	25280
28	00742	57392	39064	66432	84673	40027	32832	61362	98947	96067	64760	64584	96096	98253
29	05366	04213	25669	26422	44407	44048	37937	63904	45766	66134	75470	66520	34693	90449
30	91921	26418	64117	94305	26766	25940	39972	22209	71500	64568	91402	42416	07844	69618
31	00582	04711	87917	77341	42206	35126	74087	99547	81817	42607	43808	76655	62028	76630
32	00725	69884	62797	56170	86324	88072	76222	36086	84637	93161	76038	65855	77919	88006
33	69011	65795	95876	55293	18988	27354	26575	08625	40801	59920	29841	80150	12777	48501
34	25976	57948	29888	88604	67917	48708	18912	82271	65424	69774	33611	54262	85963	03547
35	09763	83473	73577	12908	30883	18317	28290	35797	05998	41688	34952	37888	38917	88050
36	91567	42595	27958	30134	04024	86385	29880	99730	55536	84855	29080	09250	79656	73211
37	17955	56349	90999	49127	20044	59931	06115	20542	18059	02008	73708	83517	36103	42791
38	46503	18584	18845	49618	02304	51038	20655	58727	28168	15475	56942	53389	20562	87338
39	92157	89634	94824	78171	84610	82834	09922	25417	44137	48413	25555	21246	35509	20468
40	14577	62765	35605	81263	39667	47358	56873	56307	61607	49518	89656	20103	77490	18062
41	98427	07523	33362	64270	01638	92477	66969	98420	04880	45585	46565	04102	46880	45709
42	34914	63976	88720	82765	34476	17032	87589	40836	32427	70002	70663	88863	77775	69348
43	70060	28277	39475	46473	23219	53416	94970	25832	69975	94884	19661	72828	00102	66794
44	53976	54914	06990	67245	68350	82948	11398	42878	80287	88267	47363	46634	06541	97809
45	76072	29515	40980	07391	58745	25774	22987	80059	39911	96189	41151	14222	60697	59583
46	90725	52210	83974	29992	65831	38857	50490	83765	55657	14361	31720	57375	56228	41546
47	64364	67412	33339	31926	14883	24413	59744	92351	97473	89286	35931	04110	23726	51900
48	08962	00358	31662	25388	61642	34072	81249	35648	56891	69352	48373	45578	78547	81788
49	95012	68379	93526	70765	10592	04542	76463	54328	02349	17247	28865	14777	62730	92277
50	15664	10493	20492	38391	91132	21999	59516	81652	27195	48223	46751	22923	32261	85653

Source: Interstate Commerce Commission

absolute error The absolute value of the difference between the actual measure, *x*, and the specified measure. (192)

absolute value For any number *x*, if *x* is greater than or equal to 0, $|x| = x$, and if *x* is less than 0, $|x| = -x$. (101)

Addition Principle of Counting Suppose there are *m* ways to make a first choice, *n* ways to make a second choice, and *t* ways that have been counted twice. Then there are $m + n - t$ ways to make the first choice OR the second choice. (378)

Addition Property of Equality If equal amounts are added to the expressions on both sides of an equation, the expressions remain equal. (126)

Addition Property of Inequality If equal amounts are added to the expressions on both sides of an inequality, the resulting inequality is still true. (132)

Addition Property of Zero For any number *a*, $a + 0 = 0 + a$. (107)

additive identity The number that adds to a second number to give a sum that equals the second number. Zero is the additive identity for addition. (107)

additive inverses Two numbers, such as *a* and $-a$, whose sum is 0. (102)

address Each entry in a matrix can be located by its matrix address. (309)

approximate solution A reasonable estimate for a point of intersection for a system of equations. (264)

Associative Property of Addition For all numbers *a*, *b*, and *c*, $(a + b) + c = a + (b + c)$. (104)

asymptotes The lines that a hyperbola will approach, but not touch. (427)

axis of symmetry The line along which a figure can be folded so that the two halves of the figure match exactly. (418)

base In an expression of the form x^a, *x* is the base. (456)

coefficient The number multiplied by a variable. (112)

common binomial factor A binomial factor that is common to all terms of a polynomial. (526)

common monomial factor A monomial factor that is common to all terms of a polynomial. (525)

Commutative Property for Addition For any numbers *a* and *b*, $a + b = b + a$. (114)

composite number A whole number greater than 1 that is not a prime number. (525)

compound interest Money paid or earned on a given amount of money. (60)

conjecture A statement about observations that is believed to be true. (10)

consistent system A system of equations that has one or more solutions. (279)

constant A term in an algebraic expression that represents a fixed amount. (112) In the constant function $y = b$, *b* is the constant. (241)

constant of variation In direct variation of the form $= k$ or $y = kx$, if *y* varies directly as *x*, *k* is called the constant of variation. (228)

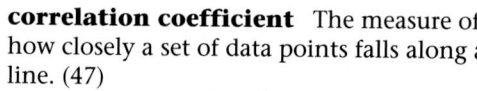

correlation coefficient The measure of how closely a set of data points falls along a line. (47)

degree of a polynomial The degree of a polynomial in one variable is the exponent with the greatest value of any of the polynomial's terms. (507)

dependent events Two events are dependent if the occurrence of the first event does affect the probability of the second event occurring. (396)

dependent system A system of equations that has an infinite number of solutions. The graph of each equation is the same line with the same y-intercepts. A dependent system is called a consistent system. (280)

descending order The order of the terms of a polynomial when they are ordered from left to right, from the greatest to the least degree of the variable. (507)

difference of two squares A polynomial of the form $a^2 - b^2$ that can be written as the product of two factors, $a^2 - b^2 = (a - b)(a + b)$. (531)

direct variation If y varies directly as x, then $\frac{y}{x} = k$ or $y = kx$. The k is called the constant of variation. (228)

discriminant The expression $b^2 - 4ac$ in the quadratic formula. (577)

Distance formula The distance, d, between two points $A(x_1, y_1)$ and $B(x_2, y_2)$ is given by $d = \sqrt{(x_2 - x_1)^2 + (y_2 - y_1)^2}$. (626)

Distributive Property For all numbers a, b, and c, $a(b + c) = ab + ac$ and $(b + c)a = ba + ca$. (113)

Distributive Property Over Subtraction For all numbers a, b, and c, $a(b - c) = ab - ac$, and $(b - c)a = ba - ca$. (153)

Division Property of Equality See Multiplication and Division Property of Equality.

Division Property of Inequality See Multiplication and Division Property of Inequality.

Division Property of Square Roots For all numbers $a \geq 0$ and $b > 0$, $\sqrt{\frac{a}{b}} = \frac{\sqrt{a}}{\sqrt{b}}$. (606)

domain The set of first coordinates, or x-values, in an ordered pair of a function. (413)

elimination method A method used to solve a system of equations in which one variable is eliminated by adding or subtracting opposites. (272)

equally-likely outcomes Outcomes of an experiment that have the same probability of happening. (389)

equation Two equivalent expressions separated by an equal sign. (23)

experimental probability Let t be the number of trials in an experiment. Let f be the number of times an event occurs. The experimental probability, P, of the event is given by $P = \frac{f}{t}$. (354)

exponent The number that tells how many times a number is used as a factor. In an expression of the form x^a, a is the exponent. (457)

exponential decay A situation in which a number is repeatedly multiplied by a number between 0 and 1. (60)

exponential growth A situation in which a number is repeatedly multiplied by a number greater than 1. (60)

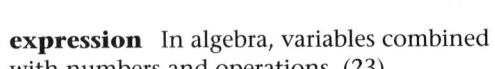

expression In algebra, variables combined with numbers and operations. (23)

extremes In the proportion $\frac{a}{b} = \frac{c}{d}$, a and d are the extremes. (684)

factor Numbers or variables that are multiplied. (112) Numbers or polynomials that are multiplied to form a product. (525)

FOIL method A method for multiplying two binomials.

1. Multiply **F**irst terms.

2. Multiply **O**utside terms. Multiply **I**nside terms. Add products from 2 and 3.

3. Multiply **L**ast terms. (520)

formula An equation that describes a numerical relationship. (127)

function A set of ordered pairs for which no two pairs have the same first coordinate. (411)

greatest common factor (GCF) The greatest factor that is common to all terms of a polynomial. (516)

horizontal translation The graph of $y = f(x - h)$ is translated horizontally by h units from the graph of $y = f(x)$. (438)

hyperbola The graph of the reciprocal function. (427)

hypotenuse The side opposite the right angle in a right triangle. (619)

identity matrix for addition A matrix that is added to a second matrix to give a sum that equals the second matrix. The identity matrix for addition is the zero matrix. (317)

identity matrix for multiplication A matrix that is multiplied by a second matrix to give a product that equals the second matrix. (327)

inconsistent system A system of equations that has no solution. The graphs of the equations are parallel lines with the same slope but different y-intercepts. (279)

independent events Two events are independent if the occurrence of the first event does not affect the probability of the second event occurring. (396)

independent system A system of equations that has one solution. The graphs of the equations are lines that intersect at one point and have different slopes. Independent systems are called consistent systems. (280)

inequality A statement containing one of the signs $<$, $>$, $\leq$, $\geq$, or $\neq$. (132)

integer function A function written in the form $\text{INT}(x)$. The integer function rounds the number x down to the nearest integer. (80)

integers Whole numbers and their opposites. (100)

inverse variation If $x \cdot y = k$, or $y = \frac{k}{x}$ and k is a constant, then y varies inversely as x. The constant k is called the constant of variation. (658)

irrational numbers Numbers whose decimal part never terminates or repeats. Irrational numbers cannot be represented by the ratio of two integers. (597)

like terms Terms that contain the same form of a variable. (112)

line of best fit Represents an approximation of the data on a scatter plot. (46)

linear equation An equation whose graph is a straight line. (37)

linear relationship A relationship in which a number grows by a fixed amount. (60)

literal equation An equation that contains a number of different letters. Many formulas are examples of literal equations. (127)

matrix Data arranged in a table of rows and columns and enclosed by brackets [ ]. The plural of matrix is matrices. (309)

matrix equality Two matrices are equal when their dimensions are the same and their corresponding entries are equal. (311)

matrix equation An equation for a system of linear equations of the form $AX = B$ where A is the coefficient matrix, X is the variable matrix, and B is the constant matrix. (338)

maximum value The y-value of the vertex of a parabola that opens down. (560)

mean A measure of central tendency in which all pieces of data are added and the sum is divided by the number of pieces of data. (368)

means In the proportion $\frac{a}{b} = \frac{c}{d}$, b and c are the means. (684)

median The middle number in a set of data arranged in order. If there are two middle numbers, the median is the mean of the two numbers. (369)

Midpoint Formula The midpoint, M, of segment PQ where $P(x_1, y_1)$ and $Q(x_2, y_2)$ is $M\left(\frac{x_1 + x_2}{2}, \frac{y_1 + y_2}{2}\right)$. (628)

minimum value The y-value of the vertex of a parabola that opens upward. (560)

mode The piece of data that occurs most often. There can be more than one mode for a set of data. (369)

monomial An algebraic expression that is either a constant, a variable, or a product of a constant and one or more variables. (464)

Multiplication Principle of Counting If there are m ways to make a first choice and n ways to make a second choice, then there are $m \cdot n$ ways to make a first choice AND a second choice. (383)

Multiplication and Division Property of Equality If both sides of an equation are multiplied or divided by equal numbers, the results are equal. (161–162)

Multiplication and Division Property of Inequality If both sides of an inequality are multiplied or divided by the same positive number, the resulting inequality has the same solution. If both sides of an inequality are multiplied or divided by the same negative number and the inequality sign is reversed, the resulting inequality has the same solution. (188)

Multiplication Property of Square Roots For all numbers a and b, such that a and $b \geq 0$, $\sqrt{ab} = \sqrt{a}\sqrt{b}$. (606)

multiplicative inverse The reciprocal of a number. (169)

multiplicative inverse matrix The matrix A has an inverse matrix, A^{-1}, if $AA^{-1} = I = A^{-1}A$, where I is the identity matrix. (331)

Multiplicative Property of −1 For all numbers a, $-1(a) = -a$. (153)

negative exponent If x is any number except zero and n is a positive integer, then $x^{-n} = \frac{1}{x^n}$. (470)

opposites Two numbers that are on opposite sides of zero and the same distance from zero on a number line. (100)

order of operations The set of rules for computation.

1. Perform all operations enclosed in symbols of inclusion (parentheses, brackets, braces, and bars) from innermost outward.

2. Perform all operations with exponents.

3. Perform all multiplications and divisions in order from left to right.

4. Perform all additions and subtractions in order from left to right. (29)

ordered pair The address of a point in the rectangular coordinate system, indicated by two numbers in parentheses, (x, y). (35)

origin The point of intersection of the x- and y-axes in the rectangular coordinate system. (35)

parabola The graph of a quadratic function. (66)

parallel lines If two different lines have the same slope, the lines are parallel. If two different lines are parallel, they have the same slope. Two parallel, vertical lines have undefined slope. (247)

parent function The most basic of a family of functions. (418)

percent A ratio that compares a number with 100. (173)

perfect square A number whose square root is a positive integer. (596)

perfect-square trinomial A trinomial of the form $a^2 + 2ab + b^2$ or $a^2 - 2ab + b^2$. The factored form of $a^2 + 2ab + b^2$ is $(a + b)(a + b) = (a + b)^2$, and of $a^2 - 2ab + b^2$ is $(a - b)(a - b) = (a - b)^2$. (529)

perpendicular lines If the slopes of two lines are m and $-\frac{1}{m}$, the lines are perpendicular. If the slope of a line is m, then the slope of a line perpendicular to it is $-\frac{1}{m}$. (250)

point-slope form The form $y - y_1 = m(x - x_1)$ is the point-slope form for the equation of a line. The coordinates x_1 and y_1 are taken from a given point (x_1, y_1), and the slope is m. (235)

polynomial The sum or difference of two or more monomials. (506)

Power-of-a-Power Property If x is any number and a and b are any positive integers, then $(x^a)^b = x^{ab}$. (464)

Power-of-a-Product Property If x and y are any numbers and n is a positive integer, then $(xy)^n = x^n y^n$. (465)

prime number A whole number greater than 1 whose only factors are itself and 1. (525)

prime polynomial A polynomial that has no polynomial factors with integral coefficients except itself and 1. (525)

principal square root The positive square root of a number that is indicated by the radical sign. (598)

product matrix The result of multiplying two matrices. (325)

Product-of-Powers Property If x is any number and a and b are any positive integers, then $x^a \cdot x^b = x^{a+b}$. (463)

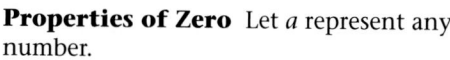

Properties of Zero Let a represent any number.

1. The product of any number and zero is zero. $a \cdot 0$ and $0 \cdot a = 0$

2. Zero divided by any nonzero number is zero. $\dfrac{0}{a} = 0$, $a \neq 0$

3. A number divided by zero is undefined. That is, never divide by zero. (148)

Property of Opposites For any number a, $-a$ is its opposite, and $a + (-a) = 0$. (102)

proportion An equation that states that two ratios are equal. (684)

"Pythagorean" Right–Triangle Theorem Given a right triangle with legs of length a and b and hypotenuse of length c, $c^2 = a^2 + b^2$ or $a^2 + b^2 = c^2$. (620)

quadrant One of the four regions in a coordinate plane. A horizontal and vertical number line divide a coordinate plane into 4 quadrants. (35)

quadratic formula The solutions of the quadratic equation $ax^2 + bx + c = 0$ are $\dfrac{-b \pm \sqrt{b^2 - 4ac}}{2a}$. (575)

quadratic inequality An inequality that contains one or more quadratic expressions. (581)

Quotient-of-Powers Property If x is any number except 0, and a and b are any positive integers with $a > b$, then $\dfrac{x^a}{x^b} = x^{a-b}$. (466)

radical expression An expression that contains a square root. (605)

radical sign The sign $\sqrt{\ \ }$ used to denote the square root. (598)

radicand The number under a radical sign. (604)

range The difference between the highest and lowest numbers in a set of data. (369) The set of second coordinates, or y-values, in an ordered pair of a function. (413)

rate of change The amount of increase or decrease of a function. (221)

ratio The comparison of two quantities. (650)

rational expression If P and Q are polynomials and $Q \neq 0$, then an expression in the form $\dfrac{P}{Q}$ is a rational expression. (653)

rational function A function of the form $y = \dfrac{P}{Q}$, or $f(x) = \dfrac{P(x)}{Q(x)}$. (653)

rational numbers A number that can be expressed as the ratio of two integers with 0 excluded from the denominator. (168)

real numbers The combination of all rational and irrational numbers. (598)

Reciprocal Property For any nonzero number r, there is a number $\dfrac{1}{r}$ such that $r \cdot \dfrac{1}{r} = 1$. (168)

rectangular coordinate system A system in which a plane is divided into four regions by a horizontal and a vertical number line and the addresses, or coordinates, of points are given by ordered pairs. (35)

reflection A transformation that flips a figure over a given line. (431)

regression line The line of best fit for a set of data. (227)

relation Pairs elements from one set with elements of another set. (410)

rise The vertical change in a line. (208)

run The horizontal change in a line. (208)

scalar The number by which each entry in a matrix is multiplied. (324)

scale factor The number by which a parent function is multiplied to create a vertical stretch. (425)

scatter plot A display of data that has been organized into ordered pairs and graphed on the coordinate plane. (40)

scientific notation A number written in scientific notation is written with two factors, a number from 1 to 10, but not including 10, and a power of 10. (476)

simplest form of a radical expression A radical expression is in simplest form if the radicand contains no perfect square factors greater than 1, contains no fractions, and if there is no radical sign in the denominator of a fraction. (605)

simulation An experiment with mathematical characteristics that are similar to the actual event. (360)

slope Measures the steepness of a line by the formula, slope $= \dfrac{\text{rise}}{\text{run}}$. (209)

slope formula Given two points with coordinates (x_1, y_1) and (x_2, y_2), the slope is

$$m = \dfrac{\text{change in } y}{\text{change in } x} = \dfrac{\text{difference in } y}{\text{difference in } x} = \dfrac{y_2 - y_1}{x_2 - x_1} \ .$$

(226)

slope-intercept form The slope-intercept formula or form for a line with slope m and y-intercept b is $y = mx + b$. (225)

square matrix A matrix with equal row and column dimensions. (310)

square root If a is greater than or equal to zero, $\sqrt{a}$ and $-\sqrt{a}$ represent the square root of a. Each square root of a has the following property. $\sqrt{a} \cdot \sqrt{a} = a \quad (-\sqrt{a}) \cdot (-\sqrt{a}) = a$

standard form An equation in the form $Ax + By = C$ is in standard form when A, B, and C are integers, A and B are not both zero, and A is not negative. (233)

standard form of a polynomial A polynomial is in standard form when the terms of the polynomial are ordered from left to right, from the greatest to the least degree of the variable. (507)

stretch A transformation in which the graph of a parent function is stretched vertically by the amount of the scale factor. (424)

substitution method A method used to solve a system of equations in which variables are replaced with known values or algebraic expressions. (267)

Subtraction Property of Equality If equal amounts are subtracted from the expressions on both sides of an equation, the expressions remain equal. (125)

Subtraction Property of Inequality If equal amounts are subtracted from the expressions on both sides of an inequality, the resulting inequality is still true. (132)

symmetric A figure is symmetric with respect to a line called the axis of symmetry if the figure can be folded along this axis and the left and right halves of the figure match exactly. (418)

system of equations Two or more equations in two or more variables. (261)

system of linear inequalities Two or more linear inequalities in two or more variables. The solution for a system of linear inequalities is the intersection of the solution sets for each inequality. (285)

tangent function In a right triangle, the ratio of the length of the leg opposite an acute angle of the triangle to the length of the leg adjacent to the acute angle. (638)

term Each number in a sequence. (10) The numbers, variables, or product or quotient of numbers and variables that are added or subtracted in an algebraic expression. (112)

theoretical probability Let n be the number of equally-likely outcomes in an event. Let s be the number of successful outcomes in the same event. Then, the theoretical probability that the event will occur is $P = \frac{s}{n}$. (389)

transformation A variation such as a stretch, reflection, or shift of a parent function. (418)

translation A transformation that shifts the graph of a function horizontally or vertically. (436)

tree diagram A diagram that is used to find all the possible choices in a situation in which one choice AND another choice need to be made. (382)

trial The number of times an experiment is conducted. (353)

variable A letter or other symbol that can be replaced by any number or other expression. (23)

variable matrix A matrix containing the variables of a system of equations. (338)

Venn diagram A diagram used to represent the relationships among different sets of data. (376)

vertex The point where a parabola or absolute value function changes direction. (66)

vertical line test A test used to determine whether the graph of a relation is a function. A relation is a function if any vertical line intersects the graph of the relation no more than once. (412)

vertical translation The graph of $y = f(x) + k$ is translated vertically by k units from the graph of $y = f(x)$. (437)

***x*-axis** The horizontal number line in the rectangular coordinate system. (35)

***x*-coordinate** A point on the x-axis in the rectangular coordinate system. (35)

***y*-axis** The vertical number line in the rectangular coordinate system. (35)

***y*-coordinate** A point on the y-axis in the rectangular coordinate system. (35)

***y*-intercept** The point where a line crosses the y-axis. (224)

zero exponent If x is any number, $x^0 = 1$. (470)

zero matrix A matrix that is filled with zeros. The zero matrix is the identity matrix for matrix addition. (317)

Zero Product Property If a and b are real numbers such that $ab = 0$, then $a = 0$ or $b = 0$. (567)

INDEX

Definitions of boldface entries can be found in the glossary.

Absolute error, 192–193
Absolute value, 101, 192
  distance and, 194
  equations, 192–194
  function, 79-80, 87, 419, 421
  inequalities, 195–196
Addition
  equations, 124–125, 164
  of expressions, 112–115
  of integers, 101–104
  of matrices, 315–318
  of polynomials, 508–509
  of rational expressions, 670–672
  principle of counting, 375–378
  properties
    associative, 104, 318
    associative, of matrices, 318
    commutative, 114, 318
    commutative, of matrices, 318
    of equality, 126
    of inequality, 132
    of zero, 107
Addition Principle of
  Counting, 375–378
Addition Property of Equality,
  126
Addition Property of
  Inequality, 132
Addition Property of Zero, 107
Additive identity, 107
Additive inverses, 102
Address, 309
Agnesi, Maria Gaetana, 488
Ahmes papyrus, 199, 277
Algebraic expression, 112, 151
Algebraic logic, 29
AND, 376–377
Applications, 25, 41–43, 148, 189
  business and economics
    accounting, 276, 580
    advertising, 33, 370, 373
    banking, 146, 150
    business, 106, 312, 348, 349,
      452, 453
    clothing manufacturing, 116
    cost of living, 43
    discount sales, 178, 518
    discounts, 176
    economics, 481

Applications *(cont.)*
    equipment rental, 257
    finance, 60, 62, 63, 94, 142,
      276, 490
    fund-raising, 27, 70, 75, 122,
      125, 172, 234, 238, 239,
      271, 291, 325, 330
    income, 276
    inventory, 123
    inventory control, 316, 319,
      324, 342, 384
    investment, 89, 127, 168,
      268, 276, 661, 663
    manufacturing, 329
    marketing, 404
    money, 293
    packaging, 447
    pricing, 329
    quality control, 192
    sales tax, 177, 178, 329, 539
    small business, 157, 521
    stocks, 348
    wages, 156
    work schedule, 336
  language arts
    communicate, 12, 19, 26, 32,
      38, 43, 48, 62, 67, 73, 81,
      87, 105, 110, 115, 121,
      128, 135, 149, 155, 164,
      171, 178, 184, 189, 197,
      212, 221, 230, 237, 243,
      250, 264, 270, 275, 281,
      289, 296, 312, 318, 328,
      334, 341, 357, 364, 371,
      379, 386, 391, 397, 415,
      421, 428, 433, 440, 446,
      460, 466, 471, 480, 487,
      493, 510, 517, 522, 527,
      532, 538, 550, 556, 561,
      571, 579, 584, 601, 609,
      616, 667, 673, 682, 687,
      692
    Eyewitness Math, 76, 158,
      216, 322, 366, 474, 564,
      676
    language, 387, 392
    language arts, 244
  life skills
    appointments, 363, 394
    auto painting, 683
    baking, 688
    budget, 244

Applications *(cont.)*
    budgeting time, 291
    career options, 283
    carpentry, 160, 165
    consumer ecomonics, 26, 89,
      172, 175, 185, 191, 337
    consumer math, 68, 82, 129,
      231
    decorating, 688
    fitness, 698
    health, 39, 197, 393
    home economics, 21, 82, 129
    home improvement, 28, 33
    house construction, 642
    investment, 491, 501
    job opportunities, 342
    landscaping, 623
    membership fees, 359
    nutrition, 462
    parenting, 399
    savings, 171, 381, 669
    scheduling, 382, 387, 394,
      399
    surveying, 634
    woodworking, 195
  miscellaneous
    age, 295
    aviation, 270
    highways, 642
    logic, 376
    mixture, 292
    number theory, 184, 283, 473
    numeration, 270, 294, 297
    opinion polls, 380
    rescue services, 630
    survey, 391
    time, 500
    transportation, 75, 89, 308,
      313, 656, 663, 688
  science
    astromomy, 476, 481, 500
    bacteriology, 459
    biology, 461, 668
    chemistry, 296, 302, 343, 387
    ecology, 24, 222, 658, 674
    genetics, 522
    meteorology, 362, 372, 374
    optics, 584
    physical science, 222, 232,
      246
    physics, 18, 20, 66, 69, 84, 94,
      105, 130, 163, 435, 478,

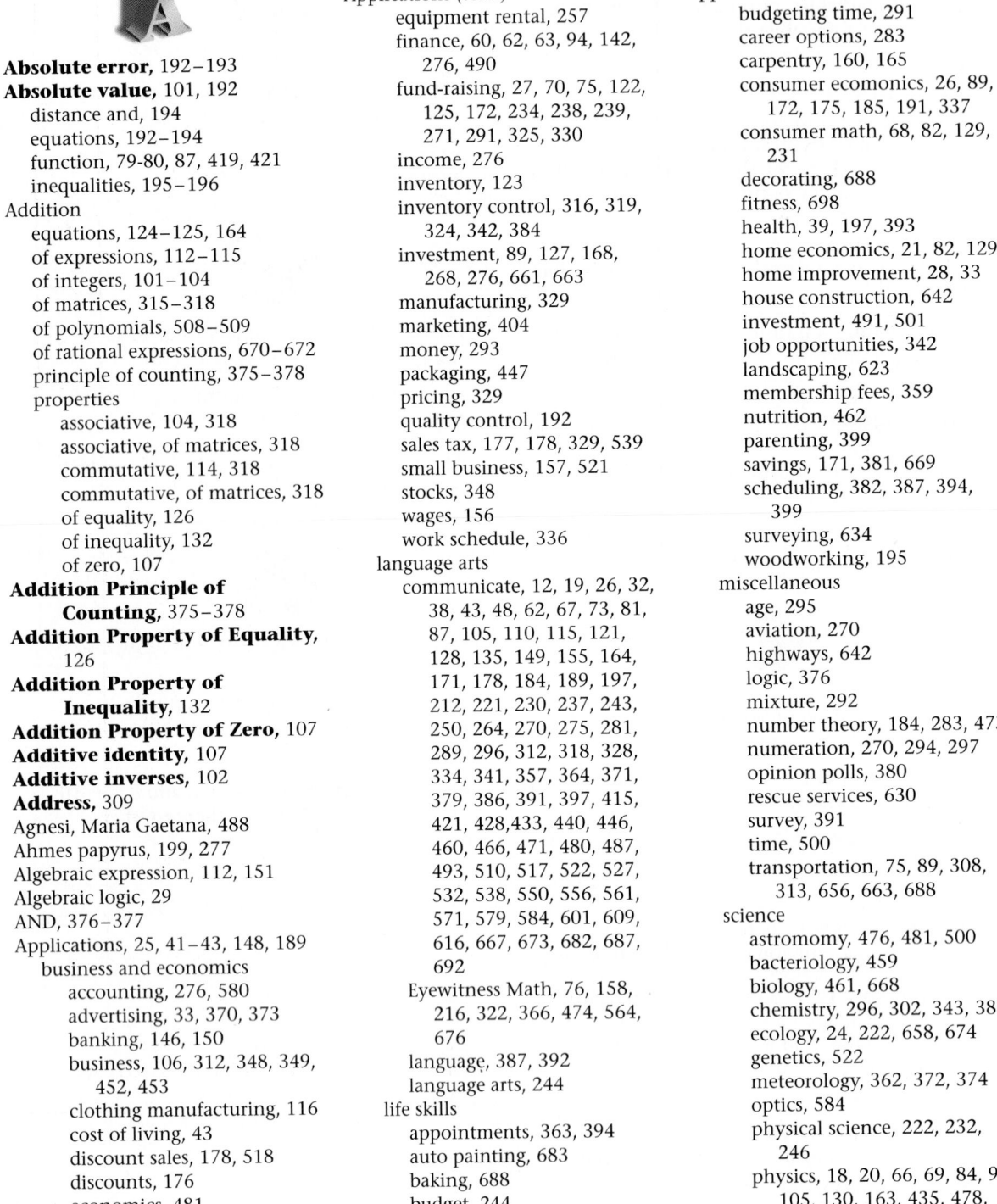

Applications *(cont.)*
528, 551, 552, 557, 562,
563, 573, 585, 611, 617,
618, 648, 660, 662
physiology, 37
science, 209, 251, 336, 392,
417, 469, 481, 489
space exploration, 313
space travel, 477
temperature, 102, 110, 136,
452
weather, 364
social studies
demographics, 42, 59, 85,
179, 283, 364, 487, 488,
492, 501
geography, 44, 313, 380, 392
government, 198
music history, 46
politics, 683
social studies, 215, 239, 480,
483, 485, 501
sports and leisure
art, 429
auto racing, 226
ballet, 189, 190
boating navigation, 621
contests, 399
crafts, 544
entertainment, 27, 54
games, 352, 358, 378, 379,
385, 386, 390, 395, 396,
397
hobbies, 142, 653
model airplanes, 642
music, 72, 74, 663
photography, 572
picture framing, 116
recreation, 213, 240, 245,
296, 398, 683
sports, 45, 88, 102, 129, 136,
162, 171, 172, 179, 202,
214, 226, 266, 291, 319,
360, 368, 373, 381, 383,
590, 617, 648, 678,
686
ticket sales, 342
travel, 129, 165, 185, 210,
256, 277, 294, 443, 468,
688, 698
Approximate solution, 264
Aristotle, 72

Assessment, Ongoing (See
*Communicate, Critical
Thinking, Try This.*)
Chapter Assessment, 55, 95,
143, 203, 257, 303, 349,
405, 453, 501, 545, 591,
649, 699
Chapter Review, 52, 92, 140,
200, 254, 300, 346, 402,
450, 498, 542, 588, 646,
696
Cumulative Assessment/
College Entrance Exam
Practice, 96, 204, 304, 406,
502, 592, 700
**Associative Property of
Addition,** 104
Asymptotes, 427
Axis of symmetry, 418, 548, 555

Banneker, Benjamin, 657
Base, 456
Binomial, 507
Boundary line, 284–286

Calculator, 24, 28, 29, 31, 109 (Also
see Graphics Calculator.)
algebraic logic, 29
constant feature, 10
EE or Exp key, 479
enter key, 11
equals key, 11
extension, 459
greatest integer function, 80
pattern exploration, 31
scientific, 30
square root, 598
TAN key, 639
TAN^{-1} key, 640
Cartesian coordinate system, 35
Chebyshev, 190

Classic applications
age, 295
mixture, 292–293
money, 293
numeration, 294
travel, 294–295
Coefficient, 112
Coefficient matrix, 338
Combinations, 400–401
Common binomial factor,
526
Common monomial factor, 525
Common solution for a system of
equations, 261
Communicate (See *Applications,
language arts*)
**Commutative Property for
Addition,** 114
Complement, 397
Completing the square, 558–561
Composite numbers, 525
Composite of a function, 449
Compound interest, 60, 490
Computers, 28, 128, 181 (Also see
Spreadsheets and *Technology.*)
Conclusion, 689
Conjecture, 10
Connections (See *Applications,
Cultural Connections, Math
Connections*)
Consistent system, 279–280
Constant, 112, 220, 241
Constant function, 241
Constant matrix, 338
Constant of variation, 228–229
Constant polynomial, 507
Converse, 627, 689
Converse of the "Pythagorean"
Right-Triangle Theorem,
627
Coordinate geometry, 632–634
Correlation, 40
little or no, 41
strong negative, 41
strong positive, 41
Correlation Coefficient, 47
Counting Principles
Addition, 378
Multiplication, 383
Critical Thinking, 10, 18, 25, 28, 35,
43, 48, 60, 65, 72, 80, 85, 104,
108, 114, 120, 124, 134, 147,

Critical Thinking, *(cont.)*
  155, 163, 168, 170, 175, 181,
  188, 194, 196, 221, 225, 235,
  243, 249, 261, 268, 279, 288,
  295, 310, 317, 333, 339, 353,
  356, 363, 370, 376, 385, 390,
  397, 411, 427, 432, 444, 458,
  459, 464, 465, 470, 471, 478,
  483, 491, 509, 510, 516, 521,
  525, 530, 532, 538, 549, 554,
  567, 576, 582, 598, 600, 608,
  616, 622, 627, 635, 655, 659,
  661, 665, 673, 679, 687,
  691
Cross products, 685
Cubic polynomial, 507
Cultural Connections
  Africa, 27, 111, 214, 252–253,
    440
  Egypt, 34, 62, 123, 185, 199,
    277, 354, 440, 635, 659
  Americas, 172
    African Americans, 657
    Guatemala, 314, 539
    Mexico, 214
    Montana, 675
    Ojibwa (Chippewa) Indians,
      352
  Asia, 297
    Babylon, 89, 574, 619
    China, 103, 271, 281, 283,
      341, 343, 434, 619
    India, 619
  Europe, 14, 35, 72, 540–541
    England, 478
    France, 473
    Greece, 624
    Italy, 488
    Russia, 190

Databases, 138–139, 375–377, 416
  files, 375
  record, 375
de Padilla, D. Juan Joseph, 314, 539
Degree of a polynomial, 507
Dense set, 598
Density, 163
Dependent events, 396

Dependent system, 280
Descartes, Rene, 35
Descending order, 507
Difference of two squares,
  531–532
Differences
  constant, 16
  finite, 586
  first, 16–18, 66–67, 84
  second, 16–18, 66–67, 84
Dimensions of a matrix, 309
Diophantus, 27, 111, 252–253
Direct variation, 228–229
Discriminant, 577–578
Disjoint sets, 376
Distance formula, 625–627
Distributive Property, 113,
  514–516, 519–520
**Distributive Property Over
  Subtraction,** 153
Division
  equations, 162–164
  of expressions, 154–155
  of integers, 147–148
  of monomials, 466
  of rational expressions, 673
Division Property of Equality,
  161
**Division Property of
  Inequality,** 188
**Division Property of Square
  Roots,** 606
Domain, 37, 413

Elimination method, 272–274
Entry in a matrix, 309
Equality
  addition property of, 126
  division property of, 161
  multiplication property of, 162
  subtraction property of, 125
Equally-likely outcomes, 389
Equation method, 176
Equations, 23
  absolute value, 192–194
  addition, 124–125
  division, 162–164
  linear, 37, 224–227, 233–237

Equations, *(cont.)*
  literal, 127
  matrix, 337–341
  multiplication, 160–161
  multistep, 180–183
  quadratic, 552–555, 558–561,
    566–570, 574–578
  subtraction, 126
  systems of, 260–264, 267–269,
    272–274, 278–281
Even numbers, 690
Events, 353
  independent, 394–397
Expanded notation, 457
Experimental probability,
  352–356
Explorations
  A Diminishing Sequence, 24
  A Fair Game With Two Number
    Cubes, 354
  A Reflection, 431
  Addition and Subtraction
    Properties, 317
  Angle and Tangents, 639
  Basketball Simulation, 361
  Changing the Slope, 229
  Counting from a Venn Diagram,
    377
  Defining a Variable, 22–23
  Defining x^0, 470
  Dependent Systems, 279
  Dividing Inequalities, 188
  Dividing Integers, 147
  Dividing Powers, 458
  Edges and Surface Area, 64–65
  Estimating Square Roots, 597
  Evaluating Your Options, 58
  Find a Correlation, 40–41
  Finding a Value, 268
  Finding Products, 515
  Finding the Distance Formula,
    626
  Finding the Formula, 633
  Finding the Midpoint Formula,
    628
  Fitting the Line to a Point, 219
  From Product Rectangles to
    Factors, 516
  Graphing, 577
  Graphing Changes, 7
  Graphing Linear Inequalities,
    284

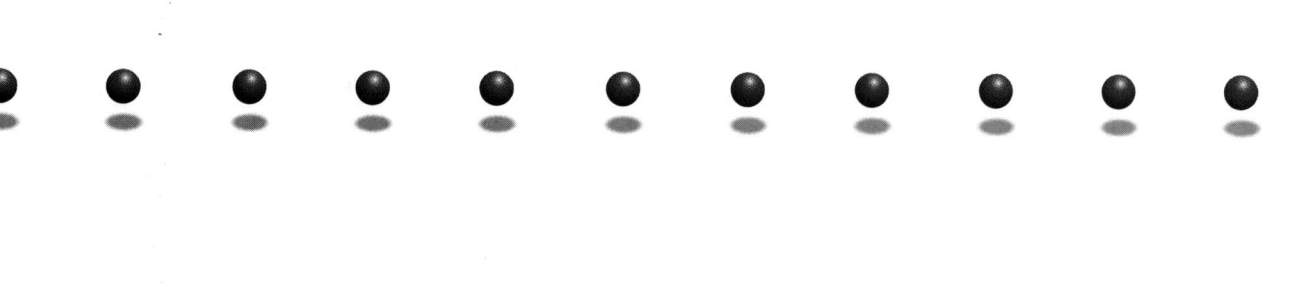

Explorations, *(cont.)*
Graphing Sequences, 66–67
Growth and Decay, 61
How Far From Zero?, 79
How Many Ones?, 559
Interpreting a Venn Diagram, 377
Introducing a Constant, 220
Margin of Error, 371
Modeling With Tiles, 514
Multiplying and Dividing Inequalities, 186–187
Multiplying Binomials, 520
Multiplying by −1, 153
Multiplying Inequalities, 187
Multiplying Integers, 146–147
Multiplying Powers, 457
Multiplying With Tiles, 519
Other Systems, 280
Parabolas and Constant Differences, 548
Parabolas and Polynomials, 549
Parabolas and Transformations, 549
Patterns in Addition, 101
Perpendicular Slope on a Calculator, 248–249
Prime Numbers, 524–525
Probability and the Coin Game, 352–353
Probability From a Number Cube, 353
Production and Profit, 581
Segments and Midpoints, 634
Slopes and Perpendicular Lines, 248
Solving an Inequality, 132
Solving Proportions, 685
Substituting Expressions, 268
Subtraction With Tiles, 108
The Diagonals of a Rectangle, 633
The Equation of a Circle, 633
The Multiplicative Identity Matrix, 327
The Power of a Power, 458
The Procedure for Multiplying Matrices, 326
Transformations of Absolute Value Functions, 419
Transformations of Quadratic Functions, 420

Explorations, *(cont.)*
Transforming the Square Root Function, 599
True or False?, 686
Understanding the Multiplication of Matrices, 325
Using Negative Tiles, 515
Using Opposites, 272
Weather Simulation, 362
What is the Cost?, 267
Exponent, 457
negative, 469–470
zero, 470–471
Exponential decay, 60, 492
Exponential form, 457
Exponential function, 58–61, 87, 483–486
applications of, 489–493
Exponential growth, 60, 492
Expressions, 23
adding, 112–115
algebraic, 112
dividing, 154–155
multiplying, 152–154
rational, 652–655
subtracting, 118–121
Extremes, 684

Factor(s), 112, 525
common, 524–526
common binomial, 526
common monomial, 525
greatest common, 516
Factored form of a polynomial, 516
Factoring
a difference of two squares, 531
a perfect-square trinomial, 530
by grouping, 526
the greatest common factor, 525–526
trinomials, 535–538
by guess-and-check, 537–538
Fermat, Pierre de, 473

Finite difference, 586–587
FOIL method, 520–521
Formula, 127
Fractals, 8
Function(s), 84–87, 410–415
absolute value, 79–80, 419, 421
composite, 449
constant, 241
domain of, 413
evaluating, 413–414
exponential, 59–60, 421, 483–486, 489–493
integer, 80–81, 421
linear, 34, 36, 37, 218–221, 421
notation, 413
parent, 418–421
piecewise, 86–87
quadratic, 64–67, 420–421
range of, 413
rational, 653–655
reciprocal, 70–73, 421
square root, 598–600
trigonometric, 638–641
vertical line test for, 412

Galileo, 552
General growth formula, 485
Golden rectangle, 644–645
Graphics calculator, 11, 86
correlation coefficient, 47
graphing exponential functions, 486
graphing linear functions, 37, 182
graphing quadratic functions, 553
graphing reciprocal functions, 71
graphing transformations on, 418–420, 445
INT, 80, 355
perpendicular slope, 248–250
RAND, 355
regression line, 227
TAN, 639
TAN^{-1}, 640
TRACE, 71, 613

Graphics calculator, *(cont.)*
    working with matrices, 334,
        339–340
Graphing method to solve systems
    of equations, 260–264
Graphs
    of absolute value functions, 80
    of exponential functions, 59, 86
    of functions, 412
    of inequalities, 134
    of integer functions, 81
    of linear functions, 59, 85,
        218–221
    of linear inequalities, 284–288
    of piecewise functions, 86
    of quadratic functions, 66, 84
    of rational numbers, 168
    of reciprocal functions, 71
    of systems of equations,
        260–264, 267
    using vertical line test on, 412
Greatest common factor (GCF),
    516
Greatest integer function (See
    integer function.)
Guess-and-check, 537–538

Han dynasty, 434
Horizontal lines, 240–243
    equation for, 242
Horizontal translation, 438
Hypatia, 27, 111
Hyperbola, 427
    asymptotes of, 427
Hypotenuse, 619
Hypothesis, 689

Identity
    additive, 107
    multiplicative, 169
Identity matrix for addition,
    317
**Identity matrix for
    multiplication,** 327

Inconsistent system, 279–280
Independent events, 394–
    397
Independent system, 280
Inequality, 131–134
    absolute value, 195–196
    linear
        graphing, 284–288
        solving , 132–133
    properties of
        addition, 132
        division, 188
        multiplication, 188
        subtraction, 132
    statements of, 132
    using a number line to represent,
        133, 196
Integer function, 80–81, 87
Integers, 100
    adding, 101–104
        like signs, 103
        more than two, 104
        unlike signs, 103
    dividing, 147–148
    multiplying, 146–147
    negative, 100
    positive, 100
    subtracting, 107–109
Intersection of sets, 376
Inverse
    additive, 102
    matrix, 331–334
    multiplicative, 169
Inverse variation, 658–661
Irrational numbers, 597

Legs, 619
Like terms, 112
Line of best fit, 46–48
Linear equation, 37
    graphing, 34–37
    point-slope form of, 235–237
    slope-intercept form of,
        224–227, 237
    standard form of, 233–235,
        237
Linear functions, 34, 36, 37
    graphs of, 59, 85, 218–221

Linear inequalities
    boundary line, 284
    graphing, 284–285
    systems of, 285–288
Linear polynomial, 507
Linear programming
    constraints, 299
    feasibility region, 299
    optimization equation, 298
Linear relationship, 60
Literal equation, 127

Margin of error, 371
Math Connections
    coordinate geometry, 80, 85,
        336, 343, 365, 374, 387, 462,
        482, 518, 627, 632
    geometry, 11, 13, 14, 15, 21, 33,
        49, 64, 68, 115, 117, 127, 129,
        130, 142, 156, 161, 164, 166,
        172, 179, 184, 191, 194, 202,
        213, 214, 232, 245, 247, 248,
        250, 251, 271, 276, 279, 282,
        283, 289, 291, 302, 321, 330,
        335, 342, 349, 381, 384, 386,
        387, 392, 394, 415, 418, 420,
        423, 435, 442, 446, 452, 462,
        467, 500, 509, 512, 518, 523,
        526, 528, 533, 534, 544, 557,
        563, 569, 572, 580, 601, 610,
        615, 622, 623, 629, 637, 642,
        662, 663, 667, 668, 686, 688,
        693, 698
    maximum/minimum, 193, 228,
        418, 549, 560, 581
    probability, 179, 417, 430, 472,
        491, 494, 500, 534, 563, 580,
        669, 675
    statistics, 27, 33, 40, 44, 78, 122,
        126, 129, 148, 150, 179, 181,
        202, 211, 266, 319, 321, 359,
        378, 379, 380, 423, 430, 435,
        439, 442, 461, 468, 482, 534,
        693
    transformations, 473
Matrices
    addition properties of, 317–318
    address, 309

Matrices *(cont.)*
  codes, and, 344–345
  columns, 309
  dimensions, 309
  entry, 309
  equal, 311
  equations, 337–341
  identity, 327, 331
  multiplication of, 325–327
  multiplicative inverse, 331
  product, 325
  rows, 309
  scalar, 324
  scalar multiplication, 324
  square, 310
  subtraction properties of, 317
  using to display and store data,
    308–311, 315–316
  zero, 317
Matrix equality, 311
Matrix equations, 337–341
  solving with technology,
    339–340
Matrix inverse, 331
Matrix solution for systems of
  equations, 338–339
Maximum value, 560
Mean, 368–369
Means of a proportion, 684
Median, 368–369
Midpoint formula, 628–629
Minimum value, 560
Mode, 368–369
Monomials, 464–465, 507
  dividing, 466
  multiplying, 464–465
Mozart, Wolfgang Amadeus, 46
Multiplication
  equations, 160–161, 164
  of binomials, 519–522
  of expressions, 152–154
  of integers, 146–147
  of monomials, 464–465
  of polynomials, 513–516
  of rational expressions, 672–
    673
**Multiplication Principle of
  Counting,** 383–385
**Multiplication Property of
  Equality,** 162
**Multiplication Property of
  Inequality,** 188

**Multiplication Property of
  Square Roots,** 606
Multiplicative inverse, 169
Multiplictive inverse matrix,
  331
Multiplicative Property of − 1,
  153

Negative direction, 210
Negative exponent, 469–470
Negative integers, 100
Neutral pair, 119
Newton, Sir Isaac, 478
Number line, 34, 113
  absolute value on, 79
  distance between points on,
    109
  graphing numbers on, 186
  graphing solutions to quadratic
    inequalities on, 582
  integers on, 100
  rational numbers on, 16
  using to represent inequalities,
    133–134, 196
  vertical, 103
Numbers
  even, 690
  integers, 598
  irrational, 598
  natural, 598
  odd, 690
  rational, 168, 598
  real, 598
  rectangular, 11
  square, 18
  whole, 598

Odd numbers, 690
Opposites, 100
OR, 376–377
Order of operations, 28–29
Ordered pair, 35
Origin, 35

Parabolas, 66, 548–549
  and constant differences, 548
  and polynomials, 549
  and transformations, 549
  axis of symmetry of, 548, 555
  vertex of, 549, 555
Parallel lines, 246–247
Parent function, 418–421
  absolute value, 79, 419, 421
  exponential, 60, 421
  integer, 80, 421
  linear, 60, 421
  quadratic, 65, 420–421
  reciprocal, 70, 421
Pascal, Blaise, 540–541
Pascal's triangle, 15
  Chinese version, 540
Patterns
  exploration, 31
  fractal, 8
  number, 8
  quadratic, 65
  repeating, 50–51
  using differences to identify,
    16–19
Percent, 173–177
  as a decimal, 173–174
  as a fraction, 173–174
  problems, 174–177
Percent bar, 174
Perfect square, 596
Perfect-square trinomial,
  529–530
Permutations, 400–401
Perpendicular lines, 248–250
Piecewise function, 86, 87
Point-slope form, 235
Polynomial, 506–507
  binomial, 507
  constant, 507
  cubic, 507
  degree of, 507
  factors of, 514
  greatest common factor (GCF)
    of, 516
  linear, 507
  monomial, 507
  prime, 525, 538

Polynomial, *(cont.)*
quadratic, 507
standard form, 507
trinomial, 507
Portfolio Activity, 3, 21, 57, 83, 99,
130, 145, 198, 207, 223, 259,
266, 271, 277, 307, 343, 351,
399, 409, 447, 455, 488, 505,
528, 547, 580, 595, 631, 651,
688
Positive direction, 210
Positive integers, 100
Power, 456–458
of 10, 476
Power-of-a-Power Property,
464, 471
Power-of-a-Product Property,
463, 471
Prime numbers, 524–525
Prime polynomial, 525, 538
Principal square root, 598
function, 598–600
Probability
and Pascal's triangle, 540–541
equally likely, 389
event, 353
experimental, 352–356
finding with the complement,
396–397
of independent events, 396
successful event, 353
theoretical, 388–391
trail, 353
Problem solving (Also see *Applications.*)
age, 295
mixture, 292–293
money, 293
numeration, 294
travel, 294–295
Problem-solving strategies
look for a pattern, 9
reason logically from the
pattern, 9
think of a simpler problem, 8–9
working backward, 18
Product matrix, 325
Product-of-Powers Property,
463, 471
Projects
Cubes and Pyramids, 90–91
Dimensional Analysis, 694

Projects, *(cont.)*
Diophantine Equations, 252–
253
Egyptian Equation Solving, 199
Find It Faster, 138–139
Minimum Cost Maximum Profit,
298–299
Pick a Number, 448–449
Please Don't Sneeze, 496–497
Powers, Pascal, and Probability,
540–541
Radical Rectangles, 644–645
Repeating Patterns, 50–51
Secret Codes, 344–345
What's the Difference?, 586–
587
Winning Ways, 400–401
Proof, 689
Properties of Zero, 148
Property(ies)
addition, of equality, 126
addition, of inequality, 132
addition, of zero, 107
associative, of addition, 104
commutative, of addition, 114
distributive, 113
distributive, over subtraction,
153
division, of equality, 161
division, of inequality, 188
division, of square roots, 606
multiplication, of equality, 162
multiplication, of inequality, 188
multiplication, of square roots,
606
multiplication, of zero, 148
multiplicative, of -1, 153
of opposites, 102
power-of-a-power, 464, 471
power-of-a-product, 465, 471
product-of-powers, 463, 471
quotient, of powers, 466, 471
reciprocal, 168
subtraction, of equality, 125
zero product, 549
Property of Opposites, 102
Proportion(s), 170, 174–177,
684–687
means, 684
extremes, 684
solving, 685
Proportion method, 176–177

**"Pythagorean" Right-Triangle
Theorem,** 619–622
converse of, 627

Quadrant, 35
Quadratic equations
discriminant, 577–578
solving
by completing the square,
558–561
by finding square roots,
553–554
using the algebraic method,
554–555
using the graphing method,
553
using the quadratic formula,
576–578
using the Zero Product
Property, 567
Quadratic formula, 574–578
discriminant, 566–578
Quadratic function, 64–67, 87, 548
maximum value, 560
minimum value, 560
Quadratic inequality, 581–584
Quadratic polynomial, 507
Quotient-of-Powers Property, 466,
471

Radical equations, 604
solving, 611–616
Radical expression(s), 605
rationalizing the denomi-
nator of, 607
simplest form of, 605
Radical sign, 598, 604
Radicand, 604
RAND, 355–356
Random integers, generating, 356
Random numbers, 355-356
Range, 37, 368–369
Range of a function, 413
Rate of change, 221

Ratio, 170, 173, 650
Rational equations
  common denominator method,
    678–679
  graphing method, 678–679
  solving, 678–681
Rational expression(s), 650,
    652–655
  operations with, 670–673
  simplifying, 664–667
Rational function, 653–655
  reducing, 665
  undefined, 653–654
Rational numbers, 168, 597
Rationalizing the denominator,
    607
Real numbers, 598
Reasoning (See *Critical Thinking*)
Reciprocal, 71, 168–170
Reciprocal function, 70–73, 87
Reciprocal Property, 168–70
**Rectangular coordinate
    system,** 35
Reflections, 431–433
  vertical, 432
Regression line, 227
Relation, 410–412
Right triangle, 619–620
  converse of the "Pythagorean"
    Right-Triangle Theorem,
    627
  hypotenuse, 619–620
  legs, 619–620
  "Pythagorean" Right-Triangle
    Theorem, 620–622
Rise, 208–212
Run, 208–212

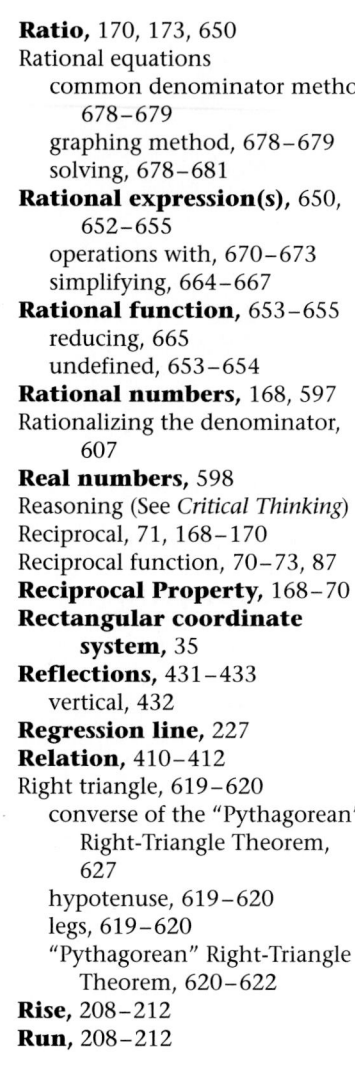

Scalar, 324
Scalar multiplication of matrices,
    324
Scale factor, 425
Scatter plot, 40–43, 46–48
Scientific notation, 476–479
Sequences, 10
  diminishing, 24
  generated by a quadratic, 66–67,
    586–587

Sequences, *(cont.)*
  number, 10
  terms of, 10, 66–67
Sets
  disjoint, 376
  intersection of, 376
  union of, 376
**Simplest form of a radical
    expression,** 605
Simulation, 360–363
  basketball, 361
  designing a, 361
  weather, 362
Slope, 208–212
  geometric interpretation of, 209
  graphic interpretation of, 209
  negative, 209–210
Slope formula, 226
Slope-intercept form, 224–227,
    237
Spreadsheets, 12, 19, 78–79, 83,
    151
Square matrix, 310
Square root(s), 553, 596–600,
    604–608
  division property of, 606
  estimating, 597
  multiplication property of, 606
  negative, 604
  positive, 604
  principal, 598
Standard form, 233–235, 237
**Standard form of a
    polynomial,** 507
Statistics
  deceptive, 370
  descriptive, 368–371
Stretches, 424–427
  scale factor, 425
Substitution method, 267–269
Subtraction, 109
  equations, 126, 164
  of expressions, 118–121
  of integers, 108–109
  of matrices, 315–317
  of polynomials, 510
  of rational expressions, 672
**Subtraction Property of
    Equality,** 125
**Subtraction Property of
    Inequality,** 132
Successful event, 353

Symbols of inclusion, 29, 30
Symmetric, 418
Systems of equations
  consistent, 279–280
  dependent, 280
  elimination method, 272–274
  graphing, 260–264
  inconsistent, 279–280
  independent, 280
  matrix solution of, 338–339
  substitution method, 267–269
Systems of linear inequalities
  boundary line, 286
  graphs of, 285–288

Tangent, 638–641
Tangent function, 538
Technology, 13, 19, 20, 28, 63, 83,
    198, 232, 238, 244, 266, 314,
    334, 335, 339–340, 343, 358,
    359, 373, 393, 461, 462, 467,
    472, 473, 482, 494, 495, 602,
    603, 631, 641 (Also see
    *Calculators, Computers,
    Graphics Calculators,* and
    Spreadsheets.)
Term, 10, 112
Theodorus, 624
Theorem, 689
Theoretical probability,
    388–391
Transformations, 418–421
  combining, 443–445
  of absolute value functions,
    419
  of quadratic functions, 420
  of square root function,
    599–600
  reflections, 431–433
  stretches, 424–427
  translations, 436–440
Translations, 436–440
  horizontal, 438
  vertical, 437
Tree diagram, 382–384, 541
Trial, 353
Trigonometric functions, 638–641
  tangent, 638–641

Trinomial(s), 507
    factoring, 535–538
    perfect-square, 529–530
Try This, 12, 19, 30, 37, 60, 71, 81,
    86, 103, 114, 121, 126, 127,
    134, 154, 155, 161, 163, 170,
    176, 182, 193, 209, 226, 237,
    241, 250, 262, 274, 287, 294,
    310, 317, 333, 339, 356, 369,
    384, 390, 395, 397, 414, 426,
    437, 438, 445, 466, 477, 478,
    485, 491, 509, 526, 531, 536,
    561, 567, 569, 570, 576, 582,
    605, 606, 614, 621, 654, 655,
    659, 666, 667, 672, 681, 691

Union of sets, 376

Variable, 23
Variable matrix, 338

Venn diagrams, 376–377
    counting from, 377
    interpreting, 377
    overlapping regions, 376
    showing disjoint sets, 376
    showing the intersection of sets,
        376
    showing the union of sets, 376
Vertex, 66, 418, 549, 555
Vertical lines, 240–243
    equation for, 242
Vertical line test, 412
Vertical stretch (See *stretches*.)
Vertical translation, 438

x-**axis,** 35
x-**coordinate,** 35

y-**axis,** 35
y-**coordinate,** 35
y-**intercept,** 224

Zero
    addition property of, 107
    division property of, 148
    multiplication property of,
        148
Zero exponent, 470-471
Zero matrix, 317
Zero Product Property, 567
Zeros of a function, 549

CREDITS

PHOTOS

Abbreviations used: (t) top, (c) center, (b) bottom, (l) left, (r) right, (bckgd) background, (bdr) border.

FRONT COVER: (l), James W. Kay/Daniel Schaefer; (c) SuperStock; (r), Viesti Associates, Inc. **TABLE OF CONTENTS:** Page v(tr), Jerry Jacka; v(br), Sam Dudgeon/ HRW Photo; vi(tr), Michelle Bridwell/HRW Photo; vi(cl), Larry Stevens/Nawrocki Stock Photo, Inc.; vi(br), Sam Dudgeon/HRW Photo; vii(tr), Tony Stone Images; vii(cl), Ron Kimball; vii(br), Nawrocki Stock Photo, Inc.; viii(tr), Dennis Fagan/HRW Photo; viii(cl), John Langford/HRW Photo; viii(br) Sam Dudgeon/HRW Photo; ix(tr), Douglas Dawson Gallery/Chicago, Illinois; ix(cl), Scott Van Osdol/HRW Photo; ix(br), Andrew Leonard/The Stock Market; x(tr), Michelle Bridwell/HRW Photo; x(b), David Madison;xi(tr), David Phillips/HRW Photo; xi(br), Dennis Fagan/HRW Photo. **CHAPTER ONE:** Page 2(bl)(tr), Jerry Jacka; 2-3(bckgd), Steve Vidler/Nawrocki Stock Photo, Inc. Inc.; 3(tr)(br), Michelle Bridwell/Frontera Fotos; 3(c), Jim Newberry/HRW Photo; 4(tr), Scott Van Osdol/HRW Photo; 5(tl), Fred Griffen; 5(cr), The Stock Market; 6(tl), Tony Stone Images; 6(br), Scott Van Osdol/HRW Photo; 7(tr), Flip McCririck; 8(cl), Dennis Fagan/HRW Photo; 8(c), Art Matrix; 9(tl), (tr),(cr), Michelle Bridwell/Frontera Fotos; 12-13(br), Dennis Fagan/HRW Photo;13 (tr), Jonathan Daniel/Allsport; 14(tl), Bettmann Archive; 15(tr), Alex Bartel/FPG International; 16(tr), Warren Faidley/Weatherstock; 16(cl), Dennis Fagan/HRW Photo; 18(c), Sam Dudgeon/HRW Photo; 19(br), Jim Newberry/HRW Photo; 20-21(bckgd), NASA; 22(c), Sam Dudgeon/HRW Photo; 24(t), Advanced Satellite Productions; 25(t), Sam Dudgeon/HRW Photo; 26(t), Dennis Fagan/HRW Photo; 28(t),(bc), Michelle Bridwell/Frontera Fotos; 29(bc), 31(t), 33(t), Sam Dudgeon/HRW Photo; 34(t), Dennis Fagan/HRW Photo; 34(tr), Sam Dudgeon/HRW Photo/Printed with permission of State of Minnesota; 36(t), Sam Dudgeon/HRW Photo; 37(cl), Jay Thomas/International Stock; 38(t), Dennis Fagan/HRW Photo; 39(t), Sam Dudgeon/HRW Photo; 39(bl), Kathleen Campbell/Allstock; 40(tr), 43(t), Sam Dudgeon/HRW Photo; 45(t), David Madison; 46(bl), (tr), (bl), Sam Dudgeon/HRW Photo; 48(b), Dennis Fagan/HRW Photo; 50-51(bckgd), The Quilt Complex Courtesy of the Espirit Quilt Collection. **CHAPTER TWO:** Page 57(r), Michelle Bridwell/ HRW Photo; 56-57(bc), Sam Dudgeon/HRW Photo; 58(tl), 59(bl), Dennis Fagan/HRW Photo; 60(br), Phil Schermeister/AllStock; 62(tr), Dennis Fagan/HRW Photo; 63(tr), Harold E. Edgerton/Palm Press; 64(tc),(bl),(bc),(b), Sam Dudgeon/HRW Photo; 65(b), Michelle Bridwell/HRW Photo; 67(br), Dennis Fagan/HRW Photo; 68(tr), (br), Michelle Bridwell/HRW Photo; 69(cr), Laurence Parent; 70(c), Michelle Bridwell/HRW Photo; 72(bl), Sam Dudgeon/HRW Photo; 73(tr), Tony Freeman/PhotoEdit; 74(tr),(bl),75(cr), Sam Dudgeon/HRW Photo; 78(t), 83(br), Michelle Bridwell/HRW Photo; 84(t), Richard T. Bryant/The Stock Source/Atlanta; 85(tr), Michelle Bridwell/HRW Photo; 88(tr), Robert E. Daemmrich/Tony Stone Images; 89(br), Babylonian Sillabary, 442 B.C., British Museum, London, England/Art Resource, NY; 90(tr), Index Stock Photography Stock; 90(cl),(bl), Sam Dudgeon/HRW Photo; 91(br), Nawrocki Stock Photo, Inc.; 91(tr), Sam Dudgeon/HRW Photo; 91(b), Index Stock Photography Stock. **CHAPTER THREE:** Page 98(bckgd), Martin Rogers/Tony Stone Images; 99(tc), Ray Massey/Tony Stone Images; 99(br), Barbara Adams/FPG International; 98(br), SuperStock; 99(bckgd), Paluan/Art Resource.; 100(r), NASA; 101(c), Michelle Bridwell/Frontera Fotos; 101(cr), Sam Dudgeon/HRW Photo; 102(tr), Michelle Bridwell/HRW Photo; 104(b), Sam Dudgeon/HRW Photo; 105(tl), Michelle Bridwell/Frontera Fotos; 107(tr), Sam Dudgeon/HRW Photo; 108(bl), Michelle Bridwell/HRW Photo; 109(br), Sam Dudgeon/HRW Photo; 110(tr), Michelle Bridwell/HRW Photo; 111(bckgd), Lobl-Schreyer/FPG International; 112(tl), Michelle Bridwell/HRW Photo; 112(b), 115-116(bckgd), Sam Dudgeon/HRW Photo; 118(t),121(b), Michelle Bridwell/HRW Photo; 123(bckgd), Trevor Wood/Tony Stone Images; 124(tr),125(tr), Michelle Bridwell/Frontera Fotos; 126(cl), 126(br), 128(r), Sam Dudgeon/HRW Photo; 129(bckgd), SuperStock; 131(c), 132(l),135(tr), Michelle Bridwell/Frontera Fotos; 136-137(bckgd),Tom McCarthy/PhotoEdit;138(bckgd), Matthew Neal McVay/Tony Stone Images; 139(tl), (cr), Bob Daemmrich; 139(cl), (br), Sam Dudgeon/HRW Photo. **CHAPTER FOUR:** Page 144(bc),145(cl,cr), SuperStock;144-145(bckgd), Scott Van Osdol/HRW Photo;146(r), Michelle Bridwell/HRW Photo; 148(cl), (cr), David Phillips/HRW Photo; 149(r), Sam Dudgeon/HRW Photo; 150(cl), Mike Valeri/FPG International; 151(tr), Michelle Bridwell/HRW Photo; 152(bl), 153(tr), Sam Dudgeon/HRW Photo; 155(br), Michelle Bridwell/HRW Photo; 156(bl),157(tl), Sam Dudgeon/HRW Photo; 157(br), David Phillips/HRW Photo; 158(bc), (br), Sam Dudgeon/HRW Photo; 158-159(bckgd), Ron Sanford/Tony Stone Images; 160(tr), Michelle Bridwell/HRW Photo; 162-163, Stock Editions; 162(cl), Robert E. Daemmrich; 164(br), Dennis Fagan/HRW Photo; 165(cr), Sam Dudgeon/HRW Photo; 165(bckgd), John W. Warden/SuperStock; 166(bckgd), Philip Habib/Tony Stone Images; 167(t), 168(tr), 169(c), Sam Dudgeon/HRW Photo; 171(tr), Dennis Fagan/HRW Photo; 171(cl), (br), 172(tl), Sam Dudgeon/HRW Photo; 172(cr), Larry Stevens/Nawrocki Stock Photo Inc.; 172(bl), Robert Wolf/Robert Wolf Photography; 173(tr), Colin Prior/Tony Stone Images; 174(br), Sam Dudgeon/HRW Photo; 175(tr), 177(tl), Michelle Bridwell/HRW Photo; 178(tl), Dennis Fagan/HRW Photo; 180(tr), Michelle Bridwell/HRW Photo; 181(tr), Dennis Fagan/HRW Photo; 183(bl), Michelle Bridwell/HRW Photo; 184(tr), Dennis Fagan/HRW Photo; 185(c), SuperStock; 186(c), John Terence Turner/FPG International; 189(tr), Maria Taglienti/The Image Bank; 190(bl), The VNR Concise Encyclopedia of Mathematics/Van Nostrand Reinhold Company; 190(br), Photri; 191(t), SuperStock; 190-191(bckgd), John Terence Turner/FPG International; 192(t), Mark Segal/Tony Stone Images; 194(t), Dennis Fagan/HRW Photo; 197(br), David Phillips/HRW Photo; 198(tc), Robert E. Daemmrich/Tony Stone Images; 199(cl), "Egyptian Equation Solving:/Image #27 Limestone/Dunasty XII. c.1900 B.C./Planet Art Classic Graphics, Ancient Egypt; 199(tr), Courtesy of The National Council of Teachers of Mathematics: The Rhind Mathematical Papyrus by Arnold Buffum Chace. **CHAPTER FIVE:** Page 206(c), Robert Harding Associates, London; 206(b), John Langford/HRW Photo; 207(bckgd) Hugh Sitton/Tony Stone Images; 208(tc), (inset) Michelle Bridwell/HRW Photo; 209(l), 212(cr), Sam Dudgeon/HRW Photo; 211(bckgd), E. Nagele/ FPG International; 212-213(bckgd), 213(cr), Birdseye, Ltd.; 213(br), Michelle Bridwell/HRW Photo; 214(tl), Tony Stone Images; 214(br), William R. Saliaz/Duomo; 215(t), Mark Reinstein/FPG International; 216,(bc)SuperStock: 216-217(bckgd), Lee Balteman/FPG International; 218(c), Joe Towers/George Hall/Check Six; 220(bl), 221(bckgd), Sam Dudgeon/HRW Photo; 222(tr), Dennis Fagan/HRW Photo; 222(bl), SuperStock; 224(r), Steven E. Sutton/Duomo; 226-227(bc), J. Zimmerman/FPG International; 227(tc), Richard Dole/Duomo; 228(bl), 229(cr), Michelle Bridwell/HRW Photo; 230(tc), James Newberry/HRW Photo; 231(br), Michelle Bridwell/HRW Photo; 232(tl), Sam Dudgeon/HRW Photo; 233(tc), Shooting Star International; 234(tl), Michelle Bridwell/Frontera Fotos; 236(cl), Sam Dudgeon/HRW Photo; 237(br), Dennis Fagan/HRW Photo; 238(cr), Michelle Bridwell/Frontera Fotos; 238-239(bckgd), Bob Daemmrich; 240(r)(l), David Madison; 241(bckgd), Chris Harvey/Tony Stone Images; 242-243(bckgd), Michael Hart/FPG International; 243(bl), Dennis Fagan/HRW Photo; 244(br), Sam Dudgeon/HRW Photo; 245(tr), 246(tr), SuperStock; 246(tc), Michael Keller/FPG International; 248(l), 248(r), Sam Dudgeon/HRW photo; 252(tr), The Bettmann Archive; 250-51(bckgd), Superstock; 252-253(bckgd), Carl Yarbrough; 253(tr), Chris Marona. **CHAPTER SIX:** Page 258(cr), John Banagan/Image Bank; 258(b),259(r), SuperStock; 258-259(bckgd), SuperStock; 260(t), David Phillips/HRW Photo; 261(c), Jeff Zaruba/The Stock Market; 262(br), Michelle Bridwell/HRW Photo; 264-265(bckgd), A.Upitis/The Image Bank; 267(tr), SuperStock; 268(br), David Phillips/HRW Photo; 270(tr), Dennis Fagan/HRW Photo; 270(bl), Robert Reiff/FPG International; 272(t), 276(cl), David Phillips/HRW photo; 276(br), Sam Dudgeon/HRW Photo; 276(bckgd), Michelle Bridwell/HRW Photo; 278(t), 279(br),Dennis Fagan/HRW Photo; 284(t),Superstock; 287(tl), (tr), Michelle Bridwell/HRW Photo; 289(tr), David Phillips/HRW Photo; 291(tr), Michelle Bridwell/HRW Photo; 292(tr), (cl), Ron Kimball; 292(bc), Dennis Fagan/HRW Photo; 293(bl), Sam Dudgeon/HRW Photo; 295(bl), Nawrocki Stock Photo, Inc.; 296(bc), Michelle Bridwell/HRW Photo; 297(tl), Joe Towers/The Stock Market; 297(c), A. Edgeworth/The Stock Market; 297(bckgd), Luis Villota/The Stock Market; 298(tl), Michelle Bridwell/HRW Photo; 298(tl), Sam Dudgeon/HRW Photo; 298-299(bckgd), Superstock. **CHAPTER SEVEN:** 306(bl), Winfield Parks/National Geographic; 306-307 (bckgd), Joseph Scherschel/National Geographic; Page 306(r)(tr), Erich Lessing/Art Resource; 307(c), Art Resource; 307(tl), Werner Forman Archive/Art Resource; 308(tc), Michelle Bridwell/HRW Photo; 309(tc), Sam Dudgeon/HRW Photo; 310(tc), Tom Van Sant/The Stock Market; 312(br), Sam Dudgeon/HRW Photo; 315(r), David Phillips/HRW Photo; 316(bl), Sam Dudgeon/HRW Photo; 318(cr), Dennis Fagan/HRW Photo; 319(br), Michelle Bridwell/HRW Photo; 320(tr), Jonathan Daniel/AllSport USA; 321(tr), Michelle Bridwell/HRW Photo; 322(tr), Mike Anich/Adventure Photo; 322(bl), D. R. Fernandez & M. L. Peck/Adventure Photo; 323(bckgd), Sam Dudgeon/HRW Photo; 324(t), David Phillips/HRW Photo; 325(br), Michelle Bridwell/HRW Photo; 328-329(bckgd), Sam Dudgeon/HRW Photo; 334(br), 336(tl), Dennis Fagan/HRW Photo; 337(t), Sam Dudgeon/HRW Photo; 340(tl), 341(c), David Phillips/HRW Photo; 343(c), Michelle Bridwell/HRW Photo; 344(t), UPI/Bettmann; 345(br), FPG International. **CHAPTER EIGHT:** Page 350-351(bckgd), Sam Dudgeon/HRW Photo; 350(c),351(tr),(c), John Langford/HRW Photo; 352(t), David Phillips/HRW Photo; 352(bc), Sam Dudgeon/HRW Photo; 353(t), John Langford/HRW Photo; 354(tl), Michelle Bridwell/HRW Photo; 354(bl), Erich Lessing/Art Resource, NY; 357(br), John Langford/HRW Photo; 358(tr), (bl), (bc), Sam Dudgeon/ HRW Photo; 360(cl), 362(l), 363(tr), 364(tr), David Phillips/HRW Photo; 366(c),Ron Chapple/FPG International; 366-367(bckgd), J. Zimmerman/FPG International; 368(t), R. Stewart/AllSport USA; 372(t), Warren Morgan/Westlight; 372(c), Dick Reed/The Stock Market; 372(b), SuperStock; 374(c), Bruce Coleman; 375(r), 379(tr), David Phillips/HRW Photo; 380(cr), Mel Digiacomo/The Image Bank; 381(b), Michelle Bridwell/HRW; 382(tl), David Phillips/HRW; 382(tc), Ken Lax/HRW Photo; 382(br), Ken Karp/HRW Photo; 383(tl), (tc), Sam Dudgeon/HRW Photo; 383(tr), Coronado Rodney Jones/HRW Photo; 384(t), 385(br), Sam Dudgeon/HRW Photo; 386(br), Michelle Bridwell/HRW Photo; 387(tr), Sam Dudgeon/HRW Photo; 388(tc), Mary Kate Denny/Photo Edit; 389(tr),(tr), 390(cl),(c), Sam Dudgeon/HRW Photo; 391(tl), Michelle Bridwell/HRW Photo; 391(br), David Phillips/HRW Photo; 392(tl), SuperStock; 392(cr), Photo Researchers, Inc.; 394(bl), David Phillips/HRW Photo; 396(tr), 398-399(bckgd), 400(c), Michelle Bridwell/HRW Photo; 401(c), Sam Dudgeon/HRW Photo. **CHAPTER NINE:** Page 408(b), 408-409(t), (l), Scott Van Osdol/HRW Photo; 410(t), Tibor Bognar/The Stock Market; 411(t), Jonathan A. Meyers/FPG International; 415(br), David Phillips/HRW Photo; 416(bckgd), Keith Gunnar/FPG International; 418(t), Michelle Bridwell/HRW Photo; 418(cl), Sam Dudgeon/HRW Photo; 421(br), David Phillips/HRW Photo; 424(r), Ron Thomas/FPG International; 428(tr), David Phillips/HRW Photo; 430(br),

Sam Dudgeon/HRW Photo; 431(t),(cl), 433(br), Scott Van Osdol/HRW Photo; 434(bl), Dennis Cox/FPG International; 435(tc), Sam Dudgeon/HRW Photo; 436(t), MC Escher Heirs/Cordon Art, Baarn, Holland; 435(tr), Tom Ives/The Stock Market; 438(t), Sam Dudgeon/HRW Photo; 440(inset), Janet Brooks, artist; 440(tl), Douglas Dawson Gallery/Chicago, Ill. c.1920, Earthenware Vessel, Bamileke Culture, Cameroon grasslands; 440(tl), Douglas Dawson Gallery/Chicago, Ill. c.1200 A.D., Earthenware slip decorated vessel, D'jenne Culture, Mali; 440(br), Michelle Bridwell/HRW Photo; 442(l), Sam Dudgeon/HRW Photo; 443(t), 444(b), 446(t), David Phillips/HRW Photo; 448(r), 449(t), Scott Van Osdol/HRW Photo. **CHAPTER TEN:** Page 454(c), Thomas Craig/FPG International; 455(tr), James King Holmes/SPL/Photo Researchers, Inc.; 454(r), Frithfoto/ Bruce Coleman, Inc; 454(b), Stan Osolinski/FPG International; 454-455 (bckgd), Lea Kuhn/FPG International; 455(b), Douglas Faulkner/ Photo Researchers, Inc; 455(r), Joe McDonald/Bruce Coleman, Inc; 456, Scott Van Osdol/HRW Photo; 459(tl), Visuals Unlimited; 459(cr), 468(tr), Sam Dudgeon/HRW Photo; 466(t), David Phillips/HRW Photo; 469 (bckgd), Dr. Mitsuo Ohtsuki/SPL/Photo Researchers, Inc.; 469(t), Telegraph Colour Library/FPG International; 472(br), David Phillips/HRW Photo; 473(br), The VNR Concise Encyclopedia of Mathematics/Van Nostrand Reinhold Company; 474(inset), Rick Friedman/New York Times; 474-475(bckgd), Sam Dudgeon/HRW Photo; 475(inset), Scott Van Osdol/HRW Photo; 475(c), Robert Friedman/New York Times; 476(l), Scott Van Osdol/HRW Photo; 476(bckgd), John Sanford/Science Photo Library/Photo Researchers; 478(t), Dr. E. R. Degginger; 480(t), Scott Van Osdol/HRW Photo; 481(c), NASA; 483(b), Jack Zehrt/FPG International; 484(b), Telegraph Colour Library/FPG International; 485(bc), NRSC LTD/Science Photo Library/Photo Researchers; 487(t), Michelle Bridwell/HRW Photo; 488(tl), Alese & Mort Pechter/The Stock Market; 488(cl), Ulf Sjostedt/FPG International; 488(br), New York Public Library; 489(t), Ken Reid/FPG International; 489(cl), James King Holmes/Science Photo Library/Photo Researchers, Inc.; 491(t), Matt Bradley/Bruce Coleman, Inc.; 493(b), Michelle Bridwell/HRW Photo; 494-495(bc), Richard Stockton; 496-497(bckgd), Andrew Leonard/The Stock Market; 497(c), Michelle Bridwell/HRW Photo. **CHAPTER ELEVEN:** Page 506(t), 510(b), Scott Van Osdol/HRW Photo; 513(t), Dennis Fagan/HRW Photo; 514-515(b), 517(tr), Sam Dudgeon/HRW Photo; 519(t), Scott Van Osdol/HRW Photo; 521(cr), 524(tc), Michelle Bridwell/HRW Photo; 527(tr), Scott Van Osdol/HRW Photo; 528(bckgd), Sam Dudgeon/HRW Photo; 532(bc), Dennis Fagan/HRW Photo; 533(t), David Phillips/HRW Photo; 535(t), Michelle Bridwell/HRW Photo; 534(bckgd), John Langford/HRW Photo; 536(b), Michelle Bridwell/HRW Photo; 539(cr), Robert Wolff/ Robert Wolff Photography; 541(bl), Michelle Bridwell/HRW Photo; 543(bckgd) John Langford/HRW Photo. **CHAPTER TWELVE:** Page 546(bckgd), Ron Kimball; 546(c), Michael Brohm/Nawrocki; W.S. Nawrocki/ Nawrocki; 548(t), Bob Burch/Bruce Coleman, Inc. 550(t), David Phillips/HRW Photo; 551(br),

Sam Dudgeon/HRW Photo; 552(cl), Shinichi Kanno/FPG International; 554(inset), Superstock; 555(tr), David Phillips/HRW Photo; 557(cl), Carlos V. Causo/Bruce Coleman; 558(tl), 558(tr), (bl), (br), Sam Dudgeon/HRW Photo; 562(bl), Michelle Bridwell/HRW Photo; 563(tr), David Phillips/HRW Photo; 564(c), Nancy Engebretson/Phoenix Gazette/Used with permission. Permission does not imply endorsement.; 565(c), Superstock; 566(br), Michelle Bridwell/HRW Photo; 568(bl), Sam Dudgeon/HRW Photo; 569(cr), David Phillips/HRW Photo; 571(tr), 572(b), Michelle Bridwell/HRW Photo; 573(tc), Robert Frerch/Tony Stone Images; 573(tr), Peter Brandt; 574(l), (r), (c), David Madison; 576(tl), David Madison; 577(tl), David Madison; 579(tr), Michelle Bridwell/HRW Photo; 581(cl), FPG International; 585(r), Dave Gleiter/FPG International; 586(inset), SuperStock. **CHAPTER THIRTEEN:** Page 594(tc), W. Warren/Weslight; 595(c), Superstock; 596(t), Michelle Bridwell/HRW Photo; 600(cl), Rhonda Bishop/Contact Press/Woodfin Camp & Associates; 603(br),609(t), Michelle Bridwell/HRW Photo; 604(t), 607(tr), 611(br), Dennis Fagan/HRW Photo; 612(tl), Comstock; 612(br), John Langford/HRW Photo; 615(bl), SuperStock; 616(b), 617(br), 619(t), Comstock; 619(bl), Dave Bartruff/Nawrocki Stock Photo, Inc.; 619(bckgd), Sam Dudgeon/HRW Photo; 621(t), Tony Stone Images; 623(cr), SuperStock; 624(b), David Phillips/HRW Photo; 625(t), Matthew McVay/Tony Stone Images; 626(tc), Nawrocki Stock Photo, Inc.; 627(tr), Michael J. Howell/SuperStock; 629(br), Dennis Fagan/HRW Photo; 630(c), Index Stock Photography, Inc.; 632(t), Comstock; 633(bckgd), Bill Ross/Tony Stone Images; 634(t), Dallas & John Heaton/Westlight; 635(bl), Alan Bolesta/Index Stock Photography, Inc.; 636(t), David Phillips/HRW Photo; 636-37(bckgd), Michael Hans/Agence Vandystadt/Allsport; 638(t), Michelle Bridwell/HRW Photo; 641(t), SuperStock; 641(br), Michelle Bridwell/HRW Photo; 642(c),Superstock; 643(bl), Dennis Fagan/HRW Photo; 644(c), Scala/Art Resource, NY; 645(tl), Paul Chelsey/Tony Stone Images; 645(cr), Dennis Fagan/HRW Photo; 645(br), Ron Rovtar/Photonica. **CHAPTER FOURTEEN:** Page 651(t), David Philips/HRW Photo; 652(t), 653(b), 654(b), 656(bckgd), Michelle Bridwell/HRW Photo; 658(t), David Phillips/HRW Photo; 660(t), Dennis Fagan/HRW Photo; 662(br), 663(tr), SuperStock; 664(t), Dennis Fagan/HRW Photo; 667(tl), Index Stock Photography Stock; 667(br), Dennis Fagan/HRW Photo; 668(bl), Robert A. Lubeck/Animals Animals; 668-669(bckgd), Nicholas DeVore/Tony Stone; 669(tr), David Phillips/HRW Photo; 670-671(t), 673(br), 674(b), Michelle Bridwell/Frontera Fotos; 675(cr), Museum of the Rockies/Montana State Univ.; 676(bl), David Phillips/HRW Photo; 677(bckgd), SuperStock; 678(t), Dennis Fagan/HRW Photo; 679(b), Sam Dudgeon/HRW Photo; 680(c), 681(t), 682(tr), David Phillips/HRW Photo; 683(tr), Robert Frerck/Tony Stone; 684-685(b), David Phillips/HRW Photo; 687(tr), Michelle Bridwell/HRW Photo; 688(l), Dennis Fagan/HRW Photo; 692(tr), 693(t), Sam Dudgeon/HRW Photo; 694(tc), Hot Rod Magazine; 695(bl), Ron Kimball.

ILLUSTRATIONS

Abbreviations used: (t) top, (c) center, (b) bottom, (r) right, (l) left.

BB&K Design pages 50, 90,138, 199, 252, 298, 344, 400, 448, 496, 540, 586, 644, 694

Bindon, John pages 158 , 159

Boston Graphics pages 12, 21, 27, 30, 42 (r)(l) 44,46, 58, 72, 84, 86, 105, 111, 125, 129, 173, 185, 191, 218, 225, 236, 239, 260 (t)(b), 266, 272, 290, 312, 313 (t)(b), 330, 342

Fisher, David pages 199 (t)(b), 377, 388, 411, 441, 461, 468, 469, 476, 483, 484, 490, 518, 522, 525, 552, 600, 620, 622, 625, 634, 638, 657,

Holcomb, Rhonda pages 129, 224, 234, 235, 238, 376, 425, 445, 457, 512, 516, 520, 599, 599, 604, 669, 684, 689

Kumar, Nishi pages 8, 146, 149, 176, 308, 327, 329, 331, 336, 342, 355, 369, 394, 394, 422, 434, 439, 483, 505, 529, 546, 547, 563, 598, 668

Longacre, Jimmy pages 14, 657

Morrow, Mike pages 10, 18, 22, 23, 25, 27, 30, 33, 34, 43, 47(t)(b), 64, 70, 79, 81, 82, 89, 103, 106, 110, 116, 117, 118, 123, 124, 147, 160, 179, 186, 192, 224, 227, 228, 245, 246, 265, 266, 267, 269, 271, 275, 277, 278, 281(t)(c)(b), 283, 283, 294, 295, 314, 319, 337, 338, 368, 370(t)(b), 371(t)(b), 373(t)(b), 392, 398, 410, 413, 414, 429, 448, 456, 462, 463, 478, 520, 584, 607 , 628, 661

Ray, Julie page 181

Rivers, Joan page 210

Wong, Ophelia pages 76 , 216, 322, 366 , 474, 564 , 676

PERMISSIONS

Grateful acknowledgment is made to the following sources for permission to reprint copyrighted material:

The Associated Press: From "Atlanta Architect Steps Up Quest for Safe Staircases" by Lauran Neergaard from the *Albuquerque Journal*, 1993. Copyright © by The Associated Press.

Hill and Wang, a division of Farrar, Straus & Giroux, Inc.: From *Innumeracy* by John Allen Paulos. Copyright © 1988 by John Allen Paulos.

Kentucky Department of Education: "Kentucky Mathematics Portfolio Holistic Scoring Guide" (Retitled: "Portfolio Holistic Scoring Guide") from *Kentucky Mathematics Portfolio*: 1994-1995. Published by the Kentucky Department of Education.

Alfred A. Knopf, Inc.: Text and charts from *Jurassic Park* by Michael Crichton. Copyright © 1990 by Michael Crichton.

Karol V. Menzie and Randy Johnson: From "Count Your Way to Stair Success" by Karol V. Menzie and Randy Johnson from "Homework" column from *The Baltimore Sun*, 1993. Copyright © 1993 by Karol V. Menzie and Randy Johnson.

The National Council of Teachers of Mathematics: From page 84 from *The Rhind Mathematical Papyrus*, translations by Arnold Buffum Chace. Published by The National Council of Teachers of Mathematics, 1979.

Newsweek, Inc.: From "Finding Order in Disorder" by Sharon Begley from *Newsweek*, December 21, 1987. Copyright © 1987 by Newsweek, Inc. All rights reserved.

The New York Times Company: From "'Hot Hands' Phenomenon: A Myth?" from *The New York Times*, April 19, 1988. Copyright © 1988 by The New York Times Company. From graph, "Getting Lost in the Shuffle," and from "In Shuffling Cards, 7 is Winning Number" by Gina Kolata from The New York Times, January 9, 1990. Copyright © 1990 by The New York Times Company.

Penguin Books Ltd.: From page 274 of *The Crest of the Peacock: Non-European Roots of Mathematics* by George Gheverghese Joseph. Copyright © 1991 by George Gheverghese Joseph. Published by Penguin Books, 1992. First published by I. B. Tauris.

Springer-Verlag New York, Inc.: From "What People Mean by the 'Hot Hand' and 'Streak Shooting'" (Retitled: "Researchers' Survey of 100 Basketball Fans") from "The Cold Facts About the 'Hot Hand' in Basketball" by Tversky, Amos, and Gilovich from *Change: New Directions for Statistics and Computing*, vol. 2, no. 1, 1989. Copyright © 1989 by Springer-Verlag New York, Inc.

Time Inc.: From "A Miraculous Sky Rescue" from *Time*, May 4, 1987. Copyright © 1987 by Time Inc.

U.S. News & World Report: From "Now, the testing question" from *U.S. News & World Report*, September 21, 1992. Copyright © 1992 by U.S. News & World Report.

The Wall Street Journal: From "Counting Big Bears" from *The Wall Street Journal*, May 2, 1990. Copyright © 1990 by Dow Jones & Company, Inc. All Rights Reserved Worldwide.

ADDITIONAL ANSWERS

Lesson 1.1, pages 8–15

Communicate

1. The table would show all possible pairs of teams A, B, C and D. To list the games, note that team A can play B, C, or D. Other distinct pairings: B can play C or D, and C can play D. The table should thus have 6 entries.
   AB   AC   AD
        BC   BD
             CD

2. Represent the sum as a triangle of dots. Form a rectangle of dots by copying the triangle of dots and rotating it. The sum is then half of the number of dots in the rectangle, 210. Another method is $\frac{20 \cdot 21}{2} = 210$.

3. Notice that each term in the sequence is 4 more than the preceding one.

4. Notice that each term in the sequence is the sum of the preceding two.

5. Four Strategies: draw a picture, think of a simpler problem, make a table, look for a pattern. One possible example: Drawing a picture or diagram helps to visualize relationships.

Lesson 1.2, pages 16–21

Communicate

1. Find the first differences by subtracting each successive term. The first differences are 3, 5, 7, 9, and 11. The second differences are found by subtracting the successive first differences. The second differences are a constant 2.

2. Notice that the first differences are a constant -12 to predict the next 2 terms. Add -12 to 40 to find the next term, 28, and add -12 to 28 to find the next term, 16.

3. The maximum height is reached after half of the flight time. So this time is $\frac{17}{2} = 8.5$ seconds.

4. C4 = B5 $-$ B4 gives the fourth first difference.

5. If a sequence of differences are constant, taking further differences would always give zero, so no new information is gained.

6. Take the second difference and add it to the last known first difference to obtain the next first difference. Now take the new first difference and add it to the last known term of the original sequence. In this way the sequence may be extended term-by-term.

Lesson 1.3, pages 22–27

Exploration 1

1. $2   2. $4   3. $8   4. 2   5. 2
6. 2h   7. $2 \cdot 5 = 10$

Exploration 2

1. 922,654 acres

2. 57,600 acres

3. $-57,600d$ acres

4. $922,654 - 57,600(10) = 346,654$ acres

5. If $f = 0$ the forest is completely destroyed.

6. After 16 days only 1054 acres would be left, so on the 17th day the forest would be completely destroyed.

Communicate

1. A variable is a letter or other symbol that can be replaced by a number or an expression. An expression is formed by combinations of numbers, variables, and operations. Two equivalent expressions separated by an equal sign define an equation.

2. For example, let x represent the number of apples Phil has. Then Jeff has $x + 3$ apples. 3 apples

3. The cost of all the pencils is 12 times the cost per pencil. Express this as $12p = \$1.92$, where p represents the cost per pencil. $p = \$0.16$

4. The cost of x number of tickets is $10x$. By solving $10x = 35$ the number of tickets may be found, $x = 3.5$. One cannot buy half a ticket, so the possible answer is 3 tickets.

5. Guess values for x and check in the equation, keeping track of the results in a table. Choose additional guesses that are closer and closer or exactly equal to the true solution, $x = 51$.

Lesson 1.4, pages 28–33

Communicate

1. Adding 3 to 2 and then multiplying by 4 gives 20, but multiplying 2 times 4 and then adding 3 gives 11.

2. 20 divided by 2, or 10, times 5 is 50, but 20 divided by 2 times 5, or 10, is 2.

3. To get the correct answer, 50, in Exercise 2, the multiplications and divisions are performed from left to right, which is the agreed upon convention.

4. Given $\{[3(8 - 4)]^2 - 6\} \div (4 - 2)$, first simplify parentheses from innermost outward. The result is $\{[12]^2 - 6\} \div 2$. Next, perform operations with exponents. Evaluate $12^2 = 144$ to find $\{144 - 6\} \div 2$. Finally, evaluate the expression in braces and then divide by 2 to get $138 \div 2 = 69$.

5. They are necessary to insure that every person will obtain the same answer.

6. Answers may vary. Round the

numbers given. $\frac{173 + 223}{151 - 21}$ is about $\frac{400}{130}$, which is close to 3.

Lesson 1.5, pages 34–39

Communicate

1. Start at the point in question. Count the number of units that the point is *above* (positive) or *below* (negative) the *x*-axis for the *y*-value. Count the number of units the point is to the *right* (positive) or to the *left* (negative) of the *y*-axis for the *x*-value.

2. No. (6, 7) and (7, 6) are ordered pairs, and thus do not specify the same point in the plane. Both the *x*-coordinate and the *y*-coordinate are different.

3. Begin at the origin. Count 7 units to the right along the *x*-axis, then count 3 units up, parallel to the *y*-axis. This locates the point (7, 3).

4. In any ordered pair, the first set of values is the domain, and the second set of values is the range.

5. A table allows one to see the function value for a given domain value, while a graph gives a picture of the whole function.

6. In the left column of the table, list several *x* values. Evaluate the expression $2x + 5$ for each listed value. Enter these *y*-values for their corresponding *x*-values in the right column.

x	2x + 5	y	(x, y)
−2	2(−2) + 5	1	(−2, 1)
−1	2(−1) + 5	3	(−1, 3)
0	2(0) + 5	5	(0, 5)
1	2(1) + 5	7	(1, 7)

Practice and Apply

19-22. The lines $y = x + 7$ and $y = x - 7$ appear to be parallel, and perpendicular to the parallel lines $7 - x$ and $-7 - x$.

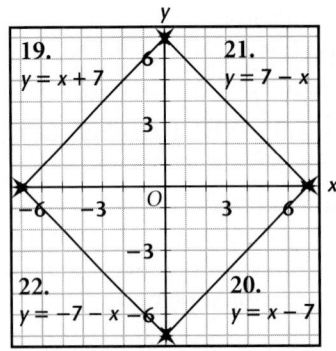

19. $y = x + 7$
21. $y = 7 - x$
22. $y = -7 - x$
20. $y = x - 7$

Lesson 1.6, pages 40–45

Exploration

1. They also tended to do well in science.

2. Most of them did not do well in science.

3. The data points cluster roughly near a line.

4. The line rises passing from left to right.

5. The data pattern shows a correlation between math and science scores. Students with high math scores tend to have high science scores, and vice versa.

Communicate

1. The data points cluster near a line that rises passing from left to right. This indicates a strong positive correlation.

2. Since the cost of the notebooks depends on the number of pages, label the horizontal axis to represent the number of pages. The vertical axis then represents cost.

3. A correlation describes the extent to which variables are related, and how closely they are related.

4. Little to none. The data points appear to be scattered randomly.

5. Strong positive. The data points cluster closely to a rising line.

6. Since the cost depends on the year, label the horizontal axis with the year, and the vertical axis with the cost.

7. Plot the ordered pairs (1982, 1.00), (1985, 1.06), (1988, 1.18), and (1991, 1.36), choosing an appropriate scale for the cost axis.

Practice and Apply

18a.

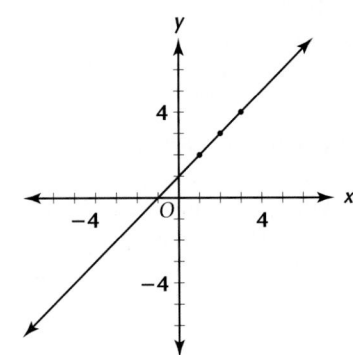

18b.

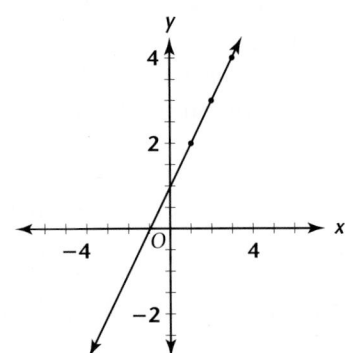

Lesson 1.7, pages 46–49

Communicate

1. Place the spaghetti so that it is close to as many points as possible.

2. A coefficient of correlation of ± 1 means that the data points fit exactly on a line.

3. The data shows a strong positive correlation.

4. The data shows a strong negative correlation.

5. 0.23 is a small coefficient of correlation, this means that the data points show little or no correlation.

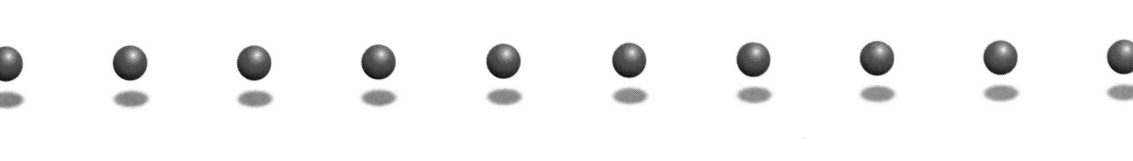

Lesson 2.1, pages 58–63

Exploration 1

1. A

2. After four days, it doesn't seem so.

3. Prize A: $2000
Prize B: $5242.88

4. Yes, day 19

5. A: $3000, B: $5,368,709.12

Exploration 2

1.

x	1	2	3	4	5
y	2	4	8	16	32

2.

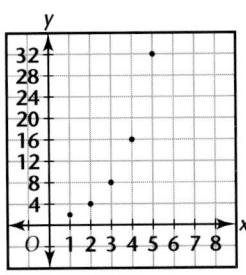

3. It rises in an upward curve.

4.

x	1	2	3	4	5
y	$\frac{1}{2}$	$\frac{1}{4}$	$\frac{1}{8}$	$\frac{1}{16}$	$\frac{1}{32}$

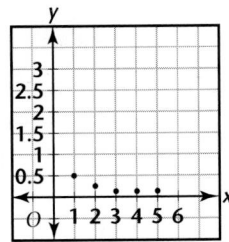

5. The graph of $y = 2^x$ is increasing and gets farther and farther away from $y = 0$. The graph of $y = \left(\frac{1}{2}\right)^x$ is decreasing and gets closer and closer to 0 but never reaches 0.

Communicate

1. Multiply the previous term by 2.

2. Divide the previous term by 2.

3. Add 10 to the previous term.

4. Subtract 20 from the previous term.

5. Sequences 1 and 2 are exponential because they increase or decrease at a fixed rate.

6. Sequences 3 and 4 are linear because they increase or decrease by adding a fixed amount.

7. Exponential decay means the amount is decreasing at a fixed rate. Exponential growth means the amount is increasing at a fixed rate.

8. The interest rate is 10%, so for the first year the principal is multiplied by 1.10. For interest to be compounded for 4 years the principal will be multiplied by $(1.10)^4$.

Lesson 2.2, pages 64–69

Exploration 1

1. It is 4 times larger.

2. It is 9 times larger.

3. 4 : 9; these numbers are the squares of the edge lengths.

4. 25; 6(25); $6(5^2)$; 150

5. $S = 6e^2$

Exploration 2

1. no; 2

2. 6th term, 36; 7th term, 49; First differences: 9, 11, 13, 15; Second differences: 2, 2, 2

3. They are constant.

4.

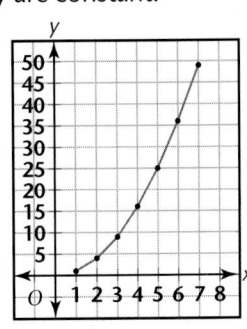

half of a parabola

5. +4, constant

6.

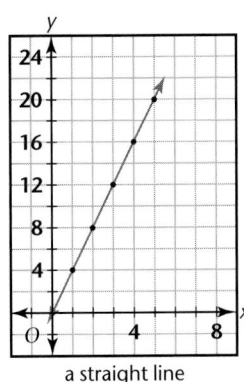

a straight line

7. In Step 4 the *second differences* are constant and the graph is a half parabola. In Step 6 the *first differences* are constant and the graph is a line.

Communicate

1. The area is 4 times larger.

2. The second differences are constant. The graph is a parabola.

3. The pattern indicates that the function oscillates between 0 and 4.

4. b; it is a curve that grows at a steady rate.

5. a; it is a straight line.

6. d; this is made up of two different linear parts.

7. c; the graph is a parabola.

8. The vertex is the point at which the parabola changes direction. It is the maximum point on the curve shown at (3, 9). In a table it will be the pair of coordinates with the greatest or least y-value, and the y-values on either side of that point will repeat.

9. If the first differences are constant, the relationship is linear. If the second differences are constant then the sequence describes a quadratic relationship. Constant differences help to determine the type of relationship.

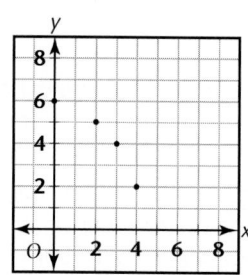

Look Beyond

53.

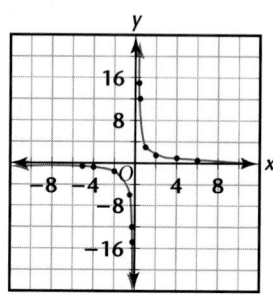

Lesson 2.3, pages 70–75

Communicate

1. As one variable increases the other decreases. For example, as your hourly rate of pay increases, the number of hours that you need to work to earn $100 decreases. As your rate of speed decreases, the time it takes to drive a fixed distance increases.

2. Divide 1 by 5. $\frac{1}{5}$

3. Divide 1 by 100. $\frac{1}{100}$

4. Divide 1 by $\frac{1}{4}$.   4

5. Divide 1 by $\frac{1}{6}$.   6

6. Divide $1000 by 4. $250

7. Divide $1000 by 200. $5

8. Let n represent the number of people contributing and a represent the amount per person, then $a = \frac{30}{n}$.

Look Back

50.

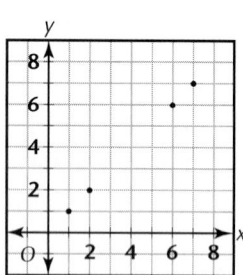

51.

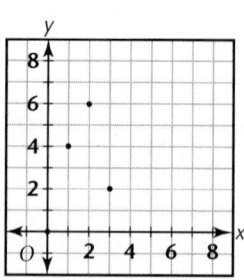

Eyewitness Math, pages 76–77

BREEDING POPULATION
Dinosaur Height

Week	A	B	C	D	E	F	G	H	I	J	K	L	M	N	O	P	Q	R	S	T
1	8	1	0	1	7	4	9	0	2	7	7	9	0	3	1					
2	5	0	9	1	2	0	9	3	9	9	2	3	5	0	1					
3	2	2	6	4	2	6	3	0	8	1	0	8	1	9	1					
4	8	9	4	2	0	6	7	8	0	0	5	5	1	3	7	5				
5	5	1	0	8	5	1	2	7	6	5	5	1	8	2	1	5				
6	1	2	5	9	7	7	4	5	2	1	6	3	0	8	6	0				
7	7	5	6	9	2	1	4	4	3	8	4	4	2	5	3	9	0			
8	0	7	0	4	6	0	6	3	8	9	0	7	5	6	0	1	4			
9	0	7	1	9	0	2	3	6	8	2	1	3	8	2	5	2	4			
10	0	4	6	0	2	6	8	8	9	3	6	8	1	9	3	8	5	5		
11	5	3	2	2	4	4	3	1	9	0	1	1	8	8	6	5	2	5		
12	5	6	4	8	3	5	4	4	9	1	9	0	5	9	4	4	5	5		
13	1	5	7	0	1	0	1	1	5	4	0	9	2	3	3	3	6	2	9	
14	0	9	0	4	3	1	2	7	3	0	4	1	4	6	1	8	5	9	4	
15	2	9	8	5	2	7	1	5	8	5	8	5	0	3	0	5	1	1	3	
16	2	0	1	9	1	5	9	2	7	4	7	6	4	9	5	1	5	2	1	6
17	2	5	3	9	1	1	4	6	2	6	9	5	8	5	8	6	2	3	2	1
18	6	1	4	5	1	3	8	3	1	4	5	9	8	7	3	6	2	3	4	9
	59	76	66	89	49	59	87	73	99	69	79	87	70	97	58	68	41	35	23	16

Height	Number of Dinosaurs
0-9	0
10-19	1
20-29	1
30-39	1
40-49	2
50-59	3
60-69	3
70-79	4
80-89	3
90-100	2

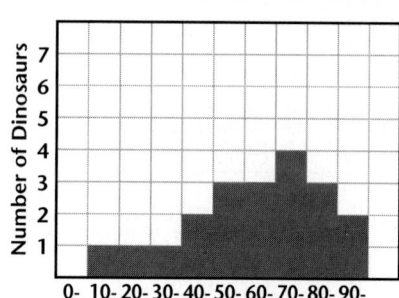

CONTROLLED POPULATION SIMULATION
Dinosaur Height

Week	A	B	C	D	E	F	G	H	I	J	K	L	M	N	O	P	Q	R	S	T
1	9	1	5	6	7	4	2	5	9	5										
2	2	7	9	5	8	3	0	1	3	4										
3	0	4	0	2	4	8	6	3	8	5										
4	2	9	8	8	0	9	9	7	3	0										
5	5	5	5	3	6	8	4	8	5	5										
6	2	9	0	8	0	0	8	2	5	0										
7	7	9	6	5	6	7	3	2	1	1										
8	1	7	9	5	5	5	6	3	4	9										
9	9	0	9	9	9	4	9	1	2	7	2	0	0	4	4	5	9	9	3	1
10	0	6	1	1	5	2	0	5	4	2	1	8	0	5	9	0	2	0	0	8
11	7	3	7	0	8	8	3	5	1	7	3	6	1	0	3	4	2	7	9	1
12	4	6	5	0	3	1	8	5	8	4	1	8	8	4	5	4	9	6	1	8
13	0	2	3	0	4	5	1	0	3	8	2	0	6	5	5	5	8	7	2	7
14	2	8	1	6	8	1	5	4	7	5	5	6	9	4	2	5	3	3	8	9
15	2	0	5	6	2	8	7	3	3	8	9	2	1	5	7	8	9	6	3	4
16	9	4	8	2	4	7	8	1	7	1	8	4	6	1	0	8	2	8	3	4
	61	80	81	66	79	80	79	55	73	71	31	34	31	28	35	39	44	46	29	42

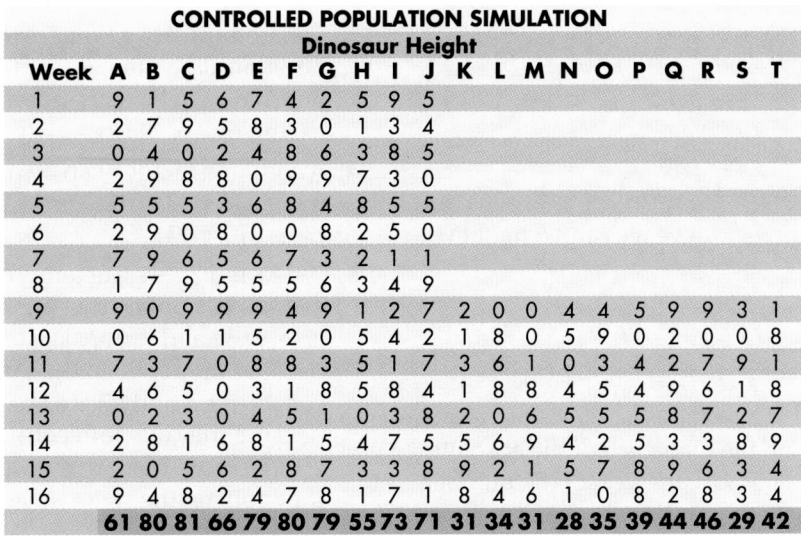

Height	Number of Dinosaurs
0-9	0
10-19	0
20-29	2
30-39	5
40-49	3
50-59	1
60-69	2
70-79	4
80-89	3
90-100	0

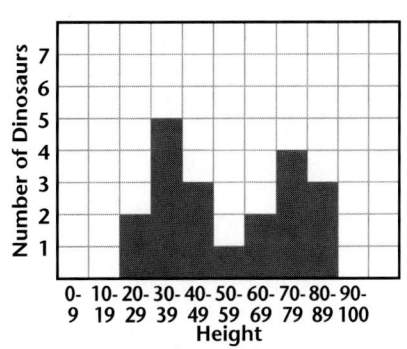

Lesson 2.4, pages 78–83

Exploration

1. 3,2,1; 1,2,3

2.

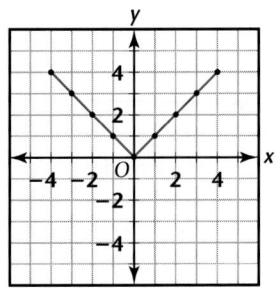

3. The graph is in the shape of a V.

Communicate

ABS is the distance of the number from 0.

1. 29 **2.** 34 **3.** 7.99 **4.** 3.44
To find INT, round the number down to the nearest integer.

5. 29 **6.** 6 **7.** 7 **8.** 3

9. The shape of a V.

10. The shape of steps.

Practice and Apply

32.

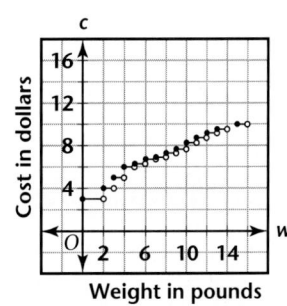

Lesson 2.5, pages 84–89

Communicate

1. e: absolute value; it has a V-shape.

2. b: exponential; it is a curve that starts out flat and then rises quickly.

3. d: reciprocal; it is a wide open curve that rises along the y-axis and decreases along the x-axis, but never touches either axis.

4. f: integer; it increases in steps.

5. a: linear; it is a straight line.

6. c: quadratic; its shape is that of a parabola.

7. Find second differences to show they are constant; the relationship is quadratic.

8. Neither first nor second differences are constant; the relationship is neither linear nor quadratic.

9. The first differences are constant; the relationship is linear.

10. The first differences are constant; the relationship is linear.

Practice and Apply

25.

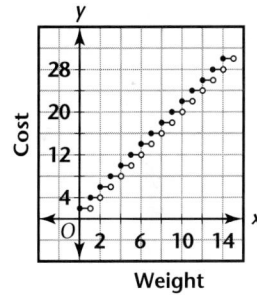

27.

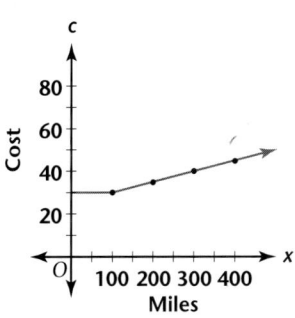

Lesson 3.1, pages 100–106

Exploration

1a. 10

1b. 9

1c. −10

1d. -9

2. Add the absolute value of the integers and use the same sign as the integers.

Communicate

1. The opposite of a number is a number the same distance from 0 on the number line, but on the opposite side of 0. For example, -6 is the opposite of 6.

2. The absolute value of a number is its distance from 0 on the number line. For example, $|-13| = 13$, $|0| = 0$, and $|16| = 16$.

3. Add their absolute values and use the sign that is the same for both integers.

4. Subtract their absolute values and use the sign of the integer with the larger absolute value.

5. Use 2 horizontal and 5 vertical red toothpicks and use 1 horizontal and 3 vertical black toothpicks. Eliminate 1 red and 1 black horizontal toothpick and eliminate 3 red and 3 black vertical toothpicks. One red horizontal toothpick and 2 red vertical toothpicks remain. The answer is 12.

6. Method 1: Since the terms have like signs, *add* the absolute values of -43 and -9. The sign of the sum will be negative because both signs are negative. Add -52 and 128. Since the terms have unlike signs, *find the difference* of the absolute values, 76. The sign of 76 is positive because 128, the number with the larger absolute value, is positive.

$$[-43 + (-9)] + 128$$
$$= -52 + 128 = 76$$

Method 2: Add 128 and -9. The sum is 119. Add 119 and -43. The sum is 76.

$$-43 + [-9 + 128] =$$
$$-43 + 119 = 76$$

734 Additional Answers

Lesson 3.2, pages 107–111

Exploration

2. 0

3. Take away 5 positive tiles.

4. Three positive tiles and 5 negative tiles are left.

5. $3 - 5 = 3 + (-5)$;
$3 + (-5) = -2$

6. $-3 - (-4) = -3 + (+4) =$
$-3 + 4 = 1$

7. Find the difference of the absolute values. Use the sign of the number with the greater absolute value.

Communicate

1. The value of the number does not change.

2. Because adding a pair of positive and negative tiles is equivalent to adding 0, the value of the problem does not change. The neutral pairs provide a model for which some of the tiles can be removed.

3. His account balance would be $-\$20$ (his account would be overdrawn by \$20).

4. Add 7 positive tiles and 7 negative tiles; the value to be subtracted is -7.

5. It means the same as subtracting the number.

6. To subtract a number, add its opposite.

Lesson 3.3, pages 112–117

Communicate

1. An algebraic expression can be numbers, variables, or combinations of the two separated by operation signs; for example, $6x$ and $13y + z$.

2. In a term such as $3n$, the coefficient of n is 3. In $4x + 3y + 3$, the coefficients of x and y are 4 and 3, respectively.

---

3. The Commutative Property of Addition allows the order of terms in addition to be rearranged; for example, $13x + 12y = 12y + 13x$.

4. A common factor x can be removed from $3x + 8x$ for $(3 + 8)x = 11x$.

5. $7x$, $2x$ and $3x$ are like terms because they contain the same variable, x. $3z$ and $-z$ are like terms because they contain the same variable, z. 5 and 23 are like terms because they are both constants. $7y$ and $3y$ are like terms because they contain the same variable, y.

6. Start with 5 positive x-tiles and 2 positive 1-tiles. Add 3 positive x-tiles and 4 negative 1-tiles. Collect the like tiles. The result is 8 positive x-tiles and 2 negative 1-tiles or $8x - 2$.

Lesson 3.4, pages 118–123

Communicate

1. Start with 4 positive x-tiles and take 3 of them away. One positive x-tile remains. The result is x.

2. Start with 3 positive x-tiles and 2 positive 1-tiles. Take away 2 positive x-tiles and 1 positive 1-tile. One positive x-tile and one positive 1-tile remain. The result is $x + 1$.

3. Start with 5 positive x-tiles and 3 positive 1-tiles. Since there are not 4 positive 1-tiles to remove, add 1 pair of positive and negative 1-tiles. Remove 2 positive x-tiles and 4 positive 1-tiles. Three positive x-tiles and one negative 1-tile remain. The result is $3x - 1$.

4. Let r represent the number of reams of paper.

$5r + 100 - (2r + 50)$ Given
$5r + 100 + (-2r - 50)$ Definition
  of subtraction
$(5r - 2r) + (100 - 50)$ Rearrange
  terms.
$3r + 50$ Combine like terms.

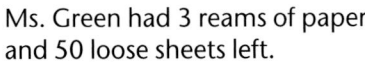

Ms. Green had 3 reams of paper and 50 loose sheets left.

5. To subtract an expression, change the signs of the terms of the expression to be subtracted to their opposites and then add the expressions.

6. $(7y + 9x + 3) - (3y + 4x - 1)$
Given
$(7y + 9x + 3) + (-3y - 4x + 1)$
Definition of subtraction
$(7y - 3y) + (9x - 4x) + (3 + 1)$
Rearrange terms.
$4y + 5x + 4$
Combine like terms.

7. $m < 0$

Lesson 3.5, pages 124–130

Communicate

1. The Addition Property of Equality states that if equal amounts are *added* to the expressions on both sides of an equation, the expressions remain equal. The Subtraction Property of Equality states that if equal amounts are *subtracted* from the expressions on both sides of an equation, the expressions remain equal. If a number is *added* to an expression, use the Subtraction Property of Equality. If a number is *subtracted* from an expression, use the Addition Property of Equality.

2a. Answers may vary. For example, John started out the week with $20 and finished the week with $50. How much money did he make during the week?

2b. Answers may vary. For example, what was Luke's inventory at the start of the game if he ended the game with 50 bags of peanuts and sold 20?

3a. Model the left side of the equation with 1 *x*-tile and 6 positive 1-tiles and the right side with 10 positive 1-tiles. Take away 6 positive 1-tiles from each side. Four

positive 1-tiles remain on the right side. $x = 4$

3b. Model the left side of the equation with 1 *x*-tile and 6 negative 1-tiles and the right side with 10 positive 1-tiles. Add 6 positive 1-tiles to each side. Simplify the left side of the equation by removing 6 pairs of positive and negative 1-tiles. Sixteen positive 1-tiles remain on the right side. $x = 16$

4. $s + x = r$ Given
$s + x - x = r - x$ Subtraction Property of Equality
$s = r - x$ Combine like terms.

5. $m + b = n$ Given
$m + b - b = n - b$ Subtraction Property of Equality
$m = n - b$ Combine like terms.

Lesson 3.6, pages 131–137

Exploration

1a. True

1b. True

1c. False

1d. False

2. 1.10

3. Subtract 1.25 from both sides of the inequality.

Communicate

1. Add 2 and 3 which makes the inequality $5 \geq 5$ true.

2. Add 2 and 3 which makes the inequality $4 \leq 5$ true.

3. Subtract 3 from 7 which makes the inequality $6 < 4$ false.

4. Subtract 2 from 10 which makes the inequality $8 > 8$ false.

5. The properties for equality and inequality are both used to solve addition equations and inequalities by using subtraction, and subtraction equations and inequalities by using addition.

6. $3x - 4 \leq 2x + 1$ Given
$3x - 4 - 2x \leq 2x + 1 - 2x$

Subtraction Property of Inequality
$3x - 2x - 4 \leq 2x - 2x + 1$
Rearrange like terms.
$x - 4 \leq 1$ Combine like terms.
$x - 4 + 4 \leq 1 + 4$
Addition Property of Inequality
$x \leq 5$ Combine like terms.

7. Start by solving the inequality.
$x - 4 \leq 9$ Given
$x - 4 + 4 \leq 9 + 4$
Addition Property of Inequality
$x \leq 13$ Combine like terms.

The value 13 represents equality and all values less than 13 represent inequality.

8. Solve the inequality.
$x + 3 < 7$ Given
$x + 3 - 3 < 7 - 3$
Subtraction Property of Inequality
$x < 4$ Combine like terms.

Draw a number line with an open dot on 4 because the inequality sign is less than x, then draw an arrow to the left. $x < 4$

$$\begin{array}{c}\longleftarrow\!\!+\!\!+\!\!+\!\!+\!\!+\!\!+\!\!+\!\!+\!\!\diamond\!\!+\!\!+\!\!\longrightarrow\\ \;-5\;\;-3\;\;-1\;0\;1\;\;\;3\;\;\;5\end{array}$$

9. Write a variable (x) and then a less than sign because the arrow is pointing to the left. Write a 4 after the less than sign because it is the number under the open circle $x < 4$.

10. Inclusive means to include numbers in between the numbers given, including the numbers given. The whole numbers between 5 and 10 inclusive are 5, 6, 7, 8, 9, and 10.

Lesson 4.1, pages 146–150

Exploration 1

1a. $-2; -4; -6$

1b. $-3; -6; -9$

1c. $3; 6; 9$

2a. positive   **2b.** negative

2c. negative

2d. positive

3a. 48

3b. -39

3c. -486

3d. 80

4a. positive

4b. negative

Exploration 2

1a. 56

1b. -15

1c. -8

1d. 8

2a. $56 \div 8 = 7; 56 \div 7 = 8$

2b. $-15 \div 5 = -3; -15 \div -3 = 5$

2c. $-8 \div -4 = 2; -8 \div 2 = -4$

2d. $8 \div -1 = -8; 8 \div -8 = -1$

If the product is positive, both divisors are negative or both are positive. If the product is negative, one of the divisors is negative when the other is positive.

3a. positive

3b. negative

3c. negative

3d. positive

4a. -24

4b. -9

4c. 8

4d. -7

5a. positive

5b. negative

Communicate

1. It would decrease by $100.

2. It would increase by $30.

3. Since the 3 is positive, an amount is added to the bank account and since the 5 is negative, the amount added to the account is negative. Therefore, the transaction is add 3 withdrawals of $5 each.

4. Divide -15 by both 3 and -5.

$-15 \div 3 = -5$ and
$-15 \div -5 = 3$

5. Find the difference between each of the numbers and 120:
$95 - 120 = -25$,
$119 - 120 = -1$,
$110 - 120 = -10$,
$130 - 120 = 10$,
$141 - 120 = 21$, and
$155 - 120 = 35$. The sum of these differences is 30. Divide 30 by 6 (the number of values) to get 5. Therefore, the average difference is 5. Add this 5 to the guess of 120 for an average number of 125; yes

Lesson 4.2, pages 151–157

Exploration

1–2. Answers may vary.

n	-5	-4.2	$-\frac{1}{2}$	0	$\frac{3}{4}$	6	7.3
$-1(n)$	5	4.2	$\frac{1}{2}$	0	$-\frac{3}{4}$	-6	-7.3

3. Their signs are the opposite of those in the first row.

4. opposites

Communicate

1. If Jan works 4 hours and earns $6 an hour, this means she will earn $6 four times or $6(4) = 24$, $24.

2. If Jan works 2.5 hours and earns $6 an hour, this means she will earn $6(2.5) = 15$, $15.

3. If Jan works h hours and earns $6 an hour, this means she will earn $6(h) = 6h$.

4. Multiply the 10 by 2 and add 3 to get 23.

5. Multiply the 3 and 4 to get 12 and y and y to get y^2 so that the result is $12y^2$.

6. Multiply $4m$ by -2 to get $-8m$ and multiply 5 by -2 to get -10. The result is $-8m - 10$. An expression is simplified when no more operations can be performed.

7. Divide $5p$ by -5 to get $-p$ and divide -15 by -5 to get 3 so that the result is $-p + 3$.

Lesson 4.3, pages 160–166

Communicate

1a. Answers may vary. For example, Benjamin makes $5 an hour at his part-time job. If his pay for one week was $100, how many hours did he work?

1b. Answers may vary. For example, when David and Marisa divided their game tokens, each received 10. How many tokens did they have originally?

2. Divide both sides of the equation by 592 for $x \approx 1.4$.

3. Add 246 to both sides of the equation for $x = 774$.

4. Divide both sides of the equation by 5 (or multiply both sides by $\frac{1}{5}$) for $x = \frac{1}{50}$.

5. Subtract 10 from both sides of the equation for $x = -5$.

6. Divide both sides of the equation by 2π for $r = \frac{C}{2\pi}$.

7. Answers may vary. For example, if Mary worked 4 math problems in 40 minutes, how long did it take her to work 1 problem?

8. Answers may vary. For example, if Mr. Johnson divided his PE class into 4 teams with 20 students on each team, how many students are in his PE class?

9. Answers may vary. For example, in 5 minutes, Adam collected $15 from people coming into the basketball game. If each person paid $2.50, how many people came to the game in that 5 minutes?

10. Answers may vary. For example, Paul organized his magazines by placing them into 4 piles of 20 each. How many magazines did Paul have?

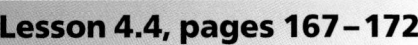

Lesson 4.4, pages 167–172

Communicate

1. A rational number is a number that can be expressed as the ratio of 2 integers with 0 excluded from the denominator. Examples will vary, $\frac{2}{3}$, $\frac{-6}{1}$, $\frac{94}{3}$.

2. An integer is a rational number because it can be expressed as the ratio of itself and 1.

3. Start at 0 and move right $\frac{1}{2}$ of a unit. Then move left 3 units. The result is $-2\frac{1}{2}$.

4. Start at 0 and move right $\frac{1}{2}$ of a unit. Then move left $\frac{5}{6}$ of a unit. The result is $\frac{-2}{6}$ or $\frac{-1}{3}$.

5. For any number r other than 0, $r \cdot \frac{1}{r}$. The reciprocal of -3 is $\frac{-1}{3}$ because $-3 \cdot \frac{-1}{3} = \frac{3}{3} = 1$.

6. Multiply both sides by 8 (the reciprocal of the coefficient of the y-term) for $y = \frac{-16}{3} \approx -5.33$. Multiply both sides by 24 (the least common denominator) for $3y = -16$. Then divide both sides by 3 for $y = -\frac{16}{3} \approx -5.33$.

Lesson 4.5, pages 173–179

Communicate

1. Place the percent over 100 and divide.

2. Draw a horizontal bar and label the lower left and right corners 0% and 100%, respectively. Label the top left corner 0. On the bottom of the bar, near the left of the half-way point, write 40% and draw a line through the bar at this point. Write x on the top of the bar at this line. At the top right corner of the bar, write 50. Shade the portion of the bar to the left of the 40% line.

3. Draw a horizontal bar and label the lower left and right corners 0% and 200% respectively. Label the top left corner 0. At the half-way point on the bottom of the bar, write 100% and draw a line through the bar at this point. Write 50 on the top of the bar at this line. At the top right corner write x. Shade the entire bar.

4. Draw a horizontal bar and label the lower left and right corners 0% and 100% respectively and the top left corner 0. At the half-way point on the bottom, write 50% and draw a line through the bar at this point. Write 30 on the top of the bar at this line. At the top right corner write x. Shade the portion of the bar to the left of the 50% line.

5. Draw a horizontal bar and label the lower left and right corners 0% and 100%, respectively. Label the top left corner 0. Label the top right corner 80. Just left of 50% of the way across the bottom of the bar, write x%. Draw a line through the bar at this point, and write 60 at the top of the line. Shade the portion of the bar to the left of the x% line.

6. less; the base (50) is multiplied by a fraction $\left(\frac{40}{100}\right)$ whose value is less than 1.

7. more; the base (50) is multiplied by a fraction $\left(\frac{200}{100}\right)$ whose value is greater than 1.

8. more; 30 is half of 60, which is more than 50.

9. more; 50% of 80 is the same as $\frac{1}{2}$ of 80, which is equal to 40, so 60 must be more than 50%.

10. Let x represent the number of ounces of pretzels that must be eaten. Since 1 ounce contains 10% of the U.S. RDA, set up the proportion $\frac{1}{10} = \frac{x}{100}$, $x = 10$.

Practice and Apply

21. 28;

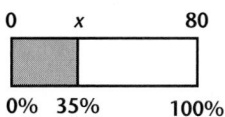

22. 20%;

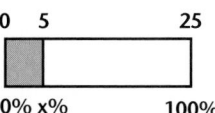

23. 0.8;

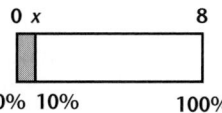

24. 60;

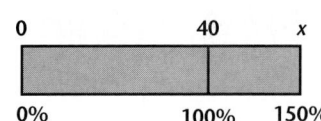

25. 44.4;

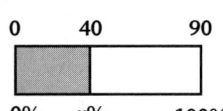

26. 90;

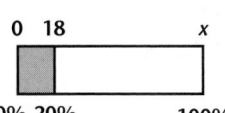

Lesson 4.6, pages 180–185

Communicate

1. Subtract p from both sides. Next subtract 57 from both sides of the equation and then divide both sides by 22.

2. Subtract 33 from both sides of the equation and then multiply both sides by 24.

3. Add $2z$ to both sides. Next add 5 to both sides and then divide both sides by 7.

4. Let s represent the score Brittany needs on her third test. The average is calculated by adding the scores and then dividing by the number of scores. In other words,

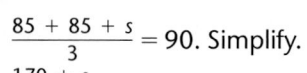

$\frac{85 + 85 + s}{3} = 90.$ Simplify.

$\frac{170 + s}{3} = 90.$ Multiply both sides by 3, $170 + s = 270.$ Subtract 170 from both sides to find that $s = 100.$

5. Set $Y_1 = 2x + 3$ and $Y_2 = 8 - 2x$. Graph Y_1 and Y_2 on the same axis. If the point of intersection is not on the screen then zoom out until it is visible on the screen. Once it is visible, use the trace function to find the approximate intersection. Zoom in on the intersection and use the trace function again to increase the accuracy of the reading. $x = 1.25$

Lesson 4.7, pages 186–191

Exploration 1

1c. =

1d. >

1e. >

2c. >

2d. >

3. The inequality sign does not change.

4. If each side is multiplied by 0, both sides are then equal to 0. If the inequality sign was >, <, or ≠, the inequality then becomes false. If the inequality sign was ≥ or ≤, the inequality remains true. Neither side can be divided by 0 because division by 0 is undefined.

5. The inequality sign is reversed.

6a. When you multiply or divide each side of an inequality by the same positive number the inequality sign remains the same.

6b. When you multiply or divide each side of an inequality by the same negative number reverse the inequality sign.

Exploration 2

2. <

3. >

4.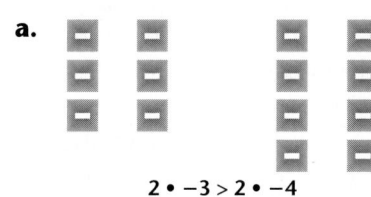

$-3 > -4$

a.

$2 \cdot -3 > 2 \cdot -4$
$-6 > -8$

b.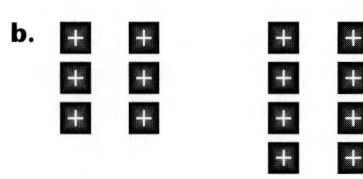

$-2 \cdot -3 < -2 \cdot -4$
$6 < 8$

Exploration 3

2. <

3. >

4.

$-4 < 2$

a.

$-4 \div 2 < 2 \div 2$
$-2 < 1$

b.

$-4 \div (-2) > 2 \div (-2)$
$2 > -1$

Communicate

1. true; 7 is less than 8.

2. false; 7 is equal to 7.

3. true; 7 is equal to 7.

4. false; 7 is equal to 7.

5. Subtract 1 from each side of the inequality to find that $x > 3$.

6. Add 3 to each side of the inequality to find that $x \leq 16$.

7. Divide both sides of the inequality

---

by -3 and reverse the inequality sign to >, $p > -4$.

8. Subtract $2x$ from each side of the inequality, $2x - 2 \geq 3$. Add 2 to each side to find that $2x \geq 5$. Divide each side by 2 for $x \geq \frac{5}{2}$.

Lesson 4.8, pages 192–198

Communicate

1. A measurement specification that falls between a minimum of 44.999 cm and a maximum of 45.001 cm.

2. Choose a variable, for example x, and subtract 45 from x. Write the absolute value of this quantity as less than 0.001, $|x - 45| \leq 0.001$.

3. Since the absolute value function measures distance from 0, the distance must be measured in both the positive and negative directions.

4. Write the absolute value of a variable x minus 7. Make this quantity less than or equal to 3 because 3 is the farthest from -7 than any of the desired numbers will be. Therefore the inequality is $|x - (-7)| \leq 3$ or $|x + 7| \leq 3$.

5. Solve the inequality to determine the boundary numbers. Choose numbers on both sides of these numbers and substitute them into the inequality. Values that make the inequality true will be inside the boundary. Values that make the inequality false will be outside the boundary.

Practice and Apply

14. $-4 < x < 10$

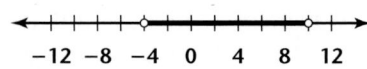

15. $x > 4$ or $x < -12$

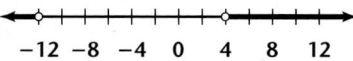

16. $4 \leq x \leq 12$

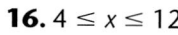

17. $x \geq 7$ or $x \leq 3$

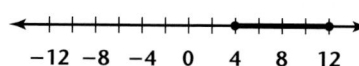

18. $x > 8$ or $x < -4$

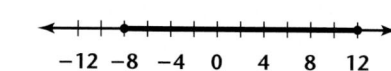

19. $-8 \leq x \leq 12$

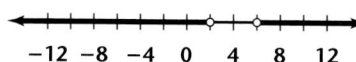

20. $-6 < x < 4$

21. $x < 2$ or $x > 6$

Lesson 5.1, pages 208–215

Communicate

1. Start at any point, go up 4 units (vertically), and then move right 3 units (horizontally). Start from this new point. Repeat the process (up 4, right 3) once more, and then draw a line through the three points.

2. The slope is negative; the line falls downward from left to right.

3. Locate two coordinates on line AC, then write the differences in y-coordinates as the numerator of a fraction, and write the difference in x-coordinates as the denominator of the fraction.

4. Points A, B, and C are on the same line.

5. $\frac{s}{t}; \frac{-s}{t}$; the steepness of the two lines is the same, but one slope is positive and one slope is negative.

---

6. Calculate the difference in the y-coordinates, divided by the difference in the x-coordinates. Subtract the x-coordinates in the same order as the corresponding y-coordinates. Slope of line $k = \dfrac{-3 - (-4)}{5 - 9}$ or $\dfrac{-4 - (-3)}{9 - 5}$, which simplifies to $-\dfrac{1}{4}$.

Practice and Apply

38.

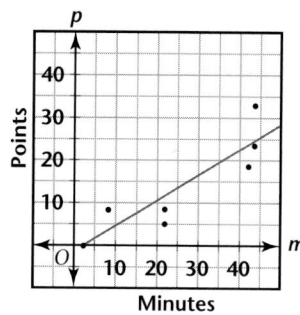

If the ruler is well-placed, it will appear to go through the data points $(2, 0)$ and $(23, 43)$. From these points the slope can be calculated to be about $\dfrac{23}{41} \approx 0.56$ or about $\dfrac{1}{2}$. A player scores about a point every minute of play. (If a calculator is used $m \approx 0.58$; $y \approx 0.58x - 1.25$; $r \approx 0.87$.)

52. The three lines have the same slope (4) but they cross the y-axis at different points.

53.

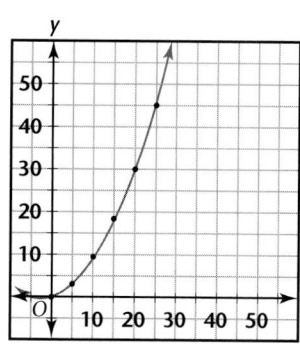

Yes, it is part of a quadratic function found in Chapter 2.

Lesson 5.2, pages 218–223

Exploration 1

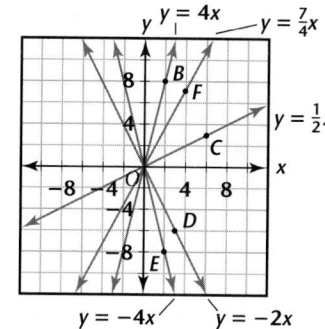

The slope, m, is the quotient of the y-coordinate divided by the x-coordinate.

Exploration 2

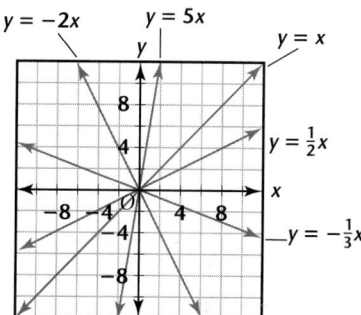

1. They all pass through $(0, 0)$.

2. Their slopes are different.

3. It would pass through $(0, 0)$ with slope 3.

4.

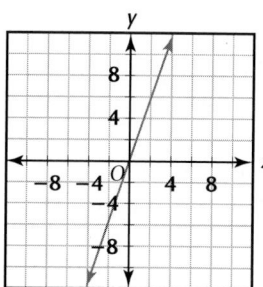

5. The larger the value of m, the steeper the line. Lines with positive slopes slope upward from left to right, and lines with negative slopes slope downward from left to right.

Exploration 3

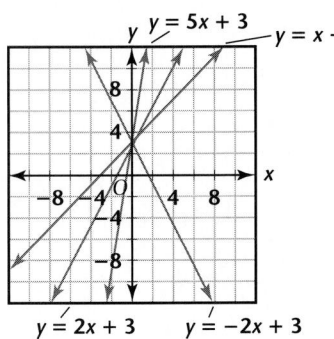

1. They all cross the y-axis at (0, 3).
2. Their slopes are different.
3. The line will pass through (0, 3) with slope 2.
4.

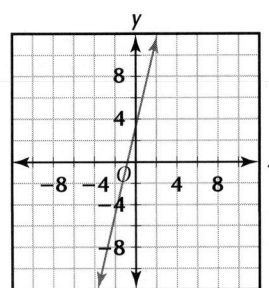

5. Adding 3 raises the graph so that the y-intercept is at (0, 3).
6.

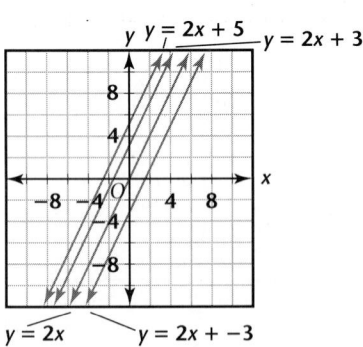

   These lines all have the same slope.
7. The lines cross the y-axis at different points.
8. The line will be parallel to the other four lines, but it will have y-intercept at (0, −4).

9. Graphs may vary.
10. The value of b is where the line crosses the y-axis. It raises or lowers the line.

Communicate

1. The slope of a line through the origin is the value of the y-coordinate divided by the value of x-coordinate. In the equation of a line $y = mx + b$, m represents the slope: the larger the value of m, the steeper the line. If m is positive, the line slopes upward from left to right, if m is negative, the line slopes downward from left to right. b represents the y-intercept of the line.

2. $y = \frac{1}{2}x$

3. $y = 2x$

4. $y = \frac{9}{4}x$

5. $y = -\frac{2}{3}x$

6. The two points are (0,0) and (3, 6). The slope of a line passing through the origin is the value of the y-coordinate divided by the x-coordinate, $\frac{6}{3} = 2$, or $m = \frac{6-0}{3-0} = 2$. Since the y-intercept is at 0, $b = 0$. The equation for the line is $y = 2x$.

7. $y = ax$

Practice and Apply

8.

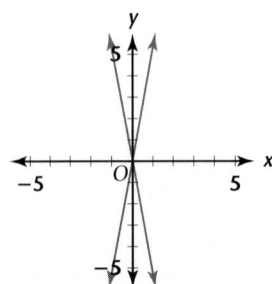

The lines both have y-intercept at (0, 0). They have the same steepness, but one has a positive

slope and the other has a negative slope.

9.

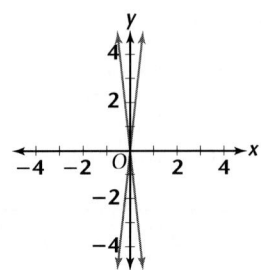

The lines both have y-intercept at (0, 0). They have the same steepness, but one has a positive slope and the other has a negative slope.

10.

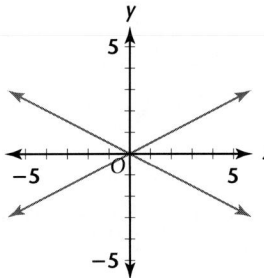

The lines both have y-intercept at (0, 0). They have the same steepness, but one has a positive slope and the other has a negative slope.

11.

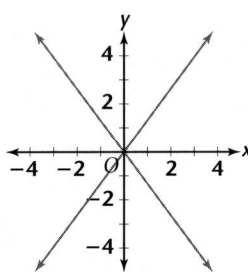

The lines both have y-intercept at (0, 0). They have the same steepness, but one has a positive slope and the other has a negative slope.

12.

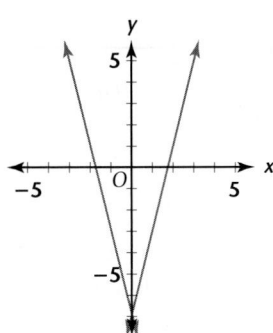

The lines both have *y*-intercept at $(0, -7)$. They have the same steepness, but one has a positive slope and the other has a negative slope.

13.

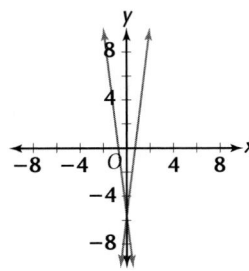

The lines both have *y*-intercept at $(0, -7)$. They have the same steepness, but one has a positive slope and the other has a negative slope.

14. The new line will have the opposite (negative or positive) slope from the one that is graphed, but the *y*-intercept will be the same.

16. $y = \frac{5}{2}x$

17. $y = \frac{8}{5}x$

18. $y = 9x$

19. $y = \frac{1}{2}x$

20. $y = \frac{3}{7}x$

26.

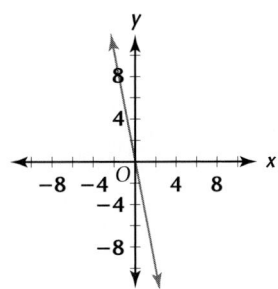

27.

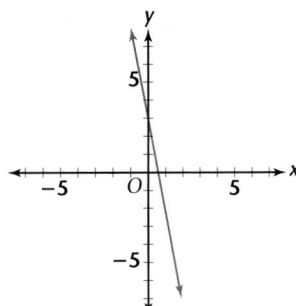

28.

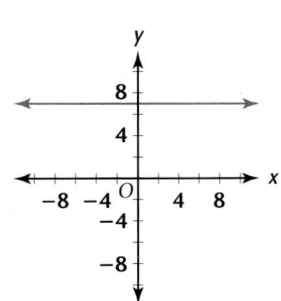

29.

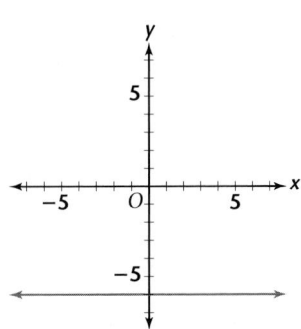

30.

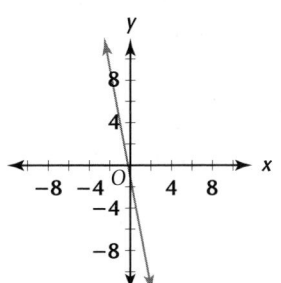

31.

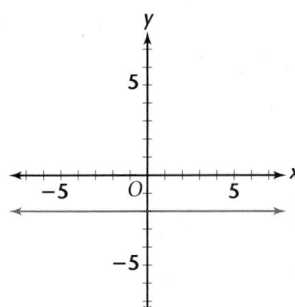

32.

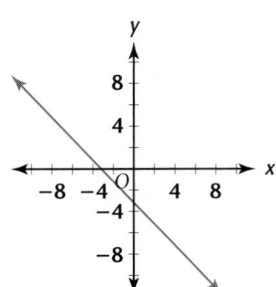

33.

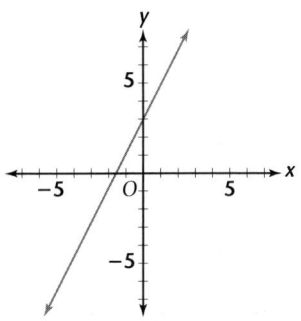

34.

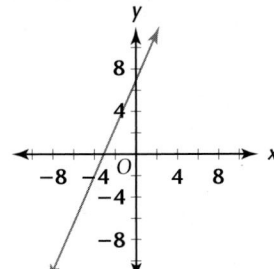

35.

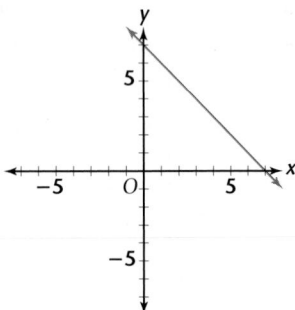

36.

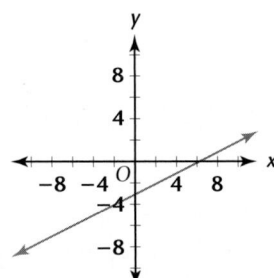

37.

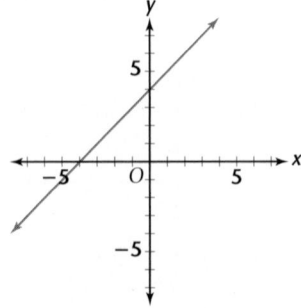

38. $m = 0.69$, if measured in cubits, the slope is 1.4; $m = 0.69$, if measured in feet, the slope is 1.4; the slopes are approximately the same.

Lesson 5.3, pages 224–232

Communicate

1. A change in b changes the y-intercept of the line.

2. A change in m changes the slope of the line.

3. Find the difference in the y-coordinates divided by the difference in the x-coordinates.

$$m = \frac{5 - 2}{3 - 7} = -\frac{3}{4} \text{ or}$$
$$m = \frac{2 - 5}{7 - 3} = -\frac{3}{4}$$

4. Substitute zero for x.

5. The slope is the numerical coefficient of x when the equation is written in the form $y = mx + b$. For the line $y = -x + 10$ the slope is -1.

6. Mark the y-intercept $(0, -3)$. Move 1 unit to the right for the run and 2 units up for the rise. Mark this point. Draw a line through the 2 points.

7. Calculate the slope, $m = \frac{4 - 5}{-2 - 1} = \frac{1}{3}$. Write the slope in the $y = mx + b$ form of the equation. Substitute one of the points to find the value of the y-intercept, b. The equation is $y = \frac{1}{3}x + \frac{14}{3}$.

Lesson 5.4, pages 233–239

Communicate

1. Add $3x$ to both sides, and then add 2 to both sides. The equation is $3x + 5y = 2$.

2. Find the x-intercept by substituting 0 for y, $3x + 6(0) = 18$. Then sub-

stitute 0 for x to find the x-intercept, $3(0) + 6y + 18$. The intercepts are $(6, 0)$ and $(0, 3)$.

3. Plot the intercepts, $(6, 0)$ and $(0, 4)$, and connect the points with a line.

4. Isolate the y-term and, and then divide both sides of the equation by the coefficient of y.

$$y = \frac{1}{3}x - 3$$

5. Use the two points given to calculate the slope, then substitute the slope value and one of the points into the equation $y - y_1 = m(x - x_1)$.

$$m = \frac{4 - (-8)}{-2 - 4} = -2$$
$$y - 4 = -2(x + 2)$$

6. Rewrite the equation in the form $y = mx + b$. The value of m is the slope.

$$y = \frac{5}{2}x + 20, \quad m = -\frac{5}{2}$$

Practice and Apply

45. October 1, 1908; 10-01-08

46. 1-31-58; 580131

47. July 20, 1969; 690720

48. Because two digit spaces are needed to allow for all 12 months

Lesson 5.5, pages 240–245

Communicate

1. Use any two points on the line, for example $(2,3)$ and $(4, 3)$, $m = 0$. The line is horizontal.

2. Use any two points on the line, for example $(1, 0)$ and $(1, 2)$, m is undefined. The line is vertical.

3. The slope is the numerical coefficient of x, 5. The line crosses the y-axis at $(0, 0)$.

4. Write the equation in the form $y = mx + b$, $y = \frac{1}{5}x$. The slope is $\frac{1}{5}$. The line crosses the y-axis at $(0, 0)$.

5. The equation can be written as $y = 0x + 15$. The slope is zero and the b value is a constant 15. The value of the y-coordinate is always the same.

6. The run of a vertical line is 0. The slope, $\frac{\text{rise}}{\text{run}}$, is undefined.

7. The rise of a horizontal line is 0. The slope, $\frac{\text{rise}}{\text{run}}$, is zero.

Lesson 5.6, pages 246–251

Exploration 1

1. $\frac{3}{5}$

2. $-\frac{5}{3}$

3. negative

4. The slope of the perpendicular line is the negative reciprocal of the original.

5. The result is the same.

Exploration 2

1. Set a square window. Graph $y = 2x$.

2. Graphs will vary.

3. $y = -\frac{1}{2}x$

Communicate

1. The slope of $y = 4x + 3$ is 4. Any line of the form $y = 4x + b$ will be parallel to the given line.

2. Write the reciprocal of $\frac{3}{2}$, with the opposite sign, $-\frac{2}{3}$.

3. The slope of the line $y = \frac{1}{3}x + 2$ is $\frac{1}{3}$ so the perpendicular slope is $-\frac{3}{1}$ or -3.

4. The slope of the line $y = 4x + 3$ is 4. Any line of the form $y = -\frac{1}{4}x + b$ will be perpendicular to the given line.

5. First write the equation in slope-intercept form, $y = 3x - 4$, then change the equation to standard form, $3x - y = 4$.

6. First write the equation in slope-intercept form, $y = -6x + 12$. Determine the slope of the perpendicular line. Write the equation of the perpendicular line, $y = \frac{1}{6}x + 12$. The equation in standard form is $x - 6y = -72$.

7. Find the slope of the line $x - 5y = 15$ by rewriting the equation in the form $y = mx + b$ ($y = \frac{1}{5}x - 3$, $m = \frac{1}{5}$). Use the negative reciprocal of this slope and y-intercept 0 to write the new equation: $y = -5x + 0$ or $y = -5x$.

Review

15–16.

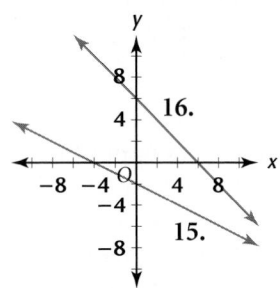

17–18

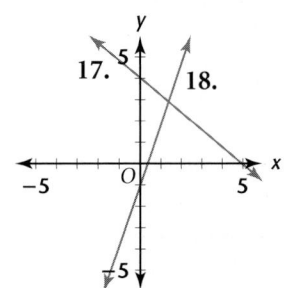

Lesson 6.1, pages 260–266

Communicate

1. Isolate y on one side of the equal sign. This form of the equation, $y = mx + b$, allows you to substitute values for x and find values for y.

2. Substitute various values for x and then determining the correspond-

ing y-values for each equation. The point of intersection is the x-value which generates the same y-values in both equations.

3. Plot the ordered pairs from the table and connect the points to form the two lines. The common solution is the point where the lines intersect, $(-1, -2)$.

4. Write the equations in slope-intercept form. Graph the equations by substituting values for x and finding values for y to plot points and graph lines. Find the point where the two lines intersect. A good estimate of the solution is, $\left(\frac{7}{2}, \frac{-1}{2}\right)$.

5. A check allows you to see if your estimate is reasonable.

Practice and Apply

26–27.

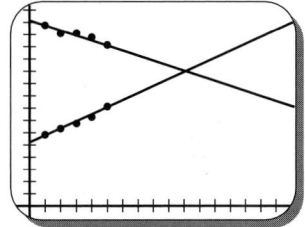

Lesson 6.2, pages 267–271

Exploration 1

1. A reasonable answer is about 3.8.

2. The y-coordinate of the intersection point is not a whole number.

3. 3.75

4. Substitute the known value into one of the equations and solve for the unknown value of the other variable.

5. Graphing or substitution

6. Substitution. Substitute 3 for x into the first equation and solve for y. $y = -\frac{5}{2}$ or -2.5

Exploration 2

1. $x = -y + 1$

1a. $3(-y + 1) + 2y = 4$

1b. $y = -1$ and $x = 2$

2. $y = -x + 1$

2a. $3x + 2(-x + 1) = 4$

2b. $x = 2$ and $y = -1$

3. Both methods yield the same solution. The method in Step 1 involved substituting an expression for x which yielded a solution for y whereas the method in Step 2 involved substituting an expression for y which yielded a solution for x.

4. Solve for either variable. Substitute the value of one variable in terms of the other variable (as an expression) into the other equation. Solve, and then substitute the known variable into the equation to find the value of the other variable.

Exploration 3

1. $y = 2x - 6$

2. $15x - 5(2x - 6) = 30; 0$

3. Solve for the variable whose coefficient is 1. Substitute this expression into the other equation to solve for the value of the unknown variable. Substitute this value into the equation to determine the value of the other variable.

Communicate

1. Substitute 42 for y and solve for x; $x = 17$.

2. Solve the first equation for y, because the coefficient of y is 1; solve for x and y. The solution is (2, 10).

3. Solve the first equation for x, because the coefficient of x is 1; solve for x and y. The solution is (10, 1).

Lesson 6.3, pages 272–277

Exploration

1. The sum is 0.

2a. $2y$ and $-2y$

2b. $8x = 16$

2c. $(2, \frac{1}{2})$

2d. $7 = 7$ True, $\quad 9 = 9$ True

3a. $8a + 4b = 36$

3b. $(4, 1)$

3c. $9 = 9$ True, $\quad 8 = 8$ True

4. Addition of opposites causes one of the variables to be eliminated when using the Addition Property of Equality. This allows the value of the other variable to be determined which in turn will lead to the determination of the first variable.

Communicate

1. y and $-y$ are opposites. Use the Addition Property of Equality to solve for x, then y. (9, 4)

2. $-3y$ and $3y$ are opposites. Use the Addition Property of Equality. (4, 0)

3. $2a$ and $-2a$ are opposites. Use the Addition Property of Equality. $\left(\frac{13}{2}, -7\right)$

4. Multiply each side of $x - 2y = 13$ by -4.

5. Multiply each side of $a - 2b = 7$ by -2.

6. Multiply each side of $3m - 5n = 11$ by 3 and $2m - 3n = 1$ by 5.

7. Multiply each side of equation 1 by -2. Solve for x, then y. $\left(11, -\frac{13}{3}\right)$

8. Multiply each side of equation 1 by -3 and each side of equation 2 by 2. Solve for x and y. $(-2, -1)$

9. Multiply each side of equation 1 by -3. Solve for x and y. $\left(\frac{1}{3}, -\frac{1}{2}\right)$

10. Solve the system $\begin{cases} x + y = 156 \\ x - y = 6 \end{cases}$ by the elimination method. (81, 75)

Lesson 6.4, pages 278–283

Exploration 1

1. The lines are the same.

2. The equation for line AB is $y = 2x - 4$ and the equation for line CD is $y = 2x - 4$; the slopes are the same.

3. $0 = 0$

4. The variables are eliminated; yes; yes; yes; they have an infinite number of points in common.

Exploration 2

1a. The slopes are not equal.

1b. The y-intercepts may or may not be equal.

1c. There is only one common solution.

2a. The slopes are equal.

2b. The y-intercepts are not equal.

2c. There is no common solution, the lines never intersect.

3a. The slopes are equal

3b. The y-intercepts are equal.

3c. There are infinitely many solutions because both equations describe the same line.

Communicate

1. Since both the slopes and y-intercepts of both lines are the same, the system is dependent.

2. Since the slopes of these two lines are not the same but their y-intercepts are, the system is independent.

3. Since the slopes of both equations are the same but their y-intercepts are not, the system is inconsistent.

4. Multiply both sides of the equation by the same number. The slope and the y-intercept remain the same.

5. Use the slope of the given equation

(−3) and choose a different y-intercept.

6. If two lines intersect at only one point, the lines have different slopes. The y-intercepts may or may not be the same.

Lesson 6.5, pages 284–291

Exploration

1.

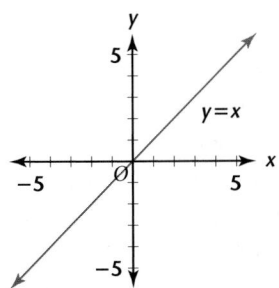

2. $x < y$

3. $x > y$

Communicate

1. If the inequality symbol includes the equal sign, then use a solid line, otherwise use a dashed line.

2. Graph the boundary line, $y = -\frac{1}{2}x + 1$. Substitute a point that is not on the line. If it satisfies the inequality, shade the side of the line containing the point. If the point does not satisfy the inequality, shade the other side of the line. Answers may vary. Try (0, 0), because 0 is a simple number to check.

3. Graph the boundary line using dashed lines. Locate a point that satisfies each inequality, and shade the half-plane that contains each point.

4. Graph the boundary line using dashed lines. Locate a point that satisfies each inequality, and shade the half-plane that contains each point.

5. Graph the boundary lines using solid lines. Locate a point that satisfies each inequality, and shade the half-plane that contains each point.

6. Graph $x + y \leq -2$ using a solid line, and graph $x + y > -2$ using a dashed line. Locate a point that satisfies each inequality, and shade the half-plane that contains each point.

7. They plan for *no more* than 20 perennials, so $p \leq 20$. Their budget has room for *at least* $30, so $5p + 1.5a \geq 30$. Use a solid line to graph the boundary line, $p = 20$. Use a solid line to graph $5p + 1.5a = 30$. Locate a point that satisfies each inequality, and shade the half-plane that contains each point. Only positive answers in the overlapping region make sense.

Practice and Apply

8.

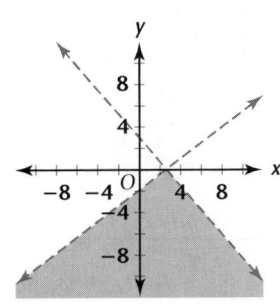

9.

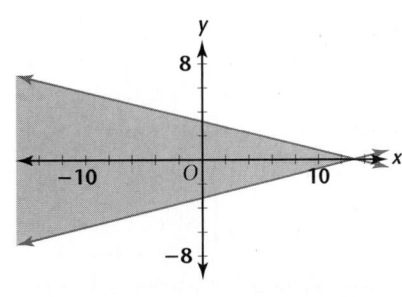

10.

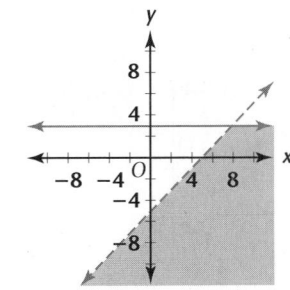

11.

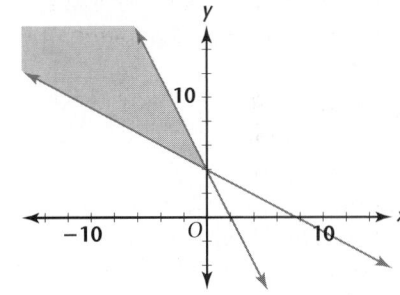

12.

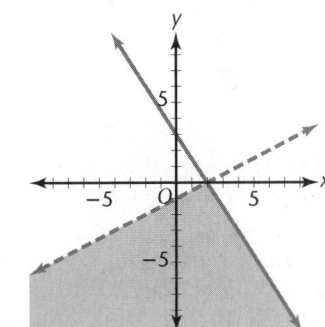

13.

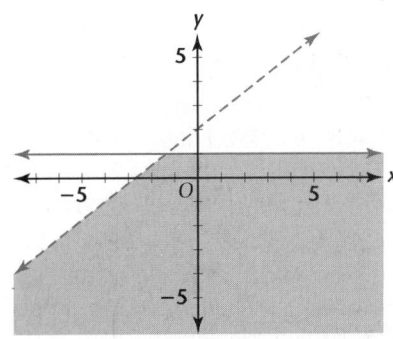

14.

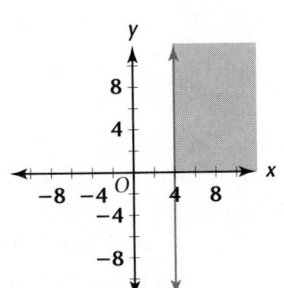

15.

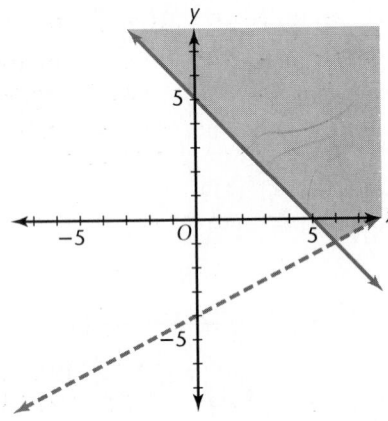

16.

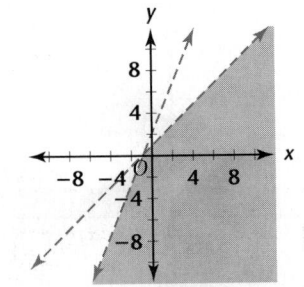

Lesson 6.6, pages 292–297

Communicate

1. There are 20 nickels and quarters, so $q + n = 20$. The money mixture totaled $2.60, so $0.25q + 0.05n = 2.60$.   8 quarters

2. There are 20 l of the final solution, so $x + y = 20$. There are 2.6 l of pure alcohol in the final solution, so $0.25x + 0.05y = 2.6$. 8 liters

3. Draw a bird and a large arrow to represent the wind. Let x represent the bird's rate and let y represent the rate of the wind.

4. Draw a bird and an arrow facing right to represent the bird flying with the wind. Draw an arrow pointing toward the bird to represent the bird flying against the wind. Let $(x + y)$ represent the bird flying with the wind, and let $(x - y)$ represent the bird flying against the wind.

5. The bird's rate is 3 times faster flying with the wind than flying against the wind. $x + y = 3(x - y)$

6. Let t represent the tens digit and u represent the units digit. The sum of the digits is 8, so $t + u = 8$. The number is 62. To check the answer, add 16 to 62 which gives 78. Multiply 26 by 3 which also gives 78.

Assessment

16.

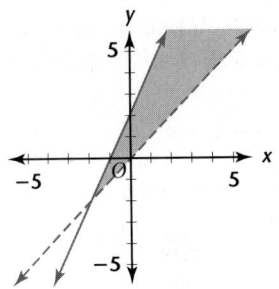

17.

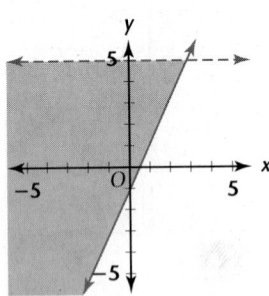

18.

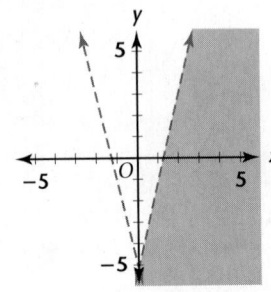

Lesson 7.1, pages 308–314

Communicate

1. Place all of the data representing sales of men's clothing items in one matrix and the data representing the sales of women's clothing items in the other matrix.

Men's Clothing Sales in thousands			
	1985	1989	1993
Suits	14.6	10.8	11.5
Shirts	16.7	16.2	17.3
Jeans	242.7	210.5	186.9

Women's Clothing Sales in thousands			
	1985	1989	1993
Suits	17.4	12.3	8.6
Shirts	25.6	21.3	16.2
Jeans	98.2	90.1	80.3

2. Count the number of rows and columns across.

$$P_{5\times2};\ Q_{5\times2};\ R_{4\times2};\ T_{6\times2}$$

3. Simplify entries in each matrix to determine if corresponding addresses contain equal entries. Simplify matrix G.

$$G = \begin{bmatrix} 5 & 2 \\ -25 & -8 \end{bmatrix}$$

Matrix A and matrix C are equal to matrix G.

Practice and Apply

10.

	Atlantis	Columbia	Discovery
1988	1	0	1
$L = $ 1989	2	1	2
1990	2	2	2
1991	3	1	3

16.

	Chula Vista	Abilene
1980	85,000	99,000
1981	89,500	100,000
1982	94,000	101,000
1983	98,500	102,000
1984	103,000	103,000
1985	107,500	104,000
1986	112,000	105,000
1987	116,500	106,000
1988	121,000	107,000
1989	125,500	108,000
1990	130,000	109,000

Lesson 7.2, pages 315–321

Exploration

1a. $\begin{bmatrix} \frac{27}{4} & -13 \\ 5 & \frac{5}{2} \end{bmatrix}$

1b. $\begin{bmatrix} \frac{27}{4} & -13 \\ 5 & \frac{5}{2} \end{bmatrix}$

1c. Matrices B and D do not have the same dimensions, therefore, they cannot be added together.

1d. $\begin{bmatrix} 2.55 & -12 \\ 5 & -4.1 \end{bmatrix}$

1e. $\begin{bmatrix} 2.55 & -12 \\ 5 & -4.1 \end{bmatrix}$

1f. Matrices C and D do not have the same dimensions, therefore, they cannot be added together. To add matrices, add the corresponding entries of the matrices.

2. In order to add matrices, the matrices must have the same dimensions. The additions in a, b, d, and e could be performed because the matrices have the same dimensions. The additions in c and f cannot be performed because the matrices have different dimensions. Addition of matrices is commutative and associative.

3a. $\begin{bmatrix} -\frac{21}{4} & -7 \\ 13 & -\frac{13}{2} \end{bmatrix}$

3b. $\begin{bmatrix} 0 & 0 \\ 0 & 0 \end{bmatrix}$

3c. $\begin{bmatrix} -10.2 & 4 \\ 4 & -11.1 \end{bmatrix}$

3d. $\begin{bmatrix} 5.25 & 7 \\ -13 & 6.5 \end{bmatrix}$ or $\begin{bmatrix} \frac{21}{4} & 7 \\ -13 & \frac{13}{2} \end{bmatrix}$

3e. $\begin{bmatrix} -4.95 & 11 \\ -9 & -4.6 \end{bmatrix}$

3f. Matrices C and D do not have the same dimensions, therefore, they cannot be subtracted. To subtract matrices, subtract the corresponding entries of the matrices.

4. In order to subtract matrices, the matrices must have the same dimensions. The subtractions in a and d can be performed because the matrices have the same dimensions. The results are the same except that the signs of the corresponding elements are opposite because the order of the matrices being subtracted is reversed. The subtraction in b involves only one matrix so the dimensions are the same. The result is a matrix of all 0s because every element is subtracted from itself. The subtractions in c and e can be performed because the matrices have the same dimensions. The subtraction in f cannot be performed because the matrices have different dimensions.

Communicate

1. Add the corresponding entries of each matrix. $\begin{bmatrix} 9 & 12 \\ 13 & -12 \end{bmatrix}$

2. Subtract the corresponding entries of each matrix. $\begin{bmatrix} -9 & -4 \\ 29 & -4 \\ 13 & -6 \end{bmatrix}$.

3. These 2 matrices cannot be added together because the first one has 2 rows and 1 column and the second one has 1 row and 2 columns.

4. Add the corresponding entries of each matrix. $\begin{bmatrix} 1 & \frac{1}{2} & -1 \\ \frac{4}{3} & \frac{8}{5} & \frac{1}{2} \end{bmatrix}$

5. Matrix subtraction is not commutative because it requires individual matrix entries, which are not commutative, to be subtracted.

6. Matrix subtraction is not associative because it requires individual matrix elements, which are not associative, to be subtracted.

7. Not necessarily; the zero matrix is simply a matrix in which every element is equal to 0 and this matrix can have any dimensions.

Practice and Apply

17. $\begin{bmatrix} 16 & 145 & 133 \\ -23 & -75 & -115 \\ -84 & 220 & 99 \end{bmatrix}$

18. Stock [June] = $\begin{bmatrix} 3498 & 1967 \\ 798 & 1994 \\ 1934 & 730 \end{bmatrix}$

20.

Total Enrollment	Mu	Ar	Te	He
[1837	1219	1822	1592]

31.

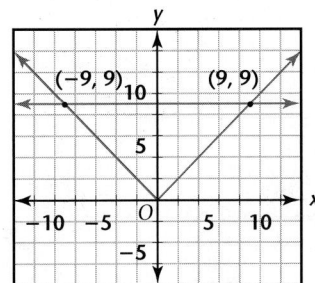

Lesson 7.3, pages 324–330

Exploration 1

1. 108 wheels

2. 191 nails

3. 174 nails

4. $\begin{bmatrix} 122 & 191 \\ 108 & 174 \end{bmatrix}$

Exploration 2

1. 4; row 1; column 2

2. -14; row 2; column 1

3. -33; row 2; column 2; p_{22}

4. You can multiply matrices if the number of columns in the first matrix are equal to the number of rows in the second matrix.

To multiply 2-by-2 matrices:

A. Multiply each entry in row 1 times each corresponding entry in column 1, then place the sum of the products at the address p_{11} in a product matrix.

B. Multiply each entry in row 1 times each corresponding entry in column 2, then place the sum of the products at the address p_{12} in a product matrix.

C. Multiply each entry in row 2 times each corresponding entry in column 1, then place the sum of the products at the address p_{21} in a product matrix.

D. Multiply each entry in row 2 times each corresponding entry in column 2, then place the sum of the products at the address p_{22} in a product matrix.

Exploration 3

1a. $\begin{bmatrix} 3 & -7 \\ 16 & 9 \end{bmatrix}$

1b. $\begin{bmatrix} 12 & 9 \\ -7 & -4 \end{bmatrix}$

1c. $\begin{bmatrix} 12 & 9 \\ -7 & -4 \end{bmatrix}$

2. The product is the matrix that does not contain 0 or 1.

3. The product matrix is identical to the original matrix when it is multiplied on both sides.

4. Matrix multiplication is not commutative.

5. The identity matrix must work on both sides. Thus, the dimensions of the original matrix and its identity matrix must be equal.

Communicate

1. three rows and one column

2. three rows and two columns

3. one row and four columns

4. Multiply each address by 3; $\begin{bmatrix} -18 & 3 \\ 12 & -3 \end{bmatrix}$.

5. $\begin{bmatrix} -14 & -22 \\ 5 & -1 \end{bmatrix}$

6. $\begin{bmatrix} 8 & 3 & 15 \\ -2 & -2 & -15 \\ 22 & 8 & 39 \end{bmatrix}$

Practice and Apply

19. not possible

20. $\begin{bmatrix} 16 & -20 \\ -14 & 60 \\ -9 & 25 \end{bmatrix}$

47.

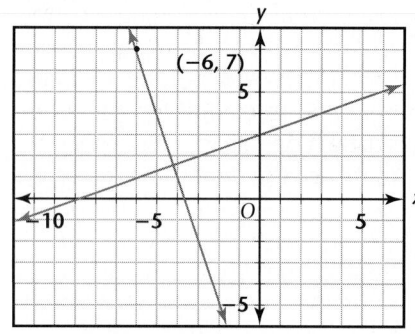

Look Beyond

50. $C = \begin{bmatrix} 1 & 0 \\ 0 & 1 \end{bmatrix}$

51. The product of the two matrices is the identity; the matrices are inverses.

Lesson 7.4, pages 331–336

Communicate

1. $\begin{bmatrix} -1 & 6 \\ 0 & 3 \end{bmatrix}\begin{bmatrix} a & b \\ c & d \end{bmatrix} = \begin{bmatrix} 1 & 0 \\ 0 & 1 \end{bmatrix}$

$-a + 6c = 1 \qquad -b + 6d = 0$

$0a + 3c = 0 \qquad 0b + 3d = 1$

Solve for a, b, c, and d. $\begin{bmatrix} -1 & 2 \\ 0 & \frac{1}{3} \end{bmatrix}$

2. not possible

3. not possible

4. If the product of the matrices does equal an identity matrix, the matrices are not inverses.

5. Matrix inverses are used to solve matrix equations.

6. The product is equal to 1 if the number is not zero.

Practice and Apply

12. $\begin{bmatrix} 10 & 5 \\ 9 & 4 \end{bmatrix}$

27.

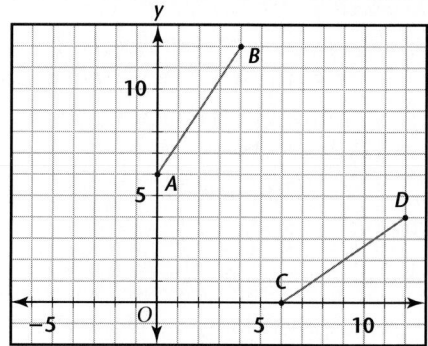

Lesson 7.5, pages 337–343

Communicate

1. Write the matrix equation $AX = B$ and solve

$\begin{bmatrix} 3 & -6 \\ 5 & 2 \end{bmatrix}\begin{bmatrix} a \\ b \end{bmatrix} = \begin{bmatrix} 9 \\ 9 \end{bmatrix}$.

2. Write the matrix equation $AX = B$ and solve

$\begin{bmatrix} 12 & 5 \\ 4 & -1 \end{bmatrix}\begin{bmatrix} x \\ y \end{bmatrix} = \begin{bmatrix} 30 \\ 6 \end{bmatrix}$.

3. Find the inverse of the coefficient matrix which is $\begin{bmatrix} 0 & 1 \\ -1 & -2 \end{bmatrix}$.

Multiply the inverse matrix times the constant matrix. Be certain that the inverse is multiplied on the left side of the constant matrix.

4. Find the inverse of the coefficient

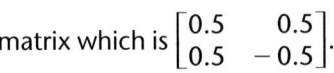

matrix which is $\begin{bmatrix} 0.5 & 0.5 \\ 0.5 & -0.5 \end{bmatrix}$.
Multiply the inverse matrix times the constant matrix. Be certain that the inverse is multiplied on the left side of the constant matrix.

Lesson 8.1, pages 352–359

Exploration 1

1–5. Answers may vary depending upon the results of the experiment.

6. $P = \frac{f}{t}$

Exploration 2

1–2. Answers may vary depending upon the results of the experiment.

3. event—when a 5 appears on the top of the cube
trial—each roll of the numbered cube

4. $\frac{\text{number of times 5 appears}}{10}$

Exploration 3

1. Player A can get a point 24 ways. Player B has only 12 ways to get a point.

2. Answers may vary. For example, player A gets a point if the sum is even. Player B gets a point if the sum is odd.

Communicate

1. Experimental probability is calculated by performing an experiment and comparing the number of times an event occurs to the number of trials in the experiment.

2. Answers may vary. For example, toss 4 coins 10 times and count the number of times that 3 or 4 heads turn up.

3. Yes. For example, if 2 pairs of players toss 4 coins each, the number of heads showing may be different for both pairs.

4. Yes. If one player tosses 4 coins

once and then again, the number of heads showing on each of the 4 coins tosses may be different.

5. integers 1, 2, 3, 4, 5, 6, or 7

6. In the 10 trials, only once were both numbers less than 50. The experimental probability is $\frac{1}{10}$.

Lesson 8.2, pages 360–365

Exploration 1

1-4. Answers will depend upon the number of heads tossed.

$$P = \frac{\text{least 4 heads in a row happen}}{20}$$

Exploration 2

1-3. Answers may vary depending upon the results of the experiment.

4. It will not rain on Saturday or Sunday.

5. It will rain on Saturday but not on Sunday.

6. 7 weekends

7. $\frac{7}{10}$

Communicate

1a. Choose a way to generate random numbers and describe what each result represents.

1b. Decide how to simulate one trial of the experiment.

1c. Carry out a large number of trials and record the results.

2. A coin toss; heads is a shot made, tails is a shot missed. A random number generator; 1 is a shot made, 0 is a shot missed. A number cube; an even number is a shot made, an odd number is a shot missed.

3. Let the numbers 1,2, and 3 represent the occurrence of rain, and the numbers 4, 5, 6, 7, 8, 9, and 10 represent no rain.

4. Answers may vary. For example: hitting a nail on the head with a

hammer, kicking a football through the goal post, and getting an answer on the phone when you call.

5. Use 5 different-colored slips of paper, where each color represents a different answer.

6. Generate random positive integers from 1 to 10 on a calculator. Let the number 1 represent a late flight, and the numbers 2 to 10 represent a flight on time. Generate 3 numbers. Let this represent 1 trial. Repeat for 20 trials. Divide the number of trials where all 3 numbers are greater than 1 by 20. This quotient is the experimental probability.

Practice and Apply

For Exercises 13-18, answers may vary.

13. Use a calculator to generate a number from 1 to 7; use INT(RAND*7)+1.

14. Use a calculator to generate a number from 1 to 365; use INT(RAND*365)+1.

15. Use a calculator to generate a number from 1 to 24; use INT(RAND*24)+1.

16. Use a calculator to generate a number from 1 to 8; use INT(RAND*8)+1.

17. Generate 100 pairs of random numbers from 1 to 100. Count the number of pairs in which both numbers are less than or equal to 20. Let that number equal n. Then the experimental probability is $\frac{n}{100}$.

18. Generate 20 sets of 5 random numbers from 1 to 10. Then count the number of sets that have exactly 2 numbers less than or equal to 3. Let that number equal n. Then the experimental probability is $\frac{n}{20}$.

21. Answers may vary. Use a calculator to generate the numbers 1 to 5. Let 1 represent a correct response on a question, and 2 to 5 represent an incorrect response. Generate 4

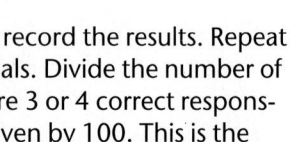

numbers; record the results. Repeat for 100 trials. Divide the number of trials where 3 or 4 correct responses were given by 100. This is the experimental probability, approximately $\frac{17}{625}$.

22. Answers may vary. Use a number cube to generate the numbers from 1 to 6. Let 1 represent a correct response, and 2 to 5 represent an incorrect response, and 6 means roll again. Generate 6 numbers; record the results. Repeat for 1000 trials. Divide the number of trials where 5 or 6 correct responses were given by 1000. This is the experimental probability, approximately $\frac{1}{625}$.

Look Back

24.

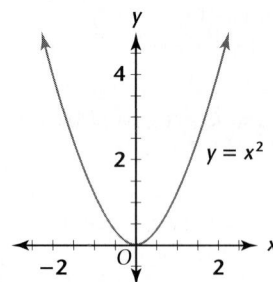

The graph of a quadratic function is a parabola.

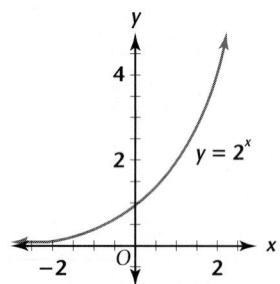

The graph of an exponential function is a curve that increases very rapidly.

27.

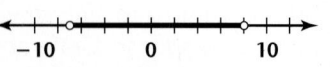

Lesson 8.3, pages 368–374

Exploration

1. 1H-1; 2H-0; 3H-5; 5H-5
2. 5
3. 95%
4. 19
5. 50%
7. 5%; outside

Communicate

1. The mean is the sum of all the data divided by the number of items. The median is the number in the middle of the ordered data. The mode is the most frequently occurring item of data.
2. The 5 numbers were placed in order from the least to the greatest; 0.342 is the middle value.
3. Bob did not put the numbers in order from the least to the greatest first.
4. Arrange the data from the least to the greatest and find the mean or average of the 2 middle numbers.
5. If lawn A is 2×1 and lawn B is 4×2, the area of A is 2 and area of B is 8. Therefore, area B is 4 times as great as area A.
6. 95%
7. Bar graphs can use misleading scales. Area or volume can be used to distort comparisons of quantity.

Lesson 8.4, pages 375–381

Exploration 1

1.

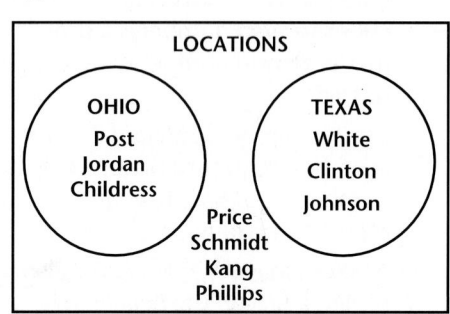

2. 3
3. 3
4. 6
5. 0
6. Ohio + Texas = Total; $3 + 3 = 6$

Exploration 2

1.

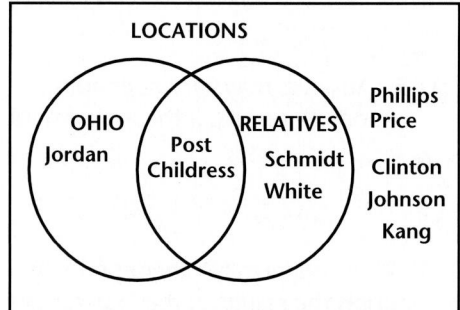

2. 3
3. 4
4. 5
5. 2
6. No, 2 people have been counted twice.
7. Ohio + Relatives − Relatives in Ohio = Total
$$3 + 4 - 2 = 5$$

Communicate

1. A field is a category; a record is each line of information.
2. Post or Childress
3. It counts once as an 8, and once as a heart.
4. There is no intersection.
5. $49 + 57 = 106$
6. $43 + 49 = 92$
7. 49
8. $49 + 57 + 43 = 149$

Lesson 8.5, pages 382–387

Communicate

1. And: you do both things
Or: you do one or the other

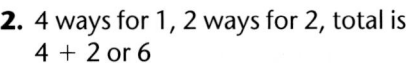

2. 4 ways for 1, 2 ways for 2, total is 4 + 2 or 6

3. 4 ways for 1, 2 ways for 2, total is 4 × 2 or 8

4. There are 2 volleyball teams, so there are initially 2 branches. Each team has 8 players, so each branch has 8 branches. The total number of branches is 2 × 8 or 16.

5. There are 4 ways to draw a three. There are 26 ways to draw a red card.   4 · 26 = 104

6. 3 ways; $\overrightarrow{AB}$, $\overrightarrow{AC}$, $\overrightarrow{AD}$

Lesson 8.6, pages 388–393

Communicate

1. experimental =
$$\frac{\text{number of times event occurs}}{\text{number of trials}}$$

theoretical =
$$\frac{\text{number of successful outcomes}}{\text{number of equally} - \text{likely outcomes}}$$

2. No, he has a 50-50 chance on one question, but with each subsequent question, he decreases the probability of getting all of the questions correct by $\frac{1}{2}$.

3. Experimental. This probability is based on the actual results of an experiment.

4. No, the probability of all heads or all tails is only
$\frac{1}{16} + \frac{1}{16}$ or $\frac{1}{8}$.

5. Theoretical probability assumes that events are randomly occurring and the situations are fair, that is, not biased towards one outcome.

Lesson 8.7, pages 394–399

Communicate

1. Use a square divided into regions for each event; use a tree diagram.

2. The outcome of one event does not affect the outcome the other.

---

3. Multiplication Principle of Counting

4. 8

5. Subtract the number of ways the event could not occur from the total number of possible outcomes. The difference represents the number of ways the event could occur. Divide the difference by the total number of possible outcomes to find the probability of an event occurring.

6. Count the squares on the left side of the vertical line. Divide by 100. 60%

7. Count the squares above the horizontal line. Divide by 100.   80%

8. Count the number of squares that are to the left of the vertical line **and** are also above the horizontal line. Divide by 100.   48%

9. Count the number of square to the left of the vertical line. Add this number to the number of squares above the horizontal line. Subtract the number of squares found in question 8 from this total. Divide by 100.   92%

Look Beyond

40. You have a $\frac{1}{2000}$ or 0.0005 chance of winning $25 from the second station. You have a $\frac{1}{100000}$ or 0.00001 chance of winning $1000 from the first station. If you call the second station 100,000 times, you would have a chance of winning 50 times, for a total of $1250. If you call the first station 100,000 times, you would have a chance of winning once, or $1000. It is more lucrative financially to call the second station.

Lesson 9.1, pages 410–417

Communicate

1. A relation pairs elements from one set with elements of a second set.

A function is a set of ordered pairs for which there is exactly one second coordinate for each first coordinate.

2. Each national park is located in only one state.

3. No vertical line will intersect the graph of a function more than once.

4. Simplify to find that $f(-2) = 2(4) + 10 = 18$. Substitute and evaluate $2x^2 - 5x$ for the integers $-2, -1, 0, 1$, and 2.

5. The first element (the x-coordinate) 6 is from the domain and the second element (the y-coordinate) -46 is from the range.

6. The -24 is the x-coordinate (the first element) and the 16 is the y-coordinate (the second element). Therefore the ordered pair is $(-24, 16)$.

Look Beyond

61.

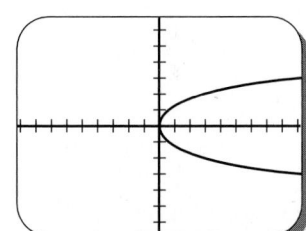

62.

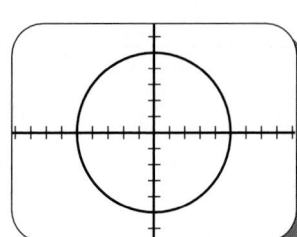

Lesson 9.2, pages 418–423

Exploration 1

1.

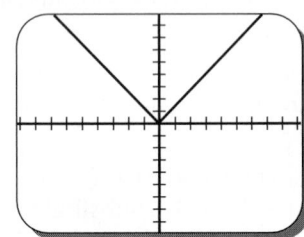

2a. reflected through the *x*-axis

2b. stretched vertically by 2

2c. shifted vertically by − 2

2d. reflected through the *x*-axis and shifted vertically by 4

2e. stretched vertically by 2 and shifted vertically by − 4

2f. stretched vertically by 3, reflected through the *x*-axis, and shifted horizontally by − 2

3. The function in 2a matches graph c; 2b matches graph a; 2c matches graph b; 2d matches graph e; 2e matches graph d, and 2f matches f.

4. To stretch the graph, multiply it by a number, the larger the number the more narrow (closer to the *y*-axis) the graph appears. To reflect *y* = |*x*|, multiply |*x*| by − 1. To shift the graph vertically, add or subtract a number to or from *x* after its absolute value has been taken. Adding shifts the graph up and subtracting shifts it down. To shift the graph horizontally, add or subtract a number from *x* and then take the absolute value of this result. Adding shifts the graph to the left and subtracting shifts it to the right.

5. If it has been stretched, then the sides of the graph would not be made up of the portions of the lines *y* = *x* and *y* = − *x* that lie above the *x*-axis. If it has been reflected, then it would open downward. If it has been shifted, then the vertex of the graph would not be at (0, 0).

6a. Stretch the graph vertically by 3.

6b. Stretch the graph vertically by 5 and reflect it through the *x*-axis.

6c. Shift the graph vertically by 2.

7. The graph will be shifted horizontally to the left by 1.

8a.

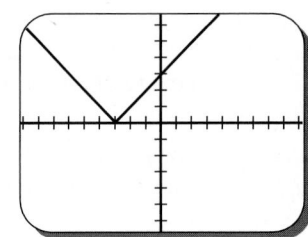

8b.

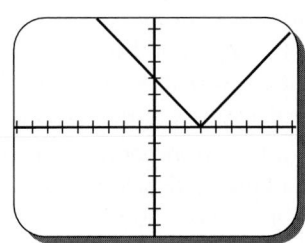

Exploration 2

1.

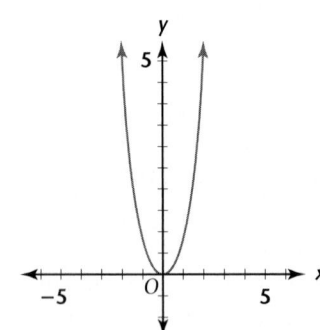

2a. stretched vertically by 2

2b. reflected through the *x*-axis

2c. shifted vertically by − 2

3. Graph 2a of Exploration 2 opens upward and is stretched vertically by 2 just as is graph 2a of Exploration 1. Graph 2b of Exploration 2 is reflected through the *x*-axis just as is graph 2b of Exploration 1. Graph 2c of

Exploration 2 is shifted down by 2 just as is graph 2c of Exploration 1.

4. To stretch the graph, you multiply it by a number. To reflect the graph, multiply x^2 by − 1. To shift the graph vertically, add or subtract a number to or from x^2. Adding shifts the graph up and subtracting shifts it down. To shift the graph horizontally, add or subtract a number to or from *x* and then square this result. Adding shifts the graph to the left and subtracting shifts it to the right.

5. Stretch the function vertically by 2, reflect it through the *x*-axis and shift it vertically by 3.

Communicate

1. The vertex is the point (0, 0). The line of symmetry is the vertical line through the vertex, that is, the *y*-axis.

2. All of the *y*-values are increased by a factor of the stretch thereby causing the graph to be more narrow vertically when compared to the parent function.

3. All of the *y*-values are negated thereby causing the graph to be upside down when compared to the parent function.

4. The graph moves either up or down if it is a vertical shift or left or right if it is a horizontal shift. All points of the graph shift by the same amount and in the same direction.

5. The amount of the shift is either added or subtracted from the *x*-coordinate of the vertex if the shift is horizontal and either added or subtracted from the *y*-coordinate of the vertex if the shift is vertical.

6. Graph each function. If there is a maximum or minimum value, the graph has a vertex. If there is a reflection through the *y*-axis, it is the axis of symmetry.

752 Additional Answers

Practice and Apply

10.

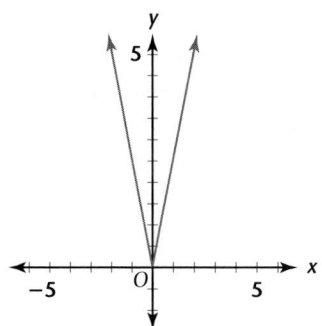

11.

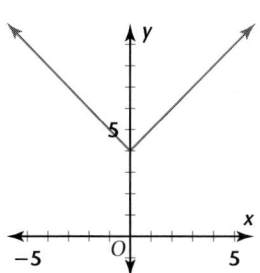

12.

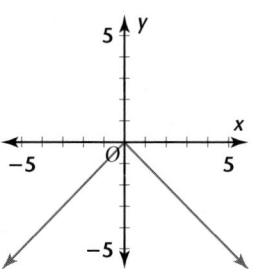

13.

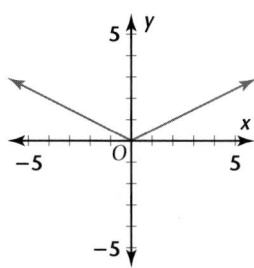

14.

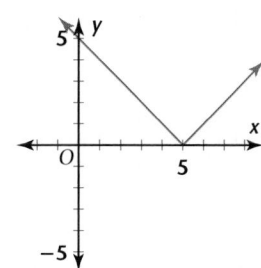

15.

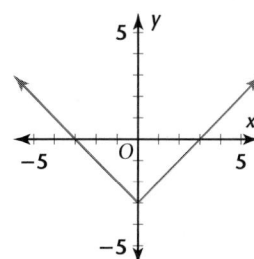

16.

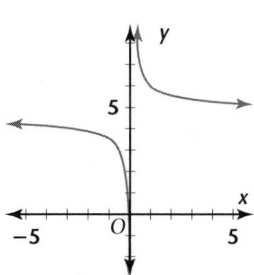

17.

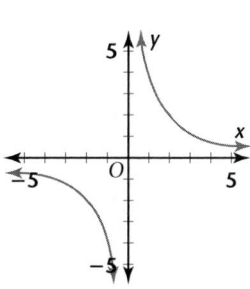

18.

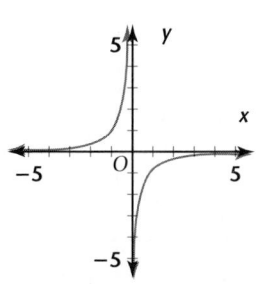

19.

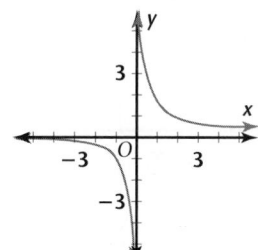

20.

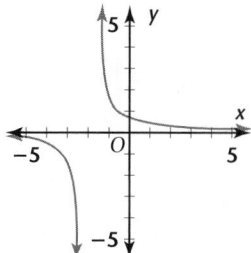

21.

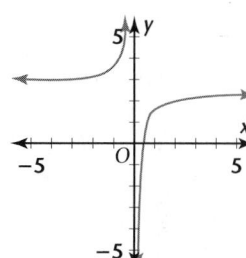

22.

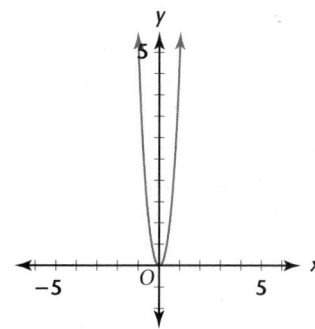

23.

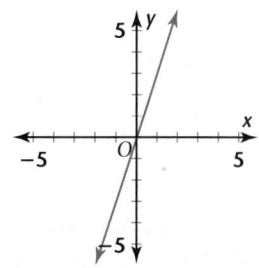

Additional Answers **753**

24.

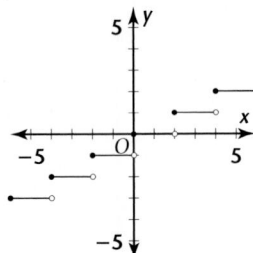

25.

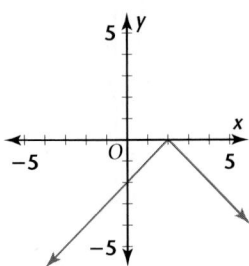

28.

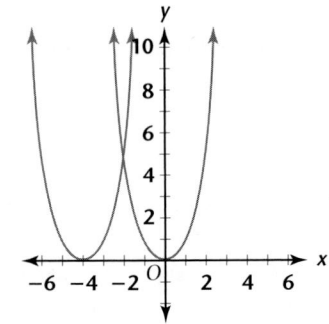

29. $A = (x + 4)^2$ is the graph of $A = x^2$ shifted 4 units to the left.

30. It is shifted up 3 pounds.

31. vertical stretch of 2

Look Back

38.

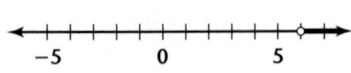

39.

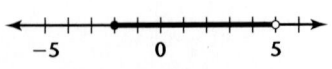

Lesson 9.3, pages 424–430

Communicate

1. When a is 1, the graph is the parent function $y = x^2$. As the value of a increases, the parabola becomes narrower (closer to the y-axis).

2. As a gets smaller and takes on values between 1 and 0, the parabola becomes wider.

3. When compared to the parent function the y-values are moved by the stretch factor a.

4. When b is 1, the graph is the parent function $y = \frac{1}{x}$. As the value of b increases, the curves of the hyperbola become flatter (move away from the origin).

5. As b gets smaller and takes on values between 1 and 0, the curves of the hyperbola become closer to the origin because the denominators of the y-values are becoming larger.

6. $\frac{m}{2}$ is the scale factor that stretches the graph vertically.

Practice and Apply

8.

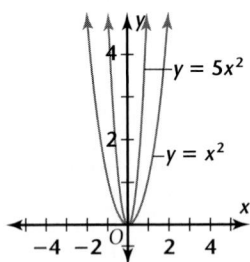

10.

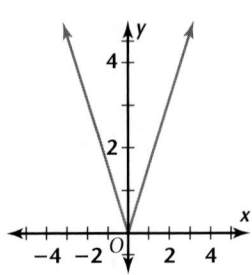

11.

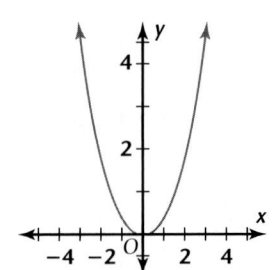

12.

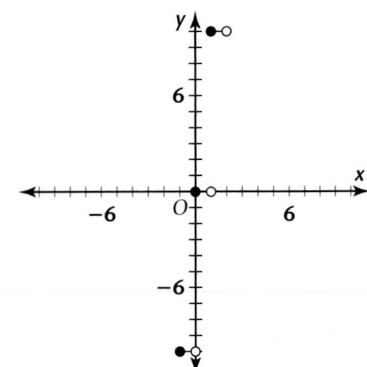

17. The scale factor of $\frac{1}{2}$ in $y = \frac{1}{2x}$ has stretched the graph $y = \frac{1}{x}$ by $\frac{1}{2}$. The graph of $y = \frac{1}{2x}$ is closer to the origin because the y-values are $\frac{1}{2}$ as large.

18. yes

19. no

20. yes

21. yes

Lesson 9.4, pages 431–435

Exploration

1.

x	-2	-1	0	1	2
y	4	1	0	1	4

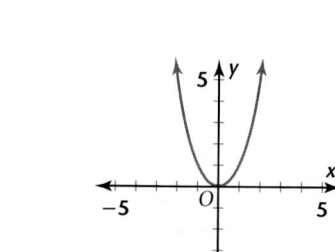

2.

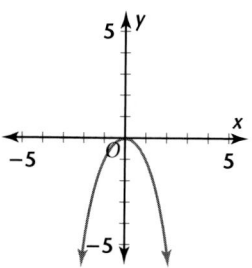

3.

x	−2	−1	0	1	2
y	−4	−1	0	−1	−4

4. −4; −9

5a. −9

5b. 0

5c. −9

5d. −a^2

6. To generate the graph reflected through the x-axis, multiply the y-values of the parent function by −1.

7. $y = -f(x)$

Communicate

1. The negative sign will reflect the graph through the x-axis.

2. The original function is multiplied by −1 giving $g(x) = -a\left(\frac{1}{x}\right)$.

3. Select points on the graph of $f(x) = 2x$ and then negate their y-values.

4. The new formula is the opposite of the original function.

5. The graph remains the same. For example, if $f(x) = x^2$ then $g(x) = (-x)^2$ is the reflection of $f(x)$ through the y-axis.

6. x-axis; y-axis; If the negative sign is outside the parentheses then the y-values are negated and this indicates a reflection through the x-axis. If the negative sign is inside the parentheses then the x-values are negated and this indicates a reflection through the y-axis.

7. A mirror image is a reflection.

Practice and Apply

12.

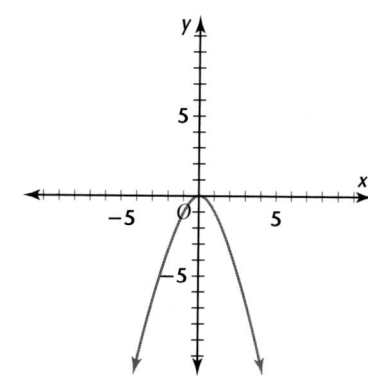

13.

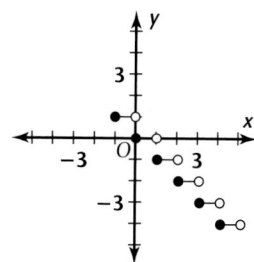

14.

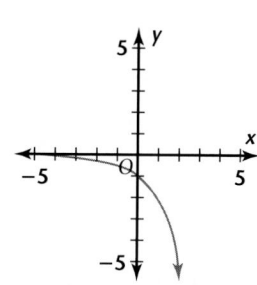

15. Same as graph

16. The V is turned upside down; y-values become the opposite. Same graph; the x-values become the opposite, but the y-values stay the same.

17.

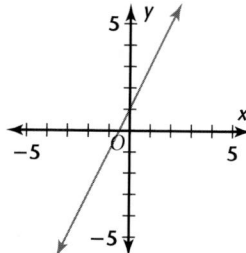

18. 7

21. For $y = x^2$, the combined transformations give $(x, y) \to (-x, -y)$.

Lesson 9.5, pages 436–442

Communicate

1. A translation shifts a graph horizontally or vertically.

2. −3

3. 5

4. h increases or decreases the x-value at each point.

5. k increases or decreases the y-value at each point.

6. The 4 shifts the graph to the right 4 units and the 3 shifts the graph up 3 units; (4, 3).

Practice and Apply

7.

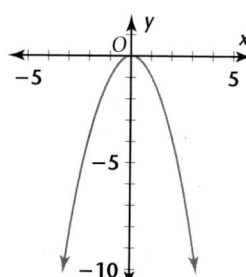

8.

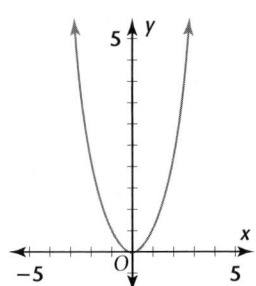

9.

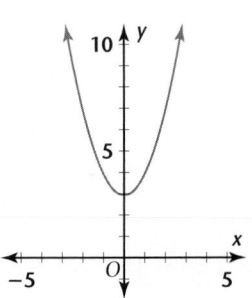

10.

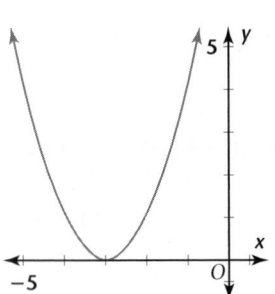

11.

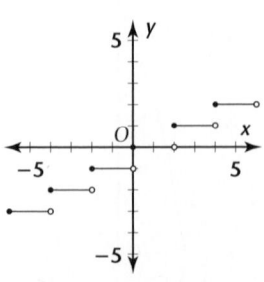

12.

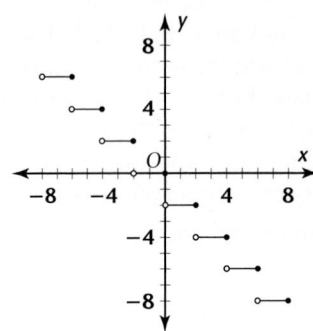

13.

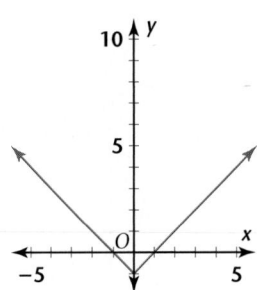

14.

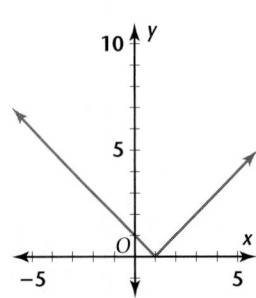

Look Back

38.

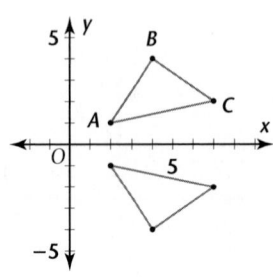

$(2, -1), (4, -4),$ and $(7, -2)$

39.

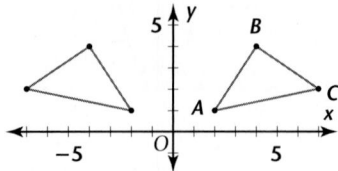

$(-2, 1), (-4, 4),$ and $(-7, 2)$

Look Beyond

40-41.

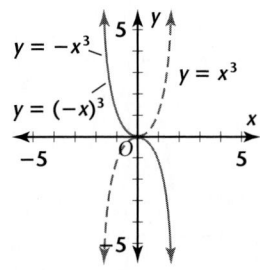

Horizontal reflection

Lesson 9.6, pages 443–447

Communicate

1. Translations occur when a constant is added or subtracted. Thus, the number -2 represents the translation of 2 to the right.

2. A scale factor stretches a graph. Since $R = \frac{100}{h}$ can be written as $R = 100\left(\frac{1}{h}\right)$, the scale factor 100 represents a vertical stretch of 100.

3. Answers may vary. For example, for the parent absolute value function the graph is a *V* with the vertex at $(0, 0)$. The graph opens upward and contains the portions of the lines $y = x$ and $y = -x$ that lie above the *x*-axis.

4. Since it is a horizontal shift, apply the translation to the *x*-value of the parent function $y = \frac{1}{x}$. It is a shift to the left so add 4 to *x* giving $x + 4$. The translated function is $y = \frac{1}{x + 4}$.

5. The function is reflected by the negative sign preceding a, stretched by a, shifted horizontally by $-h$, and shifted vertically by $+k$.

6. Derive a table of values from the function. Plot the ordered pairs from the table and graph the line.

Practice and Apply

7.

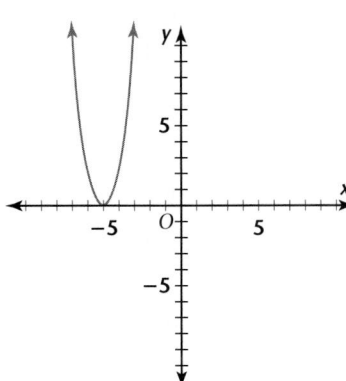

8.

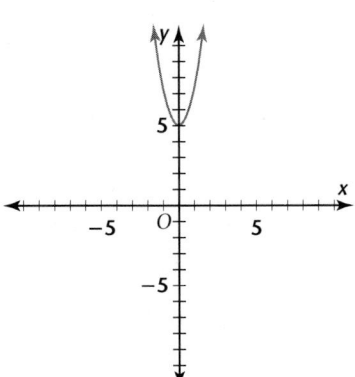

9.

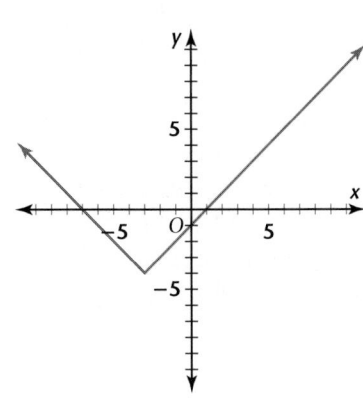

10.

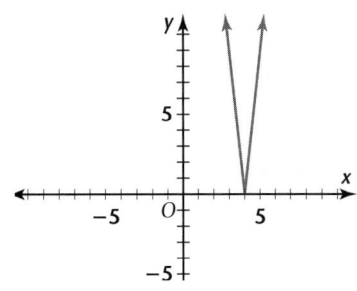

11.

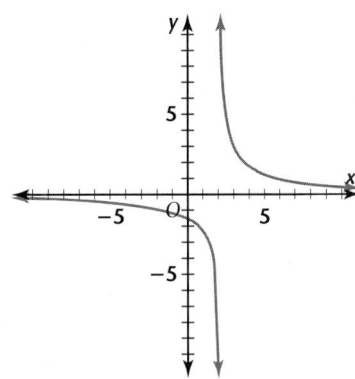

12.

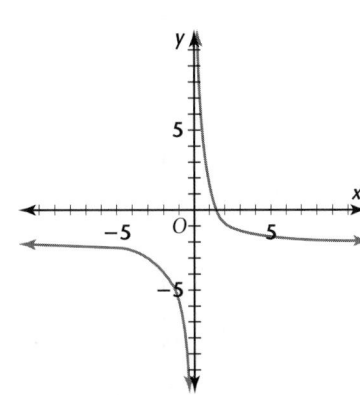

13.

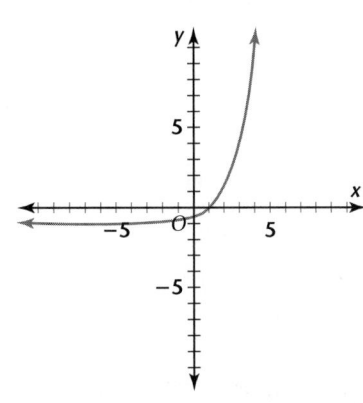

14.

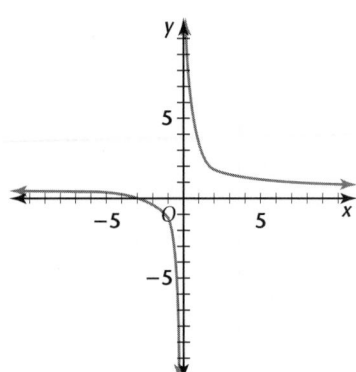

15.

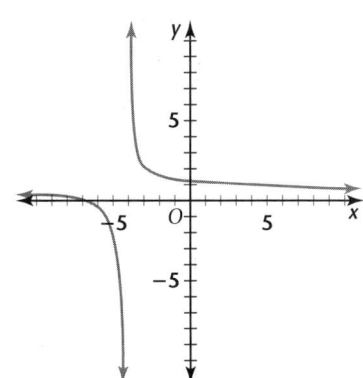

24. Designs will vary.

Look Back

25. Graphs may vary.

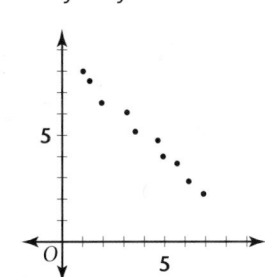

Look Beyond

33.

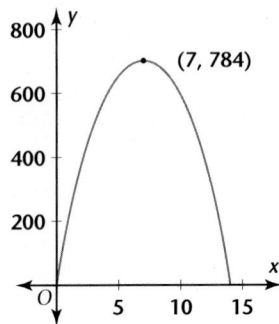

35.

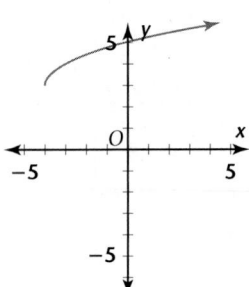

36.

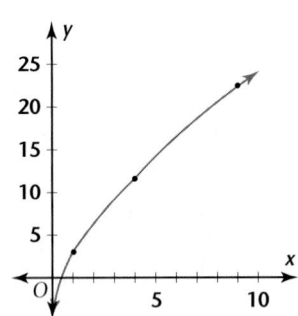

Review

10.

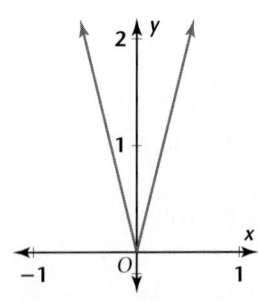

11.

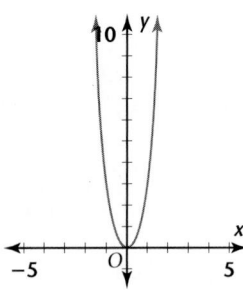

12.

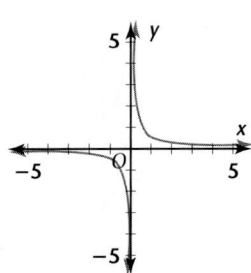

13.

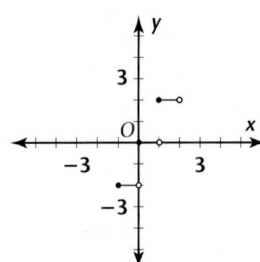

14.

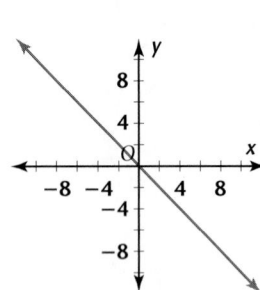

15.

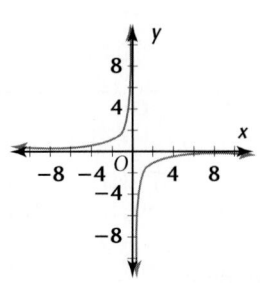

16.

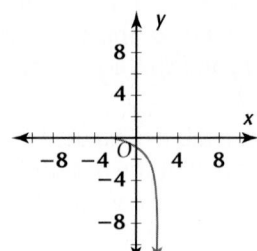

17.

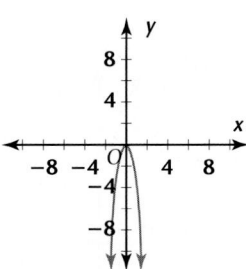

18.

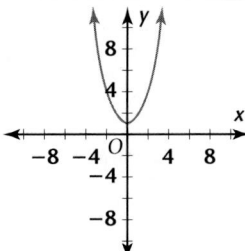

19.

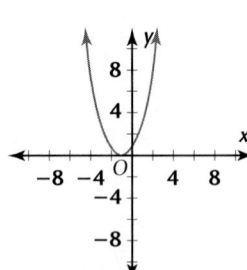

20.

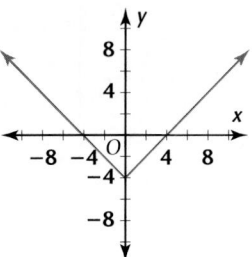

21.

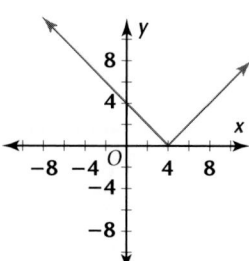

22.

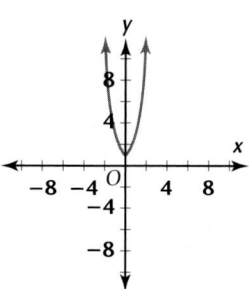

23.

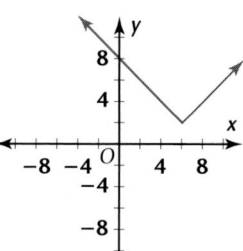

24.

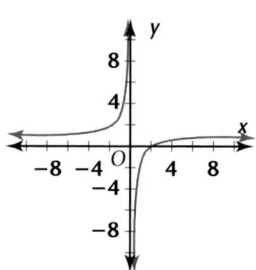

Assessment

6.

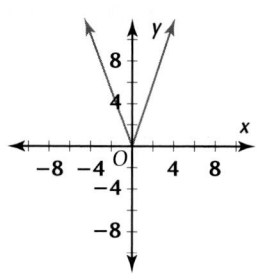

7.

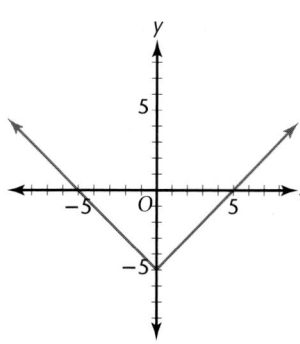

8.

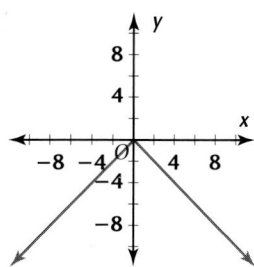

9.

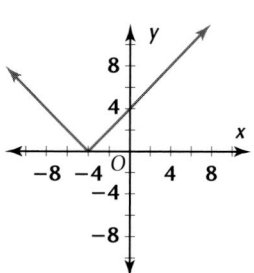

11.

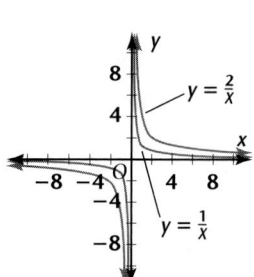

12. The vertices on both sides of the graph have moved away from their respective asymptotes, this is caused by the multiplication of all the y-values by 2.

19.

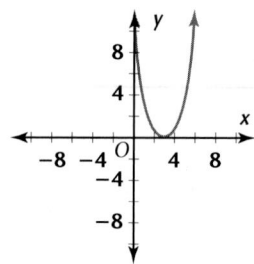

20.

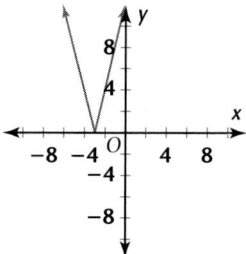

21.

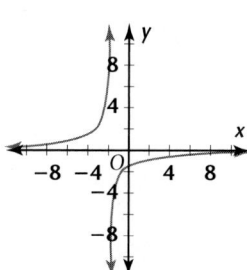

Lesson 10.1, pages 456–462

Exploration 1

1a. $10^3 \cdot 10^2 =$
$1000 \cdot 100 =$
$100,000 = 10^5$

1b. $10^3 \cdot 10^3 =$
$1000 \cdot 1000 =$
$1,000,000 = 10^6$

2. $10^4 \cdot 10^2 = 10^6$

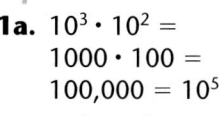

Additional Answers **759**

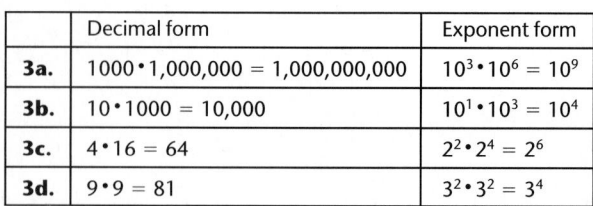

	Decimal form	Exponent form
3a.	$1000 \cdot 1{,}000{,}000 = 1{,}000{,}000{,}000$	$10^3 \cdot 10^6 = 10^9$
3b.	$10 \cdot 1000 = 10{,}000$	$10^1 \cdot 10^3 = 10^4$
3c.	$4 \cdot 16 = 64$	$2^2 \cdot 2^4 = 2^6$
3d.	$9 \cdot 9 = 81$	$3^2 \cdot 3^2 = 3^4$

4. Decimal form:

$$\underbrace{a \cdot a \cdot a \cdot \ldots}_{m \text{ times}} \cdot \underbrace{a \cdot a \cdot a}_{n \text{ times}} = \underbrace{a \cdot a \cdot a \cdot a \cdots}_{m + n \text{ times}}$$

Exponent form: $a^m \cdot a^n = a^{m+n}$

5. $a^m \cdot a^n = a^{m+n}$

Exploration 2

1. 10^4

2.

Form		Numerator	Denominator	Quotient
2a.	Decimal	100,000	100	1000
	Exponent	10^5	10^2	10^3
2b.	Decimal	64	4	16
	Exponent	2^6	2^2	2^4
2c.	Decimal	81	27	3
	Exponent	3^4	3^3	3^1

3.

Form	Numerator	Denominator	Quotient
Decimal	100,000	10	10,000
Exponent	10^5	10^1	10^4

4. $\dfrac{a^m}{a^n} = a^{m-n}$

5. Subtract the exponents:
$\dfrac{2^5}{2^3} = 2^{5-3} = 2^2$

6. No, because the bases are different. If the bases are the same then you can divide powers by subtracting the exponents.

Exploration 3

1. 100,000; 1,000,000

2. $1000 \cdot 1000 = 1{,}000{,}000$

3. 10^6

4. Multiply exponents; $(5^2)^4 = 5^8$

5. $(a^m)^n = a^{mn}$

Communicate

1. The exponent indicates the number of zeros in customary notation.

2. The exponent indicates how many times the base is used as a factor.

3. addition

4. subtraction

5. multiplication

6. 8^{17}; $8^{17} > 8^{16}$ because the exponent $17 > 16$ and $8^{17} > 5^{17}$ because the base $8 > 5$.

7. You are multiplying powers of the same base.

8. You are dividing powers of the same base.

9. You are finding a power of a power.

Practice and Apply

10. Decimal form $1{,}000{,}000 \cdot 10{,}000 = 10{,}000{,}000{,}000$
Exponent form $10^6 \cdot 10^4 = 10^{10}$

Look Back

56. $x \le 10$ and $x \ge -4$

57. $x > 2$ or $x < -6$

58. $y = \dfrac{19}{30}x - \dfrac{178}{15}$

59. $A^{-1} = \begin{bmatrix} 1 & 1 \\ 2 & 3 \end{bmatrix}$

$A \cdot A^{-1} = \begin{bmatrix} 1 & 0 \\ 0 & 1 \end{bmatrix}$

60. Experimental: it compares the number of nurses who had fewer cataracts to the number of trials.

63. $y = 2(x + 3)^2$ is a parabola shifted 3 units to the left and stretched vertically by a factor 2.

Lesson 10.2, pages 463–468

Communicate

1. $3x^2$ is a monomial because it is the product of a constant and a variable with a positive integer exponent. $3x^{-2}$ is not a monomial because the variable has a negative exponent.

2. The Product-of-Powers Property states that to multiply powers of the same base you add the exponents. The Quotient-of-Powers Property states that to divide powers of the same base you subtract the exponents.

3. The outside exponent tells how many times the inside power is used as a factor.

4. Multiplication is commutative.

5. To multiply powers of the same base, add the exponents.

Lesson 10.3, pages 469–473

Exploration

1a. 1   **1b.** 1   **1c.** 1

2a. 10^0 **2b.** 10^0 **2c.** 10^0

3. $10^0 = 1$

4. $x^0 = 1$ for any positive number x is 1.

Communicate

1. $5^{-2} = \dfrac{5^3}{5^5} = \dfrac{5 \cdot 5 \cdot 5}{5 \cdot 5 \cdot 5 \cdot 5 \cdot 5} = \dfrac{1}{5^2}$

2. $2^3 = 8, 2^2 = 4, 2^1 = 2, 2^0 = 1$ For any base except 0, the exponent zero makes the value 1.

3. No, they have different bases.

Practice and Apply

11.

Decimal Form	10,000	1000	100	10	1	0.1	0.01
Exponent Form	10^4	10^3	10^2	10^1	10^0	10^{-1}	10^{-2}

Look Back

42.

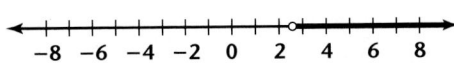

The solution set is all values greater than $2\frac{4}{7}$.

44. $y = -2(x + 3)^2 - 4$

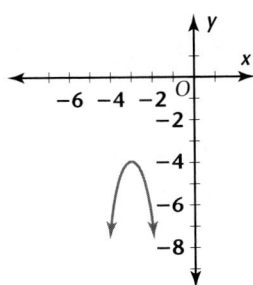

Look Beyond

47. 17; 257; $2^{16} + 1$; 65,537; $2^{32} + 1$; 4,294,967,297

Lesson 10.4, pages 476–482

Communicate

1. The numbers are easier to read and use in calculations.

2. The value is multiplied by 0.00000001; 8.

3. It is 149,000,000 km from Earth to the sun. Move the decimal point 8 places to the left: 1.49×10^8 km.

4. The value is multiplied by 1,000,000,000,000; -12.

4. The parentheses are around both 3 and a;
$$(3a)^{-2} = \frac{1}{(3a)^2} = \frac{1}{9a^2}.$$

5. It means the reciprocal of the power.

6. The power has value 1.

5. Write each number in scientific notation first, then multiply the non-power factors and multiply the powers of 10. Check that the final answer is expressed in scientific notation.

$$240,000 \times 0.006$$
$$= 2.4 \times 10^5 \times 6 \times 10^{-3}$$
$$= 2.4 \times 6 \times 10^{5+(-3)}$$
$$= 14.4 \times 10^2$$
$$= 1.44 \times 10^1 \times 10^2$$
$$= 1.44 \times 10^3$$

Look Back

68. Graphs may vary.

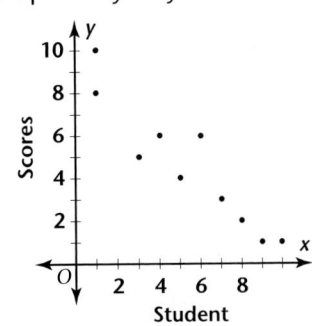

73.

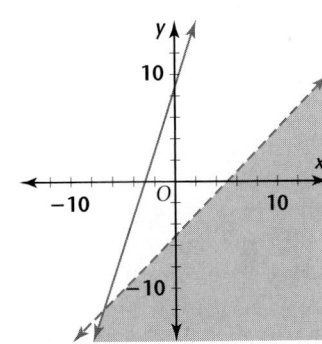

The solution is all the points below the line $y = x - 5$ and to the right of $y = 3x + 9$.

Lesson 10.5, pages 483–488

Communicate

1. Add 100% to the yearly rate of increase: 101.9% or 1.019.

2. Tables and graphs show the actual data and illustrate how quickly the population grows.

3. P = final amount, A = original amount, r = rate of increase or growth, t = time in years

4. As the base gets larger the amount grows more quickly.

5. The constant multiplier is the base to which the exponent is applied.

6. Both are increasing exponentially from an original amount at a given rate over a specified number of years.

Practice and Apply

7.

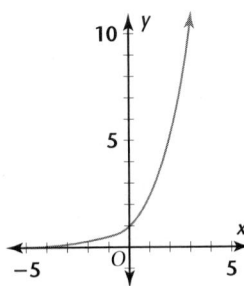

8.

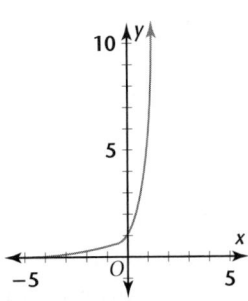

9.

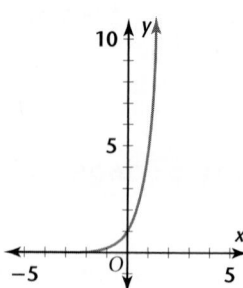

10.

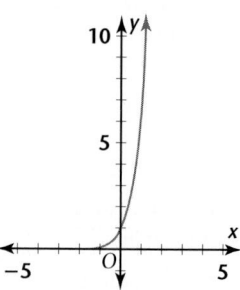

11.

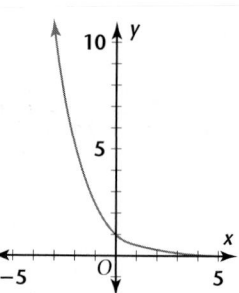

12.

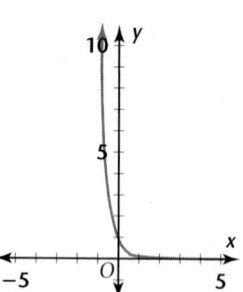

15.

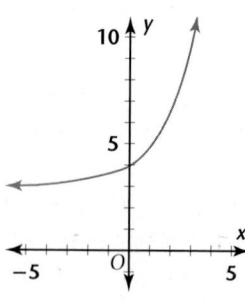

$y = 2^x + 3$ shifts the parent graph $y = 2^x$ vertically up 3 units

16.

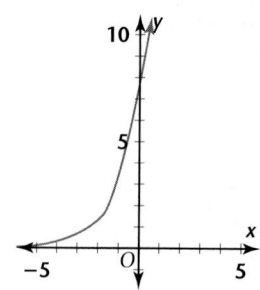

$y = 2^{(x+3)}$ shifts the parent graph horizontally 3 units to the left.

17.

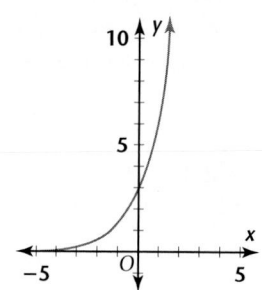

$y = 3 \cdot 2^x$ stretches the parent graph vertically by a factor of 3.

18.

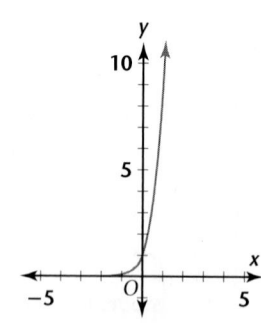

$y = 2^{3x}$ stretches the parent graph vertically by a power of 3.

Look Back

35. The first payment of 3% on an amount of $1015 is $30.45. After 12 payments, $842.01 is still owed. In approximately 9 years and 10 months, the amount paid back is $1779 (rounded to the nearest dollar).

Look Beyond

44–45.

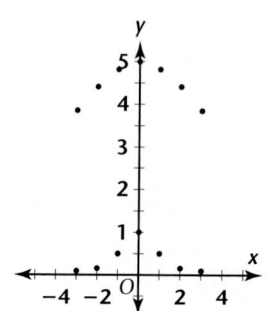

46. Both graphs have a maximum value when $x = 0$ and are symmetric about the y-axis.

Lesson 10.6, pages 489–495

Communicate

1. Examples may vary. For example, interest paid on loans, growth of bacteria.

2. The remaining amount, P, equals the original amount, A, multiplied by the constant multiplier, $1 - r$, raised to the *exponent* for the number of years of decay, t.

3. Decay is a decreasing measure.

4. For a horizontal translation, a value, h, is subtracted from the *exponent* of the parent function, $y = b^{x-h}$, $h \neq 0$.

5. For a vertical translation, a value, k, is added to the parent function, $y = b^x + k$, $k \neq 0$.

6. Answers may vary. For example: 3^1, $3^2, 3^3, 3^4, 3^5, 3^6, \ldots = 3, 9, 27, 81, 243, 729, \ldots$

First differences: 6, 18, 54, 162, 486, . . .

Each difference is 3 times the previous one.

Practice and Apply

9.

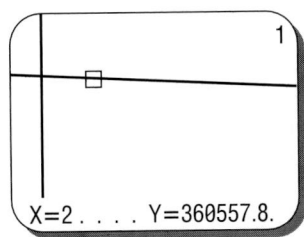

X=2 . . . . Y=360557.8.

Lesson 11.1, pages 506–512

Communicate

1. Use five x^2-tiles for $5x^2$, two negative x-tiles for $-2x$, and three 1-tiles for 3.

2. Determine the greatest exponent of all of the terms in the polynomial.

3. trinomial; cubic

4.

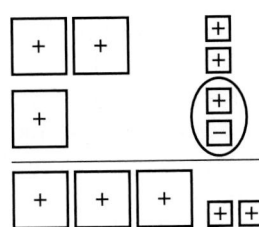

$3x^2 + 2$

5. Change the signs of the terms in the trinomial that is to be subtracted to their opposites. Then group like terms and simplify to $2b^3 - 3b + 4$.

6. Arrange the terms in descending order according to their exponents. The polynomial in standard form is $3x^4 + 5x^2 - 2x - 6$.

Lesson 11.2, pages 513–518

Exploration 1

1. $-x^2$

2. $x^2; -x^2; x^2$

3. $-x; -x$

4. $-x; x; x; x; -x; x$

Exploration 2

1. $x, x + 1$

2. $x^2 + x$

3. $x^2 + x$

4a. $2x^2 + 6x$

4b. $9x + 6$

5a. $2x^2 + 6x$

5b. $9x + 6$

6. Multiply $2x$ and x, $2x$ and 1. Add the products. $2x^2 + 2x$

Exploration 3

1. $2x, x - 1$

2. $2x^2 - 2x$

3. $2x^2 - 2x$

4a. $-4x + 8$

4b. $6x - 3$

5a. $-4x + 8$

5b. $6x - 3$

6. Multiply -3 and $-x$, -3 and 4. Add the products. $3x - 12$

Exploration 4

2. $2x + 3$

4. $x + 2$

5. Think $3x$ times what number equals $3x^2$? Think $3x$ times what number equals $6x$? $3x(x + 2)$

6. Think $5x$ times what number equals $5x^2$? Think $5x$ times what number equals $15x$? $5x(x + 3)$

Communicate

1. Use two x-tiles as one factor. Use one x-tile and five 1-tiles as the other factor. The product rectangle will be $2x^2$-tiles and 10 x-tiles.

2. Multiply each term in $x + 5$ by $2x$; $2x^2 + 10x$.

3.

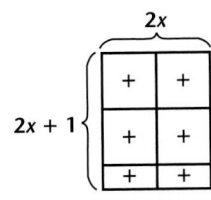

$2x$ and $2x + 1$

4. Write the factors whose product is $x^3 + x$. The factors are x and $x^2 + 1$; $x^3 + x = x(x^2 + 1)$.

Practice and Apply

5.

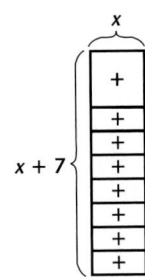

$x^2 + 7x$

6.

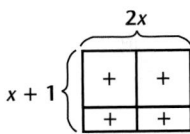

$2x$ and $x + 1$

Lesson 11.3, pages 519–523

Exploration 1

1.

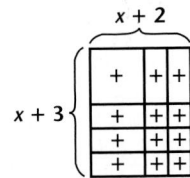

2. $x + 3$ and $x + 2$

3. $x^2 + 5x + 6$

4. $x^2 + 5x + 6$

5. One factor is one x-tile and four 1-tiles. The other factor is one x-tile and five 1-tiles. The product rectangle will have one x^2-tile, nine x-tiles and twenty 1-tiles.

Exploration 2

1. The product of the first terms of each factor is the first term of the trinomial.

2. The sum of the products of the outside terms and the inside terms is the middle term of the trinomial.

3. The product of the last terms of each factor is the last term of the trinomial.

4. Multiply $x \cdot x = x^2$, $x \cdot 5 = 5x$, $3 \cdot x = 3x$, and $3 \cdot 5 = 15$. Collect like terms, $x^2 + 8x + 15$.

5a. $x^2 - 2x - 15$

5b. $x^2 + 2x - 15$

5c. $x^2 - 8x + 15$

6. $(x + a)(x + b) = x^2 + ax + bx + ab$

Communicate

1a.

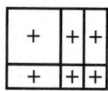

$x^2 + 3x + 2$

1b.

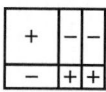

$x^2 - 3x + 2$

2a. $(x + 1)(x + 2) =$
$x^2 + x + 2x + 2 = x^2 + 3x + 2$

2b. $(x - 1)(x - 2) =$
$x^2 - x - 2x + 2 = x^2 - 3x + 2$

3. $(2x + 3)(x - 4) = 2x^2 + 3x -$
$8x - 12 = 2x^2 - 5x - 12$

4.

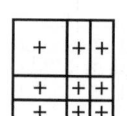 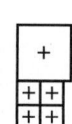

$(x + 2)^2$ has four x-tiles that $x^2 + 2^2$ does not.

Look Back

43. (4, 1)

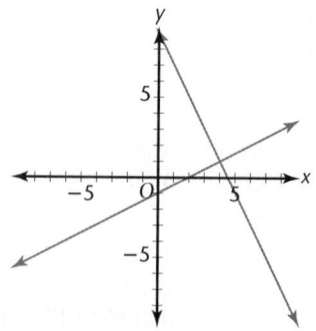

44.

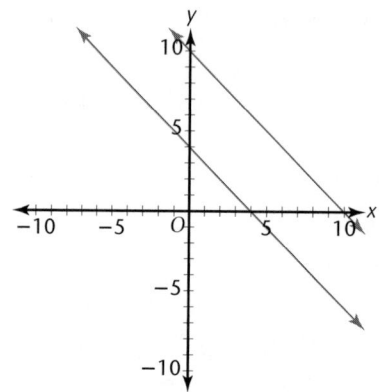

There is no solution because the lines are parallel.

Lesson 11.4, pages 524–528

Exploration

1d. 1 **1e.** 2 **1f.** 1

1g. 2 **1h.** 2 **1i.** 2

1j. 1 **1k.** 3 **1l.** 1

2. 2, 3, 5, 7, 11, 13; 2, 3, 5, 7, 11, 13, 17, 19, 23, 29

3. 4, 6, 8, 9, 10, 12, 14, 15, 16, 18, 20, 21, 22, 24, 25, 26, 27, 28

4. 1 divides every number and divides it as many times as you want it to.

5. A prime number is a number which has exactly two different factors, 1 and itself.

Communicate

1. A prime polynomial is one that has no polynomial factors with integral coefficients except itself and 1.

2. Distributive Property

3. Divide each term in the polynomial by the greatest common factor (GCF).

4. 30; Find the largest number that divides 60 and 150 evenly.

5. x^3y^2; Factor out variables that are common to both terms with the highest possible degree.

---

6. $(x + y)$; First check for a possible GCF between 25 and 39. There is no GCF, so the coefficients do not change. But both expressions have $(x + y)$ in common. Therefore, $(x + y)$ is the GCF.

7. Group terms with common variables. The result is $(y^2 + 2y) + (3y + 6)$.

Lesson 11.5, pages 529–534

Communicate

1. Determine the 2 equal factors of the first term, x^2, namely, x and x. Determine the 2 equal factors of the third term, 100, namely, 10 and 10. Since the trinomial is a perfect square the factors are $(x + 10)$ and $(x + 10)$.

2. If you assume the trinomial is a perfect square, determine the square roots of the first and last term. The square root of $4x^2$ and 9 are $2x$ and 3 respectively. Check to see if the middle term of the trinomial is equal to the twice the product of the square roots of $4x^2$ and 9. Since the values are the same, the factors are $(2x - 3)$ and $(2x - 3)$.

3. Find the positive square root of p^2 and 121. The square roots are p and 11, respectively. Since the binomial is a difference of two square, its factors are the sum and the difference of the square roots, $(p + 11)$ and $(p - 11)$.

Practice and Apply

41. The rectangle formed has dimensions $a + b$ and $a - b$, therefore, by comparing areas, $a^2 - b^2 - (a + b)(a - b)$.

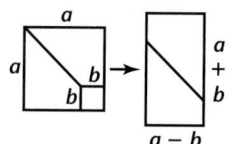

46. Size of rectangle may vary.

$$x + 4$$

$$x - 4$$

47. The area of the rectangle is $(x + 4)(x - 4)$ or $x^2 - 16$ which is the area of the original shaded area.

Lesson 11.6, pages 535–539

Communicate

1. $(x - 4)(x - 1)$

2. $(x - 6)(x + 2)$

3. $(x + 3)(x + 3)$

4. $(x + 3)(x - 2)$; the signs are opposite because the third term of the trinomial is negative.

5. $(x - 2)(x - 5)$; the signs are the same because the third term of the trinomial is positive.

6. $(x + 5)(x - 3)$; the signs are opposite because the third term of the trinomial is negative.

7. Determine the factor pairs of the third term. The factor pairs of 24 are: 1, 24; 2, 12; 3, 8; 4, 6. Check which factors can be arranged to form the value of -5, the middle term of the trinomial.

8. 1, 36; 2, 18; 3, 12; 4, 9; 6, 6

Look Back

44.

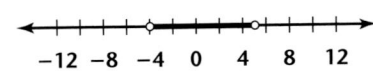

Look Beyond

48. $x = 0$

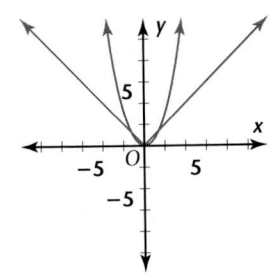

Project, pages 540–541

Activity 1

1. $a^3 + 3a^2b + 3ab^2 + b^3$

2. $a^4 + 4a^3b + 6a^2b^2 + 4ab^3 + b^4$
$a^5 + 5a^4b + 10a^3b^2 + 10a^2b^3 + 5ab^4 + b^5$

3. The number of terms is one more than the exponent of the binomial.

4. eight

5.
```
        1    6    15    20    15    6   1
      1    7    21    35    35    21   7   1
    1    8    28    56    70    56   28   8   1
  1    9    36    84   126  126   84   36   9   1
1   10   45   120   210   252  210  120  45  10  1
```

6. 1   7   21   35   35   21   7   1

7. The exponents of a decrease by 1 as you move from left to right. The exponents of b decrease by 1 as you move from left to right.

8. $a^7 + 7a^6b + 21a^5b^2 + 35a^4b^3 + 35a^3b^4 + 21a^2b^5 + 7ab^6 + b^7$

Lesson 12.1, pages 548–551

Exploration 1

1. first differences; $-10, -6, -2, 2$; second differences; 4

2. They are constant.

3. $f(5) = 0$; $f(6) = 10$

4.

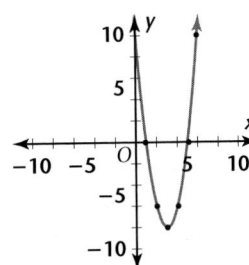

5. The graph is a parabola. If the second differences are constant the function is a quadratic function.

Exploration 2

1. $(3, -8)$

2. $(0, 10), (1, 0), (2, -6), (5, 0)$

3.

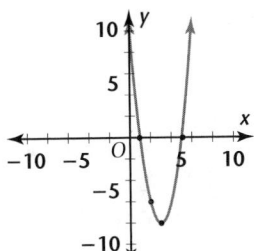

4. parabola; same graph as in Exploration 1

Exploration 3

1.

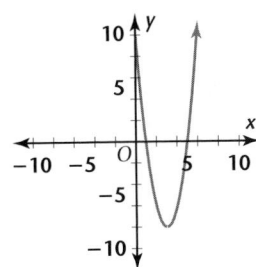

2. 5 and 1   **3.** 3

4. -8 **5.** $(3, -8)$

6. $x = 3$

7. -8

8. same graph

9. same value

Communicate

1. The second differences are constant.

2. The parent graph has been stretched vertically by a factor of 2 and moved horizontally 3 units to the right and vertically 8 units down.

3. The vertex (h, k); is $(3, -4)$.

4. The axis of symmetry $x = h$; is $x = 3$.

5. Factor the quadratic expression. Then set each linear factor equal to zero and solve these equations. $(x - 4)(x - 4) = 0$; $x = 4$, $x = 4$

6. Find the average of the two zeros. The axis of symmetry is $x = 4$.

Practice and Apply

7. Yes; the second differences are constant $+2$.

8. shift right 2 units, up 3 units

9. shift right 5 units, down 2 units; 3 times steeper

10. shift right 2 units, up 1 unit, and reflect

11. shift left 1 unit

12. shift right 2 units, up 1 unit, and reflect; 3 times steeper

13. shift right 2 units, up 3 units; half as steep

14. V($-4, -3$); $x = -4$

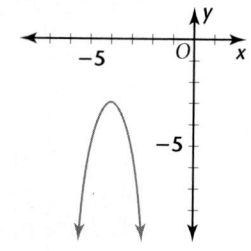

15. V(2, 3); $x = 2$

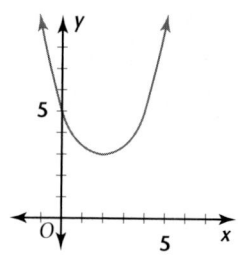

16. V(3, -7); $x = 3$

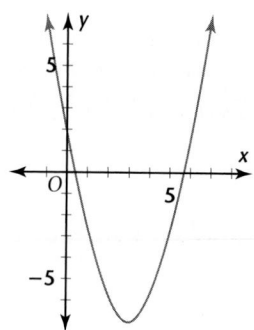

17. V(5, 2); $x = 5$

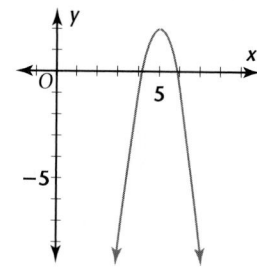

18. V($-3, -2$); $x = -3$

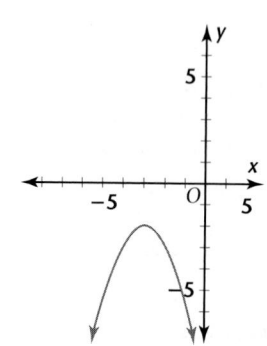

19. V($-5, 7$); $x = -5$

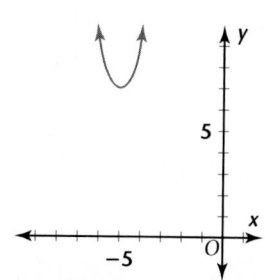

Lesson 12.2, pages 552–557

Communicate

1. The value for x for which y is 0 is the point at which the function crosses the x-axis. If the function represents a real relationship such as between time and height, negative values may have no meaning.

2. Find the numbers which when squared give 100; positive or negative.

3. Since $x^2 = 64$ and $64 \geq 0$, $x = \pm 8$.

4. Since $x^2 = \sqrt{8}$ and $\sqrt{8} \geq 0$, $x = \pm \sqrt{8}$.

5. Since $x^2 = \frac{16}{100}$ and $\frac{16}{100} \geq 0$, $x = \pm \frac{4}{10}$.

6. Add 25 to both sides of the equation. Solve the resulting equation.

$$(x + 3)^2 = 25$$
$$x + 3 = \pm 5$$
$$x = 2 \text{ or } x = -8$$

7. $(x - 8)^2 = 2$

$$x - 8 = \pm\sqrt{2}$$

$$x = 8 \pm \sqrt{2}$$

8. Plot the vertex at $(-4, -5)$. Solve the equation $(p + 4)^2 - 5 = 0$ to find the zeros. Mark these two points on the x-axis. Sketch the graph by drawing a parabola passing through the three points plotted and symmetrical about the axis of symmetry, $x = -4$.

Practice and Apply

46. V(4, −9); x = 4; zeros = 1,7

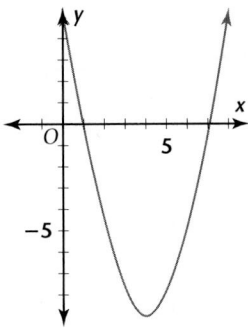

47. V(−2, −1); x = −2; zeros −3, −1

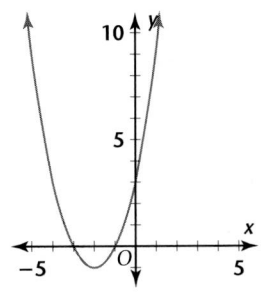

48. V(4, −3); x = 4; zeros ≈ 2.27, 5.73

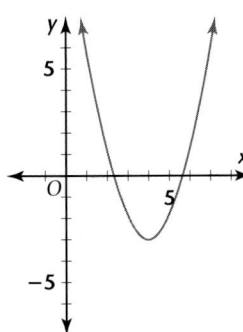

52.

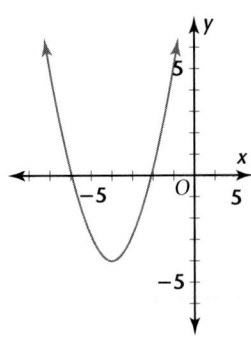

53.

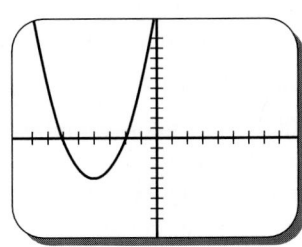

Lesson 12.3, pages 558–563

Exploration

1. 2 (5 per group)

2. 25

3. $x^2 + 10x + 25$

4a. 5

4b. 25

4c. 25

5. x + 8; the constant is half the coefficient of x

5a. $x^2 + 16x + 64$

5b. $(x + 8)^2$

6. 1369 1-tiles: half 74 to get 37 and then square 37 to get 1369.

Communicate

1. Start with one x^2-tile, arrange the four x-tiles in two groups of 2. Four 1-tiles are needed to complete the square.

2. Find half of the coefficient of x and then add its square to the expression. Factor the trinomial into a perfect square.
$$x^2 - 7x + \frac{49}{4} = \left(x - \frac{7}{2}\right)^2$$

3. The first three terms are a perfect square trinomial, so the equation can be written as $y = (x + 5)^2 - 25$.

4. Complete the square to find the vertex. The y-value, 2 is the minimum value.

5. Complete the square to find the vertex. The y-value, $-\frac{1}{4}$, is the minimum value.

6. $y = x^2 - 10x + 11$
$= x^2 - 10x + 25 - 25 + 11$
$= (x - 5)^2 - 14$
The vertex is (5, −14).

Practice and Apply

31. minimum at −25

44.

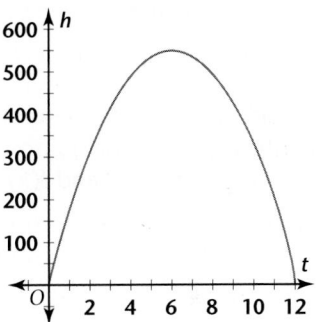

Look Beyond

60. 2.5, 10

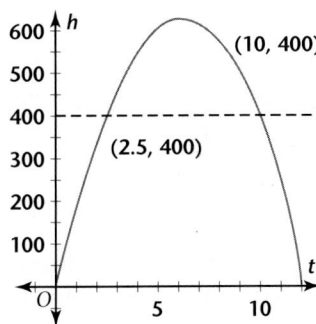

61. 2.5 seconds, 10 seconds

Lesson 12.4, pages 566–573

Communicate

1. They have the same value.

2. Complete the square on the first two terms of the trinomial. Solve the equation obtained by solving f(x) = 0. x = 2 or x = 4

3. Take half the coefficient of x and square it. Add this number to both sides of the equation. Express the perfect square on the left side as a square of a linear factor and simplify the integers on the right side. Solve the resulting equation.

4.
$$x^2 + 10x = 24$$
$$x^2 + 10x - 24 = 0$$
$$(x + 12)(x - 2) = 0$$
$$x = -12 \text{ or } x = 12$$

5. Factoring, because this is a perfect square trinomial.

6. Factoring, because this trinomial has two linear factors.

7. Solve the equation $2 = x^2 - 7x + 12$. Solve by factoring or find the intersection of the graphs of $f(x) = x^2 - 7x + 12$ and $f(x) = 2$.

8. Graph the line $y = x - 1$ and the parabola $y = x^2 - 3x + 2$ and observe where they intersect.

Lesson 12.5, pages 574–580

Exploration

1.

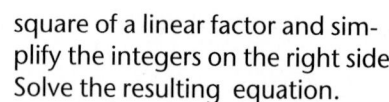

Function	a	b	c	$b^2 - 4ac$	x-intercepts
f	1	6	5	16	2
g	1	-10	25	0	1
h	1	0	1	-4	none

2. if $b^2 - 4ac > 0$ two solutions, if $b^2 - 4ac = 0$ one solution, if $b^2 - 4ac < 0$ no solutions

Communicate

1. When the equation is written in the form $2y^2 + 3y - 7 = 0$, the coefficient of y^2 is a, the coefficient of y is b and the constant term is c.

2. In $b^2 - 4ac$ substitute $a = 2$, $b = 3$ and $c = -7$.

3. Find the value of $b^2 - 4ac$ when $a = 2$, $b = -1$ and $c = -2$. Since this value is greater than zero (17), the equation has two solutions.

4. Find the value of $b^2 - 4ac$ when $a = 1$, $b = -2$ and $c = 7$. Since this value is less than zero (-24), the equation has no solutions.

5. Rewrite the equation so that it equals zero. Use $a = 4$, $b = 12$ and $c = -7$ in the quadratic formula.

6. The factors are $\left(x + \frac{1}{2}\right)\left(x + \frac{5}{2}\right)$. Multiply each expression by 2 to eliminate the fractions. The equation is $(2x + 1)(2x + 5) = 0$.

Lesson 12.6, pages 581–585

Exploration

1. 3000 and 7000 watches per week

2. when more than 3000 but less than 7000 watches are produced per week

3. when less than 3000 or more than 7000 are produced per week

4. 5000 per week

5. $4000 per week

Communicate

1. Factor the related quadratic equation and determine where the product is positive.

2. Since the inequality does not include equals to, circle the two solutions found from exercise 1. Test values for x from each of the three intervals created on the number line to see which range of values makes the inequality true.

3. The boundary is dashed since the inequality does not include *equal to*.

4. The boundary is solid because the inequality includes *equal to*.

5. The boundary is solid because the inequality includes equal to.

6. First sketch the parabola $y = x^2 - 3$ with a broken line. Test one point between the zeros of the function and if the resulting inequality is true shade that side of the parabola. If the test is false shade the other side.

7. Use the steps as in the previous answer except use a solid line for the parabola.

8. Use the same steps as in Exercise 7.

Practice and Apply

9.

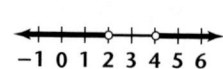

10.

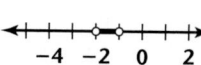

11.

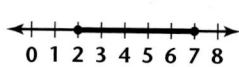

12.

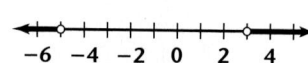

13.

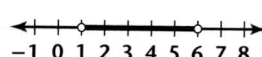

14.

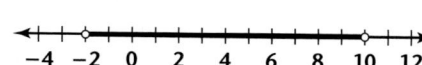

15.

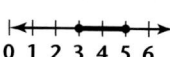

16.

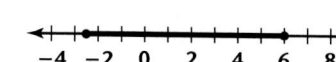

17.

18.

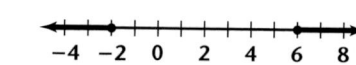

19.

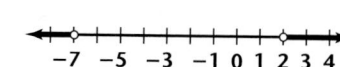

20.

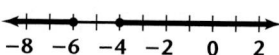

21.

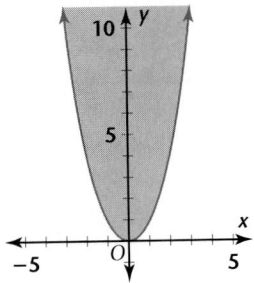

22.

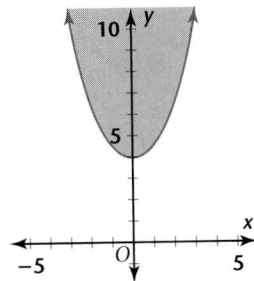

23.

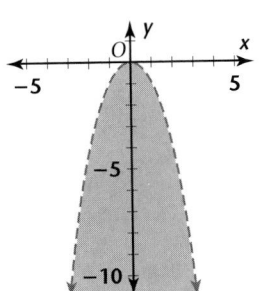

24.

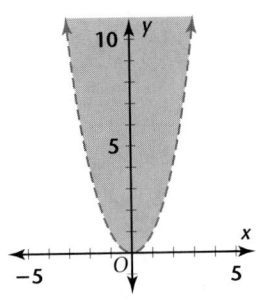

25.

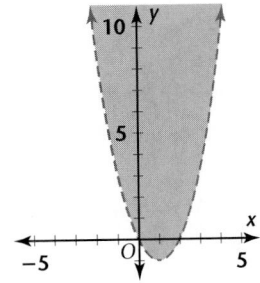

26.

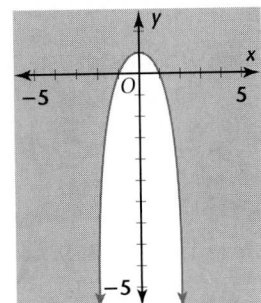

27.

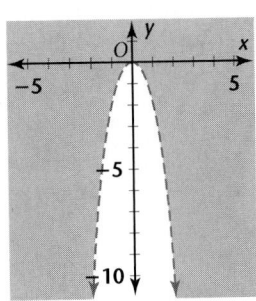

28.

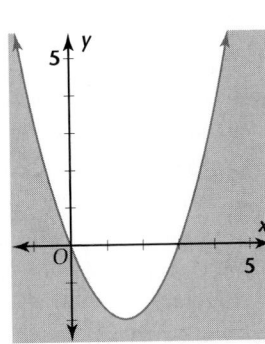

29.

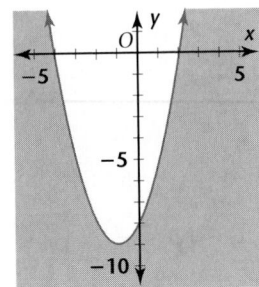

30.

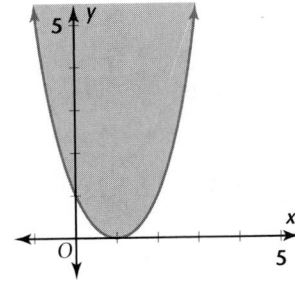

31.

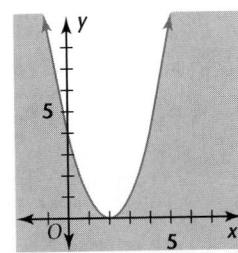

32.

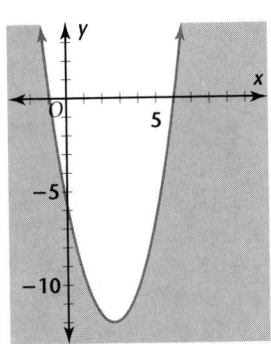

33.

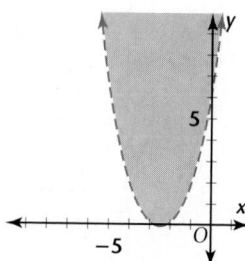

34.

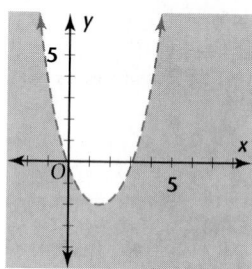

35.

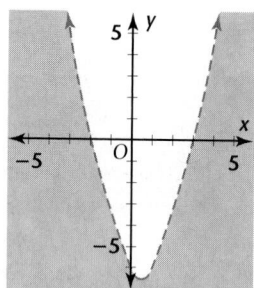

38.

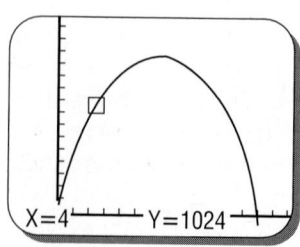

Lesson 13.1, pages 596–603

Exploration 1

1. 16 cm²

2. 9 cm²

3. Answers may vary. For example, 3.4 cm.

4. Answers may vary. For example, 11.56 cm².

5. Answers may vary. For example, the product is 0.44 cm² less than 12 cm². Try 3.45, 3.45² = 11.9025.

6. Answers may vary. An estimate of 3.46 or 3.47 is good since √12 is between 3.46 and 3.47.

7. Determine the perfect square numbers less than and greater than 20. They are 16 and 25, respectively. Determine the square roots of 16 and 25. They are 4 and 5, so the value of √20 is between 4 and 5. Use the same procedure as before to determine √20 to two decimal places. Either 4.47 or 4.48 is a good estimate since the value of √20 is between 4.47 and 4.48.

Exploration 2

1.

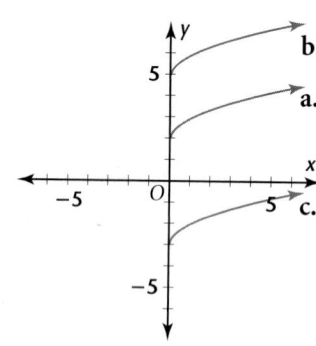

2a. $g(x)$ is the graph of $f(x)$ shifted vertically up by 2 units.

2b. $h(x)$ is the graph of $f(x)$ shifted vertically up by 5 units.

2c. $p(x)$ is the graph of $f(x)$ shifted vertically down by 3 units. When a is added to $f(x) = \sqrt{x}$, the graph of $f(x)$ is shifted vertically by

a units. A positive a value results in an upward shift, a negative a value results in a downward shift.

3.

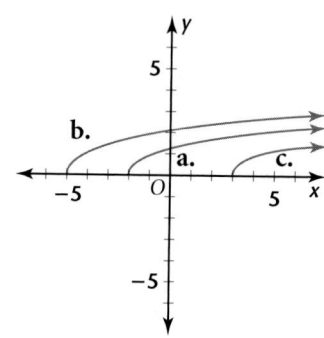

4a. $g(x)$ is the graph of $f(x)$ shifted horizontally 2 units to the left.

4b. $h(x)$ is the graph of $f(x)$ shifted horizontally 5 units to the left.

4c. $p(x)$ is the graph of $f(x)$ shifted horizontally 3 units to the right. When a is added to $f(x) = \sqrt{x}$, the graph of $f(x)$ is shifted horizontally by a units. A positive a-value results in a shift to the right, a negative a-value results in a shift to the left.

5.

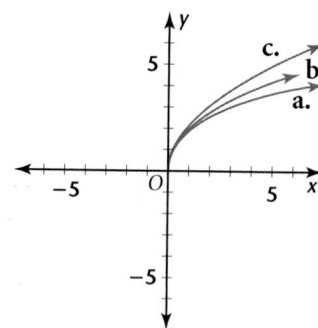

6a. $g(x)$ is the graph of $f(x)$ vertically stretched upward by a factor of 2.

6b. $h(x)$ is the graph of $f(x)$ vertically stretched upward by a factor of 3.

6c. $p(x)$ is the graph of $f(x)$ vertically stretched upward by a factor of 5. When $f(x) = \sqrt{x}$ is multiplied by a scale factor, a, the graph of $f(x)$ is stretched vertically from the vertex $(0, 0)$. When $a > 1$, then the graph of $f(x)$ is stretched upward.

When $0 < a < 1$, the graph is flattened.

7.

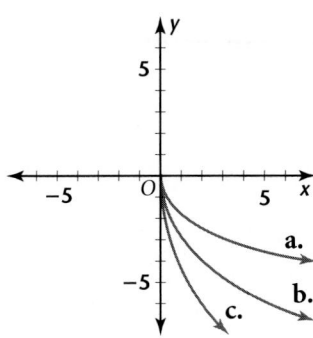

8a. $g(x)$ is the graph of $f(x)$ is reflected through the *x*-axis and stretched 2 times.

8b. $h(x)$ is the graph of $f(x)$ is reflected through the *x*-axis and stretched 3 times.

8c. $p(x)$ is the graph of $f(x)$ is reflected through the x-axis and stretched 5 times. When $f(x)$ is multiplied by $-a$, the graph of $f(x)$ is reflected through the *x*-axis and then stretched vertically by the scale factor, *a*.

9. The graph of $\sqrt{x} + a$ is a vertical shift up or down of the graph of $\sqrt{x}$ by *a* or $-a$ units, respectively. The graph of $\sqrt{x + a}$ is a horizontal shift left or right of the graph of $\sqrt{x}$ by *a* or $-a$ units, respectively. The graph of $a\sqrt{x}$ is an upward stretch of the graph of $\sqrt{x}$ by a scale factor of *a*, and a flattening stretch toward the *x*-axis, when $0 < a < 1$. The graph of $-a\sqrt{x}$ is a reflection through the *x*-axis of the graph of $\sqrt{x}$ followed by a stretch by a scale factor of *a*.

Communicate

1. Draw a square that contains the number of graph paper squares equal to the perfect square number.

2. Draw two squares, one inside the other, whose areas are equal to the perfect square numbers just less

than just greater than the number that is not a perfect square. Determine the length of the side of each square. The square root of the non-square number is between these two values. Make an estimate for the square root of the non-square number. Use the guess-and-check method to find a value such that when it is squared, it is approximately the original number.

3. An irrational number is a number that cannot be represented by a repeating or terminating decimal. Answers may vary. For example, π, $6.10100111\ldots$, $\sqrt{19}$.

4. Answers may vary. For example, use the square root key on a calculator; use the guess-and-check method with square numbers.

5. To move the square root function 2 to the right, subtract 2 from *x* before taking the square root. To move the function up 3 units, add 3 to $\sqrt{x - 2}$. The result is $f(x) = \sqrt{x - 2} + 3$.

6. A negative number multiplied by itself gives a positive result. The same is true for a positive number.

Practice and Apply

24.

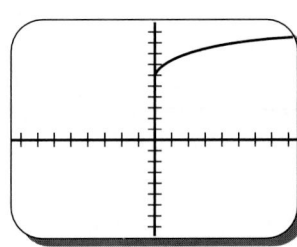

25.

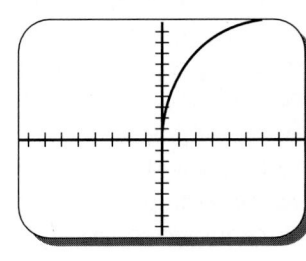

26.

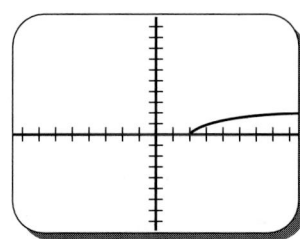

27.

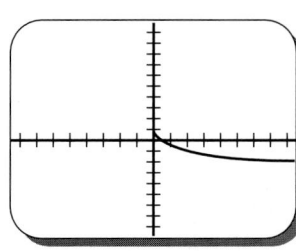

28.

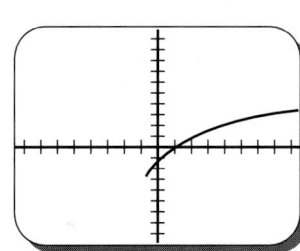

29.

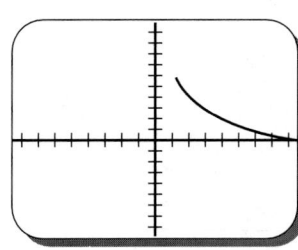

37. The graph of the parent function is shifted horizontally to the left by 3 units, then shifted vertically down 4 units, and stretched upward by a scale factor of 5.

38. The graph of the parent function is shifted horizontally to the left by 3.75 units, stretched vertically by a scale factor of 3.5, reflected through the *x*-axis then shifted upward by 2 units.

39. $y = \sqrt{x} + 3$

40. $y = -\sqrt{x + 4}$

41. $y = \sqrt{x - 2} + 6$

43.

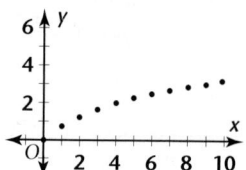

47. yes

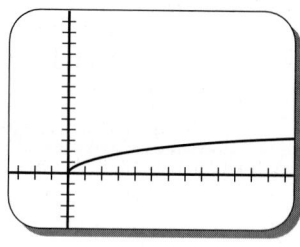

Look Back

48.

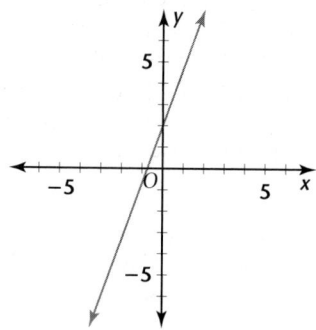

49.

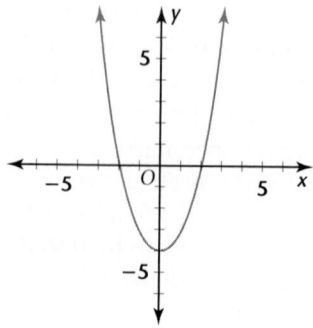

50.

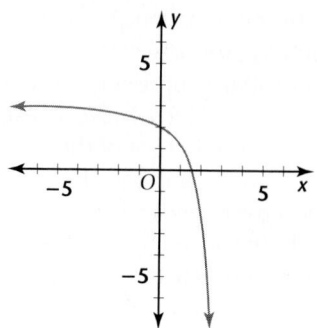

51.

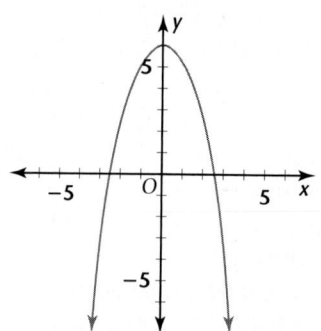

Lesson 13.2, pages 604–610

Communicate

1. when $a \geq 0$

2. When simplifying an expression perfect square factors are factored out of the radical; an expression is in simplified form when there are no perfect square factors under the radical sign, and there are no radicals in the denominator.

3. $a \leq 3$

4. Answers may vary. For example, the Multiplication Property states that the square root of a product is the same as taking the square roots of its factors separately; the Division Property state that the square root of a quotient is the same as taking the square roots of the dividend and divisor separately.

5. When the expressions under the radical signs are equal.

---

6. Answers may vary. For example, use the FOIL method.

$$(\sqrt{3} + 2)(\sqrt{2} + 3)$$
$$= \sqrt{3}\,\sqrt{2} + 3\sqrt{3} + 2\sqrt{2} + 2(3)$$
$$= \sqrt{6} + 3\sqrt{3} + 2\sqrt{2} + 6$$

Lesson 13.3, pages 611–618

Communicate

1. When both positive and negative quantities are squared, the result is positive.

2. Square both sides of the equation and then isolate and solve for the variable.

3. no; the principal square root of a number can never be negative.

4. For $a^2 = b^2 + c^2$, when you solve for a, the two values are $\sqrt{b + c}$ and $-\sqrt{b + c}$. For $a^2 = (b + c)^2$, when you solve for a, the two values are $(b + c)$ and $-(b + c)$ or $-b - c$.

5. Express the left side as a perfect square binomial, for $(x + b)^2 = c^2$. Thus, $x = c - b$ and $x = -c - b$. Answers may vary. For example, let $b = 2$ and $c = 3$. The equation becomes $x^2 + 4x + 4 = 9$. Factor both sides for $(x + 2)^2 = 9$. Thus, $x + 2 = \pm 3$. Solving for x, $x = 1$ or -5.

6. Radicals are used to identify the variable x in the equation $ax^2 + bx + c = 0$

Practice and Apply

22. $(4, 4); x = 4$

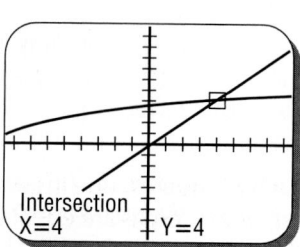

23. no solution

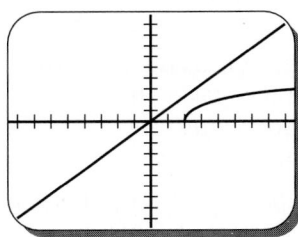

24. (1, 1) and (4, 2); $x = 1$ or 4

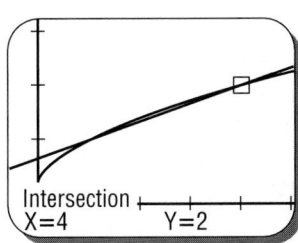

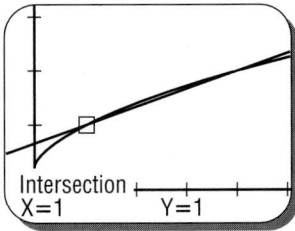

Look Back

48.

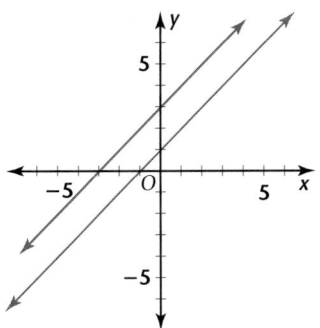

The lines are parallel. The second graph is the first translated 2 to the right.

49.

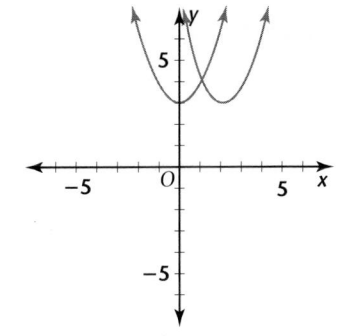

Both are parabolas that open upward with vertices of (0, 3) and (2, 3). The graphs intersect at (1, 4). The second graph is shifted 2 to the right.

50.

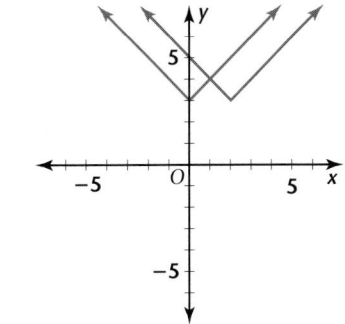

Both are V-shaped graphs that open upward one has a vertex of (0, 3) and the other has a vertex of (2, 3). They both contain the point (1, 4). The second graph is shifted 2 to the right.

Lesson 13.4, pages 619–624

Communicate

1. The side opposite the right angle is the hypotenuse. The other sides are the legs.

2. In a right triangle, the sum of the squares on the legs is equal to the square on the hypotenuse.

3. Square the lengths of the legs, add the squares together and then determine the square root of the sum.

4. Subtract the square of the known leg from the square of the hypotenuse. Then find the square root of the difference.

5. Answers may vary. For example, draw a right triangle. Measure the length of its sides, square the lengths of the legs and then add these sums. Show that the sum is equal to the square of the length of the hypotenuse.

6. Let the length of the rectangle be one leg of the right triangle, and the width be the other leg. The diagonal of the rectangle will be the hypotenuse of the right triangle. Apply the Pythagorean Right-triangle Theorem to determine the length of the diagonal.

Practice and Apply

24.

Right Triangle Number	1	2	3	4	5	6	7	8	9
Hypotenuse	$\sqrt{2}$	$\sqrt{3}$	$\sqrt{4}$ or 2	$\sqrt{5}$	$\sqrt{6}$	$\sqrt{7}$	$\sqrt{8}$ or $2\sqrt{2}$	$\sqrt{9}$ or 3	$\sqrt{10}$

  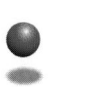

Look Beyond

39. 7 units

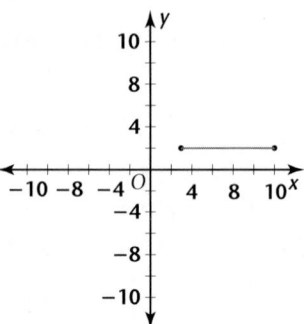

40. 12 units

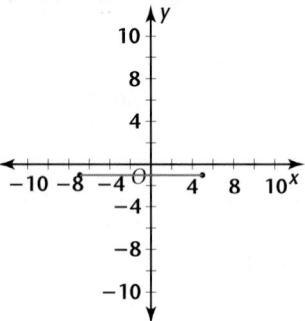

41. 7 units

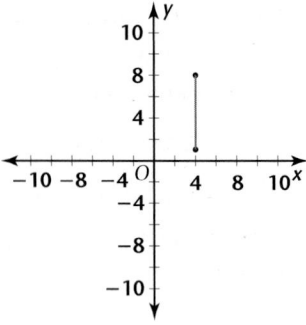

42. 3 units

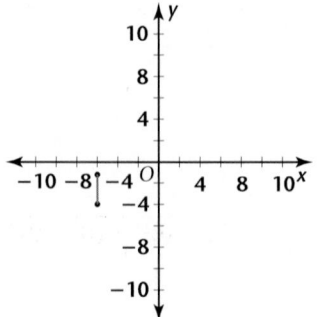

Lesson 13.5, pages 625–631

Exploration 1

1a. 5

1b. 13

1c. 10.30

1d. 6.40

2. $\sqrt{(x_2 - x_1)^2 + (y_2 - y_1)^2}$; they are the same.

Exploration 2

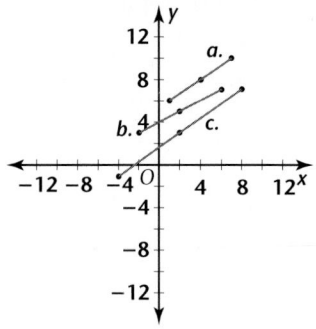

1a. (4, 8)

1b. (2, 5)

1c. (2, 3)

2a. The lengths of line segments *PM* and *MQ* are both $\sqrt{13}$.

2b. The lengths of line segments *PM* and *MQ* are both $2\sqrt{5}$.

2c. The lengths of line segments *PM* and *MQ* are both $2\sqrt{13}$

3. Average the *x*-coordinates.

4. Average the *y*-coordinates.

5. (27.5, 41.5)

6. 26.50 = 26.50

Communicate

1. A point is a position on a line. A co-ordinate is a number which is used to locate a point.

2. The Pythagorean Right-triangle Theorem can be used to find the distance between the endpoints of two perpendicular line segments.

---

3. Square the difference between the two *x*-coordinates. Repeat for the *y*-coordinates. Take the square root of the sum of the differences.

4. The *x*-coordinate of the midpoint is found by taking the average of the *x*-coordinates. The *y*-coordinate is found by the same method.

5. Interchange the *if* and *then* portions of the statement.

Practice and Apply

12.

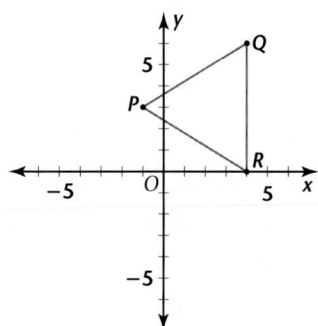

13.

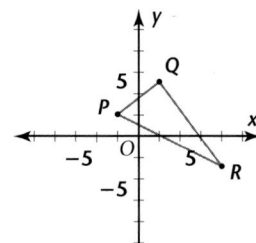

15.

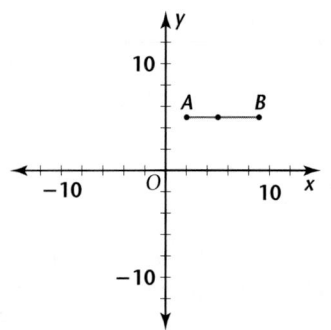

16.

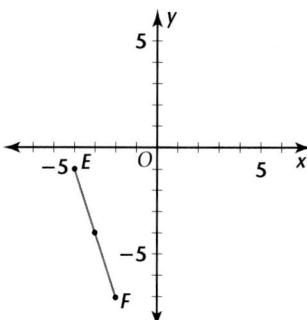

17.

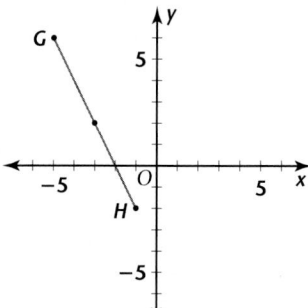

18.

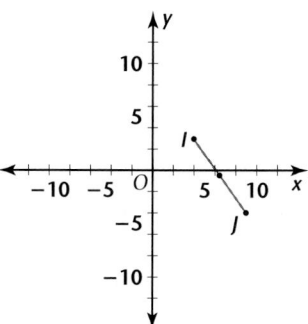

33a. $S = 4.26\sqrt{d}$.

33b. $d \approx 166.69$ ft

33c. $S = 25.56$ mph; the officer was correct.

34.

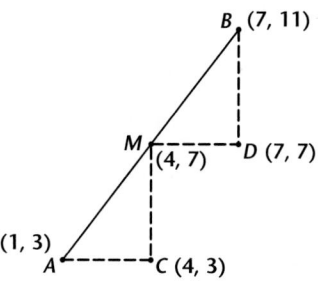

$(AC)^2 + (CM)^2 = (AM)^2$
$3^2 + 4^2 = (AM)^2$
$9 + 16 = (AM)^2$
$\sqrt{25} = AM$
$5 = AM$

$(MD)^2 + (BD)^2 = (MB)^2$
$3^2 + 4^2 = (MB)^2$
$9 + 16 = (MB)^2$
$\sqrt{25} = MB$
$5 = MB$

Since $AM = MB$, M is the midpoint of $\overline{AB}$.

Lesson 13.6, pages 632–637

Exploration 1

1.

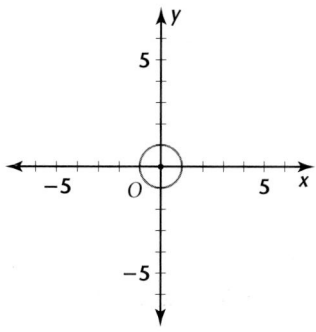

2. $x^2 + y^2 = 9$; $x^2 + y^2 = 1$
3. $x^2 + y^2 = r^2$
4. $(x - 1)^2 + (y - 1)^2 = 1$

Exploration 2

1. 5, 5
2. Answers may vary. For example, 11.31, 11.31; 17.89, 17.89
3. The diagonals of a rectangle are equal in length.

Exploration 3

1. length of a diagonal $= \sqrt{a^2 + b^2}$
2. To determine the length of the diagonal of a rectangle, square the lengths of both sides. Add the

---

squares together then take the square root of the sum. Both diagonal will have the same length.

Exploration 4

1. The length of AB is $\dfrac{1}{2}$ the length of the side to which it is parallel.
2. In an isosceles triangle, the segment connecting the midpoints of the congruent sides is $\dfrac{1}{2}$ the length of the third side.
3. It should also be true.
4. Label the vertices of the triangle as follows:

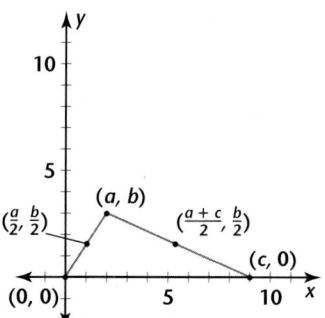

Let D be the midpoint of side AB, then the coordinates of D are $\left(\dfrac{a}{2}, \dfrac{b}{2}\right)$. Let E be the midpoint of side BC, then the coordinates are $\left(\dfrac{a+c}{2}, \dfrac{b}{2}\right)$. The length of DE and AC can be found by applying the distance formula.

length of $DE =$

$$\sqrt{\left(\frac{a}{2} - \frac{a+c}{2}\right)^2 + \left(\frac{b}{2} - \frac{b}{2}\right)^2}$$
$$= \sqrt{\left(\frac{a}{2} - \frac{a}{2} - \frac{c}{2}\right)^2 + (0)^2}$$
$$= \sqrt{\frac{c^2}{4}}$$
$$= \frac{c}{2}$$

length of $AC =$

$$\sqrt{(c - 0)^2 + (0 - 0)^2}$$
$$= \sqrt{c^2}$$
$$= c$$

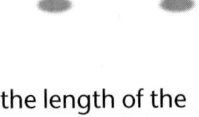

This shows that the length of the line segment connecting the midpoints of the two sides of a triangle is half the length of the third side.

Communicate

1. Answers may vary. For example, by using geometric figures to model a problem you can solve problems in algebra.

2. This is the definition of a circle.

3. Subtract the *x*-coordinates and square the difference. Repeat for the *y*-coordinates. Add these two values and take the square root of the sum.

4. Let *r* be the common distance, (x, y) the given point and (x_n, y_n) be any of the equidistant points. The distance between the given point and any of the equidistant points can be expressed as $r = \sqrt{(x - x_n)^2 + (y - y_n)^2}$. Square both sides of the equation for $r^2 = (x - x_n)^2 + (y - y_n)^2$, which is the equation of a circle.

5. Find the distance between any two opposite corners of the triangle.

Practice and Apply

11.

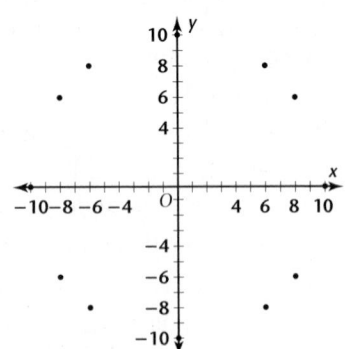

Look Beyond

44. Label the vertices of the rectangle as shown.

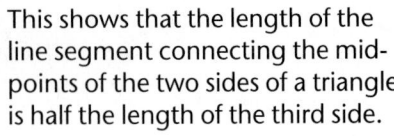

The coordinates of the diagonals' midpoint *E* are $\left(\frac{x}{2}, \frac{y}{2}\right)$. To verify that *E* is the midpoint of *AC* and *BD*, apply the distance formula for line segments, *AE* and *EC*, and *BE* and *ED*. The result is $AE = EC = BE = ED = \frac{\sqrt{x^2 + y^2}}{2}$.

45. Label the vertices of the isosceles trapezoid as shown.

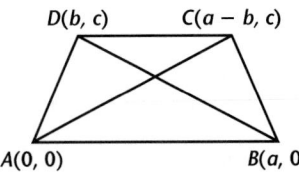

Let *a* represent the length of the longer size of the trapezoid or its base, and *b* represent the *x*-coordinate of point *D*, which is the horizontal distance *D* is from origin. Since the trapezoid is isosceles, then the horizontal distance of *C* from the origin is $a - b$. Apply the distance formula to find the length of the diagonals. The result is the length of $AC = BD = \sqrt{(a + b)^2 - c^2}$

Lesson 13.7, pages 638–643

Exploration

1a. 0.2

1b. 0.4

1c. 0.6

1d. 0.8

1e. 1.2

1f. 1.7

1g. 2.7

---

2. From 10 to 20 degrees, the value of the tangent ratio doubles. For values greater than 20, the ratio more than doubles.

3. 10°; 20°; 30°; 40°

4. They have the same length.

5. The angle is 45°. The tangent ratio is 1 and equal to the ratio of the length of the side opposite the angle *A* and the length of the side adjacent to angle *A*. Therefore, by solving for the length of the adjacent side, the result is length of the adjacent side is equal to the length of the opposite side.

Communicate

1. The slope of a line is the ratio of the change in *y* to the change in *x*. The tangent function is the ratio of the length of the side opposite an angle (the change in *y*) to the length of the side adjacent an angle (the change in *x*).

2. $\text{Tan } A = \frac{\text{length of the opposite leg}}{\text{length of the adjacent leg}}$
Solve for the length of the opposite leg = Tan *A* (length of the adjacent leg). Use a calculator to determine the value of Tan *A* and calculate the length of the opposite leg.

3. Use a calculator to apply the inverse tangent function to the ratio.

4. The angle is not measured directly but calculated using the lengths of the sides of the triangle.

Lesson 14.1, pages 652–657

Communicate

1. A rational expression is an expression that can be expressed in the form $\frac{P}{Q}$, where *P* and *Q* are both polynomials in the same variable and $Q \neq 0$. Examples are $f(x) = \frac{2x}{x^2 - 3x + 2}$, $f(x) = \frac{x^2 - 4}{x}$.

2. A rational expression $\frac{P}{Q}$ is undefined whenever the value of Q is zero.

3.

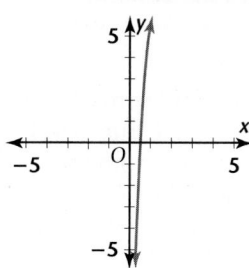

$h(x) = \frac{12x - 7}{x} = -\frac{7}{x} + 12$

4. When x is 0, $h(x)$ is undefined. The line $x = 0$ is a vertical asymptote of the graph.

5. The graph of $h(x)$ is obtained from the parent function $g(x) = \frac{1}{x}$ by a vertical stretch factor 7, a reflection in the x-axis and a vertical shift of 12 units.

Practice and Apply

20. $x \neq 0$

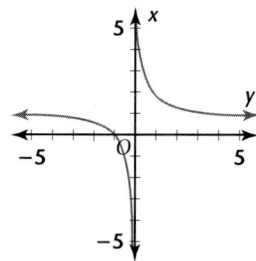

21. $x \neq -2$

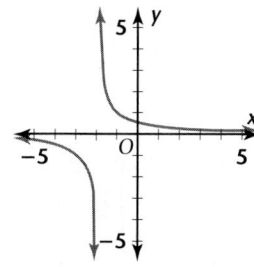

22. $x \neq 0$

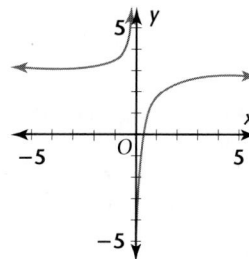

23. $x \neq 1$

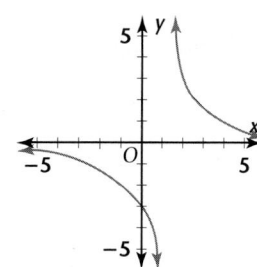

24. $x \neq 3$

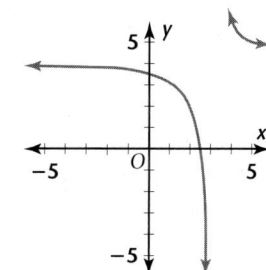

25. $x \neq 4$

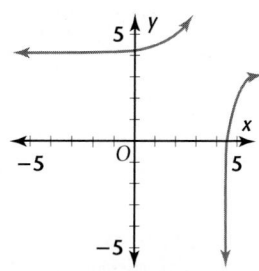

26.

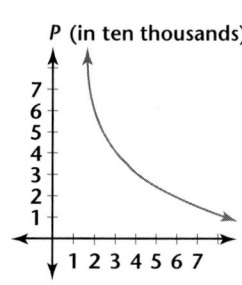

P (in ten thousands)

Lesson 14.2, pages 658–663

Communicate

1. $xy = k; y = \frac{k}{x}$

2. An inverse variation between rate and time means that as the rate increases the time decreases. For example the faster you drive the less time it takes to drive a certain distance.

3. If y varies inversely as x then $y = \frac{k}{x}$, where k is the constant of variation. Substitute $y = 3$ and $x = 8$ to find the value of k. Then use the equation of variation to find the value of y when $x = 2$.

4. The Rule of 72 is $t = \frac{72}{r}$. To find how long it would take money to double at 4%, divide 72 by 4.

5. $y = \frac{k}{x}$, where k is the constant of variation and $x \neq 0$.

Lesson 14.3, pages 664–669

Communicate

1. Any value of the variable that makes the value of the denominator 0 results in an undefined expression.

2. Set the expression in the denominator equal to zero and solve the resulting equation.

3. A common factor divides evenly into each of two or more numbers or expressions.

4. Factor the denominator to obtain $\frac{x + 1}{(x + 1)(x + 1)}$. Then remove the common factors from the numerator and the denominator. State any restrictions to avoid division by zero. So the expression simplifies to $\frac{1}{(x + 1)}$, $x \neq -1$.

5. $x - 7$ is the opposite of $7 - x$ because $-1(x - 7) = -x + 7 = 7 - x$.

Lesson 14.4, pages 670–675

Communicate

1. Multiply $x + 1$ by x or x by $x + 1$.

2. After the numerators are rewritten so the fractions have common denominators, add the numerators.

3. Follow the steps in 2, but subtract.

4. Multiply the numerators, then multiply the denominators.

5. Multiply the first expression by the reciprocal of the second expression.

Practice and Apply

6. $\dfrac{7}{3x}$, $x \neq 0$

7. $\dfrac{3}{x + 1}$, $x \neq -1$

8. $\dfrac{-3x}{y + 4}$, $y \neq -4$

9. $\dfrac{2x - 3}{a - c}$, $a \neq c$

10. $\dfrac{3a + 2b}{ab}$, $ab \neq 0$

11. $\dfrac{-5}{3t}$, $t \neq 0$

12. $\dfrac{6s^2 + t^2}{2rst}$, $rst \neq 0$

13. $\dfrac{5p - m}{p^2 q}$, $p^2 q \neq 0$

14. $\dfrac{-1}{2(x + 1)}$, $x \neq -1$

15. $\dfrac{x^2 + 3x + 4}{x + 4}$, $x \neq -4$

16. $\dfrac{d - 5}{d - 1}$, $d \neq 1$

17. $\dfrac{ad + bc}{bd}$, $bd \neq 0$

18. $\dfrac{5 + m}{a + b}$, $n \neq -3$, $a \neq -b$

19. $\dfrac{x}{x^2 + 4x + 3}$, $x \neq 0, -1, -2, -3$

20. $\dfrac{q - 1}{q}$, $q \neq 0, -1$

21. $\dfrac{2}{x^2 + x}$, $x \neq 0, -1$

22. $\dfrac{15}{x^2 - x}$, $x \neq 0, 1$

23. $\dfrac{y^2 - y + 4}{y^2 - 4y}$, $y \neq 0, 4$

24. $\dfrac{x + 3}{x^2 + x}$, $x \neq 0, -1$

25. $\dfrac{a^2 + 6a - 1}{a^2 + 4a + 3}$, $a \neq -3, -1$

26. $\dfrac{m^2 + 2m - 15}{m^2 + m - 2}$, $m \neq 1, -2$

27. $\dfrac{r - r^2}{r^2 + 6r + 5}$, $r \neq -1, -5$

28. $\dfrac{-5b - 12}{b^2 + b - 6}$, $b \neq 2, -3$

29. $\dfrac{9y - 8}{y^2 - 4y - 4}$, $y \neq 2$

30. $\dfrac{-3}{y + 6}$, $y \neq 5, -6$

31. $\dfrac{x - 1}{1 - x}$, $x \neq 1$

32. $\dfrac{7 - x}{2x^2 - 18}$, $x \neq 3, -3$

Eyewitness Math

3a. $P = \dfrac{a(1)}{(1 - a)(1 - 1) + a(1)}$

$= \dfrac{a}{(1 - a)(0) + a} = \dfrac{a}{a} = 1$

Sample: That makes sense because if $r = 1$, then *everyone* tested has the disease. So, no matter what result you get, the probability that you have the disease is 1.

3b. $P = \dfrac{ar}{(1 - a)(1 - r) + ar}$

$= \dfrac{(0.98)(0.5)}{(1 - 0.98)(1 - 0.5) + (0.98)(0.5)}$

$= \dfrac{0.49}{0.50} = 0.98$

Sample: No, because the probability would have turned out to be the same as the test accuracy.

Lesson 14.5, pages 678–683

Communicate

1. The values of the variable that makes the denominator of any rational expressions in the equation have value zero.

2. The realistic domain usually contains positive numbers because x is representing some type of measurement.

3. Find the lowest common multiple of the three denominators. The common denominator is $c(c - 1)$.

4. Graph two functions, $Y_1 =$ the left side of the equation and $Y_2 =$ the right side of the equation, and find their point(s) of intersection. The x-value(s) at the point(s) of intersection is the solution provided this value does not make the denominator of any of the terms in the equation have value 0.

5. Look at the denominator of each rational expression in the equation to find any values of the variable that make them have value 0. This equation is undefined for $c = 0$ and $c = 1$.

6. Rate $= \dfrac{1}{\text{number of hours worked}}$

Look Back

32.

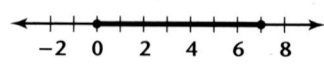

33.

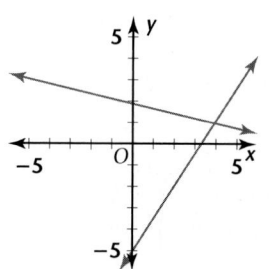

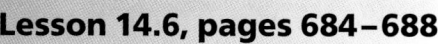

Lesson 14.6, pages 684–688

Exploration 1

	extremes	means
1a.	1, 10	2, 5
1b.	3, 12	4, 9
1c.	4, 9	6, 6
1d.	6, 2	$2\sqrt{2}, 3\sqrt{2}$

	extremes	means
2a.	10	10
2b.	36	36
2c.	36	36
2d.	12	12

3. The two products are equal.

4. The product of the extremes equals the product of the means in a proportion. Thus, the cross products are equal. **b** and **c** are proportions.

5. Cross multiply: $x \cdot 24 = 4 \cdot 42$. Then solve: $24x = 168$ and $x = 7$.

6a. 6

6b. 7, −1

7. Both proportions involve one variable and can be solved by using the cross product. The cross product in **a** results in a linear equation with one solution, while the cross product in **b** results in a quadratic equation with two solutions.

Exploration 2

1. Answers may vary. Examples are given for $a = 2$, $b = 4$, $c = 3$, and $d = 6$.

1a. $bc = 12$; $ad = 12$

1b. $c(a + b) = 18$; $c(a − b) = −6$

1c. $d(a + b) = 36$; $d(a + b) = 36$

1d. $ad = 12$; $bc = 12$

1e. $d(a − b) = −12$; $b(c − d) = −12$

1f. $a(b + d) = 20$; $b(a + c) = 20$

2. a, c, d, e, f

3a. The reciprocals of the original proportion are equal to each other and are true proportions; **c.** The sums of the numerator and the de-
nominator of the original proportion divided by the denominators of each side of the original proportion are equal to each other and are true proportions; **d.** If the means of the original proportion are transposed, the resulting proportion is a true proportion; **e.** The differences of the numerator and the denominator of the original proportion divided by the denominators of the original proportion are equal to each other and are true proportions; **f.** The left side of the original proportion is equal to the sum of the numerators of the original proportion divided by the sum of the denominators of the original proportion, and is true proportion.

Communicate

1. A proportion is an equation that states that two ratios are equal. For example $\frac{2}{3} = \frac{4}{6}$.

2. The extremes of a proportion are the numerator of the first ratio and the denominator of the second ratio. The means are the denominator of the first ratio and the numerator of the second ratio.

3. The product of the means is equal to the product of the extremes.

4. Find the cross product and solve the resulting linear equation: $4 = 4x + 2$, $x = 0.5$.

5. Let d represent the distance traveled on 75 liters. Write a proportion that involves two equal ratios comparing the number of liters used to the distance traveled: $\frac{40}{320} = \frac{75}{d}$. The car can travel 600 kilometers.

Lesson 14.7, pages 689–693

Communicate

1. A proof in algebra gives a logical justification to show that a conclusion is true.

2. definitions, postulates, and already proven theorems

3. The hypothesis, the statement that follows the word "if," is assumed to be true. The conclusion, the statement that follows the word "then," is the part that has to be proven.

4. In the converse of a statement the conclusion and the hypothesis are interchanged.

5. Start with the given equation $7x + 9 = 5$. Add −9 to both sides using the Addition Property of Equality: $7x + 9 − 9 = 5 − 9$. Since 9 and −9 are opposites, their sum is 0, so the equation simplifies to $7x = −14$. Divide both sides by 7 using the Division Property of Equality: $x = −2$.

Practice and Apply

Proofs may vary.

17. $2y + 3y = 5y$ Given
$y(2 + 3) = 5y$ Distributive Property
$y(5) = 5y$ Simplify
$5y = 5y$ Commutative Property for Multiplication

18. $5x − 3x = 2x$ Given
$x(5 − 3) = 2x$ Distributive Property
$x(2) = 2x$ Simplify
$2x = 2x$ Commutative Property for Multiplication

19. $(−3a)(−2a) = 6a^2$ Given
$(−3)a \cdot a(−2) = 6a^2$ Commutative Property
$(−3)a^2(−2) = 6a^2$ Definition of a Power
$(−3)(−2)a^2 = 6a^2$ Commutative Property for Multiplication
$6a^2 = 6a^2$ Multiplication of Integers

20. $7n − 3n = 32$ Given
$n(7 − 3) = 32$ Distributive Property
$n(4) = 32$ Subtraction of Integers

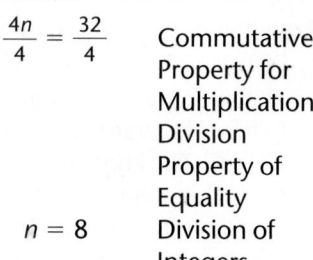

$\dfrac{4n}{4} = \dfrac{32}{4}$ Commutative Property for Multiplication, Division Property of Equality

$n = 8$ Division of Integers

21. $3(x + 2) = -15$ Given

$3x + 6 = -15$ Distributive Property

$3x + 6 - 6 = -15 - 6$ Subtraction Property of Equality

$3x + 0 = -21$ Property of Opposites

$3x = -21$ Identity for addition

$\dfrac{3x}{3} = \dfrac{-21}{3}$ Division Property of Equality

$x = -7$ Division of integers

22. $\dfrac{2}{3}x + 5 = -9$ Given

$\dfrac{2}{3}x + 5 - 5 = -9 - 5$ Subtraction Property of Equality

$\dfrac{2}{3}x + 0 = -14$ Property of Opposites

$\dfrac{2}{3}x = -14$ Identity for addition

$\left(\dfrac{3}{2}\right)\left(\dfrac{2}{3}x\right) = \left(\dfrac{3}{2}\right)(-14)$ Multiplication Property of Equality

$x = -\dfrac{42}{2}$

Reciprocal Property

$x = -21$ Division of integers

23. $\dfrac{3}{4}m + 8 = -1$ Given

$\dfrac{3}{4}m + 8 - 8 = -1 - 8$ Subtraction Property of Equality

$\dfrac{3}{4}m + 0 = -9$ Property of Opposites

$\left(\dfrac{4}{3}\right)\left(\dfrac{3}{4}m\right) = \left(\dfrac{4}{3}\right)(-9)$ Multiplication Property of Equality

$m = -\dfrac{36}{3}$ Reciprocal Property

$m = -12$ Division of integers

28. If a number is even, then it has 2 as a prime factor and is divisible by 2. When the number is squared the number 2 will appear at least twice and the number will be divisible by 2.
If n is even and k is an integer, then $n = 2k$.
$(n)^2 = (2k)^2$ and since $4k^2$ is divisible by 2, n^2 must be divisible by 2.

29. If n is odd, then $n = k + 1$, where k is an even number.
Then $n^2 = (k + 1)(k + 1)$
$= k^2 + 2k + 1.$
From question 28, if k is even then k^2 is also even. Since $2k$ is also always even, the sum $k^2 + 2k + 1$ must be odd.

---

30. $(a - b) + (b - a) = a + (-b + b) + (-a)$ Associative Property for Addition

$= a + 0 + (-a)$ Property of Opposites

$= a + (-a)$ Identity for Addition

$= 0$ Property of Opposites

31. $(xy)\left(\dfrac{1}{x}\right) = (yx)\left(\dfrac{1}{x}\right)$ Commutative Property for Multiplication

$= y(x)\left(\dfrac{1}{x}\right)$ Associative Property for Multiplication

$= y(1)$ Reciprocal Property

$= y$ Identity for Multiplication

32. $\dfrac{a}{b} + \dfrac{c}{d} = 1\left(\dfrac{a}{b}\right) + 1\left(\dfrac{c}{d}\right)$ Identity for Multiplication

$= \left(\dfrac{d}{d}\right)\left(\dfrac{a}{b}\right) + \left(\dfrac{b}{b}\right)\left(\dfrac{c}{d}\right)$ Substitution Property

$= \dfrac{da}{db} + \dfrac{bc}{bd}$ Multiplication of rationals

$= \dfrac{ad}{bd} + \dfrac{bc}{bd}$ Commutative Property for Multiplication

$= \dfrac{ad + bc}{bd}$ Addition of rationals